Third International Handbook
of Mathematics Education

Springer International Handbooks of Education

VOLUME 27

For further volumes:
http://www.springer.com/series/6189

M. A. (Ken) Clements • Alan J. Bishop
Christine Keitel • Jeremy Kilpatrick
Frederick K. S. Leung
Editors

Third International Handbook of Mathematics Education

Editors
M. A. (Ken) Clements
Illinois State University
Department of Mathematics
Normal, IL
USA

Christine Keitel
Freie University
Berlin
Germany

Frederick K. S. Leung
The University of Hong Kong
Pokfulam, Hong Kong
People's Republic of China

Alan J. Bishop
Monash University
Clayton, Victoria
Australia

Jeremy Kilpatrick
University of Georgia
Athens, Georgia
USA

ISBN 978-1-4614-4683-5   ISBN 978-1-4614-4684-2 (eBook)
DOI 10.1007/978-1-4614-4684-2
Springer New York Heidelberg Dordrecht London

Library of Congress Control Number: 2012947991

© Springer Science+Business Media New York 2013
This work is subject to copyright. All rights are reserved by the Publisher, whether the whole or part of the material is concerned, specifically the rights of translation, reprinting, reuse of illustrations, recitation, broadcasting, reproduction on microfilms or in any other physical way, and transmission or information storage and retrieval, electronic adaptation, computer software, or by similar or dissimilar methodology now known or hereafter developed. Exempted from this legal reservation are brief excerpts in connection with reviews or scholarly analysis or material supplied specifically for the purpose of being entered and executed on a computer system, for exclusive use by the purchaser of the work. Duplication of this publication or parts thereof is permitted only under the provisions of the Copyright Law of the Publisher's location, in its current version, and permission for use must always be obtained from Springer. Permissions for use may be obtained through RightsLink at the Copyright Clearance Center. Violations are liable to prosecution under the respective Copyright Law.
The use of general descriptive names, registered names, trademarks, service marks, etc. in this publication does not imply, even in the absence of a specific statement, that such names are exempt from the relevant protective laws and regulations and therefore free for general use.
While the advice and information in this book are believed to be true and accurate at the date of publication, neither the authors nor the editors nor the publisher can accept any legal responsibility for any errors or omissions that may be made. The publisher makes no warranty, express or implied, with respect to the material contained herein.

Printed on acid-free paper

Springer is part of Springer Science+Business Media (www.springer.com)

Past, Present and Future Dimensions of Mathematics Education: Introduction to the Third International Handbook of Mathematics Education

M. A. (Ken) Clements

Abstract The four major sections in this *Third International Handbook* are concerned with: (a) social, political and cultural dimensions in mathematics education; (b) mathematics education as a field of study; (c) technology in the mathematics curriculum; and (d) international perspectives on mathematics education. These themes are taken up by 84 internationally-recognized scholars, based in 26 different nations. Each of the *Handbook*'s four sections is structured on the basis of past, present and future aspects. The first chapter in a section provides historical perspectives ("How did we get to where we are now?"); the middle chapters in a section analyze present-day key issues and themes ("Where are we now, and what recent events have been especially significant?"); and the final chapter in a section reflects on policy matters ("Where are we going, and what should we do?"). An overview of the major common recurring themes and issues in the *Handbook* is presented. It is argued that mathematics education research has a vitally important role to play in improving mathematics curricula and the teaching and learning of mathematics. As a result of the expertise, wisdom, and internationalism of both authors and section editors, this *Handbook* provides an invaluable, state-of-the-art compendium of the most recent, and promising, developments in the field.

Keywords Globalization and mathematics education • History of mathematics education • *International Handbook of Mathematics Education* • Mathematics education research • Mathematics education policy • Social turn • No Child Left Behind • Technology in mathematics education

There are a number of facts about this *Third International Handbook* that should be made clear at the outset. These are:

- All 31 chapters were specifically written for this *Handbook.* There is no chapter in this *Handbook* which appeared in either the first *International Handbook of*

M. A. (Ken) Clements
Illinois State University, Department of Mathematics, Normal, IL, USA

Mathematics Education (Bishop, Clements, Keitel, Kilpatrick & Laborde, 1996) or the *Second International Handbook of Mathematics* (Bishop, Clements, Keitel, Kilpatrick & Leung, 2003).
- Although authors were expected to pay special attention to developments in scholarship, and in practice, that have occurred since the publication, in 2003, of the *Second International Handbook,* this *Third International Handbook* should not be seen merely as an update of the earlier handbooks. From the beginning, the editors aimed for a state-of-the-art compendium that identified and examined four major dimensions of contemporary mathematics education.
- The contents of this *Third International Handbook* are consistent with the inclusion of the word "International" in the title. Altogether, there are 84 authors who contributed to the 31 chapters, and at the time the chapters were written (between September 2010 and December 2011), the authors were working in a total of 26 nations: Australia, Austria, Brazil, Canada, China, Colombia, Czech Republic, Denmark, France, Germany, Hong Kong, Iran, Israel, Italy, Japan, Malaysia, Mexico, New Zealand, Portugal, Singapore, Spain, Sweden, The Netherlands, UK, USA, and Venezuela. Although we would have liked the *Third Handbook* to have been even more international than it is in its outlook, we recognize that given that there were to be only 31 chapters, it would not have been realistic, or prudent, to have attempted to have more nations represented among the authors.
- In July 2010 the editorial team met for a week to discuss the structure, likely chapter emphases, and authors for the *Third Handbook.* The first decision made, at that time, was that there would be the following four sections:
  - *Section A:* Social, Political and Cultural Dimensions in Mathematics Education;
  - *Section B:* Mathematics Education as a Field of Study;
  - *Section C:* Technology in the Mathematics Curriculum; and
  - *Section D:* International Perspectives on Mathematics Education.
  - We recognized that these sections did not cover all of the important areas of mathematics education—but we chose these major themes after reflection on what we thought offered the best follow-up potential to the *Second Handbook,* in terms of developments between 2003 and the present.
- It was also agreed that each section would be structured on the basis of past, present and future aspects. Thus, the first chapter in each of the four sections is concerned with analyses of historical antecedents ("How did we get to where we are now?"); the "middle" chapters provide analyses of present-day key issues and themes ("Where are we now, and what events since 2003 have been especially significant?"); and the final chapter in each section reflects on policy matters ("Where are we going, and what should we do?"). As far as we know, this *Handbook* is the first major mathematics education publication to adopt, consciously, this past–present–future organizational structure.
- Each author was selected, jointly by the editors, on the basis of her or his recognized excellence and experience in relation to the theme that needed to be addressed in a chapter.

Major International Developments in Mathematics Education Since 2003

I have read each chapter in this *Handbook* several times. One cannot read the chapters carefully without beginning to recognize the pervasiveness of certain influences on the field of mathematics education. It is not my intention here to comment on each chapter in the *Handbook*—the section editors will have the opportunity to do that in their own introductions, placed at the beginning of the sections. Rather, I wish to draw attention to several major developments, and sometimes associated tensions, over the last decade.

The "Social Turn" Versus Control Groups, Random Assignment, and Randomized Trials

The first major development has been in relation to what Lerman (2000) called the *social turn* in mathematics education research. Many of the authors (especially of chapters in Sections A and B of this *Handbook*) draw attention to the increasing use of socio-cultural theories in the field. Some see the selection, use and refinement of such theories as the main way by which mathematics education is developing into a discipline in its own right. This emphasis on the social, cultural and political aspects of mathematics education has resulted in many of the traditional assumptions in mathematics education, about *who* should study *what* mathematics, and *why*, being problematized. In relation to issues associated with the call for "mathematics for all," traditional concepts of "disadvantage" have been questioned and re-defined, and traditional classroom discourse patterns have been subjected to scrutiny, especially from vantage points offered by different theorists. Issues associated with the role of assessment have never been far away, and the matter of what should constitute the most appropriate forms of assessment in a given context is widely discussed. There has also been much discussion and research on the concept of teachers as researchers, and on what collaboration might mean in different areas of mathematics teaching and mathematics education research. Globalization tendencies have been, and continue to be, scrutinized from various theoretical perspectives.

An interesting feature of the last decade has been the roles and status of mathematics education researchers in the USA, where the 2001 Federal Education initiative *No Child Left Behind* (NCLB) Act heralded a series of significant reforms which sought to improve student, teacher, school, and system performance in mathematics through test-based school accountability (Learning Point Associates, 2007). The NCLB Act called for education policy to rely on a foundation of *scientifically-based research* which employed rigorous methodological designs and techniques, including control groups, random assignment, and randomized trials. National Science Foundation (NSF) grant applicants were strongly advised to strive for randomized designs, and the Department of Education's 2002–2007 strategic plan (see

Shavelson & Towne, 2002) stated that, by 2004, 75% of new research and evaluation projects funded by the Department which address causal questions should use randomized experimental designs. Mathematicians, as well as mathematics educators, were expected to be included in mathematics education research teams. The tension between those requirements and the spirit of Lerman's (2000) social turn is discussed in several chapters in this *Handbook*.

Technology

Some of the authors in this *Third International Handbook* make it very clear that these days the world of mathematics education is changing very rapidly, and that technology is a major factor influencing the directions of change. Writers in Sections C and D of this *Handbook* emphasize that recent technological developments are challenging traditional views on curriculum, teaching, learning, and assessment. What forms of curriculum, teaching, learning, and assessment are the most appropriate given the rapid technological developments? How can teachers keep up with developments and, simultaneously, cope with their often-too-heavy teaching loads? Given recent developments, what should algebra, geometry and calculus curricula look like in the future? What should proof in school mathematics look like? What technological aids should students be allowed to use in examinations, and what are the implications of that question for those responsible for developing policies with respect to assessment and evaluation?

Given the rapidity of ongoing technological developments, and the increasing reach of new technologies into even remote areas of the world, one cannot help wondering whether in 20 years time, say, there will be an agreed international mathematics curriculum. Many writers committed to the need to link curricula and teaching to social and cultural factors view such a possibility as extremely unwelcome. Issues associated with online and other distance forms of mathematics education are frequently discussed, and there is a concern that despite the socializing potential of new technology, an international mathematics curriculum would result in mathematics education becoming even more separated from local aspects of culture than it is now.

Globalization and Internationalization of Mathematics Education

When the *Handbook* editors initially met to work out the *Handbook* structure, chapter titles, authors, etc., it was agreed that it would be wise to try to avoid unnecessary repetition. In particular, it was agreed that we should try to restrict, to just a few chapters, discussion of the influence of the International Association for the Evaluation of Educational Achievement's (IEA's) "Trends in Mathematics and Science Study" (TIMSS), and the Organisation for Economic Co-operation and Development's (OECD's) "Programme for International Student Assessment" (PISA). Despite the best efforts of our editors, we failed in this regard, largely because many authors recognized the huge impact that TIMSS and PISA (and other international studies such as the Learner's Perspective Study—LPS) have had during the past decade.

There is a concern that TIMSS, PISA, and other international testing programs will have a standardizing effect on school mathematics that will cramp promising developments arising from the "social turn" in research. But some authors have argued that despite this potential danger, these international studies have drawn attention to well-performing nations like Japan, Singapore, Hong Kong, and Finland and have more or less forced researchers and policy makers to face the question: "Why have the students in such nations performed so well—and why have students in some extremely well-resourced nations performed considerably less well?" This has given rise to additional questions like: "How can we make mathematics education research more responsive to national needs, as those needs are perceived by politicians and education policy makers?" The possibility that influential policy makers do not regard the results of much mathematics education research as useful has been raised.

Who Should Read This *Handbook*?

As I read the draft chapters of this *Third International Handbook* I often found myself thinking that all mathematics educators, including mathematics teachers at all levels, should read some or all of the chapters. Then, when teaching graduate classes, I often thought that all of my graduate students would benefit from reading some of the chapters. I certainly intend to use this *Handbook* as a text for my future graduate students and, of course, I hope that other persons teaching graduate mathematics education students will do the same.

Various *Handbook* authors have drawn attention to the tendency for much mathematics education research to be carried out in teams that include school teachers and mathematics educators normally based outside of schools. Every person involved in collaborative studies of this type would likely benefit from becoming aware of what authors in this *Handbook* have said.

Chapters in this *Handbook* can provide important insights into how teachers and researchers around the world are working towards providing answers to issues that can no longer be ignored. For example, we need to answer questions such as: "What can a school do if it wants to engage all of its students actively and productively in relevant mathematics learning?" And, "What about those outside of the normal school and college system (many adults, for example) who want to learn mathematics?—What should we be doing, for them, to facilitate top-quality, and satisfying mathematics learning?"

This *Handbook* is the most-internationalized of all mathematics education handbooks that have been prepared thus far. Its chapters provide up-to-the-minute, state-of-the-art reviews on major themes; invariably, there has been an attempt to make readers aware of the international spread of opinion, methodologies, research and practice. The *Handbook* provides much insight, not only from researchers in the traditional European and North American nations but also from researchers in many other parts of the world. Throughout, any suggestion that the best mathematics education research wisdom has largely emanated from "the West" has been problematized, and basic questions—such as: "Why have Confucian-based cultures generated such productive forms of school mathematics?"—have been carefully considered. Authors charged with the responsibility of presenting historical perspectives (and authors of some of the other chapters, too) have deliberately argued from international, global, vantage points rather than from distinctly Western vantage points.

Whither Mathematics Education?

I have been privileged to work on the three Springer/Kluwer international handbooks on mathematics education. For almost all of my professional career I have worked in the field of mathematics education, and it has been a matter of principle for me to be able to say why I believe, strongly, that mathematics education is a crucially important field of endeavour.

Mathematics is one of the few areas in an individual's life in which she or he is required to spend between three and five hours per week (and, in addition, more hours on homework or with a tutor), for between 10 and 12 years (at least) studying a curriculum defined by others. What a waste of everyone's time, energy, and money, if students do not learn school mathematics as well as they possibly can, so that they develop an interest in the subject and an appreciation of its power to help them deal efficiently with important everyday problems. Furthermore, I believe that success with the subject is likely to be associated with greater satisfaction in later life (because successful students are more likely to take up vocations of their choice, or gain entry to a wider range of courses in higher education institutions). From a national perspective, the benefit of having a mathematically-competent citizenry is, it is often asserted, likely to result in strong economic performance (or, at least, stronger than would be the case if most citizens were not mathematically competent). Thus, it is important that research be conducted which will take into account students' attitudes towards mathematics, as well as their mathematical problem-solving and problem-posing performances.

But if mathematics education research is important, then how well are we doing in fostering the highest possible quality of mathematics learning as a result of our mathematics education research? Let us not put our heads in the sand on this matter. There is certainly a lot of room for improvement! The nation which has the most qualified mathematics education researchers is probably the USA—yet, many indicators (including results on international comparative studies) suggest that many US students fail to learn mathematics well. How could that be the case, considering the amount of research that has been conducted, and published within the USA, over so many years?

It is well known that many students, in most nations (perhaps all nations), experience difficulty in understanding fractions, the four operations with integers, and elementary algebra. We need to face the reality that many learners experience much difficulty in mathematizing situations for which mathematical approaches to problem solving would be informative and efficient. Why has there not been a marked improvement, given the large amount of mathematics education research conducted around the world, and over a very long period of time, with respect to such fundamentally important curriculum matters? Should our standard curricula and teaching approaches be problematized and reconceptualized? Various chapters in this *Third Handbook* consider issues such as these.

I could say much more—but perhaps, now I have succeeded in stimulating your interest and arousing an argumentative spirit within you. I should leave the core of what is said in this *Third International Handbook* to our team of very competent authors. As you read each chapter, I urge you to reflect on the basic question: Whither mathematics education?

Acknowledgments As many readers might imagine, it has been an honour and privilege to work with a team of such outstanding editors. It hardly needs to be said that Alan Bishop, Christine Keitel, Jeremy Kilpatrick and Frederick Leung are world-class mathematics educators and researchers. And I can say, sincerely, that each has expended a huge amount of effort, and time, on this project.

I would also like to thank the authors, who have worked hard to meet demanding deadlines. Together, twhe editors and authors for this Third International Handbook have provided enormous experience, expertise and wisdom on key mathematics education issues. Has any other book in the field of mathematics education ever had such a wide-ranging and authoritative international set of authors?

I would also like to thank the large number of professional reviewers of chapters. Every chapter has been read by at least four people—the appropriate section editor, at least two especially-appointed external reviewers, and myself. None of the Handbook's 31 chapters appears in the form in which it was first submitted. Always, authors were willing to respond professionally to suggestions for improvement.

During the period in which the Third Handbook was being prepared I was based at Illinois State University's Department of Mathematics. I would especially like to thank Dr George Seelinger, Head of the Department of Mathematics, for his encouragement and support.

Throughout the planning and manuscript preparation process, Harmen van Paradijs, Editorial Director, Human Sciences, for Springer, was supportive, wise and understanding. We also profited from the knowledge and cooperation of Melissa James and Natalie Rieborn, Springer's Publishing Editor, and Senior Editorial Assistant, respectively. We could not have asked for a better relationship with our publisher.

References

Bishop, A. J., Clements, K., Keitel, C., Kilpatrick, J., & Laborde, C. (Eds.). (1996). *International handbook of mathematics education* (2 vols.). Dordrecht, The Netherlands: Kluwer Academic Publishers.

Bishop, A. J., Clements, M. A., Keitel, C., Kilpatrick, J., & Leung, F. (Eds.). (2003). *Second international handbook of mathematics education* (2 vols.). Dordrecht, The Netherlands: Kluwer Academic Publishers.

Learning Point Associates. (2007). *Understanding the No Child Left Behind Act: Mathematics and science.* Chicago, IL: Author.

Lerman, S. (2000). Social turn in mathematics education research. In J. Boaler (Ed.), *Multiple perspectives on mathematics teaching and learning* (pp. 19–44). Palo Alto, CA: Greenwood.

Shavelson, R. J., & Towne, I. (Eds.). (2002). *Scientific research in education.* National Research Council, Committee on Scientific Principles for Education Research, Center for Education, Division of Behavioral and Social Sciences and Education. Washington, DC: National Academies Press.

Contents

Part I   Introduction to Section A: Social, Political and Cultural Dimensions in Mathematics Education 1
Christine Keitel

1 From the Few to the Many: Historical Perspectives on Who Should Learn Mathematics 7
M. A. (Ken) Clements, Christine Keitel, Alan J. Bishop, Jeremy Kilpatrick, and Frederick K. S. Leung

2 Theories for Studying Social, Political and Cultural Dimensions of Mathematics Education 41
Eva Jablonka, David Wagner, and Margaret Walshaw

3 Understanding and Overcoming "Disadvantage" in Learning Mathematics 69
Lulu Healy and Arthur B. Powell

4 Beyond Deficit Models of Learning Mathematics: Socio-cultural Directions for Change and Research 101
Cristina Frade, Nadja Acioly-Régnier, and Li Jun

5 Studying Learners in Intercultural Contexts 145
Yoshinori Shimizu and Gaye Williams

6 Learners in Transition Between Contexts 169
Tamsin Meaney and Troels Lange

7 Critical Perspectives on Adults' Mathematics Education 203
Jeff Evans, Tine Wedege, and Keiko Yasukawa

8 The Politics of Equity and Access in Teaching and Learning Mathematics 243
Neil A. Pateman and Chap Sam Lim

| Part II | Introduction to Section B: Mathematics Education as a Field of Study | 265 |

Alan J. Bishop

9   From Mathematics and Education, to Mathematics Education 273
Fulvia Furinghetti, José Manuel Matos, and Marta Menghini

10 Theories in Mathematics Education: Some Developments and Ways Forward 303
Bharath Sriraman and Elena Nardi

11 Research Methods in Mathematics Teacher Education 327
Uwe Gellert, Rosa Becerra Hernández, and Olive Chapman

12 Linking Research to Practice: Teachers as Key Stakeholders in Mathematics Education Research 361
Carolyn Kieran, Konrad Krainer, and J. Michael Shaughnessy

13 Teachers Learning from Teachers 393
Allan Leslie White, Barbara Jaworski, Cecilia Agudelo-Valderrama, and Zahra Gooya

14 Developing Mathematics Educators 431
Jarmila Novotná, Claire Margolinas, and Bernard Sarrazy

15 Institutional Contexts for Research in Mathematics Education 459
Tony Brown and David Clarke

16 Policy Implications of Developing Mathematics Education Research 485
Celia Hoyles and Joan Ferrini-Mundy

| Part III | Introduction to Section C: Technology in the Mathematics Curriculum | 517 |

Frederick K. S. Leung

17 From the Slate to the Web: Technology in the Mathematics Curriculum 525
David Lindsay Roberts, Allen Yuk Lun Leung, and Abigail Fregni Lins

18 Modelling with Mathematics and Technologies 549
Julian Williams and Merrilyn Goos

19 Technology and the Role of Proof: The Case of Dynamic Geometry 571
Nathalie Sinclair and Ornella Robutti

20	**How Might Computer Algebra Systems Change the Role of Algebra in the School Curriculum?** M. Kathleen Heid, Michael O. J. Thomas, and Rose Mary Zbiek	597
21	**Technology for Enhancing Statistical Reasoning at the School Level** Rolf Biehler, Dani Ben-Zvi, Arthur Bakker, and Katie Makar	643
22	**Learning with the Use of the Internet** Marcelo C. Borba, Philip Clarkson, and George Gadanidis	691
23	**Technology and Assessment in Mathematics** Kaye Stacey and Dylan Wiliam	721
24	**Technology-Driven Developments and Policy Implications for Mathematics Education** L. Trouche, P. Drijvers, G. Gueudet, and A. I. Sacristán	753
Part IV	**Introduction to Section D: International Perspectives on Mathematics Education** Jeremy Kilpatrick	791
25	**From the Local to the International in Mathematics Education** Alexander Karp	797
26	**International Collaborative Studies in Mathematics Education** Parmjit Singh and Nerida F. Ellerton	827
27	**Influence of International Studies of Student Achievement on Mathematics Teaching and Learning** Vilma Mesa, Pedro Gómez, and Ui Hock Cheah	861
28	**International Organizations in Mathematics Education** Bernard R. Hodgson, Leo F. Rogers, Stephen Lerman, and Suat Khoh Lim-Teo	901
29	**Toward an International Mathematics Curriculum** Jinfa Cai and Geoffrey Howson	949
30	**Methods for Studying Mathematics Teaching and Learning Internationally** Mogens Niss, Jonas Emanuelsson, and Peter Nyström	975

31 Implications of International Studies for National and Local Policy in Mathematics Education .... John A. Dossey and Margaret L. Wu	1009
Brief Biographical Details of Authors	1043
Names of Reviewers	1063
Author Index	1065
Subject Index	1093

Part I
Introduction to Section A: Social, Political and Cultural Dimensions in Mathematics Education

Christine Keitel

Abstract There are eight chapters in this first main section of the *Third International Handbook of Mathematics Education* and, altogether, there are 22 contributing authors, from 13 different nations. The first chapter, prepared by the five editors of the *Third Handbook*, provides historical perspectives on how far we have progressed towards the goals of mathematics for all—and also on different interpretations of that goal—over the past 200 years. The authors of the other chapters present various theoretical positions that are informing mathematics education researchers as they strive to achieve more equitable and effective environments in which the teaching and learning of mathematics occurs. Cultural, social, linguistic and political factors that not only affect views on the nature of mathematics, but also the structuring of curricula and education environments, are emphasized.

Keywords Deficit models in mathematics education • Disadvantage in mathematics education • Equity in mathematics education • Language and mathematics learning • Numeracy • Social justice in mathematics education • Social turn in mathematics education • Sociocultural directions in mathematics education • Transition between contexts

In Chapter 1, the editors argue that historically the acceptance of reckoning or mathematics as something to be taught in classrooms came rather late. Although, immediately after the invention of printing, reckoning books for independent learning appeared in Europe, and early in the 16th century private reckoning schools for bourgeois pupils were operating in central Europe, all of this happened rather slowly. At first, printed arithmetics were written in Latin, but then followed vernacular texts—like the famous arithmetic book written, in a German language, by Adam Ries. But it was mainly the children of wealthy bourgeois families in cities—and almost always boys—who attended such schools and usually the emphasis was on

C. Keitel
Freie University, Berlin, Germany

mechanistic, rule-based calculations. Although famous mathematicians like, for example, Descartes and Leibniz, were advocating that their own revolutionary mathematical discoveries (e.g., Descartes' "Cartesian Geometry" and Leibniz's "Calculus") be taught in schools and universities, those who ultimately went on to study any form of higher mathematics were few in number.

High-level schools and universities were rare and expensive, and in any case, within such institutions mathematics was rarely regarded as a subject of educational value. Not only was the number of persons capable of teaching mathematics beyond elementary arithmetic small, but also general parental attitudes to schooling, the economic circumstances of most families, and social and psychological presuppositions and prejudices about mathematical ability or giftedness, combined to condemn forms of mathematics education into a precarious position. Mathematics teaching was the domain of the private tutor or the barely mathematically literate teacher in private schools. Chapter 1 identifies a historical progression underlying the evolution of the current expectation that relevant and applicable mathematics education should be available to all people: the sequence begins with schooling for all, and proceeds to arithmetic for all, to mathematics for all, and to quantitative or mathematical literacy for all.

In mathematics education research and practice today there is a noticeable change in approaches to researching the diverse social, political and cultural dimensions of mathematics education. In Chapter 2, Eva Jablonka, Margaret Walshaw and David Wagner provide an overview of a growing number of theories that are allowing us to widen our perspectives on these dimensions. Jablonka et al. identify and discuss theoretical trends and provide critical discussion not only of the theories themselves but also of the ways they are being used to discuss and critique research and practices in mathematics education. The authors successfully summarize, compare and exploit theories and their applications from research presented at the annual meetings of the International Group for the Psychology of Mathematics Education (PME).

Past research has largely characterized disadvantage as an individual or social condition that somehow impedes mathematics learning. That approach resulted in the marginalization of individuals whose physical, racial, ethnic, linguistic and social identities were different from normative identities constructed by dominant social groups. Recent studies have consciously avoided equating difference with deficiency and instead have sought to understand mathematics learning from the perspective of those whose identities are not consistent with norms constructed by dominant social groups. With this way of thinking, traditional concepts of "disadvantage" can be interpreted as having not only been socially constructed but also as having perpetuated disadvantage among certain types of individuals. Overcoming disadvantage can be achieved by analyzing how learning scenarios and teaching practices can be more finely tuned to the needs of particular groups of learners, empowering them to demonstrate abilities beyond the limits generally set and expected within dominant discourses.

In Chapter 3, Lulu Healy and Arthur Powell consider—under the heading "Understanding and Overcoming Disadvantages in Mathematics"—theoretical and methodological perspectives associated with the search for a more inclusive mathematics education, one which generally perceives and conceptualizes the role of the teachers as active participants in the process of researching and interpreting students' learning. Drawing from examples from a diverse range of learners including linguistic, racial and ethnic minorities, as well as deaf and blind students, the authors argue that by carefully studying and trying to get a much deeper understanding of the learning processes of such students we may not only be able to design pedagogical means to allow children to learn better, but also to better understand mathematics learning in general.

Cristina Frade, Nadja Acioly-Régnier and Li Jun described the aim of Chapter 4—titled "Beyond Deficit Models of Learning Mathematics: Sociocultural Directions for Change and Research"—as providing a theoretical exposé of the inherent weaknesses of deficit models. The identification of those weaknesses only came to be recognized following major paradigmatic changes in mathematics education research which drew attention to new perspectives on learning. Whereas, previously, deficit models were foregrounded in research designs, they have now been replaced by a wide variety of theoretical directions for studying diverse approaches to learning mathematics. This has resulted in an acceptance of the need for richness and variety in research practices, so that approaches can be studied, compared, and mutually applied and improved. Psychological and quantitative approaches and methods are now increasingly complemented, or even replaced, by new directions that rely on social and anthropological theories and methods. Rather than reviving ideas about deficit research in mathematics education, Frade et al. present sociocultural perspectives of learning mathematics, and show how these perspectives demand answers to important questions that were not even considered when deficit models of learning framed research. Having placed the main traditional markers of discrimination in school mathematics—gender, social class and ethnicity—within a perspective of social justice, the chapter concludes with a reflection on equality in terms of the democratic principle of meritocracy in mathematics education.

The recognition by recent researchers that learning mathematics is a culturally-influenced activity has become increasingly more apparent as research aims, technological advances, and methodological techniques have diversified, enabling more detailed analyses of learners and what they learn. Increased opportunities for studying learners in different cultural, social and political settings have also been provided by online access to results of international benchmark testing programs. The availability of data sets from large-scale quantitative studies—like, for example, "Trends in International Mathematics and Science Study" (TIMSS) and the "Program for International Student Assessment" (PISA)—and from comprehensive qualitative studies—like the international "Learners' Perspective Study" (LPS)—have facilitated careful investigation of research questions about learners and the contexts in which they learn. In Chapter 5, "Studying Learners in Intercultural Contexts," Yoshinori Shimizu and Gaye Williams point to how results and methods from large-scale quantitative studies have stimulated questions that demand qualitative research designs for their exploration. The increasing adoption of qualitative research has raised awareness with respect to the importance of historical, social and cultural perspectives when considering the dimensions of learning. This raises questions about the roles of "local" theories in investigations involving intercultural analyses.

In Chapter 6, "Learners in Transition Between Contexts," Tamsin Meaney and Troels Lange explore conceptions of learners in transition between contexts, and evaluate pedagogical practices that have been advocated for such learners. They point out that learning occurs as learners reflect on their transition between contexts, particularly when there are differences in what content knowledge is valued, the relationships between participants, and how activities are undertaken. From this perspective, productive pedagogical practices for learners in transition are those that build and sustain relationships between learners and mathematics and between learners and others, including especially those that lead outside the mathematics classroom. Meaney and Lange look, for their inspiration, specifically at examples of pedagogical practices that draw on principles associated with ethnomathematics and critical mathematics education.

Chapter 7 provides a focussed discussion of the goals and achievements of a movement that is concerned with adults' mathematics education (AME) as a field of study and practice. Jeff Evans, Tine Wedege and Keiko Yasukawa draw attention to a broad range of settings for teaching and learning, as well as for research. AME, whose activities have developed in a dynamic context of globalization, competition, and social insecurity, has faced the same struggle for its justification, in terms of humanistic and human capital goals of education, that adult education and lifelong education have been facing over the last half-century. This struggle is well reflected in current AME practices, research and policy. Evans et al. formulate critical perspectives for examining AME in the three connected dimensions of practice, research, and policy, always with the intention of clarifying assumptions, concepts, and actions with respect to crucial areas. Thus, for example, they examine multiple and contested meanings of key terms like "numeracy," and point out that definitions vary depending on whether they seek to foreground the needs of individual learners or whether they are more concerned with particular economic imperatives (such as "needs" of the labour market). Evans et al. illuminate how variations in such definitions can affect the experiences of AME learners and practitioners. They problematize ideas associated with "the transfer of learning" of mathematics from school to work, and from formal to non-formal or informal learning situations. They argue that because a new international survey of adults' skills—the OECD-sponsored Program for International Assessment of Adult Competencies (PIAAC) is now being conducted—it is timely to question what such surveys can tell us about the development of AME as a field, and to consider which questions need to be pursued independently.

In the last chapter of Section A, Chapter 8, on "Politics of Equity and Access in Teaching and Learning Mathematics," Neil A. Pateman and Lim Chap Sam, besides clarifying definitions of equity and access, briefly contrast two philosophical positions on the nature of mathematics and speculate about the consequences of these different positions for equity and access. They raise the question "whose mathematics?" and provide a developing viewpoint on how mathematics learning depends on equity and access for students. After considering the roles of mathematics teachers and how these are related to equity and access for students, they broaden their discussion to consider political influences on both teachers and learners of mathematics. Their observations relate to the role that politics plays at different levels in influencing access and equity for teaching and learning mathematics. Pateman and Lim illustrate their position through a discussion of particular examples, some from history, and others documenting more recent events. Finally they offer a brief discussion of several international cases which, they believe, demonstrate how a form of colonization is occurring in relation to contexts in which authorities insist on an "English-first" policy whereby the language of instruction in school mathematics must be English despite the fact that English is not the pupils' first language.

Chapter 1
From the Few to the Many: Historical Perspectives on Who Should Learn Mathematics

M. A. (Ken) Clements, Christine Keitel, Alan J. Bishop, Jeremy Kilpatrick, and Frederick K. S. Leung

Abstract Today we take for granted that everybody should be offered the opportunity to learn mathematics. However, it was not until well into the 20th century that "mathematics for all" became an achievable goal. Before then, the geographical location of schools in relation to children's homes, the availability (or non-availability) of teachers capable of teaching mathematics, parental attitudes to schooling, economic circumstances of families, and social and psychological presuppositions and prejudices about mathematical ability or giftedness, all influenced greatly whether a child might have the opportunity to learn mathematics. Moreover, in many cultures the perceived difference between two social functions of mathematics—its utilitarian function and its capability to sharpen the mind and induce logical thinking—generated mathematics curricula and forms of teaching in local schools which did not meet the needs of some learners. This chapter identifies a historical progression towards the achievement of mathematics for all: from schooling for all, to arithmetic for all, to basic mathematics for all; to secondary mathematics for all; to mathematical modelling for all; and to quantitative literacy for all.

M. A. (Ken) Clements (✉)
Illinois State University, Department of Mathematics, Normal, IL, USA
e-mail: clements@ilstu.edu

C. Keitel
Freie University, Berlin, Germany

A. J. Bishop
Monash University, Clayton, Victoria, Australia

J. Kilpatrick
University of Georgia, Athens, GA, USA

F. K. S. Leung
The University of Hong Kong, Pokfulam, Hong Kong, People's Republic of China

"Mathematics for all" is the kind of goal that anticipates a world in which all people have the opportunity to learn, and benefit from learning, mathematics. This chapter offers historical perspectives, not only on who has had the opportunity to learn mathematics, but also on forms of mathematics that have been embraced by the expression "mathematics for all." We take that expression to mean a situation in which all living people, in all nations, at any particular time, will have formally studied, or are studying, or will be expected to study at least some form of mathematics. There is also an implied additional assumption that studying mathematics will bring associated benefits—personal, social, and political—for all.

Our decision to interpret "mathematics for all" in that way means that we shall not be focussing on higher-order mathematics as found in universities. We shall be more concerned with providing a historical analysis of how, gradually, during the 19th and 20th centuries, more and more people gained the opportunity to study mathematics. Our decision implies that part of our analysis needs to be concerned with the concept of "schooling for all" because progress towards mathematics for all, as we are interpreting it, presupposes schooling for all.

Towards Mathematics for All

Perhaps the best-known set of statements on "mathematics for all" is a collection of 22 papers (Damerow, Dunkley, Nebres, & Werry, 1984), published in 1984 by the United Nations Educational, Scientific, and Cultural Organization (UNESCO). Since that publication, there have been many calls for "mathematics for all" (e.g., Gates & Vistro-Yu, 2003; Krygowska, 1984), or variations or extensions of that theme, such as "algebra for all" (e.g., Viadero, 2009) and "numeracy for all" (e.g., Robinson, 1996).

One of the most stimulating papers in the UNESCO collection was jointly prepared by Peter Damerow, of the Max Planck Institute for Human Development and Education in Germany, and Ian Westbury, of the University of Illinois. Damerow and Westbury (1984) addressed the problem of designing a mathematics curriculum which meets the mathematical needs of all students in a nation. They asserted that history had shown that in such efforts the politics of the situation inevitably led to a curriculum which met the needs of only a small group of students. They maintained that that was precisely what had happened, in many nations, during the "modern mathematics" era from the mid-1950s to the mid-1970s.

In the following passage, Damerow et al. (1984) identified a major stumbling block in efforts to achieve "mathematics for all":

> Mathematics curricula were developed for an élite group of students who were expected to specialize in the subject, and to study mathematics subsequently at higher levels in a tertiary institution. As education has become increasingly universal, however, students of lesser ability and with more modest vocational aspirations and daily life requirements have entered the school system in greater numbers. A major problem results when these students are exposed to a curriculum designed for potential specialists. This same type of traditional curriculum

has frequently been transferred to developing and third world countries where, because of different cultural and social traditions, the inappropriateness for general mathematics education has only been compounded. (p. 4)

Notice how Damerow et al. assumed that students have different "abilities" with respect to mathematics, and that a curriculum for "élite" students would not be suited to the needs of other students (Kamens & Benavot, 1991).

In another paper in the 1984 collection, Ben Nebres, of the Philippines, introduced the twin concepts of vertical and horizontal curriculum relationships. He argued that education authorities in developing countries typically kept their eyes on vertical curriculum requirements in developed countries, because they not only wanted their élite students to be qualified to study in developed countries, but they also wanted their own graduates to be professionally accepted for registration purposes in those nations. Nebres (1984) pointed out that this resulted in local curricula in developing countries failing to meet the needs of the majority and, indeed, failing to provide courses that were of interest, or suitable, for, most local students.

Historically, the numbers of students permitted to study mathematics, formally, have varied from community to community, from nation to nation, and from era to era (Gates & Vistro-Yu, 2003; Li & Ginsburg, 2006; Wu & Zhang, 2006). Even within the same community, or the same school, at a particular time, there may not be agreement on which students should be allowed to study the different forms of mathematics that are offered.

If everyone is to study mathematics then should there be a "core" mathematics curriculum, and if so, what should that core mathematics curriculum look like? And, to what extent should the mathematics-for-all expectation take into account cultural factors and individual differences? Should "mathematics for all" mean that students in schools in Paris, France, be taught the same mathematics as students in schools in the remote and mountainous regions of Vietnam? If one answers no, then immediately should follow the uncomfortable but important question, why not? H. R. W. Benjamin's (1939) classic *Saber-Tooth Curriculum* helped us recognize that there are important areas of life—like, for example, sports—in which it makes little sense for everyone to be asked to learn and practise the same skills. It may not be reasonable to require all people everywhere to learn the same mathematics.

In all societies, most adults use what Bishop (1988) called "small-m" mathematics, on a daily basis. They count, reason, and use concepts like "more," "less," "the same," and so forth, to perform actions in appropriate sequences. We all estimate and measure context-relevant quantities involving money, distances, times, capacities, areas, and other quantities. In this chapter we take such ethnomathematical practices for granted and focus more on the "big-M" forms of Mathematics (Bishop, 1988) that are offered in formal education institutions.

The perspectives we provide in this chapter will mainly take account of developments over the past 200 years. The coverage provides a broad sweep, and it has not been possible to take account of changing circumstances in all nations.

From our perspective, any scholarly discussion of mathematics-for-all phenomena ought to address the following questions:

1. Should all school children be expected to study mathematics, and if so what mathematics and for how long?
2. Should different students in different cultural settings study the same mathematics?
3. Should different students in the same nation, and even at the same school or college, study the same forms of mathematics?

Although these questions appear to be straightforward, they can be interpreted in different ways.

In this chapter we identify progress towards "mathematics for all" by providing commentary on the history of the development of the concepts of "schooling for all," "arithmetic for all," "basic mathematics for all," "secondary school mathematics for all," "mathematical modelling for all," and "quantitative literacy for all" (or "numeracy for all").

Schooling for All

Table 1.1, which is adapted from the United States' Commissioner of Education's reports for 1905 and 1907, shows proportions of people, in 37 nations and states on 6 continents, who were enrolled in schools around 1900. Entries are suggestive of the acceptance, or otherwise, of formal schooling in the various countries and states that are listed.

Around 1900, many school-age children, in many parts of the world, were not enrolled in a school. In some nations—for example, in regions now known as Bhutan and Brunei Darussalam—there were no formal schools, although there were small local temple- or mosque-related arrangements in which mainly religious knowledge was taught (Horwood & Clements, 2000).

School enrolment was one thing and attendance another. Although Table 1.1 indicates, for example, that percentages of children enrolled in schools in the USA were relatively high, many boys in the north-eastern and mid-western states worked on their parents' farms for most of the year and attended local one-room schoolhouses during winter months only (Cubberley, 1920; Zimmerman, 2009). Furthermore, in the USA in 1900, "only five percent of one-room school graduates proceeded to urban high schools" (Grove, 2000, p. 75).

More generally, at the beginning of the 20th century, school mathematics beyond the most elementary forms of arithmetic was not something that most people, in most countries, had experienced, or would experience (West, Greene, & Brownell, 1930). In many nations, children did not attend school regularly, and often they received no formal instruction in mathematics at all. Thus, it would have made little sense at that time to try to create an international policy on "mathematics for all," even if someone had thought of trying to achieve that goal. Often there was no

Table 1.1
Percentage of Populations Enrolled in Schools, in Various Nations, Around 1900

State or Country	Approx. % of Population Enrolled in Schools	State or Country	Approx. % of Population Enrolled in Schools	State or Country	Approx. % of Population Enrolled in Schools
Ontario (Canada)	21	Sweden	14	Costa Rica	6
USA	21	Belgium	12	Roumania	6
Switzerland	20	Québec (Canada)	12	Mexico	5
Prince Edward Island (Canada)	20	Japan	11	Honduras	5
Victoria (Australia)	20	Cuba	10	Nicaragua	4
England and Wales	18	Cape of Good Hope	10	Portugal	4
Scotland	17	Argentina	9	Servia	4
Ireland	17	Bulgaria	9	Bombay (India)	3
German Empire	17	Italy	8	Russia	3
Norway	15	Greece	7	Egypt	2
The Netherlands	14	Puerto Rico	7	Burma	1
Austria-Hungary	14	Spain	7		
France	14	Uruguay	6		

Note. Data are taken from reports by the U.S. Commissioner of Education (1905, 1907).

school within walking distance of a child's home; often, schools were available but parents did not want their children to attend them; sometimes, teachers capable of teaching forms of mathematics beyond the four operations and simple measurement were not available (Kamens & Benavot, 1991).

Achmad Arifin (1984), an Indonesian mathematician, emphasized the need for mathematics programs to be available in all schools in developing nations like Indonesia. He added, however, that such programs needed to be related to societal needs and cultural expectations: Although mathematical correctness in school textbooks and instruction was important, and something not to be taken for granted, unless there were frequent and positive interactions between schools, mathematics educators, and mathematicians, an acceptable mathematics-for-all agenda would be difficult to develop and implement. But, Arifin argued, if well-organized school mathematics programs could be worked out, then this could have beyond-school local benefits because mathematical solutions might then be applied to social problems.

Was the ideal of schooling for all achieved during the course of the 20th century? The short answer is no. A longer answer would elaborate on the fact that although, during the 20th century, schooling for all became a reality in most nations, in many Asian, African, and Central and South American nations it has never been achieved (Freire, 1996). Nevertheless, in many nations, there was significant progress towards schooling for all.

Take, for example, the nation of Brunei Darussalam, where it was not until 1914 that the first government-supported primary school was opened, and for many years even after that, most Bruneian children—and especially girls—never attended school (Upex, 2000). It was only in the 1950s that the first government secondary school was opened. However, in Brunei Darussalam today, almost all children attend primary and secondary schools, and mathematics is a mandatory part of the curriculum that they study. Likewise, in the Malaysian states of Sarawak and Sabah, which share their borders with Brunei Darussalam, it was not until the late 19th century that government-supported schools were first established, and the value and utility of such schools were not accepted by the majority of the local populations until well into the 20th century (Abu Zahari, 1977).

Progress Towards Schooling for All

The principle of schooling for all was declared, confirmed and reconfirmed by powerful organizations at various times during the 20th century. In 1948, for example, part of the Universal Declaration of Human Rights adopted by the United Nations asserted that "everyone has a right to education." In 1990, a World Conference on "Education for All," held in Jomtien (Thailand), and sponsored by UNESCO (1990), laid down that every person—child, youth or adult—should be able to benefit from educational opportunities designed to meet his or her basic learning needs.

The Jomtien delegates set the goal that by the year 2000 every child in every country should have the chance to complete at least a primary education. However, the goal was not reached, for in 2000 UNESCO estimated that 16% of the world's children did not attend school (Skovsmose, 2006). Of those who attended school, about 20% failed to complete a primary school education (Bruns, Mingat, & Rakotomalala, 2003). A World Education Forum in Dakar, Senegal, in 2000, reaffirmed the Jomtien commitment to schooling for all and added a note about the quality of education that should be expected in schools. The following Dakar goal for universal education (UNESCO, 2000) specifically mentioned numeracy:

> Improve all aspects of the quality of education and ensure excellence of all so that recognized and measurable learning outcomes are achieved by all, especially in literacy, numeracy, and essential life skills. (Quoted in Bruns et al., 2003, p. 2)

Dakar delegates decided that strategies should be devised that would enable all children to receive instruction in elementary numeracy, and that this goal should be achieved early in the 21st century.

Although the Jomtien and Dakar meetings presented an optimistic face, at the beginning of the 21st century universal primary education was far from having been achieved. UNESCO's (1998a) *World Education Report* revealed that in some Southeast Asian nations (e.g., Cambodia, Indonesia and Laos) millions of children never attended primary school. Of those who did, many did not remain at school after Grade 5 (UNESCO, 1998a, 1998b). Towards the end of the century, UNESCO (1998b) estimated that between 100 and 140 million of the world's primary-school-aged children had never attended school.

Around 2005 there were about 860 million illiterate adults in the world, of whom about 60% lived in India, China, Pakistan or Bangladesh. Whereas middle-class families in large cities valued the processes and products of primary school arithmetic, that was not always the case with poor families—especially those in remote regions or in slum areas in large cities. Often parents of poor families found it difficult to comprehend why their children should be required to spend many years in schools being drilled on "useless" facts when the children were needed at home or in the fields (Horwood & Clements, 2000).

Harding (1995) reported that in the 1990s well over 100 million adults aged between 15 and 35 were illiterate, and of these, 62% were women. Immediately before the floods which devastated the island nation of Haiti in 2010, about 65% of school-age children in that nation had never attended school, and the country's adult literacy rate was less than 50%. In Afghanistan, the primary-school completion rate dropped from 22% in 1990 to an estimated 8% in 1999 (Bruns et al., 2003). During the 1990s, Zambia, the Republic of Congo, Albania, Cameroon, Kenya, Madagascar, Qatar, Iraq, the United Arab Emirates, Bahrain, and Venezuela, made little, if any, progress on primary-school completion rates (Bruns et al., 2003; Delors, 1996).

Even today, many children do not get the chance to complete a primary-school education because they never go to school. According to data presented at a United Nations Summit in 2010, about 30 million school-age children in sub-Saharan African nations had never attended school (UNESCO, 2010). In war-ravaged Mekong Basin nations—Cambodia and Laos, for example—many children attend school only spasmodically, if at all. Harding (1995) cited UNESCO data indicating that between 19 and 24 million children aged between 6 and 14 years in India in 1995 had never attended school, and 60% of these were girls. According to Harding (1995), almost half of the children who entered Grade 1 in India dropped out before they reached Grade 5, with the highest drop-out rate occurring immediately after Grade 1 (see also UNESCO, 1998b).

Those who have learned to value formal education can find it difficult to understand why some parents avoid sending their children to school. The educated élite tend to think that schools provide a bridge to a better world. However, those who think that way have something to learn from the following comments by Ben Nebres (2006) on education in the Philippines:

> The first impression of a visiting mathematics educator from countries with a stronger mathematics education tradition in discussions with counterparts from the Philippines might be that of similarities in situations. As solutions begin to be discussed, however, he might begin to realize that beneath these similarities are greater differences. The dominant reality in a country like the Philippines is the scarcity of resources, both human and material. Five or six students have to share a textbook. Many schools lack classrooms, so classrooms meant for 40 children are crammed with 80 students. Or schools have double sessions, in some cases triple sessions, a day. Teachers are poorly trained and have to teach in very difficult environments. (p. 278)

In school education, conditions and contexts matter.

In the first *International Handbook of Mathematics Education*, Stephen Arnold, Christine Shiu and Nerida Ellerton (1996) emphasized the potential of distance education for improving access to mathematics learning, especially, but certainly not only, in geographically remote areas. For several decades, in China, India, Indonesia and Thailand, for example, there have been large enrolments in distance courses in mathematics, especially from adults seeking to qualify for professional appointments (such as teaching). Although this movement has been accelerated by developments in information and communication technologies (hereafter "ICT"), particularly in relation to online education, too often these developments have not given sufficient credence to local cultural and societal factors (Clements & Ellerton, 1996).

Arithmetic for All

The *Abbaco* Tradition in Arithmetic

Modern scholarship has revealed that many aspects of current school mathematics curricula have descended from what has been called the *abbaco* tradition in arithmetic (Ellerton & Clements, 2012; Franci, 1992; Høyrup, 2005; Long, McGee, & Stahl, 2009; Swetz, 1987, 1992; Van Egmond, 1980). It is likely that this tradition emerged from practices associated with so-called *trattati* or *libri d'abbaco*, vernacular Italian pedagogic manuals of commercial mathematics, accounting, and geometry widely used in Italian reckoning schools from the 13th century (Long et al., 2009; Van Egmond, 1980). Sharp increases in international trade and banking in Renaissance Europe prompted city republics to form vernacular schools in which commercial mathematics, accounting and writing were taught to sons of merchants or to apprentices with important responsibilities.

In Western Europe it became common for merchant-class parents to send their sons for two-year courses at these reckoning schools, where they learned commercially-oriented *abbaco* mathematics (see, e.g., Swetz, 1987, for details of an *abbaco* text, the *Treviso Arithmetic*, an Italian arithmetic first printed in 1478). Thus, for instance, in 1522, a book by Adam Ries, the noted German *rechenmeister* (reckoning master), showed how the use of Hindu-Arabic numbers could simplify calculations. The language of the text was German, not Latin, and although the book was probably aimed at male students, Ries thought that all students should learn to use written methods for calculation. According to Karpinski (1925), 40 editions of Ries's arithmetics were published in the vernacular in the 16th century alone, and many more appeared after that.

According to the *abbaco* tradition, children were not expected to begin to study in the reckoning schools until they were about 10 years of age. Then, for several years, boys would prepare cyphering books in which they neatly made entries on a standard sequence of topics (Van Egmond, 1980). They recorded rules, cases, examples and exercises concerned with Hindu-Arabic numeration, the four operations on numbers and on quantities (e.g., distance, area, volume, capacity, time, money, angle), reduction, ratio and proportion tasks—usually associated with what was called the "rule of three"—and, perhaps, vulgar and decimal fractions and simple and compound interest (Ellerton & Clements, 2012; Fink, 1900; Franci, 1992; Høyrup, 2005).

Pestalozzi's Challenge to the *Abbaco* Tradition

In 1801 Johann Heinrich Pestalozzi, the celebrated Swiss educator, set out his ideas on teaching and learning in his book *How Gertrude Teaches Her Children* (Biber, 1831; Holman, 1913). His method was to begin with observation, to pass from observation to what he termed "consciousness," and then from consciousness to speech. This method was applied to the teaching of reading, measuring, drawing, writing, numerical notation and reckoning. Pestalozzi believed that all children could learn arithmetic if they were provided with emotionally secure learning environments in which instruction followed a process of human conceptualization that began with sensation and emphasized sensory learning. He designed object lessons in which children, guided by teachers, examined the shape and numerical aspects (such as quantity and weight) of objects, and named them.

Pestalozzi commented that thousands of children had to go begging in streets, and nobody took care of them. He thought they needed a decent job, but would only get one if they learned to read, to write and to do arithmetic. So, he took in as many "neglected" children as he could, fed and clothed them, got them helping in his gardens, and taught them how to spin and to weave. At the same time he taught them, from about the age of six, to read, to write and to do arithmetic.

He was not always successful, but royalty, politicians, and captains of industry from all over the world came to observe his efforts first hand. This meant that the idea of "arithmetic for all" was put squarely before early 19th-century European society (Holman, 1913), and it would be picked up by other educators who paid attention to the mathematical thinking of children with special needs—including Edouard Séguin (1907), the French/American educator, and Maria Montessori (1912), the Italian educator (Graves, 1914).

Pestalozzi challenged the tradition that children less than 10 years of age should not study arithmetic. For him, the *abbaco* tradition of delaying instruction on quantitative reasoning until children were about 10 years of age had resulted in many adults not being able to survive with dignity in a world in which quantitative relationships were important.

In the USA, Henry Barnard, the first US Commissioner of Education, was influenced by Pestalozzi's views (Barnard, 1859; MacMullen, 1991), as were Horace Mann and the heads of the first US normal schools (Monroe, 1907). Pestalozzi's greatest direct influence on US mathematics education, though, came through the writings of Warren Colburn (1821), whose *An Arithmetic on the Plan of Pestalozzi with Some Improvements* persuaded many American educators that all children, male or female, could and should learn arithmetic from the age of six.

One well-regarded educator, George Emerson (1842), described Colburn's classic text as "the only faultless school book that we have" (p. 442). Slowly, American educators began to accept the premise that all elementary school children could learn arithmetic in meaningful ways (see, e.g., Page, 1859).

Colonialist Assumptions

Primary school children who studied arithmetic at school did not always find it a rewarding experience. In New South Wales in the 1790s, for example, the authorities at Sydney Cove, probably the most remote of all colonial outposts, reflected on how they might civilize the "ignorant and benighted heathen" who were the indigenous inhabitants of the great southern land formerly known as New Holland. Schools were established which had similar curricula to charity schools in England, the emphasis being on reading, writing and arithmetic. These schools failed miserably because the curricula did not interest the small numbers of children who attended. Most Aboriginal families resisted efforts to get their children to attend school (Clements, Grimison, & Ellerton, 1989).

During the 19th century an arithmetic curriculum of the *abbaco* variety was directly exported from England into colonies such as New South Wales, Québec, the Cape of Good Hope, Malaya, and India. Other countries (e.g., Spain and Portugal) also prescribed sequences of *abbaco*-arithmetic topics in national or statewide arithmetic curricula (U.S. Commissioner of Education, 1891).

After 1861, state-supported schools in England and Wales were expected to follow a standard arithmetic curriculum, and this expectation was reinforced by government inspectors and a "payment-by-results" system. This same curriculum and payment-by-results system was translated into British colonies (Clements et al., 1989; Ellerton & Clements, 1988; Griffiths, 1987; Horwood & Clements, 2000; Kamens & Benavot, 1991), where government inspectors went from school to school examining children. All students in the same grade were required to be taught the same material to the same "standard" (Selleck, 1982), but often the syllabi, the imported textbooks, and the modes of assessment for primary school arithmetic directed the teachers', pupils', parents' and indeed society's attention away from what was most needed by the learners (Griffiths, 1987).

In the second half of the 19th century, colonial primary-school arithmetic curricula emphasized the four operations on whole numbers and vulgar and decimal fractions, and assessment was by externally-set written examination papers which asked questions requiring incredibly complicated calculations. In the British colonies, where payment-by-results held sway, teachers did their best to maximize their pupils' examination results because the better their students' results, the higher their pay. The following question, which was set by a primary school inspector in Victoria (Australia) in 1877, was typical:

> If 249.804 bushels of oats last 804.573 horses for 7.4 days, how many horses would 347.147 bushels feed for the same time? (Quoted in Griffiths, 1987, p. 194)

This kind of arithmetic was studied in schools across the world. Even in Vietnam, today, there is a rigid, centralized, compulsory mathematics curriculum which originated in the French and Russian systems of education (Bessot & Comiti, 2006). Although this curriculum can serve a good purpose in big cities, like Hanoi, it tends to be rejected by children and their parents in remote and mountainous provinces like Son La (Horwood & Clements, 2000).

Today's basic elementary arithmetic curriculum might be seen as the remnant of an *abbaco* curriculum that has been in place for many centuries—first in India, then in Arabic nations, then in Europe and then in European colonies. By the beginning of the 20th century, standard textbooks setting out this kind of arithmetic, usually with elementary business applications, were to be found in most of the nations listed in Table 1.1. Often, the language used in textbooks was not the first language for many learners. In the colonies, and former colonies, of European nations, textbooks were typically written in the dominant language of the "mother" country (Bessot & Comiti, 2006; Swetz, 1975; Woolman, 2001).

Colonizing powers believed that one the most important reasons for establishing colonies had been to create new export markets for the motherland, and a perceived need to generate export income by selling school textbooks written in the home country to the colonies was a by-product of that kind of thinking (Swetz, 1975; Woolman, 2001). Thus, for example, it was not until after the USA became an independent nation that most textbooks used in North American schools came to be written by American authors (Karpinski, 1980).

Problems that Arose in Relation to Efforts to Achieve "Arithmetic for All"

A century ago, the ideal of primary education for all was on the way to being achieved in some nations (e.g., in Australia, the USA, the UK, and certain Western European nations). In those nations, legislation had been introduced making attendance at primary schools compulsory for specific age-groups. Those administering the schools specified a curriculum for the primary school that emphasized basic "literacies," and in particular, the three Rs of reading, 'riting and 'rithmetic.

In US elementary schools in the second half of the 19th century, arithmetic gradually became the subject which occupied the most curriculum time (Burnham, 1911; Smith, 1911; West et al., 1930)—sometimes up to half the school day (Buswell, 1930). It also became "the chief source of non-promotion in the elementary school" (Buswell & Judd, 1925, p. 7). Despite the influence of Pestalozzi, Colburn and Séguin, arithmetic teaching increasingly relied on textbooks and pencil-and-paper quizzes and tests (Cajori, 1907; Ellerton & Clements, 2012). Many students struggled to cope with the heavy curriculum and assessment expectations, and young and inexperienced teachers were given daunting face-to-face teaching workloads.

According to William Burnham (1911), Professor of Pedagogy and School Hygiene at Clark University, in the USA, "a considerable ballast of unessential or extraneous material ... crept into [arithmetic] textbooks" (p. 207). Burnham stated that although it was claimed that the study of arithmetic developed "the habit of continuous attention," there were "exaggerated ideas of the efficiency of arithmetic in the cultivation of the mind" (p. 208). Some students had "little ability for work in mathematics," and others were "in special danger of nervous overstrain from work in this subject" (p. 208). Work in arithmetic was, according to Burnham, "a frequent cause of worry and interference with sleep" (p. 208).

Burnham (1911) called for a return to pre-Pestalozzian days when no child younger than 8 years would be asked to learn written arithmetic. After pointing out that many young children were not equal to the task of matching concrete objects through comparison, analysis and abstraction, he described some of the number relationships inherent in the standard elementary mathematics curriculum as "grotesque" (p. 209). Burnham was not the first person, and certainly would not be the last, to introduce the concept of "ability" into discussions on whether all children should study the same mathematics. His comment reflected the fact that the concept of "ability" had been defined and explored experimentally by the French psychologist Alfred Binet (Binet, Pollack, & Brenner, 1969).

Others argued that some children were "not interested" in arithmetic, or were unlikely to need to apply most of its principles in their future lives. By contrast, some educators (e.g., Brooks, 1904) maintained that since the study of mathematics strengthened mental faculties, all children should study it, irrespective of whether they liked it or would need it. These educators tended to claim that since all adults used arithmetic, in one form or another, all children should study it at school (see, e.g., Brooks, 1904; Page, 1859).

The final report of the National Education Society's (1895) "Committee of Fifteen," which investigated elementary school education in the USA, included a well-argued section supporting the idea of "arithmetic for all." But, the Committee also warned readers that requiring young children to do too much arithmetic could be psychologically damaging:

> Your Committee would report that the practice of teaching two lessons daily in arithmetic, one styled "mental" or "intellectual" and the other "written" arithmetic (because its exercises are written out with pencil or pen), is still continued in many schools. By this device the pupil is made to give twice as much time to arithmetic as to any other branch. It is contended by the opponents of this practice, with some show of reason, that two lessons a day in the study of quantity have a tendency to give the mind a bent or set in the direction of thinking quantitatively with a corresponding neglect of the power to observe, and to reflect upon, qualitative and causal aspects. For mathematics does not take account of causes, but only of equality and difference in magnitude. It is further objected that the attempt to secure what is called thoroughness in the branches taught in the elementary schools is often carried too far, in fact, to such an extent as to produce arrested development (a sort of mental paralysis) in the mechanical and formal stages of growth. The mind in that case loses its appetite for higher methods. (p. 56)

At the end of the 19th century, it appears that the quest for "arithmetic for all" had reached the point where many observers, worried about the possibility of psychological damage, maintained that too much curriculum time was being given to arithmetic.

In the USA, and in many other countries, during the 20th century, the idea that mathematics should be "for all" was *not* meant to imply that all children should study exactly the *same* mathematics. The issue of who should decide which mathematics should be studied by which children was often debated (Reisner, 1930; West et al., 1930). During the early decades of the 20th century, universal primary education, and arithmetic for all, became a reality in some nations (e.g., Australia, Canada, France, Germany, Great Britain, and the USA). It should not be assumed, though, that anything like that situation was achieved in many Asian, African, and Central American nations. Thus, for example, in the Philippines the provisions for public schooling were poor, and only a tiny proportion of children attended school for more than 2 years. In 1905 over 700 public schools in the Philippines did not have an official schoolhouse (U.S. Commissioner of Education, 1907).

It was hardly surprising that in some countries during the first decade of the 20th century the amount of curriculum time dedicated to arithmetic, especially mental arithmetic, was reduced. Nevertheless, and despite calls for change, the remnants of the *abbaco* curriculum—numeration, the four operations, vulgar and decimal fractions, percentage, and simple applications especially in the area of business arithmetic—remained in vogue wherever schooling occurred during the first half of the 20th century (Clements & Ellerton, 1996). Drill-and-practice methods were in accord with behaviourist theories advocated by psychologists such as Ivan Pavlov (Todes, 1997), Edward L. Thorndike (1921) and B. F. Skinner (1984).

Although the value of what came to known as "basic arithmetic" would be challenged in the late 1950s and throughout the 1960s, at the time of the so-called New Mathematics (Moon, 1986), it returned to centre stage in the 1970s when mastery learning became important (Block, 1974). Then, in the last two decades of the 20th century, the term *numeracy* would be coined as educators sought to recreate a Pestalozzian approach to primary school mathematics.

Basic Mathematics for All

According to Schmidt, Cogan and McKnight (2010–2011), in the 21st century most nations have education policies which posit that the intended mathematics curriculum should be the same for all primary and lower-secondary schoolchildren. Furthermore, after analyzing international curricula, school participation, and performance data, Schmidt et al. maintained that "even in countries that appear to be creating different tracks, the reality is that basic content is covered by all, with advanced students studying the same topics more deeply" (p. 19). They attributed within- and between-nation performance differences on the mathematics tests used in the Trends in International Mathematics and Science Study (TIMSS) mainly to differences in "individual student ability and effort, combined with differences in teacher quality" (p. 19).

There can be little doubt that there were major between-nation TIMSS performance differences, but the difficult question to answer is why these differences occurred. It is often assumed that quality of teaching was a significant factor. In many developing nations, teachers were poorly trained and those brave enough to face the inevitable professional challenges received inadequate remuneration. Therefore, many capable youngsters were not attracted to teaching.

The availability of suitable teachers of mathematics is a problem for education authorities in many nations, and especially, perhaps, in some Southeast Asian nations and in many African nations (see, e.g., Lao Country Paper, 1996; Tran Si Nguyen & Hoang Van Sit, 1996). According to Torres (1996), it is paradoxical that the global crusade to universalize basic education and to improve its quality coincided "with a notorious and global deterioration of teaching and of teachers' conditions," which in turn led to a "massive exodus of qualified and experienced teachers" (p. 14). It should not be imagined, however, that the quality-of-teaching factor was the only impediment to the achievement of basic mathematics for all. According to Schmidt et al. (2010–2011), the relatively poor mean scores of US students on the International Education Association's SIMS (*Second International Mathematics Study*) and TIMSS tests resulted not only from low-standard intended and implemented curricula, but also from other factors such as poverty, housing, and access to curriculum materials.

It is useful to reflect on implications of the TIMSS results (in the mid-1990s and thereafter) for the mathematics-for-all objective. The TIMSS pencil-and-paper achievement instruments comprised questions that might be described as representing "core mathematics." Most questions related to standard arithmetic or other elementary forms of mathematics. The fact that Asian students living in Confucian-heritage nations did so well on these questions suggests that most students in those nations learned their basic mathematical content well (or, at least, better than students in the nations which had lower mean scores). The implication was that if by mathematics, one means the kinds of basic, or core, mathematical concepts and skills tested on the TIMSS tests (especially those tests taken by 9- and 14-year-old students), then in those Confucian-heritage nations the goal of basic mathematics for all was well on the way to being achieved (Shin, Lee, & Kim, 2009; Stevenson, 1992).

Wong (2006, 2008) distinguished between attitudes to school mathematics in East Asian nations that had the strongest Confucian influence (he specifically mentioned China, Japan, Korea, and Vietnam) and those in East Asian nations which had what he called "a gentler, Buddhist approach to life"—he specifically included Laos and Cambodia, and might also have included Thailand in this latter category. Students from Confucian-heritage nations tended to perform uniformly well on pencil-and-paper mathematics tests, but those from more strongly Buddhist-tradition nations did less well.

Much research aimed at investigating and explaining this phenomenon has been conducted during the period 1980–2012. Variables such as approaches to learning, teaching methods, mathematics teacher education, conceptions of mathematics and mathematical problem solving, roles of memorization and repetitive learning, teacher–student relationships, achievement-orientation, especially in relation to mathematics examinations, collectivism, values, high expectations of parents, and attributions of success to effort, have been studied (e.g., Bishop, 1999; Bishop & Seah, 2008; Clarkson, Bishop, & Seah, 2010; Leu & Wu, 2006; Schmidt, Blömeke, & Tatto, 2011).

Ramakrishnan Menon (2000), who had much experience working as a mathematics educator in both Southeast Asia and North America, asked whether American families would be prepared to change their patterns of living in order that their children might do better at school mathematics. He pointedly asked whether American parents, and students, would, for example, be prepared to forego regular participation in sporting events so that there would be sufficient time for students to participate fully in after-school mathematics tutorial sessions.

Yet, Asian-American students participating in largely decentralized systems of education across the USA perform at levels comparable to those of high-performing students in Asian nations. After making this point, Clarke et al. (2006) commented that "this single illustration suggests that differences on particular measures of mathematical performance are at least as attributable to the cultural affiliation of the students as to the particular school system attended" (p. 354).

From another perspective, there are data gathered within Confucian-heritage nations that suggest that students' attitudes towards mathematics in those nations leave something to be desired. For example, in the TIMSS 1999 Repeat Study, the rank-order correlation for nations' mean-performance scores and mean-attitude scores was an amazing, and challenging, −.74 (Ellerton & Clements, 2010). There is also plenty of evidence, both quantitative and qualitative, that across the wide span of nations and cultures—from Aboriginal Australia, to Germany, to Hong Kong—many school students do not enjoy mathematics lessons (Harris, 1984; Jablonka & Keitel, 2006).

Leung (2006) conjectured that Confucian-heritage students' relatively low self concepts in mathematics, and their apparently negative attitudes to mathematics, may have something to do with their having been brought up "not to be boastful" (p. 40). Also, high expectations for student achievement, coupled with relatively low achievement when compared with peers, may have resulted in "a large number of students classified as failures in their system, and these repeated experiences of a sense of failure may have further reinforced this lack of confidence" (p. 40). Furthermore, Leung argued, "over-confidence may lower students' incentive to learn further and cause them to put very little effort into their studying, and hence result in low achievement" (p. 42). Thus, Leung conjectured, the negative correlation between Confucian-heritage students' confidence in mathematics and their achievement was something to be expected, and was not necessarily a bad thing.

Mathematics Beyond Arithmetic: For a Minority, or for the Majority?

It could be argued that in many nations, school curricula have been colonized so that they serve the needs of an élite minority, whose social backgrounds, living conditions, and education privileges have often made it appear that their children are "gifted" with respect to mathematics, when, in fact, they are not. On the other hand, there are many adults, at all levels of society, who, while agreeing that such privilege should not be allowed to dominate education opportunities, believe that all children should be taught mathematics in classes in which an uncompromised, rigorous traditional mathematics curriculum is followed. Some influential groups—including many mathematicians in colleges and universities—tend to have little patience with educators and others who try to develop more "democratic" mathematics curricula.

Keitel (2000) raised the associated issue in a challenging way:

> As long as we continue to be mainly interested in the few "gifted" students, the negative experiences of the majority will remain. Although negative presuppositions are already brought into mathematics and science classrooms, they are reinforced by contradictory goals and by a one-sided organisation of teaching and learning. But can we still indulge or afford to conceptualise mathematics and science education as a special form of education for a few, and yet still make it compulsory for all? Should we remain reconciled to a situation in which the common learning of many students is hindered or even blocked by anxiety and frustration? (p. 300)

Until 1900, most people who studied mathematics did so mainly in primary, or elementary, schools. During the 20th century, however, there was an explosion in secondary and technical school enrolments, and in the numbers of adults attending mathematics classes. It was inevitable that the question of what mathematics was most needed in such programs would be raised.

Zoltan Dienes and Abstract Mathematics for All

During the 20th century the idea that almost all children could profitably follow the same mathematics curriculum was put forward from time to time. Sometimes this viewpoint was justified by the argument that in the past many had struggled to learn mathematics because of unsatisfactory curricular offerings, or because of poor teaching. Those who argued this way tended to believe that if teachers could be trained to teach rigorous forms of mathematics well, then the subject would be understood by all learners. This was the point of view put forward by Zoltan Paul Dienes (1960, 1964).

Dienes was born in Hungary in 1916 and gained a PhD in mathematics from the University of London in 1939. During the 1960s and 1970s he became known for his advocacy of the view that an uncompromising form of mathematics, which emphasized algebraic structures, could, if well taught, revolutionize the teaching and learning of mathematics for *all* children (Clements & Ellerton, 1996; Dienes, 1960). Dienes, who worked with Jerome Bruner at Harvard University in 1960 and 1961, was given the opportunity to put his mathematics-for-all views to the test when, in the 1960s, he accepted a UNESCO invitation to implement his ideas in the community schools of Papua New Guinea (PNG).

Dienes maintained that *all* primary schoolchildren could be taught to reason logically, and that good teaching could accelerate that process. He himself was a master teacher. Bruner (1963) wrote that those "who had been privileged to see Dr Dienes at work with children" were astonished "to behold how quickly and surely his mathematical embodiments lead to insights that are at first intuitive and concrete and then more disciplined and rigorous" (p. xi). It was a matter of intense interest, then, from both theoretical and practical points of view, whether PNG community school children and teachers would be able to cope with Dienes's program.

To cut a long story short, the implementation of the Dienes program in PNG was unsuccessful. The teachers found it difficult to cope with the mathematics, and the community school children, who were expected to learn mathematics in classrooms in which English was the language of instruction (but was not their first language), made little progress mathematically. Only a few "gifted" children did well. John Hayter (1982) concluded that Dienes's curriculum and its associated teaching approaches were not in harmony with the cultures and levels of mathematical readiness of either the children or the teachers.

Secondary School Mathematics for All

In 1900 the meaning of the term *secondary school* varied from nation to nation. One definition was the type of schooling provided in schools that followed primary or elementary education. Another was the form of education offered to children aged between 12 and about 18 years. A third definition associated secondary education with courses taken in agricultural schools or manual arts colleges (Monroe, 1913).

At the beginning of the 20th century, most school students did not study mathematics beyond primary school. In some nations, though—for example, Austria, France and Germany (Smith, 1911)—a substantial core of students studied mathematics beyond arithmetic in the higher classes of primary schools and in secondary schools. Overall, though, only a small proportion of the world's population had formally studied either algebra or geometry. Less than one percent of people living at the beginning of the 20th century had studied, or were studying, or would ever study calculus (Clements & Ellerton, 1996).

Regulations set out by the Board of Education in England in 1904 specified the following minimum curriculum requirements for "secondary schools":

> The course should provide for instruction in the English language and literature, at least one language other than English, Geography, History, Mathematics, Science and Drawing, with due provision for manual work and physical exercises, and, in a girls' school, for housewifery. Not less than 4½ hours per week must be allotted to English, Geography and History; not less than 3½ to the language, where one is taken, or less than 6 hours, where two are taken; and not less than 7½ hours to Science and Mathematics, of which 3 must be for Science. (quoted in Maclure, 1971, p. 159)

Although in some nations—in France, for example—curricular sequences for post-primary education had been carefully defined by central authorities, at the beginning of the 20th century in most nations there were relatively few schools other than primary schools. Even in France and Germany, less than 10% of those who began their primary schooling stayed on to complete a secondary education, and in the USA, England, and Australia, the percentage of those who stayed on was much less than that (Stanic, 1986).

The Curriculum Revolution at the Beginning of the 20th Century

During the early years of the 20th century, many nations were prepared to reshape their education institutions in ways that would have been entertained by only a vanguard of reformers during the last few decades of the 19th century. The question "what school subjects should be studied by whom?" was a by-product of this international tendency to reform and liberalize curricula to meet the demands of a new century in which more and more students were not only attending school, but were also staying at school for more than just a few years.

In mathematics education, this international tendency towards reform expressed itself in the preparation of a series of national reports on school mathematics published by the International Commission on Mathematical Instruction (ICMI) (Reeve, 1929; Schubring, 2008). Among the national reports to appear were papers on school mathematics in Australia, Austria, Czechoslovakia, France, Germany, Holland, Hungary, Italy, Japan, Russia, the Scandinavian countries, the UK, and the USA. As can be seen from that list, most—but not all—of the reports were prepared in European nations.

Mathematics educators were not the only ones who, between 1900 and 1915, sought to create a better world through improved school curricula, instruction, and learning environments. The following seven threads could be found woven into the international curriculum revolution of the period (Dewey, 1896, 1976; McMurry, 1906):

1. Educators began to think that children of all ages should be introduced to the best literature from their own, and other, lands. Closely allied with this trend were calls for greater attention to history, and to the biographies of leading authors, poets, artists, monarchs, statesmen, pioneers, religious teachers, scholars, and scientists.
2. Many educators, in many nations, believed that nature study warranted a greater place in school curricula. Thus, for example, it was argued that children would benefit from actively investigating insects, birds, animals, and trees within their own environments.
3. The virtues of manual training and agricultural education were increasingly advocated, for all levels of schooling.
4. A demand for systematic physical training of children was increasingly expressed and funds for gymnasia and equipment were requested.
5. The view that more curriculum attention should be devoted to fine arts and music was often put forward.
6. Primary-school educators thought that many kindergarten activities would be better placed in primary grades and, at the other end of the primary school,

algebra, geometry, modern languages, and even Latin might be offered in higher primary grades.
7. Many primary school teachers believed that more curriculum time should be devoted to reading, writing, and speaking, and correspondingly less time to arithmetic.

The idea was that in each school the curriculum should cover the full range of human thought and experience (Dewey, 1976; Kilpatrick, 1992).

With respect to US school mathematics in the early years of the 20th century, Stanic (1986) identified four groups with different attitudes on the issue of what mathematics ought to be taught to students. He referred to these as (a) the *humanists*, who defended traditional rigorous mathematics, but believed that not everyone could learn it effectively; (b) the *developmentalists*, for whom all subject matter was of secondary importance because the focus of the implemented curriculum ought to be on individual learners; (c) the *social efficiency educators*, who believed that the study of secondary-school mathematics should be confined to those who would obviously need mathematics in their future employment; and (d) the *social meliorists*, who wanted all learners to have an equal chance to learn all areas of mathematics, but did not believe that everyone needed to study the same mathematics to the same level. Interestingly, all four groups thought that all students should study arithmetic, but that was the extent of their agreement as far as mathematics education was concerned (Stanic, 1986).

Around 1900, there was much confusion in the Western world with respect to theoretical bases for debate on issues concerning curriculum and pedagogy. Adherents to faculty psychology believed that the muscles of the mind needed to be exercised, and that arithmetic and elementary algebra were well suited to this purpose (Brooks, 1904). Kindergartners, on the other hand, wanted the kind of integrated curriculum that had been advocated by Friedrich Froebel.

In Europe, England, the USA, and Australia, Herbartianist philosophical and pedagogical ideas had gained at least the same level of popularity as Pestalozzian viewpoints (Connell, 1980). Herbartianists believed that the associationist psychology of Johann Friedrich Herbart (1776–1841) held the keys to quality education. But other educators preferred to reject both Pestalozzian and Herbartian thinking, and to adopt a social Darwinian perspective (Spencer, 1861).

At the beginning of the 20th century, Herbartianists greatly influenced education policy in Germany and in numerous other nations, including Australia, Bulgaria, Canada, Chile, Finland, Japan, Mexico, Roumania, Russia, South Africa, Sweden, the UK, and the USA (Connell, 1980). Selleck (1968) stated that whatever reservations commentators might have had about Herbart's theories, his work had a complexity, subtlety and coherence which made it "more impressive than the writings of comparative amateurs such as Froebel or Pestalozzi" (p. 227). During the period 1880–1900, educators from all over the world flocked to Germany to study Herbartianist theory. One of the key tenets in Herbartianist theory was the need for a "correlated" curriculum (McMurry, 1906), and the view that mathematics should not be a separate entity in school curricula was always strong among Herbartianists.

Harold Dunkel's (1970) *Herbart and Herbartianism: An Educational Ghost Story* traced the influence of Herbartianism beyond Germany, and sought to explain why Herbartianism "blazed up like a meteor and meteorlike was extinguished" (p. 4). Even non-Herbartianists, like John Dewey (1896), were influenced by Herbartian theory. Although, in the early years of the 20th century, Herbartianism quickly lost any influence it had on those with the power to change school mathematics, the idea that perhaps mathematics should not be a separate entity in school curricula would never disappear entirely. For example, in 1984 Jan de Lange, a Dutch mathematics educator, debated the proposition that all mathematical instruction should be integrated into other related disciplines. He concluded, controversially, that "the ultimate mathematics for all is no mathematics—as a separate discipline—at all" (de Lange, 1984, p. 71).

The View that All Students Should Take the Same Rigorous School Mathematics Course

At the time of the new math(s) there was considerable optimism, especially among mathematicians, that school mathematics could be brought into line with modern developments in higher mathematics. In the USA, for example, the School Mathematics Study Group (SMSG) prepared texts that were intended to be appropriate for all students in Grades K through 9. SMSG had a panel whose major task was to prepare suitable mathematics texts for non-college-bound students in Grades 7 and 8 ("Introduction to Secondary-School Mathematics") and in Grade 9 ("Introduction to Algebra"). The assumption was that by slowing the texts down—taking 2 years rather than 1—the same material could be learned by less-advanced and less-well-prepared students (Begle, 1971).

Like Dienes's over-optimistic efforts with community school children in Papua New Guinea, SMSG's efforts to get the same mathematics learned by all students were not entirely successful. Although given more time to learn, so-called slow learners were able to learn more mathematics, SMSG's approach did nothing to address the ways in which tracking students in mathematics guarantees that they will have different curriculum experiences. In all these efforts, the mathematics may not have been suited to the needs and interests of many students, and many teachers were undoubtedly not ready to teach what they were expected to teach.

Applicable Mathematics and Modelling for All

Most contemporary mathematics educators believe that the mathematics-for-all concept should extend beyond basic skills and elementary numeracy to include non-trivial mathematical modelling (Westwood, 2008; Wu, 2006). In 19th-century

France and Germany there were numerous attempts to make school mathematics curricula more "practical," on the one hand, or more "pure," on the other (Glas, 1989), and throughout the 20th century many attempts were made, around the world, to give school mathematics a stronger practical orientation. We consider just three of these attempts: Perryism, the Mathematics Applicable Project, and Realistic Mathematics Education.

As a result of ICT developments, mathematics has come to be seen, in recent decades, as providing one of the most potent means for "planning, optimizing, steering, representing and communicating social affairs" (Keitel, 2006a, p. 11). The practical and theoretical power of ICT working in tandem with mathematics has meant that models can be created which maximize the educational power, and practicality, of mathematics.

What is not so well known is that democratization as a result of developments in technology is not a recent phenomenon. The 19th century, for example, witnessed the greater availability of paper—and therefore textbooks, exercise books, and graph paper—lead pencils, pens, slates, blackboards, electricity, mechanical calculating machines, and personal slide rules. These developments made it easier for all people, including those living in remote areas, not only to gain access to, but also to learn and to apply, mathematics (Ellerton & Clements, 2012; Kidwell, Ackerberg-Hastings, & Roberts, 2008).

John Perry, Perryism, and David Eugene Smith

Around 1900, some scholars believed that there was no reason why all children should not learn a form of mathematics that had structures similar to the mathematics employed by professional applied mathematicians. One of the leading proponents of this point of view was John Perry, an Irish-born engineer and applied mathematician. When working as a mathematics instructor in Japan between 1876 and 1879, Perry developed the concept of mathematics laboratories in which problem-based approaches to mathematics teaching and learning took advantage of technological developments associated with graphical analyses and the use of personal slide rules (Brock, 1981). On returning to England, Perry achieved considerable influence over a period of 25 years as Professor of Mathematics and Mechanics at the Royal College of Science, in South Kensington, London. He worked hard to popularize a form of school and college mathematics education that emphasized links between mathematics, the physical sciences, engineering, architecture, and manual work (Perry, 1899, 1902, 1912).

Perry envisaged a form of mathematics education in which students from all levels of society constantly gathered data in laboratory sessions, graphed them, and interpreted results. He urged the regular use of squared paper, with the mathematical concept of function providing the key integrating theme (Brock & Price, 1980; Price, 1986, 1994). Much of his teaching was to artisans studying mathematics in evening classes.

In 1901, Perry delivered a major address on the teaching of mathematics to a meeting of the British Association held in Glasgow (Perry, 1902), and in 1902 his ideas were advocated in the USA by Professor E. H. Moore, of the University of Chicago—in Moore's retiring address as President of the American Mathematical Society (Moore, 1903; Roberts, 1993, 2001). During the early years of the 20th century, *Perryism* became a cause célèbre among many Chicago-based mathematicians and mathematics educators (Moore, 1903, 1906; Roberts, 2001).

Perryism was supported by groups of mathematicians and educators who saw Perry's approaches as bringing life and relevance to school mathematics. But its influence declined rapidly during the period 1910–1920, both in England and in the United States (Young, 1914). David Eugene Smith (1905, 1913) described Perryism as an extreme position held by an influential minority in England and in the midwestern states of the USA. He maintained that educators in the eastern states of the USA, in cities like New York, Philadelphia and Boston, regarded Perry's ideas as an aberration in the education process.

Smith (1905) summarized the attitudes to Perryism of teachers in the eastern states in the following way:

1. Any effort to introduce physical experiments into the classes in mathematics has no support whatever from either the teachers of mathematics or those of physics. ...
2. Any effort to seek the applications of mathematics chiefly in physics or in science generally, has not met with favor, and is not likely to find advocates. The consensus of opinion is that the number of applications of algebra to physics, for example, is exceedingly small, those to business being considerably larger, even though these are not numerous. ...
3. The attempt to have algebra and geometry appear to the pupil as having any considerable application to science or to business aside from a few special propositions will not be made. (p. 207)

Smith added that teachers in the eastern states not only wanted their students "to love mathematics for its own sake" (p. 208), but also wanted them to be well prepared for the rigid system of public examinations. He saw no inherent contradiction in those aims. According to Smith, teachers in the Eastern states would disagree strongly with the proposition that "no equation should be given without a genuine application, that no problem be assigned without a genuine application, that no problem should be assigned without a physical context, that no topic shall be considered save as it bears upon life" (p. 208). Smith added that teachers in the eastern states wanted to develop "pure mathematics" laboratories in which pupil activity took place in the mind rather than with physical apparatus.

Despite the fact that Smith was closely aligned with Felix Klein, the German mathematician who strongly advocated a function approach to algebra (Donoghue, 2008; Tobies, 1989; Young, 1914), he did little to incorporate functions fully into the numerous textbooks he wrote on school mathematics. For example, *Academic Algebra,* which he co-authored (Beman & Smith, 1902), contained no graphs and no discussion of functions, despite the fact that the preface claimed that the book would prepare students for college mathematics.

Smith (1905) said that eastern-states teachers wanted to see how much of the spirit of German gymnasium mathematics, with its pure, as opposed to applied, mathematical tradition (Jahnke, 1983), could be transplanted into American schools. He acknowledged that his position was essentially a conservative one, but emphasized that that was the position favoured by teachers of mathematics in the eastern states. For them, improving school mathematics was synonymous with writing pure mathematics textbooks that would better prepare students for examinations so that the best students would be well prepared to proceed to higher studies in mathematics.

This emphasis on written examinations for sorting students was, at least in Western nations, a 19th-century development (Ellerton & Clements, 2012; Keitel, 2006a, 2006b; Kilpatrick, 1992). Although it could be argued that it provided "bright" children in poor families with a means of accessing higher mathematics, it could also be interpreted as providing an "objective" means by which children from mainly well-to-do families would be selected for higher mathematical studies. In that sense, the bias in favour of an élite based on social class was replaced by a different, but highly related, bias based on "readiness to learn," or "giftedness" (Hansen, 2009). Expressions such as "ability to learn" came to be used to justify procedures whereby the study of higher mathematics was preserved for the few (Clements, 1992).

In London, Perry succeeded in persuading examining boards in England to offer alternative forms of examinations in which problems linked to laboratory methods were asked, and by 1910 these examinations were widely used by schools. After Perry's retirement in 1914, however, the alternative examinations were dropped.

Thus ended, at least for the moment, a promising movement in mathematics education that had begun in Japan, blossomed in the UK, and spread its wings as far as the USA, France, Italy, and Australia (Borel, 1904; Clements, 1992; Giacardi, 2009, 2010; Ruthven, 2008). The preferred approach to school mathematics across the world was, at that time, in the balance—and some thought it was tilting in the direction of applied mathematics by which most, if not all, students would become acquainted with real-world problem solving. Conservative mathematics education leaders, like David Eugene Smith, rejected this view as extreme and supported the growing influence of written examinations and printed textbooks. That influence spread quickly across the world, even to nations like Japan (Ueno, 2006), so that "mathematics for all" was increasingly interpreted as preparing all students for externally-set tests and examinations.

Christopher Ormell and the Mathematics Applicable Project

Like Perryism, the Mathematics Applicable Project, which operated in the UK between 1969 and 1978, was based on the assumption that mathematics education should take advantage of the potential of mathematics to model real-world applications (Ormell, 1972, 1991). Christopher Ormell, who led the Project, held to a concept of "applicable mathematics" that was linked with Wittgenstein's philosophy which accepted the idea that mathematical models can simulate real possibilities and thus enable situations to be explored without physical interventions (Ormell, 1972).

In the 1970s, Ormell managed to get a Mathematics Applicable examination accepted as an alternative to standard examinations for the final years of secondary school mathematics. Mathematical Applicable examinations required students to use modelling to explore possible real-world scenarios. If a candidate got "stuck," then he or she could take a "hint," but that would lead to a reduction in the mark that could be obtained.

Like Perryism, the Mathematics Applicable Project began promisingly but ultimately failed to win general acceptance from schools and examination boards. Yet, its rise and fall in the 1970s suggested that the need to make school mathematics more applicable, rather than be based purely on book learning, had never gone away during the 20th century.

Realistic Mathematics Education

The underlying philosophy of the Realistic Mathematics Education (often denoted by RME) program was originally put forward in the Netherlands in the 1960s and 1970s by Hans Freudenthal (1978). That philosophy has subsequently been refined and popularized by scholars attached to the Freudenthal Institute in the Netherlands (see, e.g., Gravemeijer, Lehrer, van Oers, & Verschaffel, 2002).

According to Gravemeijer (2002), the basic tenets of RME are as follows:

1. Instruction should start by introducing students to experientially-real contexts.
2. Initial informal mathematical activities should encourage students to abstract and construct sophisticated mathematical conceptions.
3. Increasingly, students should create and elaborate symbolic models such as drawings, diagrams, or tables.
4. The instructional method should allow for much interaction, not only between the teacher and students, but also between the students. All present should strive to justify and improve their own solutions and those of others.
5. Mathematical structures and concepts which manifest themselves within intertwining learning strands are to be sought after, identified, and analyzed.

RME emphasizes the concept of an *emergent model*—a *model of* some problem situation evolves so that it becomes a *model for* mathematical reasoning (Gravemeijer & Stephan, 2002). According to Gravemeijer and Doorman (1999), modelling and symbolizing are "an integral part of an organizing activity that aims at coming to grips with a problem situation" (p. 119). Emergent models are grounded in students' understandings of paradigmatic, experientially-real settings, and emerge more clearly as reasoning loses its dependence on situation-specific imagery, so that progressively the models become more reified (Gravemeijer, 2002). The RME curriculum is intended for *all* students (Van den Heuvel-Panhuizen & Becker, 2003). Thus, we have arrived at a point where a mathematics curriculum based on a modelling approach is no longer seen as an extreme, but as something "for all."

There have been numerous other mathematics educators who have assisted "mathematical modelling for all" to move from something regarded as an extreme to something that can form the basis of a worldwide curriculum phenomenon. Two such figures were Henry Pollak, a distinguished applied mathematician with a demonstrated interest in mathematics education (see Karp, 2007; Pollak, 2003), and John Mason, the Canadian-British mathematics educator whose textbook, *Modelling with Mathematics in Primary and Secondary Schools,* written two decades ago (Mason & Davis, 1991), showed how mathematical modelling might become the basis for curricula for primary and secondary schools.

Numeracy for All

Over the past two decades the concept of quantitative literacy, or numeracy, has been championed by governments (Westwood, 2008) and by scholars like Lynn Arthur Steen. *Numeracy*, which Steen (2001) defined as "the quantitative and mathematical requirements for contemporary work and responsible citizenship" (p. xi), is a politically powerful word because of its association with "literacy." For researchers seeking funding for projects, it is also an attractive word, because everyone recognizes the need for citizens to become "quantitatively literate."

Patricia Cline Cohen (2001) wrote that the case for quantitative literacy advanced thinking in at least four ways:

> It identifies various components ... of this style of thinking that together give us a comprehensive and appropriately complex definition of quantitative literacy. It then gives a multitude of examples of actions and behaviors ... occurring in daily life that call for this kind of thinking, from the simple to the esoteric. It next distinguishes the bundle of skills that constitute quantitative literacy as an academic subject. And finally, ... [it] makes clear that quantitative literacy and mathematics are really two quite different things. (p. 23)

But, as White and Southwell (2008) observed, different definitions of numeracy abound. Indeed, Clements (2008) argued that it is not a unidimensional, culture-free, or context-free concept, and wondered whether it should continue to be employed as a catch-all term. Nevertheless, the ill-defined concept of numeracy has provided a convenient vehicle in which governments can move to improve the quantitative literacy of citizens without having to embrace, strongly, the mathematical side of quantitative literacy. It allows the call for "mathematics for all" to be modified to "numeracy for all" (Steen, 2001).

It is an irony that in the USA between about 1910 and 1955, during what became known as the "progressive era," many elementary school mathematics textbooks (e.g., Buswell, Brownell, & Sauble, 1955; Stone, 1931) featured what might be regarded as a quantitative literacy approach to school mathematics. Attractively illustrated story shells relating to everyday situations were presented, and a series of quantitative questions relating to the story shells were asked. That approach gave way to the new math(s) period, when mathematically dense questions, not obviously related to everyday life, became the order of the day. Then followed the back-to-the-basics period of the 1970s, when teachers once again focussed on helping their students get correct answers, but often mainly for context-free, mechanical tasks.

The concept of numeracy has gradually been extended beyond purely arithmetical skills to embrace not only other elementary mathematical skills but also affective characteristics such as attitudes and confidence (Westwood, 2008). Whereas TIMSS pencil-and-paper performance instruments have been widely seen as testing basic skills, instruments developed for the Programme for International Student Assessment (PISA), which is conducted by the Organisation for Economic Co-operation and Development (OECD), are thought to be more concerned with numeracy or quantitative literacy skills. PISA aims at measuring how well students can choose and apply elementary mathematical—as opposed to strictly numerical—concepts in everyday situations (OECD, 2004; Pinxten & François, 2011).

Concluding Comments

At the beginning of the 20th century, some educators still believed that the formal study of mathematics did not stimulate higher-order thinking as much as did the study of the ancient classics, especially Latin. That kind of thinking almost disappeared during the 20th century, as many educators debated the role that mathematics should have in the curricula of primary, secondary, technical and adult forms of education. Today, most people think that everyone needs to learn mathematics in some form. The question remains: Which form?

In this chapter we have surveyed what might be called progress with respect to the development of meaning for "mathematics for all." Our analysis has revealed, however, that there is still much fuzziness about what *progress* might mean so far as "mathematics for all" is concerned. Even more serious than that, perhaps, is the development by which mathematics education is now in danger of being colonized by education administrators, politicians, literacy experts, and psychometricians, who—it could be argued—wish to exploit the term *numeracy* for their own purposes.

Given the amazingly rapid development of sophisticated online facilities for teaching and learning, there are some who believe that the world is poised to take steps that will implement mathematically and educationally sound mathematics-for-all programs (Arnold et al., 1996). However, the challenge is for mathematics educators and mathematicians to lift their eyes to the hills and be willing to work collaboratively with those who have the cultural backgrounds and professional and technical skills to give "mathematics for all" a chance. In such efforts, it will be important that "outside experts" do not seek to impose their "solutions" on the "developing" world: The expertise needs to be developed *within* groups of *local* educators (Clements & Ellerton, 1996), for only those groups are in a position to energize, monitor, and sustain progress.

It is important that mathematics educators take the high ground in their efforts to make mathematics something which everyone can study with benefit. That said, we need to make sure that the kind of mathematics put forward as suitable for everyone *is* indeed suitable for everyone. Lessons from the new math(s) era, and from experiments like the failed Dienes program in Papua New Guinea, need to be noted. Mathematics is a cultural phenomenon (Bishop, 1988), and "mathematics for all" should generate forms of mathematics that arise out of, and are obviously related to, the needs of learners and the societies in which they live.

References

Abu Zahari bin Abu Bakar (1977). *The development of education in West Malaysia during pre-independence and post-independent periods* (Unpublished M.Ed thesis). University of Sheffield.
Arifin, A. (1984). Universal mathematics education and its conditions in social interaction. In P. Damerov, M. E. Dunkley, B. F. Nebres, & B. Werry (Eds.), *Mathematics for all: Problems of cultural sensitivity and unequal distribution of mathematical education and future prospective on mathematics teaching for the majority* (pp. 36–39). Paris, France: UNESCO.
Arnold, S., Shiu, C., & Ellerton, N. F. (1996). Critical issues in the distance teaching of mathematics and mathematics education. In A. J. Bishop, K. Clements, C. Keitel, J. Kilpatrick, & C. Laborde (Eds.), *International handbook of mathematics education* (pp. 701–753). Dordrecht, The Netherlands: Kluwer Academic Publishers.
Barnard, H. (1859). *Pestalozzi and Pestalozzianism: Life, educational principles, and methods of John Henry Pestalozzi, with biographical sketches of several of his assistants and disciples.* New York, NY: F. C. Brownell.
Begle, E. G. (1971). Time devoted to instruction and student achievement. *Educational Studies in Mathematics, 4*, 220–224.
Beman, W. W., & Smith, D. E. (1902). *Academic algebra*. Boston, MA: Ginn.
Benjamin, H. R. W. (1939). *The saber-tooth curriculum*. New York, NY: McGraw-Hill.
Bessot, A., & Comiti, C. (2006). Some comparative studies between French and Vietnamese curricula. In F. K. S. Leung, K.-D. Graf, & F. Lopez-Real (Eds.), *Mathematics education in different cultural traditions—A comparative study of East Asia and the West: The 13th ICMI Study* (pp. 159–179). New York, NY: Springer.
Biber, G. E. (1831). *Henry Pestalozzi and his plan of education*. London, UK: John Souter.
Binet, A., Pollack, R. H., & Brenner, M. W. (Eds.). (1969). *The experimental psychology of Alfred Binet: Selected papers*. New York, NY: Springer.
Bishop, A. J. (1988). *Mathematical enculturation: A cultural perspective on mathematics education*. Dordrecht, The Netherlands: Kluwer Academic Publishers.
Bishop, A. J. (1999). Critical challenges in researching social, cultural and linguistic issues in science, mathematics and technology education. In M. A. Clements & Y. P. Leong (Eds.), *Cultural and language aspects of science, mathematics and technical education* (pp. 3–15). Gadong, Brunei Darussalam: Universiti Brunei Darussalam.
Bishop, A. J., & Seah, W. T. (2008). Educating values: Possibilities and challenges through mathematics teaching. In M.-H. Chau & T. Kerry (Eds.), *International perspectives on education* (pp. 118–138). London, UK: Continuum.
Block, J. H. (1974). *Schools, society and mastery learning*. New York, NY: Holt, Rinehart & Winston.
Borel, É. (1904). Les exercices pratiques de mathématiques dans l'enseignement secondaire. *Revue Générale des Sciences Pures et Appliqués, 15*, 431–440.
Brock, W. H. (1981). The Japanese connection: Engineering in Tokyo, London, and Glasgow at the end of the nineteenth century. *The British Journal of the History of Science, 14*(3), 227–243.
Brock, W. H., & Price, M. H. (1980). Squared paper in the nineteenth century: Instrument of science and engineering, and symbol of reform in mathematical education. *Educational Studies in Mathematics, 11*, 365–381.

Brooks, E. (1904). *The philosophy of arithmetic as developed from the three fundamental processes of synthesis, analysis and comparison, containing also a history of arithmetic.* Philadelphia, PA: Normal Publishing.

Bruner, J. (1963). Preface. In Z. P. Dienes. *An experimental study of mathematics learning* (pp. xi–xiii). London, UK: Hutchinson.

Bruns, B., Mingat, A., & Rakotomalala, R. (2003). *Achieving universal primary education by 2015: A chance for every child.* Washington, DC: World Bank.

Burnham, W. H. (1911). Hygiene of arithmetic. In P. Monroe (Ed.), *A cyclopaedia of education* (Vol. 1, pp. 207–209). New York, NY: Macmillan.

Buswell, G. T. (1930). A critical survey of previous research in arithmetic. In G. M. Whipple (Ed.), *Report of the Society's committee on arithmetic* (29th Yearbook of the National Society for the Study of Education, pp. 445–470). Chicago, IL: University of Chicago Press.

Buswell, G. T., Brownell, W. A., & Sauble, I. (1955). *Arithmetic we need.* Boston, MA: Ginn.

Buswell, G. T., & Judd, C. H. (1925). *Summary of educational investigations relating to arithmetic* (Supplementary Educational Monographs, No. 27). Chicago, IL: University of Chicago Press.

Cajori, F. (1907). *A history of elementary mathematics with hints on methods of teaching.* New York, NY: Macmillan.

Clarke, D. J., Shimizu, Y., Ulep, S. A., Gallos, F. L., Sethole, G., Adler, J., & Vithal, R. (2006). Cultural diversity and the learner's perspective: Attending to voice and context. In F. K. S. Leung, K-D Graf, & F. Lopez-Real (Eds.), *Mathematics education in different cultural traditions—A comparative study of East Asia and the West: The 13th ICMI Study* (pp. 353–380). New York, NY: Springer.

Clarkson, P., Bishop, & Seah, W. T. (2010). Mathematics education and student values: The cultivation of mathematical wellbeing. In T. Lovat, R. Toomey, & N. Clement (Eds.), *International research handbook on values education and student wellbeing* (pp. 111–135). Dordrecht, The Netherlands: Springer.

Clements, M. A. (1992). *Mathematics for the minority: Some historical perspectives on school mathematics in Victoria.* Geelong, Australia: Deakin University.

Clements, M. A. (2008). Australasian mathematics education research 2004–2007: An overview. In H. Forgasz, A. Barkatsas, A. J. Bishop, B. Clarke, S. Keast, W. T., Seah, P. Sullivan, & S. Willis (Eds.), *Research in mathematics education in Australasia 2004–2007* (pp. 337–356). Rotterdam, The Netherlands: Sense Publishers.

Clements, M. A., & Ellerton, N. F. (1996). *Mathematics education research: Past, present and future.* Bangkok, Thailand: UNESCO.

Clements, M. A., Grimison, L., & Ellerton, N. F. (1989). Colonialism and school mathematics in Australia, 1788–1988. In N. F. Ellerton & M. A. Clements (Eds.), *School mathematics: The challenge to change* (pp. 50–78). Geelong, Australia: Deakin University.

Cohen, P. C. (2001). The emergence of numeracy. In L. A. Steen (Ed.), *Mathematics and democracy: The case for quantitative literacy* (pp. 23–30). Northfield, MN: National Council on Education and the Disciplines.

Colburn, W. (1821). *An arithmetic on the plan of Pestalozzi with some improvements.* Boston, MA: Cummings & Hilliard.

Connell, W. (1980). *A history of education in the twentieth century world.* Canberra, Australia: Curriculum Development Centre.

Cubberley, E. P. (1920). *The history of education.* Boston, MA: Houghton Mifflin.

Damerow, P., Dunkley, M. E., Nebres, B. F., & Werry, B. (Eds.). (1984). *Mathematics for all: Problems of cultural sensitivity and unequal distribution of mathematical education and future prospective on mathematics teaching for the majority.* Paris, France: UNESCO.

Damerow, P., & Westbury, I. (1984). Conclusions drawn from the experiences of the new mathematics movement. In P. Damerov, M. E. Dunkley, B. F. Nebres, & B. Werry (Eds.), *Mathematics for all: Problems of cultural sensitivity and unequal distribution of mathematical education and future prospective on mathematics teaching for the majority* (pp. 22–25). Paris, France: UNESCO.

De Lange, J. (1984). Mathematics for all is no mathematics at all. In P. Damerov, M. E. Dunkley, B. F. Nebres, & B. Werry (Eds.), *Mathematics for all: Problems of cultural sensitivity and*

unequal distribution of mathematical education and future prospective on mathematics teaching for the majority (pp. 66–71). Paris, France: UNESCO.
Delors, J. (1996). *Learning: The treasure within: Report to UNESCO of the International Commission on Education for the twenty-first century*. Canberra, Australia: The Australian National Commission for UNESCO.
Dewey, J. (1896). Interest as related to (the training of the) will. *Second Supplement to the Herbart Year Book for 1895* (pp. 209–255). Bloomington, IL: National Herbart Society.
Dewey, J. (1976). The educational situation. In J. A. Boydston (Ed.), *The middle works, 1899–1924* (Vol. 1, pp. 257–313). Carbondale, IL: Southern Illinois University Press.
Dienes, Z. P. (1960). *Building up mathematics*. London, UK: Hutchinson.
Dienes, Z. P. (1964). *The power of mathematics*. London, UK: Hutchinson.
Donoghue, E. F. (2008). David Eugene Smith and the founding of the International Commission on the Teaching of Mathematics. *International Journal for the History of Mathematics Education, 3*(2), 35–46.
Dunkel, H. B. (1970). *Herbart and Herbartianism: An educational ghost story*. Chicago, IL: University of Chicago Press.
Ellerton, N. F., & Clements, M. A. (1988). Reshaping school mathematics in Australia, 1788–1988. *Australian Journal of Education, 32*(3), 387–405.
Ellerton, N. F., & Clements, M. A. (2010). Hidden weaknesses in mathematics education settings. In I. Hideki (Ed.), *Development of mathematical literacy in the lifelong learning society* (pp. 286–301). Hiroshima, Japan: Japan Society for the Promotion of Science.
Ellerton, N. F., & Clements, M. A. (2012). *Rewriting the history of North American school mathematics 1607–1861*. New York, NY: Springer.
Emerson, G. B. (1842). The schoolmaster. In A. Potter & G. B. Emerson (Eds.), *The school and the schoolmaster* (pp. 265–552). New York, NY: Harper.
Fink, K. (1900). *A brief history of mathematics: An authorized translation of Dr. Karl Fink* (W. W. Beman & D. E. Smith, Trans.). Chicago, IL: Open Court.
Franci, R. (1992). Le matematiche dell'abaco nel quattrocento. In *Contributi alla storia delle matematiche: Scritti in onore di Gino Arrighi* (pp. 53–74). Modena, Italy: Mucchi.
Freire, P. (1996). *Letters to Cristina: Reflections on my life and work*. London, UK: Routledge.
Freudenthal, H. (1978). *Weeding and sowing: Preface to a science of mathematical education*. Dordrecht, The Netherlands: D. Reidel.
Gates, P., & Vistro-Yu, C. (2003). Is mathematics for all? In A. J. Bishop, M. A. Clements, C. Keitel, J. Kilpatrick, & F. Leung (Eds.), *Second international handbook of mathematics education* (pp. 31–73). Dordrecht, The Netherlands: Kluwer Academic Publishers.
Giacardi, L. (2009). The school as "laboratory": Giovanni Vailati and the project to reform mathematics teaching in Italy. *International Journal for the History of Mathematics Education, 4*(1), 5–28.
Giacardi, L. (2010). The Italian school of algebraic geometry and mathematics teaching: Methods, teacher training, and curricular reforms in the early twentieth century. *International Journal for the History of Mathematics Education, 5*(1), 1–19.
Glas, E. (1989). Social determinants of mathematical change: The Ecole Polytechnique 1794–1809. In C. Keitel (Ed.), *Mathematics, education and society* (pp. 47–49). Paris, France: UNESCO.
Gravemeijer, K. (2002). Preamble: From models to modeling. In K. Gravemeijer, R. Lehrer, B. van Oers, & L. Verschaffel (Eds.), *Symbolizing, modeling and tool use in mathematics education* (pp. 7–22). Dordrecht, The Netherlands: Kluwer Academic Publishers.
Gravemeijer, K., & Doorman, M. (1999). Context problems in realistic mathematics education: A calculus course as an example. *Educational Studies in Mathematics, 39*(1–3), 111–129.
Gravemeijer, K., Lehrer, R., van Oers, B., & Verschaffel, L. (Eds.). (2002). *Symbolizing, modeling and tool use in mathematics education* (pp. 7–22). Dordrecht, The Netherlands: Kluwer Academic Publishers.
Gravemeijer, K., & Stephan, M. (2002). Emergent models as an instructional heuristic. In K. Gravemeijer, R. Lehrer, B. van Oers, & L. Verschaffel (Eds.), *Symbolizing, modeling and tool use in mathematics education* (pp. 145–169). Dordrecht, The Netherlands: Kluwer Academic Publishers.

Graves, F. P. (1914). Is the Montessori method a fad? *Popular Science Monthly, 84*, 609–614.

Griffiths, R. (1987). A tale of 804.573 horses: Arithmetic teaching in Victoria 1860–1914. *Educational Studies in Mathematics, 20*(1), 79–96.

Grove, M. J. (2000). *Legacy of one-room schools.* Morgantown, PA: Masthof Press.

Hansen, H. C. (2009). The century when mathematics became for all. *International Journal for the History of Mathematics Education, 4*(2), 19–40.

Harding, D. (1995). Teacher empowerment: An analysis of the Indian experience. In UNESCO Asia-Pacific Centre of Educational Innovation for Development (Ed.), *Partnerships in teacher education for a new Asia* (pp. 1–12). Bangkok, Thailand: Author.

Harris, P. (1984). The relevance of primary school mathematics in tribal Aboriginal communities. In P. Damerov, M. E. Dunkley, B. F. Nebres, & B. Werry (Eds.), *Mathematics for all: Problems of cultural sensitivity and unequal distribution of mathematical education and future prospective on mathematics teaching for the majority* (pp. 96–100). Paris, France: UNESCO.

Hayter, J. (1982). *The high school mathematics curriculum in Papua New Guinea: An evaluation.* Port Moresby, Papua New Guinea: University of Papua New Guinea.

Holman, H. (1913). Pestalozzi, Johann Heinrich. In P. Monroe (Ed.), *Encyclopedia of education* (Vol. 4, pp. 655–659). New York, NY: Macmillan.

Horwood, J., & Clements, M. A. (2000). A mirror to the past: Mathematics education in Australia and Southeast Asia: Past, present and future. In M. A. Clements, H. Tairab, & K. Y. Wong (Eds.), *Science, mathematics, and technical education in the 20th and 21st centuries* (pp. 133–143). Gadong, Brunei Darussalam: Universiti Brunei Darussalam.

Høyrup, J. (2005). Leonardo Fibonacci and abbaco culture: A proposal to invert the roles. *Revue d'Histoire des Mathématiques, 11*, 23–56.

Jablonka, E., & Keitel, C. (2006). Values and classroom interaction: Students' struggle for sense making. In F. K. S. Leung, K.-D. Graf, & F. Lopez-Real (Eds.), *Mathematics education in different cultural traditions—A comparative study of East Asia and the West: The 13th ICMI Study* (pp. 495–521). New York, NY: Springer.

Jahnke, H. N. (1983). Origins of school mathematics in early nineteenth century Germany. *Journal of Curriculum Studies, 18*(1), 85–94.

Kamens, D. H., & Benavot, A. (1991). Élite knowledge for the masses: The origins and spread of mathematics and science education in national curricula. *American Journal of Education, 99*(2), 137–180.

Karp, A. (2007). Interview with Henry Pollak. *International Journal for the History of Mathematics Education, 2*(2), 67–89.

Karpinski, L. C. (1925). *The history of arithmetic.* Chicago, IL: Rand McNally.

Karpinski, L. C. (1980). *Bibliography of mathematical works printed in America through 1850* (2nd ed.). Ann Arbor, MI: University of Michigan Press.

Keitel, C. (2000). Mathematics and science curricula: For whom, and for whose benefit? (or, the crisis of mathematics and science education as part of general education). In M. A. Clements, H. Tairab, & K. Y. Wong (Eds.), *Science, mathematics, and technical education in the 20th and 21st centuries* (pp. 299–316). Gadong, Brunei Darussalam: Universiti Brunei Darussalam.

Keitel, C. (2006a). Mathematics, knowledge, and political power. In J. Maasz & W. Schloegmann (Eds.), *New mathematics education research and practice* (pp. 11–23). Rotterdam, The Netherlands: Sense Publishers.

Keitel, C. (2006b). Perceptions of mathematics and mathematics education in the course of history— A review of Western perspectives. In F. K. S. Leung, K.-D. Graf, & F. Lopez-Real (Eds.), *Mathematics education in different cultural traditions—A comparative study of East Asia and the West: The 13th ICMI Study* (pp. 81–94). New York, NY: Springer.

Kidwell, A. K., Ackerberg-Hastings, A., & Roberts, D. L. (2008). *Tools of American mathematics teaching, 1800–2000.* Washington, DC: Smithsonian Institution and Baltimore, MD: Johns Hopkins University Press.

Kilpatrick, J. (1992). A history of research in mathematics education. In D. A. Grouws (Ed.), *Handbook of research on mathematics teaching and learning* (pp. 3–38). New York, NY: Macmillan.

Krygowska, A. Z. (1984). Mathematics education at the first level in post-elementary and secondary schools. In International Commission on Mathematical Instruction (Ed.), *New trends in mathematics teaching* (Vol. 4, pp. 31–46). Paris, France: UNESCO.
Lao Country Paper. (1996). Networks for the teacher upgrading programme. In I. Birch (Ed.), *Partnerships in teacher development for a new Asia* (pp. 99–106). Bangkok, Thailand: Asia-Pacific Centre of Educational Innovation for Development (UNESCO).
Leu, Y.-C., & Wu, C.-J. (2006). The origins of pupils' awareness of teachers' mathematics pedagogical values: Confucianism and Buddhism-driven. In F. K. S. Leung, K.-D. Graf, & F. Lopez-Real (Eds.), *Mathematics education in different cultural traditions—A comparative study of East Asia and the West: The 13th ICMI Study* (pp. 139–152). New York, NY: Springer.
Leung, K. S. F. (2006). Education in East Asia and the West: Does culture matter? In F. K. S. Leung, K.-D. Graf, & F. Lopez-Real (Eds.), *Mathematics education in different cultural traditions—A comparative study of East Asia and the West: The 13th ICMI Study* (pp. 21–46). New York, NY: Springer.
Li, Y., & Ginsburg, M. B. (2006). Classification and framing of mathematical knowledge in Hong Kong, Mainland China, Singapore, and the United States. In F. K. S. Leung, K.-D. Graf, & F. Lopez-Real (Eds.), *Mathematics education in different cultural traditions—A comparative study of East Asia and the West: The 13th ICMI Study* (pp. 195–211). New York, NY: Springer.
Long, P. O., McGee, D., & Stahl, A. M. (2009). *The book of Michael of Rhodes: A 15th century maritime manuscript*. Cambridge, MA: MIT Press.
Maclure, J. S. (Ed.). (1971). *Educational documents: England and Wales 1816–1968*. London, UK: Methuen.
MacMullen, E. N. (1991). *In the cause of true education reform: Henry Barnard and nineteenth century school reform*. New Haven, CT: Yale University Press.
Mason, J., & Davis, J. (1991). *Modelling with mathematics in primary and secondary schools*. Geelong, Australia: Deakin University.
McMurry, C. A. (1906). *Course of study in the eight grades*. New York, NY: Macmillan.
Menon, R. (2000). Should the United States emulate Singapore's education system to achieve Singapore's success in the TIMSS? *Mathematics Teaching in the Middle School, 5*(6), 345–347.
Monroe, P. (1913). Secondary education. In P. Monroe (Ed.), *A cyclopedia of education* (Vol. 5, pp. 312–313). New York, NY: Macmillan.
Monroe, W. S. (1907). *History of the Pestalozzian movement in the United States, with nine portraits and a bibliography*. Syracuse, NY: C. W. Bardeen.
Moon, B. (1986). *The "New Maths" curriculum controversy*. Barcombe, UK: Falmer.
Moore, E. H. (1903). On the foundations of mathematics. *Science, 17*(428), 401–416.
Moore, E. H. (1906). The cross-section paper as a mathematical instrument. *School Review, 14*, 317–338.
Montessori, M. (1912). *The Montessori method* (A. E. George, Trans.). New York, NY: Frederick Stokes.
National Education Association. (1895). *Report of the Committee of Fifteen on Elementary Education, with the reports of the sub-committees: on the training of teachers; on the correlation of studies in elementary education; on the organization of city school systems*. New York, NY: American Book Company.
Nebres, B. F. (1984). The problem of universal mathematics education in developing countries. In P. Damerov, M. E. Dunkley, B. F. Nebres, & B. Werry (Eds.), *Mathematics for all: Problems of cultural sensitivity and unequal distribution of mathematical education and future prospective on mathematics teaching for the majority* (pp. 18–21). Paris, France: UNESCO.
Nebres, B. F. (2006). Philippine perspective on the ICMI comparative study. In F. K. S. Leung, K.-D. Graf, & F. Lopez-Real (Eds.), *Mathematics education in different cultural traditions—A comparative study of East Asia and the West: The 13th ICMI Study* (pp. 277–289). New York, NY: Springer.
Organisation for Economic Co-operation and Development (OECD). (2004). *Learning for tomorrow's world—First results from PISA 2003*. Paris, France: OECD.

Ormell, C. P. (1972). Mathematics, applicable versus pure-and-applied. *International Journal of Mathematical Education in Science and Technology, 3*, 125–131.

Ormell, C. P. (1991). How ordinary meaning underpins the meaning of mathematics. *For the Learning of Mathematics, 11*, 25–30.

Page, D. P. (1859). *Theory and practice of teaching: The motives and methods of good schoolkeeping* (25th ed.). New York, NY: A. S. Barnes.

Perry, J. (1899). *Practical mathematics*. London, UK: Department of Science and Art of the Committee of Council on Education.

Perry, J. (Ed.). (1902). *Discussion on the teaching of mathematics*. London, UK: Macmillan.

Perry, J. (1912). Practical mathematics. *Nature, 90*, 34–35.

Pinxten, R., & François, K. (2011). Politics in an Indian canyon? Some thoughts on the implications of ethnomathematics. *Educational Studies in Mathematics, 78*(2), 261–273.

Pollak, H. O. (2003). A history of the teaching of modeling. In G. M. A. Stanic & J. Kilpatrick (Eds.), *A history of school mathematics* (pp. 647–672). Reston, VA: National Council of Teachers of Mathematics.

Price, M. H. (1986). The Perry movement in school mathematics. In M. H. Price (Ed.), *The development of the secondary curriculum* (pp. 103–155). London, UK: Croom Helm.

Price, M. H. (1994). *Mathematics for the multitude?* Leicester, UK: Mathematical Association.

Reeve, W. D. (1929). *Significant changes and trends in the teaching of mathematics throughout the world since 1910* (Fourth Yearbook of the National Council of Teachers of Mathematics). New York, NY: Teachers College, Columbia University, Bureau of Publications.

Reisner, E. H. (1930). *Evolution of the common school*. New York, NY: Macmillan.

Roberts, D. L. (1993). *Mathematics and pedagogy: Professional mathematicians and American educational reform, 1893–1923*. PhD dissertation, Johns Hopkins University.

Roberts, D. L. (2001). E. H. Moore's early twentieth-century program for reform in mathematics education. *American Mathematical Monthly, 108*(8), 689–696.

Robinson, S. P. (1996). With numeracy for all: Urban schools and the reform of mathematics education. *Urban Education, 30*, 379–394.

Ruthven, K. (2008). Mathematical technologies as a vehicle for intuition and experiment: A foundational theme of the International Commission on Mathematical Instruction, and a continuing preoccupation. *International Journal for the History of Mathematics Education, 3*(2), 91–102.

Schmidt, W. H., Blömeke, S., & Tatto, M. T. (2011). *Teacher education matters: A study of middle school mathematics teacher preparation in six countries*. New York, NY: Teachers College Press.

Schmidt, W. H., Cogan, L. S., & McKnight, C. C. (2010–2011, Winter) Equality of educational opportunity: Myth or reality in U. S. schooling? *American Educator,* 12–19.

Schubring, G. (2008). The origins and the early history of ICMI. *International Journal for the History of Mathematics, 3*(2), 3–33.

Séguin, E. (1907). *Idiocy and its treatment by the physiological method*. New York, NY: A. M. Kelley.

Selleck, R. J. W. (1968). *The new education, 1870–1914*. London, UK: Pitman.

Selleck, R. J. W. (1982). *Frank Tate: A biography*. Melbourne, Australia: Melbourne University Press.

Shin, J., Lee, H., & Kim, Y. (2009). Student and school factors affecting mathematics achievement international comparisons between Korea, Japan and the USA. *School Psychology International, 30*(5), 520–537.

Skinner, B. F. (1984). The evolution of behavior. *Journal of Experimental Analysis of Behavior, 41*, 217–221.

Skovsmose, O. (2006). Mathematics, culture and society. In J. Maasz & W. Schloegmann (Eds.), *New mathematics education research and practice* (pp. 6–10). Rotterdam, The Netherlands: Sense Publishers.

Smith, D. E. (1905). Movements in mathematical teaching. *School Science and Mathematics, 5*(3), 205–209.

Smith, D. E. (1911). Algebra. In P. Monroe (Ed.), *A cyclopaedia of education* (Vol. 1, pp. 90–92). New York, NY: Macmillan.

Smith, D. E. (1913). Laboratory methods in mathematics. In P. Monroe (Ed.), *A cyclopaedia of education* (Vol. 4, pp. 159–160). New York, NY: Macmillan.

Spencer, H. (1861). *Education: Intellectual, moral and physical*. London, UK: Williams & Norgate.

Stanic, G. M. A. (1986). The growing crisis in mathematics education in the early twentieth century. *Journal for Research in Mathematics Education, 17*, 190–205.

Steen, L. A. (Ed.). (2001). *Mathematics and democracy: The case for quantitative literacy*. Northfield, MN: National Council on Education and the Disciplines.

Stevenson, H. W. (1992). Learning from Asian schools. *Scientific American, 267*, 70–75.

Stone, J. C. (1931). *The Stone arithmetic primary*. Chicago, IL: Benj H. Sanborn.

Swetz, F. J. (1987). *Capitalism and arithmetic: The new math of the 15th century*. La Salle, IL: Open Court.

Swetz, F. J. (1992). Fifteenth and sixteenth century arithmetic texts: What can we learn from them? *Science & Education, 1*, 365–378.

Swetz, F. J. (1975). Mathematics curricular reform in less-developed nations: An issue of concern. *The Journal of Developing Areas, 10*, 8–14.

Thorndike, E. L. (1921). *The new methods in arithmetic*. Chicago, IL: Rand McNally.

Tobies, R. (1989). The activities of Felix Klein in the Teaching Commission of the 2nd chamber of the Prussian parliament. In C. Keitel (Ed.), *Mathematics, education and society* (pp. 50–52). Paris, France: UNESCO.

Todes, D. P. (1997). Pavlov's physiological factory. *Isis, 88*, 205–246.

Torres, R. M. (1996). Teacher education: From rhetoric to action. In I. Birch (Ed.), *Partnerships in teacher development for a new Asia* (pp. 14–55). Bangkok, Thailand: Asia-Pacific Centre of Educational Innovation for development (UNESCO).

Tran Si Nguyen, & Hoang Van Sit (1996). Retraining and fostering teachers in the multi-grade class project. In I. Birch (Ed.), *Partnerships in teacher development for a new Asia* (pp. 94–99). Bangkok, Thailand: Asia-Pacific Centre of Educational Innovation for development (UNESCO).

Ueno, K. (2006). From Wasan to Yozan: Comparison between mathematical education in the Edo period and the one after the Meiji restoration. In F. K. S. Leung, K.-D. Graf, & F. Lopez-Real (Eds.), *Mathematics education in different cultural traditions—A comparative study of East Asia and the West: The 13th ICMI Study* (pp. 65 79). New York, NY: Springer.

UNESCO. (1990). *World declaration on education for all (Jomtien report)*. Paris, France: Author.

UNESCO. (1998a). *World education report: Teachers and teaching in a changing world*. Paris, France: Author.

UNESCO. (1998b). *All human beings: A manual for human rights education*. Paris, France: Author.

UNESCO. (2000). *The Dakar framework for action: Education for all*. Paris, France: Author.

UNESCO. (2010). *The millennium development goals report, 2010*. New York, NY: Author.

Upex, S. (2000). An outline of some of the sources for the construction of history of education within Negara Brunei Darussalam. In M. A. Clements, H. Tairab, & K. Y. Wong (Eds.), *Science, mathematics, and technical education in the 20th and 21st centuries* (pp. 356–364). Gadong, Brunei Darussalam: Universiti Brunei Darussalam.

U.S. Commissioner of Education. (1891). *Report of the Commissioner of Education for the Year 1888–1889*. Washington, DC: U.S. Government Printing Office.

U.S. Commissioner of Education. (1905). *Report of the Commissioner of Education for the Year 1903*. Washington, DC: U.S. Government Printing Office.

U.S. Commissioner of Education. (1907). *Report of the Commissioner of Education for the Year 1905*. Washington, DC: U.S. Government Printing Office.

Van Egmond, W. (1980). *Practical mathematics in the Italian Renaissance: A catalog of Italian abbacus manuscripts and printed books to 1600*. Firenze, Italy: Istituto E Museo di Storia Della Scienza.

Van den Heuvel-Panhuizen, M., & Becker, J. (2003). Towards a didactic model for assessment design in mathematics education. In A. J. Bishop, M. A. Clements, C. Keitel, J. Kilpatrick, & F. K. S. Leung (Eds.), *Second international handbook of mathematics education* (pp. 689–716). Dordrecht, The Netherlands: Kluwer Academic Publishers.

Viadero, D. (2009). Algebra-for-all policy found to raise rates of failure in Chicago. *Education Week, 28*(24), 11.

West, R. L., Greene, C. E., & Brownell, W. A. (1930). The arithmetic curriculum. In G. M. Whipple (Ed.), *Report of the Society's committee on arithmetic: Twenty-ninth yearbook of the National Society for the Study of Education* (pp. 64–142). Bloomington, IL: Public School Publishing Co.

Westwood, P. S. (2008). *What teachers need to know about numeracy.* Camberwell, Australia: Australian Council for Educational Research.

White, A., & Southwell, B. (2008). Review of learning of mathematics in the middle years 2004–07. In H. Forgasz, A. Barkatsas, A. Bishop, B. Clarke, S. Keast, W. T. Seah, P. Sullivan, & S. Willis (Eds.), *Research in mathematics education in Australasia 2004–2007* (pp. 41–72). Rotterdam, The Netherlands: Sense Publishers.

Woolman, D. C. (2001). Education reconstruction and post-colonial curriculum development: A comparative study of four African countries. *International Education Journal, 2*(5), 27–46.

Wong, N. Y. (2006). From "entering the way" to "exiting the way": In search of a bridge to span "basic skills" and "process abilities." In F. K. S. Leung, G.-D. Graf, & F. J. Lopez-Real (Eds.), *Mathematics education in different cultural traditions: The 13th ICMI Study* (pp. 111–128). New York, NY: Springer.

Wong, N. Y. (2008). Confucian heritage culture learner's phenomenon: from "exploring the middle zone" to "constructing a bridge." *ZDM–The International Journal on Mathematics Education, 40*, 973–981.

Wu, M. (2006). A comparison of mathematics performance between East and West: What PISA and TIMSS can tell us. In F. K. S. Leung, K.-D. Graf, & F. Lopez-Real (Eds.), *Mathematics education in different cultural traditions—A comparative study of East Asia and the West: The 13th ICMI Study* (pp. 239–259). New York, NY: Springer.

Wu, M., & Zhang, D. (2006). An overview of the mathematics curricula in the West and East. In F. K. S. Leung, K.-D. Graf, & F. Lopez-Real (Eds.), *Mathematics education in different cultural traditions—A comparative study of East Asia and the West: The 13th ICMI Study* (pp. 181–193). New York, NY: Springer.

Young, J. W. A. (1914). *The teaching of mathematics.* New York, NY: Longmans, Green, and Co.

Zimmerman, J. (2009). *Small wonder: The little red schoolhouse in history and memory.* New Haven, CT: Yale University Press.

Chapter 2
Theories for Studying Social, Political and Cultural Dimensions of Mathematics Education

Eva Jablonka, David Wagner, and Margaret Walshaw

Abstract In mathematics education research and practice today we notice a change in the multiplicity of approaches that allow us to widen our perspectives on diverse social, political and cultural dimensions of mathematics education. This chapter provides an overview of trends and a critical discussion of the use of theories to approach, discuss and critique research and practices in mathematics education, particularly with attention to social, political and cultural dimensions.

Introduction

All research is built around a set of assumptions about the world and how it should be understood and studied. Researchers who study the social, political and cultural dimensions of mathematics education ground their work in a range of assumptions about the nature of knowledge and truth (epistemology) and being (ontology). These understandings are typically implicit, yet they inform the overarching stance of the researcher. Researchers, whether or not they acknowledge or discuss their stance, choose theories that are appropriate to their own view of the world and these, in turn, influence the kinds of projects the researchers undertake. Each perspective allows us to enrich our understandings of the diverse social, political and cultural dimensions of mathematics education. How those dimensions are conceptualized in contemporary research is the focus of this chapter.

E. Jablonka (✉)
Luleå University of Technology, Luleå, Sweden
e-mail: eva.jablonka@ltu.se

D. Wagner
University of New Brunswick, Fredericton, NB, Canada

M. Walshaw
Massey University, Palmerston North, New Zealand

The word "theory" carries with it various meanings, all of which take theory as something one sees or recognizes. The Greek roots of the word connect it to seeing. A description of a researcher's theoretical perspective, then, recognizes that the researcher looks at the researched situation from a particular vantage point. Clearly, certain vantage points that may be available to others will not be available to us. As researchers, we can choose from various vantage points and thus, ultimately, work to initiate change in what we see in the researched situation. Frameworks and models refer to conceptualizations of classes of situations, which we may compare to a situation we see in a researched situation. Thus the frameworks and models we bring with us as researchers affect the locus of our attention and affect what we see in a research context. Jablonka and Bergsten (2010) illustrate different strategies of theorizing in mathematics education in terms of their intertextuality, that is, engagement with and reference to previous work, and "relational density," that is, the extent to which relations between key concepts are established. They distinguish "ad-hoc constructions," "theory conglomerates" and "local models" from proper theories. For this chapter, we will subsume such frameworks and models that refer to previous research and make explicit their intellectual roots under the word "theory," though we are aware that there are differences among the various ways of thinking about theory.

What our exploration in this chapter seeks to do is offer an assemblage of theoretical vantage points that have been used by researchers in mathematics education in contemporary times. Arguably, among the differing perspectives, "incommensurability" (Cobb, 2007) will be a feature, which will prevent us from "providing warrants for our field's identity and intellectual autonomy within apparently broader fields such as education, psychology, or mathematics" (Silver & Herbst, 2007, p. 60). We begin from the position that the wide range of theories, characteristic of the research field today, does not symbolize a field marked by disarray, tensions and contradictions. Rather, what we wish to portray is a vibrant and diverse field, comprising influential perspectives, all of which have important things to tell us about the shape and character of mathematics education. Each perspective allows mathematics education to develop a vision of what to work toward.

Our concern, initially, is to investigate the potential of theories that have their intellectual roots outside the field of mathematics education to advance our perspectives on diverse social, political and cultural dimensions of mathematics education. We are also interested in the ways our researchers use them. In the first part of this chapter we locate trends in theorizing in mathematics education in relation to a widening of perspectives that call our attention to social, political and cultural dimensions. In locating such trends, we focus our attention on well-established theories that have been developed outside the field of mathematics education and their adoption, assimilation and potentials that are hoped for. In the second part of the chapter we review some of the work in mathematics education that has advanced our knowledge of social, political and cultural dimensions. Again, we look at work that has made use of theories developed in other fields, in particular in social linguistics and sociology as well as in postmodern analyses. We support the view of Sriraman and English (2010) to the effect that advancement in the field has often been initiated by adoption and assimilation of new theoretical vantage points that have their intellectual roots outside the field of mathematics education. However, as discussed in the third part of the chapter, there are important theories developed within mathematics education.

It becomes clear from our overview that mathematics education is no longer only concerned with the technologies of learning and teaching in institutionalized pedagogic settings. It includes researching mathematics education in sites beyond the classroom (e.g., local communities and families, workplaces, policy making, the media, textbook production) and research activities that describe and theorize these practices, including research that is directed towards studying the social, economic and political conditions and consequences of those practices.

Trends and Advances in Theorizing

Trends

Mathematics education is at the intersection of many disciplines including socio-cultural disciplines, language, mathematics, and politics. There is a smorgasbord of theories that researchers might draw upon productively from these disciplines, because each discipline also carries a variety of theories. With this diversity at our disposal, it is instructive to note which disciplines and which theories are being taken up. Tsatsaroni, Lerman, and Xu (2003), when reporting their investigation of theories in mathematics education, noticed a social turn. They noted that where once inspiration for researchers was drawn primarily from psychology, a turn to the social enabled the exploration of a broader range of research questions and issues. New perspectives, topics and methodologies arose, and influential journals (e.g., *Educational Studies in Mathematics;* the *Journal of Mathematics Teacher Education;* and the *Journal for Research in Mathematics Education*) were noticeably now more inclusive of non-traditional frames. These non-traditional frames had enabled researchers to attend to previously unseen aspects of practice.

In addition to traditional psychological and mathematics theories, a growing variety of psycho-social, sociological, socio-cultural, (social) linguistic and semiotic theories have been referred to in conference proceedings and journal articles. Also, reference to recent broader theoretical currents, such as feminism and poststructuralism has been made in the more recent publications (since the time of Tsatsaroni, Lerman and Xu's analysis). During the same period the total numbers of traditional psychological and mathematical papers did not decrease, and Jablonka and Bergsten (2010) have referred to the addition of a "social branch" rather than a social turn. Sub-fields of mathematics education grow in parallel and eventually constitute their own discourses, without one dominating or being privileged.

Inspired by Tsatsaroni et al.'s (2003) investigation, we identify expansions of theorizing in the proceedings of four recent annual conferences of the International Group for the Psychology of Mathematics Education (PME), 2007, 2008, 2009 and 2010. While three of these conferences had no special theme, PME 2009 was subtitled "In Search for Theories in Mathematics Education." We have chosen

PME principally because this is the most established organization that organizes regular conferences and thus reflects changes in what is to be considered as mainstream. We did not anticipate that all innovations in theorizing would emerge within this context, as we are well aware that such innovations take seed in edited volumes, anthologies as well as at conferences that are specifically devoted to exchanging and developing alternative views. In relation to investigating the social and political dimensions of mathematics education, the Mathematics Education and Society conferences provide such forums (see, for example, Gellert, Jablonka, & Morgan, 2010; Matos, Valero, & Yasukawa, 2008).

We compiled a list of names of theories, frameworks and authors associated with socio-linguistic, socio-cultural, sociological and postmodern theories and searched proceedings by using a global document search function. Raw numbers from this search are shown in Tables 2.1, 2.2, and 2.3. If a search term only occurred in the reference list, the paper was not included in the count. For some searches we used word roots in order to capture variations. For example, Vygotsk' captures "Vygotsky" and "Vygotskian." Similarly, sociol' captures "sociological," "sociology," and other variations. We are aware of other classifications of theories from those used to construct the tables.

Table 2.1
Number of PME Papers Mentioning Vygotskian and neo-Vygotskian Theories

Search Terms for Vygotskian and Neo-Vygotskian Theories	PME 2007	PME 2008	PME 2009	PME 2010
Vygotsk [y]	14	18	34	15
[Jean] Lave	7	8	15	9
[Etienne] Wenger	14	12	12	10
[Barbara] Rogoff	2	2	3	
Psycholinguist [ics]				
Activity theory	2	3	12	13
[Yrjö] Engeström				7

Table 2.2
Number of PME Papers Mentioning Sociological Theories

Search Terms for Sociology	PME 2007	PME 2008	PME 2009	PME 2010
Sociol [ogy/ogical]	3	33	16	8
Intellectual Roots of Contemporary Sociological Theories, by Authors:				
[Émile] Durkheim			1	
[Karl] Marx	1		3	3
[Max] Weber	1		1	

(continued)

Table 2.2
(continued)

Search Terms for Sociology	PME 2007	PME 2008	PME 2009	PME 2010
[Edmund] Husserl	1		1	2
[Alfred] Schütz			2	
[Talcott] Parson, [Louis] Althusser, [Antonio] Gramsci, [Eric Olin] Wright, [Georg] Simmel, [George Herbert] Mead, [Herbert] Blumer, [Erving] Goffmann, [Harold] Garfinkel				
Neofunctionalism:				
[Niklas] Luhmann			1	
Neofunctionalis [m/t]				
Critical Theory and Conflict Theory:				
Critical Theor [y]			2	
Conflict Theor [y]			1	1
Frankfurt School			1	
[Max] Horkheimer			1	
[Theodor] Adorno			1	
[Herbert] Marcuse			1	
[Erich] Fromm			1	
[Charles Wright] Mills				
[Pierre] Bourdieu	2	1	7	5
Analytic Sociology of Conflict:				
[Analytic] Sociology of Conflict, [Ralf] Dahrendorf, [Randall] Collins				
Theories of Evolution, Modernity and Globalization:				
[Anthony] Giddens			1	
Structuration Theory				
[Jürgen] Habermas		1	5	1
[Theory of] Communicative [Action]		12		
[Ulrich Beck]				
[Reflexive] Modernization			1	
Risk Society				1
Symbolic Interactionism and Phenomenology:				
Symbolic Interactionism		3	2	1
[Patricia Hill] Collins, [Dorothy E.] Smith				
Phenomenology	2	8	5	3
[Peter] Berger		1	1	
[Thomas] Luckmann			1	
[Max Van] Manen		1		
Rational Choice [Theories]				
Sociology of Education, of Mathematics Education:				
[Michael] Young				
[Michael] Apple			1	
[Basil] Bernstein	1	2	6	1
[Paul] Dowling	1		1	1

Table 2.3
Number of PME Papers Mentioning Literary Theory, Discourse Analysis, Social Linguistics, Positioning Theory and Postmodern Approaches

Search Terms	PME 2007	PME 2008	PME 2009	PME 2010
Literary Theory, Discourse Analysis, Social Linguistics:				
Critical Discourse Analysis			3	
Discourse Analysis	4	4	1	5
[Mikhail] Bakhtin	1	4	3	5
[Norman] Fairclough			3	
[Michael] Halliday	2		1	
[Ruqaiya] Hasan			2	
[J.R.R.] Martin		3		
[Gunther] Kress	1		4	
Positioning Theory:				
Positioning theory		2	4	
Social psychology				
[Rom] Harré	1		2	
Foucault and Postmodern Approaches:				
[Michel] Foucault, Foucauldian	1	2	4	4
Feminis [m/t]				
Psychoanaly [tic theory]	1	2	2	1
[Slavoj] Žižek			1	
[Jacques] Lacan		1	1	3
[Deborah] Britzman, [Elizabeth] Ellsworth				

There are a number of limitations we need to make explicit with respect to our use of PME proceedings. Although annual PME conferences are recognized as important international conferences for mathematics education researchers, they do not fully capture the research being undertaken by mathematics educators worldwide. Papers provided by researchers from non-English-speaking countries are published less frequently in PME proceedings relative to those of English-speaking researchers. In addition, the kinds of classrooms depicted in research reported in PME proceedings tend to reflect a prototypical mathematics classroom which is not representative of classrooms throughout the world. Skovsmose (2006) has suggested that 90% of mathematics classroom research represents only ten per cent of the classrooms in the world.

It is important for us to clarify that in our analysis of the PME volumes we were not seeking to identify papers that failed to make explicit the theory that underpinned the work. Rather, we wondered if it is possible to characterize, without explicit reference to any intellectual tradition, some research in mathematics education as adopting a sociological, political or postmodern perspective by asking research questions that bear testimony to the "spirit" of a theory. Nevertheless, we agreed that it is important to identify and make explicit one's theoretical perspective because attention to this detail makes for richer, more thoughtful interpretation. In addition, if readers outside an esoteric circle are to be addressed (which is necessary for dissemination), then the conceptual underpinning should be articulated.

We need to make clear, too, that page restrictions for papers in PME proceedings act as constraints for researchers. Theoretical frameworks were usually presented in a succinct format or not at all. Although many papers provided hints at the standpoint taken, the implicit nature of this evidence made it difficult to provide absolute characterizations of the field. There was evidence, however, from those reports which declared their positions, that the PME conference proceedings under investigation were open to a range of theoretical and methodological standpoints. That is to say, a diverse and complex array of theoretical frameworks informed inquiry. Specifically, although there are many references to cultural studies and a range of social practice theories such as symbolic interactionism, activity theory, situated learning and social constructivism, a relatively small number of studies were informed by postmodern and sociological theories.

Given that sociology challenges many assumptions of psychology, reference to sociological theories could indeed have been expected to be uncommon in PME proceedings. Tables 2.1, 2.2, and 2.3 also reveal the "white spots," that is when we did not find a reference to a theory we searched for. These white spots could indicate that the respective theories were not being integrated into the mainstream. However, in some cases, lack of such reference could also mean that although a well-established researcher from mathematics education, who has built from and elaborated a theory that has its roots outside the field, is cited, any reference to the original sources is not seen as essential anymore. However, in our investigation we were less interested in the proportions of different branches of theorizing, and more interested in how theories from sociology, linguistics, activity theory, positioning theory, situated cognition and postmodern theories were used, and to what effect.

Adoption and Assimilation of Established Theories

One difference between the work of mathematics educators and the theorists from whom we draw is that most of these theories are oriented to describing and analyzing practice, while in mathematics education there is a sense that we have to prescribe or at least identify good practice. We think that this tension is central to many of the challenges mathematics educators have when applying theories which emerge from other disciplines. As criteria for usefulness, in a technical sense, of theorizing can only be framed in relation to a given practice, there would not be any innovative or critical potential if identifying good practice were the only *raison d'être* of research. A sometimes-observed hostility towards theory in mathematics education research is based on a misreading of theory as mere contemplation and speculation. Theorizing includes systematization of and critical reflection upon practice that opens up new views. Seen in this way, theorizing is indispensable for the advancement of a field.

In considering the range of theories available to mathematics education researchers, there are a number of decisions researchers have to make with regard to theory. First, one chooses theory that enables one to address the research question, but often the theory is instrumental in formulating the question as well. Second, when choosing a well-developed theory, one chooses aspects of that theory for focus. Third, it is important to consider the "translation" of the theory that was birthed in a specific context to the context of mathematics education. Fourth, there are decisions about how much attention to give to the theory when writing about the research, including the possibility of not recognizing that the research has a perspective that is socioculturally and politically relevant. For the last three of these choices, there are continua—for example, a researcher might take one concept from a theory, more of the theory, or much of the theory. There is yet another possibility—taking two or more theories in some kind of hybridization. Moreover, in our view, importing a theory from a different tradition of research is already a form of hybridity. The recontextualization of theories from outside our field necessarily involves a change in the criteria for what counts as advancement, a shift in focus and in meaning.

In our conversations about the PME papers we considered possible ways of misusing theories. One could, for example, use a single concept from a theory and thus miss some central ideas of the theory. This could be done with intention or with naïveté (and we acknowledge that there are only degrees of naïveté, for no one can be said to know everything about a theory). We agreed that for a misread of a theory to be deemed heresy, it would have to be an intentional twisting of the theory, but then we wondered how to distinguish between heresy and "moving theory forward," both of which turn and/or move theory. Picking up on single concepts from a theory can be productive, but it may not be. Productive, deliberate re-interpretation and expansion of theory based on some principles might be called heresy or development, depending on one's point of view. Hybrids from different theoretical sources can be promising in bringing together ideas that seemed apart, but also limiting by distorting the spirit of the individual theories. The strategy might amount to a pastiche or a conglomerate, and perhaps even to an anti-theoretical bricolage. This brings forward the important question of how we might judge the qualities theories bring to our field.

When reading a selection of PME contributions, we were interested in which aspects of the theories were used, whether or not the papers included reflection on the challenges of applying these theories in particular research contexts, how the researchers described the motivation for their choice of theory, the extent to which the data interpretation drew on the theory, the extent of the description of the theory, and who the paper cites in the description of the theory—the major theorists from outside mathematics education, or mathematics educators applying the theory in our field.

From our reading, we see that "networking" theories remains a challenge for research in mathematics education. We found examples of this challenge in our reading of promising contributions that could form a starting point for moving the discipline forward. We found some innovative study designs that attempted to achieve some theoretical combinations that looked entirely novel. But in the examples of theoretical combinations, one theory often dominated. The assumptions shared by the individual theories were not elaborated. In many reported empirical studies, the motivation for theory choice is not made explicit. While the "novelty" of an approach might be mentioned, the promises of a new theory in relation to other approaches that did not carry the same promises tended not to be discussed. We also noted a geographical distribution of branches of theorizing and of innovation. This is of course due to the physical closeness of experts in a location, for example supervisors, but also to the cultural situatedness of traditions. In the PME papers we also identified some contributions that fully exploited the potential of theories and sought to advance our understanding of the field. Furthermore, some authors alerted readers to the potential of a whole branch of theories.

Examples of Providing New Terrain

Our selection of papers from the PME proceedings for further discussion was guided by the number of theories the papers connected with, by promising titles, and surprising combinations of references. We were also careful to review plenary papers and research forums. In our selection, we also have taken the number of references made in the same paper as an indication of an extended discussion of a theory, though these numbers are not represented in the tables above.

A theoretical paper by Brown (2008), for example, provides a critical analysis of Luis Radford's cultural theory of objectification. Using ideas drawn from a range of postmodern sources, Brown offered a critique of the way in which Radford conceptualized the notions of culture and subjectivity in his theoretical development. Brown drew specifically on discursive approaches to knowledge and subjectivity to develop his critique. Working from the premise that Radford's cultural theory of objectification "perhaps provides the most sustained and substantial excursion" (p. 209) into the area of cultural and historical dimensions of mathematical objects, Brown attempted to unsettle some of the foundations on which that theory is built.

Breen's plenary paper at the PME 2007 conference at the end of his term as President of PME (Breen, 2007) is suffused with ideas from enactivism and psychoanalytic theory and these provided a springboard for Breen's reflections of and hopes for mathematics education. What the paper revealed particularly is that theoretical border-crossing into enactivism and psychoanalytic ideas requires a shift in thinking and in attitude and, in that sense, the sensibilities of the theory may have been lost on some readers. Such a shift offers readers new understandings about mathematics education and its situatedness with institutions, history, and cultural fields. It also draws our attention to our ultimately compromised stance in everything we do and say within mathematics education. As a plenary paper the content could be deemed highly influential. It opened up theoretical discussion for the discipline. Of course, readers could choose to dismiss his theoretical tools or they could choose to pick up snippets of ideas that suited them. Alternatively, readers could assess his theoretical apparatus as a key resource for interrogating and understanding the dynamics and politics of mathematics education.

At the PME 2009 conference, a research forum on sociological theories in mathematics education was held. The overall agenda of the forum was researching possibilities of how more equitable outcomes may be achieved in mathematics education, as no research group in mathematics education, least of all the leading international group, can ignore the social disadvantages reproduced in mathematics classrooms in most countries of the world (Lerman, 2009). The contributors explored how research in mathematics education has made, and could make, use of sociological theories in shaping research questions and methodologies that contribute to the agenda. The forum also discussed the ideologies at work in research designs. At the PME 2010 conference the discussion group on mathematics education and democracy (Mattos, Batarce, & Lerman, 2010) also provided new terrain that is not genuinely linked with psychology. The members of the discussion group included as their theoretical underpinnings Karl Marx's concept of commodity, Jean Baudrillard's concept of sign value as well as the work of Jacques Derrida. One key issue for discussion was the constitution of mathematics knowledge as universal need in today's society and the role of mathematics education in the constitution of such an ideology.

In a plenary address at PME 2009, Morgan's (2009) account of her evolving research program provided a window into how tools and ideas from linguistics might connect with a researcher's agenda. She described how Pimm's (1987) book title connected to the questions that dominated her thinking about mathematics education, and how this connection drew her to systemic functional linguistics (SFL). From this, she became interested in Fairclough's work because it helped her move beyond description to the judgment of mathematical texts. Fairclough's work connects Halliday's to critical social theorists including Foucault and Bourdieu. Not surprisingly, Morgan was next drawn to Bernstein's theory to help her understand the social context of mathematics discourse. In her work with Evans (Evans & Morgan, 2009), she noted that discursive approaches address some of the classic dilemmas in sociology and social theory: structure versus action, order versus conflict, and official versus deviant perspectives. Morgan's path is illustrative of the connections among the three strands we are using to divide up theories that attend to social, political and cultural dimensions of mathematics education.

Opening Up New Perspectives

We now turn to include mathematics education literature beyond the recent PME proceedings to outline the way theory from outside the field has been used to move the field forward through accounting for social, cultural and political dimensions of mathematics education. We divide this work into the three broad areas, discourse analysis, sociology and postmodern approaches, though we know that these three areas are interconnected. Indeed, our work on these overviews reminded us of these intersections and the related difficulty of categorizing work. However, we are also aware of other approaches to knowledge development that provide alternatives to cognitivism and are compatible with socio-cultural learning theories.

Enactivist and complexity theories, for example, add a new twist to the influences of the social in highlighting the dynamic and interactive adaptations of the learner, and address questions of "being" rather than "knowing." An insightful application of enactivism for education can be observed in the work of Davis and Simmt (2003). A similar biological metaphor is used by Radford, Edwards, and Arzarello (2009) in their embodied theory. Ideas are not held by individuals but are embodied by human beings with normal human cognitive capacities living in a culture, situated in and productive of larger, social, cultural and historical thinking. Mathematical thinking, learning and communication involve different semiotic systems and multiple modalities of expression including gesture, speech, written inscriptions, and physical and electronic artefacts, all of which are integral to the cognitive process (Radford, 2009). The conceptualizations of one person are not assessed as a measure of "fit" or "match"; rather they are said to be viable (or otherwise) in relation to another's conceptualizations.

Discourse Analysis

Theory from linguistics has been instrumental in illuminating interpersonal interaction within the contexts of mathematics teaching and learning and, in particular, the positioning of students and teachers in relation to others and the discipline. However, discourse analysis is not limited to linguistics. As articulated by Ryve (2011), mathematics education draws on various theorizations of discourse to illuminate multiple perspectives on mathematics teaching and learning. Ryve's analysis of numerous articles in mathematics education journals shows that the concept of discourse is too-often undertheorized in research reporting. However, there are strong examples of productive use of various forms of discourse analysis in the recent years of mathematics education scholarship, which we overview below.

Halliday (1978) called the discipline-specific use of language employed in mathematics communication "the mathematics register." With the increasing use of this term the fuzzy boundaries of the register are becoming exposed, drawing attention to the goals of mathematics educators. Mathematicians speak and write differently from mathematics teachers and learners. Pimm (2007) and Barwell (2007) have commented on this distinction in response to research that seems to blur this line. Herbel-Eisenmann, Wagner, and Cortes (2010) clarified their analysis of mathematics classroom discourse as investigations of "the mathematics classroom register." Even so, any classroom or any interaction has its own peculiar forms, so it is not possible to delineate "the" register accurately.

Following Halliday's (1985) social semiotics, a powerful body of tools for understanding how people use language for various purposes and effects in discourse—called systemic functional linguistics (SFL)—has been developed. Though various scholars used SFL tools before, Morgan (2006) contributed an introduction of social semiotics to mathematics education, with the purpose of demonstrating its tools and of identifying research questions that these tools can help to answer. They are useful for analyzing transcripts to identify who or what is doing things in learning contexts, the objects of mathematics in these contexts and the relationships at work.

For example, Nachlieli and Herbst (2007) used such tools to identify the particular utterances that related to assumptions in the proof discourse they analyzed. Both Mesa and Chang (2008) and Wagner and Herbel-Eisenmann (2007) used a narrower tool set within the rubric of SFL, as they used Martin and White's appraisal linguistics to understand the engagement of mathematics learners in classrooms.

Mathematics educators are using approaches to discourse analysis in addition to social semiotics. For example, Both Mesa and Chang (2008) and Wagner and Herbel-Eisenmann (2009) complemented their SFL work with positioning theory, which is another form of analyzing discourse from a social psychology perspective. Hegedus and Penuel (2008) used Goodwin's participation frameworks to analyze discourse specific to a mathematics learning with wireless technology, aiming to document how students' identity shifts during the course of a class. Carlsen (2010) used Linell's dialogical approach to study interaction and its effect on meaning. Black et al. (2010) used Gee's approach to discourse analysis, which takes a broader view of discourse, to identify the interconnecting stories at work in students' accounts of their mathematical narratives.

In addition to using tools to identify features of language for understanding what is happening in mathematics learning contexts, linguistics and other domains provide theory for understanding in general the connections between language and thinking. This kind of theory is often called "discursive psychology." Some examples of using a linguist's observations to support one's line of attention in mathematics education include the following. Leung and Or (2007) used Michael Halliday to support their claim that language choices shape human experience. Similarly, Sakonidis and Klothou (2007) used Gunther Kress to substantiate their observation that students' writing is not necessarily read in the intended way by their assessors. De Freitas (2009) used Fairclough's "critical discourse analysis" to locate the structuring of power relationships in mathematics classrooms in the language choices.

Sfard (2008), in developing her own model to explain how communication and cognition are co-implicated in mathematics learning, invented the word "commognition" to denote this inherent connection. Barwell (2009) drew on discursive psychology to critique Sfard's use of examples to develop her model, and pointed to the general challenge of drawing inferences from excerpts of mathematics learning situations. Indeed, most analysis of mathematics education discourse works with texts identified by researchers, perhaps because these chosen texts exemplify a particular distinction or phenomenon. Alternatively, if one works from a large body of diverse classroom texts, which linguists call a "corpus" (e.g., Herbel-Eisenmann et al., 2010; Wagner & Herbel-Eisenmann, 2007), it is possible to identify features of the discourse in general. This corpus linguistics work does not undermine the importance of in-depth analysis of isolated excerpts. Nevertheless, it is important to be careful about warrants for claims made from examples of mathematics teaching and learning discourse.

Discourse analysis appears in yet other forms of mathematics education work. In researched professional development contexts, mathematics educators have been directing the attention of teachers to forms of discourse analysis. Herbel-Eisenmann supported a group of mathematics teachers in their action research projects that focussed on aspects of discourse (e.g., Herbel-Eisenmann & Cirillo, 2009). Zolkower and de Freitas (2010) guided teachers in deconstructing transcripts of their mathematics teaching to increase their awareness of semiotic choices available to them.

De Freitas (2010) used critical discourse analysis to study the classroom discourse and interaction patterns of two secondary school mathematics teachers of senior classes in Canada. She employed Fairclough's understanding that language not only produces meaning but also positions speakers in specific relations of power. The purpose was to understand the way in which teachers' subjectivity is constituted and enacted, in brief and often spontaneous and contradictory speech acts. The task demands thinking about text and context in classroom interaction as intersecting rather than separated. In the analysis de Freitas showed how one teacher, Mark, repeatedly used metaphors that signified an antagonistic relationship between students and texts, and embedded many references to sports throughout his lessons. She demonstrated how the other teacher, Roy, continuously made reference to the difficulty of learning calculus, choosing to exclude discourse from other texts that spoke calculus into existence in other ways.

Both analyses highlighted what teachers choose to say and the way in which they say it, and the power relations that descend from those linguistic decisions. In particular, the analyses provided counter-narratives about classroom discourse, pointing to the regulatory power of teacher discourse in providing access to mathematics, by shedding light on those students who were included within and those who were positioned outside of the text. Importantly, through the fine-grained reading that unpacked hidden relationships and regulatory practices operating within the classroom, de Freitas demonstrated the way in which the discursive practices of the two teachers contributed to the kind of thinking that is possible within the classroom.

The number of edited collections that have focussed on discourse in mathematics education in recent years points to the importance of discourse analysis within the field. Chronaki and Christiansen (2005) presented a collection of varied perspectives used to theorize communication and this collection also addressed associated political issues. Moschkovich (2010) likewise assembled multiple perspectives on language and mathematics education and identified new directions for research. This volume featured different authors from the Chronaki and Christiansen volume, demonstrating the depth of the field within mathematics education. Herbel-Eisenmann, Choppin, Wagner, and Pimm (2011) brought into conversation mathematics education research that focussed on equity and on discourse to show how these two are inherently connected. Barwell, Barton, and Setati (2007) edited a special issue of *Educational Studies in Mathematics* focussed on a narrower discourse-related issue—mathematics learning in multilingual contexts. There is an active group of scholars working together to focus on multilingual contexts, many of whom gathered for an ICMI study conference in 2011.

Sociology

Our study of trends in mainstream research as reflected in the PME conferences has shown that employing Vygotskian and neo-Vygotskian theories, as well as a general reference to the label "socio-cultural," has become common. Vygotsky is often cited only in the text, as it is usual with references to classical works and names that stand for an intellectual tradition (e.g., as the "Vygotskian paradigm" or "Vygotskian approach"). Similarly, activity theory is often referred to without specific references. Clearly, socio-cultural perspectives based on those theories on learning have been integrated into mainstream. These perspectives have considerably advanced the field of mathematics education by drawing attention to socially and culturally specific experiences among learners of mathematics.

However, relations to social structures remain under-theorized in those approaches. Learning within a community of practice does not occur in isolation from the power relations that operate within that practice. Conceptualizing learning through legitimate peripheral participation does not necessarily help to understand stratification of achievement in mathematics classrooms, especially in relation to social and economic class, race and gender (Ensor & Galant, 2005; Huzzard, 2004). How does social structuration come about in communities of practice? Daniels (2001) suggested that the theories of situated knowledge and learning should be related to a political analysis of power and control. A potential of productive interaction between socio-cultural perspectives on learning and sociological theories was pointed out by de Abreu (2008), who suggested that while cultural psychology allows for, if not draws attention to, the diversity among learners in their socially and culturally specific experiences, apprenticeship models are limited in conceptualizing consequences of macro-social structures on learning.

In many studies of mathematics classroom interaction, reference is made to symbolic interactionism and phenomenology, the reference sometimes being mediated through the works of mathematics educators who have followed these perspectives in their works (see, e.g., Yackel & Cobb, 1996). However, not all sociologists accept phenomenology (and its offspring ethnomethodology) and symbolic interactionism as genuine sociological theorizing. Although both share an anti-positivistic paradigm and common assumptions about the task of focussing on understanding of how meanings are developed and shared, they have been criticized for focussing merely on micro-level small group social interactions as well as for non-attention to the unintentional "hidden" consequences of actions or to the constraints of socio-political structures on people's actions.

But there are also important differences between the two traditions. Symbolic interactionism is interested in how the participants define the situation and come to make sense through the process of interaction, while ethnomethodology is interested in uncovering the taken-for-granted values, norms and rules that already operate in the interaction. Classroom studies based on symbolic interactionism illuminate how (flexible) role expectations and meanings are established through a "negotiation of meaning," with a focus on how students act in situations that demand new interpretations.

This leads to more insightful interpretations of what happens in "inquiry-based" learning situations than in classrooms with more apparent role-asymmetries (Voigt, 1996). Ethnomethodology acknowledges that the taken-for-granted rules are functional and that the participants' interpretations might be limited. Voigt (1984), for example, showed that teachers and students enact subconscious practices or "routines" when structuring the process of developing new knowledge interactively. He also pointed out that in the interactive construction of new meaning through an elicitation pattern, which starts with a teacher's open question (a new "task"), there is no shared frame of reference from the outset. For the students the new question is ambivalent, and this ambivalence is only retrospectively (reflexively) reduced when the official solution is institutionalized. This ambivalence causes a problem, especially with contextualized mathematics as a starting point for developing new concepts and methods. The issue has been taken up by researchers who are interested in the effects of "invisible" pedagogic practices on the stratification of achievement (see e.g., Jablonka & Gellert, 2011).

In order to explore the stratification of achievement in mathematics classrooms, one has to acknowledge that patterns of classroom interaction are functional in terms of the goals of the institution and are not accomplished at the initiative of the participants in a single classroom. Theorizing the reproduction of inequalities through mathematics education is the most obvious agenda of genuine sociological approaches. Advances have been made through employing the works of Bourdieu and Bernstein. In PME conferences, references to Bourdieu and Bernstein are not very common and remain often on a general level, with the exception of the research forum on sociological theories mentioned above. Bernstein's notion of visible pedagogy invites didactization, as can be seen by Sullivan's (2008) reference to Bernstein, pointing to the necessity of making explicit the criteria for evaluation when implementing non-routine tasks in classrooms. Aaron (2008) innovatively employed Bourdieu's notion of "symbolic economy" when analyzing students' views of classroom work in geometry lessons in order to conceptualize differences in the students' identities.

In an illuminative investigation of unequal achievement in mathematics secondary education in Victoria, Australia, Teese (2000) drew on Bourdieu. His analysis shows that much of the students' success at different levels of the mathematics curriculum depended on their personal characteristics, such as organizational skills, study habits, concentration and academic self-esteem. Teese argued that the discriminating potential of mathematics education is implicit in a curriculum hierarchy that raises the demands over successive levels of mathematics that call more and more on embedded scholastic attitudes and behaviours. It can be taken as a measure of the implicit cultural homogeneity of the mathematics curriculum as a whole—based on sequenced and overlapping content and shared conceptual emphasis—that the average social level of students rises at each level of performance. Teese's investigation showed the potential of data analysis from a consistent theoretical vantage point.

Bernstein's work offers a broad range of interrelated notions that are incorporated into a complex theoretical body. Most prominently in references feature the concepts of recontextualization, horizontal and vertical discourse, classification and framing, and visible and invisible pedagogy. Increasing numbers of researchers in mathematics education have extended, developed and critically engaged with that body of theory. Dowling's (1998) study of school mathematics texts and some of the methodological tools developed in that study provided a major contribution to sociological theorizing in mathematics education.

Uncovering the ideologies behind different mathematics curricula and scrutinizing the ways in which what types of knowledges are constructed for which groups of learners remains a major task for sociological approaches. Noss et al. (1990) provided a collection of sociological analyses of curriculum, Ernest (2009) attempted a critique of ideology in mathematics, science and technology education research and its globalization, Ensor and Galant (2005) reviewed studies from the South African context. Analysis of the development and effects of policy discourses in mathematics education is another domain of study to which sociological theory provides powerful tools.

Valero and Zevenbergen (2004) located approaches, sometimes also subsumed under the label "socio-cultural," which acknowledged that both mathematics education and research in the field are not only social but also political practices. Depending on the political, economic and social conditions, these practices exercise power in different forms. Institutions that contribute to the reproduction of power, as for example schooling, can be analyzed as political institutions. Perspectives explicitly sharing the acknowledgement of this fact can be described as "socio-political." Such approaches have moved beyond the tools made available within classical sociology and cultural psychology to explore the power dynamics within social interactions. They ground their investigations on the premise that the practices and processes of mathematics education are inherently political. Skovsmose (2009), for example, wrote, in his critique of mathematical rationality, of the symbolic power of mathematics.

Martin (2010) observed that race still remains under-theorized in mathematics education, as disparities in achievement are often taken as reflecting race effects rather than as consequences of the racialized nature of the students' mathematical experiences. A similar point has been made (e.g., Skovsmose, 2007) about many, mostly quantitative, studies of unequal attainment in relation to social and economic background that treat the students' background merely as an input variable. Similarly, Gutiérrez and Dixon-Román (2011) demonstrated how "gap-gazing" constructs those who do not achieve as deficitarian, while not addressing the ideological underpinnings of the goals of mathematics education. Chronaki (2011) argued that curriculum politics act as "ideological state apparatus" (Althusser, 1971), regulating the micro-level of mathematics education by creating microspaces, for example in the form of didactic innovations. She argued that hegemonic discourses of equity construct subjects with static identities as marginalized and voiceless. Chronaki observed that such discourses are underpinned by constructivist and socio-cultural approaches that overemphasize the "autonomous subject" who makes rational decisions. This points to the potential of employing psychoanalytic and poststructuralist theories, and clearly challenges psychological theorizing.

Researching unequal access to mathematical practices and discourses that provide cultural and symbolic capital might leave the conception of curriculum untouched. However, exposing the forms of mathematics privileged in a curriculum is an outcome of an analysis of the functionality of curriculum as well as of mathematics. Pais and Valero (2011) argued that mathematics education in many places must be understood within a capitalist economic and neo-liberal political setting that calls for quality and equity yet serves particular interests in these settings.

Critique and a wish to contribute to an agenda for social change is an important agenda in sociologically-oriented research in mathematics education as it was outlined by Noss et al. (1990). The fact that this is a political agenda does not mean that the research is more value-loaded than any other, but rather that the values are made more explicit.

Postmodern Approaches

Seminal edited volumes written during the last decade (e.g., Bishop, Clements, Keitel, Kilpatrick, & Leung, 2003; Boaler, 2000; Lester, 2007; Sriraman & English, 2010), although making important contributions to the discipline, did not reflect the impact and take-up of postmodern theory within mathematics education. Given that an increasing number of researchers are interested in what these social theories might mean for mathematics education, we look at the origins of and the assumptions underpinning this theoretical movement and the way in which the movement promotes local voice and critical thinking, even as it holds critical thinking itself up for scrutiny.

The specific traditions of psychology and sociology provide a bedrock of concepts and theories for the study of mathematics education from a postmodern perspective. Psychology has informed a psychoanalytical turn, designed to unsettle fundamental modernist assumptions concerning identity formations. For example, Brown and McNamara (2010) drew on the work of Lacan to investigate how preservice teachers use language to describe the world around them and how they see themselves fitting in. Sociology has helped seed poststructuralist work that aimed at drawing attention to the ways in which power works within mathematics education, at any level, and within any relationship, to constitute identities and to shape proficiencies. Walshaw (2004a), for example, built on the work of Foucault to explore the ways in which teaching practice is inherently political.

Like analyses of a modernist persuasion, at the heart of postmodern analyses lies an interest in understanding contemporary social and cultural phenomena (e.g., Brown, 2008; de Freitas & Nolan, 2008; Walls, 2009 Walshaw, 2004b, 2010). Postmodern analyses chart teaching and learning, and the way in which identities and proficiencies evolve, tracking reflections, investigating everyday classroom activities and tools, analyzing discussions with principals, mathematics teachers, students, and educators, and mapping out the effects of policy, and so forth. The point of departure from modernist narratives is derived from assumptions about the nature of the reality being studied, assumptions about what constitutes knowledge of that reality, and assumptions about what are appropriate ways of building knowledge of that reality. As a result of these specific understandings, the lived contradictions of mathematics processes and structures are able to be explored.

Poststructuralists and psychoanalysts share some fundamental assumptions about language, meaning and subjectivity. They see language as fragile and problematic and as *constituting* social reality rather than *reflecting* an already given reality. What is warranted at one moment of time, may be unwarranted at another time (see Walshaw, 2007). The claim is that because the construction process is ongoing, we do not have access to an independent reality. Hanley (2010) demonstrated that point in her exploration of the way in which teachers make sense of and enact curriculum reform. She showed that although teachers attempted to put into practice what they learned through a professional development project, what was learned and practised in professional development initiatives was never fully cashed in as educational capital within the classroom.

Objectivity is not the only concept that postmodern theorists take issue with. They debate conventional understandings of reason, insisting that rationality is always relative to time and place. They prefer to think in terms of "local" determinants, fallibility and contingency. Underwriting their projects is a "decentred self"—a self that is an effect of discourse which is open to redefinition and which is constantly in process. This point was given expression by Walls (2010) in her investigation of the "good" teacher, in a setting of compulsory standardized testing. Walls drew on the idea of teacher identity as a process embedded in discourse, to explore teachers' struggle for self and to investigate how systemic forces, in a culture of teacher accountability, are lived by teachers as individual dilemmas.

What is apparent in mathematics education research that draws on postmodern theories is a move towards exploring tentativeness and developing scepticism of the particular principles and methods that put a shine on essentialist and absolutist tendencies. What such theories also do is require researchers to consider the implicit assumptions that guide their work. The point was emphasized by Adler and Lerman (2003), who argued that there are moral obligations, and hence, ethical issues at stake in any research practice. More critical debate and evaluation of the competitive work of researchers is needed particularly at this moment of time when political influences on research are becoming deeply entrenched. In Adler and Lerman's view, such influences were "insufficiently problematized in the mathematics education community" (p. 457).

De Andrade (2008), reporting on a research project undertaken in Brazil, drew on the work of Foucault and used a form of discourse analysis to look at the relationship between research and classrooms in mathematics education. Derrida's ideas of deconstruction also informed the methodology, by providing a vehicle for keeping "the system in play," "in process," and "to set up procedures to continuously demystify the realities we create, to fight the tendency for our categories to congeal" (p. 60). Employing these ideas, De Andrade (2008) set up contradictions between model classrooms as depicted in mathematics education research, and the kinds of classrooms in which teachers in Brazil sometimes find themselves teaching. Two major themes could be discerned in De Andrade's paper: the subjectivity of learners in actual classrooms and the inherently political nature of research. These themes are in keeping with other work based in the field that draws on Foucault's framework. Such a framework provides the means to explore the relationship between power and knowledge. It is also able to signal that the views of teachers and researchers are always enmeshed in sites of knowledge production that are unavoidably political. In drawing on the work of Foucault, De Andrade dealt with meaning construction in a way that acknowledged the researcher's own complicity in the analysis.

Stentoft and Valero (2010) investigated the fragility of mathematical learning. Their discussion expressed a poststructuralist imagination that took seriously the notion that language constitutes social reality rather than reflects an already given reality. In developing an understanding of the "noise" symptomatic of everyday classrooms, Stentoft and Valero (2010) challenged interpretations of the practices within what are typically characterized as "pure mathematics classrooms." Their theoretical approach used precepts that are, in tenor, at odds with the presuppositions that ground the rational autonomous learner.

In a discussion on undermining traditional approaches to learning, Stentoft and Valero (2010) drew attention to the interrelatedness as well as the fragility of classroom discourse, identity and learning. They argued that these three elements together constitute the landscape within which a student's sense-of-self as learner is formed. In their discursive analysis, they case studied mathematics classroom interactions at a Danish teacher training college. Underlying the analysis was an intent to avoid mere descriptions of classroom life, but rather, to unpack how students and teachers were involved with constructing multiple identities over the course of a mathematics lesson. The intent was also to make clear how learning mathematics and constructing mathematical knowledge in the classroom is inextricably caught up in the discursive practices of the classroom.

Bibby (2010) used concepts from psychoanalytic theory to explore the pedagogical relation. She drew on the concepts of the oedipal family and the Oedipus complex to unpack relationships to mathematics, particularly as they are constituted in primary schools. Post-Freudian psychoanalytic theories of authority provided her with conceptual tools to investigate the way in which mathematics, with an emphasis of rules, speed and correct answers, is characterized as masculine in traditional school mathematics pedagogy. Taking care not to essentialize gender, Bibby unpacked the ideational fiction of binary characterization, and proffered, instead, masculinity and femininity, boy and girl, as "elements within gender." She drew on research data to unpack some of the potential consequences of differentiating mathematics as an unemotional, authoritative, rational, systematic and logical set of values and practices, away from so-called feminine qualities such as warmth, emotional attunement, and creativity. Specifically, she explored the tensions that result from fictions that allow for the deployment of masculinity in the discursive construction of mathematics and investigated the consequences for teachers and students living with the effects of these splits in policy and practice.

Theories from Within Mathematics Education

In addition to the theories mathematics education researchers import from other domains, there are theories that were born within mathematics education itself in order to overcome the limitations of a purely psychological paradigm. For example, as mentioned previously, Sfard (2008) developed her own model for describing the interaction between communication and cognition. This model has subsequently been used by other researchers as a theoretical perspective on cognition. Because the connection of communication and cognition could have been theorized outside of the context of mathematics education, Sfard's theory is an example of the way researchers in our field develop theory using contexts from mathematics education and theory from other domains. This is a strong example, because others have taken up her theory.

Renert and Davis (2010) proposed an integral perspective for exploring knowledge production. They developed a model that integrated self, culture, and nature, through which a plurality of perspectives could be entertained. Their proposal was towards an evolutionary perspective, one that is inclusive of the contributions of traditional, modernist and postmodern perspectives. They showed how each perspective leads to different views about the kinds of tools that mathematics uses and each makes it possible for certain understandings to be entertained and legitimated. Their integral perspective valued the enacted, creative and dynamic dimensions of mathematics and was focussed on the health and harmony of the entire system. Renert and Davis applied these ideas to their work with experienced middle-school teachers. Their work demonstrated how teachers are crucial participants in the creation of mathematical possibilities. They suggested that teachers might engage students more meaningfully with mathematics by elaborating the specific, by using active language, and by allowing them to engage with multiplicity and plurality in discourse, meaning-making and interpretation.

There are other theories that have arisen in mathematics education for which the unique context of mathematics education is a necessary aspect of the theory. These include work that is critical of mathematics education and its position and role in society. One of the more established, but also very diverse, of these approaches is ethnomathematics. In PME proceedings, the term was referred to in 10 papers in the volumes from 2010, but only in three papers from 2009 and in one paper from 2007. Ethnomathematics is concerned with practices and activities of marginalized groups, that can be identified as mathematical, but which are not institutionalized as mathematics.

The term ethnomathematics is a label used for the theoretical underpinnings as well as for the product of an analysis of the mathematical nature of such activities. It emerged from a critique of both a Eurocentric gaze in popular history of mathematics as well as an elitist pedagogic model together with a deficitarian perspective on the knowledge of students with a cultural frame not in line with the official school culture. Thus it allows seeing the political dimension of mathematics education derived from and designed for the hegemonic sectors of society. The spirit of the approach to both research and education is a commitment to inclusion. Ethnomathematics brought into attention the cultural embeddedness of mathematical knowledge and of mathematics education.

There are, of course, other forms of criticism of mathematical practices and mathematics education practices. For example, much equity work in mathematics education has criticized the way mathematics is taught and has based this criticism on sociological and political perspectives. There has also been criticism of mathematical rationality, and its effects on society. Skovsmose's work, partly inspired by critical theory and critical pedagogy, drew many to contribute to the project of critical mathematics education (see Alrø, Ravn, & Valero, 2010). His analysis of the formatting power of mathematics itself ought to be viewed as a contribution to the sociology of mathematics, a branch of sociological theorizing that has a great potential to be further developed. It is necessary for scrutinizing traditional dogmas about mathematical knowledge production and applications. Understanding the relationships between different practices that include mathematics and exhibit different knowledge structures and discourses, in school and outside school, remains a major concern (Jablonka, 2003; Jablonka & Gellert, 2007).

Though theories have emerged within mathematics education, as described above, these theories still connect with theories outside mathematics education. For example, ethnomathematics research is informed by ethnographic traditions. And critiques of mathematics use approaches that have been developed outside mathematics education. In short, there are no theories that are absolutely independent from other theories. There are no distinct theories; there are only relations among theories.

Conclusion

In this chapter we have presented a critical investigation of contemporary theoretical trends in international research in mathematics education. Our attempt at mapping the field by broad strokes has allowed us to grasp the current state of play in theory selection, to understand how particular theories gain ascendancy, and to see how differing theories are acted upon in varying research projects. What has been revealed is a vibrant international research community that validates a wide range of theoretical perspectives, each of which informs the production of new knowledge relevant to mathematics education.

Though there is vibrancy and growth in our field attributable to socio-cultural and political perspectives, it is important to recognize that the field itself is dominated by one language and also by certain cultural practices, some of which are related to that language. Our review of the field follows developments in the literature published in English because this volume is in English and because English is the primary linguistic medium for developments in our field. However, we recognize that there is good work in other languages that addresses socio-cultural and political dimensions of mathematics education. The dominance of English in our field is, of course, a characteristic that relates to social, cultural and political forces. There is some scholarship that addresses this characteristic of our field (e.g., Barton, 2008; Skovsmose, 2006), but more often this characteristic is addressed in researchers' descriptions of the limitations of their work.

As all of the perspectives we have discussed not only challenge the assumptions of psychology but are also based on partly conflicting assumptions, the question of the complementarity or juxtaposition of analyses informed by these perspectives becomes important. Research in mathematics education is diversified and the domain might be characterized as a collection of different approaches within rival discourses with little or no dialogue. If dialogue is avoided, it is then not the explanatory power of the diverse theories that prompt reception and dissemination, but the power relations among the researchers within competing discourses. However, many researchers acknowledge the potential of asking for alternative interpretations of the same empirical field from different theoretical vantage points, for these can bring tensions to the foreground. This will only happen, of course, if the theoretical underpinnings are well understood and their implications for the research design well articulated. That is particularly the case if interpretations based on distinct theories are controversial.

What can we learn from this profile? We can find out about the specific theorists who are currently influential in the field. We can learn about the way in which ideas about theory in mathematics education change. But we can do more—we can draw on the insights that our exploration offers to inform the debate about those things that are most important in mathematics education. From our interrogation we see signs of a shift away from cognitive psychology and evidence of critical questioning, of the creation of new ideas, and new ways of doing things, as well as a tolerance for multiplicity. All of these observations will contribute to the development of a body of professional knowledge in our discipline, informed by theory rather than driven by policy. We believe the international research community holds the reins of exciting potential for further development of leading edge knowledge in mathematics education.

Serious engagement with the work produced from different vantage points and openness towards different views can counteract the establishment of closed circles of academic inquiry, often labelled under a common term and declaring other projects as irrelevant to their own. We share the belief that an analysis of the situation of mathematics education needs to include critical reflection of its practices. It is only in its difference to practice as unmediated and often unconscious action, that theory transcends practice.

References

Aaron, W. R. (2008). Academic identities of geometry students. In O. Figueras, J. L. Cortina, S. Alatorre, T. Rojano, & A. Sepúlveda (Eds.), *Proceedings of the Joint Conference PME32–PMENAXXX for the Psychology of Mathematics Education* (Vol. 2, pp. 1–8). Morelia, Mexico: PME.

Adler, J., & Lerman, S. (2003). Getting the description right and making it count: Ethical practice in mathematics education research. In A. Bishop, M. A. Clements, C. Keitel, J. Kilpatrick, & F. Leung (Eds.), *Second international handbook of mathematics education* (pp. 441–470). Dordrecht, The Netherlands: Kluwer Academic Publishers.

Alrø, H., Ravn, O., & Valero, P. (Eds.). (2010). *Critical mathematics education: Past, present and future*. Rotterdam, The Netherlands: Sense Publishers.

Althusser, L. (1971). Ideology and ideological state apparatuses. In L. Althusser (Ed.), *Lenin and philosophy and other essays* (Ben Brewster, Trans.) (pp. 121–176). London, UK: New Left Books.

Barton, B. (2008). *The language of mathematics: Telling mathematical tales*. New York, NY: Springer.
Barwell, R. (2007). Semiotic resources for doing and learning mathematics. *For the Learning of Mathematics, 27*(1), 31–32.
Barwell, R. (2009). Researchers' descriptions and the construction of mathematical thinking. *Educational Studies in Mathematics, 72*(2), 255–269.
Barwell, R., Barton, B., & Setati, M. (2007). Multilingual issues in mathematics education: Introduction. *Educational Studies in Mathematics, 64*(2), 113–119.
Bibby, T. (2010). What does it mean to characterize mathematics as "masculine"? Bringing a psychoanalytic lens to bear on the teaching and learning of mathematics. In M. Walshaw (Ed.), *Unpacking pedagogy: New perspectives for mathematics classrooms* (pp. 21–41). Charlotte, NC: Information Age.
Bishop, A., Clements, M. A., Keitel, C., Kilpatrick, J., & Leung, F. (Eds.). (2003). *Second international handbook of mathematics education*. Dordrecht, The Netherlands: Kluwer Academic Publishers.
Black, L., Williams, J., Hernandez-Martinez, P., Davis, P., Pampaka, M., & Wake, G. (2010). Developing a "leading identity": The relationship between students' mathematical identities and their career and higher education aspirations. *Educational Studies in Mathematics, 73*(1), 55–72.
Boaler, J. (Ed.). (2000). *Multiple perspectives on mathematics teaching and learning*. Westport, CT: Ablex Publishing.
Breen, C. (2007). On humanistic mathematics education: A personal coming of age? In J. H. Woo, H. C. Lew, K. S. Park, & D. Y. Seo (Eds.), *Proceedings of the 31st Conference of the International Group for the Psychology of Mathematics Education* (Vol. 1, pp. 3–16). Seoul, Korea: PME.
Brown, T. (Ed.). (2008). *The psychology of mathematics education: A psychoanalytic displacement*. Rotterdam, The Netherlands: Sense Publishers.
Brown, T., & McNamara, O. (2010). *New teacher identity and regulative government: The discursive formation of primary mathematics teacher education*. New York, NY: Springer.
Carlsen, M. (2010). Appropriating geometric series as a cultural tool: A study of student collaborative learning. *Educational Studies in Mathematics, 74*(2), 95–116.
Chronaki, A. (2011). Disrupting "development" as the quality/equity discourse: Cyborfs and the subalterns in school technoscience. In B. Atweh, M. Graven, W. Secada, & P. Valero (Eds.), *Mapping equity and quality in mathematics education* (pp. 3–19). New York, NY: Springer.
Chronaki, A., & Christiansen, I. (Eds.). (2005). *Challenging perspectives on mathematics classroom communication*. Charlotte, NC: Information Age Publishing.
Cobb, P. (2007). Putting philosophy to work: Coping with multiple theoretical perspectives. In F. K. Lester (Ed.), *Second handbook of research on mathematics teaching and learning* (pp. 3–38). Charlotte, NC: Information Age/NCTM.
Daniels, H. (2001). *Vygotsky and pedagogy*. London, UK: Routledge Falmer.
Davis, B., & Simmt, E. (2003). Understanding learning systems: Mathematics education and complexity science. *Journal for Research in Mathematics Education, 34*(2), 137–167.
de Abreu, G. (2008). From mathematics learning out-of-school to multicultural classrooms: A cultural psychology perspective. In L. English, M. Bartolini Bussi, G. Jones, R. Lesh, & B. Sriraman (Eds.), *Handbook of international research in mathematics education* (2nd ed., pp. 354–386). New York, NY: Taylor and Francis.
De Andrade, S. (2008). The relationship between research and classroom in mathematics education: A very complex and multiple look phenomenon. In O. Figueras, J. L. Cortina, S. Alatorre, T. Rojano, & A. Sepúlveda (Eds.), *Proceedings of the Joint Conference PME32 – PMENAXXX for the Psychology of Mathematics Education* (Vol. 2, pp. 57–64). Morelia, Mexico: PME.
De Freitas, E. (2009). Decoding the discourse of difficulty in high school mathematics. In M. Tzekaki, M., Kaldrimidou, & H. Sakonidis (Eds.), *Proceedings of the 33rd Conference of the International Group for the Psychology of Mathematics Education* (Vol. 1, p. 365). Thessaloniki, Greece: PME.

De Freitas, E. (2010). Regulating mathematics classroom discourse: Text, context and intertextuality. In M. Walshaw (Ed.), *Unpacking pedagogy: New perspectives for mathematics classrooms* (pp. 129–151). Charlotte, NC: Information Age.

De Freitas, E., & Nolan, K. (Eds.). (2008). *Opening the research text: Critical insights and in(ter)ventions into mathematics education*. Dordrecht, The Netherlands: Springer.

Dowling, P. (1998). *The sociology of mathematics education: Mathematical myths/pedagogic texts*. London, UK: Falmer.

Ensor, P., & Galant, J. (2005). Knowledge and pedagogy: Sociological research in mathematics education in South Africa. In R. Vithal, J. Adler, & C. Keitel (Eds.), *Researching mathematics education in South Africa. Perspectives, practices and possibilities* (pp. 281–306). Cape Town, South Africa: HSRC Press.

Ernest, P. (2009). Globalization, ideology and research in mathematics education. In M. Setati, R. Vithal, C. Malcolm, & R. Dhunpath (Eds.), *Researching possibilities in mathematics, science and technology education* (pp. 3–36). New York, NY: Nova Science Publishers.

Evans, J., & Morgan, C. (2009). Sociological frameworks in mathematics education research: Discursive approaches. In M. Tzekaki, M. Kaldrimidou, & C. Sakonidis (Eds.), *Proceedings of the 33rd Conference of the International Group for the Psychology of Mathematics Education* (Vol. 1, pp. 223–226). Thessaloniki, Greece: PME.

Gellert, U., Jablonka, E., & Morgan, C. (Eds.). (2010). *Proceedings of the Sixth International Mathematics Education and Society Conference: Berlin, March 20–25, 2010*. Berlin, Germany: Freie Universität Berlin.

Gutiérrez, R., & Dixon-Román, E. (2011). Beyond gap gazing: How can thinking about education comprehensively help us (re)envision mathematics education? In B. Atweh, M. Graven, W. Secada, & P. Valero (Eds.), *Mapping equity and quality in mathematics education* (pp. 21–34). New York, NY: Springer.

Halliday, M. (1978). Sociolinguistic aspects of mathematics education. In M. Halliday (Ed.), *Language as social semiotic: The social interpretation of language and meaning* (pp. 194–204). London, UK: Edward Arnold.

Halliday, M. (1985). *An introduction to functional grammar*. London, UK: Edward Arnold.

Hanley, U. (2010). Teachers and curriculum change: Working to get it right. In M. Walshaw (Ed.), *Unpacking pedagogy: New perspectives for mathematics classrooms* (pp. 3–19). Charlotte, NC: Information Age.

Hegedus, S., & Penuel, W. (2008). Studying new forms of participation and identity in mathematics classrooms with integrated communication and representational infrastructures. *Educational Studies in Mathematics, 68*(2), 171–183.

Herbel-Eisenmann, B., Choppin, J., Wagner, D., & Pimm, D. (Eds.). (2011). *Equity in discourse for mathematics education: Theories, practices, and policies*. New York, NY: Springer.

Herbel-Eisenmann, B., & Cirillo, M. (Eds.). (2009). *Promoting purposeful discourse: Teacher research in mathematics classrooms*. Reston, VA: National Council of Teachers of Mathematics.

Herbel-Eisenmann, B., Wagner, D., & Cortes, V. (2010). Lexical bundle analysis in mathematics classroom discourse: The significance of stance. *Educational Studies in Mathematics, 75*(1), 23–42.

Huzzard, T. (2004). Communities of domination? Reconceptualising organisational learning and power. *The Journal of Workplace Learning, 16*(6), 350–361.

Jablonka, E. (2003). Mathematical literacy. In A. Bishop, M. A. Clements, C. Keitel, J. Kilpatrick, & F. K. S. Leung (Eds.), *Second international handbook of mathematics education* (pp. 77–104). Dordrecht, The Netherlands: Kluwer Academic Publishers.

Jablonka, E., & Bergsten, C. (2010). Theorising in mathematics education research: Differences in modes and quality. *Nordisk matematikkdidaktikk: Nordic Studies in Mathematics Education, 15*(1), 25–52.

Jablonka, E., & Gellert, U. (2007). Mathematisation–demathematisation. In U. Gellert & E. Jablonka (Eds.), *Mathematization and de-mathematization: Social, philosophical and educational ramifications* (pp. 1–18). Rotterdam, The Netherlands: Sense Publishers.

Jablonka, E., & Gellert, U. (2011). Equity concerns about mathematical modelling. In B. Atweh, M. Graven, W. Secada, & P. Valero (Eds.), *Mapping equity and quality in mathematics education* (pp. 221–234). New York, NY: Springer.

Lerman, S. (2009). Sociological frameworks in mathematics education research. In M. Tzekaki, M. Kaldrimidou, & H. Sakonidis (Eds.), *Proceedings of the 33rd conference of the International Group for the Psychology of Mathematics Education* (Vol. 2, p. 217). Thessaloniki, Greece: PME.

Lester, F. K. (Ed.). (2007). *Second handbook of research on mathematics teaching and learning*. Charlotte, NC: Information Age/NCTM.

Leung, A., & Or, C. (2007). From construction to proof: Explanations in dynamic geometry environment. In J. H. Woo, H. C. Lew, K. S. Park, & D. Y. Seo (Eds.), *Proceedings of the 31st Conference of the International Group for the Psychology of Mathematics Education* (Vol. 3, pp. 177–184). Seoul, Korea: PME.

Martin, D. (2010). Not-so-strange bedfellows: Racial projects and the mathematics education enterprise. In U. Gellert, E. Jablonka, & C. Morgan (Eds.), *Proceedings of the Sixth International Mathematics Education and Society Conference: Berlin, March 20–25, 2010* (pp. 57–79). Berlin, Germany: Freie Universität Berlin.

Matos, J. P., Valero, P., & Yasukawa, K. (Eds.). (2008). *Proceedings of the Fifth International Mathematics Education and Society Conference Albufeira, Portugal, February 16–21, 2008*. Lisbon: Centre de Investigacao em Educacao, Universidade de Lisboa and Department of Education, Learning and Philosophy, Aalborg University.

Mattos, A. C., Batarce, M. S., & Lerman, S. (2010). Discussion group: Mathematics education and democracy. In M. M. Pinto & T. F. Kawasaki (Eds.), *Proceedings of the 34th Conference of the International Group for the Psychology of Mathematics Education* (Vol. 1, p. 338). Belo Horizonte, Brazil: PME.

Mesa, V., & Chang, P. (2008). Instructors' language in two undergraduate mathematics classrooms. In O. Figueras, J. L. Cortina, S. Alatorre, T. Rojano, & A. Sepúlveda (Eds.), *Proceedings of the 32nd Conference of the International Group for the Psychology of Mathematics Education held jointly with the 30th Conference of PME-NA* (Vol. 3, pp. 367–374). Morelia, Mexico: PME

Morgan, C. (2006). What does social semiotics have to offer mathematics education research? *Educational Studies in Mathematics, 61*(12), 219–245.

Morgan, C. (2009). Understanding practices in mathematics education: Structure and text. In M. Tzekaki, M. Kaldrimidou, & C. Sakonidis (Eds.), *Proceedings of the 33rd Conference of the International Group for the Psychology of Mathematics Education* (Vol. 1, pp. 49–64). Thessaloniki, Greece: PME.

Moschkovich, J. (Ed.). (2010). *Language and mathematics education: Multiple perspectives and directions for research*. Charlotte, NC: Information Age Publishing.

Nachlieli, T., & Herbst, P. (2007). Students engaged in proving—participants in an inquiry process or executers of a predetermined script? In J. H. Woo, H. C. Lew, K. S. Park, & D. Y. Seo (Eds.), *Proceedings of the 31st Conference of the International Group for the Psychology of Mathematics Education* (Vol. 4, pp. 9–16). Seoul, Korea: PME.

Noss, R., Brown, A., Drake, P., Dowling, P., Harris, M., Hoyles, C., & Mellin-Olsen, S. (Eds.). (1990). *The political dimensions of mathematics education: Action & critique*. London, UK: Institute of Education.

Pais, A., & Valero, P. (2011). Beyond disavowing the politics of equity and quality in mathematics education. In B. Atweh, M. Graven, W. Secada, & P. Valero (Eds.), *Mapping equity and quality in mathematics education* (pp. 35–48). New York, NY: Springer.

Pimm, D. (1987). *Speaking mathematically: Communication in mathematics classrooms*. London, UK: Routledge.

Pimm, D. (2007). Registering surprise. *For the Learning of Mathematics, 27*(1), 31.

Radford, L. (2009). Why do gestures matter? Sensuous cognition and the palpability of mathematical meanings. *Educational Studies in Mathematics, 70*(2), 111–126.

Radford, L., Edwards, L., & Arzarello, F. (2009). Introduction: Beyond words. *Educational Studies in Mathematics, 7*(2), 91–95.

Renert, M., & Davis, B. (2010). Life in mathematics: Evolutionary perspectives on subject matter. In M. Walshaw (Ed.), *Unpacking pedagogy: New perspectives for mathematics classrooms* (pp. 177–199). New York, NY: Information Age Publishing.

Ryve, A. (2011). Discourse research in mathematics education: A critical evaluation of 108 journal articles. *Journal for Research in Mathematics Education, 42*(2), 167–198.

Sakonidis, H., & Klothou, A. (2007). On primary teachers' assessment of pupils' written work in mathematics. In J. H. Woo, H. C. Lew, K. S. Park, & D. Y. Seo (Eds.), *Proceedings of the 31st Conference of the International Group for the Psychology of Mathematics Education* (Vol. 4, pp. 153–160). Seoul, Korea: PME.

Sfard, A. (2008). *Thinking as communicating: Human development, the growth of discourses, and mathematizing.* Cambridge, UK: Cambridge University Press.

Silver, E. A., & Herbst, P. (2007). Theory in mathematics education scholarship. In F. K. Lester (Ed.), *Second handbook of research on mathematics teaching and learning* (2nd ed., pp. 39–68). Charlotte, NC: NCTM/IAP.

Skovsmose, O. (2006). Research, practice, uncertainty and responsibility. *Journal of Mathematical Behaviour, 25*(4), 267–284.

Skovsmose, O. (2007). Students' foregrounds and the politics of learning obstacles. In U. Gellert & E. Jablonka (Eds.), *Mathematization and de-mathematization: Social, philosophical and educational ramifications* (pp. 81–94). Rotterdam, The Netherlands: Sense Publishers.

Skovsmose, O. (2009). *In doubt: About language, mathematics, knowledge and life-worlds.* Rotterdam, The Netherlands: Sense Publishers.

Sriraman, B., & English, L. (2010). Surveying theories and philosophies of mathematics education. In B. Sriraman & L. English (Eds.), *Theories of mathematics education: Seeking new frontiers* (pp. 7–32). Berlin, Germany: Springer.

Stentoft, D., & Valero, P. (2010). Fragile learning in the classroom: Exploring mathematics lessons within a pre-service course. In M. Walshaw (Ed.), *Unpacking pedagogy: New perspectives for mathematics classrooms* (pp. 87–107). Charlotte, NC: Information Age.

Sullivan, P. (2008). Designing task-based mathematics lessons. In O. Figueras, J. L. Cortina, S. Alatorre, T. Rojano, & A. Sepúlveda (Eds.), *Proceedings of the Joint Conference PME32 – PMENAXXX for the Psychology of Mathematics Education* (Vol. 1, pp. 133–138). Morelia, Mexico: PME.

Teese, R. (2000). *Academic success and social power. Examinations and inequality.* Melbourne, Australia: Melbourne University Press.

Tsatsaroni, A., Lerman, S., & Xu, G.-R. (2003, April). *A sociological description of changes in the intellectual field of mathematics education research: Implications for the identities of academics.* Paper presented at the Annual Meeting of the American Educational Research Association, Chicago.

Valero, P., & Zevenbergen, R. (2004). Introduction: Setting the scene of this book. In P. Valero & R. Zevenbergen (Eds.), *Researching the socio-political dimensions of mathematics education: Issues of power in theory and methodology* (pp. 1–4). Dordrecht, The Netherlands: Kluwer Academic Publishers.

Voigt, J. (1984). *Interaktionsmuster und Routinen im Mathematikunterricht. Theoretische Grundlagen und mikroethnographische Falluntersuchungen.* Weinheim, Germany: Beltz.

Voigt, J. (1996). Negotiation of mathematical meaning in classroom processes: Social interaction and learning mathematics. In L. P. Steffe, P. Nesher, P. Cobb, G. A. Goldin, & B. Greer (Eds.), *Theories of mathematical learning* (pp. 21–50). Mahwah, NJ: Lawrence Erlbaum.

Wagner, D., & Herbel-Eisenmann, B. (2007). *Discursive tools for suppressing and inviting dialogue in the mathematics classroom.* Proceedings of the 29th Annual Meeting of the North American Chapter of the International Group for the Psychology of Mathematics Education, Lake Tahoe, USA.

Wagner, D., & Herbel-Eisenmann, B. (2009). Re-mythologizing mathematics through attention to classroom positioning. *Educational Studies in Mathematics, 72*(1), 1–15.

Walls, F. (2009). *Mathematical subjects: Children talk about their mathematics lives.* Dordrecht, The Netherlands: Springer.

Walls, F. (2010). The good mathematics teacher: Standardized mathematics tests, teacher identituy, and pedagogy. In M. Walshaw (Ed.), *Unpacking pedagogy: New perspectives for mathematics classrooms* (pp. 65–83). Charlotte, NC: Information Age.

Walshaw, M. (2004a). The pedagogical relation in postmodern times: Learning with Lacan. In M. Walshaw (Ed.), *Mathematics education within the postmodern* (pp. 121–139). Greenwich, CT: Information Age.

Walshaw, M. (Ed.). (2004b). *Mathematics education within the postmodern*. Greenwich, CT: Information Age.

Walshaw, M. (2007). *Working with Foucault in education*. Rotterdam, The Netherlands: Sense Publishers.

Walshaw, M. (Ed.). (2010). *Unpacking pedagogy: New perspectives for mathematics classrooms*. Charlotte, NC: Information Age.

Yackel, E., & Cobb, P. (1996). Sociomathematical norms, argumentation, and autonomy in mathematics. *Journal for Research in Mathematics Education, 27*(4), 458–477.

Zolkower, B., & de Freitas, E. (2010). What's in a text? Engaging mathematics teachers in the study of whole-class conversations. In U. Gellert, E. Jablonka, & C. Morgan (Eds.), *Proceedings of the Sixth International Mathematics Education and Society Conference* (pp. 508–518). Berlin, Germany: Freie Universität Berlin.

Chapter 3
Understanding and Overcoming "Disadvantage" in Learning Mathematics

Lulu Healy and Arthur B. Powell

Abstract Past research has largely characterized disadvantage as an individual or social condition that somehow impedes mathematics learning, which has resulted in the further marginalization of individuals whose physical, racial, ethnic, linguistic and social identities are different from normative identities constructed by dominant social groups. Recent studies have begun to avoid equating difference with deficiency and instead seek to understand mathematics learning from the perspective of those whose identities contrast the construction of normal by dominant social groups. In this way of thinking, "understanding" disadvantage can be discussed as understanding social processes that disadvantage individuals. And, "overcoming" disadvantage can be explored by analyzing how learning scenarios and teaching practices can be more finely tuned to the needs of particular groups of learners, empowering them to demonstrate abilities beyond what is generally expected by dominant discourses. In this chapter, we consider theoretical and methodological perspectives associated with the search for a more inclusive mathematics education, and how they generally share a conceptualization of the role of the teacher as an active participant in researching and interpreting their students' learning. Drawing from examples with a diverse range of learners including linguistic, racial and ethnic minorities, as well as deaf students, blind students, and those with specific difficulties with mathematics, we argue that by understanding the learning processes of such students we may better understand mathematics learning in general.

L. Healy (✉)
Bandeirante University of São Paulo, São Paulo, Brazil
e-mail: lulu@baquara.com

A. B. Powell
Rutgers University-Newark, Newark, NJ, USA

In this chapter, we consider research in mathematics education that has concerned itself with documenting, analyzing, and critiquing the social construction of disadvantaged mathematics learners and in investigating the participation of students from marginalized cultural and social groups. To begin, we will discuss how the notion of disadvantage as deficiency in mathematics learning further stigmatizes and marginalizes social groups whose identities are not congruent with those of dominant social groups and consider alternative approaches to understanding and interpreting issues of equity and access in mathematics education. We go on to consider developing research perspectives that aim to look beyond models in which difference is equated with deficiency and focus instead on how mathematical agency and identities are mediated by a diverse range of resources, including language, cultural artefacts and sensory experiences. Finally, we turn to attempts to engage teachers in the challenge of capitalizing upon this diversity to create more inclusive mathematics classrooms.

Difference, Not Disadvantage

Discourses of disadvantage in mathematics education parallel larger societal discourses. Being social, these discourses could have been built otherwise. The societal discourses beget disciplinary discourses and disciplinary discourses both reflect and contribute new ideas to their parent discourses. Ideas of disadvantage tend to be based on physical, racial, ethnic, linguistic, social and gendered identities that are different from normative identities constructed by dominant social groups. As researchers such as Gutiérrez (2008) and Martin (2009a) have argued, a problem with this perspective is that it treats marginal groups as static categories and runs the risk of equating group membership with connotations of innate intelligence. At the least, this perspective implies that students from particular cultural groups are deficient in something—like mathematical achievement—that those from the dominant ideal have. Hence to overcome their possibly innate disadvantage the marginalized need to become more like their more "normal" contemporaries. But, as many researchers have suggested, physical, racial, ethnic, linguistic, social and gendered identities are far from static, they are constructed in association with social, political and economic processes. Viewed in this way, identities are continually constructed and reconstructed, experienced and re-experienced. Identity is simultaneously cultural and transcends culture.

In this chapter then, we consider that particular groups come to be disadvantaged not as a result of some static characteristic that defines the group in question but by the social, political, economic and psychological practices of the wider society to which they belong. This brings us to a difficult challenge. We are faced, in this chapter, with the enormous task of considering developments in the field of mathematics education in relation to disadvantage—or, perhaps better, difference—in general. As if everyone somehow experiences difference in the same ways. This challenge is, of course, impossible to tackle in a fully adequate way, and inevitably we have had to make choices on which learners we will focus in our discussion. Indeed, any attempt to list those who society disadvantages is similarly dangerous. On the one hand, we risk excluding some groups from the list and on the other we risk assuming terminology that will be deemed unacceptable by some of those who would position themselves either inside or outside the groups we label. Yet, we cannot write this chapter without doing so.

In their chapter on issues of equity and access in the *Second Handbook of Research on Mathematics Teaching and Learning*, Bishop and Forgasz (2007) suggested that a wide range of student groups had suffered what they called "conflicts with mainstream mathematics education," including, students from racial and ethnic minorities and indigenous peoples, rural learners, non-Judeo-Christian religious student groups, working-class students, female students and students with disabilities. We would add to this list students who have been expected to learn mathematics in a language different from their first or home language and lesbian, gay or transgender students, although this last student group, as Rands (2009) has argued, is almost completely absent from the research literature.

Another impossible task for this single chapter is to consider all the fronts on which action is necessary if disadvantage is to be overcome. The very view of disadvantage to which this chapter subscribes should make it clear that disadvantage cannot be overcome at any global level without a restructuring of a society as a whole. At a more local level, however, we will argue that one way in which "overcoming" disadvantage can be explored is by analyzing how learning scenarios and teaching practices can be more finely tuned to the practices of particular groups of learners, empowering them to demonstrate abilities beyond what is generally expected by dominant discourses. We therefore focus our attention on research related to understanding the mathematical practices of students from groups marginalized by wider society, using a lens in which deviances from any documented norms are treated as differences not as deficiencies. This suggests a shift from focussing on disadvantage to moving towards equitable approaches in mathematics education. In this chapter, we concentrate in particular on school mathematics and, more specifically, we have chosen to focus mainly on research related to mathematical practices as they occur within classrooms. We begin by considering recent views on equity in the context of mathematics education.

Views on Equity

Our shift in focus from overcoming disadvantage to equity means that the latter category requires examination. Equity does not exclusively affect students positioned as disadvantaged by virtue of their linguistic, cultural, ethnic and racial, physical, sexual, and gender identities. As such, discourses on equity are not marginal issues in mathematics education policy, research and practice. Bishop and Forgasz (2007) provided details on possible research approaches to access and equity in mathematics education. Here, instead, we focus on examining critiques of the equity approaches previously identified. We attempt to define equity and to provide a deeper theoretical framework for understanding connections between macro and micro perspectives.

For some, in mathematics education, "equity" has to do with its legal denotation of "fairness" and "justice" and is made indistinct from "equality." As such, equity is thought to equate to providing the same for all. Internationally, the notion "mathematics for all" gained acceptance and emerged as a Theme Group at ICME-5 (Damerow, Dunkley, Nebres, & Werry, 1984) and continued on by a host of authors (for example, Croon, 1997; National Council of Teachers of Mathematics, 1989; Steen, 1990). Nearly two decades later, from the perspective of critical theory, this view is subject to negative appraisal. Frankenstein (2010) describes a profound concern that "mathematics for all" assumes that all students have the same social, economic context and further pointed out that Apple (1992) concluded that the NCTM *Standards* (1989) did not address "the question of *whose problem* ... by focussing on the reform of mathematics education for 'everyone,' the specific problems and situations of students from groups who are in the most oppressed conditions can tend to be marginalized or largely ignored" (Secada, 1989, p. 25). The *Standards* did not contain, for example, suggestions for mathematical investigations that would illustrate how the current US government's real-life de-funding of public education, through funding formulas based on property taxes, creates conditions in which the real-life implementation of the NCTM student-centered pedagogy is virtually impossible except in wealthy communities (Kozol, 1991).

Others, not necessarily critical theorists, have proffered a similar line of analysis: equity in mathematics education is not likely to be achievable within societies suffering from structural socio-economic inequalities. For instance, Clarke and Suri (2003) problematized cultural explanations of observed differences and similarities in international comparative research of mathematical achievement. In discussing how analyses of between countries rankings on international comparative assessments, such as TIMSS and PISA, masks within country inequities, he cited Berliner (2001) as saying,

> Average scores mislead completely in a country as heterogeneous as [the United States of America] ... The TIMSS-R tells us just what is happening. In Science, for the items common to both the TIMSS and the TIMSS-R, the scores of white students in the United States were exceeded by only three other nations. But black American school children were beaten by every single nation, and Hispanic kids were beaten by all but two nations. A similar pattern was true of mathematics scores. ... The true message of the TIMSS-R and other international assessments is that the United States will not improve in international standings until our terrible inequalities are fixed. (p. B3)

The consequence of internal social and economic variations or inequalities is missed in the aggregation of performance data for countries as socially and culturally plural, for instance, as in the cases of, Australia, Brazil, Canada, England, South Africa, and the USA.

Socio-political variations within and between countries not only skew interpretations of international comparative data but also mask possibilities for equitable access, treatment, and outcomes in mathematics education. In their discussion of social inclusion and diversity in mathematics education, Baldino and Cabral (2006) argued the need for researchers to develop a theoretical stance to examine and understand practices in mathematics education from broader social and political perspectives. Assuming this challenge, Pais and Valero (2011) constructed a theoretical perspective of equity for understanding connections between the micro views of equity in mathematics teaching and learning practices and the macro-social conditions in which those practices occur. They argued that research on equity has not fully theorized the complexity of social and political life that entails the practices of mathematics education to understand how this engulfing complex conditions possibilities for equitable access, treatment, and outcome in mathematics education.

From another theoretical perspective, progress toward equity is viewed as a tension between dominant and critical mathematics education (Gutiérrez, 2007). The latter practice of mathematics education was first theorized by Frankenstein (1983). Distinctions between dominant and critical draw attention to differences among practices of mathematics education that reflect the social–political status quo of societies and practices that admit the positioning of students as members of a society rife with issues of power and domination. Critical mathematics takes students' cultural identities and builds mathematics around them in ways that address social and political issues in society, especially highlighting the perspectives of marginalized groups. This is a mathematics that challenges static notions of formalism, as embedded in a tradition that favors the West. For us, the distinction between dominant and critical is not one of acquisition and application, but rather one of aligning with society (and its embedded power relations) or exposing and challenging society and its power relations (Gutiérrez, 2007).

Gutiérrez's (2007) point was that attitudes and practices in mathematics education that align with dominant perspectives of who can and does mathematics lead to inequity. She proposed a way to define equity that implies how both to achieve and to measure it. Borrowing from D'Ambrosio's (1999) trivium—literacy, matheracy, and technoracy—and illustrating with data from a high school that supports Latina and Latino students' participation in calculus courses while enabling them to maintain their linguistic and cultural identities, she posits three criteria for achieving and measuring equity in mathematics education:

1. Being unable to predict students' mathematics achievement and participation based solely upon characteristics such as race, class, ethnicity, gender, beliefs, and proficiency in the dominant language.
2. Being unable to predict students' ability to analyze, reason about, and especially critique knowledge and events in the world as a result of mathematical practice, based solely upon characteristics such as race, class, ethnicity, gender, beliefs, and proficiency in the dominant language.
3. An erasure of inequities between people, mathematics, and the globe.

The first of Gutiérrez's three criteria addresses the acquisition of cultural capital needed to participate fully in the economic life of dominant society. The second, points to students' abilities to use mathematics to analyze and critique injustices in society. The third criterion is clearly far-reaching and seeks to position students, now possessing both dominant and critical mathematics, as active users of mathematics in the service of eliminating social inequities. With these micro and macro criteria of equity in mind, in the next section we consider the changing views of learning that characterize research investigations into the participation of different groups of students in the practices associated with doing and learning mathematics.

Perspectives on Learning: From Individual to Socio-political Approaches

Just as perspectives on disadvantage and equity have changed substantially in mathematics education literature over the past 20 years or so, so too have views on learning. In particular, by the end of the last century, what Lerman (2000) termed the "social turn" had already begun to take place, with socio-cultural theories on learning ever more present and a shift in the balance between those who equate learning with a culture of acquisition and those who focus on the practice of understanding. Following Lave (1990), Sfard (1998) described the two poles on this balance as distinct metaphors for learning—with the metaphor of acquisition emphasizing knowledge as a commodity, as possession, and learning as coming to have, whereas the metaphor of participation posits knowledge as an aspect of the activity or discourse of a cultural domain and learning as a process of coming to belong.

The increasing prevalence of socio-cultural theories (Atweh, Forgasz, & Nebres, 2001), and attention to learning as participation in cultural practices, characterizes the field of mathematics education as a whole and is not limited to those whose research lens is focussed on those who continue to be marginalized players in practices associated with school and university mathematics. Indeed, if researchers use socio-cultural approaches to examine the extent to which students become "successful" participants only in *existing*, privileged mathematical practices or the cognitive behaviours that characterize those already included, then there is a danger of making ever more invisible those whose life experiences lead them to appropriate these practices in ways that differ from a supposed norm. That is to say, if researchers treat learning school mathematics as some kind of general process of enculturation, expecting that all learners, regardless of their differences, experience and appropriate the artefacts that currently compose school mathematics in the same ways, then once again there is a risk of failing to recognize as valid forms of appropriating and using mathematical tools which deviate from the expected, with the result that researchers reinforce discourses that see members of certain groups as somehow innately disadvantaged. This perspective aligns with central premises of the ethnomathematics research program (D'Ambrosio, 2001; Gerdes, 2007; Knijnik, 2002), critical and social justice pedagogy (Frankenstein, 1983, 1998; Gutstein, 2006; Skovsmose, 1994, 2011; Sriraman, 2008; Wager & Stinson, 2012), and culturally responsive mathematics education (Greer, Mukhopadhyay, Powell, & Nelson-Barber, 2009).

Indeed, the kinds of quantitative comparisons between student groups that tend to be used when equity is considered in terms of outcomes are usually made on the basis of high-stakes assessment instruments designed to measure achievement in relation to the current hierarchies of mathematical knowledge of existing school curricula (Gutiérrez & Dixon-Román, 2011). If we accept that learners' appropriations of the artefacts associated with the discipline of mathematics are mediated by cultural tools, then any attempt to judge learners' achievement using tools and practices associated exclusively with what Gutiérrez (2002) calls "dominant mathematics" and Bishop and Forgasz (2007) term "western mathematics" may privilege the participation of certain groups of learners at the expense of others. Perhaps more worryingly, as researchers such as Martin (2009a) and Gutiérrez (2010) have suggested, viewing learning only in terms of enculturation into the dominant culture can imply that learning to succeed is equated with learning to be like those idealized in the dominant culture. For learners who do not fit this ideal, this process, if possible, would involve a denial of their very identity. These concerns confirm that understanding different patterns of participation in school mathematics necessitate more than comparing outcomes: it also involves focussing on the mathematics learner as a cultural being and on investigating how different aspects of this being have an impact upon the particular ways that the practices of school mathematics are appropriated. This returns us to the idea of mathematics learning as a process of appropriation and especially to how the term appropriation might be interpreted. In what follows, we explore two points of view—enculturation and emancipation—as presented in the current literature.

On the one hand, Gutiérrez (2010) has pointed out that not all research that adopts a socio-cultural perspective addresses issues of power or how power relations contribute to the marginalization of certain groups of learners. She suggested that this has led researchers such as Greer, Mukhopadhyay, Powell, and Nelson-Barber (2009), Mukhopadhyay and Greer (2001), Valero and Zevenbergen (2004), and Walshaw (2001) to demarcate between socio-cultural research whose goal is that of enculturation and that research which aims for emancipation. Alluding to a second turn in mathematics education, analogous to the social turn mentioned above, she highlighted the increasing attention to theoretical perspectives and tools of an overtly socio-political nature (see, for instance, Mellin-Olsen, 1987; Valero & Zevenbergen, 2004). Perhaps not surprisingly, although the social now permeates many aspects of mathematics education researchers, it is researchers interested in equity and social justice who are most responsible for this sociopolitical turn, since any comprehensive attempt to challenge the privileges and disadvantages that currently characterize educational institutions involves a political gaze. Gutiérrez (2010) highlighted in particular work emanating from critical mathematics education, critical theory and post-structuralism. These perspectives bring conceptual tools that aim to illuminate how issues of power and identity manifest in mathematics education. They adopt methodologies which emphasize the voices and stories of students from marginalized groups (see, for example, Martin, 2006; Mendick, 2006) and they question perspectives in which cultural identities are used as static cultural markers, instead positing identity construction as an on-going dynamic perspective which reflects how senses of self are continually created and recreated:

> In mathematics education we recognize that learners, practitioners, and researchers are constantly creating themselves—writing themselves into the space of education and society as well as drawing upon and reacting to those constructions. (Gutiérrez, 2010, p. 10)

Within this emerging socio-political tradition, "narratives of self" (Mendick, 2005) and analyses of discursive positioning (Evans, Morgan & Tsatsaroni, 2006) are means to explore the complex and continuous processes by which students develop their identities as mathematics learners in relation to discursive binaries such as masculine and feminine, active and passive, black and white, and so on of the dominant culture. Counter narratives (Stanley, 2007) can serve as alternatives to the dominant discourses, offering stories of struggle, of resistance, of achievement and of success, and hence challenge views which associate deviation from the mainstream with failure and deficiency (Berry, 2008; Berry, Thunder, & McClain, 2011; Martin, 2009a; Stinson, 2006).

On the other hand, although emancipation is clearly an explicit facet on the agenda in socio-political approaches, not all researchers would necessarily agree that it makes sense to dichotomize enculturation and emancipation. It might even be asked what this dichotomization implies about the processes of "enculturation"—does it suggest a process by which all learners should develop identical senses for a particular artefact, regardless of their cognitive resources, or worse, a kind of imposition of cultural norms in which the individual is a passive recipient? On the contrary, it is also possible to view enculturation as part of emancipation and not in binary opposition to it.

At the very least, within the socio-cultural perspectives which have their roots in Vygotsky's work appropriation cannot be viewed as a one-way process (Moschkovich, 2004; Newman, Griffin, & Cole, 1989; Rogoff, 1990). And although both social meanings and personal senses play their parts (Leontiev, 1978), appropriation does not involve a gradual replacement of personal senses by culturally accepted meanings. Rather, it might be characterized as a kind of entanglement of perspectives on an activity, out of which emerges new forms of thinking about the objects in question for all—or for some—of those involved. Hence, the social is always fully present: the activities undertaken and the expressions associated with them being essentially social acts, mediated by *all* the means available to those interacting within the setting in question. This means not only the physical resources and semiotic presentations, but also the cognitive resources associated with the multiple identities which the learners bring to the setting. Hence, it is only when it is assumed that everyone will, or should, appropriate the tools and practices which comprise mathematics in the same way that enculturation becomes equated with imposition. In the following section, we consider the growing corpus of research focussed on how the mathematical agency of learners mediates and is mediated by cultural, cognitive and corporal resources.

Examinations of Multiple Resources for Mathematics Learning

Although the social turn in mathematics education began in earnest only toward the end of the last century, Vygotsky (1978/1930) was already attributing analytic primacy to the social and cultural rather than the individual in the theory he developed during the 1920s and 1930s. A central tenet of his theory is that human beings have a special mental quality which involves the need and ability both to use artefacts to mediate their activities and to encourage the appropriation of these forms of mediation by subsequent generations (Cole & Wertsch, 1996). At particular moments in the history of a given culture, artefacts are created as a response to the demands of particular practices. In turn, these artefacts modify the activities of those using them and, further, can also be modified in use.

Hence, as Cole (1996) argued, "artefacts are the fundamental constituents of culture" (p. 144): in any given setting, a multitude of coordinated artefacts mediate our attitudes and beliefs as well as our social interactions and our actions on the human and nonhuman world. From this perspective, learning mathematics can be described as coming to use artefacts that historically and culturally represent the body of knowledge associated with mathematics. It is important to add two caveats to this definition. First, in the light of the discussion of the previous section, mathematics needs to be viewed in its broadest sense and not restricted to mathematical practices associated exclusively with dominant forms of school mathematics. Second, as the research explored in the next section reveals, the ways in which learners appropriate and use different artefacts should not be expected to be identical for all.

Interplays Between the Sensory, the Material and the Semiotic

Mediation has been well documented in the mathematics education literature (e.g., Bartolini Bussi & Mariotti, 2008; Forman & Ansell, 2001; Moreno-Armella & Sriraman, 2010). The idea that all intellectual activities involve an indirect action on the world is particularly attractive given the nature of mathematics, whose objects depend for their materialization in activity on the mediating presence of some perceivable entity, be it of material or semiotic form. In the context of this chapter, it is perhaps interesting to note that Vygotsky's work on mediation has its roots in his work with blind learners, deaf learners and learners with different disabilities (Vygotsky, 1997).

Bringing arguments characteristically before his time, rather than associating disability with deficit and focussing on quantitative differences in achievements between those with and without certain abilities, Vygotsky proposed that a qualitative perspective should be adopted to understand how access to different mediating resources impacts upon development. This position became associated with his first formulations of the notion of mediation, as he began to discuss the idea that the eye and speech are "instruments" to see and to think respectively, and that other instruments might be sought to substitute the function of sensory organs (Vygotsky, 1997). For example, he argued that, for the blind individual, the eye might be substituted by another instrument. Consistent with his view that artefacts both modify the activities of those using them and become modified as a result of their use, this substitution can be expected "to cause a profound restructuration" of the intellect and of the personality of the blind individual (Vygotsky, 1997, p. 99). That is, since hands and eyes are fundamentally different tools, when one is used instead of the other, it is to be expected that different perspectives on activities they mediate will emerge.

Vygotsky's writings suggested that he was, at least implicitly, attributing to organs of the body—more specifically, to the eye, to the ear and to the skin—the role of tool. This implies that sensory tools should be included alongside material and semiotic resources as mediators in learning. And, rather than using a model that posits students with disabilities as deficient in relation to those without, Vygostsky's stance involves considering how and when the substitution of one tool by another may empower different mediational forms and hence engender different mathematical practices. In this sense, in their investigations of the practices of blind mathematics learners in Brazil, Fernandes and Healy (Fernandes & Healy, 2007a; Healy & Fernandes, 2011) have argued that to understand blind learners, it is important to identify these differences and explore how the particular set of material, semiotic and sensory tools by which blind learners seek to give sense to their activities in the world motivate different forms of participation in mathematics. Pointing to some of the differences associated with seeing with one's hands and seeing with one's eyes, Fernandes and Healy explored how tactile means of accessing visuo-spatial information became associated with the highlighting of certain mathematical abstractions by blind mathematics students. Although mathematically valid, these abstractions were not always those that the teacher was intending to highlight in the teaching situation and tended to be expressed both bodily and linguistically in accord with the dynamic manners in which the learners' hands explored artefacts used to represent mathematical objects.

This last point is illustrated in a case they report in which two blind learners explored reflective symmetry (Fernandes & Healy, 2007a). One of the learners was blind from birth while the other gradually lost his sight over a 10-year period, becoming completely blind only at the age of 15 years. There were differences between approaches to symmetry adopted by the two students. For example, the student who had never had access to the visual field tended to treat geometrical objects as dynamic trajectories and attempted to look for invariance relationships among the sets of points which defined the trajectories; the second student attempted to characterize the objects he was feeling in terms of objects he remembered from before he lost his sight. Nevertheless there were also similarities. Notably, both students tended to move their hands or corresponding fingers from each hand in a symmetrical manner over the materials they were exploring. This was not something that the researchers had anticipated in the design of the tasks—which had been developed based on research into sighted learners' understandings of symmetry and reflection. Fernandes and Healy suggested that concentrating more specifically on how blind learners use their hands to conceive mathematical objects might highlight the existence of alternative learning trajectories for those with or without visual impairment—or at least differences in preferred routes to mathematics.

A similar point was made by Nunes (2004) in relation to deaf mathematics learners. She argued that, in the light of the results of recent studies highlighting the role of visuo-spatial representations in the mathematics learning of the deaf and the hard of hearing (Bull, 2008; Kelly, 2008; Nunes & Moreno, 2002), the participation of deaf learners in mathematical activities might be prejudiced if tasks are consistently presented to them in forms that privilege serial over spatial coding. Nunes maintained that deaf learners should be given opportunities to learn to use their preferred and superior visuo-spatial abilities to represent and manipulate the sequential information within mathematical problems.

Marschark, Spencer, Adams and Sapere (2011) also stressed the need to teach to the specific strengths and needs of deaf and hard-of-hearing (DHH) learners. Their view was that there has been a general assumption in approaches to teaching that these learners are "simply hearing children who cannot hear" (Marschark et al., 2011, p. 4). This practice, they argued, is misplaced, as it does not take into account the specific cognitive and language abilities of DHH learners. As far as mathematics learning is concerned, alongside the importance of visuo-spatial representations, they pointed to other factors that have an impact on the participation of DHH in school mathematics—including early experiences with quantitative concepts (Bull, 2008; Nunes & Moreno, 1998), limited opportunities for informal, incidental mathematics learning (Nunes & Moreno, 2002; Pagliaro, 2006), and sensory and language differences in how those with or without hearing loss process information (Marschark & Hauser, 2008). This is consistent with Mayer and Akamatsu's (2003) position that in designing learning activities for DHH students, it is necessary to take into account the sensory modalities available to them and to ensure they have opportunities to appropriate and manipulate all possible mediational means at their disposal.

Much of the research related to DHH learners has focussed on language issues rather than mathematics and, even when mathematics learning is under study, as Bishop and Forgasz (2007) noted, language fluency is frequently cited as a factor which contributes to the differential engagement of DHH learners with mathematics problems (Kelly, Lang & Pagliaro, 2003; Pagliaro, 2006). Fluency in the language of instruction is an issue which has implications for participation in mathematics learning activities for many marginalized students, not only those with hearing loss. In the next section, we turn to questions addressed in the literature concerned with equity, language and mathematics learning.

Language and the Mediation of Mathematics Learning

Not surprisingly, the central stage given to language has resulted in the application of socio-cultural perspectives by researchers investigating the mathematics learning of those who are bi- or multilingual (Civil, 2009; Moschkovich, 2002, 2007). Such studies avoid a deficit view of linguistic minority students by discussing all their language options as potential cognitive resources that may contribute to their appropriation of mathematical knowledge. Setati and Moschkovich (2010) took this point a little further, arguing that rather than comparing the performances of bilinguals or multilinguals to monolinguals in situations which privilege only the language of the dominant monolingual group, research should better "focus on the multiple ways that bilingual learners might describe mathematical situations" (p. 3). Indeed, this position also applies to those who communicate using sign languages as well as to speakers of variants of a dominant language such as Black English Vernacular (BEV) speakers. With a culturally appropriate pedagogy, they as well as speakers of the dominant variant (e.g., Standard American English), could enjoy access to multiple perspectives and expressions of mathematical ideas. From the point of view of equity, this underlines how difference cannot be understood as deficiency. On the contrary, there is an implication that access to more than one language might be associated with positive benefits for the mathematics learner. Empirical support for this claim can be found in the work of Clarkson (2006).

In this vein, and drawing from the work of Grosjean (1985), Setati, Adler, Reed, and Bapoo (2002) argued that bilinguals and multilinguals have a unique and specific language configuration, and hence it makes little sense to consider their linguistic abilities as the sum of two or more complete or incomplete monolinguals. The question then arises of the impact of this unique language configuration on their mathematical practices. One difference is that when learners are bi- or multilingual, their mathematical activity is not necessarily confined to one or other of their languages. Planas and Setati (2009) and Moschkovich (2007) described how bilingual learners switched between their two languages during mathematical activities. How and when these switches occurred did not relate only, or even necessarily, to the relative proficiency in one language over another—rather they were more complex, being interweaved with the social circumstances in which the activity took place and infused with questions of power and status. That is to say, for many students, and especially those from immigrant or indigenous groups who learn mathematics in a language that is not their first, linguistic identities and activities are intertwined with cultural identity.

Multilinguals and Cognitive Resources

The multiplicity of linguistic and cultural diversity that exists in some countries challenges educational institutions and teachers to provide equitable instruction so that all students are respected and develop their intellectual potential, especially in mathematics. School children whose cultural and linguistic backgrounds differ from the institutional culture and language of schools often confront cognitive obstacles that are invisible and incomprehensible to others, and are viewed as a disadvantage in mathematics classrooms (Garcia & Gonzalez, 1995).

To understand sources of disadvantage for linguistic minorities in mathematics classrooms, Vazquez (2009) and Powell and Vazquez (2011) investigated differences in mathematical thinking of two groups of Spanish-dominant fourth graders in a poor urban community in the northeast of the USA by examining their problem-solving representations. Powell and Vazquez (2011) analyzed students who received bilingual instruction in Spanish and English and students who received instruction only in English. In the classroom, the former group of students was allowed to use Spanish and English as they wished in their discursive interactions, but the latter group of students was expected to use English only. The researchers examined how each group of students built mathematical representations in language and with inscriptions, focussing particularly on the discursive interactions as students within a group justified and attempted to persuade each other of their results. The researchers found that the group of students who communicated bilingually moved fluidly between English and Spanish, showed greater facility in solving the problem task and built more refined representations. The group that was expected to communicate in English only experienced difficulty in their discursive interactions. Compared to the English-only group, the bilingual group had greater ease in communication and construction of mathematical representations.

For some researchers, results such as these are interpreted as indicating the existence of differences in the cognitive practices of bi- and monolinguals. According to Hagège (1996), for example, bilinguals have a greater cognitive elasticity than monolinguals. Furthermore, investigating the plasticity of the bilingual brain, Mechelli et al. (2004) tested the density differences of the gray and white mass of the brain among monolingual and bilingual individuals. Their results revealed that the grey mass in bilingual individuals is larger than that of monolingual individuals. They found that the human brain undergoes structural changes in response to the environment, including the learning of new languages.

Irrespective of whether fluency in more than one language leads to structural differences in the brain, being multilingual offers advantages for learning mathematics. Internationally, mathematics education researchers have paid increased attention to how multilingualism relates positively to cognitive development, flexibility, and the promotion of academic achievement in learners (Adler, 2001; Gorgorio & Planas, 2001; Moschkovich, 1999; Setati, 2002; Setati & Adler, 2000). However, instructional environments may prejudice the participation and performance of multilinguals when they do not invite and encourage them to use their rich linguistic resources for mathematical sense making.

Taken together, these research studies suggest that those who learn mathematics in a language that is not their first may experience it in different ways than monolingual learners. Equity in participation may therefore require the recognition that particular linguistic resources support particular mathematical practices. To stress this point, we return to the case of deaf learners whose first language is a signed rather than a spoken language. Most of the research related to bilingual learners within the mathematics education literature relates to those with some or complete fluency in two spoken languages. Many deaf people have a sign language as their first language and the written version of the mainstream language within their country as a second. Although sign languages are now regarded as true, natural languages (even if this recognition only began to come about in the 1960s and 1970s after Stokoe's

(1960/2005) work), there are some differences between signed and spoken languages. In particular, sign languages are visual-gestural whereas spoken languages are serial-auditory, with simultaneity a pronounced feature of sign languages (Mayer & Akamatsu, 2003). Nunes (2004) reported on a strategy, spontaneously developed by students in a number of studies with British deaf learners, which involved simultaneous counting up through the number signs on one hand and down on the other in order to arrive at the sum of two whole numbers. Given the task of adding 8 and 7, they would sign 8 using one hand and 7 on the other, then they would count down through the signs from 7 to 0, while counting up from 8 at the same time (left hand: 7, 6, 5, 4, 3, 2, 1, 0; right hand 8, 9, 10, 11, 12, 13, 14, 15). For students using spoken language, this strategy would be rather difficult to perform purely linguistically. It could of course be modelled using concrete material, but this is not the point—the simultaneity of sign languages in this case associated with the spontaneous use of a perfectly valid strategy not usually observed among those who speak with their mouths.

The evidence from research with bi- and multilingual learners reported in this section suggested that particularities associated with the language(s) in use in mathematics learning scenarios had an impact on the mathematics practices that developed within them. Understanding these particularities is important for including students from language minorities, as is recognizing that language is a central aspect of the learner's identity both in the mathematics classroom and beyond. The research also indicated that the ways that learners feel that they can use, or not use, their various language resources, and the ways that they experience the valuing of certain languages, are likely to have consequences for their participation within the mathematics classroom. Identifying how minority languages and multilingual learning contexts empower alternative—valid—mathematical strategies represents a central research challenge, which may contribute not only to increasing the participation of groups previously marginalized or excluded, but also to understanding learning mathematics as a whole.

Power and Disadvantaging Linguistic Resources

The positive resource of language for mathematical cognition notwithstanding, extra-cultural processes can cause a syntactic or semantic resource to be or be experienced as a disadvantage. Here we discuss two examples. Students can experience difficulties learning mathematics when their linguistic heritage suffers uncritical adoption or imposition of distinct and distant cultural and linguistic conventions. In the People's Republic of China, even educated adults experience difficulties reading multi-digit numerals—for instance, 1,335,013,694—without first pointing and naming from right to left the place value of each digit before knowing how to read the "1" in the billions place and the rest of the numeral. Powell (1986) reported that this state of affairs results from an extra-cultural, syntactic convention of delimiting digits in a many-digit numeral that varies from the linguistic structure of Mandarin numeration. Conventions for delimiting digits in a many-digit numeral serve to facilitate reading it and saying it aloud. In contrast to certain Romance and Germanic languages of the West, where commas, spaces, or points are used in accordance with the linguistic structure of those languages to delimit groups of three digits, the linguistic structure of naming numerals in Mandarin is instead based on groups of four digits. A western convention for delimiting digits in a many-digit numeral does not facilitate a Mandarin speaker's reading of it. For a Mandarin speaker to read with ease China's approximate population figure—1,335,013,694—the numeral should be delimited alternatively like this: 13 3501 3694.

This example draws attention to how experience in learning mathematics arises from the interaction of language, mathematics, and power. Researchers have examined the nature and causes of mathematics learning difficulties manifested when educators adopt curricula for use in a cultural and linguistic milieu distinct and distant from the one for which the curricula were developed (see, e.g., Berry, 1985; Orr, 1987; Philp, 1973). Based on his analysis of problems in second-language mathematics learning in Botswana, Berry (1985) put forward a general theory of types of language-associated learning problems, consisting of two categories. Of interest here is his second category of problems, those that "result from the 'distance' between the cognitive structure natural to the student and implicit in [the semantics of] his mother tongue and culture, and those assumed by the teacher (or designer of curriculum or teaching strategies)" (p. 20). Adding to the notions of semantic and cultural differences by which Berry defined the term "distance," the example presented by Powell (1986) suggests that there are *syntactic* differences, as well.

The issues of power as well as semantic, syntactic, and cultural differences all figure in the second example. It examines attitudes of some educators toward the linguistic variant of English that some African Americans speak in the USA exemplified in *Twice as Less: Black English and the Performance of Black Students in Mathematics and Science* (Orr, 1987). Orr taught at a white, middle-class private high school in Washington, DC, to which a group of urban, African-American students were given places for a number of experimental years. When these students performed poorly in mathematics and science, she and her colleagues questioned why. Explanations focussed on linguistic features of the work done in class and at home. Orr and her colleagues found "explicit evidence" that African-American "students were using one kind of function word, prepositions, in a manner different from other students; their misuses [sic] were different even from the misuses with which [they] were familiar" (p. 21). That is, the semantic and syntactic use of words similar to Standard American English (SAE) by students speaking Black English Vernacular (BEV) were different from those used by students who belonged to the culture with power. Orr concluded that this linguistic difference *was* the reason why African-American students did poorly: "For students whose first language is BEV, then, language can be a barrier to success in mathematics and science" (p. 9). Furthermore, she claimed that, unlike the grammar of BEV, "the grammar of standard English [SAE] has been shaped by what is true mathematically" (p. 158). She offered no substantiation for this claim of a supposed intrinsic superiority of the language of a culture of power, and, as a result, appears to distort connections between conceptual understanding and semantic and syntactic differences. As linguists, like Labov (1972), have demonstrated, as with any other language, BEV and SAE are both capable of generating labels for concepts attended to by the culture of the speakers. The effect of Orr's viewpoint is to confer privilege on the culture and language (SAE) of the dominant power and, thereby, to deny legitimacy to other culturally-based linguistic and cognitive experiences.

These examples of how power can disadvantage the linguistic resources of students illustrate the work that mathematics education researchers and others in society have yet to accomplish in order to achieve Gutiérrez's (2007) first criterion of equity. We now turn attention to research on the mathematics learning of those diagnosed as needing "special education."

Equity and Disability: Research into Specific Difficulties in Mathematics Learning

At the beginning of our review of research documenting the mathematical agency of different groups of mathematics learners, we pointed to a general shift towards social and political perspectives in research related to the search for more equitable mathematics classrooms. To end the section we return to this theme, looking more closely at the literature concerning the mathematics learning of one group of learners: those described as having special education needs in mathematics. A first question that arises in relation to the label "special educational needs" is how to decide which students are included. Gervasoni and Lindenskov (2011), who preferred to use the expression "students with special rights for mathematics education," drew attention to this challenge and the lack of any universally accepted definition. They focussed on two groups. The first group encompassed learners with disabilities defined by the United Nations convention on the rights of persons with disabilities, as having long-term physical, mental, intellectual or sensory impairments which in interaction with various barriers may hinder their full and effective participation in society on an equal basis with others (United Nations, 2006, cited in Gervasoni & Lindenskov, 2011).

The second group they delineated was those who underperform in mathematics. Deciding and defining who should be classified as a member of this second group raises a multitude of questions for those interested in issues of equity and social justice. Referring to special education more generally, O'Connor and DeLuca Fernandez (2006) referred to the first group as a "non-judgemental" category of special education, and the second as a "judgemental" category.

In a review of the literature related to students in both groups, Magne (2003) claimed that the move toward social and cultural interpretations was only beginning to emerge in this particular area of research, with many of the studies surveyed concentrating on the search for neurological explanations. His claim referred mainly to the literature related to the "judgemental" category, the members of which are characterized in relation to some notion of low achievement. Like O'Connor and DeLuca

Fernandez, Magne's view was that low achievement is a social construct, "not a fact but a human interpretation of relations between the individual and the environment" (p. 9). However, he believed that this relativist view does not represent the dominant view in much of the research in this area. To explore his claim, we consider the literature related to "dyscalculia" as a condition associated with specific difficulties in learning mathematics as a case in point.

To a certain extent, the migration of the term "dyscalculia" from neuropsychology to education underlines the prevalence of the search for neurologically-based explanations for the low performances of learners identified as experiencing particular difficulties in participating in the practices of school mathematics (Munn & Reason, 2007). According to the neuropsychological perspective, difficulties in learning mathematics (or rather arithmetic, since the majority of studies confine their attention to this area of mathematics) are associated with a cognitive "disorder" or a specific "learning disability." Gifford (2005), in her review of the dyscalculia literature, suggested that it is still not clear that dyscalculia can be considered to be associated with a specific cognitive deficit since there is not even a robust consensus on what precisely are its defining characteristics, aside from poor recall of number facts. Although she did not discount the possibility that there may exist differences between individuals in the neurological processing of number, she concluded that there is no firm evidence linking particular brain deficits with mathematical difficulties and pointed to several criticisms of the exclusively neuropsychological approaches. One critique related to a particular view of mathematics that has been adopted by some involved in building brain-based explanations for learning difficulties. These researchers tend to determine mathematical performance in relation to mainly knowledge of arithmetic facts and procedures, and pay little attention to conceptual understanding.

Even in relation to calculating procedures, Gifford (2005) was concerned that neuropsychologists make assumptions about what procedures should be tested to diagnose learners and what calculation procedures are considered as "normal." For example, she cited Geary's (2004) study in which it was suggested that students with dyscalculia have problems in sequencing the steps in adding numbers with more than one digit in column arithmetic. In Geary's study, a strategy of adding, for example, 45 and 97, was described in terms of the paper and pencil algorithm of arranging the numbers into the correct columns and "carrying the 10." Other possible strategies, such as adjusting the numbers to 42 and 100, appear not to have been taken into account. Furthering this critique, Ellemor-Collins and Wright (2007) offered evidence that the collection-based strategies which underline the written algorithm are not necessarily the most efficient for all learners, and that for some learners sequence-based strategies (keeping the 45 whole and counting on first 7 and then 90) tend to correlate with more robust arithmetic knowledge—especially for those previously identified as low achievers.

Such critiques support the view that the nature of the mathematics involved and students' experiences of this mathematics are factors to be taken into account if we wish to further our understandings of difficulties that learners have when participating in school mathematics. Adopting this position, Magne (2003) pointed to

D'Ambrosio's (2001) work in ethnomathematics and suggested that cultural and sociological interpretations of students' reactions to particular mathematical topics should also figure in attempts to understand underperformance in mathematics. Gervasoni and Lindenskov (2011) also stressed the influence of the mathematics background against which achievement is being assessed, arguing that low mathematics achievers are those "who underperform in mathematics due to their explicit or implicit exclusion from the type of mathematics learning and teaching environment required to maximize their potential and enable them to thrive mathematically" (p. 308).

Another possible problem underlying some of the research seeking to identify neurological causes for mathematical difficulties is an assumption that all students learn the same way. This assumption can become a self-fulfilling prophecy when it translates into teaching programs based on the premise that classrooms consist of a relatively homogenous group of students who will all gain the same value from the same type of experience (Ginsburg, 1997). Neither learning difficulties nor response to teaching interventions can be expected to be homogenous, as Ann Dowker (1998, 2004, 2005, 2007) has shown in her extensive exploration of arithmetical difficulties of young mathematics learners. Dowker's view is that arithmetical ability is not unitary, but composed of a variety of components. Students who have difficulty with one component will not necessarily experience difficulty with others, although, without teaching intervention specifically aimed at the problems that an individual learner is experiencing, difficulties in different components may come to be correlated over time for a variety of reasons—not the least of which is an increasing perception by the learner that they are "no good at mathematics" (Dowker, 2007). As well as investigating the wide range of arithmetical difficulties, Dowker also considered research related to how students might be supported in overcoming such difficulties. After reviewing a number of early intervention programs, most of which were carried out in the UK, she concluded that although many learners have arithmetic difficulties, many of these can be overcome if appropriate teaching interventions are made. She wrote:

> No two children with arithmetical difficulties are the same. It is important to find out what specific strengths and weaknesses an individual child has; and to investigate particular misconceptions and incorrect strategies that they may have. Interventions should ideally be targeted toward an individual child's particular difficulties. If they are so targeted, then most children may not need very intensive interventions. (Dowker, 2004, p. 45)

Gervasoni and Sullivan (2007) analyzed data collected during more than 20,000 assessment interviews aiming to identify learners in Australia with difficulties in learning arithmetic, and arrived at a similar conclusion. They stressed that "there is no single 'formula' for describing students who have difficulty learning arithmetic or for describing the instructional needs of this diverse student group" (p. 49). Like Dowker, they too emphasized that a learner who has difficulty in one aspect of number learning will not necessarily have difficulties in all (or even any) others.

Although these findings do not rule out neurological explanations for difficulties in learning about number, they suggest that it may not be appropriate to label the difficulties experienced by many mathematics learners as learning disabilities.

Instead, the evidence indicates that under the right conditions and interventions, many of those experiencing specific arithmetical difficulties can, and indeed do, learn. Moreover, even if it is the case that different students process numbers differently, cognitive factors are not the only factors which influence student performance. Emotional and attitudinal factors as well as socially mediated factors such as curriculum and teaching approaches have also been suggested as likely to be involved.

The recognition that low achievement is a social construct and not simply an individual characteristic has contributed to the recent growth in socio-political interpretations of how students of mathematics come to be defined as underachieving. These have been largely founded on critical theories and perspectives from disability studies. For example, Borgioli (2008) presented a critical examination of learning disabilities in mathematics within the USA. Like Magne, she referred to the prevalence of brain-based explanations and argued that labelling a child as someone in need of special mathematics education involves determining "normal" or "ideal" achievement, and positioning those that deviate from this norm as problematic and in need of remediation. Her view is that the school rather than the learner benefits most from the labelling, since "locating the obstacle within the brains of the individual offers a convenient explanation for student failure" (p. 137). Reflection on the mathematics curriculum and how it is offered is avoided because it is the low achieving students who are seen as the problem. Woodward and Montague (2002) described how frequently "the solution" involves removing the "special learner" from the mainstream classroom for highly directed training with specific step-by-step problems, since the practices associated with special mathematics education in the USA have "a history of placing a considerable emphasis on rote learning and the mastery of math facts and algorithms" (p. 91).

The process of "othering," that is to say, framing students who differ from the socially and politically defined norms as outsiders, can have the effect of perpetuating inequitable practices, since it legitimizes exclusion. Indeed, in many countries, concern has been raised about the disproportionate representation of ethnic minority students, indigenous students groups and those living in poverty in Special Education programs (Artiles, Klingner, & Tate, 2006; Dyson & Gallannaugh, 2008; Mantoan, 2009; McDermott, 1993). Although this is not an issue specific to mathematics education, it is important that it is not brushed aside. Ways in which the culture and organization of schools constrain the achievement of particular groups of students, at times even pathologizing their bodies and behaviours, need to be further studied, especially in relation to the labelling of underachievement (O'Connor & DeLuca Fernandez, 2006).

As we end this section, it is worth commenting that learners with special educational needs and learners with disabilities have, until relatively recently, been largely absent from the mathematics education literature related to equity and social justice. The questions of how these learners become less peripheral participants in the micro-practices of mathematics classrooms, and what macro-social conditions are necessary for their inclusion in the sense proposed by Pais and Valero (2011), are particularly important areas that urgently need to be addressed by future researchers.

Research into Practice: Considerations of Equity in Teacher Education

Although there have been changes in discourses surrounding students whose identities do not conform to the dominant norms within the mathematics education research community, and new associated research foci have emerged, if these changes are to have an impact on mathematics classrooms it is critical that all principal actors be involved—from policymakers and researchers to teachers and students. In this section, we examine the developing strategies for involving teachers in challenging the social processes which sustain disadvantage and in preparing them to create mathematical learning scenarios based on respect, justice and equity—and by so doing make progress towards reaching, at least, Gutiérrez's (2007) first equity criterion. We should stress that the issue of preparing teachers for equity is not new. It has long been considered an issue for inclusion in preservice courses, since prospective teachers tend to have limited experience interacting with cultures outside of their own (Grant & Secada, 1990). For the professional development of inservice teachers, equity too was on the agenda in the 1990s, with Little (1993) questioning the adequacy of a *training model*, "a model focussed primarily on expanding an individual repertoire of well-defined and skilful practice" (p. 129), for preparing teaching for the aspects of teaching and schooling in a changing society. Regarding equity and diversity, she argued that a new perspective was needed for the professional education of teachers, one in which collaboration and the establishment of teaching networks played a central role.

Concerning the preparation of mathematics teachers, Matos, Powell, and Stzajn (2009) argued that the last 20 years have indeed seen a shift from models based exclusively on training to more those requiring more practice-based professional development. They associated this shift with the move discussed above from seeing learning as a process of individual acquisition of knowledge to understanding it as the appropriation of forms of participation in social practices. In their chapter, Matos et al. (2009) did not explicitly consider the move towards more practice-based models of teacher education in relation to the challenge of deconstructing disadvantage. Nevertheless, that connection could be important because any approach to preparing teachers for cultural diversity based on a model in which sensitivity is treated as something that can be trained rather than experienced would seem to be doomed to failure. However, it is recognized that the implementation of practice-based models will not necessarily guarantee more inclusive approaches to teaching.

Though not specifically related to mathematics education, Jennings's (2007) survey of how diversity was addressed in 142 public university elementary and secondary teacher preparation programs across the USA suggested that some attention was being given to diversity in all the courses surveyed, and that diversity topics were included in many different aspects of the programs (including foundation courses, teaching methods courses and teaching experiences). This study indicated similar patterns across both elementary and secondary programmes in how diversity topics were prioritized. In both cases, race and ethnicity were the most emphasized forms of diversity, followed in order by special needs, language, social class, gender, and finally sexual orientation. No information was given on whether the prioritization patterns applied equally across all areas of subject areas, but other studies have confirmed that, although gender has been, and continues to be a major point of debate within the mathematics education community, it apparently remains underexplored in the area of mathematics teacher education.

In their survey of the European literature, Hourbette, Baron, and Khaneboubi (2008) were "forced to acknowledge" the existence of relatively few contributions focussing on the field of gender issues in mathematics teacher education. Battey, Kafal, Nixon, and Kao (2007) suggested that programs which address gender in relation to the education and professional development of mathematics teachers, and teachers of other STEM subjects, lack elements essential for the effective promotion and implementation of equity principles in the classroom. Elements that Battey et al. pinpointed as central included inquiry, collaboration, a focus on classroom practice, and consideration of the larger social and political context. They stressed that inquiry, in particular, was important in professional development related to equity in mathematics learning because it can be considered with respect to the teaching institutions involved as well as to the subject matter, teaching practices, and teachers' attitudes and beliefs. The suggestion was that to achieve more equitable mathematics classrooms, the teacher needed to become an active participant in researching and interpreting their students' learning, and should engage in the processes of reflecting on their beliefs about the mathematics that different students do and how they do it. We now turn to research in which explores how teachers might be involved in such activities.

For some researchers, an important first step is to involve teachers in deconstructing disadvantage and moving away from views of differences as deficits. For example, among the concerns that Aguirre (2009) raised about privileging equity and mathematics in preservice and inservice teacher education courses, was the need to develop strategies to confront resistance among practising and future teachers to both ideological change and pedagogical change. Her work centred in particular on Latino/a learners in US classrooms. Among the resistances of an ideological nature that were identified, she focussed in particular on the need to challenge what she called the "recycling of the cultural deficit position in mathematics learning" (p. 308). In a similar vein, a recent review of European research into teacher education and inclusion concluded that any teaching is likely to be ineffective where the dominant belief system is one that "regards some students as being 'in need of fixing' or worse, as 'deficient and therefore beyond fixing'" (European Agency for Development in Special Needs Education, 2010, p. 30).

In light of results such as these, we would argue that a common factor identified among those working to understand inequity and to undermine approaches which sustain it is the need to support teachers, at the earliest possible opportunity (preferably before they start teaching), to develop positive attitudes towards the learning possibilities of students from marginalized groups and to understand larger social and political forces which support inequity and position students as disadvantaged. Some indications of the methodologies and activities by which this might be achieved can be found in the literature related to teacher education and cultural sensitivity, as well as in attempts to involve teachers in investigating the mathematics of different student groups.

Teacher Education and Cultural Sensitivity

A consensus among mathematics education researchers concerned with preparing teachers to work with diversity and for equity is that any attempt to understand disadvantage brings into play questions of social justice. This in turn implies that those entering and working in the teaching profession today should "understand the historical, socio-cultural and ideological contexts that create discriminatory and oppressive practices in education" (Ballard, 2003, p. 59). Although Ballard referred to inclusive education in general, an examination of some of the teacher education programs in mathematics education which have explicitly addressed the question of equity shows that at least some mathematics educators agree that socio-political understanding does indeed merit centre place. Gutstein (2006), for example, identified three essential knowledge bases for teaching mathematics for social justice: classical mathematical knowledge, community knowledge, and critical knowledge. Similarly, in considering the question of what teachers need to know to support learners in bilingual and multilingual classrooms, Moschkovich and Nelson-Barber (2009) stressed the importance of addressing issues related to cultural content, social organization and cognitive resources. They contended that the ways that different learners come to know often represent values and beliefs which are specific to their cultural identity, and that it is these identities that mediate their preferences for adopting forms of thinking, observing, acting and interacting in the mathematics classroom.

Unless they have knowledge of how mathematics might appear and be expressed in the practices of different cultures, teachers can believe there is only one ("western") mathematical discourse, and even that, in some multilingual contexts, if this is not expressed in the dominant tongue, then students are somehow failing to engage in mathematical discourse at all. Gay (2009) stressed that preparing teachers to work with ethnically diverse students requires a deep and broad knowledge base concerning the cultures, histories and heritages of different ethnic groups. Hughes et al. (2007) suggested that one way in which teachers can become more aware of the mathematics which their students engage in outside of school is to create knowledge exchange programs and activities which explicitly aim to make connections between learners' activities at home and at school. They described how the Home School Knowledge Exchange Project in the UK opened a channel of communication that brought teachers into contact with the variety of ways by which their students had contact with mathematics in different aspects of their life. One example they described was how this communication channel enabled teachers to understand the differences between the finger-counting strategies they were emphasizing at school and those in use in the homes of some of their students of Bengali origins—in which it was often the case that three sections on each finger were counted rather than just a single finger.

Despite these recent attempts to improve our understanding of the processes of both preparing and supporting teachers to work in cultural diverse settings, Johnson and Timmons-Brown (2009) have argued that teacher development courses that are not tailored to "generic" populations of mathematics learners remain relatively rare. This would suggest that, notwithstanding the progress we have made as an academic community to understand, interpret and challenge disadvantage, we still need research in which teachers and future teachers, as well as researchers, have opportunities to examine in more detail the mathematical practices of particular populations. Although the move towards a more practice-based education (Matos et al., 2009) can be seen as a move forward in this respect, since the contexts in which practitioners work would be central to the courses, Johnson and Timmons-Brown (2009) have warned us that there is still a danger that the focus will continue to be what works in existing practice, with the result that teacher preparation will continue to be governed by the dominant voice and prospective teachers will continue to learn to teach using strategies that advantage dominant groups.

Essentially, we interpret this as a call for new practices for teaching, fine-tuned to the lives, the strengths, and the needs of particular groups of learners. Such a call necessitates the fostering of reciprocal relationships between teacher educators, researchers, teachers and future teachers. It is to examples of research projects born out of such collaborations that we now turn our attention.

Investigating Difference Collaboratively

Critical to changing teachers' perceptions of students from marginalized groups is a change in perspective: removing the "do not" from the phrase "what students do not do" so that it becomes "what students do." That is, the focus needs to become how students' mathematical ideas develop—and the pedagogical strategies appropriate to support their development—rather than the difficulties that students experience (Jaworski, 2004; Wood, 2004). Participation in research studies and in research-based teacher development programs appears to offer possible means of promoting such a shift.

Willey, Holliday, and Martland (2007), for example, reflected on how collaboration in the Mathematics Recovery Project, a program aimed at meeting the needs of young learners experiencing difficulties in the area of numeracy, influenced teachers in a region of the UK. They reported that teachers who participated in the Mathematics Recovery Project developed an enhanced faith in students' abilities to solve problems by themselves. In addition, the teachers became more confident in their ability to assess what their students knew, and what were thinking, and to offer appropriate support to help the students learn. Thomas and Ward (2001) arrived at a similar conclusion in relation to the increased understanding of numerical concepts and principles among teachers who participated in similar intervention programs in

Australia and New Zealand. Although the intervention programs themselves were aimed primarily at children identified as underachieving in mathematics, it was reported that participating teachers developed their own understandings of numbers and children's learning about numbers. In these projects, teachers in the programs were participants in professional development courses and did not, apparently, act as researchers in their own right.

Another strategy increasingly used in a variety of research contexts involves projects based on sustained collaborations between researchers and practitioners, and the development of research methodologies which recognize not only the need to interleave the theoretical with the practical but also to make connections between the micro issues of individual learning and the macro issues related to the context in which this occurs. Methodologies that characterize these collaborations include participatory action research in which participants work together to conduct a process of co-generative inquiry (Greenwood & Levin, 2000), as well as methods associated with design research—especially in relation to what has been described as multi-tiered teaching experiments (Lesh & Kelly, 2000).

Regardless of the particularities of the research methodologies used, we can locate a number of examples in the mathematics education literature of how participation in a research project supported teachers in focussing on the inclusion of previously marginalized groups within their classrooms. Here, we mention two examples. The first is the *Informal Mathematics Learning Project* (Powell, Maher, & Alston, 2004; Weber, Maher, Powell, & Stohl Lee, 2008), a research-based, professional development model for mathematics education that aimed at engaging teachers in attending to and reflecting on the development of students' mathematical ideas and reasoning, and on using their reflections to inform their own teaching practices. The project was conducted as an after-school program in a partnership between the Robert B. Davis Institute for Learning at Rutgers University and the Plainfield School District, New Jersey, an economically disadvantaged, urban community whose school population was 98% African American and Latino. In the context of the ideological narrative of closing the racial achievement gap in US society, which embodied assumptions relating to a supposed intellectual inferiority of African-American and Latino/a students (Martin, 2009b), the project aimed at providing a counter narrative. The approach involved the teachers in documenting the students' development of mathematics ideas and forms of reasoning, and attending to these developments so that they could access these students "having of wonderful ideas" (Duckworth, 1996).

Our second example is the project *Towards an Inclusive Mathematics Education*, which began in São Paulo, Brazil, in 2002, as practising mathematics teachers enrolled in a post-graduate course expressed a desire to improve their understanding of how they might work with the students with disabilities who were beginning to join their regular classes—which was something the teachers felt neither preservice nor inservice courses had prepared them to do (Fernandes & Healy, 2007b, Healy, Jahn & Frant, 2010). Since the project began, a series of sub-projects have been carried out. Each has involved the establishing of partnerships between school- and university-based participants in designing and evaluating learning scenarios either for blind learners or for deaf learners.

Just as discourses related to gap-gazing led to students from certain racial, linguistic and social minorities being seen as lacking in mathematical ability, discourses about students with disabilities have also been infused with narratives underestimating their mathematics learning potential (see, e.g., Gervasoni & Lindenskov, 2011). Emerging from the work of those participating in the *Towards an Inclusive Mathematics Education* project have been alternatives which challenge traditional narratives. One factor that is critical to empowering students who lack access to one or other sensory field is access to communicational tools that enable them not only to access conventional forms of mathematics but also to express mathematical ideas in innovative ways which make sense to them. One outcome of the project has been the collaborative development of digital tools which support new means of expressing mathematics by capitalizing on the forms of reasoning available to the participating students, including the use of sound, touch, movement and visual–spatial representations. One of the findings associated with teachers' involvement in developing and using these classroom tools is that collaborating teachers seem open to accept the potential and legitimacy of rather unconventional expressions of mathematical objects, properties and relations. This led Healy, Jahn, and Frant (2010) to conclude that:

> Those working with the deaf and with the blind seem to come to the design process already with an acceptance that conventional mathematical expressions alone are not always accessible to their students. The need for new expression is hence legitimized from the start. As other teachers evidence the mathematical practices afforded by these tools, it may be, although this is as yet is an untested conjecture, that they judge that these practices would also be beneficial for all of their students. (p. 402)

The message from this project seems to be that when teachers become involved in researching how deaf and blind students develop mathematical ideas and reasoning, not only do they, along with the university-based researchers, become more sensitive to the value of a variety of ways of accessing and representing mathematical ideas, but they also fine-tune their understandings of the particular abilities of blind students, and deaf students. Furthermore, they begin to reflect on how the novel approaches by which the participation of these students was encouraged might also be appropriate for the rest of their students. That is, rather than seeing the minority student as disadvantaged because the ways they experience the world do not correspond to the supposed norms, when teachers attend to their students' experiences this can open new windows on what they come to recognize and value as mathematical practices. That, in its turn, may open new windows for the teachers, as they learn to interpret how a wide range of students learn mathematics.

Reflection

In this chapter, we have drawn attention to how the recent increased attention accorded to socially and politically motivated accounts of mathematics learning have contributed to a shift away from associating disadvantage with innate or static characteristics of individual students or groups. These traditional practices and societal discourses resulted in the disadvantaging, and alienating of many students. The shift has been towards socio-cultural approaches to mathematics education, by which researchers and teachers have come to recognize learners as culturally-situated and embodied beings. This new focus has enabled researchers to identify the mathematical potential of previously marginalized students. The new focus has been on identifying qualitative differences mediated by cultural, linguistic and sensory tools, rather than on measuring quantitative performance differences among and between different groups through assessment tools geared to idealized norms.

However, perhaps in part because we chose to focus our attention on research related to mathematical practices as they occurred within classrooms, on the whole these emerging counter-narratives have been mainly confined to reporting and putting forward new micro views for promoting equity in mathematics teaching and learning practices. On their own, these narratives may have insufficient power to challenge successfully inequities in the macro-social conditions in which the practices occur. Such challenges are a necessary condition for equity and social justice to be achieved in school mathematics.

References

Adler, J. (2001). *Teaching mathematics in multilingual classrooms*. Dordrecht, The Netherlands: Kluwer Academic Publishers.

Aguirre, J. M. (2009). Privileging mathematics and equity in teacher education: Framework, counter-resistance strategies and reflections from a Latina mathematics educator. In B. Greer, S. Mukhopadhyay, A. Powell, & S. Nelson-Barber (Eds.), *Culturally responsive mathematics education* (pp. 295–320). New York, NY: Routledge, Taylor & Francis Group.

Artiles, A., Klingner, J. K., & Tate, W. F. (2006). Representation of minority students in special education: Complicating traditional explanations. *Educational Researcher, 35*(6), 3–5.

Atweh, B., Forgasz, H., & Nebres, B. (2001). *Sociocultural research on mathematics education: An international perspective*. Mahwah, NJ: Lawrence Erlbaum.

Baldino, R., & Cabral, T. (2006). Inclusion and diversity from a Hegel-Lacan point of view: Do we desire our desire for change? *International Journal of Science and Mathematics Education, 4*, 19–43.

Ballard, K. (2003). The analysis of context: Some thoughts on teacher education, culture colonization and inequality. In T. Booth, K. Ness, & M. Stromstad (Eds.), *Developing inclusive teacher education* (pp. 59–77). London, UK: Routledge/Falmer.

Bartolini Bussi, M. G., & Mariotti, M. A. (2008). Semiotic mediation in the mathematics classroom: Artefacts and signs after a Vygotskian perspective. In L. English, M. Bartolini Bussi, G. Jones, R. Lesh, & D. Tirosh (Eds.), *Handbook of international research in mathematics education* (2nd ed., pp. 720–749). Mahwah, NJ: Lawrence Erlbaum.

Battey, D., Kafal, Y., Nixon, A. S., & Kao, L. L. (2007). Professional development for teachers on gender equity in the sciences: Initiating the conversation. *Teachers College Record, 109*(1), 221–243.

Berliner, D. (2001, 28 January). *Our schools vs. theirs: Averages that hide the true extremes, Washington Post*. Retrieved from http://courses.ed.asu.edu/berliner/readings/timssroped.html.

Berry, J. M. (1985). Learning mathematics in a second language: Some cross-cultural issues. *For the Learning of Mathematics, 5*(2), 18–23.

Berry, R. Q., III. (2008). Access to upper-level mathematics: The stories of African American middle-school boys who are successful with school mathematics. *Journal for Research in Mathematics Education, 39*(5), 464–488.

Berry, R. Q., III, Thunder, K., & McClain, O. L. (2011). Counter narratives: Examining the mathematics and racial identities of black boys who are successful with school mathematics. *Journal of African American Males in Education, 2*(1), 10–23.

Bishop, A. J., & Forgasz, H. J. (2007). Issues in access and equity in mathematics education. In F. K. Lester (Ed.), *Second handbook of research on mathematics teaching and learning* (pp. 1145–1168). Charlotte, NC: Information Age Publishing.

Borgioli, G. (2008). A critical examination of learning disabilities in mathematics: Applying the lens of ableism. *Journal of Thought Paulo Freire Special Issue*. Retrieved from http://www.freireproject.org/content/journal-thought-springsummer-2008

Bull, S. (2008). Deafness, numerical cognition and mathematics. In M. Marschark & P. C. Hauser (Eds.), *Deaf cognition: Foundations and outcomes* (pp. 170–200). New York, NY: Oxford University Press.

Civil, M. (2009). *A survey of research on the mathematics teaching and learning of immigrant students*. Working Group 8, Cultural Diversity and Mathematics Education, Sixth Conference of European Research in Mathematics Education (CERME 6). Lyon, France.

Clarke, D. J., & Suri, H. (2003). *Issues of voice and variation: Developments in international comparative research in mathematics education*. Retrieved from http://extranet.edfac.unimelb.edu.au/DSME/lps/assets/Issues_of_VoiceClarke,Suri.pdf.

Clarkson, P. C. (2006). High ability bilinguals and their use of their languages. *Educational Studies in Mathematics, 64*(2), 191–215.

Cole, M. (1996). *Cultural psychology: A once and future discipline*. Cambridge, MA: Harvard University Press.

Cole, M., & Wertsch, J. V. (1996). Beyond the individual-social antinomy in discussions of Piaget and Vygotsky. *Human Development, 39*, 250–256.

Croon, L. (1997). Mathematics for all students: Access, excellence, and equity. In J. Trentacosta & M. J. Kenney (Eds.), *Multicultural and gender equity in the mathematics classroom: The gift of diversity* (pp. 1–9). Reston, VA: National Council of Teachers of Mathematics.

D'Ambrosio, U. (1999). Literacy, materacy, and technoracy: A trivium for today. *Mathematical Thinking and Learning, 1*(2), 131–153.

D'Ambrosio, U. (2001). *Etnomatemática. Elo entre as tradições e a modernidade*. Belo Horizonte, Brazil: Autêntica, Coleção Tendências em Educação Matemática.

Damerow, P., Dunkley, M. E., Nebres, B. F., & Werry, B. (Eds.). (1984). *Mathematics for all: Problems of cultural selectivity and unequal distribution of mathematical education and future perspectives on mathematics teaching for the majority*. Paris, France: UNESCO.

Dowker, A. D. (1998). Individual differences in arithmetical development. In C. Donlan (Ed.), *The development of mathematical skills* (pp. 275–302). London, UK: Psychology Press.

Dowker, A. D. (2004). *What works for children with mathematical difficulties?* London, UK: Department for Children, Schools and Families.

Dowker, A. D. (2005). Early identification and intervention for students with mathematics difficulties. *Journal of Learning Disabilities, 38*, 324–332.

Dowker, A. (2007). What can intervention tell us about the development of arithmetic? *Educational and Child Psychology, 24*, 64–82.

Duckworth, E. (1996). *The having of wonderful ideas, and other essays on teaching and learning*. New York, NY: Teachers College.

Dyson, A., & Gallannaugh, F. (2008). Disproportionality in special needs education in England. *The Journal of Special Education, 42*(1), 36–46.

Ellemor-Collins, D., & Wright, R. J. (2007). Assessing pupil knowledge of the sequential structure of number. *Educational and Child Psychology, 24*(2), 54–63.

European Agency for Development in Special Needs Education. (2010). *Teacher education for inclusion—International literature review*. Odense, Denmark: European Agency for Development in Special Needs Education.

Evans, J., Morgan, C., & Tsatsaroni, A. (2006). Discursive positioning and emotion in school mathematics practices. *Educational Studies in Mathematics, 63*(2), 209–226.

Fernandes, S. H. A. A., & Healy, L. (2007a). Transição entre o intra e interfigural na construção de conhecimento geométrico por alunos cegos. *Educação Matemática Pesquisa, 9*(1), 1–15.

Fernandes, S. H. A. A., & Healy, L. (2007b) Ensaio sobre a inclusão na Educação Matemática. *Unión. Revista Iberoamericana de Educación Matemática.* Federación Iberoamericana de Sociedades de Educación Matemática, *10*, 59–76.

Forman, E., & Ansell, E. (2001). The multiple voices of a mathematics classroom community. *Educational Studies in Mathematics, 46*, 115–142.

Frankenstein, M. (1983). Critical mathematics education: An application of Paulo Freire's epistemology. *Journal of Education, 165*(4), 315–339.

Frankenstein, M. (1998). Reading the world with math: Goals for a critical mathematical literacy curriculum. In E. Lee, D. Menkart, & M. Okazawa-Rey (Eds.), *Beyond heroes and holidays: A practical guide to K-12 anti-racist, multicultural education and staff development* (pp. 306–313). Washington, DC: Network of Educators on the Americas.

Frankenstein, M. (2010). Developing a critical mathematical numeracy through real real-life word problems. In U. Gellert, E. Jablonka, & C. Morgan (Eds.), *Proceedings of the Sixth International Mathematics Education and Society Conference* (pp. 258–267). Berlin, Germany: Freie Universitat.

Garcia, E. E., & Gonzalez, R. (1995). Issues in systemic reform for culturally and linguistically diverse students. *Teachers College Record, 96*(3), 418–431.

Gay, G. (2009). Preparing culturally responsive teachers. In B. Greer, S. Mukhopadhyay, A. Powell, & S. Nelson-Barber (Eds.), *Culturally responsive mathematics education* (pp. 189–206). New York, NY: Routledge, Taylor & Francis Group.

Geary, D. (2004). Mathematics and learning disabilities. *Journal of Learning Disabilities, 37*, 4–15.

Gerdes, P. (2007). *Etnomatemática: Reflexões sobre matemática e diversidade cultural [Ethnomathematics: Reflections on mathematics and cultural diversity].* Ribeirão, Brazil: Edições Húmus.

Gervasoni, A., & Lindenskov, L. (2011). Students with "special rights" for mathematics education. In B. Atweh, M. Graven, & P. Valero (Eds.), *Mapping equity and quality in mathematics education* (pp. 307–324). New York, NY: Springer.

Gervasoni, A., & Sullivan, P. (2007). Assessing and teaching children who have difficulty learning arithmetic. *Educational & Child Psychology, 24*(2), 40–53.

Gifford, S. (2005). Young children's difficulties in learning mathematics. Review of research in relation to dyscalculia. *Qualifications and Curriculum Authority (QCA/05/1545).* London, UK: Department for Children, Schools and Families.

Ginsburg, H. P. (1997). Mathematics learning disabilities: A view from developmental psychology. *Journal of Learning Disabilities, 30*, 20–33.

Gorgorio, N., & Planas, N. (2001). Teaching mathematics in multilingual classrooms. *Educational Studies in Mathematics, 47*, 7–33.

Grant, C. A., & Secada, W. G. (1990). Preparing teachers for diversity. In W. R. Houston, M. Haberman, & J. Sikula (Eds.), *Handbook of research on teacher education* (pp. 403–422). New York, NY: Macmillan.

Greenwood, D., & Levin, M. (2000). Reconstructing the relationships between universities and society through action research. In N. K. Denzin & Y. Lincoln (Eds.), *Handbook of qualitative research* (2nd ed., pp. 85–106). Thousand Oaks, CA: Sage Publications.

Greer, B., Mukhopadhyay, S., Powell, A. B., & Nelson-Barber, S. (Eds.). (2009). *Culturally responsive mathematics education.* New York, NY: Routledge.

Grosjean, F. (1985). The bilingual as a competent but specific speaker-hearer. *Journal of Multilingual and Multicultural Development, 6*(6), 467–477.

Gutiérrez, R. (2002). Enabling the practice of mathematics teachers in context: Toward a new equity research agenda. *Mathematical Thinking and Learning, 4*(2–3), 145–187.

Gutiérrez, R. (2007). (Re)defining equity: The importance of a critical perspective. In N. S. Nasir & P. Cobb (Eds.), *Improving access to mathematics: Diversity and equity in the classroom* (pp. 37–50). New York, NY: Teachers College.

Gutiérrez, R. (2008). A "gap gazing" fetish in mathematics education? Problematizing research on the achievement gap. *Journal for Research in Mathematics Education, 39*(4), 357–364.

Gutiérrez, R. (2010). The sociopolitical turn in mathematics education. *Journal for Research in Mathematics Education, 41*, 1–32.

Gutiérrez, R., & Dixon-Román, E. (2011). Beyond gap gazing: How can thinking about education comprehensively help us (re)envision mathematics education? In B. Atweh, M. Graven, & P. Valero (Eds.), *Mapping equity and quality in mathematics education* (pp. 21–34). New York, NY: Springer.

Gutstein, E. (2006). *Reading and writing the world with mathematics: Toward a pedagogy for social justice*. New York, NY: Routledge.

Hagège, C. (1996). *L'enfant aux deux langues*. Paris, France: Éditions Odile Jacob.

Healy, L., & Fernandes, S. H. A. A. (2011). The role of gestures in the mathematical practices of those who do not see with their eyes. *Educational Studies in Mathematics, 77*, 157–174.

Healy, L., Jahn, A-P., & Frant, J. B. (2010). Digital technologies and the challenge of constructing an inclusive school mathematics. *ZDM—International Journal of Mathematics Education, 42*, 393–404.

Hourbette, D., Baron, G., & Khaneboubi, M. (2008). *Towards mathematical gender sensitive teacher education Lessons from researchers, institutions and practitioners: First elements for a state of the art*. PREMA2 deliverable. Retrieved from http://prema2.iacm.forth.gr/docs/dels/MathGenderSensitive.pdf.

Hughes, R. M., Greenhough, P. M., Yee, W. C., Andrews, J., Winter, J. C., & Salway, L. (2007). Linking children's home and school mathematics. *Educational and Child Psychology, 24*(2), 137–145.

Jaworski, B. (2004). Grappling with complexity: Co-learning in inquiry communities in mathematics teaching development. In M. J. Høines & A. B. Fuglestad (Eds.), *Proceedings of the twenty-eighth conference of the International Group for the Psychology of Mathematics Education* (Vol. 1, pp. 17–36). Bergen, Norway: Bergen University College.

Jennings, T. (2007). Addressing diversity in U.S. teacher preparation programs: A survey of elementary and secondary programs' priorities and challenges from across the United States of America. *Teaching and Teacher Education: An International Journal of Research and Studies, 23*(8), 1258–1270.

Johnson, M. L., & Timmons-Brown, S. (2009). University/K–12 partnerships: A collaborative approach to school reform. In D. B. Martin (Ed.), *Mathematics teaching, learning, and liberation in the lives of Black children* (pp. 333–350). New York, NY: Routledge.

Kelly, R. R. (2008). Deaf learners and mathematical problem solving. In M. Marschark & P. C. Hauser (Eds.), *Deaf cognition: Foundations and outcomes* (pp. 226–249). New York, NY: Oxford University Press.

Kelly, R. R., Lang, H. G., & Pagliaro, C. M. (2003). Mathematics word problem solving for deaf students: A survey of perceptions and practices. *Journal of Deaf Studies and Deaf Education, 8*, 104–119.

Knijnik, G. (2002). Currículo, etnomatemática e educação popular: um estudo em um assentamento do Movimento Sem-Terra. *Reflexão e Ação, 10*(1), 47–64.

Kozol, J. (1991). *Savage inequalities: Children in America's schools*. New York, NY: Harper Collins.

Labov, W. (1972). The logic of nonstandard English. In L. Kampf & P. Lauter (Eds.), *The politics of literature: Dissenting essays in the teaching of English* (pp. 194–244). New York, NY: Pantheon.

Lave, J. (1990). The culture of acquisition and the practice of understanding. In J. W. Stigler, R. A. Shweder, & G. Herdt (Eds.), *Cultural psychology* (pp. 259–286). Cambridge, UK: Cambridge University Press.

Leontiev, A. N. (1978). *Activity, consciousness, and personality*. Englewood Cliffs, NJ: Prentice-Hall.

Lerman, S. (2000). Social turn in mathematics education research. In J. Boaler (Ed.), *Multiple perspectives on mathematics teaching and learning* (pp. 19–44). Palo Alto, CA: Greenwood.

Lesh, R., & Kelly, A. (2000). Multitiered teaching experiments. In A. E. Kelly & R. A. Lesh (Eds.), *Handbook of research design in mathematics and science education* (pp. 197–205). Mahwah, NJ: Lawrence Erlbaum.

Little, J. W. (1993). Teachers' professional development in a climate of educational reform. *Educational Evaluation and Policy Analysis, 15*, 129–151.

Magne, O. (2003). Literature on special needs in mathematics: A bibliography with some comments. In *Educational and Psychological Interactions 124* (4th ed.). Malmo, Sweden: School of Education, Malmo University.

Mantoan, M. T. (2009). *Inclusão escolar: O que é? Por quê? Como fazer?* (2nd ed.). São Paulo, Brazil: Moderna.

Marschark, M., & Hauser, P. (2008). Cognitive underpinnings of learning by deaf and hard-of-hearing students: Differences, diversity, and directions. In M. Marschark & P. C. Hauser (Eds.), *Deaf cognition: Foundations and outcomes* (pp. 3–23). New York, NY: Oxford University Press.

Marschark, M., Spencer, P. E., Adams, J., & Sapere, P. (2011). Evidence-based practice in educating deaf and hard-of-hearing children: Teaching to their cognitive strengths and needs. *European Journal of Special Needs Education, 26*, 3–16.

Martin, D. B. (2006). Mathematics learning and participation as racialized forms of experience: African American parents speak on the struggle for mathematics literacy. *Mathematical Thinking and Learning, 8*(3), 197–229.

Martin, D. B. (Ed.). (2009a). *Mathematics teaching, learning and liberation in the lives of Black children*. New York, NY: Routledge.

Martin, D. B. (2009b). Researching race in mathematics education. *Teachers College Record, 111*(2), 295–338.

Matos, J. F., Powell, A. B., & Stzajn, P. (2009). Mathematics teachers' professional development: Processes of learning in and from practice. In D. L. Ball & R. Even (Eds.), *The professional education and development of teachers of mathematics: The 15th ICMI study* (pp. 167–183). New York, NY: Springer.

Mayer, C., & Akamatsu, C. T. (2003). Bilingualism and literacy. In M. Marschark & P. E. Spencer (Eds.), *Oxford handbook of deaf studies, language, and education* (pp. 136–147). New York, NY: Oxford University Press.

McDermott, R. (1993). The acquisition of a child by a learning disability. In S. Chaiklin & J. Lave (Eds.), *Understanding practice: Perspectives on activity and context* (pp. 269–305). Cambridge, UK: Cambridge University Press.

Mechelli, A., Crinion, J. T., Noppeney, U., O'Doherty, J., Ashburner, J., Frackowiak, R. S., & Price, C. J. (2004). Structural plasticity in the bilingual brain: Proficiency in a second language and age at acquisition affect grey-matter density. *Nature, 7010*, 757–759.

Mellin-Olsen, S. (1987). *The politics of mathematics education*. Boston, MA: Kluwer Academic Publishers.

Mendick, H. (2005). A beautiful myth? The gendering of being/doing "good at maths." *Gender and Education, 17*(2), 89–105.

Mendick, H. (2006). *Masculinities in mathematics*. Buckingham, UK: Open University Press.

Moreno-Armella, L., & Sriraman, B. (2010). Symbols and mediation in mathematics education. In B. Sriraman & L. English (Eds.), *Advances in mathematics education. Theories of mathematics education: Seeking new frontiers* (pp. 211–212). New York, NY: Springer.

Moschkovich, J. N. (1999). Supporting the participation of English language learners in mathematical discussions. *For the Learning of Mathematics, 19*(1), 11–19.

Moschkovich, J. N. (2002). A situated and sociocultural perspective on bilingual mathematics learners. *Mathematical Thinking and Learning, 4*(2–3), 189–212.

Moschkovich, J. N. (2004). Appropriating mathematical practices: A case study of learning to use and explore functions through interaction with a tutor. *Educational Studies in Mathematics, 5*, 49–80.

Moschkovich, J. N. (2007). Using two languages when learning mathematics. *Educational Studies in Mathematics, 64*(2), 121–144.

Moschkovich, J. N., & Nelson-Barber, S. (2009). What mathematics teachers need to know about culture and language. In B. Greer, S. Mukhopadhyay, A. Powell, & S. Nelson-Barber (Eds.), *Culturally responsive mathematics education* (pp. 111–136). New York, NY: Routledge, Taylor & Francis Group.

Mukhopadhyay, S., & Greer, B. (2001). Modeling with purpose: Mathematics as a critical tool. In B. Atweh, H. Forgasz, & B. Nebres (Eds.), *Sociocultural research on mathematics education: An international perspective* (pp. 295–312). Mahwah, NJ: Lawrence Erlbaum.

Munn, P., & Reason, R. (2007). Arithmetical difficulties: Developmental and instructional perspectives. *Educational and Child Psychology, 24*(2), 5–14. Extended editorial.

National Council of Teachers of Mathematics. (1989). *Curriculum and evaluation standards for school mathematics*. Reston, VA: National Council of Teachers of Mathematics.

Newman, D., Griffin, P., & Cole, M. (1989). *The construction zone: Working for cognitive change in school*. Cambridge, UK: Cambridge University Press.

Nunes, T. (2004). *Teaching mathematics to deaf children*. London, UK: Whurr Publishers.

Nunes, T., & Moreno, C. (1998). Is hearing impairment a cause of difficulties in learning mathematics? In C. Donlan (Ed.), *The development of mathematical skills* (pp. 227–254). Hove, UK: Psychology Press.

Nunes, T., & Moreno, C. (2002). An intervention program for promoting deaf pupils' achievement in mathematics. *Journal of Deaf Studies and Deaf Education, 7*(2), 120–133.

O'Connor, C., & DeLuca Fernandez, S. (2006). Race, class, and disproportionality: Reevaluating the relationship between poverty and special education placement. *Educational Researcher, 35*(6), 6–11.

Orr, E. W. (1987). *Twice as less: Black English and the performance of Black students in mathematics and science*. New York, NY: W. W. Norton.

Pagliaro, C. (2006). Mathematics education and the deaf learner. In D. F. Moores & D. S. Martin (Eds.), *Deaf learners: Developments in curriculum and instruction* (pp. 29–40). Washington, DC: Gallaudet University Press.

Pais, A., & Valero, P. (2011). Beyond disavowing the politics of equity and quality in mathematics education. In B. Atweh, M. Graven, W. Secada, & P. Valero (Eds.), *Mapping equity and quality in mathematics education* (pp. 35–48). New York, NY: Springer.

Philp, H. (1973). Mathematics education in developing countries: Some problems of teaching and learning. In A. G. Howson (Ed.), *Developments in mathematics education* (pp. 154–180). Cambridge, UK: Cambridge.

Planas, N., & Setati, M. (2009). Bilingual students using their languages in the learning of mathematics. *Mathematics Education Research Journal, 21*(3), 36–59.

Powell, A. B. (1986). Economizing learning: The teaching of numeration in Chinese. *For the Learning of Mathematics, 6*(3), 20–23.

Powell, A. B., Maher, C. A., & Alston, A. S. (2004). Ideas, sense making, and the early development of reasoning in an informal mathematics settings. In D. E. McDougall & J. A. Ross (Eds.), *Proceedings of the twenty-sixth annual meeting of the North American Chapter of the International Group for the Psychology of Mathematics Education* (pp. 585–591). Toronto, Canada: PMENA.

Powell, A. B., & Vazquez, S. C. (2011, September). *Thinking mathematically in two languages*. Paper presented at the Proceedings of the ICMI Study 21 Conference: Mathematics Education and Language Diversity, São Paulo, Brazil.

Rands, K. (2009). Mathematical Inqu[ee]ry: beyond "Add-Queers-and-Stir" elementary mathematics education. *Sex Education, 9*(2), 181–191.

Rogoff, B. (1990). *Apprenticeship in thinking: Cognitive development in social context*. New York, NY: Oxford University Press.

Secada, W. G. (1989). Agenda setting, enlightened self-interest, and equity in mathematics education. *Peabody Journal of Education, 66*(2), 22–56.

Setati, M. (2002). Researching mathematics education and language in multilingual South Africa. *The Mathematics Educator, 12*(2), 6–20.

Setati, M., & Adler, J. (2000). Between language and discourses: Language practices in primary multilingual mathematics classrooms in South Africa. *Educational Studies in Mathematics, 43*(3), 243–269.

Setati, M., Adler, J., Reed, Y., & Bapoo, A. (2002). Incomplete journeys: Code-switching and other language practices in mathematics, science and English language classrooms in South Africa. *Language and Education, 16*(2), 128–149.

Setati, M., & Moschkovich, J. N. (2010). Mathematics education and language diversity: A dialogue across settings. *Journal for Research in Mathematics Education, 41*, 1–28.

Sfard, A. (1998). On two metaphors for learning and the dangers of choosing just one. *Educational Researcher, 27*(2), 4–13.

Skovsmose, O. (1994). *Towards a philosophy of critical mathematics education*. Dordrecht, The Netherlands: Kluwer Academic Publishers.

Skovsmose, O. (2011). *An invitation to critical mathematics education*. Rotterdam, The Netherlands: Sense Publications.

Sriraman, B. (Ed.). (2008). *International perspectives on social justice in mathematics education*. Charlotte, NC: Information Age Publishing.

Stanley, C. (2007). When counter narratives meet master narratives in the journal editorial-review process. *Educational Researcher, 36*(1), 14–24.

Steen, L. A. (1990). Mathematics for all Americans. In T. J. Cooney (Ed.), *Teaching and learning mathematics in the 1990s* (pp. 130–134). Reston, VA: National Council of Teachers of Mathematics.

Stinson, D. (2006). African American male adolescents, schooling (and mathematics): Deficiency, rejection, and achievement. *Review of Educational Research, 76*(4), 477–506.

Stokoe, W. C. (1960/2005). Sign language structure: An outline of the visual communication system of the American deaf. *Studies in Linguistics, Occasional Papers 8*. Buffalo, NY: Department of Anthropology and Linguistics, University of Buffalo. Reprinted in *Journal of Deaf Studies and Deaf Education, 10*, 3–37.

Thomas, G., & Ward, J. (2001). *An evaluation of the Count Me In Too pilot project*. Wellington, New Zealand: Ministry of Education.

Valero, P., & Zevenbergen, R. (Eds.). (2004). *Researching the socio-political dimensions of mathematics education: Issues of power in theory and methodology*. New York, NY: Kluwer Academic Publishers.

Vazquez, S. C. (2009). Combina ou não combina? Um estudo de caso com alunos bilíngües e não bilíngües nos EUA. *BOLEMA: O Boletim de Educação Matemática, 55*, 41–64.

Vygotsky, L. S. (1978/1930). *Mind in society: The development of higher psychological processes* (M. Cole, V. John-Steiner, S. Scribner, & E. Souberman, Trans.). Cambridge, MA: Harvard University Press.

Vygotsky, L. S. (1997). *Obras escogidas V–Fundamentos da defectología [The Fundamentals of defectology]*. Traducción: Julio Guillermo Blank. Madrid, Spain: Visor.

Wager, A. A., & Stinson, D. W. (Eds.). (2012). *Social justice mathematics*. Reston, VA: National Council of Teachers of Mathematics.

Walshaw, M. (2001). A Foucauldian gaze on gender research: What do you do when confronted with the tunnel at the end of the light? *Journal for Research in Mathematics Education, 32*(5), 471–492.

Weber, K., Maher, C., Powell, A. B., & Stohl Lee, H. (2008). Learning opportunities from group discussions: Warrants become the objects of debate. *Educational Studies in Mathematics, 68*(3), 247–261.

Willey, R., Holliday, A., & Martland, J. (2007). Achieving new heights in Cumbria: Raising standards in early numeracy through mathematics recovery. *Educational and Child Psychology, 24*(2), 108–118.

Wood, T. (2004). In mathematics classes what do students' think? *Journal of Mathematics Teacher Education, 7*(3), 173–174.

Woodward, J., & Montague, M. (2002). Meeting the challenge of mathematics reform for students with LD. *The Journal of Special Education, 36*, 89–101.

Chapter 4
Beyond Deficit Models of Learning Mathematics: Socio-cultural Directions for Change and Research

Cristina Frade, Nadja Acioly-Régnier, and Li Jun

Abstract Major paradigmatic changes in mathematics education research are drawing attention to new perspectives on learning. Whereas deficit models were previously in the foreground of research designs, these have been replaced by a wide variety of theoretical directions for studying diverse approaches to learning mathematics. There is now an acceptance of the need for richness and variety in research practices so that approaches can be studied, compared and mutually applied and improved. Psychological and quantitative approaches and methods are now increasingly complemented, or even replaced, by new directions that rely on social and anthropological theories and methods. Rather than reviving ideas about deficit research in mathematics education, the aim of this chapter is to present some socio-cultural perspectives of mathematics learning, and to show how these perspectives go beyond the deficit model of learning. Framing the main traditional markers of discrimination in school mathematics—gender, social class and ethnicity—in a perspective of social justice, the chapter concludes with a reflection on equality in terms of the democratic principle of meritocracy in mathematics education.

C. Frade (✉)
Federal University of Minas Gerais, Belo Horizonte, Brazil
e-mail: frade.cristina@gmail.com

N. Acioly-Régnier
EAM 4128-SIS-Université Lyon 1, Lyon, France

L. Jun
East China Normal University,
Shanghai, People's Republic of China

The assumption that people of low socio-economic background, or of different genders or ethnic groups, are intellectually less capable than others has deep implications in society. In educational settings, this assumption is particularly perverse, for it strongly influences the development of policies and practices for dealing with differences as markers of segregation. Under the label of "deficit model"—sometimes labelled as deficit thinking, deficit theory, the deficit paradigm, and the deficit discourse of learning—from the early 1960s, in the USA, for example, the deficit assumption seemed to adopt a definite theoretical perspective in attempts to explain why students "failed." Social and political factors embedded in the educational system, which favoured segregation among groups of students, were ignored. In other countries—such as Australia (with Aborigines), New Zealand (with indigenous peoples), the UK (with immigrants), the Netherlands (with immigrants), South Africa (with Black and poor populations) and Brazil (with poor and indigenous people)—the same debate occurred, not so much under the label of *deficit*, but in relation to the alleged deficit transmitters under investigation: gender, social class, race, culture or familial context.

Richard Valencia (2010) described six main characteristics of the deficit model in the educational context:

- *Victim blaming.* The deficit model of learning links the school failure to a membership community. It attributes the performance of poor students, students of color, students of different genders and ethnic groups to their alleged cognitive and affective deficits.
- *Oppression.* The deficit model holds little possibility of success to these students, privileging some and oppressing others.
- *Pseudoscience.* Deficit research draws on deeply negative bias in relation to persons of color, of different genders, of low socio-economic class and minority culture, "basing their research on flawed assumptions, using psychometrically weak instruments, not controlling for key variables" (p. 95), and communicating their findings in proselytizing ways.
- *Temporal changes.* The deficit discourse varies depending on when they are made. Alleged deficits can be transmitted by low-grade genes, gender, minority culture, social class, familial context, and other related transmitters.
- *Educability.* The deficit model often goes beyond the description, explanation and prediction of elements of poor students, students of color and of different genders, classes and cultures. It is also "a prescriptive model based on educability perceptions" (p. 18) of these students.
- *Heterodoxy.* The deficit model reflects the "dominant, conventional scholarly and ideological climates of the time. Through an evolving discourse, heterodoxy has come to play a major role in the scholarly and ideological spheres in which deficit thinking has been situated" (p. 18).

Taking the USA as the scenario of his critique (which can be fairly extended to other countries), Valencia (2010) carefully scrutinized several North-American documents, dismantling the fallacy of the deficit discourse. One of his main conclusions was that:

> Students are not at risk for academic problems due to their alleged deficits. Rather, schools are organized and run in such oppressive ways (e.g., inequities in the distribution of teacher quality characteristics and inequities in the distribution of economic resources for schooling) that many students are placed at risk for school failure. (p. 125)

At the time deficit models were in the foreground of research designs, the work developed by Klineberg (1935) brought an important psychological contribution in challenging the assumption that certain racial groups are intellectually inferior to others. Klineberg studied Black American children's IQ scores, and showed that they can be directly affected by environmental circumstances. Klineberg's research did not reach to a definite conclusion about the specific role of the environment over these achievements but, as stated by Lieberson (1985), it did "present evidence that such events can occur, that IQ is at least affected by the environment, and that judgments on a rather clear-cut matter can be altered by the influence of a social group" (p. 220).

Similar reactions on intelligence tests among different ethnic groups appeared in the works of Bruner (1990), Canady (1936), Cole (1985), Gould (1995), Long (1925), Menchaca (1997), Thomas (1982), and Van der Veer and Valsiner (1991), among others. All these works share somehow the conclusion that intelligence tests measure the familiarity of certain minority groups with the culture and language proficiency of dominant groups, not intelligence. Bruner (1990), for instance, questioned some ideas concerning the relationships between school learning and development and *intellectual prowess*. Without any reflection about what exactly we want to mean by intellectual prowess, said Bruner, we decided "to use school performance as our measure for assessing 'it' and predicting 'its' development" (p. 26). For Bruner, a definition of intellectual prowess or successful performance intimately depends on which traits a culture selects to honor, reward and cultivate. So, whatever definition of these terms is used, that definition should lead us to issues concerning the use we wish to make of them in "a variety of circumstances—political, social, economic, even scientific" (p. 27). This is to say, the cognitive development of the individuals cannot be evaluated out of the culture they are inserted in, and the operatory power and limits of theoretical models of learning and development adopted by diverse research lines must be analyzed in their emergent political-historical context.

By the 1970s, deficit research designs began to be challenged by a wide variety of theoretical perspectives of learning. Since then, there has been an acceptance of the need for richness and variety in research practices so that approaches can be studied, compared and mutually applied and improved. Psychological and quantitative approaches and methods have been increasingly complemented, or even replaced, by emergent approaches that rely on social and anthropological theories and methods.

Rather than reviving ideas about deficit research in mathematics education, the aim of this present chapter is to present some socio-cultural perspectives of mathematics learning, and to show how these perspectives go beyond the deficit model of learning. The chapter has been structured in four main parts. In the first part, we provide a description on how the research field of mathematics education has been reacting to deficit assumptions. In the second part, we discuss some socio-cultural perspectives of mathematics learning, showing that their mainstream assumptions challenge any deficit discourse. The third part promotes a perspective on social justice. Here some recent research related to the main traditional markers of discrimination in school mathematics—gender, social class and ethnicity—are approached. Finally, in the summary, we present an overview of the main issues and claims discussed.

Socio-cultural Reactions to Deficit Assumptions in Mathematics Education

It is well-documented in the literature that, historically, mathematics education developed as a research field from the late 19th century under the influence of two main disciplines—mathematics itself and psychology (D'Ambrosio, 1993; Kilpatrick, 1992; Lerman, 2000; Schoenfeld, 1992). It is also well-documented that mathematics has had an important role in the intellectual selection, preparation and guidance of students to enter higher education studies. Mathematics has been used to help select those who will occupy different social positions, thereby serving as a *critical filter* (Bishop, 1999; Ernest, 2007a; Gomes, 2008; Sells, 1978). Ernest (2007a) argued that, in Western culture, this "critical social function of mathematics is exacerbated by the preconception that mathematical performance is largely inherited" (p. 2), or, put another way, determined by deficit transmitters like those we have discussed. This discrimination, apparently stronger in mathematics than in other school disciplines, is still supported, in Ernest's view, by a significant corpus of quantitative research that correlates student mathematical achievements with gender, race, class, culture, familial socialization, and other divisors of society such as special needs, disability, sexual orientation, age, creed and religion.

Socio-cultural perspectives that differed from the deficit explicative approach of causality in mathematical performance started to appear by the late 1970s. These perspectives shared the assumption that it is too restrictive to consider merely "the gaps" of a population, and argued that each culture should be examined from tasks or practices that are significant or meaningful to their members. There emerged a number of works within and around mathematics education (e.g., Bernstein, 1996; Bishop, 1988; Bourdieu & Passeron, 1977; Carraher & Schliemann, 2002; D'Ambrosio, 1985; Gardner, 1983; Geertz; 1973; Greenfield & Childs, 1977; Lave, 1977; Scribner & Cole, 1973; Sternberg, 1985) that problematized the deficit approach, in terms of both specificities of the socio-cultural groups under investigation and methods that are pertinent to study these groups. The socio-cultural variable was taken into consideration in these works (Perret-Clermont & Brossard, 1988).

This movement towards new paradigms of learning in academic communities has been described by Lerman (2000) as the *social turn*, having its peak around 1988. Lerman observed that the positive receptivity of new alternative perspectives of learning by the mathematical education community "was due more to political concerns that inequalities in society were reinforced and reproduced by [deficit assumptions] in school mathematics, than social theories of learning" (p. 24). Valero (2004) pointed out that this signalled that some researchers found support, in these socio-cultural perspectives, for their understandings of these inequalities. On the other hand, for other researchers, these perspectives offered an explanatory power for them better to understand the mathematical practices in terms of the interactions, relationships, and discourses that effectively occur in the classroom. Whatever the case, these perspectives became an intellectual commitment for many mathematics education researchers: for some it is more political, and for others more pragmatic or affective.

The social turn in mathematics education was influenced not only by emerging socio-cultural approaches, notably those originating from cultural psychology, anthropology, sociology and philosophy of mathematics, but also by ethnomathematics, issues of gender, social class and ethnicity, history of mathematics, sociolinguistics, semiotics, and other topics in the social sciences (Ernest, Greer, & Sriraman, 2009; Lerman, 2000). In particular, emergent socio-cultural views of intelligence in response to deficit discourses played a special role in the social turn within academic communities in general. Two particular alternative contributions have challenged the deficit paradigm by arguing that intelligence is a social construct that manifests, in many ways and means, different things to different social groups. One contribution came from Gardner's (1983) *Theory of Multiple Intelligence*. Gardner defined intelligence "as the ability to solve problems, or to fashion products, that are valued in one or more cultural or community settings" (p. 7). The other contribution, known as the *Triarchic Theory of Intelligence*, was introduced by Sternberg (1985). It distinguished between three contexts in which intelligence manifests itself: the first relates to successful performance in standardized school norms (e.g., appropriated ways of thinking and reasoning, tests and socio norms); the second is associated with creativity and motivation toward novelty; and the third concerns successful performances in out-of-school activities.

New conceptualizations for intelligence generated new ways of thinking about both cognition and learning, and all of these demanded the development of alternative methods to complement statistical studies, or even replace them. In relation to cognition and learning, Jean Lave's book *Cognition in Practice* (1988) had a very important influence on thinking about mathematics education. Grounded on Vygotsky's ideas, Lave demonstrated that cognition is a phenomenon that emerges in social interactions, and that learning and identity formation occur as a result of participation in social practices. This resulted in a radical shift of paradigm in relation to traditional views of cognition and learning in that meaning, thinking, and reasoning came to be seen as products of social activity (Lerman, 2000).

Alternative methods of empirical research, involving qualitative approaches (see Groulx, 2008), challenged the authority of statistical methods, which came to be seen as being relevant only to events that could be "classified, operationalized and organized" (p. 97; our translation). By contrast, qualitative methods focussed on the particularities, conditions and circumstances of the historical/socio-cultural environments in which the events occurred; the subject-participants become actors in that their voices were heard, revealing a diversity of situations in which they acted in various manners and made use of a varied resource repertoires. Yet, qualitative approaches pushed academic communities to rethink studies concerning the needs of the groups of individuals according to the socio-cultural singularities of their *forms of life*, and not as measurement indicators.

Although the works mentioned so far claimed that intelligence, cognition and learning should not be explained any more from deficit parameters—the issue still remains alive in the agenda of a number of scholars from different Western countries (e.g., Ernest, 2007a; Ford, Harris, Tyson, & Trotman, 2002; Gillborn, 2005; Gomes, 2003; Gorgorió, Planas, & Bishop, 2004; Gutiérrez, 2007; Keitel, 1998; Martin, 2009; Stevens, Clycq, Timmerman, & Van Houtte, 2009; Valencia, 2010; Weiner, 2006).

In the next section some socio-cultural perspectives of mathematics learning are presented, showing that their mainstream assumptions go beyond deficit discourses.

Socio-cultural Perspectives for the Learning of Mathematics

Socio-cultural perspectives of mathematics learning are found under different denominations and within different research foci. Some of these perspectives conform to the main research foci proposed by Bishop (1999)—mathematics learning, mathematics curricula and mathematics teaching. These three foci are described by Bishop in the following words: mathematics learning relates to the ways *cultural learners* learn and use mathematics. This includes "characteristics of learners, types of learning, attitudes, beliefs, motivations, feelings, ways of remembering, imagining, representing" (p. 4). Mathematics curricula deal with *cultural issues* involved in "aspects of content, sequences of ideas, relationship to other topics, other subjects, other contexts, both real and virtual" (p. 4). Mathematics teaching covers all that encompasses *the context of mathematics teaching*, which, at the end, converges to the classrooms in the form of "interactions, explaining, clarifying, linking with other knowledge, inspiring, leading, communicating" (p. 4).

In the analysis which follows we will show that these foci are not disjoint: each overlaps or complements the others.

Cultural Learners, Cognition and Affect

Acknowledging that learning and cognitive processes should not be analyzed outside a learner's culture led to the development of studies of beyond-school mathematical practices in culturally relevant contexts. Barton (1996) identified four bodies of literature in these studies, one of them focussing on the exploration of relationships between the thinking processes of an individual's cultural group and mathematics education. Thus, for example, Terezinha Nunes, Analúcia Schliemann and David Carraher's studies of street mathematics and school mathematics with some groups of Brazilian children analyzed data on the similarities and differences between different groups of people as they attempted to solve mathematical problems at work and in school. These data, and data from other like studies, constituted strong evidence against deficit models as they showed that, despite failing in school mathematics, children from poor economic backgrounds could understand and apply basic mathematical principles as they solved problems in familiar work contexts.

In their first analysis of the mathematics that people practise in everyday settings, Carraher, Carraher, and Schliemann (1985) found that young street vendors in Brazil correctly solved 99% of the arithmetic problems that emerged during selling transactions. However, when asked to solve similar problems presented to them as school-like computations, the percentage of correct answers dropped to 37%. Nunes, Schliemann, and Carraher's (1993) studies, together with those by other authors (e.g., Lave, 1977, 1988, 1989; Reed & Lave, 1979; Saxe, 1991), demonstrated that specific socio-cultural activities, such as buying and selling, promote the development of mathematical knowledge previously thought of as accessible only through formal instruction. These findings strongly challenged the adequacy of deficit models in relation to mathematical learning: failure to learn mathematics in school cannot be attributed to deficits, given that the same children who failed in school tasks showed mathematical understanding in other contexts. The analysis of school failure needs to focus therefore on the school itself, its values, its assessment procedures, and, above all, the different practices developed in and out-of-school contexts.

Nunes, Schliemann and Carraher, and their students—the so-called *Recife Group*—developed over more than 20 years new contexts of observation in which mathematical activities were not necessarily related to school mathematics patterns (see, e.g., Acioly, 1994; Acioly-Régnier, 1997; Acioly & Schliemann, 1987; Carraher, 1986; Carraher et al., 1985; Da Rocha Falcão, 1995; de Abreu & Carraher, 1989; Nunes, Schliemann, & Carraher, 1993; Schliemann, 1985; Schliemann & Acioly, 1989; Schliemann, Araújo, Cassundé, Macedo, & Nicéas, 1994; Schliemann, & Carraher, 2004; Schliemann & Magalhães, 1990). One of the contexts, discussed by Da Rocha Falcão (2005), referred to a specific community of Brazilian fishermen, the *jangadeiros* from Recife. Although most of these fishermen were illiterate and possessed no conceptual-vectorial schemes at all, Da Rocha Falcão showed how they were able to pilot their sailing boats conforming to vectorial principles of composition of the direction and intensity of the wind and the orientation of the sail and keel.

Refusing to accept deficit models to explain difficulties in learning school mathematics, researchers in the Recife group built upon aspects of Piaget's and Vygotsky's theoretical accounts of cognitive development and followed methods similar to those developed by Cole and Scribner (1974), Luria (1976), and Reed and Lave (1979). Thus, the group developed a conceptual and contextual analysis of empirical data, using methods from anthropology, psychology, and mathematics education, to bring out the different levels of conceptualization and representations of participants in their studies. Vergnaud's (2009) theoretical proposal of *conceptual fields* provided a fruitful background for their analysis of the invariant, symbolic, and situational aspects of concepts developed both in and out-of-school.

Although initially formed with the above-mentioned theoretical and methodological orientations, some members of the Recife group reelaborated them and incorporated others to continue their own investigations. For instance, Da Rocha Falcão pointed out that, although Brazilian fishermen—the *jangadeiros* from Recife—and amateur sailing apprentices displayed clear differences in their psychological competences of sailing, both groups of competences were semiotically and culturally mediated. Supported by Leontiev's (1994) theoretical concept of activity, Da Rocha Falcão argued that the classification of these Brazilian fishermen's sailing competences, proposed by Vergnaud (1991) as being *competences-in-action*, or else *savoir-faire* as proposed by Piaget (1974), suggested the possibility of non-semiotic, strictly practical human actions. Da Rocha Falcão stated that the fact that many people could not explain or discuss their competences should not be taken as evidence that these competences had a purely enactive character.

The systematic research program developed by the Recife group not only drew attention to the weakness of deficit models for learning mathematics to explain the academic failure of children but also demonstrated common aspects of concepts developed out-of-school and in school. In discussing analytical tools for the study of mathematical activity, Araújo et al. (2003) proposed, among other things, to take into account pre-conceptual competences characterized in two ways: First, by their effectiveness in culturally meaningful contexts; and second, by the fact that these competences are, by nature, quite difficult to express using symbolic-explicit representations (see also Frade & Da Rocha Falcão, 2008). For these authors, effectiveness and tacit quality are invariants of mathematical activity, irrespective of whether we are considering school or out-of-school mathematical practices such as those performed by tailors (Lave, 1988), carpenters (Millroy, 1992), *cambistas de jogo do bicho*—Brazilian bookmakers dealing with what is called the "animal lottery" (*jogo do bicho*) (Acioly & Schliemann, 1987), fishermen (Da Rocha Falcão, 2005) and other communities of practice (e.g., Santos & Matos, 2002).

For researchers in the Recife group, the core issue regarding predictors of selective school failure relates to particular characteristics of the semiotic interactions and concepts developed in different practices. But what are those characteristics? Are they linked to the context of learning, to students' identities, or to mathematical concepts involved in the activity? It seems that the *simultaneous* consideration of these three aspects distinguishes this group as researchers of the psychology of mathematics education inspired by the theoretical perspectives of Vygotsky, Piaget, Vergnaud, Leontiev, and Lave, among others.

Within psychology, the role of culture and contexts in the cognitive development of individuals is a fundamental issue. The difficulties of integrating cultural and conceptual aspects within works on mathematical competences can be illustrated in the analysis by Saxe and Posner (1983) of the strengths and weaknesses of transcultural research into the development of the concept of number, associated with the approaches of Piaget and Vygotsky. For Saxe and Posner, each of the theories provides a base to analyze universal or culture-specific processes on the creation of numerical concepts in children.

Piaget described how numerical operations are developed, but he did not analyze the mechanisms through which social factors contribute to the creation of numerical thought. His theory did not give us enough information about the level of conceptualization of the individual in specific domains of knowledge—a fact later recognized by Piaget (1971) himself, at least in relation to what he called formal operations in adolescence and adulthood. The lack of analysis concerning possible differences of conceptualization across contexts or situations led to the abandonment of the theoretical frame of stages of development in the sense Piaget gave to them. By contrast, Vygotsky's approach, as interpreted by a group of American psychologists (Cole, Gay, Glick, & Sharp, 1971; Cole & Scribner, 1974), considered cultural experience as a differentiated theoretical construction. With this approach, concepts are regarded as important, but conceptual development is not analyzed in depth.

Taking advantage of both perspectives, the theoretical perspective of conceptual fields proposed by Vergnaud (1991, 2009) provides a pertinent and operative frame that allows a new type of analysis of different types of conceptualizations occurring in different contexts. The core of Vergnaud's theory lies in the importance attributed to situations for the development of concepts. We recall that this theory defines a concept as a tripolar system constituted by three groups that he called *signifiers, situations, and invariant operatories*. The group of signifiers allows the representation, the communication and the treatment of a concept; the group of situations refers to situations in which the concept operates, and to the idea of reference; the group of invariant operators refers to the idea of meanings.

Using Vergnaud's tripolar system, Acioly-Régnier (2010) identified a distinction between school and non-school contexts in terms of *focus of consciousness*. In this identification, the difficulty an individual is faced with relates to the recognition of whether the concepts or representations are relevant to a given situation, be it a school or a non-school situation, or even to lack of the cultural tools to represent the situation. Acioly-Régnier showed that, within a school frame, the focus of consciousness is essentially directed to the bipolar relation signifier-signify, leaving aside the situations they may refer to. In non-school contexts the stress is mainly on the axis signify-referent. In this case, Acioly-Régnier (2010) noted that the conceptualization becomes somewhat incomplete and the equilibrium of the triple (signifier, referent, signify) is lost.

In terms of the focus of consciousness, this has been justified as everyday concepts are linked to local knowledge as opposed to universal knowledge (Rogoff, 1981). For Acioly-Régnier, this view is controversial. Her study indicated that the same lack of generalization applies to learning that takes place in school contexts. As studies of transfer show (e.g., Boaler, 2002a; Carreira, Evans, Lerman, & Morgan, 2002; Frade, Winbourne, & Braga, 2009; Greeno, Smith, & Moore, 1993; Lerman, 1999), generalization and transfer across contexts are not, in general, without mediation, automatic, or even comfortable.

It is fundamentally important to take explicit account of the contexts in which learning takes place and to study the specificities of the concepts developed by the individual in a given context. That is particularly the case when the specificities of the semiotic interactions cannot be understood by the dichotomy: street mathematics context versus school mathematics absence of context.

In this respect, the perspective taken by Lave and Wenger (1991) illustrated a way of breaking with this dichotomical model. Regarding the appropriation of certain kinds of knowledge, they expanded Lave's initial views, placing the development of learning as a matter of identity development that takes place in social relations within the situations of coparticipation. This participation not only refers to local events that set in motion certain activities with certain people, but to a more global process that integrates the active participants to the practices of social communities and leads them to build their own identities to connect with the community. Lave and Wenger illustrated their theory of situated cognition by considering previous empirical studies of different learning processes among several groups: the midwifes of Yucatec, the tailors of Vai and Gola, the quartermasters of the American marine, the carvers at slaughterhouses, and a group of alcoholics anonymous. At first the individuals who join communities remain mostly at the periphery, where they do their first learning acquisitions. As they become more competent they move to the centre of the community. Therefore, learning is not seen as a simple acquisition of knowledge by individuals, but as a process of social participation in a certain practice or situation.

Acioly-Régnier (2010) adopted the characterization of concept proposed by Vergnaud (2009) in which a concept involves a set of situations, a set of operational invariants, and a set of linguistic and symbolic representations, but took into account the context of the conceptual development. At a more general level, Acioly-Regnier proposed a framework for psychological processes and a conceptualization of reality that includes three poles: culture, cognition and affect, as depicted in Figure 4.1.

This framework considers, simultaneously, the idea that performance in a given context occurs under the triple influence of cognitive, affective and cultural factors. Empirical evaluations of this proposal require multiple methodological approaches aiming at providing different and relevant perspectives regarding the phenomenon

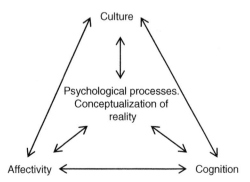

Figure 4.1. Schematization of the frame culture, cognition and affect.

under analysis. For example, one needs to consider the role given to the individual by the researcher—as holder of knowledge, as simply a source of information, or even as someone submissive to a research protocol built by the researcher to test the limits and strengths of her knowledge. The nature of the interaction proposed by the researcher to collect the data, also needs to take into account and be examined from the perspective of, for example, the classic interaction individual-researcher or the binomial interaction within a pair or a larger group (Acioly-Régnier, 1996). Acioly-Régnier (2010) argued that cognitive performance depends on the nature and the context of the question, an individual's familiarity with the situation, the kind of formulations (oral or written) that are required, etc. All these parameters play an important role in the construction of the data, as do classroom-specific factors such as the extent to which students are allowed to display their mathematical understandings.

Numerous studies into cognition and context have clarified important issues regarding individual challenges and strengths as learners develop concepts across different contexts. Questions still remain, however, so far as the relevance of these research findings and theoretical approaches to mathematics education—see Moshkovich and Brenner (2002) for a collection of studies on this issue. Also many questions need to be addressed regarding children who fail to learn mathematics in schools. For instance: What mathematical understandings do these children develop outside of schools, which are relevant to the mathematics curriculum? How are these beyond-school understandings different from the mathematics they are supposed to learn in schools? How can a teacher identify the strengths and limitations of children's previous concepts? How can the teacher create environments that will allow children to learn mathematics that is not merely a set of memorized procedures, but rather a set of meaningful, related concepts—as they seem to be capable of doing when they learn outside of schools? These kinds of questions show that psychology alone is not able to account for socially- and culturally-orientated theoretical perspectives of cognition. Possible responses to these questions can be elaborated from issues outside of psychology, however, and these will be discussed in the following sections.

Culture and Mathematics Curricula

Three particular socio-cultural perspectives of mathematics learning have had a strong influence on mathematics education research on curricula issues: ethnomathematics, Bishop's perspective of mathematical enculturation and acculturation, and situated perspectives originated from Lave's ideas on cognition in social practices.

Ethnomathematics. In his articles *The Name Ethnomathematics: My Personal View* and *Ethnomathematics: My Personal View*, D'Ambrosio (2010a, 2010b), reported on the trajectory of ethnomathematics, presenting his personal view of this already consolidated research field. In the first article, he noted that the word ethnomathematics was always used by him when he was describing the mathematics of other cultures, especially those without writing and those marginalized by the colonial process (D'Ambrosio, 2010a). D'Ambrosio (1997) recognized, however, that this word has been broadened to encompass other ethnomathematics currents, for example, critical approaches to the eurocentric character of mathematical knowledge (Powell & Frankenstein, 1997), research on educational policies and society (Gerdes, 1994), and studies of mathematical ideas of non-literate groups (Ascher & Ascher, 1997). In the second article, D'Ambrosio (2010b) explained what he meant by ethnomathematics and an ethnomathematics program: the former relates to a theoretical framework, and the latter to the empirical dimension of it. In both articles, D'Ambrosio made it clear that ethnomathematics consists of a theory of knowledge of different cultural groups, with special emphasis on both the history and philosophy of mathematics, the aim being to understand, explain, learn about, cope with and manage the natural, social and political environment of processes involving counting, measuring, sorting, ordering and inferring, of well-identified cultural groups (see also D'Ambrosio, 1988, 1997). He linked his theoretical approach to its empirical dimension by saying that his "proposal is a transcultural perception of the nature of mathematical knowledge, which demands a transdisciplinarian approach to knowledge in general" (D'Ambrosio, 2010a). In this sense, an ethnomathematics program focusses on epistemological investigations of mathematical ideas and practices of different cultures—such as those developed or used by indigenous populations, labor and artisan groups, periphery communities in urban environment, farms, and professional groups—using methodological procedures inspired by ethnography.

The paragraph above shows the wide scope of ethnomathematics studies as viewed by its best known scholar. It leads us to reflect, in particular, on some pedagogical implications of these studies as they are directed to societies, communities or groups in which education is structured by any type of formal instruction. The main challenge of ethnomathematics studies, when restricted to any type of formal education, lies with curricular issues, for their assumptions depend on a curriculum planned and developed around the specific socio-cultural needs and life history of these groups. This implies a teaching context design based on the socio-cultural environments of the learners, and a view of cognition in which reasoning reflect cultural roots (D'Ambrosio, 2010c), because meaning-making derives from the learners' socio-cultural needs and life histories. This articulation of *curriculum-context of teaching-cognition* was well illustrated in the work carried out by Knijnik (2004) with landless peasant communities in Southern Brazil. In this work, Knijnik clearly showed her careful role as researcher in the construction and planning of such articulation to attend to the social needs of the communities under investigation.

Ethnomathematics has inspired studies into the education of youths and adults (e.g., Fonseca, 2010), indigenous communities (e.g., Barton, 2008; Costa & Silva, 2010), other minority communities (e.g., Knijnik, 1999; Palhares, 2008), professional groups (e.g., Palhares, 2008) and has generated a range of socio-political approaches to research (Gerdes, 1994; Powell & Frankenstein, 1997; Valero &

Zevenbergen, 2004). Yet, ethnomathematics has brought important contributions to mathematics learning and teaching in traditional classrooms, in particular to cultural roots, interactions between mathematics and languages, human interactions and values and beliefs (Bishop, 2002, 2010). However, we do not find substantial ethnomathematics studies in the literature on mathematics education within school environments.

We suspect that this is due to two main reasons. First, the complexity of traditional classrooms in terms of the multiplicity of the students' needs and life histories are much too diverse to permit the students to be considered as well-identified cultural groups in the sense of ethnomathematics. Besides, these classrooms are inserted in a type of educational system in which the mathematics curriculum is basically the same in all countries (Bishop, 2010), generally elaborated and developed by pre-established contents, guided by national policies and mechanisms of assessment performances, and with little opening for changes. The second reason, which can be viewed as a consequence of the first, concerns a probable lack of interest in researching mathematics classrooms because ethnomathematical thinking, as Knijnik (2004) observed, emphasizes "other mathematics, usually silenced in school, as the cultural production of non-hegemonic groups" (p. 136). What we find in the literature about ethnomathematics and formal mathematics learning and teaching is a set of proposals concerning pedagogical lines of actions to incorporate the cultural diversity in the educational context (Borba, 1997; Gerdes, 1996; Palhares, 2008; Shirley, 1995).

Mathematical enculturation and acculturation. Bishop's perspective of mathematical enculturation (Bishop, 1988) and acculturation (Bishop, 2002) fills the space left open by ethnomathematics concerning school mathematics in some important aspects. It is a perspective that helps us to understand affective imbalances within mathematics classrooms, especially multi-ethnic classrooms, between students and teachers, *culturally*. Borrowing from the literature on anthropology, Bishop introduced the concepts of enculturation and acculturation into mathematics education. These concepts primarily address curricular issues in that they are strongly linked to the culture the student brings from home and the teachers' cultures, values, beliefs and choices in relation to mathematics, mathematics education and education in general (Bishop, 1988, 2002; Bishop, FitzSimons, Seah, & Clarkson, 1999; Seah & Bishop, 2000).

Frade and Faria (2008) noted that Bishop's educational analysis was initially grounded on the perspective of enculturation, where enculturation was taken to mean the induction, by a particular cultural group, of young people into their culture. This perspective presupposes the existence of a cultural consonance between school mathematics and the culture the student brings from home. Frade and Faria (2008) observed, however, that Bishop (1994) re-evaluated his premises for the purpose of reaching an understanding of cultural conflicts within multi-ethnic classrooms—moving from the assumption that mathematics education may not be a process of enculturation, but rather a process of *acculturation*, the induction into an outside culture by an outside agent. Often one of the contact cultures is dominant, irrespective of whether such dominance is intended. At this stage, according to Frade and Faria, Bishop's studies began to focus not so much on individual students, but on the acculturation process per se, and on the role of the so-called *acculturators*. After observing apprentices (in general) during their experiences with cultural conflicts, Bishop (2002) proposed a more radical hypothesis: "all mathematics education is a process of acculturation ... every learner experiences cultural conflicts in that process. However, cultural conflicts need not be conceptualized exclusively in a negative way" (p. 192).

Frade (2006) reported that for Bishop (2002) mathematics teachers are the main agents of mathematics acculturation. He considered two types of acculturator-teachers: the teacher who does not make any reference to any out-of-school mathematical knowledge; and the teacher who imposes what she wants through her privileged position and power. In both cases, Bishop claimed, although the resulting cultural conflicts contain a cognitive component, they are infused with emotional and affective traces or nuances indicating deeper and more fundamental aspects than can be accounted for from a cognitive perspective. These affective traces clearly appeared in the works of Frade and Machado (2008), and Frade and Faria (2008), who reported on two studies into teachers' mathematical culture and values, and the corresponding affective reactions of the students to both their learning and their teachers' practices.

In an attempt to humanize the imbalanced relationship between the culture of the teachers and the culture of the students, Frade and Faria (2008) suggested that Bishop (2002) proposed to reconceptualize mathematics learning environments based, to a great extent, on Gee's (1996) theoretical construct of *borderland discourse*. This would correspond to the area of intersection between the students' primary and secondary discourses. The primary discourse refers to the discourse learned and used within the family, at home or with surrounding groups. The secondary discourse, more institutional or formal than the primary one, is related to traditions passed forward to us by various generations through time, aiming at learning conducted in external environments. According to Frade and Faria, the potential oppressive character of an acculturation process led Bishop to state that the intentional mathematics acculturation of a young person is turned into some type of cultural production while schools should be the place where the primary discourse of the students' families and communities meet the secondary discourse of the mathematics community. This *turning* was explained by Bishop as he explored the idea of *transition* (see de Abreu, Bishop, & Presmeg, 2002).

More recently, Bishop (2010) revisited the evolution of his works and discussed the universality of mathematics curriculum—that is to say, the fact that school mathematics curricula are almost the same in every country, apparently disregarding the cultural diversity that characterizes the population in general. Bishop urged that "mathematics curricula be designed which deal with numeracy/ethnomathematics practices as one strand together with Mathematical theory as a separate but related strand" (p. 339). By Mathematical theory for a school's curriculum, Bishop meant "an approach which focusses on the many 'Why' questions provoked by numeracy/ethnomathematics practices" (p. 339). In this way, Bishop suggested that we would have "a balanced mathematics curriculum for all—one which respects local ethnomaths/numeracy practices, 'legitimizes' them, and accepts them within the school context, and investigates their rationales by using appropriate Mathematical theory" (p. 340). In relation to values, Bishop restated their importance in educational research. For him, although we can perceive an increased interest in culture-based research constructs, they are still insufficiently addressed, given that "shared values are a significant part of any culture" (p. 341).

Bishop's perspective has influenced not only studies focussing on school mathematics. His ideas have been taken up by those conducting general studies into the development of curricula in conflictive political and/or multicultural contexts (de Abreu et al, 2002; Civil, 2007; Gorgorió & Planas, 2001; Powell & Frankenstein, 1997; Valero & Zevenbergen, 2004), and also on teachers' values and students' affect (Clarkson, FitzSimons, & Seah, 1999; Frade & Faria, 2008; Frade & Machado, 2008).

It is clear that both ethomathematics and Bishop's perspective are unquestionably rooted in humanistic views such as respect for cultural diversity, equality and social justice, and human rights. Mathematics education researchers and teachers who have a commitment to these approaches do not accept assumptions and claims based on the cultivation of deficit practices or discrimination of any type. In particular, ethnomathematics and Bishop's perspective are very aligned with Bruner's (1990) position concerning the socio-cultural character of intellectual prowess. In the case of ethnomathematics, it is common to refer to those groups which, for some reason, are excluded from the cultural production of Western hegemonic education. In Bishop's perspective, teachers and students are the main perpetrators of cultural conflicts in which students' "primary" cultures are often oppressed by the "secondary" cultures of the teachers.

Situated perspectives of learning. With a different emphasis on culture from that given by ethnomathematics and by Bishop's perspective, but still relying on anthropology, we find the situated approaches to mathematics learning originating from Lave's (1988) perspective on cognition in practice (see also Lave and Wenger, 1991).

Lave's core idea is that cognition is a product of semiotic interactions between the individuals and the social practices in which they participate. According to Frade et al. (2009), this implies that cognition is "a phenomenon that emerges from the practice, from the fact that an essential feature of the practice is making resources available for ... involving and encouraging the individuals" (p. 16) to interact semiotically within it. For these authors, this is what it can be understood by learning as a process which does not depend on an individual only, but notably on the potential of the appeals of practice to bring individuals to participate in it (for other situated approaches, see, for example, Boaler, 2000; Brown, Collins, & Duguid, 1989; Cobb & Bowers, 1999; Engeström, 1999; Greeno, 1997; Kirshner & Whitson, 1997; Watson & Winbourne, 2008). By focussing on the practices in which individuals are expected to learn to participate, learning is then seen as a process of changing participation and identity formation within these practices (Lave & Wenger, 1991).

Lave and Wenger's (1991) social-practice perspective of learning had profound implications not only for those studying mathematics practices in and out-of-school settings but also for the interdisciplinary character of research in mathematics education. The emphasis on practices as the emergent locus for the production of meanings indicated that mathematics learning and its use by common people was no longer a matter for discussion by psychologists only. Psychology alone was not able to account for the processes involved in learning and using mathematics from a situated point of view (Boaler, 2000).

In the context of school mathematics, learning as a result of participation and identity formation is particularly challenging in two special aspects. First, it claims a refocussing of the teachers' attention away from students' cognitive differences/performances (conveyed by expressions such as "good student" and "weak student") towards the students' semiotic interactions within mathematical practices or activities, which are situated in a broader historical/socio-cultural context. This does not imply that students' individual needs are ignored. On the contrary, cognition viewed in terms of symbolic mediation means that different individuals will interact in semiotically different ways. In producing different meanings, teachers should be aware that students will have distinct needs that must be considered by the practice.

Second, and consequently, it demands a rethinking of the idea that the mathematics curriculum should be centred on a "universal" or taken-for-granted list of pre-established subjects. What is being called for, now, is a re-direction to propositions of mathematical practices within the space of signification/meaning of the students, to allow them to interact semiotically. Participation, in Lave and Wenger's sense, is not merely an act of engagement in a certain practice or activity; changing forms of participation are part of a process that shape identity formation. That is to say, by beginning to participate in new ways, participants come to see and deal with concepts or situations that either they have not seen before or, if they have, they now see and deal with them in different ways. It might be said that individuals have learned or become different persons in relation to a certain domain. Indeed, according to Lave and Wenger (1991), the concept of participation involves, above all, a contribution of the individual to the development of the practice and the contribution of the practice to the development of the individual as well.

This refocussing of the teachers' attention and the rethinking of the universal character of mathematics curricula are key aspects of situated approaches that challenge deficit assumptions. Cognition is now to be seen as a process which does not depend on an individual's "natural" attributes. And, by focussing mathematics curricula on mathematical practices that include diversity considerations as earlier proposed by Bishop (2010), for instance, while at the same time providing access to globalized knowledge, educators foster student participation and avoiding differentiation, division, qualification and disqualification among students (Popkewitz, 2004).

Some researchers (e.g., Walkerdine, 1997) have pointed out that Lave's situated perspective does not clarify how subjectivities are produced in social practices. Similarly to the discussion developed by Frade and Meira (2010) about the social nature of affective behaviours and the constitution of identity, the production of one's subjectivity can be also seen as results of internalizations that occur from the interactions between individuals and those practices in which they are involved. How an individual reacts to and internalizes what she has learnt as a result of her engagement in social practices depends not only on her previous life experiences, choices and judgments, but also—and perhaps mainly—on a combination of the contingencies, circumstances and social norms to which she is subjected at the moment. Thus, according to Frade and Meira (2010), one's subjectivity is subjected to continuous changes, and depends on the historical/socio-cultural circumstances which the individual has experienced and is experiencing at that particular moment of life.

Studies on ethnomathematics, Bishop's perspective, and situated learning perspectives clarify important issues concerning the learning and the use of mathematics by different cultural groups. However, there are many unanswered questions that need to be investigated: What effects do mathematical content and its use have on the processes of participation and identity formation? How can teachers effectively evaluate their students' mathematical developments in terms of participation and identity formation? What needs to be done to create curricula based on mathematical practices, rather than merely on content? Which kind of curricular materials would teachers need to support work with a curriculum based on mathematical practices? How can cultural specificities be incorporated into the curriculum of different cultural groups without avoiding the mere reproduction of their cultures?

The Classroom Dynamic: The End Point of Mathematics Teaching

As indicated earlier, the context of any formal teaching is configured under the influence of a number of factors—like institutional issues, curricular policies, pedagogical organization, teacher qualifications, values and beliefs, power relationships, and people's expectations and needs. In classrooms, all of these are manifested, somehow, in the form of interactions, explanations, clarifications, linking with other knowledge, inspiration, leadership, and communication. The unique combination of these in a classroom produces a classroom dynamic in terms of norms, negotiations, designs and modes of teaching. Below we provide illustrations of some socio-cultural approaches, the intention being to explore how these factors combine in classrooms to affect teaching and learning.

Social negotiations. A classic example of a proposal for the negotiation of meanings and conduct between teacher and students comes from Guy Brousseau, one of the pioneer scholars of the well-known *didactique Française*. Brousseau (1986, 2006) introduced the notion of *didactic contract* as a theoretical framework aimed at understanding certain didactic situations involving the triple relationship "teacher–students–*savoir* (knowing)." Chevallard, Bosch, and Gascón (1997) described a didactical contract as a set of generally tacit norms or clauses which regulates the reciprocal duties of teacher and students concerning a common project of study. For Chevallard et al. (1997), this set of norms or clauses is not static, for it evolves as long as the didactical process goes forward. A didactical contract is therefore a construct to illuminate studies whose objective is to understand and support the work developed in the classroom (e.g., Galligan, 2005; Novotná & Hošpesová, 2008; Passos & Teixeira, 2011; Sierpinska, 2007).

Another example of a framework for analyzing interactions in classrooms was offered by Cobb and his colleagues (Cobb, 2000; Cobb, Stephan, McClain, & Gravemeijer, 2001; Yackel & Cobb, 1996). Based on a situated approach, these authors developed a framework that was intended to link the "social" and the "individual" dimensions of classroom interactions. In doing so, they saw these interactions as a coordination between the establishment of common mathematical practices (a social perspective) and the individuals' reorganization of mathematical reasoning during the evolution of these practices (a psychological perspective). The social perspective concerns the regulation of the classroom microculture regarding three main features: classroom social norms (established jointly by the teacher and students), socio-mathematical norms (normative aspects of classroom discourse and interaction that are specific to mathematics), and classroom mathematical practices (normative ways of reasoning mathematically during specific tasks). The psychological perspective focusses on the individual students' particular ways of participating in such common mathematical practices, more precisely on the individuals' mathematical beliefs (about their own role, the role of others, and the general nature of mathematical activity in school), values, interpretation and reasoning. For Cobb (2000), "each perspective constitutes the background against which mathematical activity is interpreted from the other perspective" (p. 64). Recent explorations of this interpretative framework in classrooms can be found in the works of Levenson, Tirosh, and Tsamir (2009), Lopez and Allal (2007), Tatsis and Kolezab (2008), Yackel (2001), and Yackel, Rasmussen, and King (2000).

Classroom designs. From a critical mathematics perspective, the construct of *landscapes of investigation,* introduced by Skovsmose (2001), refers to a dialogical environment in which mathematics is discussed through thematic projects that lead the students to develop a critical position about the role of the discipline in society concerning social, political and economical interests. In proposing these landscapes of investigation, Skovsmose observed that the teacher will probably have the experience of transposing a risk zone marked by the unpredictability of some events. On the other hand, he argued that not only may the students' mathematical abilities be developed in relation to certain contents but also their competence to interpret critically and act in a social and political situation structured by mathematics. Skovsmose stated that landscapes of investigation can be set no matter how the learning processes are organized.

Complementing the illustrations provided by Skovsmose, the work of Araújo (2009) offered an example of the processes of negotiation, production and development needed to build landscapes of investigation in the classroom. Acting as teacher-researcher, Araújo reported on a whole-year term course for undergraduate geography students designed from mathematical modelling projects (see also Barbosa, 2006, for a discussion on mathematical modelling in classroom from a socio-critical and discursive perspective, and the work of Pontes, 2003, about classroom design based on activities in mathematical investigations).

Another classroom design that has been taken up by researchers and teachers came from the core ideas surrounding Lave and Wenger's concept of *community of practice*. Winbourne and Watson (1998) proposed an adaptation of such ideas so that they would become applicable to certain school settings. This adaptation was intended to account for a design for teaching, and provide an analytical tool to evaluate relationships and student participation in mathematics classrooms. Winbourne and Watson suggested that, in some school settings, we can have, or not have, what they called a *local community of practice*. Frade et al. (2009) synthesized this construct as follows:

> A local community of practice in a school setting is, amongst other things, continuing activity where the participants—teacher and students—work purposefully together towards the achievement of a common goal. In doing so they share ... ways of behaving, language, habits, values and tool-use, and can see themselves as an essential part of the regulation of their activity and progress towards the common goal. (p. 15)

Based on these ideas, Frade et al. (2009) examined the students' crossing of boundaries between some specific, apparently insulated school practices. This crossing of boundaries was the focus of research carried out by secondary mathematics and science teachers who planned and developed an interdisciplinary collaboration aimed at creating a local community of practice. The authors concluded that it was mainly the activity of these teachers that enabled the students to cross the boundaries between their disciplines: the teachers translated their specific discipline language codes, worked together to prepare and organize their collaborative work, and shared their goals and purposes with the students.

Other studies—for instance, those by Watson and Winbourne (2008) and Graven (2004)—have applied the concept of communities of practices to educational and professional mathematical settings. Thus, Graven (2004) used Wenger's (1998) concept of communities of practice to investigate the teacher learning which occurred within a mathematics senior-phase inservice program fostered by a change in the curriculum in South Africa.

Modes of teaching. Here, we offer some illustrations that share the assumption that the enhancement of "the competencies and identities of all learners, to a large extent, rests with how teachers operationalize the core dimensions of pedagogy" (Walshaw & Anthony, 2008, p. 518)

The social turn translated into new pedagogies and classroom organization demands considerable effort and commitment from teachers. They are expected to move their pedagogical actions from traditional modes and conducts of teaching, as well as their ways to organize the classroom, toward the production of a new pedagogy featuring classroom dynamics which foster mathematical and social interactions. From a socio-cultural perspective committed to equity to the students' access to globalized knowledge and to educational change, Walshaw and Anthony (2008), in discussing the teacher's role in discursive practices within the classroom, have provided a rich and comprehensive review of the literature about connections between teachers' pedagogies and desirable mathematical and social outcomes for students. Walshaw and Anthony showed how engagement in mathematical discourse can successfully develop students' understanding while at the same time fostering a respectful exchange of ideas (between the teacher and the students and among students) as well as teacher listening, attentiveness, and reflection-interaction. These authors also discussed the effectiveness of the teacher's role in building bridges between students' everyday ideas and their mathematical ideas. It is through language—in particular by acknowledging students' difficulties when attempting to use mathematical language—that teachers can build these bridges. Walshaw and Anthony argued that by teaching and involving students in mathematical language, teachers contribute to students' development of mathematical clarity through argumentations, critiques, and justification of assertions.

For David (2004), teachers have a decisive influence over the interactions that occur in the classroom. Her assumption was that all enunciations made by the teacher directly act on how the students internalize mathematics (see Blanton, Stylianou, & David (2003) for a development of patterns of these enunciations). Based on the work of Vygotsky and his colleagues (Luria, 1976; Vygotsky, 1962, 1978), David's analysis of a number of lessons taught by a group of elementary and secondary mathematics teachers, revealed how much the mediation role of the teacher's language and discourse can contribute to the development of aspects of the students' mathematical thinking and actions.

Meira and Lerman (2010) focussed on the role of language and discourse in conceptual development. They employed Vygotsky's notion of the *zone of proximal development* (ZPD) as a semiotic space when analyzing interactions between a preschool teacher and her students. When investigating the communicative moves of the teacher and a 2.5 year-old child around a plantation of beans set-up on cotton wool, Meira and Lerman demonstrated how the teacher positioned herself to be receptive to the pupil's attempts to use new words in idiosyncratic ways. In this way they sustained a shared field of attention that enhanced communication by allowing both the teacher and the child to recognize ambiguities in their own discursive contributions.

Numerous other studies grounded on socio-cultural perspectives have directly or indirectly drawn attention to the roles of teachers in classrooms. These studies addressed distinct factors such as the relationships that both the students and teachers developed with mathematics and mathematical practices as a result of *participation* (e.g., Back & Pratt, 2007; David & Watson, 2008; Frade & Tatsis, 2009; Goos, Galbraith, & Renshaw, 2002; Jaworski, 2008; Martin, Towers, & Pirie, 2006; McVittie, 2004; Williams & Clarke, 2003) and *identity* (e.g., Boaler, 2002b; Boaler & Greeno, 2000; Brown & McNamara, 2011; Frade, Roesken, & Hannula, 2010; Ingram, 2008; Sfard & Pursak, 2005). Other matters addressed by researchers have been the effects of possibilities for communication between the teacher and students (e.g., Chronaki & Christiansen, 2005; Setati & Adler, 2000; Silver & Smith, 1996), classroom culture (e.g., Seeger, Voigt, & Waschescio, 1998), teachers' mathematical knowledge for teaching (e.g., Ball, 2003; Moreira & David, 2008; Tardif, Lessard, & Lahaye, 1991), and teacher's professional development (e.g., Fiorentini, 2003; Nicol, 2002). Although these studies and those discussed above clarify many aspects concerning how socio-cultural perspectives can reach the classrooms, there is an important question which needs more attention—To what extent do findings of research on classroom interactions and teacher communication and behaviour offer guidance on how students can be *guaranteed* reasonable participation?

In completing the present section, we suggest that there is a fourth emergent focus of research in mathematics learning represented by a substantial and increasing corpus of research whose main concern is specifically with socio-political dimensions of mathematics education (Baldino, 1998; Knijnik, 2010; Mattos & Batarce, 2010; Restivo & Sloan, 2007; Valero & Zevenbergen, 2004). Although methods and approaches in this area of research need to be consolidated (Valero, 2004), the socio-political perspectives neatly go across the three foci or areas we have been discussing so far. They are intimately associated with the notion of *power* and *inclusion* in their different manifestations (social, economic, political, educational and cultural) and include critical mathematics approaches (e.g., Chronaki, 2004; Mellin-Olsen, 1987, Powell & Frankenstein, 1997; Skovsmose, 2001; Skovsmose & Borba, 2004), and equality issues concerning learner gender, ethnicity, social class, language, and other divisors of society (e.g., Barton, 1996, 2008; D'Ambrosio, 2001; Ernest, 2007a, 2007b; Ernest et al., 2009; Frankenstein, 1995; Gerdes, 1996; Gorgorió & Planas, 2001; Keitel, 1998).

The next section is dedicated to a discussion on social justice and equality in mathematics education. For equality, we selected issues of gender, ethnicity and social class.

Social Justice: An Emergent Discourse in Mathematics Education

Social justice refers to the realization of a common good to be applicable in a certain practice of a group, community or society; it is a concept associated with *praxis*, with human action. The discussion in this chapter will make it clear that social justice is an emergent discourse in mathematics education: the common good is *equality* (and all that it subsumes, like diversity, inclusion, accessibility, dignity, respect, assistance, opportunity)—applicable to *all mathematics learners and users as well* (not withstanding ethnicity, gender, social class, age, …), and to *mathematical practice in its several manifestations* (educational, social, cultural, professional, economic, political and technological).

There are many ways in which a discussion on social justice in mathematics education can be organized (see, e.g., Atweh & Keitel, 2007; Dowling, 2007). We will consider the *meritocratic model* to structuring our discussion for two main reasons: (a) it has a direct impact on the students' lives and identities inside and outside the school; and (b) mathematics teachers are very familiar with it and its mode of functioning.

In his analysis of equal opportunities and its limitations in formal systems of education, François Dubet (2004), the sociologist of education, discussed why social justice, in the sense described above, is not really found. Dubet reported that, contrary to the aristocratic societies which prioritized the "well born," democratic societies have chosen merit as the essential principle of justice in education: by giving equality of access to all, the school becomes fair because everyone can become successful as a result of their efforts and qualities. This principle, said Dubet, was progressively implemented in the modern and rich countries with the expansion of common compulsory schooling, and the opening of tertiary and secondary education. Then, gradually, the formal frame of both the equal opportunity and merit principles was globally installed in a great number of countries. For Dubet, this school, however, did not become fairer for reducing performance differences between social classes, even though all students were allowed to enter into a supposedly balanced competition—see Kariya (2011) for a discussion of this remark in the context of Japanese education.

This purely meritocratic aspect of justice in schools brought a number of difficulties in that it reinforced markers of segregation between various groups of students (Valencia, 2010). In particular:

- The accessibility of the meritocratic model to all did not eliminate the inequalities of social classes, gender and social groups. The more favored students still had decisive advantages.
- The meritocratic school especially did not adequately address the needs of the most disadvantaged students. The barriers are more rigid for the poor, and teachers' expectations are less favorable to children from disadvantaged families. In competing with others, disadvantaged students more than often lose and become the despair of their teachers. They are left aside, marginalized within a differentiated curriculum, and become increasingly weak.
- The "losers," that is to say, the students who fail, are seen as solely responsible for their failure—because, it is argued, the school gave them the same opportunities to succeed as it gave the other students. As a consequence, these students tend to lose their self-esteem and motivation, refuse to attend school, or particular classes within school, and, in many cases, become violent individuals: after all, the meritocratic school placed them in a competitive environment without giving them the support they needed to succeed. From the point of view of the teachers, the meritocratic school is also cruel as they become the major agent of social and educational selection.

Dubet (2004) stated that he doubted whether the model of justice based on merit would be abandoned because within a society that, in principle, demands equality between all, individual merit is seen, by many, as the only way of producing fair or legitimate inequalities—other inequalities, based on birth and biological attributes, for example, are recognized as unacceptable.

Yet, as Dubet noted, we cannot ignore the fact that inequalities within schools can cause social and economic inequalities. As Cole (1985), a US scholar, stated over 25 years ago:

> Our society, founded upon the principle that all men are created equal, has never lived easily with the recognition of enormous *de facto* social inequality. We need a rationale for such inequality and our traditions strongly bias us to seek the causes of inequality, in properties of individuals, not society. At the same time, we realize that social and economic inequality can be the causes of individual intellectual inequalities, as well as their consequences. (p. 218)

This issue poses a challenging question to educators: Can social justice ever be achieved in schools in which, in the name of democracy, the meritocratic model is adopted?

For Dubet, social justice in schools (and other formal educational institutions) should consist, on the one hand, of assuring education accessibility to all, and on the other hand, of using this accessibility to suppress obvious privileges and complicity between the school and certain social groups. This understanding of social justice would be measured by *the way the school treats the disadvantaged students; by recognizing them as individuals in evolution, rather than students engaged in a competition.* According to Dubet, a school committed to social justice does not humiliate and hurt the students usually identified as "losers." Instead, it values and works on those students' interests and needs, assists them in their evolution preserving their dignity and the equality of principles in relation to the others and in the fair sharing of human and material intellectual resources available.

As already suggested in this chapter, when applied to the teaching and learning of mathematics, the meritocratic principle seems to play a special role in the provision—or obstruction—of social justice. It is through it that mathematics acts as a *critical filter* in schools. And the results of several recent research studies, mentioned earlier in this chapter in the context of a discussion on deficit models have confirmed that *the fair treatment to all in schools is still far from being achieved.* Those studies mostly reacted to deficit discourses involving issues of gender, ethnicity and social class (though other markers are also claimed by social justice). They shared the conclusion that inequalities in mathematics education concerning these traditional deficit markers were consequences of socially-constructed discourses to meeting the political, social and economic interests of some groups within the society. In this sense, the meritocratic model can be seen as an efficient, but a perverse mechanism of implementation of these interests in schools.

Next, we will briefly approach issues of gender, ethnicity and social class, the aim being to present current views of researchers in mathematics education on these issues.

Gender

Recent researchers on gender have seemed to agree that inequalities between females and males in mathematics education emerged from a traditional discourse which relied on the premise that females have a fragile emotional nature in comparison to the strong rational nature of males (Walkerdine, 1998). This premise had other associated beliefs and assumptions: females tend to do better in affective matters demanding care, assistance and sensitive support, as well as in the humanities and professional areas. This explains in part why it is not surprising that many young and adult women do not hesitate to comment on their supposed incompetence in mathematics and related subjects. Males, on the other hand, are supposed to do better in objective and rational matters, in the hard sciences, and in competitive professions. Although this is a simplistic description for the presence of inequality between genders, there is a sense in which it is true: the public image *is* that mathematics is a male domain (e.g., Burton, 1986; Forgasz, 1998; Hyde, Fennema, Ryan, Frost, & Hopp, 1990; Keitel, 1998).

In terms of the meritocratic model, females and males are the actors of the competition. Females always start with a disadvantage in comparison to males, for, the public perception is that women are competing in a territory essentially meant for men. This competition appears more or less explicit in several studies. Willis (1998) reported that the Head of the Mathematics Department of a school in Australia believed that girls did not enrol in mathematics classes at the senior levels as much as boys do, not because they lack mathematical skills, but because of emotional insecurities. The school authorities, and often the girls themselves, thought that girls could not cope as well as boys with social pressures (see Sukthankar, 1998, for comments on familial pressures with respect to the career choice of females). And, if for any reason, the rules of the competition are changed so that females perform better than males, then this is blamed on "feminist initiatives" (Zevenberger, 1998). Skelton (2010), in discussing the repositioning of girls from "victims" to "victors" at school, showed how the Australian media dealt with the fact that in 1996, for the first time, girls from one State performed higher than boys in the end-of-secondary-school mathematics examinations. A similar study focussing on community reactions to this fact was reported by Coupland and Wood (1998).

The situation with respect to females' participation and performance in mathematics is showing significant signs of changing, although these signs vary according to time, nation, ethnicity, school level, and socio-economic status (Ernest, 2007b; Fennema, 1995; Forgasz, Leder, & Kloosterman, 2010; Grevholm, 2007; Nkhwalume, 2007; Rossi-Becker, 1998). In relation to the image of mathematics as a male domain, the study by Forgasz et al. (2010) indicated that most North American and Australian research student-participants see mathematics as relatively gender neutral. Seliktar and Malik's (1998) study showed that, in the USA, differences between males and females in academic choices are reducing, resulting in a fairer competition. Seliktar and Malik attributed this to a socio-economic need, typical of modern countries: the traditional occupations of women are sufficiently low in prestige, autonomy and financial compensation that they do not enable a modern and autonomous woman to support a family. So, more women are wanting to secure higher paying jobs which often require technical qualifications. Without mathematics qualifications and expertise, these jobs may not be achievable. Rossi-Becker (1998) attributed the change to extensive and diverse intervention programs aiming at increasing the participation of women in mathematics and related professions. Ernest (2007b) reported that in Latin-American countries, the Caribbean and Scandinavia, a higher proportion (at least 50%) of those taking up mathematics and science studies, and occupations, are women. Regarding imbalances in mathematical performance, Ernest analyzed data from different countries and concluded that "no unambiguous differences in achievement levels can be identified" (p. 5).

In many communities the meritocratic model has been replaced by a model based on everyday competence. For instance, studies conducted by Knijnik (1998), with landless peasant communities in Brazil, and McMurchy-Pilkington (1998), with the Maori community in New Zealand, reported data suggesting that for females, school mathematical subjects were not useful or related to their everyday lives or to their roles in their communities. In the case of the Maori community, McMurchy-Pilkington showed that although the women regarded themselves as not mathematically competent (considering the school parameters), they were able to think in mathematically complex ways, especially in family-related situations. Thus, what was valued by the Maori women was their ethnomathematical competence in everyday life. Singh (1998), who studied a group of South African Indian females, reported a strong tension between the females' aspirations for gaining better qualifications in order to gain control over their lives and be able to enter the labor market, and barriers imposed by historical, economic and social hegemonic forces.

Ethnicity

Although studies relating gender and participation in mathematics seem to point to real signs of changes, even in countries in which women have been oppressed, the same cannot be said in relation to ethnic and social class differences. Over the past four decades, much research on race and minority ethnic culture in mathematics education has continually shown the cruel effects for many students of meritocratic competition in mathematics. Here, the actors of the competition are often the socially and politically constructed individuals identified as the "non-whites" and the "whites." As mentioned previously, many teachers continue to maintain low expectations and negative images of some under-achieving groups of students, especially the "non-white" ones. Consequently, these students are often subjected to a differentiated treatment and curriculum, reinforcing inequalities and obstructing their access to quality education. In addition, in most developed and developing countries, issues of race and minority ethnic cultures are strictly linked to issues of class. This is the case, for instance, of Brazil (e.g. Costa & Silva, 2010), South Africa (e.g., Dowling, 2007), and the USA (e.g., Livingston, 2007), where the color of the skin clearly reveals separate social positions in society. In the Netherlands, research carried out by Stevens et al. (2009) called attention to the importance of considering family processes and characteristics as an essential aspect of understanding the relationship between race/ethnicity and educational inequality.

The meritocratic model relies on the democratic principle of equality and is based on the logic of a universal curriculum for mathematics: everybody should learn the same set of contents and achieve a fixed set of abilities or skills (Bishop, 2010). By effectively ignoring the cultural diversities of groups of students, those supporting this model align themselves with the continuation of educational deficit practices (Ford et al., 2002; Gillborn, 2005; Glevey, 2007; Martin, 2009; Powell, 2002; Valencia, 2010), cultural conflicts (e.g., Gorgorió et al., 2004), and conflictive communications and affective relationships between students and their teachers (e.g., Gates, 2002; Gillborn, 1990; Sewell, 1997; Wright, Weeks, & McGlaughlin, 2000). In relation to the maintenance of deficit practices in schools, the studies of Glevey (2007) and Martin (2009) are particularly noteworthy in that they not only provide complementary explanations for the persistent inequality between some groups of students, but they also propose actions to remove these inequalities within classrooms.

Taking the classroom in England as the context of his study, Glevey (2007) discussed the persistent underachievement of Black students (that is to say, students of any African heritage). Some of his conclusions were

- The lack of care, attention, teacher expectation, and consequently the non-access to education quality can be considered the major factors in the mathematical underachievement of pupils of color, minority ethnic cultures and low social classes. These pupils often develop unhealthy identities marked by painful feelings of poor self-esteem and low self-expectations in life.
- How schools succeed in providing social justice to the disadvantaged pupils depend on their appreciation of the ideological positions and tensions within which they function. We have noted that competencies, abilities, skills and motivation to participate are not innate—they result from learning which develops in healthy affective environments.
- The persistent underachievement of these pupils (all over the world) is a challenge that must be confronted and defeated. According to Glevey (2007), "while legislations are useful in persuading teachers to treat all pupils with dignity, the crucial importance of genuine care and compassion cannot be overlooked if real progress is to be made in supporting all pupils" (p. 12).

Similar illuminations are also found in the works of Martin (2006, 2009), who has discussed the learning of mathematics by Afro-American pupils in the USA. Martin (2009) called for teachers and schools to implement mathematics classroom practices that "promote the development of positive racial and mathematical identities and situate the learning of mathematics in the social (and racial) realities confronting students" (p. 299).

Regarding cultural conflicts, research carried out by Gorgorió et al. (2004) with a group of immigrant youngsters in a Catalonian school district clearly revealed the major tensions to which immigrant students are subjected. Although Gorgorió et al.'s (2004) research focussed on identifying social and political circumstances which generated conflicts in the context of the research, it indirectly revealed the impact of such conflicts on the communication and relationship between students and teachers. For instance, Gorgorió et al. showed that a teacher-participant in the research attributed these conflicts to the immigrant students' rejection of the Catalonian school culture. As a consequence, this teacher-participant had, as would be predicted by Dubet (2004), Glevey (2007), Martin (2009) and many other researchers, low learning expectations for those students, as well as a negative image of them, and a lack of awareness of the need to understand the particularities of their cultural roots. Gorgorió et al.'s proposal is "to spread the idea [in schools] that cultural and social diversity, far from being a problem, can be a source of richness if the teachers can take the advantage of it" (p. 121).

Social Class

Most of the studies on social class in the mathematics education literature frame their questionings, arguments and claims in the terms of theories and models put forward by scholars, like Freire, Bourdieu, Bernstein, Foucault, and others. Relatively few modern scholars have reported empirical studies in which social class and mathematics learning are variables (Cooper & Dunne, 2000).

Nonetheless, three empirical studies deserve our attention due to the inequalities with which they are often associated: differentiated curriculum for low-class students, poor performance of low-class children on national curriculum tests, and difficult school and life conditions for poor children. The first study, reported by Dowling (1998, 2007), provided a critical investigation of school mathematics textbooks in the UK. Dowling analyzed a series of popular mathematics textbooks for school years 7 and 8—namely, the SMP 11–16 textbooks. According to Dowling, these textbooks consisted of a large number of booklets organized for levels and topics which could be used flexibly by all students. However, at the beginning of school year 9, the format of the SMP 11–16 changed, presenting three series of textbooks for use in year 9 and the subsequent 2 years. In a careful analysis of two samples of these series (series Y and series G), Dowling concluded that the Y series and the G series were clearly distinguished in terms of the "ability" of the proposed student audience. His main findings pointed out to a strong bias concerning perceived ability and social class: the Y series was specifically directed at high-ability students, and the G series at lower-ability students. The result was that student groups reflected social class and, among other things, these were marked by differentiated content and classroom discourse.

Dowling's study demonstrated that the meritocratic model, assumed by the use of the SMP series, orientated teachers and students to the belief that ability and social class positioning walked hand in hand, that mathematical ability is an attribute somehow encapsulated in social positioning, and that ability is not changeable or achieved during school years. This same belief clearly was in evidence in the following statement of a mathematics teacher during a conversation with Gates (2006):

> You know, a lot of my bottom group really struggle with maths—and I've noticed they all come from the same part of town, and they have got similar family backgrounds. Surely that can't be a coincidence? (p. 367)

A second study, reported by Cooper and Dunne (2000), addressed the relationships between mathematics success or failure, and social class. Taking the context of the UK National Curriculum and assessment in mathematics, these authors compared a large number of test and interview data. Cooper and Dunne showed that many children failed these tests because they got confused when interpreting items that were concerned with supposedly "realistic" situations, and not because they lacked related mathematical knowledge and understanding. Drawing on Bernstein's and Bourdieu's accounts of social-class differences and cultural orientation, Cooper and Dunne explored whether the same patterns of responses occurred with male and female children and with children from different social classes. They concluded that performance on National Curriculum items in general, and what they called "esoteric" and "realistic" items—referring to Dowling's (1998) introduction of these terms—in particular, varied by both gender and social class.

For instance, in relation to the primary school context, Cooper and Dunne's (2000) results indicated that middle-class children tended to move flexibly and appropriately between and across the boundaries of the "esoteric" and the "realistic" items but working-class children did not. Cooper and Dunne also showed that the tendency of working-class children to solve esoteric items was marked by bringing to their responses considerations of their everyday lives, which were not always appropriate from the point of view of the *language games* that were being played. This indicated at least two things: (a) working-class students seemed to be subjected to a differentiated curriculum, which somehow prioritized mathematical contents drawn on the public domain, rather than on the esoteric domain; and (b) National Curriculum items seemed to be designed for middle-class students. Whatever the case, both teachers and designers of these item tests needed to be aware that inequalities between social classes were being reproduced through National Curriculum test data.

The third, study was developed by Vithal (2003, 2004). Vithal reported on the painful life and school experiences of two Black adolescents—a boy, Wiseman, and a girl, Nellie—who were identified as living in the margins of society. Wiseman and Nellie were *street children* in the city of Durban, South Africa. Both had lived in and attended shelter (usually called "home") schools. Nellie had moved on to a "normal" school. In both cases, Vithal noted that the physical and intellectual conditions of the schools were very poor, insofar as they needed more adequate physical and pedagogical resources. Nellie had attended three different schools and had had a disrupted primary schooling. Like many street children, Nellie had faced experiences of abuse, neglect and poor health while trying to cope with schooling.

The extent of the discrimination suffered by Nellie from both her teacher and her classmates was revealed in an interview, when she commented that because the other students did not understand her situation, they laughed at her, and teased her. Nellie said she liked mathematics, but her test results were very low (Vithal, 2004).

The experience of Wiseman was quite different. He was recognized not only by his teacher, but also by his peers as one of the best students, someone who would definitely be placed into one of the public schools. He was proud and confident of his mathematical ability and assisted his classmates when participating in his mathematics classes.

Nellie's and Wiseman's appreciation for mathematics, despite their harsh conditions of life and schooling, point to, as indicated by Vithal, the need for future research in mathematics education to consider more seriously why and how learners in poverty and in potentially violent situations continue to learn, and want to learn mathematics.

Results of the studies summarized above strengthen the claim that social justice is an urgent matter that needs to be more carefully considered by education policy makers, mathematics teachers, and mathematics education researchers. Social justice in mathematics education cannot be achieved without a political and affective commitment from those responsible for creating mathematics education learning environments.

Appropriate and culturally-sensitive policies, based on modern research findings, need to be devised and implemented. There has been too much rhetoric and too much deficit thinking. In terms of research, how can studies on mathematics learning, mathematics curricula and mathematics teaching effectively lead to fair treatment for disadvantaged students? Can issues like affect, education policies and actions that divide society, and the need for social justice, become central issues in mathematics teacher development courses? Can the powerful meritocratic model be tweaked, so that it becomes a mechanism for equality? If it can, then how?

We close this section with a message to all Nellies and Wisemans, adapted from a reflection of Richard Rorty (1989): "To fail as a human being is to accept somebody else's description of oneself" (p. 28).

Summary

This chapter offers a view of how various socio-cultural perspectives of learning mathematics go beyond the deficit model of learning. The chapter was not intended to revive ideas or discuss data from the deficit research. Instead, it attempted to address the issue in a broad sense, showing a variety of perspectives and reporting on a number of relevant studies, within and around mathematics education. In reporting these studies, we chose to highlight the main conclusions, rather than discuss methods, arguments and evidence used to reach conclusions.

By contrasting deficit models and socio-cultural perspectives of mathematics learning, the chapter displays an uncomfortable reality: despite all academic advances and efforts to emphasize the fundamental role of culture in any individual's learning and development, deficit thinking is still a cloud hanging over the educational context, particularly in relation to mathematics education. By looking at the results of several current research studies that generated results that challenged deficit discourses, and by providing a brief overview of recent research concerning the three traditional deficit markers in mathematics education—gender, ethnicity and social class—the chapter has shown that inequality does persist within the walls of many schools, manifesting itself in different ways and varying across time and within and between nations.

The perspective from which we addressed issues of gender matches what Ernest (2007b) called as "The Public Educator" view, which is that "the gender and mathematics problem is a product of the distorted social construction of gender roles and differences and of mathematics itself" (p. 7). The result of this distortion, said Ernest, can be explained in terms of a vicious cycle: Gender-stereotyped cultural views (mathematics=male, mathematics≠feminine)→Lack of equal opportunities to learn mathematics, plus the stereotyped self perceptions of mathematics and mathematical abilities by women → Women's lower participation rate in mathematics → Unequal opportunities to study and work: "critical filter"→ Women in lower paid jobs → Reproduction of gender inequality in society → Confirmation of gender stereotyping → Gender-stereotyped cultural views (closing the cycle). In his conclusion, Ernest stated that "only if every link in the cycle is attacked can the reproductive cycle of gender inequality in mathematics education be broken" (p. 8). It is clear that the reproduction of this cycle involves distinct factors that are associated with economic and political conditions, theories and research methodologies, and education practice. The challenge to mathematics educators might be formulated as follows: "What can mathematics educators, teachers and policy makers effectively do to reduce, or even break this cycle?"

Regarding ethnic issues, a very strong argument about the biological non-existence of human races has been provided by Birchal and Pena (2010), who stated:

> The notion of "race" was imported from the common sense to science … Recently, however, the advances of the molecular genetics and the sequencing of the human genome … showed that the labels previously used to distinguish races do not have biological importance. It may seem easy to distinguish phenotypically a European from an African or an Asian, but such ease disappears completely when we look for evidence of these racial differences in genomes. In spite of that, the concept of race persists, qua social and cultural construction, as a way of favouring cultures, languages, beliefs and emphasizing the differences between groups with different economic interests. (p. 24)

These authors analyzed some aspects of the tension between the social and the biological views of race (in connection with the philosophical question of the relation between science and ethics).

Birchal and Pena (2010) cited Relethford (1994, 2002), and Jablonski and Chaplin (2000, 2002)—to support the assumption that, from the biological point of view, human races do not exist. This evidence strongly indicated that there is an excellent correlation between levels of UV radiation and levels of skin pigmentation worldwide: "The degree of skin pigmentation is determined by the amount and the type of melanin in the skin, and these in turn are apparently determined by a small number of genes (4–6) of which the melanotropic hormone receptor appears to be the most important" (p. 24). Birchal and Pena added that external phenotypic features (e.g., nose format, thickness, hair colour and texture) most likely indicate adaptation to environmental conditions and are influenced by sexual selection. And these phenotypical features also depend on relatively few genes. For these authors, these iconic "race" features correlate well with the continent of origin, but depend on variation in an insignificantly small portion of the human genome. In this sense they argued that "race is skin deep. Yet, human societies have constructed elaborated systems of privilege and oppression based on these insignificant genetic differences" (p. 24).

Birchal and Pena illustrated their assumption by analyzing the broad admixture of genes within the three founding continental groups forming the Brazilian population—the Amerindians, Europeans and Africans. The evidence produced a weak correlation between colour (a race correlate) and ancestrality. Consequently, they concluded, that "in Brazil, the colour, as socially perceived, has little or no biological consequence" (p. 24), and raised the question: "Since race does not exist from a biological point of view, would it lead to the moral consequence that the social use of the concept of race should be banned?" (p. 25) This question offers a strong challenge not only for society in general but also for those who support, consciously or subconsciously, the deficit model of learning concerning ethnicity.

Socio-cultural perspectives of mathematics learning cannot by themselves guarantee equality in mathematics education. But they can guide and help policy makers and mathematics teachers to improve their understandings of the diversity of identities and *forms of life* that are encountered in classrooms. It is a matter of being sensitive and dealing with differences not as deficit qualities, but instead as evidence of varieties of singular human beings and familial realities, who need different levels of assistance and care. It is in this sense that this chapter makes claims for social justice.

Osler and Starkey (2010), in proposing to discuss educational inequality and discrimination in terms of human rights, stated

> These standards provide a common point of reference for teachers and educators as they engage with students from a wide diversity of cultural, [economic], ethnic and religious backgrounds. Schools can help to ensure that human rights are known and understood, not simply as normative standards for encouraging pro-social behaviour, but also as a set of principles for critically engaging with social and political realities. (p. 43)

Osler and Starkey argued that the realization of justice is at the heart of the human rights project.

The approach taken in this chapter is consistent with the view expressed by Osler and Starkey (2010). Our discussion of inequality and discrimination in terms of social justice can be viewed as a claim for human rights concerning the specific case of deficit thinking in mathematics education. Our decision to address the main issues in terms of social justice allowed us to develop the critique of the meritocratic model of justice as presented by Dubet (2004). As previously suggested, it is important to question this model since, on the one hand, it is based on the democratic principle of equality, and on the other hand, it has been used as a mechanism of discrimination and exclusion, especially in relation to inequality between social classes. Of course, our approach also has economic and political implications.

Further research on the effects of the meritocratic model on practices, and therefore on people, is needed. How can socio-cultural perspectives guide and support mathematics education researchers, policy makers and teachers to implement these perspectives in educational systems based on the meritocratic model more effectively? If mathematics education is to become a domain that features justice and equality, then responses to this question must incorporate ways of rethinking the model and its use. This is a challenge that this chapter leaves to both practice and to future research in mathematics education.

Acknowledgments We are very grateful to Ricardo Scucuglia for his generous support in directing us to important international references, to Alexandre Rodrigues for his assistance in the organization of the references, and to Steve Lerman and Simon Goodchild for helpfully reviewing this chapter.

References

Acioly, N. M. (1994). *La juste mesure: Une étude des competences mathématiques des travailleurs de la canne a sucre du Nordeste du Brésil dans le domaine de la mesure* (Doctoral dissertation). Université René Descartes, Paris, France.
Acioly, N. M., & Schliemann, A. D. (1987). Escolarização e conhecimento de matemática desenvolvido no contexto do jogo do bicho. (Schooling and knowledge of mathematics developed in the context of a "numbers game") *Cadernos de Pesquisa, 61*, 42–57.
Acioly-Régnier, N. M. (1996). Diz-me com quem resolves um problema de matemática e dir-te-ei quem és. In M. G. B. Dias & A. Spinillo (Eds.), *Tópicos em psicologia cognitiva* (pp. 195–227). Recife, Brazil: Editora Universitária da UFPE.
Acioly-Régnier, N. M. (1997). Analyse des compétences mathématiques de publics adultes peu scolarisés et/ou peu qualifiés. In F. Andrieux, J.-M. Besse, & B. Falaise (Eds.), *Illettrismes: Quels chemins vers l'écrit? Les actes de l'université d'été du 8 au 12 Juillet 1996*. Lyon, France: Magnard.
Acioly-Régnier, N. M. (2010, September). Culture et cognition: Domaine de recherche, champ conceptuel, cadre d'intelligibilité et objet d'étude fournissant des instruments pour conduire des analyses conceptuelles et méthodologiques en psychologie et en sciences de l'éducation. *Habilitation à diriger des recherches*. Université Lumière Lyon 2.
Araújo, J. L. (2009). Formatting real data in mathematical modelling projects. *IMFUFA Text, 461*, 229–239.
Araújo, C. R., Andrade, F., Hazin, I., Da Rocha Falcão, J. T., Nascimento, J. C., & Lins Lessa, M. M. (2003). Affective aspects on mathematics conceptualization: From dichotomies to an integrated approach. In N. A. Pateman, B. J. Dougherty, & J. Zilliox (Eds.), *Proceedings of the 27th Conference of the International Group for the Psychology of Mathematics Education held jointly with the 25th Conference of the North American Chapter of the International Group for the Psychology of Mathematics Education* (Vol. 2, pp. 269–276). Honolulu, HI: PME.
Ascher, M., & Ascher, R. (1997). Ethnomathematics. In A. B. Powell & M. Frankenstein (Eds.), *Ethnomathematics: Challenging eurocentrism in mathematics* (pp. 25–50). Albany, NY: SUNY Press.
Atweh, B., & Keitel, C. (2007). Social (in)justice and international collaboration in mathematics education. In B. Atweh, B. Barton, M. Borba, N. Gough, C. Keitel, C. Vistro-Yu, & R. Vithal (Eds.), *Internationalisation and globalisation in mathematics and science education* (pp. 95–112). New York, NY: Springer.
Back, J., & Pratt, N. (2007). Spaces to discuss mathematics: Communities of practice on an online discussion board. *2nd Socio-cultural Theory in Educational Research and Practice Conference*, University of Manchester, UK. Retrieved March 24, 2008 from http://www.lta.education.manchester.ac.uk/ScTIG/papers/Jenni%20Back.pdf.
Baldino, R. R. (1998). School and surplus-value: Contribution from a Third-World country. In P. Gates (Ed.), *Proceedings of the First International Mathematics Education and Society*

Conference (pp. 73–81). Nottingham, UK: Centre for the Study of Mathematics Education, Nottingham University.

Ball, D. L. (2003, February 6), *What mathematical knowledge is needed for teaching mathematics?* Remarks prepared for the Secretary's summit on Mathematics, U.S. Department of Education, Washington, DC. Retrieved from http://www.ed.gov/rschstat/research/progs/mathscience/ball.html.

Barbosa, J. C. (2006). Mathematical modelling in classroom: A critical and discursive perspective. *Zentralblatt Fur Didaktik der Mathematik, 38*(3), 293–301.

Barton, B. (1996). Making sense of ethnomathematics: Ethnomathematics is making sense. *Educational Studies in Mathematics, 31*(1/2), 201–233.

Barton, B. (2008). *The languages of mathematics*. New York, NY: Springer.

Bernstein, B. (1996). *Pedagogy, symbolic control and identity: Theory, research, critique*. London, UK: Taylor and Francis.

Birchal, T. S., & Pena, S. D. J. (2010). The biological non-existence *versus* the social existence of human races—Can science instruct the social ethos? In S. D. J. Pena (Ed.), *Themes in transdisciplinary research* (pp. 23–59). Belo Horizonte, Brazil: Editora UFMG.

Bishop, A. (1988). *Mathematical enculturation: A cultural perspective on mathematics education*. Dordrecht, The Netherlands: Kluwer Academic Publishers.

Bishop, A. J. (1994). Cultural conflicts in mathematics education: Developing a research agenda. *For the Learning of Mathematics, 14*(2), 15–18.

Bishop, A. (1999, May). *Democratising mathematics through education in the 21st century—Lessons from research*. Address to 8th Southeast Conference on Mathematics Education, Manila, The Philippines.

Bishop, A. (2002). Mathematical acculturation, cultural conflicts, and transition. In G. de Abreu, A. J. Bishop, & N. C. Presmeg (Eds.), *Transitions between contexts of mathematical practices* (pp. 193–212). Dordrecht, The Netherlands: Kluwer Academic Publishers.

Bishop, A. (2010). Directions and possibilities for research on mathematics and culture, in relation to mathematics education: A personal view. In M. M. F. Pinto & T. F. Kawasaki (Eds.), *Proceedings of the 34th Conference of the International Group for the Psychology of Mathematics Education* (Vol. 1, pp. 338–342). Belo Horizonte, Brazil: PME.

Bishop, A. J., FitzSimons, G. E., Seah, W. T., & Clarkson, P. C. (1999, December). *Values in mathematics education: Making values teaching explicit in the mathematics classroom*. Paper presented at the Australian Association for Research in Education, Melbourne.

Blanton, M., Stylianou, D., & David, M. M. (2003). The nature of instructional scaffolding in undergraduate students' transition to mathematical proof. In N. A. Pateman, B. J. Dougherty, & J. Zilliox (Eds.), *Proceedings of the 2003 Joint Meeting of PME and PMENA* (Vol. 2, pp. 113–120). Honolulu, HI: International Group for the Psychology of Mathematics Education.

Boaler, J. (2000). Introduction: Intricacies of knowledge, practice, and theory. In J. Boaler (Ed.), *Multiple perspectives on mathematics teaching and learning* (pp. 1–17). Westport, CT: Ablex Publishing.

Boaler, J. (2002a). The development of disciplinary relationships: Knowledge, practice and identity in mathematics classrooms. *For the Learning of Mathematics, 22*(1), 42–47.

Boaler, J. (2002b). Exploring the nature of mathematical activity: Using theory, research and "working hypotheses" to broaden conceptions of mathematics knowing. *Educational Studies in Mathematics, 51*(1–2), 3–21.

Boaler, J., & Greeno, J. G. (2000). Identity, agency, and knowing in mathematics worlds. In J. Boaler (Ed.), *Multiple perspectives on mathematics teaching and learning* (pp. 171–200). Westport, CT: Ablex Publishing.

Bonilla-Silva, E. (1997). Rethinking racism: Toward a structural interpretation. *American Sociological Review, 62*, 465–480.

Borba, M. C. (1997). Ethnomathematics and education. In A. B. Powell & M. Frankenstein (Eds.), *Ethnomathematics: Challenging eurocentrism in mathematics* (pp. 261–272). Albany, NY: SUNY Press.

Bourdieu, P., & Passeron, J.-C. (1977). *Reproduction in education, society and culture*. London, UK: Sage Publications. Published in *For The Learning of Mathematics, 10*(1), 39–43, in 1990.

Brousseau, G. (1986). Fondements et méthodes de la didactique des mathématiques. Résumés–*Recherches em Didactique dês Mathématiques*, 7(2). Retrieved from http://rdm.penseesauvage.com/Fondements-et-methodes-de-la.html.

Brousseau, G. (2006). Mathematics, didactical engineering and observation. In J. Novotná, H. Moraová, M. Krátká, & N. Stehlíková (Eds.), *Proceedings of the 30th conference of the International Group for the Psychology of Mathematics Education* (Vol. 1, pp. 3–18). Prague, Czech Republic: PME.

Brown, J. S., Collins, A., & Duguid, P. (1989). Situated cognition and the culture of learning. *Educational Researcher*, 18(1), 32–42.

Brown, T., & McNamara, O. (2011). *Becoming a mathematics teacher—Identity and identifications*. Dordrecht, The Netherlands: Springer.

Bruner, J. (1990). *Acts of meaning*. Cambridge, MA: Harvard University Press.

Burton, L. (1986). *Girls into maths can go*. London, UK: Holt, Rinehart and Winston.

Canady, H. G. (1936). The effect of "rapport" on the IQ: A new approach to the problem of racial psychology. *Journal of Negro Education*, 5, 209–219.

Carraher, T. N. (1986). From drawings to buildings: Working with mathematical scales. *International Journal of Behavioural Development*, 9, 527–544.

Carraher, T. N., Carraher, D. W., & Schliemann, A. D. (1985). Mathematics in the streets and in schools. *British Journal of Developmental Psychology*, 3, 21–29.

Carraher, D. W., & Schliemann, A. D. (2002). Is everyday mathematics truly relevant to mathematics education? In J. Moshkovich & M. Brenner (Eds.), *Everyday and academic mathematics in the classroom. Monographs of the Journal for Research in Mathematics Education 11*, 131–153.

Carreira, S., Evans, J., Lerman, S., & Morgan, C. (2002). Mathematical thinking: Studying the notion of transfer. In A. Cockburn & E. Nardi (Eds.), *Proceedings of 26th Annual Conference of the International Group for the Psychology of Mathematics Education* (Vol. 2, pp. 185–192). Norwich, UK: International Group for the Psychology of Mathematics Education.

Chevallard, Y., Bosch, M., & Gascón, J. (1997). *Estudiar matemáticas: El eslabón perdido entre la enseñanza y el aprendizaje*. Barcelona, Spain: I.C.E. Universitat Barcelona.

Chronaki, A. (2004). Researching the school mathematics culture of "others": Creating a self-other dialogue. In P. Valero & R. Zevenbergen (Eds.), *Researching the socio-political dimensions of mathematics education* (pp. 145–165). Dordrecht, The Netherlands: Kluwer Academic Publishers.

Chronaki, A., & Christiansen, I. M. (Eds.). (2005). *Challenging perspectives on mathematics classroom communication*. Greenwich, CT: Information Age.

Civil, M. (2007). Building on community knowledge: An avenue to equity in mathematics education. In N. Nassir & P. Cobb (Eds.), *Improving access to mathematics: Diversity and equity in the classroom* (pp. 105–117). New York, NY: Teachers College Press.

Clarkson, P. C., Fitzsimons, G. E., & Seah, W. T. (1999). Values relevant to mathematics? I'd like to see that! In N. Scott, D. Tynan, G. Asp, H. Chick, J. Dowsey, B. McCrae, J. McIntosh, & K. Stacey (Eds.), *Mathematics: Across the ages. Proceedings of the Thirty-sixth Annual Conference* (pp. 129–132). Melbourne, Australia: Mathematical Association of Victoria.

Cobb, P. (2000). The importance of a situated view of learning to the design of research and instruction. In J. Boaler (Ed.), *Multiple perspectives on mathematics teaching and learning* (pp. 45–82). Westport, CT: Ablex Publishing.

Cobb, P., & Bowers, J. (1999). Cognitive and situated learning perspectives in theory and practice. *Educational Researcher*, 28(2), 4–15.

Cobb, P., Stephan, M., McClain, K., & Gravemeijer, K. (2001). Participating in classroom mathematical practices. *Journal of the Learning Sciences*, 10, 113–164.

Cole, M. (1985). Mind as a cultural achievement: Implications for IQ testing. In E. Eisner (Ed.), *Learning and teaching the ways of knowing* (84th Yearbook of the National Society for the Study of Education) (pp. 218–249). Chicago, IL: National Society for the Study of Education.

Cole, M., Gay, J., Glick, J., & Sharp, D. W. (1971). *The cultural context of learning and thinking*. New York, NY: Basic Books.

Cole, M., & Scribner, S. (1974). *Culture and thought, a psychological introduction*. New York, NY: John Wiley & Sons.

Cooper, B., & Dunne, M. (2000). *Assessing children's mathematical knowledge—Social class, sex and problem solving*. Buckingham, UK: Open University Press.

Costa, W. G., & Silva, V. L. (2010). A desconstrução das narrativas e a reconstrução do currículo: a inclusão dos saberes matemáticos dos negros e dos índios brasileiros (Narrative deconstruction and curriculum reconstruction: The inclusion of Brazilian indians' and black people's mathematical knowledge). *Educar, 36*, 245–260.

Coupland, M., & Wood, L. (1998). What happens when the girls beat the boys? Community reactions to the improved performance of girls in final school examinations. In C. Keitel (Ed.), *Social justice and mathematics education—Gender, class, ethnicity and the politics of schooling* (pp. 238–244). Berlin, Germany: Freie Universität Berlin.

D'Ambrosio, U. (1985). Mathematical education in a cultural setting. *International Journal of Mathematics Education in Science and Technology, 16*(4), 469–477.

D'Ambrosio, U. (1988, October). A research program in the history of ideas and in cognition. *International Study Group on Ethnomathematics (ISGEm) Newsletter, 4*(1). Retrieved from http://www.ethnomath.org/resources/ISGEm/035.htm.

D'Ambrosio, U. (1993). Educação matemática: Uma visão do estado da arte. *Pro-Posições* (Revista Quadrimestral – Faculdade de Educação – UNICAMP), *4*(1[10]), 7–16.

D'Ambrosio, U. (1997). Foreword. In A. B. Powell & M. Frankenstein (Eds.), *Ethnomathematics: Challenging eurocentrism in mathematics education* (pp. xv–xxi). Albany, NY: State University of New York Press.

D'Ambrosio, U. (2001). *Etnomatemática: Elo entre as tradições e a modernidade (Ethnomathematics: Link between the traditions and the modernity)*. Belo Horizonte, Brazil: Autêntica.

D'Ambrosio, U. (2010a, January 22). Re: The name ethnomathematics: My personal view. Retrieved from http://vello.sites.uol.com.br/what.htm.

D'Ambrosio, U. (2010b, January 22). Re: Ethnomathematics: My personal view. Retrieved from http://vello.sites.uol.com.br/what.htm.

D'Ambrosio, U. (2010c, January 22). Re: Why ethomathematics? Or, what is ethnomathematics and how can it help children in schools? Retrieved from http://vello.sites.uol.com.br/what.htm.

Da Rocha Falcão, J. T. (1995). A case study of algebraic scaffolding: From balance scale to algebraic notation. In L. Meira & D. Carraher (Eds.), *Proceedings of the 19th Conference for the International Group for the Psychology of Mathematics Education* (Vol. 2, pp. 66–73). Recife, Brazil: PME and Universidade Federal de Pernambuco.

Da Rocha Falcão, J. T. (2005). *Conceptualisation en acte, conceptualisation explicite: Quels apports théoriques à offrir à la didactique des mathématiques et des sciences? Actes du Colloque Les processus de conceptualisation en debat-hommage a Gérard Vergnaud*. Paris, France: Association pour la Recherche sur le Développement des Compétences.

David, M. M. (2004). Interações discursivas em sala de aula e o desenvolvimento do pensamento matemático dos alunos (*Discursive interactions in classrooms and the development of the students' mathematical thinking*). Anais do VIII ENEM. Recife, Brazil (CD-ROM).

David, M. M., & Watson, A. (2008). Participating in what? Using situated cognition theory to illuminate differences in classroom practices. In A. Watson & P. Winbourne (Eds.), *New directions for situated cognition in mathematics education* (pp. 31–57). Dordrecht, The Netherlands: Springer.

de Abreu, G., Bishop, A. J., & Presmeg, N. (Eds.). (2002). *Transitions between contexts of mathematical practices*. Dordrecht, The Netherlands: Kluwer Academic Publishers.

de Abreu, G., & Carraher, D. W. (1989). The mathematics of Brazilian sugar cane farmers. In C. Keitel, P. Damerow, A. Bishop & P. Gerdes (Eds.), *Mathematics, education and society* (pp. 68–70). Paris, France: UNESCO Science and Technology Education Document Series, No. 35.

Dowling, P. (1998). *The sociology of mathematics education: Mathematical myths/pedagogic texts*. London, UK: Falmer.

Dowling, P. (2007). Organising the social. *Philosophy of Mathematics Education Journal, 21*. Retrieved from http://people.exeter.ac.uk/PErnest/pome21/index.htm.

Dubet, F. (2004). O que é uma escola justa? [What is a fair school?]. *Cadernos de Pesquisa, 34*, 539–555.

Engeström, Y. (1999). Situated learning at the threshold of the new millennium. In J. Bliss, R. Saljo, & P. Light (Eds.), *Learning sites: Social and technological resources for learning* (pp. 249–257). Oxford, UK: Pergamon-Earli.

Ernest, P. (2007a). Why social justice? *Philosophy of Mathematics Education Journal, 21*. Retrieved from http://people.exeter.ac.uk/PErnest/pome21/index.htm.

Ernest, P. (2007b). Questioning the gender problem in mathematics. *Philosophy of Mathematics Education Journal, 20*. Retrieved from http://people.exeter.ac.uk/PErnest/pome20/index.htm.

Ernest, P., Greer B., & Sriraman, B. (2009). Introduction: Agency in mathematics education. In P. Ernest, B. Greer, & B. Sriraman (Eds.), *Critical issues in mathematics education* (pp. ix–xvi). Missoula, MT: Age Publishing Inc. & Montana Council of Teachers of Mathematics (The Montana Mathematics Enthusiast—monograph series in mathematics education).

Fennema, E. (1995). Mathematics gender and research. In B. Grevholm & G. Hanna (Eds.), *Gender and mathematics education, an ICMI Study* (pp. 21–38). Lund, Sweden: Lund University Press.

Fiorentini, D. (Ed.). (2003). *Formação de professores de matemática: Explorando novos caminhos com outros olhares [Mathematics teachers formation: Exploring new ways with other views]*. Campinas, Brazil: Mercado de Letras.

Fonseca, M. C. F. R. (2010). Adult education and ethnomathematics: Appropriating results, methods, and principles. *Zentralblatt fur Didaktik der Mathematik, 42*, 361–369.

Ford, D. Y., Harris, J. J., Tyson, C. A., & Trotman, M. F. (2002). Beyond deficit thinking: Providing access for gifted African American students. *Roeper Review, 24*(2), 52–58.

Forgasz, H. (1998). The "male domain" of high school and tertiary mathematics learning environments. In C. Keitel (Ed.), *Social justice and mathematics education—Gender, class, ethnicity and the politics of schooling* (pp. 32–44). Berlin, Germany: Freie Universität Berlin.

Forgasz, H. J., Leder, G. C., & Kloosterman, P. (2010). New perspectives on the gender stereotyping of mathematics. *Mathematical Thinking and Learning, 6*(4), 389–420.

Frade, C. (2006). Humanizing the theoretical and the practical for mathematics education. In J. H. Woo, J. H. C. Lew, K. S. Park, & D. Y. Seo (Eds.), *Proceedings of the 31st Conference of the International Group for the Psychology of Mathematics Education* (Vol. 1, pp. 99–103). Seoul, Korea: International Group for the Psychology of Mathematics Education.

Frade, C., & Da Rocha Falcão, J. T. (2008). Exploring connections between tacit knowing and situated learning perspectives in the context of mathematics education. In A. Watson & P. Winbourne (Eds.), *New directions for situated cognition in mathematics education* (pp. 203–231). Dordrecht, The Netherlands: Springer.

Frade, C., & Faria, D. (2008). Is mathematics learning a process of enculturation or a process of acculturation? In J. F. Matos, P. Valero, & K. Yasukawa (Eds.), *Proceedings of the Fifth International Mathematics Education and Society Conference* (Vol. 1, pp. 248–258). Lisbon, Portugal: Centro de Investigação em Educação, Universidade de Lisboa, and PME.

Frade, C., & Machado, M. C. (2008). Culture and affect: Impacts of the teachers' values over the students' affect. In O. Figueras, J. L. Cortina, S. Alatorre, T. Rojano, & A. Sepúlveda (Eds.), *Proceedings of the 32nd Conference of the International Group for the Psychology of Mathematics Education* (Vol. 3, pp. 33–40). Morelia, Mexico: Cinvestav-UMSNH.

Frade, C., Machado, M. C., & Faria, D. (2008). Culture and affect: Two studies about impacts of the teachers' values over the students' affect—Topic Study Group 30. *The 11th International Congress on Mathematical Education—ICME 11*. Monterrey, Mexico: ICMI. Retrieved from http://tsg.icme11.org/tsg/show/31.

Frade, C., & Meira, L. (2010) The social nature of affective behaviors and the constitution of identity. In C. Frade, B. Roesken, & M. Hannula (Orgs.). Identity and affect in the context of teachers' professional development. In M. M. F. Pinto & T. F. Kawasaki (Eds.), *Proceedings of the 34th Conference for the International Group for the Psychology of Mathematics Education*

(Vol. 1, pp. 247–249). Belo Horizonte, Brazil: International Group for the Psychology of Mathematics Education.

Frade, C., Roesken, B., & Hannula, M. S. (2010). Identity and affect in the context of teachers' professional development. In M. M. F. Pinto & T. F. Kawasaki (Eds.), *Proceedings of the 34th Conference of the IGPME* (Vol. 1, pp. 247–279). Belo Horizonte, Brazil: PME.

Frade, C., & Tatsis, K. (2009). Learning participation and local school mathematics practice. *The Montana Mathematics Enthusiast, 6*(1–2), 96–112.

Frade, C., Winbourne, P., & Braga, S. M. (2009). A mathematics-science community of practice reconceptualising transfer in terms of crossing boundaries. *For the Learning of Mathematics, 29*(2), 14–22.

Frankenstein, M. (1995). Equity in mathematics education: Class in the world outside the class. In W. Secada, E. Fennema, & L. Adajiand (Eds.), *New directions for equity in mathematics education* (pp. 165–190). Cambridge, MA: Cambridge University Press.

Galligan, L. (2005). Conflicts in offshore learning environments of a university preparatory mathematics course. In H. L. Chick & J. L. Vincent (Eds.), *Proceedings of the 29th Conference of the International Group for the Psychology of Mathematics Education* (Vol. 3, pp. 25–32). Melbourne, Australia: International Group for the Psychology of Mathematics Education.

Gardner, H. (1983). *Multiple intelligences: The theory in practice*. New York, NY: Basic Books.

Gates, P. (2002). Issues of equity in mathematics education: Defining the problem, seeking solutions. In L. Haggarty (Ed.), *Teaching mathematics in secondary schools: A reader* (pp. 211–228). London, UK: Routledge/Falmer.

Gates, P. (2006). The place of equity and social justice in the history of PME. In A. Gutiérrez & P. Boero (Eds.), *Handbook of research on the psychology of mathematics education: Past, present and future* (pp. 367–402). Rotterdam, The Netherlands: Sense Publishers.

Gee, J. P. (1996). *Social linguistics and literacies: Ideologies in discourse*. London, UK: Taylor and Francis.

Geertz, C. (1973). *The interpretation of cultures*. New York, NY: Basic Books.

Gerdes, P. (1994). Reflections on ethnomathematics. *For The Learning of Mathematics, 14*(2), 19–22.

Gerdes, P. (1996). Geometry from Africa—Mathematical and educational explorations. In A. J. Bishop, K. Clements, C. Keitel, J. Kilpatrick, & C. Laborde (Eds.), *International handbook of mathematics education* (pp. 909–943). Dordrecht, The Netherlands: Kluwer Academic Publishers.

Gillborn, D. (1990). *Race, ethnicity and education—Teaching and learning in multi-ethnic schools*. London, UK: Unwin Hyman.

Gillborn, D. (2005). Education policy as an act of white supremacy: Whiteness, critical race theory and education reform. *Journal of Education Policy, 20*(4), 485–505.

Glevey, K. E. (2007). Pupils of African heritage, mathematics education and social justice. *Philosophy of Mathematics Education Journal, 20*. Retrieved from http://people.exeter.ac.uk/PErnest/pome20/index.htm.

Gomes, N. L. (2003). Cultura negra e educação [Black culture and education]. *Revista Brasileira de Educação, 23*, 75–85.

Gomes, M. L. M. (2008). *Quatro visões iluministas sobre a educação matemática: Diderot, d'Alembert, Condillac e Condorcet*. [Four visions on Mathematics Education from the Age of Enlightenment: Diderot, d'Alembert, Condillac and Condorcet]. Campinas, Brazil: Editora da UNICAMP.

Goos, M., Galbraith, P., & Renshaw, P. (2002). Socially mediated metacognition: Creating collaborative zones of proximal development in small group problem solving. *Educational Studies in Mathematics, 49*, 193–223.

Gorgorió, N., & Planas, N. (2001). Teaching mathematics in multicultural classrooms. *Educational Studies in Mathematics, 47*(1), 7–33.

Gorgorió, N., Planas, N., & Bishop, A. (2004). Dichotomies, complementarities and tensions: Researching mathematics teaching in its social and political context. In P. Valero & R. Zevenbergen (Eds.), *Researching the socio-political dimensions of mathematics education* (pp. 107–123). Dordrecht, The Netherlands: Kluwer Academic Publishers.

Gould, S. J. (1995). *The mismeasure of man*. New York, NY: Norton.
Graven, M. (2004). Investigating mathematics teacher learning within an in-service community of practice: The centrality of confidence. *Educational Studies in Mathematics, 57*, 177–211.
Greenfield, P., & Childs, C. (1977). Weaving, color terms, and pattern representation: Cultural influences and cognitive development among the zinacantecos of Southern Mexico. *International Journal of Psychology, 11*, 23–48.
Greeno, J. G. (1997). On claims that answer the wrong questions. *Educational Researcher, 25*(1), 5–17.
Greeno, J. G. (1998). The situativity of knowing, learning, and research. *American Psychologist, 53*(1), 5–26.
Greeno, J. G., Smith, D. R., & Moore, J. L. (1993). Transfer of situated learning. In D. K. Detterman & J. R. Sternberg (Eds.), *Transfer on trial: Intelligence, cognition and instruction* (pp. 99–167). Norwood, CT: Ablex Publishing.
Grevholm, B. (2007). Critical networking for women and mathematics: An intervention project in Sweden. *Philosophy of Mathematics Education Journal, 21*. Retrieved from http://people.exeter.ac.uk/PErnest/pome21/index.htm.
Groulx, L.-H. (2008). Contribuição da pesquisa qualitativa à pesquisa social. In J. Poupart, J. Deslauriers, L. Groulx, A. Laperrière, R. Mayer, & A. P. Pires (Eds.), *A Pesquisa Qualitativa: enfoques epistemológicos e metodológicos [The qualitative research: methodological and epistemological approaches]* (pp. 95–124). Petrópolis, Brazil: Vozes.
Gutiérrez, R. (2007). Context matters: Equity, success, and the future of mathematics education. In T. Lamberg & L. R. Wiest (Eds.), *Proceedings of the 29th Annual Meeting of the North American Chapter of the International Group for the Psychology of Mathematics Education* (Vol. 1, pp. 1–5). Stateline, Lake Tahoe, NV: University of Nevada, Reno.
Holland, D., Skinner, D., Lachicotte, W., & Cain, C. (2001). *Identity and agency in cultural worlds*. London, UK: Harvard University Press.
Hyde, J. S., Fennema, E., Ryan, M., Frost, L. A., & Hopp, C. (1990). Gender comparisons of mathematics attitudes and affect: A meta-analysis. *Psychology of Women Quarterly, 14*, 299–324.
Ingram, N. (2008). The importance of length, breadth and depth when studying student's affective responses to mathematics through the lens of identity—Topic Study Group 30. *The 11th International Congress on Mathematical Education—ICME 11*. Monterrey, Mexico: ICMI. Retrieved from http://tsg.icme11.org/tsg/show/31.
Jablonski, N. G., & Chaplin, G. (2000). The evolution of human skin coloration. *Journal of Human Evolution, 39*, 57–106.
Jablonski, N. G., & Chaplin, G. (2002). Skin deep. *Scientific American, 287*(4), 74–81.
Jaworski, B. (2008, March 24). *Theory in developmental research in mathematics teaching and learning: Social practice theory and community of inquiry as analytical tools*. Paper presented at the 2nd Socio-cultural Theory in Educational Research and Practice Conference, University of Manchester, Manchester, UK. http://www.lta.education.manchester.ac.uk/ScTIG/papers/Barbara%20Jaworski.pdf.
Kariya, T. (2011). Japanese solutions to the equity and efficiency dilemma? Secondary schools, inequity and the arrival of "universal" higher education. *Oxford Review of Education, 37*(2), 241–266.
Keitel, C. (1998). *Social justice and mathematics education—Gender, class, ethnicity and the politics of schooling*. Berlin, Germany: Freie Universität Berlin.
Kilpatrick, J. (1992). A history of research in mathematics education. In A. D. Grouws (Ed.), *Handbook of research on mathematics teaching and learning* (pp. 3–28). New York, NY: Macmillan.
Kirshner, D., & Whitson, J. A. (Eds.). (1997). *Situated cognition: Social, semiotic and psychological perspectives*. Mahwah, NJ: Lawrence Erlbaum.
Klineberg, O. (1935). *Race differences*. New York, NY: Harper.
Knijnik, G. (1998). Exclusion and resistance in Brazilian struggle for land: (Underprivileged) Women and mathematics education. In C. Keitel (Ed.), *Social justice and mathematics*

education—Gender, class, ethnicity and the politics of schooling (pp. 116–122). Berlin, Germany: Freie Universität Berlin.

Knijnik, G. (1999). Ethnomathematics and the Brazilian landless people education. *Zentralblatt fur Didaktik der Mathematik, 31*(3), 188–194.

Knijnik, G. (2004). Lessons from research with a social movement. A voice from the South. In P. Valero & R. Zevenbergen (Eds.), *Researching the socio-political dimensions of mathematics education* (pp. 125–141). Dordrecht, The Netherlands: Kluwer Academic Publishers.

Knijnik, G. (2010). Studying people's out of school mathematical practices. Research Forum: The role of context in research with digital technologies: Towards a conceptualisation. In M. M. F. Pinto & T. F. Kawasaki (Eds.), *Proceedings of the 34th Conference of the International Group for the Psychology of Mathematics Education* (Vol. 1, pp. 318–322). Belo Horizonte, Brazil: International Group for the Psychology of Mathematics Education.

Krzywacki, H., & Hannula, M. S. (2010). Tension between present and ideal state of teacher identity in the core of professional development. In C. Frade, B. Roesken, & M. Hannula (Eds.). Identity and affect in the context of teachers' professional development. In M. M. F. Pinto & T. F. Kawasaki (Eds.), *Proceedings of the 34th Conference of the International Group for the Psychology of Mathematics Education* (Vol. 1, pp. 267–271). Belo Horizonte, Brazil: International Group for the Psychology of Mathematics Education.

Kwame, E. G. (2007). Pupils of African heritage, mathematics education and social justice. *Philosophy of Mathematics Education Journal, 20*. Retrieved from http://people.exeter.ac.uk/PErnest/pome20/index.htm.

Lave, J. (1977). Cognitive consequences of traditional apprenticeship training in Africa. *Anthropology and Educational Quarterly, 7*, 177–180.

Lave, J. (1988). *Cognition in practice*. New York, NY: Cambridge University Press.

Lave, J. (1989). The acquisition of culture and the practice of understanding. In J. Stigler, R. Shweder, & G. Herdt (Eds.), *The Chicago Symposia on human development*. Cambridge, UK: Cambridge University Press.

Lave, J., & Wenger, E. (1991). *Situated learning: Legitimate peripheral participation*. New York, NY: Cambridge University Press.

Leontiev, A. N. (1994). Uma contribuição à teoria do desenvolvimento da psique infantil. In L. S. Vygotsky, A. R. Luria, & A. N. Leontiev (Eds.), *Linguagem, desenvolvimento e aprendizagem*. São Paulo, Brazil: Ícone/EDUSP.

Lerman, S. (1999). Culturally situated knowledge and the problem of transfer in the learning of mathematics. In L. Burton (Ed.), *From hierarchies to networks in mathematics education* (pp. 93–107). London, UK: Falmer.

Lerman, S. (2000). The social turn in mathematics education research. In J. Boaler (Ed.), *Multiple perspectives on mathematics teaching and learning* (pp. 19–44). Westport, CT: Ablex Publishing.

Levenson, E., Tirosh, D., & Tsamir, P. (2009). Students' perceived sociomathematical norms: The missing paradigm. *The Journal of Mathematical Behavior, 28*(2–3), 171–187.

Lieberson, G. (1985). *Making it count: The improvement of social research and theory*. Los Angeles, CA: University of California Press.

Livingston, C. V. (2007). The privilege of pedagogical capital: A framework for understanding scholastic success in mathematics. *Philosophy of Mathematics Education Journal, 20*. Retrieved from http://people.exeter.ac.uk/PErnest/pome20/index.htm.

Long, H. H. (1925). On mental tests and race psychology—A critique. *Opportunity, 3*, 134–138.

Lopez, M. L., & Allal, L. (2007). Sociomathematical norms and the regulation of problem solving in classroom microcultures. *International Journal of Educational Research, 46*, 252–265.

Luria, A. R. (1976). *Cognitive development*. Cambridge, MA: Harvard University Press.

Martin, D. B. (2006). Mathematics learning and participation in African-American context: The co-construction of identity in two intersecting realms of experience. In N. Nasir & P. Cobb (Eds.), *Diversity, equity, and access to mathematical ideas* (pp. 146–158). New York, NY: Teachers College Press.

Martin, D. B. (2009). Researching race in mathematics education. *Teachers College Record, 111*(2), 295–338.

Martin, L., Towers, J., & Pirie, S. (2006). Collective mathematical understanding as improvisation. *Mathematical Thinking and Learning, 8*(2), 149–183.

Mattos, A. C., & Batarce, M. S. (2010). Mathematics education and democracy. *Zentralblatt fur Didaktik der Mathematik, 42*, 281–289.

McMurchy-Pilkington, C. (1998). Positioning of Maori women and mathematics: Constructed as non-doers. In C. Keitel (Ed.), *Social justice and mathematics education—Gender, class, ethnicity and the politics of schooling* (pp. 108–115). Berlin, Germany: Freie Universität Berlin.

McVittie, J. (2004). Discourse communities, student selves and learning. *Language and Education, 18*(6), 488–503.

Meira, L., & Lerman, S. (2010). Zones of proximal development as fields for communication and dialogue. In C. Lightfoot & M. C. D. P. Lyra (Eds.), *Challenges and strategies for studying human development in cultural contexts* (pp. 199–219). Rome, Italy: Information Age.

Mellin-Olsen, S. (1987). *The politics of mathematics education*. Dordrecht, The Netherlands: Reidel.

Menchaca, M. (1997). Early racist discourse: The roots of deficit thinking. In R. R. Valencia (Ed.), *The evolution of deficit thinking* (pp. 13–40). New York, NY: Falmer.

Millroy, W. (1992). *An ethnographic study of the mathematical ideas of a group of carpenters*. Reston, VA: National Council of Teachers of mathematics.

Moreira, P. C., & David, M. M. (2008). Academic mathematics and mathematical knowledge needed in school teaching practice: Some conflicting elements. *Journal of Mathematics Teacher Education, 11*, 23–40.

Moshkovich, J., & Brenner, M. (2002). Everyday mathematics. *Monographs of the Journal for Research in Mathematics Education* (pp. 131–153). Reston, VA: The National Council of Teachers of Mathematics.

Nicol, C. (2002). Where's the math? Prospective teachers visit the workplace. *Educational Studies in Mathematics, 50*, 289–309.

Nkhwalume, A. A. (2007). Deconstructing the scholarly literature on gender differentials in mathematics education: Implications for research on girls learning mathematics in Botswana. *Philosophy of Mathematics Education Journal, 20*. Retrieved from http://people.exeter.ac.uk/PErnest/pome20/index.htm.

Novotná, J., & Hošpesová, A. (2008). Theory of didactical situations in mathematics. In O. Figueras, J. L. Cortina, S. Alatorre, T. Rojano, & A. Sepúlveda (Eds.), *Proceedings of the 32nd Meeting of the International Group for the Psychology of Mathematics Education* (Vol. 1, pp. 105–106). Morelia, Mexico: Cinvestav-UMSNH/PME.

Nunes, T., Schliemann, A. D., & Carraher, D. W. (1993). *Street mathematics and school mathematics*. Cambridge, UK: Cambridge University Press.

Osler, A., & Starkey, H. (2010). *Teachers and human rights education*. Stoke-on-Trent, UK: Trentham Books.

Palhares, P. (Ed.). (2008). *Etnomatemática—Um olhar sobre a diversidade cultural e a aprendizagem matemática [Ethnomathematics—A look at the cultural diversity and the mathematics learning]*. Vila Nova de Famalicão: Húmus.

Passos, C. C. M., & Teixeira, P. J. M. (2011, June). Um pouco da teoria das situações didáticas (tsd) de Guy Brousseau [Some of the theory of didactic situations (tsd) of Guy Brousseau]. *Anais da XIII Conferência Interamericana de Educação Matemática (CIAEM-IACME)*, Recife, Brazil.

Perret-Clermont, A. N., & Brossard, A. (1988). L'intrication des processus cognitifs et sociaux dans les interactions. In R. A. Hinde, A. N. Perret-Clermont, & J. Stevenson-Hinde (Eds.), *Relations interpersonnelles et développement des saviors* (pp. 441–465). Genève, Switzerland: Del Val.

Piaget, J. (1971). Intellectual evolution from adolescence to adulthood. *Human Development, 15*, 1–12.

Piaget, J. (1974). *Réussir et comprendre*. Paris, France: PUF.

Ponte, J. P. (2003). Investigar, ensinar e aprender [Investigations, teaching and learning]. *Actas do ProfMat* (CD-ROM, pp. 25–39). Lisboa, Portugal: APM.

Popkewitz, T. (2004). School subjects, the politics of knowledge, and the projects of intellectuals in change. In P. Valero & R. Zevenbergen (Eds.), *Researching the socio-political dimensions of mathematics education* (pp. 251–267). Dordrecht, The Netherlands: Kluwer Academic Publishers.

Powell, A. B. (2002). Ethnomathematics and the challenges of racism in mathematics education. In P. Valero & O. Skovsmose (Eds.), *Proceedings of the 3rd International Mathematics Education and Society Conference* (Vol. 1, pp. 17–30). Copenhagen, Denmark: Centre for Research in Learning Mathematics.

Powell, A. B., & Frankenstein, M. (Eds.). (1997). *Ethnomathematics: Challenging eurocentrism in mathematics*. Albany, NY: SUNY Press.

Reed, H. J., & Lave, J. (1979). Arithmetic as a tool for investigating relations between culture and cognition. *American Anthropologist, 6*, 568–582.

Relethford, J. H. (1994). Craniometric variation among modern human populations. *American Journal of Physical Anthropology, 95*, 53–62.

Relethford, J. H. (2002). Apportionment of global human genetic diversity based on craniometrics and skin colour. *American Journal of Physical Anthropology, 118*, 393–398.

Restivo, S., & Sloan, D. (2007). The sturm and drang of mathematics: Casualties, consequences, and contingencies in the math wars. *Philosophy of Mathematics Education Journal, 20*. Retrieved from http://people.exeter.ac.uk/PErnest/pome20/index.htm.

Rogoff, B. (1981). Schooling and the development of cognitive skills. In H. C. Triandis & A. Heron (Eds.), *Handbook of cross-cultural psychology* (Vol. 4, pp. 233–294). Boston, MA: Allyn and Bacon.

Rorty, R. (1989). *Contingency, irony, and solidarity*. Cambridge, MA: Cambridge University Press.

Rossi-Becker, J. (1998). Research on gender and mathematics in the USA: Accomplishments and future challenges. In C. Keitel (Ed.), *Social justice and mathematics education—Gender, class, ethnicity and the politics of schooling* (pp. 251–257). Berlin, Germany: Freie Universität Berlin.

Santos, M., & Matos, J. F. (2002). Thinking about mathematical learning with Cabo Verde Ardinas. In G. de Abreu, A. J. Bishop, & N. Presmeg (Eds.), *Transitions between contexts of mathematical practices* (pp. 81–122). Dordrecht, The Netherlands: Kluwer Academic Publishers.

Saxe, G. B. (1991). *Culture and cognitive development: Studies in mathematical understanding*. Hillsdale, NJ: Lawrence Erlbaum.

Saxe, G. B., & Posner, J. (1983). The development of numerical cognition: Cross-cultural perspectives. In H. P. Ginsburg (Ed.), *The development of mathematical thinking* (pp. 291–317). London, UK: Academic Press.

Schliemann, A. D. (1985). Mathematics among carpenters and carpenters' apprentices: Implications for school teaching. In P. Damerow, M. Dunkley, B. Nebres, & B. Werry (Eds.), *Mathematics for all*. Paris, France: UNESCO Science and Technology Education Document Series, No. 20.

Schliemann, A. D., & Acioly, N. M. (1989). Mathematical knowledge developed at work: The contribution of practice versus the contribution of schooling. *Cognition and Instruction, 6*(3), 185–221.

Schliemann, A. D., Araújo, C., Cassundé, M. A., Macedo, S., & Nicéas, L. (1994). Multiplicative commutativity in school children and street sellers. *Journal for Research in Mathematics Education, 29*(4), 422–435.

Schliemann, A. D., & Carraher, D. W. (2004). Everyday mathematics: Main findings and relevance to schools. In N. M. Acioly-Régnier (Ed.), *Apprentissages informels: De la recherche à l'apprentissage scolaire* (pp. 171–182). Rapport d'études sur la thématique Apprentissages informels. PIREF (Programme Incitatif à la Recherche en éducation et Formation). Ministère de la Recherche, France.

Schliemann, A. D., & Magalhães, V. P. (1990). Proportional reasoning: From shops, to kitchens, laboratories, and, hopefully, schools. In G. Booker, P. Cobb, & T. de Mendicuti (Eds.), *Proceedings of the 14th Conference of the International Group for the Psychology of Mathematics Education* (Vol. 3, pp. 67–73). Oaxtepec, Mexico: International Group for the Psychology of Mathematics Education.

Schoenfeld, A. (1992). Learning to think mathematically: Problem solving, metacognition, and sense making in mathematics. In D. A. Grouws (Ed.), *Handbook of research on mathematics teaching and learning* (pp. 334–370). New York, NY: Macmillan.

Scribner, S., & Cole, M. (1973). Cognitive consequences of formal and informal education. *Science, 182*, 553–559.

Seah, W. T., & Bishop, A. J. (2000, April). *Values in mathematics textbooks: A view through two Australasian regions*. Paper presented at the 81st Annual Meeting of the American Educational Research Association, New Orleans.

Seeger, F., Voigt, J., & Waschescio, U. (1998). *The culture of the mathematics classroom*. Cambridge, UK: Cambridge University Press.

Seliktar, M., & Malik, L. P. (1998). A study of gender differences and math-related career choice among university students. In C. Keitel (Ed.), *Social justice and mathematics education—Gender, class, ethnicity and the politics of schooling* (pp. 83–94). Berlin, Germany: Freie Universität Berlin.

Sells, L. W. (1978, February). Mathematics—Critical filter. *The Science Teacher, 28*–29.

Setati, M., & Adler, J. (2000). Between languages and discourses: Language practices in primary multilingual mathematics classrooms in South Africa. *Educational Studies in Mathematics, 43*, 243–269.

Sewell, T. (1997). *Black masculinities and schooling*. London, UK: Trentham.

Sfard, A. (2006). Participationist discourse on mathematics learning. In J. Maaß & W. Schlöglmann (Eds.), *New mathematics education research and practice* (pp. 153–170). Rotterdam, The Netherlands: Sense Publishers.

Sfard, A., & Pursak, A. (2005). Telling identities: In search of an analytic tool for investigating learning as a culturally shaped activity. *Educational Researcher, 34*(4), 14–22.

Shirley, L. (1995). Using ethnomathematics to find multicultural mathematical connections. In P. A. House & A. F. Coxford (Eds.), *Connecting mathematics across the curriculum* (pp. 34–33). Reston, VA: National Council of Teachers of Mathematics.

Sierpinska, A. (2007). I need the teacher to tell me if I am right or wrong. In J. H. Woo, H. C. Lew, K. S. Park, & D. Y. Seo (Eds.), *Proceedings of the 31st Conference of the International Group for the Psychology of Mathematics Education* (Vol. 1, pp. 45–64). Seoul, Korea: International Group for the Psychology of Mathematics Education.

Silver, E., & Smith, M. (1996). Building discourse communities in mathematics classroom: A worthwhile but challenging journey. In P. Elliott & M. Kenney (Eds.), *Communication in mathematics K–12 and beyond* (pp. 20–28). Reston, VA: National Council of Teachers of Mathematics.

Singh, S. (1998). Women's perceptions and experiences of mathematics. In C. Keitel (Ed.), *Social justice and mathematics education—Gender, class, ethnicity and the politics of schooling* (pp. 101–107). Berlin, Germany: Freie Universität Berlin.

Skelton, C. (2010). Gender and achievement: Are girls the "success stories" of restructured education systems? *Educational Review, 62*(2), 131–142.

Skovsmose, O. (2001). Landscapes of investigation. *Zentralblatt fur Didaktik der Mathematik, 33*(4), 123–132.

Skovsmose, O., & Borba, M. (2004). Research methodology and critical mathematics education. In P. Valero & R. Zevenbergen (Eds.), *Researching the socio-political dimensions of mathematics education* (pp. 207–226). Dordrecht, The Netherlands: Kluwer Academic Publishers.

Sternberg, R. J. (1985). *Beyond IQ: A triarchic theory of human intelligence*. Cambridge, MA: Cambridge University Press.

Stevens, P. A. J., Clycq, N., Timmerman, C., & Van Houtte, M. (2009). Researching race/ethnicity and educational inequality in the Netherlands: A critical review of the research literature between 1980 and 2008. *British Educational Research Journal, 37*(1), 5–43.

Sukthankar, N. (1998). Influences of parents, family, and environment on attitudes of teachers and pupils in Papua New Guinea. In C. Keitel (Ed.), *Social justice and mathematics education—Gender, class, ethnicity and the politics of schooling* (pp. 95–100). Berlin, Germany: Freie Universität.

Tardif, M., Lessard, C., & Lahaye, L. (1991). Les enseignants des orders d'enseignement primaire et secondaire face aux savoirs. Esquisse d'une problématique du savoir enseignant. *Sociologie et Sociétés, 23*(1), 55–69.

Tatsis, K., & Kolezab, E. (2008). Social and socio-mathematical norms in collaborative problem-solving. *European Journal of Teacher Education, 31*(1), 89–100.

Thomas, W. B. (1982). Black intellectuals' critiques of early mental testing: A little known saga of the 1920s. *American Journal of Education, 90*, 258–292.

Valencia, R. R. (2010). *Dismantling contemporary deficit thinking—Educational thought and practice*. New York, NY: Routledge.

Valero, P. (2004). Socio-political perspectives on mathematics education. In P. Valero & R. Zevenbergen (Eds.), *Researching the socio-political dimensions of mathematics education* (pp. 5–23). Dordrecht, The Netherlands: Kluwer Academic Publishers.

Valero, P., & Zevenbergen, R. (Eds.). (2004). *Researching the socio-political dimensions of mathematics education*. Dordrecht, The Netherlands: Kluwer Academic Publishers.

Van der Veer, R., & Valsiner, J. (1991). *Understanding Vygotsky—A quest for synthesis*. Oxford, UK: Basil Blackwell Inc.

Vergnaud, G. (1991). La théorie des champs conceptuels. *Recherches en Didactique des Mathématiques, 10*(2–3), 133–169.

Vergnaud, G. (2009). The theory of conceptual fields. *Human Development, 52*, 83–94.

Vithal, R. (2003). Student teachers and "Street Children": On becoming a teacher of mathematics. *Journal of Mathematics Teacher Education, 6*, 165–183.

Vithal, R. (2004). Researching, and learning mathematics at the margin: From "shelter" to school. In M. J. Høines & A. B. Fuglestad (Eds.), *Proceedings of the 28th Conference of the International Group for the Psychology of Mathematics Education* (Vol. 1, pp. 95–104). Bergen, Norway: Bergen University College.

Vygotsky, L. (1962). *Thought and language* (E. Hanfmann & G. Vakar, Trans.). Cambridge, MA: Massachusetts Institute of Technology.

Vygotsky, L. (1978). *Mind in society*. Cambridge, MA: Harvard University Press.

Walkerdine, V. (1997). Redefining the subject in situated cognition theory. In D. Kirshner & J. A. Whitson (Eds.), *Situated cognition: Social, semiotic and psychological perspectives* (pp. 57–70). Mahwah, NJ: Erlbaum.

Walkerdine, V. (1998). Science, reason and the female mind. In V. Walkerdine (Ed.), *Counting girls out: Girls and mathematics* (pp. 29–41). London, UK: Falmer Press.

Walshaw, M., & Anthony, G. (2008). The teacher's role in classroom discourse: A review of recent research into mathematics classrooms. *Review of Educational Research, 78*(3), 516–551.

Watson, A., & Winbourne, P. (Eds.). (2008). *New directions for situated cognition in mathematics education*. New York, NY: Springer.

Weiner, L. (2006). Challenging deficit thinking. *Teaching to Student Strengths, 64*(1), 42–45.

Wenger, E. (1998). *Communities of practice: Learning, meaning and identity*. Cambridge, UK: Cambridge University Press.

Wenger, E. (2007, September). *Learning for a small planet: Agency and structure in the constitution of identity*. Paper presented at the 2nd Socio-Cultural Theory in Educational Research and Practice Conference, University of Manchester, Manchester, UK.

Wertsch, J. V., Del Rio, P., & Alvarez, A. (1995). *Sociocultural studies of mind*. New York, NY: Cambridge University Press.

Williams, G., & Clarke, D. (2003, April). *Dyadic patterns of participation and collaborative concept creation: 'Looking In' as a stimulus to complex mathematical thinking*. Paper presented as part of the symposium "Patterns of Participation in the Classroom" at the Annual Meeting of the American Educational Research Association, Chicago.

Willis, S. (1998). Perspectives on social Justice, disadvantage, and the mathematics curriculum. In C. Keitel (Ed.), *Social justice and mathematics education: Gender, class, ethnicity and the politics of schooling* (pp. 1–19). Berlin, Germany: Freie Universität Berlin.

Winbourne, P., & Watson, A. (1998). Learning mathematics in local communities of practice. In A. Watson (Ed.), *Situated cognition in the learning of mathematics* (pp. 93–104). Oxford, UK: University of Oxford, Department of Educational Studies.

Wright, C., Weeks, D., & McGlaughlin, A. (2000). *"Race," class and gender in exclusions from school*. London, UK: Falmer Press.

Yackel, E. (2001). Explanation, justification and argumentation in mathematics classrooms. In M. van den Heuvel-Panhuizen (Ed.), *Proceedings of the 25th Conference of the International Group for the Psychology of Mathematics Education* (Vol. 1, pp. 9–23). Utrecht, The Netherlands: International Group for the Psychology of Mathematics Education.

Yackel, E., & Cobb, P. (1996). Sociomathematical norms, argumentation, and autonomy in mathematics. *Journal for Research in Mathematics Education, 27*(4), 390–408.

Yackel, E., Rasmussen, C., & King, K. (2000). Social and sociomathematical norms in an advanced undergraduate mathematics course. *Journal of Mathematical Behavior, 19*, 1–13.

Zevenbergen, R. (1998). Gender, media, and conservative politics. In C. Keitel (Ed.), *Social justice and mathematics education—Gender, class, ethnicity and the politics of schooling* (pp. 59–68). Berlin, Germany: Freie Universität Berlin.

Chapter 5
Studying Learners in Intercultural Contexts

Yoshinori Shimizu and Gaye Williams

Abstract Researchers have increasingly recognized that learning mathematics is a cultural activity. At the same time, research aims, technological advances, and methodological techniques have diversified, enabling more detailed analyses of learners and learning to take place. Increased opportunities to study learners in different cultural, social and political settings have also become available, with ease of access to results of international benchmark testing online. Large-scale quantitative studies in the form of international benchmark tests like Trends in International Mathematics and Science Study (TIMSS), the Programme for International Student Assessment (PISA), and detailed multi-source (including video) qualitative studies like the international Learners' Perspective Study (LPS), have enabled a broad range of research questions to be investigated. This chapter points to the usefulness of large-scale quantitative studies for stimulating questions that require qualitative research designs for their exploration. Qualitative research has raised awareness of the importance of socio-cultural and historical cultural perspectives when considering learning. This raises questions about uses that could be made of "local" theories in undertaking intercultural analyses.

Intercultural Contexts, Learners, and Learning Processes

There is now a greater recognition that mathematics classrooms need to be considered as cultural and social environments in which individuals participate, and that teaching and learning activities taking place in these environments should

---

Y. Shimizu (✉)
University of Tsukuba, Ibaraki, Japan
e-mail: yshimizu@human.tsukuba.ac.jp

G. Williams
Deakin University, Burwood, Victoria, Australia

be studied as such (e.g., Cobb & Hodge, 2011; Lerman, 2006; Seeger, Voigt, & Washescio, 1998). Seminal works by Bishop (1988a, 1988b) have directed researchers to the pivotal role that culture plays in both teaching and learning of mathematics. Säljö (2010) emphasized the importance of Vygotsky's (1978) work on the role of cultural tools in supporting learning. He drew attention to ways in which historical aspects of culture support student learning: "[Cultural tools] are the products of the development of practices in society over time" (Säljö, 2010, p. 499). They support communication between teacher and learner, and they also support a child thinking alone using cultural tools such as language, mathematical symbols, and concrete artefacts in doing so (Vygotsky, 2009). Thus, learning is supported through access to the knowledge, tools, rules, and ways of thinking a culture has previously developed. These realizations have led to studies of differences between learning outcomes in different countries (Stigler & Hiebert, 1999), learners learning in cultural contexts that differ from their own (de Abreu, Bishop, & Presmeg, 2002), and differences in the nature and magnitude of learning for the same child in different learning contexts (Nunes, 2010; Zang & Sternberg, 2010).

Terminology

In this chapter the terms "contexts," "intercultural," and "local," will have the following meanings:

- *Contexts* relates to the cultural settings in which learning occurs.
- *Intercultural* describes studies of learners in different cultural contexts whether they be contexts: (a) in different countries (described as: "multicultural scholastic contexts" by Favilli, Oliveras, & César, 2003, and "cross-cultural contexts" by Zang & Sternberg, 2010); or (b) of practice of learners in the same classroom; or (c) during learner "transition" from one context to another (de Abreu et al., 2002); or (d) in which the same learner participates (Jorgensen, 2010; Nunes, 2010).
- *Local* refers to those cultural practices that belong to, or those theories formulated by, researchers who belong to the culture under discussion.

Intercultural Studies and Theoretical Perspectives

Researchers in the field of mathematics education have recently drawn upon a broader array of theoretical positions and research methods to frame their inquiries. This broadening of research perspectives is closely tied to the emergence of advances technological advances that have increased the capacity for large-scale qualitative research methods and analysis applications. Qualitative methodologies have increasingly complemented quantitative methodologies. Certain methods are more appropriate to particular research questions than to others (Battista et al., 2009). Distinctions between the usefulness of the product of learning and the processes through which learning occurs can be crucial when deciding which research designs will be most appropriate for exploring particular research questions.

Among the theoretical perspectives now employed, there has been a growth in attention to sociological and socio-cultural theories, and more recently, closer connections have been made between social and cognitive theories of learning (Hershkowitz, Schwarz, & Dreyfus, 2001). Social practices have been argued to be discursively constituted whereby people become part of practices as practices become part of them (Lerman, 2002). In addition, those topics arising from and extending the notions of ethnomathematics and everyday cognition provide broader perspectives on learning mathematics (Presmeg, 2007; Watson & Winbourne, 2008). A productive area for intercultural research could be links between socio-cultural aspects of learning and learner cognition to answer such questions as: are there cultural differences in the thought processes students use in different cultural contexts? Battista et al. (2009) called this new emphasis, "Qualitative Cognition-Focused Research." According to Battista et al. (2009):

> Research that focuses on describing cognition attempts to account for individual students' and teachers' actions, reasoning, and learning (Cobb, 2007). The value of these descriptions [is] in the insights they provide researchers and teachers and curriculum/assessment developers into the nature of students' mathematical learning. (p. 222)

Through video study, Qualitative-Cognition-Focused-Research could be studied in conjunction with the learning context to learn more about connections between socio-cultural, and cognitive activity during the process of learning mathematics. Wood's (2007) abstract for her intended ICME Study Topic Group 26, Learning and Cognition in 2008 presentation made this connection:

> In the past 15 years, following Vygotsky's interest in the influence of culture on children's intellectual development research in mathematics education shifted to an interest in the *social or cultural* aspects of learning mathematics. ... Although the goal of these theoretical perspectives is to account for the social conditions of learning, they do not provide constructs to account for internal mental processes. (Wood, 2007, abstract)

Wood identified social and cultural aspects of learning mathematics as insufficient on their own to gain a deep understanding of processes of learning.

Developmental psychology aims to analyze individual psychological processes involved in learning and using mathematics in specific socio-cultural contexts (de Abreu, 2008). In contrast, a cultural psychological perspective pays attention to the interplay between the individual, society, and the culture. Learning mathematics is viewed as a function of what an individual accomplishes over time and across the various communities and practices in which he or she participates. By contrasting two perspectives in research on mathematics relating to out-of-school contexts, namely, ethnomathematics education (D'Ambrosio, 1985) and developmental psychology (Nunes, Schliemann, & Carraher, 1993), de Abreu (2008) identified different levels and foci of analysis in the studies to discuss a cultural psychological perspective.

By making closer connections between socio-cultural, historical cultural, and cultural psychological theoretical perspectives, there is potential to more fully explore Vygotsky's (1978) work on how thought processes used within communities influence the ways learners participate in their cultures.

This broadening of theoretical perspectives employed in mathematics education research reflects the growing interest in influences of different cultural settings, and different social and political contexts, and contributions of internal learner processes.

Qualitative and Quantitative Studies

Adapted from other research fields like agriculture, the quantitative research methods in education employed in and beyond the 1960s relied on experimental and quasi-experimental research designs and statistical analyses as a base for generalizable claims on teaching and learning. More recently, there has been increased interest in detailed qualitative analysis of learners and learning, rather than only the reporting of patterns or tendencies.

TIMSS and PISA used quantitative techniques to study learning outcomes (student performances), and also collected information about learner characteristics, and the contexts within which they learn. More could be done, though, to present results in ways that benefit teachers and learners (see, e.g., Andrich & Styles, 2011; Doig, 2006). Formative assessment could be provided through:

> Multiple-choice items whose distractors are based on research evidence, and open-response items whose response categories are similarly based. ... [and] reporting student results ... on item-wise indicators of performance, not global aggregations. (Doig, 2011, personal communication).

TIMSS has to some extent pursued such directions but more could be done.

The validity of ranked country performances from international benchmark tests has been brought into question. These results are frequently "accepted as fact" (Loveless, 2011, p. 12). Analyses of test rankings and historical cultural contexts demonstrated that international test scores must be interpreted cautiously, and some scholars believe that much of what one may hear or read about them is misleading. Although quantitative experimental research is considered, by some, to be inherently more scientific, rigorous, and valid than qualitative research, limitations to findings from quantitative research point to the need for caution when considering results from such studies.

Interpreting Findings from Large-Scale Quantitative Studies

A cautionary note about accepting findings from large-scale quantitative studies without interrogating these results using qualitative methods was delivered by the President of the European Association of Researchers in Learning and Instruction (EARLI) at her presidential address at the 2011 EARLI Annual Conference (Lindlom-Ylanne, 2011). Her team employed quantitative analyses in studying the stability and variability of tertiary learners' approaches to learning mathematics and preliminary findings indicated no significant differences. Subsequent qualitative analysis of subsets of data showed differences existed between two learner cohorts, but these differences had cancelled each other out when results for the two cohorts were aggregated. Separating learner cohorts has also created impressions of learner outcomes on international benchmark results. In the USA, for example, the average derived from considering all US data together hides the high performances of several learner cohorts, and the extremely low performances of other US learner cohorts (see, e.g., Loveless, 2011). Thus caution is required in interpreting statistical data on learner performance in large-scale quantitative studies.

Shimizu's (2005) analysis of learner performances on two items from PISA (Organisation for Economic Co-operation and Development, 2003) showed that aggregated performances by country hid important differences in learner performances on two test items. Shimizu raised questions about the influences of cultural practices in different countries on the accessibility of test items set within different "real-life" contexts. He found the overall high mathematical performance of Japanese students compared to their international counterparts on PISA (Organisation for Economic Co-operation and Development, 2003) was not reflected in performances on one of the two test items. Japanese students outperformed their international counterparts on the item involving spatial interpretations, and under-performed in comparison to their international counterparts on the item requiring cost-related decisions in selecting items to assemble a skateboard. Shimizu pointed out the possibility of Japanese learners' familiarity with number cubes and Origami as a cultural activity contributing to their higher performance on the spatial interpretation item. This finding fits with Vygotsky's (1978) perspective of learning supported by cultural artefacts developed within the community to which the learner belongs. Shimizu raised questions about cultural practices in other countries that could have contributed to learners from some countries achieving higher performances than Japanese learners on the skateboard item, which required the use of simple numerical procedures and decision-making.

In countries in which mathematics learning is undertaken in a language other than their own (e.g., the Philippines, South Africa), students face an additional disadvantage as well as other constraints in their learning environments (Clarke et al., 2006). Limitations associated with providing a single ranking for performances of different countries on international benchmark tests show the highly problematic nature of reporting results in this way.

In Summary

Extreme caution is necessary in interpreting results of large-scale quantitative studies associated with studying learner performances. Inequities are built into international benchmark items due to varying degrees of accessibility to different learner cohorts. As international benchmark test items tend to be contextualized using artefacts from the more "well-resourced" countries, and these contexts may have little or no meaning for students from less well-resourced countries, this raises questions about the assumptions on which international comparative studies of school mathematics are predicated. Keitel and Kilpatrick (1999) suggested that the spectre of an "idealized international curriculum" lies behind even the most sophisticated research designs, including text and document analyses and the use of video to study classroom practice:

> A pseudo-consensus has been imposed (primarily by the English-speaking world) across systems so that curriculum can be taken as a constant rather than a variable, and so that the operation of other variables can be examined. (Keitel & Kilpatrick, 1999, p. 253)

Some far-reaching consequences of publishing such results online and in print are now discussed.

Impacts of International Benchmark Rankings: Local and International

The purposes of international studies such as PISA and TIMSS include providing policy makers with information about educational systems. The primary interest of policy makers in such information is generally to see their own country's relative rank among participating countries. They welcome a simple profile of student performance. There is a close match between the objectives of PISA, in particular, and the broad economic and labour market policies of host countries. This fit naturally invites a lot of public talk about respective rankings of learner performances, and this public talk is both competitive and evaluative in emphasis. For example, the release of results of the OECD-PISA 2009 (Programme for International Student Assessment, OECD, 2010) and the TIMSS 2007 (Trends in International Mathematics and Science Study, Mullis et al., 2008) received huge publicity through the media in some countries (see, for example, Loveless, 2011).

The extent of interest in cultural practices of high-performing countries is demonstrated through the study of pedagogical practices (e.g., Japanese Lesson Study) of some high-achieving countries in other countries: by researchers, regions, and schools. Interest is also evidenced in the attention paid to cultural practices in high-achieving countries in East Asia (Li & Shimizu, 2009), and interest in cultural practices in countries not in East Asia who have achieved high-ranking performances (e.g., Finland).

Lesson Study

Partly due to the high ranking of Japan in international benchmark tests (see for example, Stigler & Hiebert, 1999), Japanese *lesson study* has become a focus of attention in countries including Australia (see White, 2004), Malaysia (see Chiew & Lim, 2003), and the USA (Fernandez, Cannon, & Chokshi, 2003; Fernandez &

Yoshida, 2004). Lesson study is an approach to developing and maintaining quality mathematics instruction through a particular form of activity of Japanese teachers (Shimizu, 2002). Generally, a lesson study consists of the following events: the actual classes taught to pupils, observation by others, followed by intensive discussion called the study discussion. Designing, enacting, and analyzing are the three stages of lesson study that evolve before, during, and after the lesson. There is extensive preparation made before the class, and extensive work to be done after the lesson study as well, including follow-up and preparation for the next lesson to be presented and studied. These events form a cyclic process that can also be iterative in nature.

The presence of Japanese teachers during the initial stages of introducing lesson study to a group of US teachers helped US teachers identify appropriate lenses through which to view the processes of teaching and learning. These "findings suggest that to benefit from Lesson Study teachers will first need to learn how to apply critical lenses to their examination of lessons" (Fernandez et al., 2003, p. 171). To develop better understandings of educational activities in local contexts, researchers need to consider the underlying values and beliefs shared by the people in the community. It should be noted, for instance, that valuing students' thinking as necessary elements to be incorporated into the development of a lesson is key to the approach taken by Japanese teachers (Shimizu, 2009). Describing anticipated students' responses is, among other activities, key to lesson planning because the whole-class discussion depends on the solution methods the students actually come up with. Having a clear sense of the ways students are likely to think about and solve a problem prior to the start of a lesson makes it easier for teachers to know what to look for when they are observing students' work on the problem. Thus, the likelihood of succeeding with integrating pedagogical practices from high-ranking countries into local situations is expected to be highly dependent on whether those in the local situation are aware of the key aspects of the practice, the purposes for which it was developed, and nuances of cultural activity that are implicit within the way in which this practice is implemented in the country of origin.

Finland: Educational, Cultural and Historical Perspectives

> Finland has achieved high rankings on international benchmark tests (e.g., PISA, 2003).
>
> Finland once again came out top in the OECD's latest PISA study of learning skills among 15-year-olds, with high performances in mathematics and science matching those of top-ranking Asian school systems in Hong Kong-China, Japan and Korea. (OECD, 2004)

Such reports focussed interest of the international community on educational policy in Finland. Sahlberg (2007) drew attention to Finland's historical and cultural perspective on education and to the way Finnish education authorities have remained "faithful" to their educational philosophies:

> Steady improvement in student learning has been attained through Finnish education policies based on equity, flexibility, creativity, teacher professionalism and trust. Unlike many other education systems, consequential accountability accompanied by high-stakes testing and externally determined learning standards has not been part of Finnish education policies. (p. 147)

Instead of demanding teacher accountability by introducing national benchmark testing for learners, Finland achieved top ranking by retaining their educational policies based upon mutual trust, and faith in their teachers to meet the needs of learners creatively and flexibly. This included autonomy for both teachers and learners:

> The academic prowess of Finland's students has lured educators from more than 50 countries in recent years to learn the country's secret, ... What they find is simple but not easy: well-trained teachers and responsible children. Early on, kids do a lot without adults hovering. And teachers create lessons to fit their students. "We don't have oil or other riches. Knowledge is the thing Finnish people have," says Hannele Frantsi, a school principal. (Gamerman, 2008, p. W1)

The Finnish education system has become an attractive and internationally-examined example of a well-performing system that successfully combines quality with widespread equity and social cohesion through reasonable public financing (Sahlberg, 2007, p. 147). The reasons for its success appear to be their willingness to retain their historically-developed focus on giving teachers the leeway to respond idiosyncratically to the needs of the learners in a country in which developing autonomy in students is a cultural practice. That said, Finland was achieving high rankings on international benchmark tests before its present educational policy was developed. "This suggests that cultural and societal factors, which predate and are intertwined with the policies in question, may be the real drivers of success" (Loveless, 2011, p. 11).

In Summary

League ladders ranking national performances are widely publicized and have influenced both research, and teaching practices—and therefore learning. Retaining confidence in teaching and learning practices developed within a local culture may sometimes be more appropriate than attempting to emulate practices that have been found to be successful in other cultures. Importing teaching practices from other countries into local contexts without also paying attention to the cultural practices within which those practices were embedded can be unproductive. Findings indicate that teaching practices that are consistent with the historical and cultural practices of learners may sometimes better support the development of learner mathematical performances. This raises questions about the types of teaching practices that could optimize student learning in a multicultural classroom.

Dichotomies that Have Focussed Mathematics Education Research

There are many dichotomies evident in intercultural research, some of which have already been identified and discussed to varying extents within this chapter. They include: (a) high-performing/low-performing, (b) affluent/not-affluent,

(c) teacher-centred/student-centred, (c) autonomous/not autonomous, (d) East/West, (e) in-school/out-of-school. This section illustrates how questions raised from quantitative research focussed on dichotomies have led to qualitative research designs developed for the purposes of answering further questions.

Beyond Labels "East" and "West": Problematizing the Dichotomy of Cultural Traditions

The dichotomy East/West has been foregrounded by international benchmark testing, and has led to a qualitative focus on learning in different geographical regions as a result. Accumulated research over the past decade has contributed to our understanding of similarities and differences in mathematics teaching and learning between East Asia and the West (e.g., Leung, Graf, & Lopez-Real, 2006) or between Eastern and Western cultures (Cai, 2007). The ICMI study reported by Leung et al. (2006), "A comparative study of East Asia and the West," is distinguished here from other studies in that it was "specifically concerned with comparing practices in different settings and with trying to interpret these different practices in terms of cultural traditions." The discussion document for the study argued that "those based in East Asia and the West seem particularly promising for comparison." In this study a comparison was made between "Chinese/Confucian tradition on one side, and the Greek/Latin/Christian tradition on the other."

Juxtaposing the two different cultures indicated that researchers wanted to examine teaching and learning in each cultural context by contrasting differences between them. The labels "East/Eastern" and "West/Western," however, could be problematic in several ways. First, the terms East and West literally mean geographical areas but not cultural regions. Needless to say, there are huge diversities in ethnicity, tools, and habits that are tied to the corresponding cultures. Further, Cobb and Hodge (2011) argue that two different views of culture can be differentiated in the mathematics education literature on the issue of equity, and that both are relevant to the goal of ensuring that all students have access to significant mathematical ideas. "In one view, culture is treated as a characteristic of readily identified and thus circumscribable communities, whereas in the other view it is treated as a set of locally instantiated practices that are dynamic and improvisational" (p. 179). With the second view, in particular, it is problematic to specify different cultures based on geographical areas.

Second, it is possible to oversimplify and mislead the cultural influence on students' learning within each cultural tradition by using the same label for different communities. For example, there are studies which suggest much child education in Japan diverges from the Confucian approach in "East Asia" (Lewis, 1995; Rohlen & LeTendre, 1996). Also, in the special issue on exemplary mathematics instruction and its development in selected education systems in East Asia, it was manifested that there is a variety of approaches to accomplish quality mathematics instruction in these different systems. Thus, any framework for differentiating cultural traditions runs the risk of oversimplifying the cultural interplay. In particular, there is a need to question whether polarizing descriptors such as "East" and "West," "Asian" and "European," are maximally useful. Perhaps we need more useful ways to examine differences, for the purposes of learning from each other and identifying ways to optimize learner practices.

In-School and Out-Of-School Understandings

Earlier studies of mathematics in out-of-school contexts have identified differences in the ways learners were able to use (or not use) the same mathematical procedures in in-school and out-of-school contexts (e.g., Jorgensen, 2010; Nunes et al., 1993; Rogoff, 2003). The study of young street sellers in Brazil showed that these young people used mathematics in meaningful ways for the purpose of selling their wares and managing their finances. In school, they were unable to give answers for the same calculations when they were presented without the real-life contexts that were personally meaningful to them (Nunes, 2010). Instead of dichotomizing in-school/out-of-school performances through quantitative analyses, Nunes (2010) undertook a qualitative analysis of the differing understandings developed by students. This study provided a powerful illustration of how students whose in-school performances would be ranked low on international benchmark tests had the capacity to think mathematically when working in meaningful contexts. The study stimulated reflection on how in-school learning could be changed to capture facets of what assisted learners in out-of-school contexts.

Further developments of the studies of cultural accounts of mathematics learning have raised alternative perspectives on the role of cultures in learning mathematics. By contrasting two perspectives in research on mathematics in out-of-school contexts, namely, ethnomathematics education (D'Ambrosio, 1985) and developmental psychology (Nunes, Schliemann, & Carraher, 1993), de Abreu (2008) identified different levels and foci of analysis in the studies to discuss a cultural psychological perspective. Although the level of analysis in ethnomathematics education relates to historical and anthropological analysis of the mathematics of different sociocultural groups (sociogenetic level), developmental psychological studies aimed to analyze the psychological processes of individuals when learning and using mathematics in specific socio-cultural contexts (de Abreu, 2008). A cultural psychological perspective pays attention to the interplay between the individual, society, and the culture. Learning mathematics is viewed as a function of what an individual accomplishes over time and across the various communities and practices in which he or she participates. In this way, research on mathematics in out-of-school contexts has shown a shift from cross-cultural comparisons to social practice within cultures. Notions such as participation and identity are keys to understanding and studying learning mathematics in cultural contexts.

5 *Studying Learners in Intercultural Contexts*   155

In Summary

Our discussion of the studies in this section pointed to the need to examine the labels used in mathematics education literature that refer broadly to dichotomies, and to consider how best to gain greater understanding of teacher and learner practices and influences upon these practices by taking a finer-grained look at what is happening. In particular, it is becoming abundantly clear that the cultural practices of the communities in which the teaching and learning takes place should be considered when making sense of classroom practices.

Beyond Dichotomies in International Comparative Research Studies

Not all international research studies of learners focus on dichotomies. International comparative classroom research is viewed as the exploration of similarity and difference in order that our understanding of what is possible in mathematics classrooms can be expanded by consideration of what constitutes "good practice" in culturally diverse settings.

> Our capacity to conceive of alternatives to our current practice is constrained by deep-rooted assumptions, reflecting cultural and societal values that we lack the perspective to question. The comparisons made possible by international research facilitate our identification and interrogation of those assumptions. (Clarke, Emanuelsson, Jablonka, & Mok, 2006, p. 3)

Analysis of video data collected in the video component of TIMSS, as reported by Stigler and Hiebert (1999), centred on the proposition that the teaching practice of a nation (at least in the case of mathematics) could be explained to a significant extent by a teacher's adherence to a culturally-based "teacher script." Central to the identification of these cultural scripts for teaching were the lesson patterns reported by Stigler and Hiebert (1999) for Germany, Japan and the USA. The contention of Stigler and Hiebert was that at the level of the lesson, teaching in each of the three countries could be described by a "simple, common pattern" (Stigler & Hiebert, 1999, p. 82).

On the other hand, one focus of the Learner's Perspective Study (LPS) has been on the form and function of recognizable activity conglomerates within lessons that LPS researchers termed "lesson events" (Clarke, Keitel, & Shimizu, 2006). A lesson event was characterized by a combination of form and function, both of which were subject to local variation, but with an underlying familiarity and frequency of use that suggested both intercultural relevance and utility. Each individual lesson event had a form (visual features and social participants) sufficiently common to be identifiable within the classroom data from each of the countries studied. In each classroom there were idiosyncratic features that distinguished *each teacher's* enactment of each lesson event, particularly with regard to the function of the particular event (intention, action, inferred meaning, and outcome). At the same time, common features could be identified in the enactment of lesson events across the entire international data set and across the data set specific to a country.

TIMSS Rankings in the USA as a Catalyst for the TIMSS Video Study

The TIMSS 1995 Video Study of mathematics teachers' practices in Japan, the USA, and Germany was designed by US researchers to interrogate differences in practices between teachers in the USA, and teachers in countries ranked differently from the USA (Germany and Japan) on mathematical performances on the Third International Mathematics and Science Study. It was the first attempt to collect and analyze videotapes from a national random sample of mathematics classrooms (Stigler, Gonzales, Kawanaka, Knoll & Serrano, 1999). According to Stigler, Gallimore, and Hiebert (2000):

> There is another more subtle reason for studying teaching across cultures. Teaching is a cultural activity. Because cultural activities vary little within a society, they are often transparent and unnoticed. ... Cross-cultural comparison is a powerful way to unveil unnoticed but ubiquitous practices. ... Comparative research invites reexamination of the things "taken for granted" in our teaching, as well as suggesting new approaches that never evolved in our own society. (pp. 87–88)

The catalyst for this video study was the US ranking on TIMSS, which was lower than expected by political and educational stakeholders in the USA. The "high-performing/low-performing" dichotomy that focussed the large-scale quantitative analysis of TIMSS data raised questions about why learners in some countries performed better than learners in other countries. The qualitative TIMSS Video Study was designed to provide answers to such questions.

The TIMSS 1999 Video Study expanded the design of the TIMSS 1995 Video Study from three to seven (Hiebert et al., 2003). These studies used a single camera focussed predominantly on the teacher because previous international studies of mathematics classroom had identified coherent sets of actions, and associated attitudes, beliefs and knowledge, that appeared to constitute culturally-specific teacher practices (Stigler & Hiebert, 1999). The LPS team from Australia, Germany, Japan, and the USA, at its inception in 1999, hypothesized that there is also a set of actions and associated attitudes, beliefs, and knowledge of students that constitute a culturally-specific coherent body of learner practices (Clarke, Keitel, & Shimizu, 2006). They considered that teaching and learning, as classroom practices should be studied together as interdependent activities within a common setting. Findings from the LPS raised questions about the "culturally-based teacher script" identified through the 1995 TIMSS Video Study. LPS findings showed lesson structure differed from lesson to lesson for some teachers, and teachers in a particular country did not generally display a common set of teaching practices.

International Comparative Study: Learner's Perspective Study

By extending its focus to learners and learning within each context, LPS addressed an identified need:

> What is absent from nearly all the rhetoric and variables of TIMSS ... is ... the notion that students themselves are agents. TIMSS makes students from 41 countries into passive subjects ... all linked to the seduction of one global economic curriculum. (Thorsten, 2000, p. 71)

Among the methodologically most interesting aspects of LPS are the collaborative negotiation of the research design, the method of data generation, the intercultural and local analyses, and the processes by which various complementary accounts can be integrated into a rich and useful portrayal of mathematics classrooms internationally. The combination of participating countries gives good representation to European and Asian educational traditions, well-resourced and less-well resourced school systems, and mono-cultural and multi-cultural societies (Clarke, Emanuelsson et al., 2006). LPS developed a common research protocol intended to capture the activity of learners, their perspective on their mathematics lessons, and their learning outcomes, in addition to the types of data on teachers and teaching collected in TIMSS video studies. By the end of 2000, the LPS community had expanded to include researchers from nine countries or regions (Australia, Hong Kong, Germany, Israel, Japan, South Africa, Sweden, The Philippines, and the USA) (Clarke, Keitel, & Shimizu, 2006).

Inclusivity as a methodological principle is pervasive within the LPS research design, with the inclination to integrate rather than segregate being at the heart of the Study:

> The inclination to integrate rather than segregate is ... at the heart of the Learner's Perspective Study (LPS), since it was intended from the project's inception that any documented differences in classroom practice be interpreted as local solutions to classroom situations and, as such, be viewed as complementary rather than necessarily oppositional alternatives. (Clarke, Keitel, & Shimizu, 2006, p. 215)

LPS data generation techniques included a three-camera approach including on-site mixing of the teacher and student camera images into a picture-in-picture video record. These mixed images were used to stimulate participant reconstructive accounts of classroom events in post-lesson interviews of students and teachers. Video records were supplemented by student written material, and test and questionnaire data from students and the teacher. These data were collected for sequences of at least 10 consecutive lessons occurring in the "well-taught" eighth-grade mathematics classrooms of teachers in participating countries. The three mathematics teachers in each country were identified for their locally-defined "teaching competence" and for their situation in demographically diverse government schools in major urban settings (Clarke, Emanuelsson et al., 2006).

LPS research teams from each participating country generate their own data, and control who uses their data and for what purposes. Half-yearly meetings and progressive publications of LPS books (Clarke, Emanuelsson et al., 2006; Clarke,

Keitel, & Shimizu, 2006; Shimizu, Kaur, Huang, & Clarke, 2010) about different aspects of LPS findings have strengthened research collaborations. The number of countries employing the LPS research design, and sharing the data they generate, grew from 9 in 2000 to 14 in 2011. This points to the value researchers internationally place on participating in a collaborative international team with access to rich video data on teaching and learning, in mathematics classrooms across the world, and the opportunities to interact with their international counterparts in analyses within this data-set.

In Summary

International benchmark rankings have prompted deeper studies of teaching and learning as a result of the patterns they show. TIMSS (1995) raised questions which qualitative studies—the TIMSS 1995 and 1999 video studies—were designed to explore. But, the design of these qualitative studies prioritized the "teacher's voice." This raised questions about the capacity of the TIMSS video studies to capture important information about how students learn. As a result, another qualitative video study, the LPS, was designed to capture in detail the reciprocal practices of teachers and learners, and the perceptions of lesson participants about this activity. Thus, although questions have been raised about the validity of the results of large-scale international studies (with regard to student performances), these tests have stimulated questions that have led to qualitative research designs employing video techniques to interrogate classroom activity. This has resulted in a deeper understanding of similarities and differences in teacher and learner practices within and across cultures.

Benefits and Limitations of International Video and Video-Stimulated Interview Studies

Methodological challenges and benefits of international "video survey" studies have been usefully discussed (Clarke, Keitel, & Shimizu, 2006; Jacobs, Kawanaka, & Stigler, 1999; Stigler et al., 2000). Advantages associated with the use of video are that it reveals classroom practices clearly, facilitates reflection on alternatives in practice within each country, and stimulates discussion about teaching and learning. Video enables the study of complex processes, enables coding from multiple perspectives, stores data in a form that allows unanticipated and novel analyses at a later time, facilitates integration of qualitative and quantitative information, and facilitates communication of results (Clarke, 2000; Hiebert et al., 2003). Limitations and benefits of video research designs can be associated with various factors including: (a) camera configuration; (b) participant response to camera/s; (c) post-lesson video-stimulated interviews; and (d) local team involvement. Limitations in research scope can be associated with cameras configuration that captures a narrow focus (e.g., teacher only) of classroom activity. Broader capture (including the teacher, the whole class, and a pair of focus students) can support the interrogation of a wider range of research questions using a rich variety of theoretical perspectives. Participant consciousness of the cameras can alter lesson activity. A familiarization period can be added to the research design to minimize such effects. Clarke (2000) introduced a familiarization period because he found that students were likely to display the same types of lesson activities with lesser frequency and intensity at the beginning of the research period (Clarke, 2001).

As marked differences have sometimes been found between teacher, student, and researcher perspectives, it is important to collect multi-perspective data. Williams and Clarke (2002) showed, for example, that the observed behaviour of a student in a lesson video did not capture their development of deep understandings reconstructed in post-lesson video-stimulated interviews. Post-lesson video-stimulated interviews in which participants reconstruct their lesson activity can provide valid data where the participant is sufficiently comfortable to respond with their own ideas rather than give responses they consider the interviewer would want. The video stimulation adds to the validity of the responses by giving the interviewee access to "memory traces" about a specific instance and thus limiting generalized responses (Ericsson & Simons, 1980). Even when an interview protocol is provided for intercultural studies, the nature of the probes has been found to differ from country to country due to the research foci of local research teams (Williams, 2005). There are both limitations and advantages to this. The locally formulated probes increase opportunity to illuminate aspects of the context valued by the local team, but may decrease opportunity to interrogate data for some intercultural studies. That said, the local probes can in themselves become a fruitful area for intercultural study.

One of the most powerful outcomes of large-scale video studies, such as the TIMSS video studies, has been the interest they have stimulated about multi-perspective video-data-capture research designs supporting multiple analyses. Clarke (2001) demonstrated this potential in his study of Australian mathematics and science classrooms where different researchers used different theoretical perspectives to analyze video of the same lesson. The design used in this Australian study was the forerunner to the research design developed for LPS (see, e.g., Clarke, Keitel, & Shimizu, 2006).

Qualitative Studies: Studying Social Interaction and Meaning-Making

Studies of learning and learners across contexts and across countries have the potential to extend our understandings of how to optimize in-school learning. It is in the examination of classrooms across a variety of cultural settings and school systems that we find educational assumptions most visible and open to challenge. The contrasts and unexpected similarities offered by research in such culturally diverse settings reveal and challenge existing assumptions and theories and make essential a reconstruction of some of our theoretical perspectives.

A major benefit of comparing classroom practices in different cultural contexts is to describe in detail the nuances of mathematics teaching and learning in the contexts, so that the stereotypical, one-size-fits-all perceptions are shattered and we can better understand the key elements of such practice and use this understanding to reflect on mathematics teaching and learning in our own culture.

Comparing learning across cultures has additional advantages (Hiebert et al., 2003; Leung et al., 2006). It allows educators to examine understandings of learners, and teaching practices that influence this from a fresh perspective by widening the known possibilities. In addition to examining how teachers across one's own country approach mathematics and what type of learning results, such research provides opportunities to use theoretical lenses developed in another local community (called "local theories" in this chapter), to examine how teachers from that community approach the same topic, and what students learned as a result. This can make one's own teaching practices and the achievements of learners from these practices more visible by contrast and therefore more open for reflection and improvement. Comparing teaching across cultures can reveal alternatives and stimulate discussion about the choices being made within a country. Although a variety of teaching practices can be found in a single country (see, e.g., Williams, 2005), it sometimes requires looking outside one's own culture to see something new and different. These observations, combined with carefully crafted follow-up research, can stimulate debate about the approaches that may make the most sense for achieving the learning goals defined within a country.

LPS Research Raising Questions for Intercultural Studies

The studies reported in this chapter differed in focus and the extent to which they explored the learning of mathematics, and learners of mathematics. They were provided to stimulate thinking about possible questions which can be answered through qualitative intercultural studies. They include single-country studies with potential to be extended to intercultural studies, and intercultural studies. Included in these descriptions are the study focus, and primary data sources accessed.

Illustrations of single-country studies are: (a) learners' responses to motivational strategies utilized in a class in the Philippines, and in video and student interviews (Ulep, 2006); (b) discrepancies between learner and teacher perceptions of lesson climaxes ("yamaba") in a Japanese-utilized lesson video, and in teacher and student interviews (Shimizu, 2005); (c) learners' points of view on mathematics lessons in a Swedish-utilized lesson video (Emanuelsson & Sahlström, 2006); and (d) learners' cognitive, social, and affective activity and psychological characteristics associated with creative mathematical activity simultaneously interrogating student interviews and lesson video (Williams, 2006). Although not undertaken for the purpose of intercultural comparison, Williams' (2005) study of creative student thinking in classrooms using LPS data showed the potential for studying learners' processes of thinking and social influences upon it in intercultural settings. Given the identified influences of culture on the ways learners learn, this could be a productive focus for further research.

Clarke, Emanuelsson et al. (2006) included various intercultural studies undertaken by LPS teams. These studies provide a window into the activities of learners in different cultural contexts, including what they attended to, ways they participated in the lesson, thinking processes they employed, and what was valued by learners and teachers in the contexts.

LPS Intercultural Studies

O'Keefe, Xu, and Clarke (2006), an Australian research team, interrogated "kikan-shido"—the Japanese word for "between-desk instruction"—using LPS post-lesson data from Australia, Germany, Hong Kong, Japan, Shanghai, and Tokyo. The same form of "between-desk walking" was found to occur in all countries, but its functions differed from teacher to teacher. Students differed in their perceptions of the purpose of this activity, and student perceptions tended to differ from the intended teacher purposes. Instead of a "country signature" for kikan-shido, with each teacher in a particular country employing kikan-shido for the same function, they found "teacher signatures" and these differed from teacher to teacher. Within-country similarities and differences tended to be as great as intercultural similarities and differences.

Jablonka's (2006) analysis of patterns of participation in classrooms in Germany, Hong Kong, and the USA used lesson video and interviews to study activity and perceptions associated with learners coming to the front of the class. The study found differences in the functions for this activity: sometimes students came to the front of the class to facilitate individual interactions between teacher and learner, and sometimes for the purpose of "public talk" to the class. Learners differed in the risks they perceived to be associated with such activity. There were across-country and within-country similarities and differences.

Khuzwayo (2006) analyzed teacher interviews from South Africa, Australia, and the USA to identify constraints associated with teachers changing practices. In each country he found a high level of commitment of teachers to the learners in their classes.

Williams (2005) examined learners' cognitive, social, and affective activity in Australia, the USA, and Japan. Although she found creative mathematical learner activity in all three countries, the interview probes for data available from Japan at that time (Japanese School 1 only) focussed on the mathematical object not the process of learning so there was insufficient data to develop Japanese case studies. Williams simultaneously interrogated student interviews and lesson video and found that the creative thinkers identified all possessed certain psychological characteristics (Seligman, 1995; Williams, 2006). Although not intended as an intercultural study, this study demonstrated the potential of the LPS Research Design to support an intercultural study of student thinking during the creative development of new mathematical ideas if local probes fit with this research focus.

Summarizing LPS Contributions to Intercultural Studies

Instead of focussing on dichotomies, LPS teams have explored similarity and difference in order to understanding what is possible in mathematics classrooms. In intercultural studies, researchers used findings from other countries to illuminate what had not previously been transparent about learners and contexts for learning mathematics in their own country. The findings are useful at a "local" level and also at the international mathematics education level. Studies that do not presently have an intercultural focus could be extended to include such a focus due to the richness of this data set.

LPS has shown that intercultural research does not have to focus around evaluation of practices in different countries according to dichotomies where one pole is judged as optimal and the other pole as non-optimal. LPS research undertaken thus far has demonstrated the richness of the data for supporting a broad range of research foci and a diversity of theoretical perspectives. By showing there can sometimes be more similarities across countries than within a country, LPS research raises questions about the validity of considering learners and learning contexts within a particular country as homogeneous.

Challenges Ahead

As this chapter has shown, the focus of intercultural research studies has been influenced by findings of other intercultural studies, and collaboration between "local" researchers from different cultural settings has focussed beyond dichotomies and deepened our understandings of learners and learning in different cultural contexts. Qualitative studies have been designed to interrogate quantitative findings, and researchers have developed appropriate theories to inform these studies. It has been through careful use of theory that detailed analyses of teaching and learning in classroom has been able to generate generalizable claims.

Research designs that harness the potential for complementarities between quantitative and qualitative approaches could be a productive focus. Quantitative studies employ numerical indicators of students' capabilities for reasoning that are linked via statistical procedures and experimental designs but do not examine thought in the complex multifaceted way it occurs. In contrast, qualitative research can investigate in depth the nature and structure of individual learners' understandings, sensemaking, and learning that are useful not only for researchers but also for teachers. In any scientific inquiry methods can only be judged in terms of their appropriateness and effectiveness in addressing a particular research question. For studying such complex phenomena as learners and their learning in mathematics classroom, multiple methodological approaches are more appropriate in various parts of a study or in different studies within a series of studies.

Challenges confronting the international research community include development of:

- Test instruments that can legitimately measure the achievement of students who have participated in different mathematics curricula,
- Research techniques by which the practices, motivations, beliefs, and thought processes, of all classroom participants can be studied and compared with sensitivity to the cultural context, including contexts in multicultural classrooms,
- Theoretical frameworks by which the structure and content of diverse mathematics curricula, their enactment, and their consequences can be analyzed and compared within the cultural contexts in which they occur,
- Increased understanding of the role of local theories in illuminating intercultural analyses, and
- Increased understandings of how local theories might inform decisions about integrating practices across cultural contexts, and inform the processes of doing so.

Concluding Remarks

Mathematics education research in recent years tends to include more international endeavours than ever before. As the globalization and internationalization of research activities has continued to increase, the field of mathematics education research has clearly shown the diversification of perspectives on teaching and learning embedded in local contexts. International comparative studies have recognized the need to focus on existing diverse voices and perspectives among members of the community. Recent research has illuminated connections between learners and learning, and socio-cultural and historical cultural influences upon these. More recently there has been some focus on connecting psychological cultural perspectives to socio-cultural and historical cultural perspectives. Most important for the future is to stop positioning mathematics learning and mathematics learners as needing to comply with some idealized "international standard," and instead to find ways to give learners opportunities to show what they know, and opportunities to build upon this knowledge.

References

Andrich, D., & Styles, I. (2011). Distractors with information in multiple-choice items: A rationale based on the Rasch model. *Journal of Applied Measurement, 12*(1), 67–95.

Battista, M., Smith, M., Boerst, T., Sutton, J., Confrey, J., White, D., Knuth, E., & Quander, J. (2009). Research in mathematics education: Multiple methods for multiple uses (NCTM Research Committee Perspective). *Journal for Research in Mathematics Education, 40*(3), 216–240.

Bishop, A. J. (1988a). *Mathematical enculturation. A cultural perspective on mathematics education*. Dordrecht, The Netherlands: Kluwer Academic Publishers.

Bishop, A. J. (1988b). *Mathematics education and culture*. Dordrecht, The Netherlands: Kluwer Academic Publishers.

Cai, J. (2007). What is effective mathematics teaching? A study of teachers from Australia, Mainland China, Hong Kong SAR, and the United States. What is effective teaching? A dialogue between East and West. *Special Issue of ZDM—The International Journal of Mathematics Education, 39*(4), 311–318.

Chiew, C. M., & Lim, C. S. (2003, October). *Impact of lesson study on mathematics trainee teachers*. Paper presented at International Conference for Mathematics and Science Education, University of Malaya, Kuala Lumpur.

Clarke, D. J. (2000). *Learners' perspective study (LPS): Research design*. Melbourne, Vic: Technical report for the LPS team, University of Melbourne.

Clarke, D. J. (Ed.). (2001). *Perspectives on practice and meaning in mathematics and science classrooms*. Dordrecht, The Netherlands: Kluwer Academic Publishers.

Clarke, D. J., Emanuelsson, J., Jablonka, E., & Mok, I. A. C. (Eds.). (2006). *Making connections: Comparing mathematics classrooms around the world*. Rotterdam, The Netherlands: Sense Publishers.

Clarke, D. J., Keitel, C., & Shimizu, Y. (Eds.). (2006). *Mathematics classrooms in twelve countries: The insider's perspective*. Rotterdam, The Netherlands: Sense Publishers.

Clarke, D. J., Shimizu, Y., Ulep, S. A., Gallos, F. L., Sethole, G., Adler, J., & Vithal, R. (2006). Cultural diversity and the learner's perspective: Attending to voice and context. In F. K. S. Leung, K.-D. Graf, & F. Lopez-Real (Eds.), *Mathematics education in different cultural traditions—A comparative study of East Asia and the West: The 13th ICMI Study* (pp. 353–380). New York, NY: Springer.

Cobb, P. (2007) Putting philosophy to work: Coping with multiple theoretical perspectives. In F.K. Lester, Jr. (Ed.) Second handbook of research on mathematics teaching and learning (pp. 3–38). Charlotte, NC: Information Age.

Cobb, P., & Hodge, L. L. (2011). Culture, identity, and equity in the mathematics classroom. In E. Yackel, K. Gravemeijer, & A. Sfard (Eds.), *A journey in mathematics education research: Insights from the work of Paul Cobb* (pp. 179–195). New York, NY: Springer.

D'Ambrosio, U. (1985). Ethnomathematics and its place in the history and pedagogy of mathematics. *For the learning of Mathematics, 5*(1), 44–48.

de Abreu, G. (2008). From mathematics learning out-of-school to multicultural classrooms: A cultural psychology perspective. In L. English (Ed.), *Handbook of international research in mathematics education* (2nd ed., pp. 352–384). London, UK: Routledge.

de Abreu, G., Bishop, A. J., & Presmeg, N. C. (2002). *Transitions between contexts of mathematical practices*. Dordrecht, The Netherlands: Kluwer Academic Publisher.

Doig, B. (2006). Large-scale mathematics assessment: Looking globally to act locally. *Assessment in Education: Principles, Policy and Practice, 13*(3), 265–288.

Emanuelsson, J., & Sahlström, F. (2006). Same from the outside, different on the inside: Swedish mathematics classrooms from students' points of views. In D. Clarke, C. Keitel, & Y. Shimizu (Eds.), *Mathematics classrooms in twelve countries: The insiders' perspective*. Rotterdam, The Netherlands: Sense Publishers.

Ericsson, K., & Simons, H. (1980). Verbal reports of data. *Psychological Review, 87*(3), 215–251.

Favilli, F., Oliveras, M. L., & César, M. (2003). Bridging mathematical knowledge from different cultures: Proposals for an intercultural and interdisciplinary curriculum. In N. A. Pateman, B. J. Dougherty, & J. Zilliox (Eds.), *Proceedings of the 27th Conference of the International Group for the Psychology of Mathematics Education* (2nd ed., pp. 365–372). Honolulu, HI: International Group for the Psychology of Mathematics Education.

Fernandez, C., Cannon, J., & Chokshi, S. (2003). A U.S.–Japan lesson study collaboration reveals critical lenses for examining practice. *Teaching and Teacher Education, 19*(2), 171–185.

Fernandez, C., & Yoshida, M. (2004). *Lesson study: A Japanese approach to improving mathematics teaching and learning*. Mahwah, NJ: Lawrence Erlbaum.
Gamerman, E. (2008, February 29). What makes Finnish kids so smart? *Wall Street Journal*. Retrieved October 26. http://online.wsj.com/article/SB120425355065601997.html.
Hershkowitz, R., Schwarz, B., & Dreyfus, T. (2001). Abstraction in context: Epistemic actions. *Journal for Research in Mathematics Education, 32*(2), 195–222.
Hiebert, J., Gallimore, R., Garnier, H., Givvin, K., Hollingsworth, H., Jacobs, J., Chui, A., Wearne, D., Smith, M., Kersting, N., Manaster, A., Tseng, E., Etterbeck, W., Manaster, C., Gonzales, P., & Stigler, J. (2003). *Teaching mathematics in seven countries: Results from the TIMSS 1999 video study*. Washington, DC: U.S. Department of Education, National Center for Education Statistics.
Jablonka, E. (2006). Student(s) at the front: Forms and functions in six classrooms from Germany, Hong Kong and the United States. In D. J. Clarke, J. Emanuelsson, E. Jablonka, & I. Ah Chee Mok (Eds.), *Making connections: Comparing mathematics classrooms around the world* (pp. 107–126). Rotterdam, The Netherlands: Sense Publishers.
Jacobs, J., Kawanaka, T., & Stigler, J. (1999). Integrating qualitative and quantitative approaches to the analysis of video data on classroom teaching. *International Journal of Educational Research, 31*, 717–724.
Jorgensen, R. (2010). Structured failing: Reshaping a mathematical future for marginalised learners. In L. Sparrow, B. Kissane, & C. Hurst (Eds.), *Shaping the future of mathematics education: Proceedings of the 33rd Annual Conference of the Mathematics Education Research Group of Australasia* (pp. 26–35). Fremantle, Australia: Mathematics Education Research Group of Australasia.
Keitel, C., & Kilpatrick, J. (1999). The rationality and irrationality of international comparative studies. In G. Kaiser, E. Luna, & I. Huntley (Eds.), *International comparisons in mathematics education* (pp. 241–256). London, UK: Falmer Press.
Khuzwayo, H. (2006). A study of mathematics teachers' constraints in changing practices: Some lessons from countries participating in the Learners' Perspective Study. In D. J. Clarke, J. Emanuelsson, E. Jablonka, & I. Ah Chee Mok (Eds.), *Making connections: Comparing mathematics classrooms around the world* (pp. 201–214). Rotterdam, The Netherlands: Sense Publishers.
Lerman, S. (2002). Cultural, discursive psychology: A sociocultural approach to studying the teaching and learning of mathematics. In C. Kieran, E. Forman, & A. Sfard (Eds.), *Learning discourse: Discursive approaches to research in mathematics education* (pp. 87–113). Dordrecht, The Netherlands: Kluwer Academic Publisher.
Lerman, S. (2006). Cultural psychology, anthropology and sociology: The developing "strong" social turn. In J. Maasz & W. Schloeglmann (Eds.), *New mathematics education research and practice* (pp. 171–188). Rotterdam, The Netherlands: Sense Publishers.
Leung, F. K. (1995). The mathematics classroom in Beijing, Hong Kong and London. *Educational Studies in Mathematics, 29*(4), 297–325.
Leung, F. K., Graf, K.-D., & Lopez-Real, F. (Eds.). (2006). *Mathematics education in different cultural traditions: A comparative study of East Asian and the West*. New York, NY: Springer.
Lewis, C. (1995). *Educating hearts and minds: Reflections on Japanese preschool and elementary education*. New York, NY: Cambridge University Press.
Li, Y., & Shimizu, Y. (Eds.) (2009). Exemplary mathematics instruction and its development in East Asia. *ZDM—The International Journal of Mathematics Education, 41*(3), 257–395.
Lindlom-Ylanne, S. (2011, August). *Presidential address*. Annual Conference of the European Association for Learning and Instruction, Exeter, UK.
Loveless, T. (2011). *How well are American students learning? The 2010 Brown Center Report on American education*. Washington DC: The Brown Center on Educational Policy. Retrieved 21 November 2011. http://www.brookings.edu/reports/2011/0207_education_loveless.aspx.
Mullis, I. V. S., Martin, M. O., Foy, P., Olson, J. F., Preuschoff, C., Erberber, E., Arora, A., & Galia, J. (2008). *TIMSS 2007 international mathematics report: Findings from IEA's Trends in International Mathematics and Science Study at the fourth and eighth grades*. Chestnut Hill, MA: TIMSS & PIRLS International Study Center, Boston College.

Nunes, T. (2010). Learning outside of school. In P. Peterson, E. Baker, & B. McGaw (Eds.), *International encyclopedia of education* (3rd ed., pp. 457–463). Amsterdam, The Netherlands: Elsevier.

Nunes, T., Schliemann, A. D., & Carraher, D. W. (1993). *Street mathematics and school mathematics*. Cambridge, MA: Cambridge University Press.

O'Keefe, C., Xu, L. H., & Clarke, D. J. (2006). Kikan-shido: Between-desks instruction. In D. J. Clarke, J. Emanuelsson, E. Jablonka, & I. Ah Chee Mok (Eds.), *Making connections: Comparing mathematics classrooms around the world* (pp. 73–106). Rotterdam, The Netherlands: Sense Publishers.

Organisation for Economic Co-operation and Development (2003). *Learning for tomorrow's world: First results from PISA 2003*. Paris, France: Author.

Organisation for Economic Co-operation and Development. (2004, June 12). *Top-performer Finland improves further in PISA survey as gap between countries widens*. [Online OECD comment]. Retrieved from http://www.oecd.org/document/28/0,3746,en_21571361_4431511 5_34010524_1_1_1_1,00.html.

Organisation for Economic Co-operation and Development. (2010). *PISA 2009 results: What students know and can do: Student performance in reading, mathematics and science* (Vol. I). Paris, France: OECD Publishing.

Presmeg, N. (2007). The role of culture in teaching and learning mathematics. In F. K. Lester (Ed.), *Second handbook of research on mathematics teaching and learning* (pp. 435–458). Reston, VA: National Council of Teachers of Mathematics & Information Age Publishing.

Rogoff, B. (2003). *The cultural nature of human development*. Oxford, UK: Oxford University Press.

Rohlen, T., & LeTendre, G. (1996). *Teaching and learning in Japan*. New York, NY: Cambridge University Press.

Sahlberg, P. (2007). Educational policies for raising student learning: The Finnish approach. *Journal of Education Policy, 22*(2), 147–171.

Säljö, R. (2010). Learning in a sociocultural perspective. In P. Peterson, E. Baker, & B. McGaw (Eds.), *International encyclopedia of education* (3rd ed., pp. 498–502). Amsterdam, The Netherlands: Elsevier.

Seeger, F., Voigt, J., & Waschescio, U. (Eds.). (1998). *The culture of the mathematics classroom*. New York, NY: Cambridge University Press.

Seligman, M.E.P. (1995). The effectiveness of psychotherapy: The Consumer Reports study. *American Psychologist*, 50, 965–974

Shimizu, Y. (2002). Lesson study: What, why, and how? In H. Bass, Z. P. Usiskin, & G. Burrill (Eds.), *Studying classroom teaching as a medium for professional development: Proceedings of a U.S.–Japan workshop* (pp. 53–57, 154–156). Washington, DC: National Academy Press.

Shimizu, Y. (2005). From a profile to the scrutiny of student performance: Exploring the research possibilities offered by the international achievement studies. In H. Chick & J. Vincent (Eds.), *Proceedings of 29th Conference of the International Group for the Psychology of Mathematics Education* (pp. 75–78). Melbourne, Australia: International Group for the Psychology of Mathematics Education.

Shimizu, Y. (2009). Characterizing exemplary mathematics instruction in Japanese classrooms from the learner's perspective. *ZDM—The International Journal of Mathematics Education, 41*(3), 311–318.

Shimizu, Y., Kaur, B., Huang, R., & Clarke, D. J. (Eds.). (2010). *Mathematical tasks in classrooms around the world*. Rotterdam, The Netherlands: Sense Publishers.

Stigler, J. W., Gallimore, R., & Hiebert, J. (2000). Using video surveys to compare classrooms and teaching across cultures: Examples and lessons from the TIMSS and TIMSS-R video studies. *Educational Psychologist, 35*(2), 87–100.

Stigler, J. W., Gonzales, P., Kawanaka, T., Knoll, S., & Serrano, A. (1999). *The TIMSS videotape classroom study: Methods and findings from an exploratory research project on eighth-grade* mathematics instruction in Germany, Japan, and the United States. Washington, DC: U.S. Government Printing Office.

Stigler, J. W., & Hiebert, J. (1999). *The teaching gap: Best ideas from the world's teachers for improving education in the classroom.* New York, NY: Free Press.

Thorsten, M. (2000). Once upon a TIMSS: American and Japanese narrations of the Third International Mathematics and Science Study. *Education and Society, 18*(3), 45–76.

Ulep, S. A. (2006). Ganas—A motivational strategy: Its influences on learners. In D. Clarke, C. Kietel, Y. Shimizu (Eds). *Mathematics classrooms in twelve countries: The insider's perspective* (pp. 131–149) Rotterdam, The Netherlands: Sense Publications.

Vygotsky, L. S. (1978). *Mind in society: The development of higher psychological processes.* M. Cole, V. John-Steiner, S. Scribner, & E. Souberman, (Eds.), (J. Teller, Trans.). Cambridge, MA: Harvard University Press.

Vygotsky, L. (2009). *Play and its role in the mental development of the child.* (C. Mulholland, Trans.). Retrieved from http://www.fhcds.org/ftpimages/436/download/Play_and_Child_Development.pdf (originally published, 1933, and republished 1966).

Watson, A., & Winbourne, P. (Eds.). (2008). *New directions for situated cognition in mathematics education.* New York, NY: Springer.

White, A. L. (2004). The long-term effectiveness of lesson study: A New South Wales mathematics teacher professional development program. In I. P. A. Cheong, H. S. Dhindsa, I. J. Kyeleve, & O. Chukwu (Eds.), *Globalisation trends in science, mathematics and technical education* (pp. 320–328). Gadong, Brunei Darussalam: Universiti Brunei Darussalam.

Williams, G. (2005). *Improving intellectual and affective quality in mathematics lessons: How autonomy and spontaneity enable creative and insightful thinking.* Doctoral dissertation, The University of Melbourne. Retrieved from http://repository.unimelb.au/10187/2380.

Williams, G. (2006). Autonomous looking-in to support creative mathematical thinking: Capitalising on activity in Australian LPS classrooms. In D. J. Clarke, C. Kietel, & Y. Shimizu (Eds.), *Mathematics classrooms in twelve countries: The insider's perspective* (pp. 221–236). Rotterdam, The Netherlands: Sense Publications.

Williams, G., & Clarke, D. J. (2002). The contribution of student voice in classroom research. In C. Malcom & C. Lubisi (Eds.), *Proceedings for the South African Association for Research in Mathematics, Science and Technology Education 2002* (pp. 398–404). Durban, South Africa: South African Association for Research in Mathematics, Science and Technology Education.

Wood, T. (2007). *Social cognitive processes in learning mathematics.* Abstract of paper intended for presentation in Study Topic Group 26, Learning and Cognition, 2008. Retrieved from http://tsg.icme11.org/tsg/show/27.

Zang, L. F., & Sternberg, R. J. (2010). Learning in a cross-cultural perspective. In P. Peterson, E. Baker, & B. McGaw (Eds.), *International encyclopedia of education* (3rd ed., pp. 450–456). Amsterdam, The Netherlands: Elsevier.

Chapter 6
Learners in Transition Between Contexts

Tamsin Meaney and Troels Lange

Abstract In this chapter, we explore, from a social justice perspective, conceptions of learners in transition between contexts and evaluate pedagogical practices that have been advocated for such learners. Learning occurs as learners reflect on their transition between contexts, particularly when there are differences in what content knowledge is valued, the relationships between participants and how activities are undertaken. From this perspective, productive pedagogical practices for learners in transition are those that build and sustain relationships between learners and mathematics and between learners and others, including those outside the classroom. We look specifically at examples of pedagogical practices that draw on ethnomathematics and critical mathematics education for their inspiration.

Transitioning between contexts, such as home and school, can be a fairly minor issue for learners if they perceive similarities in what knowledge is valued and how learners and others should interact together and with the mathematical content. However, for other learners who perceive the contexts as being very different, the transitioning process can limit the possibilities for their future. This is because transitioning between contexts affects not just what knowledge is valued, and thus learnt, but also learners' processes of becoming. In this chapter, we describe how learning is connected to transitioning between contexts, before discussing different positions on social justice in mathematics education. In so doing, we identify two pedagogical approaches, ethnomathematics and critical mathematics education, to analyze using Wenger's (1998) three modes of belonging. Using learners' opinions, we identify the features of these approaches which support learners to broaden horizons of possibilities for their futures.

T. Meaney (✉) • T. Lange
Malmö University, Malmö, Sweden
e-mail: Tamsin.meaney@mah.se

Adjusting to new contexts always involves learning. The degree of adjustment varies depending on the similarities or differences between contexts. Nevertheless, even when contexts have a connection to mathematical knowledge, learning may not include gaining school mathematical outcomes. de Abreu, Bishop and Presmeg (2002) saw the transitioning process as being part of a dynamic relationship between the learner and the contexts being transitioned. Contexts act as mediators between what is structurally possible, through schooling for example, and what actually happens, such as learning. Thus, we see contexts as the enactment of systems of knowledge within social practices, whose elements Fairclough (2003) described as: (a) action and interaction; (b) social relations; (c) persons (with beliefs, attitudes, histories, etc.); (d) the material world; and (e) discourse.

The following is an example of how social practices contribute to actual events:

> Classroom teaching articulates together particular ways of using language (on part of both teachers and learners) with the social relation of the classroom, the structuring use of the classroom as a physical space, and so forth. ... Social events are casually shaped by (networks of) social practices—social practices define particular ways of acting, and although actual events may more or less diverge from these definitions and expectations (because they cut across different social practices, and because of the causal powers of social agents), they are still partly shaped by them. (Fairclough, 2003, p. 25)

Thus, contexts are not just physical settings, but include the valuing of knowledge, the typical distribution of power within relationships that interact around that knowledge and the sorts of interactions that are expected to occur between participants with that knowledge. As such, contexts can be considered systems of knowledge enacted in social practices.

Transitions between contexts have been referred to as boundary crossings (Crafter & de Abreu, 2011). Lipka, Yanez, Andrew-Ihrke, and Adam (2009) described boundary work as education across cultures which "requires bridges between elders and schooling" (p. 267). Bishop (2004) used Gee's (1996) description of "borderland discourses" to describe the differences between home and school mathematical practices. An alternative is to consider transitions in relationship to a change in horizons of learners' possibilities for their futures. Gadamer (1996) stated "the horizon is the range that includes everything that can be seen from a particular vantage point. ... we can speak of narrowness of horizon, of the possible expansion of horizon, of the opening of horizon and so forth" (p. 302).

For example, when transitioning between contexts involves the loss of connections with the home, this can be considered a narrowing of learners' horizons of possibilities for their futures. An Indigenous group "may well recognize that schooling provides the skills necessary to survive in a technological world, but it will also blame the school for alienating students from their home culture, whether deliberately or unintentionally" (Cantoni, 1991, p. 34).

At times, transitions can be one-way, so it is not possible to return to the original context—the horizon in one direction closes while another expands. For example, Gorgorió and Planas (2003) stated, "[w]hen referring to the schooling of immigrant children, transition processes may be viewed as the gradual adaptation to societal expectations" (p. 3). On the other hand, transitions between home and school occur on a regular basis and although learning may result in a reinterpretation of each context, the knowledge valued in one context would not be replaced by that from the other context. de Abreu et al. (2002) labelled these as collateral transitions. In these situations, learners have to learn how to juggle the discontinuities between the different social practices.

> We understand the construct of *transition* not as a moment of change but as the experience of changing, of living discontinuities between cultures; in particular, discontinuities between different school cultures, and different mathematics classroom cultures, and between how the home and school culture understand, value and use mathematics. Transitions include the processes of developing both individual and social identities while coping with new social and cultural experiences. (Gorgorió & Planas, 2005, p. 93)

Transitioning is dynamic and never-ending. For example, César (2007) described a student in Portugal who is first seen as being disruptive. When a new teacher provided him with different activities, he began to engage in learning mathematics. He then had to re-negotiate his role with his Cape Verde friends and community who had come to expect that he would have a leadership role in disrupting the mathematics class. Transitioning into new contexts and between contexts results in learning. New understandings can be used to reflect on the same contexts, but in new ways, thus contributing to transitioning being an ongoing process. Learning is more than a passive interpretation of the world, as it can result in changes to the contexts themselves (Diversity in Mathematics Education Center for Learning and Teaching [DiME], 2007) and to the horizons of future possibilities.

For Radford (2008), learning involved becoming progressively conversant with the collectively and culturally constituted forms of reflection. Learning is "not just about knowing something but also about becoming someone" (Radford, 2008, p. 215). In this way, the object of learning is not only within the awareness of the learner, but the learner him/herself is part of what is to be appropriated in the learning process. Similarly, Brown (2009) stated "a person's becoming occurs through engagement in the ways of knowing, doing and valuing of a particular social group, for example, philosophers, mathematicians, lawyers, gang-members, etc." (p. 172). Thus, learning embeds the individual within the historically developed societal context. The appropriation of forms of reflection happens in the entangled relationship between the individual, the collective and forms of practice, mediated through artefacts. Reflection produces conceptions of contexts while, simultaneously, interactions within and between contexts support the appropriation of the socially constructed and culturally constituted forms of reflection.

When learners transition between two contexts that are very similar, the need for adjustments may not be large. For other learners, such as Indigenous students, learning mathematics may involve querying their perceptions of what knowledge is and how it is gained (see for example Barta & Brenner, 2009). It is likely that many Indigenous students would agree with Gorgorió and Planas (2005) that "the common understanding of the [mathematics] student is still 'monolingual,' belonging to the dominant culture, and having middle class social attitudes" (p. 92). Consequently, when participants construct relationships in the mathematics classroom that position learners as being different from "typical" mathematics learners, then transitioning is likely to be difficult. Similarly, when learners transition into outside-school contexts where school mathematics is not valued, then the transitioning can be problematic. This can occur when immigrant children have to do homework that their parents consider to have limited mathematical value (Civil, 2008). In these cases, learners are forced into reflecting on the differences, resulting in a different kind of learning from that which probably was intended by, for example, the teachers.

Although the transitioning process is never completed, it continually produces outcomes. As a result of reflecting on their learning experiences, learners may decide to adapt so they more closely resemble what they consider to be "typical" mathematics learners, or they may choose not to engage with mathematics, or they may accommodate to these new forms of reflection by doing something in between. An example of this would be when learners do not perceive that transferring knowledge across contexts is valuable. For example, Nunes, Schliemann, and Carraher (1993) described how children who operated as street vendors were able to do complex calculations as part of their jobs but could not relate these calculations to what they were required to do in mathematics classrooms. Similarly, Brenner (1998) found that first-grade children recognized that the prices given for textbook items did not reflect the money exchanges that they engaged in outside of school. For both sets of learners, an outcome was that mathematics was compartmentalized so that only one type of mathematics could be used in each context. Transitioning can result in a range of outcomes depending on the learners' reflections.

Skovsmose (2005) saw learners' perceptions of their situation as being pivotal to the sort of learning in which they engaged. Learners' backgrounds as well as their foregrounds, that is, perceived opportunities for their futures, form their dispositions to learn. Like Radford's descriptions of reflections, these perceptions are not individually formed but are collectively and culturally situated. When learners transition between contexts, different foregrounds and backgrounds come to their attention. As a result, dispositions to learn can be contradictory.

> Intentions of learning emerge out of dispositions. Dispositions are concerned with "background" as well as "foreground" and are revealed when the learner produces, creates or decides his or her intention. A situation which could raise intentions for learning does not automatically belong to the background of the student having to do with his or her situation and social or cultural heritage. It is just as much to do with the students' possibilities but the possibilities as the student perceives them. The decision of the learner to act or learn therefore has a role to play when conditions for learning are created. The student has to be involved in the learning—should want to learn—if the learning activity is to become learning as action. Furthermore, the learning has to be performed by the learner if it is to include reflections and a critical awareness. (Vithal & Skovsmose, 1997, p. 147)

If the learning situation supports the active involvement of learners, intentions for learning are formed and the resulting learning process is one of action (Alrø & Skovsmose, 2002). When learners identify with the teacher's suggested outcomes of the learning activity, then joint ownership and shared perspectives between the teacher and learners develop. When learners' intentions differ, then so will the outcomes from learning.

In this chapter, we investigate pedagogical approaches, advocated for learners whose learning seems to involve a complex transition process. Reflection on the intertwining of backgrounds and foregrounds will affect learners' decisions to engage and to a broadening or narrowing of their horizons of possibilities for their future. Consequently, we consider the impact of transitioning between contexts, in which mathematical knowledge and ways of interacting around it are perceived differently, to be an issue of social justice.

Situating Pedagogical Approaches as an Issue of Social Justice

Investigations of social justice within mathematics education have focussed on the learner–teacher relationship and the pedagogical practices used in mathematics classrooms (Atweh & Brady, 2009). However, Fairclough's (2003) description of learning suggests that as one part of the entangled relationships contributing to learning, pedagogical practices, as social practices, involve the valuing by participants of certain systems of knowledge. Learners' dispositions will be affected by relationships with mathematics held by other participants, such as teachers, students, and families. Relationships are not formed solely within an individual classroom but mirror the wider societal valuations of who and what is seen as important.

> It is through this process of drawing on the resources of the various discourses available within a given classroom that individuals construct their identities as teachers and students of mathematics, positioning themselves in relation to the mathematical and nonmathematical activity within the classroom and in relation to the other participants in the classroom and accounting—to themselves and to others—for the nature of their own participation. The privileged official discourses provide what may be constructed "natural" positions for teachers and students, although individuals may resist this discourse. (Morgan, 2009, p. 98)

Curriculum, representing official discourse, identifies what should be learnt and sometimes how it should be learnt, constrains teachers' abilities to implement pedagogical practices.

Table 6.1, from Willis (1998), summarizes four social justice approaches within mathematics education and illustrates how these are likely to channel teachers and learners into forming different kinds of relationships with each other and with mathematical knowledge. Yet, the table also illustrates that it is difficult to discuss perspectives on social justice in ways that do not position some groups as always being outside of the curriculum. This is especially the case in Perspectives 1 and 2, but even in Perspectives 3 and 4 there is an implicit comparison between "others" and the "norm." In the final perspective, the educational task is described as to "help children develop different views of who does mathematics and what it means to be good at it, to understand how they are positioned by mathematics and how to use it in the interests of social justice" (p. 15). The word "different" indicates that the current norm does not do this. It is implicit that those who are affected most by social justice inequities and so have the more complex transitioning to do are the ones who need this alternative curriculum.

In considering learners transitioning between contexts, all the perspectives acknowledge that there might be differences in how knowledge is valued between

Table 6.1
Different Perspectives on the Relationship Between Disadvantage, the Mathematics Curriculum and Social Justice (from Willis, 1998, p. 15)

	Perspective 1 Remedial	Perspective 2: Non-Discriminatory	Perspective 3: Inclusive	Perspective 4: Socially Critical
The mathematics curriculum is …	given, including what is to be learnt, how it is taught and how it is assessed.	given with respect to what is to be learnt, but how it is taught and how it is assessed are not.	a selection from all possible curricula and therefore neither given nor unchangeable.	actively implicated in producing and reproducing social inequality being one of the ways in which dominant cultural values and group interests are maintained.
The problem of "disadvantage" lies with …	the children some of whom by virtue of their race ethnicity, gender, social class or disability are less well prepared than others to get the full benefits of the curriculum.	pedagogy and assessment practices which favour or relate to the experiences, interests, and cultural practices of some social groupings of children more than others.	curriculum content and sequence which reflect the values, priorities and lifestyles of the dominant culture and match the typical developmental sequences associated with their children.	the way the mathematics learner is constructed through the curriculum and the way mathematics is used inside and outside schools to support and produce privilege.
The solution is to …	help such children become better prepared for school mathematics.	change pedagogy and assessment practices to ensure children have real equity of access both to the mathematics and to the means of demonstrating their learning.	rethink who "the typical child" is for whom our curriculum is developed, what school mathematics is, what should be learned, by whom and when.	challenge and modify the hegemony of mathematics and use mathematics explicitly in the services of social justice.

The educational task is to ...	provide children with the missing skills, experiences, knowledge, attitudes or motivations.	draw upon and extend children's experiences, provide a supportive learning environment and more valid assessment opportunities.	provide children with curricula which better acknowledge, accommodate, value and reflect their own and their social groups' experiences, interests and needs.	help children develop different views of who does mathematics and what it means to be good at it, to understand how they are positioned by mathematics and how to use it in the interests of social justice.

home and school contexts. However, the assumption is that the transitioning process is one way—into classrooms. Perspective 4 is the only one that seems to recognize that learners also transition out of school contexts and that the curriculum may have a role in easing that transitioning process.

Focussing on pedagogical approaches readjusts the emphasis from the deviant performance of particular groups of learners, where the main outcome of an appropriate transitioning would be to have them perform similarly to "normal" students. Mathematics education has for a long time documented cases of learners who underachieve but research on pedagogical practices that produce more positive results has not been so prolific (Anthony & Walshaw, 2007; Nasir & Cobb, 2007). We choose to analyze learners' views on pedagogical approaches, ethnomathematics and critical mathematics education, because they had similarities with perspectives 3 and 4. In these perspectives, the mathematics curriculum is considered to be changeable and there is an emphasis on learners using their mathematics knowledge from outside of school within the classroom. Vithal and Skovsmose (1997) stated:

> Whilst ethnomathematics seems to deal mainly with cultural and social issues, critical mathematics education has largely focused on social and political aspects. These perspectives are, of course, connected. We conceive of ethnomathematics and critical mathematics education as two important educational positions in the attempt to develop an "alternative" mathematics education which expresses social awareness and political responsibility. (p. 131)

These perspectives on the connections between social justice and mathematics curriculum provide information on the structure of the relationships that learners can forge with mathematics and other participants within mathematics classrooms.

Biddy (2009) interviewed a large number of primary school students in England about their mathematics classes. Many interviews indicated that these children did not like the relationships in which they were positioned within the classroom. Consequently, she suggested that "a definition of pedagogy needs to be founded in relationships or relationality" (p. 135). Pedagogy as relationships has two components; mutuality and being seen and valued. Mutuality meant that the learning of the group was more important than learning of an individual which was in contrast to their teachers' view of the importance of the individual. For the learners, being seen and valued involved their teacher listening to them and valuing their contributions. If this is not done, Presmeg (2002) suggested that "*symbolic violence* [will be] experienced by students in transition between contexts when their cultural capital is devalued by significant others" (p. 226). Mathematics curricula provide parts of the structures in which teachers operate, but it is how they interpret these structures that will affect the relationships that are forged both inside and outside the classroom around mathematical understandings.

As discussed in later sections, ethnomathematics or critical mathematics education do not provide details about the sorts of relationships that they should foster. Yet in accepting that learning is about becoming and thus more than gaining familiarity with knowledge and skills, then there is a need to understand learners' perspectives on the relationships that they form. In the next sections, we identify learners' perspectives on being involved in ethnomathematics and critical mathematics perspectives from previous research. We then discuss these pedagogical approaches using Wenger's (1998) three modes of belonging to investigate the relationships to which they contribute.

Our primary sources for relevant research have been conferences such as Mathematics Education in Society and the International Conferences on Ethnomathematics. From lists of papers, we identified authors working in the field and located related journal articles and book chapters. Nevertheless, we have not located all relevant material.

Each section begins with a description of the pedagogical approach and theoretical concerns. Then, learners' perceptions of the impact of these approaches are described. At times we have used learners' views as reported second-hand through quotations from teachers. Although not ideal, the paucity of research limited the available data. The final section analyzes students' perspectives using the modes of belonging to illustrate what supported their transitioning between contexts.

Ethnomathematics

Ethnomathematics began in the 1980s as "the study of mathematical ideas of non-literate people" (Ascher & Ascher, 1986) but soon broadened to the mathematical practices of specific groups, whether they be carpenters (Masingila, 1994) or cardiovascular surgeons (Shockey, 2002). The mathematics used and developed by Western mathematicians is one kind of ethnomathematics (Borba, 1990), although academic mathematics has sociological implications that other kinds of mathematics do not have (Knijnik, Wanderer, & Oliveira, 2005). D'Ambrosio (1992) described a research program in ethnomathematics as "the study of the generation, organisation, transmission, dissemination and the use of jargons, codes, styles of reasoning, practices, results and methods" (p. 1183).

An ethnomathematical research program can be traced to two complementary research agendas: to understand better the mathematical practices of different groups (Bishop, 2004), and to support the development of a more just and socially equitable society that deals with the economic and environmental problems facing the world (D'Ambrosio, 2010). The first arose from work from the 1960s and 1970s which showed that the development of mathematical understandings was culturally related (Wedege, 2010). The second came from concerns about the loss of human dignity through the continual conflicts that afflict the world and the need for mathematics and mathematics education to contribute to efforts for peace (D'Ambrosio, 2010).

Despite these worthwhile aims, ethnomathematics has not been without criticisms. For example, the valuing of a practice only if it can be labelled as Western mathematics has been questioned (Jablonka & Gellert, 2010). Barton (2004) stated that although ethnomathematics provides opportunities to reconsider how aspects of Western mathematics are perceived, labelling cultural activities as "mathematics" was problematic unless certain conditions were met. He specified that the knowledge "should be systematised, should be formalised and should relate to quantity, relationships, or space. It must also be sufficiently abstracted to be removable from its practice" (p. 23). These conditions enable practitioners to be able to discuss their ideas as mathematics. Similarly, Pais (2011) suggested that although learners may engage in a range of activities, it is not until these activities are recognized that they become mathematics. However, labelling of traditional activities as mathematics runs the risk that they are seen as having no intrinsic value, except as examples of a Western knowledge system (Roberts, 1997).

Yet, ethnomathematics "has obvious pedagogical implications" (D'Ambrosio, 2010, p. 9). In a description of the *Math in Cultural Contexts* project, developed in Alaska over several decades, Lipka et al. (2009) summarized many of the assumptions on which their ethnomathematical pedagogy is based:

> The assumptions include that students will gain increased access to the math curriculum because they can identify with the curriculum and pedagogy on multiple levels, from familiar contexts to familiar knowledge, and that they will have multiple ways of engaging with the material. ... Further it is assumed that the inclusion of local knowledge, language and culture may well have a positive effect on students' identity that will be different from the typically reported process of schooling that marginalizes so many AI/AN (American Indian/ Alaskan Native) students. (p. 266)

Researchers in other parts of the world acknowledge similar assumptions (see, e.g., Adam, 2003; Laridon, Mosimege, & Mogari, 2005). However, the diversity of aims could result in conflict, making it difficult to implement an ethnomathematical approach that supports students to transition between contexts and broaden their horizons of possibilities.

Vithal and Skovsmose (1997) suggested that, to South Africans, the aims of ethnomathematics closely resemble those of apartheid where perceptions of cultural differences were used to differentiate education opportunities. To overcome the likelihood that some learners' opportunities would be limited, they recommended that students' foregrounds should be considered when choosing mathematics activities. As well, by presenting an activity as representative of a culture, a teacher could gloss over differences within that culture (Vithal & Skovsmose, 1997). Considerations of foregrounds and backgrounds in designing of mathematics activities have been discussed as the need for "permeability."

> A serious commitment to encouraging children to use mathematics to contribute to the solution of problems drawn from everyday life (whether textually represented in texts and tests or actually experienced in their life outside school) will also need to increase the permeability of the boundary between children's everyday knowledge and experience and their more purely mathematical knowledge. (Cooper & Harries, 2002, p. 21)

If permeability is achieved then learners are likely to transition between contexts more easily. Knijnik (1998) reported that in the mathematics education program in which she worked with the Landless People Movement of Brazil "the interrelations between popular knowledge and academic knowledge are qualified, allowing the adults, youths and children who participate in it to concurrently understand their own culture more profoundly, and also have access to contemporary scientific and technological production" (p. 188). This suggested that for these learners participating in activities allowed them to transition both into and out of formal mathematical contexts.

In regard to immigrant classrooms, there have been consistent calls for teachers to know their learners better and to base their teaching on everyday mathematics that learners bring to the classroom (Civil & Planas, 2010; de Abreu & Gorgorió, 2007). Moreira (2007) suggested that the mathematics teacher should act as an ethnomathematical researcher. When ethnomathematical practices have been used in the classroom (Barta, 2002; Masingila, Davidenko, & Prus-Wisniowska, 1996), the activities have tended to be those of adults rather than children. There are few examples of children's own activities being used (see Masingila, 1996; Presmeg, 1996). Carraher and Schliemann (2002) suggested that "there seems to be relatively little mathematical activity in children's out-of-school activities, and when it does come into play, it does not seem to call for a deep understanding of mathematical relations" (p. 150).

As well, Stillman and Balatti (2001) warned that the process of bringing cultural activities into the mathematics classroom potentially "divorces the cultural practices from their context and trivializes and fragments them from their real meaning in context" (p. 325). In Papua New Guinea, the curriculum was changed to support the use of traditional knowledge in the mathematics classroom. Esmonde and Saxe (2004) suggested that the support for using vernacular languages and Indigenous counting systems in community schools may revive the use of *tok ples* counting systems but only by altering its structure so that it resembled the Hindu-Arabic system. Paraide (2005) also warned of the difficulties in trying to make connections between another Papua New Guinean counting system and the Hindu-Arabic one. Thus, although incorporation of traditional counting practices would achieve some of the aims of an ethnomathematical approach, alteration of the traditional knowledge could lead to a narrowing of learners' perceptions about the value of that knowledge in the future.

Reconciling the differences in how knowledge is valued in different contexts is often left to teachers who can struggle to do this (de Abreu, 1993). For example, a Ghanaian teacher felt that she was not able to bring in the learners' outside school knowledge about sharing according to status because she saw it as being in conflict with the fraction knowledge that the curriculum required her to teach (Davis, Seah, & Bishop, 2009). In the USA, Cahnmann and Remillard (2002) described how one teacher working in a low socio-economic area was able to make cultural connections for her students so that they would enjoy mathematics, but struggled to make the tasks mathematically challenging. Consequently, the learners had limited access to academic mathematics.

In the Funds of Knowledge project described by Civil (n.d.), teachers visited the homes of some of their learners to identify activities that then could be used in the classroom to support connections being made between school and out-of-school mathematical knowledge systems.

> Overall the money module focused on children discussing social issues in relation to money (such as welfare, food stamps, buying a car, a house) in the third grade class and on researching topics such as "money, power, and politics" or "foreign currencies," in the fifth grade class. Hence, in this class, the main academic areas emphasized through this module were

social studies, reading and writing. In the third grade class, mathematics was more present, for example through connections to children's literature that had money as the focus. But even with the third graders, I think that we only scratched the surface of the mathematics in a module around money. The very rich discussions in both classrooms showed the wealth of knowledge that these children had about everyday uses of money, budgeting, and what it means not to have enough money. Yet, in terms of our mathematical agenda, I did not feel we succeeded in exploring the potential in this module. (p. 7)

Later, working with a different teacher, a unit based on gardening seemed to be more solidly grounded in mathematics, although at times the mathematics activities were contrived. Notwithstanding, the teacher drew on parents and researchers as resources rather than the learners, Civil (n.d.) suggested that the learners cared about their plants and this contributed to their being interested in the mathematics problems.

Although there are high expectations about using an ethnomathematics approach to ease the transition between school and out-of-school contexts, some mathematics educators have queried its potential. In the next section, we describe learners' views about being involved in ethnomathematics activities.

Learners' Views on Ethnomathematical Approaches

From the perspective of learners, incorporating cultural activities into the mathematics classroom was valuable. The reasons for this varied from finding the activities interesting, to seeing these activities in a new light. This supported learners to reflect not only on what they were learning but also on how they were learning, leading potentially to a broadening of their horizons for future possibilities. In reflecting on how they were learning, learners expressed a desire to be more involved in group work and to learn by doing, through working with artefacts.

Mosimege and Ismael (2004) reported on learners' enjoyment in mathematics lessons that were based around traditional African games.

> The last sessions were very nice. The game practice was very nice. We used to play this game at home without knowing what is essential in it. (pp. 132–133)

> I liked the lessons, they were very exciting because we were taught by doing … With this way of teaching you can learn really (…) other teachers should also teach us in this way if there is a possibility. (p. 133)

In the first quotation, recognizing the game as mathematics seemed to make it more valuable for the learner. In a later article, Nkopodi and Mosimege (2009) commented on the need to ensure that the mathematics was visible and the games were not merely considered fun activities. However, as noted above, there is a risk that the inherent value of the game itself is lost and this could lead to a narrowing of horizons around traditional practices.

Getting students involved in ethnomathematics projects was considered to be a way of supporting learning. In Israel, a tenth-grade teacher in an Arab school worked with researchers to introduce a geometry unit based on traditional geometric designs

(Massarwe, Verner, & Bshouty, 2010). Comments from the learners indicated that some of their enjoyment was connected to being involved in group work but, like the previous set of learners, also doing something with the designs, rather than simply working with them abstractly, was appreciated:

Salam: First time ever that I understand geometry.
Yusof: I discovered that geometry has a special magic and that it is important.
Nimr: I very much enjoyed it. The group work drew us close.
Hanna: Not only theorems and proofs—it is an enjoyable experience of discovering and drawing.
Ranya: I would prefer to study geometry this way. (Massarwe et al., 2010, p. 17)

In other places, the implementation of ethnomathematics units evoked similar comments about the need for mutuality of learning (Biddy, 2009). In describing a mathematics lesson on the Andean flute, zampoña, Favilli and Tintori (2004) provided comments from five teachers and their students. The students' comments came from a questionnaire, with many commenting on how they worked together:

- What I really liked about the zampoña lessons was the way we all worked together and the new experience.
- The thing I liked most of all about the zampoña lessons was being able to work all together: we were a real team, just like a real family; I also liked it when we found the mathematical law, because we were all enthusiastic, we felt like … important mathematicians. (p. 44)

When ethnomathematical activities are introduced into the mathematics classrooms where there are Indigenous students, traditional interaction patterns need to be respected. Lipka et al. (2005) described how one teacher supported learners' use of gestures to describe their ideas in collaborative discussions, even to the degree that talking was replaced. Thus, "the safe learning environment, in which nonverbal communication was honored as a cultural way of 'talking' and communicating mathematically in the classroom, allowed [the learner] to take a leadership role and contribute her knowledge in a culturally congruent way" (p. 379). Mutuality of learning may be achieved in a range of ways.

Relationships with artefacts were also valued by learners. In Favilli and Tintori's (2004) project, it was clear that the artefacts used in the lesson supported learners' engagement and thus also their reflection about how they liked to learn:

- The thing I liked about the zampoña lessons was that … now I know how to make one!
- The thing I liked about the zampoña lessons was seeing something we had made working, and working well because some of our classmates even played a tune with it.
- The work was good fun and, to tell the truth, I really like manual work. (Favilli & Tintori, 2004, p. 41)

The teachers' comments suggested that the learners had gained mathematical understanding from being involved. Nonetheless, as Pais (2011) pointed out, the zampoña was not from the culture of the learners and there is little evidence that cultural considerations in which the zampoña was embedded were discussed in any detail. Thus, there was a risk of trivializing the culture of the activity and this could result in a narrowing of horizons in respect to out-of-school contexts. If the activities are not related to contexts with which the learners were familiar outside of the classroom, what contexts are learners being transitioned between? Or is it sufficient that enjoyment of making the flute supported learners' transition into the context of the mathematics classroom?

In other ethnomathematical activities, like the Arab geometric unit, the connection to the cultural background of the learners was evident. These examples suggested that learners' horizons of possibilities for futures were broadened, in regard not just to the school context but also to their home contexts. After the implementation of an ethnomathematics unit in nine Grade 5 Maldivian classrooms, Adam (2003) had learners complete a survey. Their comments indicated that they valued the connection between activities done outside of school and mathematics.

> I can understand mathematics better now ... I know how to use formulae and things better after seeing how people do things in [for example] construction of houses. (p. 47)

> Before the measurement topic was taught, I did not think of mathematics outside school. Now I see mathematics everywhere. On the street—Mum also use [sic.] measurement in cooking—to measure the rice. At the fish market to sell the fish. (p. 46)

Similarly, at a teacher education college in Israel, prospective Bedouin and Jewish teachers presented projects in which they had identified the mathematics in cultural activities (Katsap & Silverman, 2008). The following quotations from two Bedouin prospective teachers illustrate how they found the activities stimulating and supportive of their learning:

> A subject that I did not previously like, such as the theory of different symmetries, I saw suddenly in a new way in this course, after it was connected to the culture of my people. It was easy to understand and I now like it, and therefore I definitely think that the process of exposing the teacher to the cultural aspects of the mathematical ideas is one that contributes to the training of the teacher. (p. 96)

> I was very happy during the presentation ... because the material was linked to our culture, and everyone is proud when other people learn and become acquainted with their own culture... I've noticed during the presentation of the study unit that the participating teachers showed considerable interest and desire to learn about our culture... Our group prepared many examples of embroidery and in one activity Jewish teachers were asked to describe orally the transformation types revealed by this example. At the end of the lesson the teachers described their sensations during this task. The satisfaction they've felt was evident. (p. 91)

For many prospective teachers, learning about their own culture through mathematics was a surprise. It also surprised them that other prospective teachers who did not share their religion/culture were interested in the activities.

The activities seemed to broaden learners' horizons in regard to possibilities for their futures outside of formal mathematics education situations. Katsap and Silverman (2008) commented on how the activities contributed to dialogue:

> In a lesson on the concept of time presented by a Jewish group, the discussion ranged from the philosophical understanding of the concept to mathematical insight. When a Bedouin student presented his own position on the matter without referring to his culture's attitude, one of the presenters immediately said, "Why are you presenting this example? It would be better if you told us how you view time in the desert. What does your sheikh (literally, 'elder') think about the essence of time?" ... In any other class, a political argument could have easily ensued. However, in this case, the Bedouin prospective teacher began

to relate how Bedouin religious leaders saw the concept of time. Everyone listened with interest, as moments before the class had discussed the Western attitude of "time is money." Two opinions from the two respective cultures represented their emotional perceptions of time and their cognitive mathematical perceptions of time. (pp. 86–87)

Lipka and Adams (2004) summarized similar comments by learners about how they could relate to the mathematics activities which were set in cultural contexts. However, they also provided examples where the use of materials was not received as had been expected. In one instance, learners told their teacher that they did not want to do the activity because they were already familiar with it.

It would seem that learners engaged with the ethnomathematics activities because the activities were of interest to them and/or because they made connections between different contexts that they were transitioning. It may be that the value given to knowledge that originated outside of the mathematics classroom enabled them to reflect on the sorts of the connections that could be made as they transitioned both into and out of the classroom. Many activities allowed learners to interact with each other and artefacts in collaborative, dialogical ways which supported them as they developed mathematical understandings. However, more research is needed to understand better how and if learners' horizons for future possibilities were broadened as a result of working with ethnomathematics projects.

Critical Mathematics Education

There are two main geographical groups of critical mathematics educators. Notwithstanding, both groups often refer to each other's theoretical positions and the distinctions between them are not large. One group located in Europe, but with connections to Brazil (see Campos, Wodewotzki, Jabobini, & Lombardo, 2010), uses the work of Ole Skovsmose who was influenced by the Frankfurt School of Critical Theory (Skovsmose, 2004). Critical theorists aimed at "'emancipating' people from positivist 'domination of thought' through understanding their circumstances and taking action to change their situation" (Patrick, 1999, p. 86). The other group, mainly in America, uses the ideas of Marilyn Frankenstein (2010) on critical mathematical literacy. She drew her inspiration from Paolo Freire. In a re-issue of her first paper, Frankenstein (2010) wrote:

> Freire's theory compels mathematics teachers to probe the non-positivist meaning of mathematical knowledge, the importance of quantitative reasoning in the development of critical consciousness, the ways that math anxiety helps sustain hegemonic ideologies, and the connections between our specific curriculum and the development of critical consciousness. In addition, his theory can strengthen our energy in the struggle for humanization by focusing our attention on the interrelationships between our concrete daily teaching practice and the broader ideological and structural context. (p. 9)

Following the suggestions that Freire made about teaching literacy to adults, Frankenstein (2010) suggested that the problems that learners engage in should be drawn from their own experiences and this would become the starting point for the curriculum. From a philosophical perspective, Skovsmose (2004) identified critical mathematics education as "a preoccupation with challenges emerging from the critical nature of mathematics education. Critical mathematics education refers to concerns which have to do with both research and practice, and a concern for equity and social justice being one of them" (p. 1).

With Helle Alrø, Skovsmose described the need to build learners' "mathemacy" which by connecting to their own contexts empowers their mathematics learning (Alrø & Skovsmose, 2002). A slightly different interpretation, but one still with an emphasis on empowerment, is that of Ernest (2002):

> A successful critical mathematics education must succeed in empowering the learner, first to overcome internal inhibitions and perceptions of inadequacy, second to question the teacher, the subject, and the constraints of school, and third to question the "facts" and edicts of authority at large in society. (p. 1)

Similarly, Frankenstein (1998) suggested that there are four aims for a critical-mathematical literacy curriculum:

1. Understanding the mathematics.
2. Understanding the mathematics of political knowledge.
3. Understanding the politics of mathematical knowledge.
4. Understanding the politics of knowledge. (p. 1)

As Jablonka and Gellert (2010) stated, "critical mathematics literacy intends to be simultaneously a pedagogy of access and a pedagogy of dissent" (p. 43). However, it is not easy for educators to resolve the inherent tension between these two aims (Powell & Brantlinger, 2008), nor for learners to achieve them because of the high level of reflection required.

An examination of the pedagogical practices advocated by critical mathematics educators illustrates how difficult this tension is to resolve. Alrø and Skovsmose (2002) advocated the inquiry cooperation model (see Figure 6.1), which emphasizes the role that the teacher has to play in listening to learners' contributions both as they meet and then engage with a new problem. The teacher must listen respectfully to the learners but also challenge their reasoning. "Challenging good reasons, therefore, means making the students reflect upon and widen their perspective and knowledge" (Alrø & Skovsmose, 1996, p. 33).

Alrø and Skovsmose (1996) found few examples in their empirical data of teachers actually engaging learners in dialogue similar to that of the Inquiry Co-operation Model. As well, in a discussion of an project, called "Terrible Small Numbers," done with 15- and 16-year-olds about the sampling of eggs for salmonella, Alrø and Skovsmose (2002) wrote: "'Terrible small numbers' *in principle* provided topics for reflection, which *in principle* may face the challenge of critique. But as experienced by the students, the reflections were not developed into any powerful ideas of critique" (p. 229).

In his research on how four teachers incorporated culturally relevant teaching ideas into their mathematics lessons, Matthews (2003) reported that only one teacher combined developing learners' mathematical understanding with critiquing their situation with any degree of success. Matthews suggested that this teacher was successful because she built strong relationships with her students based on traditional Bermudian understandings about friendship. The pedagogical practises of this

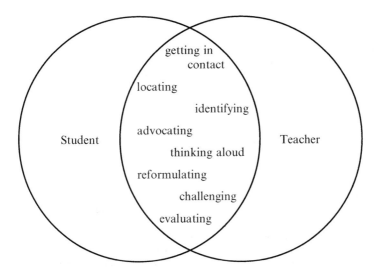

Figure 6.1. Inquiry co-operation model (from Alrø & Skovsmose, 2002, p. 63).

teacher were similar to those advocated by in the inquiry co-operation model (Alrø & Skovsmose, 2002), as the successful teacher expected learners to work together to justify the appropriateness of their answers. The inquiry co-operation model is in alignment with the conception of pedagogies as relationships as it requires teachers both to hear the learners' contributions and to value them.

Learners' Perspective on Critical Mathematics Education

As was the case with ethnomathematical pedagogical practices, learners involved in critical mathematics education activities responded positively to them. Again this seemed to be related to how learners were expected to work together as well as finding the project work interesting. One difference with ethnomathematics activities was that learners seemed to have more choices in regard to the project work that they did. Critical mathematics education projects also supported learners to make transitions both into mathematics classrooms but also to outside-school contexts, thus broadening their horizons for future possibilities in multiple directions.

Moreira and Carreira (1998) describe an activity in a calculus class for Portuguese students completing a business degree. Although the class did not include critical mathematics goals, "some of the problem situations used in teaching mathematical topics were suitable for promoting the act of acquiring a new consciousness" (p. 4). In the following extract, the students used a mathematical model to explore income distribution in an imaginary country.

Miguel: Now we just need to compute $f(0.5)$.
Paulo: Which gives 25%.

Miguel: So 25% is the income of the poorest half of the population. This means that the other half is receiving 75%. It's a striking difference!
Cristina: What are you saying? I don't get it.
Paulo: He's saying that the first half, which is the worse paid, gets 25% of the total income. Therefore, the second half is receiving all the remaining, that is 75%. These are the better paid, they're the richer people.
Cristina: Sure, according to this model ...
Eduardo: Yeah [*speaking with an ironic tone in his voice*], I doubt that the income can ever be so unfairly distributed.
Isabel: Well, you'd better not!
Eduardo: OK. Let's move on to this one: how do you interpret the fact of having $f(0)$ equal to 0 and $f(1)$ equal to 1?
Paulo: Well, if there's no population there can't be any distribution of incomes.
Eduardo: No people, no income.
Paulo: The $f(1)$ equals 1 means that 100% of the people receive the whole income.
Eduardo: Exactly.
Isabel: Which means that there is no embezzlement of money.
Cristina: There's no sense here to speak of an embezzlement of money ...
Isabel: I mean that there are no false donations, no frauds, no fake payments, no funds deviations and no tax evasions.
Eduardo: You're making a good point there ... (Moreira & Carreira, 1998, pp. 4–5)

In the exchange, learners used their mathematical understandings but also justified their responses from their own experiences: "There is no embezzlement of money." While Cristina saw this as a mathematical exercise, distinct from reality, Isabel and Eduardo interpreted the results from what they knew about the Portuguese economy. The complexity of the situation contributed to them engaging with the problem. This is in direct contrast with Esmilde, an immigrant student in Gorgorió and Planas' (2005) research, who tried hard to bring his knowledge of different housing situations to a mathematical problem about population density. Both his teacher and a non-immigrant peer rejected his suggestion, only valuing the mathematical content. Consequently, Esmilde withdrew from engaging in the problem, thus narrowing his horizons of possibilities for his future.

In reported studies, many learners expressed varying levels of dislike for mathematics in the initial stages. Gutstein (2003) quoted from one of his middle-grade students who described his learning after two years of being involved with critical mathematics education as:

> Well, I thought of mathematics as another subject in school that I hated. And I didn't bother to think too much about world issues or everyday issues. Now I know it all relates. And I've learned how powerful math can be to help us explain our decisions and help us express ourselves because, like I said before, math makes things more clear. (p. 61)

Critical mathematics education seems to have the possibility to broaden learners' horizons for future possibilities both within a school context but also in relationship to wider societal contexts that included learners' own foregrounds and considerations of the sort of world that they wanted to live in. Thus, the horizons for future possibilities seemed to be broadened in a multitude of directions.

In Andersson's (2011) research, learners in two classes worked in groups on projects—such as working out each learner's carbon footprint—that she [Andersson] deemed as being related to societal contexts and likely to support critical reflections. One of Andersson's end-of-high-school learners, Sandra, stated:

> What surprised me most though was how important a role mathematics plays when talking about environmental issues. With support of mathematics we can get people to react and stop. […] I am so interested in environmental questions and did actually not believe that maths could be important when presenting different standpoints. (p. 7)

The projects described by Moreira and Carreira (1998), Gutstein (2003) and Andersson's (2011) appear to have been designed without soliciting the learners' opinions on the sorts of projects in which they would like to engage. This lack of consultation could be in conflict with learners' preference for being seen and valued and also leaves learners as the mediators of the transitions between different contexts. Although the activities seemed to have broadened learners' horizons both for possibilities for their futures from the school context and from outside-school contexts, more research is needed to better understand this connection.

Gutstein (2003) claimed that the problems he set were from the learners' lived experiences. Dowling (2010) criticized one of Gutstein's problems, random traffic stops, because the premise on which it was based, that the police would know the ethnicity of the drivers before they stopped them, was unlikely to be a reality. One of Gutstein's (2007) own students challenged him similarly in a journal entry about whether racism was a factor in getting house mortgages.

> In my first article I said that I thought racism was not a factor; after our second discussion I thought racism was a factor, but I think that we don't really know. Even though the rate for Blacks was 5Xs higher than whites in being rejected, that does not necessarily mean it is racism. It could be because of debt, income, or maybe it could be racism. (p. 58)

Gutstein responded to the learner's comments by discussing the relationship between individual and institutional racism. At the end of the project, he also had learners adopt the perspectives of other participants in the problem, including the bank lenders. The project resulted in much discussion and Gutstein hypothesized that it was because the problem was meaningful for the learners. As one learner, Leandro wrote:

> This project was very interesting because it has happened to one of my uncles. He was looking for a house and found one. But in the end, he was turned down. This really is important to me because I will like to buy a house when I grow up, not only for me, but for my cousin and my sister. (Gutstein, 2007, p. 61)

Others have noted a similar impact on statistical learning when the topics were important to the learners. In work with student teachers in Columbia, Rojas (2010) found that "the students learned to use these statistical tools and concepts in the context of identifying and addressing the most pressing problems facing their own communities" (p. 3). In Brazil, a group of medical researchers worked with sixth-grade students in high poverty areas on how to interpret statistical data about mouth cancer so that they could then present this information to their families (Sundefeld, Homse, Prieto, & Rodrigues, 2010). By the end of the project, students had increased their understanding both about mouth cancer and also statistics. Many of their families commented that they had learnt a lot from their children about mouth cancer, although their statistical knowledge had not improved. This was the only project which seemed to investigate learners transitioning into out-of-classroom contexts explicitly.

Having learners engage in topics that investigate inequities can mean that children deal with adult topics. Stocker and Wagner (2007) discussed some of the ethical tensions arising from this. In support of his argument for using mathematics to understand social justice issues better, Stocker quoted two of his 12- and 13-year-old students:

> If we don't know about social issues, if we don't learn to be critical, we won't form views of our own. We'll grow up in a politically impotent society.
>
> Only through learning can things change. Schools need to teach this stuff. Children can't be hidden from the world. (p. 19)

Similarly, Gutstein's (2003) learners felt that being involved in considering societal issues which required them to understand mathematics was important.

> Also, I thought math was just a subject they implanted on us just because they felt like it, but now I realize that you could use math to defend your rights and realize the injustices around you. I mean you could quickly find an average on any problem, find a percentage on any solution, etc. I mean now I think math is truly necessary and I have to admit it, kinda cool. It's sort of like a pass you could use to try to make the world a better place. (p. 62)

In her work with adult learners, Tomlin (2002) requested them to bring in problems that they found relevant and which required mathematics. Although one student's problem fitted the criteria, it could be solved more sensibly, through the combined wisdom of classmates, in ways which did not involve using mathematics. Civil (n.d.) commented that in real-world problems, "mathematics is often hidden; it is not the center of attention and may actually be abandoned in the solution process" (p. 27).

Although learners could not choose their own problems in Andersson's (2011) projects, there was some room for them to make some choices, and that was appreciated. The following comment is again from Sandra, who had described herself as having maths anxiety:

> We distributed the time well, I think. [...] The group worked well. We were good at different things and helped each other. I am proud of the work I have done as I felt I could contribute a lot in the beginning when we talked about borrowing money and interest rates. To plan time and content [my]self got me to feel it related to me. I think mathematics has been a little more fun than usual. [...] I feel the project has been meaningful and to look at mathematics from different angles (vändra och vrida på matematiken) was positive. But I would like more time for explanations from the teacher, as mathematics is difficult for me. (Andersson, 2011)

For Sandra, the pedagogical approach seemed to respect both her need to be seen and valued. "I felt in that I could contribute a lot," and for mutuality in learning "the group worked well. We were good at different things and helped each other." Working in groups seemed to support her to overcome her anxiety about mathematics which can be seen as one form of transition. Nevertheless, the final cry about needing more help from the teacher suggests that she had not yet come to see herself as being a capable learner of mathematics. At the end of the project on carbon footprints, Sandra wrote:

> I have probably learnt more now than if I had only calculated tasks in the book. Now I could get use of the knowledge in the project and that made me motivated and happy! I show my knowledge best through oral presentations because there you can show all the facts and talk instead of just writing a test. To have a purpose with the calculations motivated me a lot. (Andersson, 2011)

Activities linked to critical mathematics education seemed to support learners' reflections on their learning both of content knowledge but also about themselves. These reflections enabled learners to see connections to their backgrounds, their foregrounds and their hopes for the future of the world. Thus, they could be said to support them transitioning between contexts—into and out of the mathematics classroom—thereby broadening their horizons for their future possibilities in more than one direction.

Modes of Belonging

In order to analyze learners' perceptions, we draw on Wenger's (1998) modes of belonging. Nasir and Cooks (2009) suggested that Wenger "reconceptualizes learning from an in-the-head phenomenon to a matter of engagement, participation, and membership in a community of practice" (p. 42). The modes of belonging indicate how people position themselves in relationship to communities of practice, and thus the forms of becoming that are possible. Commonly, mathematics classrooms are considered to be communities of practice.

> The recognition of the mathematics classroom as a functioning community where teacher and student activity in it is shaped by (and shapes) a set of norms and practices for learning mathematics highlights the importance of issues such as competence, ownership and alignment in engaging in this community. In particular, alignment between practices and identities of home and school has implications for whether students negotiate ways of participating that serve their individual goals (Cobb & Hodge, 2002; Hand, 2003). (DiME, 2007, p. 408)

Consequently, it would seem that the ultimate goal is to understand how learners transition into the community of practice of the mathematics classroom. Yet this view of a mathematics classroom tends to decontextualize it from the "larger social, cultural, economic and political structures" (Valero, 2010, p. LXI) that influence what occurs within those classrooms. In contrast, we see membership of the mathematics classroom community of practice as only one outcome of the never-ending transitioning process and one that occurs concurrently with learners transitioning from the classroom context into outside-classroom contexts such as the home. Therefore, we use modes of belonging to consider how pedagogical approaches support learners to transition between contexts in ways that broaden rather than narrow their horizons of possibilities for their futures in a number of directions. Figure 6.2 summarizes the three modes of belonging: engagement; imagination; and alignment.

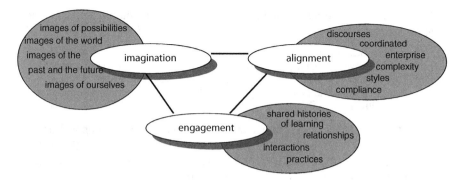

Figure 6.2. Modes of belonging (from Wenger, 1998, p. 174).

According to Wenger (1998), engagement has three parts: "the ongoing negotiation of meaning," "the formation of trajectories," and "the unfolding histories of practice" (p. 174). The distribution of power affects participants' willingness and ability to engage and conversely engagement "affords the power to negotiate our enterprises and thus to shape the context in which we can construct and experience an identity of competence" (p. 175). Although some practices are engaged in many times, each new experience will need to be made sense of again, leading to an evolution of these practices. Continual negotiation of meaning within familiar and unfamiliar situations contributes to the formation of trajectories. "As trajectories, our identities incorporate the past and the future in the very process of negotiating the present" (p. 155). However, "the understanding inherent in a shared practice is not necessarily one that gives members broad access to the histories or relations with other practices that shape their own practices" (p. 175).

Imagination is embedded within "our own experience" (Wenger, 1998, p. 173). Like Radford's view of learning, imagination arises in the interaction between the individual and in Wenger's case the community of practice. In retelling a story about two stone-cutters who gave different responses when questioned about what they saw themselves as doing when chiselling a block of stone, Wenger (1998) stated: "Their experiences of what they are doing and their sense of self in doing it are rather different. This difference is a function of imagination. As a result, they are learning very different things from the same activity" (p. 176). Imagination is closely linked to whether learners perceive the process of transitioning between contexts as leading to a narrowing or widening of their horizons of possibilities for their futures.

For Wenger (1998) "the process of alignment bridges time and space to form broader enterprises so that participants become connected through the coordination of their energies, actions, and practices" (p. 179). Although alignment is usually connected to both engagement and imagination, it may not be. Some learners in mathematics classrooms may complete textbook exercises and have a clear sense of what it means to be a mathematics learner, but when an opportunity arises to withdraw from taking mathematics classes they do so because they do not want to align themselves with this practice (Ingram, 2011). Coerced alignment may impact on the belonging that learners ultimately assume.

Modes of Belonging: Ethnomathematics and Critical Mathematics Education

From the learners' perspectives, pedagogical practices based on ethnomathematics and critical mathematics education seem to have features that support them to transition between contexts. Nevertheless, researchers considered that both sets of approaches have some inherent tensions arising from multiple and sometimes conflicting assumptions. On the other hand, learners' perceptions, as reported in research reports, suggest that they were not concerned with these tensions. Generally, learners verified that learning experiences based in ethnomathematics and critical mathematics education provided them simultaneously with mathematical skills and an awareness of outside school issues. Analyzing the comments using Wenger's modes of belonging indicated that working as a group on complex problems that the learners cared for were features that supported learners to transition between contexts.

Engagement

The learners' comments indicated that they engaged in the projects and so reflected on what mathematics was, how it could be used and how it should be learnt. Often the projects spanned several lessons and in some instances, learners worked on them at home. However, the transition into home contexts, and their engagement in activities there, was the focus of only one research study (Sundefeld et al., 2010).

Teachers and facilitators often assumed a shared history with learners when choosing an activity. Usually, the activity was seen as the vehicle for connecting the learners' backgrounds to improved school mathematics understanding, which was an important consideration in learners' foregrounds.

The negotiation between familiar and unfamiliar situations could contribute to the formation of a trajectory. However, there is a need to recognize the impact of learners' own perceptions of their situation on their dispositions to learn. Andersson (2010) commented on how labelling a day, set aside for project work about the United Nations Rights of the Child, as a Maths Day meant that many students chose to attend doctor and dentist appointments. One student, Petra, stated:

> First I thought, a whole day of mathematics, I can't do it; I just can't be there the whole day. But when I got there it was actually quite fun and now, afterwards, I read and look in the newspapers in a different way. So I actually learnt something that was really unexpected of a math-day. (p. 14)

In the critical mathematics education projects, learners were supported to engage by being able to make choices about aspects of the projects, although not usually the topic of the projects. In the ethnomathematics activities, learners commented on the value of being physically involved in an activity, such as playing games and making flutes. This increased their interest and supported them to reflect on both the activity and mathematics in different ways. It would have been valuable to know whether they also engaged in these activities outside of the classroom so that the activities could be considered as easing their transition in that direction as well.

Project work provided learners with opportunities to form supportive and collaborative relationships that contributed to developing a shared history of learning. A requirement to present to others supported learners to share more than just information about the project (e.g., Katsap & Silverman, 2008). This has links to Biddy's (2009) learners' requests for mutuality in learning.

When learners worked in groups to solve challenging problems, power was not located within the teacher but flowed between the participants. Sundefeld et al.'s (2010) project showed that school children could take on the role of informants to their families. Again, it would have been valuable to know how this eased their transition from school to out-of-school contexts.

Imagination

Imagination is closely connected to learners considering themselves as being seen and valued. For this to occur, learners need to reflect on how they are learning. Often project work supported the learners to see that mathematics could be used in situations that were important to them. In Gutstein's critical mathematics education projects, learners began to understand the power of mathematical ideas and this provided a motivation for them to want to learn. In the gardening project described by Civil's (n.d.), the children's devotion to their plants was a surprise to the teacher but supported them to participate in solving the somewhat contrived mathematical problems about the gardens.

When learners could connect mathematics to their interests, they could imagine engaging in future tasks that required them to use mathematics. However, some learners did not immediately embrace the idea that it was mathematics that provided them with these possibilities. The following quote from one of Andersson's learners, Zizzie, exemplifies this tension for the learner:

> A math-day, how fun could that be, and why did you call it a math-day? We worked on posters, we sought information, we rewrote mathematics stuff for best effect, but that is not mathematics! It was a really good day, but definitely not maths. (Andersson, 2010, p. 15)

Imagination of future possibilities was connected to the learners' experiences with mathematics in the past (Patrick, 1999). Allowing spaces for discussion of different facets of the activities meant that they were opened to discussion and reflection. This is likely to broaden horizons for imagining future possibilities for school mathematics. However, the relationship to transitions to outside-school contexts is not so clear.

Alignment

On the whole, the learners aligned themselves with the aims of the ethnomathematics and critical mathematics agendas to which they were introduced. This can be seen in the way that they combined their everyday knowledge, interests and mathematical understandings when working on the different activities. Yet, some learners had to be encouraged to align themselves with the activities by a teacher or in some cases an elder (Lipka & Adams, 2004).

It was not always clear how learners' changed perceptions of mathematics and how it could be learned were connected to their willingness to align themselves with mathematics activities in the future. As one of Gutstein's (2003) learners stated "I think that now I can understand the world better by using math, but that doesn't mean I *like* connecting math with what surrounds me. I still think that there are some 'BIG IDEAS' you can understand without using math" (p. 61, upper-case and underlining in the original). This learner may have learned to tolerate doing mathematics in the classroom but was unlikely voluntarily to choose to use mathematics in situations outside the classroom. The transition into the mathematics classroom was eased but the mathematical knowledge was likely to remain compartmentalized as something done at school.

For other students, the power that they gained from solving problems that mattered to them, with mathematics, supported their continuing engagement in mathematics classrooms. By enrolling in courses adult learners—such as those described by Tomlin (2002) and Patrick (1999)—showed that mathematics learning was an enterprise with which they wished to align themselves. This was the case even though their imagination of mathematics seemed clearly connected to what they had not been able to do when at school.

By using the modes of belonging as a lens, it is possible to see that activities based on ethnomathematics or critical mathematics education approaches provided possibilities for easing learners' transitions into mathematics classrooms. This is because it enabled learners to see possibilities about why they should learn mathematics. This was not possible when connections were not made to their backgrounds, foregrounds or to improving the world in which they lived. Nevertheless, although the activities provided opportunities for learners to engage in, imagine and align themselves to learning mathematics at school, there were still tensions when teachers'/facilitators' assumptions about the activities did not match those of the learners. The focus of activities was on easing learners' transition into the mathematics classroom and only secondary consideration, if any, was given to easing learners' transition into out-of-school contexts.

Transitioning Into and Out of School Mathematics Contexts

In this chapter, we have defined contexts as systems of knowledge. Transitioning contexts, therefore, involved coming to terms with differences in knowledge, how it was organized and valued and the interaction patterns around how it was used. Learners' reflection on these differences led to learning, both about the knowledge but also about the learners themselves. When the knowledge system of school mathematics was similar to those of outside-school contexts then the transitioning process was likely to lead to a broadening of horizons of possibilities for learners' futures, in both the school and out-of-school directions. However, when there were differences, then transitioning could leave learners as the mediators between knowledge systems. This could result in a narrowing of learners' horizons in regard to school mathematics learning and out-of-school learning.

In this final section, we first discuss why learners' perspectives are so important before revisiting Willis' table showing the relationship between disadvantage, mathematics education and curriculum. Then, we summarize those practices that seemed to ease learners' transitioning and conclude with suggestions for further research.

If we take seriously the proposition that learning occurs when learners transition between contexts and that it includes a process of becoming, as well as understanding content, then it is important to consider the pedagogical practices from the learners' own perspectives. It is their understanding of the activities that enables connections between their foregrounds and backgrounds to form their dispositions to learn (Skovsmose, 2004). As can be seen in the modes of belonging analysis, regardless of the pedagogical approaches adopted, learners' perceptions may be different from the perceptions of those who develop the mathematical activities. This can lead to a narrowing of horizons of possibilities for learners' futures, both in regard to formal mathematics and to out-of-school activities. Nevertheless, in the studies described, ethnomathematics and critical mathematics education approaches did seem to support learners to transition between in-school and out-of-school contexts, at least in one direction—towards school—and thus broadened horizons from this viewpoint.

Ethnomathematics and critical mathematics education have similarities with Perspectives 3 and 4 in Willis' (1998) table (see Table 6.1). As described earlier, the perspectives appear to position some groups as being other, so the focus remains of their differences. By investigating pedagogical approaches, we instead focus on the relationship between transitioning and the narrowing or broadening of horizons of learners' possibilities for their futures. Using learners' perspectives means that although their reflections form the heart of the analysis, it is the approaches and the outcomes of the approaches which are discussed.

The features that the learners suggested supported their transitioning process varied between what was learnt and how it was learnt. On the whole, learners suggested that activities had to be based on something that they cared for: in their past, such as poor performance in mathematics when at school; in their future, for example using ethnomathematical activities themselves as teachers; and in their present, cultivating a garden and growing plants. Learners also valued activities which were socially or politically motivating for them—such as investigating issues to do with climate change. The activities in which learners were interested generally involved integrating knowledge systems from inside and outside school.

The relevance of Biddy's (2009) pedagogy of relationship was obvious in learners' comments about how they should learn. For example, they valued being seen and valued, not only by their teachers but also by other learners. Being able to negotiate aspects of the activities with each other and with the teacher was one aspect of this, and this contributed to their aligning themselves with the present activities. Working in collaborative groups that had to resolve complex issues meant that all learners had to contribute, and those contributions were seen as valuable. The continuing negotiation within the groups, with the teacher, and in relation to the artefacts with which they had to work, meant that power was distributed between participants. Working with artefacts that learners found interesting, such as the Andean flute, supported their engagement in mathematics education activities. Many learners commented on the value of "learning by doing." However, when the tasks were not explicitly linked to the learners' own culture, the only transitioning that the tasks supported was into the classroom.

Learners have valuable insights into the sorts of pedagogical practices that contribute to their transitioning between contexts. Over the last few decades there has been much theorizing about reasons why certain groups of learners do not do well in mathematics classrooms. This has led to awareness that learners' backgrounds and their expectations about what they can learn in the classrooms will contribute to their dispositions to learn. Yet this has not led to a proliferation of research that asks learners about their experiences. In this chapter, we have presented much of the research that is available. Nonetheless, there is a need for more work if we are to learn how to ease learners' transitioning processes so that their horizons of possibilities for futures are enlarged rather than reduced. We do not know, for instance, whether enjoyment of an activity in itself eases learners' transitioning between contexts or whether there is a need to bridge different knowledge systems with which learners are familiar.

As well we would suggest that further work needs to be done in regard to improving our understanding of the range of contexts that are likely to facilitate learners' transitions. In our analysis, it was clear that both ethnomathematics and critical mathematics education have the potential to ease the transitioning of learners into out-of-school contexts. However, this has rarely been the object of mathematics education research, which has remained focussed on transitioning into mathematics classrooms. The obsession with national testing, with equity issues being strongly tied to increased test results, has blinkered much of the research, leading to a focus on the importance of easing learners into the mathematics classroom. Yet it is naïve to believe that learners would form dispositions to learn simply from their experiences in those classrooms. The reflection in which learners engage as they transition into out-of-school contexts, including the home, is likely to have an equal impact on their dispositions to learn, within those very classrooms which are deemed to be so important.

References

Adam, S. (2003). Ethnomathematical ideas in the curriculum. In L. Bragg, C. Campbell, G. Herbert, & J. Mousley (Eds.), *Mathematics Education Research: Innovation, Networking, Opportunity: Proceedings of the 26th Annual Conference of the Mathematics Education Research Group of Australasia* (pp. 41–48). Melbourne, Australia: Mathematics Education Research Group of Australasia. Retrieved from http://www.merga.net.au/publications/counter.php?pub=pub_conf&id=1336.

Alrø, H., & Skovsmose, O. (1996). Students' good reasons. *For the Learning of Mathematics, 16*(3), 31–38.

Alrø, H., & Skovsmose, O. (2002). *Dialogue and learning in mathematics education: Intention, reflection, critique*. Boston, MA: Kluwer Academic Publishers.

Andersson, A. (2010). Can a critical pedagogy in mathematics lead to achievement, engagement and social empowerment? *Philosophy of Mathematics Education, 25*, 1–25. Retrieved from http://people.exeter.ac.uk/PErnest/pome25/index.html.

Andersson, A. (2011, February). *Interplays between context and students' achievement of agency*. Proceedings from 7th Conference for European Research in Mathematics Education, Rzeszów, Poland. Retrieved from http://www.cerme7.univ.rzeszow.pl/index.php?id=wg10

Anthony, G., & Walshaw, M. (2007). *Effective pedagogy in mathematics/pangarau*. Wellington, New Zealand: New Zealand Ministry of Education. Retrieved from http://www.educationcounts.edcentre.govt.nz/publications/series/ibes/effective_pedagogy_in_pangaraumathematics.

Ascher, M., & Ascher, R. (1986). Ethnomathematics. *History of Science, 14*, 125–144.

Atweh, B., & Brady, K. (2009). Socially response-able mathematics education: Implications of an ethical approach. *Eurasia Journal of Mathematics, Science & Technology Education, 5*(3), 267–276.

Barta, J. (2002). The mathematical ecology of the Florida Seminole and its classroom implications. In W. Secada (Ed.), *Changing faces of mathematics: Perspectives on Indigenous people of North America* (pp. 167–174). Reston, VA: National Council of Teachers of Mathematics.

Barta, J., & Brenner, M. E. (2009). Seeing with many eyes: Connections between anthropology and mathematics. In B. Greer, S. Mukhopadhyay, A. B. Powell, & S. Nelson-Barber (Eds.), *Culturally responsive mathematics education* (pp. 85–110). New York, NY: Routledge.

Barton, B. (2004). Mathematics and mathematical practices: Where to draw the line? *For the Learning of Mathematics, 24*(1), 22–24.

Biddy, T. (2009). How do pedagogic practices impact on learner identities in mathematics? In L. Black, H. Mendick, & Y. Solomon (Eds.), *Mathematical relationships in education: Identities and participation* (pp. 123–135). New York, NY: Routledge, Taylor and Francis.

Bishop, A. J. (2004, July). *Critical issues in researching cultural aspects of mathematics education*. Paper presented in Discussion Group 2, 10th International Congress on Mathematics Education, Copenhagen, Denmark. Retrieved from http://www.icme-organisers.dk/dg02/.

Borba, M. C. (1990). Ethnomathematics in education. *For the Learning of Mathematics, 10*(1), 39–43.

Brenner, M. E. (1998). Meaning and money. *Educational Studies in Mathematics, 36*(2), 123–155.

Brown, R. (2009). Teaching for social justice: Exploring the development of student agency through participation in the literacy practices of a mathematics classroom. *Journal of Mathematics Teacher Education, 12*(3), 171–185.

Cahnmann, M. S., & Remillard, J. T. (2002). What counts and how: Mathematics teaching in culturally, linguistically, and socioeconomically diverse urban settings. *The Urban Review, 34*(3), 179–204.

Campos, C. R., Wodewotzki, M. L., Jabobini, O. R., & Lombardo, D. F. (2010). Statistics education in the context of the critical education: Teaching projects. In C. Reading (Ed.), *Data and context in statistics education: Towards an evidence-based society: Proceedings of the Eighth International Conference on Teaching Statistics (ICOTS8), July, 2010, Ljubljana, Slovenia*.

Voorburg, The Netherlands: International Statistical Institute. Retrieved from http://www.stat.auckland.ac.nz/~iase/publications.php?show=icots8.

Cantoni, G. (1991). Applying a cultural compatibility model to the teaching of mathematics to indigenous populations. *Journal of Navajo Education, 9*(1), 33–42.

Carraher, D. W., & Schliemann, A. D. (2002). Is everyday mathematics truly relevant to mathematics education?. In M. Brenner & J. N. Moschkovich (Eds.), *Journal for Research in Mathematics Education Monograph: Everyday and academic mathematics in the classroom (Monograph)* (pp. 131–153). Reston, VA: National Council of Teachers of Mathematics.

César, M. (2007). Dialogical identities in students from cultural minorities or students categorised as presenting SEN: How do they shape learning, namely in mathematics. In ScTIG Group (Ed.), *Second Socio-Cultural Theory in Educational Research and Practice Conference Proceedings*. Manchester, UK: University of Manchester. Retrieved from http://www.education.manchester.ac.uk/research/centres/lta/ltaresearch/socioculturaltheoryinterestgroupsctig/socioculturaltheoryineducationconference2007/conferencepapers/groupthreepapers/files/.

Cobb, P., & Hodge, L. (2002, July). *Learning, identity, and statistical data analysis.*, Paper presented at the Sixth International Conference on Teaching of Statistics, Cape Town, South Africa.

Civil, M. (2008, July). *Mathematics teaching and learning of immigrant students: A look at the key themes from recent research*. Manuscript prepared for the 11th International Congress of Mathematics Education (ICME) Survey Team 5: Mathematics Education in Multicultural and Multilingual Environments, July 2008, Monterrey, Mexico. Retrieved from http://math.arizona.edu/~cemela/english/content/ICME_PME/MCivil-SurveyTeam5-ICME11.pdf.

Civil, M. (n.d.). *Building on community knowledge: An avenue to equity in mathematics education*. Retrieved 1 June 2011 from http://cemela.math.arizona.edu/spanish/content/workingpapers/.

Civil, M., & Planas, N. (2010). Latino/a immigrant parents' voices in mathematics education. In E. L. Grigorenko & R. Takanishi (Eds.), *Immigration, diversity and education* (pp. 130–150). New York, NY: Routledge.

Cooper, B., & Harries, T. (2002). Children's responses to contrasting "realistic" mathematics problems: Just how realistic are children ready to be? *Educational Studies in Mathematics, 49*, 1–23.

Crafter, S., & de Abreu, G. (2011, February). *Teachers' discussion of parental use of implicit and explicit mathematics in the home*. Proceedings from 7th Conference for European Research in Mathematics Education, Rzeszów, Poland. Retrieved from http://www.cerme7.univ.rzeszow.pl/index.php?id=wg10.

D'Ambrosio, U. (1992). Ethnomathematics: A research program on the history and philosophy of mathematics with pedagogical implications. *Notices of the American Mathematical Society, 39*(10), 1183–1185.

D'Ambrosio, U. (2010). Ethnomathematics: A response to the changing role of mathematics in society. *Philosophy of Mathematics Education, 25*. Retrieved from. http://people.exeter.ac.uk/PErnest/pome25/http://people.exeter.ac.uk/PErnest/pome25/.

Davis, E. K., Seah, W. T., & Bishop, A. J. (2009). *Students' transition between contexts of mathematical practices in Ghana*. Proceedings of 2009 Mathematical Association of Victoria Conference (pp. 62–71). Retrieved from http://www.mav.vic.edu.au/files/conferences/2009/10Davis.pdf.

de Abreu, G. (1993). *The relationship between home and school mathematics in a farming community in rural Brazil* (Unpublished doctoral dissertation). The University of Cambridge, Cambridge.

de Abreu, G., Bishop, A. J., & Presmeg, N. C. (2002). Mathematics learners in transition. In G. de Abreu, A. J. Bishop, & N. C. Presmeg (Eds.), *Transitions between contexts of mathematical practices* (pp. 7–21). Dordrecht, The Netherlands: Kluwer Academic Publishers.

de Abreu, G., & Gorgorió, N. (2007). Social representations and multicultural mathematics teaching and learning. In D. Pitta-Pantazi & G. Philippou (Eds.), *Proceedings of the Fifth Congress of the European Society for Research in Mathematics Education, 22–26 February 2007, Larnaca, Cyprus* (pp. 1559–1566). European Society for Research in Mathematics Education. Retrieved from http://ermeweb.free.fr/CERME5b/WG10.pdf.

Diversity in Mathematics Education Center for Learning and Teaching [DiME]. (2007). Culture, race, power and mathematics education. In F. K. Lester (Ed.), *Second handbook of research in mathematics teaching and learning* (pp. 405–433). Charlotte, NC: Information Age.

Dowling, P. (2010, January). *Abandoning mathematics and hard labour in schools: A new sociology of knowledge and curriculum reform*. Paper presented at the Seventh Swedish Mathematics Education Research Seminar (MADIF-7), Stockholm, Sweden.

Ernest, P. (2002). Empowerment in mathematics education. *Philosophy of Mathematics Education, 15*. Retrieved from http://people.exeter.ac.uk/PErnest/pome15/empowerment.htm.

Esmonde, I., & Saxe, G. B. (2004). 'Cultural mathematics' in the Oksapmin curriculum: Continuities and discontinuities. ICLS '04: Proceedings of the 6th International Conference on Learning Sciences. Retrieved from http://dl.acm.org/citation.cfm?id=1149146&dl=ACM&coll=DL&CFID=46231086&CFTOKEN=37114030.

Fairclough, N. (2003). *Analysing discourse: Textual analysis for social research*. London, UK: Routledge.

Favilli, F., & Tintori, S. (2004). Intercultural mathematics education: Comments about a didactic proposal. In F. Favilli (Ed.), *Ethnomathematics and mathematics education: Proceedings of the 10th International Congress of Mathematics Education Copenhagen. Discussion Group 15 Ethnomathematics* (pp. 39–47). Pisa: Tipografia Editrice Pisana. Retrieved from http://www.dm.unipi.it/~favilli/Ethnomathematics_Proceedings_ICME10.pdf.

Frankenstein, M. (1998, July). *Reading the world with maths: Goals for a critical mathematical literacy curriculum*. Paper presented at the First International Conference of Mathematics Education and Society, Nottingham, UK. Retrieved from http://www.nottingham.ac.uk/csme/meas/measproc.html.

Frankenstein, M. (2010). Critical mathematics education: An application of Paolo Freire's epistemology. *Philosophy of Mathematics Education, 25*. Retrieved from http://people.exeter.ac.uk/PErnest/pome25/index.html.

Gadamer, H.-G. (1996). *Truth and method* (2nd ed.). New York, NY: Continuum.

Gee, J. P. (1996). *Social linguistics and literacies: Ideologies in discourse*. London, UK: Taylor and Francis.

Gorgorió, N., & Planas, N. (2003). *Transitions from backgrounds to foregrounds*. Proceedings for the Third Congress for European Research in Mathematics Education. Retrieved from http://www.dm.unipi.it/~didattica/CERME3/proceedings/Groups/TG10/TG10_Gorgorio_cerme3.pdf.

Gorgorió, N., & Planas, N. (2005). Social representations as mediators of mathematics learning in multiethnic classrooms. *European Journal of Psychology of Education, 20*(1), 91–104.

Gutstein, E. (2003). Teaching and learning mathematics for social justice in an urban, Latino school. *Journal for Research in Mathematics Education, 34*(1), 37–73.

Gutstein, E. (2007). "So one question leads to another": Using mathematics to develop a pedagogy of questioning. In N. S. Nasir & P. Cobb (Eds.), *Improving access to mathematics: Diversity and equity in the classroom* (pp. 1–9). New York, NY: Teachers College Press.

Hand, V. (2003). *Reframing participation: Meaningful mathematical activity in diverse classrooms*. Unpublished doctoral dissertation, Stanford University.

Ingram, N. (2011). *Affect and identity: The mathematical journeys of adolescents* (Ph.D. thesis). University of Otago, Dunedin, New Zealand.

Jablonka, E., & Gellert, U. (2010). Ideological roots and uncontrolled flowering of alternative curriculum conceptions. In U. Gellert, E. Jablonka, & C. Morgan (Eds.), *Proceedings of the Sixth International Mathematics Education and Society Conference, 20–25 March 2010* (pp. 31–39). Berlin: Freie Universität Berlin. Retrieved from http://www.ewi-psy.fu.berlin.de/en/v/mes6/research_papers/index.html.

Katsap, A., & Silverman, F. L. (2008). A case study of the role of ethnomathematics among teacher education students of highly diverse cultural backgrounds. *Journal of Mathematics and Culture, 3*(1), 66–102.

Knijnik, G. (1998). Ethnomathematics and political struggles. *Zentralblatt für Didaktik der Mathematik, 30*(6), 188–194.

Knijnik, G., Wanderer, F., & Oliveira, C. J. D. (2005). Cultural differences, oral mathematics and calculators in a teacher training course of the Brazilian Landless Movement. *Zentralblatt für Didaktik der Mathematik, 37*(2), 101–108.

Laridon, P., Mosimege, M., & Mogari, D. (2005). Ethnomathematics research in South Africa. In R. Vithal, J. Adler, & C. Keitel (Eds.), *Researching mathematics education? in South Africa* (pp. 133–164). Pretoria, South Africa: Human Sciences and Research Council. Retrieved from http://www.hsrcpress.ac.za/product.php?productid=2034.

Lipka, J., & Adams, B. L. (2004). Some evidence for ethnomathematics: Quantitative and qualitative data from Alaska. In F. Favilli (Ed.), *Ethnomathematics and mathematics education: Proceedings of the 10th International Congress of Mathematics Education Copenhagen. Discussion Group 15 Ethnomathematics* (pp. 87–98). Pisa, Italy: Tipografia Editrice Pisana. Retrieved from http://www.dm.unipi.it/~favilli/Ethnomathematics_Proceedings_ICME10.pdf.

Lipka, J., Hogan, M. P., Webster, J. P., Yanez, E., Adams, B., Clark, S., & Lacy, D. (2005). Math in a cultural context: Two case studies of a successful culturally based math project. *Anthropology & Education Quarterly, 36*(4), 367–385.

Lipka, J., Yanez, E., Andrew-Ihrke, D., & Adam, S. (2009). A two-way process for developing effective culturally-based math: Examples from math in a cultural context. In B. Greer, S. Mukhopadhyay, A. B. Powell, & S. Nelson-Barber (Eds.), *Culturally responsive mathematics education* (pp. 257–280). New York, NY: Routledge.

Masingila, J. O. (1994). Mathematics practice in carpet laying. *Anthropology & Education Quarterly, 25*(4), 430–462.

Masingila, J. O. (1996, July). *What can we learn from students' out-of-school mathematics practices?* Paper presented at Working Group 21, ICME 8, Seville, Spain.

Masingila, J. O., Davidenko, S., & Prus-Wisniowska, E. (1996). Mathematics learning and practice in and out of schools: A framework for connecting these experiences. *Educational Studies in Mathematics, 31*, 175–200.

Massarwe, K., Verner, I., & Bshouty, D. (2010). An ethnomathematics exercise in analysing and constructing ornaments in a geometry class. *Journal of Mathematics and Culture, 5*(1), 1–20.

Matthews, L. E. (2003). Babies overboard! The complexities of incorporating culturally relevant teaching into mathematics instruction. *Educational Studies in Mathematics, 53*(1), 61–82.

Moreira, D. (2007). Filling the gap between global and local mathematics. In D. Pitta-Pantazi & G. Philippou (Eds.), *Proceedings of the Fifth Congress of the European Society for Research in Mathematics Education 22–26 February 2007, Larnaca, Cyprus*, (pp. 1587–1596). European Society for Research in Mathematics Education & Department of Education, University of Cyprus. Retrieved from http://ermeweb.free.fr/CERME5b/WG10.pdf.

Moreira, L., & Carreira, S. (1998). *No excuses to command, not excuses to obey, no excuses to ignore. Some data to reflect upon.* Proceedings of the First International Conference of Mathematics Education and Society, 6–11 September 1998, Nottingham, UK. Retrieved from http://www.nottingham.ac.uk/csme/meas/measproc.html.

Morgan, C. (2009). Questioning the mathematics curriculum: A discursive approach. In L. Black, H. Mendick, & Y. Solomon (Eds.), *Mathematical relationships in education: Identities and participation* (pp. 97–106). New York, NY: Routledge, Taylor and Francis.

Mosimege, M., & Ismael, A. (2004). Ethnomathematical studies of Indigenous games: Examples from Southern Africa. In F. Favilli (Ed.), *Ethnomathematics and mathematics education: Proceedings of the 10th International Congress of Mathematics Education Copenhagen. Discussion Group 15 Ethnomathematics* (pp. 119–137). Pisa, Italy: Tipografia Editrice Pisana. Retrieved from: http://www.dm.unipi.it/~favilli/Ethnomathematics_Proceedings_ICME10.pdf.

Nasir, N. S., & Cobb, P. (2007). Introduction. In N. S. Nasir & P. Cobb (Eds.), *Improving access to mathematics: Diversity and equity in the classroom* (pp. 1–9). New York, NY: Teachers College Press.

Nasir, N. S., & Cooks, J. (2009). Becoming a hurdler: How learning settings afford identities. *Anthropology & Education Quarterly, 40*(1), 41–61.

Nkopodi, N., & Mosimege, M. (2009). Incorporating the indigenous game of *morabaraba* in the learning of mathematics. *South African Journal of Education, 29*(3), 377–392.

Nunes, T., Schliemann, A. D., & Carraher, D. W. (1993). *Street mathematics and school mathematics*. New York, NY: Cambridge University Press.

Pais, A. (2011). Criticism and contradictions of ethnomathematics. *Educational Studies in Mathematics, 76*, 209–230.

Paraide, P. (2005, March). *The value of Indigenous mathematical knowledge in formal learning*. Paper presented at the PNG Curriculum Reform Conference. Retrieved from http://www.pngcurriculumreform.ac.pg/research/evaluation.htm.

Patrick, R. (1999). Not your usual maths course: Critical mathematics for adults. *Higher Education Research and Development, 18*(1), 85–98. doi:10.1080/0729436990180107.

Powell, A. B., & Brantlinger, A. (2008). A pluralistic view of critical mathematics. In J. F. Matos, P. Valero, & K. Yasukawa (Eds.), *Proceedings of the Fifth International Mathematics Education and Society Conference* (pp. 424–433). Lisbon, Portugal: Centro de Investigação em Educação, Universidade de Lisboa and Department of Education, Learning and Philosophy, Aalborg University. Retrieved from http://pure.ltu.se/portal/files/2376304/Proceedings_MES5.pdf.

Presmeg, N. C. (1996, July). *Ethnomathematics and academic mathematics: The didactic interface*. Paper presented at Working Group 21, ICME 8, Seville, Spain.

Presmeg, N. C. (2002). Shifts in meaning during transitions. In G. de Abreu, A. J. Bishop, & N. C. Presmeg (Eds.), *Transitions between contexts of mathematical practices* (pp. 213–228). Dordrecht, The Netherlands: Kluwer Academic Publishers.

Radford, L. (2008). The ethics of being and knowing: Towards a cultural theory of learning. In L. Radford, G. Schubring, & F. Seeger (Eds.), *Semiotics in mathematics education: Epistemology, history, classroom and culture* (pp. 215–234). Rotterdam, The Netherlands: Sense Publishers.

Roberts, T. (1997). Aboriginal maths: Can we use it in school? In N. Schott & H. Hollingworth (Eds.), *Mathematics: Creating the future: Proceedings of the 16th Biennial Conference of the Australian Association of Mathematics Teachers (AAMT)* (pp. 95–99). Adelaide, Australia: Australian Association of Mathematics Teachers.

Rojas, Y. M. (2010). How students learn about data distribution from addressing a problem affecting their community. In C. Reading (Ed.), *Data and context in statistics education: Towards an evidence-based society: Proceedings of the Eighth International Conference on Teaching Statistics (ICOTS8), July, 2010, Ljubljana, Slovenia*. Voorburg, The Netherlands: International Statistical Institute. Retrieved from http://www.stat.auckland.ac.nz/~iase/publications.php?show=icots8.

Shockey, T. L. (2002). Etnomatematica de uma classe profissional: Cirurgiões cardiovasculares [Ethnomathematics a Professional Class: Cardiovascular Surgeons]. *Bolema, 15*(17), 1–19.

Skovsmose, O. (2004). *Critical mathematics education for the future*. Retrieved from http://www.icme10.dk/proceedings/pages/regular_pdf/RL_Ole_Skovsmose.pdf.

Skovsmose, O. (2005). Foreground and politics of learning obstacles. *For the Learning of Mathematics, 25*(1), 4–10.

Stillman, G., & Balatti, J. (2001). Contribution of ethnomathematics to mainstream mathematics classroom practices. In B. Atweh, H. Forgasz, & B. Nebres (Eds.), *Sociocultural research on mathematics education: An international perspective* (pp. 313–328). Mahwah, NJ: Lawrence Erlbaum.

Stocker, D., & Wagner, D. (2007). Talking about teaching mathematics for social justice. *For the Learning of Mathematics, 27*(3), 17–21.

Sundefeld, M. L. M. M., Homse, L. C., Prieto, A. K. C., & Rodrigues, M. A. B. (2010). The opinion of the family about the performance of the schoolchild bringing knowledge to his/her family: Statistics on prevention of mouth cancer. In C. Reading (Ed.), *Data and context in statistics education: Towards an evidence-based society: Proceedings of the Eighth International Conference on Teaching Statistics (ICOTS8), July, 2010, Ljubljana, Slovenia*. Voorburg, The Netherlands: International Statistical Institute. Retrieved from http://www.stat.auckland.ac.nz/~iase/publications.php?show=icots8.

Tomlin, A. (2002). "Real life" in everyday and academic maths. In P. Valero & O. Skovsmose (Eds.), *Mathematics education and society: Proceedings of the Third International Mathematics*

Education and Society Conference (pp. 1–9). Copenhagen, Denmark: Centre for Research in Learning Mathematics.

Valero, P. (2010). Mathematics education as a network of social practices. In V. Durand-Guerrier, S. Soury-Lavergne, & F. Arzarello (Eds.), *Proceedings of the Sixth Congress of the European Society for Research in Mathematics Education, 28th January to 1st February 2009, Lyon (France)* (pp. LIV–LXXX). Institut National de Recherche Pèdagogique. Retrieved from http://www.inrp.fr/editions/editions-electroniques/cerme6/plenary-2.

Vithal, R., & Skovsmose, O. (1997). The end of innocence: A critique of "ethnomathematics". *Educational Studies in Mathematics, 34*, 131–158.

Wedege, T. (2010). Ethnomathematics and mathematical literacy: People knowing mathematics in society. In C. Bergsten, E. Jablonka, & T. Wedege (Eds.), *Mathematics and mathematics education: Cultural and social dimensions: Proceedings of MADIF7, The Seventh Mathematics Education Research Seminar, Stockholm, January 26–27, 2010* (pp. 31–46). Linköping, Sweden: Svensk Förening för Matematikdidaktisk Forskning (SMDF).

Wenger, E. (1998). *Communities of practice: Learning, meaning, and identity*. Cambridge, UK: Cambridge University Press.

Willis, S. (1998). Perspectives on social justice, disadvantage and the mathematics curriculum. In C. Keitel (Ed.), *Social justice and mathematics education: Gender, class, ethnicity, and the politics of schooling* (pp. 1–19). Berlin, Germany: Freie Universität.

Chapter 7
Critical Perspectives on Adults' Mathematics Education

Jeff Evans, Tine Wedege, and Keiko Yasukawa

Abstract Adults' mathematics education (AME) as a field of study and practice displays a broad range of settings for teaching and learning and for research. At the same time, its activities develop in a dynamic context of globalization, competition, and social insecurity. AME is faced with the same struggle for its justification, between humanistic and human capital goals of education, that adult education and lifelong education have been facing over the last half-century. This struggle is reflected in AME practice, research and policy. In this chapter, we formulate critical perspectives for examining AME in these three dimensions with a view to helping ourselves and others to clarify and act in crucial areas. Thus, we examine multiple and contested meanings of key terms like numeracy, and how definitions vary depending on whether they seek to foreground the individual learners' needs or particular economic imperatives (for example, labour market needs). We illuminate how such variable definitions are experienced by AME learners and practitioners, and how they lead us to problematize ideas such as "the transfer of learning" of mathematics, for example, from school to work, and from formal to non-formal or informal learning situations. It is timely now, when a new international survey of adults' skills, the OECD-sponsored Programme for International Assessment of Adult Competencies (PIAAC) is being conducted, to question what these surveys can tell us for the development of AME as a field, and what alternative questions we need to be pursuing independently.

J. Evans (✉)
Middlesex University, London, UK
e-mail: J.Evans@mdx.ac.uk

T. Wedege
Malmö University, Malmö, Sweden
e-mail: tine.wedege@mah.se

K. Yasukawa
University of Technology, Sydney, Sydney, Australia
e-mail: keiko.yasukawa@uts.edu.au

Introduction

Adults' mathematics learning as a focus of research has developed in relation to a range of educational practices in formal, non-formal and informal settings, many of which have not been focussed primarily on learning mathematics. But, once established and institutionalized, a field of practice like adults' mathematics education (AME) becomes partially dependent on the development of ideas, pedagogic strategies, and solutions to problems that can be developed by research. As we will argue in the next section, the two main pillars of AME research are the fields of mathematics education and adult education. But, although the development of the field of practice is an important criterion for relevance, it is crucial for research to be able not only to solve problems in the field of practice but also to review critically and to reformulate these problems (Olesen & Rasmussen, 1996; Wedege, 2004).

In the first *International Handbook of Mathematics Education*, the chapter "Adults and Mathematics (Adult Numeracy)," by Gail FitzSimons, Helga Jungwirth, Jeurgen Maaß and Wolfgang Schlöglmann (1996), introduced the field with some of its main problems, such as adults' beliefs about and attitudes towards mathematics, and important distinctions like mathematics versus numeracy and school mathematics versus out-of-school mathematics. It also pointed to the heterogeneity of the field—in terms of learner as well as teacher diversities. The complexity of the field—in research, in practice, and in policy terms—was further emphasized in the chapter by Gail FitzSimons, Diana Coben and John O'Donoghue (2003) in the *Second International Handbook of Mathematics Education*. In that chapter the authors analyzed the field of "lifelong mathematics education" in terms of its social, cultural and economic dimensions and from the perspectives of globalization and lifelong learning (LLL).

Clarification about our use of the term "adult" is needed. Official definitions often define an adult as someone who has finished compulsory schooling, or who is 15+ or 16+ years; in labour statistics, attention may focus only on those up to 65 years (or retirement age). But the definition of an "adult" is not straightforward, as has been shown in the international research forum Adults Learning Mathematics (e.g., Safford, 1999), and in recent ICME topic study groups (e.g., Wedege, Evans, FitzSimons, Civil, & Schlöglmann, 2008). In this chapter we adopt a broad understanding of *adults* as including people of a wide range of ages, who:

- participate in a substantial range of social practices, such as working (or seeking work), parenting, caring and housework, budgeting and organizing consumption, voting; and
- are conscious of having social or political interests (cf., Wedege et al., 2008).

In particular, this definition includes adults over 65 years of age and many adolescents.

Mathematics education, too, is to be understood in a broad sense, including formal, non-formal and informal learning (UNESCO, 2000). We can, following FitzSimons et al. (2003), distinguish Formal Adult Mathematics Education (FAME), from Non-Formal Mathematics Education (NFAME), and Informal Adult Mathematics Education (IFAME). FAME normally has a well-specified curriculum, and usually results in certification. NFAME focusses on practical knowledge and skills of fairly immediate usefulness to particular learners, for example in workplace training, and may not result in assessment or certification. IFAME includes the lifelong processes by which every person accumulates knowledge, skills, attitudes and insight from experiences within their environments.

In Coben's (2006) reflection on issues highlighted by the *Second International Handbook* chapter, one task pinpointed as necessary for developing a field-specific framework for AME was a specification of the key concept of numeracy. Strictly speaking, it is not correct to talk about "the concept of numeracy" because numeracy is a contested notion and there are a number of plausible, yet different, definitions (Coben et al., 2003).

There are important reasons for this. Though introduced to policy debates in the UK in the late 1950s, the term "numeracy" only came to the fore in the 1980s. By then, policy makers had become concerned that adults should be able to use mathematics in "real contexts," especially work (see, e.g., Cockcroft, 1982).

At the same time, adult literacy tutors were looking to develop their teaching to address mathematical skills which would complement literacy skills. And progressive educators were seeking to draw on research displaying "everyday mathematics" capabilities of ordinary people, to foreground their knowledge and experience while critiquing élitist notions of education and of mathematics (e.g., D'Ambrosio, 1985; Lave, 1988; Nunes, Schliemann, & Carraher, 1993). As Johnston and Yasukawa (2001) recounted:

> Frustrated with a mathematics whose history kept it within strong disciplinary boundaries … we colonized numeracy, permitting, indeed requiring, it to be a bridge between mathematics and society. (p. 291)

It is difficult to disentangle the differing notions of numeracy proposed from the underlying (and competing) ideologies: we attempt to do this by examining different examples of studies using the concept and by problematizing their assumptions. Nevertheless, the different approaches share some concerns: for example, in most, there is a desire to address the problem of the transfer of knowledge between differing contexts by emphasizing an idea of numeracy as "bridging" mathematics and adult life in some way.

Several definitions of numeracy are given in this chapter. We can make sense of the differences, by analyzing key dimensions on which they might vary. One is whether numeracy is considered as mainly low-level "basic" skills: some views focus on lists of skills at the "lower levels" of a skills hierarchy. In contrast, other researchers (e.g., Hoyles, Noss, Kent, & Bakker, 2010; Nunes et al., 1993) describe adults' use of more advanced modes of mathematical thinking which might be called numeracy.

Some approaches see numeracy as able to be (a) defined generally within a professional group, across a country, or even transnationally and (b) measured/tested in a standardized way. We might call this a generalizing approach—starting either with claims about societal and/or labour market requirements for adults' mathematical competences, or with demands from the academic discipline. In contrast, some prefer a subjective approach—starting with adults' own perceived needs for mathematical competences and their beliefs and attitudes towards mathematics.

Another related dimension might be called contextuality: the extent to which the context of mathematical/numerate thinking is considered crucial in characterizing it. The term "basic" is sometimes taken to mean "abstracted from any context." However, most adults are interested in their use of mathematical ideas in specific contexts, as are their community and work colleagues. In AME research and in practice, the emphasis on contextuality is a strong feature of most definitions of numeracy, and it relates to the problem of transfer mentioned above.

This definition from Denmark (see also Wedege, 2010a) stresses contextuality, particularly its societal aspects:

> *Numeracy* consists of functional mathematical skills and understanding that in principle all people need to have. Numeracy changes in time and space along with social change and technological development. (Lindenskov & Wedege, 2001, p. 5)

Another dimension relates to whether numeracy is seen largely as a cognitive skill, or whether the affective aspect is also seen as crucial. Much AME research emphasizes the affective—beliefs, attitudes, emotions, and motivations.

As with numeracy, lifelong learning (LLL) is a contested concept. Adopting LLL as a focus means that the rights and the obligations concerning education do not stop with childhood and youth but include adult life. According to Rubenson (2001), in the late 1960s, UNESCO introduced LLL as a utopian-humanistic guiding principle for restructuring education. The concept reappeared in policy debates in the late 1980s in a different form. These debates were driven by interests based on an economistic worldview, emphasizing the importance of highly developed human capital, and science and technology. Thus, from the first to the second generation, LLL had changed from a utopian ideal to an economic imperative—though not without criticism. From the late 1990s, a third generation concept of LLL (sometimes called "inclusive liberalism"—Walker, 2009) was being emphasized, with both active citizenship and employability being presented as important aims. This was the view promoted by the Organisation for Economic Co-operation and Development (OECD), the European Union (EU), and other transnational organizations. A strong connection between economy and education has been promoted and maintained through the economics of education and human capital theory.

Rubenson (2008) argued that the OECD and other transnational organizations have now achieved hegemony over educational discourse in industrialized countries—with respect to the situation in less industrialized regions (see, e.g., Aitchison, 2003). In this chapter, we aim to discuss and to elucidate the dominant human capital discourse, as well as important alternative discourses. The adoption of critical perspectives (from critical theory) helps us to develop an understanding of these often conflicting points of view. There are a number of ways that we aim to open up critical perspectives in this chapter:

- We trace the multiple, ambiguous and contested meanings of key concepts such as numeracy and LLL, and relate them to the sometimes contradictory educational goals held by institutions and groups, and to these groups' exercise of power.

- We challenge dominant discourses, received ideas and conventional formulations of problems (such as "learning transfer"), sometimes bringing to bear widely used criteria (e.g., of methodological validity), sometimes deploying critical theoretical perspectives.
- We stress the ways that historical changes in society's ideology, economic organization, or political power relations may change the meaning that numeracy has for adults in society (Cohen, 1982): for example, reference is made to people's need for a greater degree of numeracy or "financial literacy" as the welfare provisions in some countries are eroded (e.g., Bond, 2000).
- We illustrate the need to document the differential effects of ongoing social and political changes on different groups within society, and aim to foreground issues of social justice and inclusion/exclusion (Rogers, 2006), in terms of various dimensions of social difference.
- We present examples of research and practice that are specific to adults and discuss the need to problematize whether and how ideas from education at the school level can be transferred to adults' mathematical learning/numeracy (cf., Coben, 2006).

Thus, the emphasis in this chapter is on the adult learner with numeracy as a key concept, and the set of research problems is related to adults, mathematics, and lifelong education in a societal context. In the next section, we consider AME as a field of study, and the variety of theoretical approaches deployed. In the two sections after that, we illustrate the wide range of educational practice in AME, and focus on policy issues, in particular those relating to international surveys of adult numeracy. The final section brings together our conclusions. Throughout, we consider the issues from our critical perspectives.

AME as a Field of Study

In this section, we consider AME as a field of study, the variety of theoretical approaches used, and some illustrations of research deploying critical perspectives from the field.

Emergence and Identity of AME as a Field of Study

The research field of AME can be seen as having been cultivated in the "borderland" between the two domains of mathematics education and of adult education from where concepts, theories, methods and findings have been imported and reconstructed. Both adult education (Olesen & Rasmussen, 1996) and mathematics education (Sierpinska & Kilpatrick, 1998) have a strong sense of their own specificity. Multidisciplinarity, the importing and juxtaposing of theories and methods from other disciplines, has been vital in building the research domain of mathematics education. However, interdisciplinarity requires a coherent framework of concepts, so the imported frameworks must be reconstructed and recontextualized (Brousseau, 1986). In AME, both in practice and in research, reconstruction of conceptual frameworks from other disciplines has been a central task (FitzSimons et al., 2003).

The institutional supports for AME research, especially internationally, in terms of journals, meetings of researchers, and reviews of research, are important. These include (a) the international organization "Adults Learning Mathematics—A Research Forum" (ALM) which brings together researchers and practitioners (Coben, O'Donoghue, & FitzSimons, 2000); and (b) Topic Study Groups at ICME congresses that have had AME as their principal focus.

The Australian-based journal *Literacy and Numeracy Studies: An International Journal in the Education and Training of Adults* has been publishing for some 20 years, and reflects the origins of the AME field in adult literacy concerns. A journal dedicated to AME research, *Adults Learning Mathematics—An International Journal*, has been published since 2005. The journal *Numeracy*, an offspring of the National Numeracy Network in the USA, began publishing in 2008.

Work done under the aegis of public agencies has also been important. For example, in England, the National Research and Development Centre for Adult Literacy and Numeracy (NRDC), established in 2002 by the UK government within its Skills for Life strategy, has sponsored many relevant projects. In the USA, the university-based National Center for Adult Literacy and the federally-funded National Center for the Study of Adult Learning and Literacy have both produced several research reports on numeracy (see, e.g., Gal, 2000). In 2004, the Brazilian government created a secretariat in the Ministry of Education dedicated to Lifelong Education, Adult Literacy, Diversity and Inclusion—aiming to reduce educational inequalities by ensuring all citizens have access to education. And there has been useful institutional work in other European countries—for example in Sweden (Gustafsson & Mouwitz, 2004) and cross-European funded networks, e.g., EMMA (see the list of Web sites at the end of this chapter). In 2010, New Zealand established a National Centre for Literacy and Numeracy for Adults.

Thus, it now makes sense to talk about a field of AME which has a legitimate and distinctive place within the mathematics education research domain. We will aim to characterize the field by considering three key aspects: (a) the aims of studies in AME; (b) the theories and research questions that provide the organizing basis for research; and (c) the range of methodologies used.

On the basis of a specific theoretical and methodological approach, research questions concerning issues in the problem field have been formulated. Any research question in AME is, for example, based on a specific conception of adult mathematical knowledge, of how adults learn mathematics, and of justifications for AME (Wedege, 2009). It has been recognized that within AME, the same key concept or term, such as "numeracy" or "adults' everyday mathematical knowledge," may be defined and construed in inconsistent or conflicting ways because of different theoretical commitments, methodologies, aims, interests, or values.

For example, generalizing and subjective approaches, discussed in the Introduction to this chapter, can be illustrated with a series of projects. The generalizing approach was assumed in an influential early British report, *Mathematics Counts* (Cockcroft, 1982), on the mathematical needs of adults. This "Cockcroft Report" asserted that the needs of adults should be taken into account when curricula in schools were being defined. In contrast, Benn's (1997) *Adults Count Too* took a subjective approach, starting with adults' perspectives. Benn argued that mathematics is not a value-free form of knowledge, but is imbued with élitist notions which can often exclude and mystify. She therefore rejected approaches where any problem with mathematics learning is located within the learner—rather than with systemic inequalities and prejudices that may themselves be reinforced by a generalizing approach, and may themselves have effects on individual adults' mathematics learning. However, to understand the affective and social conditions for people's learning processes in mathematics, one has to combine generalizing and subjective approaches, as Gail FitzSimons did in *What Counts as Mathematics?* (2002), which discussed "technologies of power" in adult and vocational education.

Why Study AME?

At the societal level, reasons for research into AME are formulated within specific discourses—within, for example, discourses on transnational policies or human capital theories which position adult numeracy learners according to the definition of numeracy formulated by the discourses.

Another related purpose is to ensure that entrants to/employees in a particular sector meet standards of knowledge or skill required in that work. This matter is taken up in one of our case studies in this section on the standards required for entry to the nursing profession.

Other studies have aimed to describe adults' learning, knowledge and use of mathematics in different settings—some of these settings, though not all of them, are related to work (see, e.g., Evans, Alatorre, van der Kooij, Noyes, & Potari, 2010). Much significant research has been done with adults in this general area. Examples include:

- Studying the distinctive mathematical practices among identifiable cultural groups (D'Ambrosio, 1985).
- Describing the mathematical thinking used by adults in everyday activities, and contrasting it with thinking in school mathematics (Lave, 1988).
- Analyzing mathematics practices of working adults in Brazil, and comparing this mathematics with school mathematics (Nunes et al., 1993).

Discussions of the context of the use of mathematics relate closely to the problem of "learning transfer"—see Chapter 6 of this *Handbook*. This remains a crucial concern for research and learning in a subject claiming wide applicability, like mathematics. The traditional use of the term "transfer" as a metaphor describing a situation where a person carries the learning product from one problem, situation, or institution, to another has been widely criticized in psychology and in adult educational circles (e.g., Hager & Hodkinson, 2009; Tuomi-Gröhn & Engestrom, 2003). Transfer has been reformulated in several ways: for example, in terms of (a) translation across contexts (Williams & Wake, 2007b) or across discourses (Evans, 2000b); (b) as consequential transition, involving a developmental change in the relationship between an individual and social activities (Beach, 1999); and (c) as boundary-crossing of a person across activities (Tuomi-Gröhn & Engestrom, 2003).

Another central aim of AME research has been to elucidate relationships between cognitive and affective aspects of adults learning mathematics. In the early studies of adults, problems were formulated within the affective domain with several foci:

- Mathematics anxiety (e.g., Buxton, 1981; Tobias, 1978/1993);
- Experience of failure and success (e.g., Burton, 1987);
- Empowerment (e.g., Benn, 1997; Frankenstein, 1989);
- Emotions (e.g., Evans, 2000a).

Because of the importance in post-compulsory education of the adult's motivation to learn in ways that are meaningful to him/her, the tendency to emphasize the study of affective factors (beliefs, attitudes and emotions towards mathematics) in AME has always been strong.

How to Study AME?

As the international research field of AME was being established many of the researchers formulated problems and constructed their research questions within theo-retical perspectives and methodologies found in mathematics education (e.g., Bishop, 1988; Ernest, 1991; Skovsmose, 1994). They also adopted perspectives found in adult education (e.g., D'Ambrosio, 1985; Freire, 1972); in developmental or social psychology (e.g., Nunes et al. 1993; Vergnaud, 1988; Walkerdine, 1988); and in social anthropology or sociology (e.g., Lave, 1988; Lave & Wenger, 1991). Studies such as these provided the concepts that formed the basis for early studies within AME.

We now describe and illustrate a range of theoretical perspectives, and related methodologies, in studies currently conducted in AME.

Utilitarian perspectives. In the UK the Cockcroft Report (1982) on school mathematics commissioned research including a street survey of adult numeracy performance and a set of semi-structured interviews - both of these including performance questions and questions about attitudes (Advisory Council for Adult and Continuing Education, 1982). This was followed by a series of studies involving home interviews, culminating in a "Skills for Life" policy (Department for Education and Skills (DfES), 2003). Each survey drew on a specific definition of numeracy as an adult skill, formulated in general terms. The recent international surveys have used a similar definition, test, and measurement approach. Other studies of this type focus on employees in a particular sector, and whether or not they meet standards of knowledge or skill required in that work.

Ethnomathematics perspectives. Within the classic ethnomathematical perspective the approach is subjective almost by definition. It emphasizes adults' "everyday knowledge" (D'Ambrosio, 1985). Most ethnomathematics research programs have focussed on these developing aims (Jablonka, 2003):

- Relating out-of-school practices with school (or college) mathematics.
- Uncovering latent mathematical content hidden or "frozen" (Gerdes, 1996) in traditional artefacts of indigenous peoples.
- Challenging a "Eurocentric" version of the historical origins of mathematical concepts and methods.

Classic ethnomethodological research (e.g., Gerdes, 1996) has tended to use ethnographic approaches. However, other approaches may be combined with ethnography, as in the activist approach of Knijnik's (1993) work with the "landless," which will be discussed later in this chapter.

Situated perspectives. We consider the work of Lave (and colleagues) and that of Nunes, Schliemann and Carraher under this heading, though their theoretical commitments are somewhat different.

Lave has studied the use of mathematics by adults in settings outside the school. Her early work championed *situated cognition*, the idea that knowing, thinking and learning depend in crucial ways on the situation in which they are done. The strong form of situated cognition argues that there is a disjunction between doing mathematics problems in school and numerate problems in everyday life, because these different contexts are characterized by different "structuring resources." And people's thinking is specific to these distinct practices and settings. Accordingly, aiming for the "transfer" of learning from school or academic contexts to outside settings is likely to be fruitless.

Methodologies used included "activity shadowing" (semi-participant observation) of adults shopping, and Weight Watchers, and Piagetian clinical interviews, which posed arithmetical problems in relevant contexts.

Lave's later work has focussed on describing learning within communities of practice, including apprenticeship as a model of situated learning (Lave & Wenger, 1991). This emphasized less the discontinuities between practices, and acknowledged that no practice could ever be completely closed. Lave argued that a community of practice must be understood in relation to other tangential, or overlapping, communities. The approach now consisted of identifying communities of practice which were interdependent (Lave, 1996; Wenger, 1998), and studying the bridges between them, particularly the social relations and identities across them.

In contrast, Terezhina Nunes, Analucia Schliemann and David Carraher have been optimistic about the possibilities of applying learning from school in outside practices—and vice versa. They studied the everyday "mathematics-containing" practices of various communities of workers around Recife in Brazil—for example, carpenters, bookies, farmers, and fishermen (see Nunes et al., 1993). These studies generally began with ethnographic descriptions of the work practices (particularly the numerate aspects), followed by studies employing experimental designs (researcher-controlled allocation of different types of problems) with Piagetian clinical interviews. Here the subject was asked to solve several sets of problems constructed by the researchers.

Nunes et al. (1993) documented clear differences in calculation methods between street mathematics and school mathematics, characterized as using oral and written procedures, respectively. Drawing on the work of Gérard Vergnaud (1988) and others, and going beyond ethnomathematical analyses, they emphasized the mathematical invariants (e.g., symmetry or commutativity) underlying the properties or relations of quantitative concepts—even when these concepts had different representations in different contexts.

Concerning transfer, Schliemann (1995) concluded:

> Mathematical knowledge developed in everyday contexts is flexible and general. Strategies developed to solve problems in a specific context can be applied to other contexts, *provided that the relations between the quantities in the target context are known by the subject as being related in the same manner as the quantities in the initial context are.* (p. 49, emphasis added)

Developing this idea, evans (2000a) emphasized *both similarities and differences between signs* in meaning-making and facilitating transfer.

Both Lave and Nunes et al. made ground-breaking contributions to understanding mathematical thinking in context—through seeking out and describing numerate thinking in a wide range of contexts relevant to adults.

Cultural–Historical Activity Theory. Cultural–Historical Activity Theory (CHAT) is an extension of the work of Lev Vygotsky (1930), Alexei Leontiev (1978), and other Soviet psychologists from the early 20th century. This approach was further developed by Michael Cole (1996) and Yrjö Engeström (2001), among others. Here the context for any action is the activity in which the subject is engaged (Engeström, 2001). Activities are oriented toward collective motives, which have arisen in the course of cultural historical development, and are organized in the triplet of activity/action/operation (Roth, 2007a).

Some "third generation" CHAT researchers see emotion as a crucial basis for motivation and identity. Identity is related to an individual's participation in collective activity; this relates to individual and collective emotional "valences" arising from face-to-face interaction with others. Cognition and emotion are seen not only as mutually influencing, but also as having "inner connections in activity" (Roth, 2007a).

The preferred research methods in the CHAT approach are ethnographic (participant observation). However other methods have been used. Thus, for example, when Roth (2007b) required more convincing indicators for his claims about emotions, he made systematic measurements of speech intensity and pitch.

Other key concepts in CHAT and related approaches are those of boundary crossing and boundary objects—see our discussion of the recent work of Hoyles et al. (2010) below.

Discursive perspectives. We group a range of approaches under this heading, including many that specify the context of adults' mathematical thinking as social practices. Most of these authors see adults' activity within social practices as regulated by discourse, and as taking form within positions made available by discourse. According to Evans, Morgan and Tsatsaroni (2006), a discourse is a system of signs that:

> provides resources for participants to construct meanings and identities, experience emotions, and account for actions. Discourses specify what objects and concepts are significant and what *positions* are available to participants in the practice ... They also provide standards of evaluation. (p. 210; emphasis in original)

This perspective problematizes the idea of context, and stresses the need to analyze an adult's positioning in multiple social practices. It aims to describe the subject's navigation of general social requirements (crystallized in the roles, resources and constraints offered by discursive practices), and "subjective" goals and impulses (Evans, 2000a).

Many discursive approaches emphasize affect and emotion, which are central constructs in AME research. Some work has aimed to show how individuals' experiences emerged from interactions between a personal history of involvement in discursive practices, and present positionings, including those in pedagogic practices (Evans et al., 2006). Researchers focussing on discursive approaches sometimes discuss affect and emotion within psychoanalytic perspectives—following Lacan's work which emphasizes desire, which, it is argued, permeates the workings of language (see, e.g., Walkerdine, 1988). The term "desire" captures the energy and intensity of emotion, and supports a unified approach to cognition and affect, seeing emotion as "attached" to (chains of) signifiers (words, symbols, etc.) representing ideas.

Another strand of discursive approaches has focussed on having "numeracy practices" and "numeracy events" as the unit of analysis (Street, Baker, & Tomlin, 2005). This approach uses parallel ethnographic approaches from the study of literacy. Street et al. examined numeracy events in the home in order to understand the relationship (or lack thereof) between the numeracy that children were learning from parents and that in the school. Gebre, Rogers, Street and Oepnjuru (2009) used similar methods to study the literacy and numeracy practices of "non-literate" adults in Ethiopia.

Some research based on discursive approaches has used the work of Bernstein (2000) on pedagogic discourse (e.g., Evans, Tsatsaroni, & Staub, 2007; FitzSimons, 2002). Others have used the work of Foucault, focussing on knowledge and plays of power (e.g., FitzSimons, 2002; Reis & Fonseca, 2008).

Case Studies Using Critical Perspectives in Research on AME

As discussed in the introduction to this chapter, critical perspectives on any theme are not only concerned with societal and individual ways of coping with economic, social and technological change, but also with related equity and inclusion considerations. Some draw on critical theories to question received notions of numeracy or other key concepts, or accepted beliefs about the functioning of a social system, especially mechanisms or groups within that system. Overall, they problematize issues about dominant perspectives, particularly issues from policy discourses of recent decades. In order to show how critical perspectives have been deployed in a range of AME research investigations we have selected four studies of adults' mathematics learning or usage within specific social practices. These studies were concerned with (a) adult numeracy students; (b) trainee nurses; (c) financial advisers and industrial engineers; and (d) landless peasants.

Motivation of adults returning to study mathematics. For Swain, Baker, Holder, Newmarch, and Coben (2005), a central research question concerned students' motives for attending—and continuing with—adult numeracy classes. The dominant view at the time was that adults would, or should, be motivated to take further basic skills classes in order to cope better with numerical challenges arising in work contexts, and in everyday life.

The researchers selected three Further Education colleges in southern England, involving 80 adult learners attending stand-alone numeracy classes. Their methodology included (a) classroom observation and semi-structured interviews by the project leader; (b) interviews, session plans and field-notes made by three teacher-researchers; and (c) analyses of diaries and photographs produced by the students to record numerate experiences outside class (Swain et al., 2005).

Theoretical resources included:

- Theories of identity about how people's beliefs about self have an influence on their perceptions of the world.
- Sociological theories of Bourdieu (e.g., 1986), whose concept of habitus (dispositions acting subconsciously to organize people's social experiences) supports the investigation of identity, and whose concepts of cultural capital (knowledge and skills acquired largely through education) and social capital (resources gained via "connections" with particular groups) explain how people gain power and status in society.
- Lave's (1988) position that out-of-school practices should not be seen as merely the application of school techniques.
- The idea of a learning career (Swain et al., 2005).

The main findings were that reasons for joining, and continuing to attend, numeracy classes were varied and complex. The students' three main motivations were (a) to prove that they could succeed in mathematics; (b) to achieve understanding and engagement; and (c) to help their children with homework. Less mentioned were the need to gain a qualification and to help their functioning in everyday life. By reflecting on data both from the whole sample and from individual case studies, the researchers were able to analyze influences on motivations from social class, ethnicity, and especially gender factors.

Numeracy for nurses. Numeracy is a key skill for professional nurses—it pervades many aspects of their professional practices, such as calculating fluid balance, drug dosages and intravenous drip rates (Hoyles, Noss, & Pozzi, 2001; Hutton, 1998). Coben (2010) and her colleagues referred to "a growing literature revealing a lack of proficiency amongst both students and registered nurses" (p. 13) when solving written tests of "relevant mathematics."

Coben (2010) pointed out that a professional registration body for nursing in the UK required nursing students to achieve 100% in a test of numeracy before they would be permitted to register as nurses. However, other than that stipulation, there was no recognized standard for numeracy for nursing, and therefore it was difficult to determine which skills required development, or to ascertain when competences had or had not been achieved. Hence, it could be argued, a multiplicity of tests, processes and criteria, which may have been neither reliable nor valid, were being developed and deployed in pre-registration nursing programs throughout the UK. In this context, a system of high-stakes testing, using instruments of doubtful quality, threatened both to undermine social justice, and to fail to solve the problem of the "safety-critical nature of nursing" (p. 13).

In current projects, Coben and her colleagues are motivated by the need to avoid a proliferation of numeracy tests, especially if based on simplistic notions of "competency." Instead, they proposed a notion of numeracy for nursing based on what Gigerenzer, Todd, and ABC Research Group (1999) called "fast and frugal heuristics," within a framework expecting bounded rationality, rather than complete rationality. This view of heuristics is aligned with Coben's (2000a) characterization of numeracy:

> To be numerate means to be competent, confident, and comfortable with one's judgements on *whether* to use mathematics in a particular situation and if so, *what* mathematics to use, *how* to do it, what *degree of accuracy* is appropriate, and what the answer *means* in relation to the context. (p. 35, emphasis in the original)

On this basis, the project team designed ICT-based simulations of practical assessment situations aiming to meet a set of conditions for measurement validity, reliability and authenticity. Team members believed that such an approach would avoid inequitable outcomes—which had disadvantaged particular social groups—and would retain an "enablement" or formative focus (Coben, 2010; Coben et al., 2010).

Techno-mathematical literacies for intermediate-level professionals. Hoyles et al., (2010) conducted a large project on mathematics in workplaces in the manufacturing and financial service sectors in the UK. Their initial focus was on the use of mathematics in these workplaces. However, using historical and sociological evidence about major shifts that have occurred in the workplace—including increased competitive constraints on UK firms, increasing customer demand for "customization," and a need for more flexible customer communication—they showed that the type of mathematical skills required for work had greatly changed. Thus, Hoyles et al. thought it appropriate to adopt a generalizing approach for the study.

According to Hoyles et al. (2010), the notion of "numeracy," as it is often used, is insufficiently linked with developing technologies in specific workplaces. On the other hand, "mathematics" is rich enough, but not perceived as relevant, to workplace tasks faced by managers and intermediate employees (see also Noss, 1998). What matters are Techno-mathematical Literacies (TmLs), abilities which allow the adult employee to understand and communicate fluently in the language of mathematical inputs and outputs to specific technologies. Hoyles et al. (2010) argued:

> Much of the discussion around "skills gaps" and the non-transferability of school mathematics misses the essential characteristics of the knowledge required in technology-mediated work, where there has been a shift in requirement from fluency in doing explicit pen-and-paper calculations, to fluency in using and interpreting outputs from IT systems in order to informate workplace judgements and decision-making. (p. 7)

Hoyles et al. (2010) proceeded to show how the concept of TmLs had been developed by analyzing the limitations of the concept of numeracy, and this resulted in the construction of a new object of study. Their research had three main aims:

1. To understand the TmLs required by employees at different levels in four different industrial and commercial sectors.
2. To identify specific cases in companies of techno-mathematical skills gaps.
3. To design, in collaboration, learning resources to help employees develop new skills in order to work more effectively, and to work with relevant sector and training organisations on policies related to forms of qualification and accreditation. (p. 19)

The methodologies used included a combination of sociological/historical analysis, plus ethnographies of workplaces for the first aim; and a design phase for the second. With respect to the third aim, the researchers designed interventions, making use of software tools that "adapt or extend symbolic artefacts [for example a graph] identified from existing work practice, that are intended to act as boundary objects, for the purposes of employees' learning and enhancing workplace communication" (p. 17).

An example of results from the ethnographic phase is given in a discussion of the problems experienced by those customer-service employees in pensions companies who did not understand the calculations involved in the predicted value at retirement of clients' pension investments. One result of this situation was that client queries were often slowly and poorly dealt with. In policy terms, the need for an educational/policy response in the workplace, and in the sector, was due to the fact that many UK workers in financial services faced the prospect of wide-scale outsourcing of their jobs by companies to other countries where employees were typically more mathematically skilled and significantly cheaper to employ—an immediate skills problem which the formal education system would not be likely to address. The researchers argued that the shift that had occurred in the type of mathematical skills required for work had not yet fully been recognized by the formal education system or by employers and managers, and was therefore not being addressed in educational and policy debates about "numeracy."

Landless peasants. The Landless Movement is the largest social movement in Latin America. It was estimated that, in 2003, the Movement comprised 1.5 million landless members organized in most states in Brazil. Research and pedagogic work with this Movement has been the basis of much of the activity of Knijnik (2007), and her research group (see Bicudo, Knijnik, Domite, & Fonseca, 2010).

In the first and second phases of her work, drawing on ethnomathematics and the work of sociologists such as Pierre Bourdieu, Knijnik (2007) problematized the dichotomy between élite academic mathematics and the "popular" mathematics practised by rural workers, but which was often not regarded as socially legitimate. From Bourdieu, Knijnik (1997) utilized concepts of cultural capital and social capital, in investigating the traditions, practices and mathematical concepts of the landless peasants. She undertook the pedagogical work needed for them to be able to interpret and decode their knowledge, to acquire the knowledge produced by academic mathematics, and to establish comparisons between the two kinds of knowledge. In this way, she was able to analyze the use of the two kinds of knowledge and the power relations involved between them.

In the latest phase of her work, she has developed an ethnomathematical approach using poststructuralist theorizations from Foucault to investigate more deeply conflictive and unstable aspects of culture, and associated power relations, differences, and eurocentrism in academic and school mathematics discourses (Bicudo et al., 2010; Knijnik, 2007). According to Knijnik (2007):

> When they come to adult education projects, their peasant culture comes with them, even when the school curriculum tries to impose a sort of "forgetfulness" about who they are, [and] the grammar they use when adding, subtracting, multiplying and dividing. When this subtle imposition of denying their culture occurs, it is not surprising to see that it brings with it a resistance process. ... When they go outside school, their peasant mathematics is revived, showing that it can survive the school conservative practices that are bound by only one kind of rationality. ... Maybe it will be possible to enlarge our adult mathematics education world, including other mathematics, other rationalities ... If so then our dreams of solidarity in our societies can be fulfilled. (p. 61)

Summary

This section has drawn attention to the diverse theoretical and methodological approaches used within AME studies. The four case studies exhibited critical perspectives in different ways. Swain et al. (2005), using theoretical resources from sociology to describe adult students' identities and learning careers, criticized received views about the motivations of adults returning to study numeracy.

In the context of nursing education, Coben (2000a) and her colleagues were challenged to combine generalizing requirements for qualifications in society with appreciation of adults' existing competences and subjective needs, including for the need for effective formative assessment as a basis for development. They challenged the conventional "deficit" characterization of nurses' numeracy, and maintained that often the high-stakes testing and assessment programs from which this characterization derived used instruments which lacked reliability, validity, and authenticity.

Hoyles et al. (2010) also focussed on policy issues, beginning from a socio-historical study of changes in the national and international business context. They questioned the usefulness of the idea of numeracy, the conventional formulation of the problem of transfer of numeracy skills, and "the language of skills policy that organises competences into divisions of skills hierarchies" (p. 13). Among their critical theoretical resources, they draw on CHAT.

Working with landless peasants in Brazil, Knijnik (2007) and her colleagues critiqued the taken-for-granted superiority of general knowledge forms over local ones. Drawing on critical theory (e.g., from Foucault), Knijnik elaborated a notion of knowledge discourses, and promoted a way of balancing knowledge of the powerful, with knowledge of the adult grounded in the setting. Putting it another way, they sought to articulate "knowledge of the powerful" into "powerful knowledge" (Young, 2010) for the landless.

AME as Educational Practice

In this section we examine examples of AME practice in different contexts. We consider examples of formal adults' mathematics education (FAME), non-formal learning (NFAME) and informal learning (IFAME), though these categories sometimes overlap. We consider several aspects of these types of AME: (a) the aims/goals of programs and learners; (b) the learners themselves; and (c) curriculum and pedagogy. We then identify and discuss some of the tensions within each of these aspects.

AME in Practice in Different Contexts

FAME in higher education. Hahn (2010) described a "service mathematics" course within a postgraduate professional management apprenticeship program. The course was offered in the context of a partnership between the education provider and a firm, within the French *alternance* system, which aimed to link mathematics to students' professional experience. The pedagogy involved teachers asking students to confront their different conceptualizations of statistical distributions and summary measures, built through multiple experiences at school and at work.

Carreira, Evans, Lerman, and Morgan (2002) reported on a first-year calculus course within an undergraduate business degree course in Portugal. Here, the lecturers focussed on achieving transfer of the mathematics taught and learned into problem contexts in the disciplinary context of economics. They concluded that although transfer can be a problematic notion, careful consideration of the roles of language and the social organization of learning can facilitate the construction of new mathematical meanings in new situations.

In Thailand, a distance-education program was designed to meet the needs of students with diverse educational, socio-economic and age profiles (Boondao &

Chantarasonthi, 2008). The program allowed the learners to choose from different media—textbooks, workbooks, and CD ROMs. The authors attributed the students' satisfaction and achievements to the program's flexibility: it catered for different learning styles, and gave learners control over their own learning. It allowed the students to use self-paced materials and their own choice of learning media. It also provided realistic contexts in problems.

FAME in basic education. Fonseca and Lima (2008) described their work in Brazil as "Youth and Adult Mathematics Education" (see also Bicudo et al., 2010). In Brazil, as in many other countries, basic education must be offered for adults aged 18 years or more, since large numbers historically have been excluded from basic schooling in their youth. The adult courses aim at students learning to handle the "texts" available in their social practices, and to produce their own texts by learning about new "mathematical genres."

In a non-vocational context, Hassi, Hannula, and Saló i Nevado (2010) examined adult basic education, including numeracy, in Finland. One study involved a Folk High School, based on the tradition of liberal education for adults in Scandinavia. There, learners could study subjects similar to those available in secondary schools— but "unlike the curriculum for regular students, adults' curriculum recommends taking into account local circumstances, local history, culture and students' living conditions" (p. 8). Like Swain et al. (2005), these researchers found varied reasons for students' attending classes: to overcome their lack of formal education, for intrinsic interest, or to facilitate access to further courses.

NFAME with parents. Díez-Palomar (2008) offered three reasons why more research focussing on parents as mathematics learners was needed:

- There is a connection between pupils' performance in school mathematics, parental engagement and family involvement.
- The desire to help children with school work provides strong motivation for parents to learn mathematics.
- There is evidence that low-income families usually have fewer opportunities than middle- or upper-class families to engage in their children's education.

The issue is how parents—particularly (but not only) those in minority and working-class communities—learn to help their children do mathematics. A crucial need is to transform parents' own perceptions about themselves as learners and doers of mathematics and to develop teaching innovations that capitalize on students' (and families') everyday knowledge and experiences (Civil, 1999; Díez-Palomar, 2008).

Chodkiewicz, Johnston, and Yasukawa (2005) reported a case study of a financial literacy program conducted in disadvantaged school communities in Australia. The program was conducted concurrently for schools and parent groups. The curriculum for the parents took account of the "taboos, the privacy around money, exploring attitudes to money, people's 'money stories' ... and learning from life experiences" (p. 38), as a way both to establish the parents' contexts for their learning, and to avoid imposing dominant discourses about what competent financial management involved. In this study the curriculum writers' and facilitators' perceptions of financial literacy also needed to be challenged.

NFAME in workplaces. Mathematics education in and for work has been studied from different angles, including preparation for the workplace (see Hahn's example above and the apprenticeship example below), issues of transfer of knowledge from academic to workplace domains (and vice versa - Wedege, 2010b), and ways in which workers learn and are supported in learning in the workplace (see, e.g., *Education Interfaces between Mathematics and Industry: EIMI 2010 Conference, Proceedings,* Araújo, Fernandes, Azevedo, & Rodrigues, 2010; Strässer, 2000).

The view put forward by Hoyles et al. (2010), that new forms of technology-mediated work rendered a lot of the mathematics in work practices invisible (see the previous section), was discussed by Williams and Wake (2007a). As "outsiders and boundary-crossers," Williams and Wake studied workers' practices in manufacturing sites with a view to understanding contradictions between the mathematics taught in college and the (often invisible) mathematics embedded in workplace practices. They described the mathematical knowledge and procedures of work practices and procedures as being "black-boxed," to capture how the history of negotiations and conflicts between different interest groups during the development of new material or symbolic artefacts was lost once they come into use. Williams and Wake's ethnographies in workplaces sought to open these black boxes. They thereby uncovered some of the politics of the workplace that influenced what mathematical knowledge workers were entitled to know, thus complicating the question of what is learnable, in formal education as preparation for work.

In a study that looked at tailoring in Senegal, Shiohata and Pryor (2008) compared experiences of the learners in an apprenticeship and those who took courses in a vocational training centre. Apprentices were first trained in the desired attitude (*comportement*), for working both within the workshop and dealing with customers (p. 192), and then in practical skills for the tailor's workshop, including cutting cloth. These latter skills the apprentices found challenging because "errors cannot be corrected" (p. 193). In contrast, vocational centre trainees attended formal classes emphasizing both theory and practice—though they never actually handled cloth because it was considered too expensive. The researchers found that the practical learning of the apprentices in the workplace appeared to be more influential in the development of vocational identities than the more theoretical focus taken in the training centre.

IFAME. Mathematics learning also takes place in informal settings. One large-scale example is the process of conversion to the euro, which occurred in 17 European countries. Each country had taken initiatives to educate the public in the likely consequences of the change, including the need to develop "price intuition" in the new currency; the several ways to do this involve different forms of numeracy. Mullen and Evans (2010) studied, in the last-but-one country to convert (the Slovak Republic in January 2009), methods used, both by the state to educate citizens and by citizens themselves to cope, during the process.

Yasukawa and Brown (2012) discuss a case study of informal learning in a trade union campaign for better pay and conditions of "casually-employed" academics. Collective learning emerged through the academics deciding to organize around shared experiences of discontent, and working with union delegates as "barefoot mathematicians" [cf., "barefoot statisticians" described by Evans and Rappaport (1998)]. The workers learned how the complex pay formula worked, and documented evidence that led to a dispute, which ultimately resulted in a win for workers. Throughout this campaign, the casually-employed academics understood at first hand the relationship between membership density of casual academics and strength in the union.

Mathematics/statistics learning is also produced by a longstanding social movement, the Radical Statistics Group, which engages its members and interested citizens in critical analysis of statistics, particularly statistics relating to public policy making. This takes place via email list debates, and through a range of publications (see the Radical Statistics Group website at http://www.radstats.org.uk/).

Aspects of AME Practice

Here we examine some of the examples introduced above, according to three aspects: (a) aims/goals; (b) the adult learner; and (c) curriculum and pedagogy. In so doing, the influences of competing interests and agendas, learners' lives and identities, and different approaches to pedagogies and curricula will reveal the political nature of AME practice.

Aims/goals. The "justification problem," the question of why we need mathematics education, has both a generalizing and a subjective dimension. Niss (1996) presented the following three types of reasons from the societal point of view:

1 Contributing to the technological and socio-economic development of society at large.
2 Contributing to society's political, ideological and cultural maintenance and development.
3 Providing individuals with prerequisites which may help them to cope with life in various spheres: education, occupation, private life, social life, life as a citizen. (p. 13).

The five spheres mentioned in the third reason relate broadly to Steen's (1997) dimensions of importance to learners.

Societal aims as expressed in policy discourses and the multi-faceted personal goals of the learner may not always be aligned. Certainly, education is often experienced by adults as a field of tension between felt needs concerning what one wants to learn, or has to learn, and various economic and societal constraints (Illeris, 2003a). "Monica," an adult numeracy student interviewed by Swain et al. (2005), was an unemployed single parent who had to respond to the government demand to go back to work, or else to go into training. Her motives for choosing a numeracy class were to obtain a mathematical qualification and to get a better job, to prove to herself that she was worth something, and to set a good example to her son. Monica's subjective needs were linked to self-esteem.

Such sets of conflicts often serve as a backdrop to adults' learning processes. To educators or policy makers, they sometimes appear as *resistance* to learning—but they can also relate to the students' perceptions of themselves as competent persons without mathematics, and to seeing mathematics as not relevant to their daily activities and life projects (Wedege & Evans, 2006).

The adult learner. In our construction of the adult learner, we consider the three dimensions of the cognitive, affective and social dimensions as important, and as always potentially in tension (Illeris, 2003b). We see the cognitive and the affective as parts of a whole, itself understood as an important aspect of the subjectivity(ies) or identity of the adult who is participating in a range of social practices. Many recent studies in AME include a strong emphasis on the affective dimension, including consideration of motivation (see Wedege & Evans, 2006). This resonates with reports of research discussed in the previous section and reports from the field of practice.

Mathematics life history interviews allow practitioners, as well as researchers, to gain a better understanding of adults' motivations and experiences with mathematics throughout life (Barton et al., 2007; Buerk & Szablewski, 1993; Coben, 2000b; Evans, 2000a; Martin, 2007; Swain et al., 2005; Wedege, 1999). For example, in Coben's (2000b) study, in England, almost all interviewees mentioned the importance of mathematics and of success on mathematics examinations. One familiar theme in these stories was:

> *the door* marked "Mathematics," locked or unlocked, through which one has to go to enter or progress within a chosen line of work or study. This image was often used, reflecting the frequency with which mathematics tests are used to filter entry into training and employment. (p. 54)

Barton et al. (2007) studied "the relationship between lives and ... learning ... over time in a range of settings" (p. 1). They concluded that although the "dominant discourse in the literacy, numeracy and language field is often one of progression to further education, higher qualifications and better jobs ... if you understand what is going on in people's lives over time it becomes clear that this is one of many possibilities" (p. 159). They illustrated this point by offering the case of "Barbara," who moved from being a self-employed business woman to working part time as a care assistant—a move that the dominant discourse might regard as "regressive" but which enabled Barbara to work where she had long wanted to be. They also showed how learning careers for adults are often non-linear: adults may be in and out of formal college learning as other aspects of their life, such as health and family, interfere in both predictable and unpredictable ways.

Thus, both research and practitioners' experiences show that adult students are motivated to study mathematics for many different reasons.

Curriculum and pedagogy. The preceding discussions about the aims for adult mathematics learning, and the goals of the learners, illustrate tensions between accounts from adult learners' perspectives, and the generalizing perspectives coming from educational providers, employers and policy makers. Thus, aiming to construct a comprehensive and meaningful framework for an adult mathematics curriculum or pedagogy may well lead to a highly contested space, with conflicting accounts of the underpinnings to programs experienced by adult learners. However, as indicated in the introduction to this chapter, by taking a critical perspective, we orient ourselves toward examining these tensions, rather than avoiding them.

In a community education setting, Oughton (2008) confronted the fundamental critical dilemma of who decides what should be the curriculum. Her study involved getting the learners, as part of their learning, to articulate what they wanted to learn. However, this exposed the dilemma faced by many who seek to enact an empowerment model of adult education. Although she was aware of the critiques of the national curricula for adult literacy and numeracy— that it was constructed around a "deficit model" of learners—she found that some learners effectively wanted their "deficits" remediated through an experience of the school mathematics they had missed earlier in their lives.

The politics around the determination of the curriculum are encountered not only within formal education but also in NFAME, in workplace numeracy training and in "citizenship training" and financial literacy programs. Hull (1997), Gallo (2004) and Bond (2000) presented arguments about who the curriculum is for, and what is needed in it—with Bond maintaining that the aim should be to develop critical understanding of power relations in workplaces and society at large, and of the ways to build greater worker/learner agency through collective action.

Researchers using ethnographic approaches to study learning needs associated with various workplaces, have often uncovered the complex interactions and tensions between changing work practices and numeracy and literacy learning (Belfiore et al., 2004; Black & Yasukawa, 2011a). In some cases, employees' resistance to replacing ways of working that have meaning to them with new practices that are not understood by them or are not in their interests, has been attributed, by employers, to literacy and numeracy "deficits."

However, not all of the ambiguities and contestations about the determination of curriculum and pedagogy are directly attributable to tensions in power relations between established curricula or employers' agendas. Different understandings of issues like "learning transfer" add to confusion about what the learner is learning and why. This is particularly pertinent in relation to the interface between formal adult education in the higher and vocational education sectors, and work. Education providers, as well as professions and industries, may claim that students studying towards a professional or trade qualification need to learn certain kinds of mathematics. However, this need is not always obvious to students, because their motivations for study are focussed on vocational qualifications, not on learning mathematics (Strässer & Zevenbergen, 1996). Hahn (2010) and FitzSimons (2002) have shown that it is the ways in which mathematics is made realistic and authentic in the curriculum, combined with the pedagogic approaches such as affording agency to learners in shaping problems, that influence the extent to which learners can make connections between the mathematics they are learning and the professional/occupational field with which they identify.

These studies suggest that the notion of learning transfer between formal education and work cannot be understood without understanding how the learners are forming (or not forming) a professional or occupational identity through which they will activate their mathematical knowledge and skills.

The influence of broader social and economic changes on the forms of mathematics and numeracy that adults might learn is played out not only in formal and non-formal education settings but also in informal learning through community development initiatives and social movements. We point to the very specific learning needs that emerged for adults with the euro (Mullen & Evans, 2010). Bond's (2000) approach to financial literacy explicitly challenged neo-liberal ideologies and their implications for lower-income adults: "the aim of any financial literacy program for adults should be to enable them, individually and collectively, to understand and question the way in which financial institutions, the state and personal and household decision-making connect to shape numerous aspects of their daily lives" (p. 76).

Critical Perspectives in AME Practice

The transfer of mathematical learning. This transfer, between school and everyday practices, and in the opposite direction, is, of course, a problem both for research and for practice—see Chapter 6.

A number of ways of "teaching mathematics for transfer" to adults have been discussed in the literature (e.g., Araújo et al. 2010; Carreira et al. 2002; Evans, 2000b; Williams & Wake, 2007a, 2007b). Wake and Williams (2010) explored how curriculum specification and classroom activity in mathematics might be informed by findings from research into the uses of mathematics in workplaces. They re-conceptualized learning "transfer" using Beach's (1999) construct of collateral transition to describe transformations of mathematics required in crossing boundaries. In considering how better to prepare students to use mathematics in different settings, they recognized the need to understand how mathematics was constituted differently in college and in workplaces and suggested how the "academic practice" of mathematics might be developed and enriched in ways suggested by its use by workers.

Social difference and different skills and needs. Addressing social justice issues in practice requires more than well-formulated curricula and well-meaning pedagogy: it also requires that larger systemic problems, and perceptions of these problems, be addressed.

Martin (2007) focussed on the construction of identities at the intersections of two areas of experience: "being African American" and "becoming a doer of mathematics" (p. 147). His narrative interviews with African American adults returning to study mathematics suggested that "issues of racial boundaries, perceived position and devalued social status, meaning-making for mathematics, and identity, assume prominence" (p. 148). The initial motives for studying mathematics for one student,

Keith, were job improvement and helping his children at school. However, Keith came to see that his status and identity, and the meanings assigned to them by "Whites," created boundaries that limited his opportunities in society, and particularly in mathematics. Thus, his struggle for mathematics literacy can be seen as a part of a general struggle for literacy and for freedom from prejudice and discrimination.

For Shiohata's and Pryor's (2008) tailoring apprentices and trainees, access to workplaces was very gendered. Although the men had access, through apprenticeships, into an authentic but very conservative way of working, the women could only train in fee-paying training centres where the focus was on theoretical knowledge designed to enable textile workers to work in the globalized and technological workplace of the future. Social differences in this case led to different future opportunities for apprentices and trainees.

Policies aiming to increase access to higher qualifications and skills do not necessarily affect people's lives in the way intended. In a study of embedded literacy and numeracy support in vocational education and training, Black and Yasukawa (2011b) interviewed a trade teacher of young Australian Aboriginal men in an animal husbandry course. This trade teacher explained that many of the learners completed the initial course, but were unwilling to proceed to the next course (which promised access to better job opportunities) because it included aspects of managing teams of workers. The young men were concerned that they could find themselves in a culturally problematic situation of having to "manage" elders in their communities who were working in the pastoral industry. Those concerned with social justice need to remember that individual achievements not valued by the individual's community may create new tensions that the individual will have to negotiate.

From a social practices perspective, AME must start from the everyday numeracy practices of learners. This means that social differences can give rise to different numeracy practices—see, for example, Jorgensen (2011) on younger learners, Houssart (2007) on older learners, and Henningsen (2008) and Reis and Fonseca (2008) on gender. Inevitably, it seems, the generalizing views of mathematics that smooth out these differences, and pedagogical practices that seek to affirm the knowledge and skills that learners bring from the particularities of their lives and culture, will co-exist in tension.

"Invisibility" of mathematics. Practitioners and researchers are often surprised to find that if one asks adults whether they use mathematics in everyday life, or in the workplace, the answer is often "No!" Some reasons for this emerged from the mathematics life history interviews described above. In those interviews, adults spoke about their mathematical experiences throughout life—both those that were explicitly mathematical (such as being taught subtraction at school, or budgeting as an adult) and those where mathematics was implicit (such as knitting or judging distances when driving). These interviews suggested a theme of "invisible mathematics—the mathematics one can do, which one does not think of as mathematics—also known as common sense" (Coben, 2000b, p. 55). Thus, for those people who have never perceived themselves as successful in mathematics,

mathematics was always what they could *not* do. The meaning of "invisible" is here subjectively invisible, that is to say, people do not recognize as mathematics the quantitative reasoning that they do.

This phenomenon appears to be widespread. If so, it is important in its effects on the beliefs and motivations of learners, and especially on their confidence. It is not surprising, then, that a number of mathematics educators working with adults have tried to counter it (see, e.g., Keogh, Maguire, & O'Donoghue, 2010).

Jungwirth, Maaß and Schlöglmann (1995) made a different, but related, distinction based on "visibility in the curriculum": the distinction was between courses in which mathematics is explicitly taught, and "mathematics-containing" courses in which mathematical concepts and methods were used implicitly but not explicitly. Both senses are different from that of "objectively invisible" mathematics, that is to say, mathematics hidden in technology. This latter sense was used by some researchers (e.g., Araújo et al., 2010; Noss, 1998).

Summary

In this section we have illustrated the many promises and tensions in adults' mathematics education practice that warn against attempts at simplistic characterizations of the field. AME takes place in formal education settings such as in higher education and further education sectors. However, within those sectors, the institutional aims of mathematics range from specialized mathematics degrees, to service teaching for other disciplines (e.g., Hahn, 2010), to numeracy courses for adults (Oughton, 2008). In NFAME, parents engage in learning numeracy, for example in a financial literacy program complementary to their children's studies (e.g., Chodkiewicz et al., 2005). Mathematics is learned at work in both non-formal and informal ways (Wake & Williams, 2010). In community settings and social movements, both non-formal and informal adult education facilitates learning such as financial literacy (Bond, 2000, Mullen & Evans, 2010) and critical citizenship (see the Radical Statistics website).

However, AME can be characterized not just in terms of the type of education provision (formal, non-formal and informal) or the institutional aims for the programs. It is critical to examine adult mathematics learning from the perspectives of the learners and their goals, and the curricular and pedagogical responses to them.

Engagement in mathematics learning at all ages can also be planned, incidental, or necessitated by larger societal changes, such as the introduction of the euro currency. But it can be difficult to nurture because of resistance engendered by difficult school experiences or by the invisibility of mathematics in many work and other social practices.

At a time when a crisis narrative is being created in many countries as a result of poor national performance on international surveys such as PISA, governments have responded with strategies that tend to smooth out, and hide, the rich and diverse realities of adult learners' lives, needs, aspirations, and constraints. Therefore, a critical perspective is needed to ensure that learners and their learning are not lost in the generalizing stories that are written and told about adults' mathematics education.

The Policy Context of AME

In previous sections, we have gained some insight into the policy contexts of AME. Here we consolidate those insights and consider the role of international adult surveys in policy developments.

The Changing Policy Context

So far, we have shown the need to problematize or contest definitions of key concepts, such as numeracy and LLL, within differing discourses; this can be extended to terms like "globalisation" (FitzSimons, 2002); and "competence" (Moore, 2007). We have also described the competing policy discourses—labelling the two main alternatives "human capital" and "humanistic" in the introduction to this chapter—and have noted changes in these definitions and discourses over time.

In particular, we have kept in perspective the generalizing versus subjective perspectives on policy, and the challenges for learners and practitioners in navigating the tensions between them. There seems to be broad consensus that numeracy, of some kind, is needed by adults in modern societies. However, we note that policy discourses in countries like Australia, England and the USA refer constantly to numeracy (and literacy) as "essential" or "foundation" skills, implying that the same types of skills are needed by all groups within society, or even across different societies. In this chapter we have problematized that point of view.

International Surveys of Adults' Skills

The formal level of literacy, published for every country worldwide (United Nations Development Programme, 2009, Table H) is a basic social indicator, often used in discussions of "level of development." Yet many governments now feel the need to monitor functional, and not only formal, literacy of a population. For example, the UK completed its first Skills for Life Survey in 2003 (Department for Education and Skills, 2003). Since 2001, the Brazilian National Functional Literacy Index (INAF), sponsored by non-governmental organizations (NGOs), has regularly monitored functional literacy.

Since 1990 there has been a series of international studies of adults' literacy and numeracy skills designed to inform governments. These have included:

- The International Adult Literacy Survey (IALS) conducted by OECD in 1994, 1996 and 1998.
- The Adult Literacy and Lifeskills Survey (ALL) conducted in 2003 and 2005.
- The Programme for the International Assessment of Adult Competencies (PIAAC) first conducted during 2011 and 2012.

IALS and ALL. The report on the first cycle of IALS elaborated on the reasons for undertaking the survey. It was stated that the production and use of knowledge was important:

> But the measurement of knowledge and skills and of their benefits is still imperfect ... We need to understand the value of competencies ... during different phases of the lifespan, so as to make informed decisions about human capital investment. ... education provides many benefits, including social cohesion ... (OECD & Statistics Canada, 1995, pp. 5–7)

The concerns of the OECD group of highly industrialized countries clearly went beyond the simple proportion of citizens qualifying as formally "literate." IALS was based on a human capital approach, which focussed on the social "return" from investing in peoples' attainment of qualifications, at all levels of education. Though the introduction referred to social benefits like social cohesion, the main concern seems to have been economic and educational efficiency.

In IALS, three measures (for dimensions of literacy) were produced for each respondent. "Numeracy," as such was not measured, but some indications of it could be inferred from results in Quantitative Literacy, and to a lesser extent, Document Literacy.

For ALL, a measure of Numeracy was created to have a much wider "breadth of mathematical skills and purposes" (Gal, van Groenestijn, Manly, Schmitt, & Tout, 2005). In contrast with IALS, which had 19 participating countries, ALL's two stages had only 7 and 5 participating nations, respectively.

Programme for the International Assessment of Adult Competencies

At the same time as it was developing the much-publicized PISA (Programme for International Student Assessment), in the late 1990s, OECD was also commissioning its PIAAC study. PIAAC's objectives were summarized by Andreas Schleicher (2008), of the Education Directorate at OECD, as helping participating countries to:

- Identify and measure differences between individuals and across countries in key competencies.
- Relate measures of skills based on these competencies to a range of economic and social outcomes of policy relevance to participating countries, including *individual outcomes* such as labour market participation and earnings, or participation in further learning and education, and *aggregate outcomes* such as economic growth, or increasing social equity in the labour market.

- Assess the performance of education and training systems, and clarify which policy measures might lead to enhancing competencies through the formal educational system—or in the work-place, through incentives addressed at the general population. (pp. 2–3, *emphasis added*)

These objectives offered a "human capital" approach, linked with social concerns. They placed heavy emphasis on comparisons between countries—which presupposed a basically competitive international economic context. PIAAC thus followed the earlier international adult surveys, IALS and ALL, but with some crucial developments (see below). Three skills or competencies were to be measured: Literacy, Numeracy, and Problem Solving in Technology-Rich Environments.

Conception of numeracy in PIAAC. In the conceptual approach used by PIAAC (and PISA), numeracy is seen as a competency, which is "an internal mental structure of abilities and dispositions," made up not only of cognitive skills and a knowledge base but also motivations, attitudes, and other non-cognitive components. It has been defined for the purposes of designing the items as "the ability to access, use, interpret, and communicate mathematical information and ideas, in order to engage in and manage the mathematical demands of a range of situations in adult life" (PIAAC Numeracy Expert Group, 2009, p. 21).

This definition represented an attempt to conceptualize a broad range of adults' mathematical thinking in context. But what is measured by a scale depends both on a conceptual scheme, and an assessment scheme, describing both the tasks used and the modes of administration and scoring (PIAAC Numeracy Expert Group, 2009).

Thus, in order to *operationalize* numeracy, it is, first of all, necessary to specify a number of dimensions of "numerate behaviour," which can be used in the construction of a set of items. PIAAC identified the following four dimensions:

- Context (four types): everyday life, work, societal, further learning.
- Response (to mathematical tasks—three main types): identify/locate/access (information); act on/use; interpret/evaluate.
- Mathematical content (four types): quantity and number, dimension and shape, pattern and relationships, data and chance.
- Representations (of mathematical/statistical information): e.g., text, tables, graphs.

Each item was categorized according to these four dimensions, and its "estimated difficulty" (or "ability level") was also given. The aim was to stipulate the proportions of the items that were from each key dimension (e.g., the proportion of "data and chance" items of moderate difficulty)—in order to try to assure the content validity of the overall set of items that was used in the test (Gal et al., 2005; PIAAC Numeracy Expert Group, 2009). (For illustrative ALL items, similar to those used in PIACC, see Gal et al., 2005.)

Numerate behaviour, in turn, was understood as "founded on the activation of enabling factors and processes," including numeracy-related experience, literacy skills, beliefs and attitudes, and "context/world knowledge" (PIAAC Numeracy Expert Group, 2009, p. 29). Thus PIAAC, more than the earlier international adult surveys, aimed at producing affective and other contextual data that could be related to a respondent's performance. Demographic and attitudinal information were gathered via a background questionnaire, and self-report indicators on the respondent's use of job-related skills at work were also obtained.

Survey design and administration. PIAAC data have been collected in 2011 and 2012. Twenty-five countries were involved, including 18 members of the European Union. In each participating nation about 5,000 adults between 16 and 65 were interviewed. Results will be released in 2013.

Adult surveys cannot rely on "captive populations" of children at school, so PIAAC (like IALS and ALL) combined household survey methods with educational testing methodology. In addition, PIAAC's "default" method of survey administration was by laptop computer. This allowed for adaptive testing, which attempted to assess the initial "level" of the respondent from a few responses, and once that had been achieved then more appropriate items (in terms of difficulty) were put to the person throughout the interview.

Aspects of survey validity. The "curriculum" that was assumed when the PIAAC numeracy instruments were being constructed was implied by the four-dimension definition of numeracy that had been developed. This was for an international survey, and a transnational definition of numeracy was used. One should question how well that definition, and the corresponding items, "fit" adults' lives in any particular country. The four types of context (everyday life, work, societal, further learning) are of course idealizations—rather than actual contexts that any particular respondent might meet in his or her everyday life. Concerning pedagogy, PIAAC related numeracy, for example, to a potentially fruitful combination of informal and formal learning (PIAAC Numeracy Expert Group, 2009). However, in the previous section, we gave several examples of tensions between what is learned formally and what is learned informally.

Computer administration of test questions should help with the reliability of administration across interviews and with reliability of marking. But it raises questions about validity. For example, it is difficult to assess the possible effects on respondents' thinking and behaviour of the on-screen presentation of tasks. These response modes contrast with the ways that many adults' numeracy practices are acted out in the participants' day-to-day lives. Similar problems arise of course for much educational assessment (e.g., assessments associated with PISA).

The validity of the concept of an adult's "level" of numeracy, used in PIAAC and in other national and international surveys, has been challenged. For example, Gillespie (2004), in reflecting on the Skills for Life survey carried out in the UK, argued: "The findings confirm that for many, being 'at a given level' is not meaningful for the individual, as levels embody predetermined assumptions about progression and relative difficulty" (p. 1). This is because many adults have different "spiky profiles," due to distinctive life experiences: thus, for example, some find items of Type A (say, "data and chance") more difficult than Type B items (say, "dimension and shape"), and others find the opposite.

When results are reported, "the minimum level of numeracy needed to cope with the demands of adult life" is sometimes stipulated—but that concept is of questionable worth. Such generalizing claims group together adults with different work, family and social situations—and further assume that the demands are the same across the countries studied.

These sorts of concerns about validity are relevant for all surveys which include assessments, especially those that aim at making comparisons across countries, or over time. Nevertheless, questions of validity must be assessed for every survey or assessment, especially if it is intended that the results will inform policy or practice.

Summary

We have aimed in this section to give an informative and balanced, yet critical, description of the design of the PIAAC survey, and have offered some reflections on methodological validity. Our concern has been to inform readers about what was planned, to alert them to issues that need to be kept in mind when interpreting the results from such studies, and to elucidate the policy context in which PIAAC has been produced. Here we elaborate on the latter.

Possible benefits and risks of international surveys for AME research and practice. These studies, despite their limitations, offer educational planners and researchers new data on some aspects of the competencies of adults, who are mostly outside of formal educational systems, and thus less accessible to researchers. They can help us to understand the effects of formal educational systems, characteristics of their graduates, and relations of performances to categories of respondents—via demographic, attitudinal, and "skills use" data.

Results of international surveys can also provide the context for other types of studies that will supplement or probe survey results. OECD policy is to make available national datasets from the surveys after publication to researchers and policy makers.

However, the results of international surveys usually lead to "high-profile" reporting, by policy-making bodies and by the media. This entails risks. Conceptions that we have seen to be highly contestable, such as numeracy, skill, and LLL, may be narrowed and fixed in public discussions and in subsequent research. This may also be true of the idea of "the adult learner," resulting in an "agreed" concept that ignores the rich diversity we have described previously.

The changing policy context. Whereas, in earlier studies, there was a multiplicity of aims for adult education, featuring both economic/vocational and humanistic goals, increasingly, "under globalisation, educational values tend to be interpreted through neo-liberal imperatives" (Rizvi & Lingard, 2010). The emerging international policy discourse uses a human capital approach—from among all of the social-scientific perspectives that could have been employed—in pursuit of economic efficiency, in a context constructed as one of international competitiveness, brought about by global factors.

In this context, the OECD, a transnational organization, has promoted the collection of data through PIAAC and other international surveys, as a way to support the achievement—and demonstration—of a country's state of competitive readiness. In addition, OECD and the European Union (with 18 of its 27 member-states participating in PIAAC) have taken on a key role in disseminating ideas and practices that will strongly influence national policy-making around the world. Such transnational organizations are already dominant players in constructing "the skills and competencies agenda" in industrialized countries at least (Grek, 2010; Rubenson, 2008).

Thus, PIAAC, and the associated technological and administrative machinery supporting it, have generated data for monitoring the acquisition and updating of adults' competencies, for facilitating international comparisons, and for assessing progress over time. Therefore PIAAC may be implicated in ongoing shifts in the meaning of LLL, and of numeracy, in the globalized environment (Evans & Tsatsaroni, 2011; Rizvi & Lingard, 2010).

Conclusions

Ways in Which AME is Specific and Different from Mathematics Education in General

What is *specific* about the teaching and learning of mathematics and numeracy for adults?

Our survey of AME suggested that associated with the practices of teaching and learning of mathematics/numeracy for adults are a number of "awkward realities" (Coben, 2006, p. 29). These include qualities of the learners:

- The diverse aims and goals that learners bring to learning may not always sit comfortably with the aims and goals of policy makers, or indeed with educators in general.
- The diverse ways in which adults' identities and engagements with mathematics learning mutually shape each other are likely to differ from individual to individual, and from nation to nation.
- The intensity of affective challenges for many adults, when mathematics is included in their educational programs, needs to be taken into consideration.

And educational and social environments can also generate awkward realities:

- There is a wide diversity of professional, vocational and community learning programs within which mathematics is a part, but not the primary focus for most learners.
- In official discourses there are many paradoxical claims made about mathematics being "basic," yet for many adults mathematics is largely invisible in their social practices.

The awkward realities also include characteristics of research peculiar to AME: while children's experiences in learning mathematics are likely to be largely dependent on formal schooling, adults' knowledge and skills in mathematical areas are likely to be more dependent on NFAME and IFAME learning. This makes a number of processes more challenging for AME research:

- It is difficult to provide empirical descriptions of an adult's mathematical knowledge, for it is encoded in different terminologies among different groups of stakeholders.
- Measurement of performance in numeracy is not a simple matter because of the likelihood that, by contrast with school-age students, adults have more "spiky profiles."
- Adults typically participate in a wider variety of practices than school students, and this can affect how "transfer" of learning (transitions or boundary crossing) between different contexts by learners and doers of mathematics is described.
- Affective aspects of adults' positioning, vis-à-vis mathematics, is likely to need greater attention because of their greater life experiences.

These awkward realities further justify the adoption of critical perspectives when examining AME because the key issues are not easily resolved and they affect different groups of adults differently. They are all linked in some ways to questions of power.

Nevertheless, the mutual interdependence between research fields in AME and mathematics education (ME) has been fruitful.

- Frameworks from ME have been adopted for AME research, straightforwardly, or in multi- or interdisciplinary fashion.
- Theoretical contributions from research in AME have been adopted in ME. That is true, for example, in the areas of (a) learning transfer and (b) the importance of affect.

Undoubtedly, cross-fertilization between these fields has been valuable for both AME and ME.

Future Developments in AME as a Field of Study

What kinds of research are needed over the next decade to ensure a balanced focus in adults' mathematics education?

In this chapter we have discussed studies that re-assert the values of alternative approaches in contesting the human-capital approach. Consider, for example:

- Hoyles et al.'s (2010) use of "Techno mathematical Literacies" (TmLs), in flexible IT-supported decision-making by professionals;
- Knijnik's (2007) and colleagues' work with the Landless Movement in Brazil, which has helped adults and researchers bridge academic and "local" knowledge; and
- Barton et al.'s (2007) description of the interaction between adults' lives, learning and identity change, within dynamic socio-economic conditions.

On-going attention to adults' needs is required, within a context of changing dimensions of social difference, e.g., differently evolving skills of those in different age groups. Emphasis ought to be given to demands for re-skilling in a rapidly changing and increasingly competitive labour market. Relationships between learning and identity formation (Swain et al., 2005) mean that the dynamics of changing identities, changing economic and cultural environments, and changing policy constructions of numeracy, will remain important foci for research and practice.

There is a range of positions for the AME researcher to take in a given setting. These include the following:

- Objective *reporter* of what is "really" going on;
- *Producer of accounts* from those engaged in activities in context;
- *Advocate* for social or educational change;
- *Activist*, working alongside those engaged in trying to bring about changes.

These roles are illustrated by the positions taken by researchers and practitioners in the accounts given in this chapter, and clearly relate to ideas of education for social justice.

Future Developments in AME as a Field of Practice

In curriculum, new areas of numeracy (and literacy) are related to emerging social and political issues. These emerging issues include:

- Financial literacy (see "AME as Educational Practice");
- Health-related decision making: interpreting expressions of risk from specific diseases, and likelihoods of success of specific medical interventions (Gigerenzer et al., 1999; O'Hagan, 2011);
- Environmental numeracy: the ability to participate skilfully in often highly quantitative public debates and decision-making, and also to implement informed "sustainability skills" in industrial, home and community settings.

In pedagogy, as larger populations in many countries gain easier access to new forms of information and communication technologies, increasing use of multimodal forms of learning are likely to emerge to support learning outside formal classroom settings. Thus, for example, there will be much more attention given to distance or "blended" learning for university students (Boondao & Chantarasonthi, 2008), and for workplace learning (Hoyles et al., 2010).

It is likely that the "crisis" discourses surrounding knowledge and skills, especially literacy, numeracy, and technology skills, in many countries, will result in numeracy coming to be seen, even more than it is now, as foundational for acquiring new occupational skills. This is related to important questions associated with issues surrounding effective pedagogy for "embedding" numeracy into vocational and professional education in FAME and NFAME. The need to facilitate transfer of learning, and to make visible the mathematics in vocational and professional practices, will increase, and such matters present a big challenge to AME researchers.

Future Developments in AME Policy

1. We aim to avoid numeracy being reduced to a narrow competency, through the dominance of human capital discourses and a narrow range of studies that monopolize the study of adults' competences. Instead we must work towards a richer notion of numeracy, building it as an element of "powerful knowledge" (Young, 2010).
2. We must maintain connections with wider currents of educational and social science research that explore issues relevant to AME research and practices. For example, notions of skill, which have been debated by Moore (2007), and Sennett (2008), need to be further problematized.
3. Numeracy tutors and adult tutors generally have been relatively neglected in most countries' educational systems. They need, and deserve, greater attention and support in responding to the generalizing trends highlighted in this chapter — especially in regard to negotiating tensions between personal philosophies of AME, the demands and needs of the students, and wider educational policies.

Acknowledgements We thank Jürgen Maaß, Iddo Gal, and Anna Tsatsaroni for comments on earlier drafts.

References

Advisory Council for Adult and Continuing Education (ACACE). (1982). *Adults' mathematical ability and performance.* Leicester, UK: Author.
Aitchison, J. (2003). Adult literacy and basic education: A SADC regional perspective. *Adult Education and Development, 60,* 161–170.
Araújo, A., Fernandes, A., Azevedo, A., & Rodrigues, J. F. (Eds.). (2010, October). *Educational interfaces between mathematics and industry: EIMI 2010 Conference Proceedings.* Lisbon, Portugal: Centro Internacional de Matemática. Retrived from http://www.cim.pt/files/proceedings_eimi_2010.pdf.
Barton, D., Ivanic, R., Appleby, R., Hodge, R., & Tusting, K. (2007). *Literacy, lives and learning.* London, UK: Routledge.
Beach, K. D. (1999). Consequential transitions: A sociocultural expedition beyond transfer in education. *Review of Research in Education, 24,* 109–139.
Belfiore, M. E., Defoe, T. A., Folinsbee, S., Hunter, J., & Jackson, N. S. (2004). *Reading work: Literacies in the new workplace.* Mahwah, NJ: Lawrence Erlbaum.
Benn, R. (1997). *Adults count too: Mathematics for empowerment.* Leicester, UK: NIACE.
Bernstein, B. (2000). *Pedagogy, symbolic control and identity: Theory, research, critique.* Oxford, UK: Rowan & Littlefield Publishers.
Bicudo, M. A. V., Knijnik, G., Domite, M. C., & Fonseca, M. C. F. R. (2010). Research in mathematics education in Brazil. In M. M. F. Pinto & T. F. Kawasaki (Eds.), *Proceedings of the 34th Conference of the International Group for the Psychology of Mathematics Education* (Vol. 1, pp. 401–419). Belo Horizonte, Brazil: International Group for the Psychology of Mathematics Education.
Bishop, A. J. (1988). *Mathematical enculturation: A cultural perspective on mathematics education.* Dordrecht, The Netherlands: Kluwer Academic Publishers.
Black, S., & Yasukawa, K. (2011a). A tale of two councils: Alternative discourses on the "literacy crisis" in Australian workplaces. *International Journal of Training Research, 9*(3), 218–233.

Black, S., & Yasukawa, K. (2011b). *Working together: Integrated language, literacy and numeracy support in vocational education and training*. Sydney: Centre for Research in Learning and Change, University of Technology. Retrieved from http://www.rilc.uts.edu.au/projects/working-together.html.

Bond, M. (2000). Understanding the benefits/wages connection: Financial literacy for citizenship in a risk society. *Studies in the Education of Adults, 32*(1), 63–77.

Boondao, S., & Chantarasonthi, U. (2008, July). What mathematical media suit adult students? Paper presented in Topic study group 8: Adult Mathematics Education, 11th international congress on mathematical education (ICME-11). Monterrey, Mexico: ICME. Retrieved from http://tsg.icme11.org/tsg/show/9#inner-documents.

Bourdieu, P. (1986). The forms of capital. In J. G. Richardson (Ed.), *Handbook of theory and research for the sociology of education* (pp. 241–258). New York, NY: Greenwood Press.

Brousseau, G. (1986). Fondements et méthodes de la didactique des mathématiques. *Recherches en Didactique des Mathématiques, 7*(2), 33–115.

Buerk, D., & Szablewski, J. (1993). Getting beneath the mask, moving out of silence. In A. White (Ed.), *Essays in humanistic mathematics* (pp. 151–164). Washington, DC: Mathematical Association of America.

Burton, L. (1987). From failure to success: Changing the experience of adult learners of mathematics. *Educational Studies in Mathematics, 18*(3), 305–316.

Buxton, L. (1981). *Do you panic about maths? Coping with maths anxiety*. London, UK: Heinemann Educational.

Carreira, S., Evans, J., Lerman, S., & Morgan, C. (2002). Mathematical thinking: Studying the notion of "Transfer." In A. Cockburn & E. Nardi (Eds.), *Proceedings of 26th International Conference of Psychology of Mathematics Education Group (PME-26)* (Vol. 2, pp. 185–192). Norwich, UK: University of East Anglia.

Chodkiewicz, A., Johnston, B., & Yasukawa, K. (2005). Educating parents: The EvenStart financial literacy program. *Literacy and Numeracy Studies, 14*(1), 33–47.

Civil, M. (1999). Parents as resources for mathematical instruction. In M. V. Groenestijn & D. Coben (Eds.), *Mathematics as part of lifelong learning. Proceedings of the Fifth International Conference of Adults Learning Maths—A research forum, Utrecht, July 1998* (pp. 216–222). London, UK: Goldsmiths University of London.

Coben, D. (2000a). Numeracy, mathematics and adult learning. In I. Gal (Ed.), *Adult numeracy development: Theory, research, practice* (pp. 33–50). Cresskill, NJ: Hampton Press.

Coben, D. (2000b). Mathematics or common sense? Researching "invisible" mathematics through adults' mathematics life histories. In D. Coben, J. O'Donoghue, & G. E. FitzSimons (Eds.), *Perspectives on adults learning mathematics: Research and practice* (pp. 53–66). Dordrecht, The Netherlands: Kluwer Academic Publishers.

Coben, D. (2006). What is specific about adult numeracy and mathematics education? *Adults Learning Mathematics—An International Journal, 2*(1), 18–32.

Coben, D. (2010). At the sharp end of education for an ethical, equitable and numerate society: Working in a safety-critical context—Numeracy for nursing. In U. Gellert, E. Jablonka, & C. Morgan (Eds.), *Proceedings of the Sixth International Mathematics Education and Society Conference* (pp. 9–22). Berlin, Germany: Freie Universität.

Coben, D., Colwell, D., Macrae, S., Boaler, J., Brown, M., & Rhodes, V. (2003). *Adult numeracy: A review of research and related literature*. London, UK: National Research and Development Centre for Adult Literacy and Numeracy.

Coben, D., Hall, C., Hutton, M., Rowe, D., Weeks, K., & Woolley, N. (2010). *Research report: Benchmark assessment of numeracy for nursing: Medication dosage calculation at point of registration*. Edinburgh, Scotland: NHS Education for Scotland.

Coben, D., O'Donoghue, J., & FitzSimons, G. E. (Eds.). (2000). *Perspectives on adults learning mathematics: Research and practice*. Dordrecht, The Netherlands: Kluwer Academic Publishers.

Cockroft, W. H. (Chairman of the Committee of Inquiry into the Teaching of Mathematics in Schools). (1982). *Mathematics counts*. London, UK: Her Majesty's Stationery Office.

Cohen, P. C. (1982). *A calculating people: The spread of numeracy in early America.* Chicago, IL: University of Chicago Press.
Cole, M. (1996). *Cultural psychology: A once and future discipline.* Cambridge, MA: Harvard University Press.
D'Ambrosio, U. (1985). Ethnomathematics and its place in the history and pedagogy of mathematics. *For the Learning of Mathematics, 5*(1), 44–48.
Department for Education and Skills (DfES). (2003). *The skills for life survey: A national needs and impact survey of literacy, numeracy and ICT skills.* Norwich, UK: HMSO.
Díez-Palomar, J. (2008). Introduction to the special issue parents' involvement in mathematics education: Looking for connections between family and school. *Adults Learning Mathematics—An International Journal, 3*(2), 6–7.
Engeström, Y. (2001). Expansive learning at work: Toward an activity theoretical reconceptualisation. *Journal of Education and Work, 14,* 133–156.
Ernest, P. (1991). *The philosophy of mathematics education.* London, UK: Falmer Press.
Evans, J. (2000a). *Adults' mathematical thinking and emotions: A study of numerate practices.* London, UK: Routledge/Falmer.
Evans, J. (2000b). Adult mathematics and everyday life: Building bridges and facilitating "transfer". In D. Coben, J. O'Donoghue, & G. E. FitzSimons (Eds.), *Perspectives on adults learning mathematics: Research and practice* (pp. 289–305). Dordrecht, The Netherlands: Kluwer Academic Publishers.
Evans, J., Alatorre, S., van der Kooij, H., Noyes, A., & Potari, D. (2010). Plenary panel: Mathematics in many settings. In M. M. F. Pinto & T. F. Kawasaki (Eds.), *Proceedings of the 34th Conference of the International Group for the Psychology of Mathematics Education* (Vol. 1, pp. 109–139). Belo Horizonte, Brazil: International Group for the Psychology of Mathematics Education.
Evans, J., Morgan, C., & Tsatsaroni, A. (2006). Discursive positioning and emotion in school mathematics practices. *Educational Studies in Mathematics, 63*(2), 209–226.
Evans, J., & Rappaport, I. (1998). Using statistics in everyday life: From barefoot statisticians to critical citizenship. In D. Dorling & S. Simpson (Eds.), *Statistics in society: The arithmetic of politics* (pp. 71–77). London, UK: Arnold.
Evans, J., & Tsatsaroni, A. (2008). Methodologies of research into gender and other social differences within a multi-faceted conception of social justice. *Adults Learning Mathematics—An International Journal, 3*(1), 13–31.
Evans, J., & Tsatsaroni, A. (2011, September). The policy relevance of OECD's Project for International Assessment of Adult Competencies (PIAAC): *Towards a "totally pedagogised society"?* Paper presented at ECER 2011 Conference, Network 23. Berlin, Germany: Policy Studies and Politics of Education.
Evans, J., Tsatsaroni, A., & Staub, N. (2007). Images of mathematics in popular culture/adults' lives: A study of advertisements in the UK press. *Adults Learning Mathematics—An International Journal, 2*(2), 33–53. Retrieved from http://www.alm-online.net.
FitzSimons, G. E. (2002). *What counts as mathematics? Technologies of power in adult and vocational education.* Dordrecht, The Netherlands: Kluwer Academic.
FitzSimons, G. E., Jungwirth, H., Maaß, J., & Schlöglmann, W. (1996). Adults and mathematics (adult numeracy). In A. J. Bishop, M. A. Clements, C. Keitel, J. Kilpatrick, & C. Laborde (Eds.), *International handbook on mathematics education* (Vol. 2, pp. 755–784). Dordrecht, The Netherlands: Kluwer Academic Publishers.
FitzSimons, G. E., O'Donoghue, J., & Coben, D. (2003). Lifelong mathematics education. In A. J. Bishop, M. A. Clements, C. Keitel, J. Kilpatrick, & F. K. S. Leung (Eds.), *Second international handbook of mathematics education* (Vol. 1, pp. 103–142). Dordrecht, The Netherlands: Kluwer Academic Publishers.
Fonseca, M. C. F. R., & Lima P. C. (2008, July). Numeracy practices for tables construction in youth and adult education. Paper presented at ICME-11, Topic Study group 8: Adult Mathematics Education. Retrieved from http://tsg.icme11.org/tsg/show/9.
Frankenstein, M. (1989). *Relearning mathematics: A different third R—Radical maths.* London, UK: Free Association Books.

Freire, P. (1972). *Pedagogy of the oppressed.* Harmondsworth, UK: Penguin.
Gal, I. (Ed.). (2000). *Adult numeracy development: Theory, research, practice.* Cresskill, NJ: Hampton Press.
Gal, I., van Groenestijn, M., Manly, M., Schmitt, M. J., & Tout, D. (2005). Adult numeracy and its assessment in the ALL Survey: A conceptual framework and pilot results. In T. S. Murray, Y. Clermont, & M. Binkley (Eds.), *Measuring adult literacy and life skills: New frameworks for assessment* (pp. 137–191). Ottawa, Canada: Statistics Canada.
Gallo, M. (2004). *Reading the world of work: A learner-centred approach to workplace literacy and ESL.* Malabar, FL: Krieger.
Gebre, A. H., Rogers, A., Street, B., & Oepnjuru, G. (2009). *Everyday literacies in Africa: Ethnographic studies of literacy and numeracy practices in Ethiopia.* Kampala, Uganda: Fountain Publishers.
Gerdes, P. (1996). Ethnomathematics and mathematics education: An overview. In A. J. Bishop, M. A. Clements, C. Keitel, J. Kilpatrick, & C. Laborde (Eds.), *International handbook on mathematics education* (Vol. 2, pp. 909–944). Dordrecht, The Netherlands: Kluwer.
Gigerenzer, G., Todd, P. M., & ABC Research Group. (1999). *Simple heuristics that make us smart.* Oxford, UK: Oxford University Press.
Gillespie, J. (2004, July). The "Skills for Life" national survey of adult numeracy in England. What does it tell us? What further questions does it prompt? Paper presented to Topic Study Group 6: Adult and Lifelong Education, 10th International Congress on Mathematical Education (ICME-10). Retrieved from http://www.icme10.dk.
Grek, S. (2010). International organisations and the shared construction of policy "problems": Problematisation and change in education governance in Europe. *European Educational Research Journal, 9*(3), 396–406.
Gustafsson, L., & Mouwitz, L. (2004). *Adults and mathematics—A vital subject.* Gothenburg, Sweden: National Center for Mathematics Education. Retrieved from http://ncm.gu.se/media/ncm/rapporter/adult-en.pdf.
Hager, P., & Hodkinson, P. (2009). Moving beyond the metaphor of transfer of learning. *British Educational Research Journal, 35*(4), 619–638.
Hahn, C. (2010). Linking professional experiences with academic knowledge: The construction of statistical concepts by sales managers' apprentices. In A. Araújo, A. Fernandes, A. Azevedo, & J. F. Rodrigues (Eds.), *Proceedings of Educational Interfaces Between Mathematics and Industry 2010 Conference* (pp. 269–279). Lisbon, Portugal: Centro Internacional de Matemática.
Hassi, M. L., Hannula, A., & Saló i Nevado, L. (2010). Basic mathematical skills and empowerment: Challenges and opportunities, Finnish adult education. *Adults Learning Mathematics—An International Journal, 5*(1), 6–22.
Henningsen, I. (2008). Gender mainstreaming of adult mathematics education: Opportunities and challenges. *Adults Learning Mathematics—An International Journal, 3*(1), 32–40.
Houssart, J. (2007). They don't use their brains, what a pity: School mathematics through the eyes of the older generation. *Research in Mathematics Education, 9,* 47–63.
Hoyles, C., Noss, R., Kent, P., & Bakker, A. (2010). *Improving mathematics at work: The need for techno-mathematical literacies.* New York, NY: Routledge.
Hoyles, C., Noss, R., & Pozzi, S. (2001). Proportional reasoning in nursing practice. *Journal for Research in Mathematics Education, 32*(1), 4–27.
Hull, G. (Ed.). (1997). *Changing work, changing workers: Critical perspectives on language, literacy, and skills.* New York, NY: State University of New York.
Hutton, B. M. (1998). Should nurses carry calculators? In D. Coben & J. O'Donoghue (Eds.), *Proceedings of ALM-4, the Fourth International Conference of Adults Learning Maths—A research forum held at the University of Limerick, Ireland, July 4-6 1997* (pp. 164–172). London, UK: Goldsmiths College.
Illeris, K. (2003a). Adult education as experienced by the learners. *International Journal of Lifelong Education, 22*(1), 13–23.

Illeris, K. (2003b). Towards a contemporary and comprehensive theory of learning. *International Journal of Lifelong Education, 22*(4), 396–406.

Jablonka, E. (2003). Mathematical literacy. In A. J. Bishop, M. A. Clements, C. Keitel, J. Kilpatrick, & F. Leung (Eds.), *Second international handbook of mathematics education* (pp. 75–102). Dordrecht, The Netherlands: Kluwer Academic Publishers.

Johnston, B., & Yasukawa, K. (2001). Numeracy: Negotiating the world through mathematics. In B. Atweh, H. Forgasz, & B. Nebres (Eds.), *Sociocultural research on mathematics education: An international perspective* (pp. 279–294). Mahwah, NJ: Lawrence Erlbaum.

Jorgensen, R. (2011). Young workers and their dispositions towards mathematics: Tensions of a mathematical habitus in the retail industry. *Educational Studies in Mathematics, 76*(1), 87–100.

Jungwirth, H., Maaß, J., & Schlöglmann, W. (1995). *Abschlussbericht zum forschungsprojekt mathematik in der weiterbildung*. Linz, Austria: Johannes Kepler Universität.

Keogh, J., Maguire, T., & O'Donoghue, J. (2010). Looking at the workplace through "mathematical eyes"—Work in progress. In G. Griffiths & D. Kaye (Eds.), *Numeracy works for life: Proceedings of the 16th International Conference of Adults Learning Mathematics—A research forum (ALM) incorporating the LLU + 7th National Numeracy Conference* (pp. 132–145). London, UK: ALM/London South Bank University.

Knijnik, G. (1993). An ethnomathematical approach in mathematical education: A matter of political power. *For the Learning of Mathematics, 13*(2), 23–25.

Knijnik, G. (1997). Mathematics education and the struggle for land in Brazil. In G. E. FitzSimons (Ed.), *Adults returning to study mathematics. Papers from working group 18, 8th International Congress on Mathematical Education (ICME-8)* (pp. 87–91). Adelaide, Australia: Australian Association of Mathematics Teachers.

Knijnik, G. (2007). Brazilian peasant mathematics, school mathematics and adult education. *Adults Learning Mathematics—An International Journal, 2*(2), 53–62.

Lave, J. (1988). *Cognition in practice: Mind, mathematics and culture in everyday life*. Cambridge, MA.: Cambridge University Press.

Lave, J. (1996). Teaching as learning in practice. *Mind, Culture and Activity, 3*(3), 149–164.

Lave, J., & Wenger, E. (1991). *Situated learning: Legitimate peripheral participation*. Cambridge, UK: Cambridge University Press.

Leontiev, A. N. (1978). *Activity, consciousness, and personality*. Englewood Cliffs, NJ: Prentice-Hall.

Lindenskov, L., & Wedege, T. (2001). *Numeracy as an analytical tool in mathematics education and research*. Roskilde, Denmark: Centre for Research in Learning Mathematics, Roskilde University.

Martin, D. B. (2007). Mathematics learning and participation in the African-American context: The co-construction of identity in two intersecting realms of experience. In N. S. Nasir & P. Cobb (Eds.), *Improving access to mathematics* (pp. 146–158). Danvers, MA: Teachers College, Columbia University.

Moore, R. (2007). *Sociology of knowledge and education*. London, UK: Continuum.

Mullen, J., & Evans, J. (2010). Adult numeracy in the context of economic change: The conversion to the Euro in the Slovak Republic. In G. Griffiths & D. Kaye (Eds.), *Numeracy works for life: Proceedings of the 16th International Conference of Adults Learning Mathematics—A Research Forum (ALM) incorporating the LLU + 7th National Numeracy Conference* (pp. 189–200). London, UK: ALM/London South Bank University.

Niss, M. (1996). Goals of mathematics education in mathematics. In A. J. Bishop, M. A. Clements, C. Keitel, J. Kilpatrick, & C. Laborde (Eds.), *International handbook of mathematics education* (pp. 11–47). Dordrecht, The Netherlands: Kluwer Academic Publishers.

Noss, R. (1998). New numeracies for a technological culture (1). *For the Learning of Mathematics, 18*(2), 2–12.

Nunes, T., Schliemann, A. D., & Carraher, D. W. (1993). *Street mathematics and school mathematics*. Cambridge, UK: Cambridge University Press.

O'Hagan, J. (2011, June). *When can we trust ourselves to think straight?* Paper presented at Adults Learning Mathematics Conference (ALM-18), Dublin, Tallaght, Ireland.

OECD & Statistics Canada. (1995). *Literacy, economy and society: Results of the First International Adults Literacy Survey*. Paris, France: OECD.

Olesen, H. S., & Rasmussen, P. (Eds.). (1996). *Theoretical issues in adult education: Danish research and experiences*. Frederiksberg, Denmark: Roskilde University Press.

Oughton, H. (2008). Mapping the adult numeracy curriculum: Cultural capital and conscientization. *Literacy and Numeracy Studies, 16*(1), 39–61.

PIAAC Numeracy Expert Group [Gal, I. (Chair), Alatorre, S., Close, S., Evans. J., Johansen, L., Maguire, T., Manly, M., Tout, D.]. (2009). *PIAAC Numeracy framework*, OECD Education Working Paper no. 35 (24 Nov 2009), OECD Publishing. http://www.oecd.org/officialdocuments/displaydocumentpdf/?cote=EDU/WKP(2009)14&doclanguage=en.

Reis, M. C., & Fonseca, M. C. (2008, July). Women, men and mathematics: A view based on data from the 4th National Functional Literacy Indicator (INAF-Brazil). Paper presented at ICME-11, Topic Study group 8: Adult Mathematics Education. http://tsg.icme11.org/tsg/show/9.

Rizvi, F., & Lingard, B. (2010). *Globalizing education policy*. London, UK: Routledge.

Rogers, A. (2006). Escaping the slums or changing the slums? Lifelong learning and social transformation. *International Journal of Lifelong Learning, 25*(2), 125–137.

Roth, W.-M. (2007a). Emotion at work: A contribution to third-generation cultural historical activity theory. *Mind, Culture and Activity, 14*(1–2), 40–63.

Roth, W.-M. (2007b). Mathematical modelling "in the wild": A case of hot cognition. In R. Lesh, J. J. Kaput, E. Hamilton, & J. Zawojewski (Eds.), *Foundations for the future in mathematics education* (pp. 7–98). Mahwah, NJ: Lawrence Erlbaum.

Rubenson, K. (2001). Lifelong learning for all: Challenges and limitations of public policies. In *The Swedish Ministry of Education and Science European Conference: Adult lifelong learning in a Europe of knowledge; Eskilstuna March 23–25, 2001* (pp. 29–39). Stockholm, Sweden: Ministry of Education and Science in Sweden.

Rubenson, K. (2008). OECD education policies and world hegemony. In R. Mahon & S. McBride (Eds.), *The OECD and transnational governance* (pp. 242–259). Vancouver, Canada: UBC Press.

Safford, K. (1999). Who is an adult? How does the definition affect our practice? In M. Groenestijn & D. Coben (Eds.), *Lifelong learning: Adults learning maths—A research forum. ALM-5. Proceedings of the Fifth Conference of ALM, 1–3 July 98, Utrecht* (pp. 96–102). London, UK: Goldsmiths University of London.

Schleicher, A. (2008). PIAAC: A new strategy for assessing adult competencies. *International Review of Education, 54*(5–6), 627–650.

Schliemann, A. (1995). Some concerns about bringing everyday mathematics to mathematics education. In L. Meira & D. Carraher (Eds.), *Proceedings of the 19th International Conference for the Psychology of Mathematics Education* (Vol. I, pp. 45–60). Recife, Brazil: PME.

Sennett, R. (2008). *The craftsman*. London, UK: Yale University Press.

Shiohata, M., & Pryor, J. (2008). Literacy and vocational learning: A process of becoming. *Compare, 38*(2), 189–203.

Sierpinska, A., & Kilpatrick, J. (Eds.). (1998). *Mathematics education as a research domain: A search for identity*. Dordrecht, The Netherlands: Kluwer Academic Publishers.

Skovsmose, O. (1994). *Towards a philosophy of critical mathematics education*. Dordrecht, The Netherlands: Kluwer Academic Publishers.

Steen, L. A. (Ed.). (1997). *Why numbers count: Quantitative literacy for tomorrow's America*. New York, NY: College Entrance Examination Board.

Strässer, R. (2000). Mathematical means and models from vocational contexts—A German perspective. In A. Bessot & J. Ridgway (Eds.), *Education for mathematics in the workplace*. Dordrecht, The Netherlands: Kluwer.

Strässer, R., & Zevenbergen, R. (1996). Further mathematics education. In A. J. Bishop, M. A. Clements, C. Keitel, J. Kilpatrick, & C. Laborde (Eds.), *International handbook on mathematics education* (pp. 647–674). Dordrecht, The Netherlands: Kluwer.

Street, B., Baker, D., & Tomlin, A. (2005). *Navigating numeracies: Home/school numeracy practices.* New York, NY: Springer.

Swain, J., Baker, E., Holder, D., Newmarch, B., & Coben, D. (2005). *Beyond the daily application: Making numeracy teaching meaningful to adult learners.* London, UK: National Research and Development Centre for adult literacy and numeracy (NRDC). http://www.nrdc.org.uk/publications_details.asp?ID=29.

Tobias, S. (1978/1993). *Overcoming math anxiety.* New York, NY: Norton.

Tuomi-Gröhn, T., & Engestrom, Y. (Eds.). (2003). *Between school and work: New perspectives on transfer and boundary crossing.* Amsterdam, The Netherlands: Pergamon.

UNESCO. (2000). *World education report 2000—The right to education: Towards education for all throughout life.* Paris, France: Author.

United Nations Development Programme. (2009). Human development report 2009. Paris, France: Author. http://hdr.undp.org/en/reports/global/hdr2009/.

Vergnaud, G. (1988). Multiplicative structures. In J. Hiebert & M. Behr (Eds.), *Number concepts and operations in the middle grades* (pp. 141–162). Hillsdale NJ: Lawrence Erlbaum.

Vygotsky, L. S. (1930/1978). *Mind in society: The development of higher psychological processes* (M. Cole, V. John-Steiner, S. Scribner, & E. Souberman, Trans.). Cambridge, MA: Harvard University Press.

Wake, G., & Williams, J. (2010). Mathematics in transition from classroom to workplace: Lessons for curriculum design. In A. Araújo, A. Fernandes, A. Azevedo, & J. F. Rodrigues (Eds.), *EIMI 2010 Conference: Educational interfaces between mathematics and industry: Proceedings* (pp. 553–564). Lisbon, Portugal: Centro International de Matemática.

Walker, J. (2009). The inclusion and construction of the worthy citizen through lifelong learning: A focus on the OECD. *Journal of Education Policy, 24*(3), 335–351.

Walkerdine, V. (1988). *The mastery of reason: Cognitive development and the production of rationality.* London, UK: Routledge.

Wedege, T. (1999). To know or not to know—Mathematics, that is a question of context. *Educational Studies in Mathematics Education, 39*(1–3), 205–227.

Wedege, T. (2004). Mathematics at work: Researching adults' mathematics-containing competences. *Nordic Studies in Mathematics Education, 9*(2), 101–122.

Wedege, T. (2009). Quality of research papers: Specific criteria in the field of adults learning mathematics? *Adults Learning Mathematics—An International Journal, 4*(1), 6–15.

Wedege, T. (2010a). Adults learning mathematics: Research and education in Denmark. In B. Sriraman, C. Bergsten, S. Goodchild, G. Palsdottir, B. D. Søndergaard, & L. Haapasalo (Eds.), *The first sourcebook on Nordic research in mathematics education: Norway, Sweden, Iceland, Denmark and contributions from Finland* (pp. 627–650). Charlotte, NC: Information Age Publishing.

Wedege, T. (2010b). Researching workers' mathematics at work. In A. Araújo, A. Fernandes, A. Azevedo, & J. F. Rodrigues (Eds.), *EIMI 2010 conference: Educational interfaces between mathematics and industry: Proceedings* (pp. 565–574). Lisbon, Portugal: Centro International de Matemática.

Wedege, T., & Evans, J. (2006). Adults' resistance to learn in school versus adults' competences in work: The case of mathematics. *Adults Learning Mathematics—An International Journal, 1*(2), 28–43.

Wedege, T., Evans, J., FitzSimons, G. E., Civil, M., & Schlöglmann, W. (2008). Adult and lifelong mathematics education. In M. Niss (Ed.), *Proceedings of the 10th International Congress on Mathematical Education, 2004* (pp. 315–318). Roskilde, Denmark: IMFUFA, Department of Science, Systems and Models, Roskilde University.

Wenger, E. (1998). *Communities of practice: Learning, meaning and identity.* Cambridge, UK: Cambridge University Press.

Williams, J., & Wake, G. (2007a). Black boxes in workplace mathematics. *Educational Studies in Mathematics, 64*(3), 317–343.

Williams, J., & Wake, G. (2007b). Metaphors and models in translation between college and workplace mathematics. *Educational Studies in Mathematics, 64*(3), 345–371.

Yasukawa, K., & Brown, T. (2012). Bringing critical mathematics to work: But can numbers mobilise? In O. Skovsmose & B. Greer (Eds.), *Opening the cage: Critique and politics of mathematics education* (pp. 249–264). Rotterdam, The Netherlands: Sense Publishers.

Young, M. (2010). Alternative educational futures for a knowledge society. *European Educational Research Journal, 9*(1), 1–12.

Websites

Adults Learning Mathematics—A Research Forum (ALM) and Adults Learning Mathematics—An International Journal. http://www.alm-online.net/.

European Network for Motivational Mathematics for Adults (EMMA). http://www.statvoks.no/emma/.

Literacy and Numeracy Studies: An International Journal in the Education and Training of Adults. http://epress.lib.uts.edu.au/ojs/index.php/lnj.

National Center for Adult Literacy (USA). http://www.literacy.org.

National Center for the Study of Adult Learning and Literacy (USA). http://ncsall.net/.

Numeracy: Advocating Education in Quantitative Literacy (journal set up by Mathematical Association of America and the National Numeracy Network). http://serc.carleton.edu/nnn/numeracy/index.html.

Programme for the International Assessment of Adult Competencies (PIAAC). http://www.oecd.org/document/57/0,3343,en_2649_33927_34474617_1_1_1_1,00.html.

Radical Statistics Group. http://www.radstats.org.uk/.

Chapter 8
The Politics of Equity and Access in Teaching and Learning Mathematics

Neil A. Pateman and Chap Sam Lim

Abstract Besides clarifying the definitions of equity and access we briefly contrast two philosophical positions on the nature of mathematics and speculate about their consequences for equity and access. We next discuss "whose mathematics," and provide a viewpoint for mathematics learning as related to equity and access for students. We also consider mathematics teachers and their teaching role as these are related to equity and access for students, and then broaden the chapter to include political influences on both teachers of mathematics and learners. Given the diverse political systems in operation throughout the world, and the range of conditions within and between countries, we are unable to frame questions that can be definitively answered. Our observations relate to the role that politics plays at different levels to influence access and equity for teaching and learning mathematics and are supported by particular examples, some from history, others documenting more recent events. Finally we offer a brief discussion of several international cases of what we believe is a form of colonization that follows from official insistence on "English first" in teaching mathematics in some states where English is a second language for students.

Equity and Access

So far as mathematics education is concerned, access and equity are mostly concerned with whether a complete range of mathematics courses is available at the school level to satisfy the needs and demands of every student and the degree

---

N. A. Pateman (✉)
University of Hawaii, Honolulu, HI, USA
e-mail: pateman@hawaii.edu

C. S. Lim
Universiti Sains Malaysia, Gelugor, Malaysia

to which that access remains open to intending students provided that student performance is satisfactory. So for mathematics learning to be equitable and accessible, all students, regardless of social and cultural background, gender, religious beliefs, ethnicity, geographical location, and family financial circumstances, should have the same "opportunity to learn" (OTL) mathematics (Husén, 1967). For Husén, OTL was the degree of overlap between content taught and content assessed. Classroom conditions, curriculum decisions, teacher beliefs about mathematics and about which students can learn mathematics, teacher preparation in mathematics, and teacher knowledge of effective teaching strategies are factors not considered in OTL, nor are factors that operate to make the content differentially accessible for different students.

A seemingly direct way to make some assessment of the access part of equity and access is to collect information on the provision of mathematics courses, particularly at the high school level. The premise that supports this approach is that without the opportunity to take courses beyond basic arithmetic and elementary level mathematics, students will find it difficult to continue on to mathematics and science courses at upper secondary school that are necessary for success at the college or university levels. Assessing participation rates of different socio-economic and ethnic groups of students within a particular school in those advanced mathematics courses that are provided should generate a second measure of access and equity. These data may be disaggregated to allow comparison not only between countries but also between subpopulations within each country and at the state or district level as well. Of course, performance within courses is an important component of access. At the secondary school level, if the performance of a particular student in required mathematics courses is assessed by teachers as being not up to some specified standard, then further access to mathematics for that student may be quickly closed-off. In almost all countries, performance in secondary mathematics courses acts as a gatekeeper, not only limiting access to further school mathematics courses, but also limiting student choices in higher education. In only a small number of countries is it the case that students who are prevented from moving on in mathematics may re-enter their studies of mathematics as adults.

Data from a variety of sources including international tests (such as PISA, TIMSS), national tests (such as SAT in the USA) and local tests (for example, the state-imposed NCLB-mandated tests in the USA), on both access and performance, are currently collected and examined by a correspondingly broad range of groups with particular interests in education: school administrators, educators, members of policy groups and politicians. For example, Akiba, LeTendre, and Scribner (2007), after reporting the 2003 TIMSS data from 46 countries on student access to qualified teachers, noted the not-surprising outcome that access to qualified teachers was positively related to student performance. However for the USA, which had similar teacher quality to other countries, there was a large gap in access to qualified teachers for low-SES students compared with high-SES students. By contrast, Korea, which had a much higher rate of qualified teachers and higher student achievement in comparison with the USA, still had a substantial achievement gap between the high and low-SES students. That suggested that qualified teachers alone may not be able to overcome the effects of low-SES.

Another example of international comparisons of mathematics performance is provided by the surveys conducted by the Programme for International Student Assessment (PISA). Comparative data have been collected, analyzed and published by the Organization for Economic Co-operation and Development (OECD), a forum of some 34 countries whose mission, according to its Web site (http://www.oecd.org/pages/0,3417,en_36734052_36734103_1_1_1_1_1,00.html), is "to promote policies that will improve the economic and social well-being of people around the world." PISA has been conducting its tests of reading literacy, mathematics literacy and science literacy every 3 years since 2000. The latest mathematics results are of data collected in 2009 from a sample of 15-year-olds selected in participating countries (OECD, 2010a). These data allow many comparisons to be made within and between countries by educational researchers and educational administrators. For example, rates of participation and performance outcomes of males in a range of mathematics tests and courses are compared with those of females; and rates and performances for those same courses and tests of different minority groups are compared with those of other minorities and of course with the rates and performances of the members of the dominant group. Measures other than achievement can also be made—with analyses of measures of variables based on ethnicity, socio-economic status, geographic location and similar characteristics providing a wealth of data allowing comparisons at international, state and local levels (OECD, 2010b, 2011).

International comparative studies (such as TIMSS, PISA) have established performance gaps of different kinds; for example, gender, Black versus Caucasian, Latino versus Caucasian. Much energy has gone into devising ways to close these gaps (NCTM, 2005). The other longer-term aspect of access and performance is their possible influence on opportunities for individuals in the future, which may lead to improvement in the economic, intellectual, and social lives of those with strong performances and a corresponding downturn in life chances for those with mediocre or poor performances.

Equity has found expression in terms of keeping track of performance within and between diverse groups identified by such considerations as, for example, gender, socio-economic status, and ethnicity (including minority language speakers). In mathematics education, a great deal of ground-breaking work, over many years, has established gender and ethnicity as attributes worthy of continuing consideration (see, e.g., Fennema & Leder, 1990; Fennema & Sherman, 1977, 1978; Forgasz, Leder, & Kloosterman, 2004; Reyes & Stanic, 1988; Secada, 1990), and the impact of poverty as a negative correlate of performance (Bracey, 2009; NCES, 2010) is also well-documented. The USA has one of the highest rates of childhood poverty among industrialized nations, a situation that raises issues concerned with equity and access for many US students.

Notice that poverty is not an attribute of individuals; it is rather a condition of their existence that leads to secondary consequences. Poverty is often accompanied by debilitating effects for young children, and these effects can compromise almost all attempts to achieve greater equity and access to education in general and to mathematics education in particular. But our lens must also bring into focus ways to decide upon the most appropriate nature of the mathematics into which young children and older students are to be inducted.

The Added Dimension of "The Politics of …"

The *Oxford Dictionaries Online* (http://oxforddictionaries.com/definition/politics) offers a set of meanings for the word "politics." In particular, there is one broad, neutral meaning—"the academic study of government and the state"—and a second meaning associated with "The politics of …." This second definition is: "The assumptions or principles relating to or inherent in a sphere, theory, or thing, especially when concerned with power and status in a society." At a more contentious level we read a third definition: "Activities within an organization that are aimed at improving someone's status or position and are typically considered to be devious or divisive." In this chapter, all three definitions will be relevant to our purposes.

The first is relevant because government-funded schools, as the designated sites for educating the majority of those soon to enter the general society, are institutions established by the state with functions and roles subject to state regulation. The state has political authority over these functions and roles, and that authority is codified in laws and statutes. At the base of these laws and statutes is a set of assumptions and beliefs about the purpose and nature of education. The school is the instrument intended to ensure that those graduating from it will in some sense be prepared to participate in the society envisioned by those in power.

In practice this is highly problematic. Assumptions about education and its purposes are varied and always contested by political groups and individuals within any citizenry. We all want our children to get a "good education," but there are within any one country very different images of what that means—negotiating which notions of the good are to guide the provision of education brings us to our second definition of politics as it applies to mathematics education—politics consists of those "assumptions or principles relating to or inherent in all *aspects of mathematics teaching and learning*, especially when concerned with power and status in a society." Each generation is inducted into a world that is adopting new layers of technological complexity; schools in the developed world are currently educating students all of whom have always known the Internet. Many of these students will find employment in fields or roles that are yet to be invented. The traditional argument that someone or some group knows what basic mathematical knowledge and skills will prepare students for their roles in society rings hollow when set against these realities.

Two commonly-held positions on education and the importance of learning mathematics may be labelled for our purposes as the *utilitarian perspective*—that only those mathematics courses that prepare the student for the world of work are necessary—and the *liberal perspective*—that all students can learn mathematics and individuals should be encouraged to pursue those mathematics courses that will allow them best to develop their own lives and careers. Of course, this is an oversimplification. Ernest (1991), for example, pointed out that each perspective will be enacted within a range of different groups with very different rationales and educational aims. The essential point is that the position held by those with responsibilities for political action may influence equity and access to mathematics, as well as the kind of mathematics education that should be supported as part of public education.

The utilitarian perspective may be described as conserving the status quo, but the political solutions offered by the liberal or humanist perspective are aimed at transforming society through the emergence of individuals who have sought their own pathway and who will bring new insights into problem solving. Somewhere between the two ideals, utilitarian and liberal, can, perhaps, be found the politics of the practical as the small numbers of the highly influential negotiate with the masses whose members each have limited individual power in the continuing struggle between different classes. How do these negotiations influence educational realities when considering the mathematical education of students and the preparation of their mathematics teachers in different societies?

What Mathematics? Whose Mathematics?

This is not the place to undertake a full discussion of the range of philosophical positions that may be taken on the nature of mathematics and the relationship of those positions to possible beliefs about mathematics education (Ernest, 1991). Our purpose here is to contrast two general perspectives and speculate about their consequences for equity and access.

Those holding a political position that we will continue to describe as utilitarian tend to perceive mathematics as neutral and uncontroversial, but those holding a more liberal perspective are likely to take an entirely different view of the nature and role of mathematics, for individuals and within society. To make the distinction clearer, let us consider the following comparison of positions on the nature of mathematics. It should be emphasized we do not intend to suggest that commentators who offer what we describe below as an example of what it means for mathematics to be neutral are therefore utilitarian in their political stance. The same disclaimer holds for those whose viewpoint exemplifies the liberal political perspective on the meaning of mathematics. In the interests of full disclosure we, the authors, declare ourselves to hold a liberal perspective.

At the Research Pre-session of the April 2010 meeting of the National Council of Teachers of Mathematics (NCTM) in the USA, a symposium entitled *Keeping the Mathematics in Mathematics Education Research* was held. This symposium came close on the heels of an editorial in the March 2010 issue of the *Journal for Research in Mathematics Education* (Heid, 2010) in which the editor stated that "*JRME* publishes research in which mathematics is an essential component rather than being the backdrop for another area of inquiry" (p. 103).

After reporting their impressions of part of the NCTM symposium, Martin, Gholson and Leonard (2010) reacted very strongly to some of the statements made during the symposium in relation to mathematics and the neutrality of the questions about the relationship between mathematics and mathematics education. Some of the words that drew their reaction were in the published symposium summary: "… the session addresses a growing concern among many mathematics education scholars regarding the lack of attention to mathematics in much of the current work in mathematics education" (NCTM, 2010, p. 60). Guershon Harel (2010) offered questions during the symposium about the role of mathematics in mathematics education research that he claimed were neutral and apolitical. This claim led Martin, Gholson and Leonard to present a strong case for the political and cultural nature of all mathematics and all mathematics education research.

The editorial comments by Heid, and the statement by Harel, had political implications for access and equity to mathematics teaching and learning—and also for what should count as research in the field of mathematics education, including mathematics teacher preparation and inservice development. Martin, Gholson and Leonard wrote:

> To whose mathematics are Heid and Harel referring? Is it the very same school mathematics that has been used to stratify students, affording privilege to some and limiting opportunities for others? ... Mathematics can also be used as a tool for understanding the work and, in the case of marginalized students, it can aid in understanding the social forces that contribute to their marginalization. (p. 14)

To rephrase: A critically aware approach to mathematics may help those who are marginalized to understand how their marginalization came about and it may also provide opportunities to resist that marginalization. Strong support for the position on mathematics outlined in the above quotation may be found in Bishop (1988), D'Ambrosio (1985), Mellin-Olsen (1987), Powell and Frankenstein (1997), Skovsmose (2010), and many others. Contributors to the edited collection, *Ethnomathematics: Challenging Eurocentrism in Mathematics* (Powell & Frankenstein, 1997) provided strong arguments for this position.

The plight of smaller nations, struggling to survive in the swirl of world-wide globalization, provides a case in point. Later in this chapter we will outline some of the unintended consequences in several nations that have taken the political decision to require that all education, including the teaching of mathematics, be conducted in English only. We believe that this is an extreme form of marginalization, bordering on a form of neo-colonization, that is taking place with the tacit agreement of local politicians and administrators. The prevailing course of action in those nations is almost universally to adopt existing textbooks from the USA or Britain, and the result is that there is little local cultural input into the mathematics that is taught. As a result, the mathematics in the curriculum can be irrelevant to much of the daily lives of students in those nations. How likely is it that students in such a situation will come to see mathematics as a tool that allows them to understand their marginalization and attempt to do something about this neo-colonization?

So we question whether students in small nations with highly-developed sets of cultural practices and long-established languages should be required by political fiat to undertake their entire education in a language other than their first language. Should teachers in those nations be required to teach only in a language that is, for many of those teachers, a second language?

As we discuss equity and access to mathematics education across nations we will assume that the mathematics with which students should interact should be in a form that is relevant and meaningful to their lives, not only in an economic sense but also in a more holistic culturally-appropriate sense (Bishop, 1988).

Equity and Access for Students: A Developing Viewpoint

Earlier, we indicated that we would refer to more recent approaches to defining equity and access in mathematics education. Nasir and Cobb (2007) reminded us that the meanings of these terms are neither fixed nor transparent to all who are interested in ensuring that students enjoy every opportunity to participate successfully in mathematics beyond rudimentary levels. Not only do the concepts continue to evolve within the mathematics education community, but so too do mathematics educators' understandings of how they relate to mathematics learning.

Nasir and Cobb (2007) sought to reframe common understandings of access and equity by pointing out that although earlier reports on the constructs were made within an environment that accepted that culture and other factors were in play there was not, at that time, a deep recognition of their effects. They were treated merely as background factors. This is not to say that those conducting the studies were not aware of cultural impacts, but rather to suggest that the analyses were insufficiently sensitive and unable to treat these impacts in a functional way.

Nasir and Cobb's (2007) perspective raised the need to understand culture and its impact, and to generate more productive ideas by applying "sociocultural theory [which] provided us with not only a common language, but also with a toolkit of ideas that potentially offered important insights into long-standing equity and diversity issues in mathematics education" (p. xi). We recognize a parallel to the concerns expressed by White, Altschuld, and Lee (2006) that are discussed later in this chapter. In particular, we recognize that those students belonging to a cultural minority, or who speak a language that is different from the language spoken by the majority, are too often treated as if they suffer from some kind of deficit. Thus minority students are offered equity and access, but to take advantage of that offer, they can be expected to move away from their home languages (or cultures), and asked to engage with their education using an unfamiliar language. For students who are studying in mathematics classrooms in which more than half of the time they are unable to understand what is said by their teachers, the "access" provided is no access at all.

Instead, we would offer a more positive view of the potential gains brought to the table by minority language students who are learning mathematics or any other subject, whether with peers only or with majority students. We shall argue that these gains could be a direct consequence of either their cultural or language differences. Such students have the capacity to enrich the classrooms in which they study, provided their teachers are suitably prepared to take advantage of their presence and their cultural and language differences.

Mathematics Teachers and Teaching Mathematics

How do equity and access interact in relation to teaching? Arguably one element of access is the competence of the teacher in terms of both mathematical knowledge and preparation to function effectively in the classroom. Ill-prepared teachers, or those with inadequate background knowledge, are likely to undermine any claims of access or equity (cf. Akiba et al., 2007). Of course, similar concerns may be raised about the preparation of students on entering the school—their access to mathematics and their claims for equitable treatment may be influenced by the degree to which they are prepared to learn in terms of attitude and knowledge.

An issue that is not often framed at all is that of access of minorities into preparation as teachers of mathematics, and their retention once they do enter the profession. In the USA, the Business-Higher Education Forum (BHEF) (2007) documented a serious situation (emphases in original):

- The USA will need more than 280,000 new mathematics and science teachers by 2015.
- Shortages are most apparent in *high-minority and high-poverty classrooms,* where students are less likely to be taught by a teacher who is well-prepared in the subject area.
- In 2002, 72% of high-minority middle school mathematics classes were taught by teachers who had not majored or minored in mathematics, compared with 55% of low-minority classes.
- There is also a *critical shortage of minority teachers*, which is outpacing the overall mathematics and science teacher shortage.
- In 2003, 42% of public school students were from minority groups—yet only 16% of their teachers were minorities. (p. 1)

The last two bullet points draw attention to a significant problem. The usual factor mentioned in support of minorities as teachers is that the minority teacher is a role model for minority students. But diversity in teaching faculty teaches all students that diversity is to be valued in everyday society. In addition, the BHEF document provided evidence that in the USA there was a serious retention problem with an annual attrition of 394,000 teachers. The attrition rate of mathematics and science teachers led all other areas and was particularly high in schools regarded as high poverty. Other statistics in the BHEF list related to the impact of poverty thereby signalling the growing importance of this factor in equity and access, even though the BHEF document, in developing its recommendations, did not mention the need to deal with poverty.

From the above statistics, we better understand the possible explanation put forward by White et al. (2006), who stated that "college retention rates for under-represented minorities (URM) in science, technology, engineering, or mathematics (STEM) are lower than other groups. One reason may be that the studies often do not view premature departure from a cultural perspective" (p. 41). That is to say, although those conducting the studies reported the data, too often they did not take account of possible explanations based on ethnicity. This lapse contributes to the continuing problem of recruitment of minority teachers in these fields—if relatively fewer minorities take mathematics courses in college, the pool of potential minority teachers of mathematics will be correspondingly reduced.

The reader may wonder why it should be important for minorities to teach mathematics to minorities. Such a reader may subscribe to the mainstream position that mathematics is value-neutral, entirely objective and essentially the product of western thought, so that the ethnic background and the first language of the teacher should be irrelevant. In response to this mainstream view (which is a political position supported by those usually described as holding the utilitarian perspective described earlier), we raise again the very different picture of the place of mathematics that may be found in, for example, D'Ambrosio (1985), Mellin-Olsen (1987), and Powell and Frankenstein (1997). For these liberal scholars, mathematics is not merely a skill to be acquired in the service of a global society envisioned by the owners and managers of the dominating multi-national corporations. Mathematics is also a tool for the enlightenment of individuals and the transformation of societies. Many smaller societies are losing their cultures because of a continuing colonization that is supported at administrative levels in many of those societies by the insistence that all instruction should take place in, for example, the English language. An examination of prevalence of this scenario, including reports of relevant studies and the effects of the acceptance within some non-English speaking societies, will form a major part of the second half of this chapter.

We would claim that most students are taught by teachers who began their lives in a very different world from that of their students in terms of everyday access to technology. The exponential growth in worldwide forms of almost instantaneous communication combined with seemingly limitless access to information of all kinds has widened the gap between the current generation and the preceding ones from which the majority of teachers of today originated. Cheap cell phones that have taken on computer-like functions, including texting and email, Netbooks, and handheld tablets, are ubiquitous and not only in the developed world. Children beginning school in many parts of the world have always known the Internet. World Internet usage is measured according to a penetration index corresponding to the percentage of a population that uses the Internet (Miniwatts Marketing Group, 2011). Even in some African countries penetration exceeds 10% and in Europe, the USA, Asia and Australia the reported penetration is in excess of 30%. Given that students have always had access to more technology than their teachers, an important question for further research may be what assumptions and principles should be established both for mathematics teaching and learning, and for the preparation of teachers of mathematics, in order that students will be prepared, and able, to take advantage of technological advances.

That the third view of politics outlined earlier in the chapter as a set of practices that are intentionally devious and divisive is relevant to equity and access to mathematics education, is evidenced in the USA by the concerted attacks on teachers and teacher unions that have occurred for the last decade (see Maher, 2002). In 2011, these attacks reached a fever pitch with calls for "value-added" measures of teacher quality. The discussion typically opens with a position that few would argue against—that all students should have access to a competent and knowledgeable teacher of mathematics. Educational administrators argue that agreement with that point of view implies that there needs to be reliable ways of identifying and supporting teachers whose performance is less than competent.

In the USA a chain of reasoning has been developed that has growing appeal. That chain goes like this:

- Many students' test scores are unacceptable.
- Teachers are directly responsible for their students' individual scores.
- Therefore many teachers are ineffective.
- It is now possible through advanced technology to link each teacher with the test scores of each student that the teacher teaches.
- As students move from year to year it is possible to measure those students' changes in test scores from one teacher to the next.
- These change scores are statistically manipulated using different models to produce value-added measures (VAM).
- These measures are then attributed to the current teacher of each student.
- Averaging out the VAM for a class provides a measure of the effectiveness of the current teacher for that class for that year.
- Collecting VAM each year allows over time the identification of "good" and "bad" teachers.

Despite the rhetoric that of course such measures should not be the only measures, in the USA they are rapidly becoming the sole measure of teacher effectiveness and are often used as the sole criterion for teachers to retain their teaching positions.

Are there any problems with the VAM approach? First, many factors independent of the teacher contribute to what and how students learn. Second, the tests used are underestimates of student knowledge and they are also not appropriate for the sophisticated statistical models needed to create the VAM for each teacher. Complex models are necessary for a variety of reasons, for example, to allow for test differences from year to year and district to district. Indeed studies of various VAM approaches show wide variation in results on the same initial data, and in some cases it is possible to draw absurd inferences from their implementation. Now contrast the VAM perspective that rests solely on student results with that of Ingvarson and Rowe (2007), who pointed out the essential difficulty of conceptualizing and evaluating "teacher quality."

Are teachers alone responsible for the scores their students produce? The answer from many countries is a resounding "No." The model of learning that assigns all responsibility to teachers is the input model, which assumes that students sit passively while the teacher fills their brains with new knowledge—the test score is then assumed to provide a direct measure of the presence of that new knowledge. Very few educationists believe this is how students learn. Furthermore, almost no-one believes that standardized test data provide a direct measure, or even a good measure, of what students know.

Politicians and educational administrators universally preface any comments on test scores with the qualification that "test scores alone are not a good indicator of student performance." However, having said that, they then abandon their own caution and arrive at important decisions about the quality of teachers and students based solely on the test scores.

This whole process objectifies students; it reduces them to ciphers. It denies the reality that students make decisions to participate (or not participate); that students have a range of motivating factors that come into play in classroom situations, some conscious, some at the level of the subconscious.

There is nothing in the statistical models which takes into account factors external to the classroom; no matter how well-documented it is that these external factors strongly influence how well individuals learn. The principal external factor is *poverty*, which is not a student attribute but rather a debilitating condition of a student's existence. Consider these mathematics results from the National Assessment of Educational Progress (NAEP) in the USA:

> In 2009, about 49% of 8th-graders from high-poverty schools performed at or above *Basic*, 13% performed at or above *Proficient*, and 1% performed at *Advanced*. In contrast, about 87% of 8th-graders from low-poverty schools performed at or above *Basic*, 50% performed at or above *Proficient*, and 15% performed at *Advanced*. (Condition of Education: Special Analysis High-poverty Public Schools, 2010, para 1.)

Many of the designers of VAM are ambivalent in relation to the use of standardized test results for high-stakes decisions. Thus, for example, Steven Rivkin (2007) recognized the many difficulties with standardized test scores as the major source of data:

> The imprecision of tests as measures of achievement, failure of some examinations to measure differences throughout the skill distribution, and limited focus of the tests on a small number of subjects further complicate efforts to rank teachers and schools based on the quality of instruction. (p. 1.)

But in the very next paragraph, we read: "Yet despite these potential drawbacks, value-added analysis may still provide valuable information to use in personnel decisions and teacher compensation structures" (p. 1). Later, Rivkin noted that it was "unlikely that available variables account for all school and peer factors systematically related to both achievement and teacher quality" (p. 3). Still later: "The myriad factors that influence cognitive growth, the purposeful sorting of families and teachers into schools and classrooms, and the imperfections of tests as measures of knowledge complicate efforts to estimate teacher effects" (p. 5).

Although the drawbacks are real, and the information is suspect, teachers on the wrong end of personnel decisions can be dealt with harshly. Even following his enthusiastic support for VAM, Rivkin advocated important direct forms of teacher evaluation such as those practised at the school level, with well-prepared supervisors available to observe and provide relevant feedback aimed at supporting those teachers who need to improve.

The political aim of these attacks from political conservatives within the USA has been to lay the blame for "poor" US student performances in international and national standardized tests at the feet of teachers and teacher unions. That these attacks are unfair and based on misunderstandings about the interpretation of test scores, deliberate or otherwise, has been made clear by writers such as Bracey (2009), and Ravitch (2010). The effects on the morale of teachers are as yet unknown but most certainly are unlikely to be positive or neutral.

Some of the effects on the teachers' work practices, in the USA at least, are becoming apparent to those working with teachers in schools. An elementary school that the first author of this chapter visits regularly is a school with a majority of students of minority status, many of whom are classified as second-language learners. Now in a "restructuring" year, because of poor test performance on a state-wide standardized test, its students take the state tests for English and mathematics three times over the course of the school year. Only the best scores are counted. The pressure to meet yearly upgraded targets, which according to statisticians must eventually become unrealistic, has caused many schools to allocate ever-increasing periods of test preparation in mathematics and language. This extra time is almost exclusively directed at those students whose performances on the first opportunity to test are not quite "satisfactory" but are "approaching satisfactory." The colloquial expression for these children is the "bubble kids." The reader is left to speculate about the impact these changed practices are having on those students who are achieving at either high or at very low levels (as determined by their test performances).

Other areas, in which similar concerns have been expressed and rapidly followed by politically-motivated attacks, are the arrangements and requirements for teacher education and provisions for inservice education for teachers. Curiously the attacks often cite Finland as a place for the USA to emulate because of its successes in international tests. Never mentioned are three critical facts about teachers and teaching mathematics in Finland; the curriculum is determined at a local level, teachers are fully-unionized and almost all have the type of masters degree that is being attacked as inappropriate for US teachers (Kupiainen, Hautamäki, & Karjalainen, 2009).

International Cases of Colonization: "English First" in Teaching Mathematics

Mathematics does not consist solely of symbols, and it is concerned with more than manipulation and computation with numbers. Despite a commonly-held belief to the contrary, mathematics requires considerable language skills if it is to be well learned. Learning mathematics involves the development of concepts and the mastering of skills. Mathematical concepts are necessarily abstract and eventually come to be recorded with concision and precision. However to develop concepts successfully, most students need to engage in a great deal of spoken discourse with teachers and fellow students. Productive discourse is only possible when students engage in interchanges involving rich language to explain their individual perspectives. Mathematics also involves logical thinking, together with deductive and analytical reasoning.

Therefore to teach and learn well in mathematics, access to the language of instruction for both teacher *and* her students, is an important factor. Both the teachers and the students must be competent in the language of instruction if their discussions and explanations are to be understood by all parties. If the students are not familiar with the language of instruction clearly they will be deprived of access into higher levels of mathematics learning. This would seem to make obvious the necessity of the language of instruction matching the language spoken by the students. Furthermore, the teacher should have a deep knowledge of that same language. Otherwise, equity and access to mathematics learning will be compromised. This is not an issue in most developed countries where teachers and students by and large share a common language. Some developing countries in Africa (such as Kenya, Malawi) have changed their language policies over the past five years to ensure that their primary school pupils are taught in their mother tongue or home languages. This policy change has raised objections from some upper- and middle-class parents who believe that if their children were to be taught in English they would be more likely to gain access to the global world than if they continue to be taught in their home language.

Other countries have changed their language policy away from an emphasis on local languages first, presumably to suit political and economic agendas, and seemingly without knowledge of the potentially negative impact of an inappropriate choice of language of instruction on both equity and access into mathematical knowledge of the students within the country. English, which has rapidly become the dominant international language, is considered the language of power—fluency in its use is regarded more and more as a pre-requisite for gaining status and prestige, particularly for countries seeking to compete within the globalized economy.

Although English is a second or third language for the children of many countries (e.g., South Africa, Malaysia, Hong Kong, The Philippines, and American Sāmoa), in a number of such countries English is now required as the sole language of instruction for many school subjects including mathematics at all levels within the public school. In many places the "English-only" edict begins at the primary school and continues through all grades of the high school. The perceived political advantages of having confident English-speaking school graduates entering their workforce have trumped the local educational aims in many of the countries that have made this choice.

One wonders if sufficient thought has been given to the social and cultural impacts on the people of the non-English speaking nations that are making this choice, almost all of which are former European colonies. Below are several international cases seeking to illustrate the consequences of political action. Almost certainly, there are other cultures being marginalized with subsequent loss of their unique ways of thought through the insistence on the use of English-only in the schools and increasingly in the mainstream society. The irony is that this marginalization is being initiated and promulgated by influential members within each culture as those members seek to position their countries as players in the globalized society.

The Case of South Africa

Setati (2005) reminded us that "language is always political, not only at the macro level of policy making but also at the micro level of classroom interaction" (p. 450).

This is because the choice of language use and the purposes for its use are not only pedagogic but also serve the political purpose of developing a work force to enable the country to compete internationally. As a result, English is further advanced as *the* international language. It is also generally observed that in many formerly colonial countries such as South Africa, Nigeria, Malaysia, and others, a "change in the language policy of a country is often linked to change in political power" (Setati, 2005, p. 450).

Setati (2005) analyzed the language used in teaching and learning mathematics in a multilingual primary mathematics classroom in South Africa. The class was taught by a qualified and experienced African teacher who was competent in both English and the home language of her pupils (Setswana). Her analysis highlighted the dilemma and tension experienced by the mathematics teacher. On one hand, she was aware of the potential power of English as a gateway to access educational and other resources in South Africa, but on the other hand, she realized the importance of using her pupils' home language as the language of mathematics in conceptual discourse. However, insisting that the teacher and pupils used only English, far too often led to a parody of discourse as a consequence of which pupils came to memorize words and symbols without a complete understanding of their meanings. The negotiation of meaning that is one of the most important outcomes of genuine discourse was simply not possible in such circumstances.

More importantly, because English was imposed, pupils learned by inference that their home language could not be very important. So, unless a teacher was extraordinarily competent the only school discussions that the pupils experienced in mathematics classes would take place in a language in which pupils were only barely functional, and the cognitive content would be presented at a very low level only. Small wonder, then, that the pupils experienced "a devaluing of conceptual discourse as valuable mathematical knowledge" (Setati, 2005, p. 462).

In fact, a similar dilemma has been experienced by Malaysian mathematics teachers who have been operating under the policy of making students' non-home language the language of instruction for teaching mathematics and science in Malaysian schools (see Lim & Ellerton, 2009; Lim & Presmeg, 2011). Related issues will be discussed later in this chapter.

The Case of Botswana

In the case of Botswana, Garegae (2007) described problems faced by teachers and students that were similar to those in South Africa. In mathematics classes, it has been mandated that English will be the language of instruction. Garegae's (2007) study was very different from that reported by Setati (2005) above in that Garegae focussed on a more linguistic analysis of the language use in the classrooms in which mathematics was being taught. As the language of instruction, English was not the learners' first language, and teachers preferred to code-switch between English (L2) and Setswana (L1) even though this was not officially acceptable.

Garagae observed three junior secondary school teachers teaching mathematics and found that they used three types of code-switching: insertion, alternation, and sentence translation. Insertion referred to when "teachers inserted a word in a sentence expressed in another code; say a Setswana code inserted into an English sentence" (p. 235). Alternation refers to those situations where, "sentences are being alternated, and a complete sentence in one code is followed by another in a different code" (p. 235). The third type of code-switching, sentence translation, was found to be the most common and for this type, "the next sentence is the translation of what was expressed in the previous code" (p. 235). Garegae (2007) observed that the first two forms of code-switching encouraged a positional simplification strategy for acquiring L2 and thus disadvantaged the pupils. This was because when students heard isolated words regularly from the teachers, they were not able to learn the meaning of the word as well as the rules of syntax and grammar.

Garegae proposed that the translation of whole sentences from one code to another is a better type of code-switching because it helps to clarify the meanings of words, expressions and sentences expressed in another code through an entire reformulation of instructions. In Botswana's school mathematics curriculum, students were expected to be given a chance to experience a change in teaching methodology from the traditional method of transmission teaching strategy to more of a problem-solving approach, by which students were asked to conjecture and formulate hypotheses about a mathematical problem. Classroom discourse was encouraged whereby learners exchanged ideas, discussed and justified their arguments.

But for this to succeed, learners needed to be well versed in the language of teaching and learning. Therefore, Garegae (2007) argued, "if teachers code-switch without helping students to be able to construct proper sentences, then classroom discourse in Botswana schools will remain an unattainable dream" (p. 236). This case again highlighted the inequity in access to certain kinds of teaching approaches due to the lack of student competency in the language of instruction, with the root cause of inequity being the politically-mandated use of English as language of instruction.

The Case of Malaysia

Viewing English as the "language of power" in meeting the challenges of globalization, in 2003 the Malaysian Ministry of Education took a bold and drastic step implementing the new language policy of Teaching Mathematics and Science in English (or better known as PPSMI). According to Choong (2004), the initial rationale was "teaching the subjects in the science disciplines in English would expedite acquisition of scientific knowledge in order to develop a scientifically literate nation by the year 2020" (p. 2). However, English was not the first language of the majority of Malaysian teachers and students in schools. In fact, those teachers who were less than 45 years old had experienced their entire education (primary to secondary to tertiary) with languages other than English as the medium of instruction. Before 2003,

English was taught as a subject, but not as the medium of instruction. Therefore, teaching mathematics in English posed great challenges, particularly to mathematics teachers in this age group—if their preparation in English was only as a stand-alone subject it was unlikely that they would have sufficient knowledge of English to conduct mathematics classes in English.

In one local study, Lim, Saleh, and Tang (2007) surveyed the perspectives of 20 primary school administrators, 443 mathematics and science teachers, and 787 primary Year 5 pupils from 20 schools in three northern states of Peninsular Malaysia, 5 years after the implementation of PPSMI. Their results showed that one-fifth of the teacher participants rated their own competency in spoken and written English as "poor." By comparison, almost all of these teachers rated their language competency in Malay and/or Mandarin as "good." Indeed, "if nearly one-fifth of the primary school teachers were incompetent in the English language, then their lack of confidence when teaching mathematics and science in English is entirely understandable" (Lim & Presmeg, 2011, p. 145).

After a review of the various related studies in Malaysia, Lim and Ellerton (2009) concluded that the overall confidence among mathematics teachers in their English language proficiency remained low enough for teaching in that language to appear as threatening. This lack of confidence might have led them to code-switch, or discouraged them from using English fully. For example, Tan, Lim, Chew, and Kor (2011) analyzed the discourse of 12 video-recorded mathematics lessons and found generally that the pattern of language use reflected the ethnicity of the pupils. Their discourse analysis showed that teachers talked more than their students, and that mathematics talk was much more common than non-mathematics talk. The use of English was greater than the use of mother tongues in all classes except the weak classes in the Chinese vernacular schools. In those schools the mother tongue dominated classroom discourse. However, the pupils' mother tongue was "the language to fall back on for the teaching of mathematics" (p. 141). The English language functioned more significantly in providing contextual discourse rather than conceptual discourse. Additionally, the pupils' mother tongue played "a major role as the language of conceptual discourse which required reflection and the articulation of one's reasoning" (p. 142).

Based on the data from the same study, Lim and Presmeg (2011) analyzed in depth the dilemma of teaching mathematics in two languages in one Malaysian Chinese primary school. Because of the complex socio-cultural demands of the Malaysian Chinese community, mathematics was taught in both Mandarin (the pupils' mother tongue) and English (the official language of instruction for mathematics) in this type of school. Both teachers in Lim and Presmeg's study emphasized that they resorted to code-switching so that those among their students who were weak in English would have a better chance of catching up with their peers. Consequently, a substantial amount of teaching time was wasted in making translations, especially of the terminology of mathematics. To expedite the teaching, these teachers sometimes opted to teach in the students' mother tongue (Mandarin) only. Hence, some of these students, particularly the weaker ones, may have been denied the opportunity to speak and express their mathematical thoughts in English.

Although the practices mentioned above were understandable from the points of view of teachers struggling to survive, they could well have created real learning problems for non-English speaking students. The source of these difficulties could lie in the decision at some administrative level to adopt a short-sighted political solution to what is a very complex educational problem. Malaysia will revert to requiring the mother-tongue to be the language of instruction in mathematics and science classes from year 2012 (Chapman, 2009). Inevitably, the debate concerning whether access and equity for some groups of pupils will be improved by the planned policy change will continue. We would maintain only that the issues of equity and access in learning mathematics are significantly related to the language of instruction for mathematics.

The Case of Aboriginal Australians in Homeland Communities

The example here is of a majority English-speaking nation with a small minority of Aboriginal people leading relatively traditional lives in remote communities now mostly in the far north and north-west of the country. Most Australia Aboriginal people have become urban dwellers and have adopted "white-fella" ways—but, as Harris (1991) reminded us, many still look to the remote groups leading traditional lives to maintain the cultural knowledge and languages of the Aboriginal people. Schooling in these communities was in theory conducted in English, but the common practice was best described in the words of an Aboriginal Australian colleague, "English is the language of instruction but *Yolngu Matha* is the language of explanation."

Two points were evident: first, the oral language was extremely and uniquely important so far as learning was concerned; and second, in *Yolngu Matha,* conceptions of space were expressed in ways that were completely unfamiliar in European languages—and hence, attempts at cross-translation mentioned earlier could not be successful. As Christie (1995) pointed out, political correctness in the 1970s "seemed to dictate that all languages are ultimately capable of communicating the meanings of all other languages" (p. 2). Christie also pointed out that languages express different epistemologies arising from different world views, and that it is likely that different mathematics is a real consequence of these different worldviews. This raised several points, not the least of which is the negative effect on Aboriginal languages and culture of being instructed in English. That, together with the added difficulty that many teachers are non-Aboriginal, and are non-native speakers of the local language, has created many very problematic scenarios.

How different are Aboriginal languages from English and other European languages? An emphasis on cardinal directions is one aspect that Aboriginal languages share with languages of other ancient cultures. English words "left" and "right" are not used, and there are not simple words for those concepts in the local language. Teaching children how to form letters in English provides an example of how different the languages can be—an English speaker might use up, down, left and right, but the

Aboriginal teacher would use the cardinal directions, north, south, west and east (in the local language) and the correct words in a situation would be relative to the child's spatial orientation.

The Case of American Sāmoa

Hunkin-Finau (2006) describes the situation in American Sāmoa thus:

> Although American Samoan society values the idea of bilingualism and biculturalism, its teachers are bound by an education system that promotes, and is heavily oriented towards, English and western values. Outside of their professional work, Samoan teachers live as Samoans in the community; inside the schools, they employ English and operate within a system that is tied to western values. (p. 49)

By adopting an English-only policy in the schools, there is loss of the traditional ceremonial forms of Samoan at the same time as there is growth of a form of Samoan contaminated by the kind of code-switching that Garegae reported in Botswana. The long-term result is that although Samoan students are losing their Samoan, they are not improving much in Standard English, and are certainly doing poorly in mathematics by most standards. A significant aspect of the problem is the limited knowledge of English of many teachers (Hunkin-Finau, 2006).

Concluding Remarks

Political decisions lead to policy formulation. A positive example of policy formulation that has had an impact on access and equity in mathematics for both students and teachers is that made, and implemented at the national government level, some 10 years ago, by Finland. Some believe that these decisions culminated in recent very strong international PISA performances by Finnish students (OECD, 2010a). Table 8.1 shows the major policy decisions taken in Finland on the right, which offer a strong contrast to the conventional model that is in effect in many countries, shown in the left column (Kupiainen et al., 2009, p. 12).

It could be argued that the key to Finnish success was the successful implementation of the outlined policies. This is to be contrasted with the unsuccessful implementation of the English-only education policy described in several cases earlier in this chapter. The usual reasons given by outside commentators for Finland's success are that it has a relatively small population, that the population is essentially homogeneous in terms of culture, with a high literacy level, and that Finland enjoys a low level of poverty among children. However two points should be made. First, 10 years previously, Finland, by its own admission, had an education system plagued with problems (Kupiainen et al., 2009), despite all the factors mentioned in the last paragraph that should have been associated with success. Second, the policies

Table 8.1
Comparison of Two Models of Policy Formulation

General Western Model	The Finnish System
Standardization Strict standards for schools, teachers and students to guarantee the quality of outcomes.	**Flexibility and diversity** School-based curriculum development, steering by information and support.
Emphasis on literacy and numeracy Basic skills in reading, writing, mathematics and science as prime targets of education reform.	**Emphasis on broad knowledge** Equal value to all aspects of individual growth and learning: personality, morality, creativity, knowledge and skills.
Consequential accountability Evaluation by inspection.	**Trust through professionalism** A culture of trust on teachers' and headmasters' professionalism in judging what is best for students and in reporting of progress.

adopted recognized and took advantage of the professionalism of teachers and administrators. The new policies removed many bureaucratic restrictions that are faced in education reform in most countries.

Perhaps at the heart of the politics and policy decisions taken within a nation, decisions that have an impact on equity and access in mathematics teaching and learning, is a general desire to develop the nation's political economy. Participation in the global marketplace is seen as very desirable as nations strive to improve the circumstances of their people. This essentially means that successful entry into the global marketplace becomes a major rationale behind many decisions, including educational, taken by the political leaders of a nation. Perceptions of success are related to producing an educated workforce. But this requirement demands a great deal of both teachers and students, particularly in countries where the political economy is unable to provide the necessary infrastructure and resources.

In such cases the implementation of the policies thought essential to joining the global market place, such as educating non-English speaking students in English only, falters and becomes counterproductive. Keady (2006) used the term "vulnerability" to remind leaders of nation states seeking to participate in the global economy that there are costs that go unrecognized associated with the supposed economic benefits. These costs include unwanted changes in social and cultural life within the state that may lead to a reduction in equity and access to mathematics learning, and indeed many other aspects of formal education.

It is impossible to close this chapter with a set of pronouncements concerning the overall situation with respect to the politics of access and equity to mathematics teaching and learning across the world. The level of complexity of such a task would be far too great. We have pointed to possible factors that make it less likely for some students to gain access to mathematics and to learn from a well-prepared teacher. Poverty remains a major factor—even in resource-rich developed countries there are pockets of citizens living below the poverty line. With few exceptions, hungry students do not learn as well, or as much, as others and are thus denied access and equity.

References

Akiba, M., LeTendre, G. K., & Scribner, J. P. (2007). Teacher quality, opportunity gap and national achievement in 46 countries. *Educational Researcher, 36*(7), 369–387.

Bishop, A. J. (1988). *Mathematical enculturation: A cultural perspective on mathematics education.* Dordrecht, The Netherlands: Kluwer Academic Publishers.

Bracey, G. (2009). *The Bracey Report on the condition of public education, 2009.* Boulder and Tempe: Education and the Public Interest Center & Education Policy Research Unit. Retrieved March 20, 2011 from http://epicpolicy.org/publication/Bracey-Report.

Business-Higher Education Forum. (2007). *An American imperative: Transforming the recruitment, retention, and renewal of our nation's mathematics and science teaching workforce: Teaching quick facts.* Bakersfield, CA: Author.

Chapman, K. (2009, July 9). It is Bahasa again but more emphasis will be placed on learning English. *The Star, N2.* Retrieved from http://thestar.com.my/news/story.asp?file=/2009/7/9/nation/4286168&sec=nation.

Choong, K. F. (2004). *English for the teaching of mathematics and science (ETeMS): From concept to implementation.* Retrieved January 1, 2009 from http://eltcm.org/eltc/resource_pabank.asp.

Christie, M. J. (1995, July). *The purloined pedagogy: Aboriginal epistemology and maths education.* Presented to the joint meeting of the Mathematics Education Research Group of Australia and the Australian Association of Mathematics Teachers, Darwin, Australia.

D'Ambrosio, U. (1985). Ethnomathematics and its place in the history and pedagogy of mathematics. *For the Learning of Mathematics, 5,* 44–48.

Ernest, P. (1991). *The philosophy of mathematics education.* London, UK: Routledge Falmer.

Fennema, E., & Leder, G. (Eds.). (1990). *Mathematics and gender: Influences on teachers and students.* New York, NY: Teachers College Press.

Fennema, E., & Sherman, J. (1977). Sex-related differences in mathematics achievement, spatial visualization, and affective factors. *American Educational Research Journal, 14*(1), 51–71.

Fennema, E., & Sherman, J. (1978). Sex-related differences in mathematics achievement and related factors: A further study. *Journal for Research in Mathematics Education, 9*(3), 189–203.

Forgasz, H. J., Leder, G. C., & Kloosterman, P. (2004). New perspectives on the gender stereotyping of mathematics. *Mathematical Thinking and Learning, 6*(4), 389–420.

Garegae, K. G. (2007). On code switching and English language proficiency: The case of mathematics learning. *The International Journal of Learning, 14*(3), 233–238.

Harel, G. (2010, April). *The role of mathematics in mathematics education research: Question for public debate.* Paper presented at the Research Pre-session of the annual meeting of the National Council of Teachers of Mathematics, San Diego, CA.

Harris, P. (1991). *Mathematics in a cultural context: Aboriginal perspectives on space, time and money.* Geelong, Australia: Deakin University.

Heid, M. K. (2010). Where's the math (in mathematics education research)? *Journal for Research in Mathematics Education, 41,* 102–103.

Hunkin-Finau, S. S. (2006). An indigenous approach to teacher education in American Sāmoa. *Educational Perspectives, 39*(2), 47–52.

Husén, T. (1967). *International study of achievement in mathematics* (Vol. 2). New York, NY: Wiley.

Ingvarson, L., & Rowe, K. (2007). Conceptualizing and evaluating teacher quality: Substantive and methodological issues. *Student Learning Processes.* Retrieved from http://research.acer.edu.au/learning_processes/8.

Keady, P. (2006). Theorising globalisation's social impact: Proposing the concept of vulnerability. *Review of International Political Economy, 13*(4), 632–655.

Kupiainen, S., Hautamäki, J., & Karjalainen, T. (2009). *The Finnish education system and PISA.* Helsinki, Finland: Ministry of Education.

Lim, C. S., & Ellerton, N. F. (2009). Malaysian experiences of teaching mathematics in English: Political dilemma versus reality. In M. Tzekaki, M. Kaldrimidou, & C. Sakonidis (Eds.),

Proceedings of the 33rd Conference of the International Group for the Psychology of Mathematics Education (Vol. 4, pp. 9–16). Thessaloniki, Greece: International Group for the Psychology of Mathematics Education.

Lim, C. S., & Presmeg, N. (2011). Teaching mathematics in two languages: A teaching dilemma of Malaysian Chinese primary schools. *International Journal of Science and Mathematics Education, 9*(1), 137–161.

Lim, C. S., Saleh, F., & Tang, K. N. (2007). *The teaching and learning of mathematics and science in English: The perspectives of primary school administrators, teachers and pupils.* Centre for Malaysian Chinese Studies Research Paper Series No. 3, Kuala Lumpur, Malaysia.

Maher, F. (2002). The attack on teacher education and teachers. *The Radical Teacher, 64*, 5–8.

Martin, D. D., Gholson, M. L., & Leonard, J. (2010). Mathematics as gatekeeper: Power and privilege in the production of knowledge. *Journal of Urban Mathematics Education, 3*(2), 12–24.

Mellin-Olsen, S. (1987). *The politics of mathematics education.* Dordrecht, The Netherlands: Reidel.

Miniwatts Marketing Group. (2011). *World Internet users and population statistics.* Retrieved from http://www.internetworldstats.com/stats.htm, July 11, 2011.

Nasir, N. S., & Cobb, P. (Eds.). (2007). *Improving access to mathematics: Diversity and equity in the classroom.* New York, NY: Teachers College Press.

National Center for Educational Statistics [NCES]. (2010). *Condition of education: Special analysis high-poverty schools.* Retrieved from http://nces.ed.gov/programs/coe/2010/analysis/section3a2.asp.

National Council of Teachers of Mathematics. (2005). *Position paper: Closing the achievement gap.* Retrieved from http://www.nctm.org/about/content.aspx?ide=6350, July 7, 2011.

National Council of Teachers of Mathematics. (2010). *Program for the research presession.* Reston, VA: Author.

OECD. (2010a). *PISA 2009 results: What students know and can do: Student performance in reading, mathematics and science* (Vol. I). Paris, France: Author. http://dx.doi.org/10.1787/9789264091450-en.

OECD. (2010b). *PISA 2009 results: Overcoming social background: Equity in learning opportunities and outcomes* (Vol. 2). Paris, France: Author. http://dx.doi.org/10.1787/9789264091504-en.

OECD. (2011). *Lessons from PISA for the United States: Strong performers and successful reformers in Education.* Paris, France: Author.

Powell, A. B., & Frankenstein, M. (Eds.). (1997). *Ethnomathematics: Challenging Eurocentrism in mathematics education.* Albany, NY: SUNY Press.

Ravitch, D. (2010). *The death and life of the great American school system: How testing and choice are undermining education.* New York, NY: Basic Books.

Reyes, L. H., & Stanic, G. M. A. (1988). Race, sex, socioeconomic status and mathematics. *Journal for Research in Mathematics Education, 19*(1), 26–43.

Rivkin, S. G. (2007). *Value-added analysis and teacher policy. Brief 1.* Washington, DC: The Urban Institute.

Secada, W. G. (1990). Needed: An agenda for equity in mathematics education. *Journal for Research in Mathematics Education, 21*(5), 354–355.

Setati, M. (2005). Teaching mathematics in a primary multilingual classroom. *Journal for Research in Mathematics Education, 36*(5), 447–466.

Skovsmose, O. (2010, October). Mathematics: A critical rationality? *Philosophy of Mathematics Education Journal, 25*, 1–23.

Tan, K. E., Lim, C. S., Chew, C. M., & Kor, L. K. (2011). Talking mathematics in English. *The Asia-Pacific Education Researcher, 20*(1), 133–143.

White, J. L., Altschuld, J. W., & Lee, Y.-F. (2006). Cultural dimensions in science, technology, engineering and mathematics: Implications for minority retention research. *Journal of Educational Research and Policy Studies, 6*(2), 41–59.

Part II
Introduction to Section B: Mathematics Education as a Field of Study

Alan J. Bishop

Abstract The eight chapters in Section *B* were written by 22 scholars from 13 different nations. The first chapter provides historical perspectives on the growth of mathematics education as a field of study, with an emphasis being on the move towards greater internationalization. The "middle chapters" discuss theoretical and practical developments, especially in relation to the trend towards greater degrees of collaboration between school teachers, based in schools, and mathematics educators not based in schools. Issues such as: (a) "How should we educate mathematics teacher educators?"; (b) "How much should practice be influenced by a perceived capacity to deliver success in terms of international competitiveness linked to economic agendas?"; and (c) "How can teachers of mathematics become effective mathematics education researchers?" are taken up. In the final chapter in the section, the issue of how much notice is being taken by curriculum developers and policy makers of the "findings" of mathematics education research is raised.

Keywords Action research in mathematics education • Mathematics education as a field of study • Teachers as researchers • Teacher education • Theory and mathematics education

The aim of this section is to document and analyze various issues concerned with developing a field of study—in this case, mathematics education. As the first chapter in this section shows, our field is a relatively new one, especially when compared with that of mathematics, and even with education. However, increasingly over the last decade, as the various practices in mathematics education have proliferated, albeit seemingly without any comparable increases in overall mathematical achievements and attitudes, researchers have taken serious steps to address the nature of our field. It is timely, therefore, in this international handbook, to analyze what has changed in the field since the last international handbooks (Bishop, Clements,

Alan J. Bishop
Monash University, Clayton, Victoria, Australia

Keitel, Kilpatrick & Laborde, 1996; Bishop, Clements, Keitel, Kilpatrick & Leung, 2003) and to ponder where mathematics education is as a field of study at present.

The Field of Study in Previous Handbooks

In earlier handbooks of mathematics education this topic has received limited and relatively narrow attention. In the first handbook produced in the field (Grouws, 1992) there was no specific section, or indeed chapter, with this focus. However an invitation came to me to contribute a chapter to that handbook under the title "International Perspectives on Research in Mathematics Education," and in that chapter I did to some extent indicate how our field of study was growing. The title of the chapter also indicated an assumed brief for the chapter, which was to document the various research activities happening in different countries.

On reflection it occurred to me that the brief given to me for the Grouws (1992) *Handbook* was an impossible brief to follow, as there was no way of establishing the validity of any claims which might be made about what was occurring within several countries. It also assumed that it was the country, the "nation state," which was somehow the determining factor in the choice of research study. Given the debates within our research field, as I saw them, due to increasing numbers of conferences, journals and collaborative research activities, it seemed much more appropriate to shed light on these perspectives by referring in some way to their historical and cultural backgrounds.

This I achieved by considering research in our field as following three distinct traditions: namely the Pedagogue tradition, the Empirical Scientist tradition, and the Scholastic Philosopher tradition (Bishop, 1992). In particular, rather than viewing research trends as particular national styles or traditions, these three traditions were shown to occur both within and across national boundaries.

Nowadays, with the growth in international contacts in research, publishing and conferences, it is less easy to see the three traditions in research practices. Nor would that necessarily be a good way to analyze the trends and developments in our field in this current *Handbook*. Research these days trawls many diverse fields for its constructs and its processes. Anthropological, socio-cultural, and political perspectives are among many being brought to bear on the complex problems of the practice of mathematics education today.

The previous Kluwer/Springer handbooks have demonstrated the growth of interest in the theoretical nature of the field itself. In the first *International Handbook on Mathematics Education* (Bishop et al., 1996), two sections contained chapters focussing on different aspects of the topic. Chapter 22 was titled "Epistemologies of Mathematics and Mathematics Education" (Sierpinska & Lerman, 1996), and Chapter 28 was titled "The Role of Theory in Mathematics Education and Research" (Mason & Waywood, 1996)—and as they introspected on the constructs and methods being used, the authors of these chapters demonstrated how researchers had begun to address the issues faced by researchers in a developing research field.

Epistemology is a fundamental part of any field of study, but what Sierpinska and Lerman (1996) showed in their chapter was that mathematics education, as a field, contains many related but also competing epistemologies. As they saw it: "There is much debate within the international community of mathematics educators about theoretical approaches and their underlying epistemological issues" (p. 867). Diversity of epistemology was echoed in the chapter by Mason and Waywood, in terms of both the range and the articulation of theory and theories relevant for mathematics education. As they stated: "Instead of asserting exclusivity, people will increasingly acknowledge that each discourse provides a way of seeing and speaking which may or may not seem appropriate to the context, issue, and participants, which may or may not prove to be informative in the future" (p. 1083).

The assumed brief for that first *Handbook* was to represent the range of activities and ideas being developed in the field up until its publication in 1996. In the *Second International Handbook of Mathematics Education* (Bishop et al., 2003) there was, once again, no particular section focussing on the field itself. Partly the reason for this was that the assumed brief for that *Second Handbook* was not to just repeat the chapters and sections of the first, but to focus on what were felt to be the growth aspects and issues and concerns in the field since the publication of the first *Handbook*. Mathematics education as a field of study did not appear to be a priority area, compared with policy issues, technological developments, issues in research, and professional practices in mathematics education, which formed the four sections. However in Section 3 "Issues in research in mathematics education" there were several chapters which showed that whenever research is undertaken, theoretical issues are always present. It was clear that in the seven years which had passed since the publication of the first *International Handbook*, more interest, and also more concern, had been expressed about the field—how it should be studied, what quality controls should be exerted on the research, and the role of ethics in research.

In introducing that section, Kilpatrick (2003) referred to some of the questions addressed by the chapter authors: "What is ethical practice in our research and how is that research to be done amid situations of social and political conflict? What impact does educational research have on mathematics education? How is our research to overcome various obstacles to dissemination? What is the role of mathematics teachers as researchers? How is the next generation of researchers in mathematics education to be prepared? These questions would probably not have been posed by most researchers a half-century ago, if only because few then were interrogating their own practice as researchers" (p. 436).

Regarding other handbooks, in 2007 the National Council of Teachers of Mathematics (NCTM) published a *Second Handbook of Research on Mathematics Teaching and Learning* (Lester, 2007), as a follow-up to its first 1992 NCTM handbook. Part 1 of the *Second Handbook* (which was one of six parts) put the nature of the field of study front and centre with three powerful and comprehensive chapters: "Putting Philosophy to Work: Coping with Multiple Theoretical Perspectives" (Cobb, 2007), "Theory in Mathematics Education Scholarship" (Silver & Herbst, 2007), and "Method" (Schoenfeld, 2007). But although these three chapters gave an overall perspective on the issues pertaining to the field and its concerns, there were no follow-up chapters on specific issues.

In the recent handbook on mathematics education, part of the Major Themes series by Routledge, I took the notion of democratizing research in mathematics education as a way of selecting the articles to be included (Bishop, 2010). This handbook was formed from already published papers, and Part 8, in Volume 4 of that work, focussed on research and theoretical analyses. It contained 17 significant papers.

The chapters and sections in this and previous handbooks demand that any researcher in our field these days cannot escape the obligation to recognize that their research is in a field which has its special professional ethics and responsibilities. Any researcher seeking to explore a specific issue these days has to cross several complex hurdles before they can even begin the study. Likewise any researcher seeking to publish research findings these days will have to satisfy strong peer reviews in which their methods and theoretical base will be carefully scrutinized.

The Structure of This Section

The chapters in this section document and analyze the diverse ways in which mathematics education is now being researched and theorized as a field. The authors have drawn on research more recently published, and show how current research builds on the work done in the past. They cannot claim to be fully comprehensive in their coverage of the literature, but they do give us an account of where the field is now.

Furinghetti, Matos and Menghini, the authors of the first chapter in this section, set the scene by giving us a historical view of the development of mathematics education as a field. They argue, in Chapter 9, that mathematics education has grown from being mostly an enterprise controlled by mathematicians to a worldwide, developing intellectual and scholarly area of activity with many professionals playing a role. They particularly focus on the role of mathematical communication as a crucial aspect of this development, and this role is analyzed through the growth of journals and research conferences. As well as developing communication about mathematics education, with the sharing of pedagogical and curricula solutions to practical problems, internationalization has also played a strong part in the growing field. Journals such as *L'Enseignement Mathématique* and *Educational Studies in Mathematics* influenced the international spread of ideas, as did organizations such as the International Commission on Mathematical Instruction (ICMI).

As well as painting a general picture of an academic field developing through the growth of organizations and international means of communication, Furinghetti et al. analyze three specific areas of mathematics education. These are presented not just as exemplars of the developing field but as pointers to three aspects of that growth: the relationships with psychology, the study of social, cultural and political dimensions, and the increasing relevance of theories for mathematics education.

This, then, is the historical backdrop against which the current approaches to the study of our field are detailed in the subsequent chapters.

Fundamental to any field of knowledge field are the theories, models and frameworks used to interpret, analyze and research the field. Since the time of the first Kluwer *Handbook* we have seen a veritable explosion in the range of theories and constructs used in mathematics education research. Some of these were explored in different chapters in the *Second International Handbook*, and Jablonka, Wagner and Walshaw draw attention to many theoretical constructs in Chapter 2 of this *Third Handbook*. Chapter 10, by Sririman and Nardi, addresses the increasing interest and concern about this proliferation. Not only does the variety confuse and challenge new researchers, it also means that the diverse analyses are more and more difficult to reconcile, synthesize, and apply.

As the theoretical field of mathematics education has grown, so too have the research methods used to study the field. In the special area of teacher education, research has moved from the psychological and the quantitative, to the social and anthropological, and increasingly we are seeing a reduction in the use of one single method, and a preference for "mixed-method" approaches. In Chapter 11, Gellert, Becerra Hernández and Chapman argue that far from dealing with the area of "which method," studies have skirted around the problem. Research nowadays relates to the fact that as well as studying teachers and teaching, researchers want to see their findings and theories applied to the teachers, their teaching, and their education. This has led to more research in which mathematics educators work "with" teachers rather than "on" teachers. Gellert et al. also report a survey that focussed on "methodologies, research methods and research techniques," drawing on journal articles published between 2005 and 2010. Chapter 11 complements Chapter 10 by analyzing the relationship between the theories and constructs being used in research studies and the research methods now available.

Mathematics education is both a professional and an academic field, and as we find in other such fields of knowledge there are both advantages and disadvantages with that situation. One advantage is that the professional context offers an important reality check on the theories coming from academic research. Another is that there is an ever-present need for increased knowledge in the professional field as society becomes yet more complex. One major "disadvantage," however, for academic researchers is that the needs of the professionals do not always sit easily with the academy's research agendas.

In Chapter 12, Kieran, Krainer and Shaughnessy explore this relationship in relation to research and the practices of teaching. The authors focus on a specific idea—namely considering teachers not as recipients of research findings but as key stakeholders in the research enterprise itself. They explore what this means and how this idea develops in five international contexts. Kieran et al.'s analyses reveal that teachers' subjective theories play a significant role in this new relationship. Their analyses also draw attention to three important dimensions of research where teachers are key stakeholders: reflective, inquiry-based activity with respect to teaching action; a significant action-research component accompanied by the creation of research artefacts by the teachers (sometimes assisted by university researchers); and the dynamic duality of research and professional development.

As has already been indicated, in the field of mathematics teacher education, there is much debate about the balance between professional and academic foci. On the one hand, there is the emphasis on the craft knowledge of the teacher which is based at the particular and local level; on the other hand is academic knowledge at theoretical and general levels. In Chapter 13, by White, Jaworski, Agudelo and Gooya, the focus is on how to blend and balance the two through the activities of teachers learning from other teachers. Demonstration lessons have been a feature of teacher preservice and inservice programs for years, but recent programs such as lesson study (LS) and the Learner's Perspective Study (LPS) have helped to raise the research to another theoretical level.

Although the issues around the theory–practice relationship in mathematics education have generally been focussed on teachers, and typically on school teachers, there is another set of issues concerning a different group of practitioners. The development of mathematics education as a field has depended to a large part on the development of a group we could refer to as "mathematics educators." This category not only includes school mathematics teachers but also, increasingly, research students, junior and senior researchers and professors, who may or may not have been school teachers previously. Chapter 14, by Novotna, Margolinas and Sarrazy addresses issues of teacher education, both preservice and inservice, as well as the growth of master's and doctoral research, from the perspective of the development of the research practitioners themselves.

Chapter 15, by Brown and Clarke, analyzes the idea and significance of institutional contexts, and examines the research benefits and drawbacks of these contexts, both for the training and education of researchers and also for the development of the field itself.

Research in mathematics education has increasingly turned to issues of context, but it too is situated in many contexts, and the researchers themselves cannot be neutral concerning their values and beliefs about mathematics education. Far from being the province of the lone university researcher, research these days takes place increasingly in small and large teams, usually but not always at universities. It is often funded by agencies and governments, each with its own agenda.

At the highest level, therefore, the future of the field of mathematics education may indeed be at stake with concerns being raised about the interaction of governmental policy with institutional practice. This is a source of much debate at present, as Hoyles and Ferrini-Mundy make clear in Chapter 16. At the heart of the debate is the accusation that researchers are too often concerned about pursuing their own "interesting," and general, research questions—instead of paying sufficient attention to relationship issues of policy and practice.

It has to be hoped that the research community recognizes and faces this challenge, by considering the policy implications of developments in researching mathematics education, by determining appropriate research questions, and by addressing the nature and implications for practice of research findings. Also implicated are the many issues of making research findings professionally and publicly available (and understandable) through conferences, research publications and other materials. Of the greatest concern, therefore, in this section of the *Handbook*, is the challenge that practical policy imperatives bring for the future development of the field itself.

References

Bishop, A. J. (1992). International perspectives on research in mathematics education. In D. A. Grouws (Ed.), *Handbook of research on mathematics teaching and learning* (pp. 710–723). New York, NY: Macmillan.

Bishop, A. J. (Ed.). (2010). *Mathematics education*. London, UK: Routledge.

Bishop, A. J., Clements, K., Keitel, C., Kilpatrick, J., & Laborde, C. (Eds.). (1996). *International handbook of mathematics education*. Dordrecht, The Netherlands: Kluwer Academic Publishers.

Bishop, A. J., Clements, M. A., Keitel, C., Kilpatrick, J., & Leung, F. K. S. (Eds.). (2003). *Second international handbook of mathematics education*. Dordrecht, The Netherlands: Kluwer Academic Publishers.

Cobb, P. (2007). Putting philosophy to work: Coping with multiple theoretical perspectives. In F. K. Lester (Ed.), *Second handbook of research on mathematics teaching and learning* (pp. 3–38). Charlotte, NC: Information Age Publishing.

Grouws, D. A. (Ed.). (1992). *Handbook of research on mathematics teaching and learning*. New York, NY: Macmillan.

Kilpatrick, J. (2003). Introduction to Section 3: Issues in research in mathematics education. In A. J. Bishop, K. Clements, C. Keitel, J. Kilpatrick, & F. K. S. Leung (Eds.), *Second international handbook of mathematics education* (pp. 435–439). Dordrecht, The Netherlands: Kluwer Academic Publishers.

Lester, F. K. (Ed.). (2007). *Second handbook of research on mathematics teaching and learning*. Charlotte, NC: Information Age Publishing.

Mason, J., & Waywood, A. (1996). The role of theory in mathematics education. In A. J. Bishop, K. Clements, C. Keitel, J. Kilpatrick, & C. Laborde (Eds.), *International handbook of mathematics education* (pp. 1055–1089). Dordrecht, The Netherlands: Kluwer Academic Publishers.

Sierpinska, A., & Lerman, S. (1996). Epistemologies of mathematics and of mathematics education. In A. J. Bishop, K. Clements, C. Keitel, J. Kilpatrick, & C. Laborde (Eds.), *International handbook of mathematics education* (pp. 827–876). Dordrecht, The Netherlands: Kluwer Academic Publishers.

Schoenfeld, A. H. (2007). Method. In F. K. Lester (Ed.), *Second handbook of research on mathematics teaching and learning* (pp. 69–108). Charlotte, NC: Information Age Publishing.

Silver, E. A., & Herbst, P. G. (2007). Theory in mathematics education scholarship. In F. K. Lester (Ed.), *Second handbook of research on mathematics teaching and learning* (pp. 39–68). Charlotte, NC: Information Age Publishing.

Chapter 9
From Mathematics and Education, to Mathematics Education

Fulvia Furinghetti, José Manuel Matos, and Marta Menghini

Abstract This chapter takes a historical view of the development of mathematics education, from its initial status as a business mostly managed by mathematicians to the birth of mathematics education as a scientific field of research. The role of mathematical communication is analyzed through the growth of journals and research conferences. Actions of internationalization and cooperation in facing instructional and educational problems are illustrated with reference to the journal *L'Enseignement Mathématique* and to ICMI. Curricular and methodological reforms in the 20th century which generated changes in school mathematics are considered. Starting from the acknowledgement that research in mathematics education demands more than the traditional focus on discussing curricular options at distinct grade levels, we identified several specialized clusters, debating specific issues related to mathematics education at an international level. We grouped the clusters into three main areas: relationships with psychology, the study of social, cultural and political dimensions, and the relevance of a theory for mathematics education.

Introduction

In this chapter we consider the evolution of mathematics education from its initial status as an enterprise mostly managed by mathematicians to the birth of mathematics education as a scientific field of research. We start our story in the 19th century

---

F. Furinghetti (✉)
Dipartimento di Matematica, Università di Genova, Genoa, Italy
e-mail: furinghe@dima.unige.it

J. M. Matos
Faculdade de Ciencias e Tecnologia, Universidade Nova de Lisboa, Lisbon, Portugal
e-mail: jmm@fct.unl.pt

M. Menghini
Dipartimento di Matematica, Università di Roma Sapienza, Rome, Italy
e-mail: marta.menghini@uniroma1.it

when old states acquired a modern organization, new states were created, and systems of education had to be updated or constructed. In this story researchers in many fields (psychology, philosophy, medicine, sociology, linguistic, anthropology, etc.) had a role, but the main players were professional mathematicians and mathematics teachers. The transition was via a lengthy pathway leading to a clarification of relationships between them and consequent autonomy from mathematicians acquired by mathematics educators. This autonomy was officially acknowledged through the new election procedure for the International Commission on Mathematical Instruction (ICMI)—adopted in Santiago de Compostela (August 19–20, 2006) by the General Assembly of the International Mathematical Union (IMU)—that transferred the election of the ICMI Executive Committee from the IMU General Assembly to the ICMI General Assembly (see Hodgson, 2009). To outline steps taken on this pathway we will focus on the following important moments:

- The attainment by mathematics education of an international dimension at the beginning of the 20th century through the journal *L'Enseignement Mathématique* and the International Commission on the Teaching of Mathematics;
- Curricular reforms; and
- The autonomous initiatives inaugurated by the new approach to mathematics education that made it an academic discipline with a new field of research.

Since the shaping of the new discipline benefited from the interaction with other domains, we will also outline the most influential of these interactions.

The movement of communication, internationalization, and solidarity that endowed mathematics education with an international dimension at the beginning of the 20th century involved countries from all around the world, but most events and people that contributed to making mathematics education an academic discipline belonged to Europe and North America. For this reason our history is mainly devoted to these two regions. We leave to the other chapters of this *Handbook* (see, for example, Chapter 26, by Singh and Ellerton) discussion of other rivulets along which mathematics education developed before and after its emancipation from mathematicians and how attention shifted to other regions of the world.

The roots of mathematics education date back to the origins of mankind. Ancient civilizations left us documents that evidence an intertwining between the development of mathematical culture and concern about the transmission of this culture (Karp & Schubring, in press; Kilpatrick, 1994). In the immense landscape of social, economic, and political events that accompanied the evolution of mathematics teaching, we put forward two important developments that affected it. First, the invention of printing in the 15th century created the possibility of universal literacy and for mathematical knowledge to be transmitted easily to large numbers of people. Over centuries, this led to the second development, the creation of schools to educate the masses. As a result, mathematics—which was an arcane subject 600 years ago—has become a subject studied by virtually all students in the world.

For many centuries the roles of mathematics teachers and researchers in mathematics were largely overlapping. Slowly, when mathematical topics reached an advanced stage far from the elementary level, this overlapping happened only in the case of university teachers who were carrying out research as part of their profession. In the primary and secondary schools, however, the division between teaching and researching mathematics became evident. Among other things, this led to a diversified production in mathematical literature: on the one hand, textbooks specifically aimed at school teaching were published; on the other hand, there was a production of materials reporting new results from mathematics research. According to Struik (1987), the process of professionalization of researchers in mathematics was strongly accelerated by the stimuli given to scientific research in the years of the Industrial Revolution, which created "new social classes with a new outlook on life, interested in science and in technical education" (p. 141). New democratic ideas generated by the French Revolution "invaded academic life; criticism rose against antiquated forms of thinking; schools and universities had to be reformed and rejuvenated" (p. 142). In the 19th century the mathematicians' "chief occupation no longer consisted in membership in a learned academy; they were usually employed by universities or technical schools and were teachers as well as investigators" (p. 142).

Around the middle of the 19th century the profession of mathematics teacher at the primary or secondary level was assuming a new shape, in connection not only with the modernization of old nations, but also with the emerging of new social pulses which manifested themselves in new associations and trade unions, political and social movements, and solidarity initiatives. The transmission of mathematical knowledge was no longer a private matter left to families or to religious bodies, but became a public business under the responsibility of the state. In the following years the establishment of modern national systems of instruction took place in the new and old countries. In this process the main concern became the development of curricula, the production of suitable textbooks, and problems associated with teacher education and recruitment. Soon the need for reflecting on problems inherent in the whole construction gave impulse to the creation of specific journals and associations. This is the setting in which our story of the transition from "mathematics and education" to "mathematics education" began.

Mathematical Communication

Mathematicians, like all scientists, have always felt the need to communicate their results. Towards that end, for a long time they mainly used private communication but after the establishment of academies and societies they began to write proceedings and reports. Following changes in the cultural and social milieu provoked by the Industrial and French Revolutions, the means of communicating became modernized and the first journals devoted specifically to mathematics appeared. Initially they were ephemeral or, like the French *Annales de Mathématiques Pures et Appliquées* lasted for a few decades (1810–1832). But soon, important periodicals, some of them still existing, were published—in 1826, for example, the *Journal für die Reine und Angewandte Mathematik*, founded by August Leopold Crelle, first appeared, and in 1836 the *Journal de Mathématiques Pures et Appliquées*, founded by Joseph Liouville, was published. These journals, and others of the same kind available around that time, contained not only original essays, but also mathematical memoirs extracted from eminent works and abstracts of important papers. In this way they contributed to the progress of mathematics by making available new results and important works not easily accessible to all their readers (among them beginning researchers). They were mainly devoted to research and had an international readership.

Around the middle of the 19th century another kind of journal, usually termed an *intermediate journal*, appeared. Between 1877 and 1881, for example, the *Journal de Mathématiques Élémentaires [et Spéciales]* (editor Justin Bourget) was published, and in 1882 it became the *Journal de Mathématiques Spéciales* (editor Gaston Albert Gohierre de Longchamps). Some of these intermediate journals were specifically addressed to teachers and students in classes preparing for admission to special schools, and mathematical themes were treated at an intermediate level between secondary and university. Earlier, in Great Britain, periodicals such as *The Ladies' Diary* or *Woman's Almanack* first issued in 1704 and *The Educational Times*, issued from 1847 to 1929, contributed in some way to the growth of mathematical knowledge by publishing mathematical questions addressed to amateurs. *The Educational Times,* which was linked to the College of Preceptors (1849), developed from the Society of Teachers (1846). This society had been established to improve the standards of secondary school teaching and, according to Howson (2010), "initially offered qualifications for pupils and teachers" (p. 43). It was from this journal that *Mathematical Questions with their Solutions. From "The Educational Times"* (editor William John Clarke Miller) originated, and this was published between 1864 and 1918.

In the panorama of journals of diverse nature appearing in the 19th century it is difficult to identify journals that specifically addressed secondary mathematics teaching. Indeed, a sign that attention would be given to secondary-level mathematics was the presence of the word "elementary" in the title—though the meaning of this term differed in different journals. Examples of the genre were three French publications *Journal de Mathématiques Élémentaires* (editor Henri Vuibert, founded in 1876), *Journal de Mathématiques Élémentaires* (editor de Longchamps, founded in 1882), *L'Éducation Mathématique* (editors Jean Griess and Henri Vuibert, founded in 1898), and the Italian publication *Rivista di Matematica Elementare* (editor Giovanni Massa, founded in 1874). Sometimes the founders and editors of these journals were schoolteachers, and indeed most of the contributors to the Italian *Rivista di Matematica Elementare* were secondary schoolteachers. The dates of foundation show that journals related to mathematics teaching at secondary level were born later than research journals. This delay is understandable if one considers that primary and secondary teachers, who were constructing their professionalism and their identity when the establishment of the systems of education in the various countries was taking place, constituted the main readership of this kind of journal.

The creation of journals devoted to mathematics teaching was often linked with associations of mathematics teachers. In some cases the periodicals provided roots for the idea of founding professional associations. For example, in 1915 in the USA the MAA (Mathematical Association of America) assumed responsibility for the *American Mathematical Monthly*, which was aimed at teachers of mathematics and, since 1894, had been published, privately. The Association of Teachers of Mathematics of the Middle States and Maryland began publishing a quarterly journal, *The Mathematics Teacher* in September 1908, which eventually was adopted as the official journal of the NCTM (National Council of Teachers of Mathematics) upon its founding in 1920. In Italy the journal *Periodico di Matematica* was founded in 1886 and became the official organ of the Italian National Association of Mathematics Teachers, Mathesis. In other cases the founding of teacher associations led to the publication of new journals for the purpose of spreading information and ideas. In Germany the Deutscher Verein zur Förderung des mathematischen und naturwissenschaftlichen Unterrichts was founded in 1891 and the journal *Unterrichtsblätter für Mathematik und Naturwissenschaften* followed in 1895. In the UK the Association for the Improvement of Geometrical Teaching (AIGT), founded in 1871, evolved into the Mathematical Association in 1897. The Association continued to publish *The Mathematical Gazette*, which had first appeared in 1894. In France the APMEP (Association des Professeurs de Mathématiques de l'Enseignement Public), begun its activities and the publication of its *Bulletin* in 1910.

Communication through journals devoted to mathematics (sometimes together with other sciences) accompanied the growth of the community of mathematicians and later of mathematics educators. Some of the 182 mathematics periodicals listed in a *Catalogue* prepared by the Mathematical Association (1913) still survive; and many new ones would be created. Some of these publications primarily addressed mathematical research, but others were devoted to mathematics teaching. The number of the latter grew considerably in the 20th century so that there were 253 in a list compiled by Schubring and Richter (1980). Some of these journals are examined in (Hanna, 2003; Hanna & Sidoli, 2002).

As we mentioned above, in the decades on either side of 1900 most important national associations of mathematics teachers were founded. These associations, and their journals, helped to promote communication and to shape mathematics teacher identity. In particular, the role of the associations was crucial in stimulating and guiding reforms which took place during that period. These reforms worked towards updating school mathematics in accordance with the new trends in research and towards making curricula suitable in an age of industrial and technological innovation. As observed by Nabonnand (2007), it is true that the spirit of reforms was often embodied by strong personalities such as John Perry in the UK, Felix Klein in Germany, and Charles Émile Ernest Carlo Bourlet in France, but the programs of reforms were discussed, worked out and spread with the teacher associations as important players.

In the United States of America the American Mathematical Society (AMS) was formed in 1888. AMS always emphasized research (and still does), whereas MAA emphasized teaching (in colleges), and still does. Eliakin Hastings Moore advocated Perry's ideas in his 1902 AMS Presidential address, see (Moore, 1903), and many AMS members were angry with him for doing so. In the UK, the Association for the Improvement of Geometrical Teaching (AIGT) and, later on, the Mathematical Association, were born with the aim of supporting reforms in the geometric syllabus. In Switzerland, new programs centred on the introduction of graphical representation of functions were introduced following the proposal of the mathematics teachers association. In 1906, after a talk delivered by Emanuel Beke at the annual meeting of the society of Hungarian teachers, a Commission charged with studying general reforms and changes in secondary mathematics teaching was instituted by the same society with Beke as its first president. In certain cases an important role of associations was to defend mathematics teaching when it was marginalized. For example, in Italy the mathematics teacher association *Mathesis*, founded in 1895, had the aim of supporting mathematics teaching against a decline which had started in the 1890s. Most of the associations are still alive and in good health; new ones have been founded. Many publish journals, bulletins, and newsletters, as well as organize national meetings and other activities.

Often, both teachers and professional mathematicians participated in these initiatives. There were also initiatives carried out by secondary teachers alone; this happened, for example, in Italy during the initial period of the teacher associations. In other cases, for example in France, academic mathematicians drove these initiatives and led reform movements. The problem of the relationship between the two communities (mathematicians and mathematics teachers) and the need to share responsibility and authority are ever-present in the background of the development of mathematics education to the status of an academic discipline.

Mathematics Education Unbounded

The national journals and teacher associations became an important tool for transmitting ideas and information among teachers within many nations, and proved to be of crucial importance in shaping the identity of mathematics teachers. Considering that the themes treated were related to the national systems of education and that the teachers of a country constituted the readership, it is not surprising that most contributors were national and that the actions of teacher associations were mainly confined to dealing with national problems. In the journals devoted to mathematics teaching the contributions by foreign authors were very few and usually translated into the local language. In spite of these national settings, we can identify some common ground in reflections, at the beginning of the 20th century, on the problems of mathematics teaching. Discussions about the organization of curricula were often based on three main themes:

- Relationship between parts of programs;
- Rigor versus intuition; and
- Relationships between mathematics and the other disciplines.

What emerged was the need to go beyond discussions on the reorganization of the curricula. It was recognized that there was a need to consider new methods of teaching that took into account the following:

- "Practical approaches to teaching," based on observation, experiments and laboratories;
- New findings about children's development; and
- A focus on applications.

Due to many common features among mathematics education problems, possible advantages of international cooperation in working towards solutions to the instructional and other educational problems were recognized in many countries. In the following we will describe two main initiatives that strongly contributed to this growing internationalism.

The Journal *L'Enseignement Mathématique*

In the second half of 19th century, internationalization was a perennial idea in many aspects of society. Transportation was becoming speedier, and technological developments facilitated long-distance communication. In this context it was not surprising to see the emergence of the idea of world exhibitions, or fairs, which provided occasions for showcasing new industrial and technological productions and sharing ideas and projects. The first world exhibition was held in London (1851), and this was followed up over the next 30 years with exhibitions in Paris, Vienna, Philadelphia and Melbourne. Internationalism invaded all aspects of life, among them mathematics. It was not by chance, then, that in 1893 a congress of mathematicians was held in Chicago, where a world exhibition was being organized. The 1893 congress of mathematicians was the cornerstone in the process of making mathematics unbounded, and heralded a tradition (started in 1897) of organizing International Congresses of Mathematicians (ICMs). One of the promoters of the tradition of having ICMs was the French mathematician Charles-Ange Laisant, who was stimulated both by his cultural view of the nature of mathematics and by social ideals of fraternity and solidarity (Furinghetti, & Giacardi, 2008).

Following the Congress of Paris, in 1900, ICMs have been held every 4 years (except for breaks due to the two World Wars). These regular forums have contributed remarkably to shaping the identity of an international community of research mathematicians. The International Mathematical Union (IMU) was founded in 1920, and although it was dissolved in 1932 it was re-established in 1951, with the first General Assembly of the new IMU being held in 1952.

The idea of internationalism was not easily transferable into the world of education for two obvious reasons: (a) issues of instruction are mainly national; and (b) mathematics teachers have a status different from that of mathematicians—in particular, they have less opportunities and financial resources for communicating and traveling together. Still, mathematics education was touched by internationalization, thanks to the foundation in 1899 of the journal *L'Enseignement Mathématique* by Laisant and the Swiss mathematician Henri Fehr. The mission and vision of this publication, explicitly declared by the editors in the first issue, was to make mathematics instruction join the movement of solidarity, internationalism and communication of the times.

This international character of *L'Enseignement Mathématique* marked the difference between this journal and the other existing journals addressed to mathematics teaching: immediately, it published surveys on the situation of mathematical instruction in different countries. The editorial board included mathematicians and historians of mathematics who had already shown a genuine interest for the problems of mathematics teaching (notably Klein), and of communication in mathematics (notably Magnus Gustaf Mittag-Leffler, founder of the mathematical journal *Acta Mathematica*).

The early years and the development of *L'Enseignement Mathématique* have been outlined by Furinghetti (2003, 2009). The journal was special not only for its international character, but also for its scope. In the sixth volume (1904) the editors claimed that for them the word "enseignement" (teaching) had the widest possible meaning: it meant teaching to pupils, as well as teaching to teachers—and, indeed, the editors made clear, one can hardly have the one without the other. For this reason they explicitly stated their intention to dedicate a wide place to questions of philosophy, methodology, and history. For them, teachers needed to enlarge their horizons beyond the program of their classrooms and their countries.

L'Enseignement Mathématique was a product of the mathematical milieu—but Fehr was teaching in Geneva, where the psychologists Édouard Claparède and Théodore Flournoy were working. They used the journal to launch a questionnaire investigating the ways of working of mathematicians. This study is important because it pointed to aspects that were not merely cognitive—using terminology that we would now say was concerned with the affective domain. On the other hand, research mathematicians, like Henri Poincaré, published articles in the journal that focussed on aspects related to the nature of the mathematical invention.

The Rise and Development of an International Project: The Early ICMI

In 1905 David Eugene Smith published in *L'Enseignement Mathématique* a paper that advocated more international cooperation and the creation of a commission to be appointed during an international conference with the aim of studying instructional problems in different countries (see Smith, 1905). This article was the seed for the establishment, during the fourth ICM (Rome, 1908), of the International Commission on the Teaching of Mathematics, with Klein as its first president. In the first decades of its life the Commission was most commonly referred to as CIEM

(*Commission Internationale de l'Enseignement Mathématique*), in French, or IMUK (*Internationale Mathematische Unterrichtskommission*), in German. Though it underwent many changes in status and scope, this Commission may be considered the first incarnation of the present ICMI.

The significance of the foundation of ICMI goes beyond the mere creation of an organizational structure. What was important was that it pointed to the existence of an international community for whom the main focus of attention would be mathematics education. Given that the initial members of ICMI were nations, and that the representatives of those nations were predominantly academic mathematicians, it was not surprising that for a long time ICMI's activities were developed inside the community of mathematicians. During ICM meetings, ICMI presented its reports and received mandates for future activities (Furinghetti, 2007; Furinghetti & Giacardi, 2008; Menghini, Furinghetti, Giacardi, & Arzarello, 2008).

The main ICMI outcomes in the early years were national reports on mathematical instruction in the various countries, and international inquiries on important themes of the teaching of mathematics. Although Klein (1923) explicitly claimed that ICMI recognized that all levels of school mathematics deserved attention, in practice attention was mainly paid to secondary and tertiary levels, and to teacher education. These priorities were evident in the following list of activities launched by ICMI between 1908 and 1915:

- Current situation of the organization and of the methods of mathematical instruction;
- Modern trends in the teaching of mathematics;
- Rigor in middle school teaching and the fusion of the various branches of mathematics;
- The teaching of mathematics to students of physical and natural sciences;
- The mathematical training of the physicists in the university;
- Intuition and experiment in mathematical teaching in the secondary schools;
- Results obtained on the introduction of differential and integral calculus into the upper years of middle school;
- The place and role of mathematics in higher technical instruction; and
- Inquiry into the training of teachers of mathematics in secondary schools in the various countries.

Like many other scientific institutions, ICMI suffered a general crisis during the First World War, and the period between the two world wars was a time of stagnation in ICMI's activities (Schubring, 2008). During the first General Assembly of the reconstituted IMU, held in Rome in 1952, ICMI became a permanent sub-commission of IMU.

However, times had changed, and in the 1950s and 1960s the old agenda based on inquiries and national reports was felt to be inadequate to face new situations. Also, relationships with mathematicians needed to be reconsidered in order to deal with educational problems efficiently.

Curricular Reforms in the 20th Century

Reforms at the Beginning of the 20th Century

At the time ICMI was born issues associated with the construction of mathematics curricula were often hotly debated in many countries. These debates not only discussed issues common to the different nations, but also nation-specific matters.

For instance in both the UK and Italy, the adequacy of Euclid's *Elements* for the teaching of geometry was a much-debated topic. In the early 1870s the AIGT (Association for the Improvement of Geometrical Teaching) had been created to consider, and to challenge, the tradition of using rote exercises for the entrance examinations to British Universities. The ensuing discussions generated numerous alternative textbooks and also led to some changes in the entrance examinations. But in 1901 the British Association for the Advancement of Science hosted an address by John Perry that would influence mathematics education throughout the world. Perry attacked the whole system of a mathematical education which, he claimed, did not take into account children's minds, their interests, the applications of mathematics, and connections between different areas of mathematics. His idea of *practical mathematics* applied to the study of geometry meant that the first work with geometry should involve students using rulers, compasses, protractors, set squares, and scissors. In England, the "Perry movement" initiated much discussion about mathematics syllabi and about the need for the reconstitution of secondary mathematics education (Howson, 1982). It also influenced many countries outside England, such as Japan, where Perry had taught for a brief period (Siu, 2009), and the USA, where, as previously mentioned, Moore accepted Perry's arguments and convictions in relation to mathematics education (Moore, 1903).

In Italy an adaptation of Euclid's *Elements* was published in 1867/1868 as the first Italian textbook after the unification. The authors were famous mathematicians who defended the idea of the purity of geometry against criticisms expressed in Italy and in the UK. The Italian reformers emphasized the importance of preparing and publishing good manuals based on the *Euclidean method*. In Italy, research in the field of geometry was flourishing, and many important researchers were engaged in authoring textbooks. For lower secondary school an intuitive geometry was introduced based on observation and on experimental activities.

Towards the end of the 19th century in the USA a "Committee of Ten" was appointed to make recommendations on the standardization, in contents and methods, of American school curricula (Kilpatrick, 1992). The subcommittee for mathematics produced a range of recommendations, for elementary to high school mathematics curricula, which can be summarized in the key words "exercise the pupil's mental activity" and "rules should be derived inductively instead of being stated dogmatically."

In France, a reform of 1902 especially directed at the lycées recognized the need for emphasis on new modern humanities, including mathematics, and to do away with the monopoly of the classical humanities. The reformers also called for school mathematics to take on a greater sense of reality, displaying more applications to the life sciences (Gispert, 2009). An important aspect of the reform was the introduction of elements of differential and integral calculus into secondary schools. We recall that during this period France was a leading country in the field of analysis.

Both the French reform and the Perry movement with its demand for increased emphasis on calculus gave impetus to the German reform movement led by Klein. This movement, whose key phrase was "functional reasoning," had among its principal aims the shifting down of some elements of differential and integral calculus from university to secondary school. However, the contents of the reform were not limited to the last school years; on the contrary, the reform started from the lower grades and involved many teachers (Schubring, 2000). The present-day emphasis given to functions as the conceptual building block for the teaching and learning of algebra and geometry is reminiscent of this German reform movement (Törner & Sriraman, 2005). In particular, the role of analytical geometry in the study of functions was stressed and thus a link between school geometry and algebra was established. Moreover, Klein's Erlanger Program, which characterized geometry as the study of invariant properties under a group of transformations, provided a stimulus for deeper work on geometric transformations in mathematics teaching.

After becoming the foundation president of ICMI in 1908, Klein promoted an international reform based on the ideas of the German reforms. An international comparison of curricula, which was part of ICMI's agenda from the start, was to serve as a key enabling element for this proposal (see Schubring, 2003). Although not all countries participated actively, many initiated significant curriculum reform activities during that period. According to Schubring (2000) these countries included Austria, Belgium, Denmark, France, Germany, Great Britain, Hungary, Sweden, and the USA.

Our analysis of the contents of the mathematics curriculum in various nations led us to concur with Howson (2003) that up to the late 1950s there was considerable agreement on what school algebra might mean. After the introduction of letters to denote numbers or variables should come the construction of algebraic formulae, followed by the formation or solution of linear equations, then quadratics, then simultaneous linear equations, and the properties of the roots of quadratic and cubic equations. In contrast, there might be notable differences in the teaching of geometry. These concerned the closeness to the original Euclid, the level of rigor, the use of algebraic or analytical means, the experimental or intuitive, the use of geometric transformations, and the attention given to space geometry. Nevertheless, it was generally agreed that in most nations attention to a small number of classical theorems in geometry was required—these theorems included the theorem of Pythagoras, the theorem of Thales or intercept theorem, the circle theorems, and congruence and similarity properties.

Modern/New Math(s)

A second international reform that occurred in the 1960s is thought to have originated from the group of mathematicians established in 1932 under the assumed name Bourbaki. The interest of the Bourbaki group in mathematics education started in the 1950s, when some of its members joined the International Commission CIEAEM (Commission Internationale pour l'Étude et l'Amélioration de l'Enseignement des Mathématiques), founded by Caleb Gattegno with the aim of studying and improving mathematics teaching (see Félix, 1985). This Commission comprised people from different backgrounds (mathematicians, pedagogists, psychologists, epistemologists, and secondary teachers).

In its initial years, CIEAEM's actions may be summarized in the following points: democratization of mathematics, active pedagogy, and actual involvement of teachers. Among the mathematicians of this research group we find the Bourbakists Jean Dieudonné, Gustave Choquet, and André Lichnerowicz, who also contributed to the text by Piaget et al. (1955), which was the first of the two books edited by CIEAEM. In that book all authors recognized the opportunities that modern mathematics offered in relation to the reform of mathematics teaching, and Dieudonné claimed that the essence of mathematics was reasoning on abstract notions.

The "modern mathematics" movement that developed in Europe had common roots with a parallel movement in the USA (see Moon, 1986)—the new math movement started in the early 1950s by Max Beberman with the creation of the University of Illinois Committee on School Mathematics (UICSM). Soon after the launch of Sputnik in 1957, the American Mathematical Society set up the School Mathematics Study Group (SMSG) to develop a new curriculum for high schools. In 1958, Edward G. Begle, then at Yale University, was appointed as its Director (see Griffiths, & Howson, 1974; Wooton, 1965). Among the many curriculum groups established in the USA during the new math period, SMSG was, perhaps, the most influential. The experiences of this group and the numerous other mathematics curriculum groups established around that time benefited from contributions of psychology (Kilpatrick, 1992).

All these streams of reform related to modern, or new, mathematics met in 1959 at an international conference held in Royaumont, near Paris. The conference was organized by OEEC (Organisation for European Economic Co-operation), and chaired by Marshall Stone, the president of ICMI. An important role was played by members of CIEAEM, particularly by Dieudonné, who gave a lecture concerning the transition from secondary school to university. According to Dieudonné, the treatment of geometry should proceed from the real numbers, establishing rules for the operations on a set of undefined objects so that a vector space structure would be created. Metric relations would then be introduced by means of a scalar product. Euclidean geometry could be dealt with in only three lessons, in which the system of axioms would be presented. The properties of triangles would not have a role in this new development (OEEC, 1961).

We note that in the same year, 1959, the Woods Hole Conference took place in the USA, with the more general aim of improving science education, and bringing together scientists, mathematicians, psychologists and others (Bruner, 1960).

The aim of the Royaumont Conference was to achieve mathematics curriculum reform in Europe—but, since both the USA and Canada had been invited to attend, it could be argued that an international reform stretching beyond European nations was desired. The conference had a more practical sequel in 1962 in Dubrovnik,

Yugoslavia, when a group of experts met to produce a modern program for mathematics teaching in secondary schools. In the geometry programs for the ages 15–18 produced by the Commission, the Cartesian plane was defined as a vector space of dimension two with a scalar product. In line with the proposals of Choquet (OEEC, 1962), these concepts were to be introduced via axioms. For children aged from 11 to 15 years, a more intuitive approach to geometry was recommended, in line with the proposals by the Belgian mathematician Paul Libois. So far as algebra was concerned, the contents listed in Dubrovnik included sets, applications and functions, the introduction to real numbers, elements of number theory, combinatorics, groups and structures, linear applications and matrices. Some of these topics would become standard in many curricula. Set theory was to be a major integrating theme, and strongly influenced the language used in textbooks written for modern mathematics.

In both Europe and the USA, the path of innovation was to start at the university and proceed down through the secondary schools to primary schools. Set theory would be present at all levels of education with, for example, cardinal and ordinal aspects of natural numbers being introduced at the beginning of elementary grades (Pellerey, 1989). Many countries officially adopted modern mathematics programs, and in France and Belgium the proposals were completely in line with Bourbakist viewpoints.

Although the modern/new math movements soon aroused strong criticisms (see, e.g., Ahlfors et al., 1962; Kline, 1973; Thom, 1973), the ample debates about changes in school mathematics provided a springboard for subsequent, more solidly based reform initiatives in the 1960s. In the UK the School Mathematics Project was launched in 1961, and the work of Edith Biggs and the "Nuffield Project" popularized the use of concrete materials and of laboratory techniques in British primary school mathematics programs. In 1967 the Nordic Committee for the Modernization of School Mathematics (Denmark, Finland, Norway, and Sweden) presented a new syllabus inspired by new math. One of the best-known members of this Committee was Bent Christiansen, of Denmark. In 1968 the Zentrum für Didaktik der Mathematik (Centre for the Didactics of Mathematics) was founded in Karlsruhe by Hans Georg Steiner and Heinz Kunle. This was followed in 1973 by the IDM (Institut für Didaktik der Mathematik), founded in Bielefeld by Steiner, Michael Otte and Heinrich Bauersfeld, whose aims combined practice in school and theoretical research. In 1969 the first IREMs (Instituts de Recherche sur l'Enseignement des Mathématiques) were established in Lyon, Paris, and Strasbourg. In the early 1970s the Collaborative Group for Research in Mathematics Education was established at the University of Southampton's Centre for Mathematics Education, with Geoffrey Howson and Bryan Thwaites as collaborators. In 1971 Hans Freudenthal founded the Institut Ontwikkeling Wiskunde Onderwijs (IOWO, Institute for the Development of Mathematics Teaching). This initiative had its far roots in the "Mathematics Working Group" founded in 1936 by Tatiana Ehrenfest-Afanassjewa. The meetings of this group were attended by Freudenthal and constituted a first step in the successive development of the "Realistic Mathematics" movement, initially led by Freudenthal (Smid, 2009).

New Bourbakist-type topics such as vectors, transformations, matrices, and set theory were included in the school mathematics curricula of numerous countries, and a greater emphasis on probability and statistics became the order of the day. The 1970s were fertile years for the creation of projects, as shown by the fact that the presentations of 15 projects were mentioned in the *Proceedings of the Third International Congress on Mathematical Education* (ICME-3), held in Karlsruhe, Germany, in 1976. These and other changes in mathematics education were outlined in a special issue of *Educational Studies in Mathematics* entitled "Change in Mathematics Education Since the Late 1950s—Ideas and Realisation: An ICMI Report" (1978).

Creeping Reforms

In addition to these strong curricular innovations there were also some creeping reforms that influenced both curriculum content and teaching and learning methods in school mathematics. The experimental work of psychologists, new teaching aids, and the reform movements of the early 20th century brought an interest among mathematicians in mathematics laboratories (Borel, 1904) in which students actively used drawing instruments, calculating machines, and manipulatives. At the beginning of the 20th century, Peter Treutlein, a German mathematician, developed more than 200 models that could assist the teaching of geometry (Treutlein & Wiener, 1912). These models were manufactured and distributed by famous manufacturers such as those of Ludwig Brill (Darmstadt) and Martin Shilling (in Halle and then Leipzig) in the middle of the 20th century, and came to be widely used in German universities and polytechnics.

After the Second World War the use of concrete materials was taken up again in many contexts. In 1945 an NCTM yearbook was devoted to measuring and drawing instruments and to the creation of three-dimensional physical models. An active promoter in this field was Gattegno, who focussed the early activities of CIEAEM on concrete materials (see Gattegno et al., 1958). This activity had an important didactical transposition in the work of the teacher Emma Castelnuovo. Gattegno, as well as the mathematician and psychologist Zoltan Dienes, strongly supported the use of manipulatives, such as Cuisenaire rods and logic blocks, in classroom activities. The presence of Dienes at ICME-1 in Lyon, France, in 1969 testified to the interest of the ICMI community in the use of concrete materials.

Other psychologists, including Jean Piaget, influenced the movement. Willmore (1972) and Price (1995) have pointed out the importance of this in changing thinking about the teaching and learning of mathematics. Libois used concrete materials at the École Decroly in Brussels, and in the UK, the Association of Mathematics Teachers (ATM) strongly supported Gattegno's initiative in promoting the use of manipulatives. Manipulatives became a vehicle for intuition and experiment in the classroom, and prepared school milieu to receive subsequent innovations with mathematical technology (Ruthven, 2008). Gattegno authored innovative software for teaching elementary numeration concepts and films for teaching geometry that extended some of the themes in Jean Nicolet's films (Powell, 2007).

In the *Proceedings* of the first ICME Congress (1969) we find reference to games, worksheets, films, overhead projectors, and to concrete materials to be used in the classroom. The use of materials is put in relation to a new methodology of classroom activities that also includes working groups and classroom discussion. At that time, computers were entering into discussions on mathematics education. An explicit reference to the role of computers in school mathematics, especially for applied mathematics, was made by Bryan Thwaites (1969) in his address at ICME-1. At the same conference, Frédérique Papy presented the "minicomputer" (Papy, 1969). The initial interest in the algorithmic aspects or in discrete mathematics created a place for programming to be considered as a means for attaining rigor (Furinghetti, Menghini, Arzarello, & Giacardi, 2008).

In the 1970s and 1980s attention turned towards learning environments, or microworlds, for example, in the form of turtle geometry as presented by Seymour Papert at ICME-2 (Howson, 1973; Papert, 1972a, 1972b). Software was developed, including forerunners to the dynamic geometry software which helped in revitalizing parts of mathematics, for example, proofs and Euclidean geometry. Technology was considered as a means for changing both the curriculum and teaching practices; mathematical activity could be enriched by modelling or processing data in statistics, by experimenting, and by visualizing. Research on the role and use of technology in the teaching of geometry was conducted, at first using a constructivist perspective in a broad sense, and later using additional theoretical perspectives, in particular, the social interactions in which learning takes place (Laborde, 2008). The use of dynamic geometry software was explored as a mediator between constructivist and other theoretical levels, highlighting the need for precise curricular construction (Borba & Bartolini, 2008).

The increasing availability of ordinary calculators, scientific calculators, and graphics calculators generated interesting experimental approaches to instruction. On the one hand, attention was directed at algorithmic aspects (see Engel, 1977), but on the other hand the ways in which some topics—functions, for example—might be dealt with in secondary schools using the new technology began to be investigated (Guin, Ruthven, & Trouche, 2005). Already, at ICME-2 in Exeter, a Working Group had explicitly addressed technology, and at ICME-3 in Karlsruhe this happened with five official activities. The survey presented by Fey (1989) at ICME-6 in Budapest described developments in the use of technology during this pioneering period. The first ICMI Study, launched in 1984, was devoted to computers and informatics (Churchhouse et al., 1986).

From Mathematics and Education to Mathematics Education

Emergence of New Approaches in Mathematics Education

In the 1950s mathematical research changed direction, and also the role of mathematics in society changed. New uses of mathematics were promoted by advances in technology, and by the political associations with the space race and the iron curtain. Mathematics instruction was perceived by governments as linked to an important potential for power among nations. In the meantime, schools were being called upon to deal with rapidly increasing populations and associated educational problems.

Given the complexity of emerging educational problems, the mere study and comparison of curricula and programs, which had been the main activities of early ICMI, were judged to be insufficient. New approaches to mathematics education suitable to the changed mathematical and social contexts were needed (Furinghetti, Menghini, Arzarello, & Giacardi 2008). Various initiatives, such as CIEAEM and the USA curricular groups, pointed to the need for cooperation among mathematicians, teachers psychologists, mathematics teacher educators and mathematics teachers. Clearly, there had emerged a need for new professional expertise featuring what Krygowska (1968) called "frontier research," which acknowledged mathematics education as a scientific discipline.

Freudenthal (1963) observed that history had shown the sterility of the problems of mere organization. By the end of the 1960s research interest shifted from curricular issues to the wider study of various dimensions of mathematics education. There emerged a trend towards widening the scope of curricular interventions, for example to pre-school and to vocational and adult education settings. There was also a call for more careful scientific research in mathematics education. A strong case for the importance of empirical research was made in the first ICME in 1969 by Begle, then at Stanford University. According to Begle (1969):

> ... the factual aspect has been badly neglected in all our discussions and ... most of the answers we have been provided have generally had little empirical justification. I doubt if it is the case that many of the answers that we have given to our questions about mathematics education are completely wrong. Rather I believe that these answers were usually far too simplistic and that the mathematical behaviours and accomplishments of real students are far more complex than the answers would have us believe. (p. 233)

Interest in empirical research in mathematics education was growing in the USA and, by the mid-1960s, several conferences discussing priorities for research for mathematics education took place.

By 1968 a Special Interest Group on mathematics education research had been formed within the American Educational Research Association (Kilpatrick, 1992). Although this kind of research was not being embraced in many other countries, the growth of international research journals and centres would change this perspective. As Fehr and Glaymann (1972) stated in the UNESCO publication *New Trends in Mathematics Teaching*:

> The curriculum reform movement of the last two decades in school mathematics was aimed primarily at improving educational practice. It was not designed to increase the number or the quality of research studies in mathematics education. Nevertheless, the reform movement did enormously stimulate such research—in part because curriculum reformers have been asked to demonstrate that their work can make a difference in the classroom; in part because these reformers have recognized that future changes can be managed better if we understand more about the teaching and learning of mathematics; and in part because the ferment in the curriculum has attracted many new scholars to the study of problems in mathematics education. (p. 127)

There was also a growing recognition of the need for the academic legitimacy of specialists in mathematics education to be recognized and respected. The "Resolutions of the First International Congress on Mathematical Education" (1969) assumed that mathematics education was becoming a science in its own right, with its own problems relating to both mathematical and pedagogical content. ICME called for the new science of mathematics education to be given a place in suitable mathematical departments of universities or research institutes.

This discussion about the identity of *mathematics education*, or *didactics of mathematics*—the preferred nomenclature in some countries—was continued at ICME-2 in 1972. Anna Zofia Krygowska, for example, in her contribution to the Working Group on teacher training for prospective secondary teachers, which was chaired by Steiner, identified four aspects of didactics of mathematics: a synthesis of the appropriate mathematical, educational, cultural and environmental ideas; an introduction to research; the nature and situation of the child; and practical experience (see Howson, 1973). Bent Christiansen (1975) distinguished between mathematics education as a process of interaction between teachers and learners in their classes and the didactics of mathematics, which was the study of this process. He recognized in didactics of mathematics the status of a new discipline and pointed out that it must be taught by specialists—"didacticians of mathematics"—and not by general education specialists.

New Initiatives in Mathematics Education

The rethinking on the role and the methods of mathematics education carried out in the 1950s and the 1960s led to a global discussion that included rethinking about the relationship between mathematicians and mathematics educators and a plan for new ways of communicating among mathematics educators. Two ICMI presidents faced these issues with particular energy—Heinrich Behnke and Freudenthal (Furinghetti & Giacardi, 2010). The former tried to settle administrative relationships, including financial issues, with mathematicians after the rebirth of ICMI in the 1950s and looked for new terms of references. But this was not enough: a cultural cut with mathematicians was necessary and this was made by Freudenthal who acted on the two main issues that were characterizing the dependence on the mathematical community, journals and conferences. Both the initiatives he took were taken independently from IMU.

L'Enseignement Mathématique, the official organ of ICMI since its foundation, was becoming a mathematical journal with little room for educational issues. On the other hand, the professional mathematics teaching journals were local and, due to their mission and vision, not suitable for publishing articles on didactic research. So in 1968 Freudenthal founded *Educational Studies in Mathematics* (ESM) (Furinghetti, 2008). According to Hanna (2003), this initiative stimulated other groups to publish mathematics education research journals: *Zentralblatt für Didaktik der Mathematik* (ZDM) (now *The International Journal on Mathematics Education*) was first published in 1969 (the first editors were Emmanuel Röhrl and Steiner) and the *Journal for Research in Mathematics Education* (JRME) was first published in 1970 (the first editor was David C. Johnson). ESM and its contemporary journals would become important vehicles in which questions, methods, and research within the discipline "mathematics education" were developed, reported and discussed.

The development of mathematics education was also accelerated by new ways of meeting at the international level. At its inception ICMI promoted important conferences, such as those in Milan (1911) and Paris (1914) but, because conferences were no longer held during and after World War I, the only scheduled places for discussing didactical issues were in those sections within the quadrennial ICMs that were devoted to the didactics of mathematics. These sections usually encompassed also philosophy of mathematics, history, and logic, and were variously put together or separated according to the inclinations of the organizers. No plenary talk was ever devoted to mathematics education.

In the 1960s the new math movement stimulated some important meetings in the USA and in Europe that focussed on mathematics education research, and ICMI collaborated with UNESCO in organizing some of these conferences. Occasionally the audience was enlarged to include teachers. Freudenthal succeeded in establishing the tradition of having an international conference—the International Congress on Mathematical Education (ICME)—with regular dates. The first of these conferences (1969 in Lyon) was organized according to a traditional pattern of presenting a sequence of talks, but already at the second of Exeter (UK), in 1972, working groups were organized, and projects presented with the aim of creating the very place for discussion of ideas. Since ICME-3 (Karlsruhe, 1976), ICME meetings have been held on a quadrennial basis.

New perspectives for looking at mathematics education also emerged from within the body of mathematicians. The concluding sentence of the talk delivered by Hassler Whitney, a mathematician who became president of ICMI in 1979, provided evidence that attention might be shifted to the learner:

> We are too used to thinking of the subject matter, and how children can learn it. We must start with the children, to see what they really are. (Whitney, 1983, p. 296)

At ICME-3, in Karlsruhe, the first affiliated study groups were established—HPM (the International Group on the relations between the History and Pedagogy of Mathematics) and PME (the International Group for the Psychology of Mathematics Education) (see Furinghetti & Giacardi, 2008). With these groups a new period began with regular meetings and proceedings. This marked the evolution of the provision of support for researchers in mathematics education.

Clusters of Specific Issues in the "Discipline" of Mathematics Education

By the middle of the 1970s new tendencies outlined in the last section were manifesting themselves more clearly at the international level. Understanding that the endeavour of searching for directions for mathematics education required more than merely discussing curricular options at the distinct grade levels, ICMI officials met with staff members of UNESCO at the end of 1974 to prepare the elaboration of the fourth volume in the series of books, *New trends in mathematics teaching*. This was an important step towards deepening discussion of issues that had already been raised. The aim was not only to identify major problems in the field of mathematics education but also to guide and monitor the direction and intensity of changes taking place in that field (Steiner & Christiansen, 1979). A methodology favouring in-depth discussion of the chapters was chosen, leading to broader approaches to the issues of mathematics education (D'Ambrosio, 2007). The results of this careful preparation of the book became visible during the third ICME that took place in Karlsruhe in 1976 and constituted a landmark in the history of mathematics education.

As a consequence of this in-depth approach, the fourth volume of *New Trends* contained chapters dedicated to the discussion of curricular issues at various levels—including adult education, university teaching, and the use of technology. These were discussed at a deeper level than ever before and a critical analysis of curriculum development and issues associated with the evaluation of students, teachers and educational materials was presented. The importance of moving on from curricular issues was noticed and appreciated: "Until recently, both research and development had focussed on only one of two main determinants of the learning process: the pupil or the curriculum. They did not consider the influence of the teacher nor of the general context of instruction" (Bauersfeld, 1979, p. 200). The book also contained a chapter on the professional life of teachers of mathematics and another discussing goals and objectives for mathematical education.

The third ICME, at Karlsruhe, and the publication of the fourth volume of *New Trends* have been acknowledged as the starting point for the formation of several specialized clusters of specific issues related to mathematics education at an international level. We will group them into three areas: (a) relationships with psychology; (b) the study of social, cultural and political dimensions; and (c) the relevance of a theory for mathematics education.

Psychology and Mathematics Education

Since the late 19th century answers to issues related to mathematics teaching and learning have been sought in fields outside of mathematics. Important contributions came from the merging of competencies within various educational sciences and other disciplines: pedagogy, psychology, philosophy, and medicine. The early works carried out in this field concerned pupils with particular needs, but the methods applied in these cases soon proved to be suitable for dealing with problems associated with the teaching and learning of normal children in the primary school. The mathematical content taken into consideration was mainly concerned with arithmetic, but the use of concrete materials affected also the teaching of geometry.

Educators carried out their work in *practice schools* founded and directed with the purpose of experimenting with new teaching methods. In these schools practice was strongly interwoven with research and two different research streams arose: one was concerned with research on teaching methods, the other with the observation of pupil behaviour. In these developments the roots of theories of learning that are concerned with what goes on in the brain of the learner (such as in Piaget's theory) can be recognized, as can theories of instruction that refer to the behaviours a child should undertake in order to learn (such as in Bruner's theory).

The influence of the work of pedagogists and psychologists in mathematics education probably started at the beginning of the 19th century through the Swiss educator Johann Heinrich Pestalozzi. Pestalozzi influenced the teaching and learning of arithmetic and geometry in primary schools in Europe (de Moor, 1995; Howson, 2010) and in the USA (Cajori, 1890). One of his followers was Friedrich Fröbel, the founder of the German kindergarten organization. Fröbel brought his pupils to learn by means of games and other activities—wooden blocks were used to teach arithmetic and concrete geometrical objects to teach geometry.

Johann Friedrich Herbart was another scholar to influence how mathematics was taught in schools. Around 1900, Herbart's ideas influenced elementary teaching and teacher education in various countries (Howson, 1982). Notwithstanding the stages of instruction that Herbart urged teachers to follow (see Ellerton & Clements, 2005), his views of the relationship between teaching and learning can be regarded as being consistent with what later became known as constructivism. It was largely based also on human and social interactions.

The interest in child education grew particularly in the USA as a result of the writings of John Dewey who, in 1896, founded a laboratory school at the University of Chicago. In 1904 Dewey moved to Columbia University, where he spent the rest of his career. Dewey framed all learning as the result of activity. As for mathematics learning, one of his leading premises was that the notion of quantity is grasped by the child as a result of solving practical problems (Stemhagen, 2008). This idea of *active learning* was also present in the work of Maria Montessori, who created a school for children in Rome, and of Ovide Decroly who created the *École de l'Ermitage* in Brussels. Both were physicians who developed their methods when working initially with children with minor disabilities. Decroly's method was based on observations of the surrounding world, but Montessori developed specific materials (materiale strutturato) that were intended to help children to learn autonomously. After that period many psychological laboratories were established in Europe, often by psychologists such as Alfred Binet—the French psychologist famous for his contributions to intelligence theory and testing—and the Swiss neurologist and child psychologist, Claparède. Children's attempts to learn mathematics were often studied in Binet's and Claparède's laboratories.

In the USA, research in the learning of mathematics was conducted by Edward Lee Thorndike, a behaviourist psychologist who had a strong interest in mathematics learning, and William Brownell, a teacher, psychologist, mathematics educator, and education psychologist. Brownell and Thorndike, although coming from different theoretical positions, were part of a broader movement to create a science of education. In 1922 Thorndike published his *Psychology of Arithmetic*, and soon after that his *Psychology of Algebra* (1923). Both were based on the theory of associations in a "connectivist" perspective, and were intended to support Thorndike's series of school mathematics textbooks. Brownell, following the ideas of his advisor Charles H. Judd, stressed the importance of "meaningful learning" with respect to "rote" methods, in contrast to Thorndike's more behaviourist views (Kilpatrick & Weaver, 1977).

Behavioural psychological theories ("behaviourism"), which had been developed via experiments with animals, were linked to school learning by Burrhus Frederic Skinner during the period 1930–1950 with an emphasis being given for what became known as operant conditioning. Skinner emphasized reinforcement processes, seen as fundamental in the shaping of behaviour. According to the corresponding instructional theory, changes in behaviour could be obtained through programmed instruction (or, later on, through mastery learning and computer-assisted learning). These ideas had a wide application in mathematics instruction (see Skinner, 1954), and in particular on theory supporting the early uses of computers in learning.

The major influence of psychology on mathematics education, however, came from the work of the Swiss psychologist, Jean Piaget. While studying the behaviour of children in a clinical manner and identifying "cognitive stages," Piaget developed methods that permitted broadening the range of mathematical topics in primary school. Piaget's stages were paralleled in the USA by the instructional stages of Jerome Bruner but, as Kilpatrick (1992) put it, only "with the arrival of cognitive psychology in 1950s and 1960s, marked by the availability of Piaget's work in English translation and the reinterpretation of that work by Jerome Bruner, [did] researchers in mathematics education begin to have a more judicious regard for psychological theory and to collaborate more frequently with psychologists" (p. 18).

Although the Russian Lev Semënovič Vygotskij was born in the same year as Piaget, it was not until the 1960s that his ideas began to have an impact on mathematics education. This delay was due to the lack of translations of his works and also to a lack of interest in a social perspective in this field. The introduction of Vygotsky's ideas, especially in relation to the crucial role of social interactions in the advancement of learners through their zone of proximal development (ZPD), would prove to be important. For Vygotsky, all knowledge was socially constructed and internalized by joint processes into which learners brought their personal experiences. It followed that close and supportive relationships played an important role in an individual's knowledge growth. In the perspective of cultural mediation, the world of meaning in the child developed by means of tools (artefacts) and signs. Over the past 25 years Vygotskian theory has been applied extensively in mathematics education, the focus being on the mathematical activities of a group of learners or a dyad rather than the individual (Berger, 2005).

An important contribution to the tie between mathematics education and educational sciences came from scholars—such as Caleb Gattegno, Zoltan Dienes, Richard Skemp, and Efraim Fischbein—whose training was both in mathematics and in educational sciences. The work of Skemp and Fischbein stimulated thinking about the role of psychological factors so far as the teaching and learning of mathematics in the higher grades were concerned. Skemp (1976) distinguished between "instrumental" and "relational understanding": Instrumental understanding is the result of a mechanic learning of rules, theorems and their immediate applications, and relational understanding is the result of a personal engagement of the learner with mathematical objects, situations, problems, ideas. We owe to Fischbein deep work on the interactions between intuition and rigor in mathematics education (Tirosh & Tsamir, 2008). Both Skemp and Fischbein were among the founders of PME. Fischbein was the first president of PME, Skemp the second.

During ICME-1, a round table discussion on the psychological problems of mathematics education was organized under the leadership of Fischbein, who also organized and led a similar discussion group at ICME-2. In the introduction to the *Proceedings* for ICME-2, Howson (1973) stressed the importance that Piagetian psychology had in relation to elementary school mathematics. He also noted that the working group on "The Psychology of Learning Mathematics" was the most attended of all working groups at the Congress. According to Howson (1973), the topic discussed "underpins the whole of mathematics education" (p. 15).

In his 1990 introductory chapter providing a research synthesis for PME of the *ICMI Studies Series*, Fischbein (1990) claimed "the psychological problems of mathematical learning and reasoning are scientifically exciting and at the same time genuinely relevant for mathematics education" (p. 4). This sentence epitomized more than a century of interaction between psychologists and mathematics educators. As a matter of fact, though many domains of knowledge have been linked to mathematics education, such as psychology, philosophy, medicine, sociology, linguistic, and anthropology, the main external conceptual support to the development of mathematics education has come from psychology.

Social, Cultural and Political Dimensions

In 1972 the chapter dedicated to research in mathematics education in the third volume of *New Trends in Mathematics Teaching* (Fehr & Glaymann, 1972) proposed three areas for research: curricula, methods and materials; learning and the learner; and teaching and the teacher. Four years later, by the time of ICME-3 in Karlsruhe, the chapter on the same issue in the fourth volume (Bauersfeld, 1979) enlarged the possibilities for research activities by listing five possibly fruitful areas for research: investigations of interactions, studies of real classroom situations, research interests of the teacher, extension of the repertoire of research methods, and a theoretical orientation. Events and perspectives presented at ICME-3 were instrumental in mathematics education adopting more comprehensive perspectives.

This widening of prospective research interests was also accompanied by a broader understanding of the dimensions involved in the place and roles of mathematics and mathematics education in society. By the late 1970s there was a growing interest in the importance of social factors either in discussing the role of mathematics in curricula or in the ways in which social and cultural factors intervened in teaching and learning mathematics. It was increasingly recognized that didactics of mathematics is (or should be) "concerned not only with the process of interaction in the classroom but also with mathematics education as a societal aspect: a process of development imbedded in the process of development of the educational system as a whole" (Christiansen, 1975, p. 28). This viewpoint was also expressed elsewhere (for example, Bishop, 1979). Preparation for ICME-3 brought into focus two early tendencies about this theme. One, championed by Ubiratán D'Ambrosio (1979), reflected on the overall objectives and goals of mathematics education; the other was outlined by Bauersfeld (1979), who advocated, among other things, the importance of the study of interactions in the teaching–learning process.

These two areas, a broad perspective of the cultural and social bases for teaching and learning mathematics and the consequent enlargement of the scope of research, saw significant developments in the 1980s. D'Ambrosio's early elaboration of the goals for mathematics education, produced for ICME-3, evolved into a broader perspective offered at his plenary session at ICME-5 in Adelaide (D'Ambrosio, 1985, 2007) when the concept of ethnomathematics was first presented in a major international event in mathematics education. He suggested that mathematics education should take into account the diversity of cultural attitudes and cultural diversity of distinct "societal groups, with clearly defined cultural roots, modes of production and property, class structure and conflicts, and senses of security and of individual rights" (D'Ambrosio, 1985, p. 5). The consideration of the diverse ways in which mathematics blends in distinct cultures and social milieux, together with a reflection of its consequences for mathematics education, prompted a flurry of investigations, many of them uncovering undervalued mathematical activities in daily practices of social groups and professions. This kind of research stimulated further study and reflection on associated educational practices.

Almost at the same time in Europe two lines of research emerged valuing the social dimensions of teaching and learning. Bauersfeld (1980) published his early work about "hidden" social dimensions in the interactions between teacher and students in the mathematics classroom. And Guy Brousseau (1986), immersed in a French tradition of research, proposed a theory accounting for the transformation (and pitfalls) of scientific mathematics into school mathematical knowledge. Both of these lines saw significant developments in further years.

By the end of the 1980s, research on the influence of social and cultural dimensions on mathematics curricula and mathematics teaching and learning was consistently being reported in mathematics education research publications. In a book published in 1987, Stieg Mellin-Olsen, after discussing the mismatch between the mathematical competencies of students in school and in daily life, argued that mathematics education researchers needed to recognize that political dimensions were inevitably at the centre of mathematics teaching and learning (Mellin-Olsen, 1987).

With the benefit of hindsight it can be seen that 1988 was a key year in the development of mathematics education research. During that year, *Educational Studies in Mathematics* dedicated a special issue to "Socio-cultural studies in mathematics education" (Bishop, 1988a, 1988b); Bishop (1988b) authored a book on the subject; a "Fifth Day Special Programme on Mathematics, Education, and Society," at ICME-6 in Budapest, was devoted to "examining the political dimensions of mathematics education" (Keitel, Damerow, Bishop, & Gerdes, 1989, p. i); and, at a plenary at the twelfth PME conference Terezinha Nunes reported on her team's work detailing the mathematical competencies of illiterate children selling small goods in the streets of Brazilian cities (Carraher, 1988). This *social turn*, as Stephen Lerman (2000) called it, signaled "the emergence into the mathematics education research community of theories that see meaning, thinking, and reasoning as products of social activity" (p. 23).

We can include in this social turn the analysis, from an educational stance, of the role of mathematics and mathematics education in society, echoing D'Ambrosio's early reflections on goals for mathematics education (D'Ambrosio, 1979). In the middle of the 1990s, Ole Skovsmose (1994) discussed the relations between mathematics, society, and citizenship. Acknowledging mathematics power in contemporary societies, he proposed the adoption of a critical stance in mathematics education that allowed for a comprehensive perspective connecting issues of globalization, content, and applications of mathematics, as a basis for actions in society, and for empowerment through mathematical literacy.

A Concern with Theory

The understanding that mathematics education should look for an adequate place in the academic field was already present at the beginning of the 20th century (Kilpatrick, 1992). One of the resolutions passed at the first ICME (1969), related to the need for a "theory of mathematics education" (p. 416). From the middle of the 1970s, in the wake of ICME-3, this push towards theory development became evident. Steiner, based at the IDM at the University of Bielefeld, led this thrust towards theory development and reflection. He formed an international study group called Theory of Mathematics Education (TME), which held five conferences until 1992, and was a regular special group at international conferences. The debate about the nature, the possibilities, the limits and the legitimacy of mathematics education as a scientific field conducted by the group (Steiner et al., 1984) enlarged earlier discussions (e.g., Begle, 1969; Christiansen, 1975) and involved prominent researchers from several countries. The relationship between mathematics education and other fields of knowledge (psychology, education, sociology, mathematics, etc.), the explanatory power of competing paradigms, the viability of home-grown theories, the relationship between theory and practice, and reflections on curriculum change were among the many contributions of this group. The most tangible productions were two books, one edited by Steiner and Vermandel (1988) on the foundations and methodology of mathematics education and another (Biehler, Scholz, Sträßer, & Winkelmann, 1994) offering a comprehensive survey of how mathematics education was viewed around the world.

Several books (Bishop, Clements, Keitel, Kilpatrick, & Laborde, 1996; Bishop, Clements, Keitel, Kilpatrick, & Leung, 2003; English, 2002; Grouws, 1992; Sierpinska & Kilpatrick, 1998; and this *Third Handbook*, in particular) have made an effort to account for the diversity of mathematics education research. In an attempt to characterize this diversity, Bishop (1992, 1998) drew attention to research traditions that were the "result of upbringing education, cultural background, and research training" (Bishop, 1992, p. 712). In 1992, he applied this construct to the characterization of three different traditions and later he used it as the background for a reflection about the relationship between research and educational practice (1998). One tradition is the pedagogue tradition, which values the role of teachers reflecting on their practice, with experiment and observation being the key components of the research. The empirical-scientist tradition was reflected in Begle's paper at the 1969 ICME-1, and "the key to knowledge, and the research process focusses attention on the methods of obtaining that evidence and of analyzing it, often quantitatively" (Bishop, 1992, p. 712). Thirdly, there is the scholastic-philosopher tradition, based on analysis, rational theorizing, and criticism. The actual teaching reality is an imperfect manifestation of these theoretical proposals.

References

Ahlfors, L. V., Bacon, H. M., Bell, C., Bellman, R. E., Bers, L., & Birkhoff, G. (1962). On the mathematics curriculum of the high school. *The Mathematics Teacher, 55*, 191–195.

Bauersfeld, H. (1979). Research related to the mathematical learning process. In H. G. Steiner & B. Christiansen (Eds.), *New trends in mathematics teaching* (Vol. IV, pp. 199–213). Paris, France: UNESCO.

Bauersfeld, H. (1980). Hidden dimensions in the so-called reality of a mathematics classroom. *Educational Studies in Mathematics, 11*, 23–41.

Begle, E. G. (1969). The role of research in the improvement of mathematics education. *Educational Studies in Mathematics, 2*, 232–244.

Berger, M. (2005). Vygotsky's theory of concept formation and mathematics education. In H. L. Chick & J. L. Vincent (Eds.), *Proceedings of the 29th Conference of the International Group for the Psychology of Mathematics Education* (Vol. 2, pp. 155–162). Melbourne, Australia: International Group for the Psychology of Mathematics Education.

Biehler, R., Scholz, R. W., Sträßer, R., & Winkelmann, B. (Eds.). (1994). *Didactics of mathematics as a scientific discipline*. Dordrecht, The Netherlands: Kluwer.

Bishop, A. (1979). Editorial statement. *Educational Studies in Mathematics, 10*, 1.

Bishop, A. J. (Ed.). (1988a). Special issue: Socio-cultural studies in mathematics education. *Educational Studies in Mathematics, 19*(2), 117–268.

Bishop, A. J. (1988b). *Mathematical enculturation. A cultural perspective on mathematics education*. Dordrecht, The Netherlands: D. Reidel.

Bishop, A. J. (1992). International perspectives on research in mathematics education. In D. A. Grouws (Ed.), *Handbook of research on mathematics teaching and learning* (pp. 710–723). New York, NY: Macmillan.

Bishop, A. J. (1998). Research, effectiveness, and the practitioners' world. In A. Sierpinska & J. Kilpatrick (Eds.), *Mathematics education as a research domain: A search for identity, An ICMI Study* (pp. 33–45). Dordrecht, The Netherlands: Kluwer.

Bishop, A. J., Clements, M. A., Keitel, C., Kilpatrick, J., & Laborde, C. (Eds.). (1996). *International handbook of mathematics education*. Dordrecht, The Netherlands: Kluwer.

Bishop, A. J., Clements, M. A., Keitel, C., Kilpatrick, J., & Leung, F. S. (Eds.). (2003). *Second international handbook of mathematics education*. Dordrecht, The Netherlands: Kluwer.

Borba, M., & Bartolini Bussi M. G., (2008). Report of the Working Group 4. Resources and technology throughout the history of ICMI. In M. Menghini, F. Furinghetti, L. Giacardi, & F. Arzarello (Eds.), *The first century of the International Commission on Mathematical Instruction*.

Reflecting and shaping the world of mathematics education (pp. 289–300). Roma, Italy: Istituto della Enciclopedia Italiana.

Borel, E. (1904). Les exercices pratiques de mathématiques dans l'enseignement secondaire. *Revue Générale des Sciences, 10*, 431–440.

Brousseau, G. (1986). Fondements et méthodes de la didactique des mathématiques. *Recherches en Didactique des Mathématiques, 7*(2), 33–115.

Bruner, J. (1960). *The process of education*. Cambridge, MA: Harvard University Press.

Cajori, F. (1890). *The teaching and history of mathematics in the United States*. Washington, DC: Government Printing Office.

Carraher, T. (1988). Street mathematics and school mathematics. In A. Borbás (Ed.), *Proceedings of the twelfth annual meeting of the International Group for the Psychology of Mathematics Education* (Vol. 1, pp. 1–23). Veszprém, Hungary: International Group for the Psychology of Mathematics Education.

Christiansen, B. (1975, January). *European mathematics education—The future*. Paper presented at the Sixth Biennial Conference of the Australian Association of Mathematics Teachers.

Churchhouse, R. F., Cornu, B., Howson, A. G. Kahane, J.-P., van Lint. J. H., Pluvinage, F., Ralston, A., & Yamaguti, M. (1986). *The influence of computers and informatics on mathematics and its teaching*. ICMI Study series. Cambridge, UK: Cambridge University Press.

D'Ambrosio, U. (1979). Overall goals and objectives for mathematical education. In H. G. Steiner & B. Christiansen (Eds.), *New trends in mathematics teaching* (Vol. IV, pp. 180–198). Paris, France: UNESCO.

D'Ambrosio, U. (1985). Sociocultural basis for mathematics education. In M. Carss (Ed.), *Proceedings of the Fifth International Congress on Mathematics Education* (pp. 1–6). Boston, MA: Birkhäuser.

D'Ambrosio, U. (2007). The role of mathematics in educational systems. *Zentralblatt für Didaktik der Mathematik, 39*, 173–181.

De Moor, W. A. (Ed.). (1995). Vormleer—An innovation that failed. *Paedagogica Historica, 31*(1), 103–113.

Ellerton, N. F., & Clements, M. A. (2005). A mathematics education ghost story: Herbartianism and school mathematics. In P. Clarkson, A. Downton, D. Gronn, M. Horne, A. McDonough, R. Pierce, & A. Roche (Eds.), *Building connections: Theory, research and practice—Proceedings of the annual conference of the Mathematics Education Group of Australasia, Melbourne* (pp. 313–321). Sydney, Australia: Mathematics Education Research Group of Australasia.

Engel, A. (1977). *Elementarmathematik vom algorithmischen Standpunkt*. Stuttgart, Germany: E. Klett Verlag.

English, L. D. (Ed.). (2002). *Handbook of international research in mathematics education*. Mahlab, NJ: Lawrence Erlbaum.

Fehr, H. E., & Glaymann, M. (Eds.). (1972). *New trends in mathematics teaching* (Vol. III). Paris, France: UNESCO.

Félix, L. (1985). Essai sur l'histoire de la CIEAEM. In J. de Lange (Ed.), *Mathematiques pour tous ... à l'age de l'ordinateur* (pp. 375–378). Utrecht, The Netherlands: OW & OC.

Fey, J. T. (1989). Technology and mathematics education: A survey of recent developments and important problems. *Educational Studies in Mathematics, 20*, 237–272.

Fischbein, E. (1990). Introduction. In P. Nesher & J. Kilpatrick (Eds.), *Mathematics and cognition: A research synthesis by the International Group for the Psychology of Mathematics Education* (pp. 1–13). Cambridge, UK: Cambridge University Press.

Freudenthal, H. (1963). Enseignement des mathématiques modernes ou enseignement modernes des mathématiques? *L'Enseignement Mathématiques, 2*(9), 28–44.

Furinghetti, F. (2003). Mathematical instruction in an international perspective: The contribution of the journal *L'Enseignement Mathématique*. In D. Coray, F. Furinghetti, H. Gispert, B. R. Hodgson, & G. Schubring (Eds.), *One hundred years of L'Enseignement Mathématique, Monographie 39 de L'Enseignement Mathématique* (pp. 19–46). Geneva, Switzerland: International Commission on Mathematical Instruction.

Furinghetti, F. (2007). Mathematics education and ICMI in the proceedings of the International Congresses of Mathematicians. *Revista Brasileira de História da Matemática Especial no 1—Festschrift Ubiratán D'Ambrosio*, 97–115.

Furinghetti, F. (2008). Mathematics education in the ICMI perspective. *International Journal for the History of Mathematics Education, 3*(2), 47–56.

Furinghetti, F. (2009). The evolution of the journal *L'Enseignement Mathématique* from its initial aims to new trends. In K. Bjarnadóttir, F. Furinghetti, & G. Schubring (Eds.), *"Dig where you stand": Proceedings of the conference on "On-going Research in the History of Mathematics Education"* (pp. 31–46). Reykjavik, Iceland: University of Iceland.

Furinghetti, F., & Giacardi, L. (2008). *The first century of the International Commission on Mathematical Instruction (1908-2008). The history of ICMI*. Retrieved from http://www.icmihistory.unito.it/

Furinghetti, F., & Giacardi, L. (2010). People, events, and documents of ICMI's first century. *Actes d'Història de la Ciència i de la Tècnica, nova època, 3*(2), 11–50.

Furinghetti, F., Menghini, M., Arzarello, F., & Giacardi, L. (2008). ICMI renaissance: The emergence of new issues in mathematics education. In M. Menghini, F. Furinghetti, L. Giacardi, & F. Arzarello (Eds.), *The first century of the International Commission on Mathematical Instruction. Reflecting and shaping the world of mathematics education* (pp. 131–147). Roma, Italy: Istituto della Enciclopedia Italiana.

Gattegno, C., Servais, W., Castelnuovo, E., Nicolet, J. L., Fletcher, T. J., Motard, L., Campedelli, L., Biguenet, A., Peskette, J. W., & Puig Adam, P. (1958). *Le matériel pour l'enseignement des mathématiques*. Neuchâtel, Switzerland: Delachaux & Niestlé.

Gispert, H. (2009). Two mathematics reforms in the context of twentieth century France: Similarities and differences. *International Journal for the History of Mathematics Education, 4*, 43–50.

Griffiths, H. B., & Howson, A. G. (1974). *Mathematics, society and curricula*. Cambridge, UK: Cambridge University Press.

Grouws, D. A. (Ed.). (1992). *Handbook of research on mathematics teaching and learning*. New York, NY: Macmillan.

Guin, D., Ruthven, K., & Trouche, L. (Eds.). (2005). *The didactical challenge of symbolic calculators*. New York, NY: Springer.

Hanna, G. (2003). Journals of mathematics education, 1900–2000. In D. Coray, F. Furinghetti, H. Gispert, B. R. Hodgson, & G. Schubring (Eds.), *One hundred years of L'Enseignement Mathématique, Monographie 39 de L'Enseignement Mathématique* (pp. 69–84). Geneva, Switzerland: International Commission on Mathematical Instruction.

Hanna, G., & Sidoli, N. (2002). The story of ESM. *Educational Studies in Mathematics, 50*, 123–156.

Hodgson, B. R. (2009). ICMI in the post-Freudenthal era: Moments in the history of mathematics education from an international perspective. In K. Bjarnadóttir, F. Furinghetti, & G. Schubring (Eds.), *"Dig where you stand": Proceedings of the conference on "On-going research in the History of Mathematics Education"* (pp. 79–96). Reykjavik, Iceland: The University of Iceland.

Howson, A. G. (Ed.). (1973). *Developments in mathematical education. Proceedings of the Second International Congress on Mathematics Education*. Cambridge, UK: Cambridge University Press.

Howson, A. G. (1982). *A history of mathematics education in England*. Cambridge, UK: Cambridge University Press.

Howson, A. G. (2003). Geometry: 1950–70. In D. Coray, F. Furinghetti, H. Gispert, B. R. Hodgson, & G. Schubring (Eds.), *One hundred years of L'Enseignement Mathématique, Monographie 39 de L'Enseignement Mathématique* (pp. 115–131). Geneva, Switzerland: International Commission on Mathematical Instruction.

Howson, A. G. (2010). Mathematics, society, and curricula in nineteenth-century England. *International Journal for the History of Mathematics Education, 5*(1), 21–51.

Karp, A., & Schubring, G. (Eds.). (in press). *Handbook on history of mathematics education*. New York, NY: Springer.

Keitel, C., Damerow, P., Bishop, A., & Gerdes, P. (Eds.). (1989). *Mathematics, education, and society*. Paris, France: UNESCO.
Kilpatrick, J. (1992). A history of research in mathematics education. In D. A. Grouws (Ed.), *Handbook of research on mathematics teaching and learning* (pp. 3–38). New York, NY: Macmillan.
Kilpatrick, J. (1994). History of mathematics education. In T. Husén & T. N. Postlethwaite (Eds.), *International encyclopedia of education* (2nd ed., pp. 3643–3647). Oxford, UK: Pergamon.
Kilpatrick, J., & Weaver, J. F. (1977). The place of William A. Brownell in mathematics education. *Journal for Research in Mathematics Education, 8*, 382–384.
Klein, F. (1923). Göttinger Professoren. Lebensbilder von eigener Hand. 4. Felix Klein, *Mitteilungen des Universitätsbundes Göttingen, 5*, 24.
Kline, M. (1973). *Why Johnny can't add: The failure of the new mathematics*. New York, NY: St. Martin's Press.
Krygowska, A. Z. (1968). Méthodologie de l'enseignement des mathématiques sujet d'étude au niveau supérieur. In N. Teodorescu (Ed.), *Colloque International UNESCO "Modernization of mathematics teaching in European countries"* (pp. 435–448). Paris, France: UNESCO.
Laborde, C. (2008). http://www.unige.ch/math/EnsMath/Rome2008/WG4/Papers/LABORD.pdf Retrieved 20 July 2011.
Lerman, S. (2000). The social turn in mathematics education research. In J. Boaler (Ed.), *Multiple perspectives on mathematics teaching and learning* (pp. 19–44). London, UK: Ablex Publishing.
Mathematical Association. (1913). *Catalogue of current mathematical journals, etc. with the names of the libraries in which they may be found compiled for The Mathematical Association*. London, UK: G. Bell & Sons.
Mellin-Olsen, S. (1987). *The politics of mathematics education*. Dordrecht, The Netherlands: D. Reidel.
Menghini, M., Furinghetti, F., Giacardi, L., & Arzarello, F. (Eds.). (2008). *The first century of the International Commission on Mathematical Instruction (1908–2008). Reflecting and shaping the world of mathematics education*. Rome, Italy: Istituto della Enciclopedia Italiana.
Moon, B. (1986). *The "new maths" curriculum controversy: An international story*. London, UK: Falmer Press.
Moore, E. H. (1903). The foundations of mathematics (Presidential Address at the Annual Meeting of the American Mathematical Society, New York, 1902). *Bulletin of the American Mathematical Society, 9*, 402–424.
Nabonnand, P. (2007). Les réformes de l'enseignement des mathématiques au début du XXe siècle. Une dynamique à l'échelle internationale. In H. Gispert, N. Hulin, & M.-C. Robic (Eds.), *Science et enseignement* (pp. 293–314). Paris, France: INRP-Vuibert.
Organisation for European Economic Co-operation (OEEC). (1961). *Mathématiques nouvelles*. Paris, France: Author.
Organisation for European Economic Co-operation (OEEC). (1962). *Un programme moderne de mathématiques pour l'enseignement secondaire*. Paris, France: Author.
Papert, S. (1972a). Teaching children thinking. *Mathematics Teaching, 58*, 2–7.
Papert, S. (1972b). Teaching children to be mathematicians versus teaching about mathematics. *International Journal of Mathematical Education in Science and Technology, 3*, 249–262.
Papy, F. (1969). Minicomputer. *Educational Studies in Mathematics, 2*, 333–345.
Pellerey, M. (1989). *Oltre gli insiemi*. Napoli, Italy: Tecnodid.
Piaget, J., Beth, E. W., Dieudonné, J., Lichnerowicz, A., Choquet, G., & Gattegno, C. (1955). *L'enseignement des mathématiques*. Neuchâtel, Switzerland: Delachaux et Niestlé.
Powell, A. B. (2007). Caleb Gattegno (1911–1988): A famous mathematics educator from Africa. *Revista Brasileira de História da Matemática Especial no 1—Festschrift Ubiratan D'Ambrosio*, 199–209.
Price, M. (1995). The AMT and the MA: A historical perspective. *Mathematics Teaching, 150*, 6–11.
Resolutions of the First International Congress on Mathematical Education (1969). *Educational Studies in Mathematics, 2*, 416.

Ruthven, K. (2008). Mathematical technologies as a vehicle for intuition and experiment: A foundational theme of the International Commission on Mathematical Instruction, and a continuing preoccupation. *International Journal for the History of Mathematics Education, 3*(1), 91–102.

Schubring, G. (2000). The first international curricular reform movement in mathematics and the role of Germany—A case study in the transmission of concepts. In A. Gagatsis, C. P. Constantinou, & L. Kyriakides (Eds.), *Learning and assessment in mathematics and science* (pp. 265–287). Nicosia, Sicily: Department of Education, University of Cyprus.

Schubring, G. (2003). *L'Enseignement Mathématique* and the first International Commission (IMUK): The emergence of international communication and cooperation. In D. Coray, F. Furinghetti, H. Gispert, B. R. Hodgson, & G. Schubring (Eds.), *One hundred years of L'Enseignement Mathématique, Monographie 39 de L'Enseignement Mathématique* (pp. 49–68). Geneva, Switzerland: International Commission on Mathematical Instruction.

Schubring, G. (2008). The origins and early incarnations of ICMI. In M. Menghini, F. Furinghetti, L. Giacardi, & F. Arzarello (Eds.), *The first century of the International Commission on Mathematical Instruction (1908-2008). Reflecting and shaping the world of mathematics education* (pp. 113–130). Rome, Italy: Istituto della Enciclopedia Italiana. See also: The origins and the early history of ICMI. *International Journal for the History of Mathematics Education, 3*(2), 3–33.

Schubring, G., & Richter, J. (1980). *International bibliography of journals in mathematics education*. Schriftenreihe des IDM 23/1980. Bielefeld, Germany: IDM.

Sierpinska, A., & Kilpatrick, J. (Eds.). (1998). *Mathematics education as a research domain—A search for identity: An ICMI Study*. Dordrecht, The Netherlands: Kluwer.

Siu, M. K. (2009). Mathematics education in East Asia from antiquity to modern times. In K. Bjarnadóttir, F. Furinghetti, & G. Schubring (Eds.), *"Dig where you stand": Proceedings of the conference on "On-going research in the History of Mathematics Education"* (pp. 197–208). Reykjavik, Iceland: University of Iceland.

Skemp, R. R. (1976). Relational understanding and instrumental understanding. *Mathematics Teaching, 77*, 20–27.

Skinner, B. F. (1954). The science of learning and the art of teaching. *Harvard Educational Review, 24*(2), 86–97.

Skovsmose, O. (1994). *Towards a philosophy of critical mathematics education*. Dordrecht, The Netherlands: Kluwer.

Smid, H. (2009). Foreign influences on Dutch mathematics teaching. In K. Bjarnadóttir, F. Furinghetti, & G. Schubring (Eds.), *"Dig where you stand": Proceedings of the "Conference on On-going Research in the History of Mathematics Education"* (pp. 209–222). Reykjavik, Iceland: The University of Iceland.

Smith, D. E. (1905). Opinion de David Eugene Smith sur les réformes à accomplir dans l'enseignement des mathématiques. *L'Enseignement Mathématique, 7*, 469–471.

Steiner, H.-G., Balacheff, N., Mason, J., Steinbring, H., Steffe, L. P., Brousseau, G., et al. (Eds.). (1984). *Theory of mathematics education (TME)*. Bielefeld, Germany: Institut für Didaktik der Mathematik der Universität Bielefeld.

Steiner, H. G., & Christiansen, B. (Eds.). (1979). *New trends in mathematics teaching* (Vol. IV). Paris, France: UNESCO.

Steiner, H.-G., & Vermandel, A. (Eds.). (1988). *Foundations and methodology of the discipline of mathematics education (didactics of mathematics). Proceedings of the 2nd TME Conference*. Bielefeld, Germany: University of Bielefeld and University of Antwerp.

Stemhagen, K. (2008). Doin' the Math: On meaningful mathematics-ethics connections. *The Montana Mathematics Enthusiast, 5*(1), 59–66.

Struik, D. J. (1987). *A concise history of mathematics* (4th ed.). New York, NY: Dover Publications.

Thom, R. (1973). Modern mathematics. Does it exist? In A. G. Howson (Ed.), Developments in mathematical education. In *Proceedings of the Second International Congress on Mathematical Education* (pp. 194–209). Cambridge, UK: Cambridge University Press.

Thwaites, B. (1969). The role of the computer in school mathematics. *Educational Studies in Mathematics, 2*, 346–359.

Tirosh, D., & Tsamir, P. (2008). Intuition and rigor in mathematics education. In M. Menghini, F. Furinghetti, L. Giacardi, & F. Arzarello (Eds.), *The first century of the International Commission on Mathematical Instruction. Reflecting and shaping the world of mathematics education* (pp. 47–61). Roma, Italy: Istituto della Enciclopedia Italiana.

Törner, G., & Sriraman, B. (2005). Issues and tendencies in German Mathematics-didactics: An epochal perspective. In H. L. Chick & J. L. Vincent (Eds.), *Proceedings of 29th Conference of the International Group for the Psychology of Mathematics Education* (Vol. 1, pp. 197–202). Melbourne, Australia: International Group for the Psychology of Mathematics Education.

Treutlein, P., & Wiener, H. (1912). Verzeichnis mathematischer Modelle. Sammlungen H. Wiener und P. Treutlein. Leipzig – Berlin, Germany: B. G. Teubner.

Whitney, H. (1983). Are we off the track in teaching mathematical concepts? In A. G. Howson (Ed.), *Developments in mathematics education* (pp. 283–296). Cambridge, UK: Cambridge University Press.

Willmore, T. (1972). The mathematical societies and associations in the United Kingdom. *The American Mathematical Monthly, 79*(9), 985–989.

Wooton, W. (1965). *SMSG. The making of a curriculum.* New Haven, CT: Yale University Press.

Chapter 10
Theories in Mathematics Education: Some Developments and Ways Forward

Bharath Sriraman and Elena Nardi

Abstract In this survey, roots of mathematics education are traced from Piaget to the current work on theorizing which utilizes sociological and commognitive frameworks. Attention is given to the critiques of Sriraman and English's (2010) edited collection, *Theories of Mathematics Education,* and productive discussions from the reviews are unpacked. The notions of "operational" versus "functional," and "models" versus "theories," are also tackled by focussing on conceptual frameworks which harmonize the terms as opposed to exemplifying their polarities.

Introduction

This chapter flows from the extensive discussion of *Theories of Mathematics Education: Seeking New Frontiers* (Sriraman & English, 2010), in book reviews and critical notices in several major outlets (Artigue, 2011; Ely, 2010; Fried, 2011; Jankvist, 2011; Schoenfeld, 2010; Umland, 2011). The chapter has been structured so the second author (Nardi) is mainly responsible for the sections entitled "Critiques of *Theories of Mathematics Education*" and "Discursive Approaches to Research in Mathematics Education: The Case of Sfard's 'Commognitive' Framework," and the first author (Sriraman) is mainly responsible for all other sections.

The dialogic form in which the community of scholars has undertaken theoretical developments in the field of mathematics education has allowed for the necessary intellectual critique needed to advance the field. Mathematics education as a research discipline occurs at the nexus of numerous other domains of inquiry

---

B. Sriraman (✉)
Department of Mathematical Sciences, The University of Montana,
Missoula, MT, USA
e-mail: sriramanb@mso.umt.edu

E. Nardi
University of East Anglia, Norwich, UK

(Sriraman, 2008; Sriraman & English, 2010). Unlike other fields, such as the natural and physical sciences, hermeneutic continuity in theory development is by and large absent in the learning sciences. Although psychological theories served as the theoretical underpinnings of the field in the 1950s and 1960s, methods from other domains of study such as sociology, anthropology, cultural historical studies, have swayed mathematics education into multidisciplinary and uncharted directions. Although this has contributed to both the complexity inherent in a field that deals with cognizing and socially situated subjects within the larger contexts of institutions and culture, it has also given cause for celebrating the multidisciplinary nature of mathematics education.

The perceived objectivity of mathematics coupled with the perceived subjectivity of the social sciences makes theory development in the field of mathematics education a particularly nuanced and complex task. In general, theories have been a troubling issue for the field of mathematics education, and today a "bewildering array of theories, theoretical models, or theoretical frameworks" (Jablonka & Bergsten, 2011) is abundantly found in the literature that characterizes mathematics education research (MER).

In this chapter we discuss briefly critiques of *Theories of Mathematics Education* (Sriraman & English, 2010) as well as address the psychological foundations of what has been called the "social turn" (Lerman, 2000), or the "social brand" (Jablonka & Bergsten, 2011) in mathematics education. We also discuss different schools of thought on what theory means in mathematics education, and assess ways in which we are progressing. Our overall observations of the field indicate that there seems to be increasingly more effort to employ theoretical frameworks with a more pronounced focus on institutional and social dimensions (e.g., the Anthropological Theory of Didactics (ATD), as derived from the work of Brousseau and Chevallard), and discursive practices (e.g., Sfard's commognitive framework). Piagetian foundations are also in need of being revisited and revitalized, given the canon of experimental studies in the last two decades (e.g., the Rational Number Project; Models and Modeling) that have developed models that can be subsumed within the larger theory. We begin with a brief presentation of critiques of *Theories of Mathematics Education* (Sriraman & English, 2010).

Critiques of Theories of Mathematics Education

Theories of Mathematics Education (Sriraman & English, 2010) grew out of a research forum at the 29th meeting of the International Group on the Psychology of Mathematics Education in 2005 and of a selection of articles published in *ZDM—The International Journal on Mathematics Education*, between 1994 and 2008, many of which were in the two special issues of *ZDM* on theories published in 2005 and 2006 (Sriraman & English, 2005, 2006). Some of the early parts set the scene through reflection on the role of theory in MER, on theory pluralism, etc. In resonance with the eclecticism of *Theories of Mathematics Education,* most parts can be read independently. Through its structure, both across and within the 19 parts, the book invited the reader to engage in a dialogue about theoretical issues.

The book has been reviewed in several key mathematics and mathematics education publications. Although many reviews have acknowledged its considerable ambition and strengths (e.g., Artigue, 2011; Ely, 2010; Fried, 2011; Jankvist, 2011; Schoenfeld, 2010; Umland, 2011), in this section we focus on those constructively critical parts of some of the reviews that have led partly to the shaping of this chapter.

A first concern (Artigue, 2011) was in regard to the presentation in the book in relation to the French tradition of *didactique des mathématiques*. Artigue noted that "seen from the outside, the [French] theoretical landscape seems homogeneous" (p. 311). She went on to say that "those who know this culture are well aware that the situation is much more complex, that different local theoretical constructions have emerged from TDS (Theory of Didactical Situations) and ATD or from the connection of these with other frameworks such as Vergnaud's theory of conceptual field, the theory of activity or cognitive ergonomy" (pp. 311–312). Furthermore, "most researchers productively combine different theoretical approaches in their research work, for instance joint action theory (Sensevy, 2009; Sensevy & Mercier, 2007), the didactic-ergonomic approach of teachers' practices (Robert & Rogalski, 2005), or the instrumental approach of technological integration (Artigue, 2002; Guin, Ruthven, & Trouche, 2004)" (p. 312).

To redress the balance a little, in this chapter we have attempted a necessarily brief and abridged account of recent developments in the areas highlighted in Artigue's comment. This only partially addresses Artigue's comment that the educational discourse in the *Theories* volume is rather "too culturally connoted" (p. 313) and that the contribution of other, non-Anglo-Saxon approaches remain largely "invisible" (p. 315) to the international community—a concern that has been echoed in historical treatments of the development of mathematics education in non-Anglo-Saxon parts of the world (e.g., Germany, France, and Italy (Sriraman & Törner, 2008).

Another concern, also expressed by Artigue (2011), was in regard to an impression that a reader may garner from the book that "investigating the relationships between … different theoretical approaches and developing networking activities" (p. 312) is somewhat a novel enterprise. As Artigue noted, the *didactique* community has been dedicating parts of its energy to this task since the early 1980s. Furthermore, in recent years, theory "networking," substantially addressed in Parts XV and XVI of the book, has been the focus of several European projects and work within ICMI. We shall comment briefly on these considerations in the final section of the chapter, where "networking" is the main focus.

A third concern raised by Artigue (2011) was that a reader may leave the book with an impression of a somewhat light-touch approach to the pitfalls of excessive eclecticism. We agree with Artigue that "theoretical diversity can only be a richness if it is not synonymous to theoretical fragmentation or eclecticism" (p. 312). We share with her the concern about "the theoretical explosion in mathematics education" (p. 312), and we recognize the need to "reflect on the way we tacitly contribute to it as a community when we value much more the creation of new theoretical entities than the hard work often required in order to appropriate and adequately exploit what has been already built by other researchers and communities" (p. 312). It is for that reason that we conclude this chapter with a call to mathematics education researchers for a more incisive approach to, and employment of, theory that goes beyond knowing the "grammar" of a theory (Lerman, 2010, p. 101)—a trend that Lerman and other contributors to the book identified as a symptom of careless eclecticism. As observed in Umland's (2011) review, "a well-articulated set of linked and nested theoretical structures" (p. 74) need not be perceived as an out-of-reach, or totally undesirable goal of MER.

A valuable observation made in several reviews of *Theories of Mathematics Education*—in, for example, Artigue (2011) and Umland (2011)—has been that any enterprise with a focus on the generation and employment of theory in mathematics education must consider the interplay between general and meta-theoretical developments as well as theoretical developments within specific areas of research in mathematics education. As Umland (2011) observed, this enterprise often entails considering "reflections on the philosophical foundations of mathematics education as a field" (meta-theory), "theoretical perspectives from other disciplines that could be brought to bear on mathematics education research," and "proto-theories" (p. 73), descriptors of the processes underlying the teaching and learning of mathematics. In this chapter our focus is on the former part of Artigue's distinction (or the second and third of Umland's distinction), even though the examples we mention originate in studies conducted within specific areas of research. Finally, as several commentators have noted (e.g., Umland, 2011), any enterprise with this focus needs to allocate some attention to what the authors mean by "theory." We do so, modestly, in the Introduction and in other sections of this chapter.

Dynamic Interactionism Between Models and Theories: The Case of Piaget's Notion of "Operational"

Piaget's theory of cognitive development in children serves as one of the barometers through which we can analyze both model and theory development in mathematics education, particularly in ensuing experimental work in North America in various longitudinal projects funded by the National Science Foundation (see, e.g., Lesh, Cramer, Doerr, Post, & Zawojewski, 2003; Lesh & Doerr, 2003; Lesh, Post, & Behr, 1987, 1988; Lesh et al., 1992). Put simply, the function of any theory is to explain phenomena. The natural and physical sciences have an established body of theories which have been validated over time through scientific experimentation and conforming data. Theory development has also been spurred by falsification (Popper, 1959). Newer theories are able to subsume older theories. Newtonian mechanics occurs as a special case in Hamiltonian mechanics. Euclidean geometry can be reduced to a special case of Riemannian geometry. Weyl's (1918) mathematical formulation of the general theory of relativity by using the parallel displacement of vectors to derive the Riemann tensor revealed the interplay between the experimental (inductive) and the deductive (the constructed object). The continued evolution of the notion of tensors in physics/Riemannian geometry can be viewed as a culmination or a result of flaws discovered in Euclidean geometry.

On the Notion of "Operational" in Mathematics Education

Mathematics education, however, is neither one of the natural/physical sciences nor mathematics per se. In mathematics education one often has to ask the question: "What are the phenomena that we are trying to explain?" (Schoenfeld, 2010). One answer to this question in relation to Piaget's body of foundational work is mathematical cognition. For instance, mathematics education has developed empirically validated models within the Rational Number Project (see Lesh et al., 1987, 1989, 1992, 2003) which explain how proportional reasoning develops in children in a way that, by and large, coheres with Piagetian stage theory. There are models that also explain how combinatorial reasoning develops (see, e.g., Sriraman & English, 2004). Some of the findings suggest that when Piaget's experiments are repeated with age-appropriate materials, the stages proposed by him are not as discrete as they might have seemed, but more porous with the possibility of children being able to reason at a more advanced level given contextual play materials. Dienes' six-stage theory of learning mathematics bears resemblance to both Piaget's and Bruner's theories, provided a somewhat *different* conceptualization of the meaning of "operational" is permitted (Dienes, 1960, 1963, 1964, 1971, 2000; Dienes & Jeeves, 1965). Similarly, the recent body of work by the models and modelling group has built on operational definitions from within the work of Dienes to develop models of student thinking in contextual problems.

When viewed from the biological perspective of neural networks, much of the post-Piagetian body of experimental work in mathematics education can be made to cohere within Piaget's stage theory—with a critical number of exceptions that warrant reconceptualization which could give rise to a more dynamic theory with the possibility of the contexts allowing children to function at a higher stage. However a foundational point of continued argument remains the working definition of "operational."

Biologists have found that methodological reductionism, that is to say going to the parts to understand the whole, which was central to the classical physical sciences, is less applicable when dealing with living systems. Analogously, the challenge confronting mathematics educators in the learning sciences who hope to create models (of the underlying conceptual systems) that students, teachers and researchers develop to make sense of complex systems occurring in their lives is the mismatch between learning science theories based on mechanistic *information processing* metaphors and recent discoveries on how complex systems work. Not everything that students know can be methodologically reduced to a list of condition-action rules, given that characteristics of complex systems cannot be explained (or modelled) using only a single function—or even a list of functions. Physicists and biologists have proposed that characteristics of complex systems arise from the *interactions* among lower-order/rule-governed agents—which function simultaneously and continuously, and which are not simply inert objects waiting to be activated by some external source (Hurford, 2010).

Given the paradigm shifts which have occurred in the physical and natural sciences, there have been proposals to view learning and the modelling of learning as analogous to the study of complex systems (Mousoulides, Sriraman, & Lesh, 2008). Piaget sought answers to fundamental questions about the nature and origins of knowledge by studying children. His focus was the child's understanding of space, time, and causality, and relations of invariance and change (Piaget, 1971, 1975). Trained as a biologist, he was also interested in exploring the "metaphors" from biology such as organization, development, and adaptation. His theory of development proposed sensorimotor, pre-operational, concrete operational and formal operational stages of cognitive development. Operations were defined as internalized actions, derived directly from the subject's physical actions as enacted in sensorimotor behaviour.

Consider the stage of generalized formal operations characterized by the organization of operations in a structural whole and the culmination of the sensorimotor, pre-operational, and concrete operational stages. Piaget (1958) suggested that at the stage of formal operations, there is a "structural mechanism" which enables students to compare various combinations of facts and decide which facts constitute necessary and sufficient conditions to ascertain truth. Those that were able to transform propositions about reality, such that the relevant variable could be isolated and relations deduced, were said to have achieved "functionality" in their structural flexibilities. Another characteristic of the stage of generalized formal operations was the relative ease with which reversibility of thought operations occurred. Piaget's essential claim was that there was a link between mathematics and biology. In other words, operationality became functionality after reversibility in order was achieved, so that the person became able to generalize mathematical structures or subsume classes of counter examples into an existing structure. This was a highly unusual claim 60 years ago, but one that Piaget tried to substantiate with decades of research. Piaget's characterization of knowledge and cognition was that there are forms of biological adaptation within which structures of *action* (evolving upwards from individual sensorimotor schemes) play a role.

An operational definition of mathematics is that it is an intellectual activity concerned with the creation of structure, with new characterizations that emphasize embodiment and anchoring in culture. Piaget (1958, 1987) conceptualized the whole of mathematics in terms of creation of structures, not in a physical or literal sense but operations carried out in the idealized world of the mathematician. The relationship between the two worlds was explained as follows: the idealized constructions emerge as a result of a series of abstractions from their literal counterparts, which are the real actions and physiological movements human beings make in the world. Piaget's psycho-genetic account of mathematics retraced this descent from actions to formal thinking as one of increasing abstraction and generalization. Being enamored by the ongoing attempt of the Bourbaki at that time to formalize all of mathematics, Piaget compared his operator structures of thinking to the structures espoused by the Bourbaki.

The Bourbaki, who aimed to write a body of work based on a rigorous and formal foundation, which could be used by mathematicians in the future, identified three fundamental structures on which mathematical knowledge rest: (a) algebraic structures; (b) structures of order; and (c) topological structures (Bourbaki, 1970). More information on the Bourbaki can be found on the Bourbaki Web site located at http://www.bourbaki.ens.fr/.

Piaget claimed that there existed a correspondence between the mathematical structures of the Bourbaki and the operative structures of thought. He felt that in the teaching of mathematics a distinctive synthesis would occur between the psychologist's operative structures of thought and the mathematician's mathematical structures. Dubinsky (1991) used the Piagetian notion of *reflective abstraction*, to develop a model using "schemas" as a way of better understanding cognitive processes in advanced mathematics. APOS theory (Asiala, Cottrill, Dubinsky, & Schwingendorf, 1997), developed in mathematics education to explain advanced mathematical thinking, is another instance of theory development cohering within a larger theory. Within the realm of mathematical cognition there are micro-theories, or what should more accurately be labelled models, that build on Piaget's work and inform it sufficiently that it is possible to re-conceptualize the theory as being more dynamic. The only contentious point in reconceptualizing Piagetian theory on cognitive development is how to gain consensus on the definition of "operational" thinking and how to put this concept on a firmer theoretical ground.

In physics, one could say that theoretical terms are invariants of operations represented by physical measurement devices. Physicists have "learned" that theoretical terms have to be defined operationally in terms of theories and that this can be supported via experimentation which can back up notions occurring within the theories (Dietrich, 2004). The question is how can this be adapted by researchers in mathematical cognition? That is to say, how can we operationally define observational terms, namely perceived regularities that we attempt to condense into theories—or as Piaget attempted to do—phylogenetically, to evolve mental cognitive operators (Dietrich, 2004)? The purpose of theoretical terms is to clarify the meaning of concepts. On the other hand the purpose of observational terms is to delineate how the concepts/constructs have been measured. Ideally there should be a perfect match between theoretical terms and observational terms. That is to say, observations should confirm theory irrespective of when or where the observations are made so long as the initial conditions of an experiment are somewhat the same. For instance, within models and modelling research, if researchers consistently report similar observations of students modelling processes when confronted with the same complex situation across age groups and locations, then these observations can be used to develop a sound theoretical construct. Conversely, when the theoretical construct is tested (or subject to experiment) at a new location, researchers should be able to predict the types of behaviours that will be observed as long as the integrity of the experiment (starting conditions, etc.) is replicated.

Having developed only slightly beyond the stage of continuous theory borrowing, the field of mathematics education currently is engaged in a period in its development replete with inquisitions aimed at purging those who don't vow allegiance to not always well-defined perspectives on mathematical learning (such as "constructivism"—which most modern theories of cognition claim to endorse, but which rarely generates testable hypotheses that distinguish one theory from another). Certain others would want to purge those who don't pledge to conform to psychometric notions of "scientific research"—such as pretest/posttest designs with "control groups" in situations where nothing significant is being controlled, where the most significant achievements are not being tested, and where teaching-to-the-test is itself the most powerful untested component of the "treatment." With the exception of small schools of mini-theory development, which occasionally have sprung up around the work of a few individuals, most research in mathematics education appears to be ideology-driven rather than theory-driven or model-driven (Lesh & Sriraman, 2005). Furthermore, as Artigue (2011) has pointed out, there is also less acknowledgement of non-Anglo Saxon approaches to conceptualizing researchable phenomena in mathematics education.

Theories are cleaned-up bodies of knowledge that are shared by a community. They are the kind of knowledge that gets embodied in textbooks. They emphasize formal/deductive logic, and they usually try to express ideas elegantly, using a single language and notation system. The development of theory is absolutely essential in order for significant advances to be made in the thinking of communities (or individuals within them). But, theories have several shortcomings. Not everything we know can be collapsed into a single theory. For example, models of realistically complex situations typically draw on a variety of theories. Pragmatists (such as Dewey, James, Peirce, Meade, Holmes) argued that it is arrogant to assume that a single "grand theory" will provide an adequate basis for decision-making for most important issues that arise in life (Lesh & Sriraman, 2005). Models are purposeful/situated/easily-modifiable/sharable/re-useable/multi-disciplinary/multi-media chunks of knowledge. They often (usually) integrate ideas from a variety of theories. They are directed toward solving problems (or making decisions) which lie outside the theories themselves. That is, they are created for a specific purpose in a specific situation. Models are seldom worth developing unless they also are intended to be: (a) sharable (with other people) and (b) re-useable (in other situations). So, one of the most important characteristics of an excellent model is that it should be easy to modify and adapt, and ideally should cohere with a larger theory and in some instances develop it further.

From Social/Institutional Branding to Bernstein and Beyond

Theory is also important as it provides the lens through which to construct and look at data. Although MER theories were originally largely psychological, recent realizations of the importance of social and cultural contexts have led to the emergence of sociological theories as well. Lerman (2010) drew on these recent traditions, largely Bernstein, to establish the claim that plurality of theories in MER is not a problem, and in fact is necessary. Below we outline his argument roughly.

Despite external pressures on educational research, MER is thriving in quantity (journals, conferences, etc.) as well as in terms of generated theories. Lerman (2010) looked at whether this is unusual and beneficial. He did so through an empirical meta-study that drew on the sociological theory of Basil Bernstein. Hierarchical discourses (as opposed to horizontal discourses) are those that require apprenticeship and involve gradual distancing and abstraction (e.g., mathematics, whether academic or in school). Horizontal discourses are acquired tacitly and concern specific contexts. Traditional pedagogies are explicit and are about performance. Authority is clearly located in the teacher. "Reform" pedagogies are often implicit, sometimes invisible and privilege those with prior linguistic wealth (Bernstein's "elaborated code"). Verticality is a concept that describes how knowledge domains grow vertically (new theories replace old ones, as more or less in science) or horizontally (new discourses and theories develop in parallel and are often incommensurable with previous ones). MER, as any other social science, develops horizontally. Because, unlike mathematics, MER has a "weak" grammar (but it does have a grammar, you need to know MER theories to do work in it), building theory across the boundaries of prior theories is possible.

Lerman's study was an analysis of papers published in *Educational Studies in Mathematics* (ESM), or in *Journal for Research in Mathematics Education* (JRME), and in the *Proceedings* of conferences of the International Group for Psychology in Mathematics Education (PME). Although a majority used theory, it was mostly traditional psychological learning theory. There was an increase, with time, towards more social theories (Vygotskian, ethnomathematical, social/critical, post-structural: belonging to what Lerman terms "the social turn") and the vast majority of authors were content with using a theory, not building upon it, refuting it, etc. A caveat was that only accepted, published research was looked at. Is there an issue of gatekeeping here? Lerman's proposition was that new theories were new voices which needed to be heard. Theory should be used appropriately in a way that also demonstrates concern for practice (unlike any other social science). In that sense it is a bit like Medicine or Computing. In mathematics education, one should not ignore social and political implications.

In their commentary on Lerman's (2010) paper, Jablonka and Bergsten (2010), although agreeing with Lerman's overall position, noted the following:

1. There are examples of "strong" grammar theories in MER—like for example, the ATD and embodied cognition.
2. "Use of theory" is weakly defined by Lerman. Many papers in ESM and JRME cite theory but do not use it in any substantial way. Some offer common sense re-workings of data that are not informed by theory at all.
3. Theoretical hybrids can achieve some level of communication across theories.
4. There is a trend towards selective acceptance of parts of a theory (e.g., accepting importance of bodily-based metaphors but not the entire embodied cognition framework).

5. What Lerman calls a "social turn" is more a "social brand."
6. MER is more focussed on the "knower" than on "knowing," and this can be dangerous.
7. Looking at other fields can be beneficial, for it looks beyond insular approaches to theorizing by mathematics education researchers.

Several explications of the so-called state-of-the-art of mathematics education in the past have been put forward by mathematicians and even mathematics educators who have done little or no research in the classroom or other learning settings in order to substantiate their claims. This has created several canons of literature which are by no means easy to differentiate, and sieve out. Research that advances theory through empirical data is what is needed, as opposed to theories advocated in the form of rhetoric in introspective articles, or theories in which quotations are the main form of data. "Theories" of dubious origins are potentially dangerous when pieces of legislation or reports from advisory panels of government bodies become the basis of curricular changes and research programs.

For instance the call of the National Mathematics Advisory Panel in the USA to make forms of algebra the panacea for curricular ills needed to be carefully scrutinized. The same should be said with respect to the stipulation that psychometric aptitude–treatment–interaction-based clinical studies were the only genre for research worthy of funding. It could be argued that that stipulation has led many mathematics education researchers into a blind following of this mode of research without regard for the fundamental problems of the field (see Greer, 2008). This swerving of research focus based on political tides does not bode well for any field of research.

One of the main points made by Jablonka and Bergsten (2011) was the need to build strong theoretical bases for the field based on ongoing research as well as to establish a rigorous framework for theorizing according to a specific research grammar—such as Basil Bernstein's internal/external languages of description. If that could be achieved, so that a sort of coherence pervaded the objects of theoretical discussion, there would be a natural and much welcomed end to the need, every decade or so, to justify the existence of mathematics education as a "research field" (see, e.g., Sierpinska & Kilpatrick, 1998; Sriraman & English, 2010).

Jablonka and Bergsten (2011) critiqued the strengths and weaknesses of four ways of theorizing and illustrated what they meant by modes and qualities of theorizing. The four modes of theorizing that they selected from the literature and presented as examples were diverse enough to cover the spectrum of existing "theoretical" trends. The first example was the PISA framework, by which a vaguely defined notion of "mathematization" became a major constituent of mathematical literacy. Despite the weak operationalization of basic notions and despite criticisms arising from a lack of cross-cultural validity for PISA test items, the framework has dangerously mutated into a basis for curricular reform in many countries of the world. The second example was the theory of authentic task situations, taken from Sweden. The arbitrariness of categories (or aspects) chosen in the operational framework was noted, and Jablonka and Bergsten (2011) claimed that relationships between categories were vague and empirically tenuous. APOS theory was used as the third example of a theory that deals with conceptual development in mathematics. This neo-Piagetian theory was appraised as having a relatively strong internal consistency in its grammar and in its specific theorizing in terms of actions, processes and objects within schemas. Fourthly, Jablonka and Bernstein tackled the French school which is the focus of the next section.

The "French" Way

Jablonka and Bergsten (2011) referred to the ATD as an example of a theory that uses a specialized language and develops hierarchical relationships between praxeologies. ATD is of particular interest to the MER community given its ecological nature and the wideness of its applicability. Simply put, ATD is an extension of Brousseau's ideas from within the institutional setting to the wider "Institutional" setting [the upper-case I is our choice]. Artigue (2002) clarified this subtlety when she stated:

> The anthropological approach shares with "socio-cultural" approaches in the educational field (Sierpinska and Lerman, 1996) the vision that mathematics is seen as the product of a human activity. Mathematical productions and thinking modes are thus seen as dependent on the social and cultural contexts where they develop. As a consequence, mathematical objects are not absolute objects, but are entities which arise from the practices of given institutions. The word "institution" has to be understood in this theory in a very broad sense…[a]ny social or cultural practice takes place within an institution. Didactic institutions are those devoted to the intentional apprenticeship of specific contents of knowledge. As regards the objects of knowledge it takes in charge, any didactic institution develops specific practices, and this results in specific norms and visions as regards the meaning of knowing or understanding such or such object. (p. 245)

Chevallard proposed a theory much larger in scope than the TDS in order to be in a position to move beyond the cognitive program of MER, with its classical concerns (Gascón, 2003) such as the cognitive activity of an individual explained independently of the larger institutional mechanisms at work which affect the individual's learning. Chevallard's (1985, 1992a, 1992b, 1999a, 1999b) essentially contended that a paradigm shift was necessary within mathematics education, one that began within the assumptions of Brousseau's work, but shifted its focus on the very origins of mathematical activity occurring in schools, the institutions which produce the knowledge (K) in the first place.

The notion of didactical transposition (Chevallard, 1985) was developed to study the changes that K goes through in its passage from scholars/mathematicians cur- riculum/policymakers → teachers → students. In other words, Chevallard's ATD was an "epistemological program" which attempted to move away from the reductionism inherent in the cognitive program (Gascón, 2003). Bosch, Chevallard, and Gascon (2005) clarified the desired outcomes of such a program of research in the following way:

> ATD takes mathematical activity institutionally conceived as its primary object of research. It thus must explicitly specify what kind of general model is being used to describe mathematical knowledge and mathematical activities, including the production and diffusion of

mathematical knowledge. The general epistemological model provided by the ATD proposes a description of mathematical knowledge in terms of mathematical praxeologies whose main components are types of tasks (or problems), techniques, technologies, and theories. (pp. 4–5)

It is noteworthy that the use of ATD as a theoretical framework by a large body of researchers in Spain, France and South America resulted in the creation of an International Congress on the ATD, which has been held biennially since 2005. The aim of these congresses has been to propose a cross-national research agenda and identify research questions which can be systematically investigated with the use of ATD as a framework.

In Sriraman and Törner (2008), several focal points were isolated via historical analysis to suggest ways in which the theoretical differences between the German, French and Italian schools of thought can be bridged (or networked) and made to interact in the present and future. In outlining the differences and similarities between the various positions and schools of thought in these three countries it became apparent that researchers were often entrenched in "ideological" perspectives. Lerman (2000) explained that these ideological tendencies were a result of the field adopting theoretical frameworks via a process of recontextualization (Bernstein, 1996). In this process "different theories become adapted and applied, allowing space for the play of ideologies" (p. 19). However, Jablonka and Bergsten (2011), by using Bernstein's sociological framework, clarified and elaborated on different modes of classification, modelling and theorizing with respect to relational densities among basic concepts within a theory, as well as levels of discursive saturation (or lack of it) in the four examples of theorizing in mathematics education chosen. In mathematics education, there is a preponderance of homegrown theories, and a lack of high relational density and intertextuality in the current modes of theorizing. More importantly there is sometimes a tendency of researchers in our field borrowing from terms and concepts fields such as sociology, social anthropology, linguistics, etc., without committing to the deeper levels of theorizing that occurs in those fields.

Discursive Approaches to Research in Mathematics Education: The Case of Sfard's "Commognitive" Framework

We close this chapter with commentary on discursive approaches to research in mathematics education, and in particular on Anna Sfard's commognitive framework.

Compared to, say, two decades ago, reports of research in mathematics education are substantially different: longer, often qualitative and no longer examining learning merely in terms of individual acquisition or construction but with ample reference to the context in which the learning occurs. Focussing on environmental factors associated with learning has often meant shifting to a focus on communication and language; in other words our examination of learning has become *discursive*. In this section we pay special attention to this opening up of the field to discursive approaches—an opening up that has been confirmed quantitatively by Ryve (2011). We do so by tracing—by necessity, rather selectively—some relevant developments, and we focus on one framework that has been attracting increasing attention in recent years, namely Anna Sfard's (2008) *commognitive framework*.

Discursive approaches to research in mathematics education are one of recent steps in a trail of attempts to describe human thinking—the work of Jerome Bruner, artificial intelligence metaphors and models and Piaget's genetic epistemology being of special importance. Although these works often had remarkable success in advancing our understanding of how human beings think, they did not always succeed in providing adequate explanations of persistent learning behaviours such as individual or collective failures in mathematics. Over the last 40 years or so, these approaches to research, which have sometimes been given as the label "acquisitionist," have been met with increasing doubt both on methodological and epistemological grounds.

With regard to methodological grounds, their clinical-experimental methods were deemed too remote from the milieu (of the classroom, the home, etc.) in which mathematical learning typically occurs. Accordingly, ethnographic approaches to studying learning began to emerge with the observation of ordinary practices (of learners, teachers, etc.), gradually replacing laboratory-based data collection.

From an epistemological perspective, doubt was cast on the appropriateness of the ways in which the older research approaches viewed learning—that human thought somewhat mirrored nature, reflected external phenomena, and that therefore learning could be described as context-invariant and universal. The sociocultural perspective, pioneered by Vygotsky, and epistemologically and philosophically bolstered by the work of Wittgenstein, Schutz and Mead, largely opposed this view. It emphasized that learning was an activity that occurred in, and was co-constituted by, situational, cultural and historical milieu. As editors of *Learning Discourse: Discursive Approaches to Research in Mathematics Education*, Kieran, Forman and Sfard (2002) noted that the discursive perspective espoused this socio-cultural tenet. Furthermore, its emphasis was firmly placed on the view of human thinking as a type of communication:

> Within the discursive framework, thinking is conceptualised as a special case of the activity of communication and learning mathematics means becoming fluent in a discourse that would be recognised as mathematical by expert interlocutors. (Kieran et al., 2002, p. 5)

Obviously, sweeping discursive approaches under an apparently unified, umbrella term such as "the discursive framework" does not do justice to the diversity of these approaches. This diversity is evident, for example, in the seven chapters of Kieran et al.'s (2002) edited collection, as well as in more recent efforts to delineate the many and varied strands of discourse-oriented research—such as Sfard's entry on Discourse in the forthcoming *Encyclopedia of the Sciences of Learning* (Seel, 2012), Jaworski and Coupland's (2005) *The Discourse Reader*, and, within mathematics education, Andreas Ryve's (2011) formidable review of 108 papers which reported research deploying a discursive approach. The 2001 *Educational Studies in Mathematics* special issue, from which the 2002 Kieran et al. volume originated, had the subtitle "Bridging the Individual and the Social" and, indeed, the impression is that the editors extended an open invitation to the reader to reflect on "the social nature of the individual" (p. 10). To do so, and to carry out research within this complex framework, is a very challenging task. For the rest of this section we focus on the recent work to this purpose of one of that volume's contributors, Anna Sfard (see, e.g., Ben-Yehuda, Lavy, Linchevski, & Sfard, 2005; Kieran et al., 2002; Sfard, 1987, 2002, 2007, 2008; Sfard & McClain, 2002; Sfard & Prusak, 2005). On the way, we aim to illustrate the potency and some of the limitations of the discursive perspective. Part of the discussion which follows has been adopted from a review-essay on the 2002 volume by one of us (Nardi, 2005).

Beyond recognizing communication in mathematical learning as an *aid to*, or a component of, thinking, Sfard's position on communication was that it is "almost tantamount to the thinking itself" (Kieran et al., 2002, p. 13). Sfard distanced herself from the perspective of learning as acquisition of entities (e.g., concepts, schemes, etc.). For her, learning was to be seen as change in one's participation in well-defined forms of activity. The shift to this perspective was necessary because earlier perspectives had failed to provide satisfactory accounts for problems of both theory and practice. Two examples from classroom data reported by Sfard (2002) drew attention to two such problems. One concerned our limited ability to explain failure, and success, in mathematical learning, despite extensive work on students' perceptions of, and difficulties with, specific mathematical topics. The other concerned our limited ability to establish pedagogical practices that warrant understanding (as opposed to merely instrumental task completion). Both highlight our limited insight into what determines the ways in which interlocutors (in this case, students and teachers) choose to proceed in a mathematical conversation. In a more recent work, Sfard (2008) provided many more examples—see, for instance, the five quandaries that she discussed in Chapter 9 of that book.

What is it, then, that the learning-as-acquisition metaphor, the Piagetian account—the equating of understanding with "perfecting mental representations," and learning-with-understanding with "relating new knowledge to knowledge already possessed" (Sfard, 2002, p. 21)—has left unaccounted? The shift from behaviourism to acquisitionism was clearly useful, Sfard stressed, but the approach that she now proposed would move on from Piaget's work. That approach, which originated in the work of Vygotsky, was intended to complement the latter in order to enrich our understanding of the issues that acquisitionism had left unresolved.

The basic difference between an acquisitionist and a participationist perspective on learning is that the latter dispenses with the belief in the existence of context-free cognitive invariants, and emphasizes the social origins of human learning. Sfard's (2002) account clearly delineates differences between the two frameworks, but also draws attention to how they complement each other. Thinking, in the participationist framework, is conceptualized as communicating with oneself, whereas "communication may be diachronic or synchronic, with others or with oneself, predominantly verbal or with the help of any other symbolic system" (p. 28). If "communicational" (a term which is explained in the 2002 paper as epistemologically distinct from but akin to "discursive," and was later, in the 2008 volume, replaced by "commognitive") psychology of human thinking posits that speech is no longer a window to thought but its determining element, then as long as thought is in language, the two—the thought and the speech—are inseparable. The interlocutor is constrained by the situation in which the communication takes place and influences it in return. In this sense, learning mathematics, or history, is initiation to a discourse, where discourse is meant as a type of communication that characterizes a particular community. From this perspective, we can talk about the discourse of science, or the discourse of a professional group, or a social class. Mathematical learning, then, is initiation into mathematical discourse. Sfard (2002) introduced four factors that determined this initiation, two of these being introduced as "mediating tools" (p. 29) and "metadiscursive rules" (p. 30).

Sfard (2002) closed her introduction to what at the time was called "the communicational framework" with a cautionary methodological remark. Wittgenstein wisely attacked mentalism—any reference to "mental states" and the inherently unobservable entities "in the mind" (p. 32). However, as "experiences, feelings and intentions are central to all our decisions" (p. 32), research must find ways of incorporating those into its accounts. Doing so is "safe," Sfard suggested, "as long as it is understood that the status of any claim about other people's intentions the researcher can make is *interpretive* [Sfard's emphasis]" (p. 32). All that research can offer is compelling, cogent, trustworthy researchers' *interpretations*.

A key contribution, in our view, of this approach has been in the research methods proposed as a means towards generating the aforementioned interpretations. These include "focal analysis" and "preoccupation analysis":

- "Focal analysis" involves an analysis of the "effectiveness of communication" between interlocutors with regard to the degree of clarity of the "discursive focus"—defined as "the expression used by an interlocutor to identify the object of her or his attention" (p. 34). There are three "focal ingredients" considered in this type of analysis: "pronounced" (what one is attending to), "attended" (how is one attending to what one is attending?) and "intended" (the collection of experiences evoked by the "pronounced focus" and the "assortment of statements that the interlocutor is now able to make on the entity identified by the pronounced focus," all the "discursive potentials" borne out of the "pronounced focus"). The constantly evolving nature of the "intended focus" is "the crux of the matter": successful communication often relies on the "attended focus" being "used as a public exponent of the intended focus."
- "Preoccupation analysis" involves an analysis of the "two types of intentions which may be conveyed through communicative actions" (p. 38): "object-level" intentions, like, for example, the intention to solve the mathematical problem in question, and "meta-discursive intentions," which are "often less visible even if not less influential" (e.g., the ways in which the interaction is managed, the relationship between the interlocutors, etc.). The former are often taken care of with the help of "focal analysis." "Preoccupation analysis" aims to explore the *interrelation* between these two types of intentions. Its principal tool is the "interactivity flowchart," a diagram characterized by "proactive" and "reactive arrows" to reveal "initiating" and "responsive" attitudes, "discourse-spurring" and "face-saving" techniques used by the interlocutors.

Both methods have the potential to illustrate learning and teaching in useful ways. To us, the ways in which the analyses generated through these methods not only acknowledged that not any communication generates learning, but also offered ways of describing the *type* of communication that does generate learning, was very important. This is a vital issue to which several other authors in the Kieran et al. (2002) volume returned, including Kieran herself, in her own chapter in the volume.

There is little doubt that the work summarized above is a well-thought through endeavor to bridge the individual with the social within research in mathematics education. Any concerns that such an elaborate analysis of mathematical conversation may miss an emphasis on the development of individual interlocutors are largely assuaged by the painstaking detail with which "focal analysis" and "interactivity flowcharts" elaborate such developments (and also generate *mathematically* rich accounts of the data).

A pragmatic concern about the employment of these methods is one of scale. To generate an account of data through any of the above approaches is a time-consuming enterprise that by definition needs to focus on only parts of the typically vast amount of data collected in naturalistic studies of learning and teaching. This focussing on fragments of the data may have an impact on the analyst's capacity to observe learning over a substantial period of time.

Finally, and this is an issue that Sfard elaborated upon in later work (e.g., Sfard, 2008), there is the issue of "ecological validity" (Seeger, 2002, p. 293) of findings generated by discourse-oriented analyses. Seeger (2002) employed this term to describe the degree to which findings "give a fairly comprehensive and typical account," particularly in cases where "experimental conditions do not match conditions in the real world" (p. 293): by focussing on collecting evidence on the observable elements of mathematical behaviour, we may miss implicit, unobservable but perhaps significant processes taking place. Later, when Sfard (2008) elaborated upon this issue, she made a strong plea for the discursive researcher to "alternate between being an *insider* and an *outsider* to the discourse under study" (p. 278).

Sfard's (2008) ambitious book, which came a few years after the 2002 volume, presented her proposition in fuller scope—and in a way that situated her perspective "at the intersection of consciousness studies, linguistics, philosophy, and mathematics education" (Sriraman, 2009, p. 541). The perspective is known as the "commognitive framework," with the hybrid term "commognition" emphasizing the interrelatedness, almost inseparability, of "cognition" and "communication." In Sfard's (2008) words, the term is meant to refer "to those phenomena that are traditionally included in the term cognition, as well as to those usually associated with interpersonal exchanges" (p. 83).

From the assumption that thinking is a special case of communicational activity it follows that mathematical learning can be seen as a particular type of communication. In Sfard's participationist view of learning, mathematical learning is seen as initiation into the discursive practices of the mathematical community. The learning of mathematics therefore involves a change of discourse. Teaching mathematics, in this sense, involves the changing of the students' discourse. Forms of communication include communication through written language, spoken language, physical objects and artefacts deployed for discursive ends. Specifically, a discourse is made distinct by a community's *word use, visual mediators, endorsed narratives* and *routines*:

- *Word use* (vocabularies, keywords and their use) includes the use of mathematical terminology as well as ordinary words with a specific meaning within mathematics (such as "limit," "open," "continuous," and "group").
- *Visual mediators* include diagrammatic and symbolic mediators of mathematical meaning (graphs, diagrams, symbols, etc.) as well as the physical objects we often employ in mathematics lessons in school settings.
- *Endorsed narratives* include definitions, theorems and proofs and generally text, either spoken or written, which describe objects and processes as well as relationships among those, and is subject to rejection or endorsement according to rules defined by the community.
- *Routines* include regularly employed and well-defined practices that are employed by, and distinctly characterize the community. Within mathematics, *routines* include conjecturing, proving, estimating, generalising, abstracting, etc.

The question might be asked: "What makes Sfard's efforts to illuminate our understanding of thinking any different from previous efforts?" (Yackel, 2009, p. 90). Sfard's efforts to go to great lengths to develop an approach that meets accepted standards of scientific rigor by providing operational definitions of keywords, such as thinking, communication, discourse, and mathematical object would be one answer to that question. And we note that her approach to developing operational definitions is Wittgenstcinian, as evident in her endorsement of Wittgenstein's statement, "the meaning of a word is its use in the language" (Sfard, 2008, p. 73). Another answer to this question is that, although Sfard cast a critical eye on acquisitionist metaphors for learning, she was also "careful to point out the benefits of objectification, namely the ways it contributes to effective and efficient communication," thereby "not decrying our propensity for objectification" (Sfard, 2008, p. 91), but simply calling for a careful scrutiny of the assumptions we make when we employ "object" metaphors.

The potency and degree of acceptance of Sfard's proposition will depend on how comfortably it will sit alongside other approaches in what we earlier described, borrowing from Jablonka and Bergsten (2010), as the "social brand" of approaches to research in mathematics education. For example, could the radical constructivist position be "subsumed as an extreme case within the commognitive framework" (Sriraman, 2009, p. 544)? For the moment it seems that the framework puts forward a "grammar by which communication can be better fostered between researchers analyzing the same discursive 'mathematical' objects in teaching and learning situations" (Sriraman, 2009, p. 544). The framework seems to be accumulating several "theoretical and methodological" elaborations (Ryve, 2011, p. 187) and seems to have attracted numerous users of its grammar. We close this section with a brief outline of a few examples of such use: one from secondary MER, specifically the award-winning *Research in Mathematics Education* paper by Natalie Sinclair and Violeta Yurita (2008); and, three from research into the teaching and learning of post-compulsory mathematics presented at CERME7, the 7th Congress of European Research in Mathematics Education.

Sinclair and Yurita (2008) investigated the impact of the introduction of a dynamic geometry environment (DGE) on the mathematical thinking of students and teachers in a secondary geometry class, by identifying changes in the discourse engendered by its introduction. The paper focussed on the teacher and it is to the credit of the authors' (commognitive) analysis that they revealed substantial differences between static and dynamic geometry—for example, "in the ways the teacher talks about geometric objects, makes use of visual artifacts and models geometric reasoning" (p. 135).

Nardi (2011) examined data from interviews with university mathematicians—reported extensively in (Nardi, 2008)—in order to outline issues related to university students' discursive shifts in the early periods after their arrival at university. The paper focussed on verbalization skills, namely skills in the employment of ordinary language to convey mathematical meaning. During the interviews the mathematicians emphasized the role of verbal expression to drive noticing to the key idea of a symbolically-formulated mathematical sentence; the importance of good command of ordinary language; the role of verbalization as a mediator between symbolic and visual mathematical expression; and the precision proviso for the use of ordinary language in mathematics. The analyses revealed that the community's discourse on verbalization in mathematics tended to be risk-averse; that wordless mathematics discourse remained alluring; and that more explicit, and less potentially contradicting, pedagogical action was necessary in order to facilitate students' appreciation of verbal mathematical expression and acquisition of verbalization skills.

Stadler (2011) examined a particular case of the transition from secondary school to university mathematics (the mathematical context was solving a parametric system of simultaneous equations) in order to discuss students' experiences of the transition as an often perplexing re-visiting of content and ways of working that seem simultaneously familiar and novel (in this case dealing with variables, parameters and unknowns when solving equations). The paper's commognitive analyses of the student observations and interviews allowed the multi-faceted (individual, institutional, social) nature of the transition from school to university mathematical discourse to emerge. Throughout, the impression was that Sfard's perspective provided a good fit to studies of transition.

Finally, Viirman (2011) traced the variation within and between three university lecturers' discourses as they introduced the concept of function. Through analysis of the observations that focussed particularly on the routines and endorsed narratives characterizing the lecturers' discourse, the variation across the three came to the fore (the variation concerned the different ways the lecturers resorted to definitions, examples, the "why" or "when" of endorsing certain narratives, etc.).

Concluding Remarks

We conclude this chapter with a call to mathematics education researchers for a more incisive approach to, and employment of, theory that goes beyond merely knowing the "grammar" of a theory (Lerman, 2010, p. 101), a trend that Lerman and other contributors (e.g., Sriraman & English, 2010) identified as a symptom of careless eclecticism. As Artigue (2011) noted:

> This is indeed hard work. Theoretical frameworks and constructs being dynamic entities shape research practices and are shaped by these, one cannot make sense of them without considering their different components and the research practices (or research praxeologies following Chevallard and ATD) they make possible and those they result from. It is not enough to know the elements of its "grammar" … for making sense of a theory and appreciating its potential. Any productive dialogue around theoretical issues cannot stay at the level of the theoretical objects themselves but needs to enable collaborative work around appropriate exemplars of research praxeologies, and this is also a real challenge for us. (p. 312)

Orchestrated efforts in this direction have been evident in the field for some time (see, for example, Parts XV and XVI of Sriraman and English (2010) that were dedicated to "networking of theories"). The call for an even more systematic communication between theories that would go beyond merely borrowing extends to developments in other fields too—neuroscience being prominent among those fields where recent developments have made this call even more topical (see, for example, Campbell's (2010) strong case for a mathematics education neuroscience). The words of Kristin Umland (2011) serve as a good indication of a way forward: after calling for a systematic survey of the big questions in mathematics education that need to be addressed, Umland added:

> This survey should include a discussion of research methods that might be appropriately used to investigate them and weaknesses in both the relevant empirical record and extant theories, many of which are still very immature and should necessarily be refined as time passes. The ultimate test of the value of the ideas … is whether they or their progeny help solve the problems that teachers, administrators, and policymakers face as they work to improve mathematics teaching and learning. (p. 74)

References

Artigue, M. (2002). Learning mathematics in a CAS environment: The genesis of a reflection about instrumentation and the dialectics between technical and conceptual work. *International Journal of Computers for Mathematics Learning, 7*, 245–274.

Artigue, M. (2011). Review of Bharath Sriraman & Lyn English: Theories of Mathematics Education—Seeking New Frontiers. *Research in Mathematics Education, 13*(3), 311–316.

Asiala, M., Cottrill, J., Dubinsky, E., & Schwingendorf, K. (1997). The development of students' graphical understanding of the derivative. *Journal of Mathematical Behavior, 16*(4), 399–431.

Ben-Yehuda, M., Lavy, I., Linchevski, L., & Sfard, A. (2005). Doing wrong with words: What bars students' access to arithmetical discourses. *Journal for Research in Mathematics Education, 36*(3), 176–247.

Bernstein, B. (1996). *Pedagogy, symbolic control and identity: Theory, research, critique.* London, UK: Taylor & Francis.

Bosch, M., Chevallard, Y., & Gascon, J. (2005). Science or magic? The use of models and theories in didactics of mathematics. In M. Bosch (Ed.), *Proceedings of CERME4 (4th Conference of European Research in Mathematics Education* (pp. 1254–1263). Sant Feliu de Guixols, Spain: CERME.

Bourbaki, N. (1970). *Théorie des ensembles de la collection éléments de mathématique.* Paris, France: Hermann.

Campbell, S. (2010). Embodied minds and dancing brains: New opportunities for research in mathematics education. In B. Sriraman & L. English (Eds.), *Theories of mathematics education: Seeking new frontiers* (pp. 309–332). New York, NY: Springer.

Chevallard, Y. (1985). *La transposition didactique. Du savoir savant au savoir enseigné.* Grenoble, France: La Pensée Sauvage.

Chevallard, Y. (1992a). Fundamental concepts of didactics: Perspectives given by an anthropological approach. *Recherches en Didactique des Mathématiques, 12*(1), 73–112.

Chevallard, Y. (1992b). A theoretical approach to curricula. *Journal für Mathematik Didaktik, 2*(3), 215–230.

Chevallard, Y. (1999a). L'analyse des pratiques enseignantes en théorie anthropologique du didactique. *Recherches en Didactique des Mathématiques, 19*(2), 221–266.

Chevallard, Y. (1999b). Didactique? You must be joking! A critical comment on terminology. *Instructional Science, 27*, 5–7.

Dienes, Z. P. (1960). *Building up mathematics.* London, UK: Hutchinson.

Dienes, Z. P. (1963). *An experimental study of mathematics learning.* London, UK: Hutchinson.

Dienes, Z. P. (1964). *The power of mathematics.* London, UK: Hutchinson Educational.

Dienes, Z. P. (1971). An example of the passage from the concrete to the manipulation of formal systems. *Educational Studies in Mathematics, 3*, 337–352.

Dienes, Z. P. (2000). The theory of the six stages of learning with integers. *Mathematics in School, 29*, 27–33.

Dienes, Z. P., & Jeeves, M. A. (1965). *Thinking in structures.* London, UK: Hutchinson.

Dietrich, O. (2004). Cognitive evolution. In F. M. Wuketits & C. Antweiler (Eds.), *Handbook of evolution* (pp. 25–77). Weinheim, Germany: Wiley-VCH Verlag GmbH.

Dubinsky, E. (1991). Reflective abstraction in advanced mathematical thinking. In D. Tall (Ed.), *Advanced mathematical thinking* (pp. 95–126). Dordrecht, The Netherlands: Kluwer.

Ely, R. (2010). Book review: Theories of Mathematics Education. *Educational Studies in Mathematics, 75*(2), 235–240.

Fried, M. (2011). Theories for, in, and of mathematics education. *Interchange, 42*(1), 81–95.

Gascón, J. (2003). From the cognitive program to the epistemological program in didactics of mathematics: Two incommensurable scientific research programs? *For the Learning of Mathematics, 23*(2), 44–55.

Greer, B. (2008) (Ed). Critical notice on the National Mathematics Advisory Panel Report. *The Montana Mathematics Enthusiast, 5*(2&3), 365–428.

Guin, D., Ruthven, K., & Trouche, L. (Eds.). (2004). *The didactic challenge of symbolic calculators.* Dordrecht, The Netherlands: Kluwer Academic Publishers.

Hurford, A. (2010). Complexity theories and theories of learning: Literature reviews and syntheses. In B. Sriraman & L. English (Eds.), *Theories of mathematics education: Seeking new frontiers* (pp. 562–589). New York, NY: Springer.

Jablonka, E., & Bergsten, C. (2010). Commentary on theories of mathematics education—Is plurality a problem? In B. Sriraman & L. English (Eds.), *Theories of mathematics education: Seeking new frontiers* (pp. 11–117). New York, NY: Springer.

Jablonka, E., & Bergsten, C. (2011). Theorising in mathematics education research: Differences in modes and quality. *Nordic Studies in Mathematics Education, 15*(1), 25–52.

Jankvist, U. T. (2011). Theories of Mathematics Education, edited by Bharath Sriraman and Lyn English: Common ground for scholars and scholars in the making. *Mathematical Thinking and Learning, 13*(4), 247–257.

Jaworski, A., & Coupland, N. (Eds.). (2005). *The discourse reader* (2nd ed.). London, UK: Routledge.

Kieran, C., Forman, E., & Sfard, A. (Eds.). (2002). *Learning discourse: Discursive approaches to research in mathematics education*. Dordrecht, The Netherlands: Kluwer Academic Publishers.

Lerman, S. (2000). The social turn in mathematics education research. In J. Boaler (Ed.), *Multiple perspectives an mathematics teaching and learning* (pp. 19–44). Westport, CT: Ablex Publishing.

Lerman, S. (2010). Theories of mathematics education—Is plurality a problem? In B. Sriraman & L. English (Eds.), *Theories of mathematics education: Seeking new frontiers* (pp. 99–109). New York, NY: Springer.

Lesh, R., Behr, M., Cramer, K., Harel, G., Orton, R., & Post, T. (1992). Five versions of the multiple embodiment principle. *ETS Technical Report*. Retrieved May 28, 2011 from http://education.umn.edu/rationalnumberproject.

Lesh, R., Cramer, K., Doerr, H., Post, T., & Zawojewski, J. (2003). Model development sequences perspectives. In R. Lesh & H. Doerr (Eds.), *Beyond constructivism: A models & modeling perspective on mathematics teaching, learning, and problem solving*. Hillsdale, NJ: Lawrence Erlbaum.

Lesh, R., & Doerr, H. (Eds.). (2003). *Beyond constructivism*. Mahwah, NJ: Lawrence Erlbaum.

Lesh, R., Post, T., & Behr, M. (1987). Dienes revisited: Multiple embodiments in computer environments. In I. Wirszup & R. Streit (Eds.), *Developments in school mathematics education around the world* (pp. 647–680). Reston, VA: National Council of Teachers of Mathematics.

Lesh, R., Post, T., & Behr, M. (1988). Proportional reasoning. In J. Hiebert & M. Behr (Eds.), *Number concepts and operations in the middle grades* (pp. 93–118). Reston, VA: National Council of Teachers of Mathematics.

Lesh, R., & Sriraman, B. (2005). Mathematics education as design science. *Zentralblatt für Didaktik der Mathematik, 37*(6), 490–505.

Mousoulides, N., Sriraman, B., & Lesh, R. (2008). The philosophy and practicality of modeling involving complex systems. *Philosophy of Mathematics Education Journal, 23*, 134–157.

Nardi, E. (2005). "Beautiful minds" in rich discourses: On the employment of discursive approaches to research in mathematics education. *European Educational Research Journal, 4*(2), 145–154.

Nardi, E. (2008). *Amongst mathematicians: Teaching and learning mathematics at university level*. New York, NY: Springer.

Nardi, E. (2011). "Driving noticing" yet "risking precision": University mathematicians' pedagogical perspectives on verbalization in mathematics. In M. Pytlak, T. Rowland, & E. Swoboda (Eds.), *Proceedings of the 7th Conference on European Research in Mathematics Education* (pp. 2053–2062). Rzeszow, Poland: Congress of the European Society for Research in Mathematics Education.

Piaget, J. (1958). *The growth of logical thinking from childhood to adolescence*. New York, NY: Basic Books.

Piaget, J. (1971). *Biology and knowledge*. Edinburgh, Scotland: Edinburgh University Press.

Piaget, J. (1975). *The child's conception of the world*. Totowa, NJ: Littlefield, Adams.

Piaget, J (1987). *Possibility and necessity: The role of necessity in cognitive development* (2 Vols.). Minneapolis, MN: University of Minnesota Press.

Popper, K. (1959). *Logik der forschung*. London, UK: Hutchinson.

Robert, A., & Rogalski, J. (2005). A cross-analysis of the mathematics teachers' activity: An example in a French 10th-grade class. *Educational Studies in Mathematics, 59*(1–3), 269–298.

Ryve, A. (2011). Discourse research in mathematics education: A critical evaluation of 108 journal articles. *Journal for Research in Mathematics Education, 42*(2), 167–198.

Schoenfeld, A. (2010). Review of Bharath Sriraman & Lyn English: Theories of Mathematics Education—Seeking New Frontiers. *ZDM—The International Journal on Mathematics Education, 42*(5), 503–506.

Seeger, F. (2002). Research on discourse in the mathematics classroom. In C. Kieran, E., Forman & A. Sfard (Eds.), *Learning discourse: Discursive approaches to research in mathematics education* (pp. 287–297). Dordrecht, The Netherlands: Kluwer Academic Publishers.

Seel, N. M. (Ed.). (2012). *Encyclopedia of the sciences of learning*. New York, NY: Springer.

Sensevy G. (2009, January). *Outline of a joint action theory in didactics*. Paper presented at the Sixth Congress of the European Society for Research in Mathematics Education, Lyon, France.

Sensevy, G., & Mercier, A. (Eds.). (2007). *Agir ensemble. L'action didactique conjointe du professeur et des élèves dans la classe*. Rennes, France: Presses Universitaires de Rennes.

Sfard, A. (1987). Mathematical practices, anomalies and classroom communication problems. In P. Ernest (Ed.), *Constructing mathematical knowledge: Epistemology and mathematics education*. London, UK: The Falmer Press.

Sfard, A. (2002). There is more to discourse than meets the ears: Looking at thinking as communicating to learn more about mathematical learning. In C. Kieran, E. Forman, & A. Sfard (Eds.), *Learning discourse: Discursive approaches to research in mathematics education* (pp. 13–57). Dordrecht, The Netherlands: Kluwer Academic Publishers.

Sfard, A. (2007). When the rules of discourse change, but nobody tells you: Making sense of mathematics learning from a commognitive standpoint. *The Journal of the Learning Sciences, 16*(4), 565–613.

Sfard, A. (2008). *Thinking as communicating. Human development, the growth of discourse, and mathematizing*. New York, NY: Cambridge University Press.

Sfard, A., & McClain, K. (2002). Analyzing tools: Perspectives on the role of designed artifacts in mathematics learning. *Journal of the Learning Sciences, 11*(2/3), 153–161.

Sfard, A., & Prusak, A. (2005). Telling identities: In search of an analytic tool for investigating learning as a culturally shaped activity. *Educational Researcher, 34*(4), 14–22.

Sierpinska, A., & Kilpatrick, J. (Eds.). (1998). *Mathematics education as a research domain: An ICMI study*. Dordrecht, The Netherlands: Kluwer Academic Publishers.

Sierpinska, A., & Lerman, S. (1996). Epistemologies of mathematics and mathematics education. In A. Bishop, M. A. Clements, C. Keitel, J. Kilpatrick, & C. Laborde (Eds.), *International handbook of mathematics education* (pp. 827–876). Dordrecht, The Netherlands: Kluwer Academic Publishers.

Sinclair, N., & Yurita, V. (2008). To be or to become: How dynamic geometry changes discourse. *Research in Mathematics Education, 10*(2), 135–150.

Sriraman, B. (2008). Let Lakatos be. *Interchange: A Quarterly Review of Education, 39*(4), 483–492.

Sriraman, B. (2009). What's all the commotion over commognition? *The Montana Mathematics Enthusiast, 6*(3), 541–544.

Sriraman, B., & English, L. (2004). Combinatorial mathematics: Research into practice. Connecting research into teaching. *The Mathematics Teacher, 98*(3), 182–191.

Sriraman, B., & English, L. (2005). Theories of mathematics education: A global survey of theoretical frameworks/trends in mathematics education research. *Zentralblatt für Didaktik der Mathematik, 37*(6), 450–456.

Sriraman, B., & English, L. (2006). Theories of mathematics education: European perspectives, commentaries and viable research directions. *Zentralblatt für Didaktik der Mathematik, 38*(1), 1–2.

Sriraman, B., & English, L. (Eds.). (2010). *Theories of mathematics education: Seeking new frontiers*. New York, NY: Springer.

Sriraman, B., & Törner, G. (2008). Political union/mathematical education disunion: Building bridges in European didactic traditions. In L. D. English, M. G. Bartolini Bussi, G. A Jones, R. A. Lesh, & D. Tirosh (Eds.), *Handbook of international research in mathematics education* (2nd edn., pp. 660–694). New York, NY: Routledge, Taylor & Francis.

Stadler, E. (2011). The same but different—Novice university students solve a textbook exercise. In M. Pytlak, T. Rowland, & E. Swoboda (Eds.), *Proceedings of the 7th Conference on European Research in Mathematics Education* (pp. 2053–2062). Rzeszow, Poland: Congress of the European Society for Research in Mathematics Education.

Umland, K. (2011). Review of theories of mathematics education: Seeking new frontiers. *The Mathematical Intelligencer, 33*(2), 73–74.

Viirman, O. (2011). Discourses of functions—University mathematics teaching through a commognitive lens. In M. Pytlak, T. Rowland, & E. Swoboda (Eds.), *Proceedings of the 7th Conference on European Research in Mathematics Education*. Rzeszow, Poland: Congress of the European Society for Research in Mathematics Education.

Weyl, H. (1918). *Raum-Zeit-Materie. Vorlesungen überallgemeine relativitätstheorie*. Berlin, Germany: Springer Verlag.

Yackel, E. (2009). Book review: "Thinking as communicating: Human development, the growth of discourses, and mathematizing". *Research in Mathematics Education, 11*(1), 90–94.

Chapter 11
Research Methods in Mathematics Teacher Education

Uwe Gellert, Rosa Becerra Hernández, and Olive Chapman

Abstract As the field of mathematics education grows so too do the research methods used to study the field. In the special area of teacher education, the last decade has witnessed a substantial increase in attention. New perspectives and new methodologies have been constituted and new research techniques established. Choosing the "right" method is not a trivial task for any researcher, and increasingly we are seeing more sophisticated research methods, including different forms of mixed methods. A main concern of research these days relates to the fact that as well as studying teachers and teaching, researchers want to see their findings applied to the professional development of teachers and to a critical modification of teacher education practices, in the frame of social changes. This has led to more research *with* teachers rather than *on* teachers. After surveying state-of-the-art of methods in research on mathematics teacher education published in renowned international journals, this chapter focusses on *participatory action research* as an example of a research method from the politicized periphery of the field.

Introduction

Prior to the establishment of the *Journal of Mathematics Teacher Education* (JMTE) in 1998 only a relatively small number of studies of mathematics teacher education were published in academic journals. This was indicated in Lubienski and

U. Gellert (✉)
Freie Universität Berlin, Berlin, Germany
e-mail: ugellert@zedat.fu-berlin.de

R. Becerra Hernández
Universidad Pedagógica Experimental Libertador, Caracas, Venezuela

O. Chapman
University of Calgary, Calgary, Canada

Bowen's (2000) survey of mathematics education research reports published in 48 major educational research journals between 1982 and 1998. Only 6% of their "general topics related to teaching and learning" (p. 630) category, among the 3,011 articles they identified as research on mathematics education, dealt with teacher education. The situation seems to have changed substantially since 1998. For example, Sfard (2005), in the course of summarizing a survey on the relation between mathematics education research and practice presented at the *10th International Congress on Mathematical Education* [ICME10], reported about a "prevalent focus on teacher and teacher practice" (p. 397) based on the responses of the 74 participants when they were asked how they would describe their "work in mathematics education over the last five years" (p. 396). The survey team found that "teacher-centeredness in research could be identified in two-thirds of the respondents who claimed to be engaged in research" (p. 397).

Although the field of research on mathematics teacher education has grown substantially and continues to grow, "research method" seems to be a peripheral aspect of this research. As a prominent example, the four volumes of *The International Handbook of Mathematics Teacher Education* (Jaworski & Wood, 2008; Krainer & Wood, 2008; Sullivan & Wood, 2008; Tirosh & Wood, 2008) were organized along the questions of the *what*, the *how* and the *who* of mathematics teacher education as well as on the knowledge and roles of teacher educators working with teachers in teacher education processes and practices. However, there was no explicit focus on the methods by which all that knowledge of mathematics teacher education had been accumulated in any of the 60 chapters. A similar absence can be observed in an editorial of JMTE in which the three articles of the issue were introduced with the words: "These three articles differ in many aspects, including scope, nature, objects of research, specific aims, research questions and suggestions for improving mathematics teacher education" (Tirosh, 2007, p. 143). Again, research method was apparently not a category to name.

This chapter intends to lessen this research deficit. It reports on a survey that exclusively focussed on methodologies, research methods and research techniques, drawing on journal articles published between 2005 and 2010. It builds on a survey by Adler, Ball, Krainer, Lin, and Novotna (2005) that included a focus on methods used in reports on research related to mathematics teacher education published between 1999 and 2003. The chapter offers examples of techniques used in the research surveyed. It also discusses issues related to what and which research gets published in the journals surveyed. In particular, it discusses work which has been done at the periphery of the field of research on mathematics teacher education. As an example of this, emphasis is placed on describing a version of *participatory action research*. The chapter does not discuss research methods in terms of which method could be used for which purposes under what conditions. Such a discussion about *doing* research in teaching or teacher education that deals with the working of different research methods, although not specific to mathematics education, is available in sources such as Linn (1986) for quantitative methods, Erickson (1986) for qualitative methods, and others noted in later sections of this chapter.

Surveying Research Methods in Mathematics Teacher Education

For the sake of conceptual precision it will be useful to distinguish between a methodology, a research method and a research technique. In this chapter these terms are differentiated conceptually as follows: Research techniques, for example videotaping, can be used for different purposes. Research techniques do not constitute a research method, although many methods draw on a specific set of techniques. Methods, for example case study, can be regarded as ways for gathering evidence. Methodologies refer to the rationale and the theoretical assumptions guiding the research. They justify why the research is proceeding the way it does. In this section, we present the results of our survey of research methods, methodologies and techniques in mathematics teacher education. We build on a survey of research on mathematics teacher education conducted for ICME10 and make comparisons to it and other previous surveys to establish possible trends.

A Survey Conducted at ICME10

At ICME10, Jill Adler, Deborah Ball, Konrad Krainer, Fou Lai Lin and Jarmila Novotná (2005) reported on an international survey of published research in mathematics teacher education for the period 1999–2003. The survey was based on publications in many international journals and some special conference proceedings and on handbook articles. The final focus of the survey was research published in JMTE (65 articles), in the *Journal of Research in Mathematics Education* [JRME] (7 articles) and in the proceedings of the annual conferences of the International Group for the *Psychology of Mathematics Education* (88 papers). The survey identified four emerging themes and Adler et al. (2005) summarized their findings by making the following claims:

"**Claim 1: Small-scale qualitative research predominates**" (p. 368). For small-scale research, Adler et al. counted studies focussing on a single teacher or on groups of size less than 20. Of the 145 studies drawing on empirical data, 98 fell into this category. Adler et al. argued that since research on mathematics teachers' education and professional development is a rather new field of study, hypotheses needed to be identified through qualitative inquiry before large-scale testing of hypotheses occurred. "It seems natural that the interest in particularization precedes generalization" (p. 369). Although Adler et al. acknowledged the "significant contributions for conceptualizing the complexity of teacher education and modelling individual teachers' learning process" (p. 370) made by small-scale qualitative studies, they also called for more large-scale studies, cross-case analyses and longitudinal studies.

"Claim 2: Most teacher education research is conducted by teacher educators studying the teachers with whom they are working" (p. 371). As more than 80% of the studies were of this type, Adler et al. called for more external research of large-scale type and/or for the development of "strong and effective theoretical languages that enable us to create a distance between us and what we are looking at" (p. 372).

"Claim 3: Research in countries where English is the national language dominates the literature" (p. 372). In JMTE, 80% of the articles published from 1998 to 2003 were from such countries. For JRME and for the *Psychology of Mathematics Education* conference proceedings the figures were somewhat lower (71% and 43% respectively), but still noteworthy. As Adler et al. argued, it seems to be troubling that the perspectives of some, and the particularities of their conditions, become the basis of the generally-accepted knowledge in the field.

"Claim 4: Some questions have been studied, not exhaustively, but extensively, but other important questions remain unexamined" (p. 375). Issues which have been studied extensively and for which there are numerous international publications are:

- the effectiveness of particular programs of teacher education;
- teachers' (re-)learning within reform processes; and
- professional communities and other institutional settings.

Issues which, according to Adler et al. (p. 376), have remained unexamined are:

- teachers' learning in contexts where reform is not the dominant issue;
- the nature of teachers' learning from experience;
- teachers learning to address issues of gender, language and socio-economic status;
- comparisons of the effectiveness of different teacher-education settings; and
- the effects of extending programs to multiple teacher-learning settings.

Survey 2005–2010

Building on the survey by Adler et al. (2005), we conducted a survey for the period 2005–2010 focussing on research methods, methodology and techniques in studies on mathematics teacher education. For such a survey, it is always difficult to decide which publications to include. Our decision to focus on JMTE, JRME and *Educational Studies in Mathematics* (ESM) was motivated by the high prestige of these journals in the field of mathematics education in general and, particularly for JMTE, in mathematics teacher education. We assumed that the publications in these journals reflected and produced accepted knowledge in the field (Dowling, 2009; Ernest, 1991). This position was similar to that of Wilson and Cooney's (2002) justification of why they reviewed JMTE, JRME and ESM for articles on mathematics teachers' beliefs or teacher change: "We have chosen to review articles from these three journals, given their international readership and the fact that they provide important outlets for researchers concerned with teaching or teacher education" (p. 128). We recognized the limitation in focussing only on these English-based journals, but given their prominence in the field, they provide a standard to understand how the field is growing in relation to the boundaries they implied, whether intentional or not.

Another challenge was deciding which articles to include and exclude. As Adler et al. (2005, p. 364) noted, the boundary between what to include and what to exclude is somewhat blurred. For the survey reported in this section, not all articles published in JMTE were included. Our focus was on the education and professional development of mathematics teachers. Like Adler et al. (2005), we excluded studies on teachers' knowledge, beliefs and practices, and those in the "mathematics teacher education around the world" category, where there was only a limited relationship to the *processes* of teacher education and professional development. We also did not include articles that focussed on the description of tasks for mathematics teacher education and professional development purposes, for these could be found in a special issue of JMTE (2007, *10*(4–6)).

A final concern could be the description of the research method, methodology and technique in the article. As Burton (2002) reminded us, journal articles are notoriously short in their discussion of methodology.

> In the majority of articles in journals and chapters in books, a description is provided of "how" the research was done but rarely is an analysis given of "why" and, more particularly, out of all the methods that could have been used, what influenced the researcher to choose to do the research in the manner described. (p. 1)

However, this shortness is not being taken as an indication of an underdeveloped methodology on which the research is based.

In the following subsections we report on selected findings from the survey. We connect our results with the results from other surveys in order to estimate trends or tendencies of development. We focus on research methods and techniques as methodologies, in our definition of the term, necessarily connect to research questions and the activated theories. In the two last subsections we move beyond the survey to argue that methodologies, research methods and techniques are constantly being developed further.

Number of participating teachers. Adler et al. (2005) claimed that small-scale research dominates the field of mathematics teacher education. This claim is still legitimate. Although our data corpus differs from that of Adler et al. in that we have focussed only on journal articles, there seems to be little progress towards an increase in the number of teachers involved. Although the percentage of studies with more than 100 participating mathematics teachers has nearly doubled, still 89% of the studies involve less than 100 teachers (see Table 11.1).

Geographic origin of the research. Research in countries where English is a national language (nearly always *the* national language) continued to dominate in

Table 11.1
Numbers of Teachers Studied in JMTE, ESM and JRME Articles

Numbers of Teachers Participating in the Study	JMTE/ESM/JRME Number of Articles ($n=151$)
1	23
2–9	46
10–19	15
20–99	45
100 or more	17
No data/No empirical study	5

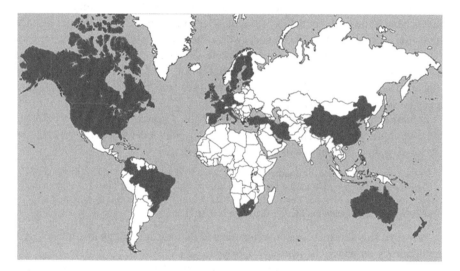

Figure 11.1. Global distribution of the surveyed research on mathematics teacher education.

these English-based journals. Seventy percent of the articles with single-nation authors came from English-speaking contexts (Québec was not included in the 70%). For only two of the 19 articles did contributing authors work in different countries, and English was not the national language in any of those countries. For 54% of the articles, at least one author worked in the USA. Figure 11.1 displays (in dark) the countries represented by at least one research institution within our data corpus.

References to standard procedures, or standard references. Established methods or techniques generally provide a standard for referencing a researcher's choice of a method or technique. However, 40% of the articles surveyed do not provide any substantive reference to such methods or techniques. For those with references, most of the sources are mentioned only once. In this survey, we accounted for only those references where the relation to the referenced work was made explicit in the 151 articles. This explicitness was in the form of a quotation or in some

Table 11.2
Most Frequently Referenced Research Volumes

Title (Author)	Frequency
Basis of Qualitative Research: Grounded Theory Procedures and Techniques (by A. L. Strauss & J. Corbin; 1st and 2nd editions)	19
The Discovery of Grounded Theory: Strategies for Qualitative Research (by B. G. Glaser & A. L. Strauss)	12
Qualitative Data Analysis: An Expanded Sourcebook (by M. B. Miles & A. M. Huberman; including earlier editions)	10
Qualitative Research and Evaluation Methods (by M. Q. Patton; including earlier editions)	9

reference to a particular feature or interpretation of the method or technique used—as was the case in the following example from Barrantes and Blanco's (2006) explanation of "discussion groups" in which both alternatives are visible.

> This is "an unmanaged technique whose purpose is the controlled production of discourse by a group of subjects who meet for a limited time to debate a topic designated by the researcher" (Gil, 1992–1993, p. 201). The idea is to establish and facilitate oral debate, not to interview the group (Watts & Ebbutt, 1987). (Barrantes & Blanco, 2006, p. 418)

In this example, both Gil (1992–1993) and Watts and Ebbutt (1987) were included in the survey as referenced sources. However, bibliographical references that appear merely as name-dropping are not included. We admit that this distinction was not always clear-cut and unambiguous. The result of the analysis of the 151 articles was a list of 80 referenced volumes, chapters and articles with 63 of these 80 mentioned only once and 7 of the 80 only twice (this happened particularly in the case of authors publishing more than one article during 2005–2010). For the remaining 10, some sources stood out by frequency (see Table 11.2).

Frequencies of 3 to 6 occurred for reference to various qualitative research methods (Bogdan & Biklen, 1998; Creswell, 1998; Erickson, 1986) and, more specifically, to work on case studies (Stake, 1995, 2000; Yin, 1989) and ethnography (Goetz & LeCompte, 1984; LeCompte, Preissle, & Tesch, 1993). A predominance of qualitative research methods is overtly visible.

It is remarkable that the two most often-cited volumes elaborate grounded theory. This is particularly interesting as any grounded-theory approach emphasizes the importance of a sensitive relation to the phenomena under study without drawing systematically on a specific theory. Either the field of research on mathematics teacher education and development is still empirically underdeveloped, or there is widespread scepticism of the usefulness of the theories that have been generated in the field. This scepticism seems to be paralleled by a dichotomy of idiosyncrasy/standardization of the literature on methods and research techniques which are most often referenced. The effect of language standardization was particularly strong given the English-based focus of the journals: only 3% of these references are work that was not originally written in English.

From a distinction between qualitative and quantitative methods to mixed methods? Although the distinguishing characteristics of qualitative and quantitative research have been, and still are, disputed, the distinction itself has been proven useful for a broad categorizing of research studies. The notion of "mixed methods" is much younger and has been made popular through a set of handbooks and articles (e.g., Creswell, 2003; Johnson & Onwuegbuzie, 2004; Tashakkori & Teddlie, 2003); since 2007 the *Journal of Mixed Methods Research* has contributed to the distribution of knowledge about possibilities for combining qualitative and quantitative methods and techniques.

Hart, Smith, Swars and Smith (2009) surveyed the research methods used in mathematics education by examining the use of qualitative, quantitative and mixed methods in articles published between 1995 and 2005 in major research journals in mathematics education. A subset of their data corpus consisted of all the articles published in JMTE, ESM and JRME, and therefore a comparison of findings is possible. Of particular interest here is how qualitative, quantitative and mixed methods research were distributed and whether a trend towards one of the options could be observed. For this purpose, we assemble in Table 11.3 the percentages of articles that used qualitative, quantitative and mixed methods in the period from 1995 to 2005 (from Hart et al., 2009) and the respective percentages from our own survey of research on mathematics teacher education and development based on the period 2005–2010. Since Hart et al. did not focus specifically on teacher education and development, we display Hart et al.'s percentages for JMTE separately.

Table 11.3 shows a decline of the percentage of mixed methods from 28% to 19%, suggesting that there was not a trend towards an increase in research that uses mixed methods. But the suggestion of a trend away from mixed methods could be a technical artefact: It is not a straightforward task to decide whether a combination of quantitative and qualitative research techniques actually results in a mixed methods approach. Hart et al. (2009) decided to "categorize an article as mixed methods if both qualitative and quantitative methods are used *in any part* of the article" (p. 38, emphasis added). They concluded that for the 10 JRME articles with mixed qualitative methods and inferential statistics, 6 did not report any findings from the interpretive analysis in the conclusions of the articles. That may explain the higher rate for mixed methods in Hart et al.'s survey.

A closer look at our survey data (see Figure 11.2) shows a slight increase in the number of studies that combined qualitative and quantitative methods at the

Table 11.3
Distribution of Quantitative, Qualitative and Mixed Methods

	Hart et al. (1995–2005)		Our Survey (2005–2010)
	ESM, JRME+JMTE	JMTE only (from Hart et al., 2009)	Research on Mathematics Teacher Education and Development
Quantitative only	14%	6%	17%
Qualitative only	58%	66%	64%
Mixed	28%	28%	19%

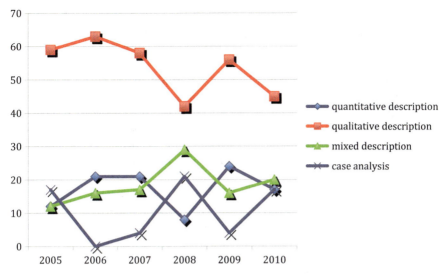

Figure 11.2. Development of the distribution of methods (2005–2010), in percentages.

expense of the number of research studies that used qualitative methods only. We distinguished between research that provided a quantitative description only of the phenomenon under study (including different-treatment designs, control-group designs), research reports that generated a qualitative description only (excluding case studies), case studies, and research that substantially combined qualitative and quantitative methods or techniques and generated both qualitative and quantitative descriptions. The trends are not considerable, and the year 2008 seems to be singular.

Techniques of data construction and data analysis. The spectrum of techniques by which the data of the surveyed research were constructed was rather broad. Most of the published research drew on a combination of two or more of the following elements: questionnaires and surveys, interviews, group conversations, videotaped classroom lessons (including copies of students' work), videotaped inservice lessons (including copies of students' work), and fieldnotes. A few studies used inservice course participants' reflective writing, mathematical autobiographies, different kinds and formats of tests (including multiple-choice, open-ended assessments, pre- and post-tests). In the 151 studies reviewed, research that drew on one single technique for the construction of data was extremely scarce.

If it was true that the techniques for data construction were rather variegated, the same was a fortiori true for the techniques of data analysis. Whether or not the techniques used could be subsumed under inferential or descriptive statistics, or as interpretive, interactionist, epistemological, psychoanalytical, critically discursive, phenomenological, etc., most of the research seemed to be such that the individual research topic or research interest determined the research methods—and not vice versa. Nevertheless, and this observation might relate to the predominance of the grounded theory literature, coding according to *emergent* categories seemed to be a very common pattern in much of the research reported. Apparently, there is not much coherence in the field in terms of a mutual enhancement or synergy of research methods and techniques.

Dynamical development of methodologies, research methods and techniques. The development of methodologies, research methods and research techniques can be linked to the interconnected nature of theory, research questions, research methods and research findings. This development can be triggered by conceptual advances, as can be witnessed in research on the professional knowledge of mathematics teachers (cf., Baumert et al., 2010; Fennema & Franke, 1992; Neubrand, Seago, Agudelo-Valderrama, DeBlois, & Leikin, 2009). It can also be generated by shifts in the perception of favourable settings for learning reflected in mathematics teachers' professional development as collaborative learning (cf., Arbaugh, 2003; Gellert, 2003; Krainer, 2003). This could partly account for the apparent increased interest in grounded theory or *emergent* methods, reflected in the studies reviewed which employ inductive approaches to interpret or evaluate such learning contexts and experiences.

Sometimes methods and methodologies develop because of a reflection on the methods and methodologies themselves. Such a development is particularly relevant here. As a case, we consider the developments of methodology and method used in the field of mathematics teachers' espoused and attributed beliefs. Speer (2005) examined the distinction that is often made—within cognitive frameworks (Frykholm, 1999; Thompson, 1992; Vacc & Bright, 1999)—between what mathematics teachers state and what is reflected in their practice:

> I assert that in some cases, reported discrepancies between professed and attributed beliefs may in fact be the result of methodological artifacts and not an accurate reflection of the phenomena researchers seek to understand. In particular, reported inconsistencies may be the result of a lack of *shared understanding* among the researchers and teachers about what descriptive terms mean. (Speer, 2005, p. 362)

Speer's critique of the distinction of professed and attributed beliefs was stimulated by researchers who investigated mathematics teachers' beliefs from non-cognitive frameworks—like, for instance, from interactionist frameworks (Skott, 2001) and discursive (psychology) frameworks (Barwell, 2003; Gellert, 2001). However, her critique focussed on the methods used in research on mathematics teachers' beliefs and the respective methodologies. She argued that "research designs should incorporate opportunities to assess and generate shared understanding in studies of beliefs and practice" and that data on beliefs should be "obtained in conjunction with data on the practices that one seeks to understand" (p. 370). For Speer (2005), one research technique that closely relates data on beliefs and practices is the video-stimulated interview:

> By using videoclips of teachers' classes in interviews, it is possible to obtain information beyond what is possible in traditional, de-contextualized interviews or in a combination of interviews and observation. (p. 377)

Video-stimulated interviews have the potential to facilitate a shared understanding of beliefs and practices among teachers and researchers. The notion of shared understanding raises questions about the very dichotomy of professed versus attributed beliefs. As Speer concluded, the premise that beliefs can be regarded as "pure" representations of teachers' cognition can no longer be taken for granted. Researchers are principally involved in the construction of data, and not the collection of data (Gellert, 2009), and the selected research techniques have an impact on the characteristics of the data.

Challenging conventional standards of research methods. What counts as legitimate methodologies, research methods and techniques seems to be, at the present moment, in a particularly dynamic development. Some researchers are deliberately going beyond mainstream interpretations of how a research process ideally is to be conducted. This dynamic development is most visible in a set of papers published as a special issue of JMTE (2010, *13*(5–6)) titled *Observing the Process of Mathematics Teacher Change*. We illustrate this shifting of research guidelines by referring to two articles published in the JMTE special issue.

One of the traditional and common central tenets of research is that it starts with one or more research questions. Research is intentional and follows a certain logic of development—at least in its presentation in journal articles, volumes or lectures. It is not dependent on chance, coincidence or accident. It identifies a research goal and elaborates precise research questions to be pursued in the course of the research process. A different position has been outlined by Liljedahl (2010):

> Working as both a mathematics inservice educator and a researcher interested in the contextual and situational dynamics of the inservice setting, I find myself too embroiled in the professional development activities to adopt the removed stance of observer. At the same time, my specific role as facilitator prevents me from adopting a stance of participant observer. As such, I have chosen to adopt a stance of *noticing* (Mason, 2002). ... This stance allows me to engage in these experiences as a researcher without the requirement of an a priori research question. (p. 412)

This pragmatic approach took the involvement of the researcher-as-facilitator in the process of mathematics teachers' professional development into account. Liljedahl would certainly not be the first and only person who did not start a research study with a clearly defined research question. By accepting the article for publication, JMTE supported this perspective as legitimate.

Along a similar line of argument, Boylan (2010) promoted an "actor-network theory" (Law, 2004) that described "the creation of reality that occurs through the process of examination" (Boylan, 2010, p. 385). In his article, he included

> an account of the ways in which the material presented in this article was gathered and produced when describing the professional development project itself. My aim here is *to blur the boundaries* between the teacher education and research aspects of this project. This reflects fluidity in my roles as project co-ordinator acting as a supporter of teacher change and researcher, who is in part a narrator (that is a constructor of narratives) of teacher change. (Boylan, 2010, p. 385, emphasis added)

Boylan's (2010) article was also accepted for publication by JMTE. This acceptance seems to have legitimized the position of the researcher-as-facilitator and the facilitator-as-researcher. It acknowledges that those who enable mathematics teachers' professional development might also be those who are able to produce significant accounts of these processes.

Examples of Techniques: Extending the Survey

In this section we extend the depth of our survey of techniques by highlighting examples of how and for what purpose they were used in the studies on mathematics teacher education in the three journals we surveyed. Although the examples of techniques were drawn from all categories of studies that were directly associated with teacher education in these journals, the intent is not to provide a complete picture but a sample of the landscape of techniques. We provide examples other than interviews to illustrate the scope and nature of techniques for the period 2005–2010. Despite the fact that interviews formed the dominant category of techniques in the qualitative and mixed method studies, the focus here is on highlighting trends with respect to other categories of techniques which were used by researchers. Five categories are presented with examples that reflect different research goals and different ways that a technique was used to show the diversity of techniques within and across categories.

Techniques Involving Narratives/Stories

Narratives or stories play an important role in some studies as in the following examples: Lloyd (2006) used narratives to gain insight into preservice secondary teachers' emerging identities as mathematics teachers. The *fictional* narratives of mathematics classrooms written by the participants were the key source of data. The primary analysis of data involved examining the participants' stories and antistories, both structurally and thematically, through narrative analysis. The initial analysis of the stories involved using the structural elements of orientation, complicating action, evaluation, and resolution to focus on *how* the preservice teachers represented and interpreted classroom issues and events. Following structural analyses, particular attention was devoted to the *meanings* of the complicating actions and resolutions in the content of the stories.

Drake (2006) used teachers' narrative descriptions of themselves as learners and teachers of mathematics to understand teachers' interpretations and implementations of a reform-oriented mathematics curriculum. The focus was on turning-point stories that captured teachers' mathematics life stories at a particular point in time. Each of the teachers participated in a mathematics story interview based on an established life-story interview protocol that prompted them to recall all of their previous experiences of both learning and teaching mathematics and to think of these experiences as a story. Analysis involved using a coding scheme for life-story interviews that focussed on "tone and specificity."

Harkness, D'Ambrosio, and Morrone (2007) used preservice elementary school teachers' mathematical autobiographies to examine why the elementary school teachers were highly motivated in a social constructivist *Problem Solving* mathematics course in which mastery goals were emphasized. The data included participants' beginning-of-the-year mathematical autobiographies and end-of-semester reflections. Analysis of the autobiographies included separating them into three categories—liked, disliked, or had mixed feelings about mathematics—based on students' descriptions of their mathematical experiences prior to the course.

Kaasila (2007) considered the use of narrative and rhetorical inquiry as research methods when constructing a mathematical biography of an elementary preservice teacher. A story was created that described how the teacher constructed her mathematical identity. Kaasila used two complementary approaches in choosing the episodes comprising biography: *emplotment* and *linguistic features*. Emplotment related events to one another by configuring them as contributors to the advancement of a plot. When searching for turning points and key episodes in a participant's story, Kaasila used linguistic features in the data.

Chapman and Heater (2010) focussed on understanding teacher change through a high school mathematics teacher's journey to inquiry-based teaching. Their data included stories that the participant was prompted to tell during interviews and chose to write based on what she considered to be significant episodes that described specific events or situations at different stages of her journey of change. The stories also included detailed descriptions of examples of her inquiry-based lessons. In the analyses of her stories, particular attention was paid to elements of their structures—their orientation, complicating actions, evaluation, and resolution. The contents of the stories were also scrutinized to identify other issues or situations that appeared to be significant in relation to the nature of, and influences on, change.

Interest in narrative has grown within the field of education over the past two decades, mostly because it "represents a way of knowing and thinking that is particularly suited to explicating the issues with which we deal" (Carter, 1993, p. 6). As Clandinin and Connelly (2000) explained, teachers' knowledge is not something fixed and static to be replaced by something else but rather it is something lifelike, something storied, something that flows forward in ever changing shapes. The sample of studies above offers examples of ways in which the use of narratives or stories is beginning to emerge in research on mathematics teacher education.

In most examples, however, stories were used mainly as a conveyer of teachers' knowledge and experiences and not as a narrative research methodology (which was what was proposed by Clandinin and Connelly, 2000). These studies used data analysis techniques that were more closely related to those of conventional qualitative methods based on text analysis, developing themes in which findings were situated within larger meanings. On the other hand, as noted by Creswell (2008), narrative researchers analyze a participant's stories by retelling or "restorying" them into a framework that makes sense (e.g., a chronology, or plot). This often involves the researcher rewriting a participant's stories to place them within a chronological sequence and/or a plot that incorporates a main character who experiences a conflict or struggle which comes to some sort of resolution. Kaasila's (2007) study, for example, reflected elements of this. So, although there seems to have been a growing acceptance of the use of narrative, its use as a research methodology will require still more consideration by researchers of mathematics teacher education and still more acceptance by the journals surveyed. The same applies to biographical methods (e.g., Smith, 1994) that embody narrative.

Techniques Involving Videos

Like audio taping, videos have been used as a tool for constructing data. For example, classroom observations are accompanied with video recordings of the lessons which are either transcribed or analyzed directly in a similar way as transcripts. In the following examples, the focus is not on this use of videos. Instead, videos are used as a basis of engaging the participants as an integral aspect of the research design and provide the data for the studies.

Morris (2006) investigated preservice teachers' abilities to collect evidence about students' learning in order to analyze the effects of instruction and to use the analysis to revise the instruction. Participants analyzed the effects of a videotaped mathematics lesson on student learning. They were randomly assigned to one of two conditions: (a) the "children's learning condition" in which they had the freedom to decide whether the lesson was successful, and to decide which instructional activities worked well and which did not; and (b) "sources of the problems condition" in which the task instructions indicated that the lesson was not successful. The data were coded using established categories to determine the number of participants who gave each type of response, and to score each participant's lesson revisions.

Santagata, Zannoni, and Stigler (2007) investigated what preservice teachers can learn from the analysis of videotaped lessons and how their analysis ability and its improvement can be measured. Each participant worked individually on a computer, watched videotaped mathematics lessons, and completed analysis tasks. For the pre- and post-tests, participants watched the videotape of an Italian eighth-grade mathematics lesson, on the area of the sector of a circle, projected on a big screen. A scoring grid was developed to evaluate their lesson analyses. Five dimensions were coded: elaboration; links to evidence; mathematics content; student learning; and critical approach.

Stockero (2008) investigated how the use of a video-case curriculum affected the reflective stance of prospective middle school teachers and the extent to which a reflective stance developed while reflecting on other teachers' practice transferred to reflecting on one's own practice. Data sources included videotapes of course sessions and participants' written work based on the video-cases. The portions of these videotapes consisting of whole-class discussions focussed on analyzing and interpreting student thinking and pedagogical issues based on the video cases were transcribed, coded, and statistically analyzed to understand changes in the participants' group reflections during the course. Four main coding categories were used: agent, topic, grounding, and level.

Star and Strickland (2008) investigated the impact of video viewing as a means for improving preservice teachers' observations of classroom practice. They utilized a pre- and post-test design to measure the quantity and type of classroom events that participants noticed before and after a methods course where improving observation skills was an explicit goal. In the pre-test, participants watched a video of one entire class period of an eighth grade mathematics class. At the conclusion of the video, they were given 60 minutes to work individually on the pre-assessment, which consisted of questions which were concerned with what they noticed about the class.

The procedure for the post-assessment at the end of the semester was identical to that at the pre-assessment. A scoring rubric was used to grade the assessments.

Llinares and Valls (2010) investigated how participation and reification of ideas about mathematics teaching were constituted in online discussions when prospective primary mathematics teachers analyzed video-cases about mathematics teaching. Data consisted of the participants' contributions to the virtual social spaces of interaction in which they had to draw on the video-clips to support and provide evidence for their ideas. Data analysis focussed on how participants provided evidence for their claims, generated alternatives to their peer's decisions or questioned their assumptions.

In these studies, videos were used as intervention tools with some form of pre- and/or post-"test" research design. Statistical analysis featured in most of the studies as a result of this design. However, videos could be used in different ways—for example, in video-stimulated interviews, or as a means of representing the findings in narrative or phenomenological studies. iMovies could also be used in similar ways as videos to support and study preservice teachers' development of knowledge of mathematics for teaching and mathematics pedagogy (Li & Chapman, 2010). Clearly, there is room for both to be further explored as research techniques.

Techniques Involving Concept Maps

Concept maps were used by some researchers to engage teachers and generate data for the studies. Chinnappan and Lawson (2005) used concept maps to investigate a framework for describing and analyzing the quality of teachers' content knowledge for teaching in one content area—such as squares, within the domain of geometry. They adopted a simple form of representation for the node-link structure for the maps, assuming that they could be used to identify teachers' knowledge of geometry and knowledge of geometry for teaching. In analyzing the maps, the researchers proposed that the connectedness of knowledge could be described both in terms of the number of knowledge components present, and in terms of the qualitative relations that existed among the knowledge components. The measures to analyze the concept maps consisted of quantity in terms of number of nodes and links and quality in terms of integrity (completeness and accuracy) and connectedness (depth, branching, cross-linking, and complexity of relationships). These formed the basis of scoring and statistical analysis.

Concept maps were also used by Hough, O'Rode, Terman, and Weissglass (2007), who explored teachers' growth in the understanding of algebra. They collected pre-course concept map data on the first day of the 10-day course on algebra in order to gauge participants' initial understanding of algebra. After being introduced to the concept map activity, the teachers were given 12 minutes to draw and complete a concept map about algebra. Towards the end of the course, without referring to their initial map, participants drew a second concept map about algebra in the allotted time of 12 minutes in order to capture their post-course understandings of algebra. Data analysis consisted of a structural/numerical analysis and a content analysis. For example, they used the notion that the number of concepts on a map can be thought of as assessing the breadth of a participant's mathematical understanding of algebraic concepts.

Given the focus on promoting conceptual understanding in mathematics education, it might be expected that the use of the concept map as a data-generating research tool would be common in studies on teacher education. But this does not appear to be the case, based on the period and journals surveyed. The two examples show how they can be used, with accompanying statistical analyses, to evaluate teachers' understanding. But their potential for unpacking teachers' sense-making could have been exploited in other ways.

Techniques Involving Tests and Tasks

Tests and mathematical tasks played a key role in some of the studies involving teachers' mathematics content knowledge. Osana, Lacroix, Tucker, and Desrosiers (2006) examined the nature of preservice teachers' evaluations of elementary mathematics problems using a model designed to discriminate among tasks according to their cognitive complexity. For the task classification, 32 mathematical problems containing eight cards at each of four cognitive demand levels (memorization, procedures without connections, procedures with connections, and doing mathematics) were created. The participants' mathematics knowledge was measured using the TerraNova (a 40-minute, 25-item, multiple-choice standardized instrument which gives measures of basic mathematical skill) and a 20-minute "doing mathematics" test that contained two open-ended items created by the research team. Scoring of the card-sorting task was based on the number of correctly sorted problems and the average distance from the agreed answer. For content knowledge, the participants' responses on the TerraNova standardized were scored for correct answers. For the open problems a rubric was created and used.

Leavy and O'Loughlin (2006) investigated how preservice teachers understood important elementary statistical concepts related to the mean. The primary data collection instrument consisted of five mathematical tasks. Individual tasks indexed different aspects of the mean and required conceptual understanding of the mean in order to arrive at accurate solutions. The participants engaged in a written think-aloud protocol when completing the tasks. Data were analyzed from several perspectives: accuracy of solution, solution strategy employed, and evidence of conceptual understanding. A number of codes were pre-established based on a pilot study.

Davis (2009) examined the influence of reading and planning from two differently organized mathematics textbooks on prospective high school mathematics teachers' content knowledge and pedagogical content knowledge of exponential functions. The participants completed three paper-and-pencil tests (pre-test, first post-test, second post-test) that were designed by examining the content knowledge and pedagogical content knowledge present in both textbooks. Test items measuring content knowledge were marked as correct or incorrect, with no partial credit being given. Questions concerning pedagogical content knowledge were analyzed for validity, with the number of different valid components being recorded.

Son and Crespo (2009) examined, through a teaching-scenario task, the reasoning and responses of prospective elementary and secondary teachers to a student's non-traditional strategy for dividing fractions. The main task used in this study presented a student's non-traditional strategy within a teaching-scenario task so as to simulate how mathematical work arises in the context of teaching. Prior to completing this main task, prospective teachers were asked to find the value of $2/9 \div 1/3$ and to create a story problem for this fraction division. The main task consisted of two prompts in the form of questions that, altogether, took about 30 minutes to complete. Analyses of responses to the first prompt focussed on examining the range of the participants' responses in terms of what these revealed about their ways of reasoning and whether they could be categorized as student- or teacher-focussed. The second part was analyzed for the influencing factors that participants identified.

Thanheiser (2010) investigated preservice elementary school teachers' responses to standard written place-value operation tasks concerned with addition or subtraction. The tasks consisted of four questions and required explanation of the regrouped digits in the addition algorithm and comparing the value of the regrouped digits in addition and subtraction. The participants completed the tasks during class time at the beginning of a methods course, before any discussion of place value or algorithms had occurred in the course. To analyze responses to the addition task, the researcher examined which values the participants assigned to the regrouped units, noting whether they considered the location of the regrouped digit, its source, or both. Those participants who correctly interpreted all digits in terms of their values on the survey were subsequently categorized as holding one of the correct conceptions.

In the above-mentioned examples, researchers used both standardized measures and researcher-created measures, with the latter dominating. The use of tests and tasks would seem to be natural choices for studying teachers' mathematics knowledge. However, with the current focus on mathematics knowledge for teaching, exploring other appropriate techniques is important.

Techniques Involving Questionnaires/Surveys

In some studies, and especially those framed in a quantitative or mixed methods perspectives, the use of questionnaires was central. Wilkins' (2008) study included a quantitative investigation of inservice elementary teachers' level of mathematical content knowledge using a 44-item mathematics content survey. Items on the survey were selected from the *Third International Mathematics and Science Study*, the *Longitudinal Study of American Youth*, the *Second International Mathematics Study*, and some developed by the author. Jones-Newton (2009) investigated prospective elementary school teachers' motivations for working with fractions before and after taking a course designed to deepen their understanding of mathematics, as well as what instructional practices might be related to any changes detected in their motivations. The motivation questionnaire included 14 items adapted from prior research, which suggested three separate scales for these items: anxiety (four items); value (four items); and self-concept of ability (six items). The domain-specific questionnaire was adapted to be topic-specific by replacing the word mathematics with fractions. A pre-test, post-test design was used.

Swars, Smith, Smith, and Hart (2009) conducted a longitudinal study that examined prospective teachers' initial pedagogical beliefs, teaching efficacy beliefs, and anxiety; changes in these constructs across two academic years in a teacher-preparation program; and the relationships between participants' beliefs and knowledge for teaching elementary mathematics at the end of the program. Four instruments were used to gather quantitative data: The *Mathematics Beliefs Instrument* and the *Mathematics Teaching Efficacy Beliefs Instrument* were administered four times; the *Mathematics Anxiety Rating Scale* was administered three times; and the *Learning Mathematics for Teaching Instrument* was administered once, at the end of student teaching.

Levenson, Tsamir, and Tirosh (2010) investigated elementary school teachers' preferences for mathematically based (MB) and practically based (PB) explanations. Using the context of even and odd numbers, they explored the types of explanations teachers generated on their own, the types of explanations they preferred after reviewing various explanations, and the basis for these preferences. The teachers filled out questionnaires based on the property of parity. The questionnaire consisted of two parts: teacher-generated explanations and teachers' choices of preferred explanations. Teachers' explanations given in the first part of the questionnaires were categorized into MB explanations, PB explanations, and "other explanations." Quantitative data were analyzed using descriptive statistical processes.

Bekdemir (2010) examined whether the worst experiences and most troublesome mathematics classroom experiences affected mathematics anxiety in preservice elementary teachers, and how the causes of their anxiety related to these negative experiences. Three different instruments were used to collect data: *Mathematics Anxiety Rating Scale*, *Worst Experience and Most Troublesome Mathematics Classroom Experience Reflection Test*, and *Interview Protocol*.

The above studies suggested that conventional quantitative techniques still hold a visible presence in teacher education in spite of the dominance of qualitative studies. However, although they use techniques that can be applied to large-scale studies, the general pattern was still to have relatively small sample sizes.

What Surveys Often Miss: An Example from the Politicized Periphery

There is a concern about an "apparent mismatch between the amount of research in mathematics education undertaken and the limited amount that filters down into teachers' classroom practice" (Cockburn, 2008, p. 344). Cockburn identified several obstacles for teachers who like to engage with research in mathematics education—such as the fragmented character of research results, and the inaccessibility of research journals. As a way of distributing research results to teachers, Cockburn mentioned inservice courses in which teachers could become influenced by research. She argued that those courses "that require teachers to undertake personalized projects which encourage them to reflect on, and criticize, their classroom practice in the light of others' research" (p. 345) are particularly suitable for getting teachers in contact with the results of latest research. However, Liljedahl (in Brown & Coles, 2010) reminded us that a change of research perspectives to become more responsive to the needs of teachers might be needed. When the structures of professional development programs do not balance the programs' objectives with the teachers' experience and personal learning plans, helpful professional development is unlikely to occur.

From this perspective, it might only be a short step towards professional development programs that are initiated and organized by the teachers themselves—though supported by researchers or teacher educators. The main aim of what has been called "action research" is not a simple acquisition of research results by teachers, but investigations by teachers into their own practices, or into the conditions of their pedagogical work. It is presumed that in carrying out these investigations the teachers are aware of, and guided by, state-of-the-art research findings related to the themes being investigated.

That action research is indeed possible in education is, of course, not a new insight. According to Erickson (1986), the inherent logic of the widely accepted interpretive paradigm in research on teaching

> leads to collaboration between the teacher and the researcher. The research subject joins in the enterprise of study, potentially as a full partner. In some of the most recent work (e.g., Florio & Walsh, 1981) the classroom teacher's own research questions—about particular children, about the organization of particular activities—become the focus of the study. (p. 157)

Erickson continued:

> It is but a few steps beyond this for the classroom teacher to become the researcher in his or her own right. As Hymes (1981) notes, interpretive research methods are intrinsically democratic; one does not need special training to be able to understand the results of such research, nor does one need arcane skills in order to conduct it. Fieldwork research requires skills of observation, comparison, contrast, and reflection that all humans possess. In order to get through life we must all do interpretive fieldwork. What professional interpretive researchers do is to make use of the ordinary skills of observation and reflection in especially systematic and deliberate ways. Classroom teachers can do this as well, by reflecting on their own practice. (p. 157)

According to Erickson (1986), fieldwork as a research method has the potential to help "researchers and *teachers* to *make the familiar strange* and interesting again …. The commonplace becomes problematic" (p. 121; original emphasis retained). It is a response to the "need for specific understanding through documentation of concrete details of practice" (p. 121). And it considers "the local meanings that happenings have for the people involved in them" (p. 121). Fieldwork is concerned with "the need for comparative understanding of different social settings" (p. 122), and a comparative understanding that goes "beyond the immediate circumstances of the local setting" (p. 122).

Action research has a rather long tradition in mathematics education and beyond (e.g., Altrichter, Feldman, Posch, & Somekh, 2008; Breen, 2003; Krainer, 1999; Zack, Mousley, & Breen, 1997; Zeichner, 2001). Two conceptualizations of action research seem to dominate: on the one hand, many professional development programs for mathematics teachers include forms of action research. In those programs, action research is mainly directed at the personal knowledge growth of mathematics teachers. The role of the teacher educator, who supports the teachers' introspections, can vary from a distant facilitator to a co-researcher who, with different expertise and different aims, participates as a full member of the teachers' action research group. Research results have suggested that such a form of action research produces results generated *through* rather than *about* professional development, but of course these *results-through* might generalize to *results-about*—a process that is not directly possible the other way around. However, as Kilpatrick (2000) noted, "teacher research has not had much impact on the larger community" (p. 87). On the other hand, researchers look for a deeper ecological validity by doing research *with* as well as *on* research participants (e.g., Planas & Civil, 2009; Scherer & Steinbring, 2006; Setati, 2000; Uworwabayeho, 2009).

A third conceptualization of action research, which has not received much attention in the area of mathematics education during the last several decades, focusses on and emphasizes the transformative potential of action research in terms of emancipatory educational, cultural and political processes. Under this perspective, the transformative moment of action research is not only directed at the personal development of the research participants or at qualities of new insights, but also at a macro-social liberation from oppressive societal forces and their local and global manifestations (Noffke, 2009). This kind of action research has been especially common in regions of the world facing structures of colonization or neo-colonization. Fals Borda (2001) cited efforts made in India (De Silva, Wignaraja, Mehta, & Rahman, 1979), Colombia (Fals Borda, 1986), Tanzania (Swantz, 1970), Brazil (Freire, 1970) and Mexico (Warman, Nolasco, Bonfil, Olivera, & Valencia, 1970) as evidence for the acceptance of the conception of *participatory* action research in nations that had previously been colonies.

Greenwood and Levin (2007) maintained that emancipatory action research projects were most likely to have occurred, and to have been successful, in "miserably poor" parts of the world (p. 154)—regions in parts of Africa, Latin America and Asia which had been most strongly affected by colonization. Emancipatory participatory action research, developed in the frame of post-colonial theories (e.g., Dussel, 1995, 2011; Spivak, 1990, 2011), is committed to affirming "solidarity with the oppressed," to working towards a "fundamental alteration in the distribution of power and money" (Greenwood & Levin, 2007, p. 154) and to problematizing "uncontested 'colonial' hegemonies of any form" (Parsons & Harding, 2011, p. 2). Kanu (1997) specifically referred to its appropriateness for the "prevalent appalling conditions of teaching and teacher education" in impoverished nations.

In short, the aim was to "de-colonize" the future (Parsons & Harding, 2011, p. 4). In this kind of action research, participation is regarded as an active engagement in a process that works towards the achievement of social justice, rather than a mere share in the research process.

In this section, we will elaborate a methodology of participatory action research of mathematics teacher education. We will suggest that participatory action research is crucially important when profound social and political changes are needed and is a particularly promising methodology for scrutinizing the social and political power structures that frame teaching practice.

Needless to say, reports of participatory action research that feature explicitly political positioning do not often appear as publications in high-prestige scientific journals. Our decision to emphasize participatory action research as a research methodology within this *Handbook* chapter is based on an important result of our survey of publications on teacher education and professional development in international journals. With some few exceptions, these journals seem to affect a standardizing of research reports and a narrowing of the scope of research methodologies reported for, and accepted in, the mathematics teacher educators' scientific community. Burton (2002) commented in the following terms:

> How knowledge is constructed is a function of values and, indeed, is also about the community that can define those values and establish the gatekeeping criteria for maintaining them. Inevitably, therefore, I see epistemology as interlocked with methodology. (p. 6)

In this chapter, participatory action research is regarded as a methodology in research on teacher education and professional development that responds to the particularities of "situations of social and political conflict" and transformation (Vithal & Valero, 2003). As an example, we focus on a Latin American context in which decolonization and transition were part and parcel of the official rhetoric and agenda, although not always part of the educational reality.

Development of Participatory Action Research in Latin America

Participatory action research has been developed in Latin America as a critical and transformative conception. The terminology was introduced by Marja-Lissa Swantz in 1970 and achieved recognition at the *World Symposium on Action Research and Scientific Analysis* in Cartagena, Colombia, in 1977. This event examined the issue of participation, noting that a central point had to do with the need to break the classic pairing of research subject and object. It is assumed that those setting the research problem, those who analyze and solve problems, are actors who are involved, called base groups, which are grouped for the "action" of researching any fact or issue that affects them, looking for its transformation.

Among the notable educators-researchers and advocates of this new focus was Orlando Fals Borda, who put forward a radical perception about research coupled with political commitment in his *Militant Sociology* (1986, 1995, 2003), a work in which he presented science as being committed to the so-called popular sectors.

Similarly, the influential work and writings of Paulo Freire (1970, 1990), Luis Bigott (1992) and by Ezequiel Ander-Egg (2003) were important in conceptualizing participatory action research to the point where it offered a strong theoretical base. Fals Borda (1986, 1995, 2003), who emphasized the necessity of social transformations, defined participatory action research as a social praxis that enabled problems of exploitation, dependence and issues associated with retention of power among the privileged few to be identified and challenged.

However, research in mathematics education in Latin America has long been dominated by a positivist paradigm and a psychological orientation (Messina, 1999). Gómez and Valero (2004), reflecting on ICME10 and the *28th Conference of the International Group for the Psychology of Mathematic Education* (PME 28) held in 2004, spoke of a shift in attention from the learning of mathematics to its teaching, and added that this shift was manifest in research in mathematics education in Latin America. According to Gómez and Valero, the knowledge and the beliefs of mathematics teachers, including their initial and inservice training, had become important foci for research. In their analysis of the research presented at ICME 10 and PME 28, Gómez and Valero (2004) noted increased interest in studying the social aspects that affect mathematics and its teaching and learning: "Topics of democracy, equality, diversity or gender appear more and more frequently in the research literature" (p. 3).

According to Gómez and Valero (2004), in modern mathematics education research "most studies use qualitative methods and tend to worry about the detailed description and interpretation of specific phenomena" (p. 3). In short, it seems as if research in mathematics education has evolved as much in its sphere of interest as in its methodologies. Different international forums have revealed the need to pay attention to the diverse social aspects that strongly influence the teaching of mathematics and the impact that these aspects have on the possibility of giving access for a growing number of students to socially valued forms of mathematical knowledge.

Action research in mathematics education in Latin America has been steadily gaining ground and importance. That has been documented by the annual meetings of the Colombian Association of Mathematics Education (ASOCOLME), at which numerous papers have been presented for which action research has been a fundamental methodological guide. In the Research and Advanced Studies Center of the National Polytechnic Institute of Mexico (CINVESTAV) a combination of four methods has predominantly been used: action research, didactic engineering, developmental psychology, and socio-epistemology (Cantoral, 2000). The 12th Interamerican Conference of Mathematics Education (IACME XII) held in 2007 in Mexico, the 6th Ibero-American Conference on Mathematics Education (CIBEM VI) held in Chile in 2009, and the Latin American Meeting of Mathematics Education (RELME), have all shown how action research can be used as a methodology that opens new sources of inquiry for mathematics education, in which historical, social, cultural and political issues are brought together to enable the process of teaching and learning of mathematics to be seen as a significant undertaking of humans (Rojas Olaya, 2011).

In Venezuela, the Group for Research and Dissemination of Mathematics Education (GIDEM) was established in 1999 and, since its creation, the sociopolitical role of mathematics education has been at the forefront of Group considerations, with any supposed neutrality of mathematics being rejected. The work of GIDEM is grounded in a critical conception of mathematics and mathematics education and in a methodology of participatory action research. The critical stance is based on the sociological tenet that the serious conflicts existent in (not only) Latin American societies are essentially class conflicts. As Skovsmose (1994) stated, "even after the belief in a Marxist definition of classes has lost supporters" (p. 12), conflicts continue to be the result of inequalities of opportunity, differential access to information, and worse, manipulation of information in order to handle large masses of people. Under such circumstances, Habermas' (1984–1987) objective that the force of the better argument should determine any dispute, cannot be reached. As a consequence, for the democratization of society it is important that citizens acquire the capacity to analyze and judge the results and consequences of social and political developments, in short, to develop "democratic competence" (Skovsmose, 1994, p. 34). The development of critical citizens is linked to an education for democracy and to John Stuart Mill's statement that "any education which aims at making human beings other than machines, in the long term makes them claim to have the control of their own actions" (Mill, 1975, p. 185; cited in Skovsmose, 1994).

A conception of mathematics education as a scientific research field devoid of values has helped to perpetuate what Giroux (1989) called "a policy of silence and an ideological amnesia" (p. 19). This silence supports the hiding of the mathematization of reality, and has transformed the relation between mathematics, technology and society, thus resulting in a lack of awareness (Jablonka & Gellert, 2007). As an example of the mathematization of our reality, witness the global financial structures that are supported by mathematical models, measures and systems. Hence trade rules depend on the use of mathematics. Examinations, grades, salaries, funding and many other subsystems that form our society are highly influenced by, and dependent on, mathematics (Davis & Hersh, 1986). Mathematics is shaping our society (Keitel, Kotzmann, & Skovsmose, 1993) and Skovsmose (1994) attributed to mathematics a formatting power.

However, one of the main perils of the mathematization of societies is that not all issues receive the same attention. Mathematical models are structuring our society, but not equally. Some aspects are highlighted, but others are ignored. Thus, models such as the gross domestic product, income tax, value added tax, the distribution of social benefits in the population, among other things, become the guidelines to design and build our world. To evaluate and scrutinize the symbolic power of mathematics, a critical stance and competence is needed that questions how mathematical models are designed, and who benefits by their application. The methodology of participatory action research in mathematics (teacher) education is grounded on a critical conception of mathematics and mathematics education.

Implementation and Examples of Participatory Action Research

Becerra and Moya (2010) conceptualized a participatory action research methodology for research in mathematics teacher education that has eight elements:

1. The *first* element is the important task of recreating and transforming teaching practice.
2. A *second* consideration is a recognition of the conceptional relationship between theory and practice, and the attempt to turn it into a single dialogic unit, where educational theory is determined by how it relates to practice, and this practice in turn adjusts our theoretical references. For Becerra and Moya (2008), the alleged dichotomy between theory versus practice is a false dilemma (see also, e.g., Bazzini, 2007; Steinbring, 1994; Steiner, 1985).
3. The *third* element is related to knowledge and understanding. It is assumed that knowledge is a process that does not end with the completion of an investigation. Knowledge is seen as something that is done, in process, in transit, en route (Bigott, 1992). Knowledge is conceptualized as a dialectical process.
4. In the *fourth* element, *dialogue* is presented as a fundamental research tool, and is understood as more than a simple conversation or a lively exchange of ideas. This dialogue involves the confrontation of different views around common interests, with the intent to understand, to learn and advance in the quest for truth shared with others (Fierro, Fortoul, & Rosas, 1999).
5. The *fifth* element is the premise that reflection and construction are not carried out in isolation, as men and women are social, historical beings. The construction of knowledge makes sense in the frame of its social relevance.
6. The *sixth* element is the relationship between epistemology and methodology, which carries with it an insistence on explanations of how insight and understanding are constructed (Becerra, 2003, 2005; Moya, 2008).
7. A *seventh* element is linked to the inalienable right to participate actively and consciously in the construction of a new form of citizenship. To enhance and to increase the active participation of others, especially of future mathematics teachers, as part of research and education, including mathematical education, means being committed to the full development of women and men as social beings.
8. According to the socio-critical paradigm the *eighth* and final element is that research cannot be regarded as a political neutral activity.

These eight elements direct GIDEM's participatory action research projects. In the concrete process of research, these methodological elements translate into methods and research techniques. According to Becerra (2003, 2010) and Lanz (1994), characteristic steps in participatory action research projects are:

- *The social framing of the topic:* Lanz suggests approaching the participants through open discussions, conducting presentations about the critical issues affecting the group or the established practice, and encouraging research into major problems that they confront.
- *Delineating the object of study:* The research problem is clarified by outlining the problematized social action, the social subjects involved in research,

both directly and indirectly, and the spatial dimension and the temporal scope of it.
- *Directionality of the investigation*: A central characteristic is the definition of the envisaged change. From the analysis and reflection of the collective social praxis, goals and objectives are formulated, and some strategies of articulation are established.

Following this scheme, a team of researchers at the Pedagogical University Experimental Libertador of Venezuela developed a research project based on its teacher education program, involving the full participation of students and teachers responsible for 1st through 6th grade of primary mathematics education (Becerra, 2003). One of the research goals was the construction of a participatory methodology for the geometry course of the teacher education curriculum. Flexible action plans were developed, which could be modified depending on work being done during the course. Participant observation and interviews were used as techniques of data collection, with some students selected at the beginning of the course as key informants for each specific activity, and some appointed as participant-observers in each group.

One of the in-class strategies used for this research was the development of conjectures by students working in small groups. This strategy consisted in the planning of workshops, which contained a series of exercises and problems in relation to mathematical activities. Each group had to submit a report describing what had been done in the workshop, including the assumptions made and justifications which supported them. Reviewing the work delivered by the end of each workshop, the most interesting observation was probably the domain of the justifications. The conjectures made in most cases were adequate, even when the teacher students might have had difficulties in resolving the problem. The results show the transformation that can occur when students are the protagonists of their learning and become aware of their potential. This educative process does not imply leaving the students to their own devices; rather, education might mean preparing the ground for them to be the protagonists and owners of their own knowledge, with the teacher intervening, with clarifications, when needed.

Silva (2010) conducted a research project entitled *From the Real to Formal Mathematics*, which applied participatory action research principles and was intended primarily to develop educational projects related to the assessment of different energy sources that enable students to acquire mathematical knowledge needed for the third year of secondary education (*Bachillerato*). Silva analyzed lessons, written tests and notebooks. Among the study's conclusions were: (a) the issue of the valuation of various energy sources enables participants to develop themes for the mathematics program, such as real functions, proportions, systems of equations and inequalities, among others; (b) the students used representations of mathematical concepts and procedures to interpret the situations under consideration, and to analyze phenomena in order to identify and understand related mathematical concepts and applications; and (c) the process of teaching and learning of mathematics that was used in the study drew on economic and ecological crises—this approach was needed in order to overcome a structure of permanent competition, individualism, and evaluation as a form of repression and uncritical thinking (cf., Gellert, 2011; Renert, 2011). Instead, the process facilitated the development of a new structure featuring teamwork, participation, cooperation, democratic evaluation and commitment to studying the world around us.

Starting from the premise that mathematics education is a *political* fact, Serrano (2010) studied the potential role of mathematics education in and for Venezuelan society. Working with high school students from Caracas, he focussed on the characteristics of mathematical literacy (cf., Jablonka, 2003) in the particular context of the group of students. He used participatory action research as a methodology that epitomized the emancipation of students. Among the key elements of such a mathematical literacy were a combination of mathematical skills, meta-mathematics, socially and axiologically characterized, and knowledge of problems and crises—such as population growth—within the students' community or region. These were necessary for transformative action, since without understanding the role of mathematics, as well as the mathematics involved, no profound understanding of these situations was possible and thus no transformation would be achieved.

Final thoughts. From the perspective of a participatory action research methodology, we cannot ignore that we live in a capitalist society, and although we do not think that the economic and class dynamics can explain everything that is of particular importance to mathematics teacher education research, we should not ignore that "their influence means putting aside some of the most insightful analytical tools we have" (Apple, 1997, p. 177). Therefore, it is our responsibility as researchers in mathematics teacher education to carry out research with the consciousness of being involved in organizations that reproduce unequal class relations in our society. Particularly in Latin America, researchers committed to participatory action research are working towards making those institutions more democratic and egalitarian, close to the evolution of the societies and the problems they confront. In this way, the proposal of a participatory action research that is critical and transformative is an important task. The characteristics of the research–action–reflection–emancipation process that we propose cannot be developed in a sudden manner. Changes and transformations that emerge will have a deep impact on the various organizations, both formal and non-formal, and the extent and forms of this impact will be influenced by the extent to which each member of a collective or group assimilates them, and the extent to which the rest of the group provides sufficient support and encouragement for the creation of a more egalitarian and democratic society.

Afterword

Claims and disputes about the "right scientific method" stand at the philosophical centre of any knowledge generation. The ground-breaking work of the French 17th century philosopher René Descartes (1637) was tellingly titled: "Discourse on the Method of Rightly Conducting the Reason, and Seeking Truth in the Sciences." Descartes' method has been characterized as objective, analytical and systematic. It has often been cited as the cornerstone for any kind of objectivist positivist enquiry.

Opposition to Descartes' fundamental work can be found in the writings of Giambattista Vico, During his lifetime in the 17th and the 18th century, Vico was not really considered influential but in the last decades of the 20th century his ideas received much attention. In his inauguratory lecture at the University of Naples, Vico maintained that modern research (at that time) paid undue attention to the geometrical method modelled on the discipline of physics, and to abstract philosophical criticism. For Vico, such a point of view undermined the importance of exposition, persuasion, and pleasure. It benumbed the imagination and stupefied the memory, both of which were central to complex reasoning and the discovery of truth. Vico's lecture was also a lecture on method—it was published in 1709 under the title "De Nostri Temporis Studiorum Ratione"—"On the Study Methods of Our Time." Disputes over the "right scientific method" have a long tradition, and it is essential for any researcher to make the methods used in research explicit. This is one of the standards for the production of any kind of research report.

Are we still disputing the "right method," specifically in research on mathematics teacher education and professional development? This chapter has given two different answers. The survey indicates that most published work has referred idiosyncratically to methods and techniques used and discussed by others. Although the works of Glaser, Corbin and Strauss seemed to effect some standardization, the research journals (personified by the editors and peer reviewers) have not insisted on a common canon of referenced methodologies, methods or techniques. As long as the methodical research procedure is explained, anything seems to go.

The other answer is that there seem to be a set of hidden values that regulate—or self-regulate—the development of mathematics teacher education as a research field. Although our survey list of the keywords associated with the research reports we reviewed contained the terms "critical thinking," "discourse," "empowerment," "equity/diversity," "Foucault," "ideology," "neoliberal," "policy issues," "social and cultural issues," and "social justice," concepts that indicate a critical-political perspective or a transformative-political perspective were noticeably scarce. We did not find other keywords that could indicate any political perspective, and the keyword entries just listed constituted only 1.4% of all keyword entries. Undoubtedly, international research journals can effect a severe channelling of visible research perspectives. Our report on research that has been characterized as participatory action research demonstrates that a politicized perspective on mathematics teacher education has been developed at the social periphery of the research field. Apparently the dispute for the "right method" continues: Does method aim at value-free description or at principled transformation?

References

Adler, J., Ball, D., Krainer, K., Lin, F.-L., & Novotna, J. (2005). Reflections on an emerging field: Researching mathematics teacher education. *Educational Studies in Mathematics, 60*(3), 359–381.

Altrichter, H., Feldman, A., Posch, P., & Somekh, B. (2008). *Teachers investigate their work: An introduction to action research across the professions* (2nd ed.). London, UK: Routledge.

Ander-Egg, E. (2003). *Repensando la investigación-acción participativa*. Buenos Aires, Argentina: Lumen.
Apple, M. (1997). *Teoría crítica y educación*. Buenos Aires, Argentina: Miño y Dávila Editores.
Arbaugh, F. (2003). Study groups as a form of professional development for secondary mathematics teachers. *Journal of Mathematics Teacher Education, 6*(2), 139–163.
Barrantes, M., & Blanco, L. J. (2006). A study of prospective primary teachers' conceptions of teaching and learning school geometry. *Journal of Mathematics Teacher Education, 9*(5), 411–436.
Barwell, R. (2003). Discursive psychology and mathematics education: Possibilities and challenges. *Zentralblatt für Didaktik der Mathematik, 35*(5), 201–207.
Baumert, J., Kunter, M., Blum, W., Brunner, M., Voss, T., Jordan, A., Klusmann, U., Krauss, S., Neubrand, M., & Tsai, Y.-M. (2010). Teachers' mathematical knowledge, cognitive activation in the classroom, and student progress. *American Educational Research Journal, 47*(1), 133–180.
Bazzini, L. (2007). The mutual influence of theory and practice in mathematics education: Implications for research and teaching. *ZDM The International Journal on Mathematics Education, 39*(1–2), 119–125.
Becerra, R. (2003). *Construyendo una estrategia metodológica participativa en el curso de Geometría del currículo de formación del docente integrador* (Doctoral dissertation). Universidad Pedagógica Experimental Libertador, Caracas, Venezuela.
Becerra, R. (2005). La educación matemática crítica: Origen y perspectivas. In D. Mora (Ed.), *Didáctica crítica, educación crítica de las matemáticas y etnomatemática* (pp. 165–203). La Paz, Bolivia: Campo Iris.
Becerra, R. (2010). Action-research in the Venezuelan classrooms. In U. Gellert, E. Jablonka, & C. Morgan (Eds.), *Proceedings of the Sixth International Mathematics Education and Society Conference* (pp. 121–129). Berlin, Germany: Freie Universität Berlin.
Becerra, R., & Moya, A. (2008). Hacia una formación docente crítica y transformadora. In D. Mora & S. De Alarcón (Eds.), *Investigar y transformar: Reflexiones críticas para pensar la educación* (pp. 109–155). La Paz, Bolivia: III-CAB.
Becerra, R., & Moya, A. (2010). Investigación-acción participativa crítica y transformadora: Un proceso permanente en construcción. *Integra Educativa, 3*(2), 133–156.
Bekdemir, M. (2010). The pre-service teachers' mathematics anxiety related to depth of negative experiences in mathematics classroom while they were students. *Educational Studies in Mathematics, 75*, 311–328.
Bigott, L. (1992). *Investigación alternativa y educación popular en América Latina*. Caracas, Venezuela: Fondo Editorial Tropykos.
Bogdan, R. C., & Biklen, S. K. (1998). *Qualitative research in education: An introduction to theory and methods*. Boston, MA: Allyn and Bacon.
Boylan, M. (2010). "It's getting me thinking and I'm an old cynic": Exploring the relational dynamics of mathematics teacher change. *Journal of Mathematics Teacher Education, 13*(5), 383–395.
Breen, C. (2003). Mathematics teachers as researchers: Living on the edge? In A. J. Bishop, M. A. Clements, C. Keitel, J. Kilpatrick, & F. K. S. Leung (Eds.), *Second international handbook of mathematics education* (pp. 523–544). Dordrecht, The Netherlands: Kluwer.
Brown, L., & Coles, A. (2010). Mathematics teacher and mathematics teacher educator change: Insight through theoretical perspectives. *Journal of Mathematics Teacher Education, 13*(5), 375–382.
Burton, L. (2002). Methodology and methods in mathematics education research: Where is "the why"? In S. Goodchild & L. English (Eds.), *Researching mathematics classrooms: A critical examination of methodology* (pp. 1–10). Greenwich, CT: Information Age Publishing.
Cantoral, R. (2000, July). La matemática educativa en Latinoamérica. Interview at the 14th Reunión Latinoamericana de Matemática Educativa (Relme 14), Panama. Retrieved from http://www.educ.ar/educar/site/educar/la-matematica-educativa-en-latinoamerica--entrevista-al-investigador-ricardo-cantoral-uriza.html.

Carter, K. (1993). The place of story in the study of teaching and teacher education. *Educational Researcher, 22*(1), 5–12.
Chapman, O., & Heater, B. (2010). Understanding change through a high school mathematics teacher's journey to inquiry-based teaching. *Journal of Mathematics Teacher Education, 13*, 445–458.
Chinnappan, M., & Lawson, M. J. (2005). A framework for analysis of teachers' geometric content knowledge and geometric knowledge for teaching. *Journal of Mathematics Teacher Education, 8*, 197–221.
Clandinin, D. J., & Connelly, F. M. (2000). *Narrative inquiry: Experience and story in qualitative research*. San Francisco, CA: Jossey Bass.
Cockburn, A. (2008). How can research be used to inform and improve mathematics teaching practice? *Journal of Mathematics Teacher Education, 11*(5), 343–347.
Creswell, J. W. (1998). *Qualitative inquiry and research design: Choosing among five traditions*. Thousand Oaks, CA: Sage.
Creswell, J. W. (2003). *Research design: Qualitative, quantitative, and mixed methods approaches* (2nd ed.). Thousand Oaks, CA: Sage.
Creswell, J. W. (2008). Narrative research designs. In *Educational research: Planning, conducting and evaluating quantitative and qualitative research* (3rd ed., pp. 511–550). Upper Saddle River, NJ: Pearson Education
Davis, J. D. (2009). Understanding the influence of two mathematics textbooks on prospective secondary teachers' knowledge. *Journal of Mathematics Teacher Education, 12*, 365–389.
Davis, P. J., & Hersh, R. (1986). *Descartes' dream: The world according to mathematics*. San Diego, CA: Harcourt Brace Jovanovich.
De Silva, G. V. S., Wignaraja, P., Mehta, N., & Rahman, M. A. (1979). Bhoomi Sena: A struggle for people's power. *Development Dialogue, 2*, 3–70.
Descartes, R. (1637). *Discours de la méthode pour bien conduire la raison et chercher la verité dans les sciences*. Leyde, Belgium: Ian Maire.
Dowling, P. (2009). *Sociology as method: Departures from the forensics of culture, text and knowledge*. Rotterdam, The Netherlands: Sense Publishers.
Drake, C. (2006). Turning points: Using teachers' mathematics life stories to understand the implementation of mathematics education reform. *Journal of Mathematics Teacher Education, 9*, 579–608.
Dussel, E. (1995). *The invention of the Americas: Eclipse of "the other" and the myth of modernity*. London, UK: Continuum.
Dussel, E. (2011). Filosofía de le liberación en la era de la globalización y la exclusión. In *Pensando el mundo desde Bolivia: I ciclo de seminarios internacionales* (pp. 259–272). La Paz, Bolivia: Vicepresidencia del Estado Plurinacional de Bolivia.
Erickson, F. (1986). Qualitative methods in research on teaching. In M. C. Wittrock (Ed.), *Handbook of research on teaching* (3rd ed., pp. 119–161). New York, NY: Macmillan Publishing Company.
Ernest, P. (1991). *The philosophy of mathematics education*. London, UK: The Falmer Press.
Fals Borda, O. (1986). *Investigación-acción participativa en Colombia*. Bogotá, Colombia: Punta de Lanza y Foro Nacional por Colombia.
Fals Borda, O. (1995). *Conocimiento y poder popular*. Bogotá, Colombia: Siglo XXI.
Fals Borda, O. (2001). Participatory (action) research in social theory: Origins and challenges. In P. Reason & H. Bradbury (Eds.), *Handbook of action research: Participative inquiry and practice* (pp. 27–37). London, UK: Sage.
Fals Borda, O. (2003). *Ante la crisis del país: Ideas/acción para el cambio*. Bogotá, Colombia: El Áncora Editores.
Fennema, E., & Franke, M. L. (1992). Teachers' knowledge and its impact. In D. A. Grouws (Ed.), *Handbook of research in mathematics teaching and learning* (pp. 147–164). New York, NY: Macmillan.
Fierro, C., Fortoul, B., & Rosas, L. (1999). *Transformando la práctica docente: Una propuesta basada en la investigación-acción*. México D.F: Paidós.

Florio, S., & Walsh, M. (1981). The teacher as colleague in classroom research. In H. T. Trueba, G. P. Guthrie, & K. H.-P. Au (Eds.), *Culture in the bilingual classroom: Studies in classroom ethnography* (pp. 87–101). Rowley, MA: Newbury House.

Freire, P. (1970). *Pedagogy of the oppressed*. New York, NY: Seabury Press.

Freire, P. (1990). *La naturaleza política de la educación: Cultura, poder y liberación*. Barcelona, Spain: Paidós.

Frykholm, J. A. (1999). The impact of reform: Challenges for mathematics teacher preparation. *Journal of Mathematics Teacher Education, 2*(1), 79–105.

Gellert, U. (2001). Research on attitudes in mathematics education: A discursive perspective. In M. van den Heuvel-Panhuizen (Ed.), *Proceedings of the 25th Conference of the International Group for the Psychology of Mathematics Education* (Vol. 3, pp. 33–40). Utrecht, The Netherlands: Freudenthal Institute.

Gellert, U. (2003). Researching teacher communities and networks. *Zentralblatt für Didaktik der Mathematik, 35*(5), 224–232.

Gellert, U. (2009). Analysing accounts, discourse and mathematics classroom interaction: Reflections on qualitative methodology. In R. Kaasila (Ed.), *Matematiikan ja luonnontieteiden* (pp. 9–34). Rovaniemi, Finland: Lapin yliopistopaino.

Gellert, U. (2011). Now it concerns us! A reaction to sustainable mathematics education. *For the Learning of Mathematics, 31*(2), 20–21.

Gil, J. (1992–1993). La metodología de investigación mediante grupos de discusión. *Enseñanza, X–XI*, 199–212.

Giroux, H. (1989). *Schooling and the struggle for public life: Critical pedagogy in the modern age*. Minneapolis, MN: University of Minnesota Press.

Glaser, B. G., & Strauss, A. L. (1967). *The discovery of grounded theory: Strategies for qualitative research*. Chicago, IL: Aldine.

Goetz, J. P., & Lecompte, M. D. (1984). *Ethnography and qualitative design in educational research*. New York, NY: Academic Press.

Gómez, P., & Valero, P. (2004). Algunas reflexiones sobre el ICME 10 y el PME 28. Retrieved from http://www.cumbia.ath.cx:591/pna/Archivos/GomezP.

Greenwood, D., & Levin, M. (2007). *Introduction to action research* (2nd ed.). Thousand Oaks, CA: Sage.

Habermas, J. (1984–1987). *The theory of communicative action* (2 vols.). Cambridge, UK: Polity.

Harkness, S. S., D'Ambrosio, B., & Morrone, A. S. (2007). Preservice elementary teachers' voices describe how their teacher motivated them to do mathematics. *Educational Studies in Mathematics, 65*, 235–254.

Hart, L. C., Smith, S. Z., Swars, S. L., & Smith, M. E. (2009). An examination of research methods in mathematics education (1995–2005). *Journal of Mixed Methods Research, 3*(1), 26–41.

Hough, S., O'Rode, N., Terman, N., & Weissglass, J. (2007). Using concept maps to assess change in teachers' understandings of algebra: A respectful approach. *Journal of Mathematics Teacher Education, 10*, 23–41.

Hymes, D. (1981). Ethnographic monitoring. In H. T. Trueba, G. P. Guthrie, & K. H.-P. Au (Eds.), *Culture in the bilingual classroom: Studies in classroom ethnography* (pp. 56–68). Rowley, MA: Newbury House.

Jablonka, E. (2003). Mathematical literacy. In A. J. Bishop, M. A. Clements, C. Keitel, J. Kilpatrick, & F. K. S. Leung (Eds.), *Second international handbook of mathematics education* (pp. 75–102). Dordrecht, The Netherlands: Kluwer Academic Publishers.

Jablonka, E., & Gellert, U. (2007). Mathematisation—Demathematisation. In U. Gellert & E. Jablonka (Eds.), *Mathematisation and demathematisation: Social, philosophical and educational ramifications* (pp. 1–18). Rotterdam, The Netherlands: Sense Publishers.

Jaworski, B., & Wood, T. (Eds.). (2008). *The mathematics teacher educator as a developing professional*. Rotterdam, The Netherlands: Sense Publishers.

Johnson, R. B., & Onwuegbuzie, A. J. (2004). Mixed methods research: A research paradigm whose time has come. *Educational Researcher, 33*(7), 14–26.

Jones-Newton, K. (2009). Instructional practices related to prospective elementary school teachers' motivation for fractions. *Journal of Mathematics Teacher Education, 12*, 89–109.
Kaasila, R. (2007). Mathematical biography and key rhetoric. *Educational Studies of Mathematics, 66*, 373–384.
Kanu, Y. (1997). Understanding development education through action research: Cross-cultural reflections. In T. R. Carson & D. J. Sumara (Eds.), *Action research as a living practice* (pp. 167–186). New York, NY: Peter Lang.
Keitel, C., Kotzmann, E., & Skovsmose, O. (1993). Beyond the tunnel vision: Analysing the relationship between mathematics, society and technology. In C. Keitel & K. Ruthven (Eds.), *Learning from computers: Mathematics education and technology* (pp. 243–279). Berlin, Germany: Springer.
Kilpatrick, J. (2000). Research in mathematics education across two centuries. In M. A. Clements, H. H. Tairab, & Wong, K. Y. (Eds.), *Science, mathematics and technical education in the 20th and 21st centuries* (pp. 79–93). Gadong, Brunei Darussalam: Universiti Brunei Darussalam.
Krainer, K. (1999). PFL-Mathematics: Improving professional practice in mathematics teaching. In B. Jaworski, T. Wood, & S. Dawson (Eds.), *Mathematics teacher education: Critical international perspectives* (pp. 102–112). London, UK: Falmer.
Krainer, K. (2003). Editorial: Teams, communities & networks. *Journal of Mathematics Teacher Education, 6*(2), 93–105.
Krainer, K., & Wood, T. (Eds.). (2008). *Participants in mathematics teacher education: Individuals, teams, communities and networks*. Rotterdam, The Netherlands: Sense Publishers.
Lanz, C. (1994). *El poder en la escuela: El método INVEDECOR como fundamento del currículo alternativo*. Caracas, Venezuela: Edit-Art.
Law, J. (2004). *After method: Mess in social science research*. London, UK: Routledge.
Leavy, A., & O'Loughlin, N. (2006). Preservice teachers' understanding of the mean: Moving beyond the arithmetic average. *Journal of Mathematics Teacher Education, 9*, 53–90.
LeCompte, M. D., Preissle, J., & Tesch, R. (1993). *Ethnography and qualitative design in educational research* (2nd ed.). San Diego, CA: Academic Press.
Levenson, E., Tsamir, P., & Tirosh, D. (2010). Mathematically based and practically based explanations in the elementary school: Teachers' preferences. *Journal of Mathematics Teacher Education, 13*, 345–369.
Li, Q., & Chapman, O. (2010, April). *iMovie in prospective secondary mathematics teacher education*. Presented at the American Educational Research Association Annual Meeting, Denver, CO.
Liljedahl, P. (2010). Noticing rapid and profound mathematics teacher change. *Journal of Mathematics Teacher Education, 13*(5), 411–423.
Linn, R. L. (1986). Quantitative methods in research on teaching. In M. C. Wittrock (Ed.), *Handbook of research on teaching* (3rd ed., pp. 92–118). New York, NY: Macmillan Publishing Company.
Llinares, S., & Valls, J. (2010). Prospective primary mathematics teachers' learning from on-line discussions in a virtual video-based environment. *Journal of Mathematics Teacher Education, 13*, 177–196.
Lloyd, G. M. (2006). Preservice teachers' stories of mathematics classrooms: Explorations of practice through fictional accounts. *Educational Studies in Mathematics, 63*, 57–87.
Lubienski, S. T., & Bowen, A. (2000). Who's counting? A survey of mathematics education research 1982–1998. *Journal for Research in Mathematics Education, 31*(5), 626–633.
Mason, J. (2002). *Researching your own practice: The discipline of noticing*. New York, NY: Routledge.
Messina, G. (1999). Investigación en o acerca de la formación docente: Un estado del arte en los noventa. *Revista Iberoamericana de Educación, 19*, n/p.
Miles, M. B., & Huberman, A. M. (1994). *Qualitative data analysis: An expanded sourcebook* (2nd ed.). London, UK: Sage.
Mill, J. S. (1975). *Three essays on liberty, representative government and the subjection of women*. Oxford, UK: Oxford University Press.

Morris, A. K. (2006). Assessing pre-service teachers' skills for analyzing teaching. *Journal of Mathematics Teacher Education, 9*, 471–505.

Moya, A. (2008). *Elementos para la construcción de un modelo de evaluación en matemática para el nivel de educación superior* (Doctoral dissertation). Universidad Pedagógica Experimental Libertador, Caracas, Venezuela.

Neubrand, M., Seago, N., Agudelo-Valderrama, C., DeBlois, L., & Leikin, R. (2009). The balance of teacher knowledge: Mathematics and pedagogy. In R. Even & D. L. Ball (Eds.), *The professional education and development of teachers of mathematics: The 15th ICMI Study* (pp. 211–225). New York, NY: Springer.

Noffke, S. (2009). Revisiting the professional, personal, and political dimensions of action research. In S. Noffke & B. Somekh (Eds.), *The SAGE handbook of educational research* (pp. 6–23). Thousand Oaks, CA: Sage.

Osana, H. P., Lacroix, G. I., Tucker, B. J., & Desrosiers, C. (2006). The role of content knowledge and problem features on preservice teachers' appraisal of elementary mathematics tasks. *Journal of Mathematics Teacher Education, 9*, 347–380.

Parsons, J. B., & Harding, K. J. (2011). Post-colonial theory and action research. *Turkish Online Journal of Qualitative Inquiry, 2*(2), 1–6.

Patton, M. Q. (2002). *Qualitative research and evaluation methods*. Newbury Park, CA: Sage.

Planas, N., & Civil, M. (2009). Working with mathematics teachers and immigrant students: An empowerment perspective. *Journal of Mathematics Teacher Education, 12*(6), 391–409.

Renert, M. (2011). Mathematics for life: Sustainable mathematics education. *For the Learning of Mathematics, 31*(1), 20–26.

Rojas Olaya, A. (2011). El GIDEM: 10 años de revolución en educación matemática. *Foro del Futuro, 4*, 10–20.

Santagata, R., Zannoni, C., & Stigler, J. W. (2007). The role of lesson analysis in pre-service teacher education: An empirical investigation of teacher learning from a virtual video-based field experience. *Journal of Mathematics Teacher Education, 10*, 123–140.

Scherer, P., & Steinbring, H. (2006). Noticing children's learning processes: Teachers jointly reflect on their own classroom interaction for improving mathematics teaching. *Journal of Mathematics Teacher Education, 9*(2), 157–185.

Serrano, W. (2010). *La educación matemática crítica en el contexto de la sociedad Venezolana: Hacia su filosofía y praxis* (Doctoral dissertation). Universidad Central de Venezuela, Caracas, Venezuela.

Setati, M. (2000). Classroom-based research: From *with or on* teachers to *with and on* teachers. In J. F. Matos & M. Santos (Eds.), *Proceedings of the Second International Mathematics Education and Society Conference* (pp. 351–363). Lisbon, Portugal: CIEAFC-Universidade de Lisboa.

Sfard, A. (2005). What could be more practical than good research? On mutual relations between research and practice of mathematics education. *Educational Studies in Mathematics, 58*(3), 393–413.

Silva, D. (2010). *De lo real a lo formal en matemática* (M.A. thesis). Universidad Pedagógica Experimental Libertador, Caracas, Venezuela.

Skott, J. (2001). The emerging practice of a novice teacher: The role of his school mathematics images. *Journal of Mathematics Teacher Education, 4*(1), 3–28.

Skovsmose, O. (1994). *Towards a philosophy of critical mathematics education*. Dordrecht, The Netherlands: Kluwer Academic Publishers.

Smith, L. M. (1994). Biographical method. In N. K. Denzin & Y. S. Lincoln (Eds.), *Handbook of qualitative research* (pp. 286–305). Thousand Oaks, CA: Sage.

Son, J.-W., & Crespo, S. (2009). Prospective teachers' reasoning and response to a student's non-traditional strategy when dividing fractions. *Journal of Mathematics Teacher Education, 12*, 235–261.

Speer, N. M. (2005). Issues of methods and theory in the study of mathematics teachers' professed and attributed beliefs. *Educational Studies in Mathematics, 58*(3), 361–391.

Spivak, G. C. (1990). *The post-colonial critic: Interviews, strategies, dialogues.* London, UK: Routledge.
Spivak, G. C. (2011). Colonidad, capitalismo y descolonización en las sociedades periféricas. In *Pensando el mundo desde Bolivia: I ciclo de seminarios internacionales* (pp. 311–320). La Paz, Bolivia: Vicepresidencia del Estado Plurinacional de Bolivia.
Stake, R. E. (1995). *The art of case study research.* Thousand Oaks, CA: Sage.
Stake, R. E. (2000). Case studies. In N. K. Denzin & Y. S. Lincoln (Eds.), *Handbook of qualitative research* (2nd ed., pp. 435–454). Thousand Oaks, CA: Sage.
Star, J. R., & Strickland, S. K. (2008). Learning to observe: Using video to improve preservice mathematics teachers' ability to notice. *Journal of Mathematics Teacher Education, 11,* 107–125.
Steinbring, H. (1994). Dialogue between theory and practice in mathematics education. In R. Biehler, R. W. Scholz, R. Sträßer, & B. Winkelmann (Eds.), *Didactics of mathematics as a scientific discipline* (pp. 89–102). Dordrecht, The Netherlands: Kluwer.
Steiner, H. G. (1985). Theory of mathematics education: An introduction. *For the Learning of Mathematics, 5*(2), 11–17.
Stockero, S. L. (2008). Using a video-based curriculum to develop a reflective stance in prospective mathematics teachers. *Journal of Mathematics Teacher Education, 11,* 373–394.
Strauss, A. L., & Corbin, J. (1990). *Basics of qualitative research: Grounded theory procedures and techniques.* Newbury Park, CA: Sage.
Strauss, A. L., & Corbin, J. (1998). *Basics of qualitative research: Techniques and procedures for developing grounded theory* (2nd ed.). Thousand Oaks, CA: Sage.
Sullivan, P., & Wood, T. (Eds.). (2008). *Knowledge and beliefs in mathematics teaching and teaching development.* Rotterdam, The Netherlands: Sense Publishers.
Swantz, M. L. (1970). *Ritual and symbol in transitional Zaramo society with special reference to women.* Lund, Sweden: Gleerup.
Swars, S. L., Smith, S. Z., Smith, M. E., & Hart, L. C. (2009). A longitudinal study of effects of a developmental teacher preparation program on elementary prospective teachers' mathematics beliefs. *Journal of Mathematics Teacher Education, 12,* 47–66.
Tashakkori, A., & Teddlie, C. (Eds.). (2003). *Handbook of mixed methods in social and behavioral research.* Thousand Oaks, CA: Sage.
Thanheiser, E. (2010). Investigating further preservice teachers' conceptions of multi-digit whole numbers: Refining a framework. *Educational Studies of Mathematics, 75,* 241–251.
Thompson, A. G. (1992). Teachers' beliefs and conceptions: A synthesis of the research. In D. A. Grouws (Ed.), *Handbook of research on mathematics teaching and learning* (pp. 127–146). New York, NY: Macmillan.
Tirosh, D. (2007). What is research in mathematics teacher education? *Journal of Mathematics Teacher Education, 10*(3), 141–144.
Tirosh, D., & Wood, T. (Eds.). (2008). *Tools and processes in mathematics teacher education.* Rotterdam, The Netherlands: Sense Publishers.
Uworwabayeho, A. (2009). Teachers' innovative change within countrywide reform: A case study in Rwanda. *Journal of Mathematics Teacher Education, 12*(5), 315–324.
Vacc, N. B., & Bright, G. W. (1999). Elementary preservice teachers' changing beliefs and instructional use of children's mathematical thinking. *Journal for Research in Mathematics Education, 30*(1), 89–110.
Vico, G. (1709). *De nostril temporis studiorum ratione.* Naples, Italy: Dissertatio.
Vithal, R., & Valero, P. (2003). Researching mathematics education in situations of social and political conflict. In A. J. Bishop, M. A. Clements, C. Keitel, J. Kilpatrick, & F. K. S. Leung (Eds.), *Second international handbook of mathematics education* (pp. 545–591). Dordrecht, The Netherlands: Kluwer.
Warman, A., Nolasco, M., Bonfil, G., Olivera, M., & Valencia, E. (1970). *De eso que llaman antropología Mexicana.* Mexico D. F: Nuestro Tiempo.
Watts, M., & Ebbutt, D. (1987). More than the sum of the parts: Research methods in group interviewing. *British Educational Research Journal, 13*(1), 25–34.

Wilkins, J. L. M. (2008). The relationship among elementary teachers' content knowledge, attitudes, beliefs, and practices. *Journal of Mathematics Teacher Education, 11*, 139–164.

Wilson, M., & Cooney, T. (2002). Mathematics teacher change and development: The role of beliefs. In G. C. Leder, E. Pekhonen, & G. Törner (Eds.), *Beliefs: The hidden variable in mathematics education?* (pp. 127–147). Dordrecht, The Netherlands: Kluwer Academic Publishers.

Yin, R. K. (1989). *Case study research: Design and methods*. Thousand Oaks, CA: Sage.

Zack, V., Mousley, J., & Breen, C. (Eds.). (1997). *Developing practice: Teachers' inquiry and educational change*. Geelong, Australia: Deakin University Press.

Zeichner, K. (2001). Educational action research. In P. Reason & H. Bradbury (Eds.), *Handbook of action research: Participative inquiry and practice* (pp. 273–283). London, UK: Sage.

Chapter 12
Linking Research to Practice: Teachers as Key Stakeholders in Mathematics Education Research

Carolyn Kieran, Konrad Krainer, and J. Michael Shaughnessy

Abstract Teachers are regarded as having a major role in the development of mathematics teaching and students' learning. Nevertheless, in much mathematics education research, teachers are viewed as recipients, and sometimes even as means to generate or disseminate knowledge, thus conserving a distinctive gap between research and practice. The theme of this chapter is to regard teachers as key stakeholders in research (i.e., as (co-)producers of professional and/or scientific knowledge) in order to make the link between research and practice more fruitful for both sides. After exploring the concept of stakeholder, the authors present five international examples, all of them involving teachers researching their own or their colleagues' practice. An analysis of the commonalities and differences among these examples reveals the presence of three important dimensions of research where teachers are key stakeholders: reflective, inquiry based activity with respect to teaching action; a significant action-research component accompanied by the creation of research artefacts by the teachers (sometimes assisted by university researchers); and the dynamic duality of research and professional development. This chapter illustrates how traditional barriers between research and practice are being replaced by synergistic interactions between the two, enabling the intersection of the two worlds.

C. Kieran (✉)
Université du Québec à Montréal,
Montréal, Canada
e-mail: kieran.carolyn@uqam.ca

K. Krainer
Alpen-Adria Universität Klagenfurt,
Klagenfurt, Austria

J. M. Shaughnessy
Portland State University, Portland, OR, USA

Introduction: Teachers' and Researchers' Diverse Worlds

In most cases, the worlds of teachers and researchers differ greatly, even if there are also cases where they work together so closely that the traditional roles begin to blur. Nevertheless, mathematics education research and mathematics teachers' practices can overlap considerably, as is reflected in the case of *students' mathematical thinking*. Let us start with a concrete example. When middle-school students deal with identities like $(a+b)^2 = a^2 + 2ab + b^2$, a variety of errors or misconceptions appear. For instance, many students come up with $a^2 + b^2$ as the result of expanding $(a+b)^2$. Several researchers have published studies on this phenomenon (see, e.g., Davis, Jockusch, & McKnight, 1978; Kieran, 2007; Kirshner & Awtry, 2004; Matz, 1982). As well, most teachers are aware of this phenomenon and have developed—consciously or subconsciously—strategies for dealing with it. They have worked out ways to support students' thinking and re-designed their introduction to the topic in order to decrease the likelihood that this error will occur. Some teachers might have been influenced by mathematics educators' research, and a few of them might have even collaborated very closely with them. In addition, some teachers are highly respected researchers in their own right. However, the picture is even more complex than this since there is considerable variety within the worlds of teachers and researchers.

The variation within the *world of mathematics teachers* can be illustrated by their ways of dealing with the identity $(a+b)^2 = a^2 + 2ab + b^2$ in their classroom. For example, teacher Anna might show on the blackboard that $(a+b)(a+b)$ just leads to the identity; then the identity is written with colors and some similar examples are given. Anna covers the topic in one hour because she believes that dealing with algebra software systems at a later grade will be much more effective. Björn, who spends four to five lessons on the topic, regards this identity as essential and aims at offering his students rich learning opportunities in order that they will remember very well the identity and its generation. He builds on links to geometry—interpreting $(a+b)(c+d)$ as expanding the size of the rectangle (a, c) to $(a+b, c+d)$—and then encourages the students to find the identity themselves. Cecile has a flexible strategy and only decides on her concrete teaching design after some repetition work where she develops a sense of her students' pre-knowledge and interests on this issue. Davido always starts a larger unit with a diagnostic test in order to know all his students' mathematical abilities. According to his findings, he forms three to five ability groups in class with different tasks and task levels, supporting in particular those who might have problems in meeting the minimum standards. And there are many other approaches, including fostering students' own ways to reach their goals (and perhaps documenting their progress in a portfolio), as well as training students by "teaching to the test."

The diversity of students' knowledge and interests in a classroom, the (subjectively) giant obstacles to overcome, and the always missing time for dealing with these challenges are some major factors that contribute to the complexity and unpredictability of teaching. "How great it would be," said Maria, an experienced mathematics teacher in a professional development course, "if I had videotapes of all interesting mathematical situations in my classroom, and the time to analyze them with colleagues; however, I have to react immediately to each error, and this causes errors on my part too."

This marks a good counterpart to the *world of mathematics education researchers* dealing with students' knowledge, be it an identity like $(a+b)^2=a^2+2ab+b^2$, the characteristics of a proof in geometry, or advanced stochastical thinking. Although the identity above is a very tiny piece within mathematics, research on it is abundant, with new ways of framing the research emerging over time. For example, the researcher, Albert, might investigate the challenges to the student who comes up with $(a+b)^2=a^2+b^2$ and then construct a mind-map about her algebraic thinking, explaining her difficulties on a theoretical basis. Bruno investigates how students' mathematical ideas are dealt with by other classmates and by the teacher, and how the negotiation of meaning takes place or the didactical contract is generated. Corinne is interested in the interplay between a particular student's mathematical abilities and interests. And there are many other perspectives that researchers might take, including large-scale investigations of students' answers on national standard tests, "design research" activities with small samples of teachers, evaluations of studies dealing with students' and teachers' mathematical growth, and systematic reflections by teacher educators on their learning processes while leading intervention projects.

We can summarize the situation as follows: Even if focussed on a very specific topic (e.g., students' thinking), *mathematics education research as well as mathematics teaching is highly diverse*. Much empirical and theoretically-based knowledge is produced by the scientific community and much, mostly unpublished, knowledge is produced by the rich experiences of thousands of teachers. From this it is also clear that the communication and possible collaboration between teachers and researchers is diverse.

The major question is: How can mathematics education research have an impact on mathematics classrooms, on students' learning, abilities, beliefs, and interests? And how can researchers benefit from the rich body of knowledge and subjective theories teachers have? *And who is responsible for dealing with this question?*

Regarding Teachers as Key Stakeholders in Research

Researchers and teacher educators neither have the role nor the capacity to influence directly mathematics teaching on a large scale. Their major impact on teaching seems to be related to the *production of relevant knowledge* and generating opportunities for teachers (and to some extent also for other relevant groups like principals) to confront this knowledge with their existing knowled*ge*. In general, teachers are regarded as key persons of educational change (e.g., Fullan, 1993). This view is largely supported by research evidence. For example, an analysis of student learning over many large-scale projects (Hattie, 2003) shows that teachers' impact on students' learning is high: identified factors that contribute to major sources of variation in student performance include students (50%) and teachers (30%) as the most important factors, whereas home, schools, principals, peer effects (altogether 20%) play a less important role (see, e.g., Pegg & Krainer, 2008). Research on "successful" schools shows that such schools are more likely to have teachers who have continual substantive interactions (Little, 1982) or that inter-staff relations are seen as an important dimension of school quality (Pegg, Lynch, & Panizzon, 2007; Reynolds et al., 2002).

The implication of this research is that approaches with the most potential to bring about genuine improvement in learning mathematics are those that resonate with teachers—with their interests, beliefs, emotions, knowledge, and practice—as well as those that encourage further collaboration among them. Krainer and Llinares (2010) have emphasized that "it is desirable to use the synergy of teachers' expertise and therefore to engage them in research activities and to support action research, among others, with the goal that some of them might develop deeper interest in research and thereby to enlarge the scientific community" (pp. 704–705). The idea of viewing teachers as experts and competent partners in research is not new at all. For example, in the literature, they are regarded as *researchers* (e.g., Altrichter, Feldman, Posch, & Somekh, 2008; Crawford & Adler, 1996; Stenhouse, 1975), *reflective practitioners* (e.g., Schön, 1983), and *experts* (e.g., Bromme, 1992). *Intervention research* with teachers as partners and *action research* by teachers or teacher educators is becoming more prominent in mathematics teacher education (see, e.g., volumes 6.2 and 9.3 of *JMTE* in 2003 and 2006). Lesson study, as a teacher-led professional development approach, has a long tradition in Japan and has begun over the last decade to spread to other countries (see, e.g., Hart, Alston, & Murata, 2011). Recently, the fourth volume of the *First Handbook of Mathematics Teacher Education* (Jaworski & Wood, 2008) drew attention to the crucial importance of activity involving *learning and self-reflection* for both teachers and teacher educators.

It is the ethical responsibility of a scientific community and at the same time a wise strategy to raise questions (see Krainer, 2011) such as: How does our knowledge get known, used, and reflected upon by relevant people and institutions? How can their experiences, which form a new kind of knowledge, be fed back to the researchers? What can be done by researchers apart from writing papers and giving talks—predominantly within the scientific community—and from teaching classes of student teachers and offering professional development courses? It cannot be taken for granted that the majority of those to whom research might possibly be addressed do in fact read the tremendously increasing number of research papers and that traditional teacher education is a viable means to link research results with the challenges of practice.

There have been efforts by individual researchers and groups to raise this issue, for example, in a conference on "*Systematic Cooperation between Theory and Practice in Mathematics Education*" (Bazzini, 1994), in papers like the "Dialogue between theory and practice in mathematics education" (Steinbring, 1994), in a special issue of *Educational Studies in Mathematics* on connecting research, practice, and theory (Even & Ball, 2003), in the chapter "Mathematics Teacher Education" in the *International Encyclopedia of Education* (Krainer & Llinares, 2010), and most recently in this chapter in the *Third International Handbook of Mathematics Education*. Not long ago, an initiative to create stronger links between researchers and practitioners was undertaken by the National Council of Teachers of Mathematics (2010) as researchers and practitioners met together to create a research agenda consisting of questions deemed most critical to conduct collaborative research. Despite these efforts and continuous claims on the importance of the role of teacher-researcher collaboration, teachers are still often seen as more or less passive recipients of researchers' knowledge production and sometimes as a means (e.g., as data supplier) to help produce knowledge. What is missing, in particular, is a systematic effort by the scientific community (such as societies, commissions, universities, research groups) to analyze and promote the potential role of teachers in research and its benefit for teachers and researchers.

In the 1980s, an interesting change of paradigm started in *management strategy* (in particular in the USA). The traditional view was the *shareholder approach*, which regarded it the duty of management to protect the interests of the shareholder, basically in order to avoid having poor social performance hurt the company financially. Management aimed at satisfying clients, consumers, society, etc., by specific strategies (e.g., public relations). In contrast, Freeman (1984) and others developed a stakeholder approach, defining "*stakeholder*" as "*any group or individual that can affect or is affected by the achievement of a corporation's purpose*" (Freeman, 2004, p. 229). The approach dealt with the practical concerns of managers—"How could they be more effective in identifying, analyzing and negotiating with key stakeholder groups?" (p. 230). The stakeholder idea is connected to ethics and values, which are regarded as equally important as the business itself (see also Krainer, 2011). The main message is that "*looking at the whole system*" *(of interests) is a benefit for all parts of the system aiming at sustainable development.*

The mathematics education research "enterprise," whose "business" includes the improvement of the teaching and learning of mathematics, is distinctly unlike a corporation in many respects. Nevertheless, the similarities between the two can be useful, including the presence of a multitude of stakeholders. It is not just researchers who have a stake in the research enterprise, even if they are generally considered to have the most expertise in research (e.g., with respect to theory, methodology, etc.) and tend to set the trajectories for research. In addition, they are assumed to form their decisions not only for the sake of the scientific community but more broadly for society too. Nevertheless, the research enterprise in mathematics education has other stakeholders: for example, students, teachers, parents, principals, superintendents, mathematicians, teacher educators, educational publishers, test-developers, firms, education policy-makers, and even the whole society can be regarded as "stakeholders" of the joint societal enterprise of promoting students' mathematical knowledge. They all have an effect on students' knowledge and at the same time they are affected by their knowledge. But of all these stakeholders, it is the teacher who *can affect to the greatest extent the achievement of one of the main purposes of the research enterprise, that is, the improvement of students' learning of mathematics.*

The scientific community needs to regard teachers not just as stakeholders, but also as "*key stakeholders*" of research. At least five aspects should be discussed when analyzing the role of teachers with regard to the production and dissemination of scientific knowledge—an activity central to the research enterprise and one by which its participants aim at contributing to the improvement of students' learning of mathematics. The first three aspects suffice if researchers are mainly attempting to optimize their own interests as researchers and seeing the production of knowledge being predominantly done within the scientific community (excluding practitioners like teachers and non-researching teacher educators). With regard to these three aspects, teachers are seen as "stakeholders" in that they have a stake in the results of research, which can inform them about elements of student learning; but they are not seen as *key stakeholders*—a term that we reserve for the fourth and fifth aspects. The fourth aspect deals with embracing teachers as experts who are principally able to contribute heavily to the quality of research, and the fifth aspect regards them as co-producers of scientific knowledge. The following presents a brief sketch of these five aspects.

Teachers as Means

For most of the research where the beliefs, knowledge, and practice of students and/or teachers are the focus, a collaboration with teachers is needed. They supply data, which are analyzed by the researchers. It should be a viable standard to provide involved teachers with a rationale for the research and its possible implications for teachers' work before the collection of data, and a summary of the research and its relevant findings after it. For example, it would be of interest to teachers to read which different ways of introducing algebraic identities (like $(a+b)^2 = a^2 + 2ab + b^2$) different teachers use, and what the rationale behind their approaches is, probably accompanied by comments, evidence, and suggestions from the authors of the study.

Teachers as Recipients

The primary responsibility of teachers is to *teach* their students, not to read research papers, and there is some evidence that most teachers don't read such papers very often (Zeuli, 1994). Strategies by members of the scientific community in order to increase teachers' interest in reading research papers are manifold (see, e.g., Debien, 2010; Shearer, Lundeberg, & Coballes-Vega, 1997). Some scholarly journals have sections that are specifically intended to share research with teachers and some teacher journals have sections devoted to "research in practice." Many researchers publish additional papers with a clear practice-oriented focus in journals widely read by teachers, write practice-focussed summaries and put them on Web sites and in teacher journals, write papers and give talks about the results of research studies that would be of interest to teachers, use pieces of research in teacher education and, for example, engage student or practising teachers in short parts of this research (e.g., having teachers construct a multiple-choice item on "$(a+b)^2 = \ldots$" with the correct answer and three other tempting answers, and having them estimate the distribution of answers of students in their class).

Teachers as Alumni

Teachers are regarded as life-long learners, having spent a considerable amount of time at teacher-education institutions. Hopefully, while there, they were confronted with a selection of interesting activities in the context of research and came to realize that research is fascinating, and that it provides insights and thereby a strong basis for understanding their own thinking and their students' thinking. Thus, they might have developed a kind of "inquiry stance" that could increase their interest in trying out small pieces of research in their classrooms, in looking for contact with teacher educators and researchers, or being open to offers from the wider scientific community. Teachers' calls to university partners like, "Do you have news about research on students' algebraic thinking?" or "Are you running another interesting project?" would be indicators of teachers' inquiry stance and former teacher educators' success at evoking such interests.

Teachers as (Co-)Producers of Professional Knowledge

Teachers deal on a daily basis with students' thinking, their beliefs and conceptions, errors and ideas, interests and fears, emotions and cognitions, views of mathematics and mathematics teaching, etc. They can be regarded as experts on students' subject-related learning. On each curricular topic they teach for a long period, they develop specific expertise; however, it varies from teacher to teacher, dependent on pedagogical, didactical, and mathematical abilities and interest. Teachers who share their experiences with peers (e.g., within the context of joint lesson study or other kinds of professional development) are more likely to intensify their abilities and interest. For example, teachers having extensively discussed their approach to the introduction of identities like $(a+b)^2 = a^2 + 2ab + b^2$ and its effect on students' learning surely develop forms of professional knowledge and subjective theories about students' algebraic thinking of interest to researchers. In particular, reflecting on the growth of students' and/or teachers' knowledge might be a beneficial endeavour for both parties. Through being involved in such projects, bridges between teachers and researchers might be built—bridges to link professional and scientific knowledge, which are not easy to separate in many cases, anyway.

Teachers as (Co-)Producers of Scientific Knowledge

There is evidence of research where teachers are equal partners or even central figures. For example, there are research projects where teachers not only help to gather data but also give advice concerning the design of research and the refinement of methods. Studies have been carried out where teachers design the research themselves, collect data, and are engaged in the analysis and interpretation of data, as well as in the process of formulating and disseminating the results. There are projects where the people involved decide intentionally to avoid the distinction between teachers and researchers since both are regarded as researchers, with differentiated roles in the research process. The presence or absence of teachers in a research project is not an indicator of research quality per se. In contrast, bringing in additional perspectives, data, and forms of communicative validation can be regarded as a feature enriching scientific research. Having teachers participating in such kinds of research, the dissemination into practice is facilitated.

In particular in the field of student- and practising-teacher education, there is a considerable amount of research where those who are educating the teachers are also those who are carrying out the research. A special kind of such research includes those projects where teachers or teacher educators investigate their own practice in order to improve it (action research). An example of research with regard to teaching algebra might be an action-research project within the framework of a professional development program where teachers try out and investigate new ways of algebra teaching (e.g., a different approach for dealing with $(a+b)^2 = a^2 + 2ab + b^2$), finally producing small case studies of their experiences. The teacher educators support teachers' innovations and investigations and probably investigate their own growth and support processes in order to improve them—a kind of second-order research (see, e.g., Altrichter et al., 2008). In addition, or alternatively, they might write a cross-case study on teachers' approaches and growth or/and investigate students' thinking together with the teachers, or/and write together a handbook for teachers with learning units based on examples and reflections from the project.

The question of how intensively researchers regard teachers and others as *key stakeholders* is an expression of the intended and/or lived relationship between teachers and researchers. This means that our view of "teachers as stakeholders" is about "us," about our beliefs and roles, about our understanding of "research."

In the following section, we aim at providing examples of research projects, where teachers are regarded as *key stakeholders* in research, in the sense that teachers and researchers (or teachers with other teachers) act as co-producers (or as producers) of professional and scientific knowledge.

Five International Examples

In reviewing the international mathematics education research literature, we sought approaches to linking research and practice that were innovative and where collaborative research partnerships had a clear focus on teachers researching their practice. Several examples presented themselves, including successful recent endeavours in both developed and developing countries (see, e.g., OECD, 2011). Space constraints, however, restricted our selection to five examples, each of which offers specific insights into the diverse ways in which teachers engage as key stakeholders in research. Two are large nation-wide programs (Japan, China), while three (USA, Norway, Canada) are initiatives that are much smaller in scale and do not claim to be widespread within their given country. We note that the terminology used for *teacher* and *researcher* is not uniform across the five examples. We have tried to respect the nomenclature adopted by the authors of the reports of each specific example by using the same terms that they have employed for teacher and researcher. We have also strived, by means of various forms of contact with individuals involved in the projects and programs described herein, to do justice to all examples and to represent their multifaceted dimensions as fairly and as accurately as possible.

The USA Example

In a 4-year project (2004–2008) led by Beth Herbel-Eisenmann, teacher researchers collaborated with university researchers in reflecting on their own teaching and in conducting cycles of action research that focussed on improving the mathematical discourse of their classrooms (Herbel-Eisenmann, 2010). Eight mathematics teachers from grades 6 to 10, whom Herbel-Eisenmann had met through her work in her university position, were interested in learning more about classroom discourse, and they agreed to be the teacher-researcher participants in the project. These teacher-researcher volunteers came from a variety of types of school settings—rural, urban, and suburban. They had teaching experiences that varied from 4 to 23 years and taught from a variety of different curricular materials. Herbel-Eisenmann, together with several graduate students, served as the university researchers over the life of the project.

At the beginning of the project, the group agreed that the primary goal of their activity, for both the university researchers and the teacher researchers, would be to learn about, reflect upon, and change mathematical discourse in classrooms. The book *Promoting Purposeful Discourse* (edited by Herbel-Eisenmann and Cirillo, 2009) provides a reflective narrative of the details of the project, including timeline, details of the study-group activities of the eight teacher researchers, data generation and analysis phases of the action-research projects, write-ups by the teachers on their own research projects, and reflections on the experience from both the university researchers and teacher researchers.

The first year of the project was spent gathering baseline data on the teachers' practices, beliefs, and patterns of discourse in their classrooms. Each teacher researcher had one of his/her classrooms videotaped for an entire week, for four different weeks over a six-month time period in the school year. When classes were not being taped, the teachers and university team met and analyzed mathematical tasks and shared artefacts from their teaching in the study group. The university researchers provided quantitative and qualitative discourse analyses of the taped classroom episodes for the teachers. Discussions were then held in which the teacher researchers reacted to the videotapes, discussed them, and had the opportunity to provide interpretations which differed from those given by the university researchers. The collaboration of university researchers and teacher researchers extended to all aspects of this project, planning, readings, data analyses, reflective writing, and developing the action-research projects themselves.

The interactions between university researchers and teacher researchers were designed to develop a community of trust and support. The main goal of the project was to give teachers the opportunity to find *their own* research voice, to tap the researchers within themselves, in order to gather evidence to help change their practice. As such there were multiple levels of "research" occurring within this project that linked research and practice, research by the teacher researchers themselves as well as research by the university researchers.

Early on in the project, the teachers were asked to create belief maps, professed beliefs about what was closest to their heart in their teaching, and then to write journal entries about these professed beliefs. Compact versions of their belief maps were created for continued reference throughout the project, so that both the university and teacher researchers could continually look for congruence between professed beliefs about teaching, and actual behavior in the classrooms by the teacher researchers. Throughout the project the teacher researchers were continually provided with prompts for creating reflective journal entries. Questions were posed after study group discussions that led to journal entries. Teacher researchers were encouraged to write journal entries on what they were learning from the discussions on their classroom videotapes. Discussions were also punctuated with commentary related to the readings on classroom discourse in which they were engaged. The habit of becoming a reflective practitioner, keeping a journal, and reflecting on their practice, was being instilled in these teacher researchers throughout this project.

As the readings, discussions, and shared classroom video segments progressed, the teacher researchers began to identify "performance gaps" that they noticed between what they claimed were their professed beliefs, and what they actually did while teaching in their classrooms. This process provided the seeds and incubation time for the teacher researchers to identify their own research questions to investigate during their cycles of action research throughout the last two years of the project. The teachers noted the importance of wait time—not only after questions are posed, but also after a student responds. They realized that wait time was critical to provide opportunities for richer, deeper student discourse about mathematical content in the classroom.

The teacher researchers found that the process of revoicing students' comments, suggestions, and questions proved to be a powerful tool for improving content discourse. Another primary focus of their work was on improving classroom discourse for social purposes. They found there was a critical need to provide a safe classroom environment for students to share their thinking, solutions, and ideas and to feel comfortable to ask questions of the teacher and of one another.

Promoting Purposeful Discourse included reflective research chapters written by each of the eight teacher researchers in which they documented and shared their passage through cycles of action research—what they did in their own research projects and what they learned throughout the four years of the project. The teacher researchers made their own choices on how they wished to approach writing up their action-research experience. In addition to their chapters, the teacher researchers had opportunities to present their work and experiences at several meetings and conferences attended by other teachers and mathematics education researchers. The topics investigated by the teacher researchers in their action-research projects included: increasing student participation in conceptual discourse; attending to particular performance gaps in their classroom practices uncovered in their belief maps; working towards giving students more ownership in the mathematical discourse in classes; revoicing student questions; addressing vagueness in classroom mathematical discourse; and improving listening to students' mathematical discourse.

Several things are of particular note in this long-term effort by Herbel-Eisenmann to link research and practice. There were multiple levels of research linked to practice that were created and continued over the entire project. Throughout the project the university researchers were investigating what moves, actions, and support structures might be helpful to create an environment where teachers could become researchers. The university researchers also kept reflective journals throughout the project, and met to discuss and plan study group meetings based on their own research observations of the group. The teacher researchers were conducting research on their classroom discourse behaviours, and on patterns of student discourse interactions in their classrooms as they developed their own action-research projects. A "linking" of research and practice occurred continually throughout this project within the discussions and reflective activity of the community meetings of the study group. Ultimately, the reflective story of this project as captured in *Promoting Purposeful Discourse* provided yet another level of research itself. It presented both a meta-reflection by the university researchers that identified the major themes, trends, and activities of the project, along with the stories told by the teacher researchers as they described their own action-research projects.

The Norwegian Example

The three-year Learning Communities in Mathematics (LCM) Project (2004–2007) was a research and development project that brought together teachers and didacticians to work together as both practitioners and researchers (Jaworski et al., 2007). It involved a team of 14 didacticians (the term that the team preferred to use for the teacher educators), which included 5 doctoral students, working with 8 schools (including primary, lower, and upper secondary) with a minimum of three teachers from each school (Jaworski, 2006). Schools volunteered to be part of the project as a result of an invitation from Agder University College in Norway where Barbara Jaworski, who led the project, held a faculty position.

The motivating principle on which the didacticians and teachers agreed to work together was the desire to develop better learning environments for mathematics students at the levels of schooling with which each teacher was associated. In fact, co-learning was central to this project. Jaworski (2011) cited Wagner (1997) to make the point:

> In a co-learning agreement, researchers and practitioners are both participants in processes of education and systems of schooling. Both are engaged in action and reflection. By working together, each might learn something about the world of the other. Of equal importance, however, each may learn something more about his or her own world and its connections to institutions and schooling. (p. 16)

Workshops at the college were an important tool in the co-learning process. During the first two years, six workshops were held per year, and four during the third year. Workshops were three and a half hours in length and consisted of both plenary and small group activity. Plenary input from both didacticians and teachers included introducing mathematical tasks (usually by the didacticians), reporting about classroom activity (mainly by the teachers), and reporting from small group activity (by all). Small group activity included working on mathematical tasks, usually followed by didactical discussions in which both teachers and didacticians participated.

The teachers in the school teams worked together on designing tasks for the classroom. Didacticians were available to discuss the ideas that the teachers had generated, as well as to observe the classroom unfolding of the activities. Three didacticians were associated with each school to discuss the planned activities, to provide support, and to collect data.

All classroom lessons related to the project, as well as the workshop sessions, were videotaped. Jaworski (2006) stated that "the data and its analysis was largely owned by didacticians, with video data also providing a source for teachers to review classroom activity and reflect on teaching" (p. 11). All data were available to all of the didacticians of the project; in addition, the teachers had access to the data for their school should they so wish. The video data also proved to be a valuable resource within both the workshops and school settings as a tool for reflecting on developing student thinking within classroom activity. The video data were not related to particular research questions; rather research questions evolved through activity and data were used according to need. Jaworski (2008) pointed out that, as the didacticians followed up initial research questions in analysis of data and writing of papers, more refined questions emerged which then fed into future activity and further research.

At the heart of this collaborative project was the resolve to frame it around an inquiry-based approach within communities of practice. Inquiry, which involved questioning, exploring, investigating, and researching within everyday practice, was conceptualized at three levels:

1. Inquiry in mathematics: (a) teachers and didacticians exploring mathematics together in problems and tasks in workshops; (b) pupils in schools learning mathematics through exploration in tasks and problems in classrooms;
2. Inquiry in teaching mathematics: teachers using inquiry in the design and implementation of tasks, problems, and mathematical activity in classrooms in association with didacticians;

3. Inquiry in developing the teaching of mathematics: teachers and didacticians researching the processes of using inquiry in mathematics and in the teaching and learning of mathematics.

This emphasis on inquiry was, in the words of those who were asked to evaluate the LCM project at its close (Skovsmose & Säljö, 2007), a "challenge to the traditional notion of school mathematics in Norway … the inquiry approach explicitly and radically breaks with this [traditional] conception of learning mathematics; the power of the [LCM] project has to do with how the inquiry approach informs and comes to be a part of reformed classroom practices" (p. 11).

Within the LCM project, an inquiry community for the project at large had been created, but it could not be separated from the established communities of which project members were a part. According to Jaworski (2008), "teachers participated in the day-to-day life of their schools and, integrally, explored the use of inquiry-based tasks in their classrooms and observed their students' mathematical activity and learning; didacticians collected and analyzed data and wrote research papers, as expected of university academics and, integrally, explored the design of tasks for workshops and their work with teachers in school environments to support teachers in their project activity" (p. 320). But even more importantly, Jaworski emphasized that the alignment of both didacticians and teachers with their respective communities was a "critical alignment." By this she meant that they did so with a critical attitude whereby they questioned, explored, and sought alternatives while engaging, so as to "have possibilities to develop and change the normal states" (p. 314). Teachers and didacticians had engaged in a research activity that yielded evidence of both teachers' learning and didacticians' associated learning. Jaworski (2008, p. 326) argued that "seeing the enterprise in terms of an activity system made it possible to pick out elements in their complexity and trace developmental patterns for participants in the project (see Goodchild & Jaworski, 2005; Jaworski & Goodchild, 2006)."

The Canadian Example

In 1989, the CIRADE research centre attached to the Université du Québec à Montréal established research links with some schools. Over the years, the research engaged in at these schools began on to take on a distinctive shape where the emphasis was clearly on collaboration between teachers and researchers—research was being conducted "with" rather than "on" teachers. The example presented herein involved a group of teachers at one of these research schools and some of the CIRADE university researchers, led by Nadine Bednarz, who collaborated with that school (Bednarz, 2004). The collaborative project that emerged was one that combined professional development with supported action research in the classroom.

A group of first-grade teachers approached the researchers because they were having difficulty conceptualizing a way in which they might implement a ministerial-mandated, problem-solving approach to the teaching of mathematics in their classes. The questions that the teachers put to themselves were the following: Is it possible to adopt a problem-solving process with young children? What does such an approach mean, and how can it be developed? These questions provided the basis for a collaborative research project that initially lasted for a year, but was extended for three more years. The team consisted, at first, of four first-grade teachers, a remedial teacher, and two researchers, but then brought in teachers from second and third grade during the following years. During the course of the project's being extended to the second and third grades, the mathematical content was also extended.

The design of problem situations, and ways in which to intervene with the children, was the central focus of the meetings that took place between the teachers and researchers. The dimension of professional development, referred to as *reflection on action* by Bednarz, was constituted by the discussions regarding the problem situations, the strategies used by the children, their approaches and ways of reasoning, and the teachers' management of the activity in the classroom context. In the process of reflecting, other questions of a more general nature arose among the teachers regarding problem solving and its integration into their practice. The research dimension was also fuelled by the joint construction of these problem situations, in particular by a reflection on the ways in which the problem situations were enhancing the mathematical learning of the children.

Over the course of the four years during which the joint process of constructing teaching situations occurred, approximately 1 day per month was given to reflection. In addition, one day of assessment was also included at the end of each year in order to review the outcomes of the project. As described by Bednarz (2004):

> The reflective activity was conducted in such a way as to encourage a planned, regular alternation between classroom experience and review of this experience. Work was performed in groups using accounts of the in-class activities, the difficulties arising in context, the records of statements by the children, and the difficulties they encountered. This review of the experience took different forms and served as a starting point for developing a new intervention sequence. This reflective activity thus developed around the meanings that the teacher developed in context and indeed imparted meaning to the situations or actions put forward. (p. 7)

Researchers and teachers interacted and jointly explored teachers' practice and engaged in the reflective review of that practice. The regular meetings of researchers and practitioners permitted, according to Bednarz (2004), the creation of an "interpretive zone" around the practice that was the subject of the exploration. This reflective activity was deemed to serve a dual function: "It is an opportunity for professional development through reflective review of the practice, with the objectives of clarifying, making explicit, and improving understanding of this practice—hence, of ultimately contributing to its restructuring; it is a research opportunity, as this meeting zone (interpretive zone) constitutes material for analysis to be used for investigating a certain object of interest to practice-related knowledge" (p. 11). In addition, Bednarz argued that, in the process of joint reflection on their action in collaboration with the researchers, the teachers were co-constructing new knowledge about their practice.

By the end of the project, several professional artefacts had been produced by the teachers: a collection of activities, observation grids, and classroom materials for the school. Jointly, the teachers and researchers produced a book containing mathematical games for first graders, as well as videos of classroom teaching and student engagement in problem solving. Scientific publications were also produced by the researchers based on analyses of video recordings of in-class situations, the records of students' statements, and audio recordings of reflection-oriented meetings between researchers and teachers. Some of these analyses dealt with teaching situations and their potential for stimulating children's learning (Bednarz, 1996; Bednarz, Dufour-Janvier, Poirier, & Bacon, 1993; Poirier & Bacon, 1996); the process of co-construction that took place and the respective contributions (Bednarz, Poirier, Desgagné, & Couture, 2001); and the structuring of a teaching situation over time and the principles that guided this restructuring (Poirier, Bourdage, & Bednarz, 1999).

To close, we note that Bednarz (2004) argued that collaborative research such as that engaged in within this project not only contributed to the growth of knowledge for the research community but also, and equally importantly, to the professional development of the teachers involved. Moreover, she emphasized that the need of the researcher to integrate the practitioner in the construction of practice-related knowledge was based on "the idea of better understanding the reasoning that supports his or her [the teacher's] practice; … the teacher is considered as a partner in the inquiry 'with' whom one looks into the practice, who contributes in joint reflection (with the researcher) to the development of the practice" (p. 6).

The Japanese Example

This fourth example—on which, there are more details in Krainer (2011)—is unique in that it is not an approach initiated by a teacher educator or researcher, but rather is a longstanding, nation-wide approach conducted by teachers for teachers: Japanese lesson study. In their brief history of Japanese *lesson study*, Fernandez and Yoshida (2004) indicated that the origins can be traced back to the early 1900s. In the 1960s, teachers started combining lesson study (*jugyokenkyu*) and school-based inservice professional development (*konaikenshu*). Recognizing the value of *konaikenshu*, in the 1970s the Japanese government started supporting these grassroots activities. This support—small financial and other incentives—still exists today. Lesson study is by far the most common *konaikenshu* activity.

There are manifold versions and sizes of Japanese lesson study. They range from small-scale in-school initiatives with from four to six teachers to large-scale nation-wide ones with hundreds of participants, many travelling long distances. However, a typical *lesson study process* (Fernandez & Yoshida, 2004; see also Hart et al., 2011) contains four to six steps, with a study lesson (*kenkyujugyo*) as the centerpiece of a lesson study (*jugyokenkyu*):

Step 1: Collaboratively planning the study lesson
Step 2: Seeing the study lesson in action
Step 3: Discussing the study lesson
Step 4: Revising the lesson (optional)

Step 5: Teaching the new version of the lesson (optional)
Step 6: Sharing reflections about the new versions of the lesson

Many schools solicit the support of an external advisor (most often instructional superintendents, sometimes experienced teachers on leave, or university staff). Schools often organize their *konaikenshu* work around a lesson study open house (*kokaijugyo*). Here well-developed ideas are shared with visitors (mostly teachers and other educators from neighboring schools). When distinguished guests take part (e.g., an external advisor), their reactions are paid considerable attention, often indicating very clear and pragmatic missions [e.g., Mr. Saeki's statement in Fernandez & Yoshida (2004): "A lesson cannot just start with giving students a problem on a sheet of paper"; teachers need to pay "attention to connecting lessons to students' prior knowledge" (p. 202)]. In many cases, lesson study open houses are followed by a joint celebration in the evening (with a mixture of relaxed socializing and exchanging opinions not articulated at the formal meeting). Some schools even produce written reports about their work (*kenkyukiyo no matome*). In the early 1990s, for example, the National Institute for Educational Research compiled every year over 4,000 reports written by teachers (see Fernandez & Yoshida, 2004, p. 213, referring to Sato, 1992).

The vast majority of elementary schools and many middle schools in Japan conduct *konaikenshu* (in all subjects). In contrast, very few high schools are engaged. In principle, *konaikenshu* activities are voluntary; in reality however, they are regarded as quasi-required. However, and most importantly, many teachers find *konaikenshu*, in particular lesson study, highly beneficial. Three mathematics teachers' opinions might give a flavor of their high regard for lesson studies:

> Developing a great lesson is an ideal thing but I think the best thing about the lesson study experience is that it gives you a chance to reflect about and rethink your own teaching. … I think even if it is a short period of time, having a place where everybody gets together and discusses instruction very seriously is an extremely valuable experience. … Anyway, lesson study can help teachers develop strong relationships, something I think is really important for all teachers. (Fernandez & Yoshida, 2004, p. 17)

It is common for individual teachers to belong to more than one lesson study group. In addition to within-school lesson study groups, autonomous cross-school study groups (regional study groups and teacher clubs) are also organized by teachers or unions (sometimes funded; in most cases membership fees are collected). A system of regular teacher rotations allows lesson study groups to learn from each other.

There are several features that are regarded as *key elements*—and at the same time as *success factors*—of lesson study. Murata (2011) highlighted five key characteristics. Lesson study: is centered on teachers' interests, is student focussed, has a research lesson, is a reflective process, and is collaborative. Further named key elements (see also Fernandez & Yoshida, 2004) are that lesson study: has its roots in strong movements (e.g., child-centred and problem-solving-based learning), regards teaching as a complex and profound enterprise (being not a one-way—and only a didactic—path, but a two-way integration of student ideas and content exploration), is part of a culture of school-based professional development, is a way of enculturing novice teachers by serious academic activity, and is a way of improving yourself by looking at others (*Hito no furi mite waga furi naose*), with no end to improving teaching (indicating a culture of life-long effort and continuous further development). In addition, it should be stressed that lesson study is an autonomous and sustained effort by the teaching profession for the teaching profession. It has a process and also a product dimension (lesson plans and books, indicating a rich body of knowledge), and has created a language of its own (indicating the status of a well-developed profession). It is supported by townships, boards of education, the ministry, etc., indicating a culture of trust in teachers.

While the lesson study movement has become very popular internationally, the way in which it is practised in Japan is quite different from its many applications in Western countries. For example, the recent book *Lesson Study Research and Practice in Mathematics Education. Learning Together*, edited by Hart et al. (2011), addressed research and practice in 16 different locations (mostly in the USA). Due to the lack of experienced lesson study teachers and teacher educators, and lacking prior participation in the whole culture of *konaikenshu* activities, teacher educators act as initiators of lesson studies and support practitioners or student teachers in the practice of lesson study. This is in contrast to the Japanese lesson study approach where teachers themselves are the initiators and school externals (e.g., teacher educators at universities) are invited. Because other countries lack the grassroots teacher movement on which the Japanese lesson-study system builds, the initiating role taken on by school externals in adaptations of lesson study should not be considered too surprising.

The Chinese Example

In China, at the turn of the millennium, the National Mathematics Curriculum Standards (NMCS) were issued, and this ushered in a new set of curriculum guidelines emphasizing creative thinking, problem solving, and mathematical exploration (Huang & Bao, 2006). That document presented a challenge to teachers, who experienced difficulty in implementing these changes, as well as to mathematics educators who wanted to be able to assist in this endeavour. To address the problem, Chinese scholars developed an innovative model of inservice teacher education, called the *Keli* approach.

According to Huang and Bao (2006), development of the new model was to include the following key features:

> First, it is necessary to have expert input in order to upgrade teacher ideas, in a context of peer support; second, it is necessary to include the whole process of action, follow-up, and reflection; and third, it is necessary to form a community, which consists of experts, researchers and teachers. Thus, the program of in-service teacher education, called Xingdong Jiaoyu (Action Education) has been created. In this program, a community consisting of teachers and experts and researchers is formed, and the teachers improve their teaching action and upgrade their professional theory through unfolding the Keli process in cooperation with the members of the community. (p. 284)

Li, Huang, Bao, and Fan (2011) emphasized that innovative approaches to teachers' professional development in China establish direct connections with teachers' practices and what they try to do in their own classrooms. The *Keli* approach is no exception.

The implementation of the *Keli* approach in a school or school district usually unfolds in three phases: (a) familiarization and focussing; (b) a cycle of teaching, reflection, and revision; and (c) disseminating the *Keli* process and the exemplary lesson.

During the first phase, "familiarization and focussing," teachers' approaches are updated and they are introduced to the procedures of developing an exemplary lesson, usually by some experts. Within the Chinese educational system, an expert or master teacher is one who holds a senior rank:

> The conditions for being a senior secondary teacher include 5 years or more serving as a secondary school teacher at the intermediate level or being the holder of a PhD and demonstrating the ability to take the responsibility of senior secondary teacher. Moreover, the candidates should (a) have either systematic and sound fundamental theory and subject content knowledge, plentiful teaching experience and good teaching effectiveness, or specialize in political and moral education and classroom management, and achieve high performance and acquire rich experience; (b) engage in education research on secondary education and teaching and write an experience summary, scientific report, or research paper on the integration of theory and practice at a certain academic level or make remarkable contributions to the improvement of other teachers' academic levels and teaching abilities. (Huang, Li, & He, 2010, p. 295)

At a certain moment a collaborative group, which consists of researchers and interested teachers, is formed. Huang and Bao (2006) provided a couple of examples: one study group consisted of two researchers (one from the District Education Institute and the other from a Teachers College/Normal University) and the teachers from one school; another group consisted of two professors, a PhD holder from the Shanghai Academy of Education Sciences, three PhD candidates from East China Normal University, and researchers from a local educational institute, together with the mathematics teachers from one secondary school. The study group members then decide on a particular research question related to one of the challenging areas of the curriculum, which thereby becomes the focus for the construction of the exemplary lesson.

During the second phase, "teaching, reflection, and revision," an exemplary lesson is developed through a cycle of three teaching stages and two reflection stages. At the outset one teacher—often someone with considerable teaching experience—is selected for all three teaching stages. The first stage involves the teacher designing the lesson by him/herself and then presenting it to a class of students, with all the members of the *Keli* group observing. This is followed by a first feedback meeting immediately after the lesson, which focusses on the teacher's rationale for the design of the lesson, commentary from the group, and suggestions for revision of the lesson. Group members may work together at developing a new and improved version of the lesson.

Following this first reflection stage and the subsequent revision of the lesson, the teacher then presents the lesson to other classes of students at the same grade level within the same school—once again being observed by the *Keli* group. After that second round of teaching, further reflection by the group takes place, which focusses on the promising features of the exemplary lesson and on the differences that remain between its design and what is considered to be effective practice according to the new curricular guidelines. An additional revision is made and a third teaching stage follows.

The third phase, "disseminating the *Keli* process and the exemplary lesson" involves writing a lesson description that can be shared with the public. According to Huang and Bao (2006), this description focusses on the following aspects: "(a) how the learning styles and teaching strategies have been changed in the classroom; (b) how the teacher's conception of teaching and ways of developing a lesson have been updated to meet the new ideas of the new NCMS curriculum; and (c) challenges faced during the process of *Keli* or the reflections occurring during *Keli*" (p. 286). Huang and Bao emphasized, as well, that teachers collaborate with the researchers and university members of the *Keli* group in the writing of the report. Once the report has been completed for publication, a video case study is produced for eventual use in teacher-education programs; it includes the main sections of the lesson, the reflections by members of the group, and an analysis of the lesson in both quantitative and qualitative terms.

In the example of the *Keli* group provided by Huang and Bao (2006), teachers were asked to keep a diary. Some of the commentary that they entered emphasized in particular the value they found in the process of reflecting on the lesson immediately afterwards, revising it, and then redelivering it. One teacher, who was interviewed on this point, stated that, "Traditionally, without follow-up action, the same content will probably be taught 4 years later, so there is only a little impression about how the content was handled before. Then the lesson will be re-designed repeatedly. Nowadays, the lesson plan was revised three times, and the lesson will be observed and reflected two times; it is definitely helpful" (p. 293).

One of the researchers from a Normal University, who had about 10 years teaching experience and held a PhD specializing in mathematics education, reflected on the role he played in this group and in others like it: "It is an important phase to summarize the particular implementation of *Keli*. … At this stage, I usually play a key role in helping them in theorizing and abstracting such as how to organize events to support the main findings, how to effectively organize a paper suitable for publication" (p. 294).

A final issue concerns the commonalities and differences between the *Keli* approach and Japanese Lesson Study. According to Huang and Bao (2006):

> The common features of both Japanese Lesson Study and the Keli process are their common concern with practical issues and the attention both pay to developing a particular lesson through collaborative lesson planning, classroom observation and post-lesson discussion to tackle the particular issues in question. However, the Chinese Keli process emphasizes the expertise stemming from experts, the revision of lesson design and the consequent new action. (p. 295)

Experts thus would seem to have a much more involved role in the *Keli* approach than is the case in Japan where it is the teachers who choose the goals they wish to pursue and the ways of achieving them within their lessons and, in fact, control the entire lesson study process—even if external experts are sometimes invited to join a given Japanese study-lesson sequence.

Discussion: Three Dimensions Central to these Examples

In the first section of this chapter, we situated our perspective on teachers as key stakeholders within two broader contexts, one related to the general notion of stakeholder and the other related to a distinction between teacher as *stakeholder* and teacher as *key stakeholder* in mathematics education research. The term "key stakeholder" was adopted in reference to research where the teacher is considered a co-producer of professional and/or scientific knowledge. In the subsequent presentation of examples drawn from the international corpus of research in mathematics education, we synthesized five cases of research where the teacher participants had a "key stakeholder" role to play. Reflecting upon these examples and focussing on their commonalities and differences allows us now to draw out some of the important dimensions of this research. These dimensions include the following: reflective, inquiry-based activity with respect to teaching action; a significant action-research component accompanied by the creation of research artefacts by the teachers (sometimes assisted by the university researchers); and the dynamic duality of research and professional development.

Reflective, Collaborative, Inquiry-Based Activity with Respect to Teaching Action

All five of the examples presented in the previous section involved sustained reflection on teaching action. Although the specific focus and form of the reflection varied from one example to the other, the importance of this dimension cannot be overemphasized. Let us look more closely at the ways in which reflection was engaged in across the example-set.

The underlying assumption of the USA example was that teachers can improve their practice by studying what they do, learning how to do it better, and sharing their experiences with others in the field. Reflection was considered an essential part of this overall process:

> From the outset of the project, the teacher researchers engaged in many kinds of reflection. Some activities that the teacher researchers cited as provoking especially meaningful reflection included creating belief mappings, juxtaposing their belief mappings with classroom videotapes [of their own teaching], and incorporating ideas from the study-group readings into their own daily practice. (Herbel-Eisenmann, Cirillo, & Otten, 2009, p. 211)

Before beginning the project work, the teachers had not yet made explicit to themselves the beliefs that they thought drove their instructional practice. They were asked to create belief maps, which were a kind of semantic net that described "what was closest to their hearts" when they practised their teaching of mathematics.

According to the university researchers, "the increased awareness gained from developing a belief mapping enabled the teacher researchers to identify what they *wanted* to happen (and why) [in their classrooms] and to continually examine whether what they *wanted* to happen was *actually* happening" (p. 212). The continuous examining and reflecting on their practice in relation to what they had described in their belief mappings, which occurred over the duration of the project, took place largely as the teacher researchers watched and reviewed videotaped lessons. Everyone in the group watched, discussed, and reflected upon the videotapes of all the teacher researchers' classroom teaching, with a particular focus on the discourse of both teacher and students. Teachers talked about how the various forms of reflection they were engaging in were enabling them to transform their thinking about their practice and described their increasing awareness as they constantly revisited their belief mappings throughout the project. In particular, the reflections that were encouraged during the project meetings helped the teachers to develop their own ideas for their action-research projects, of which more will be said shortly.

The joint reflective activity in the Canadian example, which alternated between classroom experience and review of that experience, focussed in particular on the difficulties that arose for the teachers, and for the children, as they attempted to put into practice the novel situations that they had co-constructed during the previous meeting sessions involving teachers and university researchers. This reflective activity often centred on the didactical and pedagogical principles that were underpinning the teachers' practices. For example, the teachers focussed on issues such as having the maximum number of children active, getting the children to be organized, and having the children see different ways of solving a problem and listening to different points of view (Bednarz, 2004). This kind of collective review of their practice then served as a starting point for developing new teaching sequences in the next cycle of reflective activity.

Although the Norwegian example was similar to the Canadian one in that it alternated between school activity where innovation could take place and workshops where both the design of tasks and reflective discussion occurred, the focus of the reflections was somewhat different. At the base of the Norwegian project was the principle of co-learning inquiry: people learning together through inquiry, where both didacticians and teachers were engaged in action and reflection, so as to learn not only something about the world of the other but also more about his or her own world. According to Jaworski (2008), one of the reasons for introducing inquiry as a tool was to challenge the normal state of school mathematics teaching and to question what that teaching was achieving. She emphasized that in an inquiry community, participants are not satisfied with the normal state, but approach their practice with a questioning attitude, "to start to explore what else is possible; to wonder, to ask questions, and to seek to understand by collaborating with others in the attempt to provide answers to them" (p. 314). Thus, teachers' reflections during the workshops centred on questioning, exploring, and seeking alternatives to their usual approaches to teaching mathematics.

One of the distinguishing features of the Chinese approach (sometimes also a part of Japanese lesson study) is the form that the reflection takes—one involving successive iterations of a lesson. Reflections that are based on the observation of a lesson and which focus on how the lesson could be improved, which in turn feed into the revising of the lesson and the teaching of the new version, are then followed by further shared reflections about the new version. According to Huang and Bao (2006), the reflections centre in particular on the promising features of the lesson and on the differences which remain between its design and what is considered to be effective practice according to the new curricular guidelines that emphasize creative thinking, problem solving, and mathematical exploration.

The shared reflections that take place during Japanese lesson study tend to focus on the well-developed foundational principles of Japanese mathematics teaching, such as paying attention to connecting lessons to students' prior knowledge, engaging students intellectually with important mathematics, having clear and explicit goals that address student understanding and performance, and ensuring that a given lesson fits into an overall unit within a specific grade level (Fernandez & Yoshida, 2004; for more discussion of these principles, see Corey, Peterson, Lewis, & Bukarau, 2010).

In all five examples, we noted the role of the discussions and joint activities which served to link teachers' practice to the reflective review of that practice. In some of the examples, these conversations involved teachers and university researchers; in others, teachers with teachers. But in all cases, the reflective activity was used as a vehicle for teachers' clarifying and making explicit certain aspects of teaching practice. It thereby constituted a form of professional development, which is further discussed below.

The Action-Research Dimension: Teachers as Researchers

Action research is generally defined as "systematic inquiry into one's own practice for the purpose of learning about and changing one's practice in order to better support students' learning" (Herbel-Eisenmann, 2009, p. 7; see also Altrichter et al., 2008; Benke, Hospesová, & Tichá, 2008; Krainer, 2006). Action research challenges the assumption that knowledge is separate from and superior to practice. Atweh (2004) has argued that action research serves as a conduit between theory and practice because it bridges the gap between the two. In action research, the production of local knowledge is seen as equally important as general knowledge. All of the examples that are offered in our chapter of this volume present various approaches to action research, the most significant variation being between Japanese lesson study where teachers carry out the activity autonomously with, in some cases, externals (e.g., university researchers) being invited, and the other examples where the university researchers initiate the activity and support teachers engaged in action research.

The most extensive and nationally widespread version of action research by teachers is practised in Japan within the framework of "lesson study" with its systematic reflection of practitioners on action. The teachers in a lesson study context are collaborative researchers who collect data, interpret it, and write down their experiences in papers and books. In many cases, in order to increase the effectiveness of the outcomes or the dissemination of knowledge, experienced others ("critical friends") are invited. Their role varies tremendously. They might participate in order to observe (primarily as learners), to give occasional feedback, to present an invited reaction, to give input, to (co-)investigate students' growth, or to (co-)investigate lesson-study participants' growth. However, in general, lesson study in Japan is initiated, done, reflected, and transferred to written artefacts by teachers for teachers, in an investigative attitude towards their own practice.

Jaworski (2011) in discussing teachers as researchers, distinguished between, on the one hand, research programs in which teachers research their own practice within collaborative teacher practitioner-university didactician groups and, on the other hand, research initiatives by teachers where they are the designers of the research. The example of Japanese lesson study is clearly of the latter type, with teachers designing the research, carrying it out, and producing artefacts to be shared with other teachers. However, the other examples presented within this chapter do not fall neatly into Jaworski's former category. Some traverse the two. For instance, the USA example involved a collaboration of teachers and university researchers. But it was the teachers who selected aspects of their classroom discourse that they wanted to change and then designed and carried out cycles of action research occurring over more than a year, during which time they studied the impact of the changes on students' social and mathematical experiences.

Each teacher in the project then wrote up an account of his/her action-research project in separate chapters of a book which documented the overall project (see Herbel-Eisenmann & Cirillo, 2009). In their action research, the teacher researchers collected their own videotapes and other artefacts of practice and used these to engage in systematic inquiry related to their goals. Their earlier belief-mapping schemas were used as the standards by which the teacher researchers evaluated their own teaching.

Although, within the Canadian example, it was the teachers at the research school who approached the university researchers and asked for their assistance in a project that they themselves initiated, it was not the teachers who designed the research. This was a joint collaborative venture involving both university researchers and teachers. The products of the collaborative action research described in the Canadian example consisted of a collection of activities, of observation grids, and of classroom materials for the school. In addition, several videos related to the situations tested out in class were produced by the university researchers, in collaboration with the teachers; these videos were to serve as material for preservice and inservice teacher education, as well as for a number of research publications written primarily by the university researchers.

Similarly, the research on their own practice that was carried out by the teachers of the Norwegian project was also designed in collaboration with the university researchers with whom they worked, yielding products much like the Canadian example. The Chinese example of teachers researching exemplary lessons, in collaboration with university researchers, also yielded research reports, written jointly by the teacher researchers and the university researchers, and video case studies for eventual use in teacher education programs.

The five examples thus present a picture of action research that includes the co-production of professional and scientific artefacts. The ways in which the action research was carried out, and the artefacts produced, can be characterized as a continuum ranging between two poles: one pole where the work is collaborative and shaped by input from both university researchers and teachers and where researchers and teachers together design, implement, and report findings of their research, but where the university researchers also write additional articles of a scholarly nature; the other pole where teachers collaborate with other teachers doing this work. The USA example was one that clearly straddled both poles with its teacher-initiated action-research studies of an individual nature and teacher-written publications on that research, but within a supportive collaborative framework involving other teachers and university researchers.

The Dynamic Duality of Research and Professional Development

The vision that teachers conducting research constitutes a form of professional development presents a powerful image. In the words of Cochran-Smith and Lytle (1993): "Because teacher research challenges the dominant views of staff development and preservice training as transmission and implementation of knowledge from outside to inside schools, it has the potential to reconstruct teacher development across the professional life span so that inquiry and reform are intrinsic to teaching" (cited in Herbel-Eisenmann, Cirillo, & Males, 2009, p. 219). In an interview just before the USA project came to an end, when teachers were asked how they felt about not being told what to do for their action-research studies, most responded that it was quite different from any of their other professional development experiences. "To have your ideas taken seriously and to be supported in what you think is best over a long time" was, in the words of one of the teacher participants, a foreign but rewarding experience. Although teacher action research is still quite rare in the field of mathematics education, and it is even rarer for it to be viewed as a form of professional development, especially in the USA, the examples presented in this chapter are not unique. In Australia, for example, a model of professional development, titled Improving Teaching Approaches to Mathematics (Pegg & Panizzon, 2011), has been elaborated to underpin the process whereby teachers work collaboratively, with support from university practitioners, in developing and researching strategies to address issues that they have identified and which are relevant to their own teaching contexts. In Austria, several programs have been launched where teachers are supported in carrying out action-research projects, writing reflective papers, and in forming learning communities at their schools or in their districts (see, e.g., Krainer, 2011).

Although much has already been said in this discussion section with respect to the importance of reflective activity within the five examples, its role in relation to professional development has not yet been articulated. Bednarz (2004) drew our attention to the ways in which reflection on action constitutes professional development.

In so doing, she emphasized the relevance of the knowledge that the practitioner constructs and develops throughout the course of his or her teaching experience, which then feeds into the knowledge constructed during the action-research experience. Furthermore, the shared reflection that occurs within the context of the research experience, with or without the university researcher's contribution (as is quite often the case with Japanese lesson study), renders explicit the knowledge that might otherwise remain implicit. In the group construction process, a variety of resources are brought into play, all of them nourishing the professional development that is inherent to the situation—professional development that, according to Bednarz, is as significant for the university researcher as for the teacher. More specifically and based on her experience with the Canadian project, Bednarz (2004) noted the following components of the process of collaborative research that she viewed as contributing to the teachers' professional development:

- A deeper reflection on mathematical content (learning situated in practice), where teachers have the opportunity, during the discussions around the teaching situations and the productions of children, to improve their understanding of the mathematical concepts at play;
- A new awareness of the nature of mathematical activity, where the collaborative research process is also the occasion to debate what mathematical activity means;
- New ways to look at children's statements, where teachers have the opportunity, during the discussions on the teaching situations and productions of students, to develop new ways to look at children's productions, to take some distance, to consider different ways to solve a problem;
- Reflection on the didactical variables involved in a given task and their influence, where the analysis of tasks moves away from superficial aspects and towards student reasoning, thereby encouraging the seeing of complexity;
- Teaching strategies, where the arguments underlying decision making are rendered explicit, thereby opening up other points of view;
- An evolving relationship to the teaching of mathematics, where a changing relationship with teaching "know-how" is encouraged.

Jaworski (2008) has described, in relation to the Norwegian project, the professional development that occurred both for the teacher researchers and the university researchers (didacticians): "For example, teachers suddenly came to see, through their study of students' thinking and activity in algebra, how they could explore in their school environment ways to develop teaching and learning; didacticians saw the nature of a task that could lead to teachers' effective recognition of the nature of school goals for students' development and learning in mathematics" (p. 326).

The fact that the professional development that takes place in these types of projects occurs not just for the teachers but also for the university researchers of the project is a very important point. The initiators of these projects (usually university researchers) also experience professional development and growth in these collaborative research efforts—that part is seldom carefully documented or written about. More recently, Makar and O'Brien (2012) discussed the transformative nature of collaborative research, the changes in identity, and the growth in the participation and perspectives of both teachers and researcher that developed over a 6-year, design research project on inquiry-based teaching. The teachers in Makar and O'Brien's project experienced an "identity renegotiation" as they became aware of and then acknowledged their research contribution to the project. Meanwhile, the researcher documents her own professional growth as a collaborative researcher, and what she is learning from the teachers in the project. Makar and O'Brien refer to this as *reflexivity,* the joint contributions and joint benefits of teachers and researchers engaged in collaborative research.

In the Japanese example, the research that is associated with the lesson study process goes hand in hand with professional development and is in fact part of the culture of school-based professional development. The professional development aspect of lesson-study activity is also captured by one of its key elements in that it is viewed as "a way of improving yourself by looking at others (*Hito no furi mite waga furi naose*), with no end to improving teaching (indicating a culture of life-long effort and continuous further development)" (Murata, 2011, p. 10).

In the Chinese example, the direct link between professional development and research involving teachers' practices and what they try to do in their own classrooms was an explicit focus, according to Li, Huang, Bao, and Fan (2011). More specifically, the entire research process of action, follow-up, and reflection, as well as the necessity of forming a community consisting of experts, researchers and teachers, is considered integral to the professional development approach adopted in China.

In their reflective discussions and their written research chapters, the teacher researchers in the USA project identified three major factors that transformed their own practice with regard to discourse in their classrooms and which constituted a form of professional development for them: (a) the influence of the readings and research literature, (b) the importance of reflection by the teachers—both in study-group discussions and written reflections in journal entries, and (c) the power available within a collaborative community of teachers to support one another in this kind of effort by teacher researchers. The creation of belief maps and subsequent opportunities to reflect on the videotapes they made of their practice proved to be transformative for the teacher researchers. Just seeing the data alone was not sufficient to change practice—the teacher researchers said that opportunities to reflect and to discuss with the study group whether those beliefs were actually being implemented in their classrooms was critical to making changes in their practice.

Extrapolating from the research by Herbel-Eisenmann and her colleagues suggests that, for professional development to have the potential to help teachers transform their practice, consideration of whether the following conditions are in place would be useful. Having a supportive, safe, community for the teacher researchers to share and discuss, maintained over a very long period of time, was clearly a decisive piece in this research effort. In addition to the safe harbour of the community of practice, the opportunity was provided for the teacher researchers to select from a collection of thoughtfully chosen readings that linked to the project goals and to their own practice. Open discussions and analyses of the video data were conducted jointly during group meetings of the university and teacher researchers. And finally, these teacher researchers had the opportunity to write their own stories in their own ways, supported by the university researchers in the process. The teachers' voice was crucial to the success of the work in this project.

Closing Remarks

This chapter has attempted to close the distinctive gap between research and practice that exists in much of the mathematics education research literature by viewing teachers as key stakeholders in research—stakeholders who co-produce professional and scientific knowledge—rather than as "recipients of research," and sometimes even "means" to generate or disseminate knowledge. We presented five examples, drawn from individual and nation-wide projects around the world, examples that offered the potential to link research and practice in clear and explicit ways. Our analysis of these projects revealed three salient dimensions to research where the teacher is considered a key stakeholder: (a) teacher reflection, (b) teachers in the role of researchers themselves, and (c) the multi-leveled professional development experience within the research process for both teacher researchers and university researchers. The (co-)production of professional and scientific knowledge, which cut across all three of these dimensions in the examples presented, is considered a critical aspect of the notion of the "teacher as key stakeholder" in research, an aspect to which we now briefly return.

The (co-)production of professional and scientific knowledge is clearly linked with writing papers and thus making one's findings open for public discussion and critique (Krainer, 2006). In general, this is rather more difficult for teachers than for teacher educators and researchers who live in a "culture of publishing." Despite the diversity between teachers' and researchers' worlds, discussed earlier in this chapter, all five of the approaches that were presented were able to bridge these worlds and, as well, succeeded in promoting teachers' writing down of the findings of their inquiries and investigations. This promotion was done for several reasons: systematic reflection by teachers on their own work creates new knowledge which in turn positively influences their (future) teaching and enhances the quality of teaching. Writing down is an additional opportunity to learn; written artefacts increase the opportunities for communicating and cooperating with interested people (teachers, theoreticians, administrators); written artefacts help to make teachers' professional knowledge more visible and accessible, and thus contribute to the further development of the teaching profession as a whole; these artefacts also give teacher educators and researchers an additional opportunity to learn from teachers. Teachers' own investigations increase their interest in research, in reading research papers, and in collaborating in research projects, thus building further bridges between research and practice.

The challenge now for all of us in the international mathematics education community is to consider how further to promote and systematize collaborative research work among teachers, with or without university researchers, in ways that will reflect and build upon what has been documented in the five examples presented in this chapter. Given the potential for professional growth from the expanded roles for both classroom teachers and researchers alike, and the growing documentation of the long-term benefits for researchers, teachers, and their students from such collaborative research, a case can be made that all countries should consider implementing a *systematic* integration of linked research and practice. Collaborative research with teachers has heretofore arisen on a case-by-case basis, and somewhat haphazardly, especially in the western countries where it has occurred. We feel that every country could benefit by implementing its own national commitment to linked inquiry. As has been illustrated in examples discussed in this chapter, promoting a national effort and national discourse around creating stronger links between research and practice is not only possible, but can also be rewarding for all concerned. These examples can thus serve both as inspiration and model for truly bridging the gap between mathematics education research and practice. The crucial element is to regard *researchers as key stakeholders in practice* and *teachers as key stakeholders in research*.

Acknowledgments We are grateful to Nadine Bednarz, Beth Herbel-Eisenmann, Barbara Jaworski, Minoru Ohtani, Rongjin Huang, Jiansheng Bao, and many others, for the information they have provided, either directly or indirectly, about the projects and programs presented in this chapter. We also appreciate the feedback received from the reviewers and editors on earlier versions of this chapter.

References

Altrichter, H., Feldman, A., Posch, P., & Somekh, B. (2008). *Teachers investigate their work; An introduction to action research across the professions* (2nd ed.). London, UK: Routledge [German original: Altrichter, H. & Posch, P. (1990). *Lehrer erforschen ihren Unterricht*. Bad Heilbrunn, Germany: Klinkhardt. Chinese translation 1997, Taipei, Taiwan: Yuan-Liou.].

Atweh, B. (2004). Understanding for changing and changing for understanding. Praxis between practice and theory through action research in mathematics education. In P. Valero & R. Zevenbergen (Eds.), *Researching the socio-political dimensions of mathematics education: Issues of power in theory and methodology* (pp. 187–206). New York, NY: Kluwer Academic Publishers.

Bazzini, L. (Ed.). (1994). *Proceedings of the 5th International Conference on Systematic Cooperation between Theory and Practice in Mathematics Education (SCTP 5) in Grado: Theory and Practice in Mathematics Education*. Pavia, Italy: ISDAF.

Bednarz, N. (1996). Language activities, conceptualization and problem solving: The role played by verbalization in the development of mathematical thought by young children. In H. M. Mansfield, N. A. Pateman, & N. Bednarz (Eds.), *Mathematics for tomorrow's young children: International perspectives on curriculum* (pp. 228–239). Dordrecht, The Netherlands: Kluwer Academic Publishers.

Bednarz, N. (2004). Collaborative research and professional development of teachers in mathematics. In M. Niss & E. Emborg (Eds.), *Proceedings of the 10th International Congress on*

Mathematical Education (CD version) (pp. 1–15). Copenhagen, Denmark: IMFUFA, Roskilde University.

Bednarz, N., Dufour-Janvier, B., Poirier, L., & Bacon, L. (1993). Socioconstructivist viewpoint on the use of symbolism in mathematics education. *The Alberta Journal of Educational Research, 29*(1), 41–58.

Bednarz, N., Poirier, L., Desgagné, S., & Couture, C. (2001). Conception de séquences d'enseignement en mathématiques: Une nécessaire prise en compte des praticiens. In A. Mercier, G. Lemoyne, & A. Rouchier (Eds.), *Le génie didactique: Usages et mésusages des theories de l'enseignement* (pp. 43–70). Brussels, Belgium: De Boeck Université.

Benke, G., Hospesová, A., & Tichá, M. (2008). The use of action research in teacher education. In K. Krainer & T. Wood (Eds.), *Participants in mathematics teacher education: Individuals, teams, communities and networks (International handbook of mathematics teacher education)* (Vol. 3, pp. 283–307). Rotterdam, The Netherlands: Sense Publishers.

Bromme, R. (1992). *Der Lehrer als Experte. Zur Psychologie des professionellen Wissens*. Bern, Switzerland: Huber.

Cochran-Smith, M., & Lytle, S. L. (1993). *Inside/outside: Teacher research and knowledge*. New York, NY: Teachers College Press.

Corey, D. L., Peterson, B. E., Lewis, B. M., & Bukarau, J. (2010). Are there any places that students use their heads? Principles of high-quality Japanese mathematics instruction. *Journal for Research in Mathematics Education, 41*, 438–478.

Crawford, K., & Adler, J. (1996). Teachers as researchers in mathematics education. In A. Bishop, K. Clements, C. Keitel, J. Kilpatrick, & C. Laborde (Eds.), *International handbook of mathematics education* (pp. 1187–1205). Dordrecht, The Netherlands: Kluwer Academic Publishers.

Davis, R. B., Jockusch, E., & McKnight, C. (1978). Cognitive processes in learning algebra. *Journal of Children's Mathematical Behavior, 2*(1), 10–320.

Debien, J. (2010). *Répertorier les modalités favorisant une démarche de développement professionnel chez les enseignants de mathématique de niveau secondaire* (Master's thesis). Université du Québec à Montréal, Département de mathématiques.

Even, R., & Ball, D. L. (Eds.). (2003). Connecting research, practice and theory in the development and study of mathematics education. *Educational Studies in Mathematics* (special issue), *54*(2–3).

Fernandez, C., & Yoshida, M. (2004). *Lesson study: A Japanese approach to improving mathematics teaching and learning*. Mahwah, NJ: Lawrence Erlbaum (Reprint by Routledge, 2009).

Freeman, R. E. (1984). *Strategic management. A stakeholder approach*. Boston, MA: Pitman.

Freeman, R. E. (2004). The stakeholder approach revisited. *Zeitschrift für Wirtschafts-und Unternehmensethik (zfwu), 3*(5/2004), 228–241.

Fullan, M. (1993). *Change forces. Probing the depths of educational reform*. London, UK: Falmer Press.

Goodchild, S., & Jaworski, B. (2005). Identifying contradictions in a teaching and learning development project. In H. L. Chick & J. L. Vincent (Eds.), *Proceedings of the 29th Conference of the International Group for the Psychology of Mathematics Education* (Vol. 3, pp. 41–47). Melbourne, Australia: International Group for the Psychology of Mathematics Education.

Hart, L. C., Alston, A., & Murata, A. (Eds.). (2011). *Lesson study research and practice in mathematics education: Learning together*. Dordrecht, The Netherlands: Springer.

Hattie, J. A. (2003, December). Teachers make a difference: What is the research evidence? *Building teacher quality: What does the research tell us?* Paper presented at a Conference held at the Australian Council for Educational Research. Retrieved from http://research.acer.edu.au/research_conference_2003/4.

Herbel-Eisenmann, B. (2009). Introduction to the project, the people, and the reflective activities. In B. Herbel-Eisenmann & M. Cirillo (Eds.), *Promoting purposeful discourse. Teacher research in mathematics classrooms* (pp. 3–28). Reston, VA: National Council of Teachers of Mathematics.

Herbel-Eisenmann, B. (2010). Discourse analysis: A catalyst for reflective inquiry in mathematics classrooms. In *Linking research and practice: The NCTM Research Agenda Conference report* (pp. 36–37). Reston, VA: National Council of Teachers of Mathematics.

Herbel-Eisenmann, B., & Cirillo, M. (Eds.). (2009). *Promoting purposeful discourse: Teacher research in mathematics classrooms*. Reston, VA: National Council of Teachers of Mathematics.

Herbel-Eisenmann, B., Cirillo, M., & Males, L. (2009). An argument for taking up similar work. In B. Herbel-Eisenmann & M. Cirillo (Eds.), *Promoting purposeful discourse. Teacher research in mathematics classrooms* (pp. 219–232). Reston, VA: National Council of Teachers of Mathematics.

Herbel-Eisenmann, B., Cirillo, M., & Otten, S. (2009). Synthesizing the bases of purposeful discourse: Reading, reflecting, and community. In B. Herbel-Eisenmann & M. Cirillo (Eds.), *Promoting purposeful discourse: Teacher research in mathematics classrooms* (pp. 205–217). Reston, VA: National Council of Teachers of Mathematics.

Huang, R., & Bao, J. (2006). Towards a model for teacher professional development in China: Introducing *Keli*. *Journal of Mathematics Teacher Education, 9*, 279–298.

Huang, R., Li, Y., & He, X. (2010). What constitutes effective mathematics instruction? A comparison of Chinese expert and novice teachers' views. *Canadian Journal of Science, Mathematics and Technology Education, 10*, 293–306.

Jaworski, B. (2006). Developmental research in mathematics teaching and learning: Developing learning communities based on inquiry and design. In P. Liljedahl (Ed.), *Proceedings of the 2006 Annual Meeting of the Canadian Mathematics Education Study Group* (pp. 3–16). Calgary, Canada: CMESG.

Jaworski, B. (2008). Development of mathematics teacher educators and its relation to teaching development. In B. Jaworski & T. Wood (Eds.), *The mathematics teacher educator as a developing professional (International handbook of mathematics teacher education)* (Vol. 4, pp. 335–361). Rotterdam, The Netherlands: Sense Publishers.

Jaworski, B. (2011). Situating mathematics teacher education in a global context. In N. Bednarz, D. Fiorentini, & R. Huang (Eds.), *International approaches to professional development of mathematics teachers* (pp. 2–51). Ottawa, Canada: Presses de l'Université d'Ottawa.

Jaworski, B., Fuglestad, A.-B., Bjuland, R., Breiteig, T., Goodchild, S., & Grevholm, B. (2007). *Learning communities in mathematics*. Bergen, Norway: Caspar Forlag.

Jaworski, B., & Goodchild, S. (2006). Inquiry community in an activity theory frame. In J. Novotná, H. Moraova, M. Kratka, & N. Stelikova (Eds.), *Proceedings of the 30th Conference of the International Group for the Psychology of Mathematics Education* (Vol. 3, pp. 353–360). Prague, Czech Republic: International Group for the Psychology of Mathematics Education.

Jaworski, B., & Wood, T. (Eds.). (2008). *The mathematics teacher educator as a developing professional (International handbook of mathematics teacher education)* (Vol. 4). Rotterdam, The Netherlands: Sense Publishers.

Kieran, C. (2007). Learning and teaching algebra at the middle school through college levels: Building meaning for symbols and their manipulation. In F. K. Lester Jr. (Ed.), *Second handbook of research on mathematics teaching and learning* (pp. 707–762). Greenwich, CT: Information Age Publishing.

Kirshner, D., & Awtry, T. (2004). Visual salience of algebraic transformations. *Journal for Research in Mathematics Education, 35*, 224–257.

Krainer, K. (2006). Action research and mathematics teacher education. *Journal of Mathematics Teacher Education, 9*, 213–219.

Krainer, K. (2011). Teachers as stakeholders in mathematics education research. In B. Ubuz (Ed.), *Proceedings of the 35th Conference of the International Group for the Psychology of Mathematics Education* (Vol. 1, pp. 47–62). Ankara, Turkey: International Group for the Psychology of Mathematics Education.

Krainer, K., & Llinares, S. (2010). Mathematics teacher education. In P. Peterson, E. Baker, & B. McGaw (Eds.), *International encyclopedia of education* (Vol. 7, pp. 702–705). Oxford, UK: Elsevier.

Li, Y., Huang, R., Bao, J., & Fan, Y. (2011). Facilitating mathematics teachers' professional development through ranking and promotion practices in the Chinese mainland. In N. Bednarz, D. Fiorentini, & R. Huang (Eds.), *International approaches to professional development of mathematics teachers* (pp. 72–87). Ottawa, Canada: Presses de l'Université d'Ottawa.

Little, J. W. (1982). Norms of collegiality and experimentation: Workplace conditions of school success. *American Education Research Journal, 19*, 325–340.

Makar, K., & O'Brien, M. (2012). Blurring the boundaries: The transformative nature of research participation. In W. Midgley, P. A. Danaher, & M. Baguley (Eds.), *The role of participants in education research: Ethics, epistemologies, and methods*. London, UK: Routledge.

Matz, M. (1982). Towards a process model for high school algebra errors. In D. Sleeman & J. S. Brown (Eds.), *Intelligent tutoring systems* (pp. 25–50). London, UK: Academic Press.

Murata, A. (2011). Introduction: Conceptual overview of lesson study. In L. C. Hart, A. Alston, & A. Murata (Eds.), *Lesson study research and practice in mathematics education. Learning together* (pp. 1–12). Dordrecht, The Netherlands: Springer.

National Council of Teachers of Mathematics. (2010). *Linking research and practice: The NCTM Research Agenda Conference Report*. Reston, VA: Author.

OECD. (2011). *Strong performers and successful reformers in education*. Paris, France: Author.

Pegg, J., & Krainer, K. (2008). Studies on regional and national reform initiatives as a means to improve mathematics teaching and learning at scale. In K. Krainer & T. Wood (Eds.), *Participants in mathematics teacher education: Individuals, teams, communities and networks (International handbook of mathematics teacher education)* (Vol. 3, pp. 255–280). Rotterdam, The Netherlands: Sense Publishers.

Pegg, J., Lynch, T., & Panizzon, D. (2007). *An exceptional schooling outcomes project: Mathematics*. Brisbane, Australia: Post Press.

Pegg, J., & Panizzon, D. (2011). Collaborative innovations with rural and regional secondary teachers: Enhancing student learning in mathematics. *Mathematics Education Research Journal* (special issue on "Mathematics Education in Rural Schools: Evidence-Based Approaches"), *23*(2). doi:10.1007/s13394-011-0009-0.

Poirier, L., & Bacon, L. (1996). Interaction between children in mathematics class: An example concerning the concept of number. In H. M. Mansfield, N. A. Pateman, & N. Bednarz (Eds.), *Mathematics for tomorrow's young children: International perspectives on curriculum* (pp. 166–174). Dordrecht, The Netherlands: Kluwer Academic Publishers.

Poirier, L., Bourdage, N., & Bednarz, N. (1999). Un lien possible entre la recherche en didactique des mathématiques et la pratique de classe: la recherche collaborative. In F. Jacquet (Ed.), *Les liens entre la pratique de la classe et la recherche en didactique des mathématiques. Actes de la CIEAEM 50* (pp. 193–197). Neufchâtel, Switzerland: Commission Internationale pour l'Etude et l'Amélioration de l'Enseignement des Mathématiques.

Reynolds, D., Creemers, B., Stringfield, S., Teddlie, C., & Schaffer, G. (Eds.). (2002). *World class schools. International perspectives on school effectiveness*. London, UK: Routledge.

Sato, M. (1992). Japan. In H. B. Leavitt (Ed.), *Issues and problems in teacher education: An international handbook*. New York, NY: Greenwood Press.

Schön, D. (1983). *The reflective practitioner: How professionals think in action*. London, UK: Temple-Smith.

Shearer, B. A., Lundeberg, M. A., & Coballes-Vega, C. (1997). Making the connection between research and reality: Strategies teachers use to read and evaluate journal articles. *Journal of Educational Psychology, 89*, 592–598.

Skovsmose, O., & Säljö, R. (2007). *Report on the KUL-projects: Learning Communities in Mathematics and ICT in mathematics learning*. Retrieved from http://www.navimat.dk/uploads/39600/Report-KULfinal_OSK_RS_okt07.pdf.

Steinbring, H. (1994). Dialogue between theory and practice in mathematics education. In R. Biehler, R. W. Scholz, R. Sträßer, & B. Winkelmann (Eds.), *Didactics of mathematics as a scientific discipline* (pp. 89–102). Dordrecht, The Netherlands: Kluwer Academic Publishers.

Stenhouse, L. (1975). *An introduction to curriculum research and development.* London, UK: Heinemann.

Wagner, J. (1997). The unavoidable intervention of educational research: A framework for reconsidering research-practitioner cooperation. *Educational Researcher, 26*(7), 13–22.

Zeuli, J. S. (1994). How do teachers understand research when they read it? *Teaching and Teacher Education, 10*, 39–55.

Chapter 13
Teachers Learning from Teachers

Allan Leslie White, Barbara Jaworski, Cecilia Agudelo-Valderrama, and Zahra Gooya

Abstract There is much debate within mathematics teacher education over ways in which professional and academic foci could be made to complement each other. On the one hand, teachers' craft knowledge is emphasized, mainly as this relates to the particular and local level of teaching; on the other hand, the importance of academic subject knowledge cannot be denied. In this chapter the focus will be on how to blend and balance the two through activities in which teachers learn from other teachers, particularly the co-learning of teachers and teacher educators. It will discuss professional relationships, reflective practice, community building, and research in practice. Examples of research-based programs involving *lesson study* (LS) and the *Learner's Perspective Study* (LPS) have moved the relevant research in this area to yet another level, in which theory and practice are combined. Projects such as these and others from diverse parts of the world will be presented and discussed.

Introduction

Teaching is generally regarded as a complex and demanding profession that requires a mixture of subject knowledge together with theoretical and practical knowledge, skills and understandings. Teacher learning may originate from personal

---

A. L. White (✉)
University of Western Sydney, Sydney, NSW, Australia
e-mail: al.white@uws.edu.au

B. Jaworski
Loughborough University, Leicestershire, UK

C. Agudelo-Valderrama
CONACES—Ministerio de Educación Nacional, Colombia, Bogota, Colombia

Z. Gooya
Shahid Beheshti University, Tehran, Iran

reflections on classroom experiences, professional readings, and other sources. However, the variations in teachers' learning sources have not been systematically documented and thus have had little input into the wider collective knowledge and theoretical underpinnings of teaching. Yet there exists a body of theoretical and teaching craft knowledge that is available to teachers (see, e.g., Wood, Jaworski, Krainer, Tirosh, & Sullivan, 2008). As well, focussing on a teacher's knowledge base reveals a multi-faceted, multi-sourced, highly interconnected mix that has defied the formation of widely accepted, common comprehensive frameworks. The confounding issues are whether and to what degree this knowledge is "private knowledge based on personal experience and only in the personal realm of thinking and acting," or is "knowledge coming from and staying in practice," or is "discursively generated, shared, and general knowledge" (Neubrand, Seago, Agudelo-Valderrama, DeBlois, & Leikin, 2009, p. 211).

There is a need to clarify the difference between teachers' theoretical knowledge and knowledge that arises from the teaching experience. It is common in education literatures for the term "craft knowledge" to be used to encapsulate the professional action-oriented knowledge used by teachers in their classroom teaching (Cooper & McIntyre, 1996).

> Craft knowledge describes the knowledge that arises from and, in turn, informs what teachers do. As such, this knowledge is to be distinguished from other forms of knowledge that are not linked to practice in this direct way ... Neither is it knowledge drawn from theoretical sources. Professional craft knowledge can certainly be (and often is) informed by these sources, but it is of a far more practical nature than these knowledge forms. Professional craft knowledge is the knowledge that teachers develop through the processes of reflection and practical problem-solving that they engage in to carry out the demands of their jobs. (p. 76)

In contrast, theoretical knowledge generally lays down principles and frameworks derived from research studies that are often replicable and can be generalized to other contexts. This kind of knowledge is less focussed on the individual teacher or on small practical details required for teaching. Research has sought to identify and articulate the types of professional knowledge that a successful teacher would need. The seminal work of Shulman (1986) and colleagues proposed that a basis of professional knowledge would contain: (a) content knowledge both substantive and syntactic; (b) general pedagogical knowledge including generic principles of classroom management; (c) curriculum knowledge including materials and programs; (d) pedagogical content knowledge that for a given subject area included forms of representation, concepts, useful analogies, examples and demonstrations; (e) knowledge of learners; (f) knowledge of educational contexts, communities and cultures; and (g) knowledge of educational purposes.

A number of researchers have reflected upon Shulman's work in their studies regarding teachers' learning. For instance, Even and Tirosh (2008) claimed that in coining the term pedagogical content knowledge, Shulman contributed greatly to the discussion of what teachers needed to know about students' mathematical learning. On the other hand, pedagogical content knowledge has been the subject of much debate, particularly regarding its epistemological status (Ponte & Chapman, 2008). Although Shulman's work provided a suitable beginning for the growth of a

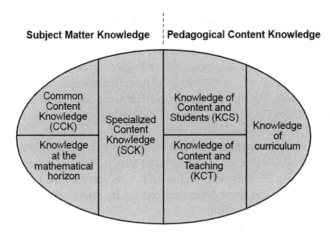

Figure 13.1. Domain map for mathematical knowledge for teaching (from Hill, Ball, & Schilling, 2008, p. 377).

framework, there has been considerable development by other researchers. For example, Hill, Ball, and Schilling (2008), in seeking to conceptualize the domain of effective teachers' unique knowledge of students' mathematical ideas and thinking, proposed the domain map for mathematical knowledge for teaching shown in Figure 13.1.

Tim Rowland and his colleagues (see, e.g., Rowland, 2008, 2009; Rowland, Huckstep, & Thwaites, 2005) suggested a framework that had four domains of knowledge: Foundation, Transformation, Connection and Contingency. This framework, which Rowland dubbed the "knowledge quartet," developed as the result of an analysis of data gathered from observations of prospective teachers, and it has now been applied to the work of practising teachers (Rowland, 2009). It drew attention to the importance of a teacher's knowledge at any given time, and also to the teacher's development of knowledge over time.

Shulman's (1986, 1987) categorization can be contrasted with the European focus on *the didactics of mathematics* (*didactique*), which is concerned with theoretical and practical issues surrounding mathematics curricula and teaching, and their relationships with learning. The European emphasis is on designing didactical situations which acknowledge and incorporate important transitions from mathematics itself to the ways in which that mathematics is brought to students in educational contexts.

Kilpatrick (2003) reported that *didactique* went beyond the art and science of teaching to include: learning and school systems; an intensive common epistemological analysis given to mathematical concepts and a shared methodology that is not to be found in U.S. research; the use of mathematics more extensively as a source of metaphors; a mode of analysis which proceeds from elaborate, a priori analyses to experimentation in the classroom. Sustained attention is to be given to classroom teaching and to the social context in which teaching and learning occur. It focusses intrinsically on the missing element for which Shulman introduced the term pedagogical content knowledge. There is considerable available research, conducted over several decades, focussing, in France, on didactical and adidactical situations (Brousseau, 1992), in Germany on the epistemological nature of teachers' thinking (Steinbring, 1998), and in the Netherlands, among researchers at the Freudenthal Institute (Gravemeijer, 1994a, 1994b, 2000; van den Heuvel-Panhuizen, 2001), on *Realistic Mathematics Education*.

The mixture of theoretical and practical learning that forms a teacher's knowledge base is open to many influences—from other teachers, friends, experiences in and out of school, subject associations, teacher-education programs, acknowledged professional experts, etc. The idea of teachers learning from teachers can conjure many different images. It is evident in the literature that there is no consensus regarding the use of terms such as teacher professional development and teacher professional learning. The terms are often used interchangeably and with little or no definition of their meanings (Even, 2008). Clements (2008) was critical of attempts to define professional development as the planned, formal activities and programs that teachers undertake to extend their professional learning, and professional learning as the individual growth of a teacher's expertise. Professional development can be the result of numerous activities that are neither planned nor ostensibly formal—such as classroom experiences, reading, and informal activities and experiences.

The term "professional development" of teachers has often, in the past, implied a deficit view of teachers, emphasizing elements of knowledge which teachers lack, or ways in which teachers need to be developed (Dawson, 1999; Hoyles, 1992; Ponte, 1994). The implication is that people who have access to theoretical knowledge (possibly teacher educators, or didacticians) are in a position to remedy the deficits in teaching by changing the practices of teachers. Such a view is simplistic, implying that those with the theoretical knowledge could translate that knowledge into classroom practice if given the opportunity to do so. It ignores the complexities of teaching practice—there are many factors which influence what teachers can do in the educational settings in which they work.

Simon (2008) wrote about two kinds of commonly available programs which influence learning and development of practising teachers—programs which focus on content *and* process, and those which are solely *process based*. According to Simon, programs which focus on content and process include "courses and workshops for teachers in which teacher educators aim to promote particular mathematical and pedagogical concepts, skills and dispositions" (p. 18). They can be considered as professional development programs in which teacher educators have an agenda for the learning of "participating teachers" (for example, for participants to become aware of research on students' strategies and errors in the teaching of algebra). With such programs it is typically assumed that participants' personal and professional learning will be stimulated. The process only category, according to Simon, includes programs such as *lesson study* (LS) and inquiry-based models, on which more will be said upon later in this chapter. In this category we would also include *developmental research* programs in which it is expected that participating teachers' engagement in research (possibly in partnership with teacher educators) will contribute to the improvement of teaching.

Teachers belong to communities situated within and around schools, and to educational systems created by the societies and cultures to which the schools belong (Wenger, 1998). The teachers construct professional identities within these sociocultural and historical settings in line with the norms and expectations which prevail. The kinds of planned professional development programs that would be expected to occur in these settings and practices would include Simon's two categories of professional learning programs. Participation would give rise to situated learning arising from everyday interactions within particular environments. In all of these cases, a teacher's development would most likely be related not only to the programs but also to that teacher's prior knowledge and experience.

The study of mathematics teachers and mathematics teaching, and associated learning outcomes, will be resumed later in this chapter with the presentation of a range of programs relating to Simon's two categories. That further discussion will highlight the concomitant learning of teacher educators who work with teachers for the purpose of developing the quality of teaching and, therefore, learning. Before resuming, it will be appropriate to briefly discuss the forces and influences that exert pressure on the nature and delivery of these programs.

Local, National, and Global Influences on Teachers, Teaching and Learning

Professional development programs and teacher professional learning are influenced in varying degrees by research across the field of education. These programs will be influenced by a mix of international, national and local research pressures and initiatives, the actual mix depending on contexts and personnel in any particular place at any particular time. An international influence could be the result of globalization; a national influence could be a perceived need to conform to a national standards document; and a local influence could be a school principal's desire to adopt an outcomes-based education approach within a school.

With the growth in communication technologies and stimulus to information flow, and the increased ease of overseas travel, it is common to hear that we live in a global world, and that the world has become a global village. Globalization has become a familiar, albeit imprecise, term associated with multiple and significant changes currently happening in all areas of social life (English, 2008; Stromquist & Monkman, 2000). Not surprisingly, education is also subject to forceful changes arising from globalization, particularly when the focus is on information flow and the possibilities for world-wide communication. Research in mathematics education is a global enterprise and as such is caught up in the wider movements that influence all educational research. Other chapters in this *Handbook* make it abundantly clear how Trends in International Mathematics and Science Study (TIMSS) and the Programme of International Student Assessment (PISA) are examples of programs whose influence has speedily transcended national boundaries.

English (2008), in her introduction to a handbook on research in mathematics education, stated: "In recent years, we have seen a major shift within the field of mathematics education from a mainly psychological and pedagogical perspective to one that encompasses the historical, cultural, social, and political contexts of both mathematics and mathematics education" (p. 4). It should be noted that globalization is not always equitable in that family and other local conditions can restrict access to information coming from, say, the Internet. This has been felt particularly in the experience of one of our authoring team who had difficulty accessing articles relating to this chapter because they appeared in books which were not available in her country.

One of the main difficulties in the dissemination of knowledge to teachers in some countries is the lack of an agreed means. We provide a concrete example from Iran, involving the quarterly journal *Roshd: Mathematics Education Journal*, which is one of 16 subject-bounded journals and 15 general magazines titled "Roshd" published by the Ministry of Education in Iran. One of the authors of this present chapter, Gooya, is the editor of the *Mathematics Education Journal*. Since 1996, a special section, titled "Teachers' Narrative," has been included in the *Journal* in order to disseminate the research findings of teacher researchers arising mainly from action-research projects conducted either locally or at district level. Teachers were also encouraged and assisted through personal communications to write scholarly papers, which were included in the *Journal*. Such publications sometimes generated workshops at annual national mathematics education conferences. The *Journal* had another section called "Viewpoints" in which teachers could share their ideas and receive feedback from their colleagues. The number of teachers communicating with this journal dropped sharply during the 2009–2010 academic year and this trend has continued. The editorial board investigated the reasons for the dramatic change and found that the formal educational system had announced that teachers could not get credit for their professional promotion by publishing in this or other similar journals. They could only get credit by publishing in university journals or journals of scientific societies approved by the Ministry of Science, Research and Technology (which is responsible for higher education and any forms of tertiary education). Thus, a single act by authorities could deny teachers the opportunities offered by the journals for disseminating their practical or craft knowledge.

It is interesting that this same kind of influence has been a reality in western academic circles for many years—where it is well known that getting a publication in a "top" journal (like, for example, *Journal for Research in Mathematics Education*, or *Educational Studies in Mathematics*) would be likely to "count" towards promotion, but a publication in a local "teaching" periodical would not. The message implicitly conveyed has been that publication in a peer-reviewed *research* journal is more important than publication in a periodical for which the readership is mainly school teachers.

Global forces should not be all powerful and should not completely mould local contexts into uniform shapes—that is because global forces do not take account for local realities. Education researchers have highlighted problems in adopting global programs because "pedagogical methods are culturally embedded, and transplanting them from one culture to another is not always feasible" (Hatano & Inagaki, 1998, p. 101). That said, there can be little doubt that, increasingly, local education contexts are being influenced by local, state and national authorities. For example, Japan, Malaysia and the UK have mandatory national curricula. Australia and the USA do not, but both may be moving towards getting one (see, e.g., Australian Curriculum Assessment and Reporting Authority (ACARA), 2010a, 2010b). Yet, independently of whether a national curriculum exists, local contexts cannot be fully understood without taking account of global influences. Stromquist and Monkman (2000) point to efforts of groups to recapture traditional values and identities as unintended effects of globalization and the reassertion of the importance of local contexts.

Is there a middle path to blend and balance the global and local forces through the activities of teachers learning from other teachers? It is within this interplay of the two forces that Robertson (1995) used the term "glocalization" to explain the process whereby the global and the local interpenetrate each other, creating a hybrid. This hybrid adapts and blends global trends with local conditions and options. In other words, global trends are contextualized into the specifics of local settings.

This interplay of global and local influences can be seen within the distinction made between formal research knowledge which is theoretical and able to be generalized across contexts and the practical knowledge of the teacher which is based at the particular and local context level (Fenstermacher, 1994). Teachers often concentrate on their own localized insights and improvements to practical—although published research can also be local in its focus. A survey of 282 research articles published between 1999 and 2003 in international journals, international handbooks of mathematics education, international mathematics education conference proceedings, and in national and regional sources revealed that more than 60% were small-scale qualitative studies of a single teacher or small group of less than 20 teachers, and that 72% were conducted by teacher educators studying teachers with whom they were working (Adler, Ball, Krainer, Lin, & Novotná, 2005). In a review of Australasian research between 2004 and 2007, Anderson, Bobis, and Way (2008) observed that "smaller-scale studies tended to rely on self-report data and that few incorporated significant amounts of observation data to help validate the self-reported findings … due to the labour-intensive and high cost involved when studies incorporate classroom observation" (p. 327).

The knowledge and results from many action research studies, conducted by teachers, have not been disseminated widely, and in such a circumstance any impact from a study is likely to have been confined within the local school or community. One result has been that teacher inquiry and practitioner research has been regarded "almost as second-level research paradigms in educational research, relevant mainly to improving professional practices rather than furthering the general field of education research and theory" (Lingard & Renshaw, 2010, p. 35). From this perspective of formal research, teachers could be seen as simply translators or interpreters of educational research completed elsewhere, or sometimes as merely the objects of formal research. One result of the fact that university- and system-based academics have often had greater access to power and resources than school-based teachers has been that many teachers have felt at liberty to ignore or reject academic research findings, which they perceive as coming from the "ivory tower." Moves towards research partnerships between teachers and teacher educators have changed this situation somewhat, however not completely, as will be seen in some of the examples which follow.

In the face of criticism, there has been increased support for the concept of "teacher-as-researcher," because of its focus on local issues and change. In the first *International Handbook of Mathematics Education,* Crawford and Adler (1996) argued that active teacher participation in research on their own professional practice was a pre-requisite to changing and improving student educational outcomes. They highlighted educational change and issues associated with the lack of dissemination of formal research findings, pointing out that often university research did not reach teachers and therefore did not have much chance of affecting teaching and learning in schools.

Since then, an International Group for the Psychology of Mathematics Education (PME) working group focussed on the "teacher as researcher in mathematics education," published a book of papers (Zack, Mousley, & Breen, 1997) germane to the teacher/teacher educator interface. And, since its first issue in 1998, the *Journal of Mathematics Teacher Education* (JMTE) has published many papers relating to teacher research, mostly written by teacher educators who work with teachers. Indeed, the first article in the first volume reported a study of the learning of teachers who explored questions relating to their own practice (Jaworski, 1998). We will briefly describe this project (the Mathematics Teacher Enquiry project) later in this chapter.

The practices of teacher research and some of the related issues for the learning and development of teachers were captured in the *Second International Handbook of Mathematics Education* in 2003, in which it was claimed that the roots of the teacher-as-researcher movement lay in a paradigm shift that focussed on teachers as knowers and thinkers. This shift grounded theory in practice and insisted that knowledge derived from research was necessarily personal. It was claimed that the value of knowledge arising from teachers' research into their own teaching "was accompanied by an explicit rejection of the authority of professional experts who produced accumulated knowledge in scientific settings for use by others in practical settings" (Breen, 2003, p. 528).

In 2005 an ICMI study conference on mathematics teacher education produced a publication focussed on teacher learning through research in practice (Even & Ball, 2009). One of the two main sections in this publication was devoted to *Teachers Learning in and from Practice.* As well, a first *Handbook of Mathematics Teacher Education* was published in four volumes, and each volume included chapters related to teacher research (Wood et al., 2008). The fourth volume was devoted to the learning of teacher educators who worked with teachers in various modes of practice-based activity.

The rise of the teacher-as-researcher movement was accompanied by a renewed focus on theory and theory development in mathematics education, evident in recent publications such as those mentioned above and in the *Second Handbook of*

Research on Mathematics Education (Lester, 2007), which devoted most of its first section to this theme. The 29th annual conference of PME held a special "Research Forum on Theories of Mathematics Education." Theories in mathematics education were emerging, not only global theories such as constructivism or socio-cultural theory, but also more localized theories in specific areas such as knowledge in teaching (cf., the knowledge quartet mentioned above), including the personal theories of teachers and teacher educators, which were mostly based on their experiences in practice. These theories often gained status through their use by members of the international community and through associated debates in scholarly publications and conferences (see, e.g., Niss, 2007). Gradually, as a result of such dissemination and debate, relationships between theory, knowledge, and practice have begun to emerge.

Teachers' theories which are tested in practice and are an influential part of that practice are often not articulated clearly. Nor are they always subjected to careful scrutiny outside a minority of theory-inclined mathematics education researchers. Teachers may develop teaching practices, and informal associations of ideas associated with their teaching, by being part of a community of teachers within a school or local area. Without the influence of more global theoretical teaching knowledge which teachers themselves embrace, both in their minds and in their professional behaviours, the teaching community may continue to perpetuate existing practices irrespective of how well, or otherwise, these practices are generating high quality student learning.

The "glocal" or balanced way was taken up by Lingard and Renshaw (2010), who entered the teacher-as-researcher debate by arguing that teaching should be both a research-informed and a research-informing profession. Not only should teachers have a "researchly disposition" but educational researchers should have a "pedagogical disposition" which entails a desire for multiple forms of dissemination. Lingard and Renshaw (2010) strongly supported the concept of co-learners and proposed the use of design research practices because, they maintained, these blend applied and theoretical positions and acknowledge teachers and academic researchers as equal partners in the production of knowledge. "Design research elevates the importance of teachers as research collaborators, not just at the local level in relation to context-specific professional practices, but in terms of developing more general insight and transferable knowledge about teaching and learning processes" (p. 36).

Jaworski (2004) made a distinction between design research and developmental research in terms of the degree of involvement of teachers. She argued that with design research, teachers often were included merely to test out designs developed by external researchers (see for example, Witmann, 1998) whereas, in developmental research, teachers were included in the decision-making process that generates a design. Cobb and colleagues, who have offered a range of activities in which the involvement of teachers can be seen to vary considerably (Cobb, Confrey, di Sessa, Lehrer, & Schauble, 2003), saw distinctions between design and developmental research as blurred.

In the *Second International Handbook of Mathematics Education,* Breen (2003) provided some examples of attempts to find connections between teacher education as a field of practice and as a field of research. One example was the spread of the Japanese process of *lesson study* (LS). Breen concluded his chapter with an appeal to mathematics education researchers to seek closer collaboration with teachers. Breen's appeal resonated with the general theoretical position emerging among mathematics educators (Even & Ball, 2009; Wood et al., 2008).

In the remainder of the chapter we explore professional relationships, reflective practices, and community building that have led to genuine learning on the part of both teachers and teacher educators. In the next section we consider relationships between research and development in mathematics teaching, focussing particularly on ways in which research can be seen to provide a basis for developing knowledge and practice in teaching.

Research as a Basis for Learning in Teaching

Earlier in this chapter, in our brief discussion of pedagogical content knowledge and *didactique,* we reported some research studies that sought to identify and articulate better the types of professional knowledge that a successful teacher or teacher educator would need. We also referred to an existing division between research and craft knowledge and to various attempts to remove it. In this section, we examine studies that seek to maximize professional knowledge creation as the practices of researching and teaching become more coordinated and knowledge conversion from one practice to the other is encouraged by educational authorities (Ruthven & Goodchild, 2008). It will be seen that it is now well recognized, both inside and outside the mathematics education research community, that there is value in minimizing the gap between the theoretical expert and the classroom teacher by using research methodologies and practices that (a) place the teacher in the genuine role of a researcher, and (b) problematize the teaching process rather than simplify it (Pritchard & Bonne, 2007).

These desirable aims must be achieved in a wider context. Thus, for example, reflecting wider global trends in the period 2004–2007, the national governments in New Zealand and Australia promoted the development of accountability measures for funding and research, and this has been reflected in the Australasian mathematics education research output. According to Forgasz et al. (2008), there has been:

- A decrease in creative and idiosyncratic research and an increase in program research;
- A decrease in individual research and an increase in group or team research;
- A decrease in funding for basic research and an increase in funding for practice-oriented projects; and,
- A decreasing concern with the quantity of research and an increasing concern with the quality of research.

During the last decade there has been a steady increase in the number of publications reporting teacher-education research from around the world, and many of the publications are making clear the value of collaborative work among mathematics teachers or between teachers and researchers (Krainer & Wood, 2008). A variety of methodologies and organizational features can be identified in these studies and the research is contributing to teaching development and the associated professional learning of teachers.

Our discussion will be informed by the use of a framework developed by Jaworski (2003) based on research with teachers in which teachers took on a practitioner-researcher role (Jaworski, 1998, 2001). She suggested that the research itself can be an important mediating tool for teaching–learning development and proposed a framework for theorizing such mediation which consisted of four paired constructs:

- knowledge and learning,
- inquiry and reflection,
- insider and outsider,
- individual and community (Jaworski, 2003).

Knowledge and learning define an epistemological dimension in which participants bring their own thinking, beliefs and expertise to the research setting and learn through interactivity and dialogue within the community. *Inquiry and reflection* form a research dimension in which questions asked about practice and reflection on engagement in practice lead to new questions and new ways of doing and being. *Insider and outsider* recognizes the roles of teachers and teacher educators in processes of teaching development, both as insiders inquiring into their own practices and as outsiders researching the practices and development in teaching related to local and general knowledge (Bassey, 1995). *Individual and community* recognizes the importance of collaborative activity to the developmental enterprise and ways in which collaboration contributes to development for individual participants.

The term "developmental research" is sometimes used to refer to research which encourages development as well as documenting the developmental process. Stenhouse (1984) suggested that research is "systematic inquiry made public" (p. 120). Consistent with this point of view, we regard as research the activity of teachers who engage in systematic inquiry into their own practices and share their thinking and outcomes with other teachers and professionals. It is hard for teachers to take on researcher roles, since the practice of teaching is extremely demanding (McIntyre, 1997), and the nature of being a researcher can be perceived as not being within the accepted roles of a teacher. However, when collaborations are formed with university researchers, or teacher educators, the knowledge that both groups bring to the collaboration can enable a research or *inquiry* process to be established (Elliot, 1991; Jaworski, 1998, 2008).

In a developmental research project, development and research act as two sides of the same coin and participants are central players collaborating in action and outcome. Teachers are insider researchers, studying aspects of their own practice and of their students' learning. Teacher educators are often outsider researchers studying the development of teaching which arises through teacher research. They can also be insider researchers if they concomitantly study aspects of their own practices in promoting teaching. The inquiry processes that are involved can result in new knowledge in practice (insider research) and new knowledge about practice (outsider research). Outsider research can lead to more generalized knowledge available for inspection and critique in the academic community (Jaworski, 2003).

Developmental research can be seen as both a democratic and a critical approach to professional enhancement and improving practice (Goodchild, 2008). It is democratic when it includes participants in collaborative engagement and respect, valuing knowledge of different kinds from different sources (Herbert, 1989). It is critical when it encourages insight into and questioning of the processes and practices of its participants by the participants themselves (Carr & Kemmis, 1986). Collaboration is a basis for democratic engagement and inquiry provides the critical dimension.

Research into the professional practice of teaching, and teachers' learning about teaching, has suggested that engagement in inquiry processes can be a strong force for teaching development (Cochran Smith and Lytle, 1999; Jaworski, 1998; Wells, 1999). Cochran Smith and Lytle (1999) referred to inquiry as "stance." Teachers taking on an inquiry stance start to think differently about teaching and through their reflections on the teaching process are able to modify teaching in critical ways. Wells (1999) reported similarly, focussing particularly on the role of dialogue in encouraging new thinking and development. The collaborative nature of an inquiry process is central to teaching development. Teachers have the opportunity not only to inquire into their own practice and to modify practice (which is extremely hard to achieve alone) but conversations with their colleagues in an inquiry community enable both the encouragement of an inquiry approach and a sustaining of inquiry activity. If the inquiry community also includes university colleagues then the outside knowledge they bring of published research and theory can provide an important additional dimension (Jaworski, 2008).

Central to such an approach is the idea of creating or developing an "inquiry community" in which practitioners reflect on their own activities and, overtly, develop knowledge in practice. In order to theorize *inquiry community*, we might start from the concept of a *community of practice* (hereafter "CoP"), drawing on Wenger (1998). The term "community" designates a group of people identifiable by who they are in terms of how they relate to each other, their common activities and ways of thinking, beliefs and values. Activities are likely to be explicit, whereas ways of thinking, beliefs and values are more implicit. Wenger (1998) described community as "a way of talking about the social configurations in which our enterprises are defined as worth pursuing and our participation is recognizable as competence," and commented that "the social configurations in which our enterprises are defined" are the basis of practice (p. 5). In our field we might think of the practice of teaching mathematics. Teachers teaching mathematics within a school setting might be seen to form a community of mathematical teaching practice with its own norms and expectations and ways of being and doing. Mathematical knowledge provides a *foundation* for such practice (Rowland, 2008), being the basis of didactical knowledge and informing pedagogy.

Wenger has suggested that *belonging* to a CoP, that is having identity within a CoP, involves *engagement*, *imagination* and *alignment*. Thus, in practices of mathematics learning and teaching, participants engage in their practice alongside their peers, use imagination in drawing on their foundational knowledge and interpreting their own roles in the practice and align themselves with established norms and values. However, the expectation that teachers will align themselves with the practices in their school environment may not promote possibilities for development. Brown and McIntyre (1993), after gathering data through observations in classrooms and conversations with teachers, talked of classroom activity settling into "normal desirable states" (p. 54) in which teacher and students were comfortable with activity and expectations. Such normal "desirable states" may run counter to the need to develop students' confidence in mathematics and strong conceptual understandings. A community of inquiry, therefore, seeks to challenge the status quo, not to change it overnight, but to start to question and to look critically at what alternatives might be possible; then to start to think and act differently. In such an inquiry approach, *alignment* becomes *critical*. This means that while aligning with the norms and expectations of the school environment, teachers might start to ask questions about ways in which teaching and learning are approached, and start to explore, and to inquire into alternative possibilities. The idea of critical alignment is central to that of an inquiry community (Jaworski, 2006).

Learning of Teachers and Teacher Educators

> Mathematics teacher education is more difficult and complex than mathematics education, because it subsumes all of the latter. Likewise, research in mathematics teacher education is more difficult and complex than research in mathematics education. (Simon, 2008, p. 27)

This quotation from Simon recognizes that research in mathematics teacher education of necessity requires attention to several layers. Study of teacher learning (of mathematics teaching) requires within it a study of the concomitant learning of students in the mathematics classrooms where teachers teach (see Figure 13.1). Without the latter, a study of teacher learning is hollow. As Pring (2004) stated, "an action might be described as 'teaching' if, first, it aims to bring about learning, second, it takes account of where the learner is at, and, third, it has regard for the nature of what has been learnt" (p. 23). Thus, to study the learning of teachers, we have to attend to how they create opportunities for the specific students with whom they work, and how they consider the associated learning outcomes.

It is possible that the issue might indeed be even more complicated than this. Although it is possible to conduct research into teacher learning in the natural settings of teachers' everyday classroom practices with their students, most often, in studying teacher learning, researchers focus on some teacher education program *designed to promote* development. Often, the people undertaking the research and reporting it in scholarly papers are themselves the teacher educators conducting the programs. As Chapman (2008) has pointed out, many such research reports focus on the nature of teaching and the learning of teachers, with no consideration given to the teacher educators' own learning from their activity for promoting teachers' learning. It is as if the practices of the teacher educators are not of critical concern.

Mathematics teacher educators ("MTEs") are themselves teachers, and in many ways their activities parallel the teaching activities of school teachers. They are professionals who work with practising teachers and/or prospective teachers to develop and improve the teaching of mathematics, just as school teachers work with students to develop students' mathematics knowledge and understanding. MTEs are often based in university settings with academic responsibilities. They are often both practitioners and researchers. They have to take account of what a teacher already knows, and does, and to have regard for the nature of what has to be learned (Pring, 2004). In their research roles, teacher educators have responsibility for conducting research into the education of teachers and such research can result in developing knowledge in practice for both groups of practitioners. Thus we might ask, *How do mathematics teachers and teacher educators learn and develop?*

- What forms of knowledge are important for teachers? For MTEs?
- In what ways does engagement in activity with teachers lead to learning and development for the MTE and vice versa?
- What programs in mathematics teacher education have been significant for the learning and development of teachers and MTEs?

Figure 13.2 suggests related aspects of teacher and MTE knowledge.

Both groups have knowledge of mathematics, pedagogy, etc., as shown in B (in Figure 13.2). This knowledge may take different forms for each group, but it nevertheless provides a basis for communication through common areas, experience and interests. In addition, each group brings its own specialist knowledge as shown in A and C. Educators do not generally have the knowledge indicated in C and teachers generally do not have that indicated in A. A surrounding rectangle (not shown) might represent the deep complexity of educational environments in which teaching development is situated.

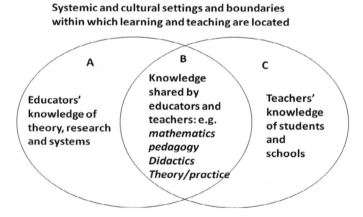

Figure 13.2. Related aspects of mathematics teacher and mathematics teacher educators' knowledge (from Jaworski, 2008, p. 336).

In the rest of the chapter we present a range of examples of projects and programs which illuminate the concepts discussed above. Our focus is on research in mathematics teacher education which has revealed and/or contributed to development in teaching mathematics and in which teachers and teacher educators have learned from each other. We draw particularly on projects with which we are familiar through our own engagement as researchers and practitioners with mathematics teachers in diverse parts of the world, seeking to develop mathematics teaching practice. We consider the development of teaching knowledge for both teachers and teacher educators.

In what follows, we use the structure of three main headings and a number of subheadings. The three main headings correspond to teachers learning from teachers as a result of participating in: (a) large-scale projects; (b) small-scale professional learning; and (c) preservice programs. These will be applied loosely as some studies could appear under more than one heading. Under the structure, in each of the modes of teachers' learning it is likely that teacher educators will be involved, sometimes as leaders in the education of teachers and sometimes as researchers. Such roles are not unproblematic and so we take up issues of teacher educators' roles and indeed teacher educators' learning alongside those relating to the learning of teachers.

Teachers Learning from Teachers in Large-Scale Projects

In this section, recent studies arising from large-scale professional development or research projects are presented. The adjective "large-scale" was considered to include those studies that drew upon systemic, state-wide, or multi-country projects as well as studies or projects in local areas that involved schools and non-school environments such as universities. The focus for these projects is their impact on the professional learning of teachers, on curriculum reform and on improved student outcomes.

All of the programs described in this section except the *Learner's Perspective Study* (LPS) possessed, to varying degrees, the following common features concerning the working process and the results (hereafter "CFPR" for common features of process and results). To avoid duplication, these common features will be assumed and only unusual or unique aspects will be highlighted. What were the common features? Firstly all programs incorporated workshops involving mathematics teachers and MTEs. These workshops were conducted at universities, schools or other institutions. All involved teachers conducting research into aspects of their own practice within their own schools, and communicating their activities and findings in the workshops. All involved MTEs who contributed, to the workshops, relevant material from research and other literature related to the teachers' own explorations, or expectations arising from mandated curriculum reform. Common features of the reported results were learning improvement of teachers in developing knowledge of theories and research, and insights into new approaches in the classroom. MTEs developed greater awareness of teachers' ways of thinking and of the challenges and limitations within schools and classrooms. Thus mathematics teachers and MTEs modified their ways of being and thinking to accommodate those of other teachers and MTEs, and this accommodation resulted in challenges to existing practices and new ways of perceiving each other. This growth of awareness led to a greater depth of understanding between mathematics teachers and MTEs which enabled them to deal with the issues that arose and to work towards productive development.

The theoretical ideas involving community of inquiry and critical alignment were the basis for two government funded, 4-year projects in Norway: "Learning Communities in Mathematics" (LCM) and "Teaching Better Mathematics" (TBM). The first involved 14 didacticians of mathematics (mathematics educators) from one university and about 30 teachers from 8 local schools in exploring the development of mathematics teaching in their schools; the second, building on the first, involved a consortium of 5 universities in different cities in Norway and schools local to each, extending the developmental process across the country. These programs followed CFPR with the workshops using collaborative inquiry-based activities between the mathematics teachers and didacticians. An uncommon feature involved the design of teaching and video-recording of innovative activities in classrooms. These video records formed part of a large bank of data from all aspects of the project which was a source of analysis for didacticians as outsider researchers in relation to a range of research questions.

The results of the program were very positive in the areas described in CFPR. Publications from the LCM project documented the learning processes in which both teachers and didacticians were engaged (e.g., Jaworski et al., 2007). The project demonstrated that learning in both groups was necessary in order to form a community of inquiry, and when there appeared to be a conflict a sincere desire to make the project work led to activity to resolve the conflict (Jaworski & Goodchild, 2006). The stakes were important for both groups and both groups felt ownership of and responsibility for the activity involved, albeit in differing ways. The implications for other programs lie in the relationships that evolved and the ways in which the program managed to foster equity. This is an important challenge for all those currently engaged, or about to be engaged, in teacher education programs.

Resonating with research in many western countries, Australasian research literature has focussed on the structures and findings of a number of large early numeracy programs—such as the Australian *Count Me In Too* (CMIT) project in the state of New South Wales, the *Early Numeracy Research Project* (ENRP) in the state of Victoria, and New Zealand's *Numeracy Development Projects* (NDP). These projects were funded by governments seeking to establish research priorities and methodological approaches. The projects aimed to deliver professional development teaching programs using a variety of strategies that included MTEs and extensive use of ICT while improving student achievement with early mathematical concepts. When these three programs were compared, researchers were able to extract common structures as well as identify the unique aspects of each project. Each featured: (a) a research-based framework for children's mathematical learning; (b) the use of individual student thinking assessment interviews; and (c) intensive whole-school professional development programs (Bobis et al., 2005). The process described in CFPR was evident in the workshops, and the results listed gains in knowledge, and improvement of relationships between teachers and MTEs.

Each of these large early numeracy programs had unique features, and we will consider just one, CMIT, as an example. The program regarded the identification, sharing and activation of knowledge of how children learn mathematics as a long-term, whole-school, classroom-based learning process. The interplay of researcher knowledge and teacher knowledge was an expectation of the CMIT program which used a design research model (Cobb, 2003) that collapsed the four groups of insiders and outsiders into one group of co-learners (the academic facilitators; the Departmental consultants (who were mostly former teachers); the teachers; and the students). The unique feature of using an on-going evaluation process conducted by external researchers (outsiders) meant that insights developed in collaboration with teachers and MTEs were used to "feed forward" into the theory development and instructional design loops that were implemented by the teachers and MTEs. Thus, theoretical knowledge was shared with teachers as active learners in their schools to be trialled and developed with their colleagues with the participation of their students. It was regarded as a factor in keeping the program dynamic and sustainable.

An extension to CMIT was the large system-wide *Counting On* (CO) program, also based in NSW. CO was designed to support the professional learning of teachers in identifying and addressing the learning needs of those students in the middle years who were having difficulties with early mathematical concepts and skills. The process and results of CFPR were recorded in a number of external evaluation studies (White, 2008, 2009, 2010). These evaluations used a framework of five critical levels (Guskey, 2000): participants' reaction; participants' learning; organizational support and change; participants' use of new knowledge and skills; and student learning outcomes. All evaluations reported positive teacher reactions and gains in organizational support, teacher learning, teacher use of new knowledge, and student achievement outcomes.

CO also had other unique features concerning the workshops, model of dissemination; and the greater autonomy given to the teachers. Each participating school sent a volunteer teacher (facilitator) to a 2-day training course. The facilitator then returned to organize and run the program in the school, supported with resources (publications, website, DVDs, and money), with mentoring being available through a Departmental consultant. Although this might first appear to be an application of a "train-the-trainer" model, the correct term is a "facilitated model," as the quality of the program was dependent on the school facilitators and their skills in leading their teams as they conducted their research and developed teaching strategies according to their needs and context. Whereas cascade models of train-the-trainer suffer from "dilution" as the process moves from level to level, by contrast the facilitated model has the potential to be better (but also worse) than what was provided with original facilitator workshops.

With CO, the school team was expected to operate using the *lesson study* (LS) model developed in Japan to enable and encourage collaborative professional learning and sharing between teachers (Stigler & Hiebert, 1999). A more detailed description of the LS model will be given later in this chapter (see Figure 13.4).

The "feed forward" in the CO process provided an excellent example of the interplay between teachers and MTEs in developing teaching that led to the improvement of student learning outcomes. For example, in response to teacher concerns involving student interactions with, and attitudes towards, mathematics word problems, Newman's diagnostic error analysis procedure (Newman, 1977; 1983; Clements, 1980) was introduced to the program. Teachers worked with MTEs to develop strategies to remedy the student difficulties revealed by this form of analysis. Initially in many classrooms, teachers displayed the diagnostic questions as a hand-made poster and the prompts were used to assist a problem-solving process. After teacher requests, a professionally designed poster was produced for dissemination throughout all schools by the NSW Department of Education Curriculum Support Directorate (see Appendix 1). Another difficulty reported by teachers was how to assist students who could not transform (or "mathematize") written mathematics problems into a suitable procedure. Teacher material involving the use of what are known as "tape diagrams" was developed by teachers and MTEs as a pedagogical strategy which assisted the teachers. Tape diagrams are visual representations (see Appendix 2) that are used extensively in Japanese schools (Murata, 2008). The success of the collaboration and co-learning between the mathematics teachers and MTEs in sharing a common goal were evident in the completed evaluation reports.

The next two cases, situated in Brunei Darussalam and Iran, shared similar CFPRs with other programs, but exhibited uniqueness in the roles of mathematics teachers, researchers and MTEs, which became blurred and interchangeable. In Brunei Darussalam, the *Active Mathematics in Classrooms* (AMIC, see Figure 13.3) was a national project designed to provide upper-primary teachers with ongoing professional learning and support (Mardiah & Shimawati, 2004; White, 2004b).

An unusual feature involved 14 practising primary teachers (called "the writers")—who were enrolled in an upgrading B.Ed program at Universiti Brunei Darussalam. These teachers adopted the roles of mathematics teachers and MTEs at different times. The writers developed, trialled, and revised AMIC workshop notes and materials for the nine topics, under the supervision of their MTE. After developing the 9 AMIC workshop units, the 14 writers then led trial workshops in which 10 future "AMIC workshop teacher leaders" (each representing a school) participated. Following these workshops the materials were revised and were then published by the Ministry of Education (Hafizah & Rosmawati, 2003; Haslina, 2003; Kamsiah, 2003; Lim & Zarinah, 2003; Maria & Ramnah, 2003; Mohammad Ariffin, 2003; Norjah, Rozaimah, & Tini, 2003; Rozina, 2003; Yunaidah, 2003). The school leaders then conducted the workshops in their schools with the help of the writers. This cycle continued, with teachers from other schools being involved, and a widening number of AMIC "graduates" becoming workshop teacher leaders.

Due to the unique geographical spread of schools, initial AMIC workshops were conducted in five schools and involved 60 upper-primary teachers. Thus a community of practice was formed in each school involving teachers, teacher leaders, writers and the MTE. Results resonated with the CFPR.

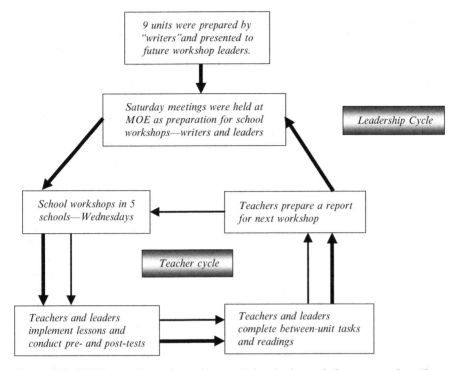

Figure 13.3. AMIC cycle for teacher writers, workshop leaders and classroom teachers (from White & Clements, 2005, p. 152).

The next example in this section relates to a national mathematics curriculum reform process in Iran that developed as a response to significant changes in secondary school education during the early 1990s. This reform process and the new curriculum challenged mathematics teachers and in particular, those who had taught geometry, and only geometry, for years. In response to teacher concerns, 11 national teacher-education sessions were planned and delivered by MTEs between 1994 and 1999. One session relating to geometry was the most controversial (Gooya, 2007). The direction and purpose of high school geometry had changed and there was an increase in the number of mathematics teachers involved in teaching geometry. Many of the new teachers were female. Previously, in Iran, geometry had been a male-dominated subject, and there was a concern that it might lose status if it became accessible to both male and female mathematics teachers and students (Gooya & Zangeneh, 2005). The most notable implication of this event for teachers' learning and their professional practices was that young mathematics teachers' views and insights about their own mathematics learning evolved and their self-confidence towards teaching geometry, in particular, was greatly improved.

Sharing the CFPR, from the outset the intention was for teachers and MTEs to work together and to integrate theoretical and practical knowledge with the teachers' craft knowledge. The results were those listed in CFPR. As well, small research projects were conducted to reveal teachers' concerns. One involved an action-research approach in which a number of graduate students, and one of the present authors (Gooya) worked with mathematics teachers from different cities. These projects involved mathematics teachers and MTEs collaborating to a degree that in some cases blurred the line between "insiders" and "outsiders," particularly when the teacher was also an MTE (Gooya, 2006). This happened quite naturally as the collaborations became more genuine and more meaningful for both groups.

The final study in this section is a large-scale study that differs from the earlier ones in the focus upon the relationships between teachers and researchers. In its original form, the *Learner's Perspective Study* (LPS) sought to document the classroom practices of competent mathematics teachers and to identify the meanings that participants held for those practices and the meanings that arose out of those practices (Clarke, 2001a, 2001b; Clarke, Keitel, & Shimizu, 2006; Clarke, Shimizu et al., 2006; Shimizu, 2002). LPS was originally a nine-country study (Australia, Germany, Hong Kong, Israel, Japan, the Philippines, South Africa, Sweden and the USA) of learner practices within the practices and meanings associated with "well-taught" Grade 8 mathematics lessons. LPS sought to uncover and to make explicit the cultural values and beliefs that framed the educational endeavours of teachers, researchers and policy makers in each country in order to contribute to the optimization of their effectiveness as sites for learning while acknowledging that optimization is shaped by the cultures of those classrooms.

LPS collected data using video and various texts such as classroom dialogue ("public" and "private"), teacher and student written material, and teacher and student post-lesson reconstructive interviews. The collaboration and sharing between MTEs and teachers through the post-lesson video-stimulated interviews contributed to accounts of the practices of classrooms and reflected teachers' intentions, actions and classroom consequences of these actions. The study challenged international comparative research practices (see, e.g., Stigler & Hiebert, 1999) by developing ways to accommodate the cultural differences through attending more closely to context and voice. The roles of teachers and learners in the examination of practice were explained using attempts to include the realities of political, societal needs and cultural plurality that were present in any particular classroom. "Teachers in Australia, Japan, The Philippines and South Africa face very different challenges with regard to cultural diversity of the communities they serve—class size, instructional resources, and societal and political priorities" (Clarke, Shimizu et al., 2006, p. 378). Many participating teachers described their participation as a powerful professional development experience. There is anticipation that value will accrue from research reports with different cultural authorship.

Although there were some differences in the last study (LPS) considered in this section, all the studies were explored in relation to the growth in learning of teachers and MTEs while they were involved in large-scale projects. The next section looks at smaller-scale studies.

Teachers Learning from Teachers in Small-Scale Professional Learning Projects

In this section, smaller studies involving learning by mathematics teachers and MTEs are considered. These studies included such things as teaching experiments, self studies, and small-group learning communities. They relied generally on self-reported data.

The structure of this section has two subsections. The first involves studies conducted within schools in which teachers worked individually or collaboratively to improve teaching practice with the aim of improving student learning outcomes. The second involves studies which focussed on the impact of teacher-development programs and activities which occurred both within and away from the school, and in which teachers from different schools worked together.

Knowledge growth within the school context. In order to explore the complexities inherent within school contexts, studies have used a range of samples of large to small numbers of mathematics teachers, sometimes involving MTEs. This then permitted the collection and analysis of rich, detailed data from multiple sources. Small sample studies can contribute to the building of a larger data set from which a synthesis across cases can form a more convincing body of evidence.

Hunter (2008, 2010) reported on a one-year-long study which involved four primary school teachers and herself, as the teacher–educator–researcher. The five participants worked together as a collaborative partnership to investigate how communication and participation patterns in the classroom might be best constituted to support student engagement in efficient and correct mathematical reasoning discourses. The study was conducted in a New Zealand primary school where the majority of students were of Pasifika or New Zealand Maori ethnic groupings. Data were collected both from study-group sessions, which took place regularly throughout the year, and from classrooms, through videotapes, done by the teachers, and researcher observations. Interviews with the teachers and reflective diaries also provided important forms of data.

Hunter (2008, 2010) described, powerfully, the gradual and, sometimes, circuitous and challenging journey through which one of the teachers, Moana, in a culturally responsive manner, shifted her positioning in the classroom culture from teacher in control of the discourse, to participant in, and facilitator of, the discourse. Inquiry of both teachers and the MTE was facilitated by the use of a specifically designed "Communication and Participation Frame." A community of inquiry focussed on how to structure and support the development of communication and participation patterns in their mathematics classrooms. The MTE inquired both into the teacher's learning and into her own practice as a colleague and supporter of the teachers in their journey of change of their pedagogical practices. Evident in this project was "a notion of teaching as learning in practice" through the overt use of "inquiry" in mathematics learning, mathematics teaching and "the development of practices of teaching in communities involving teachers and [teacher] educators" (Jaworski, 2006, p. 187).

In recent years, there has been a growing interest, particularly in western countries, in how Japanese teachers learn from each other when they are involved in the process of *lesson study* (LS). LS became highly visible beyond Japanese shores and strongly associated with mathematics education initially due to the influence of a number of American researchers and writers collaborating with Japanese counterparts (see, e.g., Fernandez, 2002; Fernandez & Yoshida, 2004; Lewis & Tsuchida, 1998; Shimizu, 1996, 1999a, 1999b; Stigler & Hiebert, 1998, 1999). This interest was stimulated by the publication of results from the Third International Mathematics and Science Study (TIMSS). The TIMSS Video Study made clear that differences did, in fact, exist not only in the mathematical achievement of American and Japanese students, but also the manner in which students were taught. One important result was a better understanding of the Japanese problem-solving teaching methods which improved student achievement on complex and novel mathematical problems. These teaching methods are now globally recognized as models for teaching that resonate with constructivist philosophical principles (Isoda, 2007).

Simon (2008) considered Japanese LS as having only process goals as there was an expectation that teachers would learn through engaging with the process and so the content was not specifically defined. LS provided a process whereby teachers could develop their professional learning and skills in order to improve classroom teaching and the learning outcomes of their students. The LS process enables and encourages collaborative professional learning and sharing between teachers and MTEs. The focus is upon the lesson instead of starting from learning theories and then trying to apply them to the classroom (Stigler & Hiebert, 1998).

LS spread throughout the world and particularly the Asia Pacific region—it has had a global influence upon the teaching of mathematics. The spread of LS has received support through the growth of information communication technologies and the ease of international travel. For example, the World Association of Lesson Studies (WALS: http://www.worldals.org/) was formed and this promoted LS at many levels from systems to individual schools across a range of countries. Another project, one which was supported by the Asian Pacific Economic Cooperation (APEC: http://hrd.apecwiki.org/index.php/Lesson_Study#Lesson_Study_in_Mathematics), was designed to encourage the spread of LS across the region. More recently LS has been promoted by the Southeast Asian Ministers of Education Organisation (SEAMEO) through their regional centres of excellence such as the Regional Centre for Education in Science and Mathematics (RECSAM) in Penang, Malaysia, and the Regional Centre for Quality Improvement of Teachers and Educational Personnel (QITEP) in Yogyakarta Indonesia (Hartono, 2010; Muchtar & Sutarto, 2010). However, different countries have adapted aspects of LS to their context (Isoda, Stephens, Ohara, & Miyakawa, 2007), so that: "The term Lesson Study has become an umbrella term for a variety of adaptations or glocal responses" (White & Lim, 2008, p. 916).

The distinctive steps and iterative nature of the LS process are illustrated in Figure 13.4. It should be noted that some "glocal" model studies do not include an iteration process. Re-teaching is a common LS feature in the USA, but is only an

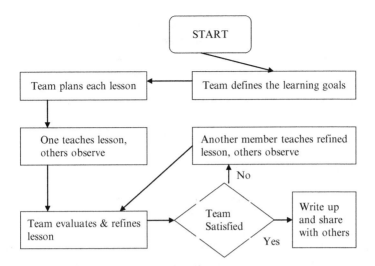

Figure 13.4. LS cycle (from White & Lim, 2007, p. 568).

occasional feature of Japanese LS (Lewis et al., 2009). Other LS studies differ over whether teachers are volunteers or are conscripted, and for other studies there are considerable differences in the composition of the teams (Hart et al., 2011). There are some "glocal" models that have a content focus and serve as counter-examples to Simon's assessment of LS as having only process goals. The following discussion will attempt to describe just three from a large number of available studies in order to highlight the growth in learning of teachers and MTEs, as well as the vast differences in the forms by which Japanese LS has been applied. We discuss studies from the USA (Lewis, Perry, & Hurd, 2009), from Australia (White, 2004a, 2006; White & Southwell, 2003; Southwell & White, 2004) and from Malaysia (Chiew, 2009; Lim, White, & Chiew, 2005; White & Lim, 2007).

The context of the US study was that it was part of a 2-week summer workshop in a North American school district. In Australia, the LS was situated within the context of a state-wide change of a mathematics syllabus, and the Malaysian context was as part of a doctoral study of a two-school professional development program.

The US study involved one team of five teachers, a teacher-coach, an MTE and two researchers. The use of an experienced and knowledgeable mathematics teacher as a coach in LS groups was regarded as important to the effectiveness and sustainability of the program (Lewis et al., 2009). The Australian study involved a large number of teams situated in schools scattered around the state, each team comprising five to six teachers, of whom one served as the facilitator. Access was limited to only one MTE and only one external researcher for all the teams. The Malaysian study consisted of one team of six and another team of eight teachers, with both teams having access to an MTE and a researcher. Membership of the US and Australian teams was voluntary, but the Malaysian participants had been directed to join by their school leaders.

In the US study the teachers chose to focus on the goal of helping primary school students indentify and express, mathematically, patterns, by working collaboratively through a sequence of activities. There were two iterations involving fourth-grade classes. The Australian teachers chose their own focus area within the new syllabus, and conducted varying numbers of iterations, depending upon the team. The Malaysian teachers were influenced by their head of mathematics in their choice of topic and iterations.

Data in the US study were collected through videotaping, field notes and artefacts such as student work samples. The Australian study collected data through questionnaires, interviews and document analysis, and the Malaysian study used observations, interviews and questionnaires.

The US study reported three types of changes produced by LS: changes in teachers' knowledge and beliefs, changes in professional community, and changes in teaching-learning resources. The Australian study reported improvements to teachers' learning and use of new knowledge, the establishment of stronger and on-going professional relationships among team members, and increased recognition and organizational support from the school leadership. The Malaysian results were mixed depending upon differences in the degree of administrative or executive support which directly affected teacher commitment in both schools.

This short discussion has highlighted the diverse range of LS "glocal" models. The studies reported improvements in the learning of teachers. Other studies, such as those reported by Arbaugh (2003) and Slavit and Nelson (2010), have generated similar benefits among teachers participating in LS groups. However, none of the three discussed here reported improvements in the learning of the MEs or researchers.

The existence of various variations to the Japanese LS model around the world has led to many other designs of professional development programs which although resonating with some of the aspects of the LS process cannot be strictly classed as an LS model. For example, in modern Iran there are examples of teachers collaborating and learning from each other within approaches and structures that resemble aspects of Japanese LS. In 1960, the teachers' council of an elementary school in a small Iran–Iraq border town called Paveh, located in the west end of Iran, began to discuss teaching methods and curriculum organization with respect to mathematics in Grades 4, 5 and 6. The council agreed to make changes and all members of the council signed an agreement. An analysis of the minutes of one of these meetings has indicated that the process of planning was similar to LS (Gooya, 2010). However, the activities that were planned and implemented would be best seen as reflecting local insights into how practice might be improved, and the aim was not to develop generally applicable findings.

Knowledge growth in and beyond school contexts. Research studies have explored the impact of professional development offered beyond the school context as teachers attended meetings, workshops or courses with teachers from other schools. In many studies there has been a synergy between out-of-school activity and related activity taking place in school.

In the Mathematics Teacher Enquiry project that ran for two years, Jaworski (1998) described teachers' learning through engaging in research into aspects of their own teaching, and the concomitant learning of the MTEs conducting the project, and studying the developmental processes involved. The project brought teachers and MTEs together in both school and university environments in which they developed mutual respect and common understandings. The program successfully used workshops in schools and university, focussing on CFPR. Thus MTEs and mathematics teachers worked together as colleagues and co-researchers in a joint professional environment which was theorized subsequently, by the MTEs as a "community of inquiry" (Jaworski, 1998, 2006).

In the nation of Colombia, a program was developed and implemented with the acronym PROMESA (Creating Science and Mathematics Connected Learning Experiences that Open Opportunities for the Promotion of Algebraic Reasoning—in Spanish, the corresponding acronym is PROMICE) (Agudelo-Valderrama & Vergel, 2009a, 2009b). In PROMESA, school mathematics and science teachers, and teacher educators, worked together as a developing community of inquiry with the shared aim of promoting students' meaningful and connected learning of mathematics and science. Eleven well-qualified mathematics and five science teachers, at three schools which served students from disadvantaged socioeconomic communities in Bogotá, and two teacher educators, worked together over a 14-month period on issues that they had identified after discussions among themselves and with their school principals. Following the teachers' participation in a series of workshops, which provided ample opportunity for analysis and discussion of ways and means of connecting science and mathematics in their schools, the teachers from each school organized themselves into sub-groups (each sub-group had one science and two mathematics teachers). These sub-groups then worked collaboratively with the teacher-education researchers on the processes of designing, implementing and documenting classroom innovations. The purpose of these innovations was always to engage the students in connected science and mathematics learning experiences which would generate opportunities for the promotion of algebraic reasoning. During the process, the teachers and the teacher educators met regularly, at each school and at the university, for whole-group-discussion and sharing sessions.

Throughout the project, data on teachers' knowledge, conceptions, beliefs and attitudes with respect to school mathematics and science, and of teaching specific concepts, were gathered using a variety of data collection methods. The students of the participating teachers were also involved in the study, and data were collected, by the teachers, in a longitudinal study of two Grade 8 groups, from different schools.

In this project, the teachers and teacher educators were both insiders and outsiders: as insiders, teachers inquired into their own thinking and their understandings of specific mathematics and science concepts. They participated in planned interviews and kept diaries in which they reflected on their own teaching practices, and expressed their feelings. As outsiders, they inquired into their students' thinking and learning in relation to the classroom work. The teacher educators were also simultaneously insiders and outsiders: as insiders, they inquired into their own practices in order to be in a better position to make informed decisions; they acted as supporters and orchestrators in the complex task of researching the various contexts and issues which emerged during the project. At the same time, they acted as MTEs in a collaborative and supportive manner. As outsiders, they inquired into the development of the teachers' thinking and teaching practices, for "in order to fulfil the task of collaborators and members of a community of inquiry, insights into the teachers' thinking processes were key" (Agudelo-Valderrama & Vergel, 2009a, p. 33). The teachers and MTEs both inquired into the students' learning, and sought to enhance and maximize student learning, which was their ultimate shared goal.

In this Colombian study there was considerable evidence of improved student learning and gain in the students' sense of purpose related to learning. The evaluations which project members carried out indicated that participants grew in their appreciation of the connections between science and mathematics knowledge and their enacted co-teaching practices during the project. Agudelo-Valderrama and Vergel (2009a) emphasized the important professional lessons MTEs learned in relation to various areas of their roles and duties as teacher educators.

Implications of the study were identified for those intending to participate in programs of initial mathematics and science teacher education, for continued professional learning, and for Local Education Authorities. As in the study by Lewis et al. (2009), there emerged issues that related not only to the sustainability of teacher professional learning, but also to possibilities and barriers affecting teacher participation in professional learning projects. For example, Agudelo-Valderrama and Vergel (2009a) drew attention to the need to establish a coherent policy in relation to the administration of school staffing and participation in inservice professional learning programs. The head teachers in the project found it difficult to allow the participating teachers to attend a weekly one-hour group meeting, citing lack of staffing and the requirements of staff management policies. As a result, the regular work sessions of teachers and teacher educators at school sites had to take place after school hours and prevented teachers from engaging in further collaborative work at the end of the 14-month period, despite this having been included as a requirement in the design of the project. Nevertheless, many of the participating teachers expressed their willingness to continue working with the researchers in order to write papers, to report on their classroom project findings and on their own learning, and to prepare these reports for publication.

Learning of Preservice Teachers and Teacher Educators

This section summarizes a collection of studies which explored teachers involved in preservice programs (including early-career teachers), focussing on the learning of all participants. It gives a brief discussion of some research models that relate the learning of teacher educators to the learning of preservice and early-career teachers with whom they worked.

In the *Second International Handbook of Mathematics Education,* the issue of teachers as mentors to early career teachers and their roles in relation to university MTEs with whom they worked was explored. Three levels of knowledge for teachers, mentors and MTEs were suggested as follows.

Level 1. Mathematics and the provision of classroom mathematical activities for students' effective learning of mathematics. This included socio-cultural mathematics education, such as the wider influences on pupils' learning, and reasons why pupils need to learn mathematics.

Level 2. Mathematics teaching and ways in which teachers think about developing their approaches to teaching.

Level 3. The roles and activities of teacher-educators in contributing to developments in (1) and (2) and including constraints on teacher education and how they can be tackled (Jaworski, 2001; Jaworski & Gellert, 2003).

Each of the first two levels incorporates those below it. Teachers operate largely (but not exclusively) at Level 1, mentors at Level 2, incorporating Level 1, and teacher educators at Level 3, incorporating Levels 1 and 2. What this framework misses is the areas of knowledge indicated earlier in this chapter in Figure 13.2—that is teachers' knowledge of students and schools and teacher educators' knowledge of theory, research and educational systems. What we recognize in considering such a different framework is the complexity of the developmental scene and the areas of knowledge on which it draws. These areas of knowledge are far from distinct and it seems important to recognize that preservice and early-career teachers and the MTEs share the knowledge in complex ways.

Perks and Prestage (2008) recognized links between their own knowledge as MTEs and the knowledge of the preservice teachers whom they taught. They offered a model for teacher learning in a teacher-education program aimed at preservice mathematics teachers, and a version of the same model aimed at teacher educators (see Figures 13.5 and 13.6). In the first case, teacher learning draws on teachers' knowledge of classroom events, professional traditions, learner knowledge and practical wisdom. The parallels for teacher educator learning draw on mathematics education sessions in the teacher-education program, professional traditions, own learner knowledge as a classroom teacher, and practical wisdom. Perks and Prestage commented particularly on the teacher educators who had experience of being teachers themselves in earlier professional practice—but of course this is not the case with all teacher educators.

An alternative way of seeing relationships between teacher educator learning and teacher learning was offered by Zaslavsky (2008), and is represented in Figure 13.7. The main idea is that the educator (or facilitator) designs activities to promote teachers' learning and then s/he learns from reflecting on the teachers' activities.

These models from Perks and Prestage and from Zaslavsky suggest that MTEs learn through engagement in and reflection on their own practice, in working with teachers, and there are parallels with teachers' learning through practice (see Even & Ball, 2009).

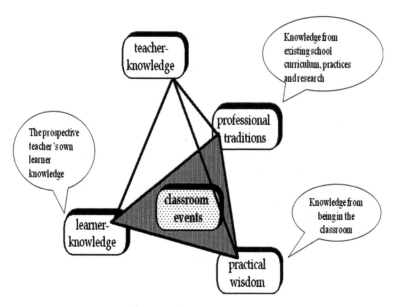

Figure 13.5. Teacher learning in a teacher-educator program with preservice mathematics teachers (from Perks & Prestage, 2008, p. 270).

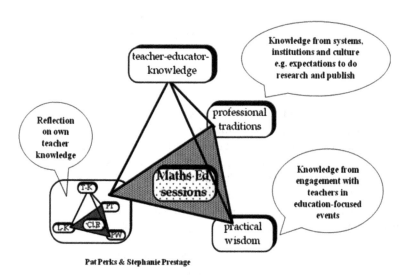

Figure 13.6. Teacher-educator learning in a teacher-educator program with preservice mathematics teachers (Perks & Prestage, 2008, p. 271).

In some parts of the world, programs have been especially designed for the learning of MTEs—an example of this was the *Manor* program in Israel (Even, 2008). This program included an introduction to research, theoretical ideas and issues related to practice, and provided professional opportunities for prospective MTEs to

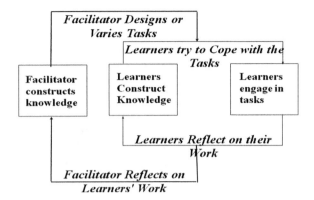

Figure 13.7. Relationships between teacher-educator learning and teachers' learning (from Zaslavsky, 2008, p. 95).

engage with teachers in professional development programs. In doing so, the program modelled ways in which those becoming teacher educators might themselves work with teachers.

These three examples related to taught programs that developed the learning of MTEs. The programs with Perks and Prestage (2008) and with Zaslavsky (2008) were for the education of teachers in which MTEs learned overtly through scrutiny of their own practice. In the third case, Even (2008) described a program for MTE learning with a model that could be adapted for teacher learning. These fit into what Simon (2008, p. 18) called teacher-education programs with *content and process* goals where there is something to be taught and teachers are expected to learn.

Concluding Comments

This chapter has attempted to convey the complexity and diversity of research focussed upon teachers learning from teachers. Our range of examples reveals the complexity of settings in which teachers learn, and the related knowledge that grows through the various developmental programs. This complexity is influenced by both global and local forces, such as the recent pressure on teachers to meet different demands imposed on them either directly by politicians and national laws such as value-added and No Child Left Behind in the United States of America, or indirectly by politicians and policy makers such as in Iran (Gooya, 2011).

Central to all the settings described were the relationships between mathematics teachers and MTEs which varied according to the nature of the program. Within three sections used to group programs of similar features we have further used the framework of Jaworski and also Simon's distinction between programs to illuminate certain important issues. In some, MTEs had a greater teaching role in guiding teachers in relation to pre-defined content, be it mathematical, didactical or pedagogical. In others, MTEs and teachers worked together in developmental roles, often in inquiry-based practices and sometimes using LS models; teachers often worked together to design their own developmental activities. Both the learning of teachers and the learning of MTEs were addressed. We can see a range of similarities and differences between the knowledge that these two groups bring to the learning interface. Importantly, neither group had all the knowledge that was needed for the development of teaching, but working together they could become a unified, powerful developmental force. Undoubtedly, both learned from each other as a result of their interactions in a research process.

Mutual respect and collaboration allow the input of critical elements of knowledge, often by MTEs, that are seen to be valuable to developmental practice. Although this input might take place in out-of-school contexts, it is within the in-school situations that knowledge can be tested and developed in practice. Here teachers' knowledge is pre-eminent and MTEs have much to learn about the systemic factors and issues that influence what can happen in schools and what is needed to put research-based knowledge into practice.

Appendix A. Appendix 1

The classroom poster (following Newman's error analysis procedure): New South Wales Department of Education and Training Curriculum Support Directorate

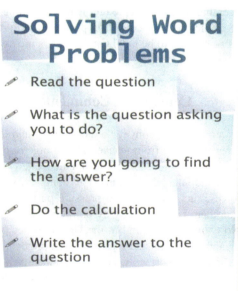

Appendix B. Appendix 2

A tape diagram for the problem:
Sue paddled 402 km along a river in her canoe over 6 days.

She paddled the same distance each day.
How far did Sue paddle each day?

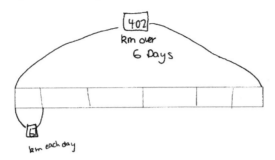

References

Adler, J., Ball, D., Krainer, K., Lin, F., & Novotná, J. (2005). Reflections on an emerging field: Researching mathematics teacher education. *Educational Studies in Mathematics, 60*, 359–381.

Agudelo-Valderrama, C., & Vergel, R. (2009a). Informe final del Proyecto PROMICE. *Promoción de un enfoque interdisciplinario y de resolución de problemas en el inicio del trabajo algebraico escolar: integrando contextos de ciencias y el uso de tecnología digital.* Bogotá, Colombia: Centro de documentación, Instituto para la Investigación Educativa y el Desarrollo Pedagógico, IDEP, Secretaría de Educación Distrital.

Agudelo-Valderrama, C., & Vergel, R. (2009b). La apertura del aula de ciencias para promover el desarrollo del pensamiento algebraico: el caso del profesor Simón, participante del Proyecto PROMICE. In L. F. Acuña & L. Zea (Eds.), *Universidad-Escuela y producción de conocimiento pedagógico: Resultados de la investigación IDEP-Colciencias* (pp. 245–258). Bogotá, Colombia: IDEP.

Anderson, J., Bobis, J., & Way, J. (2008). Teachers as learners: Building knowledge in and through the practice of teaching mathematics. In H. Forgasz, A. Barkatsas, A. Bishop, B. Clarke, S. Keast, W. T. Seah, P. Sullivan, & S. Willis (Eds.), *Research in mathematics education in Australasia 2004–2007* (pp. 313–335). Rotterdam, The Netherlands: Sense Publishers.

Arbaugh, F. (2003). Study groups as a form of professional development for secondary mathematics teachers. *Journal of Mathematics Teacher Education, 6*(2), 139–163.

Australian Curriculum Assessment and Reporting Authority (ACARA). (2010a). *The Australian curriculum: Mathematics.* Retrieved from http://www.australiancurriculum.edu.au/Mathematics/Curriculum/F-10.

Australian Curriculum Assessment and Reporting Authority (ACARA). (2010b). *The shape of the Australian curriculum version 2.0.* Sydney, Australia: ACARA.

Bassey, M. (1995). *Creating education through research.* Edinburgh, UK: British Educational Research Association.

Bobis, J., Clarke, B., Clarke, D., Thomas, M., Wright, R., Young-Loveridge, J., & Gould, P. (2005). Supporting teachers in the development of young children's mathematical thinking. Three large-scale cases. *Mathematics Education Research Journal, 16*(3), 27–57.

Breen, C. (2003). Mathematics teachers as researchers: Living on the edge. In A. J. Bishop, M. A. Clements, C. Keitel, J. Kilpatrick, & F. K. S. Leung (Eds.), *Second international handbook of mathematics education* (pp. 523–544). Dordrecht, The Netherlands: Kluwer Academic Publishers.

Brousseau, G. (1992). Didactique: What it can do for the teacher? In R. Douady & A. Mercier (Eds.), *Research in didactique of mathematics* (pp. 7–40). Paris, France: La Pensée Sauvage.

Brown, S., & McIntyre, D. (1993). *Making sense of teaching*. Buckingham, UK: Open University Press.
Carr, W., & Kemmis, S. (1986). *Education, knowledge, and action research*. Geelong, Australia: Deakin University.
Chapman, O. (2008). Mathematics teacher educators' learning from research on their instructional practices: A cognitive perspective. In T. Wood, B. Jaworski, K. Krainer, D. Tirosh, & P. Sullivan (Eds.), *The international handbook of mathematics teacher education* (Vol. 4, pp. 115–134). Rotterdam, The Netherlands: Sense Publishers.
Chiew, C. M. (2009). *Implementation of lesson study as an innovative professional development model among mathematics teachers* (PhD thesis). Universiti Sains Malaysia.
Clarke, D. J. (2001a). *Study design: Learner's Perspective Study*. Melbourne, Australia: The University of Melbourne.
Clarke, D. J. (2001b). *Perspectives on practice and meaning in mathematics and science classrooms*. Dordrecht, The Netherlands: Kluwer Academic Publishers.
Clarke, D. J., Keitel, C., & Shimizu, Y. (Eds.). (2006). *Mathematics classrooms in twelve countries: The insider's perspective*. Rotterdam, The Netherlands: Sense Publishers.
Clarke, D. J., Shimizu, Y., Ulep, S. A., Gallos, G., Adler, J., & Vithal, R. (2006). Cultural diversity and the learner's perspective: Attending to voice and context. In F. K. S. Leung, K.-D. Graf, & F. J. Lopez-Real (Eds.), *Mathematics education in different traditions—A comparative study of East Asia and the West* (pp. 353–380). New York, NY: Springer.
Clements, M. A. (1980). Analysing children's errors on written mathematical tasks. *Educational Studies in Mathematics, 11*(1), 1–21.
Clements, M. A. (2008). Australasian mathematics education research 2004–2007: An overview. In H. Forgasz, A. Barkatsas, A. Bishop, B. Clarke, S. Keast, W. T. Seah, P. Sullivan, & S. Willis (Eds.), *Research in mathematics education in Australasia 2004–2007* (pp. 337–356). Rotterdam, The Netherlands: Sense Publishers.
Cobb, P. (2003). Investigating students' reasoning about linear measurement as a paradigm case of design research. In M. Stephan, J. Bowers, P. Cobb, & K. Gravemeijer (Eds.), *Supporting students' development of measuring conceptions: Analyzing students' learning in social contexts* (pp. 1–16). Reston, VA: NCTM.
Cobb, P., Confrey, J., di Sessa, A., Lehrer, R., & Schauble, L. (2003). Design experiments in educational research. *Educational Researcher, 32*(1), 9–13.
Cochran Smith, M., & Lytle, S. L. (1999). Relationships of knowledge and practice: Teacher learning in communities. In A. Iran-Nejad & P. D. Pearson (Eds.), *Review of research in education* (pp. 249–305). Washington, DC: American Educational Research Association.
Cooper, P., & McIntyre, D. (1996). *Effective teaching and learning: Teachers' and students' perspectives*. Buckingham, UK: Open University Press.
Crawford, K., & Adler, J. (1996). Teachers as researchers in mathematics education. In A. J. Bishop, M. A. Clements, C. Keitel, J. Kilpatrick, & C. Laborde (Eds.), *International handbook of mathematics education* (pp. 1187–1205). Dordrecht, The Netherlands: Kluwer.
Dawson, S. (1999). The enactive perspective on teacher development: "A path laid while walking". In B. Jaworski, T. Wood, & S. Dawson (Eds.), *Mathematics teacher education: Critical international perspectives* (pp. 148–162). London, UK: Falmer Press.
Elliot, J. (1991). *Action research for educational change*. Milton Keynes, UK: Open University Press.
English, L. D. (2008). Setting the agenda for international research. In L. D. English (Ed.), *Handbook of international research in mathematics education* (2nd ed., pp. 3–19). New York, NY: Routledge.
Even, R. (2008). Facing the challenge of educating educators to work with practising mathematics teachers. In T. Wood, B. Jaworski, K. Krainer, D. Tirosh, & P. Sullivan (Eds.), *The international handbook of mathematics teacher education* (Vol. 4, pp. 57–74). Rotterdam, The Netherlands: Sense Publishers.
Even, R., & Ball, D. L. (2009). *The professional education and development of teachers of mathematics*. New York, NY: Springer.

Even, R., & Tirosh, D. (2008). Teacher knowledge and understanding of students' mathematical learning and thinking. In L. D. English (Ed.), *Handbook of international research in mathematics education* (2nd ed., pp. 202–222). New York, NY: Routledge.

Fenstermacher, G. D. (1994). The knower and the known: The nature of knowledge in research on teaching. *Review of Research in Education, 20*, 3–56.

Fernandez, C. (2002). Learning from Japanese approaches to professional development: The case of lesson study. *Journal of Teacher Education, 53*(5), 395–405.

Fernandez, C., & Yoshida, M. (2004). *Lesson study: A Japanese approach to improving mathematics teaching and learning*. Mahwah, NJ: Lawrence Erlbaum.

Forgasz, H., Barkatsas, T., Bishop, A., Clarke, B., Keast, S., Seah, W., & Sullivan, P. (2008). Introduction: Review of mathematics education research in Australasia 2004–2007. In H. Forgasz, A. Barkatsas, A. Bishop, B. Clarke, S. Keast, W. T. Seah, P. Sullivan, & S. Willis (Eds.), *Research in mathematics education in Australasia 2004–2007* (pp. 1–7). Rotterdam, The Netherlands: Sense Publishers.

Goodchild, S. (2008). A quest for "good" research: The mathematics teacher educator as practitioner researcher in a community of inquiry. In T. Wood, B. Jaworski, K. Krainer, D. Tirosh, & P. Sullivan (Eds.), *The international handbook of mathematics teacher education* (Vol. 4, pp. 201–222). Rotterdam, The Netherlands: Sense Publishers.

Gooya, Z. (2006). The necessity of collaborations between mathematicians and mathematics educators: Ways to have "Mathematics in the center". In J. Novotná, H. Moraová, M. Krátká, & N. Stehlikova (Eds.), *Proceedings of the 30th Conference of the International Group for the Psychology of Mathematics Education* (Vol. 1, pp. 73–76). Prague, Czech Republic: International Group for the Psychology of Mathematics Education.

Gooya, Z. (2007). Mathematics teachers' beliefs about a new reform in high school geometry in Iran. *Educational Studies in Mathematics, 65*, 331–347.

Gooya, Z. (2010). Lesson study: A souvenir from Japan, Iran, or teachers' common sense? *Roshd Mathematics Teacher Education Journal, 101*, 50–51.

Gooya, Z. (2011). How could we indicate the amount of quality in advance? *Roshd Mathematics Teacher Education Journal, 104*, 2–3.

Gooya, Z., & Zangeneh, B. (2005). How teachers conceive geometry teaching in Iran. In M. Kourkoulos, G. Roulis, & C. Tzanakis (Eds.), *Proceedings of the 4th Colloquium on the Didactics of Mathematics* (Vol. 2, pp. 247–254). Crete, Greece: The University of Crete.

Gravemeijer, K. P. E. (1994a). *Developing realistic mathematics education*. Utrecht, The Netherlands: CD-β Press, Freudenthal Institute, Utrecht University.

Gravemeijer, K. P. E. (1994b). Educational development and developmental research in mathematics education. *Journal for research in Mathematics Education, 25*(5), 443–471.

Gravemeijer, K. P. E. (2000, July). *Didactical phenomenological analysis as a design heuristic: Early statistics as an example*. Paper presented in Working Group 8 on Research, Theory and Practice, at the Ninth International Congress on Mathematical Education. Makuhari, Japan: ICME.

Guskey, T. R. (2000). *Evaluating professional development*. Newbury Park, CA: Corwin Press.

Hjh Hafizah bte Hj Salat & Hj Rosmawati bte Hj Abu Bakar (2003). *Language issues in upper primary mathematics classrooms*. Bandar Seri Begawan, Brunei Darussalam: Brunei Darussalam Ministry of Education.

Hart, L. C., Alston, A. S., & Murata, A. (Eds.). (2011). *Lesson study: Research and practice in mathematics education*. New York, NY: Springer.

Hartono, H. (2010). *Monitoring, evaluation and sustaining program of lesson study activities*. Yogyakarta, Indonesia: Regional Centre for Quality Improvement of Teachers and Educational Personnel (QITEP).

Haslina bte Hj Mahmud (2003). *Decimals: Going beyond skill and drill into reality*. Bandar Seri Begawan, Brunei Darussalam: Brunei Darussalam Ministry of Education.

Hatano, G., & Inagaki, K. (1998). Cultural contexts of schooling revisited: A review of the learning gap from a cultural psychology perspective. In S. G. Paris & H. M. Wellman (Eds.), *Global prospects for education development, culture and schooling* (pp. 79–104). Washington, DC: American Psychological Association.

Herbert, C. (1989). *Talking of silence*. London, UK: Falmer Press.
Hill, H. C., Ball, D. L., & Schilling, S. G. (2008). Unpacking pedagogical content knowledge: Conceptualizing and measuring teachers' topic-specific knowledge of students. *Journal for Research in Mathematics Education, 39*, 372–400.
Hoyles, C. (1992). Illuminations and reflections—Teachers, methodologies and mathematics. In W. Geeslin & K. Graham (Eds.), *Proceedings of the 16th Conference of the International Group for the Psychology of Mathematics Education* (Vol. 3, pp. 263–286). Durham, NH: International Group for the Psychology of Mathematics Education.
Hunter, R. (2008). Facilitating communities of mathematical inquiry. In M. Goos, R. Brown, & K. Makar (Eds.), *Navigating currents and charting directions. Proceedings of the 31st annual conference of the Mathematics Education Research Group of Australasia* (Vol. 1, pp. 31–39). Brisbane, Australia: MERGA.
Hunter, R. (2010). Changing roles and identities in the construction of a community of mathematical inquiry. *Journal of Mathematics Teacher Education, 13*(5), 397–409.
Isoda, M. (2007). Where did lesson study begin, and how far has it come? In M. Isoda, M. Stephens, Y. Ohara, & T. Miyakawa (Eds.), *Japanese lesson study in mathematics: Its impact, diversity and potential for educational improvement* (pp. 8–15). Singapore: World Scientific Publishing Co.
Isoda, M., Stephens, M., Ohara, Y., & Miyakawa, T. (Eds.). (2007). *Japanese lesson study in mathematics: Its impact, diversity and potential for educational improvement*. Singapore: World Scientific Publishing Co.
Jaworski, B. (1998). Mathematics teacher research: Process, practice and the development of teaching. *Journal of Mathematics Teacher Education, 1*(1), 3–31.
Jaworski, B. (2001). Developing mathematics teaching: Teachers, teacher-educators and researchers as co-learners. In: F.-L. Lin & T. J. Cooney (Eds.), *Making sense of mathematics teacher education* (pp. 295–320). Dordrecht, The Netherlands: Kluwer Academic Publishers.
Jaworski, B. (2003). Research practice into/influencing mathematics teaching and learning development: Towards a theoretical framework based on co-learning partnerships. *Educational Studies in Mathematics, 54*(2–3), 249–282.
Jaworski, B. (2004). Grappling with complexity: Co-learning in inquiry communities in mathematics teaching development. *Proceedings of the 28th Conference of the International Group for the Psychology of Mathematics Education* (Vol. 1, pp. 17–36). Bergen, Norway: International Group for the Psychology of Mathematics Education.
Jaworski, B. (2006). Theory and practice in mathematics teaching development: Critical inquiry as a mode of learning in teaching. *Journal of Mathematics Teacher Education, 9*(2), 187–211.
Jaworski, B. (2008). Development of the mathematics teacher educator and its relation to teaching development. In T. Wood, B. Jaworski, K. Krainer, D. Tirosh, & P. Sullivan (Eds.), *The international handbook of mathematics teacher education* (Vol. 4, pp. 335–361). Rotterdam, The Netherlands: Sense Publishers.
Jaworski, B., Fuglestad, A. B., Bjuland, R., Breiteig, T., Goodchild, S., & Grevholm, B. (2007). *Læringsfellesskap i matematikk: Learning communities in mathematics*. Bergen, Norway: Caspar Verlag.
Jaworski, B., & Gellert, U. (2003). Educating new mathematics teachers: Integrating theory and practice, and the roles of practising teachers. In A. J. Bishop, M. A. Clements, C. Keitel, J. Kilpatrick, & F. K. S. Leung (Eds.), *Second international handbook of mathematics education* (Vol. 2, pp. 829–875). Dordrecht, The Netherlands: Kluwer Academic Publishers.
Jaworski, B., & Goodchild, S. (2006). Inquiry community in an activity theory frame. In J. Novotná, H. Moraová, M. Krátká, & N. Stelikova (Eds.), *Proceedings of the 30th Conference of the International Group for the Psychology of Mathematics Education* (Vol. 3, pp. 353–360). Prague, Czech Republic: International Group for the Psychology of Mathematics Education.
Hjh Kamsiah bte Hj Ismail (2003). *Percentages, linked to decimals and fractions*. Bandar Seri Begawan, Brunei Darussalam: Brunei Darussalam Ministry of Education.
Kilpatrick, J. (2003). Twenty years of French *didactique* viewed from the United States. *For the Learning of Mathematics, 23*(2), 23–27.

Krainer, K., & Wood, T. (2008). Participants in mathematics teacher education: Individuals, teams, communities and networks. In T. Wood, B. Jaworski, K. Krainer, D. Tirosh, & P. Sullivan (Eds.), *The international handbook of mathematics teacher education* (Vol. 3, pp. 231–254). Rotterdam, The Netherlands: Sense Publishers.

Lester, F. K. (Ed.). (2007). *Second handbook of research on mathematics teaching and learning*. Charlotte, NC: Information Age and NCTM.

Lewis, C., Perry, R., & Hurd, J. (2009). Improving mathematics instruction through lesson study: A theoretical model and North America case. *Journal of Mathematics Teacher Education, 12*(4), 285–304.

Lewis, C., & Tsuchida, I. (1998). A lesson is like a swiftly flowing river: Research lessons and the improvement of Japanese education. *American Educator, Winter,* 14–17, 50–52.

Lim, G. K., & Hjh Zarinah bte Hj Jamudin (2003). *Expanding the modes of communication (including technology) in primary mathematics classrooms*. Bandar Seri Begawan, Brunei Darussalam: Brunei Darussalam Ministry of Education.

Lim, C. S., White, A. L., & Chiew, C. M. (2005). Promoting mathematics teacher collaboration through Lesson Study: What can we learn from two countries' experience? In A. Rogerson (Ed.), *Reform, revolution and paradigm shifts in mathematics education* (pp. 135–139) (Mathematics Education into the 21st Century Project: Proceedings of Eighth International Conference). Johor Bahru, Malaysia: Mathematics Education into the 21st Century Project.

Lingard, B., & Renshaw, P. (2010). Teaching as a research-informed and research informing profession. In A. Campbell & S. Groundwater-Smith (Eds.), *Connecting inquiry and professional learning in education* (pp. 26–39). Oxford, UK: Routledge.

Hjh Mardiah & Hjh Shimawati (2004). *Evaluation report: Active Mathematics in Classrooms (AMIC)*. Unpublished paper for the PS 3302 "Science and Mathematics Teachers as Researchers" course at Universiti Brunei Darussalam.

Maria bte Abdullah & Hjh Ramnah bte Pg Hj Abdul Rajak (2003). *Developing number sense throughout the primary mathematics curriculum*. Bandar Seri Begawan, Brunei Darussalam: Brunei Darussalam Ministry of Education.

McIntyre, D. (1997). The profession of educational research. *British Educational Research Journal, 23*(2), 127–140.

Mohammad Ariffin bin Hj Bakar (2003). *Teaching mathematics through a problem-solving and problem-posing approach*. Bandar Seri Begawan, Brunei Darussalam: Brunei Darussalam Ministry of Education.

Muchtar, A. K., & Sutarto, H. (2010). *Concepts of lesson study and establishing system for the proposed lesson study activities*. Yogyakarta, Indonesia: Regional Centre for Quality Improvement of Teachers and Educational Personnel (QITEP).

Murata, A. (2008). Mathematics teaching and learning as a mediating process: The case of tape diagrams. *Mathematical Thinking and Learning, 10*(4), 374–406. Retrieved from http://www.informaworld.com/smpp/title~db=all~content=t775653685, http://www.informaworld.com/smpp/title~db=all~content=t775653685~tab=issueslist~branches=10#v10, http://www.informaworld.com/smpp/title~db=all~content=t775653685~tab=issueslist~branches=10#v10, http://www.informaworld.com/smpp/title~db=all~content=g904609264.

Neubrand, M., Seago, N., Agudelo-Valderrama, C., DeBlois, L., & Leikin, R. (2009). The balance of teacher knowledge: Mathematics and pedagogy. In T. Wood (Ed.), *The professional education and development of teachers of mathematics: The 15th ICMI study* (pp. 211–225). New York, NY: Springer.

Newman, M. A. (1977). An analysis of sixth-grade pupils' errors on written mathematical tasks. *Victorian Institute for Educational Research Bulletin, 39*, 31–43.

Newman, M. A. (1983). *Strategies for diagnosis and remediation*. Sydney, Australia: Harcourt, Brace Jovanovich.

Niss, M. (2007). The concept and role of theory in mathematics education. In C. Bergsten, B. Grevholm, H. S. Masoval, & F. Ronning (Eds.), *Relating practice and research in mathematics* education: Proceedings of NORMA 05, Fourth Nordic Conference of Mathematics Education (pp. 97–110). Trondheim, Norway: Tapire Academic Press.

Norjah bte Hj Burut, Hj Rozaimah bte Hj Abdul Wahid, & Hj Tini bte Hj Sani (2003). *Word problems need not be so difficult.* Bandar Seri Begawan, Brunei Darussalam: Brunei Darussalam Ministry of Education.

Perks, P., & Prestage, S. (2008). Tools for learning about teaching and learning. In T. Wood, B. Jaworski, K. Krainer, D. Tirosh, & P. Sullivan (Eds.), *The international handbook of mathematics teacher education* (Vol. 4, pp. 265–280). Rotterdam, The Netherlands: Sense Publishers.

Ponte, J. P. (1994). Mathematics teachers' professional knowledge. In J. P. da Ponte & J. F. Matos (Eds.), *Proceedings of the Eighteenth PME Conference* (Vol. 1, pp. 195–210). Lisbon, Portugal: The University of Lisbon.

Ponte, J. P., & Chapman, O. (2008). Preservice mathematics teachers' knowledge and development. In L. D. English (Ed.), *Handbook of international research in mathematics education* (2nd ed., pp. 223–261). New York, NY: Routledge.

Pring, R. (2004). *Philosophy of education.* London, UK: Continuum.

Pritchard, R., & Bonne, L. (2007). Teachers research their practice: Developing methodologies that reflect teachers' perspectives. In J. Watson & K. Beswick (Eds.), *Mathematics: Essential research, essential practice (Proceedings of the 30th Annual Conference of the Mathematics Education Research Group of Australasia)* (pp. 621–630). Adelaide, Australia: Mathematics Education Research Group of Australasia.

Robertson, R. (1995). Globalization: Time-space and homogeneity-heterogeneity. In M. Featherstone, S. Lash, & R. Robertson (Eds.), *Global modernities* (pp. 25–45). London, UK: Sage.

Rowland, T. (2008). Researching teachers' mathematics disciplinary knowledge. In T. Wood, B. Jaworski, K. Krainer, D. Tirosh, & P. Sullivan (Eds.), *The international handbook of mathematics teacher education* (Vol. 3, pp. 273–300). Rotterdam, The Netherlands: Sense Publishers.

Rowland, T. (2009). Beliefs and actions in university mathematics teaching. In M. Tzekaki, M. Kaldrimidou, & H. Sakonidis (Eds.), *Proceedings of the 33rd Conference of the International Group for the Psychology of Mathematics Education* (Vol. 5, pp. 17–24). Thessaloniki, Greece: International Group for the Psychology of Mathematics Education.

Rowland, T., Huckstep, P., & Thwaites, A. (2005). Elementary mathematics teachers' subject knowledge: The knowledge quartet and the case of Naomi. *Journal of Mathematics Teacher Education, 8*(3), 255–281.

Rozina bte Awg Hj Salim (2003). *Finding out why children make mistakes—And then doing something to help them.* Bandar Seri Begawan, Brunei Darussalam: Brunei Darussalam Ministry of Education.

Ruthven, K., & Goodchild, S. (2008). Linking researching with teaching: Towards synergy of scholarly and craft knowledge. In L. D. English (Ed.), *Handbook of international research in mathematics education* (2nd ed., pp. 561–588). New York, NY: Routledge.

Shimizu, Y. (1996). Some pluses and minuses of "typical pattern" in mathematics lessons: A Japanese perspective. *Bulletin of the Center for Research and Guidance for Teaching Practice, 20,* 35–42.

Shimizu, Y. (1999a). Aspects of mathematics teacher education in Japan: Focusing on teachers' roles. *Journal of Mathematics Teacher Education, 2*(1), 107–116.

Shimizu, Y. (1999b). Studying sample lessons rather than one excellent lesson: A Japanese perspective on the TIMSS Videotape Classroom Study. *Zentralblatt für Didaktik der Mathematik (ZDM), 99*(6), 191–195.

Shimizu, Y. (2002, April). *Discrepancies in perceptions of lesson structure between the teacher and the students in the mathematics classroom.* Paper presented at the interactive symposium, "International Perspectives on Mathematics Classrooms," at the Annual Meeting of the American Educational Research Association, New Orleans.

Shulman, L. S. (1986). Those who understand: Knowledge growth in teaching. *Educational Researcher, 15*(2), 4–14.

Shulman, L. S. (1987). Knowledge and teaching: Foundations of the new reform. *Harvard Educational Review, 57,* 1–22.

Simon, M. (2008). The challenge of mathematics teacher education in an era of mathematics education reform. In T. Wood, B. Jaworski, K. Krainer, D. Tirosh, & P. Sullivan (Eds.), *The international handbook of mathematics teacher education* (Vol. 4, pp. 17–30). Rotterdam, The Netherlands: Sense Publishers.

Slavit, D., & Nelson, T. (2010). Collaborative teacher inquiry as a tool for building theory on the development and use of rich mathematical tasks. *Journal of Mathematics Teacher Education, 13*(3), 201–221.

Southwell, B., & White, A. L. (2004). Lesson study professional development for mathematics teachers. In M. Hoines & A. Fuglestad (Eds.), *Proceedings of the 28th Conference of the International Group for the Psychology of Mathematics Education* (Vol. 1, pp. 355–365). Bergen, Norway: International Group for the Psychology of Mathematics Education.

Steinbring, H. (1998). Elements of epistemological knowledge for mathematics teachers. *Journal of Mathematics Teacher Education, 1*(2), 157–189.

Stenhouse, L. (1984). Evaluating curriculum evaluation. In C. Adelman (Ed.), *The politics and ethics of evaluation* (pp. 77–86). London, UK: Croom Helm.

Stigler, J., & Hiebert, J. (1998). *Teaching is a cultural activity* (pp. 1–10). Winter: American Educator.

Stigler, J., & Hiebert, J. (1999). *The teaching gap: Best ideas from the world's teachers for improving education in the classroom*. New York, NY: Free Press.

Stromquist, N. P., & Monkman, K. (2000). Defining globalization and assessing its implications on knowledge and education. In N. P. Stromquist & K. Monkman (Eds.), *Globalization and education* (pp. 3–25). Lanham, MD: Rowman & Littlefield.

Van den Heuvel-Panhuizen, M. (2001). Realistic mathematics education in the Netherlands. In J. Anghileri (Ed.), *Principles and practices in arithmetic teaching: Innovative approaches for the primary classroom* (pp. 49–63). Buckingham, UK: Open University Press.

Wells, G. (1999). *Dialogic inquiry: Towards a sociocultural practice and theory of education*. Cambridge, UK: Cambridge University Press.

Wenger, E. (1998). *Communities of practice: Learning, meaning and identity*. Cambridge, UK: Cambridge University Press.

White, A. L. (2004a). The long-term effectiveness of lesson study, a New South Wales mathematics teacher professional development program. In I. P. A. Cheong, H. S. Dhindsa, I. J. Kyeleve, & O. Chukwu (Eds.), *Globalisation trends in science, mathematics and technical Education 2004* (pp. 320–338). Gadong, Brunei Darussalam: Universiti Brunei Darussalam.

White, A. L. (2004b). *Evaluation report, Active Mathematics in Classrooms: A professional development program for primary school teachers in Brunei Darussalam*. Bandar Seri Begawan, Brunei Darussalam: Brunei Darussalam Ministry of Education.

White, A. L. (2006). An examination of the international manifestations of lesson study that began as a Japanese model of teacher professional learning. In H. S. Dhindsa, I. J. Kyeleve, O. Chukwu, & J. S. H. Q. Perera (Eds.), *Shaping the future of science, mathematics and technical education* (pp. 129–138). Gadong, Brunei Darussalam: Universiti Brunei Darussalam.

White, A. L. (2008). *Counting On: Evaluation of the impact of Counting On 2007 program*. Sydney, Australia: Curriculum K-12 Directorate, Department of Education and Training. Retrieved January 11, 2011 from http://www.curriculumsupport.education.nsw.gov.au/secondary/mathematics/assets/pdf/counting_on/co_eval_2007.pdf.

White, A. L. (2009) *Counting On 2008 final report*. Sydney, Australia: Curriculum K-12 Directorate, Department of Education and Training. Retrieved from http://www.curriculumsupport.education.nsw.gov.au/secondary/mathematics/assets/pdf/counting_on/co_eval_2008.pdf.

White, A. L. (2010). *Counting On in the Middle Years 2009: Program evaluation report*. Sydney, Australia: Curriculum K-12 Directorate, Department of Education and Training. Retrieved from http://www.curriculumsupport.education.nsw.gov.au/secondary/mathematics/assets/pdf/counting_on/co_eval_2009.pdf.

White, A. L., & Clements, M. A. (2005). Energising upper-primary mathematics classrooms in Brunei Darussalam: The Active Mathematics in Classrooms (AMIC) Project. In H. S. Dhindsa, I. J. Kyeleve, O. Chukwu, & J. S. H. Q. Perera (Eds.), *Future directions in science, mathematics and technical education* (pp. 151–160). Brunei Darussalam: Universiti Brunei Darussalam.

White, A. L., & Lim, C. S. (2007). Lesson study in a global world. In C. S. Lim, S. Fatimah, G. Munirah, M. Y. Hashimah, W. L. Gan, & T. Y. Hwa (Eds.), *Meeting the challenges of developing quality mathematics education* (pp. 567–573). Penang, Malaysia: Universiti Sains Malaysia.

White, A. L., & Lim, C. S. (2008). Lesson study in Asia Pacific classrooms: Local responses to a global movement. *ZDM—The International Journal of Mathematics Education, 40*(6), 915–925.

White, A. L., & Southwell, B. (2003). Lesson study: A model of professional development for teachers of mathematics in years 7 to 12. In L. Bragg, C. Campbell, G. Herbert, & J. Mousley (Eds.), *Mathematics education research: Innovation, networking, opportunity: Proceedings of the 26th Annual Conference of the Mathematics Education Research Group of Australasia* (Vol. 2, pp. 744–751). Geelong, Australia: Deakin University.

Witmann, E. C. (1998). Mathematics education as a "design science." In A. Sierpinska & J. Kilpatrick (Eds.), *Mathematics education as a research domain: A search for identity* (pp. 87–103). Dordrecht: Kluwer Academic Publishers.

Wood, T., Jaworski, B., Krainer, K., Tirosh, D., & Sullivan, P. (Eds.). (2008). *The international handbook of mathematics teacher education*. Rotterdam, The Netherlands: Sense Publishers.

Hjh Yunaidah bte Hj Yunus (2003). *Fractions: Going beyond area models to assist student learning*. Bandar Seri Begawan, Brunei Darussalam: Brunei Darussalam Ministry of Education.

Zack, V., Mousley, J., & Breen, C. (Eds.). (1997). *Developing practice: Teachers' inquiry and educational change*. Geelong, Australia: Centre for Studies in Mathematics, Science and Environmental Education, Deakin University.

Zaslavsky, O. (2008). Meeting the challenges of mathematics teacher education through design and use of tasks that facilitate teacher learning. In T. Wood, B. Jaworski, K. Krainer, D. Tirosh, & P. Sullivan (Eds.), *The international handbook of mathematics teacher education* (Vol. 4, pp. 93–114). Rotterdam, The Netherlands: Sense Publishers.

Chapter 14
Developing Mathematics Educators

Jarmila Novotná, Claire Margolinas, and Bernard Sarrazy

Abstract This chapter addresses, from various perspectives, issues associated with teacher education and its development. Several categories of mathematics educators are characterized and their development and roles in the teaching/learning processes are summarized. Cooperation between teachers and researchers as well as the concept of teachers as researchers are discussed from different points of view. The crucial role that observations play at all levels is analyzed and illustrated by two different models of implementation of observations into teachers' and researchers' practice. Throughout the chapter the influence of the research of Guy Brousseau on mathematics education research is recognized.

Introduction

> One of the functions of didactics could be ... to contribute to the deceleration of the process of transformation of knowledge into algorithms ... To sacrifice to the god of contemporary worship to the so-called efficiency, education follows the path of algorithmic reduction and demathematization. I deeply hope that didactics will be victorious in the battle of this dispossession and dehumanization.
>
> Guy Brousseau, 1989 (translation from French, p. 68)

Our first task, in this reflection upon the development of mathematics educators, is to consider the question: "Who is a mathematics educator?" In fact, different

J. Novotná (✉)
Charles University in Prague, Prague, Czech Republic
e-mail: jarmila.novotna@pedf.cuni.cz

C. Margolinas
Laboratoire ACTé, Université Blaise Pascal, Clermont-Ferrand, France
e-mail: claire.margolinas@univ-bpclermont.fr

B. Sarrazy
Université Victor Segalen Bordeaux, Bordeaux, France
e-mail: bernard.sarrazy@u-bordeaux2.fr

answers have been given to this question by different authors. In the first part of this essay, we will discuss different ways of thinking about mathematics education and about mathematics educators, and will establish the crucial concept of "observation" in defining a mathematics educator.

The second part of the chapter will focus on the role of observation in the development of mathematics educators of various kinds. We will show how different vantage points in mathematics education can influence observational schemes and approaches to teaching mathematics. In order to illustrate different aspects of the cooperation of teachers and researchers, we will present two examples of the use of observations in mathematics education research and in the search for phenomena in mathematics. The first will be COREM (Centre for Observation and Research in Mathematics Education—*Centre d'Observation et de Recherche sur l'Enseignement des Mathématiques*), which is an example of successful cooperation of teachers and researchers; the second example will be the Learner's Perspective Study (LPS). These two projects have presented examples of different ways of observing and researching realities in mathematics classrooms. There are other perspectives that have been successfully applied in this field—for instance the ongoing comparative study of teacher education, "Teacher Education and Development Study in Mathematics" (TEDS-M) focusses on the preparation of teachers of mathematics at the primary and lower secondary levels (for more details see teds.educ.msu.edu). We do not intend to provide an exhaustive list of these other examples. Later in the chapter we will consider the role of observation within the increasingly important issues associated with the use of information and communication technologies (ICT) in mathematics education.

As mentioned by Adler, Ball, Krainer, Lin, and Novotná (2005), there is much less written about mathematics teacher educators than about teacher education itself. So in the third part of this chapter we will focus on mathematics teacher educators. The central question is "How does a person become a mathematics educator and/or a mathematics education researcher?" Based on two examples, some important aspects are identified, and these are more deeply discussed in the fourth part of the chapter. This fourth part offers a discussion of the central question of relationships between research and mathematical education, especially in didactics. This discussion will provide a synthesis of the themes covered in the chapter. We will argue that it is important to enhance teachers' didactical cultures without damaging their pedagogical beliefs.

Mathematics Education and Mathematics Educators

In this first part of the chapter, we consider different meanings of the term "mathematics education" and address the question of who mathematics educators are. In fact, various institutions involved in teacher education have their own meanings for "educator" and "mathematics educator," and if we can better understand these different meanings, then we might understand more fully what knowledge is important or required of mathematics educators. That is the main issue for this chapter.

Mathematics Education: Education to Mathematics

Although mathematics is a very old body of knowledge it is always growing. It has a history of having strong relationship with the mastery of vital aspects of reality (quantification, measures, etc.). Furthermore, the development of physics, chemistry, biology, and also economics, etc. has revealed other aspects of mathematics which offer the possibility of secure deductive reasoning. Therefore, mathematics qualifies as a body of knowledge which is universally transmitted inside various societies across the world. In this sense, "mathematics education" can be taken to mean "education to mathematics."

It can be argued that mathematics has a recursive or "Russian doll" structure: a concept that was initially constructed as a tool in order to anticipate the result of an action (e.g., integer as a tool to describe two sets of the same quantity) is considered as an object in another situation (e.g., integer as an already constructed object in the problem "what number must be added to 5 in order to obtain 22?") (Douady, 1991). Brousseau (1997) considered these two aspects as a part of the dialectic between knowledge and knowing. This aspect of mathematics is one of the reasons for the need to learn mathematics at an early stage and to continue learning it over a very long period. Mathematics is at the same time "independent of the world" (Wittgenstein, 1983) and yet something which contributes to the formation of citizens.

Therefore, if we consider "mathematics education" as the social answer to the need to educate people in mathematics, the first meaning of "mathematics educator" is "a person who is in charge of mathematics education." That meaning defines a very large category that includes parents and more generally those adults who are in charge of children's care, and teachers at all levels (from primary to tertiary education). In this chapter we consider all teachers that are in charge of teaching mathematics at any level to be mathematics teachers.

It is well known that many people have opinions, mostly based on observations of their own children, about what mathematics teachers ought to learn in order to improve their teaching. They may in fact be considered as the most basic kind of mathematics educators. However, often their point of observation is very limited, since they implicitly consider their own teaching practices as the central basis for their reflections on the nature of mathematics education.

Mathematics Education: Observing the Learning of Mathematics

Within the development of human sciences, every aspect of human activity may be subject to observation. "Learning mathematics" is therefore a legitimate field of investigation. The elements involved are the subject—child, pupil, person in general—the mathematical knowledge, and the observers of interactions.

Since mathematics is learned by children in their early years, observers of the learning of mathematics are sometimes psychologists, who consider mathematics as a system of "logic." Psychologists generally do not question the mathematics involved (which is considered to be a permanent body of knowledge) and prefer to focus on the development of children in relation to their environment (Piaget, 1985) and in relationship to parents, siblings and early childhood mathematics educators (Bruner, 1966; Vygotsky, 1962).

In school, mathematics is taught and the teachers themselves observe their own students as those students are learning mathematics. The teachers are especially concerned with whether their students are learning what they have been taught. Under certain conditions, teachers may develop further their observations by reflecting on what their students are actually learning. They might also consider which variables are involved in the learning process, and what might be the effects on learning if certain conditions were to be modified.

Since a transmission of mathematics occurs when someone learns mathematics, mathematicians may be interested in observing the learning of mathematics. Such observations would, most likely, be centred on the nature of the mathematical knowledge involved: what does this child know about the mathematics? Is the knowledge that the learner has acquired adequate with respect to my own experience and understanding of mathematics?

Therefore, if we consider mathematics education as the field of observation of the learning of mathematics, the second meaning of "mathematics educator" is "a person who observes mathematics learning." This category of mathematics educators includes teachers, mathematicians, researchers in psychology, and researchers in mathematics education. Psychology, in particular, has had a strong influence on mathematics teacher education. Often, theories from psychology have been assumed to provide satisfactory theoretical backgrounds for mathematics education, with actual mathematics teaching being regarded as an application of such theory. What has been lacking in all of this has been the teaching processes, which include both the didactic transposition that interrogates the mathematics itself (Chevallard, 1985), and some consideration of teachers' attempts to cope with the different resources and constraints within their teaching situations (Margolinas, 2002).

Mathematics Education: Observing the "Learning and Teaching" of Mathematics

As we stated above, the learning of mathematics mainly depends on the teaching of mathematics. The teaching is a conscious attempt to help learners acquire mathematical knowledge. Therefore, another field of investigation might be focussed on "learning and teaching mathematics." The elements involved are: pupil, teacher (in a broad sense: a parent, when deliberately educating his or her child about mathematics, is a teacher; university professors in mathematics are also teachers, etc.), setting, mathematical knowledge, observer.

When considering the "learning and teaching" system we can place the observer as an "outsider," someone who observes the interactions between pupil–teacher–mathematics (see Figure 14.1).

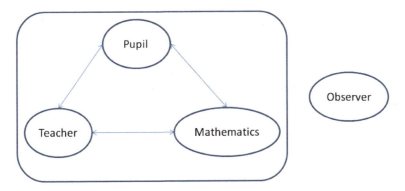

Figure 14.1. An observer outside the interactions between pupil–teacher–mathematics.

The status of the observer has some important consequences. The teacher herself or himself may be an observer, and in this case she or he is a self-observer, which is a difficult vantage point in which to be placed. Or, the "outside" observer may be another teacher, who may be inclined to identify with the teacher. Or the observer may be a mathematician, and in that case will be likely to focus on the knowledge involved and the explicit formulation of this knowledge. The observer could also be a teacher educator who wants to give advice to the teacher about how to cope with the situation, or the teacher's supervisor who has the specific task of evaluating the teacher's effectiveness.

The focus of the observer, then, is partly determined by his or her professional occupation. But the focus can also be determined by the theoretical framework of the observer or the purpose of the observation. This kind of observation may be made by a researcher who wants to increase knowledge about phenomena which occur in the learning and teaching situation. That person might be called a mathematics education researcher.

Therefore, if we consider mathematics education as the field dealing with the observation of the learning and teaching of mathematics, the third meaning of "mathematics educator" is "a person who observes mathematics learning and teaching." This category of mathematics educator can include the teacher (as a self-observer), a teacher educator, a mathematician, the teacher's supervisor, or a mathematics education researcher. What mathematics education and in particular mathematics didactics has stressed is that we need to take into account the whole didactic system (pupil–teacher–mathematics) in order to understand mathematics teaching. It is possible to focus on some of the relations (e.g., pupil–mathematics) but one should not forget the role of the teacher altogether. It is therefore crucial for mathematics teacher education that a scientific field that focusses on the phenomenon that are specific of the *entire* didactic system be developed and that mathematics educators are well informed of its main theoretical perspectives and results.

Observation as an Efficient Tool for the Knowledge Development of Mathematics Educators

What are the sources which assist a mathematics educator's development? They cannot be the same for all categories of mathematics educators mentioned in the first part of this chapter. We now summarize some of the sources associated with the different categories. The summary will not be exhaustive but it will provide some idea of the complexity of the domain.

Student Teachers

When preparing and teaching mathematics lessons, prospective teachers are profoundly influenced by mentor teachers (Cavanagh & Prescott, 2007; Vacc & Bright, 1999). Nathan and Petrosino (2003) point to the intersection between the two knowledge bases, pedagogical and mathematical; they state that preservice teachers with advanced content knowledge in mathematics have the tendency to think beyond their own content expertise when considering their students' possible reactions to the content.

Teachers

Here we draw on the burgeoning research literature on the sources of information concerning the ways teachers influence student thinking and understanding (e.g., Carpenter, Fennema, & Franke, 1996). Kinach (2002) emphasized the importance of a teacher's content knowledge when asking questions of students, anticipating likely responses, and evaluating students' responses. Feiman-Nemser (2001) drew attention to the influence on teachers of the knowledge and experiences of mentors and colleagues. Several other writers have contrasted experienced and novice teachers: when anticipating students' likely mathematical responses, experienced teachers mobilize a number of resources that novices do not have, including their past observations of students learning mathematics and their self-observations of their own teaching (Sherin, 2002). Experience in anticipating responses can help teachers identify and state learning goals embedded in a mathematical task. Research suggests that novice teachers benefit greatly from opportunities to gain experience in this domain (Morris, Hiebert, & Spitzer, 2009).

It is important for teacher educators to understand the ways in which teachers make use of available resources in their everyday teaching practices. An intermediary between research and teaching may become *journals*. In general, journals dealing with mathematics education may be classified into three groups—those aimed at (a) students and non-specialists interested in mathematics; (b) teachers of mathematics;

and (c) mathematics education researchers. When the focus is on mathematics educators, the last two categories are of the special interest.

The objective of many professional development activities is the improvement of teachers' knowledge of mathematics. But teachers often consider content knowledge as being less valuable to them than getting acquainted with the practical ideas for teaching (Wilson & Berne, 1999). Observations offer mathematics educators a wide range of both practical and theoretical information.

In the first part of this chapter we noted that different kinds of observations can be associated with different meanings for the term "mathematics educator." In this part, we show that different kinds of observations are necessary to develop the knowledge of these different kinds of mathematics educators (including mathematics education researchers). We also discuss different structures that have been used for observing mathematics education. We show that different observational vantage points can be somehow connected, even with mathematics education research. Thus, for example, when observing the educational system a researcher may adopt a position of "expert" which is very similar to the position adopted by institutional decision makers. Different vantage points can provoke different types of "observations."

Here we restrict our focus to teachers and researchers as the two main groups of mathematics educators. We show that many activities precipitate observations of different kinds. These may have the same nature and purpose, but are not based on the same knowledge and do not call into play, or monitor, the same set of variables.

The different vantage points and interests of teachers and researchers in the observation processes have been studied by authors from different perspectives. Thus, for example, Margolinas, Coulange, and Bessot (2005) focussed on teachers' learning from different situations, and Novotná, Lebethe, Rosen, and Zack (2003) focussed on differences between the roles of teachers and researchers.

In Figure 14.1, the general scheme for observation was presented. But, if we consider the different foci for the two groups of mathematics educators—teachers and researchers—we see substantial differences. Figure 14.2 represents possible perspectives for a teacher, and Figure 14.3 for a researcher.

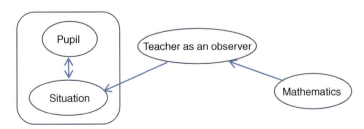

Figure 14.2. Interactions between pupil–teacher–mathematics, from a teacher's vantage point.

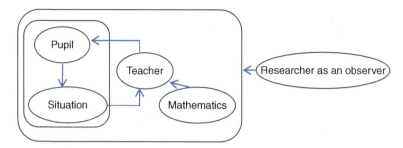

Figure 14.3. Interactions between pupil–teacher–mathematics, from a researcher's vantage point.

The main differences are in the observational role of the observer—the purpose of observation—and in the knowledge that the observer possesses. Although the teacher is interested mainly in a posteriori analysis of the teaching unit (comparison of the lesson plan and a priori analysis with the realities in the classroom and explication of differences among them leading to modifications of the unit design), the researcher's fundamental interest is in discovering general phenomena that influenced the development of the educational situation.

We also wish to focus, here, on mutual relationships between and influences of mathematics educators. We also discuss different structures that have been used in order to observe mathematics education, especially in COREM and in the LPS project.

Differences Between Teachers' and Researchers' Positions in Mathematics Education

The similarities and differences in school and research vantage points and practices were described by Brousseau (2002) in the following terms:

> When I am a *didactician*, the interpretation of every step of teaching begins with a systematic informing, a complex work of the analysis a priori and the confrontation with various aspects of contingency, of observations viewed and rejected later, etc. There is not an evident separation of what is relevant but inadequate, adequate but inadaptable, eligible but inconsistent, as well as transformations of appearance and certainties in falsifiable questions, etc. When I am a *teacher*, I have to take a number of instantaneous decisions in every moment based on the real information got in the same moment. I can use only very few of the subtle conclusions of my work as didactician and I have to fight with starting to pose myself questions which are not compatible with the time that I have and that finally have the chance to be inappropriate for the given moment. I react with my experience, with my knowledge of my pupils, with my knowledge of a teacher of mathematics which I am treating. All these things are not to be known by the didactician.

Differences between the roles of a teacher and a researcher were addressed in a panel session at the 27th annual PME conference held in Honolulu in 2003 (Novotná et al., 2003). We now consider some of the differences.

Teacher as a Researcher

In the text *Navigating Between Research and Practice: Finding My Own Way* (Novotná et al., 2003), Vicki Zack described development as a mathematics educator in the following way:

> My questions emanate from neither theory nor practice alone but from the juxtaposition of the two, and from critical reflection on the intersection between the two (Cochran-Smith & Lytle, 1993, p. 15) in areas which are of intense and enduring interest to me. There is recursiveness in the process, wherein questions are continuously reformulated, extended, re-visited, methods are revised and analysis is on-going. … I recognize the value of practical knowledge, and also respect the place research can hold in informing practice. However, I emphasize the challenge involved in understanding others' ideas. (p. 87)

Although this process in the development of a mathematics educator is individual it has common features. As Zack and Graves (2001) have emphasized, each person appropriates, reworks, re-accentuates while making her or his own way. A fundamental part of this development should be making meaning of the research and associated theoretical issues, and seeing what they might mean for the teacher's work, and for the children, who are making meaning of the mathematics as they work together with their teacher, with their peers in the classroom and, at times, with their parents at home.

Questions which become important for teachers are: How do my students proceed when asked to "prove" that they are correct? What do they consider valid arguments for proving their case and convincing others? What language do they use when presenting their arguments? What kinds of reasoning do they use: inductive, deductive, other? (Novotná et al., 2003).

Cooperation of Teachers and Researchers

The cooperation of teachers and university-based educators (in the following text we refer to them as researchers) in research teams in mathematics education is a broad and relevant topic. In most cases, the focus is on improving the quality of mathematics teaching and learning (see, e.g., Brown & Coles, 2000). Many discussions have been carried out within the last decade about the impact of this type of cooperation in mathematics education (see, e.g., Goos, 2008). Identifying and contrasting the different experiences and knowledge of teachers and researchers have been a focus of investigation in numerous studies (see, for example, contributions by Bennie, Breen, Brown, Hošpesová, Coles, Lebethe, Eddy, Macháčková, Novotná, Pelantová, Poirier, Reid, Rosen, Tichá, Zack in Novotná et al., 2006). Chris Breen (2003) drew attention to the contrasting views on the contributions that teachers are making to the field of mathematics education. Although there is a movement for more teachers to become involved in critical explorations of their practice, through such methods as critical reflection, action research, and lesson studies, some sceptics claim that these activities have done little to add to the body of knowledge in mathematics education.

Despite such controversy, there seems to be little doubt that cooperation within and between communities of practice enriches research in mathematics education. However, the components of this type of cooperation, and how the interactions of these components change teachers' opinions and approaches, are much less investigated. Without paying attention to teacher change, the results of many research activities can seem to be less significant than they actually are.

We now summarize an example of fruitful cooperation between teachers and researchers. The research project was originally designed by Guy Brousseau and Jarmila Novotná, and data collection, analysis and evaluation of the experiment were carried out in cooperation with secondary school teachers in Prague, the Czech Republic.

The experiment which was designed incorporated the following steps:

- Design of the didactical situations that were intended to change learners' approaches to solving problems.
- Development and implementation of the proposed didactical situations.
- Analysis of the implementation and, based on the experiment results, and reflections on possible modifications.

Even though the primary target group of the research comprised secondary school students, the research provided an opportunity for the participating teachers to develop their professional competences.

The influence of teacher attitudes and teaching has been formulated by Jaworski (2003):

> The action research movement has demonstrated that practitioners doing research into their own practice ... learn *in* practice through inquiry and reflection. There is a growing body of research which provides evidence that *outsider* researchers, researching the practice of other practitioners in co-learning partnerships, contribute to knowledge *of* and *in* practice within the communities of which they are a part. (p. 2)

We illustrate changes that were identified among teachers in this collaborative group exercise through the examples of two teachers, who will be referred to as Teacher *A* and Teacher *B* in the following text. The following extracts are from their self-reflections:

Teacher *A*'s reflections.

- *Experienced a "new" role as a teacher during the adidactical situation.* The teacher should rather become an observer, moderator of discussions and of the work in the classroom. This role is demanding, and from the perspective of traditional teaching, unusual. When you listen to students during group work and see that they are very close to the solution, it is not easy to answer their question without intervening in their work.
- *Gained experience in moderating students' discussion.* I learned to listen and intervene only when it was a must. If I intervene too soon there is a danger that I

divulge to students something that they could find out themselves if I had given them more space.
- **Gained experience with group work.** Before the project, I used groups very rarely. I was afraid that I would not succeed in involving all students in the activity, to be able to get all of them actively participating. The experiment showed that with an appropriate choice of activities, this is possible.
- **Gained experience with the student peer control.** The teacher is not the only one who can tell students what is correct and what is not. It proved to be more efficient when this evaluation was formulated by the students' own schoolmates.

Teacher *B*'s reflections.

- *Realized that I tended to underestimate my students' abilities.* This experiment showed me the conflicts between my expectations and what the students could really do. At the beginning, I was embarrassed that I did not manage to get from them what I wanted, but it motivated me to a deeper reflection on the ways of presenting the stages to students. At present, I find that it is not a negative if students do something differently, because we can all learn from it.
- *Benefited from gaining feedback from students.* The experiment made me want to get feedback from the students. Getting feedback should become an integral part of my work as a teacher. Before the project, I could not imagine that more fruitful discussions can took place in mathematics lessons than in lessons for other subjects.
- *Gained experience in organizing research projects.* I noticed a shift. In the beginning, I devoted myself solely to organizational items, such as the number of problems, or dividing students into groups. After gaining experience I found that I was attending to more fundamental issues, such as the definition of a mathematical model for a problem, or exploring conceptual links between aspects of the mathematics.

We observed a change in the teachers' perceptions with respect to the use of student problem posing: observing their own students in these situations broadened their knowledge about students learning mathematics. Before participation in the project, teachers were used to assigning problems to students themselves; they saw it mostly as the only appropriate way for managing the teaching/learning process. Their fears had almost certainly been influenced by their own experiences in their own schooling.

Indeed, the project considerably influenced all members of the collaborative group—the teachers as well as the researchers. It was recognized that if the work of the team was to be successful then all the participating persons needed to collaborate fruitfully. The result was changes could be observed, and not only on the teachers' side, or on the students' side, or on the researchers' side. The researchers certainly gained much from the collaboration, and the teachers' inputs helped to consolidate the experimental settings and to analyse the project results.

Interaction Between Observation and the Development of Theory in Mathematical Education: COREM and the Theory of Situations

Brousseau's ideas were successfully implemented at the Jules Michelet School, Talence, France, between 1973 and 2000. The overall project is referred to as COREM, which was created in 1973 with the following objectives (from Salin & Dreslard Nédélec, 1999):

- To conduct research necessary for the advancement of knowledge of mathematics education phenomena.
- To conceive and study new educational situations that will generate better learning of mathematics by pupils.
- To develop in this way a corpus of knowledge necessary for teacher education.

It is important to stress that Jules Michelet School was never an experimental school conceived to improve mathematics teaching or to educate the teachers of this particular school (even if it may have also this result in both cases). The Centre was conceived in order to allow a vast community of researchers to observe the real teaching process in an entire school. The scope was from the beginning a typical scientific project: to understand better didactical phenomena and not to directly implement any innovative teaching. In COREM there was always close collaboration between researchers from the university, teacher educators, elementary school teachers, pupils aged from 3 to 11, school psychologists and students of didactics of mathematics (Novotná et al., 2003). Two major data sets were generated: (a) a longitudinal collection of qualitative and quantitative information about the teaching of mathematics at the elementary level; and (b) records of two types of observations which were destined to assist in the finding and explaining of phenomena of didactics that were relevant to teaching and to research.

Michelet School consisted of 4 kindergarten and 10 elementary school classes. The school was not selective, and pupils came from a very heterogeneous population. The curricula followed in all subjects were those that applied in all other French schools. The teaching staff consisted of "ordinary" teachers without any special training. Their task was to teach, not to do research. They worked in teams, three teachers for two classes. One-third of their working hours were devoted to COREM. This time consisted of four types of activities: (a) coordinating and preparing the ordinary work of the pupils and discussing all the problems of the school (educational, administrative, social, and so on); (b) directly observing the work in the classroom, for research purposes and for normal feedback; (c) participating with the researchers in the design of sessions to be observed and collecting data about the pupils' behaviours in mathematics; and (d) participating in a weekly seminar at which themes selected by the teachers were discussed.

The daily mathematics activities were designed in collaboration with one teacher educator from a Bordeaux institute for teacher education—before 1991 this was called the Ecole Normale, but in 1991 it became the *Institut Universitaire pour la Formation des Maîtres* (IUFM). The teacher educator monitored the mathematics that the students studied during the whole school year. He was expected to make sure that the research program did not compromise the normal educational activities of the school.

There was one important rule in the decision-making processes practised in the team—specifically, in the case of consensus *not* being reached among participants on any issue, the normal teacher would have the final say about what would be done. Detailed analyses of the teaching units were carried out by the whole team, including the teachers.

The observations were of two types:

1. The first type was of observations of sequences prepared by researchers, together with teachers. In this case, the researcher was responsible for elaborating the project's teaching sequences. The researcher presented the project's sequence to the teachers, including the knowledge it was presumed the pupils would attain by the end of the teaching sequences, the problems to be presented to pupils, and a register of the expected pupils' strategies. When the project was accepted by the team, the next step was the elaboration of teaching sequences. The ideal situation was if the teacher was able to accept the scenario of the lesson directly from the project. If this was not the case, other questions were discussed—like, for example: "What vocabulary should be used in each phase and how and when?" "Should the teacher intervene in the pupils' validation of strategies, and if so, how, and when?" "What should be done if pupils do not respond as expected?" "Are the application exercises necessary?" This collective preparation was set out in the form of a written description and was distributed to the observers in advance.

    The teacher was completely responsible for what happened in the classroom. It included the right to make decisions different from those presumed.

    After the planned sequence of events had been carried out, a first analysis of what had transpired occurred immediately. In this analysis, all participants reconstructed as precisely as possible all the events of the session. Analyses proceeded according to a prescribed order: First the teacher summarized, from her or his point of view, what had been good, and what had not been good, and why. The team discussed any issues that arose, and for unusual happenings looked for explanations of why these had occurred. In such a way the observation strategy included the need for involvement. The discussions provided the researcher with a considerable amount of additional information.

2. The second type was of observations of sequences prepared by teachers themselves. Regular *weekly observation of a series of "ordinary"* lessons—that is to say, observations of lessons that had not been prepared with a researcher—served to identify and explain contingent decisions of "all" teachers. The researcher, who was interested in the overall teaching sequences and patterns during a certain period, organized the observations.

    Teachers and researchers were members of one team at least in the preparatory phase. Their roles were different. In the class, the teacher had the responsibility for pupils. Various distortions could happen: for instance, the researcher might not have formulated expectations adequately, or the teacher might not have understood what had been formulated. Sometimes, the teacher had to make important decisions in order to reach the teaching goals.

The successful functioning of COREM depended on the collaboration of all participating persons as well as much administrative and managerial work. Structures and findings were disseminated in various ways; from allowing interested persons to participate in the whole process, to presenting the organization, functioning and results at conferences and symposia in France and abroad. The teaching processes prepared for observation have never been published or given as a model for use in ordinary classroom conditions.

It is important to remark that although the functions expected of teacher and of a researcher differed, these were not differentiated so far as personal status was concerned. In COREM some persons were both teacher and researcher, but never at the same time or for the same activity. The outcomes of these interactions between an entire school and a team of researchers are enormous. The COREM research was recognized as groundbreaking—The quality and uniqueness of Guy Brousseau's work was quickly recognized, and in 2004, he was the first person to be awarded the prestigious Félix Klein medal by the International Commission on Mathematical Instruction. The importance of Brousseau's work is mainly the development of the Theory of Didactical Situation (Brousseau, 1997), that is considered by a great number of researchers as providing a paradigm for mathematics didactics. Further details about mathematics teaching in COREM have been published for the information of interested researchers or teacher educators (Brousseau, & Warfield, 1999; Brousseau, Brousseau & Warfield, 2001, 2002, 2004a, 2004b, 2007, 2008, 2009).

Researchers Observing "Ordinary Classrooms": The Learner's Perspective Study

There exist many papers describing and analysing observations of a single lesson in a single classroom. Undoubtedly, many of these provide important sources of ideas and phenomena. In this section of the chapter we will focus on another type of observation of ordinary classrooms, by researchers—the Learner's Perspective Study (LPS).

LPS methodology has been developed and applied for teaching mathematics in the eighth grade (Clarke, 2001; Clarke, Keitel, & Shimizu, 2006). The main goal of LPS has been to examine classroom practices in a more integrated and comprehensive way than in other international studies. Originally, the project was designed for in-depth analysis of mathematics classrooms in four countries (Australia, Germany, Hong Kong and the USA), but quickly other countries joined the project. In 2006 there were more than 12 countries contributing to the project materials and analyses.

The Learner's Perspective Study was designed to document the processes and events in mathematics classrooms, but not just the obvious set of events that might be recorded on a videotape. A decision was made to determine how the participants construed those events, including their memories and feelings, and the mathematical and social meanings and practices which arose as a consequence of their beliefs and conceptions. The power of the project has been greatly enhanced by the matching of LPS data from different countries.

A series of research questions were formulated in the initial phase of the project. For example: "Is there evidence of a coherent body of student practice(s), and to what extent are these practices culturally-specific?" "To what extent does an individual teacher employ a variety of pedagogical approaches in the course of teaching a lesson sequence?" "What degree of similarity or difference (both locally and internationally) can be found in the learner (and teacher) practices occurring in classrooms?" "To what extent are teacher and learner practices in a mutually supportive relationship?" "To what extent are particular documented teacher and learner practices associated with student constructions of valued social and mathematical meanings?" (Clarke, 1999).

A major characteristic of this study is its documentation of the teaching of a series of lessons instead of just one single lesson. For each participating teacher, documentation includes video from 10 consecutive lessons, obtained through three cameras in the classroom, together with post-lesson video-stimulated interviews. The common database of materials from the participating countries, with access offered to those who contribute to the project, together with their materials, represents a rich source of materials for analyses and comparative studies of classroom practices from both teachers' and learners' perspectives.

The materials obtained by LPS methodology serve as a rich source of materials for researchers. But at the same time, they represent extremely important material for teachers themselves. Combining video-recordings, the teacher's own preparation of the lessons, the real situation in the classroom and the post-lesson interviews with students provides a teacher with huge feedback and impulse for further development of her or his approaches to teaching.

Observation as a Part of Mathematics Teacher Education

The observation of classroom episodes, in both forms—observation of real classrooms or video-recordings of teaching episodes, is an irreplaceable part of teacher education (Stehlíková, 2007). In contrast to experienced teachers, student teachers usually have not obtained enough experience from real classrooms. So, when they observe lessons, their observations have a modified structure, with the mathematics content being separated from the classroom (see Figure 14.4).

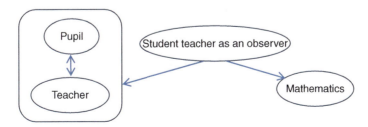

Figure 14.4. Student teacher as an observer.

It is often the case that before a student teacher observes a class she or he is asked to focus on certain features of the lesson that will be observed. Such foci could be any of the following:

Input

Scaffolding
Advance organizers and outlines
Dual code model (verbal and non-verbal representational systems)
Multiple verbal representations
Inductive approaches to learning
Textual support
Graphic organizers

Learner Differences

Varying methods according to the learners' age
Multidimensional model: something intellectual, plus something emotional
Additional time and support during writing assignments
Multiple-abilities treatment (sharing responsibilities)

Learner Processes

Strategy training (cognitive): Teaching the learners how to learn
Strategy training (social): Group-worthy tasks, cooperative strategies, peer support

Output

Support for communication
Norms of collaboration and cooperation (turn-taking; rotating roles: facilitator, materials manager, reporter, harmonizer; status treatment to equalize participation)
Inclusion of similar components in every lesson/series of lessons
Explicit evaluation criteria

The requirement of focussed observations from student teachers has obvious advantages for the development of future teachers. Student teachers will meet, and learn to recognize, a variety of teaching strategies during their study. During the observations, non-experienced student teachers will be expected to develop and interpret their theoretical knowledge and skills, linking it to real and relevant situations (Santagata, Zannoni, & Stigler, 2007).

Mathematics Educators and ICT

The use of computers in mathematics is a very up-to-date topic—see Chapter 17. Computers have become tools of motivation, and can foster comprehensible interdisciplinary links between mathematics and other subjects. However, the use of computers in teaching asks for new approaches to exposition and to mathematical content (Artigue, 2002). This might be one of the reasons why recent studies in mathematics education show that, despite many national and international actions aiming at integrating ICT into mathematics classrooms, such integration in schools remains underdeveloped.

There are several reasons for the discrepancy—ranging from the huge diversity of ICT resources (Lagrange, 2011) to the lack of experience among teachers, at all levels, in using technology in mathematics lessons. A vital part of the knowledge of mathematics educators, indeed of teacher educators, is knowledge of potential, advantages and dangers of inclusion of activities using ICT into teaching (Jančařík & Novotná, 2011).

There are many projects, seminars and conferences dealing with this topic. As a recent example, aspects of Working Group 15 ("Technologies and Resources in Mathematics Education") at the Seventh Congress of the European Society for Research in Mathematics Education (CERME 7), held in Poland in 2011, is considered. A common focus of several contributions in the Working Group was on the challenges that teachers encounter when teaching mathematics supported by ICT for developing mathematical understanding and skills. Teaching with ICT is a complex activity, requiring insight in the subject, knowledge of the ICT tools, and understanding of pupils' thinking (Fuglestad, 2011). Shulman (1986) introduced the term pedagogical content knowledge, PCK, to denote the intersection of pedagogical and content knowledge in order to consider the complex interaction between pedagogy and subject content. Mishra and Koehler (2006) extended Shulman's model to include technology and introduced the term technology pedagogical content knowledge, TPACK; Figure 14.5 (retrieved from http://tpack.org/tpck/index.php?title=TPCK_-_Technological_Pedagogical_Content_Knowledge) is a scheme indicating several areas of knowledge. Using ICT effectively in teaching requires more than just learning to handle the computers with software and other digital tools.

But what are the implications of TPACK for teacher-education programs? How can this specialized pedagogical content knowledge be best developed? When a student teacher observes a lesson within a technology-rich environment, what should she or he observe? That question, and many other like questions in the area of mathematics education and ICT, urgently need attention.

Mathematics Educators in the Position of Teacher Educators

Obviously, it would be disappointing if the results of mathematics education research did not have important implications for theory and practice in mathematics education. On the other hand, researchers often have no direct access to teachers, and vice versa; therefore, mathematics educators, viewed as a specific body of teachers (they teach teachers, and in that sense they are teacher educators) form an extremely important category influencing a great deal the spreading of theoretical

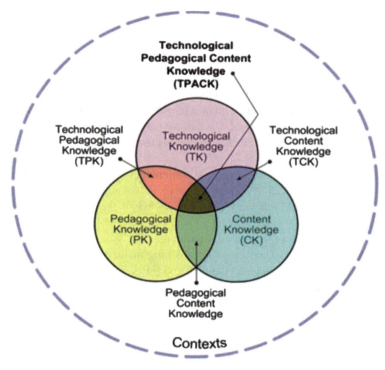

Figure 14.5. Technology pedagogical content knowledge—TPACK (Mishra & Koehler, 2006).

knowledge in the domain of mathematics education. In this part of the chapter we consider issues surrounding the development of mathematics teacher educators.

How Does Someone Become a Mathematics Teacher Educator?

There is no well-defined and unique pathway for becoming a teacher educator. Some teacher educators were originally teachers in schools and took up appointments in teacher-education institutions after years of classroom practice. Others became teacher educators immediately after completing their PhD studies (or even during their PhD studies). In cases where the PhD is completely set inside the field of education research, it has been possible to become a teacher educator without having had much experience teaching in schools. Others were originally mathematics specialists who became mathematics teacher educators without any special training in relation to psychological, pedagogical and didactical issues. The following question arises: What *basic* requirements should we expect of someone who wants to become a teacher educator (regardless of what we understand by the term "good" teacher educator), with respect to mathematics, pedagogy, psychology and mathematics education? This is a complex question, already dealt with in many papers and discussions.

It is generally accepted that mastering mathematics itself is not sufficient for successfully teaching it at any level (see, e.g., Nieto, 1996). In teacher education, it is necessary to determine the balance between the following components:

Specific knowledge. Four main areas are identified:

- Knowledge of mathematics (mathematical concepts and procedures, methodology, relationships with other areas);
- Psychological–pedagogical knowledge (general aspects of teaching/learning processes, getting to know students, planning and management of lessons, curriculum creation, knowledge of teaching contexts);
- Knowledge of learning/teaching mathematics (learning/teaching strategies for specific topics, curricular and pedagogical materials); and
- Knowledge, beliefs and attitudes towards mathematics.

Practical skills. These components are only general; they do not answer the basic question about the content and extent of knowledge required from teachers.

Teachers who become teacher educators have the experience of practice but usually lack any theoretical background. Mathematicians who wish to become teacher educators often have a tendency to overlook the importance of pedagogical–psychological components and prefer to focus on the deep and precise knowledge of the subject content; from their perspective, issues associated with the depth and extent of the mathematics to be mastered are crucially important, and other matters are much less important.

Until recently, little was known about the professional learning or development of mathematics teacher educators (Llinares & Krainer, 2006). As Chapman (2008) reported, even in cases where mathematics teacher educators have researched their own practice, not much is known about their learning, for example, how they reflected to gain self-understanding, what practical knowledge they acquired, and how this knowledge had an impact, or is likely to have an impact, on their future behaviour in working with students.

Despite the views of some sceptics, the importance of theoretical perspectives on the learning and development of university-based mathematics teacher educators is well recognized by the International Group of Psychology in Mathematics Education (IGPME). This topic arose from interactions between PME conference participants, and editors and authors of a special issue on "Teacher Change" of the *Journal of Mathematics Teacher Education*. The learning and development of mathematics teacher educators were explored in a PME discussion group in 2010, and in a PME working session in 2011 (see Goos, Brown, Chapman, & Novotná, 2010, 2011).

Instead of striving to identify a general framework, which could be a fruitless task, an example of a teacher becoming a teacher educator is presented. The following written statement was prepared by a teacher from South Africa who described the difficulties she had after she took steps to become a teacher educator (quoted in Novotná et al., 2003).

> As a teacher educator teaching teachers, my practice has often been constructed for me. Course content is sometimes prescribed and so have been the models of delivery.

During the last 2 years I have found myself strangled and twisted in a thread of tension. The Department of Education embarked on a national strategy to train and equip mathematics, science and technology teachers. They developed a five-year programme to train a substantial amount of educators in each of our provinces. The programme targeted Intermediate Phase (Grade 4 to 6) and Senior Phase (Grade 7 to 9) to ensure an early and solid foundation for learners at higher levels. The intention is that teachers will emerge with an Advanced Certificate in Education (ACE). The National Education Department set out the following outcomes for the programme and for the institutions that would deliver the programme:

- A progressive through-put of well-trained mathematics, science and technology educators per province, who can:
  - demonstrate competence and confidence in classroom practice;
  - assess teaching and learning in line with curriculum stipulations;
  - demonstrate understanding of policy imperatives impacting on teacher development; and
  - become professionally qualified educators with an ACE. (p. 78)

The course attempted to integrate theory and practice but at a very superficial level. My concerns were that as teacher educators:
- We need to think very carefully about what kind of theory is most useful and how we should teach this theory so that teachers can use it to deepen their understanding of educational processes.
- We also need to consider the educative roles of experience.
- And, how exactly should theory and practice be related when the Education authorities want well-trained maths educators?

Theories will die if they remain disconnected from me (my practice) and my practice would be lifeless if not inspired by theory.

My experience with practice has included researching my own practice. To distil the tensions I embarked on a research process that allowed me to probe my own assumptions and to investigate how these influenced the ACE course. I tried to pay attention to the voices of some of my students from the course so that this knowledge could be shared with colleagues with the possibility of reshaping the ACE programme and contributing to our understanding of professional development and teacher education. The purpose of the research was to find out from the teachers what it meant to be a mathematics teacher in their everyday, lived situations.

I do have a slight problem. I am not sure about the role that generalizability will play in the research. At this stage I remain undecided whether to use the stories (the teachers' and mine) to reflect further on the ways that individuals and institutions construct courses in teacher education in South Africa (pp. 79-80).

Theory and practice can exist separately and they can belong to the same world.

People do not stay neatly in a role: at times, setting aside the role of practitioner or of theorist. The educational theorist is a practitioner of education (a teacher); at times the teacher (as educational practitioner) is a theorist (Carr, 1995). (p. 83)

Who Teaches Mathematics Educators? How Does Research Contribute to Mathematics Education?

In the previous parts of the chapter we tried to find answers to the following questions: "Who is a mathematics educator?," "What are the most common paths for becoming a mathematics educator?," and "What is the main role of observations in mathematics educators' work and in mathematics education generally?" We have seen that, to a great extent, mathematics education is determined by "mathematics educators." The category of "mathematics educators" includes all the individuals, regardless of their status, who contribute either intentionally or non-intentionally to establishing or transforming the relationship of a subject with situations that may be modelled by mathematics. This is the place where mathematics education takes place, because knowledge of mathematics is always manifested as an expression of this relationship. But it is also the place of their establishing. As Wittgenstein (1980) stated: "Teach it to us and you established it" (p. 381).

But immediately, the paradox that Marx posed in his third thesis on Feuerbach appears: *Who will teach the educators?* Although Marx never really answered his question, Morin (1999) proposed a number of paths including that of "providing a culture that allows organizing knowledge" (p. 118). This path is promising because in fact, it allows the incorporation of the question of knowledge and its transition in the domain of educational policy and more largely in the culture: the set of ways of reacting, thinking or doing, proper to nations and communities. It is linked with considering this question in the set of strongly diverse dimensions: historical, epistemological, political, etc. These dimensions determine, but not mechanically, what pupils learn and the ways that they learn it.

In fact, although mathematics can be considered as universal, the kinds of mathematical experience pupils gain, are diverse, set in different contexts and periods, influenced by educational style (Sarrazy, 2002; Sarrazy & Novotná, 2005). Although it is possible to include questions related to mathematics education to broader discussion on education and educational policy, we can also study the specific modalities of contribution of research in social sciences and more particularly of didactics of mathematics to mathematics education. That will be our focus in the following discussion, which is a follow-up to the previous parts of the chapter. It provides a more general, more philosophical reflection on mathematics education, mathematics teachers and the education of mathematics educators. The ideas presented show the variety of possible approaches and sources. The discussion is based on the notion of didactical situation as that was introduced by Guy Brousseau in the Theory of Didactical Situations in mathematics (Brousseau, 1997).

A Necessary But Not Exclusive Specificity

From the end of the 1960s the theory of didactical situations (Brousseau, 1997) asked for mathematics education and the sciences of education to be seen in a new way. Didactical problems needed to be specific for the considered domain of education. Learning mathematics has no relationship with, for example, learning to cook or learning to play football! We will not focus on that aspect, which is largely consensual today. But if in their practice, mathematics educators (in the large sense) have no room for manoeuvre for mathematics, this room considerably increases if they examine the situations for communicating them.

This first aspect will be quickly illustrated by an anecdote. Two doctoral students were assistants in a big school in Rome; both of them were good mathematicians. The first was a perfectionist and for his lessons he always chose problems whose success was delicate and strongly clear for his pupils. The second was disordered and had no so clear and explicit vision of what he wanted his pupils to develop; he taught something because he found it interesting and useful. Despite that, the examination results of the second were regularly much better than these of the first one. A possible explanation could be that the perfect organization of the first one's teaching from the perspective of mathematics did not leave any space for interrogation with his pupils, whereas pupils of the second had to find for themselves relationships between diverse problems that looked to be entirely independent.

Fully finished mathematics (rules, algorithms, theorems etc.) might be thought of as dead mathematics. A big part of the work of teachers consists in creating specific conditions of their "resurrection" for pupils. For doing it, they do not have any other choice than to create situations enabling them to show their pupils the use, interest, and meaning of mathematics. The reason is that the concept of situations, their managing, their organization, their evaluation, their regulation, etc. have fundamentally one specific dimension. They are of an immense complexity, taking into account their multiple determinations, conscious or not, that lay stress on the structures, declared or effective functions and the dynamics of these situations: observations, evaluations, regulations. These determinants are situated at various levels of organization according to excessively complex modes of relations—political, epistemological, pedagogical, scientific, etc.—that create an ideological framework that is relatively influential in its effects. It is very difficult, if not impossible, not only to build hierarchies of the forms of determination but also to evaluate their pertinence and their course of action for mathematics education. The reason is that the theory of situations allowed isolating (in the sense of Stengers, 1995) the didactical dimension of pedagogical, social, psychological, anthropological etc. aspects; it allowed making efforts and having success in modelling properties and conditions, specific for mathematics, of pupils' interactions with the environment and thus contributing to the emergence of the didactics of mathematics. We believe that one of the conditions of mathematics education development is certainly the identification of its non-limiting specificity; this specificity is proper to the epistemology of mathematics but narrowly linked with anthropological dimensions that are not specific for mathematics but nevertheless necessary for understanding social (economical, statistical etc.) use of mathematics.

Education and Mathematical Education

It is banal to say that mathematical education does not focus merely on creating mathematicians or on communicating mathematics that is useful for social and domestic life. It is less banal to say, as many mathematicians—Bertrand Russell, for example—have said, that mathematics contributes to the creation of citizenship in its way of being in the world and of taking it into consideration.

Besides, it is sufficient to compare, in the diachrony and synchrony, forms of teaching, curricula, the roles of mathematicians in the social and school selection, for taking into account the extreme diversity of the conception of mathematics education. An equally important diversity can be found in the conceptions of mathematics by mathematicians themselves. It is not to be accepted that mathematics education could be placed under the control of one discipline or trend only. Specialists of the discipline and of its education enrich democratic discussions about the social, school and more largely political uses of mathematics from the perspective of their science. In the same way, one could imagine that researchers could clarify political decisions by their capacity for anticipating the consequences of certain political measures to the conditions of their dissemination. Unfortunately, we can confirm without much risk that the legitimate care for rationalization, efficiency and equity of education leads to the exponential development of evaluation; moreover, individualism as it appeared in the 1980s and the 1990s, together with liberalism had more impact on the ways of disseminating mathematics and mathematical culture than results of research in mathematics education accumulated during the last 40 years.

For example, Nichols and Berliner (2005) have clearly demonstrated the serious impact of evaluation policy on all the levels of the educational system. In the USA, the *No Child Left Behind* legislation envisages sanctions against teachers and institutions that do not reach the level required on mandatory high-stake tests. This policy has had serious consequences:

1. The growth of discrimination by the closing of schools in the poorest environments.
2. Teachers being forced to operate in untenable pedagogical and social environments.
3. The weakest pupils becoming frustrated, which can result in their exclusion.
4. The important development of corruption within social relationships (e.g., result fiddling).

Over two decades ago, Brousseau (1989) explained how, in such situations:

> Teachers are led to leave the objectives of high taxonomical levels for the benefit of objectives of a low level: learning algorithms and isolated facts. Each of these measures grows the teaching/learning time and presents cumulative difficulties: metadidactical shifts, repetitions and individualization swallow the collective educative time, fragmentation of knowledge cuts the comprehension and the field of its utilization, etc. This degrading form of lessons was developed since the trivialization of tests, first for the tools of information and soon as the tool for the management of educational policy. In this system, the measures of failure are a priori denounced as unsupportable and designated responsible are pupils and particularly teachers. Against all reasons, present methods are disapproved, opposed to others that are said to be forgotten, and declared better against any proof, but only for justifying the accusation of general incompetence. (Quoted in Sarrazy, 2009, p. 13)

Didactical Culture and Social Anticipation

Should education result in a "full head" or a "head well done"? Should we look for a good mastery of algorithms or allow pupils to be creative and use algorithms in new situations? This recurrent and often counterproductive debate not only has scientific overtones, it is also political because it poses questions about the type of men and women who are to be formed. If these two intentions appear together, they appear in a paradoxical relationship. In fact, the more pupils are sure of the efficiency of an algorithm, the less they authorize themselves to invent other uses than those they met originally. Like a disciple to whom a teacher shows the moon, they see the finger.

This is the place of mathematics education, between the academic dimension of knowledge and mathematical activity. The theory of didactical situations is born from the theorization and scientific study of conditions that allow exceeding this paradox. Although its recognition among the scientific community is manifest, its dissemination and use in teacher education remain strongly limited. Should we regret it?

What are the consequences for teacher and mathematics educators' education? Teacher education appears as an important lever enabling teachers to step out from the discussion between the "full head" and the "head well done." We think that it would be desirable to expand teachers' didactical culture significantly but we would make a mistake if we push them to expel their pedagogical ideas. It would be a serious mistake because teachers, as well as pupils, need a certainty and illusion at the same time. Researchers in didactics of mathematics, whose agreement on ideas is far from being unified, contribute to clarifying conditions enabling the creation of knowledge that is new for the pupil (that does not depend on the pupil but on the culture). Pedagogues are responsible for fostering such conditions under which pupils have a chance of active participation in the adventure that nobody else can experience for them, the adventure of reinventing the world by her or his activity. Pupils can hardly be expected to produce anything new unless they have had some direct experience of this process. Fostering discussions on the definition of educational policies, of clarification of the possible consequences of certain political decisions would be of much benefit for research in mathematics education in general and for teacher education.

References

Adler, J., Ball, D., Krainer, K., Lin, F.-L., & Novotná, J. (2005). Reflections on an emerging field: Researching mathematics teacher education. *Educational Studies in Mathematics, 60*(3), 359–381.

Artigue, M. (2002). Learning mathematics in a CAS environment: The genesis of a reflection about instrumentation and the dialectics between technical and conceptual work. *International Journal of Computers for Mathematics Learning, 7*, 245–274.

Breen, C. (2003). Mathematics teachers as researchers: Living on the edge? In A. Bishop, M. A. Clements, C. Keitel, J. Kilpatrick, & F. Leung (Eds.), *Second international handbook of mathematics education* (pp. 523–544). Dordrecht, The Netherlands: Kluwer Academic Publishers.

Brousseau, G. (1989). Utilité et intérêt de la didactique des mathématiques pour un professeur de collège. *Petit x, 21*, 47–68.

Brousseau, G. (1997). *Theory of didactical situations in mathematics 1970–1990.* Dordrecht, The Netherlands: Kluwer Academic Publishers.

Brousseau, G. (2002). *Cobayes et microbes* (G. Warfield, Trans.). Electronic discussion.

Brousseau, G., & Warfield, V. M. (1999). The Case of Gaël. *The Journal of Mathematical Behavior, 18*(1), 7–52.

Brousseau, G., Brousseau, N., & Warfield, V. (2001). An experiment on the teaching of statistics and probability. *The Journal of Mathematical Behavior, 20*(3), 363–411.

Brousseau, G., Brousseau, N., & Warfield, G. (2002). An experiment on the teaching of statistics and probability. *Journal of Mathematical Behavior, 23*, 1–20.

Brousseau, G., Brousseau, N., & Warfield, V. (2004a). Rationals and decimals as required in the school curriculum. Part 1: Rationals as measurement. *Journal of Mathematical Behavior, 20*, 363–411.

Brousseau, G., Brousseau, N., & Warfield, V. (2004b). Rationals and decimals as required in the school curriculum: Part 1: Rationals as measurement. *The Journal of Mathematical Behavior, 23*(1), 1–20.

Brousseau, G., Brousseau, N., & Warfield, V. (2007). Rationals and decimals as required in the school curriculum: Part 2: From rationals to decimals. *The Journal of Mathematical Behavior, 26*(4), 281–300.

Brousseau, G., Brousseau, N., & Warfield, V. (2008). Rationals and decimals as required in the school curriculum: Part 3. Rationals and decimals as linear functions. *The Journal of Mathematical Behavior, 27*(3), 153–176.

Brousseau, G., Brousseau, N., & Warfield, V. (2009). Rationals and decimals as required in the school curriculum: Part 4: Problem solving, composed mappings and division. *The Journal of Mathematical Behavior, 28*(2–3), 79–118.

Brown, L., & Coles, A. (2000). Same/different: A "natural" way of learning mathematics. In T. Nakahara & M. Koyama (Eds.), *Proceedings of the 24th Annual Conference of the International Group for the Psychology of Mathematics Education* (Vol. 2, pp. 113–120). Hiroshima, Japan: Hiroshima University Press.

Bruner, J. S. (1966). *Toward a theory of instruction*. Cambridge, MA: Harvard University Press.

Carpenter, T. P., Fennema, E., & Franke, M. L. (1996). Cognitively guided instruction: A knowledge base for reform in primary mathematics instruction. *The Elementary School Journal, 97*(1), 3–20.

Carr, W. (1995). *For education: Towards a critical educational inquiry*. Buckingham, UK: Open University Press.

Cavanagh, M., & Prescott, A. (2007). Professional experience in learning to teach secondary mathematics: Incorporating pre-service teachers into a community of practice. In J. Watson & K. Beswick (Eds.), *Mathematics: Essential research, essential practice* (pp. 182–191). Adelaide, Australia: Mathematics Education Research Group of Australasia.

Chapman, O. (2008). Mathematics teacher educators' learning from research on their instructional practices: A cognitive perspective. In B. Jaworski & T. Wood (Eds.), *International handbook of mathematics teacher education* (Vol. 4, pp. 110–129). Rotterdam, The Netherlands: Sense Publishers.

Chevallard, Y. (1985). *La transposition didactique*. Grenoble, France: La Pensée Sauvage.

Clarke, D. (1999). *The Learner's Perspective Study: Research design*. Retrieved from http://extranet.edfac.unimelb.edu.au/DSME/lps/.

Clarke, D. (2001). Complementary accounts methodology. In D. Clarke (Ed.), *Perspectives on practice and meaning in mathematics and science classrooms*. Dordrecht, The Netherlands: Kluwer.

Clarke, D. J., Keitel, C., & Shimizu, Y. (Eds.). (2006). *Mathematics classrooms in twelve countries: The insider's perspective*. Rotterdam, The Netherlands: Sense Publishers.

Cochran-Smith, M., & Lytle, S. (1993). Research on teaching and teacher research: The issues that divide. *Educational Researcher, 19*(2), 2–11.

Douady, R. (1991). Tool, object, setting, window elements for analysing and constructing didactical situations in mathematics. In A. Bishop, S. Mellin-Olsen, & J. van Dormolen (Eds.), *Knowledge: Its growth through teaching* (pp. 109–130). Dordrecht, The Netherlands: Kluwer.

Feiman-Nemser, S. (2001). Helping novices learn to teach: Lessons from an exemplary support teacher. *Journal of Teacher Education, 52*(1), 17–30.

Fuglestad, A. B. (2011). Challenges teachers face with integrating ICT with an inquiry approach in mathematics. In *CERME 7*, Working Group 15. Retrieved April 9, 2011, from http://www.cerme7.univ.rzeszow.pl/WG/15a/CERME7-WG15A-Paper02_Fuglestad.pdf.

Goos, M. (2008). *Critique and transformation in researcher-teacher relationships in mathematics education.* Symposium on the Occasion of the 100th Anniversary of ICMI. Retrieved from http://www.unige.ch/math/EnsMath/Rome2008/partWG3.html.

Goos, M., Brown, L., Chapman, O., & Novotná, J. (2010). The learning and development of mathematics teacher educator-researchers. In M. Pinto & T. Kawasaki (Eds.), *Proceedings of PME 34* (1st ed., p. 390). Belo Horizonte, Brazil: PME.

Goos, M., Brown, L., Chapman, O., & Novotná, J. (2011). The learning and development of mathematics teacher educator-researchers. In B. Ubuz (Ed.), *Proceedings of the 35th Conference of the International Group for the Psychology of Mathematics Education* (1st ed., p. 173). Ankara, Turkey: International Group for the Psychology of Mathematics Education.

Jančařík, A., & Novotná, J. (2011). Potential of CAS for development of mathematical thinking. In M. Kováčová (Ed.), *Aplimat 2011* (pp. 1375–1384). Bratislava, Slovakia: STU.

Jaworski, B. (2003). Research practice into/influencing mathematics teaching and learning development: Towards a theoretical framework based on co-learning partnerships. *Educational Studies in Mathematics, 54*(2–3), 249–282.

Kinach, B. M. (2002). A cognitive strategy for developing pedagogical content knowledge in the secondary mathematics methods course: Toward a model of effective practice. *Teaching and Teacher Education, 18*, 51–71.

Lagrange, J.-B. (2011). Working with teachers: Collaboration in a community around innovative software. In *CERME 7*, Working Group 15. Retrieved from http://www.cerme7.univ.rzeszow.pl/WG/15a/CERME7-WG15A-Paper24_Lagrange.pdf.

Llinares, S., & Krainer, K. (2006). Mathematics (student) teachers and teacher educators as learners. In A. Gutierrez & P. Boero (Eds.), *Handbook of research on the psychology of mathematics education: Past, present and future* (pp. 429–459). Rotterdam, The Netherlands: Sense Publishers.

Margolinas, C. (2002). Situations, milieux, connaissances: Analyse de l'activité du professeur. In J.-L. Dorier, M. Artaud, M. Artigue, R. Berthelot, & R. Floris (Eds.), *Actes de la 11ème Ecole d'Eté de Didactique des Mathématiques* (pp. 141–156). Grenoble: France La Pensée Sauvage.

Margolinas, C., Coulange, L., & Bessot, A. (2005). What can the teacher learn in the classroom? *Educational Studies in Mathematics, 59*(1–3), 205–304.

Mishra, P., & Koehler, M. J. (2006). Technology pedagogical content knowledge: A framework for teacher knowledge. *Teachers College Record, 108*, 1017–1054.

Morin, E. (1999). *La tête bien faite—Repenser la réforme, réformer la pensée*. Paris, France: Ed. du Seuil.

Morris, A. K., Hiebert, J., & Spitzer, S. M. (2009). Mathematical knowledge for teaching in planning and evaluating instruction: What can preservice teachers learn? *Journal for Research in Mathematics Education, 40*(5), 491–529.

Nathan, M. J., & Petrosino, A. (2003). Expert blind spot among preservice teachers. *American Educational Research Journal, 40*(4), 905–928.

Nichols, S. I., & Berliner, D. C. (2005). *Collateral damage: How high-stakes testing corrupts America's schools*. Cambridge, MA: Harvard University Press.

Nieto, L. J. B. (1996). Learning to teach mathematics: Types of knowledge. In J. Giménez, S. Llinares, & V. Sánchez (Eds.), *Becoming a primary teacher: Issues from mathematics education* (pp. 159–177). Sevilla, Spain: Gracia Alvarez.

Novotná, J., Lebethe, A., Rosen, G., & Zack, V. (2003). Navigating between theory and practice: Teachers who navigate between their research and their practice. In N. A. Pateman, B. J. Dougherty, & J. Zilliox (Eds.), *Proceedings of PME 27/PME NA 25* (Vol. 1, pp. 69–99). Honolulu, HI: University of Hawai'i.

Novotná, J., Zack, V., Rosen, G., Lebethe, A., Brown, L., & Breen, C. (2006). RF01: Teachers researching with university academics. In J. Novotná, H. Moraová, M. Krátká, & N. Stehlíková

(Eds.), *Proceedings of the 30th Annual Conference of the 30th International Group for the Psychology of Mathematics Education* (Vol. 1, pp. 95–124). Praha: Univerzita Karlova v Praze.

Piaget, J. (1985). *The equilibration of cognitive structures: The central problem of intellectual development*. Chicago, IL: University of Chicago Press.

Salin, M.-H., & Dreslard Nédélec, D. (1999). La collaboration entre chercheurs et enseignants dans un dispositif original d'observation de classes: Le centre d'observation et de recherche sur l'enseignement des mathématiques (COREM). In F. Jaquet (Ed.), *Relations between classroom practice and research in mathematics education, Proceedings CIEAEM 50* (pp. 24–37). Neuchâtel, Switzerland: IRDP.

Santagata, R., Zannoni, C., & Stigler, J. W. (2007). The role of lesson analysis in pre-service teacher education: An empirical investigation of teacher learning from a virtual video-based field experience. *Journal of Mathematics Teacher Education, 10*(2), 123–140.

Sarrazy, B. (2002). Effects of variability on responsiveness to the didactic contract in problem-solving among pupils of 9–10 years. *European Journal of Psychology of Education, 17*(4), 321–341.

Sarrazy, B. (2009, December). Les évaluations internationales et leurs effets sur l'enseignement des mathématiques. *Tangente Education* [La didactique des mathématiques], *11*, 9.

Sarrazy, B., & Novotná, J. (2005). Didactical contract: Theoretical frame for the analysis of phenomena of teaching mathematics. In J. Novotná (Ed.), *Proceedings SEMT 05* (pp. 33–44). Prague, Czech Republic: Charles University.

Sherin, M. G. (2002). When teaching becomes learning. *Cognition and Instruction, 20*(2), 119–150.

Shulman, L. S. (1986). Those who understand: Knowledge growth in teaching. *Educational Researcher, 15*, 4–14.

Stehlíková, N. (2007). Charakteristika kultury vyučování matematice. In A. Hošpesová, N. Stehlíková, & M. Tichá (Eds.) *Cesty zdokonalování kultury vyučování matematice* (pp. 13–48) [*Characteristics of the culture of mathematics teaching*, in Czech]. České Budějovice: JČU.

Stengers, I. (1995). *L'invention des sciences modernes*. Paris, France: Flammarion.

Vacc, N. N., & Bright, G. W. (1999). Elementary preservice teachers' changing beliefs and instructional use of children's mathematical thinking. *Journal for Research in Mathematics Education, 30*(1), 89–110.

Vygotsky, L. S. (1962). *Thought and language*. Cambridge, MA: MIT Press (Original work published 1934).

Wilson, S. M., & Berne, J. (1999). Teacher learning and the acquisition of professional knowledge: An examination of research on contemporary professional development. *Review of Research in Education, 24*, 173–209.

Wittgenstein, L. (1980). *Grammaire philosophique* (A.-M. Lecourret, Trans.). Paris, France: Gallimard.

Wittgenstein, L. (1983). *Remarques sur les fondements des mathématiques*. Paris, France: Gallimard.

Zack, V., & Graves, B. (2001). Making mathematical meaning through dialogue: "Once you think of it, the z minus three seems pretty weird." *Educational Studies in Mathematics, 46*(1–3), 229–271.

Chapter 15
Institutional Contexts for Research in Mathematics Education

Tony Brown and David Clarke

Abstract Mathematics has maintained an enduring image as a field of knowledge lending its resources to many intellectual pursuits and utilitarian enterprises. School mathematics, however, has increasingly learned to respond to a commonly conceived purpose of supplying the world's workforce with the resources needed to support economic wellbeing. The emergent regulation in support of this response has in some instances tempered more humanistic or idealistic conceptions of why we want to study mathematics. What had been introduced to measure school mathematics now defines and polices its boundaries. It has also privileged Western concerns in setting internationalized agenda. Mathematics, mathematics education and mathematics education research, this chapter suggests, are each conceptualized according to their location, reflecting and shaping each other, yet with each being governed by slightly different priorities. It is argued that schooling is increasingly shaped and judged by its perceived capacity to deliver success in terms of international competitiveness linked to economic agenda. This results in school mathematics being shaped to meet assessment requirements. The chapter shows how research increasingly finds its terms of reference set according to measuring delivery in these terms. It also shows how researchers become complicit in promoting particular conceptions of teaching and in constructing the field as an ideological battleground. Such complicity, it is suggested, combined with the relative insularity of the field, prevents us from occupying other worlds that might define us and serve us in different ways. The chapter concludes with a consideration of the prospects of research in mathematics education and the extent to which this activity is enabled or restricted by existing institutional contexts in re-shaping its ambitions to engage with the diversity of future needs.

T. Brown (✉)
Manchester Metropolitan University, Manchester, England
e-mail: a.m.brown@mmu.ac.uk

D. Clarke
The University of Melbourne, Melbourne, Australia

Mathematics, Mathematics Education, and Mathematics Education Research

Is mathematics defined by local conditions or can it be understood more universally as spanning nations and generations? Mathematics has maintained an enduring image as a field of knowledge lending its resources to many intellectual pursuits and utilitarian enterprises. School mathematics, however, has increasingly learned to respond to a commonly conceived purpose of supplying the world's workforce with the resources needed to support economic wellbeing. Research intended to inform the practices of mathematics classrooms has often reflected local interpretations of this fundamentally economic agenda. Since the advent of international comparisons, governments have been jockeying for a better position in the resulting league tables. The success of particular school systems in international testing programs such as the OECD *Programme for International Student Assessment* (PISA) or *Trends In Mathematics and Science Study* (TIMSS) has been variously interpreted. Good performance in these league tables has sometimes been taken as being indicative of wider economic competitiveness. Yet such comparisons can transform the content of what they compare.

> TIMSS contributes to the misrecognition of terrain where global politics motivates policy makers to apply national security responses to education. The assessment casts students as passive, nameless metaphors of national economies, whose performance in school will predict the future relations among nations. (Thorsten, 2000, p. 72)

Governments and the people they govern have been seduced by the appeal of raising standards in a statistically defined world. What had been introduced to measure school mathematics now defines what it is and polices its boundaries. This regulation has tempered more humanistic or idealistic conceptions of why we want to study mathematics. It has also done much to alter how we understand research in the area.

Howson and Mellin-Olsen (1986) documented some of the history of mathematics' evolution as a school subject for which, since the beginning of school mathematics education, the subject was stratified according to the type of student concerned, and the expectations held for them. Over the past few decades, though, the bounded vision of the measurable mathematics preferred by international testing programs (whether TIMSS or PISA), a climate of competition has been created in which nations compete for status and governments take credit or apportion blame according to these quantifications of student achievement. Among the consequences of international competition and the attendant commitment to national typification we suggest that national means of performance are given priority over the local inequalities they conceal. The success of less affluent nations in optimizing the effectiveness of their minimally resourced educational systems may go unrecognized. More informed analyses of the data generated by international testing are capable of pointing to idiosyncrasies in school systems that address, ignore or even amplify the educational difficulties experienced by particular population sectors. Our interest in this chapter is less to bewail the misuse of research in mathematics education as to examine the institutional contexts that influence the form taken by that research and explore the consequences of that influence.

Mathematics, mathematics education and mathematics education research are each conceptualized according to their location. It will be argued that they reflect and shape each other, with each being governed by slightly different agenda. For example, the assessment of school mathematics through filters such as international tests of student performance has changed the priorities of school mathematics in many countries. These changes have in turn had an impact on how the field of mathematics education research is conceived internationally. The international industry that has arisen around the assessment of student mathematics achievement has simultaneously enacted and shaped local and international conceptions of accomplished practice in mathematics and in mathematics education. Research is judged by its perceived capacity to deliver success in the prescribed terms.

For instance, the goal of comparative international measurement of student mathematics achievement is sometimes conceptualized as the raising of standards. These standards, however, result from a very specific conception of mathematical learning, often based on what US policy makers have deemed to be important through their reference to TIMSS in evaluating performance in US schools (see Bishop, 1990). Other countries have readily subscribed to these priorities, apparently with minimal questioning, or because the priorities have become the international currency to which their governments can reference their own schools' achievements in electorate-friendly terms. The assumptions about what is valuable have been encrypted into the measuring devices themselves. The results are then subject to considerations of alignment with valued mathematical performances, the affordances and limitations of the measuring devices (the test), and assumptions about levels or composition of achievements appropriate to particular age cohorts. Mathematical activity or performance, in school, workplace and other settings, is the medium by which the purposes of mathematics education are realized. Mathematics education research meanwhile draws its identity from an interest in optimizing and informing both mathematical activity and mathematics education. The focal concern of this chapter lies with those institutions that provide the context and the agenda for mathematics education research.

The scale of international research efforts and the political status of the findings have popularized a distinctive genre of mathematics education research. Accordingly, the image of the lone researcher finding out how mathematics might be taught has been eclipsed by more collective conceptions of mathematics in schools and of the research tasks developed to investigate and inform educational practice. These conceptions result from shifts in pedagogical attitudes, such as those attitudes manifest in the reform movements in the USA, China and Singapore, which combine reform zeal with very differently targeted initiatives. Associated activities can include the working through of regulative demands on curriculum definition, as in China and Australia, and the changing roles of universities in preparing teachers, as in the United Kingdom and Singapore. Researchers continue to produce knowledge and this knowledge is open to appropriation by those seeking to maintain current ideologies or by those seeking to critique and contest current ideologies. For example, Piaget and Vygotsky have been variously deployed to underwrite constructivist reforms in the USA. Freudenthal's work has been marketed as an alternative school mathematics scheme. Yet knowledge is a function of the world that produces it, which can prevent us occupying other worlds that might define us and serve us in different ways. International research has the potential to afford access to alternative visions of curriculum and practice, but filters the study of such alternatives through the normalizing demands of common measurement instruments and the use of English as the lingua franca of international education and educational research.

There are difficulties for research in exploring good practice when governments are defining what good practice is according to policy driven priorities and budgetary constraints. Research carried out according to the preferences of these governments is frequently about supporting "improvement" within the current model rather than being about producing and testing new models. Researchers can become subservient to the latest governmental vision. We suggest that such institutional contexts (a) determine the criteria by which good practice is recognized; (b) prescribe the manner in which good practice can be researched; and (c) frame and constrain the channels by which research can inform the promotion and realization of "evidence-based good practice." Central to this discussion is the determination of what constitutes evidence for the purposes of informing practice and generating policy. This shaping of the direction of research determines what mathematics and mathematics learning are considered as legitimate objects of that research.

Insistence on the universality of mathematical activity, however, represents a denial of the heterogeneity that characterizes mathematics and the way in which it is shaped to fit diverse locations. Mathematics means different things to different people, where groups may prefer particular perspectives that solidify in certain communities, according to culture, ethnicity, affluence, gender, and social class, as alternative contexts. Mathematics is held in place by its appearances in specific locations (particular pedagogical forms, representations in popular media, its use in accountancy procedures, etc.). We may ask, however, what remains if we take away these specific examples of localized cladding that at once disguise mathematics and make it recognizable and functional in those specific locations? There may be nothing left. Mathematics resides in its localized appearances addressing specific demands. Yet, not all voices or ways of life are equal on the international stage. The tension between local priorities, values and needs and the normalizing demands of international comparability make clear the sensitivity of mathematics education research to the demands of context.

Research in mathematics education has increasingly turned to issues of context, while being situated itself in many contexts. Far from being the province of the lone researcher, research these days takes place increasingly in small and large teams, usually but not always at universities, and frequently drawing membership from several educational contexts and traditions. Each stakeholder group participant in research brings its own agenda: governments, funding agencies, school systems, community groups, business, universities, research centres, research teams, teachers, students, parents, and individual researchers. But, most importantly, research takes place within communities of people governed by collective arrangements that define, regulate and normalize the practices that take place. This chapter examines the benefits and drawbacks, the affordances and the constraints, of these institutional contexts for the training and education of researchers but chiefly for the development of the field itself. Above all, it seeks to show how there are political dimensions that pertain to the practice, funding, researching and training for mathematics education, and which shape what it is. For research to be meaningful and useful it must examine the ways in which these political dynamics constitute the basic entities that make up mathematics education, namely teachers, students and mathematics itself (Otte, 1979).

The chapter commences with a preliminary account of the wider domain of mathematics education research with respect to its institutionalized contexts across and within nations, and the tools that they employ (international achievement tests; the criteria for funding deployment; conceptions of mathematics curricula). A useful approach is to examine the domain in relation to the ideological movements that legitimize mathematics as a school subject and the research carried out in this area. We have anchored this discussion on an account of "reform" mathematics as it has been conjured in the USA and, more recently, in China; as an ideology acting through the social practices in each country and beyond to produce conceptions of mathematics and its teaching. The chapter continues by examining the definition of the field of mathematics education in relation to its manifestation in specific institutional contexts: curriculum development and evidence-based policy initiatives, publication networks, academic networking and research community definition and the training and education of researchers. The chapter concludes with a consideration of the prospects of research in mathematics education and the extent to which this activity is enabled or restricted by existing institutional contexts in re-shaping its ambitions to engage with the diversity of future needs.

"Reform" as a Context for Mathematics Education Research

There is a common assumption that research in mathematics education is about informing movement towards some improved conception of teaching. But how might we conceptualize improvement? Can we agree on some set of shared aspirations? Or, alternatively, could we agree on a greater tolerance of difference? Collective movement might be harmonized towards "improvement," whether that is about being more the same, through curricular consensus or standardization of achievement measures, or more responsive to local conditions and thereby more diverse. Different goals require different approaches. What mechanisms, for example, might allow individuals to join together in such a way that a collective vision is conjured and coordinated practice is realized? What mathematics education research might inform practice within such collective arrangements? The teaching of school mathematics typically takes place within some curricular structure set for a particular community of people. The scope for individual teachers to interpret their task is tempered by their susceptibility to having their work evaluated according to local criteria. That is, teachers serve administrations aspiring to some model of teaching and take steps to align their practice with those aspirations. In turn, research is often commissioned to support or enhance practice consistent with that agenda.

Modern conceptions of "reform" as a notion within mathematics education research have developed new meanings linked to the guidelines of the US National Council of Teachers of Mathematics. Mathematics educators in the USA, the United Kingdom, and Australia have associated the term "reform" with the transition from a transmission to a constructivist pedagogical approach (Fennema & Nelson, 1997) and curricular reform in China, Korea and Singapore is now taking a similar path. By comparison, Japanese mathematics educators were making an effort in the 1960s and 1970s "to develop ways of making students discover new ideas and construct knowledge on their own" (Hino, 2007, p. 508). The result of these Japanese efforts was the development of a lesson structure called "structured problem solving" that has been the subject of much subsequent research (Hino, 2006; Sekiguchi, 2006; Shimizu, 2006). What is perceived as abrupt transformative reform in many countries is seen as the continuation of a long-term process of research and development in Japan. This contrast is important because it suggests that the reception accorded to the same instructional (or curricular) advocacy will differ according to the educational history of the community.

Constructivism, as a conception of learning, though centred in the USA, dominated international mathematics education research for some two decades (Brown, 2001; Steffe & Kieran, 1994). The pedagogy associated with constructivism involved the promotion of student agency and active engagement in advancing their own learning, through "genuine mathematical problems for students to solve" (Lloyd, 1999, p. 228) with a focus on "conceptual understanding" (Wilson & Goldenberg, 1998, p. 269). Research in the area had sometimes been conceptualized as tracking progress towards some improved state of affairs (Simon & Tzur, 1999; Tzur, Simon, Heinz, & Kinsel, 2001). Other studies focussed on how teachers responded to curriculum changes. These studies centred their analyses on individuals shaping their practice in response to the perceived reform agenda (Remillard & Geist, 2002; van Zoest & Bohl, 2002). Many of the authors identified and openly subscribed to this agenda. That is, the researchers were complicit in the promotion of a particular conception of teaching: inclined towards researching its optimization rather than towards the development of any form of critique. This is not an irrational position: if the efficacy of an instructional approach is demonstrated by research, then further research into its optimization is a logical next step. In the context of educational research, this simple rationality can be qualified by questioning: (a) the legitimacy of generalizing such instructional advocacy to all settings; and (b) the clarity and uniformity with which the advocated practice and associated theory is understood, even by those advocating its implementation. Educational advocacy—that is, reform—is always subject to contingencies of context and of consensus.

Not surprisingly, such reform did not offer a trajectory with universal appeal or applicability. There were widespread disputes within the USA itself, centred on debates that have come to be known as the "math wars." These disputes have since been replicated in other countries (in China, for example) in response to similar curricular initiatives. The "inquiry" methods associated with constructivist reform, characterized by greater learner and teacher autonomy directed at conceptual understanding, have been resisted by more traditional teachers, who preferred an emphasis on computational skills, and by some mathematicians, who saw in the new approach a loss of mathematical rigour. Similar battles continue to be fought as other countries, such as China and Korea, implement national mathematics curricula that embrace "real-life and open-ended problems" in curricular contexts dominated by examinations (Cai & Nie, 2007).

More theoretically grounded objections to constructivism pointed to the confusion caused by interpreting a theory of learning as a theory of instruction. Disputes over the effectiveness of new instructional approaches have been compounded by lack of agreement on what constitutes accomplished mathematical activity. Research in mathematics education became a weapon of the math wars, to be used (as in Andrew Lang's happy phrasing) "as a drunken man uses lampposts—for support rather than illumination" [from: http://www.brainyquote.com/quotes/authors/a/andrew_lang.html]. Since researchers in mathematics education are simultaneously members of the mathematics education community, they become complicit in the construction of the field as an ideological battleground and in the use of research as a weapon in that war. Perhaps it is inevitable that education, as a value-laden and culturally encumbered field, should be so prone to ideological division. It is not only unreasonable, but actually a misrepresentation of the nature of research, to expect educational researchers to adopt a form of ideological neutrality. The activities of mathematics education researchers are just as ideologically, politically, historically and socially situated as any other members of society: that is, just as subject to the influences of context.

Research must address not only the basic questions of teaching efficacy and learning, but also the processes and impediments by which any research-based advocacy might be actioned. For example, a few researchers sympathetic to constructivism noted resistance in some quarters, such as "veteran" or "traditional" teachers who were unable to shift so fundamentally in terms of their beliefs in what it is to be a teacher (Cohen, 1990; Lloyd, 1999; Wilson & Goldenberg, 1998). The inquiry methods would also have been less acceptable in many Eastern or Pacific cultures, where curricula, teacher/student roles and the collective good are defined differently (Brown et al., 2007). Further, the alleged autonomy understood within the "reform" agenda conflicts with the reality teachers have come to accept in many countries, assessed as they are through legislative documentation and recognized through the filter of their compliance with this. Such differences are profoundly cultural and reflect histories of educational practice that pose substantial obstacles to any reform movement predicated on autonomy, agency, dialogical reasoning and the legitimacy of contesting prevalent beliefs. The role of research and the researcher in such contested domains becomes itself the matter of debate and the authority of research and the credibility of the researcher will be equally acclaimed and decried by vested interests.

In England, for example, student-centred pedagogies emphasizing problem solving, investigations and project work dominated curriculum reform agendas some 30 years ago. The rhetoric of this tradition was largely commensurate with constructivism. A later backlash in England resulted in prescribed curricula for both teachers and students in which student-centred approaches became tightly structured. Reasons cited for this backlash included right-wing politicians—such as Kenneth Clarke, a Conservative Minister of Education—claiming that given difficulties with teacher supply the average teacher could not teach to such high-minded ideals. Left-wing commentators, meanwhile, argued that aspirations to child-centred approaches merely replaced overt regulation with a form of covert regulation (Walkerdine, 1984). The heightened status of student agency was accompanied in several Western school systems by a commensurate reduction in the importance attached to teacher agency (Chazan & Ball, 1999; Clarke, 1994; Lobato, Clarke, & Ellis, 2005). The prioritization of "higher-order thinking, self-reflection and self-regulation" in countries such as Singapore (Fan & Zhu, 2007) has been identified with the problematization of "traditional teaching" and the implicit devaluing of established tenets of teacher expertise.

The Ideological Bases for Improvement

Conceptions of "improvement" can be very localized. Trajectories of improvement do not apply across all people and all phases of development. Success depends on the criteria one uses for judging success. Many alternative criteria have been entertained in recent years, each governed by their own respective and reasonable assumptions. Relative positions on TIMSS and PISA league tables have encouraged school systems and funding agencies in the USA to adapt mathematics textbooks from Singapore for American use and to appropriate Japanese "lesson study" as a professional development tool in the hope of emulating the achievements of mathematics students in Japan and Singapore. Yet the same league tables are not interpreted in Singapore or Korea as demonstrating unequivocal educational success, where new value is being placed on creativity, imagination, and problem solving ability. Lin (2010) pointed out that Hong Kong, Korea, Japan and Taiwan, who performed well in TIMSS, "showed very poor[ly] in learning interests and self-efficacy" (p. 85). PISA has attempted to give assessment recognition to the situated nature of mathematics activity to a greater extent than TIMSS (Askew, Hodgen, Hossain, & Bretscher, 2010). The attempt within international student achievement initiatives such as PISA to honor the situatedness of mathematical activity within an international testing instrument is wholly commendable. Of course, this same situatedness renders attempts at cross-curricular measurement of student mathematical performance somewhat problematic (see Clarke, 1996). The implicit recognition that mathematics can only be assessed "in use" and that such use implies a context reflects the underlying assumptions of the Dutch *Realistic Mathematics Education* curriculum (De Lange, 1987), among others. The consequences of integrating such a perspective into an instrument intended to measure student mathematics achievement internationally can be seen in the observation that "national rankings on

TIMSS and PISA differ substantially" (Törner, Schoenfeld, & Reiss, 2007, p. 353). It is clear that "improvement" cannot be defined in absolute terms.

> In a special issue of the journal *Educational Research and Evaluation*, Cheng and Cheung (1999) provided a critique of a series of articles addressing the general theme of "TIMSS in a Western European Context." Their critique raised several concerns: (1) Challenges to the validity of country ranking; (2) Problems in relevance of TIMSS to national curriculum; (3) Methodological limitations; (4) Lack of high quality process data at classroom level; (5) Lack of contribution to theory building; and (6) Limited policy implications. The culmination of Cheng and Cheung's argument was that limitations and methodological concerns with TIMSS meant that "the policy implications for improvement of educational practices are inevitably quite limited" (Cheng & Cheung, 1999, p. 233). Given all the issues raised above, it appears that there has been sufficient consistency in the concerns raised about TIMSS to make the policy recommendations problematic. (Clarke, 2003, p. 174)

As research and the framing of policy and curriculum become more distant from the activities of the classroom, there is always a cost in the form of local preferences being suppressed resulting from one-size-fits-all suppositions. Generalized consensual aspirations, framed at the level of the state, the country or globally, lose local relevance, and alignment with them is not always so easy to grasp in the immediacy of everyday practice. There is a need to build a theoretical frame that accommodates alternatives to consensual aspiration. Utilization of such a frame would have significant impact on the way research into student achievement and instructional effectiveness was conceived and conducted. To consider alternatives to consensus is to undertake a form of ideological reconstruction. "Improvement," "success" and "quality" become pluralities contingent on context, rather than singular prescriptions.

Recent neo-Marxist theory has questioned notions of human progress being shaped by ideals relevant across all communities (Mouffe, 2005). This is hardly a radical proposal. Mathematics education, for example, might be best seen as supporting the needs of the students concerned. These needs would be culturally dependent, with each country basing its curricular aspirations on alternative conceptions of mathematics according to local need. Yet, international comparative testing has resulted in many countries teaching to those international tests, matching the style and content preferred by certain Western countries. Both curriculum content and styles of teaching have been adjusted to meet this model. For example, in the name of conformity, the United Kingdom has sacrificed its earlier facility with problem-solving approaches. Since problem solving is not assessed focally within TIMSS, this has resulted in problem solving being less common in schools (Askew et al., 2010). And recent policy has been directed towards enabling British children to be successful in the sorts of questions one finds in TIMSS. Although England succeeded in moving from 18th to 7th position on TIMSS in 2007, it dropped in its rankings from 8th to 25th on the more problem focussed PISA in 2006 (Brown, 2011; Department for Education (DfE), 2010). Tea-pickers in Sri Lanka meanwhile do not get an education suited to their local needs. The curriculum they have been obliged to follow is governed more by "internationalized" objectives than by the skills that would support the local economy. And, for those who succeed, this usually translates into a move away from their local area to work in a city, within the country, or beyond. The education intended to enable graduate mobility functions to enforce it.

Laclau (2005) has rejected the notion of the "people" as a collective actor, and, by extension, the same could apply to the possibility of a research "community" or a set of governments being able to define a common interest with regard to the purposes of school mathematics. For example, to what extent is it possible for the mathematics education research community to assume some consensus in its purposes? Examination results, facility with mathematics and enjoyment of mathematics do not always pull in the same direction (Pampaka et al., 2011). Conceptions of graduate competencies will vary from school system to school system as mathematics curricula attempt to anticipate vocational and personal capabilities likely to be required by graduates.

Instead, Laclau has examined the nature and logics of the formation of collective identities and suggested that such collectives can be seen as being held together through identification with specific populist aspirations. In mathematics education we might reference our activities to raising standards, making children happier, supporting the economy, or building richer mathematical experiences. Mathematics would then be shaped according to how it could be read against such aspirations; a quantifiable version of mathematics so that a standard can be shown to have been raised, an aesthetically pleasing version of mathematics for those more concerned with the beauty of mathematics, etc. Group affiliations might be centred on particular shared values or beliefs. Research design will reflect populist aspirations and mirror societal norms and cultural values, since society's rewards (e.g., funding) will reflect society's values. Government grants may be awarded to those promising to advise on how standards could be raised across a population. Self-elected research time might be directed at sharing with other like-minded people the intrinsic pleasures and aspirations of the individual's own teaching. The essential point is recognition of the correspondence between values and practice and the willingness to countenance and accommodate a diversity of motives to undertake research.

Some years ago, Althusser (1971) focussed on how the individual understands herself through *ideology*. Here an *ideology* is understood as a specific conception of life, a particular version of common sense. One can only inspect an ideology from the perspective of another ideology, "we are 'naturally' in ideology, our natural sight is ideological" (Žižek, 2008, p. xiii). We always occupy an ideologically derived position. We never have the luxury of speaking from outside an ideology. Althusser described schools as an instrument within the "ideological state apparatus." Here schools are seen as a hegemonic device through which the preferred ways of the state are disseminated with general consent. For many pupils and their parents, progression through school is an ideological movement to which they are readily mobilized. Sensitivity to such perspectives can focus research attention on the investigation of inequity. Mathematics and mathematics education have roles in the creation or maintenance of power differentials. These reflect societal norms or established social divisions along socio-economic lines. The role of mathematics in the entrenchment of such narratives of social reproduction has been variously studied (Anyon, 1981; Boaler, 1997; Sztajn, 2003).

Of course, the dominance of such hegemonic societal structures can act to impede any critical function that research might serve. To be published in a reputable journal a research article must typically position itself in relation to existing work and be cast in a form recognizable to a mainstream audience in the field. That is, the tools of the established order must be used to argue for anything new. There is a dynamic between the societal constraints that research might legitimately deconstruct and the action of those constraints to inhibit such critical research. This dynamic is at the heart of the dialectic whereby research becomes complicit in the structuring and maintenance of the systems it might inform. For example, research in mathematics education on gestures, teaching techniques in fractions, or the promotion of group work, may normalize the assumption that adjusting teacher classroom intervention is the main tool of mathematics education, rather than say curriculum reform, adjusting social inequalities, setting teacher education programs, etc. Research participates in constructing the boundaries of its own practice.

It is not only research as an endeavour that is seen to reflect the institutional context in which it is undertaken. Education, Mathematics, and Mathematics Education continue to evolve in ways that reflect their cultural–historical origins. The structure of a discipline such as sociology, for example, reflects its cultural–historical origins and cannot be understood without recognition that it was formed within the culture of imperialism, and embodied an intellectual response to the colonized world (Connell, 2007). Research in Mathematics Education finds itself inheritor of particular views regarding the aspirational goals of education, the legitimacy of curricular partitioning, and the role served by research to understand and optimize the realization of those goals in specific cultural settings. Within such a framework, conceptions of improvement are pre-determined to a significant extent, circumscribing the capacity of research to critique the structures from which it draws its identity.

Althusser was not persuaded by consensual aspirations where difficulties are ironed out. He saw the supposition that you could get to a consensual ideal beyond conflicting ideologies as the biggest ideology of all. The individual may recognize herself in some ideologies but not others. But, there is always a gap in this identification, a distance between the person and the story in which she sees herself. This gap stays there. For example, some American teachers may truly believe that they are subscribing to the reform agenda and following such approaches in their practice, whether or not others see it this way (Cohen, 1990). But, at the same time, some other American teachers may be sceptical about reform projecting them higher up the international league tables or they may not even agree with that ambition. Yet, both groups find their working practices defined and evaluated in line with that agenda, securing compliance at a practical level. Brown and McNamara (2011) have provided an account of how trainee and new teachers in the United Kingdom begin to include official curriculum descriptors into accounts of their own practices as they move through the accreditation process. The study demonstrated how teachers in England were *subject* to the policy framework and the terminology it employed. Their validity, professionalism and identities as teachers were understood through the filter of their compliance with this regime.

The purpose of these examples is to demonstrate that judgments of effective practice or program success, or of any other outcome that might provide a focus for educational research, are contingent on the value system structuring the construction, selection and processing of data. These value systems are determined by the context in which the research is conducted. In such circumstances, we may ask whether it is appropriate to celebrate any supposed "improvements" in the quality of mathematical learning. Such "improvements" may simply be indicative of success in the administration's project of convincing the public that the administration's understanding of mathematics is the correct one and, for example, that the content of standardized tests define what mathematics is.

American, Chinese, or any other "reform" functions as an ideology, in Althusser's sense, a specific version of common sense, insofar as it determines the key parameters shaping discussion relating to curriculum innovation. In many instances of mathematics education research, "reform" functions as a supposed consensual aspiration. However, even within each culture: "Based on their concepts of students' needs, teachers select which parts of the reform documents are appropriate for their students," which translates as "children from upper socioeconomic backgrounds get problem solving, those from lower socioeconomic backgrounds undergo rote learning" (Sztajn, 2003, p. 53). These narratives of social reproduction have been regularly revived in research studies from Anyon's (1981) seminal study to Boaler's (1997) more recent analysis. International research assists us to situate such local variation within the parameters of national boundaries, compulsory schooling infrastructure, economic status and a host of other societal assumptions. International perspectives help us guard against the temptation to over-generalize the regularities and repetitions that we find in local curriculum reform research and to recognize how the dictates of locally dominant ideologies can over-determine the processes and outcomes of our research.

Curriculum Development Initiatives and Evidence-Based Policy

Mathematical learning in schools cannot be understood fully in terms of individual students encountering idealized mathematical objects. Those objects are formed across a much broader context, and can be understood in many different ways. The "meanings circulating in the classroom cannot be confined to the interactive dimension that takes place in the class itself; rather they have to be conceptualized according to the context of the historical–cultural dimension" (Radford, 2006, p. 23). Mathematical objects in a school context are typically defined in relation to a curriculum that prescribes roles for students and teachers. The extent to which such role definitions are culturally and linguistically determined is only now becoming recognized (Brown, 2011; Clarke, 2010). The actions of teachers and students are designed, recognized and assessed according to how they conform to these definitions. This pedagogical housing of mathematics influences the objects that are studied. The housing sets the conditions for learning and the resulting apprehension of mathematics.

More generally, teacher capabilities are not merely dependent on their "delivering" mathematical ideas. The capabilities derive from a broad range of factors. The picture is much bigger. For example, the setting of policy to bring about widespread adjustment to teacher practices towards raising "standards" or national test scores is a persistent aspiration, so often disappointed (Sammons et al., 2007). Policy makers do not work to a consistent agenda in governing school mathematics, and other stakeholders, such as, advisory groups, regulators, trainers, research and development funding agencies, and potential employers and universities, work according to a variety of perspectives and priorities. At the risk of sounding repetitive, all stakeholders in the mathematics education research endeavour contribute to that endeavour in ways that are highly context-specific and mutually constitutive.

Curriculum decisions are thus divided and shared between these various stakeholder groups, which do not necessarily see eye to eye, resulting in potential disjunctions between policy formulation, implementation by teachers and the conceptualizations made of such implementations by researchers (Saunders, 2007; Whitty, 2006). In addition, much research effort is dissipated across countless small studies from which it is difficult to produce a coherent picture. As a consequence, the theoretical underpinning of such processes has been somewhat fragmentary, sometimes switching between cognitive psychology at the level of the individual student learning mathematics, to an array of policy sciences and budgetary-led political expediency at the macro level. And these various areas of work each have their own specialists, who rarely meet with specialists from other areas to swap notes. The fragmentation of the education community into specialist groups poses a challenge for the development of either an integrative or a normative narrative of curricular reform, evaluation or policy development. For the moment, the best we can hope for is that each ideologically or theoretically situated research narrative is, at least, internally coherent and transparent with respect to its underlying principles and the processes of its gestation. This gives research, evidence and evidence-based policy a contingent character unlikely to meet political demands for generalizability.

How then might we conceptualize the role of research in supporting curriculum development? Much research in the field of mathematics education is targeted at individual teachers or teacher educators, from the perspective of how they might adjust their individual practices with students, yet at the same time an array of policy interventions split between diverse stakeholders operate in the wider domain. Might alternative perspectives or points of leverage offer more effective models of curriculum change? How might we conceptualize mathematics education research having an impact on populations of teachers through affecting policy decisions?

Research is often predicated on identifying deficiencies in current practices as part of a rationale for implementing a new approach. Hargreaves (1996, p. 5) has rather optimistically suggested that educational research must demonstrate "conclusively that if teachers change their practice from x to y there will be a significant and enduring improvement in teaching and learning." Hence, a history of research would be characterized as a series of projects, papers and books, with many arguing the case for some sort of improvement against various priorities. Yet looking back at any one time it is not easy to argue how we might assess retrospectively the nature of this cumulative improvement over any given period of time. It is quite difficult to provide evidence of improvement except in narrow terms. With the introduction of any new initiative there comes an implicit assumption that it will bring improvement over the previous regime. Yet priorities are not always consensual and evaluation strategies change over time. Alternative versions of history craft their heroes, objects and time phases differently. The term "improvement" can be understood in many different ways and resists stability across time, space and circumstances. The very conceptions of progress may have moved on to be understood in different terms.

Teacher biographies are typically characterized by engagements with a number of teaching approaches throughout any one career. Each shift from one to another entails mathematics being framed in a slightly different way that perhaps results in a different teaching style and, perhaps also, in a different conception of mathematics. Elements derived from each phase feed into composite experience and contribute to that teacher's modes of practice and emergent, and perhaps convergent, professional identity. These elements might be attributed variously to fashions in school practices, learning theories, assessment preferences, career phase of the individual teacher, etc. The shifts in teaching approach would normally be locally negotiated on the basis of some supposed improvement on the previous model.

Asking teachers to move from one teaching approach to another can, it seems, never be regarded as a straightforward substitution (cf., Fullan, 2001). Nevertheless, for those charged with setting policy, there is often a perceived obligation to do something. And often this involves doing something big. In the United Kingdom, New Zealand and Australia, for example, governments have prescribed detailed curricula for students and teachers alike, along with associated industries concerned with preparing materials. Analogous to such support provision, the Chinese curriculum addresses the problem of scaffolding instructional innovation slightly differently. The mathematics curriculum itself contains sample activities, illustrative of approaches that Chinese teachers might employ in implementing the curriculum. State-orchestrated textbook construction provides Chinese teachers with an authoritative body of definitions, explanations and tasks that can be interpreted confidently as embodying the aspirations of the official curriculum.

In terms of research literature, more information is readily available about the effect of major curriculum reform in the USA, where there is also a considerable emphasis on the widespread adoption of new curriculum materials as a primary strategy for improving mathematical education (Remillard, 2005; Remillard & Bryans, 2004). Such is the extent and diversity of curriculum evaluation research in the USA that the National Research Council (USA) commissioned a meta-evaluation of mathematics curriculum evaluation studies (National Research Council, 2004; Towne, Wise, & Winters, 2005). The report of this meta-evaluation proposed clear criteria for the conduct of curricular studies employing different methodological approaches. In addition to its substantive findings, the report provides a model of effective, scholarly consideration of curricular evaluation (see also United States Department of Education, 2008).

The sheer volume of research carried out within the USA has resulted in the conceptions of teaching and curriculum implementation pertaining to this country seeping beyond its boundaries. Despite a diversity of context in the USA that defies simplistic summation, there is a sense in which it provides a context for the rest of the world. The country prescribes the parameters (through TIMSS, dominance in international research journals, setting political normalcy, promotion of the individual) whereby teaching might be classified, analyzed and informed. Ironically, American interest in Asian classrooms has stimulated a more widespread international interest in educational systems in the Asian region and encouraged researchers in Japan, China and Singapore, for example, to investigate their own practices and share the results with the international education community (Fan, Wong, Cai, & Li, 2004). The cultural specificity not only of the findings but also of the educational value systems on which the findings are predicated has perturbed the existing international acquiescence to a US-centric educational agenda. Emergent resonances of educational value and practice among European and Asian school systems may further destabilize the homogenization of international education threatened by the prominence of the international testing of student achievement and the educational imperialism of the OECD.

Conceptualizations of mathematical learning emerge through alternative curriculum models and development initiatives. Teachers, more or less, make sense of their practices adjusted in line with new descriptive lenses. They identify with successive curriculum models and the way in which these identifications frame mathematical learning. Within any curriculum implementation, both the teachers' sense of what they are doing and the curriculum itself are reconstituted through the encounter, thwarting any supposed convergence to an endpoint. This argument has implications for how we think about initiatives designed to work at creating consensus in teaching approaches. In particular, we need to question how or if research agenda encourage teachers to align with a particular model or philosophy of practice conceptualized in advance. Affinity with any particular model does not necessarily fix the mode of association or how that is viewed.

Remillard (2005, pp. 215–223) examined alternative ways in which teacher/curriculum interfaces have been understood within the research literature. She contrasted "following or subverting" a curriculum text with "drawing on" a curriculum text or "interpreting" a curriculum text. In these three alternatives, the text is present in some form and teachers respond to it. Finally, however, Remillard considered how curricula might be understood as teachers participating with the text. For a teacher "enacting" a curriculum in this mode, she suggested that teacher and curriculum might be seen as mutually constitutive. Here, curriculum use was understood as participation with the text (pp. 221–223). She identified this with "Vygotskian notions of tool use and mediation, wherein all human activity involves mediated action or the use of tools by human agents to interact with one another and the world" (cf., Cole, 1996). Such an approach is familiar within mathematics education research (e.g., Blanton, Westbrook, & Carter, 2005; Goos, 2005).

Ultimately, understood in terms of Foucault's (1989) notion of "discursive formation," both teacher and curriculum would be functions of how they are implicated in the stories that unite them. Both change as a result of curriculum development activity. Remillard (2005) identified some studies where teachers changed or learned from their use of resources (Lloyd, 1999; Remillard, 2000; van Zoest & Bohl, 2002). Yet teacher change can also be understood as being the result of increased compliance with respect to a curriculum initiative. Aspirations to consensus can suppress the specificities of alternative needs, responses, etc., and thereby serve those who are already the most powerful. We find ourselves, yet again, cautioning against the possibility that research not only reproduces values pre-determined by the institutional context of the research but also becomes complicit in the further reification of those values as universal.

Publication Networks

Journals of long-standing quality, serving different purposes and different audiences, such as *Educational Studies in Mathematics*, the *Journal of Research in Mathematics Education*, and *For the Learning of Mathematics*, continue to find a readership. Some journals, such as *ZDM—The International Journal of Mathematics Education*, successfully redefine their purpose and audience in addressing the concerns of the international research community in mathematics education. Other journals, such as the *Journal of Mathematics Teacher Education*, focus their efforts on a specialized readership within the mathematics education community. The viability of such journals is threatened by national measures that base their hierarchies on citation indices and impact factors.

Electronic publications have now established themselves within the field of recognized publication outlets. Government research productivity guidelines, such as that for the *Excellence in Research for Australia*, make no distinction among publications by mode of delivery and explicitly include e-books, for example, in the list of acceptable research publications. Such publications are subject to the same quality criteria as other forms of research output. Electronic publications have neither distorted nor diluted the quality of available outlets through which we might disseminate our research. Publication in electronic form now routinely precedes publication in hard-copy for most major journals and expedites the community's access to research.

High status conferences producing a published conference proceedings document employing a rigorous peer-review process can serve at least three essential functions: (a) Such conferences provide a forum at which the most topical issues and the most recent research can be reported and discussed; (b) The provision of an immediate publication outlet for the research reported at such conferences provides a more efficient documentation of advances in the field than that typically provided through the lengthy review and revision processes employed by journals; and (c) Provided the peer-review process is sufficiently rigorous, the resulting proceedings publication receives recognition within most measures of research productivity. The International Group for Psychology in Mathematics Education (PME) has long provided such a high-status research forum and publication outlet. Other conferences, such as the Congress of the European Society for Research in Mathematics Education (CERME), the *Commission Internationale pour l'Étude et l'Amélioration de l'Enseignement des Mathématiques* (CIEAEM, International Commission for the Study and Improvement of Mathematics Teaching) or the Research Pre-session of the annual conference of the National Council of Teachers of Mathematics (NCTM, USA), perceive their purposes differently and accord less priority to a peer-reviewed proceedings publication, placing greater emphasis on providing an interactive forum, where the contribution of research to contemporary issues in mathematics education can be critically examined. Participation by members of the mathematics education community in major international conferences of a more general nature, such as the annual conference of the American Educational Research Association (AERA) or the biennial conference of the European Association for Research in Learning and Instruction (EARLI), provides an important connection between research in mathematics education and the general field of educational research. National and regional research conferences such as the Southern African Association for Research in Mathematics, Science and Technology Education (SAARMSTE), the Mathematics Education Research Group of Australasia (MERGA) and the East Asian Regional Conference On Mathematics Education (EARCOME) all provide opportunities for the reporting and discussion of research and all produce peer-reviewed conference proceedings of high quality.

Academic Networking and Research Community Definition

As with any other professional activity, mathematics education research is undertaken within a community membership that defines itself and the field through its research activities. Advances in technology have enabled entirely new forms of international research collaboration and thereby reconstructed research communities, both in terms of their membership and the nature of their activities. Regional networks have led to the establishment of major conferences such as EARCOME and SAARMSTE, mentioned above. The availability of a regional forum where research can be reported and possibilities explored for research partnership is an essential element in the promotion and maintenance of regional research networks. Independent of participation in more global international gatherings, regional conferences provide an opportunity to develop a regional research agenda, addressing issues more immediately pertinent to school systems in the region.

Participation in international research is constrained by many factors. One of these is access to the technological resources required to generate, store and analyze large data sets. Large databases generated by projects such as the Trends in International Mathematics and Science Study (TIMSS) and the Learner's Perspective Study (LPS) are now available to participating researchers anywhere in the world through high-speed, secure, Web-mediated connection. Not only does this transform the nature of international research collaboration, by providing distributed access to storage facilities hosted within a single institution, but also less affluent research groups or institutions are saved the expense of costly storage facilities and are more able to participate in international research studies. It was previously noted that "when less affluent countries participate in international studies, it is frequently as the objects of investigation rather than as partners in the research" (Clarke, 2003, p. 177). Advances in technology and the growing emergence of international collaborative research networks are increasingly replacing such differentiated participation with true research partnership.

These emerging international research partnerships have the potential to catalyze a broadening in perceptions of the goals of research in mathematics education beyond the pragmatics of local utility. Recent curricular developments in Asian school systems, such as in China, Korea and Singapore, occur in parallel with advances through adaptation by countries such as the USA and Australia of approaches to instruction and teacher education originating in Japan and in China. These activities have been accompanied by the emergence of major research partnerships between researchers in Australia, the USA, and Europe with their counterparts in Japan, China, Korea and Singapore. This recognition of the mutual benefit afforded by international academic collaboration is an essential component in the reconceptualization of the mathematics education community as an international cross-cultural endeavour, of the manner in which research might be conducted and coordinated internationally, and of the contribution that research might make to particular school systems.

The Training and Education of Researchers

Mathematics education research is a function of the people who do it. At a local level a teacher might be concerned with doing research to teach in a more satisfying way at a personal level, or to develop or meet the demands of a school teaching scheme understood as shared guidance for a specific group of colleagues. At a national level research might be carried out by teacher educators addressing more generic issues, perhaps associated with externally defined targets or policy documentation. Or the research might be commissioned and shaped by administrators charged with managing a population of teachers and students through prescriptive curricular apparatus. At an international level, other aspirations may intervene, such as the need to speak effectively in an area of interest to a discernible group of researchers. In some countries, professional advancement in academic work is assessed by its perceived international status. Getting such an audience may be less about improving one's teaching or meeting an externally defined target through conforming to good practice, but more about learning to write or talk convincingly, even if it means neglecting one's teaching! Bordo (1999) argued that academia is often susceptible to mediatizing its image.

Academics sometimes use the accessories of theory (for example, specialised forms of jargon, predictable critical moves, references to certain authors) less in the interests of understanding the world than to proclaim themselves members of an elite club. In the process they create caricatures of themselves and of those who don't belong, peopling the scholarly world with typecast players and carving out narrow theoretical niches within which all ideas and authors are force-fit. Certain theoretical preferences, moreover, run throughout disciplines like incurable diseases, often carrying invisible racial and gender stereotypes and biases along with them. (p. 24)

A more charitable interpretation might be that academic fields get to be learnt through caricatures as it would be too overwhelming to do otherwise. Nevertheless, the impact of Bordo's comments seems to hold in educational research. The "production of educational theory and research is itself a site of ideological and political struggle" (Britzman, 2003, p. 68—citing McCarthy & Apple; see also DeFreitas & Nolan, 2008).

In parallel with the reconstruction of the international mathematics education research community, the mathematics education researcher has also undergone significant change. The contemporary researcher in mathematics education is much more likely to be well-versed in a variety of methodologies and theories than to be a doctrinaire adherent of a single theory or to engage in research restricted to a single methodological approach. In part, this ecumenical approach to research reflects the more team-driven nature of the contemporary enterprise. In many countries, such research teams combine researchers from a variety of cultural (and therefore educational) backgrounds, bringing usefully diverse perspectives to the research endeavour.

It has been changes to the institutional context of research, such as those already discussed, that have fuelled the reconstruction of the educational researcher from solitary worker to active member of a research community. It is to be hoped that the evolution of educational research (and mathematics education research, in particular) from cottage industry to international collegial enterprise will not discard cottage charm and individual creativity for a sort of industrialized and mechanical anonymity. Educational research will continue to draw many of its initiates from school settings, with a higher proportion of part-time involvement than would be found in early-career researchers in the sciences. This part-time research community brings with it a vocational situatedness that should act to the benefit of the field of mathematics education research by locating research activity in the hands of those most likely to benefit from it and best placed to implement its findings.

The argument parallels that of the action-research community and appropriately so. Nonetheless, the participation of part-time research students presents challenges for the construction of a research community that universities and research centres address with uneven success. "There are significant difficulties in influencing the professional learning of educational researchers themselves towards changing the practices of educational research" (Rees, Baron, Boyask, & Taylor, 2007). The slightly pessimistic note of this quotation should not lead us to disregard the advantages now available to the beginning researcher in mathematics education. The same technology that facilitates international networking can be exploited to create distributed research communities that integrate less and more experienced members in less and more vocationally-situated contexts. Rather, the recursiveness implicit in the research community's management of the on-going learning of its own constituents should be seen as an opportunity for continual regeneration and reflective interrogation rather than potential stagnation.

Available technologies offer the opportunity for early-career researchers to access the expertise of established researchers independent of the constraints of geography, culture or school system. Those responsible for the learning environments of beginning researchers have the opportunity to create and nurture richer, more interactive, and more diverse educational experiences for new members of the research community. The affordances provided by new connectivities and communicative networks act in the opposite direction to the constraining effects of some of the politically motivated dictates of legislation, accountability and funding provision discussed earlier.

The institutional context must be considered at least in local, national and international terms. With regard to the education of researchers, we have a tension between the local experience of improved access to the rich international diversity of theories, methodologies, issues, values, agendas, and research expertise and the potentially limiting influence of national and international political agendas (and ideological positions) that seek to channel research activity into officially sanctioned forms. In parallel with tensions in the framing of mathematics curricula, standardization in the name of accountability leads either to an impoverished curriculum offered to the beginning educational researcher or to a graduate community of mathematics education researchers, whose sophisticated research expertise is unable to be realized within the incentive schemes currently dominating the educational research landscape.

Conclusion

Mathematics education research typically seeks to inform the social interactive processes that locate but also transform teachers, students and mathematics. The task of such research can be understood from a range of perspectives that can mark out various operational levers, not just changes to teacher practice. As researchers we need to be aware of how our work is governed and formatted by a range of agencies, from employers allowing limited space between other duties, to funding agencies being specific about the perspectives they want to be depicted, to research assessment exercises or journals defining what is of value to the research community. But more generally we need to be attentive to the assumptions built into the locations of our work that restrict our scope of interest. The recommendations for practice arising from educational research are always situated recommendations, even if they are not presented as such.

Recognition of this emphasis on situated practice has implications for the sort of evidence likely to inform either educational policy or practice. Yet, the widespread enthusiasm for evidence-based policy development frequently begs the essential question as what constitutes evidence. Where this question is addressed, the answer may take the form of a prescription of valued and non-valued research paradigms. Shavelson and Towne (2002) explicitly advocated "evidence-based education" and particularly encouraged research in the social sciences to adopt if not the methods at least the principles of medical research. Subscription to such a medico-scientific standard locates research and the researcher within a discourse predicated on the identification and evaluation of educational "treatments" as the focus of the research endeavour—classifying research participants as the doing and the done-to. This leads to an inevitable emphasis on "What works?" and the implication that this can be answered in some context-free fashion. The implied parallels between physiological phenomena and socio-cognitive phenomena suggest aspirations to a misleading generalizability that educational research can seldom justify except in the reporting of trivial descriptive findings. In contrast, the practitioner research tradition has sought to emphasize how research needs to be worked into practice through time.

Either educational research accepts a responsibility to express its findings in more practical terms, so that research evidence takes the form of endorsed practices, or research itself needs to be made a part of practice (Somekh, 2006). Research also needs to attend to the mediation of teacher education so that teachers can be prepared for particular understandings of practice. What teacher education programs would need to be put in place and how would this be achieved? There is little point having a thesis on "what works" if teachers cannot access this knowledge or are insufficiently skilled to bring it about.

Structural models are often seen, through cultural bias, as ones that should be aspired to more generally or internationally. For example, any given strategy implies resource constraints and one size fits all models potentially deny key aspects of diversity. Speaking from an African context, Swanson (2010, p. 245) asked the question: What are the implications for education and mathematics education, in particular, when industrialization and economic growth are the foremost policy objectives of a nation state? We have surveyed some of the implications in Western countries and those in the Pacific Rim. This, however, is only part of the picture. "Eighty per cent of the world's children are in developing countries. Yet, much of the research in mathematics education backgrounds this reality" (Adler, 2008, p. 241). Few schools/countries could supply the teachers who could offer the sensitivities and skills required in so many proposed models of mathematical learning (cf., Skovsmose, 2005). For example, for all their rhetoric, U.S.-oriented liberal individualist constructivism and also Chinese authoritarian collectivism, support capitalism. Yet in answer to her own question, Swanson (2010) argued that this capitalism "has failed to provide the alluring 'rewards' for millions of people living in abject poverty who have little agency in relation to the hierarchy of access it has produced and which it serves to reproduce" (p. 246).

Students and teachers are not only (successful or unsuccessful) recipients of cultures but also creators of cultures insofar as their fresh perspectives on mathematical situations can be voiced, rather than being merely evaluated with respect to existing registers. Knijnik (2010) insisted on the intrinsic connection of mathematics education to culture. In discussing her work with the Landless Peasant Movement in Brazil, she described culture as a "conflictive, unstable and tense terrain, undermined by a permanent dispute to impose meanings through power relations" (p. 413), where the very concept of a unit of land remains contested. We need to ask what mechanisms might enable populations of teachers to support student creativity in challenging and renewing the cultures or contexts they occupy. As we have shown many facets of these cultures derive from externally imposed prescription, perhaps derived from norms that favour those in power.

Students and teachers are not things in themselves but are consequential to educational situations being read against specific discursive frames that shape the political domain and the priorities that domain confers. The term "teacher," for example, is constituted with respect to a particular social construction of that term and the expectations or aspirations that go with it, expectations and aspirations that differ markedly across schools and countries. As an individual teacher, I may have all sorts of personal optimism, but if I want a government job I have to fit in with the regulative structures pertaining to the context I am in, and understand myself through the terms of that regulation. Mathematics education research has a duty to enable teachers to assert a professionalism that meets yet transcends local regulative demands. To meet this duty we must reach beyond the context-specific meanings that research is obliged to service. Research might be seen as the task of rethinking mathematical teaching and learning with a view to changing them to meet or resist emerging demands. Through considering how teachers, teacher educators, trainees, pupils and researchers themselves make sense of their worlds, research can support work on how linguistic and socio-cultural contexts link to prevalent conceptions of mathematics education. Research itself can be seen as participation in cultural renewal, where the very worlds it encounters are becoming something new. This contemplates trajectories of change into fresh ways of being for teachers, teacher educators and researchers.

To represent mathematics as universal, spanning nations and generations, comes at a price. TIMSS and PISA were introduced to measure and compare school mathematics in different countries on a singular scale. Yet the resultant conceptions of school mathematics now define and police the boundaries of school mathematics. At a conference in 2011, a Mexican delegate spoke of how the exercises made her country subservient to American priorities for school mathematics (Garcia, Saiz, & Rivera, 2011). An Ethiopian educator depicted a situation in which teachers and students were obliged to engage with a form of mathematics encased in pedagogical formations largely unrecognizable in their country situation (Gebremichael, 2011). As seen, the United Kingdom has sacrificed its earlier facility with problem-solving approaches in order to meet newly understood "mathematical" objectives. Meanwhile, a Finnish commentator indicated that her country's high performance in the exercises did not release her colleagues from having to reevaluate their practices in terms of the newly dominant international discourse (Krzywacki, Koistinen, & Lavonen 2011). School mathematical knowledge has come to be a function of this newly described world, backed up by governments using these conceptions of mathematics to set their policies.

Educational research distinguishes itself from research in the sciences by its tendency to recommend the replacement rather than the augmentation of existing practice. These new ways of understanding mathematics education that throw the baby out with the bath water deflect us from occupying alternative worlds, which might define us and serve us in different ways, according to priorities that may vary from one location to another. Excessive belief in unified objectives can simultaneously disregard more localized needs and corrupt the truly universal. Researchers have become complicit in promoting and reifying the values that support these particular conceptions of teaching and thereby restrict the trajectories for change that we are able to conceive. Also, research itself in many locations is increasingly obliged to follow formal regulation, setting the ways in which educational practices can be legitimately described. Since researchers in mathematics education are simultaneously members of the mathematics education community, they have become complicit in the construction of the field as an ideological battleground, in a terrain with features falsely identified as universal.

References

Adler, J. (2008). ICMI in Africa and Africa in ICMI: The development of AFRICME. In M. Menghini, F. Furinghetti, L. Giacardi, & F. Arzarello (Eds.), *The first century of the International Commission on Mathematical Instruction (1908–2008): Reflecting and shaping the world of mathematics education* (p. 236). Rome, Italy: Biblioteca dell'Enciclopedia Treccani.

Althusser, L. (1971). Ideology and ideological state apparatuses. In *Lenin and philosophy and other essays* (B. Brewster, Trans. & Ed.). London, UK: New Left Books.

Anyon, J. (1981). Social class and school knowledge. *Curriculum Inquiry, 11*(1), 3–42.

Askew, M., Hodgen, J., Hossain, S., & Bretscher, N. (2010). *Values and variables: A review of mathematics education in high-performing countries.* London, UK: The Nuffield Foundation.

Bishop, A. (1990). Western mathematics: The secret weapon of cultural imperialism. *Race and Class, 32*(2), 51–65.

Blanton, M., Westbrook, S., & Carter, G. (2005). Using Valsiner's zone theory to interpret teaching practices in mathematics and science classrooms. *Journal of Mathematics Teacher Education, 8*(1), 5–33.

Boaler, J. (1997). *Experiencing school mathematics: Traditional and reform approaches to teaching and their impact on student learning.* Buckingham, UK: Open University Press.

Bordo, S. (1999). *Twilight zone: The hidden life of cultural images from Plato to O.J.* London, UK: University of California Press.

Britzman, D. (2003). *Practice makes practice: A critical study of learning to teach.* Albany, NY: State University New York Press.

Brown, T. (2001). *Mathematics education and language: Interpreting hermeneutics and post-structuralism* (2nd ed.). Dordrecht, The Netherlands: Kluwer Academic Publishers.

Brown, T. (2011). *Mathematics education and subjectivity: Cultures and cultural renewal.* Dordrecht, The Netherlands: Springer.

Brown, T., Devine, N., Leslie, E., Paiti, M., Sila'ila'i, E., Umaki, S., & Williams, J. (2007). Reflective engagement in cultural history: A Lacanian perspective on Pasifika teachers in New Zealand. *Pedagogy, Culture and Society, 15*(1), 107–119.

Brown, T., & McNamara, O. (2011). *Becoming a mathematics teacher: Identity and identifications.* Dordrecht, The Netherlands: Springer.

Cai, J., & Nie, B. (2007). Problem solving in Chinese mathematics education: Research and practice. *ZDM—The International Journal of Mathematics Education, 39*(5–6), 459–474.

Chazan, D., & Ball, D. (1999). Beyond being told not to tell. *For the Learning of Mathematics, 19*(2), 2–10.

Cheng, Y. C., & Cheung, W. M. (1999). Lessons from TIMSS in Europe: An observation from Asia. *Educational Research and Evaluation, 5*(2), 227–236.

Clarke, D. J. (1994). Why don't we just tell them? In C. Beesey & D. Rasmussen (Eds.), *Mathematics without limits* (pp. 11–19). Brunswick, Victoria, Australia: Mathematical Association of Victoria.

Clarke, D. J. (1996). Assessment. In A. J. Bishop, M. A. Clements, C. Keitel, J. Kilpatrick, & C. Laborde (Eds.), *International handbook of mathematics education* (Vol. 1, pp. 327–370). Dordrecht, The Netherlands: Kluwer.

Clarke, D. J. (2003). International comparative studies in mathematics education. In A. J. Bishop, M. A. Clements, C. Keitel, J. Kilpatrick, & F. K. S. Leung (Eds.), *Second international handbook of mathematics education* (pp. 145–186). Dordrecht, The Netherlands: Kluwer.

Clarke, D. J. (2010). The cultural specificity of accomplished practice: Contingent conceptions of excellence. In Y. Shimizu, Y. Sekiguchi, & K. Hino (Eds.), *In search of excellence in mathematics education: Proceedings of the 5th East Asia Regional Conference on Mathematics Education (EARCOME5)* (pp. 14–38). Tokyo, Japan: Japan Society of Mathematical Education.

Cohen, D. (1990). A revolution in one classroom: The case of Mrs. Oublier. *Educational Evaluation and Policy Analysis, 12*, 327–345.

Cole, M. (1996). *Cultural psychology: A once and future discipline*. Cambridge, MA: Belknap Press.

Connell, R. (2007). *Southern theory: The global dynamics of knowledge in the social sciences*. Sydney, Australia: Allen and Unwin.

De Lange, J. (1987). *Mathematics, insight and meaning*. Utrecht, The Netherlands: Freudenthal Institute.

DeFreitas, E., & Nolan, K. (2008). *Opening the research text: Critical insights and in(ter)ventions into mathematics education*. New York, NY: Springer.

Department for Education (DfE). (2010). *The importance of teaching*. London, UK: Author.

Fan, L., Wong, N.-Y., Cai, J., & Li, S. (2004). *How Chinese learn mathematics: Perspectives from insiders*. Hackensack, NJ: World Scientific.

Fan, L., & Zhu, Y. (2007). From convergence to divergence: The development of mathematical problem solving in research, curriculum, and classroom practice in Singapore. *ZDM—The International Journal of Mathematics Education, 39*(5–6), 491–502.

Fennema, E., & Nelson, B. (Eds.). (1997). *Mathematics teachers in transition*. Mahwah, NJ: Lawrence Erlbaum.

Foucault, M. (1989). *The archaeology of knowledge*. London, UK: Routledge.

Fullan, M. (2001). *The new meaning of educational change* (3rd ed.). New York, NY: Teachers College Press.

Garcia, R., Saiz, M., & Rivera, A. (2011). Cognitive cultural analysis of low achievement in TIMSS: Evaluating wrong answers in 8th grade. In B. Ubuz (Ed.), *Proceedings of the 35th Annual Conference of the International Group on the Psychology of Mathematics Education* (1st ed., p. 382). Ankara, Turkey: International Group on the Psychology of Mathematics Education.

Gebremichael, A. (2011). Perceptions of relevance of prior experiences of mathematics in an Ehiopian preparatory school. In B. Ubuz (Ed.), *Proceedings of the 35th Annual Conference of the International Group on the Psychology of Mathematics Education* (1st ed., p. 302). Ankara, Turkey: International Group on the Psychology of Mathematics Education.

Goos, M. (2005). A sociocultural analysis of the development of pre-service and beginning teachers' pedagogical identities as users of technology. *Journal of Mathematics Teacher Education, 8*(1), 35–59.

Hargreaves, D. (1996). Teaching as a research-based profession: Possibilities and prospects. In *The Teacher Training Agency lecture*. London, UK: Teachers Training Agency.

Hino, K. (2006). The role of seatwork in three Japanese classrooms. In D. Clarke, C. Keitel, & Y. Shimizu (Eds.), *Mathematics classrooms in twelve countries: The insider's perspective* (pp. 59–74). Rotterdam, The Netherlands: Sense Publishers.

Hino, K. (2007). Toward the problem-centred classroom: Trends in mathematical problem solving in Japan. *ZDM—The International Journal of Mathematics Education, 39*(5–6), 503–514.

Howson, G., & Mellin-Olsen, S. (1986). Social norms and external evaluation. In B. Christiansen, A. G. Howson, & M. Otte (Eds.), *Perspectives on mathematics education* (pp. 1–48). Dordrecht, The Netherlands: Reidel.

Knijnik, G. (2010). Gelsa Knijnik's research on ethnomathematics. In M. M. F. Pinto & T. F. Kawasaki (Eds.), *Proceedings of the 34th Annual Conference of the International Group for the Psychology of Mathematics Education* (Vol. 1, pp. 413–415). Belo Horizonte, Brazil: International Group for the Psychology of Mathematics Education.

Krzywacki, H., Koistinen, L., & Lavonen, J. (2011). Assessment in Finnish mathematics education: Various ways, various needs. In B. Ubuz (Ed.), *Proceedings of the 35th Annual Conference of the International Group on the Psychology of Mathematics Education* (1st ed., p. 340). Ankara, Turkey: International Group on the Psychology of Mathematics Education.

Laclau, E. (2005). *On populist reason*. London, UK: Verso.

Lin, F.-L. (2010). Mathematical tasks designing for different learning settings. In M. F. Pinto & T. F. Kawasaki (Eds.), *Proceedings of the 34th Annual Conference of the International Group for the Psychology of Mathematics Education* (Vol. 1, pp. 83–95). Belo Horizonte, Brazil: International Group for the Psychology of Mathematics Education.

Lloyd, G. M. (1999). Two teachers' conceptions of a reform-oriented curriculum: Implications for mathematics teacher development. *Journal of Mathematics Teacher Education, 2*(3), 227–252.

Lobato, J., Clarke, D. J., & Ellis, A. B. (2005). Initiating and eliciting in teaching: A reformulation of telling. *Journal for Research in Mathematics Education, 36*(2), 101–136.

Mouffe, M. (2005). *On the political*. London, UK: Routledge.

National Research Council. (2004). *On evaluating curricular effectiveness: Judging the quality of K-12 mathematics evaluations*. Washington, DC: National Academies Press.

Otte, M. (1979). The education and professional life of mathematics teachers. In B. Christiansen & H. G. Steiner (Eds.), *New trends in mathematics teaching IV* (pp. 107–133). Paris, France: UNESCO.

Pampaka, M., Williams, J., Hutcheson, G., Wake, G., Black, L., Davis, P., & Hernandez-Martinez, P. (2012) The association between mathematics pedagogy and learners' dispositions for university study, *British Educational Research Journal, 38*(3), 473–496.

Radford, L. (2006). *Elements of a cultural theory of objectification*. Sudbury, Ontario, Canada: Université Lauentienne.

Rees, G., Baron, S., Boyask, R., & Taylor, C. (2007). Research-capacity building, professional learning and the social practices of educational research. *British Educational Research Journal, 33*(5), 761–781.

Remillard, J. (2000). Can curriculum materials support teachers' learning? *Elementary School Journal, 11*(4), 331–350.

Remillard, J. T. (2005). Examining key concepts in research on teachers' use of mathematics curricula. *Review of Educational Research, 75*(2), 211–246.

Remillard, J., & Bryans, M. (2004). Teachers' orientations towards mathematics curriculum materials: Implications for teacher learning. *Journal for Research in Mathematics Education, 35*(5), 352–388.

Remillard, J. T., & Geist, P. K. (2002). Supporting teachers' professional learning by navigating openings in the curriculum. *Journal of Mathematics Teacher Education, 5*(1), 7–34.

Sammons, P., Day, C., Kington, A., Gu, Q., Stobart, G., & Smees, R. (2007). Exploring variations in teachers' work, lives and their effects on pupils: Key findings and implications from a mixed-method study. *British Educational Research Journal, 33*(5), 681–701.

Saunders, L. (Ed.). (2007). *Educational research and policy-making: Exploring the border country between research and policy*. London, UK: Routledge.

Sekiguchi, Y. (2006). Mathematical norms in Japanese mathematics lessons. In D. J. Clarke, C. Keitel, & Y. Shimizu (Eds.), *Mathematics classrooms in twelve countries: The insider's perspective* (pp. 289–306). Rotterdam, The Netherlands: Sense Publishers.

Shavelson, R. J., & Towne, L. (Eds.). (2002). *Scientific research in education*. Washington, DC: National Academies Press.

Shimizu, Y. (2006). How do you conclude today's lesson? The form and functions of "Matome" in mathematics lessons. In D. J. Clarke, J. Emanuelsson, E. Jablonka, & I. Mok (Eds.), *Making connections: Comparing mathematics classrooms around the world* (pp. 127–145). Rotterdam, The Netherlands: Sense Publishers.

Simon, M., & Tzur, R. (1999). Explicating the teacher's perspective from the researchers' perspectives: Generating accounts of mathematics teachers' practice. *Journal for Research in Mathematics Education, 30*(3), 252–264.

Skovsmose, O. (2005). *Travelling through education: Uncertainty, mathematics, responsibility*. Rotterdam, The Netherlands: Sense Publishers.

Somekh, B. (2006). *Action research: A methodology for change and development*. Maidenhead, UK: Open University Press.

Steffe, L., & Kieran, T. (1994). Radical constructivism and mathematics education. *Journal for Research in Mathematics Education, 25*(6), 711–733.

Swanson, D. (2010). The paradox and politics of disadvantage. Narratizing critical moments of discourse and pedagogy within the "glocal." In M. Walshaw (Ed.), *Unpacking pedagogy: New perspectives for mathematics classrooms* (pp. 245–263). Greenwich, CT: Information Age.

Sztajn, P. (2003). Adapting reform ideas in different mathematics classrooms: Beliefs beyond mathematics. *Journal of Mathematics Teacher Education, 6*(1), 53–75.

Thorsten, M. (2000). Once upon a TIMSS: American and Japanese narrations of the Third International Mathematics and Science Study. *Education and Society, 18*(3), 45–76.

Törner, G., Schoenfeld, A. H., & Reiss, K. M. (Eds.). (2007). Problem solving around the world: Summing up the state of the art. *ZDM—The International Journal on Mathematics Education, 39*(5–6), 353–561.

Towne, L., Wise, L. L., & Winters, T. M. (Eds.). (2005). *Advancing scientific research in education*. Washington, DC: National Academies Press.

Tzur, R., Simon, M., Heinz, K., & Kinsel, M. (2001). An account of a teacher's perspective on learning and teaching mathematics: Implications for teacher development. *Journal of Mathematics Teacher Education, 4*(3), 227–254.

United States Department of Education. (2008). *Foundations for success: The final report of the National Mathematics Advisory Panel*. Washington, DC: United States Department of Education.

van Zoest, L. R., & Bohl, J. V. (2002). The role of reform curricular materials in an internship: The case of Alice and Gregory. *Journal of Mathematics Teacher Education, 5*, 265–288.

Walkerdine, V. (1984). Developmental psychology and the child-centred pedagogy. In J. Henriques, W. Holloway, C. Urwin, C. Venn, & V. Walkerdine (Eds.), *Changing the subject: Psychology, social regulation and subjectivity* (pp. 153–202). London, UK: Methuen.

Whitty, G. (2006). Education(al) research and education policy making: Is conflict inevitable? *British Educational Research Journal, 32*(2), 159–176.

Wilson, S. M., & Goldenberg, P. (1998). Some conceptions are difficult to change: One middle school mathematics teacher's struggle. *Journal of Mathematics Teacher Education, 1*, 269–293.

Žižek, S. (2008). *The plague of fantasies* (2nd ed.). London, UK: Verso.

Chapter 16
Policy Implications of Developing Mathematics Education Research

Celia Hoyles and Joan Ferrini-Mundy

Abstract Researchers often pursue their own interesting and specific mathematics education research questions without engaging with the practical and policy issues that may have considerable bearing on mathematics education. The final chapter of this section deals with this situation by considering three interrelated themes: developments in education policy that have implications for mathematics education research; the potential for engaging the mathematics education community in pursuing research questions that have implications for policy; and the relevance, utility, and accumulation of mathematics education research findings to support policy and practice. In particular, questions are raised about the role of standards in the specifics of mathematics teaching and learning, and the challenges of making research professionally and publicly available in ways that might be used to inform the decisions and the practices of policy makers and teachers.

Introduction

The teaching and learning of mathematics occur largely within classrooms, schools, and universities that are influenced far more strongly by educational policies—"rules and regulations promulgated in state capitals and the federal government" (Sykes, Schneider, & Ford, 2009, p. 1)—than by mathematics education research. In most countries, the importance of mathematics education is judged

---

The second-named author is now at the National Science Foundation. This work was based in part on work supported by the National Science Foundation.

C. Hoyles (✉)
Institute of Education, University of London, London, UK
e-mail: C.Hoyles@ioe.ac.uk

J. Ferrini-Mundy
Michigan State University, East Lansing, MI, USA

as critical. It is presumed to be "a vehicle toward social and political progress" (Gates & Vistro-Yu, 2003, p. 62), and central to the development of a well-trained workforce that can advance the economic standing of a country. Governments face a range of distinct but related policy challenges that include providing universal mathematical literacy for all, ensuring a mathematical foundation to support the study of other subjects that are increasingly demanding higher levels of mathematics, and stimulating the most able to continue with mathematics study after it is no longer compulsory and into university.

At the same time, mathematics education research is largely conducted to take forward theory and knowledge of the domain, although impact on teaching and learning practice is a distinct purpose (e.g., Lester & Wiliam, 2002). Yet, there remains often a mismatch between questions pursued by researchers and questions facing policy makers and practitioners. It seems unlikely that most mathematics education researchers have the potential impact of their work in mind on, for example, major national economic debates or workforce capability. This has tended to mean that if mathematics education research has had rather little significant influence on practice, its influence on policy has been even less.

However, Smith and Smith (2009) have argued that policy research does influence practice but maybe not directly and obviously. As one example, Welch (1979) (cited in Smith & Smith, 2009) made a case that research on science and mathematics learning indirectly influenced the US-based K–12 curricular reforms of the 1960s and 1970s, resulting in their emphasis on hands-on instruction and inquiry-oriented approaches. A similar case can be made for comparable reforms in UK and Europe over the same period, where more investigative approaches were promoted and the need for appropriate teacher interventions recognized. Research in design experiments repeatedly reported that in such contexts, scaffolds and guidance for the teacher were needed (Noss & Hoyles, 1996). Thus, history would suggest that there is considerable untapped potential for productive interaction between the mathematics education community globally and those concerned with the development and implementation of policy that affects mathematics teaching and learning.

Education policy is defined in various ways. Wikipedia uses: "the collection of laws and rules that govern the operation of education systems" (retrieved from http://en.wikipedia.org/wiki/Education_policy). Education policies are established at the country, region, state or province, and local levels, and they are guided and communicated by documents such as national curriculum frameworks, required assessments and examinations, curriculum materials, and non-statutory guidance for use in schools. The institutions involved in setting policy "include, but are not limited to, legislatures, courts, nonprofit agencies, and national, state, and local governmental agencies" (William T. Grant Foundation, 2011). Ferrini-Mundy and Floden (2007) provide additional discussion of this area.

Policies in many countries span the range of areas of schooling (e.g., compulsory schooling policies, or assessment and examination policies that determine higher education pathways), and some are quite specific to mathematics education. In both cases—generic policies and mathematics-specific policies—there is little evidence that the mathematics education research community has engaged consistently and systematically in research that is used to formulate the policies. Nor is there a strong body of policy implementation or impact research that has examined policies that are particularly germane to issues in mathematics education. A research-like activity, policy analysis, has been undertaken in recent years by some mathematicians and mathematics educators: this might involve, for instance, assigning "grades" to standards in the USA, which often invokes comparison to standards around the world (e.g., Klein et al., 2005) and could be construed as a policy analysis activity (Clarke, 2003).

In this chapter, we explore the policy implications of developments in mathematics education research: the potential for engaging the mathematics education research community in pursuing questions that have relevance for policy, and the relevance, utility, and accumulation of mathematics education research findings to support policy and practice. The chapter will be grounded in two elaborated examples where the potential for intersection of mathematics education research and policy appears particularly fruitful, and where policy has been developed, and is developing, that is directly relevant to mathematics education. The first example is the story of the K–12 mathematics standards and related standards-based accountability in the USA. The second example traces the evolution of a national infrastructure for evidence-driven mathematics teacher professional development in England. These examples are presented as windows to illustrate how mathematics education research might relate to policy and are used to raise questions, such as: Who is involved in determining, implementing, and tracing the impact of policy? How might these stakeholders be more fully engaged with the mathematics education community? What is the role of research in these areas of policy?

With respect to these questions, we will also discuss what is available, in the research literature and elsewhere, about how policies are formed and used, focussing on the types of policies that are particularly relevant for mathematics education. In our conclusion we will discuss directions of policy, the prospects for research funding, and offer commentary on how mathematics education research agendas might embrace the possibility that mathematics education research results can inform and improve mathematics teaching, learning, and policy.

The Case of National Mathematics Standards in the USA

Efforts by the mathematics education and mathematics communities over the past two-and-a-half decades in the USA to create and implement curriculum standards as a strategy for improving K–12 mathematics education have stimulated the most vigorous policy debates and, more recently, the most widely coordinated policy incentive systems, possibly ever seen in US K–12 education within a particular discipline. The story of US mathematics standards, consistent in concept with the work of Smith and O'Day (1991) about systemic reform, illustrates a number of key policy issues that relate to research in mathematics education. In particular, these are: How does research on teaching and learning intersect with the development, implementation, and assessment of such policy levers as standards? What does research tell us about the most effective means of designing and implementing standards? What new questions become more salient when there is a lively national environment in mathematics education in the standards context? How have mathematics education researchers played key roles in this arena, and what are the prospects?

In the USA, responsibility for education is constitutionally delegated to the 50+ states and territories, which comprise about 14,000 school districts and almost 99,000 K–12 schools (see http://nces.ed.gov/pubs2011/pesagencies09/findings.asp and http://nces.ed.gov/pubs2011/pesschools09/findings.asp). Different states have different policy approaches, ranging from states with highly directive statewide curriculum standards whose adoption is expected by all districts, to states with more general standards that are then interpreted and adapted widely across school districts. Policies about such relevant matters as the required mathematical preparation of teachers, the number and nature of required mathematics courses in secondary school, and the selection of textbooks, are left to the discretion of states and vary widely.

The No Child Left Behind Federal legislation of 2001 imposed stronger Federal accountability requirements than the country had previously had, including requirements about annual assessment of students for each of Grades 3 through 8 and high school in mathematics, using instruments developed by states and aligned with state standards, and also introducing new requirements about teacher qualifications. At the same time, there have been policy influences that have emanated from the Federal level. The US Department of Education administers several billion dollars that pass directly to states, in some cases where use is highly specified. Currently the Department of Education sponsors the Mathematics and Science Partnerships program, which is heavily focussed on teacher professional development. And, the current state-led Common Core State Standards Initiative is an option that states can use in response to US Department of Education incentives to adopt standards.

A Brief History of Mathematics Education Standards in the USA

In 1989 the National Council of Teachers of Mathematics (NCTM) issued the first set of standards for curriculum guidance produced by a professional organization in the USA. The *Curriculum and Evaluation Standards for School Mathematics* (NCTM, 1989) not only specified the details of what should be taught in mathematics within grade bands, but also provided substantial guidance about instructional approaches, and offered examples and illustrations to guide teachers. The perspective reflected in this document was consistent with a constructivist view of knowledge, with a strong emphasis on "meaningful" engagement with mathematics, the use of "real-world" examples, and the role of technology. Although the 1989 NCTM standards document is not replete with references to research, a number of its authors were active researchers, and have commented that the development of the document was influenced by research findings at the time. A history of that development is recounted in McLeod, Stake, Schappelle, Mellissinos, and Gierl (1996).

The document was developed over several years, with an elaborate public reaction and comment process. NCTM leaders enlisted the endorsements of key professional organizations in mathematics. The standards were hailed by teachers and mathematics educators as a major step forward in guiding school mathematics instruction and placing issues of student engagement and understanding in the foreground. NCTM followed these initial curricular standards with three additional versions: the *Professional Standards for Teaching Mathematics* (1991), the *Assessment Standards for School Mathematics* (1995), and *Principles and Standards for School Mathematics* in 2000. Various ancillary materials were developed by the organization, including resources for teachers and instructional support materials. And standards development in other fields followed, including the *National Science Education Standards* developed by the US National Academy of Sciences (1996).

The US National Science Foundation, a Federal agency that funds grants in science and education through competitive processes, issued a call for proposals in 1990 to produce comprehensive instructional materials at grades K–6, 6–8, and 9–12 that would reflect national standards. Some of the programs developed under this call were commercially distributed. During this same period, states developed their state curriculum standards in mathematics. Anecdotal evidence suggests that many states attempted to align their standards with the NCTM document, and a series of policy-related tools to assess alignment of standards and curriculum were developed (Ferrini-Mundy, 2004). Notable among these were the curriculum framework analysis tools developed by Schmidt and colleagues for the Third International Mathematics and Science Study (TIMSS) for examining curriculum and standards around the world (see Schmidt, McKnight, Valverde, Houang, & Wiley, 1997). Following a careful comparative analysis, in which NCTM's (1989) *Standards* were considered, Schmidt and his colleagues dubbed the US mathematics curriculum as being "a mile wide and an inch deep" (Schmidt, McKnight, & Raizen, 1997, p. 62).

The convergence of many factors, perhaps including the visibility brought to the *Standards* by the funding of curricula to instantiate them, the international comparisons, the groundswell of activity from the NCTM teacher constituency, and the designation in 1999 of some of the NSF-funded and standards-based instructional materials as exemplary in a US Department of Education report (see http://www.k12academics.com/education-reform/us-department-education-exemplary-mathematics-programs) drew the attention of several prominent US mathematicians to the messages of the NCTM document. The concern of the mathematicians reached a high point in 1999, when an open letter to the US Secretary of Education, Richard Riley (see Klein et al., 1999), protested against the Department's designation of the materials as exemplary (http://www.mathematicallycorrect.com/riley.htm).

Thus the pathway of *Standards*, developed by the professional association for mathematics teachers, led to significant policy debates at the state and national level, engaging mathematicians, mathematics educators, local policy makers at the school district level, and state and federal leaders, in a new era of discussion about what school mathematics education should be. Despite the significance of the policy decisions—about standards, curriculum, and assessment—throughout this period, the definitive positions that were visible came largely from experts in mathematics, or in mathematics education, and represented professional judgment and opinion. Mathematics education research appears to have had little place or role in these debates and activities. In part, this was because the mathematics education research community's interests and inclinations in research—in the two decades spanning the release of the 1989 standards—were focussed in deep ways on important questions about student learning and understanding. Those concerned with policy were willing to use NCTM's (1989) *Standards* as an interesting site for understanding policy change (e.g., Fuhrmann, 2001), but were not necessarily driven by particular questions about the role of standards in the specifics of mathematics learning.

These circumstances, along with widespread US concern about international competitiveness and the science and mathematics education achievement of the nation's youth (articulated in *Rising Above the Gathering Storm*, 2007, a National Academies report) led in part to an Executive Order by the US President George Bush in 2006, establishing a National Mathematics Advisory Panel, charged to produce a report that contained

> … recommendations, based on the best available scientific evidence, on the following: (a) the critical skills and skill progressions for students to acquire competence in algebra and readiness for higher levels of mathematics; (b) the role and appropriate design of standards and assessment in promoting mathematical competence; (c) the processes by which students of various abilities and backgrounds learn mathematics; (d) instructional practices, programs, and materials that are effective for improving mathematics learning; (e) the training, selection, placement, and professional development of teachers of mathematics in order to enhance students' learning of mathematics; (f) the role and appropriate design of systems for delivering instruction in mathematics that combine the different elements of learning processes, curricula, instruction, teacher training and support, and standards, assessments, and accountability; (g) needs for research in support of mathematics education; (h) ideas for strengthening capabilities to teach children and youth basic mathematics, geometry, algebra, and calculus and other mathematical disciplines; (i) such other matters relating to mathematics education as the Panel deems appropriate; and (j) such other matters relating to mathematics education as the Secretary may require.
> (Retrieved from http://georgewbush-whitehouse.archives.gov/news/releases/2006/04/20060418-5.html)

The goal of this panel was to produce a report that could guide policy makers, and to employ a high standard of evidence for the inclusion of results from any research studies. The panel members represented a range of perspectives, and focussed on several aspects of mathematics education, including curricular content, learning processes, instructional practices, teachers and teacher education, instructional materials, and assessments. The report concluded that the research base for making policy decisions was not adequate:

> Systematic reviews of research on mathematics education by the task groups and subcommittees of the Panel yielded thousands of studies on important topics, but only a small proportion met standards for rigor for the causal questions the Panel was attempting to answer. The dearth of relevant rigorous research in the field is a concern. First, the number of experimental studies in education that can provide answers to questions of cause and effect is currently small. Although the number of such studies has grown in recent years due to changes in policies and priorities at federal agencies, these studies are only beginning to

yield findings that can inform educational policy and practice. Second, in educational research over the past two decades, the pendulum has swung sharply away from quantitative analyses that permit inferences from samples to populations. Third, there is a need for a stronger emphasis on such aspects of scientific rigor as operational definitions of constructs, basic research to clarify phenomena and constructs, and disconfirmation of hypotheses. Therefore, debates about issues of national importance, which mainly concern cause and effect, have devolved into matters of personal opinion rather than scientific evidence. (National Mathematics Advisory Panel, 2008, p. 63)

In summary, perhaps the most important message to come from this report was that there was not enough evidence from research in mathematics education to inform or guide some of the most pressing policy areas in the USA relevant to mathematics education.

The Higher Education Opportunity Act of 2008 included consistent emphasis on scientifically-based research, scientifically-valid research, and empirically-based practice. Earlier, in 2002 the US Department of Education had launched the "What Works Clearinghouse" (http://ies.ed.gov/ncee/wwc/), which was charged with the task of identifying instructional materials for which suitably rigorous effectiveness studies had been conducted and had resulted in positive evidence. Only a small number of mathematics instructional programs, however, were judged to have met the What Works Clearinghouse standard.

So, in the space of two decades, the paths of policy, mathematics education research, and curriculum standards had crossed and become intertwined. And, in 2009, with Federal policy support for of the NCTM standards waning, with the ascendency of "evidence-based" practices and policy, and with legislation in effect requiring high-stakes frequent assessment of K–12 mathematics learners in all states, a new phase in the US standards movement was initiated—the Common Core State Standards Initiative.

Common Core State Standards: A Policy Effort Led by States for National Impact

Over the past 15 years there have been efforts in the USA for states to build coalitions for the improvement of K–12 STEM (i.e., "Science, Technology, Engineering, and Mathematics") education. In 1996 a group of governors founded Achieve, Inc., a bipartisan organization that "helps states raise academic standards and graduation requirements, improve assessments and strengthen accountability" (http://www.achieve.org/files/AboutAchieve-Feb2011.pdf). In 2006–2007, then-Arizona governor Janet Napolitano, as President of the National Governors Association, addressed the importance of STEM education as an issue for states in the document "Innovation America" (http://www.nga.org/Files/pdf/0707INNOVATIONPOSTSEC.PDF). A related report, *Building a Science, Technology, Engineering and Math Agenda* (http://www.nga.org/Files/pdf/0702INNOVATIONSTEM.PDF), though falling short of advocating national standards, set the stage for the introduction of a national curriculum with its very strong focus on the importance of standards and international benchmarking. These discussions about standards reached the highest US policy levels when President Obama, in March 2009, outlined his education plan and discussed the need for "Encouraging better standards and assessments by focussing on testing itineraries that better fit our kids and the world they live in" (see http://www.whitehouse.gov/blog/09/03/10/Taking-on-Education/). By this time, a partnership between Achieve and the National Governors Association had been established to launch the Common Core State Standards Initiative (CCSSI).

The following, which is taken from the CCSSI Web site (http://www.corestandards.org/about-the-standards), provides a sketch of the development process used in preparing the CCSSI:

> The Common Core State Standards Initiative is a state-led effort, launched more than a year ago by state leaders, including governors and state commissioners of education from 48 states, 2 territories and the District of Columbia, through their membership in the National Governors Association Center for Best Practices (NGA Center) and Council of Chief State School Officers (CCSSO).
>
> The process used to write the standards ensured they were informed by:
> - The best state standards;
> - The experience of teachers, content experts, states and leading thinkers; and
> - Feedback from the general public.
>
> To write the standards, the NGA Center and CCSSO brought together content experts, teachers, researchers and others.
>
> The standards have been divided into two categories:
> - College and career readiness standards, which address what students are expected to learn when they have graduated from high school; and
> - K–12 standards, which address expectations for elementary through high school.
>
> The NGA Center and CCSSO received nearly 10,000 comments on the standards during two public comment periods. Comments, many of which helped shape the final version of the standards, came from teachers, parents, school administrators and other citizens concerned with education policy.
> - The draft college and career ready graduation standards were released for public comment in September 2009; and
> - The draft K–12 standards were released for public comment in March 2010.
> - The final standards were released in June 2010.
>
> An advisory group has provided advice and guidance to shape the initiative. Members of this group include experts from Achieve, Inc., ACT, the College Board, the National Association of State Boards of Education and the State Higher Education Executive Officers. (Retrieved from: http://www.corestandards.org/about-the-standards/process).

Using Policy to Incentivize Adoption of Common Core State Standards in Mathematics

The USA faces an interesting juncture in the standards trajectory, in that there is powerful momentum growing to support the use of the common core mathematics across states. Perhaps the first signal that the Federal government was supportive of this state-led effort appeared in the summer of 2009 when the US Department of

Education launched a competitive grants program among states called "Race to the Top," by which $4.35 billion dollars were made available to states to reform K–12 education. Although there was no specific focus on mathematics, the application for funding awarded points for states that were "developing and adopting common standards." The following information is from the application form.

Race to the Top
(Race to the Top Application for Initial Funding
CFDA Number: 84.395A
(http://www2.ed.gov/programs/racetothetop/application.doc)

(B) **Developing and adopting common standards** (*40 points*)
  (1) The extent to which the State has demonstrated its commitment to adopting a common set of high-quality standards, evidenced by (as set forth in Appendix B):

  (i). The State's participation in a consortium of States that— (*20 points*)
   (a). Is working toward jointly developing and adopting a common set of K–12 standards (as defined in this notice) that are supported by evidence that they are internationally benchmarked and build toward college and career readiness by the time of high school graduation; and
   (b). Includes a significant number of States; and

  (ii). — (*20 points*)
   (a). For Phase 1 applications, the State's high-quality plan demonstrating its commitment to and progress toward adopting a common set of K–12 standards (as defined in this notice) by August 2, 2010, or, at a minimum, by a later date in 2010 specified by the State, and to implementing the standards thereafter in a well-planned way; or
   (b). For Phase 2 applications, the State's adoption of a common set of K–12 standards (as defined in this notice) by August 2, 2010, or, at a minimum, by a later date in 2010 specified by the State in a high-quality plan toward which the State has made significant progress, and its commitment to implementing the standards thereafter in a well-planned way.

Common set of K–12 standards means a set of content standards that define what students must know and be able to do and that are substantially identical across all States in a consortium. A State may supplement the common standards with additional standards, provided that the additional standards do not exceed 15% of the State's total standards for that content area.

At the time of writing this chapter, As of summer 2012, 18 States and the District of Columbia had been awarded Race to the Top grants (http://www2.ed.gov/programs/racetothetop/awards.html). The Federal Department has launched a competition for two major assessment consortia to "develop a new generation of tests." The new tests will be aligned to the higher standards that were recently developed by governors and chief state school officers (http://www.ed.gov/news/press-releases/us-secretary-education-duncan-announces-winners-competition-improve-student-asse). The standards have been adopted by 45 States and 3 territories (http://www.corestandards.org/in-the-states). It would appear that the USA is on the verge of having widely used, yet voluntary, national standards in mathematics. This is a remarkable opportunity for a wide range of policy research endeavours in which the mathematics education community could take the lead.

The Role of Research

Following on the lessons of the "math wars" and the findings from the National Mathematics Advisory Panel, it seems that the organizers of the Common Core State Standards Initiative were sensitive to the need for research and evidence to provide validation for the standards. Indeed, there was, as mentioned above, a Validation Committee whose major task was to examine the evidence used to support each set of standards. Confrey (2010) summarized the types of evidence used: "Data from ACT and SAT scores and performance in 1st-year college courses; analysis of college syllabi and surveys; surveys with business members; examination of college level math and math-client fields; whether the standards are benchmarked to international standards; and evidence from student learning studies" (p. 11). Student learning studies that may prove useful in the continuing standards implementation may include work on learning progressions, though it appears that there remain important research questions needing the attention of policy makers and mathematics education researchers.

Mathematics Education in England: Policy and Research

The example from England traces some recent efforts to transform practice by brokering partnerships among mathematics education researchers, mathematicians, policy makers and teachers. It touches on similar issues to the US case study in relation to the research and the standards agenda but also considers the role of research more broadly in promoting the teaching and learning of mathematics in the country. The theoretical basis underpinning the case study— although this was rarely made explicit—is learning design that involves valuing the need for all parties to build their solutions to problems at hand together, to reflect on them together and, crucially, to allow all the groups to feel empowered to shape any innovation to fit their own goals and purposes. Cobb and Jackson (in press) noted that the learning design perspective directs us to "analyze the soundness of the intended learning supports prior to implementation" (p. 10), and policy implementation must take account of these planned supports and how they are effectively operationalized.

Mathematics presents a challenge for policy makers. The subject is highly regarded. Tests are high stakes. In addition, mathematics is widely conceived as hard and procedural by those outside the mathematics community. Mathematics is a subject that offers diverse and unique ways by which students can express themselves in creative ways. Yet this broad agenda for teaching and learning mathematics is often invisible to those outside the community, especially, it is conjectured, policy makers, who most likely only value test results and performance measures. Yet progress in improving mathematics education can only be achieved when teachers do not narrow the mathematical diet of their students to procedures to pass tests. Rather, teachers must have the confidence to introduce a broader range of tasks and activities.

To achieve this goal in England, leaders in mathematics education have struggled over many years to set up a national infrastructure for mathematics continuing professional development (CPD) in order to confer status, priority and obligation for evidence-based professional learning that is recognized by all "layers of the system" and beyond: head teachers, mathematicians, politicians at national and regional levels, as well as teachers themselves. Thus the goal was that mathematics professional development would become an expectation and a responsibility for all those involved in teaching the subject with politicians, local leaders and head teachers in schools all supporting this agenda.

This agenda for mathematics inevitably raises the question whether mathematics has a special place in schools—because of the widespread uses of mathematical knowledge, but mainly because mathematics is a core part of the "standards agenda": an agenda that monitors student performance, schools and the system over time. Measures used for this monitoring exercise included results from national tests for all students in England at the ages of 7, 11, 14 and 16 (http://en.wikipedia.org/wiki/National_Curriculum_assessment), although the national testing at 14 years was ended in summer 2009. Performance of English students in international comparative studies, such as Trends in International Mathematics and Science Study (TIMSS) (http://nces.ed.gov/timss/index.asp), the Program for International Student Assessment (PISA) (http://nces.ed.gov/surveys/pisa/), and data about adult numeracy (see the Leitch Report, 2006, *Prosperity for All in the Global Economy—World-Class Skills*), were also to be taken into account (http://www.dius.gov.uk/publications/leitch.html). The agenda was driven by what was called National Strategies (primary and secondary), alongside a system of school inspection.

The challenge that educators in England have faced is how to support children to perform better at mathematics, that is, to achieve success in tests and examinations, without sacrificing creativity and inquiry and without exerting so much pressure on students that they are put off the subject. Too much pressure can result in teaching and learning procedural rituals for getting right answers, which bypasses the need to appreciate the structure and pattern of the subject. Teachers and researchers alike have worked hard to develop among students a mathematical way of thinking while not neglecting to support them to succeed in public examinations and high-stakes tests. This balance between learning and performance is difficult to achieve. It requires teachers who focus on teaching and learning, who know their subject and its pedagogy, and are confident enough to focus on longer-term subject appreciation alongside short-term performance outcomes. One cause of imbalance can be traced to policies that have meant that the subject agenda for teaching/learning/curriculum and the standards agenda may not have been appropriately aligned due to their different goals and management structures.

In contrast to many other countries, students in England are only allowed to drop mathematics at the age of 16 years, at the end of compulsory schooling (Hodgen, Pepper, Sturman, & Ruddock, 2010). However, it is increasingly accepted that there is a need for more engagement with mathematics, so the numbers who choose to study mathematics post-16 have been added as another government target for schools alongside the standards agenda. This has been one result of the general push to work for more success in mathematics, across the policy agenda and with a better alignment of the needs of practitioners with the realities of policy makers. We now document these policy initiatives in slightly more detail.

Some History: Giving Mathematics a Policy Voice

The Advisory Committee on Mathematics Education (ACME) was established in 2002 by the Joint Mathematical Council of the United Kingdom and the Royal Society (RS), with the explicit support of all major mathematics organizations. It comprises seven members, including teachers at different phases, and has a part-time Chair, who is a Fellow of the RS, to act as a single voice for the mathematical community. Its goal is to seek ways of improving the quality of education in schools and colleges (http://www.acme-uk.org). ACME was formed after a period of many years during which there had been no conduit through which the mathematics community could have dialogue with government, despite a standards agenda that included mathematics. Like the former Mathematical Sciences Education Board within the US National Academy of Sciences, ACME's membership includes mathematicians, teachers in different phases, mathematics advisers at local or government level, and a member of the mathematics education research community.

At the time of its formation, ACME had to acquire the commitment of government to provide appropriate contacts, as well as secure some funding for meetings to pay for the time of committee members. ACME now advises government on issues such as the curriculum, assessment, and the supply and training of mathematics teachers through face-to-face meetings and a series of highly influential reports (see http://www.acme-uk.org/the-work-of-acme/publications-and-policy-documents/policy-reports). In 2011/2012, there is to be new national curriculum for mathematics and ACME will play a leading advisory role in its development and formation, thus providing a mediating layer for mathematics education research.

Over a period of two decades, a number of significant education reports of relevance to mathematics have been commissioned by the UK government to inform and drive the policy agenda. Most were in fact about science, which of course impinged on mathematics but only in a secondary way. In fact, a major breakthrough in policy circles was the transformation of a SET agenda (science, engineering and technology) in which mathematics was largely invisible, to a STEM agenda (science, technology, engineering and mathematics) in which mathematics was acknowledged as playing an important part. Some reports specifically focussed on mathematics, with *Making Mathematics Count* (2004) and the *Review of Teaching in Early Years Settings and Primary Education* (2008) being pivotal. The latter's main recommendation called for a major policy change—that there should be a trained specialist in mathematics in every primary school, a recommendation that was accepted and led to agreement about a program of training to be delivered by consortia of universities. However, later financial constraints caused this program to be tapered, with funding being shifted away from Government to schools over a period of 3 years.

Making Mathematics Count (2004), which will be abbreviated to "MMC," was particularly significant not least because it received almost universal support from all the diverse stakeholders that comprise the mathematics community, including researchers in mathematics education and mathematicians. The government of the time accepted most of the recommendations of the report, possibly because the Secretary of State was a strong supporter of mathematics and, as a result of his university background in mathematics, appreciated that mathematics was much more than arithmetic and procedural technique. This placed the mathematics community in a strong position, at least in the short term.

The MMC report underlined the need for a strategy and strong focus for mathematics. Its recommendations included issues around stimulating the supply of specialist mathematics teachers, the designation of different mathematics pathways for the 14–19-year-old age range depending on career aspirations, and support for teaching and learning. At the time, there was also considerable concern about the numbers who were opting for specialist study in advanced mathematics (A-level), following a dramatic drop in student numbers in 2001. This decline was largely due to a new policy leading to an overarching shift in curriculum structure at A-level, which had a particularly negative effect on mathematics results. The change was bought in too quickly with students examined too soon after they had met new mathematical ideas. Many students failed the new modules leading to a general loss in confidence among students and teachers alike, and a move away from taking what was perceived as a high-risk subject. Numbers entering A-level fell from over 70,000 to just over 50,000 in a matter of years. A government target of 56,000 A-level entries in 2014 was set in 2006, a target that was judged to be quite ambitious at the time, but was in fact reached well before that date (see Figure 16.3).

One recommendation of MMC was that a post of Chief Adviser for Mathematics to the UK Government should be established to provide Ministers with direct advice on the needs and requirements of the subject. This was not a political appointment but rather involved advising the Secretary of State and relevant ministers (and their civil servants) about mathematics across all phases, performance, participation and the curriculum, drawing on all available evidence—thus providing reports verbal and written that served to mediate results and "research wisdom." The first author of this chapter was selected to take up this position in 2004 and served (part-time) until 2007 when her secondment ended. At this point, the post was discontinued, mainly as a result of a shift in policy context to STEM with a new Secretary of State in charge, combined with the fact that the overall situation in mathematics had improved quite dramatically, and that ACME had been established as a voice for policy.

Another recommendation in MMC was that there should be a better alignment of the standards agenda with the mathematics curriculum and teaching agenda. This was to be achieved by merging the existing standards team, that is the National Mathematics Strategy for the Lower Secondary School and its funding, into a new national infrastructure, the National Centre (see below), with serious consideration to be given to similarly incorporating the national numeracy strategy for primary schools, into the proposed Centre. As already mentioned, these National Strategies had substantial budgets and huge political influence within the standards agenda, and this recommendation proposed quite a radical policy shift. However, it was not accepted. The two structures, one around teaching and learning mathematics generally, and the other around mathematics as part of the standards agenda, remained distinct, and with distinct roles for the Strategies and for the Chief Adviser for Mathematics. Nevertheless, during the period 2004–2007 the two structures became better aligned due to efforts from both communities.

Another focus of the MMC was on the potential role for university mathematics departments in providing enrichment in and out of school as part of the policy drive for more mathematics. This enrichment might involve organizing national competitions, mathematics clubs, and master classes, and included the promotion of mathematics careers. In addition, at the time of the report, it was also becoming evident that Further Mathematics (an optional course of post-16 mathematics that is more advanced than A-level mathematics) was a "dying subject" as fewer and fewer schools had the capacity to offer it. There were two main reasons for this: (a) many schools did not have the specialist staff needed; and (b) schools could not afford to teach the small groups who selected it. A pilot initiative to address this challenge was supported for role-out by the Government, and this was to set up a Further Mathematics (FM) Network (http://www.fmnetwork.org.uk/), a national network of FM Centres to enable every student who would benefit from it to have the opportunity to study for Further Mathematics qualifications through distance learning and mentoring. Forty-six FM Centres came into operation across England.

Along with these larger developments, a variety of smaller initiatives were also put in place, all to promote mathematics. We only mention a couple that appeared to have widespread support in the mathematics community and relevance to the thrust of this case study: a range of extra-curricular activities for gifted and talented students which provided links to universities and to employment; a national program of one-on-one tutoring for students of all ages who were falling behind in mathematics, with a particularly well-funded program, for children under 5 years, called Every Childs Counts (see http://www.everychildachancetrust.org/counts/).

Thus, during this period, expert practitioners, mathematicians, and mathematics education researchers were able to influence policy direction together and were able to communicate across the boundary of policy/practice largely through government-sponsored boards set up to work with the Chief Adviser, specifically to take forward the recommendations of the MMC. As part of this endeavour, the importance of effective teaching of mathematics in England was not only recognized, but also what this actually meant in practice was widely agreed. In addition, the country had long suffered (and still does) from an overall shortage of mathematics teachers, limited specialist capacity among mathematics teachers at every level, constant turnover, and difficulties of retention. There was therefore a manifest and distinct need for an agenda for professional development of teachers of mathematics throughout their careers, so not only could expertise be bought into the profession through changes in entry standards, but also through promoting professional learning for those already teaching. And, because of structures that had been established to align the goals and policies of government with the knowledge and expertise of the mathematics and mathematics education communities, it was possible to move forward and agree to a new agenda of professional development to support effective teaching in the subject.

In England, professional development for teachers of mathematics had existed but had tended to be rather ad hoc and geographically patchy. It was decided at a policy level that what was needed was an infrastructure that monitored and coordinated the provision nationwide. This was a recommendation of the MMC and which led to the establishment of the National Centre for Excellence in the Teaching of Mathematics (NCETM).

The National Centre for Excellence in the Teaching of Mathematics (NCETM)

The NCETM was set up in 2006 by the UK government and continues to the time of writing (November 2011). The Centre has a clear and ambitious vision. It aims to meet the professional aspirations and needs of all teachers of mathematics so that they can realize the potential of learners. It is a constant struggle to encourage teachers to see professional learning, not as a threat or a punishment for not doing well or being in some way deficient according to a standards agenda, but as something that is geared to their needs, and inspiring.

To this end, the NCETM's objectives were formulated as follows:

- To stimulate demand for mathematics-specific continuing professional development (CPD), contributing to the strengthening of the mathematical knowledge of teachers;
- To lead and improve the coordination, accessibility and availability of mathematics-specific CPD;
- To enable all teachers of mathematics to identify and access high quality CPD that will best meet their needs and aspirations.

The NCETM set out to meet these aims through a sustainable national infrastructure for mathematics-specific CPD that starts from the needs and goals of teachers. As such, it provided a counterbalance to the top-down constraints of the much more politically powerful standards agenda, which monitored student performance in the country. It is possible that these concurrent initiatives, as they gradually became more aligned, had a surprisingly positive and synergistic impact.

The NCETM provides and supports a wide variety of mathematics education networks in the country, which include universities, subject associations and the whole range of CPD providers. At the same time, the National Centre encourages schools and colleges to learn from their own best practice through collaboration among staff and by sharing good practice locally, regionally and nationally. These collaborations take place face-to-face at national and regional events and in local

Figure 16.1. Overall structure of the framework that underpins the NCETM portal.

network meetings across England, or virtually, through interactions on the NCETM portal, http://www.ncetm.org.uk.

Figure 16.1 shows the overall structure of the professional learning framework that underpins the portal, and Figure 16.2 provides a snapshot of the portal's homepage. Any portal has to be regularly updated and improved to introduce new functionality, including Web 2.0, new design, and improved tools so as to meet the needs of teachers. The portal is concerned to help teachers meet virtually in professional communities to discuss issues facing them (e.g., how to ask open questions in mathematics, how to design good formative assessments). It also implements "behind-the-scenes" speed increases and improved search facilities. The aim has been to make the portal experience user-friendly and above all useful. The statistics for NCETM portal continue on an upward trend with over 85,000 regular users in July 2012. Another statistic of interest is that, at that time, only eight countries had not visited the NCETM portal—French Guiana, Western Sahara, Mauritania, Chad, Congo Brazzaville, Guinea, North Korea and Turkmenistan.

Figure 16.2. A snapshot of NCETM's portal homepage.

Framework that Underpins the Portal

The NCETM signposts high quality resources usually organized into microsites that support the professional development of teachers. Microsites include departmental workshops that help secondary school teachers examine together a range of mathematical topics that "are hard to teach," and sector-based magazines that offer monthly articles that are stimulating and timely. The site also points to useful CPD opportunities and courses offered by a range of providers in a constantly updated Professional Development Directory, which also identifies providers that hold a quality standard for CPD that is regularly monitored. There is also the NCETM *Mathemapedia,* a wiki designed by and for mathematics education. This acts as a vehicle for improving teachers' awareness of research issues in teaching mathematics, of sharing ideas, as well as providing easy access to a range of references and interesting ideas, both theoretical and practical. The range of topics—written by NCETM portal users and moderated by the NCETM—is huge. Almost 400 articles exist, accessed over 30,000 times per month.

A later innovation, the result of teachers' requests, was to find ways to support teachers in accessing research by supporting the production of Research Study Modules (https://www.ncetm.org.uk/enquiry/35990). Each study module is based on a particular, carefully chosen, and annotated research paper which was written by a collaborative group of researchers and teachers to present a structure that would support teachers more generally to think about the ideas and findings reported, reflect on their own views and practice, and consider the implications for their own practice. The starting point for the production of each module is to present questions raised by teachers when reading the papers and to support and frame their interpretations.

A complementary approach has been used by the Institute of Effective Education, which produces research articles across different areas on "what works"—in a journal named *Better: Evidence-Based Education*. These articles are concise and written for a teacher and policy audience. However, at the time of writing only one such journal has been produced for mathematics in England and that was in 2009 (Hoyles, 2009).

A Web presence for CPD is a relatively new development—at least when the NCETM began in 2006—and one that needs to be the object of research and development in its own right as new functionalities become available. The NCETM portal is not simply a provider of online learning activities, but also provides a record of a personal learning journey. Once logged in, teachers can access their own personal learning space, in which they can store a snapshot of their own CPD experiences and reflections. Research suggests that self-evaluation is a powerful and productive way to catalyze professional development. This self-evaluation can be undertaken in the privacy of home, or as part of a professional development group in a school—anywhere, in fact, where there is time to think and reflect on what a piece of mathematics might mean, how it might be represented, or how it might be taught and assessed in new ways. The NCETM has developed self-evaluation tools (SETs) in each of the following areas: Mathematics Content Knowledge, Mathematics-specific Pedagogy, and Embedding in Practice. There are many hundreds of pages of self-evaluation steps structured in age-related phases based in the English National Curriculum. If teachers record limited confidence in any area, they are sign-posted to possible activities, on and off the portal, with which they might wish to engage to help them make progress.

One policy implication is clear and is not widely recognized by policy makers, and that is that professional development is not only about courses. Teachers can and do, with appropriate tools, learn from each other and from research about effective mathematics pedagogy and practice. The policy environment for mathematics education has made it possible in England to implement such new tools and approaches. The challenge remains for mathematics education researchers to develop the research methodologies and evidence to help improve this teacher learning system and ensure its continued growth on the basis of what elements are most effective.

The NCETM has attempted to take forward into practice research that has indicated that involving teachers in collaborative reflection and enquiry pays dividends in producing real results in the classroom, and thus is an evidence-based initiative ripe for the policy arena. Four international reviews of evaluations of CPD over a 10-year period have consistently shown that the CPD that makes a difference is: collaborative and sustained, draws on evidence from research and practice, and involves participants in experimenting with new approaches and observing effects (for a review of this research see EPPI systematic reviews of evidence about CPD: http://eppi.ioe.ac.uk and Best Evidence Synthesis *BES; Teacher Professional Learning and Development; http://www.educationcounts.govt.nz/publications). Almost all of the research reported in both reviews is generic and not subject-specific, although mathematics was not excluded. Another obvious instance of this type of teacher enquiry is Japanese lesson study methodology, which has been undertaken in mathematics classrooms and shown to be effective (Krainer, 2011).

To attempt to take this teacher enquiry agenda forward, the Centre has provided a range of opportunities and frameworks through its NCETM Funded Projects Scheme. Over 300 projects have been funded and their reports can be accessed at http://www.ncetm.org.uk/enquiry/funded-projects/view-all). The Funded Projects Scheme provides resources to scaffold the research teachers may wish to carry out in collaborative groups within or across schools and colleges. Teachers bid for funds to pursue an enquiry and are provided with useful research "starting points" and references to try to promote building on previous work in the research community. The teachers have to write a report on their work and reports and findings of the projects are posted on the portal and disseminated at NCETM events. Thus, learning is shared, and the impact maximized. Teacher groups are expected to present the results of their work and are supported in doing this (if they wish). Most, if not all, find the experience of the research and the communication to others valuable. The projects usually include a member who is an "outside catalyst" or mentor—for example a researcher from a university—who supports the team of teachers, brings a broader perspective to the work, and helps the teacher group to plan the enquiry and summarize the findings in project reports. The NCETM also produces highlights from several projects describing their impact on teachers and learners for wider dissemination in annual Teacher Enquiry Bulletins, which are widely read by teachers and researchers alike. Further reading, and the full reports and bulletins, can be found on the portal under Teacher Enquiry (http://www.ncetm.org.uk/enquiry).

The 300 or so reports from the funded projects are a tribute to the diversity of the endeavour, although many topics were in fact revisited by different groups—inevitably as selections were shaped by the policy landscape. Topics have included, for example, how to support rich mathematical questions in the classroom (that is, more open-ended investigative work); using digital tools for sharing practice or to support mathematical learning; how children's play can enrich early mathematical experience; assessment for learning; and the impact on teaching and learning of collaborative planning and review.

Independent evaluation studies of the Centre contribute to the evidence base outlining the importance of developing and supporting the practice of guided teacher enquiry. One study, in particular, documented the impact of the NCETM-funded networks on teachers, on their knowledge and practice, on their schools/colleges, and on their colleagues, pupils and students (Gouseti, Noss, Potter, & Selwyn, 2011). Another study noted that the success of the Centre stemmed from its local focus, its collaborative nature and the fact that it was driven by evidence (Sheffield Hallam University, 2010).

The findings of these evaluation studies have broad significance for policy. First the authors noted the distinct "added-value" of an external independent organization supporting the activities that take place in individual schools and colleges. The modest amounts of funding provided by NCETM could have been provided using internal school funds. However, the researchers found clear benefits of having an external organization providing the funding as a lever on school and district management and to confer status on the teachers' work. Thus, funds and the recognition and validation of the process and outcomes through conferences, accreditation and award schemes together proved a powerful incentive for professional learning. The importance of the role of the "leading" and "coordinating" teachers was recognized as fundamental to the success of the networks and projects. This pointed to the need for a policy strategy to develop the organizational and inter-personal skills-sets required to guide groups of teachers successfully. Mentoring a group of teachers in research requires specialized skills over and above those needed in teaching, as does supporting teachers to report to audiences beyond immediate colleagues. There is also the constant challenge in the research community as well as in teacher research to work out how to ensure findings are, to some extent at least, cumulative. It is clear that making research reports more accessible through careful tagging and easy availability is helpful, but although this might be necessary, it is in no way sufficient.

Several other countries have either set up or are in the process of setting up similar national centres, the most recent being in the Federal Republic of Germany, where a national centre for mathematics teacher education has been established, funded by Deutsche Telekom Foundation. An important research effort for the international mathematics education community might be to assess the impact of these centres and identify factors underpinning any successes that transcend national boundaries. Each country has different goals, strategies, funding regimes and expected outcomes but if meta-analysis pulls out overarching research findings that document the successes and challenge, they would have powerful implications for policy.

The question remains: what type of evidence is needed to convince policy makers about needed resources or infrastructure in any one country, and can research form part of this evidence and, if it can, what form should it take and how can the findings be mediated so as to be meaningful for policy makers? In England, the picture of participation in mathematics shows quite dramatic improvement. Figures 16.3 and 16.4 display the number of entries in A-level and Further Mathematics A-level over a number of years. They show the significant downturn in 2000 and 2003 mentioned earlier and the continuous and significant upward trend since 2003 in the number of entries and the proportion of the cohort opting for mathematics. But which of the many initiatives were crucially important in this upturn? Or, was it a matter of a cumulative effect? Those are important questions, worthy of investigation by future research.

Policy development processes are often "top down," coming from levels of government for implementation at the school and classroom levels. Yet, the two examples provided above—the professional society and state-led standards movements in the USA, and the collaborative community-led CPD structure in England—provide

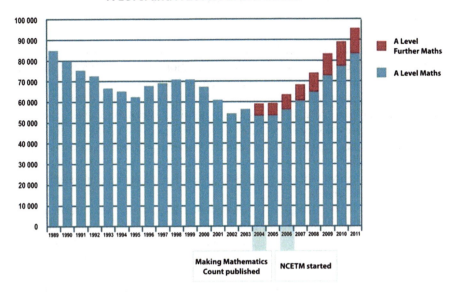

Figure 16.3. Number of entries in A-level and Further Mathematics A-level (in England).

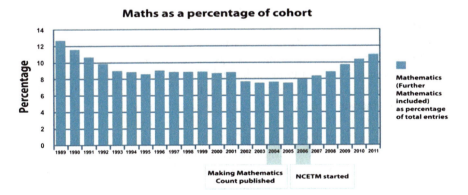

Figure 16.4. Proportion of the cohort opting for mathematics.

evidence that significant policy change can occur from a bottom-up perspective. Jacobsen (2009) discussed how the "voices of the people" are essential in development of policy. In both the US and UK examples the development of the policy has had varying levels of engagement of stakeholders and key constituencies. In contrast to these highly collaborative and inclusive processes, we offer two abbreviated examples where the approaches to policy reform have especially interesting, and different, characteristics.

Curriculum Reform in Portugal and Educational Reform in Mexico

In a fascinating account of reform in mathematics education in Portugal, Abrantes (2001) provided a description of a process of educational reform driven by a national debate about curriculum, in which schools were invited to participate. The "ultimate goal of the movement was to support the gradual creation of a new curricular organization based on a more autonomous and responsible role of the teachers and their collective structures in school" (p. 127). Given the flexibility to propose their own curricular programs, schools and teachers collaborated and formed networks over a period of years, and the activity culminated in legislation in 2001 relaxing the previously prescriptive government directions about curriculum and content, and leaving great flexibility to schools.

In contrast, the Organisation for Economic Co-operation and Development (OECD) (2011) has described an interesting partnership between the country of Mexico and OECD, an instance of an apparent trend for countries to seek collaboration from international resources to improve their educational activity. The report noted: "International organizations such as the OECD are increasingly being asked by member countries and partners to provide an analysis of state-of-the-art education policies and reform processes" (p. 30). The Mexico-OECD partnership focussed on the evaluation of schools and teachers, with efforts to draw on OECD resources, considering local issues, in developing a continual improvement strategy. The OECD team reported drawing on material in international comparative studies, on international best practices, on results of research that focussed on the specific topics of interest, and on a variety of country-based areas.

This "customized" approach to policy reform, bringing together local policy makers with teams that can bring additional research and policy evidence to the discussion, is similar to the model used in the US-based Strategic Education Research Partnership (SERP), originally grounded in work of the National Academies (2003). SERP's mission is "to conduct a program of "use-inspired" research and development, with a goal of developing, testing, and mobilizing effective programs and practices" (see http://www.serpinstitute.org/about/overview.php). The SERP partnerships involve local leaders and policy makers in school districts along with researchers concerned with the challenges faced by individual districts. Such models may offer a promising approach for more productive and influential connections between mathematics education policy needs and researchers.

Influences in the Policy Process: Considerations for Mathematics Education Researchers

How can members of the mathematics education research community internationally play a more influential role in the shaping of policy that affects mathematics education? Using the examples presented above, we will discuss some of the considerations that might be relevant as researchers become interested in undertaking studies that can intersect more directly with the world of policy formulation and implementation.

Stakeholders in the Policy Process

For mathematics education researchers contemplating how their work might be more influential in policymaking and implementation, an important context is awareness of the points of interaction by various stakeholders in the policy development and implementation process. As the previous examples illustrate, a clear understanding of the national policy context is essential in framing research agendas that will be most likely to inform future directions. Part of that context involves understanding the "intermediaries." We expand on Osborne (2011), who noted that "individuals who act as intermediaries between researchers, on the one hand, and policy makers and teachers of science on the other" (p. 27) can be important in the ways in which research might influence policy. Osborne included developers of instructional materials, local education leaders, teacher educators, and other science educators in this list. We note that in the UK, ACME serves this role. Peterson (2011) additionally suggested that advocates and lobbyists (some of whom come from professional societies) are also key intermediaries. In addition, in many countries the most important influences on policy are central ministries and departments of education.

Other entities outside of university academe have key roles—"think tank" organizations such as the RAND Corporation, the International Association for the Evaluation of Educational Achievement (IEA), and OECD provide substantial research and analysis for policy makers, and are especially skilled at the formats of policy briefs that can appeal to policy makers who are attempting to become informed quickly about a range of issues. Fowler (2004, cited in DeBoer, 2011, pp. 3–4) highlighted the importance of "issue definition," something often accomplished by intermediary groups.

In ongoing work funded by the William T. Grant Foundation, Tseng (2010) pointed out that Daly and Finnigan are studying the role of intermediary organizations in bringing research directly to policymakers. Interestingly the authors have found that grantees report that relationships have been more influential than written materials for making policymakers and practitioners aware of the results and implications of research. It seems that many policymakers and practitioners prefer to seek out information from trusted but knowledgeable personnel who are aware of comparable situations.

Prestigious national academies and high-level government panels provide authoritative reports aimed at policy makers, and international groups that engage in assessments and international comparative studies figure prominently in the directions of policy in many countries (DeBoer, 2011). Advocacy groups, professional organizations, and other interest groups also strive to be influential with policy makers.

In the USA, professional societies help to provide this function; Mexico is working with an international intermediary, OECD. And, as we have seen in the UK examples, and in Portugal, citizens and teachers can greatly influence policy makers by assembling evidence and examining key questions emerging from policies. Relationships and personal contact with those who have access to policymakers are important; indeed, in the research of Finnigan, Daly, and Che (2012) and Palinkas et al. (2011), the ways in which such relationships work in shaping policy, using social networking and other approaches, has been a subject of study. Thus, for research to inform policy, it is important that the research be useful to these "intermediaries." Using the two cases presented earlier, we explore how this happened, or could have happened in the two examples.

In the case of the US standards movement, the policy makers who in the end will either ensure successful implementation or not of the Common Core State Standards Initiative will be state leaders—governors, state boards of education, and legislative bodies—as well as local district officials, including superintendents, principals, and curriculum coordinators. Indeed, the development team was something of a microcosm of the appropriate intermediary groups. The team was headed by a mathematician with a history of working collaboratively with mathematicians and mathematics educators at both K–12 and with the undergraduate curriculum at the national level. Throughout the process there was substantial engagement of mathematicians, along with teachers and mathematics educators. This process is relatively well aligned with development processes used in the NCTM *Standards* activities, so it remains to be seen whether or not these efforts will have a role in translating to effective implementation at the state and local level—this would be an important subject of research that would require collaboration between policy makers and mathematics education experts.

In the UK, for the CPD infrastructure to be sustained, the Government and Ministers will need to be convinced of its utility, not only in terms of building a professional teacher community but also ultimately in relation to its impact on pupil learning, and the standards agenda. In this case, the "indirect" approach of engaging teachers in undertaking action research to examine the questions of interest to them in their classrooms, or even questions shaped by the policy context, is ambitious. It aims to build a network of evidence that is drawn from use-inspired research. But will the data prove convincing in the face of new political priorities? Its potential for informing future policy is as yet untested. A new contract was awarded by the Government for the NCETM to continue until 2015.

Meeting Policy Makers' Needs

There is considerable literature available indicating that if researchers better understood both the needs of policy makers and the characteristics of research and evidence that render it useful to policy makers, then their research efforts might have more impact. What mathematics education researchers might count as research and evidence are indeed only components of the various types of evidence that policy makers will use. According to Honig and Coburn (2008), school district staff were prepared to take into account evidence from social science research, from student achievement data, from practitioners, and from expert testimony, including parent and community input. In related work, Nelson, Leffler, and Hansen (2009) found that policymakers tended not to use research evidence as a primary source of guidance. They reported:

> The study revealed a surprising absence of interest by policymakers and practitioners in using research evidence. In fact, focus group members and interviewees exhibited a high degree of skepticism about the value of research. And, they did not draw a distinction between evidence based on empirical findings and "research findings" derived from the media, popular professional journals, the experiences of others, gut instinct, and their personal experience. In looking at both the research literature and the study findings, we found five common types of evidence used to inform educational policy and practice: research evidence, local data, public opinion, practice wisdom, and political perspectives. (pp. 50–51)

There are a number of factors under the control of researchers that might help ensure more visibility and usability of their work. Several authors call for framing the issue in a broader policy context (Smith & Smith, 2009; Gates & Vistro-Yu, 2003). For the UK situation, this might mean reconsidering both theoretically and practically the relationship of work in CPD to the broader standards requirements. Others call for attempting to describe causal links (Smith & Smith, 2009); in the US standards efforts, this would mean finding ways to relate student achievement to implementation of standards. Still others call for including stories to ground the claims (Smith & Smith, 2009). The evidence from this chapter suggests that both systematic evidence along with rich and interpretative narrative are needed. McDonnell (2009), suggested that researchers develop more sophisticated survey research techniques in order to address the needs of policy makers. That approach might be especially useful in mathematics, where there is a need to develop a stronger grasp of public attitudes to the importance of mathematics, including its influence on employment opportunities.

Policymakers are often forced into the situation of creating policy despite the fact that the evidence, one way or another, is inconclusive. They need tools to justify their proposed policies to other decision makers (legislators, or school board members, for example) who may not have deep familiarity with the issues. Within the educational research literature there is guidance about the needs of policymakers. For instance, Beaton and Robitaille (1999, p. 30, cited in Clarke, 2003), observed: "Educational policymakers around the world recognize the need for more and better information about the effectiveness of schools." Clarke (2003) speculated that this was a reason for the great interest of school policymakers internationally in international comparative studies such as TIMSS and PISA. In the USA, legislators have sometimes conveyed interest in identifying factors which positively influence practice. In a recent example, the US Congress requested that the National Science Foundation commission a study that would examine the characteristics of US K–12 schools that are especially effective in the areas of science, technology, engineering, and mathematics (STEM). The resulting report, *Successful STEM education: Identifying effective approaches in science, technology, engineering, and mathematics*

(National Research Council, 2011), was aimed at policy makers at the local level, and represented a synthesis of available research about effective practices (see http://www.stemreports.com/wp-content/uploads/2011/06/NRC_STEM_2.pdf). This is a current example of an intermediary entity responding to a direct request from a government policymaking body.

Tseng et al. (2008), in writing about the various ways in which policymakers use research, drew attention to the following five categories (p. 13):

1. *Instrumental* use occurs when research evidence is directly applied to decision-making.
2. *Conceptual* use refers to situations in which research evidence influences or enlightens how policymakers and practitioners think about issues, problems, or potential solutions.
3. *Tactical* use, also called political and symbolic use, occurs when research evidence is used to justify particular positions such as supporting a piece of legislation or challenging a reform effort.
4. *Imposed* use refers to situations in which there are mandates to use research evidence, as when government funding requires that practitioners adopt programs backed by research evidence.
5. *Process* use differs from the preceding terms; it does not refer to how research evidence is used but rather to what practitioners learn when they participate in conducting research.

In the case of the development of the US Common Core Curriculum Initiative, it seems that there is evidence of both conceptual use (e.g., the development of the standards using knowledge gained from research investments in learning progressions) and tactical use (e.g., the components of the validation activity calling on experts to validate whether the research cited for inclusion of particular standards was adequate). In the UK example, concerning the policy initiatives generally, and the CPD and NCETM examples, in particular, it seems that instrumental, conceptual, and process uses are all in play.

Concluding Discussion

In order for mathematics education research to be more likely to influence policy, scholars may need to consider several notions. First, deriving research questions from larger contextual circumstances that transcend mathematics education could be more important than presenting results that are directly attractive to teachers and to the mathematics education research community. As Smith and Smith (2009) noted:

> Studies designed to provide information about how to teach a specific, important concept in elementary mathematics will not be useful to policy makers in federal and state governments or even in most district offices, though they may be useful to teachers, principals, and publishers. (p. 376)

Second, the methodological preferences that are often used in mathematics education in order to address the questions of interest to researchers are dominated by descriptive work, design studies, teaching experiments, and implementation studies, which do not provide direct evidence about the potential effectiveness of innovations at scale. That limits the potential for the studies to influence policy, unless they are interpreted and seen to be valid by powerful intermediaries.

US government agencies, through policies about K–12 educational change as well as research funding policies, have placed greater emphasis on assembling research results of large-scale interventions than of small-scale studies as a source of policy guidance. Both the What Works Clearinghouse and US National Math Advisory Panel examples provide indications about what might be needed: is the methodological bar the "right one," and then what are the directions for mathematics education research that will meet the evidence standards that are put in place for influencing policy makers?

Third, it must be recognized that the particular educational challenges that a particular country is facing are essential context for framing the more specific mathematics education research questions for which an accumulation of research might well guide policy. For instance, Gates and Vistro-Yu (2003) observed that in developing countries, transforming the mathematics education system from one that was modeled originally on a system to "serve the European elite" to a system that offers universal access to mathematics education, is a key challenge faced by policy makers. Addressing both ambitious mathematics and equity is a crucial challenge in the USA. It relates closely to the global mathematics education policy challenge of how to formulate mathematics education to meet the needs of all subcultures in a society and to build on the mathematical assets inherent in those subcultures.

Finally, most Governments acknowledge the need to prepare the next generation for a world that is very different from ours. That world will innovate in mathematics teaching and learning specifically around the use of digital technology. Education in general and mathematics education in particular has been slow to grasp and exploit the findings of technology-related research into teaching and learning. This area is ripe for innovation and research with promising avenues to pursue emerging in the international scene (see, e.g., Hoyles & Lagrange, 2009).

In a world facing global challenges of unprecedented seriousness, the importance of scientific and mathematical literacy and expertise has never been more central. Around the world, nations have recognized that the mathematical education of their young people is critical to personal, societal, and economic well-being. The policies that govern education, and mathematics education in particular, have enormous relevance and implications for the effectiveness of the mathematical education of our students. Research in mathematics education stands to contribute to the shaping, implementation, analysis, and revision of these policies, and is doing so in many cases. Through strong collaborations among researchers, practitioners, and policy makers, it is possible to achieve convergence and synergy so that policies, research, and practice can address similar problems in mutually synergistic ways. The international mathematics education community has collective experience and is beginning to accumulate policy-relevant research, and the opportunities to do so more systematically and to achieve more impact in the future should be a focus in the years to come.

Acknowledgment Joan Ferrini-Mundy thanks Ed Corcoran and Eryn Stehr for their assistance in the preparation and development of the part of this manuscript concerned with US policy.

References

Abrantes, P. (2001). Mathematical competence for all: Options, implications and obstacles. *Educational Studies in Mathematics, 47*(2), 125–143.

Beaton, A. E., & Robitaille, D. E. (1999). An overview of the Third International Mathematics and Science Study. In G. Kaiser, E. Luna, & I. Huntley (Eds.), *International comparisons in mathematics education* (pp. 19–29). Philadelphia, PA: Falmer Press.

Cobb, P., & Jackson, K. (In press). Educational policies as designs for supporting learning. *The Journal of Learning Sciences*.

Clarke, D. (2003). International comparative research in mathematics education. In A. J. Bishop, M. A. Clements, C. Keitel, J. Kilpatrick, & F. K. S. Leung (Eds.), *Second international handbook of mathematics education* (pp. 143–184). Dordrecht, The Netherlands: Kluwer Academic Publishers.

Confrey, J. (2010, May). *Implications of the Common Core State Standards for mathematics*. Presentation at the North Carolina Council of Teachers of Mathematics Annual Conference, Greensboro, North Carolina. Retrieved from http://gismo.fi.ncsu.edu/confrey2010ncctm.pdf.

DeBoer, G. E. (2011). *Introduction to the policy terrain in science education*. Charlotte, NC: Information Age Publishing.

Donovan, M. S., Wigdor, A. K., & Snow, C. E. (Eds.). (2003). *Strategic education research partnership*. Washington, DC: Committee on a Strategic Education Research Partnership. National Research Council, Division of Behavioral and Social Sciences and Education. The National Academies Press.

Ferrini-Mundy, J. (2004). What does it mean to be standards-based? Issues in conceptualizing, measuring, and studying alignment with standards. In F. K. Lester Jr. & J. Ferrini-Mundy (Eds.), *Proceedings of the NCTM Research Catalyst Conference* (pp. 25–32). Reston, VA: NCTM.

Ferrini-Mundy, J., & Floden, R. E. (2007). Educational policy research and mathematics education. In F. K. Lester Jr. (Ed.), *Second handbook of research on mathematics teaching and learning* (pp. 1247–1279). Charlotte, NC: Information Age Publishing.

Finnigan, K., Daly, A., & Che, J., (2012, April). *The acquisition and use of evidence district-wide*. Paper presented at the Annual Meeting of the American Education Research Association, Vancouver, BC.

Fuhrman, S. H. (2001). *From the Capitol to the classroom: Standards-based reform in the States*. Chicago, IL: National Society for the Study of Education.

Gates, P., & Vistro-Yu, C. (2003). Is mathematics for all? In A. J. Bishop, M. A. Clements, C. Keitel, J. Kilpatrick, & F. K. S. Leung (Eds.), *Second international handbook of mathematics education* (pp. 31–73). Dordrecht, The Netherlands: Kluwer Academic Publishers.

Gouseti, A., Noss, R., Potter, J., & Selwyn, N. (2011). *Assessing the impact and sustainability of networks stimulated and supported by the NCETM*. London, UK: Institute of Education, University of London.

Hodgen, J., Pepper, D., Sturman, L., & Ruddock, G. (2010). *Is the UK an outlier? An international comparison of upper secondary mathematics education*. London, UK: Nuffield Foundation.

Honig, M. I., & Coburn, C. (2008). Evidence-based decision making in school district central offices: Toward a policy and research agenda. *Educational Policy, 22*(4), 578–608.

Hoyles, C. (2009). Understanding maths learning. *Better: Evidence-Based Education, 2*(1), 12–13.

Hoyles, C., & Lagrange, J.-B. (Eds.). (2009). *Mathematics education and technology—Rethinking the terrain*. New York, NY: Springer.

Independent Review of Mathematics Teaching in Early Years Settings and Primary Schools by Sir Peter Williams. (2008). London, UK: Department for Children, Schools and Families.

Jacobsen, R. (2009). The voice of the people in education policy. In G. Sykes, B. Schneider, D. N. Plank, & T. G. Ford (Eds.), *Handbook of education policy research* (pp. 307–318). New York, NY: Routledge and AERA.

Klein, D., Braams, B. J., Parker, T., Quirk, W., Schmid, W., & Wilson, W. S. (2005). *The state of state math standards: 2005.* Washington, DC: Thomas B. Fordham Foundation.

Klein, D., Askey, R., Milgram, M. J., Wu, H.-H., Scharlemann, M., & Tsang, B. (1999, November). An open letter to United States Secretary of Education, Richard Riley. Retrieved from http://www.mathematicallycorrect.com/riley.htm.

Krainer, K. (2011). Teachers as stakeholders in mathematics education research. In B. Ubuz (Ed.), *Proceedings of the 35th Conference of the International Group for the Psychology of Mathematics Education* (Vol. 1, pp. 47–62). Ankara, Turkey: International Group for the Psychology of Mathematics Education.

Lester, F. K., Jr., & Wiliam, D. (2002). On the purpose of mathematics education research: Making productive contributions to policy and practice. In L. D. English (Ed.), *Handbook of international research in mathematics education* (pp. 489–506). Mahwah, NJ: Lawrence Erlbaum.

Making Mathematics Count: The report of Professor Adrian Smith's Inquiry into Post-14 mathematics education. (2004). London, UK: The Stationery Office.

McDonnell, L. M. (2009). Repositioning politics in education's circle of knowledge. *Educational Researcher, 38*(6), 417–427.

McLeod, D. B., Stake, R. E., Schappelle, B. P., Mellissinos, M., & Gierl, M. J. (1996). Setting the standards: NCTM's role in the reform of mathematics education. In S. A. Raizen & E. D. Britton (Eds.), *Bold ventures: U.S. innovations in science and mathematics education. Vol. 3: Cases in mathematics education* (pp. 13–132). Dordrecht, The Netherlands: Kluwer.

National Academy of Sciences. (1996). *National science education standards.* Washington, DC: Center for Science, Mathematics, and Engineering Education.

National Academy of Sciences, National Academy of Engineering, & Institute of Medicine. (2007). *Rising above the gathering storm: Energizing and employing America for a brighter economic future.* Washington, DC: The National Academies Press.

National Council of Teachers of Mathematics. (1989). *Curriculum and evaluation standards for school mathematics.* Reston, VA: Author.

National Council of Teachers of Mathematics. (1991). *Professional standards for teaching mathematics.* Reston, VA: Author.

National Council of Teachers of Mathematics. (1995). *Assessment standards for school mathematics.* Reston, VA: Author.

National Council of Teachers of Mathematics. (2000). *Principles and standards for school mathematics.* Reston, VA: Author.

National Mathematics Advisory Panel. (2008). *Foundations for success: The final report of the National Mathematics Advisory Panel.* Washington, DC: Author.

National Research Council. (1999). *Improving student learning: A strategic plan for education research and its utilization.* Committee on a Feasibility Study for a Strategic Educational Research Program. Washington, DC: National Academies Press. Available online http://books.nap.edu/books/0309064899/html/R1.html#pagetop.

National Research Council. (2010). *Preparing teachers: Building evidence for sound policy.* Committee on the Study of Teacher Preparation Programs in the United States, Center for Education. Washington, DC: National Academies Press.

National Research Council. (2011). *Successful K–12 STEM education: Identifying effective approaches in science, technology, engineering, and mathematics.* Washington, DC: National Academies Press.

NCES. (n.d.-a). *Numbers and types of public elementary and secondary local education agencies from the common core of data: School year 2009–10.* Retrieved from http://nces.ed.gov/pubs2011/pesagencies09/findings.asp.

NCES. (n.d.-b). *Numbers and types of public elementary and secondary schools from the common core of data: School year 2009-10.* Retrieved November 20, 2011 from http://nces.ed.gov/pubs2011/pesschools09/findings.asp.

Nelson, S. R., Leffler, J. C., & Hansen, B. A. (2009). *Toward a research agenda for understanding and improving the use of research evidence.* Portland, OR: Northwest Regional Educational Laboratory. Retrieved from http://educationnorthwest.org/webfm_send/311.

Noss, R., & Hoyles, C. (1996). *Windows on mathematical meanings: Learning cultures and computers.* Dordrecht, The Netherlands: Kluwer Academic Publishers.

OECD (Organisation for Economic Co-operation and Development). (2011). *Establishing a framework for evaluation and teacher incentives: Considerations for Mexico*: Author. Retrieved from http://dx.doi.org/10.1787/9789264094406-en.

Osborne, J. (2011). Science education policy and its relationship with research and practice. In G. E. DeBoer (Ed.), *The role of public policy in K-12 science education* (pp. 13–46). Charlotte, NC: Information Age Publishing.

Palinkas, L. A., Finno, M., Fuentes, D., Garcia, A., & Holloway, I. W. (2011, August). *Evaluating dissemination of research evidence in public youth-serving systems.* Paper presented at the National Child Welfare Evaluation Summit, Washington, DC.

Peterson, J. (2011). How can science educators influence legislation at the State and Federal levels? The case of the National Science Teachers Association. In G. E. DeBoer (Ed.), *The role of public policy in K-12 science education* (pp. 241–274). Charlotte, NC: Information Age Publishing.

Prosperity for all in the global economy: World class skills: The Leitch review of skills. (2006). London, UK: Department for Employment and Learning.

Schmidt, W. H., McKnight, C. C., & Raizen, S. A. (Eds.). (1997). *A splintered vision: An investigation of U.S. mathematics and science education.* Dordrecht, The Netherlands: Kluwer.

Schmidt, W. H., McKnight, C. C., Valverde, G. A., Houang, R. T., & Wiley, D. E. (1997). *Many visions, many aims: Vol. 1. A cross-national investigation of curricular intentions in school mathematics.* Dordrecht, The Netherlands: Kluwer.

Sheffield Hallam University. (2010). *Mathematics for a stronger society: A review of the work of the NCETM in relation to the priorities of the Government.* Sheffield, UK: Centre for Education and Inclusion Research, Sheffield Hallam University.

Smith, M. S., & O'Day, J. (1991). Systemic school reform. In S. H. Fuhrman & B. Malen (Eds.), *The politics of curriculum and testing: The 1990 yearbook of the Politics of Education Association* (pp. 233–267). London, UK: Falmer Press.

Smith, M. S., & Smith, M. L. (2009). Research in the policy process. In G. Sykes, B. Schneider, D. N. Plank, & T. G. Ford (Eds.), *Handbook of education policy research* (pp. 1–14). New York, NY: Routledge and AERA.

Sykes, G., Schneider, B., & Ford, T. (2009). Introduction. In G. Sykes, B. Schneider, D. N. Plank, & T. G. Ford (Eds.), *Handbook of education policy research* (pp. 1–14). New York, NY: Routledge and AERA.

Tseng, V. (2010). Learning about the use of research to inform evidence-based policy and practice: Early lessons and future directions. *William T. Grant Foundation 2009 Annual Report* (pp. 12–17). New York, NY: William T. Grant Foundation. Retrieved from http://www.wtgrantfdn.org/File%20Library/Annual%20Reports/2009%20AR%20essay%20Learning%20about%20the%20Use%20of%20Research.pdf.

Tseng, V., Granger, R. C., Seidman, E., Maynard, R. A., Weisner, T. S., & Wilcox., B. L. (2008). Studying the use of research evidence in policy and practice. *William T. Grant Foundation 2007 Annual Report* (pp. 12–17). New York, NY: William T. Grant Foundation. Retrieved from http://www.wtgrantfdn.org/File%20Library/Annual%20Reports/2007%20AR%20essay%20Studying%20the%20Use%20of%20Research%20Evidence.pdf.

U.S. Department of Education. (n.d.). *U.S. Department of Education exemplary mathematics programs*. Retrieved from http://www.k12academics.com/education-reform/us-department-education-exemplary-mathematics-programs.

Welch, W. W. (1979). Twenty years of science curriculum development: A look back. *Review of Research in Education, 7*, 282–306.

Wikipedia. (n.d.). *Education policy*. Retrieved from http://en.wikipedia.org/wiki/Education_policy.

William T. Grant Foundation. (2011). *Request for research proposals: Understanding the acquisition, interpretation, and use of research evidence in policy and practice*. Retrieved from http://www.wtgrantfdn.org/funding_opportunities/research_grants/rfp_for_the_use_of_research_evidence.

Part III
Introduction to Section C: Technology in the Mathematics Curriculum

Frederick K. S. Leung

Abstract The eight chapters in Section *C* were prepared by 23 scholars from 13 different nations. The first chapter provides historical perspectives, not only on the use of technology through the ages, but also in relation to the rapid development of digital technologies over the past decade. The "middle chapters" discuss theoretical and practical developments with respect to the implications of emerging technologies for curriculum development, teaching, learning, and researching mathematics education. For example, what new things in algebra and geometry education can—and should—we do now, given the availability of Computer Algebra Systems, and dynamic geometry software? Will the position of statistical reasoning in the mathematics curriculum change? Will a modelling approach to mathematics problem solving be facilitated by technological developments? What about equity issues? Are we at the beginning of a new era when online and distance forms of mathematics education will become dominant? From a mathematics education perspective, how can students, teachers, schools and researchers take advantage of facilities provided by the Internet? Do technological developments profoundly affect conceptions of proof, and approaches to dealing with proof in school mathematics? How should the ready availability of technology affect the assessment of mathematics learning? The section closes with a provocative discussion of policy implications for mathematics education arising from technological developments.

Keywords Internet and mathematics education • Technology and modelling • Technology in mathematics education • Tools in mathematics education

Since the publication of the *Second International Handbook*, in 2003, it has become increasingly evident that technology is slowly changing the way school mathematics is taught and learned, at least in some countries. There can be no doubt

F. K. S. Leung
The University of Hong Kong, Pokfulam, Hong Kong, People's Republic of China

that the increasing availability of more and more powerful and portable computers, and versatile interactive software, have enabled many complex mathematics tasks to be accomplished much more easily. Furthermore, the Internet allows nearly instant access to knowledge and information worldwide. The result is that students of the modern technological era have the possibility of learning more exciting forms of mathematics more effectively than ever before. Whether this is the case or not can be traced in the literature reviewed and discussed in some of the chapters of this section.

But the more significant impact of technology on mathematics education, as will be discussed in most of the chapters in this section, is that technology is changing the very nature of the mathematics we are teaching, learning and assessing. This may be frightening for some people who perceive mathematics as eternal truth and hence a stable entity. But if we look at the issue from the perspective of history provided in the first chapter in this section, this idea that the nature of mathematics is changing because of the changing technological tools that have become available, not all that alien. In the past, compasses, the chalk board, and the abacus all changed the nature of the mathematics being represented. But, in the past few decades technology has been developing at such a fast pace that we have been able to observe the changing nature of mathematics within a relative short span of time. So, in essence, modern technology is changing both the way we learn "traditional" mathematics and, simultaneously, the nature of the mathematics that we learn. The chapters in this section portray these changes from different perspectives.

In the *Second International Handbook*, we dealt with a selected group of topics to "provide a good profile of the kinds of issues ... that researchers have been tackling" on the responses in mathematics education to technological developments (Leung, 2003, p. 233). These topics included technology and research in mathematics education, the school and undergraduate mathematics curriculum, and teacher education. In this *Handbook*, instead of selecting the same or another group of topics for this section, we decided to focus on the impact of technology on the school mathematics curriculum—in the last *Handbook* we only devoted one chapter specifically to that theme (Wong, 2003).

Although all the chapters in this current section obviously draw on research, and the issues discussed have clear implications for research, we have not devoted any chapter to address specifically the issue of technology and mathematics education research. That differs from the situation with the last *Handbook,* in which two chapters (Lagrange et al., 2003; Hoyles & Noss, 2003) were directly related to research. In the last *Handbook* there was also a chapter on technological tools for teaching undergraduate mathematics (Thomas & Holton, 2003). In this *Handbook*, Chapter 20, by Heid, Thomas and Zbiek, touches slightly on tertiary mathematics, but the main focus of that chapter, as well as that of all other chapters, is on school mathematics. Also, although the discussions in the majority of the chapters in this section of the *Handbook* have implications for mathematics teacher education, we have not devoted a chapter specifically to mathematics teacher education—although we did have such a chapter in the last *Handbook* (Mousley, Lambdin & Koc, 2003). By focussing on the school curriculum, this section of the *Third* International *Handbook* covers the impact of technology on the school mathematics curriculum comprehensively and in some depth.

In his Introduction to this *Third International Handbook* Ken Clements pointed out that all sections of the *Handbook* follow a pattern of starting with a chapter that provides a historical analysis, and ending with a chapter that deals with policy implications. For this section it starts with a chapter that surveys the evolution and curricular influence of technology in mathematics instruction, and ends with a chapter on the implications of technology-driven developments for policies on the school mathematics curriculum. In between these two chapters, there are chapters which deal with the major strands of school mathematics: Geometry (dynamic geometry and geometric proof); algebra (computer algebra systems and the role of school algebra); and statistics (technology and statistical reasoning). There are also chapters on specific technology-related issues in the school mathematics curriculum. In addition to the chapter on proof, there is a chapter on technology and mathematical modelling, one on learning mathematics with the use of the Internet, and one on technology and assessment. These chapters raise matters of central importance with respect to school mathematics education.

Although there is a chapter that deals specifically with the impact of the Internet on school mathematics, a topic of emerging importance which has not been covered in this *Handbook* is that of the impact of mobile technology or devices on mathematics teaching and learning. In the past few years, the ubiquitous presence of mobile technology and devices (smart phones, tablet PC, etc.) is affecting the lives of people tremendously, especially among the younger generation. As early as 2003, there were attempts to study the potential of mobile technology for mathematics teaching and learning (Roschelle, 2003), and scholars argued that the potential was immense. Experiments with wireless hand-held technology reported some success in Australia (Roschelle et al., 2010; Main & O'Rourke 2011) and in Israel (Daher, 2010), and there are a number of interesting ongoing projects (e.g., Geogebra Mobile: http://www.geogebraorg/trac/wiki/GeoGebraMobile; Multitouch Interactive Cinderella: http://www.youtube.com/watch?v=qraL4nIfkbI; Geometer's Sketchpad iPad version: http://www.dynamicgeometry.com/General_Resources/Sketchpad_Explorer_for_iPad.html).

However, the use of mobile technology in mathematics education is still in its infancy (Rismark, Sølvberg, Strømme & Hokstad, 2007) and has not yet reached the everyday classroom. It would be a somewhat premature to review the development yet, and so we have not devoted a chapter to the topic. But it is definitely a new area of research that is extending the current bounds of research on technology and mathematics education. My prediction is that in the years to come, as mobile technology permeates more and more into the everyday lives of members of various societies, education institutions will be bound to capitalize on the power of mobile technology and exploit its use for mathematics education.

In the first chapter in this section (Chapter 17), Roberts, Leung and Lin, provide historical perspectives on the interaction between technology and mathematics education. When talking about technology in the mathematics curriculum, one might immediately think of the Internet and computer educational software, etc. The Internet and computer software are indeed part of technology, and much of what is covered in this section of the *Handbook* deals with these two important components of technology. However, technology is much more than these, and it is instructive to look at technology from a historical perspective. Roberts et al. enable us to become aware of the roots of modern technology and the role it plays in the contemporary classroom. Their chapter provides a historical survey of technological tools, from interesting historical calculation devices to the modern virtual tools, and discusses how technology has shaped mathematics and mathematics education in the different stages of history, and the potentials technology have offered in different ages for mathematics teaching and research. Roberts et al. discuss technology as tools for information storage, information display, demonstration, and calculation, and argue that all these functions are combined in the modern computer. But their chapter also reminds us that technology is not just about the computer, and the historical perspective sets the stage for the rest of the chapters in the section.

Chapter 18, by Williams and Goos, looks at the interaction between technology and modelling—a contemporary aspect of mathematics teaching and learning that is assuming greater importance as it becomes obvious it can harness the support of new technologies. Adopting an activity theory perspective, Williams and Goos offer a theoretical framework integrating mathematical modelling and technology within a social and cultural-historical context. The authors argue that when learning through modelling, students should be taught within their zones of proximal development (ZPDs) to solve problematic and authentic mathematics tasks within a socio-cultural context. Technology allows for an expanded concept of ZPD. The relationship between technology and mathematical modelling is then illustrated with two examples where there is a "breakdown" of the modelling and/or technology within the mathematical-technological context, showing that mathematics learning is in essence an amalgamation of technology, mathematics and the social-cultural context.

While modelling is considered an important characteristic of mathematics learning, proof is often considered *the* primary characteristic of mathematics. Technology is sometimes conceived of as contributing only to the calculation and manipulation aspect of mathematics, and has little or nothing to do with mathematics proof, which is reserved for the human brain. But the development of dynamic geometry tools has raised serious questions about such a view, and has forced us to re-examine the notion of mathematical proof. Chapter 19, by Sinclair and Robutti, reviews developments over the past decade within dynamic geometry environments (DGEs) and suggests implications for the process of proving. Most relevant, are the dragging and measuring functions in DGE. Both the potentials and challenges of DGE for students' learning and understanding of proof and proving are discussed, and illustrated with examples. The discussion highlights the fact that technology is not merely an aid for studying mathematics. The use of technology affects the very nature of the mathematics we are studying, and in this case, it challenges our traditional notion of proof in mathematics.

While DGE is most relevant to the learning of geometry, the software tools that have had the greatest impact on the teaching and learning of algebra are computer algebra systems (CAS). In Chapter 20, Heid, Thomas and Zbiek offer a brief review of the history of CAS and then discuss three issues arising from CAS—namely, the effects on algebraic concepts, effects on algebraic procedures (or skills), and effects on algebraic thinking and reasoning. One pertinent aspect of mathematical thinking enhanced by CAS is generalization, which is considered to be one of the most important characteristics of algebra. It is argued that CAS not only helps students to understand symbols and to reason, but also facilitates their development of algebraic concepts and number sense. It helps to connect symbolic and graphical reasoning through its ability to link, dynamically, multiple representations of concepts. Most importantly, CAS changes the role of algebra in the curriculum. It helps to make symbolic work more achievable as students blend different areas and processes of mathematics. Some implementation issues on the curriculum and teacher training are then discussed. The latter includes the knowledge teachers need for teaching in a CAS environment (the notion of Pedagogical Technology Knowledge or PTK is introduced), and the educational significance of communities of practitioners linked through online networks, etc. The chapter ends by discussing a possible research agenda in this area.

Another major area of the school mathematics curriculum on which technology is having an impact is that of statistics. Chapter 21, by Biehler, Ben-Zvi, Bakker and Makar, focusses on the impact of technology on statistical reasoning. The chapter starts by reviewing the different technological tools for the teaching and learning of statistics. Two specific tools, *Fathom* and *TinkerPlots 2.0*, are then further discussed as examples of how technological tools can support statistical reasoning. Concrete examples are given to illustrate how the dynamic and visual nature of the software aids statistical reasoning through data exploration, connecting data and chance, and statistical inference. The chapter ends with a "wish list" for future development and research in the area of technology for statistical reasoning. To what extent that wish list will be fulfilled in the near future is perhaps something that we should look forward to in the next *Handbook*!

Chapters 19, 20 and 21 testify to how technology is slowly impacting all main domains of school mathematics. Outside of school, the most pervasive technology in the past decade has been undoubtedly the Internet, and Chapter 22, by Borba, Clarkson and Gadanidis, discusses how the Internet is interacting with mathematics education. In the last *Handbook*, we only briefly alluded to the role of the Internet in mathematics education, and with the rapid growth in the number of Internet users in the past decade, and in the power of Internet facilities, it is pertinent to ask if there has been a corresponding growth in the application of the Internet in mathematics teaching and learning contexts. Borba et al. argue that the emergence of the Internet has enormous potential for improving all aspects of mathematics education, with radical changes to curriculum and method not only becoming possible, but also feasible.

In Chapter 22 Borba et al. review the history of the emergence of the Internet and its impact on education. It offers examples illustrating education affordances of the Internet in three main areas: collaboration, multimodality and performance. The examples show how the Internet has enhanced collaborative learning, how the multimodality that the Internet affords can provide rich opportunities for learning, and how performance facilitated by and broadcasted through the Internet are ushering in new realms of possibility in mathematics learning.

As Borba et al. point out, the examples they provide draw attention to what might be possible. But Chapter 22 also discusses the limitations and problems associated with the use of the Internet for mathematics teaching and learning. The authors emphasize that the limitations and problems must be dealt with if educators are to capitalize on the potential of the Internet for mathematics teaching and learning. Although the authors do not explicitly identify the influence of the Internet on the nature of mathematics that students learn, the examples given in this chapter suggest that the Internet has ushered in an era when a much more creative approach to mathematics curricular design will become possible. But, teaching and learning approaches are likely to change drastically, and this will challenge mathematics educators to identify the essence of mathematics teaching and learning. What is disposable, what is possible, and what must remain?

The chapter by Stacey and Wiliam, Chapter 23, discusses the important issue of assessment, and addresses issues associated with how technology is not only having an impact on assessment of "traditional" mathematics but is also ushering in an era when serious attention will have to be given to the kind of new and important mathematics knowledge that can now be learned and assessed. Although technology is being employed to assess traditional mathematics efficiently—through enhancing item presentation, allowing more convenient and reliable scoring, and providing immediate and personal feedback—it must be recognized that there are still obstacles and challenges with computer-based assessment of mathematics (e.g., limitations related to a reliance on keyboard input). But, as educational technology matures, it is envisaged that such problems will be resolved, and that should happen in the near future.

The more important question is, since technology changes (or has the potential to change) the learning and teaching of mathematics, are there corresponding developments in the technology for assessment so that it will be possible for new learning skills and approaches to be appropriately assessed? As assessment usually has a strong backwash effect on teaching, developments in assessment technology will in turn lead to innovations in teaching and learning. Indeed, technology is changing, and will continue to change, the very nature of mathematics itself. This echoes the theme above that technology is not merely something which can enhance the teaching, learning and assessment of standard mathematics. It has the potential to change the nature of the mathematics taught, learned and assessed.

Although many of the chapters discussed above explicitly or implicitly touch on policy implications associated with the various subfields of technology in the mathematics curriculum, the last chapter in this section, Chapter 24, by Trouche, Drijvers, Gueudet and Sacristán, focusses on policy implications of technology-driven developments in mathematics education. A three-dimensional model is put forward for analyzing different aspects of policy. The chapter traces the historical development of policies on technology in the mathematics curriculum in different countries, and in so doing introduces the concept of learning and teaching space. Technology brings about a new paradigm of learning, but a full exploitation of available technological resources is not yet evident among most teachers. With the profusion of technological resources, policy implications for the teaching, learning and assessment of mathematics, as well as for teacher education, urgently need to be identified and acted upon. Some examples of the influence of policies on preservice and inservice teacher education, especially those that stress collaborative work in technological environments, are described. The chapter ends with some discussion on the issues concerning how new technologies are influencing curricular policies.

Altogether, the chapters in this section portray the state of the art in terms of technology in the school mathematics curriculum. The potential or affordance of modern technology for different areas of mathematics curriculum development, and for mathematics teaching and learning, is discussed. It is clear that what is happening at the moment is still a mere exploration of the affordance. The potential of modern technology for effective teaching and learning of mathematics is far from being realized in any large scale. Nevertheless, it is increasingly being recognized that not only has technology great potential for helping students learn "traditional" mathematics, it also opens up new fronts on mathematics learning which were not possible before the advent of the new technologies.

Most importantly, as pointed out in the beginning of this introduction, in many cases, the use of technology to study mathematics has changed the very nature of the mathematics we are studying. So technology in the mathematics curriculum should not be characterized by how the evolving technology will have an impact on the learning and teaching of mathematics from the curricula of previous eras. Rather, curriculum and teaching and learning methods will need to be regularly reconceptualized to take advantage of the power of modern technology to improve mathematics education in, possibly, spectacular ways.

References

Daher, W. (2010). Building mathematical knowledge in an authentic mobile phone environment. *Australasian Journal of Educational Technology, 26*(1), 85–104.

Hoyles, C., & Noss, R. (2003). What can digital technology take from and bring to research? In A. J. Bishop, M. A. Clements, C. Keitel, J. Kilpatrick, & F. K. S. Leung (Eds.), *Second international handbook on mathematics education* (pp. 323–349). Dordrecht, The Netherlands: Kluwer Academic Publishers.

Lagrange, J., Artigue, M., Laborde, C., & Trouche, L. (2003). Technology and mathematics education: A multidimensional study of the evolution of research and innovation. In A. J. Bishop, M. A. Clements, C. Keitel, J. Kilpatrick, & F. K. S. Leung (Eds.), *Second international handbook on mathematics education* (pp. 237–269). Dordrecht, The Netherlands: Kluwer Academic Publishers.

Leung, F. K. S. (2003). Responses in mathematics education to technological developments: Introduction. In A. J. Bishop, M. A. Clements, C. Keitel, J. Kilpatrick, & F. K. S. Leung (Eds.), *Second international handbook on mathematics education* (pp. 233–236). Dordrecht, The Netherlands: Kluwer Academic Publishers.

Main, S., & O'Rourke, J. (2011). "New directions for traditional lessons": Can handheld game consoles enhance mental mathematics skills. *Australian Journal of Teacher Education, 36*(2), Article 4.

Mousley, J., Lambdin, D., & Koc, Y. (2003). Mathematics teacher education and technology. In A. J. Bishop, M. A. Clements, C. Keitel, J. Kilpatrick, & F. K. S. Leung (Eds.), *Second international handbook on mathematics education* (pp. 395–432). Dordrecht, The Netherlands: Kluwer Academic Publishers.

Rismark, M., Sølvberg, A. M., Strømme, A., & Hokstad, L. M. (2007). Using mobile phones to prepare for university lectures: Students' experiences. *The Turkish Online Journal of Educational Technology, 6*(4), 86–91.

Roschelle, J. (2003). Unlocking the learning value of wireless mobile devices. *Journal of Computer Assisted Learning, 19*(3), 260–272.

Roschelle, J., Rafanan, K., Bhanot, R., Estrella, G., Penuel, B., Nussbaum, M., & Claro, S. (2010). Scaffolding group explanation and feedback with handheld technology: Impact on students' mathematics learning. *Education Technology Research and Development, 58*, 399–419.

Thomas, M., & Holton, D. (2003). Technology as a tool for teaching undergraduate mathematics. In A. J. Bishop, M. A. Clements, C. Keitel, J. Kilpatrick, & F. K. S. Leung (Eds.), *Second international handbook on mathematics education* (pp. 351–394). Dordrecht, The Netherlands: Kluwer Academic Publishers.

Wong, N. Y. (2003). The influence of technology on the mathematics curriculum. In A. J. Bishop, M. A. Clements, C. Keitel, J. Kilpatrick, & F. K. S. Leung (Eds.), *Second international handbook on mathematics education* (pp. 271–321). Dordrecht, The Netherlands: Kluwer Academic Publishers.

Chapter 17
From the Slate to the Web: Technology in the Mathematics Curriculum

David Lindsay Roberts, Allen Yuk Lun Leung, and Abigail Fregni Lins

Abstract The employment of physical tools to assist teaching and learning of mathematics did not begin with electronic devices, and has a much longer history than is often recognized. At times, technology has functioned as the inventive embodiment of mathematical ideas, progressing somewhat in step with the evolution of mathematics itself. At other times, technology has entered mathematics from outside, notably from commerce and science. This chapter surveys the evolution and curricular influence of technology in mathematics instruction in the Eastern and Western worlds from ancient times to the present day, with the primary focus being on the last 200 years. Past technology is categorized into tools for information storage, tools for information display, tools for demonstration, and tools for calculation. It is argued that today's computing technology offers teachers and students the potential to move beyond these categories, and to experience mathematics in ways that are different from traditional school mathematics curricula. A window is opened through which mathematics teaching and learning might enter into a new epistemological domain, where knowledge becomes both personal and communal, and in which connective and explorative mathematical knowledge becomes vastly more accessible.

D. L. Roberts (✉)
Prince George's Community College, Largo, MD, USA
e-mail: Robertsdl@aol.com

A. Y. L. Leung
Hong Kong Baptist University, Hong Kong SAR, China
email: aylleung@hkbu.edu.hk

A. F. Lins
State University of Paraiba, Paraiba, Brazil
e-mail: bibilins2000@yahoo.co.uk

Introduction

Since the advent of the electronic calculator it has become customary for discussion of "technology" in mathematics education to refer almost exclusively to use of electronic devices. However, this represents a manifestation of historical amnesia. The employment of physical tools to assist teaching and learning of mathematics has a much longer history, and this history provides a valuable perspective on current proposals and debates. At times technology has functioned as the inventive embodiment of mathematical ideas, progressing somewhat in step with the evolution of mathematics itself. But technology also enters mathematics from the larger world outside, notably from commerce and science. Moreover, technological tools used by mathematical practitioners need not translate immediately into mathematics education, and tools useful in an educational setting need have little appeal for professional users of mathematics. Educational use of technology is also subject to overarching educational philosophies prevailing at any given time and place; some would call these fads and fashions. The interactions among technology, mathematics, and education are thus unavoidably complex, and cannot be described by any simple model of historical progress over time.

The historical record suggests that the use of tools always has been inseparable from expressing and doing mathematics. In the ancient Western world the Babylonians carved solutions to geometric problems on small pieces of round clay. Possibly students did these as assessment tasks—for instance to find the length of a diagonal of a square using the square root of two. The ancient artefact depicted in Figure 17.1a might have been the work carried out by such a student. Another Babylonian student may have used a "calculator" to work out a rather complex arithmetic problem. In this case his tool was a counting board made from a slab of stone with groups of markings (parallel lines, semi-circles) on it. The student put pebbles on it to work out his answer. A version of this counting board, which dated

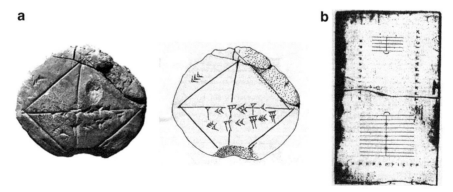

Figure 17.1. (**a**) Mathematical exercise to find diagonal of square, using the square root of 2 [Yale Babylonian Collection http://www.yale.edu/nelc/babylonian.html], (**b**) The Salamis Tablet: The oldest counting board. It is made of marble. Photo from the National Museum of Epigraphy, Athens.

17 *From the Slate to the Web: Technology in the Mathematics Curriculum*   527

Figure 17.2. (**a**) *Nuwa (left)* and *Fuxi (right)* with *Nuwa* holding a *guī* and *Fuxi* holding a *ju* [http://sunrise.hk.edu.tw/~planning/sm/images/exect-1/book-j002.JPG]; (**b**) A Chinese set-square.

back to 300 BCE, was found on the Greek island of Salamis in 1846 (Figure 17.1b). In these ancient artefacts mathematics seems to have been embodied and was being preserved under the inventiveness of ancient craft.

Looking to the Eastern world, there was a different type of embodiment. In ancient Chinese mythology, there were demigods *Nuwa* and *Fuxi* who were the progenitors of mankind and shapers of human society. Legends say that *Nuwa* and *Fuxi* invented *guī* (compasses) and *ju* (set-square) to shape the world. On an ancient stone carving found inside a tomb from the East Han dynasty (25 to 220 CE) there is engraved an intertwined image of *Nuwa* and *Fuxi* with *Nuwa* holding a *guī* and *Fuxi* holding a *ju* (Figure 17.2a).

For the ancient Chinese, the basic concept of the world was "heaven is round, earth is square" and there was an ancient motto saying that "without *guiju*, there are no square and circle." This geometrical intuition about the physical world became metaphoric in the human world. The connotative usage of the word *guiju* refers to orderliness according to underlying rules, and even applies to human affairs. Hence, for the Chinese, circle and square were elemental shapes and rules of the universe and they were embodied and symbolized by the tools that produced them. Notice that the two arms of the Chinese set square were not of the same length (Figure 17.2b). This might indicate that the ancient Chinese were already familiar with a Pythagorean-type relation about right-angled triangles. (The Chinese version of Pythagoras' Theorem was *Gougu*: Chapter 9 of the ancient Chinese mathematics treatise *The Nine Chapters*). Thus, behind the design of *ju* there lay an embodiment of a piece of mathematical knowledge. This kind of knowledge mediation, using tools embodying mathematics, was even more deep-seated in another Chinese traditional knowledge system mediated by symbolic visual tools. Ancient Chinese used dot and line pattern diagrams to represent and interpret the phenomenological world. In Figure 17.3 there are three elemental number pattern diagrams that constituted the root of Chinese thought and culture. Chinese used these diagrams (and derivations of them) as coding tools to decipher the hidden laws of the universe.

Luò Shū (The Luò River Writing) and *He Tu* (The River Map) were two different but related arrangements of 1, 2, 3, 4, 5, 6, 7, 8 and 9 using black and white dots.

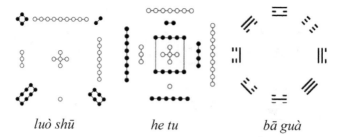

luò shū   *he tu*   *bā guà*

Figure 17.3. Three fundamental symbolic tools that form the basis of Chinese culture.

There were mythical stories about their origins signifying that these patterns were indeed very ancient and sacred. *Luò Shū* is a three-by-three magic square. It has intriguing mathematical properties and has had a deep influence in Chinese culture (Berglund, 1990). *He Tu* is a derivation of *Luò Shū:* it emphasizes the concept of duality (even and odd, *yin* and *yang*). *Bā guà* is the kernel of a binary coding system that classifies natural and human phenomena and is intimately connected to *Luò Shū*. These were the fundamental symbolic tools by which the ancient Chinese derived their concept of the world. They are supposed to embody numerical and geometrical information that guided the development of Chinese civilization. In particular, these diagrams were instrumental in facilitating mathematical calculations to predict occurrences of human affairs and natural phenomena.

The above examples from ancient Babylon and China illustrate that humans invent tools, symbols, and technology that embody mathematics. By this we mean that an object has been created, possibly simple, possibly very complex, which in some sense contains a mathematical idea or procedure. The object is capable of illustrating the idea for an observer, of facilitating the procedure, or of providing some combination of these services. Such tools can in turn endow users of the tools with enhanced ability to deepen their mathematical experiences. Mathematical experience can be thought of as "the discernment of invariant pattern concerning numbers and/or shapes and the re-production or re-presentation of that pattern" (Leung, 2010). Moreover, mathematical concepts are often developed in the process of using tools, whether the tools were designed for mathematical purposes or not. Tools used for the general betterment of social conditions, or for encapsulating features of a cultural worldview, often carry with them indigenous mathematical knowledge. In ancient India (800–500 BCE), notions of geometric shape and measuring techniques emerged in Sanskrit texts on ritual practices, such as prescriptions for constructing fire altars:

> The footprints for the altars were laid out on leveled ground by manipulating cords of various lengths attached to stakes. The manuals described the required manipulations in terse, cryptic phrases—usually prose, although sometimes including verses—called *sūtras* (literally "string" or "rule, instruction"). The measuring cords, called *śulba* or *śulva*, gave their name to this set of texts, the *Śulba-sūtras*, or "Rules of the cord." (Plofker, 2008, p. 17)

The Mayan calendar wheels (1000 BCE) in Central America, based on a vigesimal (base 20) number system, formed a complete philosophy of cyclic time that was believed to guide human destiny (Coe, 1993). The Incas, in the 1400s and 1500s in what is now Peru, used a complex system of knotted strings (*quipus*) as a data collecting and recording device which in effect served as a numerical calculator (Ascher & Ascher, 1997). The Marshall Islanders of the South Pacific used palm ribs and coconut fiber to construct navigation stick charts to represent the behaviour of wave fronts (refraction, reflection and diffraction) as they approach land (Ascher, 2002).

It must be acknowledged that our understanding of the educational practices associated with the above examples is very sparse. We see also from these examples that "technology," if interpreted broadly, can encompass a vast range of human activities, including mathematical notation and language in general. To make our discussion manageable, we therefore define technology in education more narrowly, confining ourselves to physical devices used with the aim of enhancing or amplifying the abilities of the teacher or the student in the mathematics classroom. Thus, although for our purposes we will not count a tool such as logarithms as a technology, the slide rule, a physical device embodying logarithms, will fall under our purview. Electronic devices, and algorithms realized on electronic devices, digital or analog, are also within our scope, inasmuch as there is a physical object involved. In the remainder of this chapter we offer brief histories of several representative devices that have been used in classrooms around the world. To make this survey more relevant to the present day, we furthermore focus mainly on the last 200 years, when mathematics education began to become (haltingly and unevenly across the globe) not merely an acquirement of a small elite, but a mass phenomenon.

We introduce a simple categorization to provide a framework for discussing these tools: tools for information storage, tools for information display, tools for demonstration, and tools for calculation. These categories are admittedly not entirely distinct, and we will see that they become less useful as we move into the electronic era—but they serve well for setting the stage.

Tools of Information Storage

The quintessential information storage tool is the book, which retains a powerful presence in worldwide mathematics education to the present day. The book has a history almost as old as civilization itself, from clay tablets, to the papyrus scroll, to the handwritten codex, to the printed book, and on to the modern e-book (Hobart & Schiffman, 1998). But the history of the mathematics textbook is much shorter, and falls almost entirely within the 200-year window mentioned above, especially if we neglect advanced monographs in favour of books actually used in schools. Certainly, for many centuries individuals learned mathematics independently from books, and likewise tutors used books to teach mathematics to individuals and small-groups, but a new era began with the advent of mass schooling and the mass-produced textbook. These interconnected phenomena did not become prominent until the 19th century in Europe and the Americas, and were materially aided by both political and economic developments. On the political side there was rising support for providing education for a larger proportion of children. On the economic side, there were increasing efficiencies in the production of paper and books, and increasing facilities for transporting goods over long distances, resulting in the ability to manufacture and distribute large numbers of books relatively cheaply (Cordasco, 1976).

When books were scarce, if a class had a book at all it would frequently be the exclusive possession of the teacher. If the class was of any appreciable size this led to the recitation method of teaching, which often meant the teacher simply reading aloud from the book and the pupils attempting, through writing or sheer memorization, to retain what was read, and then to recite it back to the teacher. Notable attempts to scale this system up were made in England and its colonies in the late 18th and early 19th centuries with the so-called monitorial system, in which the teacher would first teach a group of more advanced students, who would in turn teach less advanced students. In mathematics, in particular, the recitation method and the monitorial system primarily supported a curriculum centred on the rote learning of the rudiments of arithmetic (Butts, 1966).

But with cheaper books came the possibility (though still often not the reality) that not merely the teacher but also many students would have individual access to a textbook. A student with a book could now be asked to read that book both inside and outside of class and to work problems assigned from the book. More sophisticated mathematics instruction for a classroom of pupils was now far more feasible than previously. Thus the rising presence of algebra and geometry in addition to arithmetic in the curriculum of 19th-century schools surely owes a good deal to the proliferation of textbooks. The use of textbooks could also serve to hide problems arising from inadequate teacher preparation. This was certainly the case in the 19th-century USA (Tyack, 1974).

Moreover, the system amplified itself: a greater supply of books produced a greater demand for books, which in turn produced yet more books, and so on. In mathematics this resulted not merely in the creation of individual textbooks, but entire series of textbooks covering the whole range of the curriculum from the lowest grades to the colleges: basic arithmetic to the differential and integral calculus. In Europe and North America by the end of the 19th century there was a well-established textbook industry, and there were specialist authors who became wealthy writing textbooks. In the USA, notable 19th-century authors of mathematics textbooks included Charles Davies, Joseph Ray, and George Wentworth (Kidwell, Ackerberg-Hastings, & Roberts, 2008). Seymour and Davidson (2003) asserted that "until the late 1960s, the textbook was virtually the exclusive curricular and pedagogical approach to the teaching and learning of mathematics in the United States and Canada" (p. 990). A study at the close of the 20th century concluded that in the USA the textbook remained the main source used by mathematics teachers to plan daily classroom instruction (Harel & Wilson, 2011).

One effect of textbook proliferation should be especially noted: the assistance provided to standardization of the curriculum, and the difficulty of dislodging curriculum topics once they were printed in widely distributed textbooks. This is especially striking in the USA, which despite a long tradition of local control of schools, and avoidance of an official national curriculum, rapidly converged on a de facto standard curriculum in mathematics, as a relatively small number of textbooks began to dominate the market. Genuinely innovative mathematics textbooks have never fared well in the US market. Even during the "New Math" era of the 1950s and 1960s, supposedly a time of major upheaval, there was substantial continuity in high school textbooks from earlier decades (Dolciani, Berman, & Freilich, 1965; Freilich, Berman, & Johnson, 1952). Many students today have access to textbooks in electronic form, as a supplement to or instead of the traditional paper book. Whether this transition will have a marked effect on the mathematics curriculum is unclear.

Tools of Information Display

The book of course functions as a display device for individuals, as well as a storage device, but with mass education came a pressing need for multiple individuals to view the same display simultaneously. Here the representative tool is the blackboard or chalkboard and its offshoots. Prior to the wall-mounted blackboard, there had been a slow evolution of handheld writing surfaces, culminating in the slate, which could be written on with chalk. In Europe and North America this was often a facet of the recitation method of instruction. The teacher could read a problem from the book and the students could copy and display their solutions on their slates (Burton, 1850; Cajori, 1890).

Prior to the emergence of both the textbook and the blackboard, it was also common practice in many schools in Europe and North America for each student to produce a "copybook" or "cipherbook." Beginning with a collection of blank pages (paper and binding quality could vary widely, depending on economic circumstances) the student would copy out the material spoken aloud by the teacher. In the case of a teacher reading from a printed book this could often mean that the student was almost literally producing a handwritten copy of the book, or the problems from the book. Here again the use of copybooks primarily supported arithmetic instruction, but in some cases this could be fairly elaborate, including square and cube roots and complicated problems from commerce and business. The teacher could periodically inspect the copybooks, so that they could have functioned as what more recent educators would term a "portfolio." But how rigorously 18th- and 19th-century copybooks were evaluated for mathematical correctness is unclear, and some may have been assessed more on aesthetic grounds, such as penmanship (Clements & Ellerton, 2010; Cohen, 1982).

The erasable blackboard, written on with chalk, spread quietly into schools in the early 1800s and was well established by the end of that century (Kidwell et al., 2008). It allowed the teacher to display complicated verbal or pictorial details with far more exactitude than merely reading aloud from a book. Moreover, it allowed students to work out problems on the board themselves, displaying their efforts for both the teacher and other students to see and comment on, thus changing the personal dynamics of the classroom. In mathematics the blackboard worked in conjunction with the textbook to promote the rise of both algebra and geometry in the curriculum.

Blackboards have continued in use in mathematics classrooms to the present time. In many cases the chalkboard has been replaced by the "dry-erase" or "whiteboard," but with no essential change in functionality. The interactive whiteboard, developed in the late 20th century, represents a major innovation, allowing the material displayed on the board to be connected directly to a computer. Opinions vary widely on the value of this technology in the classroom (Smith, Higgins, Wall, & Miller, 2005; Wood & Ashfield, 2008). Tablet personal computers offer similar functionality, including handwriting recognition, whereby the computer is able to interpret handwriting drawn on the screen, not merely type entered via a keyboard (Anderson, 2011).

Another significant classroom display technology is the overhead projector. It came to classrooms in the USA after World War II (Kidwell et al., 2008). Much more than the blackboard, this technology usually remained the exclusive domain of the teacher. It had two primary attractions. First, it allowed the teacher to continue to face the students while displaying materials to them. Second, it allowed the teacher to display elaborate transparencies created before class. For example, a teacher of solid geometry could prepare complicated diagrams with an exactitude that could never be hoped for in hand-drawn diagrams quickly improvised while watched by the students. On the other hand, reliance on prepared slides sometimes encouraged a too rapid succession of material that overloaded the students' ability to assimilate the information presented.

Overhead projectors have continued in use to the present, but in many cases have been superseded by new technologies allowing greater ease of use and a greater range of display functionality. Computer projection systems permit the display of any image, static or moving, available to the host computer, and in particular allow slide shows formerly done via transparencies on an overhead projector to be accomplished via software such as PowerPoint. Another new technology is the document camera (also known as an image presenter or visualizer), which permits any document, or even a three-dimensional object, to be displayed on the overhead screen without any prior preparation of the document or object (Ash, 2009).

Many classrooms in the 21st century provide not only a computer and projector for the teacher but also a computer for each student, networked with the teacher's computer. In some ways this is a return of the handheld slate, with a vast increase in functionality. Its potential for mathematics instruction is just being tapped.

Tools of Demonstration

By tools of demonstration we refer to objects to be handled (physically, or, in more recent times, virtually) by either the teacher or the student, with the aim of conveying increased understanding of a concept or procedure. Rather than being tools of education in general, such tools have usually been more unique to mathematics than the tools of information storage and display. However, bringing new demonstration tools into the classroom has often only occurred in conjunction with some larger movement in educational philosophy that has affected more than mathematics alone.

The history of demonstration tools has been strikingly uneven. A few have been deeply imbedded for millennia, while others have come and gone with little trace. We have already noted the important place of the compass in Chinese thought, and it is well known that the classical geometric drawing instruments in the European tradition are the straightedge and the compass (often referred to as a pair of compasses). The Greek mathematician Euclid, in his *Elements* (ca. 300 BCE), gave priority to constructions based on these instruments. Probing the limits of such constructions (squaring the circle, trisecting the angle, etc.) was a spur to mathematical researchers from antiquity to the 19th century. Indeed, although other instruments were often used for various practical purposes, such uses were long considered illegitimate for mathematical demonstration (Knorr, 1986). Since Euclid served as the basis of geometry instruction in Europe and its colonies for centuries, the straightedge and the compass became regular features of this instruction.

In the 17th century, René Descartes, the great French philosopher and mathematician, strenuously challenged the straightedge-compass tradition, and made free use of more complicated mechanisms for geometric constructions. However, this had little influence on education. The discovery of linkages capable of producing exact straight lines in the 1870s produced a brief flurry of interest among mathematicians, and even prompted some to propose a refashioning of geometry education. In 1895 the mathematician G. B. Halsted unsuccessfully called for the Hart inversor (see Figure 17.4) to be a standard part of every elementary geometry course. Such devices have periodically created excitement among mathematics teachers and teacher educators in more recent years, but they have never become more than an enrichment topic (Kidwell et al., 2008).

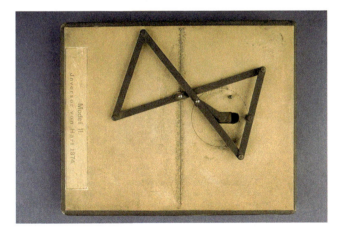

Figure 17.4. The Hart Inversor, a linkage which translates rotary into straight line motion [National Museum of American History collections, gift of Department of Mathematics, University of Michigan. Smithsonian Negative no. 2006–3].

In Europe and North America, there has been a discernable increased use of demonstration tools from the beginning of the 19th century, driven by greater emphasis on using sense data, especially visual, to convey the abstract concepts of mathematics. This has remained a feature, at least in theoretical pronouncements, of much mathematics education to the present day (Bartolini Bussi, Taimina, & Isoda, 2010). The empirical side of the 17th-century scientific revolution appears to have been crucial, with knowledge coming to be understood to depend not only on reason but also on careful sifting of material evidence; induction in addition to deduction.

But although there were some precursors, it was not until the 19th century that this stimulus was widely felt in education. Swiss educator Johann Pestalozzi and his follower Friedrich Froebel were especially influential in bringing material objects into the classroom to be seen or touched by the students. These included objects associated with mathematics, such as geometric solids. Froebel, teaching in Swiss and German towns in the 1830s and 1840s, pioneered the concept of kindergarten for very young children. He recommended organized play with blocks, which would introduce the child to geometric shapes and to arithmetic ideas up to simple fractions. Froebel's ideas spread across Europe and to the USA in the late 19th century (Allen, 1988; Butts, 1966).

One 19th-century educational tool which may have benefited from Froebel's influence was the cube root block, now little remembered. It is based on a method of extracting cube roots based on the binomial expansion of $(a+b)^3$, which can be illustrated with a cube of side $a+b$. (There is a better-known corresponding method for extracting square roots which can be illustrated with a diagram of a square of side $a+b$). Illustrations of this cube can be found in English arithmetic texts from the 17th century (e.g., Recorde, 1632), but it was not until the middle of the 19th century that it became an actual classroom device (see Figure 17.5). With the aim of helping students understand the aforementioned cube root algorithm, scientific

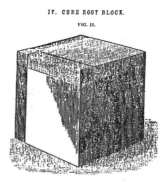

The extraction of the cube root can be explained most easily by the use of the Cube Root Block. In fact, no person who is unacquainted with Algebra or Geometry can know the reason for this rule without the aid of some such illustration.

From *The Teacher's Guide to Illustration: A Manual to Accompany Holbrook's School Apparatus*, (Hartford, 1857), 34.

Figure 17.5. Illustration of a cube root block.

instrument companies in the USA began to produce and market wooden cube root blocks that could be dissected into constituent parts.

These blocks, for advanced arithmetic students, were often advertized with other classroom objects, such as cones for displaying conic sections, and Froebel's blocks for kindergarten children. Diagrams based on the blocks were a staple of school arithmetic textbooks for many years, but the approach had detractors. The cube root block algorithm never gained any favour with engineers and other users of mathematics for practical purposes, since the efficiency of the algorithm is low compared to other methods, such as logarithms or Newton's method. Moreover, how often did mathematical practitioners even need to compute cube roots? By the 1890s many mathematics educators in the USA were campaigning against cube root extraction, but it persisted in the curriculum well into the 20th century. Cube root blocks were still being sold in the 1920s (Kidwell et al., 2008). Since no studies of the effectiveness of the cube root block as a teaching technique are known, it must be judged a demonstration tool of unclear benefit to support an algorithm of dubious value. Nevertheless for a time it was well ensconced in the curriculum.

The end of the 19th century and the beginning of the 20th saw another surge of interest in concrete instructional methods, at both the highest and lowest levels of the curriculum. For advanced instruction this was strongly influenced by a felt need to better align mathematics with science and engineering. In France, the mathematician Émile Borel, concerned that mathematics might lose its place in education due to a public perception that it was useless, called for more practical instruction, including augmenting geometry teaching with surveying exercises. He recommended "laboratoires de mathématiques," which would make many connections with physics (Borel, 1904). In the United Kingdom, the engineer John Perry promoted a more concrete and visual approach to mathematics education, helping to break the unquestioned dominance of formal Euclidean geometry in British education. His influence extended to both Japan (where he worked for a time in the 1870s) and the USA (Brock, 1975; Brock & Price, 1980). In the USA, Perry's most prominent disciple was pure mathematician Eliakim Hastings Moore of the University of Chicago, who championed a "laboratory method" of teaching mathematics at both the secondary and college levels. This involved strong emphasis on developing intuition in the student through physical models, weighing and measuring, and drawing on squared paper (an uncommon classroom item up to that time). Moore saw Perry's ideas as helping students aiming to be scientists and engineers, while at the same time supporting future teachers of mathematics and research mathematicians. His curricular program was briefly significant in the USA, but other than an increased use of graphs in algebra instruction, its long-term stimulus was slight (Roberts, 2001).

Moore was also greatly influenced by the German mathematician Felix Klein, who likewise sought to make mathematics education more supportive of engineering. Klein championed the use of geometric models in classroom instruction. This built on a tradition originating in France in the early 19th century, especially with mathematician Gaspard Monge. Models made of plaster, string, wood, and paper were developed in France and Germany. These went beyond the simple solids of Pestalozzi and Froebel to include hyperboloids and other more advanced structures, all the way to objects at the forefront of mathematical research, such as Riemann surfaces. Some of the string models were even dynamic; that is, they could be manipulated to change shape. With Klein's instigation, German models, mainly of plaster, were manufactured and sold worldwide. Colleges and universities in the USA were among the buyers, but there is little evidence to support extensive classroom use of these models; more likely they were treated more as museum pieces. There were also isolated enthusiasts at the secondary school level in the USA, who enjoyed training students to create geometric models, but their effectiveness is very hard to gauge (Committee on Multi-Sensory Aids, 1945; Kidwell et al., 2008).

Meanwhile in Italy, Maria Montessori inherited Froebel's emphasis on teaching young children through tactile experience, buttressing her theories by appealing to more recent developments in psychology and anthropology. She advised that beginning students be given the opportunity to handle objects of various shapes—such as cylinders of varying heights and diameters—continually. Colored cubes and rods were a central feature of her approach to arithmetic. Montessori schools were opened in Italy and Switzerland. After an initially rapid growth of interest in her work in the USA in the 1910s, her influence declined, in part due to criticism from American educational theorists such as William Heard Kilpatrick of Columbia University (Kramer, 1976; Whitescarver & Cossentino, 2008).

The USA experienced a Montessori revival beginning in the 1950s, and this closely coincided with, and perhaps helped to support, renewed interest in both the USA and Europe in using physical objects specifically in teaching mathematics. Other sources of support were found in the work of educational psychologists whose influence extended well beyond mathematics, such as the Swiss, Jean Piaget, and the Russian, L. S. Vygotsky. Among those in the 1960s who helped popularize what came to be called "manipulatives" in mathematics instruction were the Belgian educator Emile-Georges Cuisenaire, the Egyptian-born British educator Caleb Gattegno, and the Hungarian-born educator Zoltan Dienes, who worked in Great Britain, Australia, Canada, and elsewhere (Jeronnez, 1976; Seymour & Davidson, 2003). This period also saw the rise of the "New Math," a conglomeration of curriculum reform programs initially centred in the USA but eventually extending well beyond. Some would see manipulatives such as Cuisenaire rods as incongruous with the emphasis on axiomatics and abstraction characteristic of many of the New Math programs, although Dienes (1960, 1971), for one, saw no contradiction. In any case, the popularity of certain manipulatives to some extent rose and fell with public perceptions of the New Math as a whole. Nevertheless, while New Math programs often experienced severe backlash, the use of manipulatives never went into total eclipse.

The presence of manipulatives in classrooms during the last 50 years is testified to by the fact that the topic has been an active subject of empirical research from the 1960s to the present (Karshmer & Farsi, 2008; McNeil & Jarvin, 2007; Moyer, 2001; Sowell, 1989). This research has painted a mixed picture of the effectiveness of manipulatives. Although some studies have detected very positive effects, others have found that these effects were negated by poor teaching techniques. Some research even suggested that manipulatives could harm students by burdening them with the problem of "dual representation." According to McNeil and Jarvin (2007), "a given manipulative needs to be represented not only as an object in its own right, but also as a symbol of a mathematical concept or procedure" (p. 313).

The computer, especially as connected to the Internet, makes readily available to students and teachers all of the objects mentioned above, and many more, in virtual form. Whether this will prove to have a significantly more positive influence on the mathematics curriculum than physical models that students can hold in their hands remains to be seen. We will note some recent efforts in this direction in the last section of this chapter.

Tools of Calculation

To the consternation of many mathematicians and mathematics educators, calculation is often considered to be synonymous with mathematics by many members of the general public, so these tools naturally loom large in public discussion of mathematics education. Here we briefly discuss the history of three devices—the abacus, the slide rule, and the calculator—that have had a global impact in mathematics education, as it evolved from mechanical to electronic. It should be noted that the slide rule, though intermediate chronologically, is in no sense intermediate conceptually between the abacus and the calculator. This shows the difficulty of imposing any straightforward conception of linear progress in the use of technology in mathematics education.

The abacus. The abacus depicts numbers by means of beads on wires. It apparently evolved from marks in sand or counters on a board. The device seems to have developed somewhere in the eastern Mediterranean world in antiquity, moved east to Asia, then moved back west via Russia into Europe and thence to the Americas. The transmission to Asia is conjectural, and it is possible that it originated there independently. What is clear is that whereas the abacus became a widely used tool of calculation in China and Japan, without a serious competitor until very recent times, it never attained the same level of popularity in this role in Europe and North America. Instead, in the last-named regions, it was primarily confined to use as a demonstration tool for teaching elementary arithmetic to young children.

The Chinese abacus (*suanpan*) appears to have been in substantial use by 1200 and probably much earlier. Transmission to Japan, seems to have occurred via Korea. The Japanese modification of this instrument (called the *soroban*) was in use by 1600 (Smith, 1958). Although the abacus has been a part of education in both Japan and China for centuries, in the decades after World War II major efforts were undertaken in both nations to modernize and formalize this instruction (Hua, 1987; Shibata, 1994). The device has continued to be part of the mathematics curriculum in many East Asian nations to the present day. In Malaysia, for example, although abacus use in schools declined for a time after handheld calculators became widely available, the abacus (*sempoa* in Malay) has more recently experienced an educational resurgence in connection with an increased emphasis on mental arithmetic (Siang, 2007).

In China and Japan the beads move on vertical wires, but the version of the abacus that became common in Russia featured horizontal wires. This would prove advantageous for using it as a display device for young children, since the teacher could hold the abacus up in front of the class and the beads would remain in place. It was used in Russia for early education until recent decades. The French mathematicians Jean Victor Poncelet encountered the abacus while imprisoned in Russia following Napoleon's invasion of 1812 and introduced it to France on his return. It spread widely across France as a teaching tool in the 19th century (Gouzévitch & Gouzévitch, 1998; Régnier, 2003).

A similar teaching device began to appear in the USA in the 1820s, likely inspired at least in part by the French version. Here it meshed well with the Pestalozzian object-teaching philosophy that was gaining in popularity, and by the 1830s it was being sold under various names, including "numeral frame," by companies catering to the growing education market. These teaching abaci were not without detractors, however, some of whom felt they might even stifle the imagination of the child. They remained as a tool for only the youngest learners of arithmetic (Kidwell et al., 2008). In more recent years, some educators (e.g., Ameis, 2003), apparently reacting to the perceived success of Asian students in mathematics, have advocated more use of the Asian abacus in Western schools.

The slide rule. The slide rule was a direct embodiment of the theory of logarithms pioneered by Scottish mathematician John Napier and English mathematician Henry Briggs in the early 1600s. By marking two straightedges with logarithmic scales and sliding one with respect to the other it was possible to calculate approximate answers to multiplication problems quickly. Even more complicated problems could be handled with sufficient ingenuity, although the fact that the slide rule was an analog instrument meant that it always provided only approximate answers, and thus was not appropriate for most business applications of mathematics or for accounting. Variations involving circular rules were also possible, and both possibilities had been explored by the middle of the 17th century in England. These slide rules were slowly improved over the next century, and became a tool used by engineers, such as James Watt, in the UK. By the early 1800s they had spread to the European continent and to the USA (von Jezierski, 2000).

It was not until the late 19th century that the slide rule became an educational tool, beginning first with colleges featuring an engineering curriculum, such as Rensselaer Polytechnic, the US Military Academy, and the Massachusetts Institute of Technology. In the early 20th century the slide rule began to filter down into the secondary schools, helped by the movement to establish mathematical "laboratories" which emphasized the mathematics of measurement and applications to the physical sciences. Instrument makers were selling slide rules to the high school market by the 1920s and some were also selling oversized models that could be displayed in front of a classroom for all students to see. The slide rule remained a recognized feature, although in most cases not a central one, of many mathematics and science classrooms until the advent of cheap electronic calculators in the 1970s (Kidwell et al., 2008).

The calculator. Unlike the slide-rule, the calculator is fundamentally a digital instrument, which seems to have given it a decided advantage in achieving a place in mathematics instruction. Its fate in the classroom is still being written. European development of mechanical calculators dates from the 17th century, with such notable mathematicians as Pascal and Leibniz prominently involved (Goldstine, 1972). But it was not until the middle of the 19th century that industrial processes were sufficiently advanced to allow construction of calculating devices on a commercial basis, both in Europe and the USA. By the 1920s they had become a standard feature of many office settings. But it appears that it was not until after World War II that they received much consideration as educational devices. In the 1950s there was some minor experimentation in classrooms with mechanical calculators, or mechanical calculators with electrical assistance, but the size of these machines made them inconvenient as personal devices (Kidwell et al., 2008).

The major breakthrough occurred in the 1970s, with the arrival of inexpensive, fully electronic calculators. Initially these calculators were still relatively bulky, and were able to perform little beyond the familiar four operations of arithmetic. But by the 1980s calculators had become readily portable, and were able to compute trigonometric and other transcendental functions and to display graphs, thus far surpassing the functionality of mechanical calculators and slide rules. Classroom use became practical, and although very uneven, soon became widespread enough to create disputes between enthusiasts and detractors. Calculators greatly increased the range of feasible problems that could be given to students, but concern was expressed about the effect on basic arithmetic skills, and doubts were raised about the readiness of teachers to use calculators effectively (Kelly, 2003; Waits & Demana, 2000). By the mid-1990s computer algebra systems (CAS) were available on hand-held devices, leading to further debate. Now, in the 21st century, although the generic name persists, high-end devices referred to as "calculators" in fact provide a huge range of information storage, information display, and demonstration capabilities, in addition to pure calculation (Aldon, 2010; Trouche, 2005). Some controversy has persisted, but in recent years the use of calculators has been increasing around the world in secondary and elementary schools, and at the college level as well.

The Virtual World: The Potential of 21st-Century Technology for Mathematics Education

During the past two decades, pedagogical theories in mathematics education, such as instrumental genesis and semiotic mediation, have placed tools, artefacts, and technology at the centre stage of discussion on mathematics knowledge acquisition (see, e.g., Artigue, 2002; Bartolini Bussi & Mariotti, 2008). Studying the pedagogical potential of technology is a major research field of study in mathematics education (see, e.g., Blume & Heid, 2008; Heid & Blume, 2008). The question arises, regarding the plethora of electronic devices now available to mathematics teachers and students, and the evident integration of these devices into what appears is becoming a comprehensive technology platform: is this something fundamentally new for mathematics education or does it merely provide the means for delivering the services of the older technologies more quickly and efficiently? It would certainly appear that the distinctions made earlier in this chapter among classes of technologies are increasingly irrelevant. The computer can function simultaneously as an information storage device, an information display device, a demonstration device, a super calculator, and much more. In the remainder of this chapter we describe some indications that the new technology environment does indeed provide unprecedented opportunities.

Tools from the past are far from irrelevant to the new environment, since the Web can function as a window to access information on historical mathematical tools instantly. This provides the potential to construct mathematical knowledge via simultaneous attention to the multifarious facets in the evolution of that knowledge, as reflected in the tools, thereby creating a virtual thematic museum of mathematical artefacts. One could, if one wished, virtually go back in time, by constraining students to use only the tools available in a certain era in a specific geographic locale. This powerful capability for integrating history, pedagogy and mathematics opens a vast range of intriguing possibilities in conceptualizing the mathematics curriculum.

Research into integrating the history of mathematical tools with the school mathematics curriculum, by having students visit and study historical mathematical tools via present day accessible technology, has been carried out in teacher education and in mathematics classrooms (Bartolini Bussi et al., 2010; Maschietto & Trouche, 2010). On the one hand, this can assist students to acquire mathematical understanding in a techno-cultural context, which raises the relevance of school mathematics as a part of social development. On the other hand, students can re-visit and re-think (even re-conceptualize) familiar mathematical concepts in an old-meets-new context. This simultaneity may bring about awareness of invariants that constitute the core of abstract mathematical concepts. This looking back to *re*-interpret and *re*-present the mathematics embodied in historical tools somewhat echoes Hans Freudenthal's (1991) idea of mathematization, in which mathematical concepts are re-invented using tools that are more powerful than our predecessors possessed. According to Freudenthal, "children should repeat the learning process of mankind, not as it factually took place but rather as it would have been done if people in the past had known a bit more of what we know now" (p. 48).

There have been substantial recent efforts to study classroom use of historically significant tools, both as originally conceived and in a digital form. Maschietto and Trouche (2010) have revisited the idea of the mathematics laboratory in classroom practice, explicitly citing Borel's early 20th-century proposal. They studied the use of both "old" technology (the mechanical calculator of Blaise Pascal, the abacus) and "new" technology (networked electronic calculators) in such laboratory situations, while exploring notions of good contexts and good teaching practices. Cornell University (USA) has digitized and enhanced its collection of kinematic models, in what they call the Kinematical Models for Design Digital Library (KMDDL). These models (including linkages generating straight lines, mentioned earlier in this chapter), were originally created as physical models in the 1870s by the German engineer Franz Reuleaux. At Cornell they are being used to teach the mathematics of machine design. In the 1990s at the Centre for Research on International Cooperation in Educational Development (CRICED) at Tsukuba University (Japan), there was a rebirth of interest in using mechanical instruments in mathematics instruction, facilitated by LEGO blocks and dynamic geometry software. The project has also made use of e-textbooks to weave together historical books and interactive dynamic simulations. And the University of Modena (Italy) has established a Laboratory of Mathematical Machines, which provides digitizations of familiar mathematical instruments, including the compass. Dynamic simulations are available on the Web as a source for teaching and learning activities with prospective mathematics teachers (Bartolini Bussi et al., 2010).

There are several key research questions for this historical pedagogy. How can this re-invention process be best realized in a pedagogic process? Will the re-invention embody "more" or "less" mathematical knowledge? How can this pedagogical perspective be integrated into the curriculum? We illustrate and discuss an example in geometry.

As just noted, the Laboratory of Mathematics of the University of Modena in Italy holds a large collection of replicated mechanical "geometrical machines" from different historical periods—where by geometrical machine is meant a tool that forces a point to follow a trajectory or to be transformed according to a given law (Bartolini Bussi & Maschietto, 2008). These geometrical machines were re-constructed based on old scientific and technical literature, and after experimentation on their possible pedagogical potential. In the Museum's Web site, beside the pictures of some of the replicated geometrical machines, there are corresponding virtual animations, constructed by dynamic geometry software, showing what the machines do. Such a parallel representation is depicted in Figure 17.6, which shows a replica of a Scheiner pantograph, a device invented in Germany in 1603 by Christoph Scheiner for making a scaled copy of a given figure.

This juxtaposition of old and new technology (wooden craft and virtual craft) provides a good context for implementing historic re-invention pedagogy in the mathematics classroom. Figure 17.7 has four equal rods hinged by adjustable pivots at A, B, C and P with OA=AP and PC=P′C=AB. It is fastened by a pivot at O. Placing a pencil at P (or P′) to trace a figure, a dilated image is obtained at P′ (or P). Note that APCB is a parallelogram, O, P and P′ are collinear, and OP′/OP=OB/OA=constant. Antonini and Martignone (2011) have studied the didactical potential of the pantograph in proof and argumentation for geometrical transformations.

In a mathematics classroom, students can construct a make-shift pantograph using geometry sticks, appropriate fastening pivots and writing implements (Figure 17.8), which can be used as an explorative tool to investigate the geometry of similarity (homothety/dilation). The pivot points of this tool can be readily adjusted, which enables students to access, easily, different ratio variations between the sides of parallelogram ABCP. Since all pivot points are free, students can choose

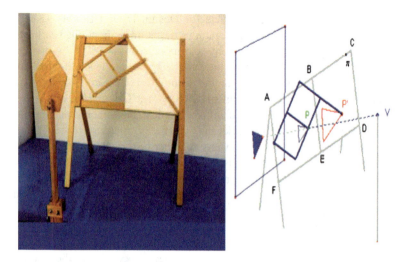

Figure 17.6. A wooden replica of a Scheiner pantograph and a dynamic animation of how it works [Source: http://www.museo.unimo.it/theatrum/macchine_00lab.htm].

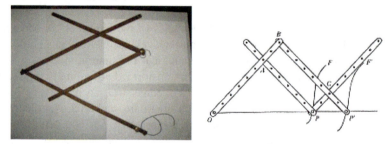

Figure 17.7. A modern day pantograph [Sources: http://www.isaacwunderwood.com/gallery2/displayimage.php?album=4&pos=0 http://www.datavis.ca/milestones/index.php?group=1600s].

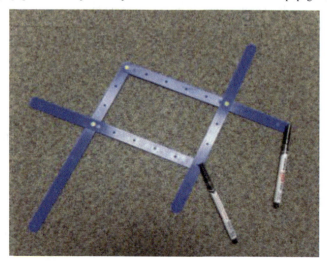

Figure 17.8. A classroom make-shift pantograph constructed using geometry sticks.

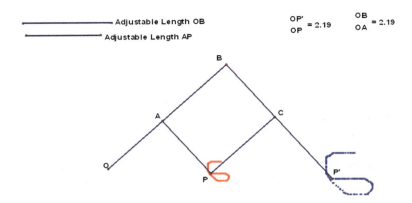

Figure 17.9. A dynamic geometry pantograph constructed in Sketchpad™.

which one to be the fastened one, and which to be where the pens are. Furthermore, the shape of ABCP can be changed to other shapes. These different degrees of freedom of the tool open up a vast pedagogic space for teachers and students.

For more advanced lessons, students can construct dynamic geometry pantographs and use the construction activity to explore the mathematics that can be embodied in a pantograph—an example of which is depicted in Figure 17.9.

For this virtual pantograph, the lengths of OB and AP are adjustable variables and points O and P are free. These features facilitate students experiencing the variation that this virtual tool can offer, providing opportunities for them to discover geometrical properties (Leung, 2008). We have here an example of old and new technologies meeting together in the mathematics curriculum, enabling meaningful mathematics teaching and learning. Such examples suggest that by utilizing the multi-functional nature of the computer, and the connectivity power of the evolving virtual technology, mathematics pedagogy could take on a new paradigm that supports connective and explorative knowledge building in a powerful way. By "connective knowledge building" is meant the ability of teachers and students to (re)construct mathematical knowledge connectively and collectively, and in particular, through the idea of "webbing." Webbing refers to "the presence of a structure that learners can draw upon and reconstruct for support—in ways that they choose as appropriate for their struggle to construct meaning for some mathematics" (Noss & Hoyles, 1996, p. 108).

Thus, webbing can be interpreted as an affordance in the virtual world to facilitate mathematics pedagogy, where connective structures that empower mathematical experience can be built by teachers and students, utilizing multi-functional tools present in the virtual environment. As Web technology advances in terms of speed, accessibility and information content, one can easily surf the Web to connect to information on mathematical artefacts, ancient or new, like those described earlier in this chapter.

The virtual platform can be designed to collect students' perception of mathematical concepts, thus forming a "knowledge database" that serves as a source to connect students' different ways of understanding. This collective understanding via a virtual environment can then be used pedagogically for developing mathematical concepts in the classroom. Leung and Lee (Lee, Wong, & Leung, 2006; Leung & Lee, 2008) have been conducting research on such a platform in an ambient dynamic geometry environment to categorize visually students' perceptions of geometrical concepts. This kind of platform may be extended to become a virtual forum (or community of practice) where teachers and students co-construct mathematical knowledge and even formulate curriculum decisions.

By explorative knowledge building is meant students engaging in explorative activities in specific virtual environments like spreadsheets, dynamic geometry software, computer algebra systems, and other purpose driven software that support mathematics knowledge construction. Students are empowered in these environments to develop tool instrumentation schemes, to discern mathematical patterns and to develop situated discourses. In this connection, Leung (2011) has proposed a framework of *techno-pedagogic task design* that aims to organize and capture trajectories of learning in a technology-rich pedagogical environment by a sequence of progressively inclusive epistemic modes: establishing practice mode, critical discernment mode, and situated discourse mode. This technology-dependent cognitive sequence can empower learners to see mathematics in situated abstract ways and hence enlighten their understanding of traditional mathematics by providing alternative passages to mathematical knowledge (Leung, 2011).

Using the mathematics knowledge embodied in computing technology, teachers and students can potentially experience mathematics in ways that are different from traditional school mathematics curricula. A window is opened through which mathematics teaching and learning might enter into a new epistemological domain, where knowledge becomes both personal and communal, and in which connective and explorative mathematical knowledge becomes vastly more accessible.

How soon or how fully this vast potential might be utilized for mathematics education is a difficult question. The historical examples given earlier in this chapter suggest that we should be cautious about predicting revolutionary changes. Moreover, it is entirely possible that the most profound effects will come not from explicit efforts to design technologies for mathematics education, but rather from the side effects of technologies adopted by the wider society. This has certainly been the case with the book, which did not originate as a special tool of mathematics education, but became ubiquitous both inside and outside mathematics classrooms. And while there is little inherently mathematical about the blackboard, its influence on the mathematics curriculum has been substantial. The computer, with its offshoots and allied technologies, represents an especially intriguing case, and a huge challenge for those who attempt to forecast the future. The computer surely does explicitly embody mathematical concepts and processes (e.g., base two arithmetic), but it does not follow that the primary applications in education will come from this direction, especially as the computer is such a versatile device. As we have indicated, mathematics educators are proposing exciting pedagogical innovations based on the newest technologies, but meanwhile the pace of technological evolution may be changing the overall place of mathematics within education and within society in ways that we cannot yet foresee.

References

Aldon, G. (2010). Handheld calculators between instrument and document. *ZDM—The International Journal on Mathematics Education, 42*, 733–745.

Allen, A. T. (1988). "Let us live with our children": Kindergarten movements in Germany and the United States, 1840–1914. *History of Education Quarterly, 28*, 23–48.

Ameis, J. (2003). The Chinese abacus: A window into standards-based pedagogy. *Mathematics Teaching in the Middle School, 9*(2), 110–114.

Anderson, M. H. (2011). Tablet PCs modernize your lectures. *MAA Focus, 31*(1), 29–30.

Antonini, S., & Martignone, F. (2011, February). *Pantograph for geometrical transformations: An explorative study on argumentation*. Paper presented in WG1, CERME7, Rzeszow, Poland. Retrieved from http://www.cerme7.univ.rzeszow.pl/WG/1/CERME7_WG1_antonini_martignone.pdf.

Artigue, M. (2002). Learning mathematics in a CAS environment: The genesis of a reflection about instrumentation and the dialectics between technical and conceptual work. *International Journal of Computers for Mathematical Learning, 7*, 245–274.

Ascher, M. (2002). *Mathematics elsewhere: An exploration of ideas across cultures*. Princeton, NJ: Princeton University Press.

Ascher, M., & Ascher, R. (1997). *Mathematics of the Incas: Code of the Quipu*. Mineola, NY: Dover Publications.

Ash, K. (2009, January 21). Projecting a better view. *Education Week's Digital Directions, 2*(3), 34–35. http://www.edweek.org/dd/articles/2009/01/21/03project.h02.html.

Bartolini Bussi, M. G., & Mariotti, M. A. (2008). Semiotic mediation in the mathematics classroom: Artifacts and signs after a Vygotskian perspective. In L. English (Ed.), *Handbook of international research in mathematics education* (2nd ed., pp. 746–783). New York, NY: Routledge.

Bartolini Bussi, M. G., & Maschietto, M. (2008). Machines as tools in teacher education. In T. Wood, B. Jaworski, K. Krainer, P. Sullivan, & D. Tirosh (Eds.), *International handbook of mathematics teacher education* (Vol. 2, pp. 183–208). Rotterdam, The Netherlands: Sense Publisher.

Bartolini Bussi, M. G., Taimina, D., & Isoda, M. (2010). Concrete models and dynamic instruments as early technology tools in classrooms at the dawn of ICMI: From Felix Klein to present applications in mathematics classrooms in different parts of the world. *ZDM—The International Journal on Mathematics Education, 42*(1), 19–31.

Berglund, L. (1990). *The secret of Luo Shu: Numerology in Chinese art and architecture*. Södra Sandby, Sweden: Tryckbiten AB.

Blume, G. W., & Heid, M. K. (2008). *Research on technology and the teaching and learning of mathematics* (Cases and perspectives, Vol. 2). Charlotte, NC: Information Age Publishing.

Borel, E. (1904). Les exercices pratiques de mathématiques dans l'enseignement secondaire. In Conference at Musée Pédagogique in Paris, *Gazette des Mathematiciens*, Juillet 2002, *93*, 47–64.

Brock, W. H. (1975). Geometry and the universities: Euclid and his modern rivals, 1860–1901. *History of Education, 4*, 21–35.

Brock, W. H., & Price, M. H. (1980). Squared paper in the nineteenth century: Instrument of science and engineering, and symbol of reform in mathematical education. *Educational Studies in Mathematics, 11*, 365–381.

Burton, W. (1850). *The district school as it was, by one who went to it*. Boston, MA: Phillips, Sampson & Co.

Butts, R. F. (1966). *A cultural history of western education: Its social and intellectual foundations* (2nd ed.). New York, NY: McGraw-Hill.

Cajori, F. (1890). *The teaching and history of mathematics in the United States*. Washington, DC: Government Printing Office.

Clements, M. A., & Ellerton, N. F. (2010, October). *Rewriting the history of mathematics education in North America*. Paper presented to a meeting of the Americas Section of the International Study

Group on the Relations Between History and Pedagogy of Mathematics, Pasadena, California. Retrieved from http://www.hpm-americas.org/wp-content/uploads/2010/12/Clements.pdf.

Coe, M. D. (1993). *Breaking the Maya code*. New York, NY: Thames & Hudson.

Cohen, P. C. (1982). *A calculating people: The spread of numeracy in early America*. Chicago, IL: University of Chicago Press.

Cordasco, F. (1976). *A brief history of education* (Revisedth ed.). Totowa, NJ: Littlefield, Adams, & Co.

Dienes, Z. P. (1960). *Building up mathematics*. London, UK: Hutchinson.

Dienes, Z. P. (1971). An example of the passage from the concrete to the manipulation of formal systems. *Educational Studies in Mathematics, 3*, 337–352.

Dolciani, M. P., Berman, S. L., & Freilich, J. (1965). *Modern algebra: Structure and method, book one*. Boston, MA: Houghton Mifflin Company.

Freilich, J., Berman, S. L., & Johnson, E. P. (1952). *Algebra for problem solving, book one*. Boston, MA: Houghton Mifflin Company.

Freudenthal, H. (1991). *Revisiting mathematics education*. Dordrecht, The Netherlands: Kluwer Academic Publishers.

Goldstine, H. H. (1972). *The computer from Pascal to von Neumann*. Princeton, NJ: Princeton University Press.

Gouzévitch, I., & Gouzévitch, D. (1998). La guerre, la captivité et les mathématiques. *SABIX: Bulletin de la Société des Amis de la Bibliothèque de l'École Polytechnique, 19*, 31–68.

Harel, G., & Wilson, W. S. (2011). The state of high school textbooks. *Notices of the American Mathematical Society, 58*(6), 823–826.

Heid, M. K., & Blume, G. W. (2008). *Research on technology and the teaching and learning of mathematics* (Research syntheses, Vol. 1). Charlotte, NC: Information Age Publishing.

Hobart, M. E., & Schiffman, Z. S. (1998). *Information ages: Literacy, numeracy, and the computer revolution*. Baltimore, MD: Johns Hopkins University Press.

Hua, Y. (1987). *Zhongguo zhusuanshi gao [A survey on the history of the abacus in China]*. China: Beijing.

Jeronnez, L. (1976). Hommage à Georges Cuisenaire. *Mathématique et Pédagogie, 6*, 75–81.

Karshmer, A., & Farsi, D. (2008). Manipulatives in the history of teaching: Fast forward to AutOMathic blocks for the blind. In K. Miesenberger, J. Klaus, W. Zagler, & A. Karshmer (Eds.), *Computers helping people with special needs* (5105th ed., pp. 915–918). Berlin/Heidelberg: Springer.

Kelly, B. (2003). The emergence of technology in mathematics education. In G. M. A. Stanic & J. Kilpatrick (Eds.), *A history of school mathematics* (Vol. 2, pp. 1037–1084). Reston, VA: National Council of Teachers of Mathematics.

Kidwell, P. A., Ackerberg-Hastings, A., & Roberts, D. L. (2008). *Tools of American mathematics teaching 1800–2000*. Baltimore, MD: Johns Hopkins University Press.

Knorr, W. (1986). *The ancient tradition of geometric problems*. Boston, MA: Birkhäuser.

Kramer, R. (1976). *Maria Montessori: A biography*. New York, NY: G. P. Putnam's Sons.

Lee, A., Wong, K. L., & Leung, A. (2006). Developing learning and assessment tasks in a dynamic geometry environment. In *Proceedings of the ICMI 17 Study Conference: Technology revisited* (Part 2, pp. 334–335), Hanoi, Vietnam.

Leung, A. (2008). Dragging in a dynamic geometry environment through the lens of variation. *International Journal of Computers for Mathematical Learning, 13*, 135–157.

Leung, A. (2010). Empowering learning with rich mathematical experience: reflections on a primary lesson on area and perimeter, *International Journal for Mathematics Teaching and Learning* [e-Journal]. http://www.cimt.plymouth.ac.uk/journal/leung.pdf.

Leung, A. (2011). An epistemic model of task design in dynamic geometry environment. *ZDM—The International Journal on Mathematics Education, 43*, 325–336.

Leung, A., & Lee, A. (2008, July). *Variational tasks in dynamic geometry environment*. Paper presented at the Topic Study Group 34: Research and development in task design and analysis, ICME 11, Monterrey, Mexico.

Maschietto, M., & Trouche, L. (2010). Mathematics learning and tools from theoretical, historical and practical points of view: The productive notion of mathematics laboratories. *ZDM—The International Journal on Mathematics Education, 42*(1), 33–41.

McNeil, N. M., & Jarvin, L. (2007). When theories don't add up: Disentangling the manipulatives debate. *Theory into Practice, 46*(4), 309–316.

Moyer, P. S. (2001). Are we having fun yet? How teachers use manipulatives to teach mathematics. *Educational Studies in Mathematics, 47*(2), 175–197.

Noss, R., & Hoyles, C. (1996). *Windows on mathematical meanings*. Dordrecht, The Netherlands: Kluwer Academic Publishers.

Plofker, K. (2008). *Mathematics in India*. Princeton, NJ: Princeton University Press.

Recorde, R. (1632). *The grounds of arts*. London, UK: Tho. Harper.

Régnier, J.-C. (2003). Le boulier-numérateur de Marie-Pape Carpantier. *Bulletin de l'Association des Professeurs de Mathématiques de l'Enseignement Public, 447*, 457–471.

Roberts, D. L. (2001). E. H. Moore's early twentieth-century program for reform in mathematics education. *American Mathematical Monthly, 108*, 689–696.

Seymour, D., & Davidson, P. S. (2003). A history of nontextbook materials. In G. M. A. Stanic & J. Kilpatrick (Eds.), *A history of school mathematics* (Vol. 2, pp. 989–1035). Reston, VA: National Council of Teachers of Mathematics.

Shibata, R. (1994). Computation in Japan from the Edo era to today: Historical reflection on the teaching of mental computation. In R. E. Reys & N. Nohda (Eds.), *Computational alternatives for the twenty-first century: Crosscultural perspectives from Japan and the United States* (pp. 14–18). Reston, VA: National Council of Teachers of Mathematics.

Siang, K. T. (2007). *The modality factor in two approaches of abacus-based calculation and its effects on mental arithmetic and school mathematics achievements* (PhD dissertation). Universiti Sains Malaysia, Malaysia.

Smith, D. E. (1958). *History of mathematics*. New York, NY: Dover Publications.

Smith, H. J., Higgins, S., Wall, K., & Miller, J. (2005). Interactive whiteboards: Boon or bandwagon? A critical review of the literature. *Journal of Computer Assisted Learning, 21*, 91–101.

Sowell, E. J. (1989). Effects of manipulative materials in mathematics instruction. *Journal for Research in Mathematics Education, 20*(5), 498–505.

The Committee on Multi-Sensory Aids of the National Council of Teachers of Mathematics. (1945). *Multi-sensory aids in the teaching of mathematics*. New York, NY: Teachers College, Columbia University.

Trouche, L. (2005). Instrumental genesis, individual and social aspects. In D. Guin, K. Ruthven, & L. Trouche (Eds.), *The didactical challenge of symbolic calculators: Turning a computational device into a mathematical instrument* (pp. 197–230). New York, NY: Springer.

Tyack, D. B. (1974). *The one best system: A history of American urban education*. Cambridge, MA: Harvard University Press.

von Jezierski, D. (2000). *Slide rules: A journey through three centuries* (Rodger Shepherd, Trans.). Mendham, NJ: Astragal Press.

Waits, B. K., & Demana, F. (2000). Calculators in mathematics teaching and learning: Past, present, and future. In M. J. Burke & F. R. Curcio (Eds.), *Learning mathematics for a new century* (pp. 51–66). Reston, VA: National Council of Teachers of Mathematics.

Whitescarver, K., & Cossentino, J. (2008). Montessori and the mainstream: A century of reform on the margins. *Teachers College Record, 110*(12), 2571–2600.

Wood, R., & Ashfield, J. (2008). The use of the interactive whiteboard for creative teaching and learning in literacy and mathematics: A case study. *British Journal of Educational Technology, 39*(1), 84–96.

Chapter 18
Modelling with Mathematics and Technologies

Julian Williams and Merrilyn Goos

Abstract This chapter seeks to provide an integrating theoretical framework for understanding the somewhat disparate and disconnected literatures on "modelling" and "technology" in mathematics education research. From a cultural–historical activity theory, neo-Vygtoskian perspective, mathematical modelling must be seen as embedded within an indivisible, molar "whole" unit of "activity." This notion situates "technology"—and mathematics, also—as an essential part or "moment" of the whole activity, alongside other mediational means; thus it can only be fully understood in relation to all the other moments. For instance, we need to understand mathematics and technology in relation to the developmental needs and hence the subjectivity and "personalities" of the learners. But, then, also seeing learning as joint teaching–learning activity implies the necessity of understanding the relation of these also to the teachers, and to the wider institutional and professional and political contexts, invoking curriculum and assessment, pedagogy and teacher development, and so on. Historically, activity has repeatedly fused mathematics and technology, whether in academe or in industry: this provides opportunities, but also problems for mathematics education. We illustrate this perspective through two case studies where the mathematical-technologies are salient (spreadsheets, the number line, and CAS), which implicate some of these wider factors, and which broaden the traditional view of technology in social context.

J. Williams (✉)
The University of Manchester, Manchester, UK
e-mail: Julian.williams@manchester.ac.uk

M. Goos
The University of Queensland, St. Lucia, QLD, Australia

Introduction

Most experienced mathematics educators probably believe they know what is meant by mathematical modelling and how this relates to problem solving, and perhaps even how it situates or is mediated by "technology." Yet, Lesh, and Zawojeski (2007) reported that there was no consensus on this issue among authors and we agree with that reading of the wider literature.

The literature on mathematical modelling is already huge, and is growing in extent, touching on almost the whole of mathematics education and its concerns: epistemology, learning sciences, curriculum, pedagogy, assessment, teacher development, innovation and change, and so on. Several attempts to help the newcomer to this literature must be mentioned. For example, the review by Lesh and Zawojewski (2007) addressed modelling with problem solving, and that by Kaiser and Sriraman (2006), among others, provided an overview and categorization of perspectives on modelling, especially as related to the literature from the International Conference on Teaching Mathematics and its Applications (ICTMA) (Kaiser, Blum, Ferri, & Stillman, 2011).

Blum, Galbraith, Henn, and Niss (2007) set out to present a state-of-the-art review on modelling in mathematics education, but their volume revealed even less convergence, suggesting the diversity of views is ever growing. There are those who see modelling as a new name for Deweyan "inquiry" (Confrey & Maloney, 2007), those from the Freudenthal tradition who see modelling as an emergent, dialectical process (e.g., Gravemeier, Lehrer, van Oers, &Verschaffel, 2002, whose approach is close in spirit to that of this chapter), and others who more or less define modelling "traditionally" through its heuristics and the modelling process, often schematized in a cyclic diagram. Those in this third category are generally guided by modelling as a metacognitive process, as a set of coordinated heuristics in the fashion of Polya (1957), as a tool for categorizing competences and thus assessment of various kinds, as an analytical tool for examining learning, and/or as a guide to teacher intervention. But then there is also a significant literature in the learning sciences, much of which is inspired as we are by cultural historical literatures, including Freudenthal, but also by Vygotskian activity perspectives (typified by authors such as Cobb, van Oers, and Gravemeier).

We will consequently certainly not try here to provide a state-of-the-art summary of mathematical modelling as a whole, but rather begin to develop an integrative, theoretical perspective (with examples and "cases" to help make "sense") that we believe can help conceptualize this field, particularly as regards the topic of this section, that is to say, "technology." Methodologically, because this approach aims for generative insight, it involves "theory and case study" of the phenomenon rather than "sampling and survey." This chapter, in this section of the *Handbook*, will mainly provide theory (with exemplification) while those that follow will likely provide deep "case studies."

We will take a risk here and define a model and modelling in a broad way that builds on a definition put forward by Lesh but also includes most perspectives discussed by Lesh, Blum, Kaiser and others: actually it was inspired by Wartofsky (1979). "A model (or modelling) is a means of seeing a situation (the target domain, sometimes called the 'real') through the lens of another situation (the source domain or 'model,' sometimes the 'mathematics')." Then modelling activity will be "activity" (a concept to be developed below) that involves modelling in a significant way.

Note that this may include all forms of re-presentation, akin to the metaphorical use of language, for instance, and tends to be "two-way", as most mathematical modellers say. Thus, just as the brain can be said to be modelled as a "computer", in computer science the computer is modelled as a brain, and our modern cultural model of computers and brains actually emerges from this two-way dialectic. For an introduction to "cultural models" see Holland and Quinn (1987), and on metaphorical modelling, see Black (1962) and Lakoff and Johnson (1980). This view of course also includes the representation of mathematics by physical models (e.g., counting-beads or the abacus as a model for arithmetic). It even includes much pure mathematical work, even proof, as invoking "modelling" (Hanna & Jahnke, 2007). Importantly, it allows for emergent modelling, and modelling within mathematics, in the sense of those such as Gravemeier and Cobb (see, e.g., Cobb, Yackel, & McClain 2000; Gravemeier, 2007; Gravemeier et al., 2002; Van Oers, 2002) as well as modelling in real problem solving in the continental European and British "trends" (e.g., Blum et al., 2007; Burkhardt, 1981; Pollak, 1969).

Similarly, the term "technology" is often taken for granted and is ill-defined and ill-theorized in the mathematics education literature, though most who address this issue argue that new technologies can be a powerful aid to enriching modelling and provide many examples and innovative approaches in mathematics education. An approach we will find fruitful comes from the analysis of mathematics in the workplace, where mathematics is found embedded or black-boxed in technological artefacts and tools, and mathematical competence may be better described as a form of techno-mathematics or techno-mathematical literacy put forward by Hoyles, Kent, and Noss (e.g., Kent, Guile, Hoyles, & Bakker 2007; Noss, Bakker, Hoyles, & Kent 2007; Noss & Hoyles, 2011).

We can define technological knowledge broadly as practical or scientific "knowledge of tools, machines, techniques, crafts, systems or methods of organization in order to solve problems" (a Wikipedia definition). Thus, technology includes the instruments, techniques and organisation that often embed mathematics "materially" in tools and methods involved in practical activity. In a sense, the "technology" available in a given context is a combination of the tools and the know-how to use them; these may embed the "ideal" mathematics in various forms, as a pair of compasses embeds the mathematics of "locus" of a circle. We will argue that mathematics in practice is always mediated by such technology, and indeed generally becomes fused with technology through such practice (such an argument was attributed by Vygotsky and others to Spinoza, who suggested that in a deep sense a circle really *is* that which is made by a pair of compasses or the equivalent).

While the literature on modelling and technology has to date emphasized the use of technical *instruments*—usually computer technology—in mathematical modelling, it usually sees the infrastructure including the "forms of organization" in schooling as something separate, a matter of learning and assessment, or pedagogy and teacher education, etc., rather than part of the technology. We will refer to these aspects as part of *educational technology*, or the technology of the industry we call "schooling" or "academe", which the modelling literature has become increasingly concerned with in recent years. In recent volumes of ICTMA proceedings one finds increasing concern for these aspects of modelling: teaching, teacher education, organisation of assessment, etc. These all centrally confront educational technology, the institution of schooling, and even politics of assessment and cultural reproduction.

In this situation, we seek to develop a theoretical perspective integrating modelling and technology in its educational, *essentially social and cultural–historical*, context. We aim thereby to help researchers to see the role of technology and mathematical modelling within activity "as a whole." We try to see how they relate to the development of youth, and to see how they essentially relate to educational institutions and systems in wider contexts. We suspect that the perspective of this chapter might challenge many readers from the field of mathematical modelling. Therefore, we will provide some examples, so as to make our proposed perspective more concrete, and perhaps more palatable.

Thus, if we provoke some to see "modelling" and "technology" in a new, broader theoretical perspective we will have succeeded in our aim. Language and mathematics, for instance, in this view, could be understood as the supreme modelling tools (Bruner, 1960), while "writing/inscribing", "sitting in rows in classrooms and copying the scribe," and later "paper-and-pencil mathematics" were perhaps historically humanity's most important technological evolutions in mathematics education—and still seem even today to be remarkably resilient.

Cultural–Historical Perspectives on Modelling and Technology

We want to conceive "mathematical modelling" as a kind of "activity", in the activity-theoretical sense. We draw on the revolutionary thinking of Vygotsky—said to be the Mozart of educational psychology—and his followers and contemporaries, especially Leontiev and Bakhtin, and those more modern, such as Cole, Engeström, and Wertsch (see the review by Roth and Lee, 2007). The unit of cultural life is "activity", prototypically that of culturally-historically situated and mediated "human labor." Labor and activity are understood to be constituted socially by a collective of joint actions on "objects", with the goal to produce previously idealized (and so planned, envisaged, initially "ideal") outcomes that fulfil a human "need." The "motive" and the "object" of activity ensure that activity is meaningful, and integrate both emotional and cognitive aspects. Activity is always mediated by the cultural artefacts that have been produced by prior generations of cultural production. Thus, mathematical work is mediated by artefacts that were produced historically by "old" mathematical technologies, and in turn produce new artefacts that embed this mathematical work in new ways. Thus, for example, one sees on the most modern computer screen an icon that looks like a pair of scissors for "cutting", a brush for "pasting," and a pair of compasses for constructing a circle.

We would like to be able to take Vygotsky's legacy—which we will call cultural–historical activity theory or CHAT—for granted. But, although Vygotsky (usually 1978, 1986) is widely cited and "well-known" in the educational literature, even in mathematical modelling literature, this whole corpus of activity theory seems to be often treated somewhat simplistically or superficially, and sometimes degraded to the trivial (there are certainly exceptions such as Bartolini Bussi, van Oers, Cobb and colleagues, etc.). Yes—Vygotsky thought that intellectual functions arise on the social plane first, and the intra-mental plane second; so, yes, the sociality of the classroom is fundamental to learning–teaching activity. But, even if he was inconsistent, so also thought Piaget, if we read his later work on children's development of logic with any care. Yes, Vygotsky explained that internalization was of fundamental importance to development, and revealed some of its essential transformations. But activity theory has much more to offer, especially regarding educational psychology, culture, history, technology, and even modelling.

For Vygotsky, the task was to formulate an educational, social-psychology, along dialectical materialist principles. This is indeed a social or cultural psychology; it invoked Marx at least as the founder of the concept of sociology and social practice in its modern sense. Thus, when Vygotsky referred to scientific concepts (sometimes translated as "academic" concepts) in contrast to "everyday" concepts, he was pointing to the specific cultural–historical, and even institutional conditions in which academe grew. Schools and academies were the source of a specific and very formal-abstract way of practising, talking and thinking that he contrasted with the "everyday" language and work of production and consumption. The leisured classes in academies escaped the immediate concerns of the poor populace (for a fascinating account of leisure and academic cultures, see the new edition of *Crest of the Peacock*, Joseph, 2010). This allowed the academy to engage in lengthy periods of scientific study, to develop and explore formal concepts and codes, and so uncover the scientific essence of things that was not superficially visible or tied to everyday practice and its associated pragmatic language use (see also Bernstein, 2000).

But, said Vygotsky (1986) and Leontiev (1981), let it be noted how this kind of academic study can lead to teaching that is excessively "verbal" and indeed "senseless" to learners. Only by "ascending to the concrete" can these academic concepts become "true", scientific concepts (for more on this theme, see Blunden's preface to Hegel in Wallace, 2008). Only through the resolution of the dialectical contradiction between everyday and academic practices can the truly scientific-yet-practical conceptions (and so new more advanced forms of social practice) emerge. As we perceive the sun "going down" in a glorious blaze of pink and orange over the blackening ocean horizon, we might still conceptualize this experience in its "academic" scientific model, and appreciate that the sun is not moving, but rather the earth is rotating, and that the light from the sun is not changing much, but rather the depth of atmosphere it must penetrate is slowly changing, leading to parts of the spectrum (blue, indigo and violet actually) being more absorbed than when the incidence is normal (thus having a more than usual proportion of red, orange and yellow). Thus, one might integrate subjective, concrete experience of the everyday with academic physics learning of planetary motion and light, and achieve a synthesis of "scientific" analysis and concrete, subjective, embodied grasp of this experience. The subjective experience gives "sense" to the academic theory and concepts; yet, the physics extends experience and potentially allows one to "see" beyond the immediate. Because it penetrates deeper into the objective reality, it tells one that the experience of sunset would be different if one were observing this phenomenon on Mars or the moon; it extends the imagination of reality far beyond the immediate perceptions and surface knowledge of the "everyday."

When the first moon-landers conducted the experiment of dropping a feather and a spanner simultaneously, they knew and we knew, in a scientific, abstract-formal way, that the two should, against all intuitive, everyday experience, fall together. This is why we watched this experiment and perceived this theoretical knowledge with such joy: we "saw" it for the first time and made this scientific knowledge both cognitively and intuitively, practically "true," in Vygotsky's (Hegelian) sense.

This then is what "modelling" means in its most general, scientific activity-theoretical, sense and this implicates what appropriate technology might do for the construction of true, scientific concepts. According to Davydov (1990), mathematics has a special role in this process: mathematics provides the formal language that distances a model theoretically from its everyday content, and allows a domain of investigation where everyday intuition can be helpfully set aside. The scientific essence of a situation or task can thus be investigated without—for the moment—the interference of the surface, and potentially dangerously misleading contents. Thus the mathematical model of the falling spanner/feather may be given by a simple table of data, or a set of related equations or their graphs: $dV/dT = g$; $V = gT$; $S = \frac{1}{2} gT^2$, which in turn relate to the similar model for the parallel situation on earth, with the appropriate modification of g. But, then, the model works less well here, where we often require a modification such as $dV/dT = g - f(V)$ or the like, allowing us—if we have the mathematical technologies to solve such equations—to explain why the feather and spanner fall differently here. Thus mathematical-technologies provide the means for modelling in problem solving in just the way that Vygotsky's highest level of scientific, or "theoretical" thinking specifies, though Vygtosky most often used formal language as the technology of choice in his own examples.

Notice in this developing formulation that the term "mathematics" is here and there substituted by "mathematical-technologies"—we could have said technomathematics which is not far off in meaning (Noss et al., 2007). But also in some cases we might say just "mathematics", as if mathematics itself *is* the technology for solving the problem. The danger is that we forget that mathematics is always mediated by the technology, even though in the most extreme case this is, as in Erdos' fine formula for pure mathematical activity, just "paper + pencil + coffee = mathematics." To this we will shortly add the educational technology, which often remains invisible in the accounts of mathematics in schools and universities.

Davydov (1990), in particular, developed the mathematical side of Vygotsky's argument, claiming that the goal of mathematics education should be to teach theoretical thinking to all children as the central goal of schooling. He believed that the gifts that talented mathematicians demonstrated in Krutetskii's (1976) studies were exactly those of good "theoretical thinking" in mathematics, available potentially to all; and Davydov's work went some way to showing this.

An example: Wason's reasoning task has come to be widely known in the psychology literature. It involves deciding which cards to turn over to test a hypothesis. Each card is said to have a number on one side, and a letter on the other side. The hypothesis to be tested is: "Every vowel has an even number on the reverse side." Which cards, out of "*A*," "*D*," "4," and "7", *must* be turned over to check if this hypothesis is true for all these cards? Very few adults, even those with training in mathematics and science, can answer this question as put (though when presented in more obvious everyday contexts its equivalent proves much easier.) Why is this such a difficult problem? One reason, we suspect, is that few apply a mathematical model to the problem. The hypothesis has the form "X implies Y", and its truth table is the same as Not [X and Not (Y)] which is always true unless both X is true (i.e., the letter is a vowel, e.g., "*A*") and Y is false (i.e., the number is not even, e.g., "7"). Those that don't produce such an argument, then, either do not know logic, do not consider mathematical modelling with truth tables relevant to logic, or are not disposed to use this knowledge in such a task—although it must be admitted that the problem can be solved perhaps more easily analogically, especially by those who have been taught empirical scientific methods for testing hypotheses; however, our solution here is the most powerful, formal, mathematical solution to this general class of problems, and arguably underpins the whole scientific logic of empirical hypothesis testing.

But let us look a bit closer—we have addressed the notion of scientific conceptions, as this pertains to the advancement of society and culture, but not really its developmental, psychological content in schooling activity. As Engeström (1991) explained, Vygotsky and Leontiev understood that schooling was an artificial institutional activity that always tended towards empty, pre-conceptual, or pseudo-conceptual "verbalism." Yet this emptying of everyday knowledge is also what makes academia essential for the specialist development of academic, scientific concepts. Thus, the social context of school is apparently historically essential, but always dangerous: what is the solution to this contradiction? In practice, the answer to this is that school must always be directed to real, problematic situations. Vygotsky and Leontiev's experiments, and Davydov's curriculum, were always directed to tough problems, just beyond the immediate grasp of the learner, in a zone of proximal development (hereafter "ZPD") where problems required the new conceptual tools or signs that the teacher (or other more advanced peers, or even research and study perhaps) could offer. Much of the best in the mathematical modelling literature and practice over the last half century has been in this mould. Thus, we conclude, new mathematics should be taught in such a zone of proximal development, where the mathematics is necessary for the learner to solve genuinely engaging, problematic, "authentic" and "meaningful" tasks (thankfully terms common in the modelling literature). This, then, is what learning through mathematical modelling should mean.

Technology may allow, however, an expanded ZPD in various ways, as case studies in the literature show. Technological instruments that embed mathematics include calculators of all kinds (from times-tables and Napier's bones to electronic and algebraic calculators and computers) that can make historically-produced mathematics "present" in all kinds of learning–teaching activity. The usual argument is that all learners might then find some task that truly motivates them, but also one that becomes accessible.

Implicit in this view is the consideration of the learner as engaging in "activity", that is defined as joint, collective activity on "objects" with substantial social "motives." In activity theory, schooling is considered to be activity in which the students may engage to please the teacher, to pass examinations, and so on, and so dangerously cut off from socially important and useful adult motives. But if the curriculum is properly directed and managed, the activity has a potential for a more advanced motive: thus, Leontiev explains, a student studying a history book, if told that it is no longer on the syllabus, may throw it aside in disgust—in which case they are clearly motivated by schooling, and examinations. But they may, perhaps, put the book aside reluctantly, or decide to read it anyway, perhaps out of a more developed "interest." In this case Leontiev considers the student to be developing adult motives, interests and capabilities—see Black, Williams, Hernandez-Martinez, Davis, and Wake (2010) for a fuller discussion. The most advanced theoretical thinking which arises in activity, then, is motivated by highly adult motives, to understand the deepest challenges of the scientific and social world. In this view, mathematical modelling is not just "intellectual" but involves social motives, affect, passion, and dispositions to act theoretically on the world.

This, then, is what mathematical modelling means, at least for adolescents (Davydov argues that it remains true for the whole of schooling after the age of seven). Or rather, we argue, this is what it might ideally mean; the implications for educating and developing youth for the school curriculum, and for pedagogy, are quite profound, we think. It involves viewing mathematics as the soft side of technology (in the sense of a semiotic tool) as well as a real theoretical world of its own, but one which is made concrete and material through the use of mathematical-technologies in socially meaningful activity. This view will be recognizable by those regarded as being in the emancipatory, critical trend in mathematical modelling and mathematics education generally. To our knowledge the literature recognizes only one serious critic of this position—Badiou argues that it is the mathematics that is "material" and the "real world" is that of "appearance." We will leave this philosophy to one side—but see Brown (2011).

But then, there are many social and political reasons why this ideal vision may not be realizable or realistic in practice: we discuss some of these below (and see Williams, 2011). We claim only that such an ideal view can provide us with a basis from which to examine and critique practice.

Reviews of Research on Problem Solving and Modelling from an Activity Perspective

The modern problem-solving literature in mathematics education really began with Polya (1957), and became a researched endeavour in the modern sense with Schoenfeld [see his review, Schoenfeld (1992)]. Research on modelling then followed this pattern: modelling being guided by heuristics that may make applied/ real problems accessible, while affective issues arise from the social context and context of curriculum. The whole genre of research and curriculum development in ICTMA conferences has represented this development well. Recent conference proceedings from, say, ICTMA-13 and ICTMA-14, offer a history and bibliography— see, e.g., Kaiser et al. (2011), Lesh, Galbraith, Haines, and Hurford (2010).

Rigorous educational research was slow to catch up with practice, but Schoenfeld's (1992) review of educational research concluded that problem-solving strategies must be made concrete in specific classes of problems to become intelligible, and so of any practical value in problem solving. Very general heuristics were also believed not to be instrumentally useful to problem solvers in the flow of practice, but might be more salient in metacognitive reflection on problem solving with classes of problems. Schoenfeld (after Lampert, 1990) also raised the issue of beliefs about the nature of problem solving, and the hidden curriculum of problem solving: like, for example, the belief that a "mathematics problem" is one that has one answer, one best method, and can usually be completed alone without lengthy working (thus revealing how significant is the institutional aspect of schooling, the educational technology). Bartolini Bussi (1998) raised this also in her development of substantial, culturally-historically based mathematical project practices in classrooms, such as the exploration of perspective in history and art. This approach is typical of many in Italy (such as within Boero's group and that of Arzarello) and elsewhere in ethnomathematics and the history of mathematics traditions. In the case of the Italians, this is usually done explicitly as part of an attempt to make mathematics classrooms social and culturally "mathematical", following a Vygotskian perspective; texts, tools and technologies, often in historical contexts, have an important place.

So, we argue the traditional genre of research may make a crucial mistake in isolating the "modelling processes or heuristics" for research and evaluation, much less teaching: removing processes from the substantive mathematics on the one side and the contexts of practical activity in which they make "sense" on the other, may leave the metacognitive aspect high-and-dry as a new mathematical "verbalism." As we argued in our previous section, the actual mathematics provides a language for theoretical thinking, a crucial "point" of schooling in the development of the learner. But concretely, heuristics like "set up a simple model" may be too general to mean much except through the study of specific mathematical theory on one hand and a space of useful activity contexts on the other. Start with a simple, linear function as a model for a relationship before being more "realistic with a non-linear function" makes lots of sense only when it is attached to practical experience in activity. However, this concretization of the general heuristic of "choosing a simple model first" implies a certain depth of understanding and expertise of the mathematics of functions themselves.

Thus the relation of heuristic and mathematics with the context, or contextual range, is also pertinent: modelling in physics in general, and kinematics in particular, is perhaps rather special and even ideal for certain pedagogic purposes. But this is very different from modelling the economy, in which even basic constructs of money supply are disputed. Additionally, this is all crucially sensitive to the technical and cultural tools at hand; the way that processes become objects (reification through automation) has a long history in activity theory itself (Leontiev, 1978) but has entered science and mathematics education through work by Latour (1987) and Sfard (1998, 2008). The dangers involved are that conscious awareness of what is hidden in the black box may become crucial at certain moments—see, for example, the literature on breakdown, but also Sfard's work, and Strässer (2007).

As we also argued, the "context" may provide a societal need, and so a "motive" that allows school study to expand beyond the traditional confines of "schooling" as an activity, because it can provide a social motivation for the student, especially but not solely the adolescent student (see, e.g., Engeström, 1991; Ryan & Williams, 2007). As such, the kind of problem solving or modelling research which isolates heuristics, while making sense in its time, represents a serious limitation in terms of understanding modelling activity within the whole mathematics-educational developmental process. It elevates metacognition but detaches it from the context and the affective (i.e., the motives and emotions).

More recently, Lesh and Zawojeski (2007) similarly summarized the field of research and called for another paradigm shift: based on Lesh and Doerr (2003), they proposed a new way of implementing modelling activity, one which incorporated traditional problem solving but engaged with a broader class of open, engineering- or design-type activities. These invoke complexity, fuzzy problems and can confront instability and inconsistency, which they regard as an essential component of modern life. The argument is that problem solving in practice, as revealed by anthropological studies of situated cognition, for instance, show that real problem solving in practice is unlike the most "realistic" and "authentic" school problems (e.g., Lave, 1988). Furthermore, they suggested that the engagement of students—following the social learning perspective of Lave and Wenger (1991) and Wenger (1998)—requires that students engage in learning via "communities of practice": arguably very difficult to simulate and perhaps impossible to realize in schooling institutions (but see studies in Watson and Winbourne, 2007).

Then there is the Freudenthal tradition which has emerged in (mainly and originally) Dutch schools, influenced by cultural–historical theory: this genre of developmental research makes explicit that the structure of the mathematics at issue is crucial: the point is to provide contexts and problems that are "realistic" (i.e., experientially real to the learners and so engaging) but which "beg to be organized" with the appropriate mathematics to be learnt (Freudenthal, 1983; Streefland, 1991; Treffers, 1987). The emphasis here is most obviously appropriate in the early years, and has the virtue of proven realizability—ecological validity. From our perspective, the notion of "realistic" is about developing "activity" in a schooling context that engages learners: inherent in this is the contradiction inherent in all schooling that tends to get cut off from "life" (Engeström, 1991; Williams, 2011). The question of societal motivation especially during adolescence seems underplayed in the Freudenthal perspective (though there are clear signs in Freudenthal-inspired practice that this has a place, as witness their various texts and materials). The argument laid at the door of socio-cultural theory by Cobb (2007) is worth considering here, i.e., there may indeed be too much "internalisation" and not enough "emergence."

In the next section we will look at a case of modelling with the empty number line in a social context where school mathematics was deployed in trying to understand a workplace mathematical practice (for more examples see Wake, 2007).

Modelling the Workplace with College Mathematics: An Illustration

In this mini case study we explore the relation between technology, mathematical modelling and education in an expansive setting. The aim is to illustrate modelling-technologies in activity as a whole, in particular how they are both shaped by and are shaping the workplace "knowledge" and the educational experience of the visitor.

Williams and Wake (2007a, 2007b) described an engineer called Dan who was trying to explain a spreadsheet formula to a researcher and two students who were visiting his plant. The formula is designed to compute an estimate of the gas a worker would need to order for the plant to use over the night shift. It is important he gets this right, or as near as possible, since there will be penalty charges from the gas supplier for drawing more or less than the amount ordered. The mysterious formula is shown in Figure 18.1.

The formula is based on a forward projection of how much gas was used (the difference between the 1st and 2nd integrating readings, taken at times which are T2 apart) in the last period of the day before the worker goes off shift, on the assumption that the rate of consumption overnight (a period of time T4) will be the same (a crucial assumption that only became clear later). A simple enough mathematical model ... it therefore uses two "readings" to calculate the rate of consumption, then multiplies the rate of consumption by the time period remaining for the shift. Here we have a not untypical mathematical-technology model in daily use, that had been produced quite some time before by Dan, the engineer, and one that is shaped by the history of workplace technology in the sense of its instruments, but also its form of organization (the times of day, etc.). But the formula is so cluttered—by the "everyday" signs that connected the formula to "practice"—that the mathematics, and the theoretical thinking behind it, are opaque to the visiting students (and the research team, and indeed to the workers themselves, and even its author!).

Dan feels obliged to explain: in order to do so he sketches a timeline, an intuitive model but an excellent pedagogical choice (Figure 18.2). He marks in the salient times on the line, then starts to mark the gas readings at each pertinent moment in time; the number line thus emerges in his explanation as a double number line. At this point the light dawns on the researcher (and the reader, perhaps?), that there is

---

$\{\{\{\{2^{nd}\text{IntegratingReading} - 0600\text{IntegratingReading}\} +\{\{\{ 2^{nd}\text{IntegratingReading}\} - \{1^{st}\text{IntegratingReading}\}/T2\}*TIME4\}\}/3.6*CALCV*1000000/29.3071\}$

Figure 18.1. Dan's formula for estimating the gas needed overnight (adapted from Williams and Wake, 2007a, 2007b).

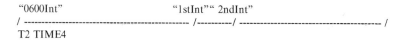

Figure 18.2. The double number line sketch of gas used (*above the line*) and times elapsed between 0600 and the same time next day (*below the line*) (adapted from Williams and Wake, 2007a, 2007b)

an assumption of linearity, and the double number line represents an appropriate ratio model. For us outsiders this linearity was counter-intuitive, as we expected gas consumption might decline when the workers go off shift for the night.

In a later episode the researcher was able to recapitulate the explanation Dan gave in a discussion in which she made sure the students "followed" the argument. The students commented, and we too found this interesting, that the assumption of linearity had not been mentioned by Dan, but they had been left to discover this for themselves. Presumably in his working life this fact of work-process knowledge was too obvious to need explanation. In fact, much mathematics that has been produced historically disappears like this in artefacts and remains hidden from conscious attention there, unexposed until for some reason there is a "breakdown" (e.g., in nursing and drug dosages—see Hoyles, Noss, & Pozzi 2001). The breakdown arose here because of our research "archaeology"—digging up this formula and seeking to understand it.

Despite the workers there present, the key element of the model lies implicit, too obvious in the practice to be spoken of. Thus mathematics, as Strässer (2000, 2007) has pointed out, disappears from conscious attention in the workplace, but actually is hidden everywhere in technological artefacts, in the work process, and of course also within mathematics itself. In activity theory this feature of the automation of processes is known as fossilization, or sometimes crystallization: we see it also in the artefacts of "schooling" (in the curriculum, in assessment, etc.) that make curriculum development and change so difficult.

This explains perhaps the difficulty in motivating mathematics: one can apparently get by "everyday" without any but the most minimal mathematics, until the everyday "breaks down," the historic mathematical work that went into the production of the everyday is suddenly required to be understood, by someone at any rate. This often involves quite "high-level" mathematics (when the reactor overheats, we call in specialists with "advanced qualifications"); but not always, even in the everyday workplace, we find examples of mathematical work done by workers like Dan.

This kind of breakdown was constructed artificially by a social situation where students and researcher were situated as questioners, and the workers felt obliged to try and explain their systems. Thus, it put a premium on mathematical communication and, indeed pedagogical discourses (informal: worker with team, formal: researcher-teacher with students). In such contexts pedagogical models such as the double number line were, perhaps naturally, prominent. But it might be argued that the model was useful to Dan's explanation for us because he already used such a model in constructing the formula in the first place. We will never know for sure, of course, but this is plausible and consistent with our theoretical framework: here Dan externalized the "mathematical thought" for our benefit, and the group understanding, insofar as it constituted group understanding, was an emergent property of the group's questions and Dan's—and then again later the researcher's own—explanations.

We argue that this kind of communication is not just "internalisation" by students in a zone of proximal development, but actually is a collective work in which emergence is constituted by internalisations *and* externalisations—in just the sense that Cobb argued is not synergetic with socio-cultural, activity theory (Cobb, 2007).

It is not a coincidence, we argue, that the double number line emerged as a powerful explanatory tool alongside the symbolic mathematics (albeit mediated by the spreadsheet, we call this a "genre" of mathematics, in the linguistic, Bakhtinian sense). As Lakoff and Núñez (2000) argued, the number line itself provides powerful affordances pedagogically in building up mathematics: these types of models are especially powerful when they allow the user to insert their body into the space the model occupies, even if only in imagination. In this case, Dan and the teacher were able to indicate segments of the timeline gesturally, and we assume the students could thereby identify the different points and intervals in time necessary to make sense of the formula.

Finally we note that the social and cultural context in the case seems vital to the students', the researcher's, and the workers' motivation and to their joint sense of the mathematical work as well. Regarded as a pedagogical episode, it broke with all the norms of schooling as an activity. Additionally, even in a narrow sense the spreadsheet formula broke all the norms about appropriate school mathematics and use of advanced technology. Yet, it is consistent with our Vygotskian perspective: making sense of adults' working practices and how mathematics is embedded there, constructing the relation with school mathematics, and perhaps even allowing for some discussion of its peculiar idiosyncrasies (the sorcery of the engineers' mathematics that kept all the other workers, including management, in the dark!). All this seems well suited to our activity perspective on modelling and technology. We argue that this adds a critical social context, an often missing element, to the case for mathematical modelling. Can this kind of expansive learning occur within the confines of schooling?

In the next example the integration of new technology in a manner more consonant with chapters later in this section will be described; and the expansive nature of mathematical-technology for mathematics education is exemplified (Figure 18.2).

Modelling and Expansive New Technology: Mathematical Technology

This case comes from a year-long study of the use of CAS-enabled technologies (TI *Nspire*) in senior secondary school mathematics classrooms (see Geiger, Faragher, & Goos 2010). The teacher had some experience with CAS but had not used it previously in his teaching. His students had begun to make use of CAS from the beginning of the school year, about 2 months before the vignette outlined below.

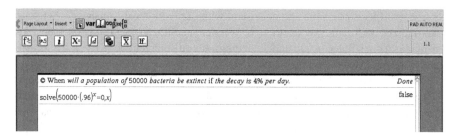

Figure 18.3. The CAS responds to a request to "solve" $50{,}000 \times (0.96)^x = 0$: "false."

The students were working on the following question: "When will a population of 50,000 bacteria become extinct if the decay rate is 4% per day?"

One pair of students developed an initial exponential model for the population y at any time x days after the initial population calculation: $y = 50{,}000 \times (0.96)^x$. They then equated the model to zero in order to represent the point at which the bacteria would be extinct, with the intention of using CAS to solve this equation. When they entered the equation into their CAS calculator, however, it unexpectedly responded *"false"* (see Figure 18.3).

The students thought this response was a result of a mistake with the syntax of their command. When they asked their teacher for help, he confirmed their syntax was correct, but said they should "think harder" about their assumptions. Eventually, when he realized that the students were making no progress, the teacher directed the problem to the whole class and one student commented: "You can't have an exponential equal to zero." This resulted in a whole-class discussion of the assumption that "extinction" should be represented by a population equal to zero. It was decided to modify the original assumption by representing extinction as "any number less than one." Students then used their CAS calculators to solve this resulting equation and obtain a numerical solution.

In a follow-up interview, directly after the lesson, the researcher asked the teacher (Teacher 1) about the episode.

Researcher: I saw an element of what we just talked about today when conflict was generated by an interpretation of the question about bacteria. Students developed an equation and then, because no bacteria were left, they equated it to zero. The calculator responded with a false message. In some ways you could think it was a distraction and that the procedure didn't work; some kids might just give up. But on the other hand, what it provoked in your class was an opportunity to discuss. "Did you push the wrong buttons? Oh, you think you did—let's look at the maths. Well your maths is right! Do you understand why it couldn't be? Let's talk about the assumption."

Teacher 1: Simon was one of those, he said—"No way you could get that to equal zero," without necessarily understanding why. Not that he couldn't solve it when it equalled zero, it was that concept he couldn't see; that population couldn't become zero.

Researcher: Yes, they didn't need CAS to understand that, they just understood it because they knew their maths well enough.
Teacher 1: Yeah we actually use the CAS to create the confrontation.

In this episode the teacher exploited the "confrontation" created by the CAS output to promote productive interaction among the class and develop a broader understanding of the role of assumptions in the mathematical modelling process.

In a later focus group interview, all teachers who participated in the project confirmed that similarly productive discussion arose from instances where technology produced unexpected, problematic results or responses. This is seen in the following transcript where Teacher 1 commented on events during the lesson on the decay of bacteria.

Teacher 1: It was pretty obvious to me why it didn't work but I deliberately made a point of that with a student to see what their reaction would be. And it was a case of pretty much what I expected. That they just grasped this new technology *Nspire* and were so wrapped up in it that they believed it could do everything and they didn't have to think too much. And so suddenly, when it didn't work, it took a fair amount of prompting to get them to actually go back and think about the mathematics that they were trying to do and why it did not give a result.

Researcher: … Interestingly you didn't just go over and tell them what to do. You just looked at it and said the syntax is all right—go and have a think about it. And they did for quite a while, and I don't know if anyone sorted it out. They may have but they didn't say. You then brought it back to the whole class and said, "What's gone wrong here?" Someone eventually said that you can't have an exponential equal to zero. What happened out of that—you might want to fill in more—is that there was quite a protracted discussion about what happened. Extinction is zero isn't it? So there is a little bit of a conflict between the way students think about it mathematically and the way it works in context. The context implies zero but there are other answers that could still make it work. So, you have to do this bit of a fudge and say the equation has to be equal to anything less than one—if it is a bacteria.

Teacher1: Even if the kids were solving that by traditional methods, they would still need to have that discussion. It was an issue with CAS that they were just expecting an instant answer and they didn't want to go and think about what was really going on.

Researcher: What is it about CAS-enable(d) technologies that would be different to ordinary technology, in this instance?

Teacher 1: I'll just reiterate and say with CAS that kids are looking for the quick solution, the immediately obvious without looking at what is underlying the discussions and the decisions that they are making. And they assume—like I did—that the machine can handle it.

In this discussion, the teacher identified a "blackbox" use of CAS (Drijvers, 2003) as the source of the impasse that the students experienced when attempting to determine when the bacteria would become extinct. Interestingly, then, students' expectation of technology's ability to produce "an answer" can potentially undermine any expected benefits of technology making challenging problems more accessible. As the teacher noted, students would have had to think carefully about their assumptions regarding extinction, whether or not they used CAS technology to tackle this task. A traditional approach might have led equally problematically to the logarithm of zero. What matters most is how the teacher responded. Such instances can be used to the advantage of students' learning if the teacher has the disposition, mathematical expertise, technological competence and confidence to manage such serendipitous opportunities.

The CAS black box here may usefully be thought of as a "mathematical-technology" which was instrumental in their modelling of bacteria-decay; but the zero value in the model here causes a "breakdown," a problematic, one which required the black box to be re-opened. As such we can argue the students had a problem in their zone of proximal development. When the students tried to enter an illegal value, the machine's response could be diagnosed as either a technology breakdown or a mathematical breakdown—the technology and the mathematics were here "fused"! They initially opted for a technology breakdown, that they had the wrong "code/syntax." The teacher said "think again/harder," because he saw the mathematical, conceptual issue, and this helped create a zone of proximal development from which, through joint exploration, there emerged a way forward, arguably a solution.

In contrast to the previous workplace case, this case revealed a naturally-occurring breakdown moment in a classroom, caused apparently by the mathematical-technology which declined to cooperate with the students and give them a solution to the equation: $50,000 \times (0.96)^x = 0$. It is interesting that the students' first thought was to question their own CAS technical competence, and this is probably quite general (cf., dividing 1 by zero and getting "error" on a numeric calculator, or sketching $y = \sin(x)$ on a graphics calculator and getting a straight line through the origin).

Teacher 1, who happened to have acquired a reasonable technical mastery of CAS, was able to see that the syntax is valid, but also had the mathematical competence to see a mathematical reason for CAS's resistance. He was thus competent to diagnose this as a moment to "think again." It seems the students and the teacher reached opposite diagnoses: the students looked to a technical fault, the teacher to a mathematical fault, and between the two there was "joint" problem-solving activity.

But actually things were a little more complex: the mathematics here was arguably not "wrong," in that the equation itself has no answer (except, perhaps, infinity). Rather, it was the mathematical modelling of the real situation that was problematic. The teacher persuaded the class—in what was (in the above account) called "whole-class discussion"—that a more sensible estimate would have been obtained by finding the time at which the model would predict a population value of one (or less) which resolved the problematic for the time being (getting a number that would be more satisfying than "infinity").

But actually, even this was questionable: one might rather ask whether the model was valid as a description of what happens to a single bacterium, and in what sense the problem of "extinction" is a "real problem" for which we need to formulate a model "fit for the purpose." A critical mathematics educator might like to run with this broader issue, and consider the purpose of such population models and the problems they can usefully address. Often, arguably, such exponential models arise at a population level (large number of particles/bacteria) of what is thought of as a probabilistic model at the micro-level (actually the probability of a bacterium dying or radioactive particle decaying in a given time is modelled as $p=0.04t$) and so when a large population becomes small one needs properly perhaps to switch back to the probabilistic model, predicting a range of time over which the last particles are likely to decay (and then maybe the time for the last few particles to decay becomes a Poisson approximation to the binomial). But whether an analyst is pushed to such model refinements really depends on the "real problem"—which is not specified in this case. A satisfactory "critical" endpoint to the class discussion might best have been "why do they want to know?" or "what really is the real problem at issue?"

The point here is that the fusion of mathematics with technology generated a problematic which was not entirely technical, not entirely mathematical, not entirely contextual, but an amalgam of all three. As such, the activity of "mathematical modelling with technology" can be trebly rich in complexity, i.e., when it is three-dimensional (mathematics, technology, activity-or-problem-context). In this case it was not just the pupils/learners who were challenged, and this case shows how such "joint problem solving" or "joint study" can become joint activity of learning and teaching, and maybe even research.

We noted here the demands this kind of work places on the teacher, to which we could add also strains on the curriculum and assessment, and the school organisation. In the ZPD both the students and the teacher were working hard at the problem from two distinct points of view; that of learning (and of engaging with the teacher) on the one hand and that of teaching (and modelling the learner) on the other (see Roth & Radford, 2011). This is truly a joint activity.

We explain this by suggesting that what is involved in "breakdown" is not so much the mathematics but the breakdown of modelling with mathematical-technology in context. Activity theory insists that "activity" is an indissoluble whole, and that any change in or neglect of one of its "moments" implies a change in all the other moments and transformation of the whole: thus an apparently innocent change in the "tools" may induce a treacherous change in the mathematics, in the subject's consciousness (the teacher's and learners' perceptions of the mathematics) and the relations and norms of behaviour in the activity system (the educational technology, curriculum/assessment, etc.) In complex systems of activity, small changes in one apparently "distant" moment can induce treacherous hurricanes downstream.

Conclusion

We have argued that mathematical modelling should ideally be conceived as adding "theoretical thinking" to real, practical problem-solving activity, and that this should have developmental consequences for students. We have used this ideal conceptualisation to situate modelling and technology within a Vygtoskyan, CHAT theoretical frame, and thus to criticize—or at least to develop a perspective from which to criticize—previous and contemporary research and practice.

It also provides a vantage point from which to see mathematics itself as a reflexive "tertiary" modelling artefact (Wartofsky, 1979, also adopted and developed by Cole, 1996), and hence as a problem-solving technology itself. We have suggested the term mathematical-technology to remind ourselves that activity tends to fuse the two in practice, and often in black boxes, and how these can provide expansive opportunities at breakdown moments. We argued that mathematics inevitably, as part of productive activity, appears alongside and even fused with, technologies in the solution of problems, producing new objects (that also may in turn hide mathematics) as outcomes. These mathematical-technological objects typically become instrumental in their turn, and provide new tools for future actions, which tend to new breakdowns. This cultural cycle fuses and re-fuses mathematics with technology, perhaps helping to solve but also causing contradictions and problems in new contexts of activity.

Particularly powerful new technologies have arisen lately (many described in the following chapters) which expand the language of mathematics, and allow learners wider scope for theoretical thinking and modelling in practice. Potentially, these may allow a wider appreciation of theoretical thinking in practical work than has been common previously, in part through the breakdowns and problems they introduce into activity. But we must not ignore the wider social context which also mediates change in educational technology, and which so often has provided the key obstacles to progress. We have only begun to touch on these here, hinting at the demands that working with mathematical-technology make of teachers and researchers, and so implicitly curriculum, assessment, and educational technology generally.

References

Bartolini Bussi, M. G. (1998). Joint activity in mathematics classrooms: A Vygotskian analysis. In F. Seeger, J. Voigt, & U. Waschescio (Eds.), *The culture of the mathematics classroom* (pp. 13–49). Cambridge, UK: Cambridge University Press.

Bernstein, B. (2000). *Pedagogy, symbolic control, and identity*. Oxford, UK: Rowman and Littlefield.

Black, M. (1962) (Ed.). *Models and metaphors: Studies in language and philosophy*. Ithaca, NY: Cornell University Press.

Black, L., Williams, J., Hernandez-Martinez, P., Davis, P., & Wake, G. (2010). Developing a "leading identity": The relationship between students' mathematical identities and their career and higher education aspirations. *Educational Studies in Mathematics, 73*(1), 55–72.

Blum, W., Galbraith, P. L., Henn, H.-W., & Niss, M. (Eds.). (2007). *Modelling and applications in mathematics education*. New York, NY: Springer.

Blunden, A. (2008). *Forward to Hegel's logic (being Part 1 of the Encyclopaedia of the Philosophical Science of 1830)*. In W. Wallace (translation, original 1873) accessed from Marxist Internet Archive, online, 2009.

Brown, A. M. (2011). Truth and the renewal of knowledge: The case of mathematics education. *Educational Studies in Mathematics, 75*(3), 329–343.

Bruner, J. S. (1960/1977). *The process of education*. Cambridge, MA: Harvard University Press.

Burkhardt, H. (1981). *The real world and mathematics*. Glasgow, Scotland: Blackie.

Cobb, P. (2007). Putting philosophy to work: Coping with multiple theoretical perspectives. In F. Lester (Ed.), *Second handbook of research on mathematics teaching and learning* (pp. 1–38). Reston, VA: National Council of Teachers of Mathematics.

Cobb, P., Yackel, E., & McClain, K. (Eds.). (2000). *Communicating and symbolizing in mathematics classrooms: Perspectives on discourse, tools, and instructional design*. Mahwah, NJ: Lawrence Erlbaum.

Cole, M. (1996). *Cultural psychology: A once and future discipline*. Cambridge, UK: Cambridge University Press.

Confrey, J., & Maloney, A. (2007). A theory of mathematical modelling in technological settings. In W. Blum, P. L. Galbraith, H.-W. Henn, & M. Niss (Eds.), *Modelling and applications in mathematics education* (pp. 57–68). New York, NY: Springer.

Davydov, V. V. (1990). *Types of generalization in instruction*. Reston, VA: National Council for Teachers of Mathematics.

Drijvers, P. (2003). *Learning algebra in a computer algebra environment. Design research on the understanding of the concept of parameter* (Doctoral dissertation). Utrecht University, The Netherlands. Retrieved from http://igitur-archive.library.uu.nl/dissertations/2003-0925-101838/inhoud.htm.

Engeström, Y. (1991). Non scolae sed vitae discimus: Toward overcoming the encapsulation of school learning. *Learning and Instruction, 1*(3), 243–259.

Freudenthal, H. (1983). *Didactical phenomenology of mathematical structures*. Dordrecht, The Netherlands: Kluwer.

Geiger, V., Faragher, R., & Goos, M. (2010). CAS-enabled technologies as "agents provocateurs" in teaching and learning mathematical modelling in secondary school classrooms. *Mathematics Education Research Journal, 22*(2), 48–68.

Gravemeier, K. (2007). Emergent modelling as a precursor to mathematical modelling. In W. Blum, P. L. Galbraith, H.-W. Henn, & M. Niss (Eds.), *Modelling and applications in mathematics education* (pp. 137–144). New York, NY: Springer.

Gravemeier, K., Lehrer, R., van Oers, B., & Verschaffel, L. (2002). *Symbolizing, modelling and tool use in mathematics education*. Dordrecht, The Netherlands: Kluwer.

Hanna, G., & Jahnke, H. N. (2007). Proving and modelling. In W. Blum, P. L. Galbraith, H-W. Henn, & M. Niss (Eds.), *Modelling and applications in mathematics education* (pp. 142–152). New York, NY: Springer.

Holland, D., & Quinn, N. (Eds.). (1987). *Cultural models in language and thought*. Cambridge, UK: Cambridge University Press.

Hoyles, C., Noss, R., & Pozzi, S. (2001). Proportional reasoning in nursing practice. *Journal for Research in Mathematics Education, 32*, 4–27.

Joseph, G. G. (2010). *Crest of the peacock: Non-European roots of mathematics* (3rd ed.). Princeton, NJ: Princeton University Press.

Kaiser, G., Blum, W., Ferri, R., & Stillman, G. (Eds.). (2011). *Trends in teaching and learning of mathematical modelling*. New York, NY: Springer.

Kaiser, G., & Sriraman, B. (2006). A global survey of international perspectives on modelling in mathematics education. *ZDM—Zentralblatt fur Didaktik der Mathematik, 38*(3), 302–310.

Kent, P., Guile, R., Hoyles, C., & Bakker, A. (2007). Characterising the use of mathematical knowledge in boundary-crossing situations at work. *Mind, Culture, and Activity, 14*(1–2), 64–82.

Krutetskii, V. (1976). *The psychology of mathematical abilities in schoolchildren*. Chicago, IL: University of Chicago Press.

Lakoff, G., & Johnson, M. (1980). *Metaphors we live by*. Chicago, IL: Chicago University Press.

Lakoff, G., & Núñez, R. (2000). *Where mathematics comes from*. New York, NY: Basic Books.

Lampert, M. (1990). When the problem is not the question and the solution is not the answer. *American Educational Research Journal, 27*, 29–63.

Latour, B. (1987). *Science in action*. Milton Keynes, UK: Open University Press.

Lave, J. (1988). *Cognition in practice*. Cambridge, UK: Cambridge University Press.

Lave, J., & Wenger, E. (1991). *Situated learning: Legitimate peripheral participation.* Cambridge, UK: Cambridge University Press.
Leontiev, A. N. (1978). *Activity, consciousness, and personality.* Englewood Cliffs, N.J.: Prentice-Hall.
Leontiev, A. N. (1981). *Problems of the development of mind.* Moscow, Russia: Progress Publishers.
Lesh, R., & Doerr, H. (2003). *Beyond constructivism: Models, and modeling perspectives on mathematics problem solving, learning, and teaching.* Mahwah, NJ: Lawrence Erlbaum.
Lesh, R., Galbraith, P. L., Haines, C. R., & Hurford, A. (Eds.). (2010). *Modeling students' mathematical modelling competences.* New York, NY: Springer.
Lesh, R., & Zawojewski, J. (2007). Problem solving and modelling. In F. Lester (Ed.), *Second handbook of research on mathematics teaching and learning* (pp. 763–804). Reston, VA: National Council of Teachers of Mathematics.
Noss, R., Bakker, A., Hoyles, C., & Kent, P. (2007). Situating graphs as workplace knowledge. *Educational Studies in Mathematics, 65*(3), 367–384.
Noss, R., & Hoyles, C. (2011), Modeling to address techno-mathematical literacies in work. In G. Kaiser, W. Blum, R. B. Ferri, & G. Stillman (Eds.), *Trends in teaching and learning of mathematical modelling* (pp. 75–78). New York, NY: Springer.
Pollak, H. (1969). How can we teach applications of mathematics? *Educational Studies in Mathematics, 2*(2–3), 393–404.
Polya, G. (1957). *How to solve it.* Princeton, NJ: Princeton University Press.
Roth, W. -M., & Lee, Y. (2007). "Vygotsky's neglected legacy": Cultural-historical activity theory. *Review of Educational Research, 77,* 186–232.
Roth, W.-M., & Radford, L. (2011). *A cultural-historical perspective on mathematics teaching and learning.* Rotterdam, The Netherlands: Sense.
Ryan, J., & Williams, J. S. (2007). *Children's mathematics 4–15.* Milton Keynes, UK: Open University Press.
Schoenfeld, A. (1992). Learning to think mathematically: Problem solving, metacognition, and sense making in mathematics. In D. A. Grouws (Ed.), *Handbook of research on mathematics teaching and learning* (pp. 334–370). Reston, VA: National Council of Teachers of Mathematics.
Sfard, A. (1998). Two metaphors for learning and the dangers of choosing just one. *Educational Researcher, 27*(2), 4–13.
Sfard, A. (2008). *Thinking as communicating.* Cambridge, UK: Cambridge University Press.
Strässer, R. (2000). Mathematical means and models from vocational contexts: A German perspective. In A. Bessot & J. Ridgway (Eds.), *Education for mathematics in the workplace* (pp. 65–80). Dordrecht, The Netherlands: Kluwer.
Strässer, R. (2007). Everyday instruments: On the use of mathematics. In W. Blum, P. L. Galbraith, H.-W. Henn, & M. Niss (Eds.), *Modelling and applications in mathematics education* (pp. 171–178). New York, NY: Springer.
Streefland, L. (1991). *Fractions in realistic mathematics education.* Dordrecht, The Netherlands: Kluwer.
Treffers, A. (1987). *Three dimensions: A model of goal and theory description, the Wiskobas project.* Dordrecht, The Netherlands: Kluwer.
Van Oers, B. (2002). The mathematization of young children's language. In K. Gravemeier, R. Lehrer, B. van Oers, & L. Verschaffel (Eds.), *Symbolizing, modelling and tool use in mathematics education* (pp. 29–58). Dordrecht, The Netherlands: Kluwer.
Vygotsky, L. S. (1978). *Mind in society: The development of higher psychological processes.* Cambridge, MA: Harvard University Press.
Vygotsky, L. S. (1986). *Language and thought.* Cambridge, MA: MIT Press.
Wake, G. (2007). Considering workplace activity from a mathematical modelling perspective. In W. Blum, P. L. Galbraith, H.-W. Henn, & M. Niss (Eds.), *Modelling and applications in mathematics education* (pp. 395–402). New York, NY: Springer.
Wartofsky, M. (1979). *Models, representations and the scientific understanding.* Dordrecht, The Netherlands: Reidel.

Watson, A., & Winbourne, P. (Eds.). (2007). *New directions for situated cognition in mathematics.* New York, MA: Springer.

Wenger, E. (1998). *Communities of practice.* Cambridge, UK: Cambridge University Press.

Williams, J. S. (2011). Towards a political economy of education. *Mind, Culture, and Activity, 18,* 276–292.

Williams, J. S., & Wake, G. D. (2007a). Black boxes in workplace mathematics. *Educational Studies in Mathematics, 64,* 317–343.

Williams, J. S., & Wake, G. D. (2007b). Metaphors and models in translation between college and workplace mathematics. *Educational Studies in Mathematics, 64,* 345–371.

Chapter 19
Technology and the Role of Proof: The Case of Dynamic Geometry

Nathalie Sinclair and Ornella Robutti

Abstract This chapter brings together two intersecting areas of research in mathematics education: teaching and learning with dynamic geometry environments (DGEs) and the teaching and learning of proof. We focus on developments in the literature since 2001 and, in particular, on (a) the evolution of the notion of "proof" in school mathematics and its impact on the kinds of research questions and studies undertaken over the past decade—including increasing use of DGEs at the primary school level; and (b) the epistemological and cognitive nature of dragging and measuring as they relate to proof.

Section A: Introduction

This chapter brings together two intersecting areas of research in mathematics education: teaching and learning with dynamic geometry environments (DGEs) and the teaching and learning of proof. Given that both of these areas of research have been predominantly related to the study of geometry, our chapter focuses mainly on this strand of the curriculum.

We begin by making a comment on terminology, and some brief remarks on recent history. In this chapter we use the term "dynamic geometry" broadly. The phrase was originally invented (and trademarked by Key Curriculum Press) to describe *The Geometer's Sketchpad* (Jackiw, 1991). *Sketchpad* and *Cabri-Géomètre* (Baulac, Bellemain, & Laborde, 1988) were independently invented in the late

---

N. Sinclair (✉)
Simon Fraser University, Burnaby, BC, Canada
e-mail: nathsinc@sfu.ca

O. Robutti
Università di Torino, Turin, Italy

1980s, and were the first software systems to offer the dragging capacity described by the term. There now exist many software packages that use these seminal ideas. Our paper loosely targets all of them, but given the importance of intellectual attribution in academic writing, we feel it important to acknowledge similarly the sources of software innovation within our field. We have opted for the term *dynamic geometry environment* (which has been used at least since 1996) over *dynamic geometry software* to underscore the fact that we are dealing with microworlds (including pre-existing sketches and designed tasks) and not just a software program.

The *Second International Handbook of Mathematics Education* (Bishop, Clements, Keitel, Kilpatrick, & Leung, 2003) included several chapters focussing on the use of digital technologies. Within these, Chapter 9 (Hoyles & Noss, 2003) considered DGEs in particular and documented the way in which research was focussing less on geometric constructions and more on the way DGEs mediate explanation, verification and proof. The authors commented on the "commonplace" understanding that students often spontaneously articulate when using DGEs, but opined that it was "not obvious that the facility to drag and conjecture will necessarily encourage an engagement with proof" (p. 335). They cited research showing how novel activities aimed at generating surprise and uncertainty can strengthen students' need for deductive proof (Hadas, Hershkowitz, & Schwartz, 2000); how 12-year-old students can move from everyday to mathematical explanations through DGE interactions (Jones, 2000); and, how DGEs can help 14–15-year-old students connect their informal geometric explanations with logical, deductive arguments as long as the tasks used are undertaken with teacher support, including a teacher-introduction to writing proofs (Healy & Hoyles, 2001).

The present chapter thus focusses on developments in the literature since 2001 and, in particular, on (a) the evolution of the notion of "proof" in school mathematics and its impact on the kinds of research questions and studies undertaken over the past decade—including increasing use of DGEs at the primary school level; and (b) the epistemological and cognitive nature of dragging and measuring as they relate to proof—thus taking up the Hoyles and Noss issue.

In preparing this chapter, we found several other overview studies in each of our primary areas of focus. Mariotti (2006) provided an overview of the past 30 years of research reports presented at the annual conferences of the International Group for the Psychology of Mathematics (PME) on "Proof and Proving in Mathematics Education," and this included a significant section on the role of DGEs. Similarly, Hollebrands, Laborde, and Sträßer (2008) offered a chapter on "Technology and the Learning of Geometry at the Secondary Level," which focussed primarily on DGEs and included a significant section on proof and proving. This chapter also includes a detailed account of the main theoretical approaches currently being used in research on the role of DGEs in proving. The two main approaches are instrumental genesis, which draws on the work of Rabardel (1995), and Verillion and Rabardel (1995), and has been elaborated by Mariotti (2002), and semiotic mediation, which draws on the work of Vygotsky (1978). These two approaches are discussed in detail in Drijvers, Kieran et al.'s (2010) overview chapter in the ICMI Study on Technology.

Briefly, a DGE can be seen, in an instrumental framework, as an "artefact" in the hands of the users. The DGE artefact can be transformed into an "instrument" by the user, according to the "schemes of use" that are activated. And the instrument obtained thus differs among users and contexts according to the particular schemes of use activated. Several studies have documented how this works in the contexts of DGE (see, e.g., Olivero, 2006; Baccaglini-Frank & Mariotti, 2010). If the students "internalize" (Vygotsky, 1978) the use of a DGE, the artefact DGE becomes a means of semiotic mediation (Mariotti, 2010) that gives rise to problem resolution.

In Section B below, we outline the perspective on proof that has emerged in the mathematics education literature during the past decade. We connect this literature to the previous theorizing on the epistemological and cognitive status of DGE diagrams and tools. While some of this theorizing dates back to the late 1990s, it was not taken up in the *Second International Handbook* and its connection to emerging perspectives on proof is new.

In section C, we identify and connect previous studies on the use of DGEs in proving in terms of the particular and distinctive tools shared by all the more widely used DGEs. We have chosen this somewhat unusual approach in order to emphasize, as well as better understand, the epistemological, cognitive and didactic implications of these tools—implications that have been studied in much more depth over the past decade than in previous research. This section will enable us to investigate the effects of DGEs on the proving process.

Having analyzed the tools in this way, we turn in section D to two case studies in which we illustrate instances of the proving process in which these tools play a significant role. Our goals are threefold: at one level, we aim to exemplify the most common theoretical perspectives, methodologies and protocols currently used—instead of merely reporting results from existing studies, we would like to show the context in which these results have emerged, as they are central to the findings proffered. At a second level, these case studies can permit us to enter into a level of detail about the proving process using DGEs that would otherwise not be possible. At a third level, given our focus on the particular tools of section C, we hope that the case studies more authentically illustrate the way in which these tools are often used in concert by learners.

In the final section, we highlight some of the major themes of sections C and D, and point to emerging questions and theoretical approaches that are likely to take centre stage in the next decade, with a particular focus on the role of teachers in classrooms using DGEs.

Section B: Proof as a Process

Policy makers, curriculum designers and researchers concurrently agree with the necessity of proof not only at the high levels of schooling, where it has traditionally occupied an important position, but also, for better continuity, at the primary level (see Bartolini Bussi, 2009; Stylianides, 2007; Stylianou, Knuth, & Blanton, 2009).

In many countries, teachers are being encouraged to offer activities favouring exploration, conjecture, argumentation, discussion and also proof (see, e.g., Centre de Recherche sur l'Ensignement des Mathématiques, 1995; Japanese Society of Mathematics Education, 2000; Ministry of Education and People's Republic of China, 2001; National Council of Teachers of Mathematics, 2000; Unione Matematica Italiana, 2004).

The notion of proof in school mathematics has changed enormously during the past century—see Herbst (2002) and Sinclair (2003, 2008) for overviews. The shift is due in part to a greater awareness of the way in which proving is done in the mathematics discipline (see Borwein & Bailey, 2008; Hanna, Jahnke, & Pulte, 2010) and, in part, we believe, to the affordances of new digital technologies, which greatly facilitate experimentation (de Villiers, 2010).

In this chapter, we have opted to frame our discussion about proof and proving in terms of the notion of the "proving process," which has been defined as a process of constructing and entering the relation "B follows from A" until the agents are satisfied with the explanation for the truth of the statement (see Rav, 1999). This explanation is a proof if it satisfies the rules of logical consequence.

The proving process consists of two phases: the formulation of a conjecture and the construction of a proof (Arzarello, Olivero, Paola, & Robutti, 1999; Leung & Or, 2007). In the first phase, a person explores the situation and formulates a conjecture, and searches for elements (e.g., properties) that will be organized later on. In the second phase, these elements are to be put in order according to the rules of logical consequence. As we emphasize below, the shift to the focus on the "proving process" is especially relevant to DGEs in the sense that the first phase of the process differs radically from pencil-and-paper environments, which, in turn, affects the way in which the second phase evolves.

The proving process also involves back-and-forth moves between the spatio-graphical field and the theoretical field (Laborde, 2004). Given the back-and-forth nature of the proving process, it is not viable to abandon visual representations in order to encourage the move toward the theoretical field—instead, learners need to be able to coordinate different semiotic resources (Duval, 2006). Once again, as we elaborate below, the status of the dynamic diagrams supported in DGEs, as well as their insistence on the spatio-graphical field, presents both opportunities and challenges when working with learners on the proving process.

The Proving Process in DGE

In terms of the back-and-forth nature of the proving process, we focus on the status of geometric objects in DGEs. Laborde (2000) considered dynamic diagrams (such as a dynamically draggable parallelogram) as scaffolding the drawing/figure divide in the sense that it remains a material object (albeit virtual on the screen), but the invariances it carries in dragging can represent its basic properties.

But Jones (2000) argued that learners in DGEs "can get 'stuck' somewhere between a drawing and a figure" (p. 58). This led Battista (2008) to conclude both that the drawing/figure distinction might not be sufficient in working with geometry objects and that more attention needs to be paid to the ways in which learners *perceive* and think about DGE objects. We focus here on the latter issue: Battista drew attention to two subtly different perspectives. One was captured by Marrades and Gutierrez (2000) when they wrote: "The main advantage of DGS learning environments" is that "students have access to a variety of examples that can hardly be matched by non-computational or static computational environments" (p. 95). This theoretical perspective tends to treat examples as *representations* of the figure. The second theoretical perspective, articulated by Laborde (1992) and Battista (2008), focusses on the continuous transformation of the draggable object rather than on the set of characteristics that can be abstracted from a given set of examples. This second perspective treats the object as an entity whose behaviour can be investigated and described—not as an entity that can be represented by particular diagrams.

It is still unclear whether learners somehow naturally see the draggable diagrams as a series of examples or as one continuously changing object, and whether this depends on their previous exposure to the static geometric discourse of the typical classroom (see also Lehrer, Jenkins, & Osana, 1998). Whatever the case may be, there is strong evidence to show that dynamic diagrams support students' transition from an exclusively spatio-graphical field to a more theoretical one by helping them attend to the visual invariance of the dynamic diagram, which can be verbally mediated through classroom discussion or teaching intervention (Battista, 2008; Laborde, Kynigos, Hollebrands, & Sträßer, 2006; Sinclair, Moss, & Jones, 2010).

Dynamic draggable diagrams can strongly affect the proving process in the sense of mediating what kinds of conjectures will be made (first phase), and what kinds of properties will be identified and then organized (second phase). In this sense, the diagram constructed in a DGE and the schemes of use activated on it by the students can be considered an actual mediator between the first and second phases, thus providing continuity—despite the epistemological discontinuity—between a conjecture (hypothetic statement) and the corresponding proof (logical deductive sequence of statements). The internalization of the schemes of use in the artefact and its transformation into an instrument help them in building the proof as final product (Arzarello, Micheletti, Olivero, Paola, & Robutti, 1998; Boero, Garuti, Lemut, & Mariotti, 1996; Olivero, 2002).

The role of DGEs in the proving process also depends on the underlying goal of proof, as perceived by teachers and students. Some have suggested that the use of DGEs could inhibit proof since students may think that a fact is empirically evident, and they do not feel the necessity of proving it (Frant & de Costra, 2000). However, as many studies have shown, exploratory experiences with DGEs do not necessarily jeopardize the development of deductive proof (Bruckheimer & Arcavi, 2001; Christou, Mousoulides, Pittalis, & Pitta-Pantazi, 2005; Guven, Cekmez, & Maratas, 2010; Hadas et al., 2000; Healy & Hoyles, 2001; Mariotti, 2006; Oner, 2008, 2009). The work of de Villiers (1990, 1997, 1998), in articulating the different functions of proof, has brought some nuance to the question of whether or not DGE's inhibit proof. He listed the following functions: verification, explanation, systematization, discovery, communication and intellectual challenge. The explanatory function of proof has been particularly prevalent in the DGE literature. In particular, proving through surprise involves appealing to the explanatory function of proof, as described by de Villiers (1990, 1997, 1998). For example, instead of questioning the conviction empirical methods can give, de Villiers invited students to accept the evidence, but then to ask *why* the relationships they were seeing must hold. In answer to that question they were expected to find a logical chain for going from hypothesis to thesis.

DGEs are used not only in Euclidean geometry but also in non-Euclidean geometry as a means of offering representations and interactive models (such as the Poincaré disc model of hyperbolic geometry) that are more powerful than paper (Jones, Mackrell, & Stevenson, 2010). Hollebrands, Conner and Smith (2010) examined the structure of arguments that students created using a DGE in light of Toulmin's theory and presented evidence of the supporting value of DGEs in such a context.

This section has focussed on the general role of DGEs in the proving process and, in particular, the important role of open problems and the teacher's insistence on explanation in supporting this proving process. In the next section we examine in more detail the nature of the dynamic geometry tools that relate to the process of proving.

Section C: Roles of Different Instruments Typical in DGEs

The role of the tools students have at their disposal in a DGE (as for example dragging, measures, locus, trace, and so on) can be that of a bridge between the spatio-graphical and the theoretical way of looking at a diagram. This point of view is supported by data from teaching experiments at different school levels (Arzarello et al., 1998; Jahn, 2002; Laborde, 1998, 2004; Olivero & Robutti, 2007; Sinclair et al., 2010; Vadcard, 1999). In the next three sections, we focus on two central tools of DGEs, that of dragging and measuring, isolating each of them (even if they can be used together in exploration), in order to provide a more fine-grained analysis of the epistemological, cognitive and didactical implications of each one.

Focus on the Dragging Tool

Dragging is the most central and distinctive tool of DGEs and allows users to select one or more objects and to move them continuously on the screen. We are concerned with the way the dragging may mediate the process of proving, in particular focussing on the epistemological and cognitive implications of it. (There are also didactical implications, but these will be considered in section D.)

Epistemological implications of dragging. Dragging changes the figural aspect of a construction such as an equilateral triangle (the representation changes) but not the conceptual one (since all the properties of the equilateral triangle are being maintained). This duality does not arise in a static pencil-and-paper environment, since the figural aspects are handled in a visual register and the conceptual one in the discursive register.

Since geometric proofs are meant to concern theoretical objects—and not just specific drawings—the role that dragging can play in managing the figural/conceptual duality is of particular interest. Making a conjecture about any configuration depends on the expectation of some sort of invariance under figural change. And any conjecture about an equilateral triangle must assume that the conjecture will hold true for any configuration of an equilateral triangle.

Taken as a tool of semiotic mediation, Mariotti (2006) wrote that dragging fosters students' access to the world of Geometric theory, citing Jones' (2000) study of students working with quadrilaterals, with the task of constructing a quadrilateral and modifying it into a special case with dragging. By dragging a rectangle, say, the students could observe that "A rectangle ... becomes a square" and "You can make a rectangle into a square by dragging the side shorter ..." At the end of the unit, when asked to classify quadrilaterals into a family tree, the students were also able to answer questions of the form "why is one quadrilateral a special case of another?" However, they persisted in saying that the set of more general quadrilaterals to which they were referring excluded the special case. This reveals an epistemological gap between the pragmatic, inductive argument enabled by the dragging (a square is a special case of a rectangle because you can turn the rectangle into a square) and the property-based, deductive inference required for a proof (a square is a special case of a rectangle because a square has four right angles).

Erez and Yerulshalmy's (2006) study showed that with some guidance from the teacher, drawing attention to the relevant properties, 5th grade students (aged 10 and 11) can, with dynamic geometry, view a square as a rhombus. In his study of Grade 5 children using the *Shape Maker* microworld, Battista (2007) theorized the effectiveness of dragging in terms of a two-folding assumption: first, subconscious visual transformations "are one of the mental mechanisms by which we spatially structure shapes" (p. 150). For example, the opposite sides of a parallelogram are subconsciously seen as parallel through a translation of one side onto another.

Battista's second assumption, which he called the *transformational-saliency hypothesis*, related more centrally to dragging. This hypothesis essentially stated that people notice invariance. As students drag the rhombus maker, they notice what stays invariant, namely the fact that all four sides are equal. For Battista, it was not just that one *might see* invariance in dragging but that one cannot help but notice it. He thus conjectured that "investigating shapes through *Shape Maker* transformations make the essence of the properties more psychologically salient to students than simple comparing examples of shapes as in traditional instruction" (p. 152). Dragging thus changes the way shapes are perceived, moving from a static visual apprehension to a temporal attention to what remains invariant. This hypothesis might explain the tendency students have to compare shapes using transformations, as reported both in DGE environments (Jones, 2000; Sinclair et al., 2010) and in non-DGE ones (Lehrer et al., 1998).

Leung's (2008) approach to theorizing dynamic geometry experiences—and particularly the temporal and spatial changes involved in dragging—also accorded a central role to invariance. Leung (2008) used Marton & Booth's (1997) theory of variation to analyze the different ways in which dragging might help a learner come to know mathematics. According to Leung (2003), "in DGE, it is possible to define a way of seeing (discernment) in terms of actually seeing invariant critical features (a visual demarcation or focussing) under a continuous variation of certain components of configuration" (p. 198). He elaborated Marton and Booth's notion of *simultaneity*—experiencing different temporal or spatial "instances" of a phenomenon at the same time—which he saw as a promising agent to help bridge the gap between experimental and theoretical mathematics. Leung described two different types of simultaneity: (a) the spatial type involves experiencing different features of the same thing simultaneously; and (b) the temporal type involves experiencing instances that have previously been encountered at different points in time, at the same time (for example, seeing three animations that had been viewed one after the other, all at the same time). Interestingly, these loosely correspond to Battista's two perspectives on dynamic diagrams: the spatial type corresponds to the multiple example perspective and the temporal type to the continuous transformation perspective. In Leung's example of the temporal simultaneity, two things were varying simultaneously and the observer was trying to identify some kind of invariance. The spatial simultaneity will involve comparing and contrasting the similarities and differences among the four types dragging experiences. Dragging is central to both types of simultaneity, as is the experience of invariance. Indeed, simultaneity can only enable discernment and awareness if invariances can be seen, an assumption that relates strongly to Battista's *transformational-saliency hypothesis*.

Leung's (2008) case study showed how these two types of simultaneity not only lead to the generation of a conjecture but also to the construction of what he calls a "DGE proof" (p. 146). Although Leung did not dwell on the transition from conjecture to "DGE proof," it is worth attending to the epistemological nature of the transition, particularly in terms of the role of dragging. The conjecture identifies an invariance. In Leung's case study, very briefly, the problem is as follows: For a quadrilateral *ABCD*, under what conditions is $\angle ABC = 2\angle ADC$ (see Figure 19.1a). Through explorations in dragging, Leung saw that if *A, B, C* are left fixed and *D* is dragged, then $\angle ABC = 2\angle ADC$ when *D* seems to be on a circle passing through *A* and *C*. Moving to a DGE proof then, involves first constructing the circle c_4 through *A, B* and *C*, then constructing the perpendicular bisector of *AC*, then finding the intersection *E* of this perpendicular bisector with c_4 (see Figure 19.1b). If *D* is merged to the circle passing through *A* and *C*, centred at E, then $\angle ABC = 2\angle ADC$.

The role of dragging has shifted from interrogation to declaration. The construction process is *declarative*, in that it asserts the circle, as well as *demonstrative* in that it shows how the geometric configuration can be put into action (much like Euclid's first proposition, which shows how an equilateral triangle can be constructed). As Jackiw (2006) wrote, the DGE proof follows a long tradition of mechanical demonstration in the history of science and mathematics, one that is replete with personal agency and aesthetic satisfaction. While not providing the propositional sequence of deductions required in a formal, written proof, the "DGE proof" is epistemologically much closer to the explanation than to verification and,

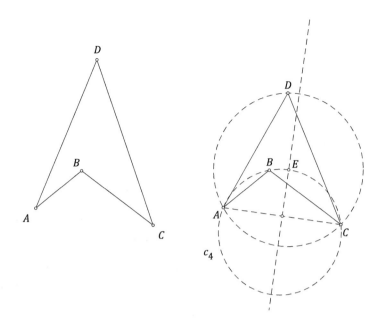

Figure 19.1. (**a**) The general configuration; (**b**) Under what conditions is ∠ABC=2∠ADC.

as such, may fruitfully become an object of study in further research on the role of DGEs in the proving process.

Cognitive implications of dragging. Dragging can be seen as semiotic tool used to express meanings and solve problems (Mariotti, 2002). It can also reveal students' thinking processes as they work on tasks. By observing students solving open problems, Arzarello, Olivero, Paola, and Robutti (2002) identified several distinct dragging modalities that were used, and that revealed specific features of their thinking. They have been widely cited and used by researchers trying to understand students' problem solving and proving activity. We highlight some of them here in order to stress the links to proving. Consider the blackbox situation given in Figure 19.2. If students are asked to identify the transformation leading from A to A', they may begin by wandering dragging in which they drag point A more or less randomly on the screen. They have no plan yet, and no conjectures about how A and A' are related. In their wandering dragging, they may notice that there are places where the two points meet. They may therefore engage in "dummy locus" dragging in which they attempt to drag A along a path that keeps A and A' together; here they are following a hidden path (the dummy locus, in this case a line) even without being aware of it. In order to keep track of these places where A and A' overlap, they may engage in line dragging, in which they draw new points on the spots of overlap, or in guided dragging, in order to obtain and observe a particular situation (for example moving A towards A'). They will thus see that the spots form a line, at

Figure 19.2. Moving from wandering to dummy dragging.

which point they may construct a line and use linked dragging to merge the point to the line and move it along that line.

These drag modalities structure a situated hierarchical scheme that, as suggested in the above example, are indicative of different modes of thinking such as exploring, forming conjectures, testing conjectures, and verifying conjectures.

The final type of dragging identified by these researchers is called the *dragging test*, and involves dragging points in order to verify whether a figure keeps its initial properties: if a student constructs an equilateral triangle, he/she then uses the drag test to see whether moving one vertex maintains the triangle as equilateral.

These modalities can be grouped in two broad categories, characterizing the moves between the spatio-graphical field and theoretical field: dragging for exploring/conjecturing and dragging for validating a conjecture or proof. In the first case, students drag objects observing the figure in search for regularities and invariants (e.g., wandering dragging): once they have found one and they express it through a conjecture, they shift from the spatio-graphical field to the theoretical field. And vice versa, they shift backward if they drag objects to check a conjecture already discovered (e.g., guided dragging).

To this list of drag modalities, Leung (2008) added one that relates to the parametric colour tool, which enables users to make the colour of an object depend on a measured quantity. For example, one could have the colour of a circle depend on its distance from a fixed point: dragging the circle, its colour would thus change. While this *spectral dragging* often involves moving the mouse (as in the example given above), it is not fundamentally about motion; rather, its power is more in the chromatic feedback offered. Leung's study focusses less on the cognitive dimension of dragging than, as he says, on the mathematical relevance of dragging strategies. He identified different types of dragging that are based on different ways of attending to variation: dragging for contrast, for separation, for generalization and for fusion. As with the dragging modalities of Arzarello et al. (2002), these are all related to conjecturing and proving.

Focus on the Measurement Tool

Measuring is a powerful tool of dynamic geometry environments, and also a complex one that requires appropriate management and interpretation, and may be used by students with varying degrees of confidence and awareness. Actually, students using a DGE in geometry tasks can utilize measures in different ways, according to the description given for dragging: in the spatio-graphical field, if they are more concerned with the perception of the drawing, or in the theoretical field, if they are more concentrated with the properties of the figure. According to their use of measurements, students can construct mathematical meanings, formulate conjectures, check them or use them to construct a proof, in a continuous process from exploration to the final product of a formal proof.

Epistemological implications of measurements. One of the specific features available in a DGE, the measuring tools, within the context of open geometry problems, can offer deep insight on the figures and ideas for the formulation of conjectures, especially if used with the dynamicity offered by dragging. However, from an epistemological point of view, the double meaning of measuring (in mathematics and in physics) can also cause misunderstanding or conflict in the proving process, in the interplay between theoretical and graphical aspects of figures. Actually, if we consider the meaning of measures in mathematics, we have a function that gives a unique real number associated to a quantity (that can be rational or irrational): for example, in the case of diagonal of a square measured with its side, it is irrational (the square root of 2). Measuring in physics and experimental sciences has a different epistemological status, because it depends on the kind of tool and its sensibility, on the number of measurements, and on calculation of uncertainty. In this case, measuring never provides an irrational number, but always a rational number, with a precision (number of significant digits) that depends on the tool and on the process, and gives as result a number in an uncertainty interval.

When using a DGE, things are more complex, and go further than this double meaning, because measurements are given as in physics (using a virtual ruler), but the software simulates a geometric environment (the Euclidean one) where the measure is mathematics (Olivero & Robutti, 2001). In a DGE, users can construct geometric figures according to Euclidean properties, but it is not the Euclidean plane, because there is no continuity, but a finite number of pixels. In this plane, computations do not have infinite precision, but they are approximate. For these reasons, coordinates of points, amplitudes of angles, and measurements in general are written with a finite number of decimal digits. The user can increase that number of digits, however the precision of the tool does not increase, because there is a technological limit to the precision of the measuring tools: the dimension of the pixel.

The outcome of these features is that many different things may happen when students are measuring in a DGE (length of segment, amplitude of an angle, and so on). For example, one might see on the screen two equal lengths corresponding to segments that are known to be not equal from a theoretical point of view, or the additive property of measures is not always satisfied (Figure 19.3). These examples show that some geometric properties may appear not to be satisfied when looking at measurements in a DGE.

When students approach problems in a DGE and find a situation like the one in Figure 19.3, they may remain stuck and attribute the mistake to the software, or may re-consider again their work from the beginning, thinking they have made a mistake.

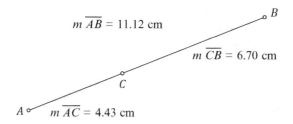

Figure 19.3. The length of AB is not equal to the sum of the lengths of AC and CB.

Otherwise, if they are aware of the epistemological implications of the use of measures in a DGE (approximation, pixels, calculation, and so on), they can look at their figure in a theoretical way, in order to interpret it with the constraints of software use. From a didactic point of view, de Villiers (1999) showed that one can also use the finite limitations of a DGE to design specific situations where students are not absolutely certain in order to motivate a need for proof as a means of further verification.

Cognitive implications of measures. The epistemological implications described above imply a complexity where measure interpretation and management surely creates opportunities for construction and exploration, but it can also be a source of conflict. Actually, the double meaning of measure in a DGE is not always transparent to students, and can sometimes act as a black box, where students accept as "true" a value of a measurement they read on the screen, without treating it contextually with some uncertainty or approximation. Students' competence in managing measures depends on their awareness of their epistemological status, for which the teacher's role is fundamental.

As for dragging, also for measure in a DGE, the different uses of the tool can correspond to different cognitive activity, from a spatio-graphical level to a theoretical one, or backwards. In fact, students may use the measure tool at times for discovering and conjecturing and at times for validating a conjecture. This was shown by Vadcard (1999), in his study on the use of measuring in *Cabri*, which distinguished between *mesure exploratoire*, used mainly as a heuristic tool and *mesure probatoire*, used as a checking tool. Furthermore, it is possible to split each category in different modalities (Olivero & Robutti, 2007). The first category (*mesure exploratoire*), related to the shift from the spatio-graphical field to the theoretical field, can be divided into different modalities. One of them is *wandering measuring*, when students do not have any precise ideas about the configuration—so they explore the situation randomly: they take measurements of some elements of the configuration, in the same way as they might use wandering dragging, in order to identify quantitative relations, invariants, congruencies, etc. Of course, it seems to be important *which* quantitative relationships will be identified: as Sinclair (2002) argued, only certain ones will be deemed interesting enough—for example, in wandering dragging that involves measuring angles, certain values (such as 90°, 120°, etc.) will seem much more relevant than others, because of an aesthetic component in conjecturing, in the sense that only some configurations, invariances, or measures are worth attending to. If students trust measurements and they believe measurements are absolutely exact, they stay at a spatio-graphical field. Instead, if they use the information provided by the reading of the measurements to formulate a conjecture in the standard form "if … then …" and in a general case ("generic example," in the sense of Balacheff, 1987), they connect the spatio-graphical field with the theoretical one.

Another modality is *guided measuring* (echoing guided dragging), when students do a guided exploration of the configuration, examining particular cases one after the other: measurements are used in order to obtain a particular figure from a generic configuration; for example a generic quadrilateral can be transformed into a parallelogram by looking at the length of the opposite sides. This modality can be useful to put in order (Guala & Boero, 1999) a set of different cases, with the aim of exploring them—from, for example, the most particular to the most general.

A third is *perceptual measuring*, when students use measurements as a means of checking the validity of a perception: having the intuition on a property, but not being sure of it, they use measurements to validate the perception, by transforming a qualitative relationship into a quantitative one, and remain in the spatio-graphical field. They may jump to the theoretical field if they then transform that perception into a conjecture.

On the opposite side, the second category is related to the shift from the theoretical field to the spatio-graphical field, when students need to check the validity of an intuition, a conjecture, even a proof, and use *mesure probatoire*. In the *validation measuring* modality, students use measures to check whether a conjecture should be accepted or refuted. This use is very similar to the dragging test, for checking a construction. Another modality is *proof measuring*, not frequently used, where students go back to the DGE after having constructed a proof, in order to check it or gain conviction. In this case, new experiments are made in the DGE and measurements are used to "validate" the proof from an experimental point of view, or to have data against the proof itself.

Students' competencies in using measurements discriminate their approach to proof. If they read values in the DGE, giving them a proper approximation and range of uncertainty, it could be useful both in the passage from spatio-graphic to theory and backwards, otherwise some problems may occur (one is described in the data below). In terms of the use of the measuring, researchers have studied the ways students use measures in the proving process (perceptually or theoretically?), with their awareness in doing so (knowing its epistemological status and its role in a proof or not?), and the relation of their use of measures to the construction of a proof (support or against?). We exemplify some findings of this research in section D.

Other DGE Tools

The dragging and measurements tools are central in DGEs, and have been studied extensively, but they are not the only ones: there are many other tools at a learner's disposal including trace, locus, transformation, animation, parametric colours, macro-constructions or custom tools, iteration and graphing. Mariotti (2002) described the way in which tracing and locus can work together with the former providing exploratory evidence of a relationship and the latter a confirmatory construction. While trace guides a student to the solution of the problem, locus supports investigation on specific characteristic of the figure or relationships among its components. Locus and tracing have been shown to be effective in problems where students construct and use transformation, and are engaged in finding invariants (Jahn, 2002), or in problems where they are involved in guessing what kind of transformation is at the basis of a given construction—as in Laborde's (2001) black boxes. While dragging and measurement will remain the focus of attention, we anticipate more work in the next decade on the use of these other tools in the proving process.

Section D: Protocols in DGE Research Around Proving

In this section we offer two case studies of student interactions with a DGE. In section C we focussed on the two tools of dragging and measuring, charting out their cognitive and epistemological implications across a range of topic areas and grade levels—particularly as they relate to the proving process. Here we consider particular classroom situations dealing with specific content, tasks and grade levels, as well as specific theoretical lenses that can be used to study geometric thinking. The first example, reported by Sinclair et al. (2010), involved primary grade children engaging in geometric argumentation in the context of a whole classroom exploration around identifying triangles. Dragging is a central feature of this example but the analysis focusses less on the modality of dragging used than on the discursive changes in students' talk about triangles. The second example, reported by Olivero and Robutti (2007), involved high school students working at a more sophisticated level of the proving process, in the context of a teaching experiment in which pairs of students solved an open problem that centrally involved the use of the measure tool.

Dragging at the Elementary School Level

This case study involves children 4–5 years old who had never seen the software before, and had not engaged in any formal work on geometry at that point in the year (January). The task was designed in such a way to promote exploration, conjecturing and argumentation. Although the episode is not directly about proof, we include it because we see it as belonging to the initial stages of the proving process, including the move to the theoretical and the use of definitions. The episode involved students engaged in identifying triangles. We focus attention on the different ways that students talked about the shapes they see and their developing discourse as they encountered a broader range of dragged triangles.

The teacher began by asking "What is a triangle?" The children use their hands and bodies to form triangles. There was no reference to specific properties of the shape. The teacher then constructed a triangle with its vertex facing down using the segment tool. Most of the children did not think this was a triangle. Some turned around, and said it was a triangle only if they looked at it upside down. The teacher then dragged one of the vertices of the triangle.

34.	T:	I'm going to ask you to tell me what you see when I'm doing this.
35.	Ss:	Ooh. Wow. [*some laughing*]
36.	T:	What do you see?
37.	Abigail:	That's actually a triangle.
38.	Leah:	A triangle.
39.	John:	It is.
40.	T:	Why is it a triangle? [*continuing to drag*]
41.	Robert:	You can stretch it out.
42.	T:	And that's what makes it a triangle?
43.	Robert:	Yeah.
44.	Dasia:	Every triangle
45.	John:	You're going to make
46.	T:	Just wait. Dasia, what are you saying?
47.	Dasia:	Every triangle could be, um, a different shape but it just has three corners.

Although brief, this exchange was long enough to show a process of discourse change in real time. The change was signalled with a collective expression of surprise (see "Ooh. Wow" in [35]). Abigail's statement "That's actually a triangle" [37] signalled that what so far was considered to be a triangle only *potentially*, pending a person's position with respect to the shape, now became a triangle as-is, unconditionally. Since it was now obvious, even if not actually shown, that this animated thing could easily reincarnate into the canonical shape she had always considered as a triangle, that canonical shape became but one of the possible images of the *moving object* called triangle. The follow-up comments by Leah [38] and John [39] suggested the same interpretation. Most notably, Robert explicitly signalled having changed his mind when he explained that whatever he could see on the screen, including the original "upside-down" shape, was a triangle simply because it could be obtained from one basic triangular shape by "stretching it out" [41]. A significant aspect of the children's talk was that it shifted toward the hypothetical and the abstract: so far, a triangle used to be something that was present visually, but the children now spoke of it as an object that changes over time.

Dasia's talk stood out in three respects. First, she spoke in plural, mentioning many triangles rather than just one (see her expression "every triangle" in [44]). Second, she referred explicitly to properties of a triangle (see her mention of "three corners" in [47]; this is similar to, but still different from Abigail's earlier mention of three sides). Finally, she turned Robert's statement around into a more general description of a triangle. On the basis of [47] we can conclude that having three corners was, for her, the only condition a shape had to fulfil to be called "triangle." With the word "just" in her requirement of "just has three corners" Dasia communicated the idea that a triangle can have many other characteristics (be upside down, stretched, etc.), but having three corners was the only property that was really indispensable.

Wishing to elicit the difficulty that children might have in identifying stick-shaped triangles as triangles, the teacher asked the students to close their eyes so that she could drag the triangle into a more "monstrous" (very long and skinny) shape. When asked whether it is a triangle, some students say "no" and others "yes." After some discussion, Morgan remained the only dissenter.

82.	Dasia:	It's a triangle for me though.
83.	Robert:	It's a triangle for me too.
84.	Other Ss:	Me too.
85.	T:	But we have to try and explain why it's a triangle for Morgan. You have to convince her. Why do you think it's a triangle because she's not sure?
86.	Dasia:	Because it has three corners.
87.	T:	What else does it have?
88.	Nadia:	Three lines.
89.	T:	We need to come up with a definition we all agree on and so far your definition is three vertices, or you can call them corners, and three sides.
90.	Michael:	Well it's a triangle as long as it has a point and as long as it had three dots and as long as it has three lines.
91.	Leah:	Yeah whatever you do with it, it's still a triangle.

Dasia, whose utterance [86] showed once again that for her deciding whether a drawing was a triangle required just checking whether it was showing three corners. In contrast, Morgan's use of triangle included a broader class of transformations of the canonical one, the routine of identification was still exclusively visual, with no appeal to properties. The rest of the class, however, could now speak about a *family* of triangles rather than a single one, and the family was unified by common properties. Dasia's and Michael's utterances [82] and [90], respectively, showed an interesting discursive gap: whereas Dasia appealed only to properties that were both necessary and sufficient, Michael added to the "three dots" the requirement that there must also be "a point" and "three lines," thus making his statement more of a description than a definition.

The children's talk bore clear evidence of its being developed in a DGE. With the phrases "as long as," (Michael, [90]) and "still" (Leah, [91]) the children communicated their reliance on transformability in trying to identify a triangle—thus supporting Battista's hypothesis. Leah, in particular, used a dynamic, material language in which a triangle was something you could do to things. Her "whatever you do" referred to what could be done with one of the triangle's corners. Unlike Michael, she did not draw on the properties; her statement spoke to the variability she thought the triangle had.

We see this episode as relevant to the proving process in that the development of the notion of triangle is not achieved through memorization of vocabulary but, instead, through a back-and-forth between the visual and discursive registers. In prompting the students to use words, to compare their visual identifications with their verbal statements, and to consider counter-examples, the teacher supported the use of explanation. The students moved toward deductive statements such as "that shape is a triangle *because* it has three sides." Further, this excerpt revealed the way in which the work on identifying shapes as simple as triangles, in the primary grades, can be vectored toward the development from description to definition—a crucial one in the proving process.

Measurement in a High School Setting

Students in the 10th grade were involved in a teaching experiment aimed at introducing the proving process as a continuing activity from construction of a figure in a DGE and its exploration until the formulation of a proof, via some conjectures. The teacher emphasized the need for proof as the explanation of *why* a statement is valid within some hypothesis. The students, adopting this didactic contract, knew that once they have found one or more conjectures in the geometric situation, they had to justify and explain them. The research study started from the hypothesis (Arzarello et al., 1998, 2002) that the DGE can support a cognitive continuity from conjecturing to proving, whereas conjecture and proof have two different epistemological statuses

In the classroom sessions (each of them lasted about two hours) the students were engaged in solving an open problem working in pairs with *Cabri* and sharing their solution at the end of the group session. Discussion was coordinated by the teacher. Data collected were field notes and videotapes of the work, made in *Cabri* and in paper and pencil. With these data, researchers can pay attention to students' approaches to measurements and to their cognitive role during the activity. The data showed students' frequent passages from the spatio-graphical to the theoretical field.

In the example we present here, two students (Alessandra and Tiziana) in their working group explored a geometric situation, made conjectures and proved one of them. After having done this, they used measurements to check their proof. So, they did a passage from their theoretical approach (doing the proof) to a spatio-graphical approach again (checking the proof with measures), which showed the use of the measurement tool in modality *proof measuring* of the list above. At this point, students were surprised to find that the two results, coming from theory and experiment, were not coherent (Olivero & Robutti, 2002, 2007). These students were medium achievers; they had not used *Cabri* very often before, and had no substantial experience with open problems before this one.

The problem given to the students is "Varignon's problem," formulated as follows: Draw any quadrilateral *ABCD*. Draw the midpoints *L, M, N, P* of the four sides.

1. Which properties does the quadrilateral *LMNP* have?
2. Which particular configurations does *LMNP* assume?
3. Which hypotheses on the quadrilateral *ABCD* are needed, in order for *LMNP* to assume those particular configurations? Make conjectures and prove them.

At the beginning the two students explored the situation in *Cabri* and formulated the following conjecture: if the external quadrilateral is a square, the internal quadrilateral is a rhombus. After that, they proved the conjecture, based on Figure 19.4 drawn on paper. Note that it is not surprising that the students first call the "inside"

Figure 19.4. Proving that "If ABCD is a square then HKLM is a rhombus."

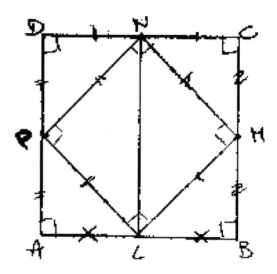

shape a rhombus, because their shape identification strategies were visually oriented, namely based on a visual association with a prototypical shape—they do not call it a "square" because its sides are not vertical and horizontal.

181	Tiziana	All this stuff…these…they are congruent (the halves of the sides of *ABCD*). [*Then Tiziana writes down the thesis*: *LM* equals *MN*, equals *NP*, equals *PL*. *Meanwhile Alessandra uses a ruler to measure* the sides of *LMNP*.)]
189	Tiziana	So *NC* is congruent to *MC PAL* (angle) is congruent to *NCM* (angle) because *ABCD* is square [*she writes down all this*]. And so it has got right angles, right? [*She marks the right angles in ABCD*]. We didn't write this in the hypothesis, but shall I explain this?
193	Tiziana	So *PL* equals *MN*. The same for *PDN* triangle and *LBM* triangle then *PN* equals *LM*. Should I do a cross comparison? *PDN* triangle and *PAL* triangle then *PN* equals *PL*. What's missing? These two are done, these two are done
195	Tiziana	*NCM* triangle and *MBL* triangle so *NM* congruent to *ML*
197	Tiziana	They all have equal sides. So it is a rhombus! OK!

The students constructed a proof with all the logical passages required at their school level. They used the congruence of triangles in which the figure was decomposed (Figure 19.4) and the congruence theorem SAS. At the end of their proving process, they were sure of the validity of their conjecture, because they had proved it, i.e., they had deduced the thesis (the quadrilateral inside the square is a rhombus) from the hypothesis (the starting quadrilateral is a square)—thus appealing to the verification function of proof.

To be convinced of their theoretical approach, at this point, after the construction of their final proof, the two students went back to *Cabri* to look for validation—that is, empirical evidence. So, a new episode started up, marking a passage from the theoretical to the spatio-graphical field:

207	Tiziana	Try to make *ABCD* a square
208	Alessandra	[*Drags A and B trying to obtain a square with a side of 6.17 cm*]
209	Tiziana	Is it a rhombus?
210	Alessandra	Well ... [Alessandra points at the sides of *LMNP*]. A rhombus has got equal sides oh no! It's not a rhombus! (see Figure 19.5)

The two students dragged *ABCD* into a square (using guided dragging, along with a guided use of measurements), to check if *LMNP* is a rhombus. And to do this, they dragged *ABCD* not randomly, but checking measures of its sides, in order to obtain equal values. They were transforming a generic quadrilateral *ABCD* into a square just by dragging (and looking at the side measures), not by constructing it. Moreover, they used the *proof measuring* modality in that they wanted to find the empirical evidence of their theoretical finding. This passage from theoretical to spatio-graphical was not so unusual (even for mathematicians!) and could give different results. The two students were surprised when they found that such numbers were not equal (*PN* and *PL*, for example, in Figure 19.5). These students had more confidence in the measurements taken by *Cabri* than in the correctness of their proof. In fact, they immediately checked each passage of the proof but concluded that the software was right—and so they rejected their proof:

257	Tiziana	Because if this is the midpoint [*she points at P*] then it divides this side in two equal parts.[*She points at AD and AP and PD*].
263	Tiziana	So it should be: if it is a square, the quadrilateral inside is a square too.
264	Teacher	Right!
265	Tiziana	Why doesn't the figure show that?
274	Teacher	I'll try to pose the problem from a different perspective: this figure. [Points at the Figure 19.4 on the screen] Is it a square? Really?

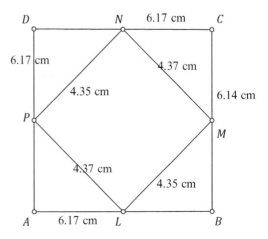

Figure 19.5. Measuring the sides of the inner quadrilateral.

The teacher's intervention focusses students' attention on the link between the spatio-graphical and the theoretical, helping them to realize that their figure (Figure 19.5) was *not* properly a square. Furthermore she suggested that they *construct* a square, and check if it was in accordance with the proof. In the sense of Laborde, these students were looking at a square as a sketch, not at a *Cabri*-square (a dynamic construction). After the teacher's suggestion, the students constructed a square and thus overcome the conflict. The use of measurements described above seemed to detach the students completely from the theoretical field and made them dependent on the spatio-graphical one.

For these students, the intervention of the teacher was crucial for overcoming the conflict and using the figure in a theoretical way, accepting the proof and going further in solving the problem. This case study thus illustrates the epistemological and cognitive implications of the measuring tool in relation to the proving process and offers a specific example of the kind of work that a teacher must do to support this process.

Section E: Conclusion and Implications

Didactic Implications of DGEs

As observed in Hoyles and Noss's (2003) *Second Handbook* chapter, research has focussed not just on the affordances of the tool, but also on the role of the teacher in the classroom. In both of the case studies shown above, the teacher played a crucial role in emphasizing relationships between objects, while trying to introduce students to the theoretical world of geometry, which is similar to Mariotti's (2000) thesis. However, as shown in Sinclair and Yurita (2008), it can be challenging for teachers to play this role, particularly when their way of thinking and talking about geometry is static in nature. Although it is clear that teachers must help students move towards explanation and deduction, there is still a great deal of research to be done on how teachers should deal with the epistemological and cognitive implications of DGE tools such as dragging and measuring. Should, for example, the various dragging modalities be modelled by the teacher? How might a teacher choose tasks that can scaffold dragging modalities in such a way to better support the proving process? Do teachers need to be more careful about how and when they introduce measure tools? How can teachers carry out in class an instrumentation of the different tools present in a DGE, in the sense of giving students schemes of use of these tools and awareness of their use according to the kind of problem (construction, exploration, modelling, and so on)?

The orchestration of different materials and technologies (Trouche, 2004), and the use of new teaching practices (working groups, discussions, distance education), are difficult and can create confusion for teachers. However, many national and international projects of teacher education are directing their efforts at supporting teachers in changing (see Sinclair, Arzarello, Gaisman, & Lozano, 2009). More research, along the lines of Laborde's (2001), needs to focus on developing better teacher integration, including supporting the development of new tasks that are suitable to DGE use and new modes of assessment that do not simply fall back to pencil-and-paper testing.

Research Implications of DGEs

Hoyles and Noss (2003) stated that "attention is turning away from the investigation of the process of construction and conjecturing with DGS, and towards consideration of how the new tools mediate the nature of explanation, verification and even proof" (p. 335). Since the early 2000s, we have seen a strong emphasis on how those two processes are related, which has led to an increasing range of contexts of study, across the grade levels (from primary school to university level, including pre-service teacher education—see Lassak, 2009). This may lead to more longitudinal research on the way in which the proving process might develop over the school curriculum—such longitudinal studies have been quite rare in the field of technology, but they would be particularly helpful in elucidating the nature of the proving process. In particular, as anticipated in the first case study, we need new models, or learning trajectories, that can account for the impact of DGE use in the development of student thinking.

Attention is also turning to the ways by which DGEs can mediate the objects of geometric investigation and the ways students can think about these objects, as discussed in section C. With the fine-grained analyses of dragging and measuring already undertaken, we tried to show that the dynamicity and the measure give students the opportunity to pass from the exploration phase, characterized by conjectures, to the proving phase, characterized by the construction of proof, with continuity.

Over the past decade, researchers have identified some of the more subtle aspects of learning to use and teach with DGEs. For example, the literature suggests that teachers need to help students develop "schemes of use" (Rabardel, 1995)—not just how to drag a point, but why a point or an object is not draggable, how and when to use a measure to infer a property, how to read a measure in a proper way, and how to use dragging to obtain a particular configuration. This awareness of the schemes of use enables students to master the problem, from the hypothesis to the thesis, during the various solution phases (Sträßer, 2009). More research is needed to study how (and how much) teachers should "instrument" their students in the use of technologies. A delicate balance is needed since, on the one hand, it is important to introduce schemes of use in a cognitive and metacognitive way, but on the other, this activity should not become a sequence of instructions and rules, but an investigation methodology. In fact, tools are active elements of the culture in which they are used as they involve cognitive structures (schemes) and also introduce cultural systems that are products of cultural evolution. For this reason, we have to examine the closeness of the notion of "scheme" to that of "situated abstraction" (Noss & Hoyles, 1996) as a complex product of activity, context, history and culture. And we have to examine the complexity of the relation between humans and technology (Borba &

Villarreal, 2005)—not only as one subject interacting with one tool, but as *more subjects* interacting together and with *more tools*.

For what concerns *more subjects*, Trouche (2004) notes that: "an instrument is the result of a construction by a subject, in a community of practice, on a basis of a given artifact, through a process, the instrumental genesis" (p. 289). The social dimension of the tool can be analyzed according to different theoretical perspectives, such as semiotic mediation (Mariotti, 2010) or the human, embodied, cultural and multimodal perspectives put forward by Radford (2010) and Arzarello and Robutti (2010). These frames consider all the signs introduced by students (verbal utterances, gestures, bodily movements, actions on the computer, etc.), during a classroom activity to be relevant for understanding cognitive processes.

In terms of *more artefacts*, we again point to Trouche's emphasis on the importance of "instrumental orchestration," which involves the design of tasks, the guidance of students' instrumental genesis, and the environmental organization (Trouche, 2004). Instrumental orchestration is achieved by "didactic configurations" (layout of available artefacts in the environment), "exploitation modes" of these configurations (decisions for carrying out the activities with artefacts) and "didactical performance" (ad hoc decisions taken while teaching on how actually to perform in the chosen didactic configuration and exploitation mode). Recent trends in research study the types of instrumental orchestration that emerge in technology-rich classroom teaching and to what extent are teachers' repertoires of orchestrations related to their expressed views on mathematics education and the role of technology (Drijvers, Doorman, Boon, Reed, & Gravemeijer, 2010). They should inform research on the use of DGEs in the proving process.

References

Arzarello, F., Micheletti, C., Olivero, F., Paola, D., & Robutti, O. (1998). A model for analysing the transition to formal proofs in geometry. In A. Olivier & K. Newstead (Eds.), *Proceedings of the 22nd Conference of the International Group for the Psychology of Mathematics Education* (Vol. 2, pp. 24–31). Stellenbosh, South Africa: International Group for the Psychology of Mathematics Education.

Arzarello, F., Olivero, F., Paola, D., & Robutti, O. (1999). I problemi di costruzione geometrica con l'aiuto di Cabri. *L'insegnamento della matematica e delle scienze integrate, 22B*(4), 309–338.

Arzarello, F., Olivero, F., Paola, D., & Robutti, O. (2002). A cognitive analysis of dragging practises in Cabri environments. *ZDM—Zentralblatt fur Didaktik der Mathematik, 34*(3), 66–72.

Arzarello, F., & Robutti, O. (2010). Multimodality in multi-representational environments. *ZDM—The International Journal on Mathematics Education, 42*(7), 715–731.

Baccaglini-Frank, A., & Mariotti, M. A. (2010). Generating conjectures in dynamic geometry: The Maintaining Dragging Model. *International Journal of Computers for Mathematical Learning, 15*(3), 225–253.

Balacheff, N. (1987). Processus de prevue et situations de validation. *Educational Studies in Mathematics, 8*, 147–176.

Bartolini Bussi, M. (2009). Proof and proving in primary school: An experimental approach. In F.-L. Lin, F.-J. Hsieh, G. Hanna, & M. De Villiers (Eds.), *Proceedings of the ICMI Study 19 Conference: Proof and proving in mathematics education* (Vol. 1, pp. 53–58). Taipei, Taiwan: National Taiwan Normal University.

Battista, M. T. (2007). The development of geometric and spatial thinking. In F. Lester (Ed.), *Second handbook of research on mathematics teaching and learning* (pp. 843–908). Reston, VA: National Council of Teachers of Mathematics.

Battista, M. T. (2008). Representations and cognitive objects in modern school geometry. In G. Blume & M. K. Heid (Eds.), *Research on technology and the teaching and learning of mathematics: Cases and perspectives* (Vol. 2, pp. 341–362). Charlotte, NC: Information Age Publishing.

Baulac, Y., Bellemain, F., & Laborde, J. M. (1988). *Cabri-Géomètre, un logiciel d'aide à l'apprentissage de la géométrie: Logiciel et manuel d'utilisation.* Paris, France: Cedic-Nathan.

Bishop, A. J., Clements, M. A., Keitel, C., Kilpatrick, J., & Leung, F. K. S. (Eds.). (2003). *Second international handbook of mathematics education.* Dordrecht, The Netherlands: Kluwer.

Boero, P., Garuti, R., Lemut, E., & Mariotti, A. M. (1996). Challenging the traditional school approach to theorems: A hypothesis about the cognitive unity of theorems. In L. Puig & A. Gutiérrez (Eds.), *Proceedings of the 20th Conference of the International Group for the Psychology of Mathematics Education* (Vol. 2, pp. 113–120). Valencia, Spain: International Group for the Psychology of Mathematics Education.

Borba, M. C., & Villarreal, M. E. (2005). *Humans-with-media and the reorganization of mathematical thinking.* New York, NY: Springer.

Borwein, J., & Bailey, D. (2008). *Mathematics by experiment: Plausible reasoning in the 21st century.* Wellesley, MA: A. K. Peters/CRC Press.

Bruckheimer, M., & Arcavi, A. (2001). A Herrick among mathematicians or dynamic geometry as an aid to proof. *International Journal of Computers in Mathematics Learning, 6*, 113–126.

Centre de Recherche sur l'Ensignement des Mathématiques (CREM). (1995). *Les mathématiques de la maternelle jusqu'à 18 ans. Essai d'élaboration d'un cadre global pour l'enseignement des mathématiques.* In L. Grugnetti & V. Villani [Italian edition, *La matematica dalla scuola materna all'Università*]. Bologna, Italy: Pitagora Editrice.

Christou, C., Mousoulides, N., Pittalis, M., & Pitta-Pantazi, D. (2005). Proofs through exploration in dynamic geometry environments. *International Journal of Science and Mathematics Education, 2*, 339–352.

de Villiers, M. (1990). The role and function of proof in mathematics. *Pythagoras, 24*, 17–24.

de Villiers, M. (1997). The role of proof in investigative, computer-based geometry: Some personal reflections. In J. King & D. Schattschneider (Eds.), *Geometry turned on* (pp. 15–24). Washington, DC: Mathematical Association of America.

de Villiers, M. (1998). An alternative approach to proof in dynamic geometry? In R. Lehrer & D. Chazan (Eds.), *Designing learning environments for developing understanding of geometry and space* (pp. 369–393). Mahwah, NJ: Lawrence Erlbaum.

de Villiers, M. (1999). *Rethinking proof with the Geometer's Sketchpad.* Emeryville, CA: Key Curriculum Press.

de Villiers, M. (2010). The role of experimentation in mathematics and mathematics education. In G. Hanna & H. N. Jahnke (Eds.), *Explanation and proof in mathematics: Philosophical and educational perspectives* (pp. 205–222). Basel, Switzerland: Springer Books.

Drijvers, P., Doorman, M., Boon, P., Reed, H., & Gravemeijer, K. (2010). The teacher and the tool: Instrumental orchestrations in the technology-rich mathematics classroom. *Educational Studies in Mathematics, 75*(2), 213–234.

Drijvers, P., Kieran, C., Mariotti, M., Ainley, J., Andresen, M., Chan, Y., Dana-Picard, T., Gueudet, G., Kidron, I., Leung, A., & Meagher, M. (2010). Integrating technology into mathematics education: Theoretical perspectives. In C. Hoyles & J. B. Lagrange (Eds.), *Mathematics education and technology—Rethinking the terrain* (pp. 89–132). New York, NY: Springer.

Duval, R. (2006). A cognitive analysis of problems of comprehension in the learning of mathematics. *Educational Studies in Mathematics, 61*, 103–131.

Erez, M. M., & Yerulshalmy, M. (2006). "If you can turn a rectangle into a square, you can turn a square into a rectangle …": Young students experience the dragging tool. *International Journal of Computers for Mathematical Learning, 11*, 271–299.

Frant, J. B., & de Costra, R. M. (2000, July). *Proofs in geometry: Different concepts build upon very different cognitive mechanisms*. Paper presented at the ICME 9, TSG12: Proof and Proving in Mathematics Education, Tokyo, Japan.

Guala, E., & Boero, P. (1999). Time, complexity and learning. *Annals of the New York Academy of Sciences, 879*, 164–167.

Guven, B., Cekmez, E., & Maratas, I. (2010). Using empirical evidence in the process of proving: The case of dynamic geometry. *Teaching Mathematics and Its Applications, 29*(4), 193–207.

Hadas, N., Hershkowitz, R., & Schwartz, B. (2000). The role of contradiction and uncertainty in promoting the need to prove in dynamic geometry environments. *Educational Studies in Mathematics, 44*, 127–150.

Hanna, G., Jahnke, H. N., & Pulte, H. (Eds.). (2010). *Explanation and proof in mathematics: Philosophical and educational perspectives*. New York, NY: Springer.

Healy, L., & Hoyles, C. (2001). Software tools for geometrical problem solving: Potentials and pitfalls. *International Journal of Computers for Mathematical Learning, 1*(3), 235–256.

Herbst, P. (2002). Engaging students in proving: A double bind on the teacher. *Journal for Research in Mathematics Education, 33*(3), 176–203.

Hollebrands, K. F., Conner, A. M., & Smith, R. C. (2010). The nature of arguments provided by college geometry students with access to technology while solving problems. *Journal for Research in Mathematics Education, 41*(4), 324–350.

Hollebrands, K., Laborde, C., & Sträßer, R. (2008). Technology and the learning of geometry at the secondary level. In K. Heid & G. Blume (Eds.), *Research in technology and the teaching and learning of mathematics: Research syntheses* (Vol. 1, pp. 155–203). Charlotte, NC: Information Age Publishing.

Hoyles, C., & Noss, R. (2003). What can digital technologies take from and bring to research in mathematics education? In A. J. Bishop, M. A. Clements, C. Keitel, J. Kilpatrick, & F. Leung (Eds.), *Second international handbook of mathematics education* (pp. 323–349). Dordrecht, The Netherlands: Kluwer Academic Publishers.

Jackiw, N. (1991, 2001). *The Geometer's Sketchpad* [Computer Program]. Emeryville, CA: Key Curriculum Press.

Jackiw, N. (2006). Mechanism and magic in the psychology of dynamic geometry. In N. Sinclair, D. Pimm, & W. Higginson (Eds.), *Mathematics and the aesthetics: New approaches to an ancient affinity* (pp. 145–159). New York, NY: Springer.

Jahn, A. P. (2002). "Locus" and "Trace" in Cabri-géomètre: Relationships between geometric and functional aspects in a study of transformations. *ZDM—Zentralblatt fur Didaktik der Mathematik, 34*(3), 78–84.

Japanese Society of Mathematics Education. (2000). *Mathematics programme in Japan*. Tokyo, Japan: Author.

Jones, K. (2000). Providing a foundation for deductive reasoning: Students' interpretations when using dynamic geometry software and their evolving mathematical explanations. *Educational Studies in Mathematics, 44*, 55–85.

Jones, K., Mackrell, K., & Stevenson, I. (2010). Designing digital technologies and learning activities for different geometries. In C. Hoyles & J.-B. Lagrange (Eds.), *Mathematics education and technology—Rethinking the terrain New ICMI Study Series* (Vol. 13, pp. 47–60). Dordrecht, The Netherlands: Springer.

Laborde, C. (1992). Solving problems in computer based geometry environments: The influence of the features of the software. *ZDM—Zentrablatt für Didaktik des Mathematik, 92*(4), 128–135.

Laborde, C. (1998). Relationship between the spatial and theoretical in geometry: The role of computer dynamic representations in problem solving. In J. D. Tinsley & D. C. Johnson (Eds.), *Information and communications technologies in school mathematics* (pp. 183–195). London, UK: Chapman & Hall.

Laborde, C. (2000). Dynamical geometry environments as a source of rich learning contexts for the complex activity of proving. *Educational Studies in Mathematics, 44*(1–2), 151–161.

Laborde, C. (2001). Integration of technology in the design of geometry tasks with Cabri-Géomètre. *International Journal of Computers for Mathematical Learning, 6*, 283–317.

Laborde, C. (2004). The hidden role of diagrams in pupils' construction of meaning in geometry. In J. Kilpatrick, C. Hoyles, & O. Skovsmose (Eds.), *Meaning in mathematics education* (pp. 1–21). Dordrecht, The Netherlands: Kluwer Academic Publishers.

Laborde, C., Kynigos, C., Hollebrands, K., & Sträßer, R. (2006). Teaching and learning geometry with technology. In A. Gutiérrez & P. Boero (Eds.), *Handbook of research on the psychology of mathematics education: Past, present and future* (pp. 275–304). Rotterdam, The Netherlands: Sense Publishers.

Lassak, M. (2009). Using dynamic graphs to reveal student reasoning. *International Journal of Mathematical Education in Science and Technology, 40*(5), 690–696.

Lehrer, R., Jenkins, M., & Osana, H. (1998). Longitudinal study of children's reasoning about space and geometry. In R. Lehrer & D. Chazan (Eds.), *Designing learning environments for developing understanding of geometry and space* (pp. 137–167). Mahwah, NJ: Lawrence Erlbaum.

Leung, A. (2003). Dynamic geometry and the theory of variation. In N. A. Pateman, B. J. Doughherty, & J. T. Zillox (Eds.), *Proceedings of PME 27: Psychology of Mathematics Education 27th International Conference* Volume 3 (pp. 197–204). Honolulu: University of Hawaii.

Leung, A. (2008). Dragging in a dynamic geometry environment through the lens of variation. *International Journal of Computers in Mathematics Learning, 13*, 135–157.

Leung, A., & Or, C. M. (2007). From construction to proof: Explanations in dynamic geometry environments. In J. H. Woo, H. C. Lew, K. S. Park, & D. Y. Seo (Eds.), *Proceedings of the 31st Conference of the International Group for the Psychology of Mathematics Education* (Vol. 3, pp. 177–184). Seoul, Korea: International Group for the Psychology of Mathematics Education.

Mariotti, M. A. (2000). Introduction to proof: The mediation of dynamical software environment. *Educational Studies in Mathematics, 44*(1–2), 25–53.

Mariotti, M. A. (2002). Influence of technologies advances on students' math learning. In L. English (Ed.), *Handbook of international research in mathematics education* (pp. 695–723). Mahwah, NJ: Lawrence Erbaum.

Mariotti, M. A. (2006). Proof and proving in mathematics education. In A. Guttiérrez & P. Boero (Eds.), *Handbook of research on the psychology of mathematics education: Past, present and future* (pp. 173–204). Rotterdam, The Netherlands: Sense Publishing.

Mariotti, M. A. (2010). Proofs, semiotics and artefacts of information technologies. In G. Hanna, H. N. Jahnke, & H. Pulte (Eds.), *Explanation and proof in mathematics: Philosophical and educational perspectives* (pp. 169–190). Dordrecht, The Netherlands: Springer.

Marrades, R., & Gutierrez, A. (2000). Proofs produced by secondary school students learning geometry in a dynamic computer environment. *Educational Studies in Mathematics, 44*, 87–125.

Marton, F., & Booth, S. (1997). *Learning and awareness*. Mahwah, NJ: Lawrence Erlbaum.

Ministry of Education, People's Republic of China. (2001). *Full-time obligatory education mathematics curriculum standards (experimental version)* [in Chinese]. Beijing, China: Beijing University Press.

National Council of Teachers of Mathematics. (2000). *Principles and standards for school mathematics*. Reston, VA: Author.

Noss, R., & Hoyles, C. (Eds.). (1996). *Windows on mathematical meanings—Learning cultures and computers*. Dordrecht, The Netherlands: Kluwer Academic Publishers.

Olivero, F. (2002). *The proving process within a dynamic geometry environment* (Doctoral thesis). University of Bristol, UK

Olivero, F. (2006). Students' constructions of dynamic geometry. In C. Hoyles, J.-B. Lagrange, L.-H. Son, & N. Sinclair (Eds.), *Proceedings of the 17th International Congress on Mathematical Instruction Study Conference "Technology Revisited"* (pp. 433–442). Hanoi, Vietnam: Hanoi Institute of Technology and Didirem University.

Olivero, F., & Robutti, O. (2001). Measuring in Cabri as a bridge between perception and theory. In M. van den Heuvel-Panhuizen (Ed.), *Proceedings of the 25th Conference of the International Group for the Psychology of Mathematics Education* (Vol. 4, pp. 9–16). Utrecht, The Netherlands: International Group for the Psychology of Mathematics Education.

Olivero, F., & Robutti, O. (2002). An exploratory study of students' measurement activity in a dynamic geometry environment. In J. Novotńa (Ed.), *Proceedings of CERME 2* (Vol. 1, pp. 215–226). Prague, Czech Republic: Charles University.

Olivero, F., & Robutti, O. (2007). Measuring in dynamic geometry environments as a tool for conjecturing and proving. *International Journal of Computers for Mathematical Learning, 12*(2), 135–156.

Oner, D. (2008). A comparative analysis of high school geometry curricula: What do technology-intensive, standards-based, and traditional curricula have to offer in terms mathematical proof and reasoning? *Journal of Computers in Mathematics and Science Teaching, 27*(4), 467–497.

Oner, D. (2009). The role of dynamic geometry software in high school geometry curricula: An analysis of proof tasks. *International Journal for Technology in Mathematics Education, 16*(3), 109–121.

Rabardel, P. (1995). *Les hommes et les technologies—Approche cognitive des instruments contemporains*. Paris, France: A. Colin.

Radford, L. (2010). The anthropological turn in mathematics education and its implication on the meaning of mathematical activity and classroom practice. *Acta Didactica Universitatis Comenianae, 10*, 103–120.

Rav, Y. (1999). Why do we prove theorems? *Philosophia Mathematica III, 7*, 5–41.

Sinclair, N. (2002). The kissing triangles: The aesthetics of mathematical discovery. *International Journal of Computers for Mathematical Learning, 7*(1), 45–63.

Sinclair, M. P. (2003). Some implications of the results of a case study for the design of pre-constructed dynamic geometry sketches and accompanying materials. *Educational Studies in Mathematics, 52*(3), 289–317.

Sinclair, N. (2008). *The history of the geometry curriculum in the United States*. Charlotte, NC: Information Age.

Sinclair, N., Arzarello, F., Gaisman, M. T., & Lozano, M. D. (2009). Implementing digital technologies at a national scale. In C. Hoyles & J.-B. Lagrange (Eds.), *Digital technologies and mathematics teaching and learning: Rethinking the terrain* (Vol. 13, pp. 61–78). New York, NY: Springer.

Sinclair, N., Moss, J., & Jones, K. (2010). Developing geometric discourse using DGS in K-3. In M. M. F. Pinto & T. F. Kawasaki (Eds.), *Proceedings of the 34th Conference of the International Group for the Psychology of Mathematics Education* (Vol. 4, pp. 185–192). Belo Horizonte, Brazil: International Group for the Psychology of Mathematics Education.

Sinclair, N., & Yurita, V. (2008). To be or to become: How dynamic geometry changes discourse. *Research in Mathematics Education, 10*(2), 135–150.

Sträßer, R. (2009). Instruments for learning and teaching mathematics. An attempt to theorise about the role of textbooks, computers and other artefacts to teach and learn mathematics. In M. Tzekaki, M. Kaldrimidou, & H. Sakonidis (Eds.), *Proceedings of the 33rd Conference of the International Group for the Psychology of Mathematics Education* (Vol. 1, pp. 67–81). Thessaloniki, Greece: International Group for the Psychology of Mathematics Education.

Stylianides, A. J. (2007). The notion of proof in the context of elementary school mathematics. *Educational Studies in Mathematics, 65*, 1–20.

Stylianou, D., Knuth, E., & Blanton, M. (Eds.). (2009). *Teaching and learning proof across the grades: A K-16 perspective*. New York, NY: Routledge.

Trouche, L. (2004). Managing complexity of human machine interactions in computerized learning environments: Guiding student's command process through instrumental orchestrations. *International Journal of Computers for Mathematical Learning, 9*(3), 281–307.

Unione Matematica Italiana. (2004). G. Anichini, F. Arzarello, L. Ciarrapico & O. Robutti (Eds.), *Matematica 2003. La matematica per il cittadino. Attività didattiche e prove di verifica per un nuovo curricolo di Matematica (ciclo secondario)*. Lucca, Italy: Matteoni Stampatore.

Vadcard, L. (1999). La validation en géomètrie au collège avec Cabri-Géomètre: Mesures exploratoire et mesures probatoires. *Petit X, 50*, 5–21.

Verillion, P., & Rabardel, P. (1995). Cognition and artifacts: A contribution to the study of thought in relation to instrumented activity. *European Journal of Psychology of Education, 10*(1), 77–101.

Vygotsky, L. S. (1978). *Mind in society. The development of higher psychological processes*. Cambridge, MA: Harvard University Press.

Chapter 20
How Might Computer Algebra Systems Change the Role of Algebra in the School Curriculum?

M. Kathleen Heid, Michael O. J. Thomas, and Rose Mary Zbiek

Abstract Computer Algebra Systems (CAS) are software systems with the capability of symbolic manipulation linked with graphical, numerical, and tabular utilities, and increasingly include interactive symbolic links to spreadsheets and dynamical geometry programs. School classrooms that incorporate CAS allow for new explorations of mathematical invariants, active linking of dynamic representations, engagement with real data, and simulations of real and mathematical relationships. Changes can occur not only in the tasks but also in the modes of interaction among teachers and students, shifting the source of mathematical authority toward the students themselves, and students' and teachers' attention toward more global mathematical perspectives. With CAS a welcome partner in school algebra, different concepts can be emphasized, concepts that are taught can be done so more deeply and in ways clearly connected to technical skills, investigations of procedures can be extended, new attention can be placed on structure, and thinking and reasoning can be inspired. CAS can also create the opportunity to extend some algebraic procedures and introduce and assist exploration of new structures. A result is the enrichment of multiple views of algebra and changing classroom dynamics. Suggestions are offered for future research centred on the use of CAS in school algebra.

Developing an understanding of algebra is central to school mathematics, and the teaching and learning of algebra is receiving increasingly greater attention in a range of national settings. In the USA, for example, the President convened a National Math Panel, a central purpose of which was to provide the best advice on preparing children for the study of algebra. As nations attack the issue of enhancing

---

M. K. Heid (✉) • R. M. Zbiek
The Pennsylvania State University, University Park, PA, USA
e-mail: mkh2@psu.edu

M. O. J. Thomas
The University of Auckland, Auckland, New Zealand

students' understanding of algebra, it becomes important to define what is meant by algebra. Textbooks dedicated to algebra identify topics to be covered, and national, state, and local goals identify the algebraic skills that students must master. But algebra is more than a list of topics. It is also a way of thinking and reasoning. In this chapter, we consider algebra to consist not only of a set of mathematical topics but also ways of thinking.

Just as it is important to examine strategies for improving students' understanding of algebra, it is important to do so in the context of the technological resources available for assisting the learning of algebra. As algebra-specific software becomes increasingly available in school mathematics classrooms, it becomes more and more important to examine the ways in which such software can affect the teaching and learning of algebra and the part that algebra plays in developing students' understanding of mathematics. One configuration of software that is particularly relevant to the learning of algebra is what has come to be known as a Computer Algebra System (CAS). The CAS can, on command, perform symbolic manipulations that often comprise much of a student's algebra skill set. The basic utilities of a CAS are enhanced by the linking capability among its components. A CAS links graphical, numerical, and tabular utilities with that of a symbolic manipulator, and interactive symbolic links to spreadsheets and dynamical geometry programs are becoming a more common part of CAS configurations. Communication among these latter components opens the possibilities for decreasing barriers that have at times separated the study of algebra from the study of other areas of the mathematical sciences. The capability of linking symbolic mathematics capabilities to graphical and dynamic geometry, for example, opens the possibility of symbolic experimentation supplemented by graphical parametric exploration and corroborated through geometrical construction and measurement. Networking with the capacity to collect and display results from a large group of students allows experimentation more easily to become a group project instead of an individual investigation. The ever-increasing possibilities for connections and interactions open the door for an algebra that links traditional notation systems and representations to new ones. The myriad current possibilities for CAS encourage substantial changes in the role of algebra in the school curriculum. This chapter discusses those potential changes.

Brief History of CAS in Mathematics Education

To provide a context, before examining how CAS might affect the role of algebra in the school curriculum, it seems useful to review how the use of technology in mathematics education has evolved. The evolution has been threefold. The type of technology available has evolved, the ways in which that technology is used have evolved, and research and theories about teaching and learning in the context of that technology have evolved. These evolutions have been interdependent with limitations on the available technologies constraining the ways that they could be used, and limitations on uses constraining the field's ability to investigate and explain learning in the context of technology. For each phase of this evolution and each general type of technology, there was initially a time during which there was experimentation with what could be done. This was generally followed by the development of curricula and instructional approaches and investigation of the effects of the technology's use in the consequent range of settings. Finally, often after periods of experimentation and development, theory was developed or expanded to explain the use of that technology. For CAS, the initial work was limited by the platforms on which the CAS was built. Early versions consisted of only symbolic manipulation programs.

Work in use of technology in mathematics education has evolved in the areas of curriculum and instruction. Neither of these foci replaced the others, but the consideration of CAS in each enriched the field's perspective on what was involved with the incorporation of technology in mathematics education. This evolution has been reflected in work with CAS as well, and the development of theories about technology use was accelerated by CAS-related work. Initial curriculum work focussed on development of CAS approaches to algebra by students as exemplified early in small trials of Computer-Intensive Algebra (CIA) (early versions of Fey & Heid, 1995), later in widespread use of CAS calculators in Austrian (Böhm, 2007) and Australian schools (see http://extranet.edfac.unimelb.edu.au/DSME/CAS-CAT/publicationsCASCAT/Publications.shtml#2009 for an extensive publication list related to Australian CAS-CAT work), and finally in the incorporation of CAS work in widely used curricula such as those of the University of Chicago Mathematics Project (Usiskin, 2004). Various configurations were tried in the course of experimentation with CAS in school algebra, ranging from supplements to an entire curriculum. Theory related to instruction has evolved from characterizing the nature of technical work with CAS (Artigue, 2002; Lagrange, 1999), to describing the work methods of students using CAS (Guin & Trouche, 1999; Trouche, 2005a), and to developing theory describing the relationships between the instructor, students, and CAS (Trouche, 2005b). Attention is now turning to the networking and connectedness possibilities with the advent of the TI Navigator for TI-Nspire with CAS (see Roschelle, Vahey, Tatar, Kaput, & Hegedus, 2003, for a discussion of networking and connectedness in mathematics instruction).

In spite of the long history of work with CAS in educational settings, the impact of technology on school mathematics has to date been marginal, and the incorporation of CAS in classrooms has been even slower. Some would attribute this slow movement to the time it takes to implement fully any change (Drijvers & Weigand, 2010). Others would attribute it to the difficulty of making such radical change in the nature of school mathematics or to the difficulties involved in preparing teachers to work effectively with such changes (Zbiek & Hollebrands, 2008). Barriers to incorporation of CAS in school mathematics, however, could also have been related to the nature of the tool itself and its potential uses in school mathematics. The prospect of incorporating CAS as a constant resource in students' algebra experiences has been regarded with trepidation by those who imagined students replacing by-hand facility with symbolic manipulation resulting in a need to depend on technology for transformation of symbolic expressions and equations. They may have suspected that what had been the essence of school algebra would, in

CAS-enabled classrooms, be set aside. On one side, the debate regarding the nature of the change needed in fully integrating the CAS into school mathematics curricula is fuelled by the supposition that a curriculum that does not focus primarily on by-hand symbolic manipulation would deprive students of the insights that could be gained from refined by-hand symbolic manipulation. On the other hand, Dick (1992) pointed out that "to realize the savings in time and to harness the power of computation that a symbolic calculator can provide, students need to pay more, not less, attention to understanding the meaning of the symbols and notation they use" (p. 2).

Throughout its history in school mathematics classrooms, CAS has offered a range of new opportunities for the teaching and learning of algebra and the resultant effects on the nature and depth of mathematical content as well as on the nature of assessment. Researchers have investigated the effects of CAS on the content, teaching, and assessment of school algebra. With constant access to CAS, the nature of tasks, classroom interactions, and views of mathematics could be transformed. Pierce, Stacey, and Wander (2010) illustrated, and richly conveyed, pedagogical opportunities in classrooms that have constant access to CAS (see Figure 20.1). Because of the CAS capacity to execute symbolic procedures rapidly and accurately, time is available to engage students regularly in an expanded range of task types. The symbolic manipulation capacity of the CAS allows for exploration of different mathematical ideas in ways that were either not possible or not feasible without such technological help. These new opportunities involve exploration of mathematical invariants, active linking of dynamic representations, and engagement

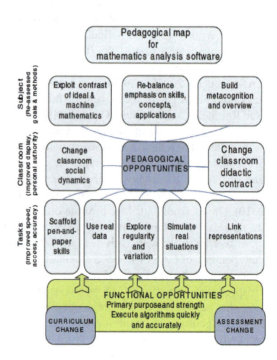

Figure 20.1. Pedagogical opportunities in classrooms that had constant access to CAS (from Pierce, Stacey, & Wander, 2010).

with real data and simulations of real and mathematical relationships. With the welcoming of CAS in school classrooms, changes can occur not only in the tasks but also in the modes of interaction among teachers and students. With powerful tools in students' hands, the source of authority can shift toward the students themselves and teachers and students can engage in a newly defined relationship that includes not only the teacher, the tasks, and the students but also the technology. Students' and teachers' attention can turn toward more global mathematical perspectives, such as recognizing the affordances and constraints of work with technology and maintaining a balance of procedural and conceptual knowledge.

Our examination of literature across the history of CAS in mathematics education suggests three topics that are central to discussions of research, theory, or practice: the interaction of concepts and procedures; new concepts, extended procedures, and structures that can be approached with CAS; and the thinking and reasoning that CAS use inspires or requires. In the following sections, we undertake each of these three topics before we come to terms with the role of algebra in the school curriculum and address associated issues and needed research.

The Role of CAS in Calibrating the Conceptual–Technical Balance of Algebra Instruction

In considering the potential for CAS to affect the role of algebra in the school curriculum, it is the symbolic manipulation capacity of CAS that has drawn the most attention. Initial concern was directed at what was perceived to be the imbalance of procedures and concepts in the algebra curriculum, even though subject matter content may not be readily categorizable into either of these subject matter types. Researchers recognized that often the classroom focus was on procedures with little attention to concepts that would signal when those procedures were called for. They experimented with relegating large parts of the symbolic manipulation to the CAS and concentrating attention on understanding fundamental concepts and when particular symbolic manipulations were appropriate. They investigated whether such a shift in focus would result in atrophy or failure to develop by-hand symbolic manipulation skills, whether such re-balancing could result in a more in-depth development of conceptual understanding, and whether a re-balanced approach would result in improved success in problem solving that required execution of particular procedures.

As described in research syntheses focussed on technology in mathematics instruction (e.g., Heid, 1997; Heid & Blume, 2008), early studies examined the effects of various approaches to using CAS on the balance of mathematical procedures and concepts in the curriculum. Studies by Heid (1984, 1988), Palmiter (1991), and Judson (1990) provided evidence that calculus courses at the collegiate level could be designed to use symbolic calculation programs to foster the development of concepts and understanding regarding when to use particular procedures without harming the development of students' skill at transforming and using symbolic forms. Similar results were obtained in early studies of students' learning of algebra using CIA—see early versions of Fey and Heid (1995)—a functions-based algebra curriculum used at the school and college levels that gave students constant access to some form of CAS (Boers van Oosterum, 1990; Heid, 1992; Heid, Sheets, Matras, & Menasian, 1988; Matras, 1988; O'Callaghan, 1998; Sheets, 1993). Early research on CAS use centred on using symbolic manipulation programs, sometimes supplemented by graphing and spreadsheet programs. In these and other studies of CAS use in algebra instruction (e.g., Hollar & Norwood, 1999; Mayes, 1995), a fairly consistent result was that, in a curriculum that prioritized concepts and applications of algebra, fundamental concepts of algebra could effectively be learned without detriment to symbol manipulation procedures.

Researchers have experimented with using CAS in a variety of curricular configurations, ranging from supplements for an existing curriculum to replacement of all or some of the existing curriculum. For example, the CIA project investigated a completely reconceptualized introductory algebra curriculum. An investigation by Edwards (2001) studied effects of regularly supplementing the traditional algebra curriculum with CAS activities, and Kieran and colleagues (Kieran & Drijvers, 2006; Kieran & Saldanha, 2008) studied the effects of specifically designed CAS activities on students' work with symbolic investigations. It should be noted that each of these studies occurred in the context of a curriculum designed to capitalize on the opportunities provided by the CAS. The question was not whether the incorporation of CAS in and of itself made a difference, but whether the CAS could enable the design of algebra curricula that exemplified particular perspectives on the teaching and learning of algebra. Although these studies gave evidence that a different type of learning could occur in the context of CAS-intensive algebra classrooms, analysis of the specific nature of the learning in those settings was largely unexplored. Not every study resulted in superior performance by the CAS group (e.g., Thomas & Rickhuss, 1992), and it became evident that one of the factors that mattered was the particular way in which CAS was integrated into the curriculum. Developers and mathematics educators became wary of the potential for CAS to obscure the symbolic work and popularized a white box–black box analogy to describe the projected role of CAS in school mathematics (Buchberger, 1989). Soon thereafter, the focus of the debate shifted from the question of what effects CAS would have on understanding of concepts and procedures to the nature of the interactive balance of concepts and skills fostered in CAS-intensive environments.

Analysis of the types of mathematical knowledge involved in use of CAS in school mathematics led to the consideration of *computational transposition*. Computational transposition refers to the formation of additional mathematical knowledge that the use of a particular computational artefact involves (Artigue, 2002; Balacheff, 1994; Hoyles & Noss, 2009). Concerned about the danger of considering technical work and conceptual understanding as separable, French researchers shifted the attention of CAS research to the construct of *technique*, which accentuated the development of integral links between procedures and conceptual reflection (Artigue, 2002; Lagrange, 2003). These researchers pointed out that, within a

CAS-enhanced setting, concepts and techniques are intertwined and embedded within a context. In a landmark book based on the work of this research team, Guin, Ruthven, and Trouche (2005) provided a language to describe how the relationship of user to tool played out in the integration of CAS into school mathematics. Drawing on the field of ergonomics (Vérillon & Rabardel, 1995), the authors of chapters in that book explained that CAS was an artefact that needed to develop into an instrument for teachers and students. They used the phrase *instrumental genesis* to describe the development of an artefact into an instrument, and noted that this genesis involves the transformation of the individual (*instrumentalization*) as well as the transformation of the artefact (*instrumentation*). This attention to the development of the relationship between the CAS and the CAS user accentuated the importance of recognizing that the nature of the use of a tool such as the CAS was not independent of the activity and experience of the user. These constructs hold considerable promise in explaining the range of effects in individual settings and situations for CAS-enhanced instruction.

As Artigue (2002) noted, "any technique, if it has to become more than a mechanically-learned gesture, requires some accompanying theoretical discourse" (p. 261). In the case of tool-assisted procedures, an additional participant in the discourse is the tool itself, and the tool brings with it its own mathematical system. The challenge for students and teachers is to account for the mathematics of the tool as well as the mathematics that students are intended to learn. At the elementary algebra level, for example, the user of a CAS needs to be aware of how the particular CAS being used handles extraneous roots and expressions that are undefined for particular input values.

The question raised by Artigue is how to determine the theoretical discourse needed for adequate student control of the artefact. Hasenbank and Hodgson (2007) suggested that the development of procedural understanding, presumably in the style of technique, can be aided through the implementation of a meta-analytical approach to procedures. They suggest that students engage in a series of questions about their procedural work:

> What is the goal of the procedure?
> What answer should I expect?
> How do I carry out the procedure?
> What other procedures could I use?
> Why does the procedure work?
> How can I verify my answers?
> When is this the "best" procedure to use?
> What else can I use this procedure to do?

Questions about the role of CAS in developing mathematical knowledge and about the nature of the balance of technical and conceptual understanding has permeated research on CAS-assisted mathematics, yet such research needs both theory that could inform the development of those approaches and venues for trying those different approaches. Empirical advances have been made with the creation and testing of CAS-intensive approaches, and the development of theoretical perspectives and frameworks have refined the field's approach to research on the effects of

CAS-assisted approaches to the learning of algebra. Yet, progress has sometimes been slowed by a general reluctance to welcome CAS into the regular school mathematics curriculum. Nevertheless, the field is positioned to engage in theory-based research with the potential for making significant advances in its understanding of the ways in which CAS can affect the balance and interplay of procedural and conceptual knowledge.

CAS Effect on Changing Emphasis on Concepts, Extending Procedures, and Attending to Structure

Incorporation of CAS in school algebra has the capacity to affect both the content of school algebra and how that content is developed. Different concepts can be emphasized, concepts that are taught can be studied more deeply, investigations of procedures can be extended, and new attention can be placed on structure. In this section we provide illustrations of each of these potential changes.

Changing Treatment of Concepts

The subgroup of the ICMI algebra study that focussed on the use of CAS in algebra learning suggested that one of the crucial questions to ask when considering implementation of CAS was "How does CAS use influence student conceptualization?" (Thomas, Monaghan, & Pierce, 2004, p. 166). One possibility is that CAS offers the opportunity to investigate concepts more deeply and to emphasize concepts that might not otherwise be prominent. In reality, in classrooms where CAS has been used by teachers themselves (rather than by researchers who involved teachers in their work), some research (e.g., Thomas & Hong, 2005b) has suggested that student activity with CAS rarely involves investigating a conceptual idea but is mostly used to obtain procedural answers and check work completed by-hand. This is an example of what Artigue (2002) called "the transmission of the bases of mathematical culture" (p. 246), passing on the socially constructed norm of what constitutes mathematical activity, which has traditionally been primarily by-hand procedural work. In this section we consider some possible activities in which CAS might be used to extend student engagement with mathematical conceptualization.

One of the keys to accessing mathematical concepts with CAS is the set of techniques that is promoted in the classroom. For many teachers these techniques are often perceived and evaluated in terms of their *pragmatic value* (Artigue, 2002), or how much can be efficiently accomplished using them. Artigue (2002) described the *pragmatic value* of techniques as their "productive potential (efficiency, cost, field of validity)" (p. 248) and the *epistemic value* as their contribution "to the understanding of the objects they involve" (p. 248). She stressed that techniques are most often considered and appreciated for their pragmatic value. An example would be the formula for solving quadratic equations, which has high pragmatic value in schools. However, in addition to this value for producing answers, drawing graphs, and other activities, a CAS instrument also has an *epistemic value*; that is, it has the capability to be used to produce knowledge of the object under study and to give rise to new questions that in turn promote new knowledge (Lagrange, 2002, 2003). It is particularly this area of how CAS can assist in construction of knowledge of mathematical concepts that is the subject of this section. We consider three main areas: how the CAS can allow some concepts in the current algebraic content in the curriculum to take on a different emphasis and importance, while emphasizing others that might not otherwise be prominent; how the CAS can create the opportunity to extend some algebraic procedures; and how the CAS can be used to assist exploration of new structures from outside the immediate curriculum.

There are two overarching principles that guide the examples presented here. One is that of using the CAS to assist in generalization. Mason, Graham, and Johnston-Wilder (2005) claim that expressing generality lies at the heart of mathematics and hence "a lesson without the opportunity for learners to express a generality is not in fact a mathematics lesson" (p. 297). They maintain that every page of a textbook should not only contain such opportunities but should clearly signal the need for generalization. This aim lies at the heart of the following examples.

The second principle used here is that, as teachers and researchers, we need to look for ways to use the epistemic value of CAS to improve students' mathematical understanding. Employing it as a "black box" in the context of which the student has little or no idea how the outputs relate to the inputs does little for students' learning of mathematics. In contrast, using the CAS as a tool for investigation can lead students to engage to some extent with the essential core of mathematical thinking. In this manner students will be encouraged to develop both mathematical *ways of thinking* and *ways of understanding* (Harel, 2008).

Delving More Deeply into Concepts

Understanding forms a crucial part of the mathematical experience for a number of fundamental, ubiquitous algebraic concepts. Examples of these concepts are variable, function, expression, and equation. CAS can offer an opportunity to engage with these concepts in a more comprehensive and deeper way than has often been the case.

One manner in which algebraic concepts can be explored more deeply is through a consideration of how they relate to other representations. In this regard Duval (2006) reminded us of two important classes of cognitive activity involving representational transformations (transformations within or between registers or representation systems). Duval designated transformations that happen within the same register as *treatments*, and those that consist of changing a register without changing the object as *conversions*. Although Duval (2006) recommended prioritizing conversions over treatments for those studying mathematical learning, and especially when analyzing student difficulties, CAS environments are capable assistants in both treatments and conversions. Important conceptual aspects arise from relating, through conversions, corresponding elements of conceptual representations. In the context of algebra, the manipulation of expressions or formulas and algebraic solution of equations would be treatments, whereas drawing a graph or producing a table of values for a given algebraic representation of a function would be conversions. CAS environments in which representation systems are linked and interactive are capable of conversion actions in which students need only to choose or enter appropriate commands and then observe the effects of the conversions. Opportunities for student engagement with conversion actions in CAS settings must be carefully crafted. From conversion activity, important aspects of epistemology, and understanding, of a mathematical object can arise, contributing to the goal of helping students attain *versatile thinking* in mathematics (Thomas, 2008a, 2008b), which involves at least three abilities:

- to switch at will in any given representational system between a perception of a particular mathematical entity as a process and the perception of the entity as an object;
- to exploit the power of visual schemas by linking them to relevant logico/analytic schemas; and
- to work seamlessly within and between representations, and to engage in procedural and conceptual interactions with representations.

This third component of the framework for versatile thinking, called *representational versatility* (Thomas, 2008a), incorporates more than Duval's treatments and conversions. The idea of conceptual interactions with representations is one that is highly relevant to CAS use and is exemplified in the following paragraphs.

Algebraic transformations. In a CAS environment the technology can help students to engage with novel (to them) mathematics through conversions. One example of a task that engages students with novel mathematics is the task of asking what algebraic form a function would take when its graph is reflected in the line $y=k$, for some real k. Applying the aforementioned principles by approaching the general through the specific we might ask students to reflect the graph of, say, $y=x^2+3x$ in the line $y=2$. The CAS can be used to draw the graphs (see Figure 20.2). A number of routes and their associated techniques are then possible to attempt to answer the problem. For example, we know that the points of intersection of $y=x^2+3x$ and $y=2$ are invariant under reflection, so we can start by determining these points. Likewise the vertex remains at the same x-value, and this may give ideas for an approach. However, students may develop a strategy involving translating the graph vertically by -2, then reflecting in the x-axis, and then translating vertically by $+2$. This nicely links the graphical transformations, such as a translation and reflection, with algebraic concepts $f(x)+k$ and $-f(x)$, and can be accomplished with the CAS (Figure 20.2). The correct answer of $y=-x^2-3x+4$ is seen in Figure 20.3, along with the graph(s) in Figure 20.2 to check that it works.

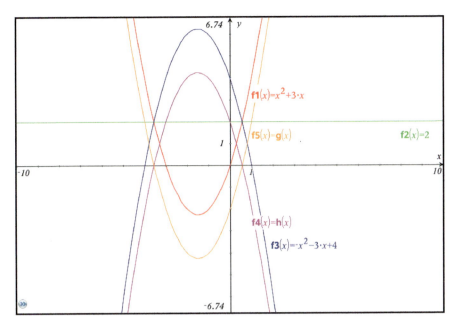

Figure 20.2. Using graphs in CAS to confirm the reflection of function in $y = 2$.

Of course the key question is whether one can generalize this, both graphically and, more importantly, algebraically. The key idea here is shown in Figure 20.4. Since $g(x)$ is a reflection of $f(x)$ in $y = k$ *every* point of the plane is reflected. Thus for a general point $(x, f(x))$, distance n above the line, $n = f(x) - k$, and so $g(x) = k - n = k - (f(x) - k) = 2k - f(x)$. Hence, the result of reflecting the graph of a continuous, well-behaved function $f(x)$ in the line $y = k$ is to obtain a function $g(x) = 2k - f(x)$. For example, the reflection of the graph of $f(x) = x^3 - 2x$ in the line $y = -1$ gives the graph of the function $g(x) = -2 - (x^3 - 2x) = 2x - x^3 - 2$. Involving students in a few examples with the CAS might serve as a model for them to engage with mathematics at this deeper level.

Equation and equivalence. The constructs of number, symbolic literals, operators, the "=" symbol itself, and the formal equivalence relation, as well as the principles of arithmetic, all contribute to building a deep understanding of equation. However, there is evidence (Godfrey & Thomas, 2008) that many students have a surface structure view of equation (Laborde, 2002), looking at the equation rather than through it (Mason, 1995), and hence failing to integrate the properties of the object with that surface structure (Thomas, 2008a). An example of this provided by Godfrey and Thomas (2008) is the way in which an embodied input–output, procedural or operational view of equation persists for approximately 25% of secondary school students, even when they reach the university level. In addition, charting student progress through the concepts, Godfrey and Thomas (2008) point

Define $f(x)=x^2+3\cdot x$	Done
solve$(f(x)=2,x)$	$x=\dfrac{-\left(\sqrt{17}+3\right)}{2}$ or $x=\dfrac{\sqrt{17}-3}{2}$
Define $g(x)=f(x)-2$	Done
Define $h(x)=-g(x)$	Done
Define $j(x)=h(x)+2$	Done
$j(x)$	$-x^2-3\cdot x+4$

Figure 20.3. Using algebra in CAS to reflect a function in $y=2$.

Figure 20.4. Generalizing a reflection in $y=k$.

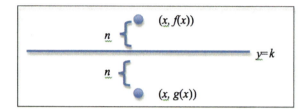

out that equivalence is not well understood, and that the reflexive, symmetric, and transitive properties forming an equivalence relation are rarely considered in schools, even though they are often assumed.

For example, when solving an equation we may go from $x+6=3x+1$ to $2x+1=6$, rather than $6=2x+1$, using the symmetric property applied to the *conditional* equation. Or we may reason along the lines that if $y=2x+1$ (*identical* equation, defining y), then when $y=0$ (*conditional* equation), $2x+1=0$ (*conditional* equation), employing the transitive property to do so. Note that *identical* equations are ones that are true for all values of the variable(s) and conditional equations are ones that are true for certain values only. However, we may not explicitly highlight these properties, or the kinds of equations employed, leaving students to abstract these themselves (Godfrey & Thomas, 2008, p. 89).

One study that addressed the issue of CAS use for equivalence, equality, and equation in algebra is that of Kieran and Drijvers (2006). As they comment about equivalence, "On the one hand, equivalence of two expressions relates to the numeric as it reflects the idea of 'equal output values for all input values.' On the other hand, the notion of equivalence of expressions from an algebraic perspective means that the expressions can be rewritten in a common algebraic form" (Kieran & Drijvers, 2006, p. 214). This is another way of describing the proceptual nature of the symbols (Gray & Tall, 1994) as having the dual faces of process (input and output) and object (expression) (Tall, Thomas, Davis, Gray, & Simpson, 2000). As part of Kieran and Drijvers' experiment, 10th-grade students (15-year-old students) considered the equivalence of the expressions in Figure 20.5 and used by-hand techniques to test

	CAS technique task		
Given expression	Result produced by the Enter button	Result produced by Factor	Result produced by Expand
1. $\dfrac{8x^2-10x-3}{6}$			
2. $\dfrac{(x-3)^2+(x-3)(7x-1)}{4}$			
3. $(3-x)(1-2x)$			
4. $\dfrac{(2x-3)(4x^2-7x-2)}{6x-12}$			

Figure 20.5. Using CAS to consider equivalence of expressions (adapted from Kieran & Drijvers, 2006, p. 216). After they carry out CAS techniques, students compare the results. The purpose is to develop understanding of equivalence.

their conclusions and tried to "reconcile the techniques in the two media." The researchers describe the different techniques arising from each and arrive at several conclusions:

> Two notions of the equivalence of two expressions can be distinguished: an algebraic view as having a common form, and a numerical view on equivalence as having—always, in most cases, or even just in some cases – the same numerical output values. The latter view is related to the previous item, and is reflected in the language issue related to the words equivalent and equal.
>
> … The issue of restrictions on equivalence is an important theoretical aspect of the concept of equivalence. It involves both the particularities of the way the CAS deals with restrictions, and the somewhat strange definition—at least possibly strange in the eyes of the students—of equivalence involving a set of admissible values.
>
> … The relation between solving an equation and the notion of equivalence of expressions, and between restrictions on equivalence and solutions of the equation, could be confusing for students. Both restrictions and solutions have a sense of "exceptions," but in a kind of complementary way. This issue needs coordination…. (p. 220)

The following activity, from Thomas (2009) was designed to assist students to distinguish equivalent equations.

> Which of the following equations have the same solutions? Explain how you worked out your answers and write down reasons for your answers. Use a graphic calculator to help you work out and support your answers with an explanation.
>
> (a) $x^2+x+1=2x^2-x-3$
> (b) $x^2+x+5=2x^2-x+1$
> (c) $x^2-x+1=2x^2-3x-3$
> (d) $x^2+2x+1=2x^2-2x-3$
> (e) $2x^2+3x-1=3x^2+x-5$ (p. 153)

Thomas maintained that the theory underpinning this task is to understand the difference between *legitimate transformations* of an equation—those that are mathematically correct and preserve the solutions—and *productive transformations*—those that also move rapidly towards finding the solutions. This distinction is often not understood by students. Linking to the graphical representation can support the students' understanding of the invariance of solutions under legitimate transformations.

Continuity. CAS can also help to use algebraic representations to make concepts such as limits and connecting limits to continuity (and possibly differentiability) more prominent in the curriculum. If we consider, for example the function $f(x) = \begin{cases} x^2 & x \leq 3 \\ x+6 & x > 3 \end{cases}$, then the question arises whether the function is continuous at $x=3$. We can define the function piecewise in the CAS using "Define $f(x)$=piecewise($x^{\wedge}2$, $x \leq 3$, $x+6$, $x>3$)" and get the CAS to draw the graph of the function (see Figure 20.6). Looking at the left and right limits provides corroborating evidence that the limit exists and is equal to 9, which is also clearly $f(3)$ [which is equal to 3^2]. If the students know about derivatives, and we are beginning to discuss their existence, then getting the CAS to draw the graph of the derived function shows clearly the discontinuity in the derived function at $x=3$. Finding the limits confirms this (see Figure 20.7).

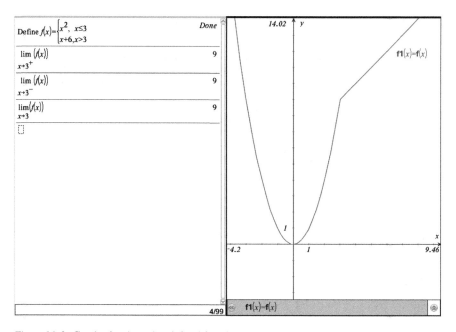

Figure 20.6. Graph of a piecewise-defined function.

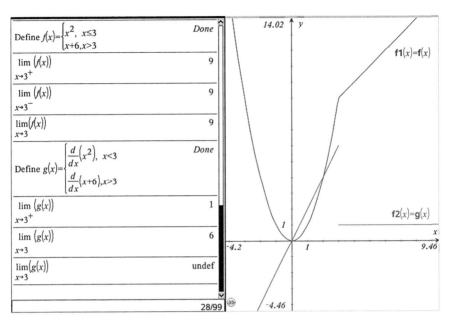

Figure 20.7. Symbolic and graphical confirmation of discontinuity of the derived function at $x=3$.

One area in which the CAS output needs careful scrutiny involves the continuity of functions such as $f(x) = \dfrac{1-x^2}{x^2-2x-3}$. Here the graph (see Figure 20.8) does not show the discontinuity at $x=-1$, although the CAS generates a warning that the "Domain of the result may be larger than the domain of the input." Encouraging students to use the CAS to link representations provides the opportunity for further insight. The table of values shows that the function is not defined at $x=-1$, and this is then confirmed by attempting to generate a value for $f(-1)$. The continuity of other interesting functions can be similarly investigated.

Extending Procedures

In mathematics one of the most important ideas that students need to develop is an understanding that all mathematical processes and constructs have conditions or limitations that influence their use. For example, consideration of the domain of a function is a vital part of its study. One way to build appreciation of this is to extend student knowledge by engaging them in areas of mathematics that lie just beyond their current understanding. In this section we consider some algebraic examples for which CAS may assist with extending procedures to objects beyond those they have experienced or by encouraging generalization of procedures.

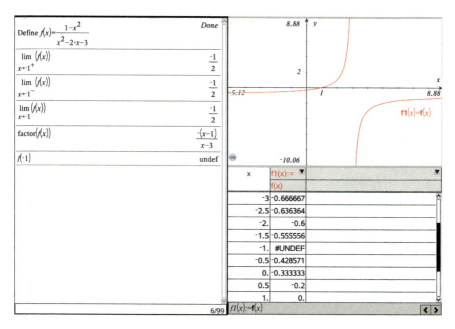

Figure 20.8. Confirmation that the function is not defined at $x = -1$.

Moving toward generalization through extension of factoring. A task used by Kieran and colleagues (Kieran & Drijvers, 2006; Kieran & Saldanha, 2008) considered the use of the factoring command in CAS to get students to move towards a generalization regarding the factorization of $(x^n - 1)$. Students worked in both directions, factoring expressions of the form $(x^j - 1)$, for $j = 2, \ldots 6$, and expanding $(x-1)(x+1)$, $(x-1)(x^2+x+1)$, and so on. The outcomes suggested that:

> The notion of complete factorization can come to the fore as soon as students attempt to factor an expression with a non-prime even exponent, such as $x^4 - 1$, according to the general rule [using only a factor of $x - 1$], and are confronted with a CAS factorization that they do not anticipate [e.g., $(x-1)(x+1)(x^2+1)$]. (Kieran & Drijvers, 2006, p. 243)

Thus by-hand techniques are helpful in reconciling these differences. In turn this can elicit further conjectures, such as $(x+1)$ is always a factor of $(x^n - 1)$ for even n, which then requires proof. Kieran and Drijvers proposed that this CAS-based approach led to theoretical development for the students in at least four areas:

1. Resolution of the conflict between by-hand and CAS results led to enhanced theoretical perception of the structure of expressions of the form $(x^n - 1)$.
2. Noticing in CAS output structure that they had not noticed in prior examples.
3. Improved reflection through tentative conjectures based on the examples they generated, and testing the conjectures by means of CAS techniques.
4. Deepening of theoretical thinking involving the coordination and integration of several discrete pieces of theory.

The researchers conclude that technique and theory emerge in mutual interaction, with CAS playing a crucial epistemic role.

Extending polynomial investigations. One general question in engaging students in investigations with CAS is whether activities should start with a general case or not. Since the CAS allows one to consider such cases, for example a cubic $x^3 + ax^2 + bx + c = 0$, it is tempting to make this a starting point. However, there appears to be a stronger case for beginning with specific examples, encouraging students to form conjectures and gradually to motivate them to move their thinking towards the general cases, as seen in the previously described example from the research of Kieran and Drijvers. This again relates to the Task–Technique–Theory (TTT) framework that Kieran and Drijvers (2006) espoused, based on ideas from Artigue (2002) and Lagrange (2002, 2003), namely that it is through the construction of techniques required to perform tasks that the understanding of mathematical objects arises, often through the production of new questions. This deepening of understanding may also arise through reflective comparison of the technique with other techniques (Lagrange, 2003). This is precisely the epistemic role of techniques.

Most school students will at some time be shown the formula for the solutions of a quadratic equation. However, if we are thinking about using CAS to extend what may be considered, then the zeros of a cubic function (or the solutions of a cubic equation) should be a topic for investigation. Careful structuring of the process of considering the Tartaglia-Cardano method of solution may be needed, but this investment would allow for a valuable extension of algebraic thinking and capability. For example, given the cubic equation:

$$x^3 + 3x^2 - 6x + 9 = 0$$

(with some discussion of why the coefficient of x^3 is 1) one could ask how a general method to solve such an equation could be derived (rather than using a black-box approach), and what mathematics would arise from doing so.

Using the CAS we can define the function f such that $f(x) = x^3 + 3x^2 - 6x + 9$. Then our first task is to remove the term in x^2. This can always be done and the resulting production of a depressed cubic is the first fundamental idea in the Tartaglia-Cardano method of solution. This draws nicely on the mathematical idea of composite function, which is usually introduced in school but may often find few applications. Here we want to find a k such that $f(z+k)$ avoids a term in z^2. Students could experiment until they find one that works (see Figure 20.9). Trying other cubics they will be asked to generalize and find a "rule" for a substitution that works. In fact for $f(x) = x^3 + ax^2 + bx + c$, making the substitution $x = z - \dfrac{a}{3}$ (which can be done relatively easily with the CAS to confirm the generalization) gives

$$f\left(z - \frac{a}{3}\right) = \left(z - \frac{a}{3}\right)^3 + a\left(z - \frac{a}{3}\right)^2 + b\left(z - \frac{a}{3}\right) + c.$$

Define $f(x)=x^3+3\cdot x^2-6\cdot x+9$	Done
$f(z-1)$	$z^3-9\cdot z+17$
Define $g(z)=z^3-9\cdot z+17$	Done
$g(u+v)$	$u^3+3\cdot u^2\cdot v+u\cdot(3\cdot v^2-9)+v^3-9\cdot v+17$
factor$(g(u+v))$	$u^3+3\cdot u^2\cdot v+3\cdot u\cdot(v^2-3)+v^3-9\cdot v+17$
factor$(3\cdot u^2\cdot v+3\cdot u\cdot(v^2-3)-9\cdot v)$	$3\cdot(u+v)\cdot(u\cdot v-3)$
Define $u=\dfrac{3}{v}$	Done
$g(u+v)$	$\dfrac{v^6+17\cdot v^3+27}{v^3}$
solve$(g(u+v)=0,v)$	$v=\dfrac{-(\sqrt{181}+17)^{\frac{1}{3}}\cdot 2^{\frac{2}{3}}}{2}$ or $v=\dfrac{(\sqrt{181}-17)^{\frac{1}{3}}\cdot 2^{\frac{2}{3}}}{2}$

Figure 20.9. TI-Nspire computer screen of the Tartaglia–Cardano method of solving cubic equations.

And this can be seen to result in an equation of the form $z^3+mz+n=0$, as shown in the example in Figure 20.9.

Then we may ask how do we solve this equation? Why is it easier than the original one? Here is where the beauty of the method comes in. If we let $z=u+v$ then, as Figure 20.9 shows, $g(u+v)$ does not, at first sight look very useful, and trying to factor with the CAS does not work. But factoring the terms other than u^3, v^3, and 17 is the key to the method (although seeing why it would be useful requires a leap of insight in the original formulation), since it gives a "nice" factorization. It is this that suggests the idea of setting $uv=3$ to remove these terms (but why?). Doing so we can reduce the cubic to a quadratic and hence find the solution. At each stage of a number of examples the student is encouraged to ask "Is this a special case or will it always happen?" and to find evidence to support their conclusions.

One may ask, why bother to do this when the original cubic can be solved on the CAS in an easy step? We remind the reader who thinks this way of our second principle above. Using CAS to investigate a method such as the one just described will lead students to engage in mathematical thinking and reasoning and will divert attention away from a purely answer-driven approach to mathematics.

Another area whereby known procedures can be extended is that of solving Diophantine equations. Of course, Pythagoras' theorem could be the springboard for this since it is often studied and there are readily accessible integer solutions to $x^2+y^2=z^2$. Although, as has been proved by Andrew Wiles (and as was stated in Fermat's Last Theorem), there are no other integer values of $n>2$ for which any triple (x, y, z) of non-zero integers, gives a solution for $x^n+y^n=z^n$, there are similar looking equations that do have positive integer solutions. One of these, $x^n+y^n=z^{n+1}$,

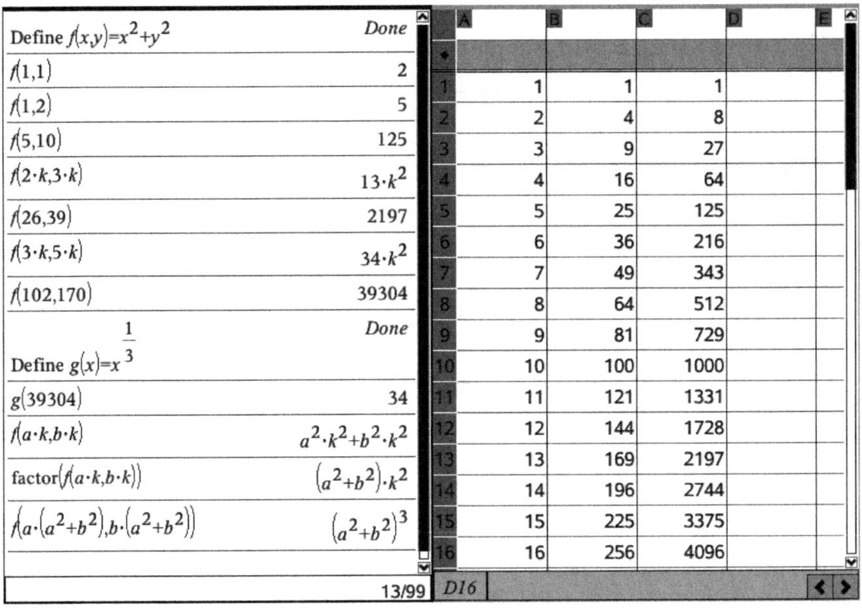

20 Computer Algebra Systems, and Algebra in the Curriculum

Define $f(x,y)=x^2+y^2$	Done
$f(1,1)$	2
$f(1,2)$	5
$f(5,10)$	125
$f(2 \cdot k, 3 \cdot k)$	$13 \cdot k^2$
$f(26,39)$	2197
$f(3 \cdot k, 5 \cdot k)$	$34 \cdot k^2$
$f(102,170)$	39304
Define $g(x)=x^{\frac{1}{3}}$	Done
$g(39304)$	34
$f(a \cdot k, b \cdot k)$	$a^2 \cdot k^2 + b^2 \cdot k^2$
factor$(f(a \cdot k, b \cdot k))$	$(a^2+b^2) \cdot k^2$
$f(a \cdot (a^2+b^2), b \cdot (a^2+b^2))$	$(a^2+b^2)^3$

	A	B	C	D
1	1	1	1	
2	2	4	8	
3	3	9	27	
4	4	16	64	
5	5	25	125	
6	6	36	216	
7	7	49	343	
8	8	64	512	
9	9	81	729	
10	10	100	1000	
11	11	121	1331	
12	12	144	1728	
13	13	169	2197	
14	14	196	2744	
15	15	225	3375	
16	16	256	4096	

Figure 20.10. CAS screens showing a method of solving $x^2 + y^2 = z^3$.

which is accessible with CAS, was described by Hoehn (1989). Once again we might start with a particular equation, say $x^2 + y^2 = z^3$, and ask students to try to find a solution using the CAS. A function of two variables could be defined (see Figure 20.10), introducing a new mathematical construct. After a few trial-and-error attempts using $x=1$ or 2, the use of a spreadsheet with values of n^2 and n^3 could help to find two of the squares that add up to a cube (for example $x=2$, $y=2$ and $z=2$ may be seen immediately). In this way $x=5$ and $y=10$ can also easily be found. Hence, there is at least one solution. If students start to flounder, then some teacher direction could suggest trying something of the form $f(ak, bk)$ for given integers a and b. However, the teacher might aim for this conjecture to come from the class.

In Figure 20.10 we can see examples with $a=2$ and $b=3$, and $a=3$, $b=5$. Now in each case we get an answer of the form ck^2 and since we are looking for something of the form z^3 the idea is to set $c=k$, giving k^3. We soon get some large values and the spreadsheet could be extended to check $\sqrt[3]{39304}$, and so on, or the CAS will do it even better. So now the generalization question comes into play. Will this always work? With the CAS we can try general a and b of course, as seen in Figure 20.10. In this case it still works if we set $a^2+b^2=k$, and the final step shown in the CAS screen shows that this gives z^3, with $z=a^2+b^2$.

The final step of a complete generalization to the solution of $x^n + y^n = z^{n+1}$ is likely to be a step too far for all but the most able school students, but we comment on it here for the sake of completeness and the principle of generalizing results. Figure 20.11 shows an attempt to use the TI-Nspire to apply the same method as above.

Define $h(x,y)=x^n+y^n$	Done
factor$(h(a \cdot k, b \cdot k))$	$(a \cdot k)^n + (b \cdot k)^n$
$h(a \cdot (a^n+b^n), b \cdot (a^n+b^n))$	$(a \cdot (a^n+b^n))^n + ((a^n+b^n) \cdot b)^n$
factor$(h(a \cdot (a^n+b^n), b \cdot (a^n+b^n)))$	$(a \cdot (a^n+b^n))^n + ((a^n+b^n) \cdot b)^n$
$h(3 \cdot (3^5+7^5), 7 \cdot (3^5+7^5))$	$341^n \cdot (350^n + 150^n)$
$341^5 \cdot (350^5 + 150^5)$	24566670447125640625000000
$24566670447125640625000000^{\frac{1}{6}}$	17050
$3 \cdot (3^5+7^5)$	51150
$7 \cdot (3^5+7^5)$	119350

Figure 20.11. CAS screens showing a method of solving $x^n + y^n = z^{n+1}$.

Defining a function $h(x, y) = x^n + y^n$ and considering $h(ak, bk)$ with $k = (a^n + b^n)$ leads to the expression $(a(a^n + b^n))^n + (b(a^n + b^n))^n$, which by hand can readily be seen by an experienced eye to factor to $(a^n + b^n)^n (a^n + b^n)$ and hence equal $(a^n + b^n)^{n+1}$. However, the TI-Nspire program does not seem to be able to cope with this factorization, making this a good example to help the students to see that CAS has its limitations and to realize that they cannot rely on it to do everything for them. Thus, in the above manner, for a given n, we can construct solutions of $x^n + y^n = z^{n+1}$. One example with $n = 5$, $a = 3$ and $b = 7$ is shown in Figure 20.11, where we see evidence that $51150^5 + 119350^5 = 17050^6$.

The previous examples focussed on determining solutions to given equations. Tasks that require the generation of equations with particular features, including given solutions, are another way in which work with polynomial functions might be extended. Relatively early in their experience with factoring polynomials and solving equations, students might be asked the following task, from Böhm (2007):

> Given is a set of solutions L = {3, −1, 1/2}
> Find two equations of degree 5 with L = set of solutions. (p. 3)

Although a CAS Solve command or graphical means could be applied in the hope of determining solutions for an equation of degree 5, the CAS work needed to generate an equation from information about the solutions is not obvious, especially to beginning algebra students. Figure 20.12 shows what we might do as starting points for symbolic, tabular, graphical approaches.

We know other things that are possible or not possible in each approach. For example, the complete symbolic form is $(x-3)(x+1)(x-1/2)(x-\square)(x-\square) = 0$ where each box represents one of 3, −1, and 1/2. Choosing one of the solutions for each of the boxes produces an equation that satisfies the conditions.

The question of producing two equations that meet the conditions then allows for generalization at a level appropriate for students. For example, we could see how

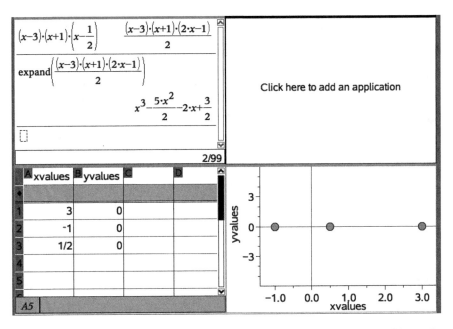

Figure 20.12. Initial symbolic, graphical, and table attempts to produce an equation of degree 5.

many distinct equations are possible when the equation $(x-3)(x+1)(x-1/2)(x-\square)(x-\square)=0$ is expressed in expanded form. The results of testing all nine combinations of two solutions and looking for distinct results could be done with nine CAS Expand commands or, as shown in Figure 20.13, with a CAS-generated table.

The task provides an opportunity for predetermining two or more distinct equations but also to characterize the number and nature of possible equations by reasoning symbolically. Filling both boxes with one of the three solutions yields three distinct quintic expressions. Filling the two boxes with different solutions yields three more distinct quintic expressions. So, there are six possible equations of the form $x^5 + bx^4 + cx^3 + dx^2 + ex + f = 0$ that satisfy the given conditions.

To this point, an underlying assumption might be that the equation is in the form of a polynomial of degree 5 with leading coefficient 1 set equal to 0. Students familiar with factoring might produce additional equations by using a constant factor with the quintic polynomial. Infinitely many more are possible when any nonzero real number, k, is used as a factor, as in the expression $k(x-3)(x+1)(x-1/2)(x-\square)(x-\square)=0$ or $kx^5 + kbx^4 + kcx^3 + kdx^2 + kex + kf = 0$.

Graphically, as in Figure 20.14, we could think about the situation in terms of behaviour at each of the three points. If it touches the x-axis at one point, then it must touch without crossing at another point and simply intersect at the third point; there are three ways in which this can happen. If the graph has an inflection point at one point, it simply crosses at the other two, which happens in three ways. If the graph simply crosses at one point, we find it falls into one of the other two cases.

	A box1	B box2	C result	D alternative	E
♦			=f(a[],b[])	=expand(c[])	
1	3	3	(x−3)^3*(x−0.5)*(x+1)	x^5−8.5*x^4+22.*x^3−9.*x^...	one
2	3	-1	(x−3)^2*(x−0.5)*(x+1)^2	x^5−4.5*x^4+13.*x^2+3.*x−...	two
3	3	1/2	(x−3)^2*(x−0.5)^2*(x+1)	x^5−6.*x^4+8.25*x^3+4.75*...	three
4	-1	3	(x−3)^2*(x−0.5)*(x+1)^2	x^5−4.5*x^4+13.*x^2+3.*x−...	repeat
5	-1	-1	(x−3)*(x−0.5)*(x+1)^3	x^5−0.5*x^4−6.*x^3−5.*x^2...	four
6	-1	1/2	(x−3)*(x−0.5)^2*(x+1)^2	x^5−2.*x^4−3.75*x^3+1.75*...	five
7	1/2	3	(x−3)^2*(x−0.5)^2*(x+1)	x^5−6.*x^4+8.25*x^3+4.75*...	repeat
8	1/2	-1	(x−3)*(x−0.5)^2*(x+1)^2	x^5−2.*x^4−3.75*x^3+1.75*...	repeat
9	1/2	1/2	(x−3)*(x−0.5)^3*(x+1)	x^5−3.5*x^4+0.75*x^3+2.87...	six

Figure 20.13. Testing nine symbolic options using a CAS-generated table to determine six distinct results.

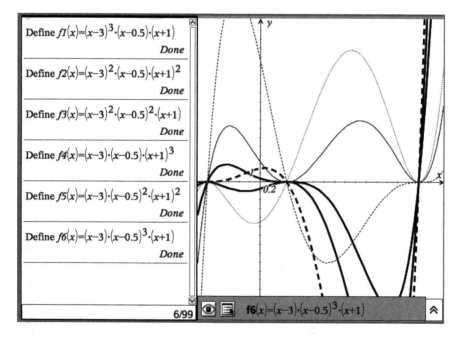

Figure 20.14. Graphs representing quintic functions which lead to six different equations.

As with the symbolic form, we have six general patterns and the graph can draw attention to the meaning of the solution set. Taking amplitude into consideration, we have the effects of the constant factor and infinitely many choices.

Böhm's task requires students to think about characteristics of equations and their solutions. Extending the task with a question about the number and nature of possible equations yields a generalizing experience in elementary algebra.

Exploring "New" Structures

Using CAS there is an opportunity to investigate the structure of other "abstract" algebras where the "rules" or axioms governing the structure of the algebra of generalized arithmetic no longer apply. It can demonstrate that the rules that we take for granted do not extend to all systems. In introducing the following examples we employ some of the appropriate mathematical language describing the structures, although teachers may not want to use this language with students. Some examples are:

1. Students expect $AB=BA$; that is, that multiplication is commutative;
2. Students expect $AB=0$ if and only if $A=0$ or $B=0$, since there are no non-zero divisors of zero.
3. Extending 2 we can see we expect that if $AB-AC=0$ then $A(B-C)=0$ and $A=0$ or $B=C$.

Using CAS it is easy to set up a situation for which this can be investigated. For example we may consider the following 2 by 2 matrices:

$$A = \begin{pmatrix} 1 & 2 \\ 3 & 6 \end{pmatrix}, \quad B = \begin{pmatrix} 3 & -8 \\ 2 & 3 \end{pmatrix}, \quad C = \begin{pmatrix} 5 & 2 \\ 1 & -2 \end{pmatrix}, \quad D = \begin{pmatrix} 2 & -1 \\ 3 & 3 \end{pmatrix}, \quad E = \begin{pmatrix} -4 & -4 \\ 2 & 2 \end{pmatrix}$$

Using a CAS, students can generate the products, AE, BD, DB, AB, and AC, and can find that $BD \neq DB$, $AE=0$ even though $A \neq 0$ and $E \neq 0$, and $AB=AC$ even though $A \neq 0$ and $B \neq C$.

$$BD = \begin{pmatrix} 3 & -8 \\ 2 & 3 \end{pmatrix}\begin{pmatrix} 2 & -1 \\ 3 & 3 \end{pmatrix} = \begin{pmatrix} -18 & -27 \\ 13 & 7 \end{pmatrix}$$

$$DB = \begin{pmatrix} 2 & -1 \\ 3 & 3 \end{pmatrix}\begin{pmatrix} 3 & -8 \\ 2 & 3 \end{pmatrix} = \begin{pmatrix} 4 & -19 \\ 15 & -15 \end{pmatrix}$$

$$AE = \begin{pmatrix} 1 & 2 \\ 3 & 6 \end{pmatrix}\begin{pmatrix} -4 & -4 \\ 2 & 2 \end{pmatrix} = \begin{pmatrix} 0 & 0 \\ 0 & 0 \end{pmatrix}$$

$$AB = \begin{pmatrix} 1 & 2 \\ 3 & 6 \end{pmatrix}\begin{pmatrix} 3 & -8 \\ 2 & 3 \end{pmatrix} = \begin{pmatrix} 7 & -2 \\ 21 & -6 \end{pmatrix}$$

$$AC = \begin{pmatrix} 1 & 2 \\ 3 & 6 \end{pmatrix}\begin{pmatrix} 5 & 2 \\ 1 & -2 \end{pmatrix} = \begin{pmatrix} 7 & -2 \\ 21 & -6 \end{pmatrix}$$

Then students can be asked to state a conjecture and continue their investigation, possibly considering a proof of it, using, for example, $A = \begin{pmatrix} a & b \\ c & d \end{pmatrix}$. For instance, they may find that in the ring of 2 by 2 matrices the zero divisors are singular, that is, with determinant 0. Questions arise about whether the order matters for the zero

$$\begin{bmatrix} -1 & -1 \\ 2 & 2 \end{bmatrix} \cdot \begin{bmatrix} -2 & -3 \\ 2 & 3 \end{bmatrix} \qquad \begin{bmatrix} 0 & 0 \\ 0 & 0 \end{bmatrix}$$

$$\begin{bmatrix} -2 & -3 \\ 2 & 3 \end{bmatrix} \cdot \begin{bmatrix} -1 & -1 \\ 2 & 2 \end{bmatrix} \qquad \begin{bmatrix} -4 & -4 \\ 4 & 4 \end{bmatrix}$$

$$\det\left(\begin{bmatrix} -1 & -1 \\ 2 & 2 \end{bmatrix}\right) \qquad 0$$

$$\det\left(\begin{bmatrix} -2 & -3 \\ 2 & 3 \end{bmatrix}\right) \qquad 0$$

Figure 20.15. TI-Nspire computer screen showing left/right zero divisors with determinant zero.

divisors, and they may find that there are left and right zero divisors (e.g., we can ask whether we can find two non-zero matrices P and Q such that $PQ = 0$ but $QP \neq 0$). Figure 20.15 shows that this is possible.

Thinking and Reasoning that CAS Use Inspires or Requires

A striking feature of the examples in the previous section of how CAS allows students to engage with new concepts is the extent to which the mathematical work involves generalization, including generalization of properties, strategies, and other relationships. As Arcavi (1994) observed, CAS is "a tool for understanding, expressing, and communicating generalization, for revealing structure, and for establishing connections and formulating mathematical arguments" (p. 24). The impact of CAS on thinking about connections and formulating arguments can be considered in terms of the objects about which students reason and the tools they employ in their reasoning.

Objects About Which to Reason

Reasoning opportunities with CAS seem to be related to the tool's multiple representation capacity. We begin with perhaps the most enticing CAS aspect—possibilities in the symbolic register.

Symbolic representations. Arguably the most documented type of CAS-generated opportunity for reasoning about symbols is the resolution of unanticipated symbolic results. Reasoning stems from the need to compare CAS-produced results to by-hand results or to a desired informative equivalent symbolic form. Alonso and colleagues (2001) provided several examples of unexpected results and their use to encourage students to reason about the results and about how they are using CAS.

The duality of reasoning about mathematics and about CAS functions is a common theme in CAS literature.

A related though less frequently mentioned reasoning opportunity is conjecturing and justifying theorems that underlie CAS procedures. Dana-Picard (2007) drew attention to CAS commands that are implementations of theorems that do not typically appear in course syllabi. Her examples include Derive's use of the following theorem when computing $I_n = \int_0^{\pi/2} \sin^n x \, dx$,

$$\int \text{SIN}(a \cdot x + b)^p \, dx \rightarrow -\frac{\text{SIN}(a \cdot x + b)^{p-1} \cdot \text{COS}(a \cdot x + b)}{a \cdot p}$$
$$+ \frac{p-1}{p} \cdot \int \text{SIN}(a \cdot x + b)^{p-2} \, dx$$

(p. 223)

Supported by evidence of student symbolic reasoning, Dana-Picard contended that the user needs to learn new mathematics in order to understand well the CAS process. She referred to these situations as *motivating constraints*, and she contended, despite the connotation of "constraint," that these situations can be used to push the user towards mathematical insight. Her construct of motivating constraint is an addition to Guin and Trouche's (1999) extension of Balacheff's (1994) ideas regarding *internal constraints* of the hardware, *command constraints* of the software, and *organization constraints* of the interface. Dana-Picard's example illustrates how CAS features can motivate identification and justification of theorems beyond the standard syllabi.

Other uses of CAS can help develop student understanding of symbols and symbolic reasoning. Cedillo and Kieran (2003) detail an experiment in which beginning algebra students generated the algebraic code needed for a CAS to produce given numerical patterns (e.g., input numbers 1, 4, 6, 9 with corresponding output numbers 1, 7, 11, 17). Students tested the code and used CAS results to revise it. Results of the study indicate that students developed the notion of "a letter as 'serving to represent any number'" (p. 231). In this case, reasoning about symbols while using CAS was the means by which concepts were developed.

As another example of reasoning about symbols while using CAS, consider the following task from McMullin (2003):

> Use the sequence operation to produce the sequence 3, 6, 9, 12, 15 as many different ways as you can. (p. 268)

Multiple possibilities, including several suggested by McMullin, appear in Figure 20.16. The reasoning for a beginning algebra student that produces each of the options could include simply replicating the terms, attending to a linear pattern, and considering multiples of three—as exemplified in the first three lines of Figure 20.16. Subsequent examples indicate how the task could be differently handled with additional mathematics experience.

Similar to activities used by Cedillo and Kieran, this task engages students' understanding of equivalence through the production of CAS code. The concept

seq$(n,n,3,15,3)$	$\{3,6,9,12,15\}$
seq$(n+3,n,0,12,3)$	$\{3,6,9,12,15\}$
seq$(3 \cdot n,n,1,5,1)$	$\{3,6,9,12,15\}$
seq$((\sqrt{3 \cdot n})^2,n,1,5,1)$	$\{3,6,9,12,15\}$
seq$\left(\dfrac{n^2-n}{n-1},n,3,15,3\right)$	$\{3,6,9,12,15\}$
seq$(20-n,n,17,5,-3)$	$\{3,6,9,12,15\}$
seq$(n \cdot \sin(0.5 \cdot \pi),n,3,15,3)$	$\{3.,6.,9.,12.,15.\}$

Figure 20.16. Sequence commands that yield 3, 6, 9, 12, 15.

under consideration in this case is not only sequence but also equivalence. The CAS seq expressions are equivalent because they represent the same finite sequence, although the defining expressions (e.g., n, $3n$, $n \sin(0.5\pi)$, $20-n$) are not necessarily equivalent. These examples underscore the need to understand symbols both as algebraic expressions and as CAS code. They also highlight the importance of distinguishing among the mathematical objects being represented (in this case, sequences and expressions).

Attention to symbolic understanding and the symbolic capacity of CAS foregrounds consideration of symbolic sense. According to Arzarello and Robutti (2010), who built on Arcavi's (1994) notion of symbol sense as they described students working with handheld CAS,

> Students have symbol sense if they are able, for example: to call on symbols in the process of solving a problem and, conversely, to abandon a symbolic treatment for better tools; to recognize the meaning of a symbolic expression; and to sense the different roles symbols can play in different contexts. (p. 720)

Arzarello and Robutti claimed that the symbolic power of a CAS-empowered spreadsheet supports the development of symbol sense in a way that tables of numerical examples cannot. Examples of student work—including the spontaneous use by two students—supported their claim. In generating a table of numerical values for second differences of $y = ax^2 + bx + c$ for integer values of x from 0 to 15, students could see a constant numerical second difference (e.g., −4) for a specific quadratic case. However, a table of symbolic results for second differences for x-values of x_0, $x_0 + h$, $x_0 + 2h$, …, $x_0 + 15h$ showed that the constant difference in the general case was $2ah^2$. CAS results made it easier for students to see symbolic patterns and then reason about them.

Reacting to CAS results that are produced in intended or spontaneous ways appears useful in helping students to develop meaning for symbols as they reason with and about these results. Some of the observations in the symbolic register seem to have parallels in other registers. For example the potential of immediate feedback has long been acknowledged in other registers, such as its impact in graphical tasks (e.g., Ruthven, 1990) and geometric environments (e.g., Hillel, Kieran, & Gurtner, 1989). We turn now to consideration of how CAS facility with graphical representations generates opportunities and supports reasoning.

Graphical representations. Graphical reasoning can be an alternative to symbolic reasoning, but connecting graphical and symbolic actions and results is one way in which CAS use provides opportunities that transcend affordances of simpler graphing utilities. For example, recall the reasoning with transformations of functions in the example of reflecting a quadratic about a horizontal line. Students could reason graphically about translating the graph vertically by −2 then reflecting the result in the x-axis and then translating that result vertically by +2. Application of this reasoning to the graph as a set of points using three points to generate a quadratic expression connects graphical and symbolic images in a solution that crosses registers.

A second example of integrated graphical and symbolic reasoning involves solving equations by graphical intersection. Such methods generalize to equations for which symbolic methods are not available. Zbiek and Heid (2011) illustrated the reasoning process that draws on characteristics of functions to reason through a solution for $\ln x = 5 \sin x$ that required manipulating graphical images, acknowledging approximate nature of values, and reasoning about the behaviour of the logarithmic and trigonometric functions. Reasoning graphically allows students to expect and identify intersection points beyond those that are produced by a direct solve command (see Figure 20.17a) or that appear in a typical viewing window

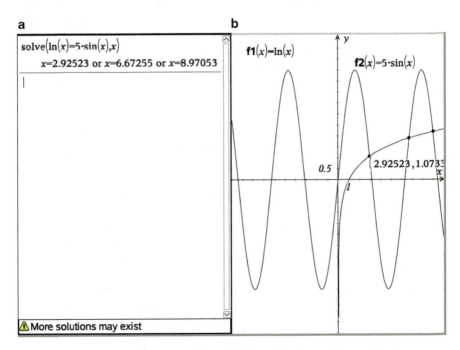

Figure 20.17. Typical direct solve results (**a**) and viewing window image (**b**) suggesting three approximate solutions for $\ln x = 5 \sin x$.

(see Figure 20.17b) and to justify why there is a finite number of solutions. By reasoning about the monotonic behaviour of the logarithmic function in contrast to the bounded values of the sine function, students concluded that, although there are many solutions that they can illustrate by scrolling to see what happens for larger values of x, there are not infinitely many solutions. They also came to terms with the difficulty of representing the solutions in compact symbolic forms due to their non-periodic values. Although it might seem that reasoning about graphs overshadows symbolic reasoning in this example, there are two important elements that symbolic forms offer. First, reasoning about properties of functions requires the symbolic forms. Unlike graphs that provide only approximate values and convey a function relationship for only a subset of a domain, symbolic forms provide the needed specificity for confidence in the argument. Second, examples like this provide opportunities for students to experience instances in which symbolic forms (or graphical forms) fall short as they coordinate among different techniques.

Graphical reasoning related to equation solving might be done not only to identify solutions but also to make sense of how properties of real numbers and properties of equality are used to make sense of steps in symbolic procedures. For example, Zbiek and Heid (2011) assumed a beginning algebra context and use the equation $6x+3=12+3x$ to illustrate how these two types of properties differently affect the values of the two expressions but not the solution of the equation. Figure 20.18a contains a set of steps executed with CAS. The sequence of graph pairs of the members of each equation appears in Figure 20.18b–f.

Figure 20.18b, c shows that application of properties of real number operations does not change the graphs, as it does not change the values of the expressions for any value of x. In contrast, Figures 20.18d–f illustrate that application of properties of equality leave the solutions unchanged but expression values changed. A comparison of these two types of graphical situations illustrates differences as well as the relationship between equivalent expressions (produced by application of properties of real numbers) and equivalent equations (produced by application of properties of equality and properties of real numbers).

CAS-supported reasoning across graphical and symbolic domains can target aspects of student understanding other than equation solving and problem solving. Kidron (2010) shared an example of a discussion of resolving a definition of horizontal asymptote in a calculus course. Nathalie, who previously offered examples and rules but not a definition for asymptote, was asked what an asymptote is. The college calculus student then worked through a specially designed set of tasks to challenge her concept image of asymptote. Kidron described how Nathalie's understanding progressed beyond her initial notion of asymptote as "some kind of a line" such that the "function tends to it—not touching it, but approaching it." Tasks provided instances in which a graph intersected a horizontal asymptote and in which there were infinitely many such intersections. As a result, Nathalie revised her concept definition to acknowledge that "'tending to' is not only when the graph of the function looks like a line which approaches steadily the asymptote, but when the value of the function at infinity equals some number, approaches some specific value." From this example, we suggest that tasks that challenge concept images

20 Computer Algebra Systems, and Algebra in the Curriculum

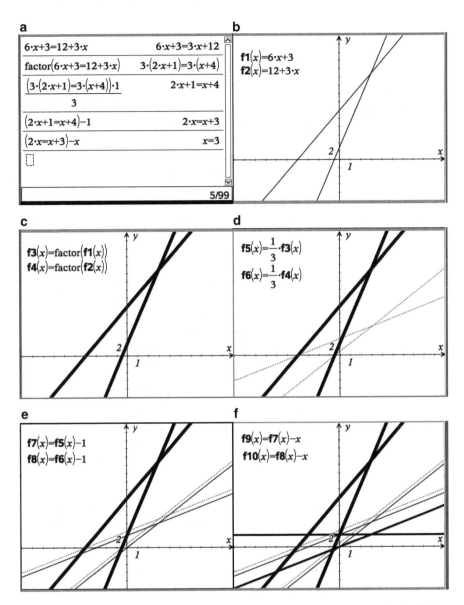

Figure 20.18. Symbolic (**a**) and graphical (**b–f**) representations of steps in solving $6x+3=12+3x$.

through the use of graphical representations can help students develop and understand rich, symbolically stated definitions in addition to common and generalized symbolic solution methods. Reasoning supports symbol sense while capitalizing on CAS multiple representation capacity in developing techniques.

Tools for Reasoning

Although CAS, with its symbolic emphasis and multiple representation capacity, has potential as a tool for reasoning, recent technology developments raise new questions. Use of the previously mentioned CAS-generated tables whose elements can be symbolic algebraic expressions is one way in which students have expedient ways to generate multiple instances within and across registers.

Dynamically linked representations. CAS environments feature not only multiple representations but also dynamically linked representations in ways that allow users to progress quickly through multiple examples by clicking or dragging an element of one representation and seeing corresponding changes in other registers. Scholars working outside of CAS environments (e.g., Hegedus & Kaput, 2007) have emphasized the potential of dynamically linked representations to allow students to see how a phenomenon in one representation might not be apparent in another. Duncan (2010) indicated that teachers believe that linked dynamic representations provide students with evidence to support their reasoning. As Kieran (2007) noted, research on effects of controlled change on dynamically linked representation is an underdeveloped research domain.

Relating both dynamically linked representations and reasoning about results come into play as users can generate multiple values of a parameter by manipulating a "slider." Zbiek and Heid (2001) provided an example with a task that was initially developed in a dynamic geometry setting and was subsequently moved to a CAS slider environment. Students used sliders to explore the family of functions generally represented by $f(x) = a/(1+be^{cx})+d$, where a, b, c, and d are real numbers. When students dragged a slider to change the value of b (as represented by the sequence of graphs in Figure 20.19), they observed a sudden "break" in the graph. The surprise was not as striking when produced with static selection of particular values for the parameter in the absence of a slider. Spurred by the sudden event in the dynamic setting, students reasoned symbolically to justify why such a break would occur. Although empirical research is not extensive, dynamically linked representations have promise as tools to elicit and support reasoning that links the symbolic register to other registers.

Integrated technology environments. Dynamic elements underlie questions that might be leading CAS-focussed researchers to work in broader technology environments. Lagrange and Chiappini (2007) describe the work of two research groups with digital tools that blend CAS with other dynamic elements. A promising feature of one of the artefacts, Cassyopée,[1] is its inclusion of geometry and a connection of algebra to other domains. The integrated or linked nature of representations with current CAS leads to the question of how one reasons within and across different representations. Lagrange and Gelis (2008) describe two lesson

---

[1] Cassyopée is the spelling used in the referenced paper.

Figure 20.19. Sequence of graphs representing dragging slider to change the value of b in $f(x) = a/(1+be^{cx})+d$.

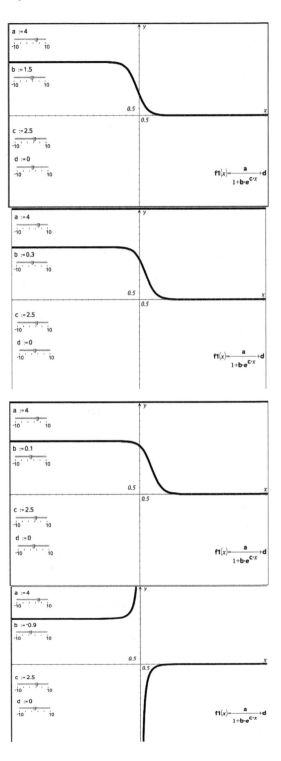

sets from the Casyopée project, a project involved in adapting or altering CAS to allow students a way to access mathematical symbols. The lessons target difficulties that students have with function ideas (e.g., notation, covariation, linked representations). CAS allows for geometrical calculations and parameter manipulation and supports conjecturing and proving, allowing symbolic work to go with graphical work. Lagrange and Gelis not only find the CAS connection to dynamic geometry in Casyopée important but they also note that a notepad feature—which is a communication medium rather than a mathematical one—allows users to give an account of their work, which is particularly useful for proof work.

As illustrated in these last instances, current research on the nature and potential of CAS is now conceptualized in terms of broader technology environments. Given the evolution of CAS technology, we question what to call tools that include CAS capability among a more extensive suite of tools. Holton, Thomas, and Harradine (2009) use *collection of technologies* (COT) rather than CAS to label calculators and computer software with symbolic manipulation in addition to other capabilities. Pierce and Stacey (2010) refer to calculators or computer software that perform algorithms necessary to execute routine procedures from any branch of mathematics, including but not limited to algebra, as *mathematics analysis software* (MAS). The examples of reasoning in CAS environments that appear in the literature suggest the potential of COT or MAS to support reasoning across registers about algebraic entities and their counterparts in other areas of mathematics.

Role of Algebra in the School Curriculum

We described three foci central to CAS research, theory, and practice: the interaction of concepts and skills, the concepts that can be approached with CAS, and the thinking and reasoning that CAS inspires or requires. With these themes from the literature and issues around teachers and other factors as background, we turn to the question of how CAS change the role of algebra in the school curriculum. Multiple perspectives, approaches, and conceptions of algebra are represented in the literature, including algebra as: generalization (Lee, 1996; Mason, 1996), a study of function (Chazan & Yerushalmy, 2003; Fey & Heid, 1995; Heid, 1996; Mayes, 2001; Yerushalmy & Chazan, 2002), a problem-solving tool (Bednarz & Janvier, 1996; Rojano, 1996), a study of structure (Cuoco, 2002), and a modelling tool (Nemirovsky, 1996).

Introducing CAS into algebra seems to have a direct effect on a functions approach to algebra. Multiple and now dynamically linked symbolic forms, graphs, and tables facilitate the study of functions. The ease of sliders and other tools to study parameter effects facilitates exploration of function families. Most CAS work, like the examples previously reported, involves functions and clearly enriches a functions approach to algebra. However, CAS also enriches other views of school algebra. The capability to construct and alter different symbolic expressions yields modelling possibilities. The ability to build and manipulate complex expressions and the new concepts introduced encourage generalization. Symbolic results to interpret and control provide a venue for algebra as a study of structure. In short, CAS allows each of the views of algebra that we have identified to be enriched.

Many of the examples we have provided have focussed on school mathematics that is likely beyond the capability of beginning algebra students. However, entire curricula have been constructed for beginning algebra students based on the premise of availability of CAS. The aforementioned CIA (Fey & Heid, 1995) curriculum is an example. In the case of the CIA curriculum, integration of CAS allowed the development of a curriculum that took as its central theme the construct of function. For example, solutions of linear equations were taken as the input value, x, for the point of intersection of the functions defined by $f(x)$ and $g(x)$. Equations in two unknowns were viewed as statements about the relationship between two functions of two variables. [See Heid, 1996, for results regarding student learning in the context of the CIA curriculum.] Through attention to blended concepts and procedures, techniques, and new concepts, CAS supports more seamless thinking across arithmetic, algebra, and calculus. Newer CAS-inclusive technologies allow other areas of mathematics, such as geometry and data analysis, to be more closely tied to the symbolic power of algebra. The impact of CAS on the role of algebra in the school curriculum seems to be as a means to make symbolic work more prevalent as students blend procedures and old and new concepts and reason symbolically across the mathematics curriculum and within the sciences.

Issues Related to Implementation of CAS

In this section we briefly address some of the issues that may arise when teachers consider implementation of CAS in their classroom. These include unfavourable attitudes of students, their parents, and society in general regarding the use of CAS calculators in mathematics teaching; the influence of external assessment practice on CAS use; the problems inherent in integration of CAS into current practice; and especially, the attitude and capabilities of the teachers themselves and the changing dynamics of the didactic contract when CAS is present. This last issue covers a number of aspects that must converge to enable the kinds of conceptual use of CAS previously described.

One issue with regard to CAS use relates to student attitudes, which in turn may tend to reflect those of parents and of society in general. The common misconception that use of any calculator is detrimental to the acquisition of mathematical skills appears widespread and persistent. A number of studies have demonstrated that a significant minority of students show some resistance to CAS use, often because they are satisfied with by-hand methods, or believe that this is the only proper way to do mathematics (Ball & Stacey, 2005; Pierce, Herbert, & Giri, 2004; Stewart, 2005). In a study of university students using computer-based CAS, Stewart, Thomas, and Hannah (2005) categorized student attitudes toward CAS, describing one group whose members are openly opposed to computers and believe strongly in the superiority of by-hand work for doing and understanding mathematics. They also described students who use CAS primarily for checking by-hand answers, a practice that has also been noticed among school students (Stewart & Thomas, 2005; Thomas & Hong, 2004, 2005b).

Researchers have identified a number of factors that influence teacher adoption and implementation of technology in mathematics teaching. These include, for example, previous experience in using technology, time, opportunities to learn, professional development, access to technology, availability of classroom teaching materials, support from colleagues and school administration, pressures of curriculum and assessment requirements, and technical support (Forgasz, 2006a; Goos, 2005; Thomas, 2006). Hence, although teachers may acknowledge that technology such as CAS may be used to improve students' learning, many teachers perceive a variety of barriers to the use of the technology (Pierce & Ball, 2009). Forgasz (2006a) lists access to computers and/or computer laboratories as the most prevalent inhibiting factor, with lack of professional development and technical problems, including lack of technical support next. Thomas (2006) agrees, citing availability of technology as the major issue, followed by a lack of resources, training, and confidence. There is also some evidence that a teacher's personal beliefs, values, and attitudes related to mathematics and technology, what Schoenfeld calls *orientations* (Schoenfeld, 2008, 2011) could influence perspectives on obstacles to CAS use. Positive orientations include a strong belief in the value of technology in learning mathematics, confidence in using technology to teach, enjoyment of technology, and an openness to personal learning (Forgasz, 2006a; Hong & Thomas, 2006; Pierce, Stacey, & Wander, 2010; Thomas & Hong, 2005a). Schoenfeld (2011) holds that the teachers' orientations not only shape the goals that they set but also the priority attached to the goals. Schoenfeld further posits that, once the teacher has oriented herself and set goals for the current situation, she then decides on the direction necessary to achieve the goals, and calls on the resources, including technology, to meet them. Goals can emerge in the process of teaching, and Monaghan (2004) claims that the presence of technology can influence goals that emerge during a lesson. Once the goals have been set decisions are made in order to meet them, and it is the quality of this decision making that affects how successful a teacher is in attaining the goals. Since the whole process is underpinned by teacher beliefs as a major part of their orientations, there is a need to focus on what teachers believe about technology use, and how this may change over time (Lagrange et al., 2003). Whereas beliefs are generally stable, and so attempts to influence them have to be long term, appropriate, targeted professional development may be able to shift beliefs about technology, leading to more positive use, as has been noted in other areas (Paterson, Thomas, & Taylor, 2011).

The pressure teachers are under to have their students perform well on external assessment has a strong influence on what they do, or do not do, in the classroom. Many feel that there is a time burden associated with adding technology to their already overcrowded lessons. This perspective is unlikely to change unless CAS use in examinations is sanctioned by educational authorities. Two issues that come to the fore with regard to using CAS in examinations are, first, the effect on what is actually being assessed, given the capability of the calculators, and second, the perceived lack of equality of access caused by the cost of handheld CAS. The latter was reported by Thomas and colleagues (2008) to be of only minor concern to teachers surveyed in New Zealand, but the same research showed that the former does worry teachers. There has been research on the use of CAS in examinations, much of it emerging from Victoria, Australia, where VCE Mathematical Methods (CAS), a CAS-permitted examination, has been in place for some years. The research from Victoria suggests that CAS scaffolds students, helping them engage with extended response analysis examination questions and achieve relatively good success (Evans, Norton, & Leigh-Lancaster, 2005; Norton, Leigh-Lancaster, Jones, & Evans, 2007). In addition there is support for the claim that students who use CAS develop at least the same level of skills as those who use graphic calculators, countering the loss of skills argument. However, to achieve this positive outcome, Ball and Stacey (2004, 2005) concluded that since new mathematical practices and processes of learning emerge when CAS is employed, communicating this to students requires active participation of teachers and a different curriculum emphasis. One aspect of this is the rubric RIPA (*Reasons–Inputs–Plan–(some) Answers*) proposed as a guide for teaching students how to record their solutions when they use CAS. The integration of CAS in the curriculum, including assessment practice, is a crucial issue impinging on CAS use. Research by Oates (2004, 2009), although focussed on tertiary mathematics, pointed out the need for a refined taxonomy to describe what is really meant by such a technology-integrated curriculum.

To use CAS in teaching to its full potential requires a particular set of skills and attitudes on the part of teachers, and so addressing teacher-related issues is crucial. One of these is that while many teachers claim to support the use of technology in their teaching (Forgasz, 2006a; Thomas, 2006) the degree and type of use in the classroom are variable (Zbiek & Hollebrands, 2008). There is also a sizeable minority of teachers who are either not convinced of its value (Forgasz, 2006b) or actively oppose its use (Thomas, Hong, Bosley, & delos Santos, 2008). This latter study reported that 60.5% of teachers disagreed with the statement that "All types of calculators should be allowed in examinations," with only 21.7% in favour, and that 27% of teachers thought that using calculators can be detrimental to student understanding of mathematics. There are many intrinsic factors that may influence a teacher's decision to use (or not to use) technology. These include their orientations; their instrumental genesis of the tools (Artigue, 2002; Guin & Trouche, 1999; Rabardel, 1995; Vérillon & Rabardel, 1995); their perceptions of the nature of mathematical knowledge and how it should be learned (Zbiek & Hollebrands, 2008); their mathematical content knowledge; and their mathematical knowledge for teaching (Ball, Hill, & Bass, 2005; Hill & Ball, 2004; Zbiek, Heid, Blume, & Dick, 2007), which includes Shulman's pedagogical content knowledge (PCK) (Shulman, 1986). PCK refers to understanding not only the mathematical ideas in a particular topic but also how these relate to the principles and techniques required to teach and learn the topic, including appropriate structuring of content and relevant classroom discourse and activities.

Considering these factors led Thomas (Hong & Thomas, 2006; Thomas, 2009; Thomas & Chinnappan, 2008; Thomas & Hong, 2005b) to propose the notion of *pedagogical technology knowledge* (PTK) as a useful way to think about what teachers need in order to use technology, such as CAS, when teaching mathematics. He also suggests that the level of a teacher's PTK may be a key driver of CAS use. The teacher development of PTK for mathematics involves adding a number of attributes to mathematical PCK. The most important of these, enabled by a strong mathematical content knowledge, is a shift in focus, from seeing the technology as simply something added to the teaching of mathematics to putting the mathematics at the centre of activity, and asking how the CAS can enable students to understand the mathematical concepts better. To attain this may require a change in orientations with regard to mathematics and CAS technology. Hence, the affective domain is also involved, with personal confidence in teaching with CAS one dimension of PTK (Thomas et al., 2008). Another aspect of PTK is instrumental genesis of CAS (comprising both instrumentation and instrumentalization), by which CAS tools are transformed into epistemic instruments. Guin and Trouche (1999) argue that instrumental genesis and conceptualization should occur concurrently in the classroom, and, in order for this to happen, teachers need to have developed their PTK sufficiently to be able to focus CAS activity on specific mathematical conceptions, such as those suggested in this chapter. It seems reasonable that teachers who have strong PTK are likely to feel comfortable in accessing CAS when designing mathematical learning experiences. Pierce, Stacey, and Wander (2010) report that initially teachers principally regarded the CAS as a tool for doing, rather than exploring mathematics. However, they believe that this may change as teachers grow in confidence and skills with the CAS. According to Pierce (2005) a teacher who can discern strategic use of CAS and model its effective use to students will make qualitative progress in technology use. One way in which strong PTK may influence teachers is in the use of CAS to mediate student learning through development and use of innovative mathematical tasks and approaches (Clark-Wilson, 2010). In turn, teacher privileging of the technology (Kendal & Stacey, 1999, 2001) has been shown to have a positive impact on students' uptake of technology in exploring mathematics.

How can teachers be assisted to develop PTK further? One critical element in the promotion of teacher PTK, which might lead to improved use of CAS for development of activities that encourage conceptual thinking, is focussed preservice training and inservice professional development (PD) of mathematics teachers (Fitzallen, 2005; Forgasz, 2006b). One suggestion by Goos and Bennison (2005) for improving PD is to employ online discussion by teachers to build a community of practice. It also appears that giving teachers personal experience of using CAS in their own classroom as a component of PD may help them develop their PTK (Ball & Stacey, 2006).

Even when teachers have a high level of PTK, studies show that there are issues involving the didactic contract that arises in classrooms when technology is introduced. Monaghan (2004) suggests that there is no common structure for teacher–student interactions in CAS classrooms, and this can lead to a disconnect between students and teachers with regard to the didactic contract (Pierce, Stacey, & Wander, 2010). While both students and teacher agree that the teacher has a responsibility to teach technology skills, students may see these skills as the main point of the lesson, while teachers view the lesson as primarily about teaching mathematics. An example of how things may change is seen in Duncan's (2010) study, in which teachers recognized that when using CAS they changed the didactic contract, moving from a general class teaching style to greater use of student investigation and discussion. It has also been shown that CAS technology can play a role in the conceptualization of mathematical models rather than simply being a tool that is used to solve a mathematical problem after it has been abstracted, and this can also provoke a change in student–student and student–teacher interactions (Geiger, Faragher, Redmond, & Lowe, 2008). In the light of these and other influences on classroom dynamics and relationships there is likely to be a need for negotiation to adapt didactic contracts.

Needed Research

As we examined the empirical and theoretical literature on the use of CAS, we found promising strands of research. We also realized that there is much yet to be learned about how the incorporation of CAS can affect the teaching and learning of school algebra. We end with a few suggestions for what we see as promising directions for future research centred on the use of CAS in school algebra.

Each of these suggestions requires developing school settings in which CAS technologies are welcome and available. In these environments, we need to know more about how CAS can affect the ways in which students reason about mathematics:

> What does research across COT or MAS suggest about student reasoning, such as the role of representations and moving across registers?

> How does use of dynamically linked representations motivate reasoning, facilitate reasoning, and contribute to the development of a capacity to reason?

> How does prolonged experience with CAS (COT or MAS) affect how students understand and use algebraic symbols?

> How can CAS be used to influence student conceptualization? What factors can improve the epistemic value of CAS?

> Are there long-term conceptual benefits from CAS use? If so what are they?

We need to know more about instructors and instructional strategies in CAS-present classrooms.

> Can we improve the student construction of CAS-related schemes through classroom presentation and discussion of techniques, and, if so, how?

> What is the relationship between teacher confidence and pedagogical technology knowledge (PTK)? Along what trajectories does PTK develop? Can PTK be validly and reliably measured, and, if so, how?

> How does the introduction of CAS change student–student and student–teacher interactions? Can these changes be captured by descriptions of the didactic contract?

We need to know more about CAS-intensive mathematics curricula.

What does it mean to have a CAS-integrated curriculum? What would it look like? How can we describe what is really meant by a CAS-integrated curriculum at any level?

References

Alonso, F., Garcia, A., Garcia, F., Hoya, S., Rodriguez, G., & de la Valla, A. (2001). Some unexpected results using computer algebra systems. *International Journal of Computer Algebra in Mathematics Education, 8*, 239–252.

Arcavi, A. (1994). Symbol sense: Informal sense-making in formal mathematics. *For the Learning of Mathematics, 14*(3), 24–35.

Artigue, M. (2002). Learning mathematics in a CAS environment: The genesis of a reflection about instrumentation and the dialectics between technical and conceptual work. *International Journal of Computers for Mathematical Learning, 7*, 245–274. doi:10.1023/A:1022103903080.

Arzarello, F., & Robutti, O. (2010). Multimodality in multi-representational environments. *ZDM–International Journal of Mathematics Education, 42*, 715–731. doi:10.1007/s11858-010-0288-z.

Balacheff, N. (1994). La transposition informatique. Note sur un nouveau problème pour la didactique. In M. Artigue, R. Gras, C. Laborde, & P. Tavignot (Eds.), *Vingt ans de didactique des mathématiques en France: Hommage à Guy Brousseau et à Gérard Vergnaud* (pp. 364–370). Grenoble, France: La Pensée Sauvage.

Ball, D. L., Hill, H. C., & Bass, H. (2005). Knowing mathematics for teaching: Who knows mathematics well enough to teach third grade, and how can we decide? *American Educator, 29*(1), 14–17, 20–22, 43–46.

Ball, L., & Stacey, K. (2004). A new practice evolving in learning mathematics: Differences in students' written records with CAS. In P. Clarkson, A. Downton, D. Gronn, M. Horne, A. McDonough, R. Pierce, & A. Roche (Eds.), *Building connections: Theory, research, and practice*. Proceedings of the 28th Annual Conference of the Mathematics Education Research Group of Australasia, Melbourne (Vol. 1, pp. 177–184). Sydney, Australia: Mathematics Education Research Group of Australasia.

Ball, L., & Stacey, K. (2005). Students' views on using CAS in senior mathematics. *Building connections: Theory, research, and practice*. In P. Clarkson, A. Downton, D. Gronn, M. Horne, A. McDonough, R. Pierce, & A. Roche (Eds.), *Building connections: Theory, research, and practice*. Proceedings of the 28th Annual Conference of the Mathematics Education Research Group of Australasia, Melbourne (Vol. 1, pp. 121–128). Sydney, Australia: Mathematics Education Research Group of Australasia.

Ball, L., & Stacey, K. (2006). Coming to appreciate the pedagogical uses of CAS. In J. Novotná, H. Moraová, M. Krátká, & N. Stehlíková (Eds.), *Proceedings of the 30th Conference of the International Group for the Psychology of Mathematics Education* (Vol. 2, pp. 105–112). Prague, Czech Republic: International Group for the Psychology of Mathematics Education.

Bednarz, N., & Janvier, B. (1996). Emergence and development of algebra as a problem-solving tool: Continuities and discontinuities with arithmetic. In N. Bednarz, C. Kieran, & L. Lee (Eds.), *Approaches to algebra: Perspectives for research and teaching* (pp. 115–136). Boston, MA: Kluwer.

Boers van Oosterum, M. A. M. (1990). *Understanding of variables and their uses acquired by students in their traditional and computer-intensive algebra* (PhD dissertation). University of Maryland, College Park, MD.

Böhm, J. (2007, June). *Why is happening with CAS in classrooms? Example Austria*. CAME 2007 Symposium: Connecting and Extending the Roles of Computer Algebra in Mathematics Education, Pécs, Hungary. Retrieved from http://www.lkl.ac.uk/research/came/events/CAME5/CAME5-Theme2-Boehm.pdf.

Buchberger, B. (1989). Should students learn integration rules? *SIGSAM Bulletin, 24*(1), 10–17.
Cedillo, T., & Kieran, C. (2003). Initiating students into algebra with symbol-manipulating calculators. In J. Fey, A. Cuoco, C. Kieran, L. McMullin, & R. M. Zbiek (Eds.), *Computer algebra systems in secondary school mathematics education* (pp. 219–239). Reston, VA: National Council of Teachers of Mathematics.
Chazan, D., & Yerushalmy, M. (2003). On appreciating the cognitive complexity of school algebra: Research on algebra learning and directions of curricular change. In J. Kilpatrick, D. Schifter, & G. Martin (Eds.), *A research companion to the Principles and Standards for School Mathematics* (pp. 123–135). Reston, VA: National Council of Teachers of Mathematics.
Clark-Wilson, A. (2010). Emergent pedagogies and the changing role of the teacher in the TI-Nspire Navigator-networked mathematics classroom. *ZDM–International Journal of Mathematics Education, 42*, 747–761. doi:10.1007/s11858-010-0279-0.
Cuoco, A. (2002). Thoughts on reading Artigue's "Learning mathematics in a CAS environment". *International Journal of Computers for Mathematical Learning, 7*, 245–274. doi:10.1023/A:1022112104897.
Dana-Picard, T. (2007). Motivating constraints of a pedagogy-embedded computer algebra system. *International Journal of Science and Mathematics Education, 5*, 217–235. doi:10.1007/s10763-006-9052-9.
Dick, T. P. (1992). Symbolic-graphical calculators: Teaching tools for mathematics. *School Science and Mathematics, 92*, 1–5. doi:10.1111/j.1949-8594.1992.tb12128.x.
Drijvers, P., & Weigand, H.-G. (2010). The role of handheld technology in the mathematics classroom. *ZDM–International Journal of Mathematics Education, 42*, 665–666. doi:10.1007/s11858-010-0285-2.
Duncan, A. G. (2010). Teachers' views on dynamically linked multiple representations, pedagogical practices and students' understanding of mathematics using TI-Nspire in Scottish secondary schools. *ZDM–International Journal of Mathematics Education, 42*, 763–774. doi:10.1007/s11858-010-0273-6.
Duval, R. (2006). A cognitive analysis of problems of comprehension in a learning of mathematics. *Educational Studies in Mathematics, 61*, 103–131. doi:10.1007/s10649-006-0400-z.
Edwards, M. T. (2001). *The electronic "other": A study of calculator-based symbolic manipulation utilities with secondary school mathematics students* (PhD dissertation). The Ohio State University, Columbus, OH.
Evans, M., Norton, P., & Leigh-Lancaster, D. (2005). Mathematical methods Computer Algebra System: 2004 pilot examinations and links to a broader research agenda. In P. Clarkson, A. Downton, D. Gronn, M. Horne, A. McDonough, R. Pierce, & A. Roche (Eds.), *Building connections: Theory, research, and practice* (Proceedings of the 28th Annual Conference of the Mathematics Education Research Group of Australasia, Vol. 1, pp. 329–336). Melbourne, Australia: Mathematics Education Research Group of Australasia.
Fey, J. T., Heid, M. K., Good, R. A., Sheets, C., Blume, G., & Zbiek, R. M. (1995). *Concepts in algebra: A technological approach*. Dedham, MA: Janson Publications, Inc.
Fitzallen, N. (2005). Integrating ICT into professional practice: A case study of four mathematics teachers. In P. Clarkson, A. Downton, D. Gronn, M. Horne, A. McDonough, R. Pierce, & A. Roche (Eds.), *Building connections: Theory, research, and practice* (Proceedings of the 28th Annual Conference of the Mathematics Education Research Group of Australasia, Vol. 1, pp. 353–360). Melbourne, Australia: Mathematics Education Research Group of Australasia.
Forgasz, H. J. (2006a). Factors that encourage and inhibit computer use for secondary mathematics teaching. *Journal of Computers in Mathematics and Science Teaching, 25*(1), 77–93. doi:10.1007/s10857-006-9014-8.
Forgasz, H. J. (2006b). Teachers, equity, and computers for secondary mathematics learning. *Journal of Mathematics Teacher Education, 9*, 437–469. doi:10.1007/s10857-006-9014-8.
Geiger, V., Faragher, R., Redmond, T., & Lowe, J. (2008). CAS-enabled devices as provocative agents in the process of mathematical modeling. In M. Goos, R. Brown, & K. Makar (Eds.), *Navigating currents and charting directions* (Proceedings of the 30th Annual Conference of the Mathematics Education Research Group of Australasia, Vol. 1, pp. 219–226). Brisbane, Australia: Mathematics Education Research Group of Australasia.

Godfrey, D., & Thomas, M. O. J. (2008). Student perspectives on equation: The transition from school to university. *Mathematics Education Research Journal, 20*(2), 71–92. doi:10.1007/BF03217478.

Goos, M. (2005). A sociocultural analysis of the development of pre-service and beginning teachers' pedagogical identities as users of technology. *Journal of Mathematics Teacher Education, 8*, 35–59. doi:10.1007/s10857-005-0457-0.

Goos, M., & Bennison, A. (2005). The role of online discussion in building a community of practice for beginning teachers of secondary mathematics. In P. Clarkson, A. Downton, D. Gronn, M. Horne, A. McDonough, R. Pierce, & A. Roche (Eds.), *Building connections: Theory, research, and practice* (Proceedings of the 28th Annual Conference of the Mathematics Education Research Group of Australasia, Vol. 1, pp. 385–392). Melbourne, Australia: Mathematics Education Research Group of Australasia.

Gray, E. M., & Tall, D. O. (1994). Duality, ambiguity and flexibility: A proceptual view of simple arithmetic. *Journal for Research in Mathematics Education, 26*, 115–141. doi:10.2307/749505.

Guin, D., & Trouche, L. (1999). The complex process of converting tools into mathematical instruments: The case of calculators. *International Journal of Computers for Mathematical Learning, 3*, 195–227.

Guin, D., Ruthven, K., & Trouche, L. (2005). *The didactical challenge of symbolic calculators: Turning a computational device into a mathematical instrument*. New York, NY: Springer.

Harel, G. (2008). DNR perspectives on mathematics curriculum and instruction, Part I: Focus on proving. *ZDM–International Journal of Mathematics Education, 40*, 487–500. doi:10.1007/s11858-008-0104-1.

Hasenbank, J. F., & Hodgson, T. (2007, February). *A framework for developing algebraic understanding & procedural skill: An initial assessment*. Paper presented at the tenth conference on Research in Undergraduate Mathematics Education, Special Interest Group of Mathematical Association of America (SIG-MAA), San Diego, CA.

Hegedus, S., & Kaput, J. (2007). *Lessons from SimCalc: What research says* (Research Note 6). Dallas, TX: Texas Instruments.

Heid, M. K. (1984). *An exploratory study to examine the effects of resequencing concepts and skills in an applied calculus curriculum through the use of a microcomputer* (PhD dissertation). University of Maryland, College Park, MD.

Heid, M. K. (1988). Resequencing skills and concepts in applied calculus using the computer as a tool. *Journal for Research in Mathematics Education, 19*, 3–25. doi:10.2307/749108.

Heid, M. K. (1992). *Final report: Computer-intensive curriculum for secondary school algebra*. Final report NSF project number MDR 8751499. University Park: The Pennsylvania State University, Department of Curriculum and Instruction.

Heid, M. K. (1996). A technology-intensive functional approach to the emergence of algebraic thinking. In N. Bednarz, C. Kieran, & L. Lee (Eds.), *Approaches to algebra: Perspectives for research and teaching* (pp. 239–256). Boston, MA: Kluwer Academic Publishers.

Heid, M. K. (1997). The technological revolution and the reform of school mathematics. *American Journal of Education, 106*, 5–61. doi:10.1086/444175.

Heid, M. K., & Blume, G. W. (2008). Algebra and function development. In M. K. Heid & G. W. Blume (Eds.), *Research on technology and the teaching and learning of mathematics* (Research syntheses, Vol. 1, pp. 55–108). Charlotte, NC: Information Age Publishers.

Heid, M. K., Sheets, C., Matras, M. A., & Menasian, J. (1988, April). *Classroom and computer lab interaction in a computer-intensive algebra curriculum*. Paper presented at the annual meeting of the American Educational Research Association, New Orleans.

Hill, H., & Ball, D. L. (2004). Learning mathematics for teaching: Results from California's mathematics professional development institutes. *Journal for Research in Mathematics Education, 35*, 330–351. doi:10.2307/30034819.

Hillel, J., Kieran, C., & Gurtner, J. (1989). Solving structured geometric tasks on the computer: The role of feedback in generating strategies. *Educational Studies in Mathematics, 20*, 1–39. doi:10.1007/BF00356039.

Hoehn, L. (1989, December). Solutions of $x^n+y^n=z^{n+1}$. *Mathematics Magazine*, 342. doi:10.2307/2689491

Hollar, J. C., & Norwood, K. (1999). The effects of a graphing-approach intermediate algebra curriculum on students' understanding of function. *Journal for Research in Mathematics Education, 30,* 220–226. doi:10.2307/749612.

Holton, D., Thomas, M. O. J., & Harradine, A. (2009). The excircle problem: A case study in how mathematics develops. In B. Davis & S. Lerman (Eds.), *Mathematical action & structures of noticing: Studies inspired by John Mason* (pp. 31–48). Rotterdam, The Netherlands: Sense.

Hong, Y. Y., & Thomas, M. O. J. (2006). Factors influencing teacher integration of graphic calculators in teaching. In *Proceedings of the 11th Asian Technology Conference in Mathematics* (pp. 234–243). Hong Kong: Asian Technology Conference in Mathematics.

Hoyles, C., & Noss, R. (2009). The technological mediation of mathematics and its learning. *Human development, 52,* 129–147. doi:10.1159/000202730.

Judson, P. T. (1990). Elementary business calculus with computer algebra. *Journal of Mathematical Behavior, 9,* 153–157.

Kendal, M., & Stacey, K. (1999). Varieties of teacher privileging for teaching calculus with computer algebra systems. *The International Journal of Computer Algebra in Mathematics Education, 6,* 233–247.

Kendal, M., & Stacey, K. (2001). The impact of teacher privileging on learning differentiation with technology. *International Journal of Computers for Mathematical Learning, 6,* 143–165. doi:10.1023/A:1017986520658.

Kidron, I. (2010). Constructing knowledge about the notion of limit in the definition of the horizontal asymptote. *International Journal of Science and Mathematics Education* [published online]. Retrieved from http://www.springerlink.com/content/b4757j83v8826527.

Kieran, C. (2007). Learning and teaching algebra at the middle school through college levels. In F. K. Lester Jr. (Ed.), *Second handbook of research on mathematics teaching and learning* (pp. 707–762). Charlotte, NC: Information Age Publishing.

Kieran, C., & Drijvers, P. (2006). The co-emergence of machine techniques, paper-and-pencil techniques, and theoretical reflection: A study of CAS use in secondary school algebra. *International Journal of Computers for Mathematical Learning, 11,* 205–263. doi:10.1007/s10758-006-0006-7.

Kieran, C., & Saldanha, L. (2008). Designing tasks for the co-development of conceptual and technical knowledge in CAS activity: An example from factoring. In G. W. Blume & M. K. Heid (Eds.), *Research on technology and the teaching and learning of mathematics* (Cases and perspectives, Vol. 2, pp. 393–414). Charlotte, NC: Information Age Publishing.

Laborde, C. (2002). The process of introducing new tasks using dynamic geometry into the teaching of mathematics. In B. Barton, K. C. Irwin, M. Pfannkuch, & M. O. J. Thomas (Eds.), *Mathematics education in the South Pacific* (Proceedings of the 25th Annual Conference of the Mathematics Education Research Group of Australasia, pp. 15–33). Auckland, NZ: Mathematics Education Research Group of Australasia.

Lagrange, J.-B. (1999). Complex calculators in the classroom: Theoretical and practical reflections on teaching pre-calculus. *International Journal of Computers for Mathematical Learning, 4,* 51–81. doi:10.1023/A:1009858714113.

Lagrange, J.-B. (2002). Étudier les mathématiques avec les calculatrices symboliques. Quelle place pour les techniques? In D. Guin & L. Trouche (Eds.), *Calculatrices Symboliques. Transformer un outil en un instrument du travail mathématique: Un problème didactique* (pp. 151–185). Grenoble, France: La Pensée Sauvage.

Lagrange, J.-B. (2003). Learning techniques and concepts using CAS: A practical and theoretical reflection. In J. T. Fey, A. Cuoco, C. Kieran, L. McMullin, & R. M. Zbiek (Eds.), *Computer algebra systems in secondary school mathematics education* (pp. 269–283). Reston, VA: National Council of Teachers of Mathematics.

Lagrange, J.-B., Artigue, M., Laborde, C., & Trouche, L. (2003). Technology and math education: A multidimensional overview of recent research and innovation. In A. J. Bishop, M. A. Clements, C. Keitel, J. Kilpatrick, & F. Leung (Eds.), *Second international handbook of mathematics education* (Part 1, pp. 237–270). Dordrecht, The Netherlands: Kluwer.

Lagrange, J.-B., & Chiappini, J. P. (2007). Integrating the learning of algebra with technology at the European level: Two examples in the ReMath Project. In *Proceedings of Conference of European Society of Research in Mathematics Education* (CERME 5) (pp. 903–912). Larnaca, Cyprus: European Society of Research in Mathematics Education.

Lagrange, J.-B., & Gelis, J.-M. (2008). The Casyopee project: A computer algebra systems environment for students' better access to algebra. *International Journal of Continuing Engineering Education and Life-Long Learning, 18*(5/6), 575–584. doi:10.1504/IJCEELL.2008.022164.

Lee, L. (1996). An initiation into algebraic culture through generalization activities. In N. Bednarz, C. Kieran, & L. Lee (Eds.), *Approaches to algebra: Perspectives for research and teaching* (pp. 87–106). Boston, MA: Kluwer.

Mason, J. (1995). Less may be more on a screen. In L. Burton & B. Jaworski (Eds.), *Technology in mathematics teaching: A bridge between teaching and learning* (pp. 119–134). London, UK: Chartwell-Bratt.

Mason, J. (1996). Expressing generality and roots of algebra. In N. Bednarz, C. Kieran, & L. Lee (Eds.), *Approaches to algebra: Perspectives for research and teaching* (pp. 65–86). Boston, MA: Kluwer.

Mason, J., Graham, A., & Johnston-Wilder, S. (2005). *Developing thinking in algebra*. London, UK: Paul Chapman Publishing and The Open University.

Matras, M. A. (1988). *The effects of curricula on students' ability to analyze and solve problems in algebra* (PhD dissertation). University of Maryland, College Park, MD.

Mayes, R. (1995). The application of a computer algebra system as a tool in college algebra. *School Science and Mathematics, 95*, 61–68. doi:10.1111/j.1949-8594.1995.tb15729.x.

Mayes, R. (2001). CAS applied in a functional perspective college algebra curriculum. *Computers in the Schools, 17*(1&2), 57–75. doi:10.1300/J025v17n01_06.

McMullin, L. (2003). Activity 8. In J. Fey, A. Cuoco, C. Kieran, L. McMullin, & R. M. Zbiek (Eds.), *Computer algebra systems in secondary school mathematics education* (p. 268). Reston, VA: The National Council of Teachers of Mathematics.

Monaghan, J. (2004). Teachers' activities in technology-based mathematics lessons. *International Journal of Computers for Mathematical Learning, 9*, 327–357. doi:10.1007/s10758-004-3467-6.

Nemirovsky, R. (1996). Mathematical narratives, modeling, and algebra. In N. Bednarz, C. Kieran, & L. Lee (Eds.), *Approaches to algebra: Perspectives for research and teaching* (pp. 197–220). Boston, MA: Kluwer.

Norton, P., Leigh-Lancaster, D., Jones, P., & Evans, M. (2007). Mathematical methods and mathematical methods computer algebra systems (CAS) 2006—Concurrent implementation with a common technology free examination. In J. Watson & K. Beswick (Eds.), *Mathematics: Essential research, essential practice* (Proceedings of the 30th Annual Conference of the Mathematics Education Research Group of Australasia, Vol. 2, pp. 543–550). Hobart, Australia: Mathematics Education Research Group of Australasia.

O'Callaghan, B. R. (1998). Computer-intensive algebra and students' conceptual knowledge of functions. *Journal for Research in Mathematics Education, 29*, 21–40. doi:10.2307/749716.

Oates, G. (2004). Measuring the degree of technology use in tertiary mathematics courses. In W.-C. Yang, S.-C. Chu, T. de Alwis, & K.-C. Ang (Eds.), *Proceedings of the 9th Asian Technology Conference in Mathematics* (pp. 282–291). Blacksburg, VA: Asian Technology Conference in Mathematics.

Oates, G. (2009). Relative values of curriculum topics in undergraduate mathematics in an integrated technology environment. In R. Hunter, B. Bicknell, & T. Burgess (Eds.), *Proceedings of the 32nd Annual Conference of the Mathematics Education Research Group of Australasia* (Vol. 2, pp. 419–427). Palmerston North, New Zealand: Mathematics Education Research Group of Australasia.

Palmiter, J. (1991). Effects of computer-algebra systems on concept and skill acquisition in calculus. *Journal for Research in Mathematics Education, 22*, 151–156. doi:10.2307/749591.

Paterson, J., Thomas, M. O. J., & Taylor, S. (2011). Reaching decisions via internal dialogue: Its role in a lecturer professional development model. In B. Ubuz (Ed.), *Proceedings of the 35th Conference of the International Group for the Psychology of Mathematics Education*

(Vol. 3, pp. 353–360). Ankara, Turkey: International Group for the Psychology of Mathematics Education.

Pierce, R. (2005). Using CAS to enrich the teaching and learning of mathematics. In S.-C. Chu, H.-C. Lew, & W.-C. Yang (Eds.), *Enriching technology and enhancing mathematics for all. Proceedings of the 10th Asian Conference on Technology in Mathematics* (pp. 47–58). Blacksburg VA: Asian Technology Conference in Mathematics.

Pierce, R., & Ball, L. (2009). Perceptions that may affect teachers' intention to use technology in secondary mathematics classes. *Educational Studies in Mathematics, 71*, 299–317. doi:10.1007/s10649-008-9177-6.

Pierce, R., Herbert, S., & Giri, J. (2004). CAS: Student engagement requires unambiguous advantages. In I. Putt, R. Faragher, & M. I. McLean (Eds.), *Mathematics education for the third millennium: Towards 2010* (Proceedings of the 27th Annual Conference of the Mathematics Education Group of Australasia, pp. 462–469). Townsville, Australia: Mathematics Education Research Group of Australasia.

Pierce, R., & Stacey, K. (2010). Mapping pedagogical opportunities provided by mathematics analysis software. *International Journal of Computers for Mathematical Learning, 15*, 1–20. doi:10.1007/s10758-010-9158-6.

Pierce, R., Stacey, K., & Wander, R. (2010). Examining the didactic contract when handheld technology is permitted in the mathematics classroom. *ZDM–International Journal of Mathematics Education, 42*, 683–695. doi:10.1007/s11858-010-0271-8.

Rabardel, P. (1995). *Les hommes et les technologies, approche cognitive des instruments contemporains*. Paris, France: Armand Colin.

Rojano, T. (1996). Developing algebraic aspects of problem solving within a spreadsheet environment. In N. Bednarz, C. Kieran, & L. Lee (Eds.), *Approaches to algebra: Perspectives for research and teaching* (pp. 137–146). Boston, MA: Kluwer.

Roschelle, J., Vahey, P., Tatar, D., Kaput, J., & Hegedus, S. J. (2003). Five key considerations for networking in a handheld-based mathematics classroom. In N. A. Pateman, B. J. Dougherty, & J. T. Zilliox (Eds.), *Proceedings of the 2003 Joint Meeting of PME and PMENA* (Vol. 4, pp. 71–78). Honolulu, Hawaii: University of Hawaii.

Ruthven, K. (1990). The influence of graphic calculator use on translation from graphic to symbolic forms. *Educational Studies in Mathematics, 21*(5), 431–450. doi:10.1007/BF00398862.

Schoenfeld, A. H. (2008). On modeling teachers' in-the-moment decision-making. In A. H. Schoenfeld (Ed.), *A study of teaching: Multiple lenses, multiple views* (Journal for Research in Mathematics Education Monograph No. 14, pp. 45–96). Reston, VA: National Council of Teachers of Mathematics.

Schoenfeld, A. H. (2011). *How we think. A theory of goal-oriented decision making and its educational applications*. New York, NY: Routledge.

Sheets, C. (1993). *Effects of computer learning and problem-solving tools on the development of secondary school students' understanding of mathematical functions* (PhD dissertation). University of Maryland, College Park, MD.

Shulman, L. (1986). Those who understand: Knowledge growth in teaching. *Educational Researcher, 15*(2), 4–14. doi:10.3102/0013189X015002004.

Stewart, S. (2005). Concerns relating to the use of CAS at university level. In P. Clarkson, A. Downton, D. Gronn, M. Horne, A. McDonough, R. Pierce, & A. Roche (Eds.), *Building connections: Research, theory and practice* (Proceedings of the 28th Annual Conference of the Mathematics Education Research Group of Australasia, Vol. 2, pp. 704–711). Melbourne, Australia: MERGA.

Stewart, S., & Thomas, M. O. J. (2005). University student perceptions of CAS use in mathematics learning. In H. L. Chick & J. L. Vincent (Eds.), *Proceedings of the 29th Conference for the International Group for the Psychology of Mathematics Education* (Vol. 4, pp. 233–240). Melbourne, Australia: The University of Melbourne.

Stewart, S., Thomas, M. O. J., & Hannah, J. (2005). Towards student instrumentation of computer-based algebra systems in university courses. *International Journal of Mathematical Education in Science and Technology, 36*, 741–750. doi:10.1080/00207390500271651.

Tall, D. O., Thomas, M. O. J., Davis, G., Gray, E., & Simpson, A. (2000). What is the object of the encapsulation of a process? *Journal of Mathematical Behavior, 18*, 223–241.

Thomas, M. O. J. (2006). Teachers using computers in the mathematics classroom: A longitudinal study. In J. Novotná, H. Moraová, M. Krátká, & N. Stehlíková (Eds.), *Proceedings of the 30th Conference of the International Group for the Psychology of Mathematics Education* (Vol. 5, pp. 265–272). Prague, Czech Republic: Charles University.

Thomas, M. O. J. (2008a). Developing versatility in mathematical thinking. *Mediterranean Journal for Research in Mathematics Education, 7*(2), 67–87.

Thomas, M. O. J. (2008b). Conceptual representations and versatile mathematical thinking. In *Proceedings of ICME-10* (CD version of proceedings, pp. 1–18). Copenhagen, Denmark, Paper available from http://www.icme10.dk/proceedings/pages/regular_pdf/RL_Mike_Thomas.pdf.

Thomas, M. O. J. (2009). Hand-held technology in the mathematics classroom: Developing pedagogical technology knowledge. In J. Averill, D. Smith, & R. Harvey (Eds.), *Teaching secondary school students mathematics and statistics: Evidence-based practice* (Vol. 2, pp. 147–160). Wellington, New Zealand: NZCER.

Thomas, M. O. J., & Chinnappan, M. (2008). Teaching and learning with technology: Realising the potential. In H. Forgasz, A. Barkatsas, A. Bishop, B. Clarke, S. Keast, W.-T. Seah, P. Sullivan, & S. Willis (Eds.), *Research in mathematics education in Australasia 2004-2007* (pp. 167–194). Sydney, Australia: Sense Publishers.

Thomas, M. O. J., & Hong, Y. Y. (2004). Integrating CAS calculators into mathematics learning: Issues of partnership. In M. J. Høines & A. B. Fuglestad (Eds.), *Proceedings of the 28th Annual Conference of the International Group for the Psychology of Mathematics Education* (Vol. 4, pp. 297–304). Bergen, Norway: Bergen University College.

Thomas, M. O. J., & Hong, Y. Y. (2005a). Teacher factors in integration of graphic calculators into mathematics learning. In H. L. Chick & J. L. Vincent (Eds.), *Proceedings of the 29th Annual Conference of the International Group for the Psychology of Mathematics Education* (Vol. 4, pp. 257–264). Melbourne, Australia: The University of Melbourne.

Thomas, M. O. J., & Hong, Y. Y. (2005b). Learning mathematics with CAS calculators: Integration and partnership issues. *The Journal of Educational Research in Mathematics, 15*(2), 215–232.

Thomas, M. O. J., Hong, Y. Y., Bosley, J., & delos Santos, A. (2008). Use of calculators in the mathematics classroom. *The Electronic Journal of Mathematics and Technology (eJMT)* [On-line Serial] 2(2). Retrieved from https://php.radford.edu/~ejmt/ContentIndex.php and http://www.radford.edu/ejmt.

Thomas, M. O. J., Monaghan, J., & Pierce, R. (2004). Computer algebra systems and algebra: Curriculum, assessment, teaching, and learning. In K. Stacey, H. Chick, & M. Kendal (Eds.), *The teaching and learning of algebra: The 12th ICMI study* (pp. 155–186). Norwood, MA: Kluwer.

Thomas, P. G., & Rickhuss, M. G. (1992). An experiment in the use of computer algebra in the classroom. *Education & Computing, 8*, 255–263. doi:10.1016/0167-9287(92)92793-Y.

Trouche, L. (2005a). An instrumental approach to mathematics learning in symbolic calculator environments. In D. Guin, K. Ruthven, & L. Trouche (Eds.), *The didactical challenge of symbolic calculators* (pp. 137–162). New York, NY: Springer.

Trouche, L. (2005b). Instrumental genesis, individual and social aspects. In D. Guin, K. Ruthven, & L. Trouche (Eds.), *The didactical challenge of symbolic calculators* (pp. 197–230). New York, NY: Springer.

Usiskin, Z. (2004). A K-12 mathematics curriculum with CAS: What is it and what would it take to get it? In W. C. Yang, S. C. Chu, T. de Alwis, & K. C. Ang (Eds.), *Proceedings of the 9th Asian Technology Conference in Mathematics* (pp. 5–16). Blacksburg, VA: Asian Technology Conference in Mathematics.

Vérillon, P., & Rabardel, P. (1995). Cognition and artifacts: A contribution to the study of though [sic] in relation to instrumented activity. *European Journal of Psychology of Education, 10*, 77–101.

Yerushalmy, M., & Chazan, D. (2002). Flux in school algebra: Curricular change, graphing, technology, and research on student learning and teacher knowledge. In L. D. English (Ed.), *Handbook of international research in mathematics education* (pp. 725–755). Mahwah, NJ: Lawrence Erlbaum.

Zbiek, R. M., & Heid, M. K. (2001). Dynamic aspects of function representations. In H. Chick, K. Stacey, J. Vincent, & J. Vincent (Eds.), *Proceedings of the 12th ICMI on the future of the teaching and learning of algebra* (pp. 682–689). Melbourne, Australia: The University of Melbourne.

Zbiek, R. M., & Heid, M. K. (2011). Using technology to make sense of symbols and graphs and to reason about general cases. In T. Dick & K. Hollebrands (Eds.), *Focus on reasoning and sense making: Technology to support reasoning and sense making* (pp. 19–31). Reston, VA: National Council of Teachers of Mathematics.

Zbiek, R. M., Heid, M. K., Blume, G. W., & Dick, T. P. (2007). Research on technology in mathematics education: A perspective of constructs. In F. Lester Jr. (Ed.), *Second handbook of research on mathematics teaching and learning* (pp. 1169–1207). Charlotte, NC: Information Age Publishing.

Zbiek, R. M., & Hollebrands, K. (2008). A research-informed view of the process of incorporating mathematics technology into classroom practice by inservice and prospective teachers. In M. K. Heid & G. W. Blume (Eds.), *Research on technology and the teaching and learning of mathematics* (Research syntheses, Vol. 1, pp. 287–344). Charlotte, NC: Information Age Publishing.

Chapter 21
Technology for Enhancing Statistical Reasoning at the School Level

Rolf Biehler, Dani Ben-Zvi, Arthur Bakker, and Katie Makar

Abstract The purpose of this chapter is to provide an updated overview of digital technologies relevant to statistics education, and to summarize what is currently known about how these new technologies can support the development of students' statistical reasoning at the school level. A brief literature review of trends in statistics education is followed by a section on the history of technologies in statistics and statistics education. Next, an overview of various types of technological tools highlights their benefits, purposes and limitations for developing students' statistical reasoning. We further discuss different learning environments that capitalize on these tools with examples from research and practice. Dynamic data analysis software applications for secondary students such as *Fathom* and *TinkerPlots* are discussed in detail. Examples are provided to illustrate innovative uses of technology. In the future, these uses may also be supported by a wider range of new tools still to be developed. To summarize some of the findings, the role of digital technologies in statistical reasoning is metaphorically compared with travelling between data and conclusions, where these tools represent fast modes of transport. Finally, we suggest future directions for technology in research and practice of developing students' statistical reasoning in technology-enhanced learning environments.

R. Biehler (✉)
University of Paderborn, Paderborn, Germany
e-mail: biehler@math.upb.de

D. Ben-Zvi
The University of Haifa, Haifa, Israel

A. Bakker
Utrecht University, Utrecht, The Netherlands

K. Makar
The University of Queensland, St. Lucia, Australia

The pervasiveness of data in everyday life is in part due to global advances in technologies. Citizens are bombarded with statistics in the media and must be savvy consumers of data (Watson, 2002). In the workplace, data are vital for quality control, monitoring and improving productivity, and anticipating problems (Bakker, Kent, Noss, & Hoyles, 2009). With data increasingly being used to add or imply credibility, there are new pressures for schools to prepare both citizens and professionals to be able to create and critically evaluate data-based claims (Garfield & Ben-Zvi, 2008). Progress in the understandings of teaching and learning of statistical reasoning and the availability of high quality technological tools for learning statistics have enabled the relatively young field of statistics education to integrate and readily capitalize on these advances (Garfield & Ben-Zvi, 2008).

The purpose of this chapter is to provide an updated perspective of advances in digital technologies and to summarize what is currently known about how these new technologies can support the development of students' statistical reasoning. We confine this chapter to the school level as other recent publications have provided discussions of technologies in statistics in tertiary settings (Chance, Ben-Zvi, Garfield, & Medina, 2007; Everson & Garfield, 2008; Garfield, Chance, & Snell, 2000; Gould, 2010), the workplace (Bakker et al., 2009; Hoyles, Noss, Kent, & Bakker, 2010; Noss, Bakker, Hoyles, & Kent, 2007), public sector (Gal & Ograjenšek, 2010; Sandlin, 2007; Trewin, 2007), assessment (Garfield et al., 2011) and the development of teachers' teaching and learning of statistical reasoning using technologies (Ben-Zvi, 2008; Burrill & Biehler, 2011; Lee & Hollebrands, 2008; Madden, 2011).

This chapter is divided into five sections. First, we provide a brief literature review of trends in statistics education relevant to the topic of statistical reasoning. Next, the historical role of technologies will be addressed, with an emphasis on how technological tools have changed the evolution of the field. Third, an overview of various types of technological tools will be covered by highlighting their benefits, purposes and limitations for developing students' statistical reasoning. In the fourth section, we will discuss different learning environments that capitalize on these tools in unique ways with exemplars from research and practice. Finally, we suggest future directions for technology in research and practice of developing students' statistical reasoning in technology-enhanced learning environments.

Trends in Statistics Education

The study of statistics provides students with competencies, tools, ideas and dispositions to use in order to respond intelligently to quantitative information in the world around them. These are considered important for all citizens to have and therefore learn as part of their education (Watson, 2006). "Statistics is a general intellectual method ... because data, variation, and chance are omnipresent in modern life" (Moore, 1998, p. 134). It has therefore become a worldwide standard for statistics to be incorporated in school education (e.g., Australian Curriculum,

Assessment and Reporting Authority, 2010; cTWO, 2007; KMK, 2004; National Council of Teachers of Mathematics, 2000; New Zealand Ministry of Education, 2007; Qualifications and Curriculum Authority, 2007).

Many research studies over the past several decades, however, indicate that most students and adults do not think statistically about important issues that affect their lives (Garfield & Ben-Zvi, 2007). Gal (2002) argued that understanding, interpreting, and reacting to real-world messages that contain statistical elements go beyond simply learning statistical content. He suggested that these skills are built on an interaction between several knowledge bases and supporting dispositions. Statistical literacy skills must be activated together with statistical, mathematical and general world knowledge. Unfortunately, traditional approaches to teaching statistics have focussed almost exclusively on mathematical skills and procedures that are insufficient for students to reason or think statistically (Ben-Zvi & Garfield, 2004). If students equate statistics with mathematics, they may expect the focus to be on a single correct outcome. However, statistics offers distinctive and powerful ways of reasoning that are distinctly different from mathematical reasoning (Cobb & Moore, 1997; delMas, 2004). Unlike mathematics, where the context can obscure the underlying abstractions, context provides meaning for the data in statistics and data cannot be meaningfully analyzed without paying careful consideration to their context, including how they were collected and what they represent (Cobb & Moore, 1997). To improve their statistical reasoning, Moore (1998) recommended that students gain multiple experiences with the messy process of data collection and exploration, discussions of how existing data are produced, experiences which ask them to select appropriate statistical summaries and draw evidence-based conclusions (delMas, 2002).

In their landmark paper, Wild and Pfannkuch (1999) provided an empirically based comprehensive description of the processes statisticians use in the practice of data-based enquiry from problem formulation to conclusions. Their paper provided the field with important research on which to build students' key experiences in learning statistics. These processes now form the foundation of much of the current research in mathematics education by focussing on the nature and development of statistical literacy, reasoning, and thinking (e.g., Ben-Zvi & Garfield, 2004). Moore (1997) summarized new recommendations in terms of changes in *content* (more key concepts, and data analysis, less probability), *pedagogy* (fewer lectures, more active learning), and *technology* (for data analysis and simulations). Garfield and Ben-Zvi (2008) provided a comprehensive background on this "reformed" approach, the history that led to this change and specific examples of ways to implement this change.

The key concepts of statistics include data, distribution, centre, variability, comparing groups, sampling, statistical inference, covariation, and statistical models (Burrill & Biehler, 2011; Pfannkuch & Ben-Zvi, 2011; Watson, 2006). Statistical reasoning (Ben-Zvi & Garfield, 2004) can then be developed through inquiry-based pedagogies and data-based activities (Garfield & Ben-Zvi, 2009) that integrate and elicit students' active engagement with these statistical concepts through the investigative processes described by Wild and Pfannkuch (1999). Throughout this chapter, we provide examples of how technology can be used to develop students' statistical reasoning. Since advances in computer software have had a major influence on how students can learn to reason statistically, we reflect historically on these advances in the next section.

A Brief History of Technology in Statistics and Statistics Education

The development of statistics has always been intertwined with the development of technology. Tukey (1965), Yates (1971) and Chambers (1980) were among the early visionaries describing the nature of future changes in doing statistics that would emerge as a result of technological advances. Moreover they envisioned the nature of new tools that were to be developed and required to support statistical practice. We will show that statistics education has also met the challenge of developing tools that are suitable to support the learning of statistics better than already existing tools.

The Emergence of New Styles of Data Analysis

A key milestone in the development of statistical reasoning was the reinterpretation of statistics into separate practices comprising exploratory data analysis (EDA) and confirmatory data analysis (CDA) (Tukey, 1977). EDA emphasizes the practical strategies of analyzing, representing and interpreting data and developing hypotheses while interacting with data. CDA is the next step, where findings are checked before the court of statistical inference and where the extent of uncertainty of inferences is quantified. This is in contrast to the more theoretical approaches to statistical inference under ideal and simplified conditions emphasized in traditional statistics. Tukey's EDA (1977) was already a step towards a more realistic account of statistical practice as was conceptualized by Wild and Pfannkuch (1999) decades later. This dramatic change in conceptualizing separate purposes for interpreting statistical information was made possible by the affordances of the new technologies which allowed for the interactive style of EDA and the practical application of graphical tools such as box plot and scatter plot, while the technological tools were concurrently re-shaped and revised by the growing popularity of EDA practices.

In the late 1970s and early 1980s, the statistical programming language S (which has evolved into the now widely used R) was initially created in parallel with developing the "new statistics" at Bell Labs in the USA (Becker, 1994). S was created specifically to support new graphical, interactive and experimental styles of EDA and designed to be able to support the development and exploration of new statistical methods by being an extensible programmable system. Another example that highlights the interaction between completely new statistical methods and new technological tools was the development of computer-intensive statistical methods, such as the bootstrapping (Diaconis & Efron, 1983). Simulation capabilities likewise played a central role as a new means of theoretical exploration of statistical methods. Following this, randomization and permutation tests became a practical option as the simulation capabilities of the new technologies expanded.

Inferential statistics under the heading of CDA also found itself adapting to new technological opportunities. Previously, model assumptions (independence, linearity, normality) had to be oversimplified because alternative methods were impractical to use without computers. New technological tools supported the careful checking of more complex assumptions of traditional procedures, for instance by residual analysis or graphical data exploration. More robust methods could therefore be developed and implemented by the support of technology. The method of least squares, for example, was relatively straightforward and computationally simple for fitting lines. Fitting curves by minimizing the sum of absolute deviations is computationally much more complex and then became a computationally feasible alternative. Moreover, the new technological support allowed the user more easily to locate patterns in the association of two variables, select an appropriate functional model, and then check the residuals for deviation from the model. This is a much more challenging process of statistical reasoning than just applying the algorithm of least squares. Moreover, this more complex process is more adequate to solve real problems and thus technology indirectly contributed to the empowerment of statistics to solve such problems.

Access to Statistical Practices Through Statistical Software

In parallel to these developments, statistical "packages" such as *SPSS* (http://www.spss.com) and *BMDP* (http://www.statistical-solutions-software.com) were developed for supporting the statistical practitioner. For many decades these two tools were characterized as a "black box" with a collection of statistical methods, where the user analyzed the statistical problem, selected the appropriate method (predominantly numerical), and obtained the corresponding results. However, neither interactive working styles nor statistical graphs were very much supported with these packages at that time. A third line of development of professional tools started in 1985 with *DataDesk* (http://www.datadesk.com; Velleman, 1998), an early prototypical tool that used the graphical user interface of Apple Macintosh creating a new quality of interactivity and user friendliness. *DataDesk* was designed to support the interactive, heavily graphical working style of EDA, replacing the command-driven interface of *S*, although with a loss in adaptability (Biehler, 1993). From the 1980s until today, borders between these three types of tools (*R*, *BMDP* and *DataDesk*) have become permeable. For instance, *R* can be also used with graphical user interfaces. And, *SPSS* comes with an interface to *R*, so that newly developed methods can be incorporated.

Technologies for Learning Statistics: Beyond the "Black Box"

Eventually, discussions took place to update statistics education that would take into account the changes in statistics (content), pedagogy and technology (Biehler, 1993; Moore, 1997). Two perspectives arose from these discussions. One perspective is based on the view that technologies for learning statistics should mirror the theory and practice of professional statistics packages to keep the gap between learning statistics and using statistical methods professionally as small as possible. Another perspective is to use technology to improve the learning of statistics. The focus in this second perspective is on other affordances of technology, such as making statistics visual, interactive and dynamic, focussing on concepts rather than computations, and offering the opportunity to experiment with data to make it engaging for students (Olive & Makar, 2010).

These two perspectives still exist today. A typical statistics text book at the tertiary level from the first perspective would be packaged with data files for different statistical tools or packages. The exercises would require students to use one of the supported software tools to apply the methods learned to real data sets. A typical text book from the second perspective may use a set of applets independent of a software tool, with which the students can interactively explore properties of statistical concepts and methods as visual–experimental support for learning. Of course, some textbooks would situate themselves somewhere between these two perspectives, including both applets for exploration and data files specific to a particular statistical package (e.g., Rossman & Chance, 2008 and their various editions for specific software such as *Fathom*, *Minitab* and graphic calculators). Some adaptable and programmable professional tools such as *R*, *Minitab*, and also spreadsheets provide the option to create interactive visualizations and experiments with and within the tool itself. The advantage is that learners and teachers can adapt these experiments to their own needs if they know how to use the tool, so that they do not need knowledge for programming applets (see Verzani, 2005 for one of the examples for using *R*). One tool can support a whole course in doing and learning statistics (see Biehler, 1997 for more details).

New Technological Tools Designed for Children's Learning

A group of statistics educators developed the vision that to realize the potential of technology at school level would require the creation of specific tools adapted to inexperienced students' needs that could also grow up with them as they gained expertise (bottom-up design vs. top-down design; Konold, 2010). In the early 1990s, *ProbSim* and *DataScope* were developed by Cliff Konold's team for doing probability simulations and data analysis, respectively, that were easy to learn and simple enough for students. The drawback was that they did not support conventional statistical experiments or the creation of new methods (Ernie, 1996) and that they were two separate tools. Chris Hancock developed *TableTop* (Hancock, 1995) with a new interface, representing data on a virtual table where the data can be rearranged. The current successor is *InspireData* (http://www.inspiration.com) which is a commercial extended version of *TableTop* that also focusses on visual representations in helping grade 4–8 students create meaning as they explore data in a dynamic inquiry software environment. Another example of software for inexperienced students is the series of *Minitools* (Cobb, Gravemeijer, Bowers, & Doorman, 1997; an updated version was designed by Bakker, Gravemeijer, & van Velthoven, 2001; see also Bakker & Gravemeijer, 2004). These *Minitools* were designed as limited experimental environments to support the development of ideas for data display and comparison.

Requirements of Software Tools

Many technological tools are available for statistics instruction. Choosing technology or a combination of technologies that is most appropriate for the student learning goals, could involve a complex set of considerations and decisions about how to best choose and use these tools, how often to use them, and for what purposes and activities (Chance et al., 2007). Therefore, we next discuss—from an educational perspective—issues related to the requirements for software to help students learn and reason about statistics.

Based on a conception of professional statistical practice and of educational tools for learning statistics, Biehler (1993, 1997) developed requirements for a more flexible tool that would support both doing and learning statistics. The vision was to have one tool that would support students in doing exploratory and confirmatory data analysis and in exploring statistical methods. Moreover it should provide features by which teachers can define exploratory interactive experiments, visualizations, simulations, and applets (defined as "microworlds") that support students in active learning processes. This is similar to using pre-prepared spreadsheet files for exploratory learning. These requirements can be used to evaluate tools for supporting doing and learning statistics. The following types of student activities and related requirements have to be distinguished:

1. *Students can practise graphical and numerical data analysis by developing an exploratory working style.* An exploratory working style would require the software to support the collection of these graphical and numerical results in an organized workspace in a multi-window system. Annotations, selection of results for a report and the ability to interact with and modify results should be possible.
2. *Students can construct models for random experiments and use computer simulation to study them.* Whereas the first requirement is increasingly met by professional software tools, this second requirement is often supported only by programmable tools such as *R* or *Excel*. At that time, this was seen as a major challenge of software design, to provide a simulation tool that would support the kind of simulation and modelling that are relevant in introductory statistics education, with a much simpler interface than what *Excel* and *R* provide, and at the same time supporting elementary notions of model building and simulation.

3. *Students can participate in "research in statistics," that is to say they participate in constructing, analyzing and comparing statistical methods.* Although many tools provide a collection of ready-made methods, the third important requirement implied that learning statistics bears similarities to creating new *methods in* statistical research, not just using methods made by others. Biehler (1997) therefore evaluated statistical software by the extent to which it allowed for the construction and analysis of new methods. However, defining new methods typically requires the input of formal notation, formulas or micros, which is challenging for students.
4. *Students can use, modify and create "embedded" microworlds in the software for exploring statistical concepts.* We use "microworld" as a notion that comprises exploratory interactive experiments, visualization, and simulations, and applets. This fourth requirement concerns the capacity of the tool to function as a meta-tool and meta-medium for teachers by supporting the teacher's construction of microworlds within the software. We then speak of "embedded microworlds" that can be used by students on various levels:
   (a) Use the microworld as a ready-made black box;
   (b) Understand the construction of the microworld by means of their knowledge of the software (transparency of the microworld);
   (c) Modify the microworld according to own needs by means of their knowledge of the software;

If teachers have software competence on level (a) and level (c) they will be able to adapt the microworlds to their teaching goals.

The software *Fathom* (Finzer, 2001; http://www.keypress.com) was later developed for secondary and tertiary statistical learning and realized and extended many of the envisioned features. Soon after, *TinkerPlots* (Konold & Miller, 2005; http://www.keypress.com) was developed by Cliff Konold's team as a kind of little sibling of *Fathom* for younger children. *TinkerPlots* 1.0 was built on ideas from *Minitools*, *DataScope* and *TableTop* in that all these tools for learning statistics can be "emulated" in *TinkerPlots*, yet at the same time allow students to address more general statistical problems. *TinkerPlots* 2.0 (Konold & Miller, 2011) added a simulation feature that built on *ProbSim's* and *Fathom's* capabilities by integrating simulation and data analysis as an overall tool for younger children. We will elaborate on *Fathom* and *TinkerPlots* in the next two sections.

Types of Technological Tools for Teaching of Statistics

Characterizing Technological Tools for Teaching of Statistics

As hinted at in the previous section, there are many types of technological tools and resources to support the learning and teaching of statistics. These include: (a) statistical software packages, (b) spreadsheets, (c) applets/stand-alone applications,

(d) graphing calculators, (e) multimedia materials, (f) data repositories, and (f) educational software. The goal of this section is to provide an overview of these types of tools and common examples of each, and to highlight the requirements for software from an educational perspective. This will be followed by thorough description of two specific educational tools, *Fathom* and *TinkerPlots*. For more comprehensive reviews of the capabilities of these tools and their possible educational uses, we refer readers to Chance et al. (2007). Some older reviews which are still instructive can be found in Ben-Zvi (2000), Biehler (1997), and Garfield et al. (2000).

Statistical software packages. Statistical packages are computer programs designed for performing statistical analyses. Several packages are commonly used by statisticians, including *SAS* (http://www.sas.com), *SPSS* (http://www.spss.com), and *Minitab* (http://www.minitab.com). Although these packages were mainly designed for use by science and industry, they have evolved into statistics learning tools for students and are increasingly used in introductory statistics classes. For example, the statistical package *Minitab* allows student exploration of statistical ideas, e.g., writing "macros" for repeated sampling and graphics that update automatically as data values are added or manipulated. The *DataDesk* package (first version in 1985, cf., Velleman, 1998) focusses on data exploration and interactive graphing by providing unique tools that allow students to look for patterns, ask more detailed questions about the data, etc. *DataDesk* supports the construction of microworlds by means of multiple linked representations (see Biehler, 1997). Several stand-alone statistical packages are also now available for free or at minimal cost, online. The increasingly used *R* package (Verzani, 2005; http://www.r-project.org) is freely accessible and provides a wide variety of statistical and graphical techniques and its programmability and extensibility makes it also an ideal tool for creating microworlds for students and for supporting students' method exploration (Requirement 3 and 4 from above). However, the command-driven interface is a major obstacle for secondary students. *StatCrunch* (West, 2009; http://www.statcrunch.com) is a fully functional, inexpensive, Web-based statistical package with an easy-to-use interface and basic statistical routines suited for educational needs.

Spreadsheets. The widely available spreadsheets (such as *Excel*, office.microsoft.com or *Google Spreadsheet*, http://www.google.com/apps) are frequently used for financial information crunching and reporting in business. Spreadsheets are frequently used as a statistical educational package to help students learn to organize and represent data, and use "automatic updating" of calculations and graphs (Hunt, 1996). There is a possibility to create microworlds within spreadsheets, the Internet is full of examples for this. However, care must be exercised due to their poor calculation algorithms and choice of graphical displays (Cryer, 2001; McCullough & Wilson, 1999). For example, it is still very difficult to make a box plot in *Excel*. Moreover, spreadsheet interfaces are less well adapted to statistical concepts (Requirement 1 and 2 are not deeply met). A recent new development is the combination of *Excel* with *R* that tries to combine advantages of both tools (Heiberger & Neuwirth, 2009).

Applets and stand-alone applications. Free on-line applets can help students explore concepts in a visual, interactive and dynamic environment. Although many of the applets are easy for students to use and often present an interesting challenge for students, they are not often accompanied by detailed documentation to guide teacher and student use. There are exceptions to that typical drawback of applets; for example, the Web-application *GapMinder* (http://www.gapminder.org) is an innovative visualization tool displaying time series of development statistics for all countries with animated and interactive graphics that offer Web-based tools and guides for use in classrooms. In addition, a large number of computer programs can be downloaded from the Internet and run without an Internet connection that allow students to explore a particular concept (e.g., *Sampling SIM,* http://www.tc.umn.edu/~delma001/stat_tools/). The *Consortium for the Advancement of Undergraduate Statistics Education* (CAUSE, http://www.causeweb.org) provides a peer-reviewed annotated list of such tools.

Graphing calculators. A graphing calculator is used in statistics education as a learning tool for analyzing and exploring data, performing statistical procedures and calculations, including inference procedures and probability distributions (Kuhn, 2003). Many graphing calculators can function as data loggers by downloading data from the Web or collecting data from sensors like electronic thermometers, decibel and light meters. Simulations can also be run allowing students to explore concepts such as sampling distributions (Flores, 2006; Koehler, 2006). Graphing calculators are used in secondary statistics classrooms in some countries but are not a full substitute for statistical packages beyond the introductory statistics course and have problematic pitfalls (Lesser, 2007). In addition, the output given by the graphing calculator often does not provide sufficient communication of statistical results (e.g., graphs with no labels and scales). Some recent graphic calculator incorporate features of educational software, such as *TI-Nspire,* which incorporates features of *Fathom.*

Multimedia materials. Multimedia materials combine several different types of content forms (such as, text, audio, still images, animation, video, and interactive media) to teach statistics (Alldredge & Som, 2002; Mittag, 2002; Sklar & Zwick, 2009). For example, *ActivStats* combines videos of real-world uses of statistics, mini-lectures accompanied by animation, links to applets, and the ability to launch a statistical software package and analyze a data set instantly. *ActivStats* (http://www.datadesk.com/products/mediadx/activstats/) was originally designed to work with *DataDesk* but has now interfaces to other tools as well. An advantage of such a learning environment is that students only need to learn one type of technology. In fact, more and more, entire lessons and even textbooks are written around these types of embedded technology to make them "living" textbooks, e.g., *CyberStats* (Symanzik & Vukasinovic, 2006). In addition, an ever-growing format of statistics teaching today is over the Internet, in the form of a Web-based course with video-taped lectures, interactive discussions, collaborative projects, and electronic text and assessment materials (e.g., Everson & Garfield, 2008).

Data and materials repositories. Pedagogically rich data sets and explorative student activities are widely available in the Internet (e.g., Schafer & Ramsey, 2003). Data repositories include data sets with "stories" outlining their background and classroom uses. Examples include *The Data and Story Library* (*DASL*, lib.stat.cmu.edu/DASL), the *Journal of Statistics Education* (JSE) *Dataset and Stories* feature (http://www.amstat.org/publications/jse/jse_data_archive.htm), *CAUSE* (http://www.causeweb.org), as well as some of the official statistical agencies Websites (e.g., Statistics Canada, http://www.statcan.gc.ca).

Educational software. Different kinds of statistical software programs have been developed exclusively for helping students learn statistics. *Fathom* (see below) is a flexible and dynamic tool designed to help students understand abstract concepts and processes in statistics. *TinkerPlots* was developed to aid younger students' investigation of data and statistical concepts (see below). *TinkerPlots* has been widely field tested in mathematics classes in grades 4–8 in the USA and in other countries (e.g., Ben-Zvi, 2006) with very positive results. Some of these educational packages make it easier for students to access large data sets (e.g., census data) and for teachers to access pre-developed classroom exercises.

Fathom and *TinkerPlots 2.0*

In this section, we briefly summarize important points and argue why we think that *Fathom* and *TinkerPlots 2.0* are tools that have met the above requirements in an exemplary way. This explains why quite a number of researchers in statistics education, including the authors of this chapter, frequently use these tools in their research and development work.

Fathom

Fathom Dynamic Statistics is software for learning and doing statistics in secondary and tertiary levels that enables students to explore and analyze data both visually and computationally. It has a menu-driven and drag-and-drop computational environment with a general formula editor incorporated in a central place. This supports the creation of new attributes and numerical procedures that open new possibilities for model and method construction. Its strengths are in the opportunities it provides students to:

- Quickly *drag-and-drop* variables into a graph to visualize distributions and relationships between variables;
- Through *dragging*, visualize how dynamically changing data and parameters affect related measures and representations in real time;
- *Link multiple representations* of data to informally observe statistical tendencies;
- Create *simulations* to investigate and test relationships in the data.

First, we discuss theory that emphasizes the opportunities and insights that dynamic learning software can provide over the static statistical measures, tests and representations used by most statistical software programs. We then provide a short overview of each of the four features above in the context of learning statistics with *Fathom*.

Drag-and-drop graphing. A key deterrent when learning new software is the time it takes to become competent to use a tool productively, which is based on perceived usefulness, ease of use and attitude (Premkumar & Bhattacherjee, 2008). One feature of *Fathom* that makes it easy to learn and become productive is the ability to *drag-and-drop* variables into a graph. This provides an easy way, through synchronous interaction, to check the distribution of a variable or its association with another variable quickly. This action of dragging and dropping is familiar to most computer users, even young children (Agudo, Sanchez, & Rico, 2010), making it more user-friendly and natural than menu-driven systems. For example, to check the amount of time students were spending on their mathematics per week, a dot plot of this variable can be examined in a few seconds. In Figure 21.1a, the number of hours per week spent working on their mathematics (variable: Hrs_Wk) is selected and dragged to the horizontal axis of a blank graph (Figure 21.1b).

Dragging points. A second key feature of *Fathom* is its ability to drag. Dynamic dragging is a feature of multiple mathematical learning software programs developed over the past two decades, primarily in geometry (e.g., *Geometer's Sketchpad, Cabri Geometry, GeoGebra*). Finzer and Jackiw (1998) highlighted three key benefits that "dynamic manipulation" (dragging) enables:

1. *Direct manipulation*. In dynamic dragging, the user is able to click on a point and manipulate it directly. If a point is dragged left, the user *sees* it move to the left on the screen. Even though this action is not directly applied to the pixels on the screen, to the user, who is unaware of computer systems running in the background, it appears that they are dragging the point directly. This allows

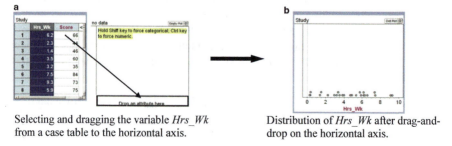

Selecting and dragging the variable *Hrs_Wk* from a case table to the horizontal axis.

Distribution of *Hrs_Wk* after drag-and-drop on the horizontal axis.

Figure 21.1. Drag-and-drop graphing in *Fathom*. (**a**) Selecting and dragging the variable *Hrs_Wk* from a case table to the horizontal axis. (**b**) Distribution of *Hrs_Wk* after drag-and-drop on the horizontal axis.

the user to focus on the effect of dragging rather than the systems which enable them to drag.
2. *Continuous motion.* The relationships between the points being dragged, whether they are vertices of a triangle (as in *Geometer's Sketchpad*) or data in a box plot (as in *Fathom*), are invariant. This continuity of dragging means that changes in objects or related measures change as the dragging takes place. This provides immediate feedback to the user on the effects of their dragging.
3. *Immersive environment.* The objects being dragged are not isolated but part of an environment in which the user is exploring. The interface is minimal to allow the focus to be on the exploration rather than the technology.

For example, suppose in *Fathom* the learner is exploring the relationship between the location of points in a scatterplot and the resulting correlation and regression line. If the learner wonders, "if the points near the centre of the scatterplot had been further out, how would this affect the correlation and regression line?," then they can test this theoretical "I wonder" question by dragging the points and observing the effect of their location on the strength (correlation coefficient) and direction (regression line) of the association (Figure 21.2).

In the example in Figure 21.2, when the student *directly* drags a point (Z), they can immediately see the equation and correlation change *continuously* and *immediately* as the point is dragged. This allows the feedback to support immersion in the exploration environment and not require overt focus on the software interface. Finzer (2006) discussed the effect of direct and continuous dragging of data and statistical objects (e.g., sliders, movable lines and axes) and of observing its effect on measures (e.g., mean, median), statistical tests (e.g., *p*-value), graphs and distributions.

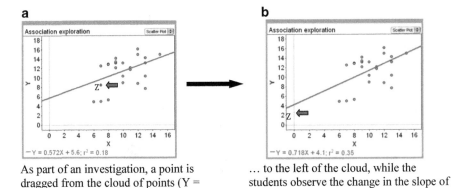

As part of an investigation, a point is dragged from the cloud of points (Y = 0.572X + 5.6; r^2 = 0.18) ...

... to the left of the cloud, while the students observe the change in the slope of the line and correlation coefficient (Y = 0.718X + 4.1; r^2 = 0.35.)

Figure 21.2. Using dragging in *Fathom* to investigate the relationship between the location of points and the direction and strength of the association. (**a**) As part of an investigation, a point is dragged from the cloud of points ($Y = 0.572X + 5.6$; $r^2 = 0.18$) ... (**b**) to the left of the cloud, while the students observe the change in the slope of the line and correlation coefficient ($Y = 0.718X + 4.1$; $r^2 = 0.35$).

He noted three opportunities that *Fathom* provides through dragging that elaborates on three elements of dragging more generally (see, also, Finzer & Jackiw, 1998). First, that the ability to drag provides multiplicity of examples in a few seconds of dragging. Second, the learner can observe the direction and magnitude of their dragging and of its effect in real time. This allows them to build a "cause–effect" relationship between the elements being dragged and their related measures or objects. Important intermediate stages can be "frozen," so that a multitude of graphs can be simultaneously compared as well. With an emphasis on the invariance of these dynamic relationships, students can build stronger foundations of generalization and abstraction. Third, the act of dragging "draws the learner in," engaging them in the exploration and facilitating them in following up on "I wonder" questions. These features are also important for constructing embedded microworlds in *Fathom*, which we will discuss later.

Linked multiple representations. A third key feature of *Fathom* that makes it a productive tool for learning statistics is its ability to *link multiple representations* dynamically. We can use this feature in data analysis, as is shown in Figure 21.3, where the relation between three variables is studied. The scatterplot shows the relation between height and weight. If we select the females in the bar graph on the right side, the females are also highlighted in the scatterplot and the somewhat different relations between height and weight in the two subgroups are shown.

Simulations. Another fairly elementary use of *Fathom* is to simulate a random experiment, in order to estimate an unknown probability by relative frequencies. Maxara and Biehler (2007) analyzed the typical stepwise structure of elementary simulations (Table 21.1).

We judge software with regard to how easily it supports this standard process of simulation. Let us take the following problem as an example.

Students have to pass a test with 10 yes or no questions. The test is passed when 70% of the answers are correct. What is the probability of passing the test just by guessing?

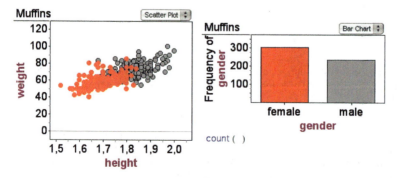

Figure 21.3. Linking multiple representations in *Fathom*.

Table 21.1
Four-Step Design as a Guideline for Stochastic Modelling (Maxara & Biehler, 2007)

Step	Probabilistic Concepts	*Fathom* Objects and Operations
1	Construct the model, the random experiment	Choose type of simulation; define a (randomly generated) collection representing the random experiment
2	Identify events and random variables of interest (*Events and random variables as bridging concepts*)	Express events and random variables as "measures" of the collection
3	Repeat the model experiment and collect data on events and random variables	Collect measures and generate a new collection with values of the measures
4	Analyze data: relative frequency (events); empirical distribution (random variables)	Use *Fathom* as a data analysis software

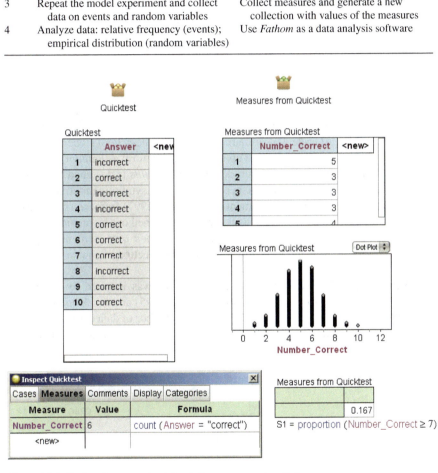

Figure 21.4. Performing a standard simulation process using *Fathom*.

For Step 1 (Table 21.1; Figure 21.4), we use the function randomPick ("correct," "incorrect") and define a column "Answer." The 10 rows stand for the 10 responses. In Step 2 we define a "measure"—a specific functionality provided by *Fathom*—by the formula count (Answer="correct"). We name it *Number_Correct*. A "collect measures" functionality allows to collect as many (*N*) measures as one wants (Step 3). *Fathom* then can be used to display and analyze the distribution of the random variable *Number_Correct*. Figure 21.4 shows a screen display of this simple simulation. The scope of simulations that can be realized depends on the available functions for defining measures. The ease depends on how simply these steps can be executed and how easily students can make sense of the various objects on the screen and their linkage structure.

We have seen how easily the four steps can be implemented in *Fathom* in this example. The broad potential of *Fathom's* simulation capabilities including its strengths and limitations were analyzed by Maxara (2009). Maxara and Biehler (2006) pointed out difficulties students had with performing simulations along these lines.

TinkerPlots

TinkerPlots is a data analysis tool with simulation capabilities (since version 2.0) that has especially been designed for supporting young students' development of statistical reasoning (Grade 4 of primary school to middle-school students, students from the age of 9 onwards). It is built upon the same platform as *Fathom*, which may smooth a transition to *Fathom* at an older age. *TinkerPlots* is designed for creating many simulation models without the necessity of using symbolic input. In addition, *TinkerPlots* meets the third requirement of Biehler's (1997) framework by making students participate in the construction and evaluation of methods by providing a graph construction tool for young students who can invent their own elementary graphs, whereas most other tools provide only a readymade selection of standard graphs.

Organizing and representing data. *TinkerPlots* was inspired by features of older technology, in particular *TableTop* (Hancock, 1995) and the *Minitools* (Cobb et al., 1997). What it shares with *TableTop* is the metaphor of data cards—providing a case by case dynamic view of a data set (Figure 21.5)—scattered on a table, to be

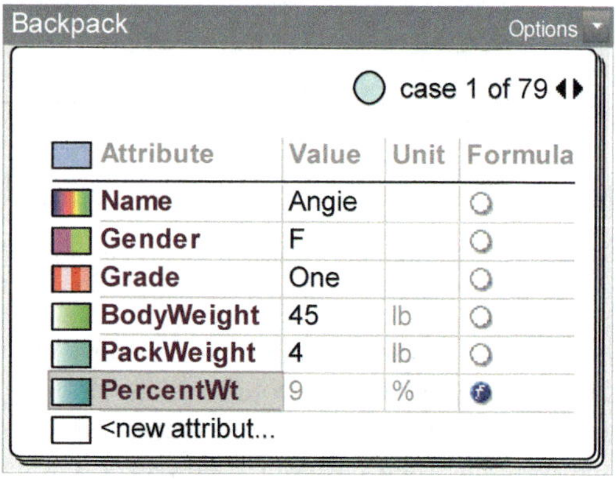

Figure 21.5. Data cards in *TinkerPlots* provide a case by case view of a data set. The student can add or change data, drag attributes into plots, and change the color scheme of attributes.

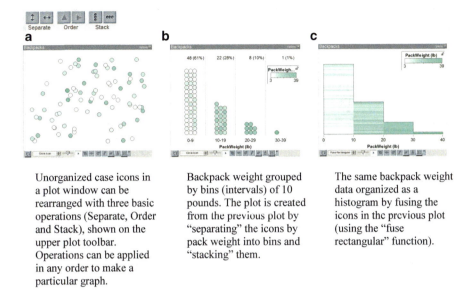

Figure 21.6. Organizing data in *TinkerPlots*. (**a**) Unorganized case icons in a plot window can be rearranged with three basic operations (Separate, Order and Stack), shown on the *upper plot toolbar*. Operations can be applied in any order to make a particular graph. (**b**) Backpack weight grouped by bins (intervals) of 10 pounds. The plot is created from the previous plot by "separating" the icons by pack weight into bins and "stacking" them. (**c**) The same backpack weight data organized as a histogram by fusing the icons in the previous plot (using the "fuse rectangular" function).

organized, changed, displayed in a table or a plot and analyzed by students. The data table metaphor eases a transition from working with real data cards to reorganizing virtual data cards (Harradine & Konold, 2006).

What *TinkerPlots* shares with the *Minitools* are many of its representations. However, *TinkerPlots* is an application (rather than a series of applets) with much more flexibility and possibilities to support the construction of students' statistical knowledge in a bottom-up manner (Konold, 2010). For example, it helps to visualize the transition from the naturally unorganized virtual data cards (Figure 21.5) represented as unorganized data icons (Figure 21.6a) to graphical representations such as dot plot (Figure 21.6b) and histogram (Figure 21.6c). By using multiple plot windows, various plots (including unconventional ones invented by children) can also be compared with other plots of the same data set. Like in *Fathom*, plots can be easily adapted (e.g., making bins narrower for histograms) and data values can be changed to see the effect on the resulting plot or measures of center or variation.

Like most other educational software, *TinkerPlots* offers computational options such as arithmetic mean, median, mode, counts, and percentages in numerical and graphical modes. What is typical of *TinkerPlots* is the possibility to transform any plot into almost any other plot using the basic actions on data of Order, Stack, and Separate (Figure 21.6a). Only some of these plots are conventional ones (bar graph, pie chart,

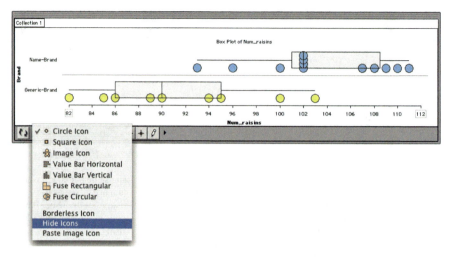

Figure 21.7. Adding box plots to dot plots and the option of hiding the individual data values in *TinkerPlots*.

histogram, box plot, etc.). Figure 21.7 shows how a dot plot can be overlaid by a box plot. This can help students to see for example that larger "boxes" of the boxplot do not necessarily represent more data points (Bakker et al., 2005). The dots can also be hidden.

Multivariate representations and analysis. Another feature that is typical of *TinkerPlots* is the use of color gradients: When clicking on a variable all data icons are color graded according to the value of this variable. This allows students to compare three variables simultaneously, for example when having arm span on the *x*-axis, height on the *y*-axis and foot length color graded (Figure 21.8).

Typically students produce different graphs for a data set that can be compared in the multi-window environment (Figure 21.9). There may not be an optimal graph, but several graphs may show different aspects of the data. Sometimes it may be clear that one graph is superior to another when a certain purpose is given. For instance, the top left (unsorted) graph is inferior to the bottom left graph if we are interested in finding out what properties animals with high (low) heart rates have in common. By providing such a richness of standard and non-standard elementary graphs, students can actively participate in selecting most suitable graphs, and can better understand to which extent the conventions in standard graphs are reasonable ones.

Graphical two-way tables can be easily produced using the commands Separate and Stack. In Figure 21.10 the level of interest in certain computer games depending on gender is displayed.

First steps in reasoning from such two-way tables can start in early grades. *TinkerPlots* makes pie charts accessible for children without any explicit knowledge about fractions by assigning circle sectors to individual cases. Figure 21.11 shows

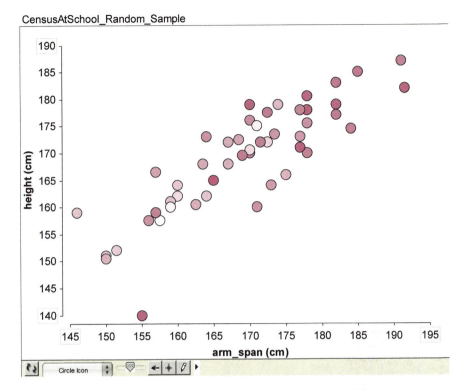

Figure 21.8. The use of *color gradients* in *TinkerPlots* to compare three variables simultaneously, arm span, height and color-graded foot length.

the display before (Figure 21.11a) and after ordering according to gender (Figure 21.11b). Such intuitive pie chart displays are also useful when young students are required to compare theoretical probabilities and empirical relative frequencies (see, for instance, Pratt, 1998).

Simulations. *TinkerPlots* 2.0 includes a simulation interface to define a random experiment by using a "sampler." The sampler is used by students for modelling probabilistic processes and for generating their data. For example, you can build a sampler to model flipping a fair (or biased) coin; run the sampler to collect data in a results table; and analyze the data in a plot to explore questions about the probability of various events.

For example, in Figure 21.12 we simulate the same example that we used for *Fathom* above. We start with ten random draws of "c(orrect)" and "f(alse)" answers that simulate guessing in a ten question test: The resulting table (Figure 21.12b) contains a column with a compound outcome ("join") of 1,000 repetitions. We apply a *TinkerPlots* function ("matchCount") by choosing from a menu to the "join" column in order to count the number of "c"s and represent this count in a new column

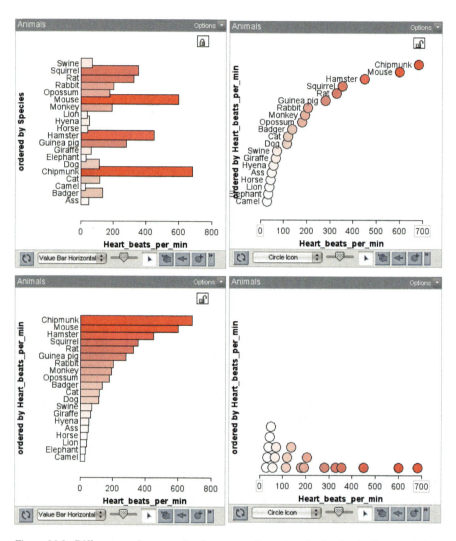

Figure 21.9. Different graphs presenting heart rates (per minute) of animals. Source of data: Ogborn and Boohan (1991, p. 35).

(Figure 21.12c). With drag-and-drop, this column can be represented as a dot plot. The percentage of those passing the test can be found graphically by marking the respective cases (Figure 21.12d).

Creating this simulation is obviously easier than in *Fathom*. The example shows the kind of detailed support (visual sampler, "join"-outcome, matchCount command) that we would need from a simulation tool for students (Requirement 2).

21 Technology for Enhancing Statistical Reasoning at the School Level 663

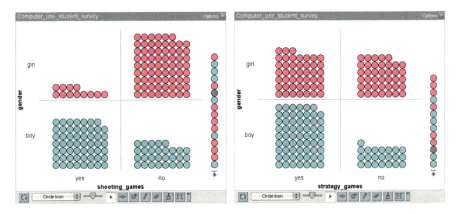

Figure 21.10. Two-way table of a student survey for German students, grades 3–8: Playing shooting games vs. playing strategy games.

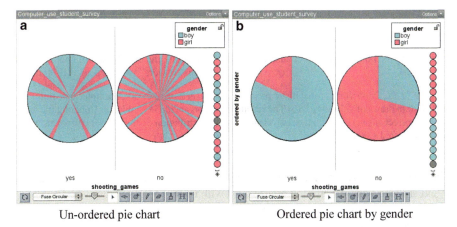

Un-ordered pie chart   Ordered pie chart by gender

Figure 21.11. Interest in shooting games by gender. *Dots* on the side represent missing values. (**a**) Un-ordered pie chart. (**b**) Ordered pie chart by gender.

Microworlds

A microworld is a term that means, literally, a tiny world inside which a student can explore alternatives, test hypotheses, and explore facts about that world. In statistics, it comprises interactive experiments, exploratory visualizations, and simulations. Students can use these microworlds on various levels: (a) use it as a ready-made black box, (b) study the construction of the microworld, and (c) modify it according to own needs (Biehler, 1997). If teachers have software competence on levels 2 and 3 they can adapt the microworlds to their teaching goals. If a microworld is embedded

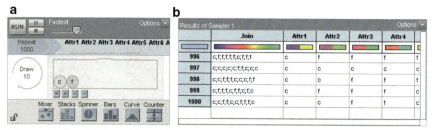

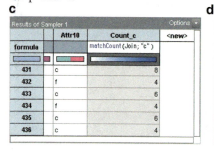
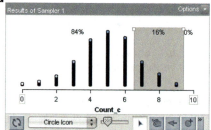

Figure 21.12. Modelling probabilistic processes in *TinkerPlots* 2.0. (**a**) A sampler that simulates 10 random draws of "*f*" and "*c*." (**b**) The resulting data organized in a table. (**c**) Counting the number of correct answers in each test. (**d**) Visualizing the distribution of "number of correct answers" when repeated 1,000 times.

in *Fathom*, *TinkerPlots*, *Excel* or *R* instead of being programmed in Java, teachers (and students) can modify the microworld with their tool knowledge.

To illustrate these points, we present a simulation microworld developed with *Fathom*. Let us assume as an example that we wish to estimate the unknown proportion p of mp3-player owners in a student population. We take a random sample n and observe a relative frequency f of mp3 owners. A 95% confidence interval for the unknown p can be calculated by using the approximate formula $\left[f - 1.96\sqrt{\frac{f(1-f)}{n}}, f + 1.96\sqrt{\frac{f(1-f)}{n}} \right]$. This interval depends on the observed frequency f, which itself is a random outcome. When we repeat the calculation of confidence intervals many times for different samples, the procedure will capture the true probability 95% of the time.

For understanding this idea we construct a simulation microworld developed with *Fathom* for exploring the random nature of confidence intervals and how their width depends on the sample size n. This is a very common visualization found in most applet collections for introductory statistics. Therefore our example also illustrates an important use of technology for understanding a central concept in inference statistics. We assume a certain true probability p, draw a random sample

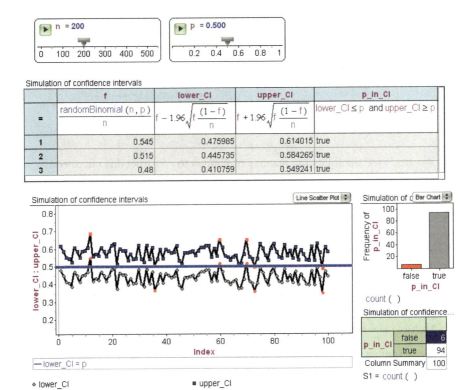

Figure 21.13. Stepwise simulation of the random nature of confidence intervals.

of n, observe the relative frequency of success, calculate the respective confidence interval and check how often our random interval captures the true probability p.

We start the simulation in Figure 21.13 with two variables, n for sample size (e.g., 200) and p for the probability (e.g. 0.5), which can be changed using sliders. In the data table, the first column simulates the observed relative frequency (f) of success in n repetitions by means of binomial distributions. Columns 2 and 3 are used to calculate the lower and upper confidence bounds according to the above-mentioned rule. Column 4 checks whether the random confidence interval contains the "true p." After 100 repetitions, the bottom graph shows the 100 confidence bounds as two line graphs, with a comparison line at p inserted for reference.

The resulting table and bar graph to the right show the distribution of the attribute *p_in_CI*. We would expect 5 out of 100 not to be correct, and in this simulation 6 out of 100 were not correct. The false cases are selected (highlighted) in the right graph and therefore highlighted in all other displays as well. In the bottom left graph we see that the highlighted cases do not contain p. The learner can re-randomize to observe the effect of chance variation on the confidence intervals, and can change the sample size n to observe that the confidence intervals get smaller as n increases; likewise, p can be varied to visually explore other true probabilities.

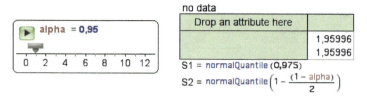

Figure 21.14. Making the confidence level a variable in the microworld.

We have used the following features of *Fathom* to construct this microworld:

1. Sliders for easy change of parameters;
2. Formula input for defining derived variables of interest;
3. Random number generator for statistical distributions with easy re-randomization options;
4. Standard displays (line graphs, bar graphs, frequency tables);
5. Possibility of enhancing graphs, e.g., plotting more than one variable in one plot; inserting reference lines depending on parameters;
6. Multiple linked displays: highlighting cases in one display will be propagated into all other displays; and
7. Multiple windows that let us see the data from various perspectives.

These features are important elements for tools that are to support the construction of microworlds for learning statistics.

We can also illustrate the advantage of microworlds being embedded in a software environment by this example. Imagine that a teacher wants to modify the microworld so that he can illustrate, how the widths of the confidence interval depend on the confidence level alpha. So far we had set alpha=0.95, which has resulted into the coefficient 1.96, which is calculated from the standard normal distribution as the 97.5% percentile. The teacher can adapt the microworld by introducing a new slider *alpha* and replacing 1.96 by a formula depending on alpha as is shown in Figure 21.14.

Supporting Statistical Reasoning

In this section we focus on three main areas of statistical reasoning and how they can be supported by computer tools: (a) data exploration, (b) connecting data and chance, and (c) preparing for statistical inference. We use classroom episodes from recent empirical research to illustrate each area.

Data Exploration with TinkerPlots

TinkerPlots 1.0 was mainly used for data exploration. Research at the middle-school level has been carried out in many countries, for example in Australia (Fitzallen & Watson, 2010; Harradine & Konold, 2006; Ireland & Watson, 2009;

Watson & Donne, 2009), Canada (Hall, 2011), Cyprus (Paparistodemou & Meletiou-Mavrotheris, 2008), Germany (Biehler, 2007c), Israel (Ben-Zvi, Gil, & Apel, 2007; Gil & Ben-Zvi, 2011), and the USA, where it was developed and first tested (Bakker, 2002; Bakker & Frederickson, 2005; Friel, 2002; Konold, 2002; Konold & Lehrer, 2008). However, it has also been used at the elementary level (Ben-Zvi et al., 2007) and in the workplace (Bakker, Kent, Noss, & Hoyles, 2006; Hoyles, Bakker, Kent, & Noss, 2007).

One type of statistical reasoning that is difficult for students is to make conclusions about differences between groups. It is well documented that students who know how to compute the arithmetic mean and median are mostly not inclined to use such measures when comparing groups (Konold & Higgins, 2002). However, Bakker and Derry (2011) showed how sixth-grade students who gained experience with *TinkerPlots* when exploring data sets came to use means and medians to compare groups, even though the teacher never asked them to do so. In their learning trajectory, students repeatedly compared groups while making different plots that helped them answer the question. They gradually developed a richer language to note many aspects of data sets in different representations.

In one of the latest lessons, they had to check a fish farmer's claim that genetically engineered (GE) fish grew twice as long as normal fish. Using *TinkerPlots*, one student, Tom, first separated the dots representing the different fish types vertically, ordered the lengths horizontally, stacked the dots, and used the mean button and reference lines to compare the types of fish (Figure 21.15). Like his fellow students, he used the term "clump" for the majority of the data. He explained further:

I clicked the mean value and the reference line, because it shows kind of where the clump is. And that helps me because it is easier for me to see where most of them are. And this one

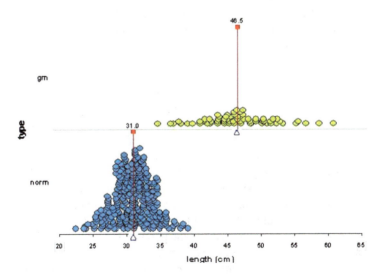

Figure 21.15. The graphs Tom used to explain his statistical inference that the GE fish were about 1.5 times the normal size. The blue triangles give the position of the mean values, and the movable vertical red lines are called reference lines and are used to indicate the means.

[the normal fish] there is a lot more, there is 292 [using the Count option]. And this one [GE], there is 67, so there is about a fourth, a little under a fourth. This one [GE] is a lot more gradual, it is spread out, but they grow a lot bigger and this one [normal] is very steep [points with the mouse along the slope of the distribution shape] and then is really steep, going down. And you can see that it is not really twice. These ones [GE] aren't twice the size of these ones. It is more like one and a half times.

Note that Tom included many observations in his reasoning. For example, it is valuable that Tom used the means to indicate where the clumps were while taking into account the different types of variation—variation between and within data sets as well as variation in sample size and perhaps even variation around an ideal shape.

A large research project in which *TinkerPlots* was used for data exploration is the *Connections* Project, which took place in a science-focussed primary school in Israel during the period 2005–2011 (Ben-Zvi et al., 2007). The project extended for 5 weeks of 6 hr per week each year (Grades 4–6, age 10–12) during which time students actively experienced some of the processes involved in experts' practice of data-based inquiry. Students conducted authentic data investigations through peer collaboration and classroom discussions using *TinkerPlots*.

Students who participated in the *Connections* Project gained a considerable fluency in techniques common in EDA (e.g., generating a research question to drive their investigation, organizing, analyzing and interpreting data, and drawing conclusions based on data evidence), use of statistical concepts (such as, data, distribution, statistical inquiry, comparing groups, variability and centre, sample and sampling), statistical habits of mind, inquiry-based reasoning skills, norms and habits of inquiry, and *TinkerPlots* as a tool to extend their reasoning about data (Makar, Bakker, & Ben-Zvi, 2011). In the inquiry-based learning environment, statistical concepts are initially problematized—that is, rather than first teach students directly about these concepts and then ask them to apply in investigations, the investigations themselves were designed to raise the need to attend to these concepts, hence deepening students' understandings of both their relevance and application.

The tasks undertaken by students in the *Connections* Project were typically a combination of semi-structured data investigations which provided students with rich and motivating experiences in inquiry, including meaningful use of statistical concepts assisted by *TinkerPlots* followed by autonomous, open-ended and extended data investigations. A key idea behind the design of activities is that of growing samples (Bakker, 2004; Ben-Zvi, 2006; Konold & Pollatsek, 2002), to support coherent reasoning with key statistical concepts. Starting with small data sets (e.g., $n=10$), students are expected to experience the limitations of what they can infer from them about the population. They are next asked to draw conclusions by resampling additional small random samples or by increasing the sample size while speculating on what can be inferred about the population.

Students were highly motivated to present their findings in short presentations during the project and at the *Statistical Happening*, a final festive event with their teachers and parents. From this event we cite three boys (aged 12) to illustrate

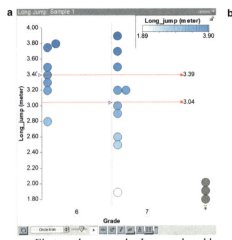
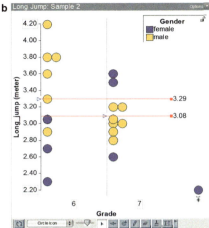

| First random sample: Icons colored by long jump values. | In the second random sample, the boys colored the icons by gender to test their hypothesis that the surprising difference in means is related to gender. |

Figure 21.16. Dot plots representing long jump results by grade of two random samples created by the boys in *TinkerPlots*. The means of the two groups are represented by *blue circles* and horizontal reference lines. The icons on the right represent missing values. (**a**) First random sample: Icons colored by long jump values. (**b**) In the second random sample, the boys colored the icons by gender to test their hypothesis that the surprising difference in means is related to gender.

what kind of statistical reasoning was supported by the software and the learning experiences. They started their presentation as follows:

Eli:	Our research questions are: What are the long jump results in grades 6 and 7?
Asi:	And does the favourite sport affect these results?
Eli:	Is there an association between favourite sport and long jump results?
Odi:	Our hypothesis was that seventh-graders jump farther because they are apparently stronger and bigger.
Asi:	And they are also more experienced than us. Therefore, we thought they'll jump farther.
Odi:	And we also hypothesized that favourite sports that include jumps, like basketball and gymnastics, will have a greater effect on long jump results.

After talking the audience through what they have learnt from analyzing two random samples ($n=20$) (Figure 21.16), the boys concluded:

Odi:	Well, in light of the results of the two samples (Figure 21.17), we discovered as a matter of fact that the inference was similar, and there was not really a change [*between the samples*], and we also found that sixth-graders as a matter of fact jumped farther than seventh-graders, and that basketball and gymnastics really influenced [*the long jump results*].
Asi:	OK, Inferences. From these two samples, we infer that the physical fitness in sixth grades is probably higher than in seventh grades or that more sixth-graders are engaged in sport subjects that can support long jump. We are certain about our inferences since… due to the reason that the two samples have revealed almost the same thing, so our inference is probably correct and we are confident in our inference [in the level of] something like 9 of 10.

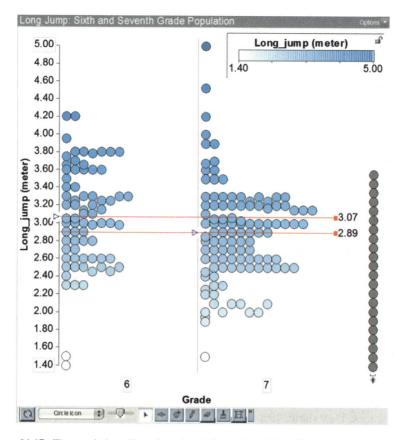

Figure 21.17. The population of long jump results by grade in *TinkerPlots*.

Makar and her colleagues (2011) characterized Asi's conclusion as an informal statistical inference. Note that the boys have both generalized from the findings and speculated on a causal explanation for their findings (that if students play basketball or do gymnastics, it influenced their long jump results). The latter does not necessarily come from the data, but may assist students in making meaning of their findings and seek further explanations (Gil & Ben-Zvi, 2011). After the boys presented their investigation process and inferences, the population graph was exposed for the first time (Figure 21.17). The sixth grade mean turned out bigger than the seventh grade mean, but the mean difference was smaller than in the samples. The boys' responses:

Asi: We see a smaller difference [*in means*], but still a difference. One explanation can be the number of boys compared to the number of girls.
R.: Is your previous conclusion reinforced or weakened by what you see in the population?
Eli: In fact, it reinforces it, but it also weakens it. It reinforces it since we see that the sixth-grade average is really bigger than the seven-grade average, but on the other hand it weakens it since the gap between the two averages here [*in the population*] is not so big.
R.: What do you think about this whole process of sampling and inference that you went through?
Odi: I think that perhaps the [*first*] sample [Figure 21.17a] was biased since we got different numbers of boys and girls.

By playing a major role in helping students learn new ways to organize and represent data, and to develop statistical reasoning, *TinkerPlots* gradually became a thinking tool for these students; it scaffolded their on-going negotiations with data, statistical ideas, inferences, and their meanings.

Connecting Data and Chance Through Modelling

For a long time, researchers (e.g., Biehler, 1994) have argued that the connection between statistics and probability should be rethought, because the emergence of EDA has led to a looser connection between probability and statistics. Tukey (1972) stated for instance, that "data analysis," instead of "statistics," is a name that allows us to use probability where it is needed and avoid it where we should (p. 51).

Probability can be taught "data-free." Under the assumption of equiprobable single events the probability of compound events can be calculated using combinatorial reasoning or tree diagrams. Data analysis can be taught "probability and model free," just doing EDA on given data sets. Data and chance, probability and statistics were traditionally brought together, when inferential statistics was taught at upper secondary level or tertiary level. Inferential statistics can also be taught nearly data-free as an ideal inference process under idealized model assumption, far away from the processes of statistical reasoning of practising statisticians that were described by Wild and Pfannkuch (1999).

In recent discussions about statistics education, a stronger early connection between data and chance have been suggested, in order to achieve a better foundation of inferential statistics and to make students aware how probability is used to model real data generating processes (Burrill & Biehler, 2011; Konold & Kazak, 2008). Other researchers argue for the introduction of informal statistical reasoning (Pratt & Ainley, 2008) or for more accessible ways to transition to statistical inference (Wild, Pfannkuch, Regan, & Horton, 2011).

Technology can have various roles in connecting data and chance. Students can construct simulations of probability models that produce pseudo-real data (that behave similar to data drawn from real processes), and study how these processes behave. A further step is to compare data from real experiments with predictions from probability models in order to validate these models. Both approaches require tool software for doing simulations and for comparing models and real data.

TinkerPlots 2.0, which has a sampling unit, supports students to make such connections between data and chance. Konold, Lehrer, and their colleagues (Konold, Harradine, & Kazak, 2007; Konold & Kazak, 2008; Lehrer, Kim, & Schauble, 2007) have chosen a modelling approach in which they help seventh- and eighth-grade students see the "data in chance" and the "chance in data." Their teaching focussed on four main ideas: model fit, distribution, signal-noise, and the law of large numbers. Their teaching materials capitalized on three main activities: repeated measures, production processes, and different individuals. Such situations are helpful when introducing students to the idea of data as comprising signal and noise. Here we highlight a well-reported activity of repeated measures to show how *TinkerPlots* 2.0 capabilities can support students to make connections between data and chance.

This activity is to model a distribution of measurements around a "true" foot length. Lehrer et al. (2007) engaged students to reason about measures of centre and spread, and think about what might cause the variation in the data sets that students produced. Next they used the *TinkerPlots* Sampler to create a distribution that matched the one they found empirically. The "data factory" comprised several causes of variation such as random error, reading angle, and rounding error, where each spinner models −.1, 0, or .1 mm deviation with particular probabilities. The model is then run 60 times to check the match with the empirical data set, after which the model is adjusted. Konold and Kazak (2008) argued that in this way students are engaged in making models fit while discussing distribution features and distinguishing signal (true value) and noise (measurement errors). The law of large numbers comes into play when students start to use larger numbers of repetitions and reflect on the effect of the distribution shapes (which become smoother with larger n).

Activities can even start with the simple question of throwing two dice. Students can make different models. Some students suggest a model where they do not distinguish between the cases of the type 1+5 and 5+1. In that way they suggest that there are 21 equally probable cases (model 1). The standard model distinguishes 36 equally probable case (model 2). Experiments and simulations can be used to judge between these models. Here we take an activity from Biehler (2007a, 2007b).

Figure 21.18a shows the deviations between the probabilities and relative frequencies from model 1 with 5,000 iterations, Figure 21.18b shows the deviations between probabilities and relative frequencies from model 2 with 5,000 iterations.

Randomization can be used to see that this structure of the residuals is typical, which speaks in favour of model 2. This approach however emphasizes the signal+noise approach, and students can gather experiences about the size of "legitimate" random fluctuations. This activity shows that model 2 is better in predicting the simulated dice. We have to analyze real data from dice experiments in a similar way to validate the model.

We will show two further examples, where technology is used to compare real data with model predictions. The data are taken from the German federal state North Rhine-Westfalia (NRW) in 2002. For every community the number of children born was recorded as well as the number of boys and girls. The proportion of males was calculated from the data and plotted against the number of children born. A line was inserted with the overall proportion of males in NRW (0.5141). The data are compatible with the law of large numbers: the higher the number of children born (sample size), the lower is the deviation from the expected proportion of males. We could have used simulated data for this purpose, but it is most important to use patterns in real data to show the reality of the law of large numbers and that the deviation from the expected value decreases with sample size (Figure 21.19).

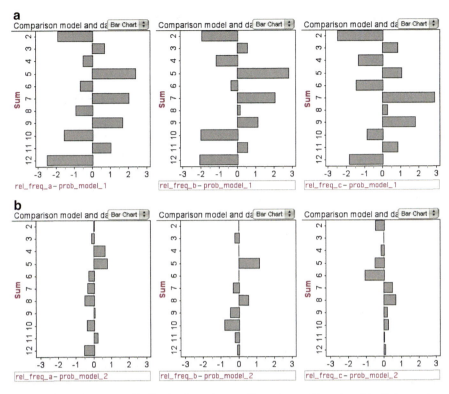

Figure 21.18. (**a**: top graph) Graphs from the two-dice microworld produced in *Fathom* using model 1 with 5,000 cases. (**b**: bottom graph) Graphs from the two-dice microworld produced in *Fathom* according to the correct model 2 with 5,000 cases.

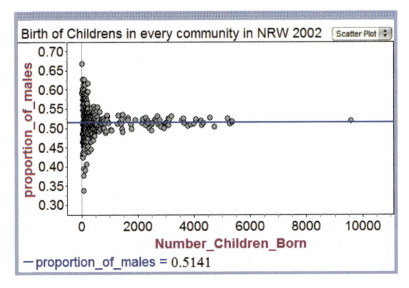

Figure 21.19. Proportion of male babies born in a community against (absolute) number of babies born in that community. Every point stands for a different community of the federal state NRW.

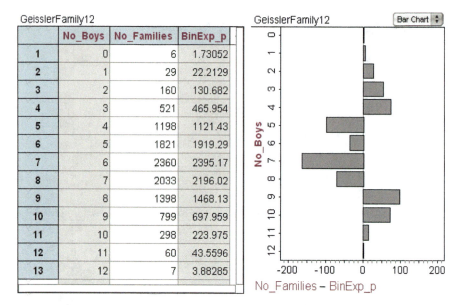

Figure 21.20. Family data; residuals from binomial fit.

We briefly refer to another example taken from Biehler (2005). We start from data on the gender distribution in German 19th century families who had exactly 12 children. We can estimate the probability of a male birth from the data as being .5168. Fitting a binomial model with these parameters leaves systematic residuals that show that the probability model is only an approximate fit and that there are fewer families with "non-extreme" gender distributions than expected. This can be partly, but not fully, explained by the occurrence of identical twins.

Simulation could be used to visualize the random fluctuation that would be expected if the model were true. This can substantiate the judgment that Figure 21.20 shows non-random deviations and can be an informal intermediate step towards formal inference supported by goodness-of-fit tests.

Pathways to Statistical Inference

Statistical inference is an area of particular difficulty for students when they reach secondary and tertiary study. *Fathom Dynamic Statistical* software has been used as a way to support students in their transition to formal hypothesis testing. Building on the discussion above, we focus in this section on the kinds of informal inferential reasoning that can be supported with *Fathom*. In particular, we discuss the use of simulations to build students' experiences with concepts of sampling distributions and confidence intervals.

Most traditional approaches to teaching statistical inference move from descriptive statistics over probability directly into formal methods of statistical inference such as

> Casino: With a particular sort of gambling machine the winning probability is 30%. At such a machine about 50 games are performed daily in a smaller casino and about 200 games in a larger casino. If more than 40% of the games will be won the gambling machine must be refilled. In which casino is the probability larger that on one day more than 40% of the games will be won?
> I. In the smaller casino
> II. In the larger casino
> III. The probability is equal

Figure 21.21. Casino problem (Maxara & Biehler, 2010).

hypothesis testing. Several studies have suggested the difficulty that students have with this transition (e.g., Garfield & Ben-Zvi, 2008). For example, students often oversimplify the leap from sample statistics as a single value (e.g., mean) to population estimates (confidence interval) as a point estimate through direct proportional scaling, ignoring the important influences of sample size and sampling variability (Burrill & Biehler, 2011; Chance, delMas, & Garfield, 2004). Alternatively, several researchers have suggested that students need to build on their experiences with descriptive statistics through exploratory data analysis, comparing distributions, and developing informal inferential reasoning (Makar et al., 2011; Wild et al., 2011).

As an example of how *Fathom* can assist students in developing inferential reasoning in a less formal setting, we draw on a research project which uses the classic "The Hospital Problem" in statistics formulated by Kahneman, Slovic, and Tversky (1982) and widely explored by Sedlmeier (1999). Researchers often use this problem to exemplify the counterintuitive nature of the relationship between sample size and level of confidence (Garfield, 2003). Maxara and Biehler (2010) used modified versions of this problem (Figure 21.21) with their students in building a learning trajectory to better support students in transitioning to statistical inference.

Initially, they found that about half of students struggled with this problem, answering incorrectly that either the larger casino was more likely to record over 40% of games won (arguing, for example, that it would have more games played, so more chance of exceeding 40% winning) or that the two casinos were equally likely to produce this result (arguing, for example, that the probability of winning is the same for both casinos and the higher number of games in the larger casino is taken into account by taken the percentage of 40% in both casinos).

A simulation built in *Fathom* can be used to illustrate the outcomes in the two casinos. Figure 21.22 shows a simulation for a single day's results for each casino. Although each casino has a 30% chance of producing a win in any single game, secondary students are usually comfortable with the idea that on any given day the total number of wins and losses will not be exactly 30%. What they don't have a good feel for, however, is by how much this figure will vary and what range of values is reasonable to expect from each casino. By first building the single case using a random number generator, students can generate what a single day's return may look like in the two casinos. They can re-randomize this data to observe data on several days and develop a sense of the variability that might occur from day to day. This kind of experience is helpful to begin to build students' underlying beliefs

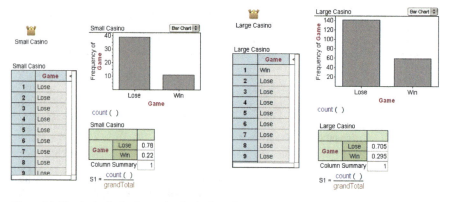

Figure 21.22. A single day's result of winnings from the small (*left*) and large (*right*) casino. The win/lose data are represented in a table, bar graph and summary table for each casino.

about which casino may more likely produce 40% winning games on a given day. Once students are comfortable with the kinds of values that might be expected, they are often ready to move on to quantify their findings. The single case is important to assist them initially, but does not help them to develop more sophisticated processes for quantifying this likelihood, as is needed in formal hypothesis testing.

Once students develop a sense of how the data change from day to day in each casino, they can collect these outcomes, thereby creating a sampling distribution, where each data point represents a single day's result. Students first record the outcomes by hand before automating this process to understand the process. To automate the recording of the winnings from day to day, students can build a data set out of the daily results. This action is straight-forward in *Fathom* defining a "measure" *PropWin* (proportion of wins) and then using the *collect measures* button. For example, in Figure 21.23, the graph shows the proportion of wins recorded in the small casino over 1,000 days. Using this distribution, students can estimate the likelihood that the casino will generate more than 40% wins by counting the number of days that *PropWin* is at least 0.4 (7.6% in the small casino and 0.2% in the large casino). Students can repeat simulating 1,000 days or increase the number of days sampled (e.g., to 5,000 days) to see how stable this figure is and from this to generate an estimate of how probable it is that the small/large casino will have to refill their machines on any given day.

For comparing the two distributions it is important to use exactly the same scales in both graphs. If this is done as in Figure 21.23, students can see that the reason for the lower probability in the large casino: The sampling distribution of the proportion concentrates around the expected value much more than in the small casino. The spread is smaller in the large casino. From this insight students can progress in the direction that the standard deviation of the sampling distribution of the proportion decreases with $1/\sqrt{n}$, where n is the sample size (see Biehler & Prömmel, 2010, who used this progression of ideas in their introductory course to statistics).

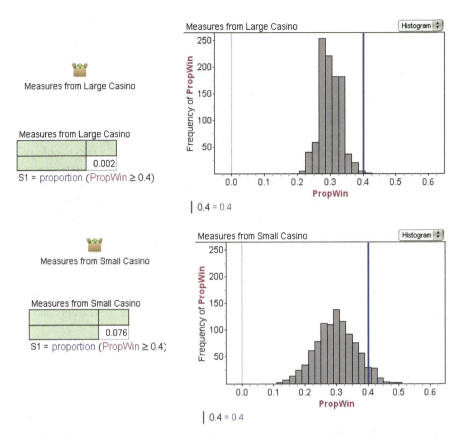

Figure 21.23. Simulations recording the proportion of winning games for each casino for 1,000 days. These proportions are represented in a histogram. The proportion of days for which at least 40% of games were winning is calculated.

These experiences repeated multiple times allow students to begin to develop a better understanding of the relationships between sample size and sampling variability. Through generating and interpreting these simulations in different contexts and under different conditions, the confusion that students often experience between sample size and number of samples drawn, between data distributions and sampling distributions, between interval estimates and confidence levels can be avoided.

A deep understanding of sampling distributions is the basis for making sense of statistical inference. Basic ideas of hypothesis testing with *p*-values can be introduced on top of this. The historical example of the tea tasting lady, who claimed that she can distinguish whether milk was poured in the cup before the tea or vice versa (Salsburg, 2001), can be presented to students in various setting such as whether they can distinguish Pepsi Cola from Coca Cola, or milk with normal milk vs. low fat milk, music with CD vs. music with MP3 quality, etc. If we set up an experiment with 20 trials, the question is whether a person does better than "just guessing."

If someone gets 15 correct out of 20, we can ask "What is the probability that someone gets 15 or more (75% or more) correct just by guessing?" Technology can be used to simulate a guessing person with the varying results of guessing. We can use the same simulation and graphs as in the casino example, display the sampling distribution and estimate the probability of having at least 75% correct (this will be about 2%). Such a result throws deep doubts on the assumption that the person is just guessing. Simulation makes the set of hypothetical cases to which the concrete person is compared to much more real than when such probabilities are just calculated by using the binomial distribution. The fact that 20 persons out of 1,000 people just guessing can pass the test became a reality for the students as in the experimental course of Meyfarth (2006, 2008) and the related empirical study of Podworny (2007). From the casino problem, students can learn that to exceed a boundary set at 75% just by guessing becomes more difficult the higher the sample size n (instead of 20) is chosen.

Using technology in such ways aims at improving students' intuitive understanding. Of course, on the basis of this understanding, software can then be used to apply statistical procedures in the process of statistical work as described by Wild and Pfannkuch (1999), where many practical steps can effectively be performed only by the support of statistical tools, which can serve both supporting the learning and the doing of statistics.

Conclusions and Discussion

In this chapter we intended to summarize what is currently known about how new digital technologies can support students' statistical reasoning. In this last section we address some of the main themes, limitations of software and speculate on what might be future directions for the field.

Statistical Reasoning as Travelling: A Metaphor

To summarize some of the main issues of using technology to support student reasoning, we use a metaphor. Like any metaphor it only highlights some issues and has its limitations. We compare statistical reasoning metaphorically with travelling from particular points (statements) to other destinations (conclusions based on data) while staying aware of the environment (e.g., uncertainty, variation, lurking variables). Our travelling is never ending: reaching a conclusion can raise further questions; conflicting conclusions can raise doubts, caveats, or even rebuttals. In short, inquiry involves a lot of metaphorical travelling—not only from A to B but also back, further, and going round in circles. The role of a computer tool is to make travelling (whichever way) easier and faster, inevitably with some "black box" effect: when travelling by plane or train we see fewer details along the road than when walking or cycling.

Main Themes

Hands-on or computer use? One of the recurring themes in the literature is when students should use the *software* and when they should do something by *hand*. Several researchers have argued that manually organizing data cards (Harradine & Konold, 2006; Watson & Callingham, 1997) or drawing data cards from hats (Bakker & Frederickson, 2005) can be a useful experience for students to understand what the software is doing for them. Once a student knows how to draw a dot plot, box plot, or histogram, making these by hand becomes boring and tedious. Metaphorically, students should know what it is to walk or cycle, stimulated to think about minor issues (which way to take) before they really appreciate what it is to arrive somewhere fast without knowing about all the decisions taken for them (when travelling by train, the direction can be chosen but the rails have been laid down somewhere by other people).

We are not recommending that hands-on experience should always precede the use of a computer tool. Metaphorically, travelling slowly can also hinder the flow of reasoning. It can be exciting to get somewhere quickly, see a lot and get to know about the world—even if only superficially. Getting somewhere quickly, and back, and further, can support both *exploratory data analysis (EDA)* and *statistical inquiry* (Ben-Zvi & Sharett-Amir, 2005).

Dynamic, visual, and personal nature of software. The *dynamic* nature of the educational software such as *TinkerPlots* and *Fathom* allows students to switch easily between many types of plots. Only some of these are conventional ones (bar graph, pie chart, histogram, box plot, etc.). One powerful advantage is that students have control over their plots and can transform any plot into almost any other. We argued that this might help them find a plot that makes sense to them and to gain insight into how data can be organized. This experience provides a good basis for meta-representational skills (diSessa, Hammer, Sherin, & Kolpakowski, 1991).

The *visual* nature of the tools discussed above is in line with Wild et al.'s (2011) advice for students "never to take their eyes off the graphs" (p. 252). However, there is one pitfall: students may be busy making graphs but not reasoning. This reasoning can be lost in the process of graphing, but has to be promoted by useful tasks and classroom discussion. Leading good whole-class discussions is well known to be difficult; however, with a data projector and the students making the plots, it becomes easier to keep their attention (Ben-Zvi, 2006).

In the context of *TinkerPlots* use, students typically name particular plots after their inventors (the "Ryan plot") or after their shapes ("snake plot"), which helps to give them a sense of ownership of their representations (Bakker & Derry, 2011; Bakker & Frederickson, 2005). Teachers can then sometimes tell them that their plot is also used by famous statisticians, who call it a histogram or value-bar graph. This could be an example of guided reinvention (Freudenthal, 1991). The many options to *personalize* icons and working documents give students a sense of agency that is favoured, for example in the games literature (Rosa & Lerman, 2011). Metaphorically, dynamic, visual and personal features of educational software packages seem to help students find their own way in the landscape of statistical reasoning.

Reasoning with aggregates and statistical key concepts. One of the main themes in promoting statistical reasoning is how to help students develop an *aggregate* view on data sets beyond their initial point-value views (Konold, Higgins, Russell, & Khalil, 2003). *TinkerPlots*, for example, offers ample opportunities to enhance students' initial point-value views of data by aggregate views. As shown in Figure 21.6c, dots in bins can be fused to make a histogram; dot plots can be combined with box plots so that students can still see the distribution of individual data points; and students can draw distribution curves with the drawing tool to indicate the shape of a data set represented in a dot plot. Of course, it is not sufficient to have the software affordances: such technical capabilities also need to be incorporated into cognitive habits. The teaching and learning process in the classroom has to be orchestrated such that an adequate instrumental genesis can take place (see, e.g., Trouche, 2004).

In *TinkerPlots*, the small steps between plots and the option to combine individual and aggregate representations support students to become tangibly aware of the *variation* inevitably involved in data. In fact, attention to *key statistical concepts* seems to be easier if the cognitive load required for computation and graph drawing is minimized by software (cf. Chance et al., 2007). We discuss a few key concepts: distribution, association, and sampling.

Distribution shapes easily emerge if we grow sample sizes with a slider (cf., work by Bakker, 2004, and Ben-Zvi, 2006). Thereby students can step forward to a deeper understanding of empirical and theoretical (probability) distributions. *Association* (e.g., correlation) typically comes rather late in the curricula, but as Ridgway, Nicholson, and McCusker (2007) have argued, real questions often involve *multivariate* data—and some educational software packages allow young students to make plots that help them study multivariate data with a minimal baggage of formal techniques (e.g., in *TinkerPlots*, by using colors and their gradients instead of numerals). Konold (2002) for example, argued that even two value-bar graphs side by side can help students see a relationship between, say, length of brushing your teeth and the amount of plaque on them (cf., Biehler, 2007c). And by using color gradients, students can explore a third variable in a scatterplot (Makar et al., 2011). These points emphasize that even without learning formal techniques, students are able to explore key statistical questions that they care about through many travelling options. Moreover, informal approaches allow teachers and designers to have students make connections between concepts from an early stage onward—something desirable from an epistemological view (Bakker & Derry, 2011).

New educational software is also handy when drawing random *samples* of predefined sizes. In *Fathom* and *TinkerPlots*, sliders can be made for changing sample size so as to explore effects on stability of statistical measures and to gain a feel for the *law of large numbers*. Because *computations* of mean, median, mode and other statistics are available at the click of a button, they require little attention. Instead students can focus on the meaning and utility of an arithmetic mean within the problem context. This seems to help them use means for comparing groups rather than see them as computations on data only—something that proves to be hard for many students (Konold & Higgins, 2002).

Technologies such as *TinkerPlots*, *Fathom*, and *Excel* allow students to see the effect of changing data (e.g., outliers) on statistics such as mean and median. Metaphorically speaking, students can go back to the data and see where they would have gone if they had taken a different route—and what different conclusions they might have reached. This possibility of exploring *"what–if questions"* is a major advantage of travelling fast, even if this means that some details of the statistics are bypassed in the black box.

Not all "what–if reasoning" should be carried out with a computer tool. For example, Bakker (2004) noticed that inventing data to match a particular hypothetical situation helped students better understand the relation between statistical concepts and patterns in data. Moreover, he noticed that the quality of reasoning was better in the discussions without computers around than when students were clicking buttons at a computer.

Tool for learning or learning the tool? One question teachers often ask is whether it takes long for students to learn to use the software at stake. In our metaphor, some modes of transport require very little learning investment (walking, cycling, taking a train) whereas others do (e.g., driving a car). The *instrumentation* process (Trouche, 2004) seems to be quite different for different tools such as applets, spreadsheets and educational applications. Applets typically require very little learning investment but are confined in their usage and adaptability. An advantage of *Excel* is that students typically encounter it often and in many situations, but as mentioned before, such spreadsheet programs have limited capabilities and do not offer the dynamic and visual features of educational software such as *Fathom* or *TinkerPlots*. One of the considerations when selecting an application such as *Fathom* is whether instruction time for statistics is long enough to justify learning about the tool. Results from implementation studies of *Fathom* in Germany led to the creation of *eFathom* (http://efathom.math.uni-paderborn.de/), a multimedia environment for self-regulated learning of basic features of *Fathom* for data analysis and simulation (Biehler & Hofmann, 2011). In our experience, middle-school students typically learn to use *TinkerPlots* without much effort, but we have also observed superficial use of this tool when students are not well scaffolded.

Future Directions: A Wish List

Following Shaughnessy's (1992) example, we end with a wish list for future development and research in the area of technology for statistical reasoning.

1. *More insight into task design when using new technologies.* It is clear that new tools require new didactics (domain-specific pedagogy). One dilemma could be categorized as route versus landscape. Bakker (2002) distinguished between route-type software such as the series of *Minitools* and landscape-type software such as *TinkerPlots*. He conjectured that many teachers prefer a more controlled

learning progression that they can both predict and oversee. They do not want students to be "all over the place," because it makes leading classroom discussions challenging and learning outcomes unpredictable and harder to assess (Makar & Fielding-Wells, 2011). Yet, there seem to be ways when using *TinkerPlots* to steer the plots used by students through the instructional materials and classroom discussions (Ben-Zvi, 2006).

2. *More insight into how teachers can be supported to use new statistics technology in their classrooms.* First of all, there are barriers to using computer laboratories (networked computers, password protection, dependency on an IT person, private folders, etc.). Computer problems at school typically require many staff to solve problems in the network. Web applets are one nice way to get around some of these issues. Connecting a dynamic plot on a computer with its static printed version is hard; something gets lost in the transformation from dynamic to static. Another challenge is that teachers have to learn the software, know how to organize a classroom for computer-based tasks, and fluidly switch between whole-class and individual computer work. Then there are the students' multiple approaches to tasks. Working with computers definitely adds a layer of complexity to organizing learning. With landscape-type software especially, teachers often feel uncomfortable with the many options and the variety of plots and conclusions that students might create or reach. This emphasizes the importance of supporting their professional development, and in some cases, helping them to find ways to limit possibilities and steer students along some trajectory.

3. *Dissemination and implementation at larger scales.* Almost all recent research on the use of technology in statistics education has been design-based, typically in close collaboration with excellent teachers in mostly favourable conditions. The growing body of research provides insights into what and how relatively young students can learn to reason statistically. However, if we want larger groups of students to enjoy the use of such dynamic tools, we also need more research on curricular issues, assessment, teacher professional development, with larger numbers of students and teachers in various contexts. We also need more insights into how successes in small-scale research projects can be replicated at larger scales.

Our personal wish is that our overview will contribute to understanding the use of technology in statistics education for both purposes: supporting meaningful learning of statistical concepts and procedures and supporting students in doing authentic statistical inquiries.

References

Agudo, J. E., Sanchez, H., & Rico, M. (2010). *Playing games on the screen: Adapting mouse interaction at early ages.* Proceedings of the 10th IEEE International Conference on Advanced Learning Technologies (pp. 493–497), Sousse, Tunisia.

Alldredge, J. R., & Som, N. A. (2002). Comparison of multimedia educational materials used in an introductory statistical methods course. In B. Phillips (Ed.), *Proceedings of the Sixth*

International Conference on Teaching Statistics: Developing a Statistically Literate Society. Voorburg, The Netherlands: International Statistical Institute. Retrieved from http://www.stat.auckland.ac.nz/~iase/publications/1/6c4_alld.pdf.

Australian Curriculum, Assessment and Reporting Authority (ACARA). (2010). *The Australian curriculum: Mathematics.* Sydney, Australia: Author.

Bakker, A. (2002). Route-type and landscape-type software for learning statistical data analysis. In B. Phillips (Ed.), *Proceedings of the Sixth International Conference on Teaching Statistics: Developing a statistically literate society.* Voorburg, The Netherlands: International Statistical Institute. Retrieved from http://www.stat.auckland.ac.nz/~iase/publications/1/7f1_bakk.pdf.

Bakker, A. (2004). *Design research in statistics education: On symbolizing and computer tools.* Utrecht, The Netherlands: CD Beta Press. Retrieved from http://www.stat.auckland.ac.nz/~iase/publications/dissertations/04.Bakker.Dissertation.pdf.

Bakker, A., Biehler, R., & Konold, C. (2005). Should young students learn about box plots? In G. Burrill & M. Camden (Eds.), *Curricular development in statistics education. International Association for Statistical Education (IASE) Roundtable, Lund, Sweden, 28 June–3 July 2004* (pp. 163–173). Voorburg, The Netherlands: International Statistical Institute. http://www.stat.auckland.ac.nz/~iase/publications.php.

Bakker, A., & Derry, J. (2011). Lessons from inferentialism for statistics education. *Mathematical Thinking and Learning, 13*, 5–26.

Bakker, A., & Frederickson, A. (2005). Comparing distributions and growing samples by hand and with a computer tool. In W. J. Masalski (Ed.), *Technology-supported mathematics learning environments: Sixty-seventh Yearbook of the National Council of Teachers of Mathematics* (pp. 75–91). Reston, VA: National Council of Teachers of Mathematics.

Bakker, A., & Gravemeijer, K. P. E. (2004). Learning to reason about distribution. In D. Ben-Zvi & J. Garfield (Eds.), *The challenge of developing statistical literacy, reasoning, and thinking* (pp. 147–168). Dordrecht, The Netherlands: Kluwer Academic Publishers.

Bakker, A., Gravemeijer, K. P. E., van Velthoven, W. (2001). *Statistical Minitools 1 and 2* [revised version based on Cobb, Gravemeijer, Bowers, & Doorman 1997]. Nashville, TN & Utrecht, The Netherlands: Vanderbilt University & Utrecht University.

Bakker, A., Kent, P., Noss, R., & Hoyles, C. (2006). Designing statistical learning opportunities for industry. In A. Rossman & B. Chance (Eds.), *Proceedings of the Seventh International Conference on Teaching Statistics.* Voorburg, The Netherlands: International Statistical Institute.

Bakker, A., Kent, P., Noss, R., & Hoyles, C. (2009). Alternative representations of statistical measures in computer tools to promote communication between employees in automotive manufacturing. *Technology Innovations in Statistics Education, 3*(2), Article 1. Retrieved from http://escholarship.org/uc/item/S3b9122r.

Becker, R. (1994). A brief history of S. In P. Dirschedl & R. Osterman (Eds.), *Computational statistics* (pp. 81–110). Heidelberg, Germany: Physica Verlag.

Ben-Zvi, D. (2000). Toward understanding the role of technological tools in statistical learning. *Mathematical Thinking and Learning, 2*, 127–155.

Ben-Zvi, D. (2006). Scaffolding students' informal inference and argumentation. In A. Rossman & B. Chance (Eds.), *Proceedings of the Seventh International Conference on Teaching Statistics.* [CD-ROM]. Voorburg, The Netherlands: International Statistical Institute. Retrieved from http://www.stat.auckland.ac.nz/~iase/publications/17/2D1_BENZ.pdf.

Ben-Zvi, D. (2008). Partners in innovation: Helping teachers to integrate technology in the teaching of statistics. In C. Batanero, G. Burrill, C. Reading, & A. Rossman (Eds.), *Proceedings of the Joint ICMI/IASE Study on Statistics Education in School Mathematics: Challenges for Teaching and Teacher Education.* Monterrey, Mexico: ITESM.

Ben-Zvi, D., & Garfield, J. (Eds.). (2004). *The challenge of developing statistical literacy, reasoning, and thinking.* Dordrecht, The Netherlands: Kluwer Academic Publishers.

Ben-Zvi, D., Gil, E., & Apel, N. (2007). What is hidden beyond the data? Helping young students to reason and argue about some wider universe. In D. Pratt & J. Ainley (Eds.), *Reasoning about informal inferential statistical reasoning: A collection of current research studies. Proceedings of the Fifth International Research Forum on Statistical Reasoning, Thinking, and Literacy (SRTL-5)*. Warwick, UK: University of Warwick.

Ben-Zvi, D., & Sharett-Amir, Y. (2005). How do primary school students begin to reason about distributions? In K. Makar (Ed.), *Reasoning about distribution: A collection of current research studies. Proceedings of the Fourth International Research Forum on Statistical Reasoning, Thinking, and Literacy (SRTL–4)*, [CD-ROM]. Brisbane, Australia: University of Queensland.

Biehler, R. (1993). Software tools and mathematics education: The case of statistics. In C. Keitel & K. Ruthven (Eds.), *Learning from computers: Mathematics education and technology* (pp. 68–100). Berlin, Germany: Springer.

Biehler, R. (1994, July). *Probabilistic thinking, statistical reasoning, and the search for causes. Do we need a probabilistic revolution after we have taught data analysis?* Revised and extended version of a paper presented at the Fourth International Conference on Teaching Statistics (ICOTS 4), Marrakech, Morocco, 25–30 July 1994. Retrieved from http://lama.unipaderborn.de/fileadmin/Mathematik/People/biehler/Homepage/pubs/BiehlerIcots19941.pdf.

Biehler, R. (1997). Software for learning and for doing statistics. *International Statistical Review, 65*(2), 167–189.

Biehler, R. (2005). Authentic modelling in stochastics education: The case of the binomial distribution. In G. Kaiser & H.-W. Henn (Eds.), *Festschrift für Werner Blum* (pp. 19–30). Hildesheim, Germany: Franzbecker.

Biehler, R. (2007a). *Skriptum Elementare Stochastik*. Kassel, Germany: Universität Kassel.

Biehler, R. (2007b, August). *Challenging students' informal inferential reasoning by means of smoothly introducing p-value based hypothesis testing*. Paper presented at the Fifth International Forum for Research on Statistical Reasoning, Thinking and Literacy. University of Warwick, Warwick, UK.

Biehler, R. (2007c). TINKERPLOTS: Eine Software zur Förderung der Datenkompetenz in Primar- und früher Sekundarstufe. *Stochastik in der Schule, 27*(3), 34–42.

Biehler, R., & Hofmann, T. (2011, August). *Designing and evaluating an e-learning environment for supporting students' problem-oriented use of statistical tool software*. Paper presented at the 58th ISI Session, Dublin, Ireland.

Biehler, R., & Prömmel, A. (2010). Developing students' computer-supported simulation and modelling competencies by means of carefully designed working environments. In C. Reading (Ed.), *Proceedings of the Eighth International Conference on Teaching Statistics*. Voorburg, The Netherlands: International Statistical Institute. Retrieved from http://www.stat.auckland.ac.nz/~iase/publications/icots8/ICOTS8_8D3_BIEHLER.pdf.

Burrill, G., & Biehler, R. (2011). Fundamental statistical ideas in the school curriculum and in training teachers. In C. Batanero, G. Burrill, & C. Reading (Eds.), *Teaching statistics in school mathematics: Challenges for teaching and teacher education (A joint ICMI/IASE Study)* (pp. 57–69). New York, NY: Springer.

Chambers, J. M. (1980). Statistical computing: History and trends. *The American Statistician, 34*(4), 238–243.

Chance, B. L., Ben-Zvi, D., Garfield, J., & Medina, E. (2007). The role of technology in improving student learning. *Technology Innovations in Statistics Education,1*(1). Article 2. Retrieved from http://escholarship.org/uc/item/8Sd2tyrr.

Chance, B. L., delMas, R., & Garfield, J. (2004). Reasoning about sampling distributions. In D. Ben-Zvi & J. Garfield (Eds.), *The challenge of developing statistical literacy, reasoning, and thinking* (pp. 295–323). Dordrecht, The Netherlands: Kluwer Academic Publishers.

Cobb, P., Gravemeijer, K. P. E., Bowers, J., & Doorman, L. M. (1997). *Statistical Minitools* [applets and applications]. Nashville & Utrecht: Vanderbilt University, Freudenthal Institute, & Utrecht University.

Cobb, G., & Moore, D. (1997). Mathematics, statistics and teaching. *American Mathematical Monthly, 104*(9), 801–823.

Cryer, J. (2001, August). *Problems with using Microsoft Excel for statistics*. Presented at the American Statistical Association (ASA) Joint Statistical Meeting, Atlanta, GA.

cTWO (2007). *Rich in meaning. A vision on innovative mathematics education.* Utrecht, The Netherlands: Commissie Toekomst WiskundeOnderwijs.

delMas, R. (2004). A comparison of mathematical and statistical reasoning. In D. Ben-Zvi & J. Garfield (Eds.), *The challenge of developing statistical literacy, reasoning, and thinking* (pp. 79–95). Dordrecht, The Netherlands: Kluwer Academic Publishers.

delMas, R. (2002). Statistical literacy, reasoning and learning. *Journal of Statistics Education, 10*(3). Online.

Diaconis, P., & Efron, B. (1983). Computer-intensive methods in statistics. *Scientific American, 248,* 96–110.

DiSessa, A. A., Hammer, D., Sherin, B., & Kolpakowski, T. (1991). Inventing graphing: Meta-representational expertise in children. *Journal of Mathematical Behavior, 10,* 117–160.

Ernie, K. (1996). Technology reviews: *DataScope* and *ProbSim*. *Mathematics Teacher, 89,* 359–360.

Everson, M. G., & Garfield, J. (2008). An innovative approach to teaching online statistics courses. *Technology Innovations in Statistics Education, 2*(1), Article 3. Retrieved from http://escholarship.org/uc/item/2v6124xr.

Finzer, W. (2001). *Fathom Dynamic Statistics* (v1.0) [Current version is 2.1]. Key Curriculum Press.

Finzer, W. (2006). What does dragging this do? The role of dynamically changing data and parameters in building a foundation for statistical understanding. In A. Rossman & B. Chance (Eds.), *Proceedings of the Seventh International Conference on Teaching Statistics* [CD-ROM]. Voorburg, The Netherlands: International Statistical Institute. Retrieved from http://www.stat.auckland.ac.nz/~iase/publications/17/7D4_FINZ.pdf.

Finzer, W., & Jackiw, N. (1998). *Dynamic manipulation of mathematical objects*. White paper presented to the NCTM 2000 Electronic Format Group. http://www.dynamicgeometry.com/documents/recentTalks/s2k/DynamicManipulation.doc.

Fitzallen, N., & Watson, J. (2010, July). Developing statistical reasoning facilitated by *TinkerPlots*. In C. Reading (Ed.), *Proceedings of the Eighth International Conference on Teaching Statistics, Ljubljana, Slovenia*. Voorburg, The Netherlands: International Statistical Institute.

Flores, A. (2006). Using graphing calculators to redress beliefs in the "law of small numbers". In G. F. Burrill (Ed.), *Thinking and reasoning with data and chance: Sixty-eighth Yearbook* (pp. 291–304). Reston, VA: National Council of Teachers of Mathematics.

Freudenthal, H. (1991). *Revisiting mathematics education: China lectures*. Dordrecht, The Netherlands: Kluwer Academic Publishers.

Friel, S. N. (2002). Wooden or steel roller coasters: What's the choice? *New England Mathematics Journal, 34,* 40–54.

Gal, I. (2002). Developing statistical literacy: Towards implementing change. *International Statistical Review, 70*(1), 46–51.

Gal, I., & Ograjenšek, I. (2010). Qualitative research in the service of understanding learners and users of statistics. *International Statistical Review, 78,* 287–296.

Garfield, J. (2003). Assessing statistical reasoning. *Statistics Education Research Journal, 2*(1), 22–38. http://www.stat.auckland.ac.nz/~iase/serj/SERJ2(1).

Garfield, J., & Ben-Zvi, D. (2007). How students learn statistics revisited: A current review research on teaching and learning statistics. *International Statistical Review, 75,* 372–396.

Garfield, J., & Ben-Zvi, D. (2008). *Developing students' statistical reasoning: Connecting research and teaching practice*. New York, NY: Springer.

Garfield, J., & Ben-Zvi, D. (2009). Helping students develop statistical reasoning: Implementing a statistical reasoning learning environment. *Teaching Statistics, 31*(3), 72–77.

Garfield, J., Chance, B., & Snell, J. L. (2000). Technology in college statistics courses. In D. Holton (Ed.), *The teaching and learning of mathematics at university level: An ICMI study* (pp. 357–370). Dordrecht, The Netherlands: Kluwer Academic Publishers.

Garfield, J., Zieffler, A., Kaplan, D., Cobb, G. W., Chance, B. L., & Holcomb, J. P. (2011). Rethinking assessment of student learning in statistics courses. *The American Statistician, 65*(1), 1–10.

Gil, E., & Ben-Zvi, D. (2011). Explanations and context in the emergence of students' informal inferential reasoning. *Mathematical Thinking and Learning, 13,* 87–108.

Gould, R. (2010). Statistics and the modern student. *International Statistical Review, 78,* 297–315.

Hall, J. (2011). Engaging teachers and students with real data: Benefits and challenges. In C. Batanero, G. Burrill, & C. Reading (Eds.), *Teaching statistics in school mathematics: Challenges for teaching and teacher education* (pp. 335–346). Dordrecht, The Netherlands: Springer.

Hancock, C. (1995). *Tabletop.* Cambridge, MA: TERC.

Harradine, A., & Konold, C. (2006). How representational medium affects the data displays students make. In A. Rossman & B. Chance (Eds.), *Proceedings of the Seventh International Conference on Teaching Statistics* [CD-ROM]. Voorburg, The Netherlands: International Statistical Institute. Retrieved from http://www.stat.auckland.ac.nz/~iase/publications/17/7C4_HARR.pdf.

Heiberger, R. M., & Neuwirth, E. (2009). *R through Excel: A spreadsheet interface for statistics, data analysis, and graphics.* New York, NY: Springer.

Hoyles, C., Bakker, A., Kent, P., & Noss, R. (2007). Attributing meanings to representations of data: The case of statistical process control. *Mathematical Thinking and Learning, 9,* 331–360.

Hoyles, C., Noss, R., Kent, P., & Bakker, A. (2010). *Improving mathematics at work: The need for techno-mathematical literacies.* London, UK: Routledge.

Hunt, D. N. (1996). Teaching statistics with Excel 5.0. *Maths & Stats, 7*(2), 11–14.

Ireland, S., & Watson, J. (2009). Building an understanding of the connection between experimental and theoretical aspects of probability. *International Electronic Journal of Mathematics Education, 4,* 339–370.

Kahneman, D., Slovic, P., & Tversky, A. (Eds.). (1982). *Judgment under uncertainty: Heuristics and biases.* New York, NY: Cambridge University Press.

KMK (Eds.). (2004). *Bildungsstandards im Fach Mathematik für den Mittleren Schulabschluss - Beschluss der Kultusministerkonferenz vom 4. 12. 2003* [National standards for mathematics in Germany]. München, Germany: Wolters Kluwer.

Koehler, M. H. (2006). Using graphing calculator simulations in teaching statistics. In G. F. Burrill (Ed.), *Thinking and reasoning with data and chance: Sixty-eighth Yearbook* (pp. 257–272). Reston, VA: National Council of Teachers of Mathematics.

Konold, C. (2002). Teaching concepts rather than conventions. *New England Journal of Mathematics, 34*(2), 69–81.

Konold, C. (2010, July). *The virtues of building on sand.* Paper presented at the Eighth International Conference on Teaching Statistics, Ljubljana, Slovenia. Retrieved from http://www.stat.auckland.ac.nz/~iase/publications/icots8/ICOTS8_PL6_KONOLD.html.

Konold, C., Harradine, A., & Kazak, S. (2007). Understanding distributions by modeling them. *International Journal of Computers for Mathematical Learning, 12,* 217–230.

Konold, C., & Higgins, T. L. (2002). Working with data: Highlights of related research. In S. J. Russell, D. Schifter, & V. Bastable (Eds.), *Developing mathematical ideas: Collecting, representing, and analyzing data* (pp. 165–201). Parsippany, NJ: Dale Seymour Publications.

Konold, C., Higgins, T., Russell, S. J., & Khalil, K. (2003). *Data seen through different lenses* (Unpublished Manuscript). University of Massachusetts, Amherst, MA.

Konold, C., & Kazak, S. (2008). Reconnecting data and chance. *Technology Innovations In Statistics Education, 2*(1), Article 1. Retrieved from http://escholarship.org/uc/item/38p7c94r.

Konold, C., & Lehrer, R. (2008). Technology and mathematics education: An essay in honor of Jim Kaput. In L. D. English (Ed.), *Handbook of international research in mathematics education* (2nd ed., pp. 49–72). New York, NY: Routledge.

Konold, C., & Miller, C. D. (2005). *TinkerPlots: Dynamic Data Exploration™ (Version 1.0)* [Computer software]. Emeryville, CA: Key Curriculum Press. Retrieved from http://www.keypress.com/x5715.xml.

Konold, C., & Miller, C. (2011). *TinkerPlots (Version v2.0)* [Computer software]. Emeryville, CA: Key Curriculum Press.

Konold, C., & Pollatsek, A. (2002). Data analysis as the search for signals in noisy processes. *Journal for Research in Mathematics Education, 33*(4), 259–289.

Kuhn, J. R. D. (2003). Graphing calculator programs for instructional data diagnostics and statistical inference. *Journal of Statistics Education, 11*(2). http://www.amstat.org/publications/jse/v11n2/kuhn.html.

Lee, H. S., & Hollebrands, K. (2008). Preparing to teach data analysis and probability with technology. In C. Batanero, G. Burrill, C. Reading, & A. Rossman (Eds.), *Proceedings of the Joint ICMI/IASE Study on Statistics Education in School Mathematics: Challenges for teaching and teacher education.* Monterrey, Mexico: ITESM.

Lehrer, R., Kim, M., & Schauble, L. (2007). Supporting the development of conceptions of statistics by engaging students in modeling and measuring variability. *International Journal of Computers for Mathematics Learning, 12*, 195–216.

Lesser, L. (2007). Using graphing calculators to do statistics: A pair of problematic pitfalls. *Mathematics Teacher, 100*, 375–378.

Madden, S. R. (2011). Statistically, technologically, and contextually provocative tasks: Supporting teachers' informal inferential reasoning. *Mathematical Thinking and Learning, 13*, 109–131.

Makar, K., Bakker, A., & Ben-Zvi, D. (2011). The reasoning behind informal statistical inference. *Mathematical Thinking and Learning, 13*, 152–173.

Makar, K., & Fielding-Wells, J. (2011). Teaching teachers to teach statistical investigations. In C. Batanero, G. Burrill, & C. Reading (Eds.), *Teaching statistics in school mathematics: Challenges for teaching and teacher education* (pp. 347–358). New York, NY: Springer.

Maxara, C. (2009). *Stochastische Simulation von Zufallsexperimenten mit Fathom—Eine theoretische Werkzeuganalyse und explorative Fallstudie.* Kasseler Online-Schriften zur Didaktik der Stochastik (KaDiSto) Bd. 7. Kassel: Universität Kassel. Retrieved from http://nbn-resolving.org/urn:nbn:de:hebis:34-2006110215452.

Maxara, C., & Biehler, R. (2006). Students' probabilistic simulation and modeling competence after a computer-intensive elementary course in statistics and probability In A. Rossman & B. Chance (Eds.), *Proceedings of the Seventh International Conference on Teaching Statistics* [CD-ROM]. Voorburg, The Netherlands: International Statistical Institute. Retrieved from http://www.stat.auckland.ac.nz/~iase/publications/17/7C1_MAXA.pdf.

Maxara, C., & Biehler, R. (2007). Constructing stochastic simulations with a computer tool—students' competencies and difficulties. *Proceedings of CERME 5.* http://www.erme.unito.it/CERME5b/WG5.pdf#page=79.

Maxara, C., & Biehler, R. (2010). Students' understanding and reasoning about sample size and the law of large numbers after a computer-intensive introductory course on stochastics. In C. Reading (Ed.), *Proceedings of the Eighth International Conference on Teaching Statistics.* Voorburg, The Netherlands: International Statistical Institute. http://www.stat.auckland.ac.nz/~iase/publications/icots8/ICOTS8_3C2_MAXARA.pdf.

McCullough, B. D., & Wilson, B. (1999). On the accuracy of statistical procedures in Microsoft Excel 97. *Computational Statistics and Data Analysis, 31*, 27–37.

Meyfarth, T. (2006). *Ein computergestütztes Kurskonzept für den Stochastik-Leistungskurs mit kontinuierlicher Verwendung der Software Fathom—Didaktisch kommentierte Unterrichtsmaterialien. Kasseler Online-Schriften zur Didaktik der Stochastik (KaDiSto) Bd. 2.* Kassel, Germany: Universität Kassel. Retrieved from http://nbn-resolving.org/urn:nbn:de:hebis34-2006092214683.

Meyfarth, T. (2008). *Die Konzeption, Durchführung und Analyse eines simulationsintensiven Einstiegs in das Kurshalbjahr Stochastik der gymnasialen Oberstufe—Eine explorative Entwicklungsstudie. Kasseler Online-Schriften zur Didaktik der Stochastik (KaDiSto) Bd. 6.* Kassel, Germany: Universität Kassel. Retrieved from http://nbn-resolving.org/urn:nbn:de:hebis:34-2006100414792.

Mittag, H. J. (2002). Java applets and multimedia catalogues for statistics education. In B. Phillips (Ed.), *Proceedings of the Sixth International Conference on Teaching Statistics: Developing a statistically literate society.* Voorburg, The Netherlands: International Statistical Institute. Retrieved from http://www.stat.auckland.ac.nz/~iase/publications/1/7a1_mitt.pdf.

Moore, D. (1997). New pedagogy and new content: The case of statistics. *International Statistical Review, 65,* 123–137.

Moore, D. (1998). Statistics among the liberal arts. *Journal of the American Statistical Association, 93*(444), 1253–1259.

National Council of Teachers of Mathematics. (2000). *Principles and standards for school mathematics.* Reston, VA: Author.

New Zealand Ministry of Education. (2007). *The New Zealand curriculum.* Wellington, New Zealand: Learning Media Ltd.

Noss, R., Bakker, A., Hoyles, C., & Kent, P. (2007). Situating graphs as workplace knowledge. *Educational Studies in Mathematics, 65,* 367–384.

Ogborn, J., & Boohan, D. (1991). *Making sense of data: Nuffield Exploratory Data Skills Project. Mini-Course 5: Scatterplots. Student book.* London, UK: Longman.

Olive, J., & Makar, K. (2010). Mathematical knowledge and practices resulting from access to digital technologies. In C. Hoyles & J.-B. Lagrange (Eds.), *Mathematics education and technology revisited: Rethinking the terrain* (pp. 133–177). New York, NY: Springer. doi: 10.1007/978-1-4419-0146-0_8.

Paparistodemou, E., & Meletiou-Mavrotheris, M. (2008). Enhancing reasoning about statistical inference in 8 year-old students. *Statistics Education Research Journal, 7*(2), 83–106.

Pfannkuch, M., & Ben-Zvi, D. (2011). Developing teachers' statistical thinking. In C. Batanero, G. Burrill, & C. Reading (Eds.), *Teaching statistics in school mathematics: Challenges for teaching and teacher education (A joint ICMI/IASE Study)* (pp. 323–333). New York, NY: Springer.

Podworny, S. (2007). *Hypothesentesten mit P-Werten im Stochastikunterricht der gymnasialen Oberstufe—Eine didaktische Analyse konkreten Unterrichts.* Master's thesis: Universität Kassel.

Pratt, D. (1998). *The construction of meanings in and for a stochastic domain of abstraction* (Unpublished Ph.D. thesis). University of Warwick, Warwick. Retrieved from http://fcis1.wie.warwick.ac.uk/~dave_pratt/papers/thesis.rtf.

Pratt, D., & Ainley, J. (Eds.) (2008). Introducing the special issue on informal inference. *Statistical Education Research Journal, 7*(2), 3–4

Premkumar, G., & Bhattacherjee, A. (2008). Explaining information technology usage: A test of competing models. *Omega: The International Journal of Management Science, 36,* 64–75.

Qualifications and Curriculum Authority. (2007). *The national curriculum.* London, UK: Author.

Ridgway, J., Nicholson, J., & McCusker, S. (2007). Reasoning with multivariate evidence. *International Electronic Journal of Mathematics Education, 2,* 245–269.

Rosa, M., & Lerman, S. (2011). Researching online mathematics education: Opening a space for virtual learner identities. *Educational Studies in Mathematics, 78,* 69–90.

Rossman, A. J., & Chance, B. L. (2008). *Workshop statistics: Discovery with data* (3rd ed.). Emeryville, CA: Key College Publishing.

Salsburg, D. (2001). *The lady tasting tea: How statistics revolutionized science in the twentieth century.* New York, NY: W.H. Freeman.

Sandlin, J. A. (2007). Netnography as a consumer education research tool. *International Journal of Consumer Studies, 31,* 288–294.

Schafer, D. W., & Ramsey, F. L. (2003). Teaching the craft of data analysis. *Journal of Statistics Education, 11*(1). Retrieved from http://www.amstat.org/publications/jse/v11n1/schafer.html.

Sedlmeier, P. (1999). *Improving statistical reasoning: Theoretical models and practical implications.* Mahwah, NJ: Lawrence Erlbaum.

Shaughnessy, J. M. (1992). Research in probability and statistics: Reflections and directions. In D. Grouws (Ed.), *Handbook of research on mathematics teaching and learning* (pp. 465–494). New York, NY: Macmillan.

Sklar, J. C., & Zwick R. (2009). Multimedia presentations in educational measurement and statistics: Design considerations and instructional approaches. *Journal of Statistics Education, 17*(3). Retrieved from http://www.amstat.org/publications/jse/v17n3/sklar.html.

Symanzik, J., & Vukasinovic, N. (2006). Teaching an introductory statistics course with *CyberStats*, an electronic textbook. *Journal of Statistics Education, 14*(1). Retrieved from http://www.amstat.org/publications/jse/v14n1/symanzik.html.

Trewin, D. (2007). The evolution of national statistical systems: Trends and implications. *Statistical Journal of the International Association for Official Statistics, 24*, 5–33.

Trouche, L. (2004). Managing the complexity of human/machine interactions in computerized learning environments: Guiding students' command process through instrumental orchestrations. *International Journal of Computers for Mathematical Learning, 9*, 281–307.

Tukey, J. W. (1965). The technical tools of statistics. *The American Statistician, 19*(2), 23–28.

Tukey, J. W. (1972). Data analysis, computation and mathematics. *Quarterly of Applied Mathematics, 30*, 51–65.

Tukey, J. W. (1977). *Exploratory data analysis*. Reading, MA: Addison-Wesley.

Velleman, P. (1998). *Learning data analysis with Data Desk*. Reading, MA: Addison-Wesley.

Verzani, J. (2005). *Using R for introductory statistics* [Online]. Boca Raton, FL: Chapman & Hall/CRC.

Watson, J. M. (2002). Doing research in statistics education: More than just data. In B. Phillips (Ed.), *Proceedings of the Sixth International Conference on Teaching Statistics: Developing a Statistically Literate Society*. Voorburg, The Netherlands: International Statistical Institute.

Watson, J. M. (2006). *Statistical literacy at school: Growth and goals*. Mahwah, NJ: Lawrence Erlbaum Associates.

Watson, J. M., & Callingham, R. A. (1997). Data cards: An introduction to higher order processes in data handling. *Teaching Statistics, 19*, 12–16.

Watson, J. M., & Donne, J. (2009). *TinkerPlots* as a research tool to explore student understanding. *Technology Innovations in Statistics Education, 3*(1), Article 1. Retrieved from http://escholarship.org/uc/item/8dp5t34t.

West, W. (2009). Social data analysis with *StatCrunch*: Potential benefits to statistical education. *Technology Innovations in Statistics Education, 3*(1). Retrieved from http://escholarship.org/uc/item/8dp5t34t.

Wild, C. J., & Pfannkuch, M. (1999). Statistical thinking in empirical enquiry. *International Statistical Review, 67*, 223–265.

Wild, C. J., Pfannkuch, M., Regan, M., & Horton, N. J. (2011). Towards more accessible conceptions of statistical inference. *Journal of the Royal Statistical Society: Series A (Statistics in Society), 174*(2), 247–295.

Yates, F. (1971). The use of computers for statistical analysis: A review of aims and achievements. *Proceedings of the ISI, Session 36*, 39–53.

Chapter 22
Learning with the Use of the Internet

Marcelo C. Borba, Philip Clarkson, and George Gadanidis

Abstract In this chapter we discuss how the Internet is interacting with mathematics education. After briefly discussing the rise of the Internet and its impact on education, we suggest that it has the potential to disrupt mathematics teaching and learning. Moving far beyond its used as a data resource, we suggest the Internet will provide on-demand access to mathematics knowledge through the collaborative, multimodal and performative affordances of the media that it supports. We note that such affordances will not come to fruition until pedagogical practices have adapted to the rapid pace of this technological change. We conclude by noting that such fundamental change in the teaching of mathematics does have many obstacles, not least that approximately two-thirds of the world's population does not have sufficient access to the Internet— and in societies where access is available, access to the Internet often remains limited in classroom settings, particularly for students in low socio-economic areas.

Introduction

Imagine a mathematics classroom before the widespread use of the Internet. Mathematics knowledge was the property of teachers and textbooks and mathematics teaching happened in formal classroom settings under the control of teachers and a mandated curriculum. Now imagine a mathematics classroom where students and teachers have constant access to the Internet. What changes might we see?

M. C. Borba (✉)
Sao Paulo State University, Sao Paulo, Brazil
e-mail: mborba@rc.unesp.br

P. C. Clarkson
Australian Catholic University, Melbourne, Australia

G. Gadanidis
The Western University, London, ON, Canada

Consider a parallel. Imagine society before the widespread use of the Internet. Information was the property of governments and news media and for the most part it was disseminated through their control. Today (in 2012) governments and news media still control and disseminate information, but they no longer have a monopoly. Every person with a cell phone can connect to some aspect of the Internet, not only to access information but also to share information with others who have Internet access. What changes as a result?

The Internet has facilitated the emergence of information sharing that is for the most part beyond the control of governments and traditional news media (Khine & Salleh, 2010). Wikileaks is one example of this, where government records and communications have been made public in unprecedented ways, by individuals posting them on the Internet. Such public sharing of typically secret information adds a level of transparency to government. But there is something else that is at play here that is more than just who controls and disseminates information. Schrage (2001) suggested that the commonly-used label of *information revolution* misses the essence of the paradigm shift due to new media. He suggested that a more accurate description of the paradigm shift is *relationship revolution*. For example, in the case of the Middle East and North Africa, it was the creation and organization of new communities through Internet tools like Facebook and Twitter that played a significant role in challenging existing government structures over the past three years.

Returning now to our initial question of what changes might we see in mathematics classrooms where students and teachers have ready access to the Internet, we can imagine some of the following occurring which in some respects are analogous with the above examples of socio-political developments at large:

- Mathematics knowledge in all its enormity is no longer just the property of teachers and textbooks, nor is it constrained by the communication forms of traditional textbooks. It also exists in publicly available information sites such as Wikipedia and the numerous mathematics education sites that offer textual, multimodal and interactive mathematics content.
- Mathematics teaching is not limited to formal classroom settings. The Internet has become a vast resource of information. For example, a student can search on YouTube for "factoring" and find numerous videos that "teach" mathematics content related to this topic.
- Online mathematics courses have created a new form of "classroom," in which no physical space exists as the classroom. The *new classroom* is a combination of the place where each student-computer is a virtual environment where messages, videos, drawings are posted synchronously or asynchronously. In this sense, the classroom is in the Internet. Thus pedagogical designs need to take into account affordances of the Internet such as collaboration, multimodality and performance (which we discuss later in this chapter).

We suggest that these three fundamental foci within mathematics education—mathematics knowledge, teaching and the context of classrooms—can all undergo, individually and together, radical change with the emergence and use of the Internet. We have noted above some recent (2010–2012) actions in society that most likely would not have occurred in the way they did without the Internet. Events such as these have prompted us to speculate on the impact of the Internet on mathematics education. In doing so, we are mindful that classrooms do function differently from society as a whole, but clearly being an artefact of society there are overlaps. Rather than using the three foci outlined above as our organizing structure for this chapter, we use a structure that incorporates possibilities that are not being practised widely as yet, a structure that the authors believe offers possibilities for mathematics education in the 21st century. Our approach will be based on three key affordances of the Internet: collaboration, multimodality and performance. But we will first start with a general discussion about the Internet.

The Internet

In the *Second International Handbook of Mathematics Education* of this series, Atweh, Clarkson, and Nebres (2003) acknowledged the international nature of mathematics and mathematics education, picking up threads of an argument they had mounted some years earlier (Atweh & Clarkson, 2001). They also detailed some aspects of the impact of globalization on mathematics education which they argued had both advantages and disadvantages, although often it seemed that this multi-pronged process seems overwhelming, unstoppable and often associated with forces that were "impersonal and beyond the control and intentions of any individual or groups of individuals" (Waters, 1995, p. 2). Later, Clarkson (2011) noted that the impact of globalization is not always easy to identify in real time, but often only becomes apparent on reflection. Within this argument, clearly the use and power of the Internet was formidable—both useful and at times overpowering of local initiatives and thinking.

Much of the hardware that is utilized in education was developed for other areas of society. Education is forever playing catch-up. Film, television, audio recording, video and then digital recording, overhead projectors, all of which have been used to varying degrees in schools, were developed first for business, and then later marketed as valuable resources for education. Some, such as video recordings, proved to be useful, but others such as television and film proved far more problematic.

When it comes to information and communication technologies (ICT), again they were invented for business and some for scientific/engineering applications, with education a secondary market. The Internet in particular was originally invented for military purposes. Hence, unlike resources that have from the start been developed for education purposes, these technologies are being utilized as best-fit possibilities in education. It is, therefore, no surprise that there are unexpected occurrences along the way. But the same is true in business. For example, a report from India on the utilization of ICT in micro-businesses shows that it is the cheap digital phones that are the most used and adaptable to that situation, not the far more powerful desk-top computer technology (Ilavarasan & Levy, 2010). Hence, in working through how to use ICT in mathematics education, researchers and curriculum developers should employ investigative techniques that do not lack rigor, but at the same time are designed to capture unexpected outcomes.

It is starting to become difficult to think of schooling, including doing mathematics, without the Internet. The Internet seems to be present when students do work at home, when they communicate with colleagues, and so on. The 2011 worldwide estimate of the number of Internet users was at more than 2.25 billion people, and rising (Internet World Stats, 2011). This is a significant growth since 1995 when there were "only" 16 million Internet users. The popularization of the Internet, which offers new popular and specialized forms of representation and communication of ideas, has an impact on mathematics education.

The Internet and Education

The use of computer technology in schooling has a long history. In the late 1960s and early 1970s enthusiastic teachers found ways to introduce students to the use of computers. This meant collecting hand-punched cards and sending or taking them to some central main-frame computer for processing (Clarkson, 1980). However, the question of whether this technology advanced the quality of teaching and learning for students was never far away. One issue was whether students' performances on assessment tasks increased over time with access to this technology, but this proved to be a very hard and not always productive type of question. It was also recognized at a social level that students needed to know about this technology and its impact, since it was seen to be the start of a revolution in our society.

Throughout the mid-1980s computers themselves began to change. They became smaller and therefore more portable. They became relatively far less expensive and hence, affordable by many more people in many societies. Their power grew exponentially meaning that small laptops could compute faster than the old giant main frames of the 1960s. A laptop now has far more computing power than the computer at Houston, in Texas, that had control of the moon landing in 1969. This rise in computer power allowed the rise of multi-function computers that not only complete mathematical calculations, but also easily handle numerical databases and alpha databases. They also became a facility for playing games. Game playing took off with the interactive screen which allowed for point and click, utilizing high quality graphics, rather than having to remember specific code to type in from the keyboard. When, in the early- to mid-1990s, easy access to the Internet using the World Wide Web (WWW) became available, anyone with a computer and a modem that connected it to the copper wire telephone cable system could have access to virtually unlimited information, and contact anyone who had an email address.

An immense amount of research has focussed on the use of computers (without the Internet). The two ICMI studies (Churchhouse, 1986; Hoyles & Lagrange, 2010) and PME proceedings (e.g., Pinto & Kawasaki, 2010) provide a representative collection of papers on the subject. Interestingly, these collections do not make clear how widespread the use of these computers is in everyday classrooms. This rather fundamental issue of the place and use of computers in everyday school education worldwide is a project that is still to be undertaken.

The popularization of the Internet which offers new popular and specialized forms of representation and communication of ideas has an impact on mathematics education and education in general. DeBell and Chapman (2006) suggested that "children and adolescents commonly use computers for playing games, completing school assignments, word processing, email, and connecting to the Internet. The most frequent online activities for students are using email, playing games, using social network sites, and finding news and product information" (p. 37) (see also Smith & Caruso, 2010).

When it comes to education, Head and Eisenberg (2010) found that "college students use *Wikipedia*. But, they do so knowing its limitation. They use *Wikipedia* just as most of us do—because it is a quick way to get started and it has some, but not deep, credibility" (para. 4). The role of Facebook in education has also been discussed by researchers (Ellison, Steinfield, & Lampe, 2007; Idris & Wang, 2009; Lampe, Ellison, & Steinfield, 2008; Tay, Tan, & Tan, 2009). Selwyn (2007), looking at the cohort of middle-class university students, saw:

> Facebook as being a highly significant but also an unremarkable means of social networking and communication in the everyday lives of the young people. ... The Internet has become enmeshed into daily lives and the social interactions of this generation ... We have seen how students were using Facebook to communicate with friends in the same house, library or computer lab in an asynchronous and sometimes quasi-synchronous manner. Conversations appeared to skip across Facebook walls, text messaging, MSN and face-to-face contact, leaving the wall postings as just one part of a seamless, multimodal exchange. (p. 17)

The use of short text messages and images through mobile technologies and social network has also become a very popular medium for communication among adolescents and college students (Nanyang Technological University, 2010), and represents a shift away from communication through email. One thing that is consistent about student use of Internet-based resources is an uncertainty about what the next popular mode of communication might be. There is no doubt whatever that when it arrives it will have an impact on education, and mathematics education in particular.

The Internet and Mathematics Education

It is more than 25 years since the interface between information technology (IT) and mathematics education started to become an issue for research. This became more important since personal computers first became available (for a few) in the mid-1980s. Nevertheless it is still not clear in terms of research whether, and if so how, information communication technology (ICT) transforms the teaching and learning of mathematics. It was with the rise of the Internet that the IT changed to ICT. We do know that access to computers is very uneven in schools worldwide. Not surprisingly, ICT is even more unevenly present in education than the presence of computers, since many schools that have computers have limited or no access to the Internet.

Software that allows students to investigate features of functions or geometric figures has become popular in mathematics education conferences, as has the exploration of using spreadsheets in the teaching of algebra. But there is no account, to our knowledge, about how widespread their use is in classrooms (Borba & Villarreal, 2005; Hoyles & Lagrange, 2010; Kieran & Yerushalmy, 2004). There is some suggestion in the literature that the widespread use of scientific and then graphical calculators from the late 1990s led to a reduction in the use of computers in mathematics classrooms compared to their use in other subject areas (Clarkson & Toomey, 2001). We do know they have not been used in international comparative assessments, even though there are movements for international surveys like PISA to introduce computer-based items in their assessment tasks.

We suspect that research on software development in mathematics education has helped to shape mathematics education technology that is available on the Internet, for example in the form of applets. However "could the Internet be fully accepted in (mathematics education)?" is a question posed by Borba (2009). At that point in time he had no comprehensive answer. But it seems that some practices, other than using it as source of reference, have been developed which have the potential to transform the way mathematics is taught and learned.

In the previous section we briefly discussed some of the research related to computers (without the Internet) and mathematics education. We also reviewed very briefly some research in education in general, regarding the use of social networks and other affordances of the Internet to provide learning and to enhance teaching. From this sampling of the research it is clear that ICT, and the Internet, in particular, are changing society, and hence there are radical implications for education, including mathematics education. However, Maltempi and Malheiros (2010) in a survey showed that until 2007 there were few studies published in English text journals, conference proceedings and books about online mathematics education, although they suggested the situation was slightly better in countries like Brazil.

There is a wide variety of free mathematics education resources that students and teachers can use for developing mathematical understanding. For example:

1. The National Council of Teachers of Mathematics (NCTM) maintains the Illuminations Web site (http://www.illuminations.nctm.org) which offers activities, lessons and interactive content for grades K-12.
2. Utah University has developed the National Library of Virtual Manipulatives for mathematics education (http://www.nlvm.usu.edu/en/nav/vlibrary.html).
3. Drexel University runs the Mathematics Forum (http://www.mathforum.org) which offers a bank of math questions and answers, and a free online math help service.
4. Other sites, such as the following, are not run by institutions or professional organizations, but are also of interest for the discussion we will develop in this chapter:
    - http://www.ted.com/talks/salman_khan_let_s_use_video_to_reinvent_education.html, and
    - http://www.wolframalpha.com/.

In addition, Engelbrecht and Harding (2005) identified a number of other online resources that are likely to benefit students: math dictionaries, libraries of puzzles and other enrichment content, online learning or extension material to support face-to-face courses—online material made available by textbook publishers and supplementary notes made available by the teacher—and exploration and demonstration sites with interactive animations.

From the early days when the Internet was beginning to be utilized in classrooms, there were issues in students' learning that were new but still remain on today's research agenda. Gerber, Shuell, and Harlos (1998) noted that when using the Internet "students did not seem to have a clear cut plan for their projects or for locating data prior to using the Internet" (p. 123). They added: "students approached the task of searching in different ways. ... [but they] did not search the Internet with a clear plan in mind. ... Most of them needed a good deal of scaffolding to focus their searches and find relevant data" (p. 127). A similar comment might also be appropriate for any project which demands students collecting data, whether this involves the Internet or not. But if they are to utilize the Internet, then peculiar issues come into play. Pritchard and Wilson (1999) alluded to this when they noted:

> The Web's very popularity is becoming one of its major weaknesses. To go about looking for data on a particular topic is fairly easy—the difficult part of sifting through the often thousands of documents a search has generated for an article which will contain something which is genuinely helpful or interesting. The fact that the authenticity or veracity of the data or information provided cannot be guaranteed is another failing. (p. 44)

Nevertheless, Herrera (2001) and Engelbrecht and Harding (2005) asserted that the Internet's hands-on environment enables students to see and explore mathematical concepts. Martinovic (2005) suggested that there are a number of potential benefits to students of mathematics using online help sites. According to Martinovic, the Internet

- has a greater potential for students to develop questions that will engage them in a process of self-diagnosis and reflection;
- provides students with answers that may provide models of *thinking through* problems;
- through online help sites offer vicarious benefits even for visitors that do not ask questions, by helping them learn the language of mathematics, how to ask questions, and how to answer them; and
- provides different approaches in answering similar mathematics questions which may help students realize that there is more than one way to solve problems.

Although there are many claims for online learning, those who are teaching such courses have not indicated that the learning of students is without difficulty. Guberman-Glebov, Baruch and Barabash (2003), reflecting on their teaching in this environment, suggested that "students in such a course left on their own, do not manage to make a sufficient progress and need permanent instruction, which renders the distant learning approach (online) in this case time consuming and not efficient" (p. 161). Wadsworth, Husman, Duggan and Pennington (2007) later noted:

> Although students in online courses are implementing many of the same strategies as their counterparts in traditional classrooms, there has been little evidence to show what strategies are most useful in this new environment and how some strategies may translate to a new learning environment. (p. 13)

The role of teacher and the form of teaching when the Internet is utilized to any degree is also of interest. Again there is much in the literature that asserts that the context has changed for the better, but there seems to be little hard research evidence on which these conclusions are based.

Stahl (2009b) called for a new way of teaching when using the Internet because:

> Students learn math best if they are actively involved in discussing math. Explaining their thinking to each other, making their ideas visible, expressing math concepts, teaching peers and contributing proposals are important ways for students to develop deep understanding and real expertise. There are few opportunities for such student-initiated activities in most teacher-led classrooms. (p. 24)

Although the Internet does afford new pedagogical possibilities, "the teacher's role in the use of the Internet is one of significant importance and not to be taken lightly" (Loong & White, 2003; p. 2). As Guberman-Glebov, Baruch, and Barabash (2003) noted:

> The computer and Internet provide some unquestionable advantages as a learning environment, if one learns to use them properly. We assert by that the technology usage is not self-evident for every course and every context, and one needs tools and skills for decision-making as to the choice of teaching methods and strategies involving these techniques. (p. 160)

There is another potential affordance offered by the Internet that may help teachers explore the new roles that are open to them. In a unique way, not available previously, the Internet affords the creation of networks of teachers. Some researchers see it as a venue for developing ideas to improve mathematics teaching. Chinnappan (2006) suggested that "by sharing the problems and concerns of their own school context, teachers can better understand, anticipate, and develop potential solutions to the learning demands of children in their classroom" (p. 357). And yet this assertion hardly needs the Internet for this to occur.

But even in large cities with many schools, teachers have often found it hard to meet and share professionally in a manner that is frequent and continuous over a long period of time. Most inter-school professional meetings of teachers only occur when there is a specific task to be accomplished. Changes may be possible with the Internet.

> Through the Internet, teachers can share expertise, offer one another their ideas on lesson plans and projects, even chat across continents about common problems and interests. Lessons made for one cultural setting may not be suitable for another, but they may still suggest ideas that can be revised and molded for your classroom. (Herrera, 2001, p. 26)

Thus building a professional community of support without having to leave your office, which can meet asynchronistically if necessary, becomes a possibility with the Internet. But how often this occurs, and the gain teachers have from such a community, has not been made at all clear in the research literature.

A critical aspect of teaching is the utilizations of resources that will help develop a useful context for student learning. The traditional resource for mathematics teaching has been the textbook. Unfortunately, many of the resources for mathematics teaching on the Internet essentially are just a reproduction of practices which are based on a paper-and-pencil medium such as downloading books or downloading exercises, a practice that does not take full advantage of possibilities of the Internet (Engelbrecht & Harding, 2005). Herrera (2001) suggested that there are alternatives:

> An in-depth treatment of a topic in this medium can include interactive animation, links to related material, video clips, and opportunities to email experts on the topic. Not all these elements are necessary, and certainly you do not want them included for their "glitz" value, but used properly they enrich the learning experience. (p. 28)

Borba and Villarreal (2005) discussed how there are new forms of communication in an online course taught via chat (see also Beatty & Geiger, 2010). Cazes, Gueudet, Hersant, and Vandebrouck (2006) used the Web to post exercises that they claimed changed the didactical contract in the classroom. However it is not clear whether their exercises just reproduced paper-and-pencil exercises, or whether they took advantage of alternative Internet possibilities. Hoyles et al. (2009), when discussing the Internet, emphasized a notion that they had developed in previous work on microworlds—how *connectivity* within a regular classroom changes the nature of collaboration.

It was recognized some time ago that "using the Internet would allow the children to locate 'real-world' data, and perhaps promote a greater understanding of instances in which one encounters such data, thereby fostering an appreciation for the use of mathematics in the real world" (Gerber et al., 1998, p. 116). The Internet has developed beyond the point where it represents merely a huge accessible database, although that advantage has not changed. Now the availability of dynamic geometry software can transform the types of tasks that can be developed in the classroom (Arzarello & Edwards, 2005; Arzarello, Olivera, Paola, & Robutti, 2002; Ferrara, Pratt, & Robutti, 2006; Laborde, Kynigos, Hollebrands, & Strasser, 2006; Mariotti, 2002; Marrades & Gutierrez, 2000). The relatively new Interactive Whiteboards (IWBs), although used in some countries (e.g., England and the USA) for more than a decade, have only now come to be used in classrooms more widely; they offer exciting opportunities to explore the use of such applications in conjunction with the Internet. Although the use of IWBs have rightly been criticized in general, as well as in mathematics teaching (Clarkson, in press; Zevenbergen & Lerman, 2008), their facility of being able to archive the records of a class's group thinking, including any use made of the Internet, and to display this quickly and easily in subsequent lessons, will be something to watch for the future.

Even with the many advantages of the Internet, there are some issues that are beyond the control of the teacher. For example, there have been critiques of the design and pedagogical quality of online interactive mathematics content. Gadanidis,

Sedig, and Liang (2004) noted that designing online mathematical investigations as pedagogical tools is not a simple undertaking. In their opinion many "do not appear to be well designed, neither from a pedagogical nor from an interface design perspective" (p. 294). They suggest that good design becomes possible when mathematics education and human–computer interaction design experts work together, rather than in isolation, simultaneously taking into account pedagogical goals and interface design principles.

Rather than analyzing in detail work such as the above, we have chosen another path. Technologies and modes of communication are rapidly changing, as we have alluded to in earlier sections, making the study of their impact on mathematics education both challenging and exciting. In the next section we discuss some of the themes that appear in the literature regarding the affordances of using the Internet in mathematics education. We have chosen not to report on studies that are predominately text based and/or use rapid response modes aimed mainly at testing students' abilities. Rather, we briefly report on studies that seem to push the boundaries of how the Internet can be used creatively and with worth in mathematics education.

Collaboration, Multimodality and Performance

Collaboration, multimodality and performance are the three new affordances that we have identified and discuss briefly using some case studies in this section. These features are not affordances only of the Internet. But we claim that the Internet transforms them. Hence in one sense they are all objective capabilities of the Internet.

Collaboration has changed with the use of the Internet not only because people who are in different geographic location can interact, but because even when they are face-to-face, collaboration involving the use of the Internet changes its nature.

Multimodality, understood as the combination of different kind of texts, has definitely been changed by the Internet. It is easy to combine video, drawings and music with regular text. Hence with the Internet one is able to bring information to online courses or to face-to-face courses in ways undreamed of in pre-Internet days.

The third subsection deals with performance. We characterize here all kind of performances (such as YouTube videos) that can be found on the Internet that are directly connected to mathematics education.

Clearly there is overlap between these three issues which we recognize. We are not trying to set out a classification system with these headings. Rather, we are identifying labels through which we can discuss what we believe are affordances arising through the Internet for mathematics education. Before we go into a more detailed discussion of collaboration, multimodality and performance in the following subsections, we note that our own teaching experiences with the Internet have significantly altered our notion of *classroom*. First, all authors have been teaching online courses for at least six years. In online courses, all the interaction, or most of it, takes place in virtual environments. Normally teacher and students never meet face-to-face. Nevertheless, often some of the students from each course have mentioned that they "feel close" even without meeting face-to-face (Borba & Gadanidis, 2008; Borba, Malheiros, & Zulatto, 2010; Engelbrecht & Harding, 2005). The second type of teaching environment that we have experienced for much longer and has relevance to this discussion is that of the *blended learning* environment in which, for the most part, the use of the Internet is combined with face-to-face regular interactions. Lin and Ponte (2008) discussed different ways of how this can help in communities of prospective mathematics teachers. Recently there has been Working Groups on *online teacher education* at PME conferences (Borba & Llinares, 2008; Borba, Llinares, Clay, & Silverman, 2010). Overall, it seems that both online courses and the blended courses seem to suit both continuing teacher education and preservice teacher education programs (Maltempi & Malheiros, 2010).

Clearly, our own experiences of teaching in various ways with the Internet, colours the following discussion. As noted above, many of the practices that involve the use of the Internet are not taking advantage of the changing possibilities that it offers. They are simply mimicking practices of the paper-and-pencil medium. Hence, as we discuss collaboration, multimodality and performance, we will also note some of the reactions from students and explore possibilities for teaching mathematics—such as Math and Science Performance Festival (see http://www.MathFest.ca)—that we believe are offering new perspectives regarding how students and teacher can express their mathematical ideas.

Collaboration

What does online collaboration look like in the case of mathematics education? The two cases we present below illustrate how new technologies can help foster collaboration in online mathematics education settings.

Case 1: "Pass the pen, please". Online mathematics teaching and learning can be in synchronous, asynchronous as well as hybrid environments. In a synchronous environment, all students and the teacher are present using video, text, and/or audio. But how does one explore mathematics in such an environment? Rather than reverting to traditional modes when the instructor simply lectures and the students listen, it is possible for a synchronous environment to provide a shared collaborative workspace, where the teacher and students work together on mathematics problems. One such possibility we call "pass the pen please." The first author has developed and used a platform that allows the screen of any of the participants to be shared with everyone else. For example, we could start by showing a screen of *Geometricks* on our computer. At the same time the class of students, no matter their geographic location, could see the dragging that is performed on a given geometrical construction. To this point there is nothing of real interest. In many ways we as the instructors

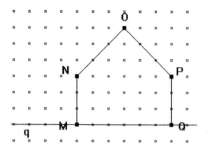

Figure 22.1. Geometricks.

have control of what is happening with the students simply watching. However a special feature of the application, which is important for the theme of collaboration, is the capability to "pass the pen" to another participant who could then add to what was done on the *Geometricks* construction. In this case, technology transforms the nature of the interaction and enables a form of collaborative problem solving to happen (see Borba & Gadanidis, 2008, for more details). This example illustrates how an online environment can support the convergence of different ideas and generate the collective construction of knowledge about geometry.

A particular example involved consideration of symmetry. A *Geometricks* file had already been given to students with the figure MNOPQ (see Figure 22.1) and they were asked to find the symmetric figure, in relation to *axis-q*.

Borba and Zulatto (2010), the professors from the university teaching the courses in which this example arose, report on how they began to learn mathematics from the interaction with the students. That is, once the authors "passed the pen" to the students and let them take the lead in the online activity, both groups, students and professors, became learners in a joint collaborative act.

However this collaborative online mathematics learning environment of "pass the pen, please" involved more than collaboration between teacher and students, and more than collaboration between students. It also involved collaboration between humans and digital mathematics tools. Borba and Villarreal (2005) have developed the theoretical notion of *humans-with-media*, as a means of stressing the idea that knowledge is constructed by collectives which involve humans and different technologies of intelligence such as orality, paper-and-pencil, and ICT (Lévy, 1993).

Hence it is hypothesized that different combinations of teachers, students and technologies result in different kinds of knowledge production. Although we do not at this stage want to make the case that new medium, such as the Internet incorporated into collectives of humans-with-media, enhance student learning, we have evidence that suggests that the Internet is a media that transforms practices of learners and teachers involved with mathematical educational practices. The research group GPIMEM (http://www.rc.unesp.br/gpimem) has documented some of these changes. For example, in online courses that use chat rooms, it is not easy to use mathematical symbolism. Participants have to resort to writing "integral of $2x \, dx$" instead of using the normal concise mathematical symbolism for such an expression. Santos (2006) in discussing research into such phenomena suggests that

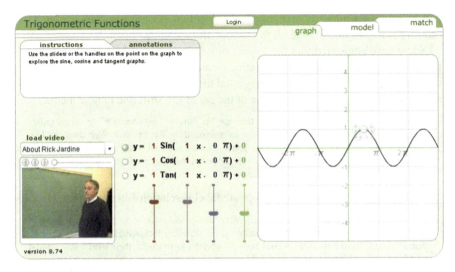

Figure 22.2. Digital Windows into Mathematics.

this change in writing such expressions online may change the nature of students' mathematical thinking.

Case 2: Annotating learning objects. Learning objects are typically viewed as "read-only" interactive content. That is, a user can explore the content but there is typically no method for annotating a particular state with ideas or questions, and then sharing these states and annotations with others. That would be a more difficult to do, as it would require more sophisticated programming and also the use of a database. Through a project called *Digital Windows into Mathematics* (http://www.edu.uwo.ca/dwm) the third author has developed learning objects (see Figure 22.2) that allow for remote collaboration (Gadanidis, Jardine, & Sedig, 2007).

Users have the ability to add their personal annotations (after obtaining a username and a password), and to incorporate personal metadata into the mathematics content. That is, the user can mark-up a learning object using text or freehand drawings and then save these annotations for later reference or for sharing with others. When saving annotations, the current state of the interactive environment is also saved (for example, the current values of the coefficients of the function being plotted, as well as the matching graph, will be saved along with the annotation). Saved annotations may be shared with others. Thus a student can share his/her ideas or questions about a certain state of the learning object, or a teacher may draw student attention to a particular aspect of a concept being explored.

Discussion. Collaboration for the purpose of learning is a prominent goal in mathematics education. Lerman (2000) has noted an emergence of a social-perspective on teaching and learning mathematics, and in particular an emphasis on collaborative learning, in mathematics curriculum documents such as NCTM's (2000) *Principles and Standards for School Mathematics*.

Some have suggested that the impact of new media, of which the Internet is an integral part, is less about the information it carries and more about the relationships that can be built. Schrage (2001) suggested that the commonly used label of *information revolution* misses the essence of the paradigm shift due to new media.

> *In reality, viewing these technologies through the lens of "information" is dangerously myopic.* The value of the Internet and the ever-expanding World Wide Web does not live mostly in bits and bytes and bandwidth. To say that the Internet is about "information" is a bit like saying that "cooking" is about oven temperatures; it's technically accurate but fundamentally untrue. (p. 1; original emphasis)

Schrage argued that a more appropriate label is relationship revolution. Hence:

> *The so-called "information revolution" itself is actually, and more accurately, a "relationship revolution."* Anyone trying to get a handle on the dazzling technologies of today and the impact they'll have tomorrow, would be well advised to re-orient their worldview around relationships ... *When it comes to the impact of new media, the importance of information is subordinate to the importance of community.* The real value of a medium lies less in the information that it carries than in the communities it creates. (pp. 1–2; original emphasis)

Lankshear and Knobel (2006) argued that the relatively recent "development and mass uptake of digital electronic technologies" represented changes on an "historical scale," which "have been accompanied by the emergence of different (new) ways of thinking about the world and responding to it" (pp. 29–30). These new ways of thinking can be characterized as more "participatory," "collaborative," and "distributed" and less "published," "individuated," "author-centric," or "expert-dominated" (Knobel & Lankshear, 2007, p. 9).

In this same vein, online mathematics learning is beginning to be associated with collaboration, suggesting a definite (which may be causal) relationship between the collaborative affordances of new media and the new emphasis on collaboration in mathematics education. For example, Stahl (2009a) noted:

> We found that participants in virtual math teams spontaneously began to explore their problems together, discussing problem formulations, issues, approaches, proposals and solutions as a group. Moreover, students generally found this interaction highly engaging, stimulating and rewarding. (p. 13)

Likewise, Sarmiento-Klapper (2009) stated:

> In our study of mathematics collaboration online we observe collective creative work as manifested in a wide range of interactions extending from the micro-level co-construction of novel resources for problem solving to the innovative re-use and expansion of ideas and solution strategies across multiple teams. (p. 227)

Another way of approaching the emerging association of online mathematics learning with an increased level of collaboration is to look at an online mathematics course that is taught asynchronously. In such a course, there is no set class time, and the instructor and students can join the course at their convenience. Two aspects of such an asynchronous course may increase online collaboration. First, the instructor needs all students to actively participate online if only to show that they are "present."

In contrast, in a typical face-to-face class (or for that matter a synchronized meeting online), many students do not have to participate actively to be "present." Second, when students participate online in an asynchronistic manner, chances are that the first person to read and possibly respond to another student's contribution or to offer assistance to their question will be another student. The teacher-centred communication norms of face-to-face classrooms are disrupted in an online asynchronous environment and there is an increased potential for student-to-student interactions.

A number of researchers have suggested that there are positive implications associated with the collaborative affordances of such an environment. Charles and Shumar (2009) stated:

> The social action that is encouraged is creative and draws upon the participants' imaginations to see knowledge production as an enjoyable, stimulating activity that is accessible by ordinary people. Understanding how to harness this agentic behavior and to leverage it for scalable, sustainable learning will be a next step for this research. (p. 224)

Sarmiento-Klapper (2009) reported:

> Group remembering and the bridging of interactional discontinuities allowed the teams to expand the referential horizon so that the objects created by themselves or by other teams could be expanded, reconsidered, or challenged. These methods allowed the teams to evolve a sense of collectivity engaged in building new knowledge and made it possible for them to interlink their collaborative interactions with those of other teams. (p. 235)

Cakir, Zemel, and Stahl (2009) also noted the benefits of collaborative online learning:

> The coordination of visual and textual realizations of the mathematical objects that the students co-construct provides a grounding of the algebraic formulas the students jointly derive using the line drawings that they inspect visually together. As the students individualize this experience of group cognition, they can develop the deep understanding of mathematical phenomena that comes from seeing the connections among multiple realizations. (p. 147)

Annetta, Folta, and Klesath (2010) suggested that young people in today's world

> … are competing and collaborating on a global scale. New technologies, or at least new to education, provide the opportunity to rebuild the collaborative social structures that we have begun to lose in our educational communities. … it is high time to rethink learning. (p. 21)

However, the concept of collaboration in online environments is complex. Issues surrounding the design of online mathematics learning require more research on how best to use and support collaboration. For example, Stahl (2009a) noted that "group size has an enormous impact on the effectiveness of different media" (p. 13). He added that most research on online mathematics learning had focussed on individual learning and commented that "there is not much research on, for instance, math collaboration by different size groups" (p. 13).

Kotsopoulos (2010) noted that when we look closely at student interactions in what appears on the surface to be collaborative learning, we find instances that are "predominantly non-collaborative despite the pedagogical efforts and intentions of the teacher and the task" (p.129). Kotsopoulos identified instances of non-collaboration while students work in groups (in a classroom setting) where "non-collaborative learning sent a message of incompetence and exclusion" to some of the students in the group (p. 138). The author continued:

> [Some] students ... received little support from their peers during collaborative learning. Moreover, efforts by these students to collaborate were thwarted by one or more members of the group. The group served to sustain a particular normalized way of collaborating that was exclusionary. (p. 138)

Kortsopoulos concluded that:

> Schools are public places of learning that ought to ensure safe and accessible learning for all students. Consequently, pedagogical strategies should work towards neutralizing the effects of power relations that restrict some learners. (p. 138)

This recent report suggested that in the classroom setting care needs to be taken with assumptions made regarding collaborative learning. It may be seen as a warning that students may not benefit from all online collaborative settings. For example, there, one can find problematic dynamics, such as bullying, occurring in group settings. These dynamics take on new forms in online settings. Cyber bullying is not uncommon among adolescents (Agatston, Kowalski, & Limber, 2007). Weigel, Straughn, and Gardner (2010) drew attention to the possibility that "bullying, which may have been limited to a small cadre of perpetrators and victims, can now spread more quickly and easily to a larger population" (pp. 17–18).

Dewey (1938) noted a long time ago that not all school experiences are educative. Some experiences are mis-educative. Similarly, we cannot assume that online interactions among students are necessarily collaborative in the positive sense. Again we note that this is an issue that needs to be worked through in the relatively new online environment for learning mathematics.

Multimodality

A challenge in teaching and learning mathematics online has been that in its initial manifestation; Internet communication was text-based. Not being able to use graphs and diagrams limited the possible representations of mathematical ideas. Although this problem has not been fully solved, as the support for communication using mathematical symbols and diagrams varies widely among e-learning platforms, the cases below point to developments that help incorporate multimodal elements to online mathematics.

Case 1: "Pass the pen" and *Digital Windows into Mathematics*. The two cases shared in the previous subsection on collaboration are also examples of how multimodal content may be used in online mathematics education. In the case of "pass the pen," the shared, collaborative geometric construction space allowed for communication using text, audio and geometric figures that could be manipulated. In the case of the *Digital Windows into Mathematics* project, the learning objects communicated mathematics ideas using text, diagrams, interactive content and videos of mathematicians talking about the mathematics explored.

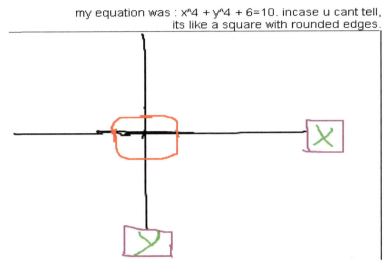

Figure 22.3. Sketch of graph of $x^4+y^4+6=10$.

Case 2: A multimodal online learning platform. For the purpose of offering online mathematics courses, the third author developed a learning platform called *Idea Construction Zone* (Gadanidis, 2007; Gadanidis & Geiger, 2010). This had the following multimodal features:

1. A rich text editor similar to ones used in word processors like *Microsoft Word*;
2. Users can embed the following within postings: video and audio recordings; graphics; *Flash* (swf); diagrams (using the built-in *Draw Tool*); hyperlinks (files and Web pages).

In addition, users have the option of making their posting *Peer Editable*, allowing other users to edit their ideas. Figure 22.3 shows what one Grade 8 student shared in the online discussion environment using the *Draw Tool* about one of the graphs he discovered while exploring an online graphing program.

Gadanidis, Hughes, and Cordy (2011) studied the nature of student learning in a classroom setting where students had ongoing access to the Internet while in a mathematics class and access to an online discussion board between classes using *Idea Construction Zone*. While exploring the graphs that were generated for missing number equations like __+__=10 and __+__=7, they wondered if they could create their own equations that would make the graphs point in a different direction or make the graphs curve. Using function plotters freely available on the Internet they investigated graphs that were well beyond the grades 7–8 mathematics curriculum: polynomial, trigonometric and even implicit, parametric and polar equations.

One example is shown in Figure 22.3. Gadanidis, Hughes, and Cordy (2011) suggested that:

> There was evident energy in the computer lab when students were creating and sharing graphs, as depicted by their eagerness to share ideas within and among groups and their willingness to take up and explore the ideas of others. Students seemed to enjoy working with equations that they initially did not understand, exploring their graphs and trying to make sense of the relationships between the equations and the graphs. Students also used Google and Wikipedia to find information about the various new equations they were encountering. (p. 418)

However, on a less positive note, this study also noted that there was a challenge in maintaining online discussion between classes. Although part of the reason had to do with poor pedagogical planning rather than the affordances of the online environment, this experience drew attention to the fact that classroom use of the Internet is not necessarily simply a positive or a negative. Rather, it also depends on how it is used pedagogically.

Discussion. Some research suggests that the collaborative aspects of online mathematics learning are supported by the multimodal online environments that are becoming increasingly available. For example, Cakir, Zemel, and Stahl (2009) stated:

> Multimodal interaction spaces—which typically bring together two or more synchronous online communication technologies such as text chat and a shared graphical workspace—have been widely used to support collaborative learning activities of small groups. ... Engaging in forms of joint activity in such online environments requires group members to use the technological features available to them in methodical ways to make their actions across multiple spaces intelligible to each other and to sustain their joint problem-solving work. (p. 140)

Horstman and Kerr (2010) suggested that multimodal content adds a further level of complexity when designing online learning environments. They stated:

> Perhaps the biggest conceptual transition for e-learning designers is to envision the content and learning objectives through graphical imagery and user interactions rather than by explaining content through text. (p. 196)

Despite the fact that the Internet is increasingly filled with multimodal content, the original text-based communication still persists for many online math courses. Martinovic (2005) noted:

> Text-based communication has little means for presenting graphs, diagrams, and tables. Both tutors and students suffered from an inability to use proper mathematical symbols and sometimes had to put in extra effort to use text editing capabilities for visual presentations. (p. 34)

Because of the original limitations posed by text-only communication, Engelbrecht and Harding (2005) suggested that "at the most basic level of mathematics on the Web is the practice of what has become known as *computerese*, using a text equivalent for formulae such as $sqrt(x)$ for the square root function" (p. 237). Clearly this formulation was needed in the early days of the Web, but nevertheless it did build another layer of complexity for communicating mathematics.

However, despite the *computerese* limitations of mathematics communication on the Internet, online communication is generally becoming increasingly multimodal in nature. This stands in contrast with many school-based experiences, especially in mathematics, which continue to rely on discourses that are monomodal or bimodal (in cases where diagrams or graphs are employed). Kress and van Leeuwen (2001) pointed out that in a digital environment "meaning is made in many different ways, always, in the many different modes and media which are co-present in a communicational ensemble" (p. 111).

The shift from text-based communication to multimodal communication is not simply a quantitative change. It is not just a case of having more communication modes. It can be seen as a qualitative shift, analogous to the change that occurred when we moved from an oral to a print culture. In the case of mathematics, we are seeing an emergence of online resources that combine text, symbols, animation, interactivity and videos. Such communication, which mirrors what young people are expecting in their overall Web-based interactions, will also be needed in their online mathematics experiences. Much is still to be done in researching this development.

Performance

Kress and van Leeuwen (2001) and Hughes (2008) noted that the multimodal nature of new media offers performative affordances. This is evident in the multimedia authoring tools used to create online content, such as *Flash*, which often use performance metaphors in their programming environments. For example, one programs on what is referred to as the *stage*, one uses *scenes* to organize *actors* or *objects* and their relationships, and one controls the performance using *scripts*. The Web as a performative medium is evident in the popularity of portals like YouTube. Hughes suggested that new media that has infused the Web draws us into performative relationships with and representations of our *content*. To use new media is to, in part, adopt a performative paradigm. Below we present two cases of Internet-based mathematics performance.

Case 1: Performing new images of mathematicians. The images of mathematicians performed in the media are typically narrow and negative. Picker and Berry (2000) have found that mathematicians are essentially invisible for students, and students rely on stereotypical images from media for their images of mathematicians. How might the Internet be a venue for offering students new views of mathematicians? The *Windows into Elementary Mathematics Project* of the Fields Institute (Gadanidis, 2010; Gadanidis & Scucuglia, 2010) uses new media tools to make mathematicians visible and offers a more positive image of mathematics and mathematicians (see Figure 22.4). In the videos, mathematicians spoke of their feelings about mathematics. Lindi Wahl stated that "One of the things that I really love about ... mathematics ... is that I'm creating something new all the time." Peter Taylor talked about choosing "the problems I do based on beauty."

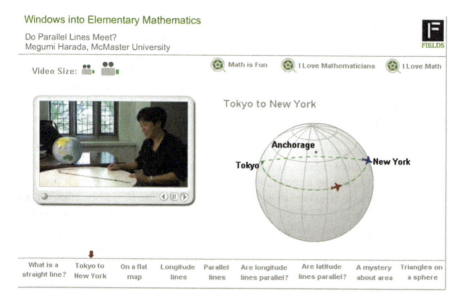

Figure 22.4. Windows into elementary mathematics.

> When one is doing mathematical biology, there are a lot of things to pay attention to, and there are a lot of papers to read, and a lot of ideas to think about, but the things I choose to work on, and the things I give to my graduate students, are the things where the structure fills me with a sense of beauty, where the aesthetics speak to me and lead me on.

Megumi Harada noted that:

> I love mathematicians. I can say that without any doubt that the math students were the most fun to be around, and I think it's because, as a group, mathematicians love what they do more than many, many other groups of people I know.

This online resource disrupts stereotypical images of mathematics as cold and abstract (Ernest, 1996) and views mathematics as a fun, interesting, imaginative, aesthetic and fully human activity (Sinclair, 2001; Sinclair, Pimm, & Higginson, 2006; Upitis, Phillips, & Higginson, 1997).

There is a little evidence that the new images of mathematicians do have some effect. The third author teaches fully online math-for-teachers courses for teachers who self-identify as "fearing or disliking mathematics." In these courses, teachers explore some of the mathematics problems explored by the mathematicians in the *Windows into Elementary Mathematics* project discussed above, and also view the video interviews with the mathematicians. It is interesting that teachers with initial negative outlooks towards mathematics end up making unsolicited positive comments about the mathematicians. For example, here are two teacher comments about mathematician Lindi Wahl:

> It is evident that she truly loves her job. She enjoys the challenge of creating brand new formulas to explain concepts. She loves collaborating with others who are specialists in their respective fields.

I love the way she talks about math! It's great to hear someone talk so passionately about it for once!

This engagement of teachers who "fear or dislike" mathematics with mathematicians who are passionate about their subject, and the resulting positive impact on teacher attitudes, has been made possible by the Internet.

Case 2: Performing classroom mathematics. In traditional mathematics classrooms, students communicate their ideas to fellow students and to their teacher. It is rare that the classroom mathematics experience spills over beyond the classroom walls. Our informal surveys of students and parents have suggested that when a student is asked "What did you do in math today?" the typical response is "Nothing" or "I don't know." In some of our work we have been exploring the idea of students as performance mathematicians, where the audience for their learning is expanded to include students in other classes, family and friends, and the wider world through the use of the Internet (Gadanidis & Borba, 2008; Gadanidis, Hughes, & Borba, 2008). An example of this is available at http://www.edu.uwo.ca/mpc/bigideas/bbw (see Figure 22.5). Here, a Grade 2 teacher relates the experience of his students:

(a) Scripting dialogues of mathematics conversations they might have at home when someone asks, "What did you do in math today?,"
(b) Performing their mathematics learning for a Grade 7 class,
(c) Performing their learning as a song and music video posted on the Internet, at the *Math and Science Performance Festival* (see http://www.MathFest.ca).

Another example of a performance from the online *Math and Science Performance Festival* in Canada that has been supported by the Fields Institute, the Imperial Oil Foundation and the Canadian Mathematical Society is *Now I'm a Trapezoid* (available at http://www.edu.uwo.ca/mathscene/geometry/geo1.html). This is a

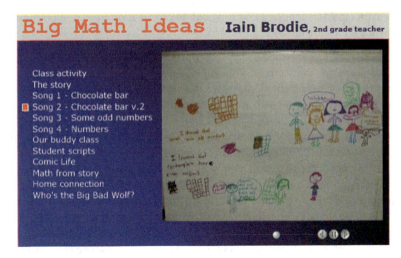

Figure 22.5. Students as performance mathematicians.

Figure 22.6. A performance from the *Math and Science Performance Festival*.

song by a fifth-grade student about a triangle that has lost its head. Saddened by this loss, the triangle laments that it is now a trapezoid (see Figure 22.6). The creation of such performances involves pedagogical shifts for mathematics teachers, putting a greater emphasis on mathematics communication through the arts and mathematics communication for a public audience.

Such pedagogical shifts are supported by the assertion of Gadanidis and Borba (2008) and Gadanidis, Hughes, and Borba (2008) that students might be viewed as *performance mathematicians* and that a performance (as in the Arts) lens might be useful in framing the teaching, learning and doing of mathematics, especially in a technology-rich setting. Such a lens helps us see and judge mathematics activity as we would see and judge a film. For example, if a mathematics activity was to be judged as we might judge a film, then Gadanidis and Borba (2008), using the work of Boorstin (1990), suggested that it would "work" if it offered us opportunities to experience the following pleasures:

- the pleasure of seeing the new and the wonderful in mathematics;
- the pleasure of being surprised mathematically;
- the pleasure of feeling emotional moments in doing and learning mathematics; and
- the pleasure of sensing mathematical beauty.

Discussion. Borba and Villarreal (2005) suggested that humans-with-media form a collective where new media serves to disrupt and reorganize human thinking. Likewise, Lévy (1993) saw technology not simply as a tool used by humans, but rather as an integral component of a *cognitive ecology* of the humans-with-technology. He added that technologies *condition* thinking. Can we imagine *what might be* if students and teachers, through their immersive experiences with performative affordances of new media, were *conditioned* to think about learning and teaching in performative ways? Lévy (1998) also claimed "as humans we never think alone or without tools. Institutions, languages, sign systems, technologies of communication, representation, and recording all form our cognitive activities in a profound manner" (p. 121).

Pineau (2005) suggested that "[t]he claim that teaching is a performance is at once self-evident and oxymoronic" (p. 15). However, as a theoretical claim, it is highly problematic. Pineau maintained that the typical interpretations of teaching-as-performance as (a) *teacher-as-actor* and (b) *teacher-as-artist* are weak, as the former reduced teaching to "teaching like an actor," and the latter equated it with "intuition, instinct, and innate creativity" (pp. 18–21). As an alternative, Pineau raised issues of power and authority and saw performance as political struggle and resistance.

Performance as a form of political struggle and resistance has been the centrepiece of the work of Boal (1985), namely his book *Theatre of the Oppressed*. In one of Boal's Forum Theatre performances, a person in poverty shopped for groceries and was confronted by the cashier as he did not have the money with which to pay for the food his family needed to survive. As the play unfolded, members of the audience (spect-actors) could at any time replace an actor and navigate the play in directions they deemed to be appropriate. There were at least two important things *at play* in such a performance. First, the common script of "shop, pay, take home" was disrupted. A second important thing at play was the agency of the audience. A spect-actor had the same right as the actor to be a part of the play.

Viewing students as performance mathematicians helps disrupt the traditional hierarchy of knowledge and authority in the mathematics classroom. Internet-based performances of mathematics help bring to public light the mathematical thinking of students themselves, who have traditionally been silenced outside the confines of mathematics classrooms. Just as importantly, seeing public performances of student mathematics raises the question, "What makes for a good mathematical performance?"

Boorstin (1990) identified three pleasures that we derive from performances such as at the movies: (a) the new and the surprising; (b) emotional moments; and (c) visceral sensations. It is interesting that Norman (2004) stated that his principles for technological design "bear perfect correspondence" (p. 123) to the principles of what make movies work, identified by Boorstin. These principles have been used in Canada and in Brazil to research how they might be used as a basis for pedagogical design in mathematics education and how they might help us see teachers and students as performance mathematicians (Gadanidis, Borba, Hughes, & Scucuglia, 2010).

Our focus on performance in mathematics parallel our immersive work with Internet-based new media. Although we cannot make a strong claim of effect, anecdotal records of our experience suggests that the performative affordance of Internet-based media helped influence and support our thinking; or, as Borba and Villarreal (2005) suggested, disrupt and reorganize our thinking in this direction.

Final Reflections

The most recent information will be easily and directly available through online databases and the World Wide Web. Students will be able to participate in deterritorialized virtual conferences, where the best researchers in the field will be present. The primary role of education will no longer be the distribution of knowledge that can now be obtained more efficiently by other means. It will help provoke learning and thinking. Education will become a driving force of the collective intelligence for which it is responsible. It will focus on managing and monitoring learning: encouraging people to exchange knowledge, relational and symbolic mediation, personalized guidance for apprenticeship programs, and so forth. (Lévy, 2001, p. 151)

Philosophers such as Pierre Lévy have made several predictions about the world with the Internet. In the above quote, from a book originally published in French in 1997, Lévy foresaw some of the transformations powered by the Internet that have already occurred, such as the availability of databases with almost any information. It is still not quite clear how this will transform education overall as the Internet shapes more and more of our world. It is not quite clear either what consequences it will have for schooling. As already noted in this chapter, traditionally teachers and books were the main source of information for students. School could be seen as the place where information would possibly become knowledge, collective knowledge. As the Internet plays an increasing role in education, including mathematics education, classrooms will be transformed or "dissolved" in the Internet (Borba, 2009).

However at present it is not clear how widespread the use of the Internet in schooling has become. But with its ever-growing presence, critical questions arise. It is fair to say that most of what is asked in mathematical examinations is easily found with the help of a search device on the Internet. How schools deal with this issue, given that all students have been born into the Internet-world, is still an open question. Will textbooks and regular lectures disappear, or just continue to be replicated online, as authors such as Engelbrecht and Harding (2005) have documented? It is too early to know.

We have tried to show how some practices are already being developed, merging arts, and particularly performance, in a way that students can post their work with little expense and can change the usual way they participate in mathematical studies. But the change in places of teacher and students is not the only result of the participation of the Internet in the production of mathematical knowledge in schools. Multimodality seems to be another key word. Students have the possibility of expressing mathematics using simulators, animations and pictures, combined with usual text and mathematical formulas. We still do not know the place that this kind of activity will have in regular schooling. The observation by Castells (2009) is worth noting, as he reminded us that advances in communication systems can not only generate possibilities but also create problems:

Each one of the components of the great communication transformation represents *the expression of the social relationships, ultimately power relationships that underlie the evolution of the multimodal communication system*. This is most apparent in the persistence of the digital divide between countries and within countries, depending on their consumer power and their level of communication infrastructure. … Even with growing access to the Internet and to wireless communication, abysmal inequality in broadband access and educational gaps in the ability to operate a digital culture tend to reproduce and amplify the class, ethnic, race, age, and gender structures of social domination between countries and within countries. (p. 57; original emphasis)

Although, in this chapter, we have tried to show possibilities of the use of the Internet, we have also hinted at various points the disproportionate spread of

Internet access. As in early 2011, still two-thirds of the world's population does not have access to the Internet. Hence although the Internet is accessible by two billion people, this also means that it is not available for between four and five billion people! In this sense the Internet can be seen as a double-edged sword: opening possibilities for some, but increasing the gap between those who have access and those who do not. In this sense, the Internet creates a new educational divide in the world. There are now the "haves" and "have nots" related to their educational access to the Internet.

Different countries have come up with different policies to include all or most of its citizens, in a time that having an electronic address seems to be as important as having a street address. But this divide is not just in terms of countries. Castells (2001) predicted that the Internet could increase the creation of a *fourth world* in many big cities. He developed the idea that the old division between first, second and third world was being modified and that we could actually have all the different worlds in almost every country. New York would have areas with high Internet access and others with low or not at all. This would coincide with the first and the fourth World respectively, in terms of economic power. We suspect this is happening and it does have implications for mathematics education.

This requires public policies that help all to be able to take advantage of such technology. The *market forces* on their own can take too long to reach the "do-not-have-Internet" since they are for the most part the ones with very little economic power. In addition, just as one aspect of Internet technology seems to become popular, another quickly and sometimes unexpectedly takes its place. For example, although most adults in developing countries continue to rely on the use of email for person-to-person communication, many students are keeping track with friends using social networking sites, such as Facebook or Twitter. All of this rapid and unpredictability makes the adoption of current Internet technology a daunting task. Nevertheless, as we have noted in this chapter, these developments, with all their hopes and confusions, do have an impact on mathematics education.

The case studies we have presented in this chapter represent not what is typical mathematics learning in today's classrooms but what might be possible. Will the Internet help transform mathematics education and enhance student learning? Past experiences with *new* technologies (such as television) indicate that the promises that they held for enhancing student learning were not fulfilled, at least not on a broad scale. Will it be different with the Internet?

We finish the chapter with one dimension of the changes brought by the Internet that has only been noted in passing but could have profound ramifications for mathematics education: assessment. The Internet has brought multimodality, which we noted may transform the nature of how we express mathematical ideas. If that occurs, what will this do to the manner in which we assess mathematics in the future? Furthermore, if most students ultimately have access to the Internet, and most answers for most mathematical problems are published on the Internet, what then becomes a challenging mathematical problem with which we can assess students' knowledge? Again, we have no answers. Nevertheless, elaborating problems that, as yet, have no answers may make us think more clearly about the potential worlds that may open before us. We will have to wait a few more years yet before we can see clearly the place the Internet will occupy in the educational scenario, including within mathematics education.

Acknowledgments We would like to thank our reviewers and editors for their comments in earlier versions of this chapter. We would also like to thank Ricardo Scucuglia Rodrigues da Silva (Western University) for his comments on earlier versions of this chapter and for helping us with the literature review. We would also acknowledge suggestions given by members of the Brazil-based research group GPIMEM: Marcus Maltempi, Silvana Santos, Felipe Heitmann, Fernando Trevisan, Nilton Domingues and Debora Soares—all of whom read at least one of the draft versions of this chapter.

References

Agatston, P. A., Kowalski, R., & Limber, S. (2007). Students' perspectives on cyber bullying. *Journal of Adolescent Health, 41*(6), S59–S60.

Annetta, L. A., Folta, E., & Klesath, M. (2010). *V-Learning: Distance education in the 21st century through 3D virtual learning environments*. New York, NY: Springer.

Arzarello, F., & Edwards, L. (2005). Research forum PME 29: Gesture and the construction of mathematical meaning. *Proceedings of the 29th Conference of the International Group for the Psychology of Mathematics Education* (Vol. 1, pp. 123–154). Melbourne, Australia: International Group for the Psychology of Mathematics Education.

Arzarello, F., Olivera, F., Paola, D., & Robutti, O. (2002). A cognitive analysis of dragging practices in *Cabri* environments. *ZDM—The International Journal of Mathematics Education, 34*(3), 66–72. doi:10.1007/BF02655708.

Atweh, B., & Clarkson, P. C. (2001). Internationalisation and globalisation of mathematics education: Towards an agenda for research action. In B. Atweh, H. Forgasz, & B. Nebres (Eds.), *Sociocultural research on mathematics education* (pp. 77–94). Mahwah NJ: Lawrence Erlbaum.

Atweh, B., Clarkson, P., & Nebres, B. (2003). Mathematics education in international and global contexts. In A. J. Bishop, M. A. Clements, C. Keitel, J. Kilpatrick, & F. Leung (Eds.), *Second international handbook of mathematics education* (pp. 185–232). Dordrecht, The Netherlands: Kluwer.

Beatty, R., & Geiger, V. (2010). Technology, communication and collaboration: Re-thinking communities of inquiry, learning and practice. In C. Hoyles & J.-B. Lagrange (Eds.), *Mathematics education and technology—Rethinking the terrain* (pp. 251–284). New York, NY: Springer.

Boal, A. (1985). *Theatre of the oppressed*. New York, NY: Theater Communications Group.

Boorstin, J. (1990). *The Hollywood eye. What makes movies work*. New York, NY: Cornelia & Michael Bessie Books.

Borba, M. C. (2009). Potential scenarios for Internet use in the mathematics classroom. *ZDM—The International Journal of Mathematics Education, 41*(4), 453–465. doi:10.1007/s11858-009-0188-2.

Borba, M. C., & Gadanidis, G. (2008). Virtual communities and networks of practicing mathematics teachers: The role of technology in collaboration. In K. Krainer & T. Wood (Eds.), *International handbook of mathematics teacher education: Participants in mathematics teacher education: Individuals, teams, communities, and networks* (Vol. 3, pp. 181–209). Rotterdam, The Netherlands: Sense Publishers.

Borba, M. C., & Llinares, S. (2008). Online mathematics education. In O. Figueras, J. L. Cortina, S. Alatorrw, & A. Mep lveda (Eds.), *Proceedings of the Joint Meeting of PME 32 and PME-NA XXX* (Vol. 1, p. 191). Morelia, Mexico: Cinvestav-UMSWH.

Borba, M. C., Llinares, S., Clay, E., & Silverman, J. (2010). Online mathematics teacher education. In M. M. F. Pinto & T. F. Kawasaki (Eds.), *Proceedings of the 34th Conference of the International Group for the Psychology of Mathematics Education* (Vol. 1, pp. 396–396). Belo Horizonte, Brazil: International Group for the Psychology in Mathematics Education.

Borba, M. C., Malheiros, A. P. S., & Zulatto, R. B. A. (2010). *Online distance education*. Rotterdam, The Netherlands: Sense Publishers.

Borba, M. C., & Villarreal, M. E. (2005). *Humans-with-media and the reorganization of mathematical thinking: Information and communication technologies, modeling, experimentation, and visualization*. New York, NY: Springer.

Borba, M. C., & Zulatto, R. A. (2010). Dialogical education and learning mathematics online from teachers. In R. Leikin & R. Zaskis (Eds.), *Learning through teaching mathematics: Developing mathematics teachers' knowledge and expertise in practice* (Vol. 5, pp. 111–125). New York, NY: Springer.

Cakir, M. P., Zemel, A., & Stahl, G. (2009). The joint organization of interaction within a multimodal CSCL medium. *International Journal of Computer-Supported Collaborative Learning, 4*(2), 155–190.

Castells, M. (2001). *The Internet galaxy: Reflections on the Internet, business, and society*. New York, NY: Oxford University Press.

Castells, M. (2009). *Communication power*. New York, NY: Oxford University Press.

Cazes, C., Gueudet, G., Hersant, M., & Vandebrouck, F. (2006). Using e-exercise bases in mathematics: Case studies at university. *International Journal of Computers for Mathematical Learning, 11*(3), 327–350.

Charles, E. S., & Shumar, W. (2009). Student and team agency in VMT. In G. Stahl (Ed.), *Studying virtual math teams* (pp. 207–224). New York, NY: Springer.

Chinnappan, M. (2006). Using the productive pedagogies framework to build a community of learners online in mathematics education. *Distance Education, 27*(3), 355–369.

Churchhouse, R. F. (1986). *The influence of computers and informatics on mathematics and its teaching*. London, UK: Cambridge University Press.

Clarkson, P. C. (1980). In-service training and curriculum. *Com-3, 21*, 31–33.

Clarkson, P. C. (2011). Thinking back: Becoming aware of the global and its impact on choice. *Cultural Studies of Science Education, 6*, 153–163.

Clarkson, P. C. (in press). Using interactive whiteboards in school settings: A resource for future pedagogies. In J. Zajda (Ed.), *International handbook on globalisation, education and policy research*. Dordrecht, The Netherlands: Springer.

Clarkson, P. C., & Toomey, R. (2001, December). *Teachers not using computers to teach mathematics*. Paper presented at the annual conference of the Australian Association of Research for Education. Retrieved from http://www.aare.edu.au/index.htm.

DeBell, M., & Chapman, C. (2006). *Computer and Internet use by students in 2003 (NCES 2006–065)*. Washington, DC: National Center for Education Statistics, U.S. Department of Education.

Dewey, J. (1938). *Experience and education*. New York, NY: Collier Books.

Ellison, N. B., Steinfield, C., & Lampe, C. (2007). The benefits of Facebook "friends": Social capital and college students' use of online social network sites. *Journal of Computer-Mediated Communication, 12*(4), article 1. Retrieved from http://jcmc.indiana.edu/vol12/issue4/ellison.html.

Engelbrecht, J., & Harding, A. (2005). Teaching undergraduate mathematics on the Internet. *Educational Studies in Mathematics, 58*(2), 235–252.

Ernest, P. (1996). Popularization: Myths, mass media and modernism. In A. J. Bishop, M. A. Clements, C. Keitel, J. Kilpatrick, & C. Laborde (Eds.), *The international handbook of mathematics education* (pp. 785–817). Dordrecht, The Netherlands: Kluwer Academic.

Ferrara, F., Pratt, D., & Robutti, O. (2006). The role and uses of technologies for the teaching of algebra and calculus: Ideas discussed at PME over the last 30 years. In A. Gutierrez & P. Boero (Eds.), *Handbook of research on the psychology of mathematics education: Past, present and future* (pp. 237–273). Rotterdam, The Netherlands: Sense Publishers.

Gadanidis, G. (2007). Designing an online learning platform from scratch. In C. Montgomerie & J. Seale (Eds.), *Proceedings of World Conference on Educational Multimedia, Hypermedia and Telecommunications 2007* (pp. 1642–1647). Chesapeake, VA: AACE.

Gadanidis, G. (2010). Digital Windows into elementary mathematics: Performing images of mathematicians using new media tools. In J. Herrington & B. Hunter (Eds.), *Proceedings of World Conference on Educational Multimedia, Hypermedia and Telecommunications 2010* (pp. 2701–2708). Chesapeake, VA: AACE.

Gadanidis, G., & Borba, M. C. (2008). Our lives as performance mathematicians. *For the Learning of Mathematics, 28*(1), 42–49.

Gadanidis, G., Borba, M. C., Hughes, J., & Scucuglia, R. (2010). "Tell me a good mathstory": Digital mathematical performance, drama, songs, and cell phones in the math classroom. In M. M. F. Pinto & T. F. Kawasaki (Eds.), *Proceedings of the 34th Conference of the International Group for the Psychology of Mathematics Education* (Vol. 3, pp. 17–24). Belo Horizonte, Brazil: International Group for the Psychology of Mathematics Education.

Gadanidis, G., & Geiger, V. (2010). A social perspective on technology enhanced mathematical learning—from collaboration to performance. *The International Journal on Mathematics Education, 42*, 91–104.

Gadanidis, G., Hughes, J., & Borba, M. C. (2008). Students as performance mathematicians. *Mathematics Teaching in the Middle School, 14*(3), 168–175.

Gadanidis, G., Hughes, J., & Cordy, M. (2011). Mathematics for gifted students in an arts- and technology-rich setting. *Journal for the Education of the Gifted, 34*(3), 397–433.

Gadanidis, G., Jardine, R., & Sedig, K. (2007). Designing Digital Windows into Mathematics. In C. Montgomerie & J. Seale (Eds.), *Proceedings of World Conference on Educational Multimedia, Hypermedia and Telecommunications 2007* (pp. 3532–3537). Chesapeake, VA: AACE.

Gadanidis, G., & Scucuglia, R. (2010). Windows into elementary mathematics: Alternate mathematics images of mathematics and mathematicians [Janelas para Matemática Elementar: imagens públicas alternativas de matemáticae matemáticos]. *Acta Scientiae, 12*(1), 24–42.

Gadanidis, G., Sedig, K., & Liang, H. N. (2004). Designing online mathematical investigation. *Journal of Computers in Mathematics and Science Teaching, 23*(3), 273–296.

Gerber, S., Shuell, T. J., & Harlos, C. A. (1998). Using the Internet to learn mathematics. *Journal of Computers Mathematics and Science Teaching, 17*(2–3), 113–132.

Guberman-Glebov, R., Baruch, R., & Barabash, M. (2003). Decision-making in construction of courses combining classroom-based and Internet-based learning and teaching strategies in mathematics teachers' education. *Journal of Educational Media, 28*(2–3), 147–163.

Head, A. J., & Eisenberg, M. (2010). How today's college students use Wikipedia for course-related research. *Mendeley Computer and Information Science, 15*(3), 1–9.

Herrera, T. A. (2001). A valid role for the Internet in the mathematics classroom. *Australian Mathematics Teacher, 57*(1), 24–27.

Horstman, T., & Kerr, S. (2010). An analysis of design strategies for creating educational experiences in virtual environments. In M. S. Khine & I. M. Saleh (Eds.), *New science of learning. Cognition, computers and collaboration in education* (pp. 183–206). New York, NY: Springer.

Hoyles, C., Kalas, I., Trouche, L., Hivon, L., Noss, R., & Wilensky, U. (2009). Connectivity and virtual networks for learning. In C. Hoyles & J.-B. Lagrange (Eds.), *Mathematical education and digital technologies: Rethinking the terrain* (pp. 439–462). New York, NY: Springer.

Hoyles, C., & Lagrange, J.-B. (Eds.). (2010). *Mathematics education and technology—Rethinking the terrain. The 17th ICMI Study*. Dordrecht, The Netherlands: Springer.

Hughes, J. (2008). The "screen-size" art: Using digital media to perform poetry. *English in Education, 42*(2), 148–164.

Idris, Y., & Wang, Q. (2009). Affordances of Facebook for learning. *International Journal of Continuing Engineering Education and Life-Long Learning, 19*(2–3), 247–255.

Ilavarasan, P., & Levy, M. (2010). *ICTs and urban microenterprises: Identifying and maximizing opportunities for economic development*. New Delhi, India: Canada's International Development Research Centre. *Internet World Stats (2010)*. Miniwatts Marketing Group.

Internet World Stats. (2011). *Internet world stats: Usage and population statistics.* Retrieved from www.internetworldstats.com/stats.htm.

Khine, M. S., & Salleh, I. M. (2010). New science of learning: Exploring the future of education. In M. S. Khine & I. M. Saleh (Eds.), *New science of learning. Cognition, computers and collaboration in education* (pp. 593–604). New York, NY: Springer.

Kieran, C., & Yerushalmy, M. (2004). Working group on technological environments. In K. Stacey, H. Chick, & M. Kendal (Eds.), *The future of the teaching and learning of algebra.* Boston, MA: Kluwer Academic Publishers.

Knobel, M., & Lankshear, C. (Eds.). (2007). *A new literacies sampler.* New York, NY: Peter Lang.

Kotsopoulos, D. (2010). When collaborative is not collaborative: Supporting peer learning through self-surveillance. *International Journal of Educational Research, 49,* 129–140.

Kress, G., & Van Leeuwen, T. (2001). *Multimodal discourse: The modes and media of contemporary communication.* London, UK: Arnold.

Laborde, C., Kynigos, C., Hollebrands, K., & Strasser, R. (2006). Teaching and learning geometry with technology. In A. Gutierrez & P. Boero (Eds.), *Handbook of research on the psychology of mathematics education: Past, present and future* (pp. 275–304). Rotterdam, The Netherlands: Sense Publishers.

Lampe, C., Ellison, N. B., & Steinfield, C. (2008). Changes in use and perception of Facebook. In *Proceedings of the 2008 Conference on Computer-Supported Cooperative Work (CSCW 2008).*

Lankshear, C., & Knobel, M. (2006). *New literacies: Everyday practices and classroom learning* (2nd ed.). New York, NY: Open University Press.

Lerman, S. (2000). The socio-cultural turn in studying the teaching and learning of mathematics. In H. Fujita, Y. Hashimoto, B. Hodgson, P. Y. Lee, S. Lerman, & T. Sawada (Eds.), *Proceedings of the ninth international congress on mathematical education* (pp. 157–158). Tokyo, Japan: Kluwer.

Lévy, P. (1993). *Tecnologias da inteligência: O futuro do pensamento na era da informática. [Technologies of Intelligence: the future of thinking in the informatics era].* Rio de Janeiro, Brazil: Editora 34.

Lévy, P. (1998). *Becoming virtual: Reality in the digital age.* New York, NY: Plenum Press.

Lévy, P. (2001). *Cyberculture.* Minneapolis, MN: University of Minnesota Press.

Lin, F. L., & Ponte, J. P. (2008). Face-to-face learning communities of prospective mathematics teachers: Studies on their professional growth. In K. Krainer & T. Wood (Eds.), *Participants in mathematics teacher education: Individuals, teams, communities and networks* (pp. 111–129). Rotterdam, The Netherlands: Sense Publishers.

Loong, E., & White, B. (2003). *Teaching mathematics using the Internet.* Retrieved from http://www.merga.net.au/documents/RR_loongwhi.pdf.

Maltempi, M. V., & Malheiros, A. P. S. (2010). Online distance mathematics education in Brazil: Research, practice and policy. *ZDM—The International Journal of Mathematics Education, 42*(3–4), 291–304. doi:10.1007/s11858-009-0231-3.

Mariotti, M. A. (2002). Influence of technologies advances on students' math learning. In L. D. English, M. Bartolini Bussi, G. A. Jones, R. A. Lesh, & D. Tirosh (Eds.), *Handbook of international research in mathematics education* (pp. 695–723). Hillsdale, NJ: Lawrence Erlbaum.

Marrades, R., & Gutierrez, A. (2000). Proof produced by secondary school students learning geometry in dynamic computer environment. *Educational Studies in Mathematics, 44,* 87–125. doi:10.1023/A:1012785106627.

Martinovic, D. (2005). What are the characteristics of asynchronous online mathematics help environments and do they provide conditions for learning? In R. Luckin & S. Puntambeker (Eds.), *Proceedings from the 12th International Conference of Artificial Intelligence in Education, Representing and analyzing collaborative interactions: What works? When does it work? To what extent are methods re-usable?* (pp. 32–39). Amsterdam, The Netherlands: AIED. Retrieved from http://hcs.science.uva.nl/AIED2005/W6proc.pdf.

Nanyang Technological University. (2010). *Student IT background survey: Usage of Internet communication and Web 2.0.* Centre for IT Services (CITS), Centre for Excellence in Learning and

Teaching (CELT) and NIE Computer Services Centre (CSC). Retrieved from http://enewsletter.ntu.edu.sg/itconnect/2010-04/Pages/Survey2009-InternetCollaboration.aspx.

National Council of Teachers of Mathematics. (2000). *Principles and standards for school mathematics*. Reston, VA: Author.

Norman, D. A. (2004). *Emotional design: Why we love (or hate) everyday things*. New York, NY: Basic Books.

Picker, S., & Berry, J. (2000). Investigating pupils' images of mathematicians. *Educational Studies in Mathematics, 43*(1), 65–94.

Pineau, E. L. (2005). Teaching is performance: Reconceptualizing a problematic metaphor. In B. K. Alexander, G. L. Anderson, & B. P. Gallegos (Eds.), *Performance theories in education: Power, pedagogy, and the politics of identity* (pp. 15–39). Mahwah, NJ: Lawrence Erlbaum.

Pinto, M. M. F., & Kawasaki, T. F. (Eds.). (2010). *Proceedings of the 34th Conference of the International Group for the Psychology of Mathematics Education*. Belo Horizonte, Brazil: International Group for the Psychology in Mathematics Education.

Pritchard, C., & Wilson, G. (1999). Sources of data on the Internet. *Mathematics in School, 28*(1), 44–46.

Santos, S. C. (2006). *A produção matemática em uma ambiente virtual de aprendizagem: o caso da geometria euclidiana especial. [Mathematical production in a learning virtual environment: the case of Spatial Euclidian Geometry]* (Master's thesis). Sao Paulo State University (UNESP), Brazil.

Sarmiento-Klapper, J. W. (2009). Group creativity in VMT. In G. Stahl (Ed.), *Studying virtual math teams* (pp. 225–235). New York, NY: Springer.

Schrage, M. (2001). *The relationship revolution*. Retrieved from http://web.archive.org/web/20030602025739/, http://www.ml.com/woml/forum/relation.htm.

Selwyn, N. (2007). "Screw blackboard ... do it on Facebook!": An investigation of students' educational use of Facebook. Retrieved from http://www.scribd.com/doc/513958/Facebookseminar-paper-Selwyn.

Sinclair, N. (2001). The aesthetic is relevant. *For the Learning of Mathematics, 21*(1), 25–32.

Sinclair, N., Pimm, D., & Higginson, W. (Eds.). (2006). *Mathematics and the aesthetic: Modern approaches to an ancient affinity*. New York, NY: Springer-Verlag.

Smith, S. D., & Caruso, J. B. (2010). *The ECAR study of undergraduate students and information technology, 2010*. Boulder, CO: EDUCAUSE.

Stahl, G. (2009a). A chat about Chat. In G. Stahl (Ed.), *Studying virtual math teams* (pp. 7–16). New York, NY: Springer.

Stahl, G. (2009b). The VMT vision. In G. Stahl (Ed.), *Studying virtual math teams* (pp. 17–29). New York, NY: Springer.

Tay, K., Tan, D., & Tan, M. (2009). Using Facebook as a multi-functional online tool for collaborative and engaged learning of pre-university subjects. In M. Kim, S. Hwang, & A.-L. Tan (Eds.), *Proceedings of the International Science Education Conference 2009* [CD-ROM] (pp. 2082–2107). Singapore: National Institute of Education.

Upitis, R., Phillips, E., & Higginson, W. (1997). *Creative mathematics: Exploring children's understanding*. London, UK: Routledge.

Wadsworth, L. M., Husman, J., Duggan, M. A., & Pennington, M. N. (2007). Online mathematics achievement: Effects of learning strategies and self-efficacy. *Journal of Developmental Education. 30*(3), 6–8, 10, 12–14.

Waters, M. (1995). *Globalization*. London, UK: Routledge.

Weigel, M., Straughn, C., & Gardner, H. (2010). New digital media and their potential cognitive impact on youth learning. In M. S. Khine & I. M. Saleh (Eds.), *New science of learning: cognition, computers and collaboration in education* (pp. 3–22). New York, NY: Springer.

Zevenbergen, R., & Lerman, S. (2008). Learning environments using interactive whiteboards: New learning spaces or reproduction of old technologies? *Mathematics Education Research Journal, 20*(1), 107–125.

Chapter 23
Technology and Assessment in Mathematics

Kaye Stacey and Dylan Wiliam

Abstract This chapter reviews the way that the decreasing cost and increasing availability of powerful technology changes how mathematics is assessed, but at the same time raises profound issues about the mathematics that students should be learning. A number of approaches to the design of new item types, authentic assessment and automated scoring of constructed responses are discussed, and current capabilities in terms of providing feedback to learners or supported assessment are reviewed. It is also shown that current assessment practices are struggling to keep pace with the use of technology for doing and teaching mathematics, particularly for senior students. The chapter concludes by discussing how a more principled approach to the design of mathematics assessments can provide a framework for future developments in this field. Specifically, it is suggested that assessment in mathematics should: (a) be guided by the mathematics that is most important for students to learn (the mathematics principle); (b) enhance the learning of mathematics (the learning principle); and (c) support every student to learn important mathematics and demonstrate this learning (the equity principle).

Introduction

This chapter addresses the use of technology in the assessment of mathematics. Using technology calls for new emphases in the learning of mathematics and the goals of the curriculum which, in turn, require different kinds of assessment to probe students' anticipated new skills and capabilities. New technology can also

K. Stacey (✉)
Melbourne Graduate School of Education, The University of Melbourne,
Melbourne, Australia
e-mail: k.stacey@unimelb.edu.au

D. Wiliam
Institute of Education, University of London, London, UK
e-mail: dylanwiliam@mac.com

provide new assistance for the work of assessment both for the teacher within the classroom and for monitoring standards at the system level. This chapter reviews the challenges and opportunities for mathematics assessment posed by the use of technology. It examines issues concerning what should be assessed under these new modes of learning; the potential for deeper, more informative assessment; and how assessment might be conducted. Throughout this chapter, the term mathematics is used to refer to all of the mathematical sciences, including statistics.

There is a large literature on research and development in computer-based testing, which identifies many different approaches to all components of testing. In this literature, distinctions are sometimes made between testing for summative and formative purposes, between fixed and adaptive item presentation (where the items presented to students depends on their success on previous items), between Web-based and other delivery systems which differ in the nature and timing of feedback to the student (if any), according to the measurement theory employed (if any), and on many other features. In this broad literature, mathematics is often selected as the content domain for research. In the present article, all forms of testing using electronic technology are included (and referred to) as "computer-based" and issues are chosen for discussion because of their relevance to mathematics teaching, learning and assessment rather than to general issues of assessment practices or measurement theory. Computer-based testing is also at the heart of intelligent tutoring, since it links the "student model" and the "tutor model," but again this is not considered beyond the issues that arise specifically in mathematics.

The Potential of Technology

Technology has potential to alter all of the aspects of the assessment process. There are new possibilities for the ways in which tasks are *selected* for use in assessments, in the way they are *presented* to students, in the ways that students *operate* while responding to the task, in the ways in which evidence generated by students is *identified*, and how evidence is *accumulated* across tasks (Almond, Steinberg, & Mislevy, 2003). Technology can improve the ways we assess the traditional mathematics curriculum, but it can also support the assessment of a wider "bandwidth" of mathematical proficiency to meet the changes in emphases of learning for the future.

Computer-based testing allows the automated generation of different items with similar psychometric properties. This allows different students to take different items or students to take the same test at different times without giving them access to items before taking the test (Irvine, 2002).

Acting as a communications infrastructure, computer-based platforms enhance item presentation, as will be demonstrated below. For the student, there may be a dynamic stimulus, three-dimensional objects may be rotated, and flexible access to complex information from multiple sources can be provided. A particularly important feature of computer-based testing is that it can ensure students comply with constraints in a problem to ensure engagement with the desired mathematics. A wider range of response types is now possible. For example, "drag-and-drop" items or the use of "hotspots" on an image may allow students to respond to more items nonverbally, giving a more rounded picture of mathematical literacy. In paper-based assessment, the validity of assessment in mathematics for some individuals has been limited by the necessity to decode written instructions for mathematical items and to express mathematical answers and ideas clearly. The software may also take into account the steps taken by a student in reaching a solution, as well as the solution itself. Computer-based platforms also support the presentation of problems with large amounts of (possibly redundant) information, mimicking the real-world scheduling and purchasing problems that are common in everyday life in the Internet age.

Automated scoring of responses has been possible for multiple-choice items for 80 years (Wiliam, 2005), but in recent years there have been significant advances in the automated scoring of items where students have to construct an answer, rather than just choose among given alternatives (see, e.g., Williamson, Mislevy, & Bejar, 2006). Computer-based assessment offers possibilities for providing more detailed information to students and teachers at lower cost, including profile scoring and other forms of diagnostic feedback that can be used to improve instruction. Automated scoring is also increasingly used in online learning systems, both "stand-alone" instructional packages and supplements to classroom instruction with integrated assessments. Such systems can give diagnostic feedback to the student during the instructional activity, as well as providing information about the final outcomes, as a single final score, or a detailed breakdown. Some interactivity may also be possible. Automated scoring also makes it easier to supply reports showing trends in performance over time. For the assessor and teacher, sophisticated reports on the assessment enable ready tracking of progress of individuals, classes and systems. Unobtrusive measurement of new aspects of student–task interaction may also be reported. Features of student-constructed drawings, displays and procedures that are impractical to code manually, can be efficiently assessed, and strong database facilities are available for statistical analysis.

Acting as a computational and representational infrastructure, the computer-based platform can enable students to demonstrate aspects of mathematical literacy that benefit from the use of the mathematics analysis tools embedded in computer and calculator technology. Without the "burden of computation," student attention can be focused on problem-solving strategies, concepts, and structures, rather than mechanical processes. They can work with multiple representations that are "hot-linked" so that a change in one representation automatically produces a change in another (e.g., a change in a data table produces a change in a chart).

Chapter Outline

The first major section of this chapter examines assessment in situations where the technology is principally used for the purpose of assessment, rather than by students in an open way for solving the mathematics items. There are subsections on items and item types, increasing the bandwidth of assessment, scoring, feedback to students, and reporting to teachers, and the comparison of computer-based and paper-based assessment. As Threlfall, Pool, Homer, and Swinnerton (2007) note, "the medium of pen-and-paper has been an inseparable part of assessment, and a change to the medium of presentation threatens that highly invested arrangement, and seems to risk losing some of what is valued" (p. 335). Most of the studies reviewed in this section assume that the mathematics curriculum and approved mathematical practices are unchanged, and what changes are the opportunities to assess these.

The second major section of this chapter considers assessment when students can use the mathematical capabilities of technology in the mathematical performance that is being assessed. This section responds particularly to the advent of mathematically-able calculators and computer software and the need to accommodate them in learning, teaching and assessment. From such a perspective, it is generally accepted that both curriculum goals and accepted mathematical practices will change.

The themes of both the major sections (the ability to use new tools for mathematics, and the changing nature of mathematical tasks) are being reflected in mathematics assessment at all levels. For example, the OECD's 2012 international PISA survey of mathematics will include an optional computer-based assessment of mathematics (Programme for International Student Assessment Governing Board, 2010). Some of the computer-based items would be suitable for paper-based delivery but the presentation will be enhanced by computer delivery. Most of the items in the computer-based assessment, however, will test aspects of mathematical literacy that depend on the additional mathematical tools that are provided by information technology, and the whole PISA assessment is now on a trajectory towards computer delivery. The intention is to move "from a paper-based assessment towards a technology-rich assessment in 2015 as well as from the traditional items to the innovative assessment formats which computer-delivery would enable" (p. 6).

The chapter concludes with reflections on the state of the art and presents some principles that can be used to guide future work in this field.

Using Technology to Assess Mathematics

This major section examines changes technology is making to assessment, organized under the various components of assessment. The first subsection examines the new possibilities for items. The following section looks at developments in automated scoring of responses. The third subsection examines progress in providing feedback to students, especially in the context of formative assessment, which has been shown to be a major strategy for improving learning (Black & Wiliam, 1998). In this first main section, technology is principally being used for enhanced item presentation, more convenient and reliable scoring, and for immediate and personalized feedback to students. In the subsequent section, attention is focused on assessments where the technology is being used by the student as a mathematical assistant, with the associated issues of changed goals for the curriculum in addition to changed procedures. As will be seen in both main sections, computer-based assessment can serve traditional goals as well as providing new opportunities to assess aspects of mathematical proficiency that relate to higher-order thinking and greater real-world relevance.

Before beginning the section proper, we note that computer-based assessment is often adopted because such test administration provides multiple points of convenience for students, teachers and educational systems. Students can often take tests at a time and place to suit themselves, and may receive immediate feedback. Teachers (and even school systems) may be freed from the burden of grading, and can receive well-designed reports by class, student or item. The expansion of online learning systems has also encouraged the use of computer-based assessment and the major commercial products have teacher-friendly tools for constructing straightforward computer-based assessment within them. Many reports in the literature discuss these features. For example, Pollock (2002) reported on a change of the teaching and assessment of "basic mathematics skills" in a course for prospective teachers. The course already used a computer-aided learning system and so adopted an associated computer-based assessment system to enable a switch from assessing with examinations to continuous assessment. Previously, such a system had been regarded as too demanding of staff time. Since the aspects of computer-based assessment related to test administration are for the most part not specifically related to mathematics, they are not discussed further.

Similarly, although access to the substantial infrastructure required for computer-based assessment is certainly a barrier to its use (by individual students, classes within schools, schools as a whole, and systems) because this does not specifically relate to mathematics, the difficulties of access are recognized but not further discussed here.

Expanding Assessment: Items and Solutions

New possibilities for computer-based items. Consider Figure 23.1 below, which shows part of two versions of an item on estimating with percentages, taken from the developmental work on "smart-tests" (see Stacey, Price, Steinle, Chick, & Gvozdenko, 2009).

The paper-based item is multiple choice. The pom-pom tree in year 1 is shown and students have to select A, B, C D, or E to indicate the height of the pom-pom tree in year 2, when its top has blown off and it is 35% shorter. This item is easily scored by hand or by computer. On the right hand side, a new version only feasible in computer-based assessment is shown. Students indicate their estimate of the height by pulling up a slider. In the figure, a student has pulled up the slider for the fir tree, but has not yet started on the pom-pom tree. The handle of its slider is visible near ground level. There are at least three advantages to the computer-based item. First, estimation is tested in a direct and active way, without guessing from alternatives (and, possibly, with less cognitive load because the choices do not need to be processed). Second, whereas such

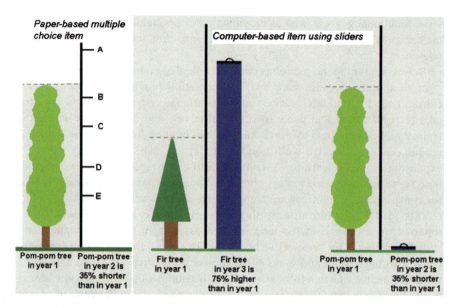

Figure 23.1. Computer-based assessment allows a wider range of item types.

an item would be very tedious to mark by hand, it is easily marked by computer, and partial credit based on the accuracy of the estimate can easily be allocated. Third, the image can be in colour, so the presentation is more attractive to students, without the substantial cost of colour printing.

Figure 23.2 shows an online mathematics question for 12–14 year olds from the example items for the "World Class Tests" (World Class Arena, 2010). These tests are designed to challenge able students, requiring creative thinking, logic and clear communication of thought processes. Solving the item in Figure 23.2 requires using the interface first in an exploratory way, gradually coming to understand the effect of certain moves (e.g., rotating twice around one point) and finally assembling a strategy to make the required shift in less than 12 moves. The computer provides the dynamic image, and itself counts the number of moves (other features of the solution could also be tracked). The item stem requires many fewer words than would be required in a paper-based version, and the item response is entirely non-verbal, which means that the mathematical proficiency of students with less developed verbal skills can be better assessed. It is hard to imagine a feasible paper-based version of this item, although it could be the basis for a mathematical investigation producing a report for teacher assessment.

As noted above, a computer-based assessment platform offers an infrastructure for communication that can enhance item presentation, the range of mathematics assessed, interaction between the student and the item, the way in which the response is provided by the student and the information that is extracted from the response.

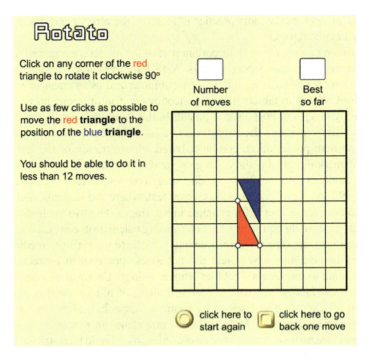

Figure 23.2. Computer screen for "Rotato," an example item from the *World Class Arena*.

There is great potential for creatively expanding the nature of assessment items and students' experience of engaging with assessment.

Authentic Assessment

We live in a society "awash in numbers" and "drenched with data" (Steen, 2001), where "computers meticulously and relentlessly note details about the world around them and carefully record these details. As a result, they create data in increasing amounts every time a purchase is made, a poll is taken, a disease is diagnosed, or a satellite passes over a section of terrain" (Steen, 2001). Knowledge workers need to make sense of these data and citizens need to be able to respond intelligently to reports from such data. This requires a change in the mathematics being learned. Full participation in society and in the workplace in this information-rich world, therefore requires an extended type of mathematical competence. For this reason, there has been increased interest in recent years in the development of "authentic" assessment in mathematics—assessment that directly assesses the competence of students in performing valued mathematics rather than relying on proxies such as multiple-choice tests that may correlate well with the desired outcomes, but may create incentives for classroom practice to focus on the proxy measures, rather than the valued mathematics.

Medication calculation is an important part of the numeracy required for nurses, since patients' lives can depend on this. *NHS Education for Scotland* funded the development by Coben et al. (2010) of a computerized assessment of medication calculations related to tablets, liquids, injections and intravenous infusions, using high-fidelity images of hospital equipment. In this "authentic assessment," the task for the student replicated the workplace task as faithfully as possible. As well as facilitating item presentation, computer-based administration of the test included automatic marking, rapid collation of group and individual results, error determination and feedback. A concurrent validity study compared the computer-based test with a "gold-standard" practical simulation test, where the students also prepared the actual dose for delivery (for example in a syringe). The two methods of assessment were essentially equivalent for determining calculation competence and ability to select an appropriate measurement vehicle (e.g., syringe, medicine pot). However, the computer assessment did not assess practical measurement errors, such as failing to displace air bubbles from a syringe. Coben et al. concluded that medication calculation assessment can be thought of in two parts: computational competence (which is best assessed by computer, especially since the whole range of calculation types can be included) and competence in practical measurement. Performance assessment, being very labour-intensive, should be restricted to assessing practical measurement.

In many cases, authentic assessment is undertaken through setting investigative projects. This is a longstanding practice, for example, in statistics education and in mathematical modelling. Since these assessments usually involve the use of mathematically-able software, they are discussed in the second main section of this paper.

Assessment with Support

A standard paper-based assessment generally aims to measure what a student can do alone and with a very limited range of tools. In the second main section, we discuss the changes when students have access to mathematically-able software when they are undertaking assessment. However, there are many other possibilities for including tools in computer-based assessments. Two educational concepts are particularly relevant here. The first is the idea of distributed cognition. Pea (1987) and others have pointed out that much cognitive activity is not carried out "in the head" but is distributed between the individual and the tools that are available for the task. The obvious consequence is that assessment of what a person can do should acknowledge tool use. The second important idea is Vygotsky's distinction between the psychological processes an individual can deploy on his or her own, and those that can be deployed when working with a teacher or a more advanced peer (see, e.g., Allal & Pelgrims Ducrey, 2000). These two ideas raise the possibility of using technology to create very different kinds of educational assessments—those that are focussed on the supports that are needed for successful performance rather than the degree of success when unsupported (Ahmed & Pollitt, 2004).

For example, Peltenburg, van den Heuvel-Panhuizen, and Robitzsch (2010) were concerned to improve the assessment of students with special education needs. Traditional assessments of these students indicated that they were operating several years below grade level, but the researchers were keen to investigate what the students might do with support. The study compared a standardized assessment with a computer-based "dynamic assessment" (Lidz & Elliott, 2001), which provided digital manipulatives that students could use to assist with subtraction questions. Students' results were better when the manipulatives were available, because the assessment showed more of what the students knew than could be inferred simply from an incorrect answer. Software running in the background also captured data on how the students used the manipulatives. Interestingly, in several instances these were not the methods that had been predicted when designing the tools.

Scoring and Gathering Other Data on Performance

In this subsection, we first examine progress in automating the work that a teacher does in evaluating the work of a student. Then, we look at non-traditional measures of the interaction between students and items that may contribute to a fuller assessment of student performance and learning.

Scoring constructed response mathematics items. Computer-based assessment, since its inception in the 1970s, has been limited by the nature of responses that can be scored reliably. The dominance of the multiple-choice format and single entry number answers, which still persists today, highlights the problem. Yet there is much more to mathematics than producing such simple responses: ideally, assessment across the full bandwidth of mathematics should deal with multiple-step calculations, checking each step as a teacher might, analyzing arguments and explanations, and certainly, as will be illustrated below, providing full credit for all solutions that are mathematically correct but differ in mathematical form. Although automated scoring that is as good as the best human scorers, if it can ever be achieved, is many years away, considerable progress has been made in recent years on assessing certain kinds of constructed-response mathematics items.

An advertisement for the commercial product *WebAssign* in the March 2011 edition of the *Notices* of the American Mathematical Society showed grading by two automated assessment systems of a student's response to a constructed response item. The item was "Find the derivative of $y = 2\sin(3x - \pi)$," and the response given was $\frac{dy}{dx} = -6\cos 3x$. The expected pen-and-paper response (by applying the chain rule) to this item would be $6\cos(3x-\pi)$, which is of course equivalent to the given response of $-6\cos 3x$. The advertisement made the point that the online assessment system *WebAssign* correctly graded this "unexpected" simplified response, whereas many other online grading systems would have graded it as incorrect (see WebAssign, n.d.). The difference lies in the computational engine (if any) being used for scoring complex constructed response mathematical answers. A powerful computer algebra system (CAS) can create items fitting specified criteria, compute the correct answer, and check students' responses.

Within the limited realm of school mathematics, less powerful mathematical software is effective. The equivalence of different algebraic expressions can be established by numerically evaluating the correct response (supplied by the item setter) and the student response at a number of points. The "m-rater" scoring engine developed by the Educational Testing Service does just this, by choosing the points to be evaluated at random, but also allows item creators to specify additional points to be evaluated. This approach has roughly the same level of accuracy as symbolic manipulation (Educational Testing Service, 2010).

In the report of the 17th ICMI Study on technology in mathematics education, only one paper specifically focussed on assessment. Sangwin, Cazes, Lee, and Wong (2010) considered the use of technologies such as CAS and dynamic geometry to generate an outcome from a student response that is a mathematical object (e.g., an algebraic expression, a graph, or a dynamic geometry object). The outcome may be right/wrong feedback to the student, a numerical mark along with automated written feedback to the student, or statistics for the teacher about the cohort of students.

Sangwin et al. first made the point that a CAS needs a range of additional capabilities to support good computer-aided assessment (CAA). As a simple example, they noted that a mainstream CAS recognizes x^2+2x+1 and $x+1+x+x^2$ as algebraically equivalent (and hence can mark either as correct), but for useful feedback to a student, a CAA system should be able to recognize the incomplete simplification and provide appropriate feedback to the student. Another simple example was an item where students needed to rotate one point about a central point. The resulting dynamic geometry diagram could be analyzed to see if the student has the correct distance and the correct angular position, opening possibilities for both partial credit and informative feedback. Drawing on examples of classroom observations the article described the development of quality feedback, useful cohort data for teachers, and new styles of mathematical tasks for which informative feedback can be given. It also described the pitfalls when a system can only examine the end product instead of examining the strategies that students use. Technology in this area is developing rapidly, and product development cycles often overtake educational research. Sangwin et al. concluded that new CAA tools require new modes of thought and action on the part of institutions, teachers and students alike.

Interest in assessment of constructed responses has been given further impetus by the shift towards integrated online learning and assessment systems, especially in tertiary education. For example, the *WebAssign* system mentioned earlier identifies its strongest features as convenience, reliability and security, compatibility with popular learning management systems, automated and customizable reporting to teachers by student or item, and easy creation or selection of assessment items. Partnerships with major textbook companies provide prepared databases of practice and assessment items and tutorial materials linked to popular textbooks, and questions can also be selected from open resources or those created by the teacher.

Another example is Maple T.A. (Maplesoft, 2011), which, being powered by the long-standing computer algebra system Maple, is specifically designed for technical courses that involve mathematics. Advertised strengths include the capacity to use conventional mathematical notation in both questions and student responses, the comprehensive coverage of mathematics and its capacity to support complex, free-form entry of mathematical equations and intelligent, automated evaluation of student responses graded for true mathematical equivalence with feedback available for the student. Maple T.A. can support open-ended questions with infinitely many answers, flexible partial credit scoring, and offers the assessment designer a high degree of mathematical control over randomly generated items, so that different students see different items testing the same content or to provide virtually unlimited on-demand practice. Maple visualization tools such as 2D and 3D plots are available to test creators and test takers.

Reports on the use of Maple T.A. and other systems are now appearing. For example, Jones (2008) reported on its ability to provide regular feedback and practice questions to engineering students. The article discussed how partial credit may be awarded, how account had been taken of techniques for designing good questions that incorporate randomly generated parameters, the coding required by the instructor, and strategies for reducing cheating in the on-line environment. Students' difficulties with the syntax for entering mathematics into the computer are commonly reported across much of the computer-based mathematics literature. Jones (2008) recommended the use of practice questions at the beginning of the course to reduce this. In this way, some of the barriers to a more expert computer-based scoring of constructed mathematical responses are now being overcome.

Awarding of partial credit is an important feature of human scoring in mathematics, but this presents significant challenges for automated scoring (Beevers, Youngson, McGuire, Wild, & Fiddes, 1999). In view of the difficulty of replicating the judgments made by humans in awarding partial credit, designers of computer-based assessments have explored a range of ways of approximating partial-credit scoring with simple dichotomous scoring. Ashton, Beevers, Korabinski, and Youngson (2006) trialled two methods of awarding partial credit in automatically-scored high-stakes pass/fail examinations. In the "steps method," some questions required the student to choose whether to enter a single response, which would be scored as correct or incorrect, or to opt to answer a series of sub-questions that led to the full answer, each of which would be assessed individually. For example, students asked to find the equation of a tangent to a curve could either choose to input the equation (for which they would either get full credit or no marks), or they could answer a series of sub-questions, requiring the coordinates of the point of tangency, the general form of the derivative, the slope of the tangent at the point and then its equation. Fewer marks were awarded for the structured approach because students did not demonstrate the ability to plan a solution strategy for themselves. The second method of approximating human-scored partial credit assessment explored by Ashton et al. simply informed students whether their submitted answer was correct or incorrect and gave them the opportunity to resubmit. The logic here is that partial credit is commonly awarded when students make small slips and so feedback would enable students to correct these small slips, bringing their score up closer to a human-assessed score. Although the total marks awarded in both methods were statistically indistinguishable from standard partial credit marking, Ashton et al. recommended adoption of the "steps" method because the correct/incorrect feedback method appeared to promote guessing rather than careful review.

The choice of a digital tool as a mathematical assistant depends on many aspects of the teaching context. For example, the Digital Math Environment (http://www.fi.uu.nl/wisweb/en/) has been designed to help secondary school students as they learn pen-and-paper algebra. It provides students with a facility to solve problems (e.g., to solve a quadratic equation) step by step, with the program providing feedback on accuracy at each step. In this way, it is primarily a learning tool, providing immediate formative assessment as the student works through problems, but summative assessment is also available.

Unobtrusive Measurement of Student–Task Interaction

Computer-based testing allows the collection and analysis of a range of data beyond a student's response, including response time and number of attempts. In cognitive psychology, response time has for many years been regarded as an important measure in the investigation of mental processing (Eysenck & Keane, 2005), and computer-based testing allows data on response times on a larger scale, and in naturalistic settings. Response time has been used for many purposes, including to inform item selection by complementing accuracy data, to identify cheating, to monitor test takers' motivation (for example, by flagging rapid guessing), and to track the development of automaticity, which is especially relevant to consolidating mathematical skills.

Gvozdenko (2010) studied the uses that teachers and test designers can make of information about student response times, using data from preservice primary teacher education mathematics courses. He found that response–time measurements provide a valuable supplement to performance data for: (a) evaluating difference in cognitive load of items; (b) identifying the presence of multiple solution strategies; and (c) monitoring the impact of teaching on specific cohort sub-groups across a teaching period.

Figure 23.3 gives an example from Gvozdenko (2010) of three versions of a test item that were intended to be classically parallel (i.e., the items should be interchangeable). The facility (percentages of students correct) and mean question

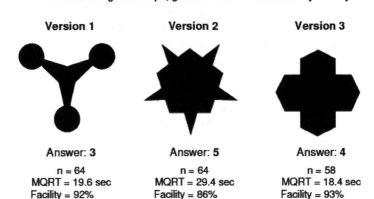

Figure 23.3. Three versions of a task, and associated mean question response times (MQRT) and facility (Adapted with permission from Gvozdenko, 2010).

response times (MQRT) of versions 1 and 3 were both similar. Version 2 looks similar from the facility data (only 6% lower) but it has a substantially greater MQRT. The 50% greater MQRT draws attention to the greater cognitive load in version 2, probably due to having two different rotated elements and a higher order of rotational symmetry.

Gvozdenko's (2010) study of preservice primary teacher education students also showed how response time can provide a supplementary measure of learning. Many students in such a course are able to solve primary-school level problems on entry, but their knowledge is not sufficiently automatic, robust, and strongly founded for flexible use in the immediacy of teaching a class. Measuring response time can give an additional indicator of developing competence for teaching. Another item from Gvozdenko (2010) involved the conversion of square metres to hectares. Conversion of 12,560 m^2 to hectares (answer 1.256) had a facility of 77%, but conversion of 690 m^2 to hectares (answer 0.069) had a facility of 72%. This group of students seems equally competent at these items. However, the MQRT of the first was 44 s, and for the second 62 s. This reveals a difference in the robustness of the knowledge that may show up in the pressured environment of the classroom.

Providing Feedback

The provision of feedback that is focussed on what a learner needs to do to improve, rather than on how well the individual compared with others, has been shown to impact significantly on learning (Wiliam, 2011). Indeed, over the last quarter century, a number of reviews of research have demonstrated that there are few interventions that have such a great impact on student achievement (Hattie, 2008). It is not surprising, therefore, that a major priority in the development of computer-based assessment software has been providing detailed feedback to test takers. Traditionally, assessment has been concerned with placing a student at a particular point on a scale. Although this may be adequate for many of the functions that assessments serve, it does not give feedback to students about what to do next. Rather, a feedback system needs to focus less on measurement, and more on classification—the assessment should indicate that the student has a particular state of knowledge that is likely to benefit from a specific intervention.

Livne, Livne, and Wight (2007) developed an online parsing system for students preparing to take college-level courses in mathematics designed to classify errors in student numerical answers, mathematical expressions, and equations as either structural (indicating the possibility of a conceptual difficulty) or computational (for example, the kinds of errors that would result from transcription errors). In terms of overall scoring, correlation between the automated scoring system and human scoring was very high (0.91). However, the automated scoring system appeared to be considerably better than human scorers at identifying patterns of errors in students' responses.

Shute, Hansen, and Almond (2008) investigated how summative and formative assessment could be linked by examining how an assessment system might be modified to include some elements of instruction for 15-year-old students learning algebra. They investigated the impact on student learning when feedback was added to an assessment system and when the presentation of items in the assessment was adaptive (responding to student answers) rather than in a fixed sequence. They found that the validity, reliability and efficiency of the summative assessment was unaffected by the provision of feedback, even when the feedback was elaborated (i.e., showing detailed solutions immediately after the item was completed). Students who received adaptive items learned as much as students who received items in the fixed sequence and students who received the elaborated feedback learned more than those who received no feedback or received feedback only on the correctness of their answers. The results suggest that it may be possible, in the near future, to derive data for summative purposes (e.g., for accountability) from experiences primarily designed to promote learning. In the authors' phrase, it may be possible to fatten the hog with the same instrument used to weigh it.

A particularly fruitful area for such research in recent years has been the development of Bayesian inference networks, or Bayesian nets for short. The basic idea is that for a particular domain, a proficiency model is specified that details the elements needed for successful performance in that domain. For each individual, a student model is constructed by observing the student's performance on a number of tasks, and using Bayes' theorem to update the likelihood that the student does indeed possess particular knowledge given the performance evidence. Such models are widely used in intelligent tutors, both to track student competence (the assessment task) and also to make decisions on what tasks a student should tackle next

(Korb & Nicholson, 2011; Stacey, Sonenberg, Nicholson, Boneh, & Steinle, 2003; VanLehn, 2006).

Diagnostic Feedback for Teachers

Although diagnostic feedback direct to students has proven educational benefits, there is also a case for providing detailed diagnostic feedback to teachers, especially when it is able to enhance their pedagogical content knowledge. Stacey et al. (2009) described a system, now in use in schools, of "Specific Mathematics Assessments that Reveal Thinking" (SMART, 2008). These "smart-tests" are designed to provide teachers with a simple way to conduct assessment to support learning. Using the Internet, students undertake a short test that is focussed narrowly on a topic selected by their teacher. Students' stages of development are diagnosed, and are immediately available to the teacher.

The programming behind the diagnosis links individual student's answers across questions to pool the evidence for particular misconceptions or missing conceptions in a way that would be impractical for teachers to do manually. Where possible, items have been derived from international research and then adapted for computer-based delivery. Online teaching resources (when available) are linked to each diagnosis, to guide teachers in moving students to the next stage. Many smart-tests are now being trialled in schools and their impact on students' and teachers' learning is being evaluated.

Comparing Computer-Based and Paper-Based Assessment

When an important goal of an assessment is to compare results over time with an unchanged content expectation, the question of how a computer-based assessment compares with a paper-based assessment for mathematics is important. One common example of such a context is when governments monitor achievement standards in schools from year to year. In response to such concerns, the European Commission Joint Research Centre commissioned a report (Scheuermann & Björnssen, 2009) on the transition to computer-based assessment for a wide range of purposes.

Research studies comparing effects of modes of assessment have shown mixed results (Hargreaves, Shorrocks-Taylor, Swinnerton, Tait, & Threlfall, 2004; Threlfall et al., 2007). There were differences in student performance in both directions and also no differences. Kingston (2009) conducted a meta-analysis of studies for 10 years up to 2007 and found that the comparability between traditional mathematics tested with computer-based and paper-based formats is slightly less than the comparability between tests of reading and science in these two formats. This difference was attributed to the need, in many items in the mathematics test, for students to shift their focus between the computer screen and writing paper. The difficulty of typing mathematics into a computer means that students undertaking computer-based mathematics assessment still usually need to do a lot of the work on paper, and transfer this to the computer when it is complete.

Hargreaves et al. (2004) found no significant difference between computer-based and paper-based testing for 10-year-old children and no advantage for students with greater familiarity with computers. In a study of complex problem solving involving fractions content, Bottge, Rueda, Kwon, Grant, and LaRoque (2009) found no difference by mode of presentation in the results of the assessment for any ability group. In general, computer-based testing creates both constraints *and* affordances for students; computer-based presentation can limit the strategies that students can use for solving problems, but can also afford more interesting and dynamic approaches to assessment. Items often change when converted from paper-based to computer-based assessment, but it does not seem possible to predict, in general, whether such conversion is likely to make items easier or more difficult.

Threlfall et al. (2007) explored how changing items designed originally for paper-based tests into a computer-based form altered what students do, and therefore what the items assess. The study examined only a narrow range of computer-based items, created by transferring paper-based items to the screen as closely as possible and marking as similarly as possible. Overall results were similar but some items showed large differences in facility. Computer-based items that supported exploratory solutions, and which enabled a solution to be adjusted, generally had higher facilities than the paper-based equivalent. For example, students ordering 4 lengths by size could drag the symbols into position and then check all of the pairwise comparisons, rearranging if necessary. Students placing circles to make a figure symmetric could drag them into position, and then check if the result looked symmetric, whereas on the paper-based item this approach was not possible. The computer-based presentation for such items enabled more sequential processing and hence effectively reduced cognitive load. However, some items where the computer allowed exploratory activity were less well done than in the paper-based version; an example was given of how the computer program did not provide exploration that was well controlled. Items where performance was better in the paper-based mode included those in which students did written calculation on scrap paper but where students tried to work mentally in the computer-based assessment. Students often do not use paper in a computer-based assessment even if it is available. Threlfall et al. concluded that each item needs to be examined to see which of the solution methods afforded by the media most closely correspond to the behaviours that the item is designed to assess. Using different item presentation media can affect performance, but the relationship with validity is complex—higher scores do not necessarily indicate greater validity.

The awkwardness of using the computer palette or other input device to construct mathematical expressions remains a potential source of construct-irrelevant variance for assessing mathematics by computer. A study of beginning tertiary quantitatively-able students by Gallagher, Bennett, Cahalan, and Rock (2002) found that ability to use the entry interface did not affect performance on a test where all answers were symbolic mathematical expressions. However, examinees overwhelmingly expressed a preference for taking a paper-based rather than computer-based test, because inputting mathematical objects was so cumbersome. The difficulties arising from the sharp contrast between hand written mathematics and keyboard-entered mathematics is a recurring theme in reports of computer-based assessment of all types and at all levels of education. Written mathematics is two-dimensional rather than strictly linear, there are symbols that are not standard on a keyboard, and different representations such as equations, graphs, diagrams, text, and symbols are used together in presenting a solution. All of these features mean that even the best of the current systems is far from ideal. Keyboard input remains a major barrier to computer-based assessment of mathematics.

In addition to whether the mode of presentation affects performance overall, it is also important to examine whether certain kinds of student are disadvantaged, or advantaged, by particular modes of presentation. Martin and Binkley (2009) suggested, for example, that the presentation of dynamic stimuli will advantage boys. Other groups of concern (see, for example, Scheuermann & Björnssen, 2009) include students with disabilities, members of different ethnic groups and students with certain cognitive characteristics. It is likely that there is too much variation in styles of computer-based assessment to obtain simple answers to such questions.

Assessing Mathematics Changed by Technology

The advertisement for *WebAssign* mentioned above appears to assume that the student differentiates the given expression using pen and paper, then enters the answer into a computer system. However the computer into which the student enters the response has the capacity to carry out the differentiation itself. If the online assessment system has access to a CAS for grading the work, it seems odd that access to this system should be denied to the student. Indeed, the widespread availability of powerful software for *doing* mathematics, rather than just checking the correctness of mathematics done on paper, raises fundamental issues about what mathematics is valued, how it should be taught and how it should be assessed. This has been a major preoccupation in many countries in recent years, and is the theme of this second major section of the chapter.

There are several reasons why assessment should take into consideration the tools that are used for mathematics outside school. As noted earlier, Pea (1987) has pointed out that tools that assist students in undertaking cognitive tasks have knowledge embedded within them, so the most meaningful unit for assessing competence is the user with the tool, rather than the user artificially working alone for the purpose of assessment. Another argument for the use of technology in formal assessment arises from the principle of validity—the context of the assessment should not differ significantly from the context of instruction. Indeed, where the context of assessment differs greatly from the context of instruction, assessment results are uninterpretable.

The College Board (2010), in the USA, explicitly made the point that the limitations of the use of technology in examination-based assessment should not limit the use of technology in classrooms, but the examination remains a powerful driver of what happens in schools. As will be demonstrated below, assessing mathematics when students are allowed to use technology has been shown to require substantial experimentation, research and a critical examination of values. Specifically, it requires clarity about the constructs to be assessed (Wiliam, 2011). Among other reasons, this is because research has led to a growing realization that mathematical thinking is almost impossible to separate from the tools with which it is learned and practised (Trouche & Drijvers, 2009). Doing mathematics with new tools leads to different ways of thinking about mathematical problems, and, indeed, to somewhat different mathematics.

Mathematical Competence and Computer Technology

Mathematics has a special relationship with computer technology, as its origins lay in the need to deal with extensive computation. An important part of mathematics has always been to develop algorithms for solving problems, and the design of effective algorithms has always had a two-way relationship with the technology of the day, from the abacus, to Napier's "bones," to ready reckoners, logarithm tables and slide rules to today's calculators and computers. Working with electronic technology, whether packaged as calculators, computers or special purpose machines, is now an essential component of doing and using mathematics in everyday life and in the workplace.

The impact of electronic technology on the ways in which individuals use mathematics, and consequently should learn it, has long been discussed, and continues to change rapidly. Thirty years ago, the Cockcroft enquiry into mathematics in UK schools (Committee of Inquiry into the Teaching of Mathematics in Schools, 1982) pointed to a change in the relative importance of methods of arithmetic calculation for personal and occupational use. Pen-and-paper algorithms had diminished in importance, being replaced by mental computation and estimation wherever appropriate and backed up by computer/calculator use when an exact answer to a difficult computation was required. This was an early indication of the need for mathematical competence to be redefined, in relation to electronic technology, with consequent impact on assessment. As Trouche and Drijvers (2009) pointed out, whereas the introduction of computers into mathematics education appears to have had limited impact on classroom practice, the use of handheld technology rapidly overcame the infrastructure limitations in schools and has made a greater difference to practice in mathematics classrooms. In the hands of students, for use at home and school when required rather than housed in a distant computer laboratory, handheld calculators (now with considerable mathematical and statistical power) are now used routinely in assessment in many countries. Much of the research reviewed in this section is therefore centred on the role of handheld technology for senior school mathematics.

The mathematical functionality of mathematically-able software such as graphics calculators, CAS, and statistics programs (especially those focussed on exploratory, rather than confirmatory data analysis) render many of the questions asked in the pen-and-paper era obsolete when looked at from a purely functional point of view. The availability of mathematically-able software shifts significant parts of the work from the student to a machine. For example, a student may decide a problem can be answered by solving two simultaneous equations and so inputs the equations to a graphics calculator using appropriate syntax, requests the graph with a suitable range and domain, examines the output and interprets the coordinates of intersection in terms of the original question. The machine does the graphing and zooming as requested, supported by a myriad of hidden numerical calculations. The student selects the method, establishes the equations, and interprets the output. This example demonstrates that assessment with technology tests very different skills from assessment without technology. Routine calculations and routine graphing can be by-passed by the student, who is left in charge of the strategic plan of solution. Hopefully, with the burden of calculation removed, emphasis can then shift to assessing more than routine skills to encompass a much broader bandwidth of mathematical proficiency, including reasoning, problem solving, modelling and argumentation. Some expansion of the range of assessable mathematical content might also be predicted. For example, non-linear equations can be treated similarly to linear models when graphical, rather than algebraic, methods are used.

Applying Three Principles for Assessment

In the USA, the National Research Council Mathematical Sciences Education Board (1993) published a conceptual guide for assessment which emphasized that assessment should make the important measurable rather than making the measurable important. To this end, they proposed the following three principles for the assessment of mathematics that are relevant at the personal, class and system level.

- *The mathematics principle:* Assessment should reflect the mathematics that is most important for students to learn. (This was called the "content principle" by MSEB)
- *The learning principle:* Assessment should enhance mathematics learning and support good instructional practice.
- *The equity principle:* Assessment should support *every* student's opportunity to learn important mathematics. (p. 1)

While these three principles are statements of values, rather than the more familiar principles of educational measurement, they do, in effect, subsume traditional concerns such as validity. The main value in the three principles presented above is that the focus was shifted from measurement to education (Carver, 1974).

These three principles do, of course, have implications for assessing traditional mathematics with technology, discussed in the first major section of this chapter. However, the major implications of the three principles, and the interactions between them, are more significant for the kinds of mathematics that can be assessed.

The Mathematics and Learning Principles

The issues at the heart of the mathematics principle and the learning principle are evident when school systems grapple with how to introduce technology into examinations. What mathematics is valued and how can good learning of mathematics be promoted? Drijvers (2009), for example, reported on the use of mathematically-able software (principally graphics calculators and CAS calculators) in 10 European countries. Consistent with earlier studies, he found four policies: technology not allowed; technology allowed but with examination questions designed so that it is of minimal use; technology allowed and useful in solving questions but without any reward for such work; and technology use allowed and rewarded in at least some components of the assessment. Drijvers concluded that the 10 countries he studied were probably moving towards consensus on the policies allowing the use of technology: (a) including some questions where it is definitely useful, and (b) ensuring pen-and-paper algebra/calculus skills are tested in some way, either by not rewarding certain technology-assisted work, or by including a special component of assessment without technology. This is consistent with the policy of several university-entrance examinations, including AP Calculus (College Board, 2010) and some Australian examinations (Victorian Curriculum and Assessment Authority, 2010).

The mathematics principle states that assessment should focus on the mathematics that is most important for students to learn, but of course exactly what this is may be strongly contested. A review of the policies above confirms that there are divided opinions on the use of technology to "do mathematics," so that compromises (e.g., to have separate components some of which allow and some of which disallow technology) are common. The learning principle is also significant here. The need for students to have basic pen-and-paper competence is widely recognized, even among strong advocates for the use of technology. It is essential, for example, to recognize equivalent algebraic forms when the technology generates an unexpected result. Having a separate component of an examination that does not allow technology is defended by some to ensure that these basic skills are not overlooked in schools. Exactly what skills should be tested and whether such a component is necessary, however, is also a contested matter. It is an interesting contrast that in the statistics education literature, the question whether students should use statistics software is rarely debated (see for example Garfield et al., 2011).

Given the enhanced computational power in the hands of students, one might hypothesize that end-of-school and university-entrance examinations allowing mathematically-able software would show a shift from mechanical questions (requiring students to perform some standard procedure that is cued in the wording of the question) towards questions requiring application in new situations and more complex construction of solutions. This might be seen as a natural outcome of the mathematics principle. However, Brown (2010) observed that the introduction of mathematically-able tools does not necessarily change the character of mathematics being assessed (and hence taught). Brown compared six end-of-school examinations in three jurisdictions, first at a point in time when students could use only a standard scientific calculator and later when students were permitted to use graphics calculators. He found that there was less emphasis on mechanical questions in two of the later examinations, but not in the other four.

Mechanical questions dominated all of the examinations before and after, even in examinations that were supplemented by an additional component where graphics calculator use was not permitted. Brown attributed the general lack of change to the unchanging mathematical values of the question writers, many of whom continue to place a high value on the accurate performance of pen-and-paper procedures. This may not however be the whole reason. For example, Flynn (2003) demonstrated that designing new questions that take advantage of technology requires creativity and experimentation, and it takes time for teachers and assessors to develop the necessary expertise. In a case study of "problems to prove," Flynn analyzed many sample examination questions, and identified difficulties that arose when the solution tools changed. With symbolic manipulation software (CAS), the key issue is what Flynn called "gobbling up" steps. For example, a student without CAS who shows that $(\sin x + \cos x)^2 = 1 + \sin 2x$ demonstrates knowledge of the identities $\sin^2 x + \cos^2 x = 1$ and $2\sin x \cdot \cos x = \sin 2x$. For the student with CAS, these steps are "gobbled up" by the CAS, and the result is given immediately. Flynn's paper provided some ways forward for assessing complex reasoning. However, there is much to be done to improve all assessment of the full bandwidth of mathematical proficiency. Having new technologies provides an extra dimension to this challenge as well as new but still embryonic opportunities.

Flynn (2003) also provided a case study of the way in which the symbolic manipulation facility of CAS calculators can actually be used in examinations that permit their use. He analyzed the two first such examinations in Victoria, Australia. Flynn found that questions yielding 12% of the total marks could not be answered with CAS features. These questions typically tested knowledge of features and properties of unspecified mathematical functions such as identifying the graph of $f(-x)$ from multiple-choice options, given the graph of a function with an *unspecified rule* for $f(x)$. This style of question came to prominence when graphics calculators were first permitted, to test understanding of the fundamental relationship between the graphs of $f(x)$ and $f(-x)$. Previously, this understanding may have been assessed by asking students to sketch the graph for a specified $f(x)$, but graphics calculators changed the cognitive demand of this task from mainly mathematical knowledge to mainly syntax and button pushing because they can automatically graph $f(-x)$ where $f(x)$ is given.

> On a particular day, the temperature y, in degrees Celsius, can be modelled by the function whose rule is $y = 9 - 5\sin(\pi t/12)$, where t is the time in hours after midnight. The maximum temperature for this particular day occurs at
> A. 3.00 pm
> B. 6.00 am
> C. 12.00 noon
> D. 6.00 pm
> E. 12.00 midnight

Figure 23.4. VCAA 2002 Mathematical Methods (CAS) Examination 1, Part I, Question 3.

Flynn found that symbolic manipulation would have advantaged students in questions worth 31% of the total marks. Most of these questions were similar to those that Brown (2010) termed mechanical questions, requiring rehearsed procedures such as factoring or differentiation—with CAS they require little more than button pushing. Perhaps surprisingly, these questions were generally well within the pen-and-paper algebraic skills of most students and hence many students would have completed them most efficiently without CAS. In fact, since examiners had probably derived the answers by hand, it was sometimes the case that multiple-choice questions presented answers in algebraic forms that favour pen-and-paper methods. There were no clear examples of questions that required algebra skills beyond expected pen-and-paper competence and in this sense took full advantage of the CAS, although subsequently this has occasionally occurred.

In questions leading to 56% of the marks, Flynn judged that a CAS calculator would give no advantage to a good student, although for a large proportion of such questions, the symbolic capability offered an additional solution or checking method, a phenomenon known as "explosion of methods." Figure 23.4 illustrates an examination question of this type.

For the question in Figure 23.4, the following methods are available:

1. Locating when the maximum temperature occurs from the graph of the function;
2. Solving $\sin(\pi t/12) = -1$ (the known minimum value of sine) using either the symbolic capabilities of CAS, with pen-and-paper, or directly from knowledge that the sine function has a minimum value at $3\pi/2$;
3. Solving $dy/dt = 0$ for t either with pen-and-paper or by using the symbolic capabilities for differentiation and/or solving;
4. Using a built-in facility on some calculators to find the maximum of a function;

For a student without technology, only the pen-and-paper versions of methods 2 and 3 are feasible; having a graphing facility adds methods (1) and (4), whereas with symbolic manipulation as well, all of the algebraic work is supported, as it would be in a question with parameters instead of specific values, when algebra would be the only viable solution method.

The large proportion of marks for questions where the newly permitted CAS facility had little or no impact demonstrate a continuity in examination practice, a continuity in what mathematics is valued, and the need for time and experience to develop a range of new question types. A broadening of the range of available solution methods is a main effect of the introduction of CAS into this examination system. Other effects of having CAS available are that it can compensate for some students' algebraic weakness, or enable them to check their own work, or simply be a strategic decision to save time.

Equity Principle

The purpose of assessment is to allow valid and reliable inferences about student learning to be made. For this reason, it is imperative that all students be given a fair chance to show what they have learned. In assessment with technology, there are many dimensions where the equity principle is relevant, including socio-economic circumstances, and certain physical disabilities. The College Board (2010) makes the point that teacher professional development is an important equity issue, as is convenient access to calculators or computers and the ancillary equipment (e.g., data projectors, calculator view screens, networks, etc.) to make the most of the technology in class. Education systems have tended to manage the latter issues by slowing the pace of change that might otherwise be desirable.

Gender is a potential equity issue, since boys are often said to be more "technically minded" than girls, and there are numerous research studies which confirm this "digital divide." Pierce, Stacey, and Barkatsas (2007) showed that although secondary school boys and girls (on average) approach learning mathematics with technology differently, this does not seem to affect their school use of technology for learning. Others, however, proposed that examinations with advanced technology disadvantage girls. Forgasz and Tan (2010), for example, proposed, on the basis of results from a special sample, that girls are disadvantaged when the more advanced CAS calculators are used instead of graphics calculators: this proposal awaits confirmation with a well-constructed sample, and a theoretical explanation of why the addition of symbolic manipulation to an already powerful technology might have such an effect.

One of the most important questions facing assessment with technology is how it can be conducted fairly if students use equipment of different quality or different brands or models with different capabilities. Of course, this is hardly a new issue. When fountain pens were first available, some worried that students rich enough to afford one would be at an advantage compared to those who had to dip the pen repeatedly in the ink-well. The examinations in Australia discussed above require students to have a calculator from a list of approved models (Victorian Curriculum and Assessment Authority, 2010), and the list is created with students' economic circumstances in mind. Any capability of the calculator can be used. Because modern calculators have the ability to store text (some more than others, and with different ease of access), students are permitted to bring notes into examinations. In other settings such as AP Calculus (College Board, 2010), any calculator can be used but only a restricted range of their capabilities can be used, with pen-and-paper working required for other processes.

As with the mathematics principle and the learning principle, the equity principle requires that assessors have a strong knowledge of the capabilities of the permitted technologies. Even when there is a list of approved calculators which have the same broad capabilities, assessors need to be certain that students are not advantaged by using one calculator over another, certainly over the whole examination and preferably in individual questions. Differences between brands and models can occur in architecture (e.g., ease of linking of representations or accessing commands, menus and keys), user-friendliness of syntax, capabilities (e.g., operations and transformations) and outputs (e.g., privileged forms and possible inconsistencies). The study of Victorian Certificate of Education questions by Flynn (2003) cited above found that 20% of available marks were affected by differences between the three permitted calculators, although when the examination was considered as a whole, these differences cancelled out. A major source of differences is that a symbolic manipulation package auto-simplifies mathematical expressions. A good example from Flynn and Asp (2002) is provided in Figure 23.5. To solve part (c) (ii), $a=\tan^{-1}(3/4)$ can be substituted into the expression for the derivative. One CAS calculator produces the answer nearly as required, but another gives an answer that is disconcerting to both students and teachers (see Figure 23.6).

In fact, the CAS2 solution can be simplified to give the same answer, but few students (or for that matter, teachers) are likely to be confident that the initial answer

---

The diagram [*not reproduced here*] shows part of the graph of the curve with equation $y = e^{2x}\cos x$.

(a) Show that $\dfrac{dy}{dx} = e^{2x}(2 \cos x - \sin x)$.

(b) Find $\dfrac{d^2y}{dx^2}$.

(c) There is an inflexion point at $P(a, b)$. Use the results from (a) and (b) to prove that
   (i) tan a = 3/4 and
   (ii) the gradient of the curve at P is e^{2a}

Figure 23.5. International Baccalaureate Mathematical Methods Standard Level 2000 Paper 2, Question 7.

CAS1: $f'(\tan^{-1}(3/4)) = e^{2\tan^{-1}(3/4)}$

CAS2: $f'(\tan^{-1}(3/4)) = e^{2\tan^{-1}(3/4)}(2\cos(\tan^{-1}(3/4)) - \sin(\tan^{-1}(3/4)))$

Figure 23.6. Different answers from different CAS calculators.

is on the correct path. This interesting example raises another issue related to the Mathematics Principle: does this technology-assisted solution constitute the proof required for this question?

After noting that users of different brands and models of technology may have "unfair" advantages on some questions, Flynn (2003) concluded that the most important goal is a fair examination, where small advantages to some on some questions balance out, thereby providing a fair overall result. This requires examinations to be rigorously scrutinized by assessors knowledgeable about all the technologies in use and about how students are likely to use them.

Assessing Project Work that Is Supported by Technology

In classroom projects and investigations, students can use technology to explore mathematical ideas for themselves, undertake more substantial work than is possible in a timed examination and deal with complex data sets, including real data, or undertake mathematical modelling of real problems, formulating relationships and interpreting results. For example, dynamic geometry programs provide excellent assistance for students to experiment, make hypotheses and test them, before creating formal proofs. In this way, students can demonstrate a wide range of abilities. Spreadsheets and statistics programs similarly enable students to search for relationships in authentic data and provide excellent graphical representations of datasets, and are ideal tools to use in project work. These are important aspects of mathematics and statistics that are difficult, if not impossible, to assess validly in traditional examinations. Since both the mathematics principle and the learning principle invite us to ensure that these "higher-order" skills do indeed feature in assessments, assessment of students using technology in investigations is important.

Rijpkema, Boon, van Berkum, and Di Bucchianico (2010) described how the program *StatLab* can be used to teach and assess engineers about the design of experiments. The *StatLab* program assists in assessing application of theoretical knowledge to practical situations by providing part of the grading and feedback to students. Bulmer (2010) described a course based around a virtual island with many inhabitants who were used by his students as subjects in virtual experiments. He described how this provided support for rich tasks that engaged students in realistic scientific practice where they confronted statistical issues, and he also described how Internet technology facilitated the assessment of project work for a large number of students by providing ready access to peer and tutor feedback. Bulmer commented that students could carry out the virtual experiments without access to statistical software, although the realism and modelling of good statistical practice would suffer from the necessarily limited samples. Callingham (2010) surveyed assessment of statistics using technology, giving examples of technology used in various phases of the assessment process, including an instance where Grade 9 students used technology to create graphs of data. Callingham concluded that more research is needed, especially on the assessment of statistical concepts.

The lack of research is surprising, given that for several decades many practitioners have expected students to use technology in statistics assignments, as a tool for calculation and for handling data. For example, the Victorian Certificate of Education, which combines both timed written examinations and school-based assessments, requires students to use statistical analysis systems in relevant topics and has done so for over 20 years (Victorian Curriculum and Assessment Authority, 2010).

Assessment from Classroom Connectivity

The vision of a connected classroom where teachers and students can exchange electronic information instantaneously and in a usefully collated form has been around for many years. In 1990, a software package called *Discourse* enabled teachers to set tasks for students, for students to respond, for teachers to monitor students' responses as they were generated, and, in later versions, to select an individual student's work and display it for the whole class, either with or without attribution (Heller Reports, 2002). This provides substantial opportunities for immediate formative assessment. However, the promise of such "classroom aggregation technologies" (Roschelle, Abrahamson, & Penuel, 2004) is still to be fully realized.

There have been several studies of the use of classroom aggregation technology for mathematics, such as the wireless-based Texas Instruments Navigator system, which has features like *Discourse* along with CAS and graphics calculator capabilities. Clark-Wilson (2010) reported on her own and other studies which found more opportunities for students to peer-assess other work and self-assess their own. They found that teachers used student responses to make decisions about the direction of subsequent work. In her study of seven teachers, Clark-Wilson found that all teachers reported new opportunities for formative assessment. By providing better opportunities to monitor students' work as entered into calculators, teachers gained additional insights, which enabled them to provide thoughtful interventions. They reported various mechanisms by which the discourse in the classroom was enhanced (e.g., discussing an interesting approach by a student to a problem), and in turn this enriched the teacher's awareness of student thinking. Additionally teachers reported many instances where students changed their opinions and moderated their responses when they saw other students' work: this provided additional opportunities for peer-assessment and self-assessment. However, learning to teach well with data arriving throughout the lesson appeared to challenge some teachers.

King and Robinson (2009) found that the use of electronic voting systems (which can also be used for immediate formative assessment providing information to teachers and students) in undergraduate mathematics classrooms was viewed positively by most students, and did increase student engagement—even for those students who did not view the electronic voting systems as positive. However, they found no relationship between increased use of electronic voting systems and student achievement.

Reflections

This review of the ways in which technology is changing assessment in mathematics was organized around two broad themes. First, the increasing sophistication and power of technology has supported five main categories of changes in the ways that assessment is conducted:

1. *Item preparation and selection*: better understanding of what makes items difficult has enabled the automated generation of items with predictable psychometric properties that reduce the cost of assessments, and make it easier to produce practice tests for students to prepare for high-stakes assessments. Technology also permits adaptive testing where the items are selected according to student responses to earlier items, thus increasing test reliability (or, equivalently, reducing test length).
2. *Item presentation*: technology allows items to be presented to students in ways that would not be possible with paper alone—for example, through the use of assessment models that focus not on how far through an item a student progresses, but the amount of support needed for successful completion of the task, thus improving the assessment experience for the student.
3. *Operation*: technology allows students to engage in tasks in different ways, and can also ensure that students adhere to constraints imposed on solutions, thus improving the validity of the assessment and expanding characteristics that are assessed, especially by reducing the reliance on verbal communication. Possibilities for authentic assessment are expanded.
4. *Evidence identification*: technology allows automatic scoring of some responses constructed by students, thus reducing the cost of scoring and supporting automated diagnostic analysis of response patterns. It allows different types of evidence (e.g., response time) to be collected unobtrusively, analyzed and reported.
5. *Evidence accumulation*: technology supports the development of models of student proficiency that go beyond simple unidimensional scales measuring competence to multidimensional models that allow the provision of detailed feedback to students and teachers.

These changes are blurring the boundaries between teaching and assessment, allowing assessment to become better integrated with instruction, and ultimately offer the prospect of integrated systems of assessment that can serve both formative and summative functions. However, several major obstacles still exist. What is possible now is a promise rather than a reality even in rich countries, not least because existing assessment systems tend to be well accepted in the contexts in which they operate, so change tends to be slow (Black & Wiliam, 2005). Furthermore, moving from pen-and-paper, human-scored systems to technology-based systems involves substantial initial investment costs. Perhaps most significantly, most current human–computer interfaces for mathematics require non-intuitive keyboard-based inputs, and students' solution processes need to combine paper-based work with computer input.

The second major theme of this chapter has been that technology prompts significant changes in the nature of mathematics that is assessed, and this creates new challenges for teachers and examiners. Creativity is needed to design assessment items which show what mathematical values are held important, and to design systems that are equitable, encourage good learning and focus the attention of teachers and students on mathematical knowledge that is important for the future.

Assessment should focus on the mathematical knowledge and skills that are most valuable. Technology, including dynamic geometry, spreadsheets, and calculators, enables students to explore mathematical ideas for themselves, formulating and testing and resolving hypotheses, so some assessment with technology needs to be without time pressure so that students can show these abilities. Similarly, some extended assessment can look at the whole modelling cycle, from formulating a problem mathematically, to solving it and interpreting the results; a process which technology assists at a number of points. Since technology takes over much of the routine work of solving, even examinations now need to look beyond assessing a narrow bandwidth of mathematical activity. Good assessment practices which permit technology use will be powerful in ensuring that systems achieve the higher-order thinking benefits that educators seek from technology in schools. Designing good assessments with technology also needs to pay attention to equity. High performance in school mathematics is often associated with social advantage, so it is important that use of technology in class or in assessment does not operate to limit further the achievement of socially and economically disadvantaged students. To accomplish all of these goals, assessors need to be very familiar with the capabilities of the technologies permitted and the sometimes unexpected ways in which students might use them.

In summary, new technologies offer considerable potential to provide the capability to support authentic assessments of complex mathematical activity, and to monitor unobtrusively how students interact with the tasks, thus supporting the development of sophisticated models of student proficiency that support the provision of high-quality feedback. Although recent developments in assessment with technology seems to have focussed primarily on the delivery of rich audio–visual content, the real power of computerized assessment is likely, in the future, to be in the creation of learning environments in which students use a range of information resources, engage with powerful software for problem solving, and collaborate with other students.

References

Ahmed, A., & Pollitt, A. (2004, June). *Quantifying support: Grading achievement with the support model*. Paper presented at the Annual Conference of the International Association for Education Assessment, Philadelphia, PA.

Allal, L., & Pelgrims Ducrey, G. (2000). Assessment of—or in—the zone of proximal development. *Learning and Instruction, 10*(2), 137–152.

Almond, R. G., Steinberg, L. S., & Mislevy, R. J. (2003). *A four-process architecture for assessment delivery, with connections to assessment design*. Los Angeles, CA: University of California Los Angeles Center for Research on Evaluations, Standards and Student Testing (CRESST).

Ashton, H. S., Beevers, C. E., Korabinski, A. A., & Youngson, M. A. (2006). Incorporating partial credit in computeraided assessment of mathematics in secondary education. *British Journal of Educational Technology, 37*(1), 93–119.

Beevers, J., Youngson, M., McGuire, G., Wild, D., & Fiddes, D. (1999). Issues of partial credit in mathematical assessment by computer. *ALT-J (Association for Learning Technology Journal), 7*, 26–32.

Black, P. J., & Wiliam, D. (1998). Assessment and classroom learning. *Assessment in Education: Principles, Policy and Practice, 5*(1), 7–74.

Black, P. J., & Wiliam, D. (2005). Lessons from around the world: How policies, politics and cultures constrain and afford assessment practices. *Curriculum Journal, 16*(2), 249–261.

Bottge, B., Rueda, E., Kwon, J., Grant, T., & LaRoque, P. (2009). Assessing and tracking students' problem solving performances in anchored learning environments. *Educational Technology Research & Development, 57*(4), 529–552.

Brown, R. (2010). Does the introduction of the graphics calculator into system-wide examinations lead to change in the types of mathematical skills tested? *Educational Studies in Mathematics, 73*(2), 181–203.

Bulmer, M. (2010). Technologies for enhancing project assessment in large classes. In C. Reading (Ed.), *Data and context in teaching statistics. Proceedings of Eighth International Conference on Teaching Statistics.* Voorburg, Netherlands: International Statistical Institute. Retrieved from http://icots.net/8/cd/pdfs/invited/ICOTS8_5D3_BULMER.pdf

Callingham, R. (2010). Issues for the assessment and measurement of statistical understanding in a technology-rich environment. In C. Reading (Ed.) *Data and context in teaching statistics. Proceedings of Eighth International Conference on Teaching Statistics.* Voorburg, Netherlands: International Statistical Institute. Retrieved from http://icots.net/8/cd/pdfs/invited/ICOTS8_5D2_CALLINGHAM.pdf

Carver, R. (1974). Two dimensions of tests: Psychometric and edumetric. *American Psychologist, 29*, 512–518.

Clark-Wilson, A. (2010). Emergent pedagogies and the changing role of the teacher in the *TI-Nspire Navigator* networked mathematics classroom. *ZDM—The International Journal of Mathematics Education, 42*(7), 747–761.

Coben, D., Hall, C., Hutton, M., Rowe, D., Weeks, K., & Wolley, N. (2010). *Benchmark assessment of numeracy for nursing: Medication dosage calculation at point of registration.* Edinburgh, UK: NHS Education for Scotland.

College Board. (2010). *Calculus AB/Calculus BC course description (effective Fall 2010).* Retrieved from http://apcentral.collegeboard.com/apc/public/repository/ap-calculus-course-description.pdf

Committee of Inquiry into the Teaching of Mathematics in Schools. (1982). *Report: Mathematics counts* [The Cockcroft Report]. London, UK: Her Majesty's Stationery Office.

Drijvers, P. (2009). Tools and tests: Technology in national final mathematics examinations. In C. Winslow (Ed.), *Nordic research on mathematics education: Proceedings from NORMA08* (pp. 225–236). Rotterdam, The Netherlands: Sense Publishers.

Educational Testing Service. (2010). *ETS automated scoring and NLP technologies.* Princeton, NJ: Educational Testing Service.

Eysenck, M., & Keane, M. (2005). *Cognitive psychology: A student's handbook* (5th ed.). Hove, UK: Psychology Press.

Flynn, P. (2003). Adapting "problems to prove" for CAS-permitted examinations. *International Journal of Computer Algebra in Mathematics Education, 10*(2), 103–121.

Flynn, P., & Asp, G. (2002). Assessing the potential suitability of "show that" questions in CAS-permitted examinations. In B. Barton, K. Irwin, M. Pfannkuch, & M. Thomas (Eds.), *Proceedings of 25th Annual Conference of the Mathematics Education Research Group of Australasia* (pp. 252–259). Sydney, Australia: MERGA.

Forgasz, H., & Tan, H. (2010). Does CAS use disadvantage girls in VCE Mathematics? *Australian Senior Mathematics Journal, 24*(1), 25–36.

Gallagher, A., Bennett, R., Cahalan, C., & Rock, D. (2002). Validity and fairness in technology-based assessment: Detecting construct-irrelevant variance in an open-ended, computerized mathematics task. *Educational Assessment, 8*(1), 27–41.

Garfield, J., Zieffler, A., Kaplan, D., Cobb, G., Chance, B., & Holcomb, J. (2011). Rethinking assessment of student learning in statistics courses. *American Statistician, 65*(1), 1–10.

Gvozdenko, E. (2010). *Meaning and potential of test response time and certainty data: Teaching perspective* (Doctoral dissertation). The University of Melbourne. Retrieved from http://repository.unimelb.edu.au/10187/11051

Hargreaves, M., Shorrocks-Taylor, D., Swinnerton, B., Tait, K., & Threlfall, J. (2004). Computer or paper? That is the question: Does the medium in which assessment questions are presented affect children's performance in mathematics? *Educational Research, 46*(1), 29–42.

Hattie, J. (2008). *Visible learning.* London, UK: Routledge.

Heller Reports. (2002, September). *Discourse finds a home in ETS acquisition.* Retrieved from http://findarticles.com/p/articles/mi_hb5695/is_11_13/ai_n28944586/

Irvine, S. (Ed.). (2002). *Item generation for test development.* Mahwah, NJ: Lawrence Erlbaum Associates.

Jones, I. (2008). Computer-aided assessment questions in engineering mathematics using Maple T.A. *International Journal of Mathematical Education in Science and Technology, 39*(3), 341–356.

King, S., & Robinson, C. (2009). "Pretty lights" and maths! Increasing student engagement and enhancing learning through the use of electronic voting systems. *Computers and Education, 53*(1), 189–199.

Kingston, N. (2009). Comparability of computer- and paper-administered multiple-choice tests for K-12 populations: A synthesis. *Applied Measurement in Education, 22*(1), 22–37.

Korb, K., & Nicholson, A. (2011). *Bayesian artificial intelligence* (2nd ed.). Boca Raton, FL: CRC/Chapman Hall.

Lidz, C. S., & Elliott, J. G. (Eds.). (2001). *Dynamic assessment: Prevailing models and applications.* Oxford, UK: Elsevier.

Livne, N. L., Livne, O. E., & Wight, C. A. (2007). Can automated scoring surpass hand grading of students' constructed responses and error patterns in mathematics? *MERLOT Journal of Online Learning and Teaching, 3*(3), 295–306.

Maplesoft. (2011). *Testing solutions from Maplesoft.* Retrieved from http://www.maplesoft.com/products/testing_solutions/

Martin, R., & Binkley, M. (2009). Gender differences in cognitive tests: A consequence of gender dependent preferences for specific information presentation formats? In F. Scheuermann & J. Björnssen (Eds.), *The transition to computer-based assessment* (pp. 75–82). Luxembourg: Office for Official Publications of the European Communities.

National Research Council Mathematical Sciences Education Board (Ed.). (1993). *Measuring what counts: A conceptual guide for assessment.* Washington, DC: National Academy Press.

Pea, R. (1987). Practices of distributed intelligence and designs for education. In G. Salomon (Ed.), *Distributed cognitions: Psychological and educational considerations* (pp. 47–87). Cambridge, MA: Cambridge University Press.

Peltenburg, M., van den Heuvel-Panhuizen, M., & Robitzsch, A. (2010). ICT-based dynamic assessment to reveal special education students' potential in mathematics. *Research Papers in Education, 25*(3), 319–334.

Pierce, R., Stacey, K., & Barkatsas, A. (2007). A scale for monitoring students' attitudes to learning mathematics with technology. *Computers and Education, 48*(2), 285–300.

Pollock, M. (2002). Benefits of CAA. *International Journal of Technology & Design Education, 12*(3), 249–270.

Programme for International Student Assessment Governing Board. (2010, November). Report of the 30th meeting of the PISA Governing Board. Retrieved from http://www.oecd.org/officialdocuments/publicdisplaydocumentpdf/?cote=EDU/PISA/GB/M(2010)2/REV1&docLanguage=En

Rijpkema, K., Boon, M., van Berkum, E., & Di Bucchianico, A. (2010). *Statlab*: Learning DOE by doing! In C. Reading (Ed.), *Data and context in teaching statistics. Proceedings of Eighth International Conference on Teaching Statistics.* Voorburg, The Netherlands: International Statistical Institute. Retrieved from http://icots.net/8/cd/pdfs/invited/ICOTS8_9C3_RIJPKEMA.pdf

Roschelle, J., Abrahamson, L., & Penuel, W. R. (2004, April). *Integrating classroom network technology and learning theory to improve classroom science learning: A literature synthesis.* Paper presented at the annual meeting of the American Educational Research Association, San Diego, CA.

Sangwin, C., Cazes, C., Lee, A., & Wong, K. L. (2010). Micro-level automatic assessment supported by digital technologies. In C. Hoyles & J.-B. Lagrange (Eds.), *Mathematics education and technology: Rethinking the terrain* (pp. 227–250). Dordrecht, The Netherlands: Springer.

Scheuermann, F., & Björnssen, J. (Eds.). (2009). *The transition to computer-based assessment.* Luxembourg: Office for Official Publications of the European Communities.

Shute, V., Hansen, E., & Almond, R. (2008). You can't fatten a hog by weighing it—Or can you? Evaluating an assessment for learning system called ACED. *International Journal of Artificial Intelligence in Education, 18*(4), 289–316.

Specific Mathematics Assessments that Reveal Thinking (SMART). (2008). *How to choose a quiz.* Retrieved from http://www.smartvic.com/smart/samples/select_preset.html

Stacey, K., Price, B., Steinle, V., Chick, H., & Gvozdenko, E. (2009). *SMART assessment for learning.* Retrieved from http://www.isdde.org/isdde/cairns/pdf/papers/isdde09_stacey.pdf

Stacey, K., Sonenberg, E., Nicholson, A., Boneh, T., & Steinle, V. (2003). A teacher model exploiting cognitive conflict driven by a Bayesian network. In P. Brusilovsky, A. Corbett, & F. de Rosis (Eds.), *Lecture notes in artificial intelligence. Proceedings of the 9th International Conference on User Modelling UM-03* (Vol. 2702, pp. 352–362). Berlin, Germany: Springer-Verlag.

Steen, L. (Ed.). (2001). *Mathematics and democracy: The case for quantitative literacy.* Washington, DC: The National Council on Education and the Disciplines.

Threlfall, J., Pool, P., Homer, M., & Swinnerton, B. (2007). Implicit aspects of paper and pencil mathematics assessment that come to light through the use of the computer. *Educational Studies in Mathematics, 66*(3), 335–348.

Trouche, L., & Drijvers, P. (2009). Handheld technology for mathematics education: Flashback into the future. *ZDM—The International Journal of Mathematics Education, 42*(7), 667–681.

VanLehn, K. (2006). The behavior of tutoring systems. *International Journal of Artificial Intelligence in Education, 16*(3), 227–265.

Victorian Curriculum and Assessment Authority (VCAA). (2002). *Mathematical Methods (CAS). Examination 1, Part I.* Melbourne, Australia: Author.

Victorian Curriculum and Assessment Authority (VCAA). (2010). *Mathematics. Victorian Certificate of Education Study Design.* Retrieved November 1, 2011, from http://www.vcaa.vic.edu.au/vce/studies/mathematics/mathsstd.pdf

WebAssign. (n.d.) *Online homework and grading.* Retrieved from https://www.webassign.net/index.html

Wiliam, D. (2005). Assessment for learning: Why no profile in US policy? In J. Gardner (Ed.), *Assessment and learning* (pp. 169–183). London, UK: Sage.

Wiliam, D. (2011). What is assessment for learning? *Studies in Educational Evaluation, 37*(1), 2–14.

Williamson, D. M., Mislevy, R. J., & Bejar, I. (Eds.). (2006). *Automated scoring of complex tasks in computer-based testing.* Mahwah, NJ: Lawrence Erlbaum Associates.

World Class Arena. (2010). *Example questions: 12 to 14 year-old questions.* Retrieved from http://www.worldclassarena.org/files/en/sample/12-14M_eng/ICM1300172.html

Chapter 24
Technology-Driven Developments and Policy Implications for Mathematics Education

L. Trouche, P. Drijvers, G. Gueudet, and A. I. Sacristán

Abstract The advent of technology has done more than merely increase the range of resources available for mathematics teaching and learning: it represents the emergence of a new culture—a virtual culture with new paradigms—which differs crucially from preceding cultural forms. In this chapter, the implications of this paradigm shift for policies concerning learning, curriculum design, and teacher education will be discussed. Also, the ubiquitous possibility of emergence of ever-new forms of technology brings about both new opportunities for learning and collaborative work (involving students and teachers), as well as potential dangers. Policy measures may give priority to technological access and developments, over the intellectual growth of learners and the professional development of teachers—which should be more demanding goals of mathematics education. Such policy issues will be discussed.

Introduction

The previous chapters in this section of the *Third Handbook* suggest that the emergence and dissemination of digital technology provides opportunities for mathematics education and affects teaching and learning practices in different ways.

L. Trouche (✉)
Institut Français de l'Education, Ecole Normale Supérieure de Lyon, Lyon, France
e-mail: luc.trouche@ens-lyon.fr

P. Drijvers
Freudenthal Institute, Utrecht University, Utrecht, The Netherlands

G. Gueudet
Brest University, Rennes, France

A. I. Sacristán
Center for Research & Advanced Studies (Cinvestav), Mexico, Mexico

The influence on different mathematical fields has been discussed: for example geometry (Chapter 19), algebra and calculus (Chapter 20), statistics (Chapter 21); as well as on different aspects of mathematics education, such as curriculum design (Chapter 17), modelling (Chapter 18), proving (Chapter 19), the use of interactive resources (Chapter 22), and assessment (Chapter 23).

The impact on mathematics education, however, is not just a matter of individual teachers and students finding their ways to use and benefit from the affordances offered by technological means; the integration of technology in mathematics education involves setting standards (International Society for Technology in Education, 2011) and is also a matter of institutional and national policies with regard to educational reform (UNESCO, 2008). Therefore, this final chapter of this section in the *Handbook* addresses technology-driven developments and policy implications for mathematics education.

Let us begin by clarifying how we understand the expressions used in the chapter's title. By *technology-driven developments* we refer to two levels of developments. At a first level, we consider the developments of digital technology that can be used in mathematics education. For example, interactive whiteboards have been integrated to many mathematics classrooms nowadays. Students have handheld technological devices at their disposal such as calculators, netbook or laptop computers, in the classroom as well as at home. Through the Internet, both students and teachers have access to online content and resources, to communication facilities and to student management systems which monitor student progress. These first-level developments foster second-level developments, namely individual students and teachers learning to work in new technological contexts. For example, students may change the way they work on tasks and in preparing for tests. Teachers may be tempted to develop new teaching and/or assessment practices. The availability of technology confronts both teachers and students with questions on the relation between paper-and-pencil work and work with technological tools, and on the approach to mathematics—as an experimental science or as a more structural, formal science.

These types of technology-driven developments have repercussions initially at local and individual scales. However, they also have an impact on more global, institutional and national policy levels. Therefore, *policy implications* need to be considered. For example, a school, a group of schools, or a regional school board may decide to abandon textbooks and to use—and eventually co-design—online resources that cover the curriculum. Also, national authorities may decide to allow specific types of technology in centralized assessments. As a third example, teachers may benefit from online collaboration with their colleagues, so as to share, and collectively develop, resources and practices.

Two dimensions seem to be of particular interest in describing policies related to the development of educational technology, namely the top-down/bottom-up dimension and the access/support dimension. The top-down/bottom-up dimension refers to the differences between policies that, on the one hand, may emerge from the needs expressed by students, teachers, parents and other persons involved in mathematics education, and on the other hand may be imposed on the mathematics education community as a result of political choices made by top-level administrations and, thus, at a distance from educational reality. For example, a top-down policy could be a national directive to impose access to graphing calculators during national examinations; whereas support for teachers who start to design their own online resources can be seen as a bottom-up policy.

The access/support dimension refers to the difference between, on the one hand, policies which focus primarily on providing teachers and students with access to technology, and leave the implementation up to the educational field itself; and, on the other hand, policies that focus on supporting teachers and students in the process of integrating technology. For example, providing schools with high-speed Internet connections is typically an access policy, whereas measures for professional development and guidelines for implementation may be more supportive. This access/support dimension is manifest in different statements on the integration of technology in mathematics education. For example, in the USA, the National Council of Teachers of Mathematics (NCTM), in a 2008 Position Statement, claimed that "all schools must ensure that all their students have access to technology" and that "programs in teacher education and professional development must continually update practitioners' knowledge of technology and its classroom applications" (NCTM, 2008).

The two policy dimensions are depicted in the left part of Figure 24.1. We believe that policies are more effective if they emerge from, and respond to, bottom-up developments rather than resulting from top-down initiatives, as will be illustrated in this chapter.

Merely providing access to technology is not enough for promoting educational change; support for teachers' professional development is a necessary precondition for a thoughtful and fruitful integration of technology. In line with this position, the right part of Figure 24.1 shows a potential trajectory towards effective policies, and

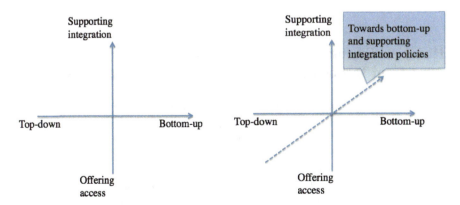

Figure 24.1. The two policy dimensions (*left*), with potential orientation towards bottom-up and supporting policies (*right*).

as such represents a policy shift. Whether these types of shifts can be observed in national developments will be discussed in this chapter. Policy shifts do not fall out of the blue, but reflect or intend to support underlying views on learning, and are mediated by new paradigms of teaching and learning. Therefore, we cannot address policy shifts without discussing, as well, shifting paradigms of learning.

The issue of educational policies and learning paradigms related to technological developments is addressed through the next four sections of this chapter, each offering a different view and illustrated through related national experiments or *windows* on experiences. Part 2 addresses some challenges of policy, curricula and assessment implementation. The shifting learning paradigm that underpins policy changes is addressed in Part 3, and illustrates how new spaces for learning can be opened. Part 4 then describes the role of digital resources in policy making, questioning the two articulated issues of design and quality. Then, since teachers are of crucial importance in mathematics education policies, Part 5 delves more deeply into teacher education, and highlights the new opportunities—such as through networking possibilities— that technology may offer for this. Finally, in the conclusion, we propose an extension to the two-dimensional top-down/bottom-up and access/support model.

Part 2: Policy, Curricula and Assessment Implementations: Evolution and Challenges

In this section we discuss how policies and curricula have tended to integrate technologies for mathematical education and their evolution linked to developments in technologies. We present some cases that illustrate the two-dimensional model discussed above; the policy tendencies in different regions; and how different policies (even within a same region) have different emphases. Finally we address the issue of technological assessment policies.

Historical Evolution of Technology Integration and the Shift Away from Technologies for New Educational Paradigms

The incredibly rapid development and dissemination of technology in society has led to a demand for policies for incorporating technologies into education—such as was proposed in UNESCO's (2005) World Report or in the *Bento Gonçalves Declaration for Action* (Carvalho, Kendall, & Cornu, 2009)—and of setting standards at national and international levels (International Society for Technology in Education, 2011; UNESCO, 2008). However, though there is a generalized political discourse that emphasizes the need to incorporate technology, there seems to be limited visions on *how* to carry this out (Fonseca, 2005). A comprehensive meta-review on research on the integration of technologies (ICT) into education in general (LeBaron & McDonough, 2009) pointed to a gap between educational practice and policies with background theory and research; and calls for research strategies that will support educators to make the best use of the resources that are emerging. It also calls for policies that will help teachers go beyond a technical focus and think of technologies as a means for improving teaching and learning. In fact, as we will illustrate here, it would seem that many policies focus on *digital power*, rather than contemplating a rethinking of educational paradigms in the light of what technologies can bring and change.

Historically, mathematics education was one of the first fields to glimpse the potential of digital technologies, and consider them for mathematics education curricula: For instance, in the 1980s, following the publication of *Mindstorms* (Papert, 1980), the Logo programming language was introduced into mainstream schools and programs, particularly for developing mathematical thinking, in many countries, including the USA and UK (Agalianos, Noss, & Whitty, 2001). Other technologies, such as calculators, spreadsheets and dynamic geometry, also were seen early on as having great potential (as evident in the first ICMI Study—Churchhouse et al., 1986).

Pimm and Johnston-Wilder (2004) provided an interesting historical account, from a UK perspective, of the evolution of the inclusion, policies and relationship in and with school mathematics of technology—from the first calculators and computer programming, to the recent interactive whiteboards. They narrated that, even before the advent of microcomputers, computer programming was part of the UK's mathematics syllabus because of the *special relationship of computers and mathematics*. In fact, technologies and computer programming (e.g., with Logo) were used as a means to develop mathematical thinking (e.g., through *construction and expression*) and to seek deep educational transformations (as inspired by the Logo philosophy). But as computer science evolved, school mathematics distanced itself from it, as explained by Ruthven (2008):

> The rise of Logo ... was facilitated by an educational climate receptive to progressive educational ideas ... the majority of classrooms took up Logo as part of an incremental view of educational change and were quick to absorb it into existing modes of work ... In terms of *disciplinary congruence*, during the period of Logo's rise the *algorithmic thinking* associated with computer programming was being proposed as a modern equivalent of Klein's *functional thinking* ... However, this position ... lost ground as a wider range of software became available with new types of user interface which pushed programming into the background ... In terms of *adoptive facility*, ... the lack of a viable platform suited to conventional classroom use was an important barrier ... Finally, in terms of *educational advantage*, the perceived value of Logo diminished as the place of more open and extended work in school mathematics was downplayed. (p. 99)

Thus, with the evolution of the nature of the technologies involved, and mathematics increasingly hidden in the software used (Pimm & Johnston-Wilder, 2004), there has been a shift in the past 15 years in how technology and its role is conceived in policy and curricula. In many cases, rather than harnessing the potential of technologies for creating new paradigms of thinking about mathematics and/or of school mathematical practices, technologies are often used to assist in existing traditional mathematical practices (used as tools for visualization, presentation, or for their computational power—see Julie et al., 2010).

Also with the increasing availability of hardware and the development of online resources, Web sites, and the possibilities of networking, there is a focus—at least at top or national levels—to *access,* seeking to provide schools and pupils with technologies (both in terms of equipment and resources). In the case of many developing countries, as discussed in the next section, access seems to be the priority, together with developing computer "literacy," which in some countries implies developing technical competencies for the use of pervading software (e.g., office suites). In fact, as some of the general research reviewed by LeBaron and McDonough (2009) pointed out, there has been a lack of sufficient technological resources in classrooms, as well as of professional development. We will now discuss some cases of national policies with regard to the incorporation of technologies in mathematics education.

Some National Curricula Recommendations and Policy Implementations

In developed countries, technology has been part of national mathematics education policies for several decades. For example, in the USA, as far back as 1980, the NCTM had as one its main recommendations that "mathematics programs must take full advantage of the power of calculators and computers at all grade levels," and that access to those tools should be provided in classrooms (NCTM, 1980); in 2000 it claimed, boldly: "Technology is essential in teaching and learning mathematics; it influences the mathematics that is taught and enhances students' learning" (NCTM, 2000, p. 24; see also Ferrini-Mundy & Breaux, 2008).

In France, mastering common information and communication technologies is considered one of the major seven competencies of the curriculum (Ministère de l'Éducation Nationale, de l'Enseignement Supérieur et de la Recherche, 2006). At the end of 2009, reforms were announced proposing to offer two weekly hours of computer science in the last-year of high school (*Terminale S*) to science and mathematics students (Ministère de l'Éducation Nationale, 2009).

Julie et al. (2010) described some developments of access to and implementation of technologies in mathematics education in various countries or regions—for example, government initiatives in Hong Kong and South Africa were described, as well as three types of integration in Latin America—the first two, bottom-up, and the third top-down: (a) due to the initiative of individual teachers and/or schools; (b) privately-funded projects (IBM, Microsoft, Intel, etc.); and (c) government-sponsored projects. The paper offered a vision of large-scale projects in several countries (such as those expanded below in Windows 1 and 2, for the case of

Mexico), highlighting the difficulty of such projects, and the problem of the digital divide. It concluded:

> The outstanding similarity is the acceptance at political and bureaucratic level of the use of digital technologies for mathematics teaching and learning in all the countries. However, the translation of policy into practice is a much more daunting task. ... Even under massive government implementation, there remain unequal access, unequal resources, and sporadic use of the digital technologies in schools. Political decisions and administrative issues also affect the implementations, the quality of the training of teachers as well as its continuity and that of the projects themselves. (Julie et al., 2010, p. 380)

More recently, many developing countries have ordered hundreds of thousands of *One Laptop per Child* (OLPC) computers, particularly Peru, Uruguay, Argentina and Rwanda (OLPC Foundation, 2011). Though some early reports (Australian Council for Educational Research, 2010) pointed to some positive results, careful evaluations of the effects of activities with these machines—on teacher training, and on mathematics teaching and learning in schools—still need to be carried out.

It is worthwhile taking up the case of Mexico in terms of its national top-down policies for the integration of technologies for mathematical teaching and learning. Between 1997 and 2007, the Mexican Ministry of Education (SEP) launched, in this respect, two very different initiatives with opposite pedagogical and implementation strategies (as explained below): The *Teaching Mathematics with Technology* (EMAT) program (Window 1) and *Enciclomedia* (Window 2). These examples offer insight into the dimensions discussed at the beginning of the chapter, with *Enciclomedia* having a top-down and access nature, whereas *EMAT* conceived as a bottom-up implementation, supporting integration. With government changes in 2007, federal support for both *EMAT* and *Enciclomedia* was discontinued, though *EMAT* continues at regional levels. In 2003, there were 731 schools officially participating in the *EMAT* program.

The availability of *Enciclomedia* resources is limited nowadays (and is no longer available from *Enciclomedia's* official Web site, http://www.enciclomedia.edu.mx/). However, some teachers still use them. The government has now conceived a program called *Habilidades Digitales para Todos* (Digital Abilities for All) with very different aims from those of past projects: this program aims to provide all

---

Window 1: A First Case of Mexico's National Implementations: The EMAT Project

EMAT, which began in 1997 (together with parallel sciences programs—*ECIT-ECAMM*) aimed to incorporate technologies in middle schools (for students from 12 to 15 years) in order to transform educational practices from the traditional teacher-to-student, top-down approach towards student-centred, exploratory, bottom-up practices. An international team of mathematics education researchers designed a constructivist, pedagogical model

(continued)

Window 1: (continued)

and activities. Universal open tools (that allowed different objectives) were preferred, such as spreadsheets, dynamic geometry (*Cabri-Géomètre*), the TI-92 algebraic calculator and, later, Logo. Emphasis was put on changes in the classroom structure, on collaborative exploratory work, and on a teaching model based on mediation and guidance (Ursini & Rojano, 2000). *EMAT* was designed to be implemented gradually—beginning with eight schools in 1997, and gradually expanding over the course of several years—so that adjustments and support could be provided, and the quality of teacher education and implementation in classrooms would be optimized.

Though the implementation in schools was not as straightforward as planned (for example, preservice and inservice education were limited in scope— see Trigueros & Sacristán, 2008)—*EMAT* was groundbreaking in the ways it opened doors to integrate technologies in schools. Its use was recommended in the official national mathematics curriculum, and has extended beyond the originally-conceived policies. Some teachers who have been working with *EMAT* over many years, have been able to integrate the use of diverse tools and develop their own long-term projects—like, for example, the series of long-term *Painless Trigonometry* projects (Jiménez-Molotla & Sacristán, 2010), which was developed by a couple of teachers on the basis of EMAT's triangle activities. In *Painless Trigonometry* projects, students participated in activities which helped develop their trigonometric concepts and ideas through complementary explorations and constructions with the *EMAT* tools and other software (Figure 24.2). This led, in one case, to the construction, by the students themselves, of 3D computer models of triangle-based figures (such as pyramids).

In some regions, local officials still coordinate and support teachers' communities of practice for *EMAT*, hold monthly workshops, develop new materials, and have developed anthologies of *EMAT* activities for different tools—see Figure 24.3 (Sacristán & Rojano, 2009).

Figure 24.2. Complementary trigonometrical explorations with Cabri, Excel and Logo.

(continued)

Window 1: (continued)

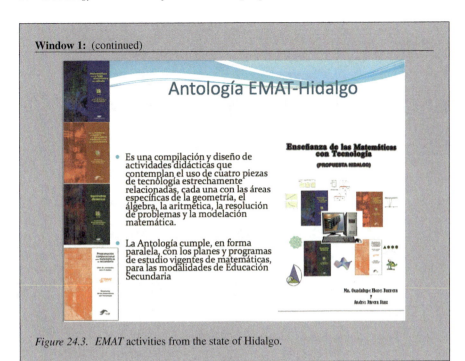

Figure 24.3. EMAT activities from the state of Hidalgo.

Window 2: A Different National Implementation in Mexico: Enciclomedia

Unlike *EMAT*, *Enciclomedia* was the result of an ambitious political decision, made in 2004, to implement digitalized versions of official Grade 5 and Grade 6 textbooks in all subjects in all primary schools in Mexico. It included accompanying digital resources and interactive whiteboards. For this project, a huge number of ad hoc interactive resources (applets) were produced in a very short time. However, the use of open universal tools, such as those used in EMAT, did not occur (Rojano, 2011). A view of this production of interactive resources for the mathematics curriculum (such as the one illustrated Figure 24.4) has been presented by Trigueros and Lozano (2007).

One of the most successful (and popular) mathematics resources from *Enciclomedia* was *La Balanza* ("The Scale," see Figure 24.4), for which users input numbers (e.g., fractions, decimals) and, using the scale metaphor, investigate notions such as equivalent fractions. Trigueros and Lozano (2007) found that this applet gave students and teachers freedom to explore mathematical situations through interesting mathematical activities and challenges.

Despite some successes, the haste with which *Enciclomedia* was implemented resulted in shortcomings. Rojano (2011) explained that there was an

(continued)

Window 2: (continued)

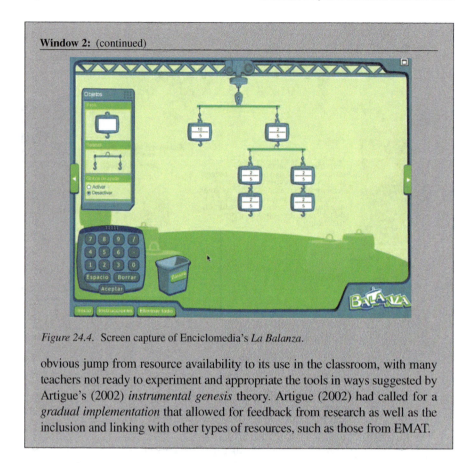

Figure 24.4. Screen capture of Enciclomedia's *La Balanza*.

obvious jump from resource availability to its use in the classroom, with many teachers not ready to experiment and appropriate the tools in ways suggested by Artigue's (2002) *instrumental genesis* theory. Artigue (2002) had called for a *gradual implementation* that allowed for feedback from research as well as the inclusion and linking with other types of resources, such as those from EMAT.

students with laptops (the access dimension) and promotes the view that all teachers should have competencies in basic software, specifically in MS Office (Bernáldez, 2011).

The case of Mexico draws attention to an issue that arose in many implementations—specifically, a lack of continuity in policies. Mexico is an example of a country where policies increasingly shifted towards *access*, and away from meaningful and supportive *integration* for mathematical learning. It also points to how social, adoptive, practical and other factors can affect policy implementation with respect to technology. Many other factors come into play (for examples, see Julie et al., 2010) and these can create gaps between political will and school and teacher implementation (Ruthven, 2007). Assude, Buteau, and Forgasz's (2010) classification into levels of factors influencing this contradiction included the social and political levels, the mathematical and epistemological level, the school and institutional level, and the classroom and didactical level. Difficulties arising from the need to develop technological competencies among teachers, and associated pedagogical difficulties, proved to be especially important (Trigueros & Sacristán, 2008).

A further consideration, related to policy and curricular changes, is that of the role of technologies for assessment. This is now discussed.

Assessment Policies

Assessment is an important and widely debated aspect of national policies, with respect to the use of technology in mathematics education. It is beyond any doubt that assessment drives teaching and affects educational reform. This particularly holds in countries where national, externally-set final examinations are used as a main form of assessment. Meanwhile, research findings on this topic are limited. In the frame of the ICMI Study 17, Sangwin, Cazes, Lee, and Wong (2010) focussed on computer use for automatic feedback during online assessment, but did not discuss policy aspects of the use of technology in assessment—issues related to the kinds of tasks that might be appropriate, and implications for pedagogy, were not considered in depth.

Leigh-Lancaster (2010), by studying how CAS technology has been incorporated into upper secondary mathematics curriculum and examinations since the year 2000 in Victoria (Australia), offered a broad perspective of the challenges and experiences of assessment that is congruent with technology integration in mathematics programs. One issue is that standard models of assessment seem to be incompatible with new educational paradigms that are promoted by the use of technologies (Stroup & Wilensky, 2000). The rationale for assessment related to these new paradigms perhaps needs further elaboration which takes into account the learner's development (Lesh, Hoover, Hole, Kelly, & Post, 2000). Some research (e.g., Hernandez-Sánchez, 2009) has delved into the issue of how to evaluate students' work and learning in classrooms in which contemporary technology tools are being used (Window 3).

As mentioned in the last chapter (Chapter 23) of this *Handbook*, concerning the role of technology in national mathematics examinations, Drijvers (2009) distinguished between four assessment policies:

1. Technology is (partially) not allowed;
2. Technology is allowed, but offers no advantage;

Window 3: A Search for Developing Assessment Methodology for Work with Technology

Hernandez-Sánchez (2009) identified three areas to assess: (a) development of abilities and mathematical content knowledge, (b) use of resources, and (c) collaboration and participation. In order to observe the work in progress

(continued)

Window 3: (continued)

with technology in a classroom, she developed a series of instruments for her own assessment and for students' self-assessment and co-assessment (with student in teams evaluating each other—see Figure 24.5).

Coevaluación:			
Criterio de coevaluación	siempre	a veces	casi nunca
	Itzel	Diana	José
Contribuyó con ideas en la realización del problema			
Participó físicamente en la solución del problema			
Trabajo en el desarrollo matemático del problema			
Habilidad técnica del software			

Figure 24.5. A co-assessment form to be used by a team of students.

3. Technology is recommended and useful, but its use is not rewarded; and
4. Technology is required and its use is rewarded.

With the fourth of these policies, conceptual skills, such as interpretation, reasoning, mathematization, justification and modelling, are examined. However, designing appropriate examination tasks for such goals is not trivial. Brown (2010) developed a similar scheme of analysis that identified four categories for technology in assessment: namely active required, active optional, active neutral and active excluded.

Drijvers (2009) investigated policies in some countries in Western Europe, and concluded that although many countries have Type 3 and Type 4 policies, they nevertheless concentrate on assessing paper-and-pencil skills, either through a non-technology part of the examination (consistent with a Type 1 policy) or through the use of specific vocabulary in the wording of items that indicates that paper-and-pencil methods are required. If technology is allowed during the assessment, a common limitation concerns communication facilities. An exception to this can be found in experimental examinations in Denmark, in which students have Internet access during the session. However, in France, after attempting to organize an "experimental test" for the *baccalauréat* in mathematics, the national authorities finally decided it was too difficult to organize both the assessment itself and the class preparation (Sur l'épreuve pratique, 2007).

Some Closing Remarks to Part 2

In this section we have presented part of the evolution of technology integration into mathematics education and related policies, which shows a shift-away from the early tendencies where technologies and computer programming were viewed as means to innovate education towards constructivist—and *constructionist* (Papert, 1991)—educational paradigms. Social and implementation difficulties, as well as the profusion of technological resources (to be discussed in Part 3 of this chapter) have brought about a change in these tendencies (Agalianos et al., 2001; Ruthven, 2008; Pimm & Johnston-Wilder, 2004).

We have also presented some examples of national policies. The contrast between the Mexican *EMAT* and *Enciclomedia* policies, not only illustrated some of the support/access, and bottom-up/top-down dimensions but also highlighted the contrast between focus on individual learning versus a collective approach, a dimension which we will discuss in the concluding section of this chapter.

Finally, concerning assessment, we claim that this is an important, but underestimated, aspect of policies on the integration of technology in mathematics education. As Kaye Stacey and Dylan Wiliam have pointed out in Chapter 23, issues relating to technology and assessment deserve more attention from the research community.

Part 3: Mathematics Learning and Teaching Spaces

The impact of national policies and strategies finally come down to teachers, either individually or collaboratively, getting involved in the design of digital resources, and facing the challenge of how to turn the available resources into effective education. Such design and integration processes, however, are not neutral, in the sense that they reflect views on learning and teaching. These views may be affected by the new opportunities technology offers. In the present section, therefore, we elaborate on this by considering relationships between the integration of technology in mathematics education, and the paradigms of its learning and teaching.

Let us first focus on learning. Technology offers opportunities to enlarge students' learning spaces. As such, it potentially extends the scope of learning, the repertory of forms of learning, and offers opportunities for new paradigms for learning. But what do we mean when we speak about "enlarging learning spaces" for mathematics? We now address some aspects of this multi-faceted concept.

Mathematical Learning Spaces

What are potential dimensions of an enlarged technology-supported learning space? A first obvious, but non-trivial, dimension that technology may bring about, concerns the learning space, in the literal sense of distance and time: technology offers new means for *ubiquitous learning*, in which students can access resources at every moment, in every place, and in a variety of synchronous as well as asynchronous modes. As an anecdotal example, it is not uncommon, these days, to see students sitting in the bus to the university campus watching video recordings of last week's class on their smart phones. Learning becomes independent from time and location, becomes *mobile*, and this is indeed an extension of the learning space. Thanks to technology, and to online resources in particular, distant learning has become quite common. The learner decides on what, where and when to learn.

A second, related aspect of the enlarged learning space concerns the opportunities for organized forms of *out-of-the-classroom* or *out-of-school learning*. Students equipped with handheld devices can go outside classrooms to gather real-life data that inform their biology or chemistry lessons. More specifically for mathematics, students can use GPS technology for a mobile geometry game in the school-yard (Window 4).

A third and more subtle aspect of the extended learning space brought about by technology, concerns what we would like to call the student's *mental learning space*. The use of technology may, on the one hand, invite mental activity, and on the other, free students from basic mental activities that may distract them from

---

Window 4: MobileMath Game with Handheld GPS Technology

In this example, taken from Wijers, Jonker, and Drijvers (2010), teams of Grades 7 and 8 students used handheld GPS devices to play an outdoor game in which they had to construct parallelograms and try to destroy other groups' geometrical shapes. The aims were to make students experience properties of geometrical figures in a lively, embodied game context.

Student actions while playing the game include looking at the map to imagine where they want to make a shape, walking to the location for the first vertex to enter this location in the mobile device, which generates a dot on the map, walking again to the location of the second vertex of their imagined shape which provides a line on the screen connecting the first vertex with the current (moving) location, etc.

The map in Figure 24.6 illustrates some student constructions. The results of the pilot experiments suggest high student engagement and motivation. Students learned how to use the GPS, to read a map, and to construct quadrilaterals. The study suggested mathematical learning opportunities that need further investigation.

(continued)

Window 4: (continued)

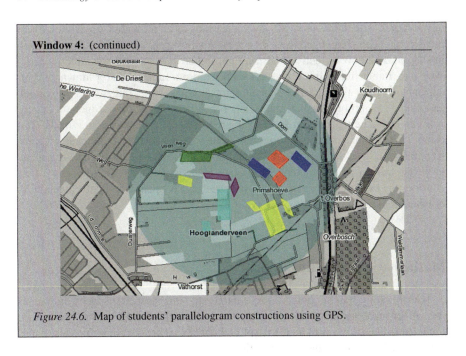

Figure 24.6. Map of students' parallelogram constructions using GPS.

higher goals. Depending on the task, technology may provide space for exploration, for discoveries in microworlds, for dynamical investigation of variance and invariance, for design of—and links between—representations; in short, for knowledge construction. Through technology, students can have early access to advanced mathematical ideas in a non-structured or nonlinear way (see Sacristán et al., 2010), as expressed by the *webbing* idea proposed by Noss and Hoyles (1996). A point of concern here, however, is that these challenging potentials are not easy to exploit in every-day mathematics teaching. For example, the seemingly trivial techniques for using technological tools are often interrelated to conceptual aspects (Lagrange, 2000).

A fourth, interesting aspect of how technology can enlarge the learning space, concerns the opportunities technology offers for *collaborative learning*. Thanks to online connectivity and social media, communication, exchange, and collaborative work are not limited to face-to-face meetings but can take place at a distance. This affects the paradigm of learning as an individual activity and widens the horizon to more intensive online collaborations (Hoyles et al., 2010).

A fifth and final aspect is that technology also enlarges the *learning space for teachers*, who are confronted with challenging questions on how to exploit the opportunities technology offers, how to organize the learning, and how to learn to organize the learning. This aspect is addressed in more detail later in the chapter.

To summarize all of the above, a new paradigm for learning has emerged, one which is influenced by the seemingly unlimited learning spaces generated by new technologies. The more classical view on learning as an individual, in-school, linear process has been challenged. Learning is now being seen: as ubiquitous, rather than in-school; as involving active construction, rather than passive reproduction; as a Web-like, rather than a linear process; as bottom-up, rather than top-down; as self-dependent, rather than teacher-dependent; as collaborative, rather than individual; and, finally, as aiming at conceptual, rather than procedural knowledge.

Even if this new paradigm for learning may sound very appropriate for the 21st century, as well as appealing in the light of new demands for workers and citizens, its realization in classroom practice—within its institutional constraints—turns out to be far from a trivial matter (Ruthven & Hennessy, 2002). Therefore, we now consider the exploitation of the teaching space as it is opened up by the availability of educational technology.

Mathematical Teaching Spaces

If technology has the potential to enlarge students' learning spaces, how does this affect teaching practice? How can teachers manage the learning spaces and *orchestrate* classroom situations to exploit them? What are the consequences of new paradigms for learning and for educational formats, classroom organization, pedagogical approaches and teaching strategies?

As a means to address these questions, Trouche and colleagues developed the notion of *instrumental orchestration* (Drijvers & Trouche, 2008; Trouche, 2004). An instrumental orchestration is a teacher's intentional and systematic organization and use of the various artefacts available in a—in this case computerized—learning environment for a given mathematical task; it includes setting up the scene, exploiting it and taking ad hoc decisions. Other models are available. For example, Ruthven and Hennessy (2002) designed a *practitioner model* for the use of technology in mathematics teaching. Pierce and Stacey (2010) offered a *pedagogical map*, which may guide teachers in their articulation of tools, task and teaching techniques. Finally, the notion of *Technological Pedagogical and Content Knowledge* (TPACK, Koehler, Mishra, & Yahya, 2007) identifies different types of knowledge that teachers need, as well as their interactions, and as such may help teachers to position their knowledge and identify possible weaknesses. Whether these models really can help teachers in their professional development on the issue of teaching with technology, is still to be investigated.

Earlier, we claimed that technology offers opportunities for ubiquitous and out-of-school learning, for widening students' learning spaces and for collaborative learning. How can these opportunities be dealt with in teaching? The idea of *ubiquitous and out-of-school learning* challenges the traditional teaching formats, as it is difficult for the teacher to know what students do and learn. Learning trajectories may take different directions at different speeds. However, technology also offers solutions to this through the availability of student monitoring systems, which allow teachers to access online students' computers or devices. This allows for the preparation of face-to-face

24 Technology-Driven Developments and Policy Implications

teaching that takes into account the students' proceedings and benefits from the different approaches they could have developed during their out-of-class work. Window 5 sketches such an approach, in what was called a *Spot-and-Show* orchestration (Drijvers, Doorman, Boon, Reed, & Gravemeijer, 2010). It illustrates the way in which the availability of technology can enlarge the mathematical teaching space, by offering the opportunity to access students' work and monitor students' progress through digital means, and fine-tune the face-to-face teaching to that.

Window 5: The "Spot-and-Show" Orchestration

In this example, taken from Drijvers et al. (2010), we imagine a teaching situation in which ICT allows a teacher to access digital student work while preparing his lesson. As he does that, he notices something special in the work of one of the students—such as a remarkable mistake, a misconception, or a surprisingly original solution. The teacher decides to exploit this during the lesson and shows the student's work to the whole class by means of a projection. Next, he may ask the student to explain his approach or reasoning. Peers can comment and the teacher can explain why he considered that this particular solution was worthy of special attention.

As an example of Spot-and-show, Grade 6 students had compared dot graphs of the square and the square root function (Figure 24.7). One pair of students typed in the digital environment: "And the square of a number is

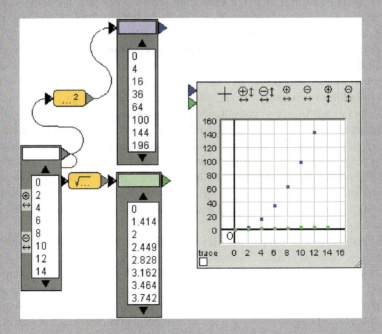

Figure 24.7. Comparing the square and the square root.

(continued)

> **Window 5:** (continued)
>
> always right above the root." The teacher wanted to draw attention to the fact that the value of the dependent variable is always positioned vertically above the value of the independent variable, and that this has nothing to do with the type of function involved. Therefore, she projected this answer to the classroom. After a whole-class discussion, one of the students said: "That's because the line underneath, that's got a number on it, which you take the square root of and square, so it's on the same line anyway."

Concerning the widening of *students' mental learning space*, the question of how best to exploit this is not an easy one to answer. Of course, students' mental activity is not stimulated by the availability of technology in itself, but largely depends on the task, the affordances and constraints of the tool, and orchestration of all this by the teacher. As a teacher, one needs to be aware of the subtle interaction between techniques for using the tool and mental activity, as it is reflected in the notion of instrumental genesis (Artigue, 2002). To enhance this, new organizational forms of teaching might be designed. Some studies suggest that teachers are less drawn to whole-class teaching in technology-rich education than they are in regular lessons (Drijvers, 2012). We strongly believe, however, that interactive forms of whole-class teaching are crucial for exploiting, making explicit and reflecting on students' individual hands-on experiences. For enhancing such whole-class interactive teaching formats, classroom connectivity tools are available, such as the *TI-Navigator*, voting boxes or different types of digital pen technology (Hoyles et al., 2010).

Technology opens new horizons for addressing *collaborative learning* in teaching. Collaborative work can be part of assessment and students could be encouraged to use online chat while working on their mathematical tasks at home or to have other types of online peer interaction. The teacher himself may be engaged in these types of collaboration. An online consultation hour for students might increase student–teacher interaction. As will be explained later in this chapter, collaborative learning also applies to teachers' collaborative work and their professional development. Technology may support teacher education through the sharing of experiences and the collaborative design and use of online resources. In this sense, technology also enlarges the teachers' own learning space. Results from the nationwide evaluation of *EMAT*, discussed earlier, showed that teachers' learning is enhanced (Trigueros & Sacristán, 2008).

To summarize this section, we claim that, on the one hand, the availability of technology enlarges students' learning spaces in several aspects and leads to new paradigms of learning. On the other hand, ways by which teachers can fully exploit the potential of these resources are not yet evident. Nevertheless, the design and diffusion of teaching resources is a major issue within educational policies. This relationship between the dissemination of resources and educational policy is the main theme of the next section.

Part 4: A Profusion of Resources, Opportunities and Questions

We now turn to a central issue for educational policy: as in the case of the examples in Mexico, provision of resources has often been seen as a way to influence what happens in the classroom (see, e.g., Ball & Cohen, 1996; Pepin, 2009). Although traditional textbooks remain central, digital textbooks are becoming much more prevalent, and there is a profusion of other available digital resources: Web sites, interactive applications, online videos, forum discussions, etc. The devisers of these resources and participants in these online exchanges may be professional designers, teachers, educators and educational researchers.

This situation raises new policy questions, such as the following:

- What are the key design modes of these new resources? Who designs and what do the design processes look like?
- How to assess the quality of the resources? Which criteria are set for linking quality and design mode, and by which assessing authority?

In the course of discussing these questions, we draw, in particular, on two examples of innovative projects in Europe: *Sesamath* and *Intergeo*.

Towards New Design Modes

From a technical point of view, designing and broadcasting online resources is within the scope of most teachers. The networking possibilities foster the development of online communities, designing resources. For example, the Geogebra community [http://www.geogebra.org/] (Lavicza, Hohenwarter, Jones, Lu, & Dawes, 2010) gathers teachers and researchers all over the world, designing resources, organizing training sessions, and conferences around this educational software. In France, an example of such an online community is the *Sesamath* association (see Window 6), whose Web site records more than 1.3 million visitors each month.

Window 6: From Drill-and-Practice to Virtual Environment: Sesamath
Sesamath [http://www.sesamath.net/], a French online association of mathematics teachers (most of them teaching in Grades 6–9), started in 2001. Its spirit is summarized on its Web site as "Mathematics for all." It offers several kinds of free resources: online exercises, dynamic geometry software, online textbooks, etc.

(continued)

Window 6: (continued)

Sesamath started with a gathering of some 20 mathematics teachers, who shared their personal Web sites and then designed together a drill-and-practice piece of software called *Mathenpoche* (Gueudet & Trouche, 2012a). *Mathenpoche* was immediately very successful, in the sense that it was used by many teachers and students. In some regions, the local educational or political authorities supported its development by offering dedicated servers.

Several changes took place between 2005 and 2006. The association started to collaborate with researchers (Kuntz, Clerc, & Hache, 2009) and the designed resources integrated results of these collaborations. For example, a virtual abacus [http://cii.sesamath.net/lille/exos_boulier/boulier.swf] was developed for primary school, and new exercises, with several solutions, were added in *Mathenpoche*. At the same time, *Sesamath* decided to develop textbooks and, through the use of an online platform, involved others teachers—outside of the association—as authors. The resulting textbooks, freely available online, were also published on paper, and sold for half of the price of regular textbooks. Some commercial publishers attempted legal action. Due to the importance acquired by *Sesamath* resources, some educational authorities started to question their quality.

The development of the association's activities continued with a Web site, *Sesaprof*, allowing users to contribute to the design of resources (Sabra, 2009). The main current *Sesamath* product is *LaboMEP* (see Figure 24.8), a virtual environment where teachers can choose various kinds of activities: online exercises, dynamic figures, extracts of textbooks. They can, among a range of possibilities, combine some of them, or assign them to specific pupils.

Explaining the reasons for the success of *Sesamath* requires specific research. The existence in France of the IREMs (Institutes for Research on Mathematics Education), a national network that involves many mathematics teachers, has played an important role. A similar project could perhaps not succeed in countries were such a network, linked with mathematics education, did not exist.

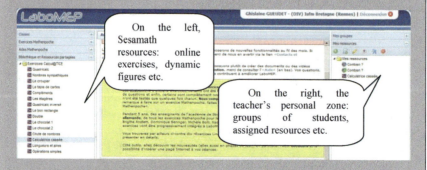

Figure 24.8. LaboMEP, a virtual environment for the teacher.

In France, no "official" online resources exist. Though files can be downloaded from several institutional Web sites (such as those of the Ministry of Education or regional academic authorities), they only concern specific topics. This is different from the *Enciclomedia* project in Mexico (Window 2), directed by the government, and providing ad hoc resources to support the official textbook; or the *Enlaces* project [http://www.enlaces.cl] in Chile that has similar features to *Enciclomedia*. Although the *Sesamath* example shows how new, bottom-up modes of design and collaboration can emerge, the examples in Latin America show that traditional centralized modes of expert production for system-wide dissemination also exist.

The availability of free resources is of economic importance, as it raises the issue of competition with commercial resources (in countries where commercial teaching resources are allowed). In some countries, governmental institutions themselves design resources, or offer opportunities for teachers to engage in the creation of resources, competing with the commercial productions (for example Wikiwijs in the Netherlands [see http://www.wikiwijs.nl/sector]).

Design issues should not be seen merely as a simple bottom-up versus top-down, or private versus public confrontation; they are more complex, involving a variety of agents. Communities of designers and users of resources include members with different positions: including regular teachers, expert teachers (with the status of teacher trainers, in some countries), and researchers.

The collaborative design of online resources is important for educational research. That is not only because research is needed to enlighten the new design modes, but also because many researchers are actively involved in the design process. This involvement is rooted in a long tradition, both in the field of research on technologies and in the field of task design (Watson & De Geest, 2005). Digital networks offer new possibilities for large projects associating teachers and researchers. Below we discuss the case of the *Intergeo* project (Window 7). Another important example of such collaboration is the UK's National Centre for Excellence in the Teaching of Mathematics (NCETM—see Chapter 16). Joint work for the design of online resources can enhance relations between researchers and teachers.

Window 7: Quality of Dynamic Geometry Resources: The Intergeo Project

Intergeo [see http://i2geo.net/] (Kortenkamp et al., 2009) is a European project that began in 2007. It has three aims: (a) inter-operability of the main existing DGS (Dynamic Geometry Systems); (b) sharing pedagogical resources; (c) quality assessment of resources (Trgalová, Jahn, & Soury-Lavergne, 2009, p. 1162).

Any user logged on the *Intergeo* platform can propose a resource, which will be immediately published online (more than 3,500 resources were published

---

(continued)

Window 7: (continued)

in January 2011). This feature makes the resource quality assessment essential. This quality assessment in *Intergeo* draws on the users' opinion, considering that the quality of a resource can only be defined in relation with a given teaching context.

The main assessment tool is a questionnaire (Figure 24.9) proposed on the user's platform (Trgalovà et al., 2009). This questionnaire takes into account nine different dimensions: metadata, technical aspect, mathematical content, instrumental content, added-value of dynamic geometry, didactical implementation, pedagogical implementation, integration in a teaching sequence, ergonomic aspects.

A user can choose to answer only a simple version of the questionnaire (giving an opinion on each dimension) or to give more details. For each dimension there are several precise statements. For example: "The activities are appropriate, given curricular and institutional constraints" (mathematical content); "The DG provides an experimental field for the learner's activity" (added-value of DG); and, "The resource describes possible students' strategies and answers" (didactical implementation). The answers are automatically collected and treated, and this treatment leads to a label (a number of *stars*) associated to the resource on the Web site.

The authors can freely modify their resources. If a participant, who is not the original author, wants to modify a resource, he/she has to copy it. The system allows following and connecting all the versions. Modifications can help improve the resource's quality; moreover, the questionnaire itself also contributes to this improvement, by raising the awareness of designers (who completed the questionnaire as users) on important dimensions of the resources.

In June 2011, *Intergeo* gathered 1,200 registered members. It contains around 3,500 resources; and altogether 700 evaluations have been proposed. This amount might seem to be limited; but the evaluation process only started in 2009.

Radio buttons: more on the left side to say that I don't agree, more on the right side to say that I agree	
○ ○ ○ ○	I found easily the resource, the audience, competencies and themes are adequate
○ ○ ○ ○	The files are technically sound and easy to open
○ ○ ○ ○	The content is mathematically sound and usable in the classroom
○ ○ ○ ○	Translation of the mathematical activity into interactive geometry is coherent
○ ○ ○ ○	In this resource, Interactive Geometry adds value to the learning experience
○ ○ ○ ○	This activity helps me teach mathematics
○ ○ ○ ○	I know how to set my class for this activity
○ ○ ○ ○	I found easily a way to use this activity in my curriculum progression
○ ○ ○ ○	The resource is user friendly and adaptable

Figure 24.9. Intergeo questionnaire on the platform, short version.

The evolution of design modes also has an impact on the articulation of design/ use as well as on the very notion of authorship. Users send their comments and suggestions; and designers modify the resources according to these contributions. A given initial resource can lead to many different versions, and identifying the contributors of one of these versions is often impossible. Moreover, teachers naturally adapt resources to their own use. This process is not new: teachers have always selected parts of textbooks, extracts from students' productions, etc. Nevertheless, the technical possibilities foster this process: teachers download files, and can easily copy and paste parts of these to produce their own files. This *documentation work* (Gueudet & Trouche, 2009) views teachers as designers of their own resources; and generally points to a need to reconsider borders between design and use.

These evolutions introduce a paradigm shift for the design of resources: the resources are never complete, but always involved in design processes. Directing this permanent move towards an increased *quality* is an essential policy issue that we discuss in the next section.

Assessing and Improving Resources Quality

Choosing a resource, for a given teaching or learning objective, is a difficult task. It is, firstly, linked to the issue of *indexation*, investigated by many computer scientists and also educational researchers (Lee, Tsai, & Wang, 2008). But the choice problem is not restricted to indexation; the metadata cannot certify the resource's *quality* that is considered both in terms of *intrinsic* quality and for its *adequacy* with respect to a user's expectations.

Defining the intrinsic quality of an online resource, for the teaching of mathematics, is not straightforward. Which criteria can guarantee this quality? Naturally, such criteria have to take into account three dimensions: *mathematical*, *didactical*, and *ergonomic* (ease of use). But even these dimensions do not fully take into account the *appropriation* by a user. Quality also encompasses the *potential* of a resource: potential for uses in class, for further design, and even for teacher professional development (see later in this chapter). In fact this question cannot have a general, unique answer. With the *Intergeo* project (Window 7) quality criteria were defined, with a focus on the added-value of dynamic geometry, particularly in terms of investigation possibilities for the students. Other criteria could be used for other foci.

Beyond the choice of criteria, the issue of *who assesses the quality* can also be delicate. In some countries educational authorities have developed certifications (in France, a national label, attributed by the Ministry of Education, indicates a resource of "Recognized Pedagogical Interest"). Different kinds of agents can intervene in the assessment process: like, for example, stakeholders such as teachers (expert or not) and researchers. In some cases, the Ministry of Education calls for researchers to intervene as experts in quality assessment tasks (as in the *Pairform@nce* program—see Window 10).

Answering the "who assesses the quality?" question drives us back to the bottom-up versus top-down confrontation and to all the intermediate possibilities.

In the *Intergeo* project, the quality assessment is grounded on the users' opinions (as these opinions are expressed by a carefully designed questionnaire). Quality and design issues are intertwined. The involvement of users in the design of a resource and the organization of *design loops* (design-use-feedback-new design) are presented by several authors (see, e.g., Hegedus & Lesh, 2008) as likely to contribute to quality, in particular by fostering the resource's appropriation potential.

Resources, Policies and Practices

We developed, in the previous sections, two important—and articulated—aspects of educational policies, concerning digital resources: their design and their quality (assessment, and improvement of quality).

These aspects can help to situate a given policy in our 2D system of axes. Indeed the design of resources can be more top-down, linked with official resources, designed by experts; or bottom-up, with a support for communities of teachers designing resources. Web sites (whoever the designers are) can propose ready-made resources, expecting the users' alignment, or can take into account the complexity of the appropriation processes, offering possibilities of adaptation. The quality assessment can be in the hands of experts; it can also be entrusted to the resources users (as in *Intergeo*).

A new important dimension appears here, concerning the production paradigm: the design of resources seems to be an increasingly collective process. We could thus complement the initial two axes displayed in Figure 24.1 with a third one, representing an individual/collective evolution, and could figure the paradigm's shift, concerning the production of resources for teaching, as a move in this 3D system of axes.

This third axis, individual/collective, is also very important for characterizing the teacher education aspects of a policy, an issue that will now be discussed.

As a final remark on designing and integrating resources, we notice that, whereas at the present time, students can be considered *digital natives*, most teachers are learning to speak *technological language* as their second, third, fourth, ..., language. This brings us to the issue of teacher education and pre- and inservice professional development.

Part 5: Teacher Education Strategies, Policies and Practices

Technology opens the horizon for new forms of orchestrations, but "the process of orchestrating technology-integrated mathematics learning is neither a spontaneous nor a rapid one" (Healy & Lagrange, 2010, p. 288). This certainly requires new resources and new competencies for teachers. To what extent do the resources for such a development exist? To what extent do new teacher education programs help teachers build such competencies? In this final section of the chapter we shall examine these questions, drawing special attention to two examples of innovative programs.

Teacher Education: Back to the Future

In the *Second International Handbook of Mathematics Education*, Mousley, Lambdin and Koc (2003) anticipated some major features of the present situation:

> There are many ways of using technology in teacher education. Generally, these meet three different purposes: ... the creation and use of videotape, videodisc and multimedia resources ...; varied facilities such as the Internet and communication software packages, ...; the use of computers, calculators and other electronic resources for doing mathematics. ... It is now not difficult to foresee a time when today's tools for meeting all three of the purposes outlined above will be able to be attended to in one apparently Internet-based seamless, interactive technological environment. (p. 396)

The time, mentioned by these authors, has apparently come (Window 8), providing resources freely, guaranteed ... or not.

Window 8: Video Resources for Helping Teachers to Integrate Technology

Figure 24.10, below, shows iTunes U, a guaranteed repository of videos linked to the results of research (videos from Universities, well-known institutions, etc.). Figure 24.11, on the other hand, shows a video obtained from the Google "jungle," via a search using as keywords "teacher education for mathematics with technology." One resource is *supporting integration* (cf., the introduction to this chapter), and the other is *offering* (magic) *access*...

Figure 24.10. Screen capture of iTunes U.

Figure 24.11. Video capture from a source obtained via Google.

More generally, looking at the mathematics teacher education landscape, we can now observe a wide range of resources, situations and devices: individual versus collective, associative (*Sesamath*, see Window 6) versus institutional (*Enciclomedia*, see Window 2), with various content–strategy privileging. Grugeon, Lagrange, and Jarvis (2010, p. 344) pointed out different strategies focussing on: mathematical knowledge, teaching skills, technology potentialities, virtual communication or dialectic old/new tasks. Throughout this diversity, some new trends appear:

- After a time of *institutional injunctions* ("teachers *have to* integrate technologies, to change their way of teaching"), there emerges a *consciousness of the complexity* of the technology integration into mathematics teaching. The perpetual and rapid technological and social changes impose the idea of *lifelong learning* by which teacher education becomes an *ongoing process*. These evolutions push a metamorphosis of *teacher training* to *teacher supporting* along deep evolutions of mathematics teacher work.
- The question is no more to privilege content, or pedagogy, or technology, but to articulate these three components: "Good teaching with technology requires understanding the mutually reinforcing relationships between all three elements taken together to develop appropriate, context specific strategies and representations" (Koehler et al., 2007, p. 741)
- The *Second Handbook* underlined a dominant point of view on teacher education as *introducing*, in a relevant way, *resources to* teachers:

    How technological resources are introduced to teachers and used in teacher education is just as important as what they are designed to do and how well they are constructed … Most authors stress the need to use the resources in the same way as one would expect teachers to use them with children (Mousley et al., 2003, p. 401).

- The idea of *supporting* teacher work implies, not only *providing* resources, but *helping them to design* their own resources. This is in line with the tendency towards *supportive policies* discussed in the introduction to this chapter.
- Helping teachers, as "instructional designers" (Visnovska, Cobb, & Dean, 2012), to design their own resources, is in line with the tendency towards bottom-up approaches presented in our introduction. It leads to conceive new devices for continuous exchanges (via Web sites or platforms) and to take into account different agents of resource design: existing resources available, particularly via the Web; student and classroom interactions, as well as teachers' interactions.

It is this new landscape that we want to illustrate now, through two contexts, one about preservice teacher education; the other, concerning inservice teacher education. Even if the border between both, in the context of lifelong learning, is vanishing, there remain some specificities: entering, and moving within, a profession, are not the same "thing."

Preservice Teacher Education: Towards New Modes of Articulating Classroom Practice and Training

In this section we want to draw attention to the role of technology for supporting teachers at the beginning of their career. This theme raises important questions that need to be faced at this time when, for economic reasons, in some countries (in France, since 2010) persons intending to be teachers are "dropped" into classrooms at the end of their academic studies, before completing their education in the field. In these conditions, new forms of training emerge, often driven by researchers, where video can have a major place, in forms of training that aim collectively to work on *cases* and to develop a reflective stance (see Window 9).

There is therefore a move "from videotape to interactive multimedia," as anticipated by Mousley et al. (2003, p. 398). The use of video is combined with the potentialities of an interactive platform, and carefully orchestrated by teacher educators. As Santagata, Zannoni, and Stigler (2007) emphasized: "The responses pre-service teachers gave to the analysis task prior to the course confirm the need for a framework to guide their observations" (p. 138). The use of video can be found in both preservice and inservice teacher education. In this case it seems to be efficient for supporting discussions, through excerpts of video, on each other's practice—see, for example, the experience of *video clubs* related by Van Es and Sherin (2010).

Window 9: Teacher Education Through Online Discussions

Llinares and Valls (2010) relate an experiment of integrating video-clips from videotaped mathematics lessons, and asynchronous, computer-mediated discussion groups (online discussions and workshops) for prospective primary teachers.

By using resources of an interactive environment (Figure 24.12), video cases and excerpts of interviews with the teacher who was "in the video," these teachers—prospective or already practising—have to: (a) notice aspects of teaching that might influence the development of primary pupils' mathematical competence; and (b) design a mathematical task to foster mathematical understanding by taking into account primary pupils' thinking. The task is realized through online discussions and online workshops, with the help of a tutor, providing the young teachers with questions and theoretical information on demand.

The authors underline the efficiency of this program, enabling the prospective teachers to reflect on, and integrate, multiple aspects of teaching. For them, this success results from the structure of the learning environment, articulating video-clips of actual mathematics lessons, providing structured guidance (task and discussion questions), participating in online debates, collaborating for designing a task; and providing theoretical background.

(continued)

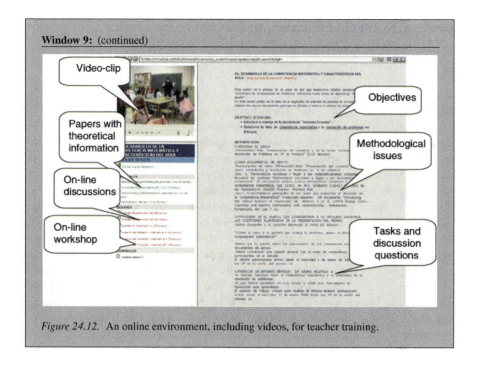

Figure 24.12. An online environment, including videos, for teacher training.

Inservice Teacher Education: Teachers as *Actors* of Their Own Development

After examining the use of digital resources for teacher education, in this section we study *teacher education for technology integration*. To illustrate this, we choose the French Pairform@nce program (Window 10), because it relies on two principles, characteristic of what we consider as new trends in teacher education:

1. *Collaboration* among teachers: Professional development, especially related to technology; results from collective activity and experience with peers, that is in line with the importance of teams; communities and networks as participants in mathematics teacher education (Krainer & Woods, 2008).
2. *Resources design and implementation in class*: A development program for teachers necessarily implies experimentation of resources in the field and, afterwards, a shared reflection that is in line with the strategy. As emphasized by Fugelstadt, Healy, Kynigos, and Monaghan (2010), "centre activities around the process of elaborating and experimenting with new instruments aimed to support new mediations of mathematics and/or teaching practices" (p. 308).

A program such as Pairform@nce would be in line with the evolutions in the field of teacher education, by considering teachers as *actors* of their own development.

Window 10: Pairform@nce, Promoting Teachers Collaborative Work on Resources

Pairform@nce is a French national inservice teacher education program featuring paths available on an online platform [http://national.pairformance.education.fr/] (Gueudet & Trouche, 2012b). Each path is structured in seven stages, combining face-to-face sessions and distance work: (1) Introduction to the training session; (2) Selection of teaching contents and organization of teams; (3) Collaboration and self-development; (4) Collaborative design of a lesson; (5) Trial of the lesson in each teacher's class; (6) Shared reflection about feedbacks of class experience; (7) Evaluation of the session. This organization seems to be close to what Fugelstadt et al. (2010, p. 297) describe as an "*inquiry cycle* … seen as consisting of the main steps: plan, act, observe, reflect and feedback."

Each stage comes with specific resources, suggestions for teacher activities, and collaboration tools. On the program's platform (see Figure 24.13), the seven stages are accessible on the left side; and some collaborative tools, like chat or forum, are accessible on the right side. Depending on the designer's choices, the tools may be specific to each stage of the path. The middle of the page displays path contents, and guidelines for the work of the participants.

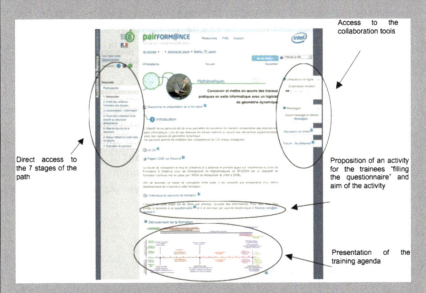

Figure 24.13. Presentation of the first stage of a training path on the Pairform@nce platform.

Analyzing the development and effects of a teacher education program constituted a "burning question," according to Mousley et al. (2003), who stated that:

> Most reporting of uses of technology in mathematics teacher education—as in teacher education more generally and school and adult education—is descriptive; such reporting, however, generally concentrates on how specific tools were used, rather than on how learning took place and the broader question of how teachers learn. (p. 425)

Ten years after, it seems that new projects are looking more carefully at the effects of what is actually done in the programs (see, e.g., Sacristán, Parada, Sandoval, & Gil, 2009; Soury-Lavergne, Trouche, Loisy, & Gueudet, 2011). Analyses of recent programs indicate that the following approaches can assist in providing valid feedback mechanisms.

- The importance of the *collective* teamwork of teacher education students for fostering their involvement in the process of designing and implementing resources in their own classrooms.
- The importance of the work on resources for supporting evolution of practices, confirming the importance of what Koehler et al. (2007) called *design talk*—that is to say, "the kinds of conversations that occur in design teams as they struggle with authentic problems of technology integration in pedagogy" (p. 741).
- The complexity of designing a development pathway that needs to be strong enough to support teachers' work, yet open enough to allow for teacher creativity.
- The necessity of conceiving a teacher education program as a "lived" entity that needs to be permanently renewed by the actors involved (both the teacher educators and the preservice and inservice teacher education students).
- The necessity of accompanying such lived entities by hybrid teams which associate researchers, designers and teacher educators and teachers with the program at stake.
- The importance of tracking the work of teacher educators and teacher education students for long enough to be able to catch real changes (a) during a program, (b) immediately after the program, and (c) one or more years later.
- Another way of monitoring the effects of a program is for outsiders to keep in touch with the continuing work of participants by means of questionnaires, interviews, "visits" of resources, and classroom observations, and for insiders to become reflective practitioners through the use of *logbooks* or diaries (prepared by teacher educators and the teachers themselves).

Networking and Professional Geneses

We agree with Grugeon et al. (2010), that "research about teacher development courses in technology and mathematics is still in infancy" (p. 343). For us, it is more than a matter of merely developing "appropriate" courses—it is a matter of *supporting* the course of teacher development. The move seems to be clearly from "teacher education for technology integration," to "teacher (co)-education in/to designing–appropriating resources (integrating technologies under various forms) for teaching mathematics." From this point of view, there has been certainly a profound evolution since the *Second Handbook*. The institutional recognition of the complexity of teaching in complex environments (continuous evolution, abundance of resources) has led to emergent forms of teacher education programs where task design, development of reflexivity (e.g., via case studies) and collaborating, play a crucial role.

New technological means have been part of these metamorphoses: for example, the role of videos for sharing and analyzing practices; or the role of distant platforms for collaborating and continuing work. The possibilities of networking appear as a major support for such evolutions, with this networking involving teachers and trainers, and also researchers in many experimental contexts.

It seems to be a time of blending: face-to-face with distance; communities involving teacher education students–teacher educators–researchers, etc. These metamorphoses renew the regard for teacher education, considered more as a *professional genesis*, resulting in teachers (individually and collectively) acting with/on resources.

Conclusion

We have come here to the end of our journey through the "mathematics education with technology" universe. We made four stops, successively visiting policies (including curricula and assessment); available resources; learning and teaching spaces; and finally teacher education strategies. It is time to close our journey's logbook, keeping in mind the main impressions.

The first impression is that the landscape we discovered through the opened windows is a *complex* one. Technology represents a deep change in mathematics learning and teaching conditions; educational policies can draw on it, but also must face associated evolutions. The two dimensions that we have distinguished in this chapter, namely the *top-down/bottom-up* dimension and the *access/support* dimension, helped us to analyze these policies. But we found that it was not always possible to characterize policies according to these dimensions. In the same country, the official institution can support the design and/or availability of resources by communities of teachers, and at the same time develop and/or provide "official resources." Also, the involvement in the design and in the provision of resources can lead some teachers to make career switches, (for example, taking on responsibilities in a district). Thus, policies do not seem to move along the neat straight lines sketched in our model.

The second impression is that technology could enlarge the digital divide between developed and developing countries. It is certainly naïve to imagine that the worldwide profusion of resources solves the essential problem of access. Access includes access to machines and access to Internet; and that is not the case in many regions. Moreover, access is dependent on official recommendations: if

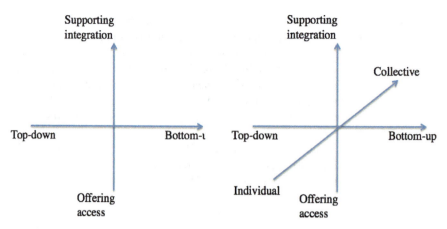

Figure 24.14. From two policy dimensions to three policy dimensions.

policies offer access to *poor* resources, this access naturally leads to dead-ends for mathematics education.

The third impression came from considering the mathematics and technology education universe as a 3D space—adding to the previous two dimensions a third axis positing an *individual/collective* dimension (Figure 24.14). For instance, the *EMAT* activities, in Mexico, were designed with the aim of developing individual learning, while most *Enciclomedia* resources were meant to be presented to the collective classroom. Our journey reveals an evolving landscape where work on resources (designing as well as offering, using or adapting) seems to be increasingly collective. We could thus complement the initial two axes with a third one, representing an individual/collective evolution, and can represent the paradigm's shift concerning the production of resources for teaching as a move in this 3D space.

Our fourth impression is that resources are never *finished*, but always involve, in the design processes, an appropriation process—individual, as well as collective—leading to a renewal of resources. Monitoring this permanent move towards increased *quality* is an essential policy issue.

Finally our journey evidenced a need for a deep reflection on what *initial* resources are required to learn and teach mathematics in technology-rich environments. How can we best give access to and support the appropriation of such critical resources? Which are the missing resources, and how can we initiate and support their design? Such reflections, which may guide future policies, do not seem to exist yet.

Each of mathematics, education, and technology is a rich world. The combination of these three worlds constitutes a very complex universe. We have tried to explore this universe. A single journey always gives a limited access to the visited universe. We are conscious of this limitation. Other chapters in this *Handbook* have enlarged this visit, and supported our reflection about what was, what is, what could be, and what should be.

References

Agalianos, A., Noss, R., & Whitty, G. (2001). Logo in mainstream schools: The struggle over the soul of an educational innovation. *British Journal of Sociology of Education, 22*(4), 479–500. doi:10.1080/01425690120094449.

Artigue, M. (2002). Learning mathematics in a CAS environment: The genesis of a reflection about instrumentation and the dialectics between technical and conceptual work. *International Journal of Computers for Mathematical Learning, 7*, 245–274. doi:10.1023/A:1022103903080.

Assude, T., Buteau, C., & Forgasz, H. (2010). Factors influencing implementation of technology-rich mathematics curriculum and practices. In C. Hoyles & J.-B. Lagrange (Eds.), *Mathematics education and technology—Rethinking the terrain. The 17th ICMI study* (Vol. 13, New ICMI Study Series, pp. 405–419). New York, NY: Springer. doi: 10.1007/978-1-4419-0146-0_19.

Australian Council for Educational Research (ACER). (2010). *Evaluation of One Laptop per Child (OLPC) trial project in the Solomon Islands.* Retrieved from http://www.box.net/keydox/1/31970050/418415076/1%20#keydox/1/31970050/601916856/1.

Ball, D., & Cohen, D. K. (1996). Reform by the book. What is—or might be—the role of curriculum materials in teacher learning and instructional reform? *The Educational Researcher, 25*(9), 6–14. doi:10.3102/0013189X025009006.

Bernáldez, M. (2011, February 18). *Habilidades digitales para todos: Los retos de democratizar la tecnología en las escuelas mexicanas.* Address presented at 4o Seminario Internacional de Educación Integral: Habilidades digitales, retos para el aprendizaje, la enseñanza y la gestión educativa, Hotel Hilton, Mexico, D.F.

Brown, R. (2010). Does the introduction of the graphics calculator into system-wide examinations lead to change in the types of mathematical skills tested? *Educational Studies in Mathematics, 73*(2), 181–203. doi:10.1007/s10649-009-9220-2.

Carvalho, A., Kendall, M., & Cornu, B. (2009, December 18). *The Bento Gonçalves declaration for action.* WCCE 2009 IFIP TC3. Retrieved from http://www.ifip-tc3.net/IMG/pdf/BGDeclaration.pdf.

Churchhouse, R. F., Cornu, B., Howson, A., Kahane, J., Van Lint, J., Pluvinage, F., & Yamaguti, M. (Eds.). (1986). *The influence of computers and informatics on mathematics and its teaching* (ICMI Study Series, Vol. 1). Cambridge, UK: Cambridge University Press.

Drijvers, P. (2009). Tools and tests: Technology in national final mathematics examinations. In C. Winslow (Ed.), *Nordic research on mathematics education, Proceedings from NORMA08* (pp. 225–236). Rotterdam, The Netherlands: Sense Publishers.

Drijvers, P. (2012). Teachers transforming resources into orchestrations. In G. Gueudet, B. Pepin, & L. Trouche (Eds.), *From text to "lived" resources: Mathematics curriculum materials and teacher development* (pp. 265–281). New York, NY: Springer.

Drijvers, P., Doorman, M., Boon, P., Reed, H., & Gravemeijer, K. (2010). The teacher and the tool: Instrumental orchestrations in the technology-rich mathematics classroom. *Educational Studies in Mathematics, 75*(2), 213–234. doi:10.1007/s10649-010-9254-5.

Drijvers, P., & Trouche, L. (2008). From artefacts to instruments: A theoretical framework behind the orchestra metaphor. In G. W. Blume & M. K. Heid (Eds.), *Research on technology and the teaching and learning of mathematics* (Vol. 2, pp. 363–392). Charlotte, NC: Information Age.

Ferrini-Mundy, J., & Breaux, G. A. (2008). Perspectives on research, policy, and the use of technology in mathematics teaching and learning in the United States. In G. W. Blume & M. K. Heid (Eds.), *Research on technology and the teaching and learning of mathematics* (pp. 427–448). Charlotte, NC: Information Age Publishing.

Fonseca, C. (2005). *Educación, tecnologías digitales y poblaciones vulnerables: Una aproximación a la realidad de América Latina y el Caribe. Consulta Regional del Programa Pan Américas.* Montevideo: IDRC. Retrieved March 15, 2011, from http://www.idrc.ca/uploads/user-S/117776589014_Paper_TIC_EDU__Fonseca_FOD.pdf.

Fugelstadt, A. B., Healy, L., Kynigos, C., & Monaghan, J. (2010). Working with teachers. In C. Hoyles & J.-B. Lagrange (Eds.), *Mathematics education and technology—Rethinking the* terrain. *The 17th ICMI study* (Vol. 13, New ICMI Study Series, pp. 293–310). New York, NY: Springer. doi: 978-1-4419-0145-3.

Grugeon, B., Lagrange, J.B., & Jarvis, D. (2010). Teacher education courses in mathematics and technology: Analyzing views and options. In C. Hoyles & J.-B. Lagrange (Eds.), *Mathematics education and technology—Rethinking the terrain. The 17th ICMI study.* (Vol. 13, New ICMI Study Series, pp. 329–345). NY: Springer. doi: 10.1007/978-1-4419-0146-0_15.

Gueudet, G., & Trouche, L. (2009). Towards new documentation systems for teachers? *Educational Studies in Mathematics, 71*(3), 199–218. doi:10.1007/s10649-008-9159-8.

Gueudet, G., & Trouche, L. (2012a). Communities, documents and professional geneses: Interrelated stories. In G. Gueudet, B. Pepin, & L. Trouche (Eds.), *From text to "lived" resources: Mathematics curriculum materials and teacher documentation* (pp. 305–322). New York, NY: Springer.

Gueudet, G., & Trouche, L. (2012b). Mathematics teacher education advanced methods: An example in dynamic geometry. *ZDM—The International Journal on Mathematics Education, 43*(3), 399–411. doi:10.1007/s11858-011-0313-x.

Healy, L., & Lagrange, J.-B. (2010). Introduction to section 3. In C. Hoyles & J.-B. Lagrange (Eds.), *Mathematics education and technology—Rethinking the terrain. The 17th ICMI study.* (Vol. 13, New ICMI Study Series, pp. 287–292). New York, NY: Springer. doi: 10.1007/978-1-4419-0146-0_12.

Hegedus, S., & Lesh, R. (Eds.). (2008). Democratizing access to mathematics through technology: Issues of design, theory and implementation—In memory of Jim Kaput's work. Special issue of *Educational Studies in Mathematics, 68*(2), 81–193.

Hernandez-Sánchez, M. (2009). *Incorporación de herramientas tecnológicas a la enseñanza de las matemáticas: Cambios en el aula y búsqueda de nuevas formas de evaluación* (Master's thesis). Cinvestav-IPN, Mexico.

Hoyles, C., Kalas, I., Trouche, L., Hivon, L., Noss, R., & Wilensky, U. (2010). Connectivity and virual networks for learning. In C. Hoyles & J.-B. Lagrange (Eds.), *Mathematics education and technology—Rethinking the terrain. The 17th ICMI study* (Vol. 13, New ICMI Study Series, pp. 439–462). New York, NY: Springer. doi: 10.1007/978-1-4419-0146-0_22.

International Society for Technology in Education. (2011). *Standards for global learning in the digital age.* Retrieved June 26, 2011, from http://www.iste.org/standards.aspx.

Jiménez-Molotla, J., & Sacristán, A. I. (2010). Eight years of journey with Logo leading to the Eiffel tower mathematical project. In J. Clayson & I. Kalas (Eds.), *Constructionist approaches to creative learning, thinking and education: Lessons for the 21st century—Proceedings Constructionism 2010 (12th EuroLogo conference)* [CD] (pp. 1–11). Paris, France: AUP/Comenius University.

Julie, C., Leung, A., Thanh, N., Posadas, L., Sacristán, A. I., & Semenov, A. (2010). Some regional developments in access and implementation of digital technologies and ICT. In C. Hoyles & J. B. Lagrange (Eds.), *Mathematics education and technology—Rethinking the terrain. The 17th ICMI study* (Vol. 13, New ICMI Study Series, pp. 361–383). New York, NY: Springer. doi: 10.1007/978-1-4419-0146-0_19.

Koehler, M. J., Mishra, P., & Yahya, K. (2007). Tracing the development of teacher knowledge in a design seminar: Integrating content, pedagogy and technology. *Computers & Education, 49*(3), 740–762. doi:10.1016/j.compedu.2005.11.012.

Kortenkamp, U., Blessing, A. M., Dohrmann, C., Kreis, Y., Libbrecht, P., & Mercat, C. (2009). Interoperable interactive geometry for Europe: First technological and educational results and future challenges of the Intergeo project. In V. Durand-Guerrier, S. Soury-Lavergne, & F. Arzarello (Eds.), *Proceedings of the Sixth European Conference on Research on Mathematics Education* (pp. 1150–1160). Lyon, France: INRP. Available from http://www.inrp.fr/editions/cerme6.

Krainer, K., & Wood, T. (Eds.). (2008). *Participants in mathematics teacher education: Individuals, teams, communities and networks* (Vol. 3). Rotterdam, The Netherlands: Sense Publishers.

Kuntz, G., Clerc, B., & Hache, S. (2009). Sesamath: Questions de praticiens à la recherche en didactique. In C. Ouvrier-Buffet & M.-J. Perrin-Glorian (Eds.), *Approches plurielles en didactique des mathématiques* (pp. 175–184). Paris, France: Laboratoire de didactique André Revuz, Université Paris Diderot.

Lagrange, J.-B. (2000). L'intégration d'instruments informatiques dans l'enseignement: une approche par les techniques. [The integration of technological instruments in education: an approach by means of techniques.] *Educational Studies in Mathematics, 43*, 1–30. doi: 10.1023/A:1012086721534.

Lavicza, Z., Hohenwarter, M., Jones, K. D., Lu, A., & Dawes, M. (2010). Establishing a professional development network around dynamic mathematics software in England. *International Journal for Technology in Mathematics Education, 17*(4), 177–182.

LeBaron, J., & McDonough, E. (2009). *GeSCI meta-review research report on ICT in education and development*. Dublin, Ireland: Global e-School and Communities Initiative. Available from http://www.gesci.org/publications.html.

Lee, M. C., Tsai, K. H., & Wang, T. (2008). Practical ontology query expansion algorithm for semantic-aware learning objects retrieval. *Computers & Education, 50*(4), 1240–1257. doi:10.1016/j.compedu.2006.12.007.

Leigh-Lancaster, D. (2010). The case of technology in senior secondary mathematics: Curriculum and assessment congruence? In C. Glascodine & K.-A. Hoad (Eds.), *ACER Research Conference Proceedings 2010* (pp. 43–46). Camberwell, Australia: Australian Council for Educational Research. Retrieved from http://research.acer.edu.au/cgi/viewcontent.cgi?article=1094&conte xt=research_conference.

Lesh, R., Hoover, M., Hole, B., Kelly, A., & Post, T. (2000). Principles for developing thought—Revealing activities for students and teachers. In A. Kelly & R. Lesh (Eds.), *Research design in mathematics and science education* (pp. 591–646). Mahwah, NJ: Lawrence Erlbaum Associates.

Llinares, S., & Valls, J. (2010). Prospective primary mathematics teachers' learning from on-line discussions in a virtual video-based environment. *Journal of Mathematics Teacher Education, 13*(2), 177–196. doi:10.1007/s10857-009-9133-0.

Ministère de l'Éducation Nationale, de l'Enseignement Supérieur et de la Recherche (MENESR). (2006). *Le socle commun des connaissances et des compétences*. Retrieved from http://media.education.gouv.fr/file/51/3/3513.pdf.

Ministère de l'Éducation Nationale. (2009, November 19). *Vers un nouveau lycée en 2010. Conférence de presse*. Retrieved from http://media.education.gouv.fr/file/11_novembre/06/8/Conference_de_presse_lycee_127068.pdf.

Mousley, J., Lambdin, D., & Koc, Y. (2003). Mathematics teacher education and technology. In A. J. Bishop, M. A. Clements, C. Keitel, J. Kilpatrick, & F. K. S. Leung (Eds.), *Second international handbook of mathematics education* (pp. 395–432). Dordrecht, The Netherlands: Kluwer Academic Publishers.

National Council of Teachers of Mathematics. (1980). *An agenda for action: Recommendations for school mathematics of the 1980s*. Reston, VA: Author. Retrieved from http://www.nctm.org/standards/content.aspx?id=17278.

National Council of Teachers of Mathematics. (2000). *Principles and standards for school mathematics*. Reston, VA: Author.

National Council of Teachers of Mathematics. (2008, March). *The role of technology in the teaching and learning of mathematics. A position of the National Council of Teachers of Mathematics*. Retrieved from http://www.nctm.org/uploadedFiles/About_NCTM/Position_Statements/Technology%20final.pdf.

Noss, R., & Hoyles, C. (1996). *Windows on mathematical meanings*. Dordrecht, The Netherlands: Kluwer Academic Publishers.

OLPC Foundation. (2011, January). Deployments. *OLPC*. Retrieved from http://wiki.laptop.org/go/Deployments.

Papert, S. (1980). *Mindstorms: Children, computers, and powerful ideas*. New York, NY: Basic Books.

Papert, S. (1991). Situating constructionism. In I. Harel & S. Papert (Eds.), *Constructionism* (pp. 1–11). Norwood, NJ: Ablex.

Pepin, B. (2009). The role of textbooks in the "figured world" of English, French and German classrooms—A comparative perspective. In L. Black, H. L. Mendick, & Y. Solomon (Eds.), *Mathematical relationships: Identities and participation* (pp. 107–118). London, UK: Routledge.

Pierce, R., & Stacey, K. (2010). Mapping pedagogical opportunities provided by mathematics analysis software. *International Journal of Computers for Mathematical Learning, 15*, 1–20. doi:10.1007/s10758-010-9158-6.

Pimm, D., & Johnston-Wilder, S. (2004). Technology, mathematics and secondary schools: A brief, UK, historical perspective. In S. Johnston-Wilder & D. Pimm (Eds.), *Teaching secondary mathematics with ICT* (pp. 3–17). Maidenhead: Open University Press. Retrieved from http://www.mcgraw-hill.co.uk/openup/chapters/0335213812.pdf.

Rojano, T. (2011). Recursos multimedia y el libro de texto gratuito: entre las herramientas universales y los desarrollos ad-hoc. In R. Barriga (Ed.), *Entre paradojas: A 50 años de los libros de texto gratuitos* (pp. 627–643). Mexico: Colegio de México-SEP-Conaliteg.

Ruthven, K. (2007). Teachers, technologies and the structures of schooling. In D. Pitta-Pantazi, & G. Philippou (Eds.), *Proceedings of the Fifth Congress of the European Society for Research in Mathematics Education* (pp. 52–67). Larnaca, Cyprus: CERME 5.

Ruthven, K. (2008). Mathematical technologies as a vehicle for intuition and experiment: A foundational theme of the International Commission on Mathematical Instruction, and a continuing preoccupation. *International Journal for the History of Mathematics Education, 3*(2), 91–102.

Ruthven, K., & Hennessy, S. (2002). A practitioner model of the use of computer-based tools and resources to support mathematics teaching and learning. *Educational Studies in Mathematics, 49*(1), 47–88. doi:10.1023/A:1016052130572.

Sabra, H. (2009). Entre monde du professeur et monde du collectif: Réflexion sur la dynamique de l'association Sesamath. *Petit x, 81*, 55–78.

Sacristán, A. I., Calder, N., Rojano, T., Santos, M., Friedlander, A., & Meissner, H. (2010). The influence and shaping of digital technologies on the learning—and learning trajectories—of mathematical concepts. In C. Hoyles & J.-B. Lagrange (Eds.), *Mathematics education and technology—Rethinking the terrain. The 17th ICMI study* (Vol. 13, New ICMI Study Series, pp. 179–226). New York, NY: Springer. doi 10.1007/978-1-4419-0146-0_6.

Sacristán, A. I., Parada, S., Sandoval, I., & Gil, N. (2009). Experiences related to the professional development of mathematics teachers for the use of technology in their practice. In M. Tzekaki, M. Kaldrimidou, & H. Sakonidis (Eds.), *Proceedings of the 33rd Conference of the International Group for the Psychology of Mathematics Education* (Vol. 5, pp. 41–48). Thessaloniki, Greece: International Group for the Psychology of Mathematics Education.

Sacristán, A. I., & Rojano, T. (2009). The Mexican national programs on teaching mathematics and science with technology: The legacy of a decade of experiences of transformation of school practices and interactions. In A. Tatnall & A. Jones (Eds.), *WCCE 2009, IFIP Advances in information and communication technology: Education and technology for a better world* (pp. 207–215). Boston, MA: Springer. doi: 10.1007/978-3-642-03115-1_22.

Sangwin, C., Cazes, C., Lee, A., & Wong, K. L. (2010). Micro-level automatic assessment supported by digital technologies. In C. Hoyles & J.-B. Lagrange (Eds.), *Mathematics education and technology—Rethinking the terrain. The 17th ICMI study* (Vol. 13, New ICMI Study Series, pp. 227–250). New York, NY: Springer. doi: 10.1007/978-1-4419-0146-0_10.

Santagata, R., Zannoni, C., & Stigler, J. W. (2007). The role of lesson analysis in pre-service teacher education: An empirical investigation of teacher learning from a virtual video-based field experience. *Journal of Mathematics Teacher Education, 10*(2), 123–140. doi:10.1007/s10857-007-9029-9.

Soury-Lavergne S., Trouche, L., Loisy, C., & Gueudet, G. (2011). *Parcours de formation, de formateurs et de stagiaires: Suivi et analyse*. Rapport à destination du Ministère de l'Education Nationale, INRP-ENSL. Available from http://eductice.inrp.fr/EducTice/equipe/PRF-2010/.

Stroup, W. M., & Wilensky, U. (2000). Assessing learning as emergent phenomena: Moving constructivist statistics beyond the bell curve. In A. E. Kelly & R. Lesh (Eds.), *Handbook of research design in mathematics and science education* (pp. 877–912). Mahwah, NJ: Lawrence Erlbaum Associates.

Sur l'épreuve pratique de mathématiques au baccalauréat en France (2007, September 21). *Educmath*. Retrieved from http://educmath.inrp.fr/Educmath/en-debat/epreuve-pratique/.

Trgalová, J., Jahn, A.-P., & Soury-Lavergne, S. (2009). Quality process for dynamic geometry resources: the Intergeo project. In V. Durand-Guerrier, S. Soury-Lavergne, & F. Arzarello (Eds.), *Proceedings of the Sixth European Conference on Research on Mathematics Education* (pp. 1161–1170). Lyon, France: INRP. Available from www.inrp.fr/editions/cerme6.

Trigueros, M., & Sacristán, A. I. (2008). Teachers' practice and students' learning in the Mexican programme for Teaching Mathematics with Technology. *International Journal of Continuing Engineering Education and Life-Long Learning (IJCEELL), 18*(5/6), 678–697. doi:10.1504/IJCEELL.2008.022174.

Trigueros, M., & Lozano, M. D. (2007). Developing resources for teaching and learning mathematics with digital technologies: An enactivist approach. *For the Learning of Mathematics, 27*(2), 45–51.

Trouche, L. (2004). Managing complexity of human/machine interactions in computerized learning environments: Guiding students' command process through instrumental orchestrations. *International Journal of Computers for Mathematical Learning, 9*, 281–307. doi:10.1007/s10758-004-3468-5.

UNESCO (2005). *Towards knowledge societies* (UNESCO World Report). Retrieved from http://www.unesco.org/en/worldreport.

UNESCO (2008). *Policy framework: ICT competency standards for teachers*. Retrieved from http://cst.unesco-ci.org/sites/projects/cst/The%20Standards/ICT-CST-Policy%20Framework.pdf.

Ursini, S., & Rojano, T. (2000). *Guía para integrar los talleres de capacitación, EMAT*. Mexico: SEP-ILCE.

Van Es, E., & Sherin, M. G. (2010). The influence of video clubs on teachers' thinking and practice. *Journal of Mathematics Teacher Education, 13*, 155–176. doi:10.1007/s10857-009-9130-3.

Visnovska, J., Cobb, P., & Dean, C. (2012). Mathematics teachers as instructional designers: What does it take? In G. Gueudet, B. Pepin, & L. Trouche (Eds.), *From text to "lived" resources: Mathematics curriculum materials and teacher development* (pp. 323–341). New York, NY: Springer.

Watson, A., & De Geest, E. (2005). Principled teaching for deep progress: Improving mathematical learning beyond methods and materials. *Educational Studies in Mathematics, 58*(2), 209–234. doi:10.1007/s10649-005-2756-x.

Wijers, M., Jonker, V., & Drijvers, P. (2010). MobileMath: Exploring mathematics outside the classroom. *ZDM—The International Journal on Mathematics Education, 42*(7), 789–799. doi:10.1007/s11858-010-0276-3.

Part IV
Introduction to Section D: International Perspectives on Mathematics Education

Jeremy Kilpatrick

Abstract International perspectives are presented by the 17 scholars—based in 9 different nations—who prepared the 7 chapters in Section D. As an academic field, mathematics education treats a universal school subject situated in vastly different local and national contexts. One can look at the field internationally from at least three perspectives: that of practice, policy, or profession. The chapters in this section take these perspectives, placing special emphasis on the last and demonstrating how the concerns of mathematics educators drive the approach they take to international studies. The chapters show dramatically not only how the international dimension of mathematics education research has developed since 1908—and especially during the last five decades—but also how far it has to go in studying the influence these studies are having on the teaching and learning of mathematics around the world.

Keywords Local • National • International • Internationalization • Globalization • Practice • Policy • Profession

Mathematics has commonly been seen as the one school subject that is universal and therefore unaffected by local circumstances. Whether schoolchildren are in North Korea, East India, South Africa, or West Germany, 23 is a prime number, 24 composite, and 25 a perfect square. The theorem bearing the name of Pythagoras was as true for learners in ancient China as it is in Brazil today. The ratio of the circumference to the diameter of a circle remains constant over time and across borders. In contrast, a nation's education system is embedded in its history and culture. Schooling operates locally, and education policy is made at the local or national level but not beyond. Consequently, mathematics education always ranges between universality and singularity.

During the last half of the 20th century, mathematics began to lose some of its presumed universality. Philosophers of mathematics began to advance fallibilist and quasi-empiricist views of the subject (Tymoczko, 1998); views, however, that were

J. Kilpatrick
University of Georgia, Athens, GA, USA

certainly not accepted by all mathematicians. At about the same time, mathematics educators began to recognize that although their field had been among the first to hold international meetings, school mathematics—as a cultural artifact—ought to be seen as embedded in the practice of a particular social group (Bishop, 1988; D'Ambrosio, 1985); again, a view that not all accepted. Today, when one looks across national boundaries, one sees practices and policies based on assumptions ranging from "school mathematics is the same everywhere" to "school mathematics is different everywhere."

The chapters in this section of the handbook survey mathematics education from various international perspectives that can be grouped into three categories: the perspective of practice, the perspective of policy, and the perspective of profession. The perspective of *practice* is the teacher's view, the ways in which mathematics education can be seen in action by looking in classrooms around the world. Observations yielding field notes and videographic records have allowed activities in mathematics classrooms to be analyzed in detail and compared so that generalizations can be made and unique features celebrated. The perspective of *policy* takes a different view; it is the policymaker's angle of vision, the perspective of one who wants to improve mathematics education not merely by learning how it is being conducted more effectively elsewhere but also by importing or adapting those characteristics that seem to make a difference. Finally, the perspective of *profession* operates along a different dimension; it is the view of the researcher in mathematics education who attempts to see beyond the local and the national. What characterizes the ways we in our country teach and learn mathematics? And how does our professional enterprise resemble or differ from what others do? Over the past century or so, as mathematics education has developed as an academic field (Kilpatrick, 2008), mathematics educators in every country have continually attempted to understand simultaneously what is being taught and learned in nearby school classrooms and how mathematics educators in other countries might be organizing and conducting their work.

As Alexander Karp notes (Chapter 25), 1908 was the year that the *Commission Internationale de l'Enseignement Mathématique* (reincarnated later as the International Commission on Mathematical Instruction) was established, and that year proved to mark a crucial point for the establishment of the field as well as the beginning of collaborative international research. The last five decades, especially, have seen enormous growth in the scope and sophistication of international research studies, beginning with the International Study of Achievement in Mathematics (later known as the First International Mathematics Study; Husén, 1967). In recent years, the Trends in International Mathematics and Science Study (TIMSS) and the Programme for International Student Assessment (PISA) have come to dominate much of the discourse about international studies in mathematics education. In fact, every chapter in this section mentions both. In particular, the chapters by Parmjit Singh and Nerida Ellerton (Chapter 26); Vilma Mesa, Pedro Gómez, and Ui Hock Cheah (Chapter 27); Mogens Niss, Jonas Emanuelsson, and Peter Nyström (Chapter 30); and John Dossey and Margaret Wu (Chapter 31) contain, with different emphases, detailed treatments of TIMSS and PISA. But many other studies, large and small, with an international dimension are also cited in the chapters in the section.

The Perspective of Practice

The development of video recording has enabled researchers to capture and study events in a variety of mathematics classrooms in different countries. The best known of these studies are the TIMSS Video Studies of 1995 and 1999, which are discussed by Singh and Ellerton (Chapter 26), Mesa et al. (Chapter 27), and Niss et al. (Chapter 30). Opinions have differed on whether teachers in a given country are following something like a national script, but all agree that the video records open a new window that allows teachers and researchers to view mathematics teaching in their own country and elsewhere. Another study that makes use of video records is the so-called Learner's Perspective Study (LPS), discussed by Singh and Ellerton and by Niss et al. The latter authors note that the LPS is different from other large-scale international studies in that it is driven by the interests of researchers in mathematics education and is not conducted under the auspices of an international organization.

Mesa and her colleagues (Chapter 27) attempted to gauge the effects at the classroom level of international studies, whether or not they included a video component. They were unable to find evidence that international studies were influencing classroom teaching and learning, and they concluded that the international mathematics education research community has not done enough to study possible influences. Studies such as TIMSS and PISA are extensively reported in the media in many countries, but any effects on school mathematics in those countries have yet to be documented.

The Perspective of Policy

The International Association for the Evaluation of Educational Achievement (IEA), which conducts TIMSS, and the Organisation for Economic Co-operation and Development (OECD), which conducts PISA, undertake those studies to help promote their goals. In the case of the IEA, the goal is to understand the effects of policies and practices within and across systems of education. The TIMSS results in mathematics are intended to help policymakers identify strengths and weaknesses in their education systems. In the case of the OECD, the goal is to stimulate economic progress and world trade, and the PISA results in mathematics provide policymakers with an indicator of how well their countries are doing in promoting mathematics literacy.

TIMSS and PISA, as the international studies relevant to mathematics education receiving the most attention in the mainstream media, are not surprisingly the studies that apparently have had the most influence on education policy. Dossey and Wu (Chapter 31) report on the effects of a variety of international studies at the policy level, with TIMSS and PISA at the forefront. They present case studies from the USA, Germany, Finland, and Singapore of differing reactions to reports of the results of these studies and consequent policy decisions. Dossey and Wu conclude that these international studies have had both positive and negative effects in policy arenas, and they raise several concerns about the manner in which survey results have been translated into policy changes. They see a fruitful agenda for research by policy researchers as well as mathematics education researchers.

The Perspective of Profession

One might easily conclude that any international study involving school mathematics can be seen from the perspective of profession. Researchers in mathematics education should presumably see any such study as speaking to their concerns. Not every study qualifies, however. For example, the IEA's First International Mathematics Study (Husén, 1967) was a study in comparative education and not mathematics education (Kilpatrick, 1971). The researchers were interested in the productivity of various national systems of education. They wanted to address such issues as the effects of school organization (e.g., selective vs. comprehensive education, class size, and school leaving age) on mathematics achievement, and mathematics was essentially taken as a black box. In subsequent IEA studies, including TIMSS, that box was opened up, and the results had many more implications for the profession.

One might also conclude that any international study involving school mathematics that can be seen from the perspective of profession can also be seen from the perspectives of practice or policy. For example, the LPS discussed by Singh and Ellerton (Chapter 26) clearly deals with practice while simultaneously addressing a variety of research topics of interest to professionals in the participating countries (see the first table in Chapter 26). But not every international study can be seen from multiple perspectives. As one example, a comparative study of teachers' knowledge of elementary mathematics in the USA and China (Ma, 1999) need not address classroom practice or have direct implications for education policy in order to be of value. As another example, an analysis of how the concept of function is treated in textbooks from 18 countries (Mesa, 2009) can address concerns of professionals in mathematics without examining teaching practice or promoting policy change.

One of the topics of principal concern to professionals in mathematics education is the school mathematics curriculum. That topic is addressed in the chapters by Karp (Chapter 25), Mesa et al. (Chapter 27), Niss et al. (Chapter 30), and Dossey and Wu (Chapter 31), and it is the theme of the chapter by Jinfa Cai and Geoffrey Howson (Chapter 29). Like Karp, Cai and Howson discuss the processes of globalization (which integrates economies, societies, and cultures to link people around the world) and internationalization (which develops products and services so that they can be adapted to local conditions). Cai and Howson see both processes as powerful influences on curriculum development although not necessarily leading toward an international curriculum. They argue for the benefits of individual countries having the freedom to experiment with the curriculum and undertake new initiatives within it.

In their chapter, Bernard Hodgson, Leo Rogers, Stephen Lerman, and Lim-Teo, Suat Khoh (Chapter 28) survey a great variety of international and multinational organizations, showing how those organizations provide niches within which mathematics education professionals can pursue their diverse interests. Hodgson et al. focus on organizations connected to research in mathematics education, and they present case studies to illustrate the various bases on which groups have been organized. They are especially struck by the proliferation of disparate subcommunities within the field, and they expect that proliferation to continue.

Conclusion

In Chapter 25, Karp looks at how the process of internationalization in mathematics education has developed over the decades. He sees the local and national as still developing, and as he notes, internationalization has its limits. Just as education systems are in flux, so international perspectives on mathematics education are shifting as well. Nonetheless, one inevitably concludes that the field has not done enough to study the effects international research studies are having on the teaching and learning of mathematics.

References

Bishop, A. J. (1988). *Mathematical enculturation: A cultural perspective on mathematics education*. Dordrecht, The Netherlands: Kluwer.

D'Ambrosio, U. (1985). Ethnomathematics and its place in the history and pedagogy of mathematics. *For the Learning of Mathematics, 5*(1), 44–48.

Husén, T. (Ed.). (1967). *International Study of Achievement in Mathematics: A comparison of twelve countries* (2 vols.). Stockholm, Sweden: Almqvist & Wiksell.

Kilpatrick, J. (1971). Some implications of the International Study of Achievement in Mathematics for mathematics educators. *Journal for Research in Mathematics Education, 2*, 164–171.

Kilpatrick, J. (2008). The development of mathematics education as an academic field. In M. Menghini, F. Furinghetti, L. Giacardi, & F. Arzarello (Eds.), *The first century of the International Commission on Mathematical Instruction (1908–2008): Reflecting and shaping the world of mathematics education* (pp. 25–39). Rome, Italy: Istituto della Enciclopedia Italiana.

Ma, L. (1999). *Knowing and teaching elementary mathematics: Teachers' understanding of fundamental mathematics in China and the United States*. Mahwah, NJ: Lawrence Erlbaum.

Mesa, V. (2009). *Conceptions of function in textbooks from eighteen countries*. Saarbrücken, Germany: VDM Publishing.

Tymoczko, T. (1998). New directions in the philosophy of mathematics: An anthology (Rev. & expanded ed.). Princeton, NJ: Princeton University Press.

Chapter 25
From the Local to the International in Mathematics Education

Alexander Karp

Abstract This chapter is devoted to a historical overview of the process of internationalization in mathematics education. The development of mathematics education is analyzed as a part of social history, and therefore the discussion inevitably touches on history and even politics. The concepts "international," "internationalization," and "globalization" themselves may be understood in different ways, and this is also discussed in the chapter. The chapter sequentially, albeit very briefly, analyzes various stages of the development of international collaboration, wherever possible identifying similar processes in the development of mathematics education in different regions that have facilitated such collaboration. The problem of the growth of scholarly articles from different regions is examined, as is the appearance and development of various international conferences and organizations of mathematics educators. The chapter also considers criticisms of internationalization as well as its limits.

Introduction

This chapter is devoted to a historical overview of the internationalization of mathematics education. Contemporary mathematics educators usually relate their work to a considerable degree to what is happening beyond the borders of their countries, and examples of the way in which education is constructed in other countries are usually a standard part of the professional discourse. This was not the case, or not exactly the case, at the beginning of the 20th century, when Americans were unlikely to become particularly interested in algebra lessons in East Asia, and even

---

A. Karp (✉)
Teachers College, Columbia University, New York, NY, USA
e-mail: apk16@columbia.edu

if any did develop such an interest for whatever reason, it would have been difficult for them to obtain information about the subject—different parts of the world were far more isolated from each another than they are now, and there were no international organizations, or any of the other institutions or instruments that are so familiar to mathematics educators in the 21st century, by which one could gather the knowledge required.

The process of the formation of increasingly close ties across national boundaries is often called *globalization*. By using this word, we inevitably enter into a thicket of controversies about globalization, which sometimes assume forms that are quite contentious. The aim of this chapter, of course, is not to offer support to this or that side or perspective. Nevertheless, it must be recognized that the problems discussed below are politicized, and that even when they concern mathematics and mathematics education, this politicization cannot be avoided. Thus, it is important to analyze what occurred in mathematics education against the background of what occurred in the world in general. The Iron Curtain separated different systems of mathematics education as much as it did everything else, although naturally its fall did not imply an automatic convergence between these systems.

Atweh, Clarkson, and Nebres (2003) wrote that they had discovered "very few references to globalization in mathematics education" (p. 187). In the ensuing years, in addition to the work just cited and the chapters in the monograph *Sociocultural Research on Mathematics Education: An International Perspective* (Atweh, Forgasz, & Nebres, 2001), we have seen the appearance of a monograph by Atweh et al. (2008), as well as such publications as Baker and LeTendre (2005), in which considerable attention has been devoted to the teaching of mathematics. We have also seen the appearance of historical works, which rely on completely different methodologies and which shed light on the formation of the international education community (e.g., Coray, Furinghetti, Gispert, Hodgson, & Schubring, 2003). Finally, a number of books about mathematics education in the non-English-speaking world have come out (e.g., books from the Series on Mathematics Education, published by World Scientific). My remarks are not intended in any way to be a full overview of the current literature which, especially as far as it concerns international comparative studies, is quite large. And yet, I would argue that the movement *from the local to the international in mathematics education* has not yet been sufficiently investigated at the level of different countries, which inevitably limits the general overview as well.

The present chapter, of course, cannot offer a complete description of what has occurred in different countries—I will confine myself merely to a general sketch, especially since many of the organizations, events, and studies that are important for international collaboration and interaction are discussed in greater detail in other chapters of this *Handbook*. A crucial watershed in the development of such collaboration was the year 1908—the year that saw the establishment of ICMI, the International Commission on Mathematics Instruction. In this chapter I will use the now familiar acronym, ICMI, but when the Commission was founded it was much more common to refer to it in French—as *Commission Internationale de l'Enseignement Mathématique* (CIEM)—or in German—as *Internationale Matematik Unterricht Kommission* (IMUK). In English, it was known as the

International Commission on the Teaching of Mathematics (ICTM). The name ICMI became standard after the Commission's "rebirth" in 1952, which will be discussed below.

In what follows, I will sequentially analyze the periods before 1908, from 1908 until the end of the Second World War, from 1945 until the collapse of the Soviet Union and the "official" end of the Cold War and finally the current period.

Did International Collaboration in Mathematics Education Exist Before 1908? Defining the Terminology

Schubring (2003) noted that "communication between different countries regarding the teaching of mathematics had been practically non-existent up to the end of the 19th century" (p. 49). At first glance, this statement may appear false. Indeed, it is not difficult to cite examples that would seem to contradict this statement. Long before 1908, the influence of foreign materials for teaching mathematics could be very great, and cases of the direct borrowing of such materials—above all, textbooks—were not few in number. In Japan, translations of foreign textbooks began appearing after the Meiji Restoration. The Russian writer Herzen recalled using the French textbook of Francoeur as a student in the 1820s, and this was by no means the only foreign textbook used in Russian mathematics education (Karp, 2007b). And American textbooks were initially based on British and French models, which were themselves also widely used. American textbooks' foreign origins were even emphasized at times in their titles, as in Colburn's famous textbook, *An Arithmetic on the Plan of Pestalozzi* (Cohen, 2003).

Foreign travelers not infrequently took an interest in the educational systems of the countries they visited, including approaches to mathematics education (one such traveler, for example, was Francisco de Miranda, the forerunner of the Latin American liberation movement, who in his diaries described his visits to educational institutions). Moreover, teachers of mathematics not infrequently moved from one country to another—Euler, who among other things created several basic sections of the traditional school course in mathematics, taught both in Russia and Prussia. Euler's experience can in no way be considered local, and there were many hundreds of less brilliant foreign teachers in many countries. Furthermore, going beyond the boundaries of specifically mathematical education, one might recall that the history of comparative education begins at the very latest in the 1820s (Brickman, 1960).

Nonetheless, I would argue that on the whole Schubring is right. And not only because there really were no international mathematics education organizations or international studies in mathematics education at the time, but also because the professional contacts just mentioned were part of a completely different organization of education. Jesuit educational institutions, widespread from Latin America to Russia, could be regarded as international, but it would be more natural to consider them extra-national—the national element, that which is distinctly characteristic of specific countries, did not play a role of any importance in them. Much else that took place in the 19th century, and in the 18th or 17th centuries and earlier, could be characterized as similarly extra-national. The evolving national systems of mathematics education naturally looked for models in other countries, and although these models were by no means always simply appropriated, and frequently went through a kind of adaptation, this process should still be distinguished from a process of communication among already-developed national educational systems (Schubring, 1989).

The national systems of mathematics education developed differently in different countries, not at the same time and not at all quickly. Consequently, in talking about the history of mathematics education communication, the theoretical distinction between what went on among already developed national systems and what went on among national systems that were not yet fully formed, to which I have just referred, is not so easy to discern in practice. Elements of different historical processes may be present simultaneously. For example, when in 1892 Andrey Kiselev's textbook in geometry, which subsequently became a kind of icon and symbol of Russian mathematics education (Karp, 2002), was published in Russia, the reviewers ingenuously praised it for following French models, and their very tone implied that Russia still had a long way to go until it reached the level of Western European countries with their more developed and established national systems of mathematics education. The Russian national system of mathematics education was still in its formative stage, although a great deal had already been done and much had already taken shape in this system which could be communicated in international professional relationships.

Recognizing the great differences between the processes that could hide behind the term *internationalization,* Atweh et al. (2003) even considered it necessary to elucidate that they did not consider "internationalization" to be synonymous with "homogenization" or "universalization" (p. 189). To put it another way, the movement from the "local" to the "international," to which this chapter is devoted, in principle presupposes that the "local" has already largely taken shape, and that its distinctive characteristics and originality are recognized and respected. Again, for the sake of precision, I should point out that in reality the local (national) systems of mathematics education are not today equally fully developed everywhere—in some places, this process is still going on and may be helped by collaboration with other countries and international organizations. This development only happens, however, if the collaboration does not turn into a kind of cultural imperialism that fails to take into account the local characteristics present at every stage, and instead imposes general, "international" schemas.

The experience of interactions among mathematics educators prior to the end of the 19th century developed along the lines indicated above—textbooks were read and translated; foreign educational institutions were studied by educators or administrators who were sometimes even sent abroad specifically for this purpose; finally, even an international labour market existed to some degree, in the sense that individuals could find work teaching in other countries as well as their own. This experience was not unimportant, since it gave at least some number of mathematics educators some notion of what was happening outside the borders of their countries.

However, what went on in mathematics education in different countries, even in Western Europe, differed widely. It is precisely these differences that Schubring (2003) sees as the reason for the lack of more developed forms of international communication. In fact, even in Germany, the teaching of mathematics differed substantially from one federal state to another, while in Britain comparable differences might be observed between schools that were located not far from one another—according to one report from 1868, five schools in Yorkshire offered wide-ranging mathematics courses while two others limited their teaching to Euclid Book I only (Howson, 2010). For all the variety of teaching models that existed within each country by itself, the differences across national borders were still more substantial. In discussing the differences between Britain and Germany, one cannot omit the differences in college preparation, as we would say today, of technical experts and engineers, and thus in the understanding of the role of mathematics in such preparation, including the role played by mathematics at the school level. The British universities of Oxford and Cambridge were radically different from German institutions of higher learning, remaining effectively theological seminaries. It is true that, from the second half of the 19th century on, a greater and greater number of educational institutions devoted specifically to engineering began to appear in the UK (Howson, 2010); but it would nonetheless not be incorrect to say that within the rigidly structured British educational system, with its pronounced differentiation between education in the higher and lower social classes, the significance of mathematics was not especially great.

Differences with other European countries—let alone non-European ones—were usually not smaller, but even greater. In some sense, it was indeed true that at some time there was not all that much to communicate across national borders in mathematics education—the conditions and problems in each country were simply too different. Nonetheless, at the very end of the 19th century, in 1899, the journal *L'Enseignement Mathématique* appeared; and in 1908, at the International Mathematical Congress in Rome, David Eugene Smith, a professor at Teachers College, Columbia University, proposed the creation of the International Commission on Mathematics Instruction (ICMI).

International Collaboration in Mathematics Education from the Beginning of the 20th Century Until the Second World War

The appearance of *L'Enseignement Mathématique* and ICMI represented a turning point in the development of communication in the international mathematics education community. For the first time, organizations and a periodical appeared that had such communication as their goal. To provide a detailed history of their appearance is not the purpose of the present chapter (see Furinghetti, 2003; Schubring, 2003, 2008). Let me merely note that this event was quite in keeping with the spirit of the time (although the path to the formation of the Commission was not without obstacles—see Schubring, 2008).

The roles played by D. E. Smith, Henry Fehr, Felix Klein, and other figures of the age were, of course, extremely large, but the movement beyond the national borders was at that time by no means characteristic of mathematics education alone, and therefore it would be misguided to reduce the collaboration that evolved—in spite of all the differences between the various national systems—to the initiative of specific individuals. Several decades prior to the appearance of ICMI, the slogan "Workers of the world, unite!" had already become popular. In spite of all the obvious differences between those who were urged to unite in these cases and the reasons for their unification, this slogan illustrates a recognition of the fact that life—economic life and cultural life alike—was now being built on new, no longer purely national foundations. The world was by this time settled and divided, and interaction with neighbors near and far—whether collaborative or hostile—was becoming unavoidable. The list of international organizations and regular international conferences established from the early 1870s on is quite long—the Universal Postal Union was founded in 1874, the International Committee for Weights and Measures in 1875, the International Statistical Institute in 1885, and the Olympic Committee in 1894. Starting in 1900, congresses of historians began to be held on a regular basis, and so on.

It should not be forgotten that contact—in the simple, technical sense of the word—became far easier at the time than it had been earlier. D. E. Smith crossed the ocean dozens of times (Donoghue, 2008) and, although this travel certainly says something about Smith's own personality, it also demonstrates the fact that travel had become much easier to do. Indeed, there was often even no need to travel in person—the telegraph already existed, and the telephone was coming.

Among the numerous international unions and alliances that appeared at this time, I cannot fail to mention the International Mathematical Union (IMU) (Lehto, 1998). It was out of the IMU that ICMI was born, and the overwhelming majority of its members were originally individuals who did not work in schools and had never done so, but were familiar with higher education only. Informal contacts among mathematicians had been maintained since ancient times—mathematics knew no borders, whether in the provinces of the Roman Empire or in the nation states of Europe. The first official international mathematics congress took place in 1893 in Chicago as part of the World's Fair, and at this congress Felix Klein emphasized the importance of cooperation: "What was formerly begun by a single mastermind, we now must seek to accomplish by united efforts and cooperation" (quoted in Albers, Alexanderson, & Reid, 1987, p. 2).

Mathematics education was seen by those who were laying the foundations for international collaboration in it as something far more closely connected with national and state administrative systems than mathematics was. That was why plans were initially made to establish the Commission and to select its participants with the help of the states and governments involved (Schubring, 2003). But the processes taking place in the world did not bypass education. As Brickman (1960) stated, at the beginning of the 20th century, after important publications had appeared, a new era of scientific comparative education commenced. At that time, certain fundamental problems in mathematics education that different countries shared, despite all of their obvious differences, began to be recognized.

The most important of these problems was the need to reform the school curriculum. The fact that many topics were obsolete, the fact that the school curriculum was disconnected from practical needs, the need to include new mathematical ideas in the school curriculum—above all, ideas connected with the concept of function—and the importance of new pedagogical and methodological approaches were discussed in different countries, both under the influence of the development of mathematics and under the influence of the development of education. These discussions were stimulated by the need to find solutions to the new problems of mass education (or, at least, education on a much larger scale than anything that had existed previously).

I am not concerned here with exactly when, where, and how this or that reformist idea was first expressed (much later, in the years of the Cold War, educators argued passionately about such issues, usually defending their country's priority—Karp, 2007a). What is important is that these ideas were heard and received in different countries.

What is noteworthy, therefore, is not simply the fact that representatives of different countries began coming together to discuss the problems of mathematics education and to publish relevant materials, but also what influence these discussions and materials had on different countries. Because it is impossible to trace how and when this process unfolded in different countries, I will confine myself to a description of these developments in just one country—Russia. The Russian representatives in ICMI prepared and published the reports that ICMI called for, but what is important is that these and other ICMI reports were also translated and published in Russia. Thus, for example, the preliminary report, which was published in *L'Enseignement Mathématique* in November 1908, was reprinted in four Russian periodicals, including such leading journals as *Zhurnal Ministerstva Narodnogo Prosveshcheniya* (*Journal of the Ministry of Education*) and *Moskovskiy matematicheskiy sbornik* (*Moscow Mathematics Compendium*), and in addition the report was sent "to all societies and organizations that had a relation to the teaching of mathematics"[1] (Sintsov, 1913, p. 4). In this way, very broad sections of the population became informed both about the work of international organizations and about the view of Russia that was expressed in international documents.

Even more importantly, the development of international contacts spurred the development of national education and discussions about it—in 1911–1912 and in 1913–1914, countrywide congresses of mathematics teachers took place in Russia. The proceedings of the first of these congresses began with an account of the creation of ICMI, and stress was laid on the fact that the "international movement aimed at investigating methods for teaching mathematics has also found an echo in Russia" (Maksheev, 1913, p. vi). Indeed, during the course of both congresses, discussions of specifically Russian problems—whether school textbooks or teacher preparation or the use of more visually-oriented manuals—took place against a

---

[1] This and later translations from Russian in this chapter are by the present author.

background of comparisons with what was being done in other countries and what could be appropriated. The participants of the congresses put together plans for new research and studies in accordance with international models (Sintsov, 1913).

To be sure, the responses to ICMI's activities were not identical in all of the countries that participated in ICMI, and indeed only relatively few countries took part in its work at first (Schubring, 2003); nonetheless, it appears plausible to conclude that evolving international cooperation developed and enriched national education.

The war of 1914–1918 interrupted this incipient development. Even D. E. Smith (1918) was forced at that time to prove his patriotism and defend himself against accusations of excessive connections with citizens of enemy states. The defeated states remained in isolation for several more years (e.g., German scholars were not invited to mathematics congresses in 1920 and 1924). To this isolation was added the gradually emerging new dividing line that walled off the Soviet Union.

Even so, attempts to develop what had been done continued, at the organizational level (the renewal of ICMI), at the level of personal contacts—Smith conducted a vast correspondence with mathematics educators around the world—and through publications about what was taking place in different countries [e.g., the articles about teaching in Germany published in the USA by Lietzmann (1924) and Malsch (1927)]. The most important of such publications, perhaps, was the fourth NCTM *Yearbook* (Reeve, 1929), which offered descriptions of the state of mathematics education in 13 countries. Nor should it be forgotten that seeds planted earlier continued to yield fruit. For example, in Brazil reformist ideas came into use and were realized after a lag by comparison with Europe (Pitombeira de Carvalho, 2006). The impending Second World War prevented efforts to develop international collaboration from coming to fruition.

From 1945 Until the Collapse of the Soviet Union

During the period after the Second World War the movement from the local to the international effectively began anew. International organizations and mechanisms for international collaboration sprung up once more. Even more significantly, however, national education systems underwent substantial changes. It was at this time that the colonial system collapsed, and new states appeared—and new school systems along with them. Around the world, education became much larger in scale. And even in the countries of Western Europe and North America, mathematics education began to be understood and implemented in new ways. The processes of change taking place in different countries often resembled one another, both because these changes were brought about by similar causes, and because what had been done in one country became an asset to others. Bass (2008) has noted that the period of the 1950s–1960s (part of what he called the "Freudenthal Era") already "witnessed the emergence of mathematics education (didactics) as an international academic discipline" (p. 10). An enormous role in this process was played by ICMI, which was re-established at the 1952 meeting of the IMU general assembly in Rome

(Kilpatrick, 2008), and which from 1969 on began holding International Congresses on Mathematics Education on a regular basis.

All around the world, however, the movement from the local to the international came up against the Iron Curtain, which separated opposed groups of countries. At certain moments during the Cold War and in certain countries, this opposition was especially emphasized in mathematics education, and what was international, that is, what came from abroad, was invariably regarded as erroneous and hostile (Karp, 2007a). Even during more peaceful periods, no common perspective on existing problems was possible, if only because of the difficulty of establishing contacts and communication.

This observation does not mean, however, that there was no movement across the Iron Curtain at all. Books from the countries of the Soviet bloc were published in the West (Kilpatrick, 2010), and something from the West did manage to penetrate into the countries of the Soviet bloc as well. Moreover, the opposition between the two blocs to some extent stimulated the development of mathematics education in both (recall the Sputnik). Below, I offer brief descriptions of the processes that unfolded during this period in five regions—Western Europe and North America, Eastern Europe, Latin America, Africa, and Asia/Australia.

Western Europe and North America. Along with the rebirth of ICMI, this period witnessed the appearance of organizations, conferences, and seminars which subsequently went on to become international in the full sense of the word, that is, encompassing the whole world, but which initially had a narrower audience of Western Europe and North America. A typical example was the Commission for the Study and Improvement of Mathematics Teaching (CIEAEM), which is officially considered to have been founded in 1952, although it first appeared in 1950. International collaboration and, in particular, the coming together of different countries in mathematics education, was naturally stimulated by growing collaboration in other fields, but they had their own causes as well, stemming above all from a recognition of the similarity of the problems being faced and their solutions (given all the existing differences).

Freudenthal (1978) began his overview of the changes taking place in education in the Netherlands with the observation that after the Second World War this predominantly agricultural country greatly changed economically, which required changes in education. New economic and social living conditions, even if they were somewhat different from those found in the Netherlands, spurred changes both in education in general and in mathematics education in particular in other countries as well (Keitel, Damerov, Bishop, & Gerdes, 1989). Howson (1978) stressed that even in the UK, with its formerly rigidly structured system which catered only to the academic child, a system intended for the average child began to develop. The idea of "mathematics for all" (Damerow, Dunkley, Nebres, & Werry, 1984) turned out to be consonant with what was taking place in many countries, but it led to changes in the traditional curriculum.

Another powerful and general movement in the direction of changing the curriculum was stimulated by the development of mathematics. As early as 1956, the well-known mathematician G. Choquet compared mathematics teachers to museum guards demonstrating archaic objects that were of no use to anyone (Charlot, 1989). A great deal in the traditional program appeared merely to obscure the essence and structure of mathematics, leaving no room for clarifying it and for introducing important new knowledge. In France, the seminar held in Royaumont in 1959, at which the famous slogan "Euclid must go!" was voiced, became an important landmark in this movement. In the USA, the new math appeared on the scene, supported and developed by the School Mathematics Study Group (Fey & Graeber, 2003); in the UK, reforms were conducted within the School Mathematics Project and by other groups of experimenters (Cooper, 1985); in Spain, analogous changes were implemented on the orders of the Ministry of Education (Ausejo, 2010), and so on. Similar (despite all their differences) movements also took shape beyond the boundaries of the region being examined here, and I shall come back to them below. Here, one should just note that the reform movement, despite the fierce and often justified criticism directed against it, significantly and universally enriched the school program: Fey (1978) noted that even at the time when his article was written, it was "hard to imagine that so many ingredients of the 'New Math' proposals were completely foreign to most mathematics programs and teachers in 1960" (p. 341).

Another movement, which is quite often mistakenly considered to have been necessarily opposed to the one just described, and which also found support in many countries, may be characterized as a struggle for "realistic mathematics." This terminology appeared in the Netherlands (Gravemeijer, 1994), but attention to modelling, and to real-world problems in general, continued to grow in many other countries also (Pollak, 2003). Again, the reasons for this growth—shared by different countries—included, on the one hand, the fact that the curriculum was now being adapted to all students, including those who had no need of mathematics for college but who needed mathematics for everyday life and, on the other hand, the fact that it was recognized how often mathematics (and mathematics of a rather high level!) turned out to be indispensable for everyday life.

Similar conceptions of the psychological–pedagogical foundations of mathematics education were developed in different countries which were based on similar changes in the understanding of the value of the individual human personality. The ideas of the Swiss Jean Piaget or of the American Jerome Bruner were accepted in different countries and consequently, at least in educational rhetoric, the value, for example, of active discovery learning came to be universally acknowledged. An understanding of the more complex nature of education, which could not be reduced to the simple transmission of knowledge from teacher to student, also became widespread. Among the publications that influenced the formation of views shared by mathematics educators in different countries, the works of the Russian Lev Vygotsky (1962) must be mentioned, although only about half of the book written by Vygotsky came to be available in its first translation, and quite an imperfect translation at that. But here I must turn to what went on in the countries of the Soviet bloc.

Eastern Europe. Howson (1980) once posed the question, in his review of Swetz's (1978) book: "Does socialist mathematics education exist?" Indeed, as Howson rightly noted, the similarities between the curricula of countries belonging to different geopolitical camps were, at least at first glance, very great. In fact, it is natural to regard the term socialist itself merely as a technical designation of the countries of the Soviet bloc. Still, mathematics education in these countries deserves to be addressed separately if only because of their separation from the rest of the world, whose repercussions could be seen even in mathematics itself (see, e.g., Lohwater, 1957), and all the more so in education.

In the Soviet Union, after the Second World War, there was a return to the style of mathematics education that had existed at the beginning of the 20th century—both in terms of the pedagogical system and in many respects in terms of the content—with the difference, however, that now this education was offered to an incomparably greater number of children than had been the case previously (Karp, 2010). During the 1960s and 1970s, a campaign of reforms spearheaded by Andrey Kolmogorov, which in many ways resembled the reform movements in the West, transformed education in the country (Abramov, 2010; Karp & Vogeli, 2010, 2011). No analogue to "mathematics for all" appeared, however. The educational system strove to become less selective (especially by the late 1970s and into the 1980s) and, in particular, to ensure a much higher pass rate in mathematics, but that was achieved less by restructuring the curriculum than by simply lowering the actual requirements.

Some works of Western mathematics educators reached Soviet readers. Thus, for example, Polya's *How to Solve It* was translated and published as early as 1959. Such works, however, were few in number. Contacts with foreign, especially Western, colleagues were also extremely limited.

Matters were more complicated with mathematics education in the other so-called socialist countries, where mathematics education developed as a result of exposure to various influences. First, many of these countries had their own strong national schools, to whose traditions educators strove to adhere even after the Second World War; second, the influence of the "older brother"—the Soviet Union—was inevitable (Alonso, Fried, & Pardala, 2010); and third, to one degree or another, the influence of what was taking place in the West also made itself felt, for example, through participation in CIEAEM and other international organizations (Ehrenfeucht, 1978), in which representatives from these countries usually participated to a greater extent than Soviet mathematics educators did. The scale of mathematics education, in terms of the number of students who received it, grew substantially in these countries (see, e.g., Halmos & Varga, 1978), but its structure and the directions of its development often turned out to be different from both many Western countries and the USSR. Halmos and Varga, for example, emphasized how different Hungarian reforms were from the new math.

Latin America and the Caribbean. In many countries of Latin America and the Caribbean, poverty and extreme social inequality have reigned for decades, preventing the development of education in general and mathematics education in particular. For this reason, Lluis (1986) wrote that, with regard to Latin America, it was necessary to pose such specific questions as: "What kind of geometry should be taught in primary schools, since 90% of pupils do not pursue their education any further?" (p. 38). Thus, for example, Colombia acquired its first group of graduates in mathematics only in 1934, and it was not until 1956 that Costa Rica had a single institution for the preparation of secondary school teachers—and those cases represented the rule for the region rather than the exception (Sangiorgi, 1962). Interactions with foreign and international influences in countries of this region therefore possessed a somewhat different character from that in the countries discussed above.

This does not mean that countries in this region had no educational traditions of their own. Such traditions did arise, taking shape in part through the adaptation of borrowings from abroad (Lluis, 1986, mentioned that "the admirable [textbook] of Wentworth and Smith" (p. 32) was used in the region 40 and 50 years after its publication). Assistance for mathematics education from abroad, which was actively offered during the years in question, did not always take these traditions into account, however (although assistance both from such international organizations as UNESCO and from individual countries—the USA, the UK, the USSR, and many others—was undoubtedly substantial). Arguments to the effect that Latin American countries had no "national culture" did not at that time appear dubious even to mathematics educators engaged in fruitful work in the countries of the region (Rosenberg, 1989). The study of ethnomathematics arose, to some degree, as a counterweight to these arguments (D'Ambrosio, 1977, 2006); it stressed the importance of the cultural context in mathematics education and quickly won recognition around the world.

Foreign trends, including the reform movement, not infrequently reached the countries of Latin America after a delay and, even more importantly, the changes to which they gave rise sometimes amounted to nothing more than "the introduction of long and unfamiliar names for simple ideas." In addition, sometimes these changes went no further than articles or academic discussions, while in actual schools everything remained as before (Wilson, 1978). At the same time, although the role of international agencies was indeed considerable, it would be wrong to attribute everything to their doings (Brito, 2008). Reformist ideas resonated with the mathematics educators of Latin America, who were developing their own teaching materials and curricula for teacher preparation (Búrigo, 2009). The region, although constrained by its own specific characteristics, nevertheless participated in the worldwide movement to renew and reform school curricula.

Africa. Until the late 1950s and the first half of the 1960s, almost the entire African continent was divided among European powers. Although European formal education began to penetrate into the continent virtually from the moment of the appearance of European missionaries there, it usually amounted to religious propaganda with only very minor additions—even during the first half of the 20th century, the ability to perform the four arithmetical operations was usually the summit of any education that was carried out on a large scale. The European powers established more advanced educational institutions, if only to prepare functionaries for colonial administrations (Cross, 2001), but the number of such institutions was very small—even at a considerably later date, when African countries had become independent. In Sudan in 1975–1976, for example, 81.3% of all students were enrolled in elementary schools, but so-called general secondary and higher secondary schools accounted for only 13.8% and 4.9% of all students, respectively (El Sawi, 1978). A similar situation existed in other countries on the continent (see, e.g.,

Mwakapenda, 2002). Nor should it be forgotten that even elementary education was by no means received by all children.

The system of education left behind by the colonial powers could in no way be considered local. To a very great extent, it reproduced the system of the colonial power, using, for example, British curricula, textbooks, and systems of examinations (Doku, 2003). In many ways and in many places, the system inherited from the colonial period continued to function after the colonized countries acquired independence. The subsequent development of education in African countries relied considerably on international aid. An important step forward was the UNESCO-sponsored Conference of African Ministers of Education in 1961. As Ohuche (1978) noted: "Rapid expansions of primary, secondary and teacher education followed. But the push for quality and appropriateness fell behind and more stress came to be put on minor modifications of foreign curricula than on bold and innovative educational experiences dictated by the African environment" (p. 272).

Through the joint efforts of a number of African countries and economically more developed states, quite ambitious projects were implemented in order to improve mathematics education (such as the African Mathematics Program, also known as the Entebbe Project, to name just one example). Aid in education came from both Western countries and the countries of the Soviet bloc—the struggle for influence which was taking place around the world did not bypass African mathematics education. In some countries, one could find American or British textbooks; in others, textbooks from the USSR.

The workshops and seminars that were conducted or the textbooks that were prepared within the framework of these projects were undoubtedly beneficial (Ohuche, 1978), although some participants in such projects noted that they could have been more effective (Karp, 2008). Although political and economic circumstances were largely to blame for this ineffectiveness, criticisms of methodological approaches were also heard.

El Sawi (1978) noted that while Western countries were going through a technological revolution, African countries were only approaching an industrial revolution. Consequently, attempts to introduce the latest Western developments—new math—into African schools met with difficulties. In curricula, the "over-dose of abstraction has been too much for the African child to assimilate" (p. 318). The living conditions of African children, quite different from the living conditions of their Western peers (and their peers in the USSR, too), caused them to develop in a different direction, and those differences could not be ignored.

To counterbalance foreign or international influences, educators voiced a desire "to create curricula based on the local culture or society" (Davis, 1992, p. 31). The study of such cultures achieved certain results (e.g., Zaslavsky, 1973); nonetheless, Davis (1992) expressed a quite skeptical view concerning achievements in "Africanizing" mathematics education.

Asia/Australia. This most densely populated region is likely also the most diverse in terms of the characteristics of its systems of education. Some of the region's countries (e.g., Australia) resemble, albeit with certain differences, the countries examined above in the section on Western Europe and North America, and consequently they actively participated in the processes described in that section. On the other hand, many of the region's countries acquired independence only after World War II, and the legacy of a colonial or semi-colonial past could be fully felt in them for the duration of the period that I am examining. In particular, certain countries in the region were plagued by mass poverty, which is inevitably destructive for mathematics education.

Perhaps the only process that was to some degree common to practically all of the region's countries was the growth of the population that was being taught mathematics. In China, the region's most populous country, approximately 80% of the population was illiterate in 1949 (Ziqiang & Monroe, 1991). Kapur (1978) noted that in India "after independence in 1947, enrollments in schools have increased 30–40 times" (p. 245). Gunawardena (1978) noted that in Sri Lanka, by the end of the colonial period, only 10% received any school mathematics education at all, whereas by the end of the 1950s this number had doubled.

Nonetheless, although mathematics education continued to spread, not all children were encompassed by it even by the end of the period discussed here. Nor are all children encompassed by it at present. Moreover, certain countries in the region went through a dramatic period of regress and destruction in their educational systems, the Chinese Cultural Revolution being one important example (Ziqiang & Monroe, 1991).

The literature (admittedly, starting at a somewhat later date, the mid-1990s) often emphasizes the resemblances between the countries if not of the entire region, then of a large part of it—East Asia (Leung, 2001; Leung, Graf, & Lopez-Real, 2006). According to the studies cited, many of these countries are united by their shared Confucian culture, which stretches far beyond the borders of the local and the national. In mathematics education, it manifests itself, for example, in the preference given to hard school work over the principle of pleasurable learning, which the authors consider typical of the West.

Over the last decade and a half, international attention to East Asian countries has noticeably grown, if only because of their high results in the Third International Mathematics and Science Study (TIMSS) and the subsequent Trends in International Mathematics and Science Studies (TIMSS, also). The culture referred to by the researchers, however, evolved long before this. Its appearance may be explained in different ways. For example, it seems reasonable to suppose that an important role is played by the fact that education in the region's countries is very selective, whereas the wealth gap between those who do and those who do not receive it is quite considerable. In general, while recognizing the deep cultural connections between, say, Taiwan and mainland China, one cannot but notice substantial political and economic differences between them during the period examined here, which also found expression in mathematics education.

Interactions with other countries took place in different ways across the region. The former colonial powers continued to exert a considerable cultural influence. This influence was exercised both through special organizations, such as the British Council (Gunawardena, 1978), and simply because the colonial powers' universities and publications remained the most natural sources of learning in the former colonies.

Consequently, many of the region's countries became involved, even if often with a certain lag, in broad international movements, including reform movements and the fight for modern mathematics (see, e.g., Kapur, 1978; Purakam, 1978).

Aid given by Western countries was not infrequently criticized in this region, too, and Western countries were rebuked, for example, for their lack of attention to cultural distinctions and their ill-considered application of Western theories in a region where corresponding findings might not apply (Clements & Ellerton, 1996). It is also clear that when graduates of Oxford or the Sorbonne returned to their homelands, they often encountered problems that were quite different from those found in Britain or France. What is noteworthy, however, is that when reforms began to be carried out in China and an open-door policy was established, Chinese mathematics educators tirelessly emphasized the importance of disseminating in their countries knowledge about the achievements of international educational theory and familiarity with the best foreign (American, British, Russian, etc.) practices (Quan, 1992; Ziqiang & Monroe, 1991).

The region's countries collaborated with the countries of the Soviet bloc, too. "Aid" for countries that had "embarked on the path of socialism" included assistance in organizing mathematics education. Ziqiang and Monroe (1991) noted that in China "older teachers are grounded in Russian educational theory" (p. 206). Russian influences on the Chinese secondary school mathematics curriculum were also evident (Leung, 1987).

One must also note collaboration within the region itself. Regional organizations and centres, such as the Mathematics Education Research Group of Australasia (MERGA) or the Regional Centre for Education in Science and Mathematics (RECSAM), have functioned effectively (and continue to do so to this day). Consequently, regional conferences, workshops, and seminars are held, and journals that bring together the region's mathematics educators are published.

Some Conclusions

The period examined above was extremely important for the formation of the international mathematics community. The degree of involvement in common or similar activity, as well as the nature of that involvement, differed from country to country and from region to region.

Strangely, although the period in question is still relatively recent, the channels of communication that existed among mathematics educators at the time have not been fully determined. In particular, it is not entirely clear how and to what extent the exchange of ideas developed in general during the period of reforms corresponding to the American new math. For example, the director of the School Mathematics Study Group in the USA, Edward G. Begle, was a guest speaker at the seminar in Royaumont, and Andrey Kolmogorov in the USSR was also well aware of it. But the degree to which these and other educators analyzed this or that specific idea—born and spread in this or that specific country—remains uninvestigated.

The existence of mutual influences, however, cannot be denied, which is why this movement as a whole may be considered international.

Examples of the international exchange of ideas are not few in number. Thus, for example, Isaak Wirszup, who did much to popularize the works of the Netherlands' van Hieles in the USA, himself learned about them from Soviet literature (Roberts, 2010). On a more fundamental level, I must mention the contribution of such international organizations as UNESCO, which facilitated such exchanges and international collaboration in mathematics education in general (Hodgson, 2009; Jacobsen, 1993).

Although noting "a growing mutual influencing of ideas, methods, practices and expectations," Bishop (1992) emphasized that they did not lead to "unification and conformity of research, although there are similarities in approach to be seen in different countries" (p. 710). He saw the international perspective as being embodied in "increased researchers' awareness of a number of key issues arising from the historically, culturally, and socially different approaches to mathematics education seen around the world" (p. 711). Bishop identified the basic types of research and to some degree even indicated which types were more widespread in various countries, although he added that this did not preclude other types of research from being carried out in these countries.

The growth of the international element in research in mathematics education in the sense indicated above is evident at the very least simply because research in mathematics education in general underwent a rapid development during these years (Clements & Ellerton, 1996; Kilpatrick, 1992). The number of researchers, studies, and publications grew along with the belief in "the role of research in the improvement of mathematics education"—as Begle (1969) titled his influential address—and with growing expectations concerning the scientific qualifications of those who were involved in mathematics teacher preparation. At the same time, educators in different countries sometimes studied similar topics, sometimes different ones; sometimes employed similar methodologies, sometimes different ones; sometimes knew about research being done abroad, and sometimes did not. It is possible to identify groups of countries, based on social-economic, linguistic, or cultural-historic principles, such that within each group awareness about what was taking place in other countries was higher. The growth of such awareness, however, has accelerated in the contemporary period, which we now consider.

The Contemporary Period

I date the contemporary period from the collapse of the Soviet Union, although of course this date is not always precise—many processes that are discussed below were already underway, in one form or another, even before this event. Still, since the end of the 1980s, since the fall of the Berlin Wall, the world has been perceived as unified to a much greater extent than it was previously, and this view could not fail to influence mathematics education. Another important feature of the new period is the technological revolution which has occurred. The Internet has gradually reached, if not every household, then at least every university. Atweh and Clarkson (2001) painted a picture of collaboration among Spanish, American, and Canadian educators helping Salvadoran students from home—and that picture, at least from a technical point of view, is absolutely plausible and realistic. If in times past information about what was happening in one or another country could be tracked down only in special reports prepared by special commissions—if it could be tracked down at all—then now this, too, can be done without leaving the house, provided one has the desire, knowledge of the language, and ability to understand a foreign culture.

Against the background of greater openness and interconnectedness among different countries, researchers have been conducting increasing numbers of various international studies, beginning with TIMSS, PISA (Programme for International Student Assessment) and ICMI studies, all of which have attracted an enormous amount of attention. International studies and their results have to some degree become a part of mass culture, inheriting all of its positive and negative aspects. An article by Romberg (1971) about an international comparative study carried out over 40 years ago pointed out certain phenomena and tendencies which have since that time only continued to grow.

In developed countries, mathematics education has clearly become organized along more similar lines. In Western countries, tendencies toward centralization had risen even at the end of the last period that I examined above—a national curriculum had appeared in the UK, and the standards movement had been gaining strength in the USA. On the other hand, in countries with a rigid curriculum designed by a centralized authority (such as the USSR), certain requirements had begun to be eased.

The idea of assessing performance in mathematics education (by no means new) has gained new momentum virtually everywhere since computer technologies opened up substantial new possibilities in this respect. The formation of new international organizations of mathematics educators, a process that had begun earlier, continued during these years. (I might mention as examples the recently formed World Federation of National Mathematics Competitions, WFNMC, and the International Group for Mathematical Creativity and Giftedness, IGMCG). Research in mathematics education, as an academic field, has been generally speaking more international than mathematics education itself, which preserves close ties to local traditions. Below, I attempt briefly to trace certain aspects of the processes and discussions which have occurred.

Exchange of People and Ideas

The borders which opened up (or even slightly opened up) in the early 1990s made easier the exchange of people and ideas in mathematics education as well as other fields. Below, I have occasion to address scholarly publications, participation in conferences, and so on. However, there is another development that must be addressed first.

The "brain drain," which is discussed below, is on the whole a phenomenon with negative consequences for the development of those countries which scientists are

Table 25.1
Size of Foreign-Born Population in the USA in 1980 and 2008

Born in …	1980	2008
China	289,079	1,339,131
Mexico	2,199,221	11,451,299
Nigeria	25,528	200,001
USSR[a]	500,728	1,096,905
World Total	14,079,906	38,016,102

[a] The figures opposite "USSR" in both columns represent the total number of immigrants from all of the now-independent countries which in 1980 were part of the Soviet Union.

leaving; at the same time, however, it provides new possibilities for the growth of international awareness concerning the achievements and problems of those countries. The number of mathematics educators who live in English-speaking countries or countries where the English language has long been used as a language of scholarship and science, and who speak and read fluently in Chinese, Spanish, or Russian, has grown dramatically over the past 20 years. Moreover, even emigrants who seemingly have no connection with mathematics education bring with them traditions and views that are widespread in their countries of origin, thus facilitating deeper understandings of the organization of mathematics education in the countries from which they came.

As an example illustrating the changes taking place, consider the statistics in Table 25.1 relating to the foreign-born population of the USA in 1980 and 2008 (Pew Hispanic Center, 2008; U.S. Census Bureau, 1999).

From another perspective, more foreigners have started visiting or working in countries that were closed off previously (An, 2008; Watson, 1993). Western countries have seen an increasing number of students, including doctoral students, from Asia, Eastern Europe, and Latin America. Although I have no data about the numbers of students specializing in mathematics education, it will be useful to examine Table 25.2, which shows the number of foreign students in all fields in the USA for the years 1984–1985 and 2009–2010 (see Institute of International Education, 2010; Zikopoulos, 1985) and offers an illustration of the processes taking place.

To this must be added the fact that contacts which are established today are incomparably easier to maintain than they were in the past, thanks once again to new technologies.

The Process of Internationalization in Mathematics Education Research

Mathematics education research as a scholarly field has developed rapidly over recent decades at least in terms of numbers of publications. The number of mathematics education journals has grown (Hanna, 2003) and continues to grow, which

Table 25.2
Number of Foreign Students in the USA

Nation of Origin	1984–1985	2009–2010
Africa	33,778	37,062
China	8,637	127,628
Europe	28,508	85,084
Latin America	41,519	65,632
USSR	196	12,707[a]
World total	292,479	690,923

[a]This figures opposite "USSR" represent the total number of students from all of the independent countries that formerly made up the Soviet Union. For 2009–2010, the number of students from the largest of these countries, Russia, was 4827.

undoubtedly is made easier by the possibility of publishing online. At the same time, the distribution of these publications among different countries is by no means proportional to their population.

Adler, Ball, Krainer, Lin, and Novotná (2005), in studying scholarly articles about mathematics teacher education, noted that "research in countries where English is the national language dominates the literature" (p. 372). More precisely, the share of such articles about the topic in question between 1998 and 2003 was 71% in the *Journal for Research in Mathematics Education* (JRME), 80% in the *Journal of Mathematics Teacher Education* (JMTE), and 43% in Psychology of Mathematics Education (PME) proceedings.

Adler et al. (2005, p. 373) provided more detailed data (see Table 25.3).

Naturally, one might ask how representative these statistics are and whether a different period or a different area of mathematics education research would yield similar figures (note, too, that in such comparisons among numbers of articles, etc., and their distribution across regions—and even more so, across countries—one must always bear in mind that these regions' population sizes may be considerably different). In certain journals which have recently appeared, the proportion of articles from English-speaking and non-English-speaking countries differs somewhat from that in the journals analyzed above. Thus, of the 39 articles published in the "Research Papers" section of the *International Journal for the History of Mathematics Education*, only 31% came from English-speaking countries. Of the 50 articles published in 2010 in the *International Journal of Science and Mathematics Education*, articles such that all authors resided in English-speaking countries constituted merely 32% of the total.

Generally speaking, the fact that in a journal published in English many of the authors are from English-speaking countries is not surprising. What is interesting is the change over time in the share of articles from non-English-speaking countries. Consider, for example, the data pertaining to *JRME,* which is not officially an international publication even though its sponsor, the National Council of Teachers

Table 25.3
Distribution of Published Articles About Mathematics Teacher Education, by Region Where Research Was Conducted (from Adler et al., 2005)

Region	JMTE ($n=65$)	PME ($n=88$)	JRME ($n=7$)
North America	68% (65% USA)	30% (24% USA)	57% (all USA)
Oceania	8%	9%	0
Europe	15% (5% UK)	25% (6% UK)	14%
Africa	3% (all South Africa)	8% (6% South Africa)	14% (all South Africa)
Asia	5%	9% (7% Taiwan)	0
South and Central America	0	3% (all Brazil)	0
Inter-continental	0	0	0
Middle East	2% (all Israel)	14% (all Israel)	14% (all Israel)

Table 25.4
Statistics on Research Articles Published in JRME

Category	1975	1985	1995	2005
Number of articles per year in which at least one author is not from the USA	2	4	3	7
Number of articles per year in which at least one author is not from an English-speaking country	1	4	2	5
Total number of research articles analyzed	23	17	19	12

of Mathematics, encompasses the USA and Canada. These data indicate that even in this journal a tendency toward internationalization is present (evidently, varying from year to year) (Table 25.4).

The number of conferences is likewise growing. "Specialized" conferences are being organized—that is, conferences devoted to this or that specific topic (often such conferences are organized by international organizations, or conversely, they serve as foundations out of which international organizations subsequently arise—the former case may be illustrated by PME, the latter by IGMCG). General subject conferences are organized as well, including regional conferences such as Inter-American Conferences or the Congresses of the European Society of Research in Mathematics Education (regional conferences are also often conducted by organizations of mathematics educators, which frequently receive support from ICMI).

The spectrum of countries represented at conferences is usually quite broad, although countries and regions are usually represented unequally. Consider, for example, the following data about presentations at two CIEAEM conferences held in Italy and Hungary (in the latter case, I do not include data about poster presentations). With respect to Table 25.5, it should be recalled that CIEAEM initially arose first and foremost as a Western European organization. As we can see, the conference remains predominantly European (at least, when it is held in Europe), although other regions are also represented. It is noteworthy that the differences between Eastern and Western Europe are no longer very significant.

Table 25.5
Statistics Relating to Countries with Which the Authors of the Presentations at the CIEAEM Conferences Are Affiliated

	CIEAEM 57 (2005)[a]	CIEAEM 59 (2007)[a]
Africa	1	0
Asia	1	1
Australia and New Zealand	1	2
Eastern Europe	23	13
Western Europe	30	27
Latin America	1	4
North America	5	13
Middle East	6	2

[a] See CIEAEM (2005, 2007). In those very few cases in which an author gives two countries of residence or in which an article has authors from two countries, both countries are counted.

Whereas the expansion of the geography of scholarly work in the field of mathematics education may be demonstrated without much difficulty (although, to repeat, the representation of different regions is still far from proportional), the growing similarity among styles and methods of scholarly studies (described by Bishop, 1992), as well as their topics, in different countries is more difficult to show, although there is evidence of that, too. At the same time, the differences between the approaches accepted in different countries remain quite considerable (see, e.g. Karp & Leikin, 2011).

It would be interesting to attempt to measure the growth of awareness about what is taking place in research in mathematics education in other countries. In particular, one would like to assess the degree to which the achievements of researchers are known in those countries in which the research was *not* conducted. As a first step toward such a study, I point out that in the 92 summaries of Russian doctoral dissertations in mathematics education mentioned by Karp and Leikin, 2011 (each of these summaries is in the order of 80,000 characters), I identified only three references to scholars who received ICMI's top awards, the Freudenthal and Klein Medals. For the sake of comparison, the names of outstanding Western mathematics educators of earlier periods, such as Polya and Freudenthal, were mentioned far more frequently—45 and 28 times, respectively, whereas the names of leading Russian scholars were usually mentioned several times in each work. An analysis of the Russian-language Google Scholar gives similar results. At the time of the writing of this chapter, there were three references to scholars who received ICMI's top awards, 828 references to Polya, and 90 references to Freudenthal (many of the references to Polya and Freudenthal, however, appeared in purely mathematical works). A similar picture emerges from an analysis of the Chinese-language Google Scholar, in which there were only 27 references to scholars who received ICMI's top awards.

Naturally, the data above may be explained in different ways. One explanation actually has already been given: the Russian-language Google Scholar in the field of mathematics education (as separate from mathematics) is still very poorly developed. Nonetheless, one may conclude that there is a degree of delay in adopting the achievements of other countries—due, in the very least, to the fact that they require translation.

Is Internationalization Useful?

Atweh and Keitel (2008) posed the question: Can international collaboration be unjust? They offered examples in which greater openness in the world led to a brain drain from less industrialized countries to industrialized countries, resulting in the loss of the best teachers. They also discussed the dominant role played by certain groups in international collaboration and consequently in the marginalization of other groups. The literature also points out the complicated role played by various international institutions that were designed to offer assistance in developing national education and indeed invest considerable resources in such development. For example, Atweh and Clarkson (2002) have written about the not-always-positive role played by the World Bank in education in Latin America. There have been presentations about the harm caused to Mongolian mathematics education by subsidies from Asian Bank, which spurred educators to restructure the system in a foreign and, in the critics' opinion, ineffective manner (Shevkin, 2010). Ernest (2008) compared such aid with what was once done by the Great Powers in their colonies, rebuking the education system thus being organized with harboring a "hidden curriculum in the form of views of knowledge, values and ideologies" (p. 32).

Evidently, these negative influences are felt most strongly in those places where national systems of mathematics education have not yet fully taken shape. However, even in countries with firm and old traditions of mathematics education, complaints about the negative influence of internationalization can also be heard. Thus, Hungarian mathematics educators have reported on the negative impact of transformations in higher education, carried out in order to bring Hungarian education in line with general European requirements, and organizational changes in the system of school education, also clearly influenced by foreign models (Connelly, 2010).

Probably the harshest criticism of globalization in mathematics education came from Igor Sharygin (2004), a former member of the ICMI executive committee. In his view, globalization has led to social polarization and to a division of mathematics education into several levels, from the serious and fundamental level, which is accessible to very few, to the mass level oriented at carrying out highly specialized and simple work. Sharygin denounced the destruction of the Russian system of mathematics education and pinned the blame for what is going on in Russia on the USA, which he sees as carrying out an ideological occupation of Russia.

In such rhetoric, one can detect the influence of Cold War propaganda, which always reduced all problems to the enemy's underhanded schemes. Nonetheless, there can be no doubt that the internationalization of mathematics education not only addresses existing problems but also creates new ones. As Atweh and Keitel (2008) noted, from the other side one hears voices that warn against investing too much in a fight "for the preservation of national culture," which according to this view can turn into a fight against the teaching of the new, thereby condemning the population of a country to falling behind in the future. While acknowledging the existence of tensions between traditional and imported cultures, Ernest (2008) pointed out that such tensions may also be quite productive, an example of this being English-language fiction, which has undoubtedly been enriched by the work of writers from former colonies.

How great are national differences really?

Baker and LeTendre (2005) commented: "Whether you find them in Mexico City, a small town in Pennsylvania, or in rural Kenya, schools all over the world appear to run in much the same way everywhere" (p. 3). And indeed, it is impossible not to agree that "widespread understanding repeatedly communicated across nations, resulting in common acceptance of ideas, leads to standardization and similar meaning, all happening in a soft, almost imperceptible, taken-for-granted way" (p. 10). No matter how much one might deplore the fact that schoolchildren in Uruguay, Poland, and South Korea are watching the same, low-quality films, this is a fact that cannot be denied, just as it is impossible to think that those films do not exert a similar influence on them. Consequently, Baker and LeTendre maintained that "emphasis on 'national cultures' of teaching is too simplistic" (p. 105).

More precisely, Baker and LeTendre (2005) did not deny certain differences, but it was differences within countries that they viewed as being the far more significant ones; as for differences between countries, these derived, in their view, not so much from national cultures, which, as they rightly point out, are not stable across time, as from various organizational details. Thus, analyses of TIMSS video recordings of classes taught by American, German, and Japanese mathematics teachers (Stigler & Hiebert, 1999) suggest that the differences "are not just the product of a different culture of teaching, they reflect basic contrasts in educational policy and school organization" (pp. 107–108), and in particular, different systems of organizing the work of the teacher.

Putting that idea somewhat more directly, one might conclude that if the Japanese system were to be introduced in the USA, mathematics lessons in the USA would become the same as mathematics lessons in Japan. Without even discussing whether this is in fact true, I should point out that the Japanese system, however, is not being introduced in the USA. I believe that the principal reason for this is the cultural differences between the two countries. The culture of education is a complex notion, which cannot be reduced to a list of the topics studied or to a system of scheduling classes that is common to practically all countries, and so on.[2] The culture of educational policy and school organization is a part of the culture of education.

---

[2] On that matter it is pertinent to reflect on Leo Tolstoy's description of Natasha Rostova, one of the characters in *War and Peace*, who presumably had been educated in an education system that was incomparably more similar to foreign education systems than what we observe today, even given all the communication across nations: "Where, how, when had she—this little countess, educated by an émigrée French governess—imbibed from the Russian air she breathed that spirit and obtained these mannerisms, which the *pas de chale* was supposed to have supplanted long ago? But the spirit and the mannerisms were the very same ones, inimitable, unlearned, Russian" (Tolstoy, 1980, p. 277).

It would be extremely interesting, although beyond the aims of this chapter, to investigate the formation of national cultures (see, e.g., Miliukov, 1960), including the national cultures of mathematics education, and the role played in their formation by foreign influences—Egyptian influences in Ancient Greece, Italian influences in French culture, Chinese or American influences in Japanese culture, and so on. For all of these influences, however, the differences between the cultures are not hard to spot. I would agree with Leung (2001) and Leung et al. (2006) that the differences between the East Asian and, say, the American classroom are quite substantial. These differences may be seen in the way lessons are structured, in the behaviours of the teacher and the students, in what is considered important in examinations, and in the aims of education in general, in the role allocated to mathematics among other subjects, and in a great number of other details.

Undoubtedly, national cultures (including national cultures of mathematics education) are not stable. If today the characteristic trait of Western education, including British education, is considered to be the striving to make learning pleasurable (Leung, 2001; Leung et al., 2006), then in the educational institutions described by Dickens or Charlotte Bronte this trait is impossible to find. On the contrary, teachers from classic British novels are often remembered as vicious sadists. The changes in values, goals, and objectives in education that have taken place were brought about by a complex combination of existing traditions and social-economic, political, and psychological changes. Changes in values, goals, and objectives in education are possible now, too, in all countries, but it would be misguided to think that they could take place merely as a result of students watching certain movies.

One may talk about national differences that are peculiar to mathematics education and hence about distinctive difficulties that give rise to the danger of oversimplification in comparative studies (Keitel & Kilpatrick, 1999). One may inquire what it is that sustains national differences in mathematics education and does not let them fade away (e.g., one may ask about the role played by "shadow education"—all kinds of conceivable additional extracurricular activities, whose significance was noted by Baker and LeTendre, 2005). At this point, however, one would be going beyond the bounds of mathematics education itself and coming up against a broader problem. At a certain time—during the euphoria that followed the end of the Cold War—it seemed that the same values, and hence the same organization of economic, political, and cultural life would triumph all over the world (Fukuyama, 1992). Speculatively, one might suppose that in such a relatively homogeneous world, mathematics education would also be relatively homogeneous. Without making futurological claims, I merely say that so far this has obviously not happened. Consequently, the internationalization of mathematics education has its limits as well.

Conclusion

International collaboration in mathematics education over the past 100 years has grown immeasurably. There is no reason to think that this process is completed or that mathematics educators in China, the Caribbean, Denmark, and Iran necessarily read the same books or value the same things in lessons. But the very rise in the number of countries whose representatives take part in international meetings, conferences, and congresses is telling in itself, pointing to the fact that the concept "international" now implies a far broader geographic range than it did 100 years ago.

Different countries are involved in international collaboration in different ways, have come to it along different paths, and expect different things from it. Mathematics is taught differently in different countries first and foremost because the need for mathematics education differs from country to country—it would be strange to prepare people for a future that they could only have abroad. For this reason alone, there is no cause to expect a mechanical transfer of what has been developed in one country into another. What we are looking at is a complex process, in which education facilitates the development of the country, but itself develops only to the extent that this development allows.

It would be naïve to think, therefore, that the growth of collaboration in mathematics education can be explained merely by the good will and inquisitiveness of mathematics educators. Mathematics education is developing in this direction following general processes taking place in the world. And yet, not a little depends on mathematics educators when it comes to discovering what is taking place abroad and making it accessible to those who are interested.

Much depends on genuine interest and the ability to go beyond the bounds of narrow frameworks that are suitable for only one or a few countries, and that do not take into account the existence of different understandings of the very same terms. Also important are relatively simple practical steps: Deborah Ball (in Adler et al., 2005) has spoken of the importance of studying foreign languages for future researchers in the field of mathematics education, for that might help them to become better acquainted with what is being done in other countries. There is also a need, I would argue, for a far wider abstracting, indexing and translating of scholarly articles—mathematics remains a model in this respect: there are widely available databases of a broad range of mathematics articles in different languages, but nothing of the kind yet exists in mathematics education. It is noteworthy that school textbooks in mathematics from different countries, of which vast numbers of copies are printed, are usually far less available in libraries than, say, collections of lyric poetry from the same countries, let alone books on mathematics.

And yet, let me stress once again that the path covered over the past century has been enormous. To an incomparably greater degree than before, the mathematics education community is now conceived of as being international, and more and more one turns to international work with the same hope that spurred David Eugene Smith to say that examining questions from an international point of view "would result in some very useful suggestions, without bringing about a uniformity in the organization of studies" (cited in Donoghue, 2008, p. 37).

Acknowledgements The author wishes to express his gratitude to Nerida Ellerton for a useful discussion of the content of this chapter and to the chapter's reviewers, Kristín Bjarnadttir and Patricia Cline Cohen, for their comments and suggestions.

References

Abramov, A. (2010). Toward a history of mathematics education reform in Soviet schools (1960s–1980s). In A. Karp & B. Vogeli (Eds.), *Russian mathematics education: History and world significance* (pp. 87–140). Hackensack, NJ: World Scientific.

Adler, J., Ball, D., Krainer, K., Lin, F.-L., & Novotná, J. (2005). Reflections on an emerging field: Researching mathematics teacher education. *Educational Studies in Mathematics, 60*(3), 359–381.

Albers, D. J., Alexanderson, G. L., & Reid, C. (1987). *International mathematical congresses. An illustrated history 1893–1986*. New York, NY: Springer Verlag.

Alonso, O. B., Fried, K., & Pardala, A. (2010). Russian influence on mathematics education in the socialist countries. In A. Karp & B. Vogeli (Eds.), *Russian mathematics education: History and world significance* (pp. 325–357). Hackensack, NJ: World Scientific.

An, S. (2008). Outsiders' views on Chinese mathematics education: A case study on the United States teachers' teaching experience in China. *Journal of Mathematics Education, 1*(1), 1–27.

Atweh, B., Barton, A. C., Borba, M. C., Gough, N., Keitel, C., Vistro-Yu, C., & Vithal, R. (Eds.). (2008). *Internationalisation and globalisation in mathematics and science education*. Dordrecht, The Netherlands: Springer.

Atweh, B., & Clarkson, P. C. (2001). Internationalization and globalization of mathematics education: Toward an agenda for research/action. In B. Atweh, H. Forgasz, & B. Nebres (Eds.), *Sociocultural research on mathematics education: An international perspective* (pp. 77–94). Mahwah, NJ: Lawrence Erlbaum.

Atweh, B., & Clarkson, P. C. (2002). Globalization and mathematics education: From above and below. In *Proceedings of the Annual Conference of the Australian Association of Research in Education*. Brisbane, Australia: University of Queensland, AARE.

Atweh, B., Clarkson, P. C., & Nebres, B. (2003). Mathematics education in international and global contexts. In A. J. Bishop, M. A. Clements, C. Keitel, J. Kilpatrick, & F. K. S. Leung (Eds.), *Second international handbook of mathematics education* (pp. 185–229). Dordrecht, The Netherlands: Kluwer Academic Publishers.

Atweh, B., Forgasz, H., & Nebres, B. (Eds.). (2001). *Sociocultural research on mathematics education: An international perspective*. Mahwah, NJ: Erlbaum.

Atweh, B., & Keitel, C. (2008). Social (in)justice and international collaborations in mathematics education. In B. Atweh, A. C. Barton, M. C. Borba, N. Gough, C. Keitel, C. Vistro-Yu, & R. Vithal (Eds.), *Internationalisation and globalisation in mathematics and science education* (pp. 95–111). Dordrecht, The Netherlands: Springer.

Ausejo, E. (2010). The introduction of "modern mathematics" in secondary education in Spain (1954–1970). *International Journal for the History of Mathematics Education, 5*(2), 1–14.

Baker, D. P., & LeTendre, G. K. (2005). *National differences, global similarities: World culture and the future of schooling*. Stanford, CA: Stanford University Press.

Bass, H. (2008). Moments in the life of ICMI. In M. Menghini, F. Furinghetti, L. Giacardi, & F. Arzarello (Eds.), *The first century of the International Commission on Mathematics Instruction (1908–2008). Reflecting and shaping the world of mathematics education* (pp. 9–24). Rome, Italy: Instituto della Enciclopedia Italiana.

Begle, E. G. (1969). The role of research in the improvement of mathematics education. *Educational Studies in Mathematics, 2*(2/3), 232–244.

Bishop, A. J. (1992). International perspectives on research in mathematics education. In D. Grouws (Ed.), *Handbook of research on mathematics teaching and learning* (pp. 710–723). New York, NY: Macmillan.

Brickman, W. (1960). A historical introduction to comparative education. *Comparative Education Review, 2*(3), 6–13.

Brito, A. (2008). Case study about how Bourbakism became implemented via international agencies in a key region of Brazil. *International Journal for the History of Mathematics Education, 3*(2), 65–72.

Búrigo, E. Z. (2009). Modern mathematics in Brazil: The promise of democratic and effective teaching. *International Journal for the History of Mathematics Education, 4*(1), 29–42.

Charlot, B. (1989). Institutional and socio-economic context of the "modern mathematics" reform in France. In C. Keitel, P. Damerov, A. Bishop, & P. Gerdes (Eds.), *Mathematics, education and society* (Science and Technology Education Document Series, Vol. 35, pp. 58–59). Paris, France: UNESCO.

CIEAEM. (2005). *Changes in society: A challenge for mathematics education.* (Proceedings of CIEAEM 57). Piazza Armerina, Italy: Author.

CIEAEM. (2007). *Mathematical activity in classroom practice and as a research object in didactics: Two complementary perspectives* (Proceedings of CIEAEM 59). Dobogoko, Hungary: Author.

Clements, M. A., & Ellerton, N. F. (1996). *Mathematics education research: Past, present and future.* Bangkok, Thailand: Asia-Pacific Centre of Educational Innovation for Development/UNESCO.

Cohen, P. C. (2003). Numeracy in nineteenth-century America. In G. M. A. Stanic & J. Kilpatrick (Eds.), *A history of school mathematics* (pp. 43–76). Reston, VA: National Council of Teachers of Mathematics.

Connelly, J. (2010). *A tradition of excellence transitions to the 21st century: Hungarian mathematics education, 1988–2008* (Doctoral dissertation). Teachers College, Columbia University. Retrieved from http://www.proquest.com/en-US/.

Cooper, P. (1985). *Renegotiating secondary school mathematics. A study of curriculum change and stability.* London, UK: Falmer.

Coray, D., Furinghetti, F., Gispert, H., Hodgson, B., & Schubring, G. (Eds.). (2003). *One hundred years of L'Enseignement Mathématique: Moments of mathematics education in the twentieth century.* Geneva, Switzerland: L'Enseignement Mathématique.

Cross, A. E. (2001). *The study of mathematics teaching and learning in selected Eritrean secondary schools* (Doctoral dissertation). Tennessee State University. Retrieved from http://www.proquest.com/en-US/.

D'Ambrosio, U. (1977). Science and technology in Latin America during discovery. *Impact of Science on Society, 27*(3), 267–274.

D'Ambrosio, U. (2006). *Ethnomathematics: Link between traditions and modernity.* Rotterdam, The Netherlands: Sense Publishers.

Damerow, P., Dunkley, M. E., Nebres, B. F., & Werry, B. (1984). *Mathematics for all. Problems of cultural selectivity and unequal distribution of mathematical education and future perspectives on mathematics teaching for the majority* (Science and Technology Education Document Series, Vol. 20). Paris, France: UNESCO.

Davis, J. C. (1992). *Young children's mathematical knowledge in Benin and the United States* (Doctoral dissertation). Teachers College, Columbia University. Retrieved from http://www.proquest.com/en-US/.

Doku, P. A. (2003). *Ghanian senior secondary school mathematics curriculum: Professors', teachers' and students' perceptions* (Doctoral dissertation). Teachers College, Columbia University. Retrieved from http://www.proquest.com/en-US/.

Donoghue, E. F. (2008). David Eugene Smith and the founding of the International Commission on the Teaching of Mathematics. *International Journal for the History of Mathematics Education, 3*(2), 35–46.

Ehrenfeucht, A. (1978). Change in mathematics education since the late 1950s—Ideas and realization: Poland. *Educational Studies in Mathematics, 9*(3), 283–295.

El Sawi, M. (1978). Change in mathematics education since the late 1950s—Ideas and realization: Sudan. *Educational Studies in Mathematics, 9*(3), 317–330.

Ernest, P. (2008). Epistemological issues in the internationalization and globalization of mathematics education. In B. Atweh, A. C. Barton, M. C. Borba, N. Gough, C. Keitel, C. Vistro-Yu, & R. Vithal (Eds.), *Internationalisation and globalisation in mathematics and science education* (pp. 19–38). Dordrecht, The Netherlands: Springer.

Fey, J. (1978). Change in mathematics education since the late 1950s—Ideas and realization: U.S.A. *Educational Studies in Mathematics, 9*(3), 339–353.

Fey, J., & Graeber, A. (2003). From the new math to the agenda for action. In G. M. A. Stanic & J. Kilpatrick (Eds.), *A history of school mathematics* (pp. 521–558). Reston, VA: National Council of Teachers of Mathematics.

Freudenthal, H. (1978). Change in mathematics education since the late 1950s—Ideas and realization: The Netherlands. *Educational Studies in Mathematics, 9*(3), 261–270.

Fukuyama, F. (1992). *The end of history and the last man.* New York, NY: Free Press.

Furinghetti, F. (2003). Mathematical instruction in an international perspective: The contribution of the journal *L'Enseignement Mathématique*. In D. Coray, F. Furinghetti, H. Gispert, B. Hodgson, & G. Schubring (Eds.), *One hundred years of L'Enseignement Mathématique: Moments of mathematics education in the twentieth century* (pp. 19–46). Geneva, Switzerland: L'Enseignement Mathématique.

Gravemeijer, K. P. E. (1994). *Developing realistic mathematics education.* Utrecht, The Netherlands: Freudenthal Institute.

Gunawardena, A. J. (1978). Change in mathematics education since the late 1950s—Ideas and realization: Sri Lanka. *Educational Studies in Mathematics, 9*(3), 303–316.

Halmos, M., & Varga, T. (1978). Change in mathematics education since the late 1950s—Ideas and realization: Hungary. *Educational Studies in Mathematics, 9*(2), 225–244.

Hanna, G. (2003). Journals of mathematics education, 1900–2000. In D. Coray, F. Furinghetti, H. Gispert, B. Hodgson, & G. Schubring (Eds.), *One hundred years of L'Enseignement Mathématique: Moments of mathematics education in the twentieth century* (pp. 67–84). Geneva, Switzerland: L'Enseignement Mathématique.

Hodgson, B. R. (2009). ICMI in the post-Freudenthal era: Moments in the history of mathematics education from an international perspective. In K. Bjarnadóttir, F. Furringhetti, & G.Schubring (Eds.), *"Dig where you stand": Proceedings of the conference "On-going research in the history of mathematics education"* (pp. 79–96). Reykjavik, Iceland: The University of Iceland.

Howson, G. (1978). Change in mathematics education since the late 1950s—Ideas and realization: Great Britain. *Educational Studies in Mathematics, 9*(2), 183–223.

Howson, G. (1980). Socialist mathematics education: Does it exist? *Educational Studies in Mathematics, 11*(3), 285–299.

Howson, G. (2010). Mathematics, society, and curricula in nineteenth-century England. *International Journal for the History of Mathematics Education, 5*(1), 21–52.

Institute of International Education. (2010). *Open doors data.* Retrieved from http://www.iie.org/.

Jacobsen, E. C. (1993). The cooperation between ICMI and UNESCO. *ICMI Bulletin, 34,* 11–12.

Kapur, J. N. (1978). Change in mathematics education since the late 1950s—Ideas and realization: India. *Educational Studies in Mathematics, 9*(3), 245–253.

Karp, A. (2002). Klassik real'nogo obrazovaniya [Classic of genuine education]. *Matematika v shkole, 8,* 7–11.

Karp, A. (2007a). The Cold War in the Soviet school: A case study of mathematics. *European Education, 38*(4), 23–43.

Karp, A. (2007b). "We all meandered through our schooling …": Notes on Russian mathematics education during the first third of the nineteenth century. *British Society for the History of Mathematics Bulletin, 22,* 104–119.

Karp, A. (2008). Interview with Geoffrey Howson. *International Journal for the History of Mathematics Education, 3*(1), 47–67.

Karp, A. (2010). Reforms and counter-reforms: Schools between 1917 and the 1950s. In A. Karp & B. Vogeli (Eds.), *Russian mathematics education: History and world significance* (pp. 43–85). Hackensack, NJ: World Scientific.

Karp, A., & Leikin, R. (2011). On mathematics education research in Russia. In A. Karp & B. Vogeli (Eds.), *Russian mathematics education: Programs and practices* (pp. 411–486). Hackensack, NJ: World Scientific.

Karp, A., & Vogeli, B. (Eds.). (2010). *Russian mathematics education: History and world significance.* Hackensack, NJ: World Scientific.

Karp, A., & Vogeli, B. (Eds.). (2011). *Russian mathematics education: Programs and practices*. Hackensack, NJ: World Scientific.

Keitel, C., Damerov, P., Bishop, A., & Gerdes, P. (Eds.). (1989). *Mathematics, education and society* (Science and Technology Education Document Series, Vol. 35). Paris, France: UNESCO.

Keitel, C., & Kilpatrick, J. (1999). The rationality and irrationality of international comparative studies. In G. Kaiser, G. Luna, & I. Huntley (Eds.), *International comparisons in mathematics education* (pp. 241–256). London, UK: Routledge.

Kilpatrick, J. (1992). A history of research in mathematics education. In D. Grouws (Ed.), *Handbook of research on mathematics teaching and learning* (pp. 3–38). New York, NY: Macmillan.

Kilpatrick, J. (2008). The development of mathematics education as an academic field. In M. Menghini, F. Furinghetti, L. Giacardi, & F. Arzarello (Eds.), *The first century of the International Commission on Mathematics Instruction (1908–2008): Reflecting and shaping the world of mathematics education* (pp. 25–39). Rome, Italy: Instituto della Enciclopedia Italiana.

Kilpatrick, J. (2010). Influences of Soviet research in mathematics education. In A. Karp & B. Vogeli (Eds.), *Russian mathematics education: History and world significance* (pp. 359–368). Hackensack, NJ: World Scientific.

Lehto, O. (1998). *Mathematics without borders: A history of the International Mathematical Union*. New York, NY: Springer-Verlag.

Leung, F. K. S. (1987). The secondary school mathematics curriculum in China. *Educational Studies in Mathematics, 18*, 35–57.

Leung, F. K. S. (2001). In search of an East Asian identity in mathematics education. *Educational Studies in Mathematics, 47*, 35–51.

Leung, F. K. S., Graf, K. D., & Lopez-Real, F. J. (Eds.). (2006). *Mathematics education in different cultural traditions—A comparative study of East Asia and the West. The 13th ICMI Study*. New York, NY: Springer.

Lietzmann, W. (1924). New types of schools in Germany and their curricula in mathematics. *Mathematics Teacher, 17*, 148–153.

Lluis, E. (1986). Geometry teaching in Latin America. In R. Morris (Ed.), *Studies in mathematics education: Teaching of geometry* (pp. 31–42). Paris, France: UNESCO.

Lohwater, A. J. (1957). Mathematics in the Soviet Union. *Science, New Series, 25*(3255), 974–978.

Maksheev, Z. (1913). [Introduction]. In *Trudy 1-go vserossiiskogo s'ezda uchiteley matematiki* [Proceedings of the 1st All-Russian Congress of Teachers of Mathematics] (Vol. 1, pp. iii–xiv). St. Petersburg, Russia: Sever.

Malsch, F. (1927). The teaching of mathematics in Germany since the war. *Mathematics Teacher, 20*, 355–368.

Miliukov, P. N. (1960). *Outlines of Russian culture*. New York, NY: A. S. Barnes.

Mwakapenda, W. (2002). The status and context of change in mathematics education in Malawi. *Educational Studies in Mathematics, 49*, 251–281.

Ohuche, O. (1978). Change in mathematics education since the late 1950s—Ideas and realization: Nigeria. *Educational Studies in Mathematics, 9*(3), 271–281.

Pew Hispanic Center. (2008). Statistical portrait of the foreign-born population in the United States, 2008. Retrieved from http://pewhispanic.org/files/factsheets/foreignborn2008/Table%205.pdf.

Pitombeira de Carvalho, J. (2006). A turning point in secondary school mathematics in Brazil: Euclides Roxo and the mathematics curricular reforms of 1931 and 1942. *International Journal for the History of Mathematics Education, 1*, 69–86.

Pollak, H. O. (2003). A history of the teaching of modeling. In G. M. A. Stanic & J. Kilpatrick (Eds.), *A history of school mathematics* (pp. 647–672). Reston, VA: National Council of Teachers of Mathematics.

Polya, G. (1959). *Kak reshat' zadachu* [How to solve it]. Moscow, USSR: Uchpedgiz.

Purakam, O. (1978). Change in mathematics education since the late 1950s—Ideas and realization: Thailand. *Educational Studies in Mathematics, 9*(3), 331–337.

Quan, W. L. (1992). Chinese advancements in mathematics education. *Educational Studies in Mathematics, 23*, 287–298.

Reeve, W. D. (Ed.). (1929). *Significant changes and trends in the teaching of mathematics throughout the world since 1919. The fourth NCTM yearbook.* New York, NY: Teachers College, Columbia University.

Roberts, D. (2010). Interview with Izaak Wirszup. *International Journal for the History of Mathematics Education, 5*(1), 53–74.

Romberg, T. A. (1971). Publicity and educational research: A case study. *Journal for Research in Mathematics Education, 2*(2), 132–135.

Rosenberg, D. (1989). Knowledge transfer from one culture to another: HEWET from the Netherlands to Argentina. In C. Keitel, P. Damerov, A. Bishop, & P. Gerdes (Eds.), *Mathematics, education and society* (Science and Technology Education Document Series, Vol. 35, pp. 82–84). Paris, France: UNESCO.

Sangiorgi, O. (1962). *The present status of mathematics teaching in secondary schools in Argentina, Brazil, Chile, Colombia, Costa Rica, Peru, Uruguay and Venezuela.* Paris, France: UNESCO.

Schubring, G. (1989). Theoretical categories for investigations in the social history of mathematics education and some characteristic patterns. In C. Keitel, P. Damerov, A. Bishop, & P. Gerdes (Eds.), *Mathematics, education and society* (Science and Technology Education Document Series, Vol. 35, pp. 6–8). Paris, France: UNESCO.

Schubring, G. (2003). *L'Enseignement Mathématique* and the First International Commission (IMUK): The emergence of international communication and cooperation. In D. Coray, F. Furinghetti, H. Gispert, B. Hodgson, & G. Schubring (Eds.), *One hundred years of L'Enseignement Mathématique: Moments of mathematics education in the twentieth century* (pp. 47–66). Geneva, Switzerland: L'Enseignement Mathématique.

Schubring, G. (2008). The origins and the early history of ICMI. *International Journal for the History of Mathematics Education, 3*(2), 3–33.

Sharygin, I. F. (2004). Obrazovanie i globalizatsiya [Education and globalization]. *Novyi mir, 10,* 110–125.

Shevkin, L. (2010). *Vserossiiskii syezd uchiteley matematiki nachal rabotu* [All-Russian Congress of Mathematics Teachers begins]. Retrieved from http://www.shevkin.ru/?action=ShowTheFullNews&ID=522.

Sintsov, D. (1913). Mezhdunarodnaya komissiya po prepodavaniyu matematiki [International Commission on Mathematics Instruction]. In *Trudy 1-go vserossiiskogo s'ezda uchiteley matematiki* (Vol. 3, pp. 1–19). St. Petersburg, Russia: Sever.

Smith, D. E. (1918, December 10). To the editor of the *New York Times. New York Times*, p. 3.

Stigler, J. W., & Hiebert, J. (1999). *The teaching gap: Best ideas from the world's teachers for improving education in the classroom.* New York, NY: Free Press.

Swetz, F. (1978). *Socialist mathematics education.* Southampton, PA: Burgundy Press.

Tolstoy, L. N. (1980). *Voina i mir* [War and peace]. Moscow, USSR: Khudozhestvennaya literatura.

U.S. Census Bureau. (1999). *Region and country or area of birth of the foreign-born population.* Retrieved from http://www.census.gov/population/www/documentation/twps0029/tab03.html.

Vygotsky, L. S. (1962). *Thought and language.* Cambridge, MA: M.I.T. Press.

Watson, A. (1993). Russian expectations. *Mathematics Teaching, 145,* 5–9.

Wilson, B. J. (1978). Change in mathematics education since the late 1950s—Ideas and realization: West Indies. *Educational Studies in Mathematics, 9*(3), 355–379.

Zaslavsky, C. (1973). *Africa counts: Number and pattern in African culture.* Boston, MA: Prindle, Weber & Schmidt.

Zikopoulos, M. (Ed.). (1985). *Report on international educational exchange.* New York, NY: Institute of International Education.

Ziqiang, Z., & Monroe, E. E. (1991). Mathematics education in China today: Four problem areas. *Educational Studies in Mathematics, 22,* 205–208.

Chapter 26
International Collaborative Studies in Mathematics Education

Parmjit Singh and Nerida F. Ellerton

Abstract This chapter focusses on the concept of "collaboration," with particular reference to mathematics education research in which the participating scholars are from different nations. After commenting that collaboration involves more than sharing, uniting, or cooperating, the concept is discussed in the light of the work of ICMI, IEA, PISA, RECSAM, MERGA, PME, LPS, and international aid programs. After providing summaries of the work undertaken in these programs and organizations, the following seven dimensions that influence the quality of "collaboration" within a program or organization were informally identified: (a) clear statement of raison d'être, (b) consistency of actions with raison d'être, (c) level of democratic governance, (d) whether wider international discussion is stimulated, (e) the extent of influence on policies, (f) the extent of influence on practices, and (g) the extent of influence on research directions. Using these dimensions as criteria, we assessed the quality of collaboration in the work of each of the above-named programs or organizations. Our conclusion is that, whereas the early work of ICMI did not feature high-quality collaboration, the ongoing work of most aspects of the other programs and organizations does feature high-quality collaboration.

Introduction

Mathematics education, and research related to mathematics education, lie at the crossroads of many well-established knowledge domains such as mathematics, psychology, sociology, epistemology, and linguistics and may be concerned with

---

P. Singh (✉)
University Technology MARA Malaysia, Selangor, Malaysia
e-mail: parmjitsingh7@hotmail.com

N. F. Ellerton
Illinois State University, Normal, IL, USA

problems imported from one or more of those fields (Sierpinska et al., 1993). Each of the disciplines mentioned typically employs a range of research paradigms and methodologies, and each has well-established traditions governing research question formulation, research design and development, and the manner in which research results are reported.

Research in mathematics education can be distilled to mean a disciplined inquiry into the structure and content of mathematics curricula and the teaching and learning of mathematics. The ultimate goal of all research in mathematics education is to improve the quality of mathematics curricula and the quality of the teaching and learning of mathematics. In order to achieve those ends a strong research base is needed.

Throughout most of the 20th century, it was accepted that, of the disciplines mentioned in the first paragraph, two were of paramount importance for those wishing to conduct research in mathematics education. The first was mathematics, which in relation to mathematics education is concerned with the kind of content that is taught and learned; the second was psychology, which deals with cognitive and affective factors influencing mathematics learning.

During the last 25 years of the 20th century, mathematics education researchers became increasingly convinced that the research agenda of the international mathematics education community needed to be concerned with far more than mathematical content and personal factors affecting learning. In order to learn mathematics well, for example, researchers recognized that an understanding of relationships between symbols, signs, and abstract entities was required. Linguistic factors therefore needed to be examined. In addition to quantitative research methodologies, qualitative methodologies were needed since research in mathematics education inevitably involved groups of people, within different societies, and different cultures.

Various postmodern approaches to research, which took into account critical theory, feminist scholarship, Vygotsky-inspired activity theories, and anti-colonialist theories (Sefa Dei & Kempf, 2006) also began to make their presence felt as a "mathematics for all" mentality swept across the planet. Indeed, within the span of the last 25 years, it has been increasingly recognized that in the past, traditional methods and traditional thinking about mathematics education resulted in the virtual exclusion from serious mathematical learning of many "marginalized" groups. Political dimensions of mathematics were taken up, replete with questions like: "What mathematics should be learned by whom, and in what education settings?" These questions had been studied in the past, but now research designs extended beyond studies that would use well-established inferential statistical approaches greatly favoured in the 1960s and 1970s. Research which teased out the implications for mathematics education, at different levels, of the incredible advances in technology was being called for. Questions associated with the educational implications of simple electronic calculators, graphing calculators, computer algebra systems, and the Internet, suddenly loomed large.

Research can be undertaken by individuals, or through collaboration with others, and such collaborations can be with others in the same institution or country, or with others in different countries. Given the diverse intersections of research into the teaching and learning of mathematics, collaborative research becomes a potentially important approach. In particular, international collaboration has the potential to bring together a rich diversity of perspectives. Although international collaboration can originate because two or more individuals of complementary interests make contact and plan research, the focus of this chapter is on structures that nurture and support collaborative research that crosses national borders, and what these structures have meant (and might mean in the future) for the products of that research.

In this chapter we first discuss the concept of collaboration and what this has involved for mathematics education research, both in terms of organizations that have the potential to nurture cooperation between researchers, and through examples of research studies that have involved collaboration. Through this discussion, criteria for the quality of collaboration emerge. The chapter concludes with a summary of the effectiveness of the various forms of international collaboration cited, and with a brief projection of what the future might hold for collaboration across international borders.

Collaborative Research in Mathematics Education

When invited to write this chapter, we were entrusted with the responsibility of exploring the meanings and importance of collaborative research in mathematics education. Early in our deliberations we began asking ourselves: What does the term *collaborative research* embrace? We found that common dictionary definitions of *collaboration* typically referred to a group of like-minded individuals working together to achieve a common goal. Such definitions triggered the reflection that over the last 30–40 years the field of mathematics education has witnessed an escalating interest in the place of mathematics education in education policy decisions and in the notion of collaborative research, especially in comparative studies in mathematics.

Mathematics education researchers around the world have become aware that planning, conducting, and evaluating mathematics education research is not a simple matter. Although researchers sometimes struggle to admit it, even to themselves, the fact is that individual researchers often lack the comprehensive knowledge and range of skills needed to conduct mathematics education research projects effectively. And, even when they do, their own pre-suppositions can stand in the way of well-triangulated research findings. There is, therefore, a well-founded demand for collaborative research in which teams of researchers combine their talents, knowledge, and beliefs, for the good of all. Terms like *action research*, *design research*, and *mixed-methods research*, as well as creative research designs associated with concepts like *lesson study*, have become commonplace. Researchers are learning how to operate, together, in new ways, on new challenges.

Collaboration in itself can take various forms ranging from offering general advice and insights, to active participation in specific pieces of research. Several reasons, and accrued benefits, have been attributed to escalating levels of research cooperation in mathematics education during the 20th century. Intuitively, collaborative research should facilitate and promote communication and cooperation among all interested parties, including education researchers and practitioners. An inherent part of any collaborative research effort is the process of reaching agreement on who should be involved in the research team, and of defining the roles of individual members. Usually, it will be decided that different team members will specialize according to their fields of expertise. Decisions about the research methodologies adopted will arise through collaborative discussion.

Collaborative research that involves participants from different nations provides a platform from which researchers can forge ongoing networks; it should link issues in mathematics education to those in the wider field of educational research. It should also explore issues in mathematics education that are of scientific interest, particularly those that benefit from the combined expertise available within the collaborative research team. Finally, collaborative research should increase the visibility of mathematics education research within the wider education research community (Katz & Martin, 1997).

However, historical antecedents and tensions within the broad fields of mathematics and mathematics education have often made it difficult for appropriate research teams to be formed. Mathematicians, for example, still believed that issues associated with mathematical content should take precedence in any thinking about mathematics education research, and they were concerned that too many mathematics education researchers lacked adequate mathematical competence. Psychometricians who have conducted research in mathematics education have been particularly interested in the research design, the validity and reliability of instruments, and analyses of data. Education administrators, for their part, have focussed on the mathematical performances of their students from one year to another, and in comparisons with students in other systems. Although mathematics teachers have also been concerned about their students' performance, they have often concentrated on their own teaching approaches, on textbooks, and on the design and implementation of curricula. Frequently, there was a gulf between researchers concerned with pre-school and elementary school mathematics, on the one hand, and mathematicians and those wishing to research the teaching and learning of secondary school and college mathematics, on the other. The former tended to believe that they had little to learn from anyone who did not have a strong background in working with young children, and the latter believed that it was dangerous for those without strong mathematical content knowledge to conduct mathematics education research at any level.

Such disparate emphases have left the way open for psychologists, psychometricians, and government education officials, who have often known little about mathematics education, to call for the development of higher quality mathematics education research. *Higher quality* has been defined in terms of investigations which feature random trials, coupled with "objective" statistical evaluations of data carried out by persons remote from the teacher and the classroom. Such calls have usually been bolstered by reference to medical research (Kaiser, Luna, & Huntley, 1999; Mosteller & Boruch, 2002). For, unless such "higher quality" investigations are designed and conducted, they claimed, results from mathematics education research would likely be untrustworthy, and often invalid.

The fundamental criterion for the success of any research in mathematics education is that ultimately its findings should enhance the quality of mathematics curricula, or a general understanding of mathematics teaching and learning. Like research communities operating in other domains, mathematics education researchers must take up the challenge of finding ways to expand the channels of communication, of improving the ways in which they communicate what has been learned from their own research and from the research of others, in order to affect practice in positive ways. Although no one is better qualified than a team of researchers to explain the practical implication and theoretical significance of their own research, such explanations still need to be suitably presented.

International Comparative Studies in Mathematics

At first glance it might be assumed that all international comparative studies in mathematics education must be fundamentally collaborative in nature, for they involve like-minded researchers working together as a collective unit. Such studies are usually large-scale studies which reach across national boundaries in order to achieve the common goal of generating scientifically-validated knowledge in mathematics education that is independent of geographical and cultural boundaries. In this chapter, we argue that not all aspects of the best-known international comparative studies conducted in mathematics education have been positive. It is instructive to identify and reflect on the positive as well as the negative features of such research, and to ponder the implications of these reflections for the concept of collaborative research.

Some of the major comparative studies in mathematics education include the early ICMI comparison studies (conducted between 1908 and 1915), the First International Mathematics Study (FIMS) conducted in the 1960s, the Second International Mathematics Study (SIMS) conducted in the late 1970s, and the Third International Mathematics and Science Study (TIMSS) conducted in the 1990s. In the present century we find Trends in Mathematics and Science Studies, the Programme for International Student Assessment (PISA), and the Learner's Perspective Study (LPS).

Although we analyze components of each of these studies, we do not confine our attention to those studies only. Put simply, our aim is to identify features of high-quality "collaborative research" in mathematics education, rather than elaborate details of the methods, results or recommendations of such studies.

Reports on international collaborations—whether these are in the form of publications, presentations, discussion groups, or meetings—are not ends in themselves. In this chapter, we accept the premise that the term *collaboration*, as it is used in mathematics education research literatures, implies more than a mere coming together of bodies and minds. We accept a point of view put forward by Schwarz, Dreyfus, and Hershkowitz (2009), that "unity comes from the fact that researchers belong to a common adventure—changing school practices and norms. This adventure is moved by societal ideals of reason and equity" (p. 1).

Thus, in this chapter we look at the work of teams of researchers who have cooperated across national borders for the purpose of constructing new entities and concepts in the field of mathematics education. These teams recognized that their work could take unexpected turns. In other words, they assumed that collaboration in the field of mathematics education involved a conscious coming together of minds, to work out new ways of thinking or operating.

One final clarifying comment on the concept of collaboration is in order. The concept embraces several familiar related concepts—for example, sharing, uniting, and cooperating—but in this chapter it outstrips each of those sub-concepts. As we have come to see it, the term *collaborative research* implies that participants share an agreed aim or mission. This aim may or may not be tightly defined from the outset, but it should be consistent with the name of the organization and with its raison d'être. The research workload should be shared by participants, but there does not have to be an equal division of labor. The term *collaborative* does not imply that all participants are in total agreement with what takes place within the agreed collaborative structure. Just as a successful marriage has its ups and downs, so too a successful collaborative venture may often have its differences of opinion, and its successes and failures.

The Scope of This Chapter

The editors of the *Third Handbook* requested that we survey, analyze, and critique collaborative studies conducted across national borders by teams of researchers who had worked together in examining and illuminating facets of mathematics education as a social practice. We were asked to examine collaborative practices that have permitted the interplay of different perspectives in mathematics education, and to tease out the implications of how the cross-national character of many of the studies have permitted the examination of trends and special cases.

It has not been possible for us to examine every international study that has contributed to the field's present understandings of how social and cultural forces impinge upon mathematics education locally, nationally and internationally. In this chapter we look, necessarily briefly, at case studies relating to the establishment and ongoing work of the following seven programs or organizations, each of which is still in existence and would claim to work in collaborative ways.

1. The *Commission Internationale de l'Enseignement Mathématique* (also known as the International Commission on Mathematical Instruction, or ICMI);
2. The International Association for the Evaluation of Educational Achievement (IEA);
3. The Programme for International Student Assessment (PISA);
4. The Regional Education Centre for Science and Mathematics (RECSAM), in Penang, Malaysia;
5. The Mathematics Education Research Group of Australasia (MERGA);
6. The International Group for the Psychology of Mathematics Education (PME);
7. The Learner's Perspective Study (LPS).

In addition, we briefly consider forms of collaborative research carried out in international aid programs. Following our brief surveys of, and commentary on, the work of the programs and organizations, we close the chapter with some comments on the type of "collaboration" that each features.

The Commission Internationale de l'Enseignement Mathématique

The Early ICMI as an Exclusive European/North American, Male, Mathematicians' Club

It has become received tradition among some historians of mathematics education that the first formal moves toward international cooperation in mathematics education came with the launching of the journal *L'Enseignement Mathématique* in 1899, and with the establishment of the International Commission on Mathematical Instruction (ICMI) at the Fourth International Congress of Mathematicians held in Rome in 1908 (Schubring, 2008a). Although we acknowledge this tradition, we feel it is important to raise serious questions, here, about the quality of collaboration involved in this early example of international co-operation.

From its creation in 1908, the *Commission Internationale de l'Enseignement Mathématique* has also been known by its English name, the International Commission on Mathematical Instruction, or ICMI, and in this chapter we henceforth refer to it as ICMI. The establishment of ICMI occurred at a meeting of the International Congress of Mathematicians (hereafter ICM) held in Zurich in 1908. Although the conference organizers had hoped to gather in Zurich "mathematicians from all countries on earth" (Curbera, 2009, p. 9), the 208 who actually attended (204 males, 4 females) came from just 16 nations, 15 of which were European. The non-European nation was the USA (pp. 15–16), which had seven conference attendees.

Much has been written about the set of events that took place at Zurich by which ICMI was established and about how a small "Central Committee" was chosen to oversee the work of ICMI (see, e.g., Donoghue, 1987; Schubring, 2008b). Here it suffices to say that much power was placed in the hands of a handful of people on the initial Central Committee, and this committee decided that ICMI's decisions would need to be ratified by 36 voting delegates: three each from Austria, France, Germany, Great Britain, Hungary, Italy, Russia, Switzerland, and the USA, and one each from Belgium, Denmark, Greece, The Netherlands, Norway, Portugal, Romania, Spain, and Sweden. Thus, 33 of 36 delegates (i.e., persons with voting rights at ICMI meetings) were to be from European nations, and the other three from the USA (Schubring, 2008b).

What is striking is how the early governance structure of ICMI was dominated by European mathematicians and educators, and subject to the control of ICM. It is easy to understand how this happened, of course, but that does not hide the fact that although ICMI identified eight "associated countries" (Australia, Brazil, Bulgaria, Canada, Cape Colony, Japan, Mexico, and Serbia), the voting nations for ICMI had a total population of about 480 million, or only about 30% of the world's population. Voting considerations aside, the following nations were among those not represented at ICMI meetings: China, India, Indonesia, Iran, Korea, Nigeria, The Philippines, Thailand, and Viet Nam—nations that accounted for more than half the world's population.

From the outset, ICMI was not representative of the worldwide body of mathematics teachers. Despite the best intentions, the Central Committee and its voting members were largely white, European, and male, and mathematics teachers in schools were not represented. Although ICMI intended that its vision would be *international*, its structure belied that vision. The principal form of collaboration in ICMI operations was little more than a sharing of information between its members. It was not truly global in its outreach, and its restricted mission arose out of an unfortunate internal belief that only certain mathematically advanced people in certain mathematically advanced nations possessed sufficient knowledge and wisdom about mathematics and mathematics education to be worthy of being admitted to ICMI's "inner circle." This belief would permeate ICMI for more than 50 years (Furinghetti, 2008).

We do not dwell on the early work of ICMI except to say that ICMI's initial mandate was to analyze the teaching of mathematics in secondary schools in various countries (Schubring, 2008b). One of the most tangible outcomes of the early work of ICMI was the production of a series of reports on the state of mathematics education in a number of countries. These reports would subsequently be published through various vehicles, including the French-language journal *L'Enseignement Mathématique* and the National Council of Teachers of Mathematics' *Fourth Yearbook* (Reeve, 1929). Among the national reports to appear were papers on school mathematics in the following countries: Australia, Austria, Czechoslovakia, England, France, Germany, Holland, Hungary, Italy, Japan, Russia, the Scandinavian countries, and the USA. The inclusion of reports from Australia and Japan testify to a widening vision of ICMI.

In 1907, David Eugene Smith, the North American mathematics educator who was significantly involved in the establishment of ICMI, had visited Japan with the aim of adding Japanese texts to his collection of rare books. Smith would subsequently have a large influence on what would be included in NCTM's *Fourth Yearbook*, and therefore the inclusion of a report on what was happening in Japan was not surprising. With the exception of Japan and the USA, all the other nations represented in the *Yearbook* were European. The impression given was that, from a mathematics education perspective, the USA had most to learn from European nations, although, perhaps, the Japanese mathematics education scene might also be of some interest.

When we studied these early summary national reports we were particularly struck by the report on the teaching of mathematics in Japan (Fukisawa, 1912). Whereas the other national reports often comprised tedious summaries of recent curricular and examination changes (see, for example, Young, Osgood, Smith, & Taylor, 1915), the Japanese report concentrated on pedagogical matters. It offered rich commentary on methods for teaching difficult topics in mathematics (e.g., fractions, ratio, elementary algebra), and for us the comments continue to have freshness and relevance even a century after they were written. These reports came from school teachers, mathematicians, school and college administrators, and government officials from all over Japan. The English version of the final Japanese report comprised 15 chapters, prepared by 17 educators and mathematicians, and occupied 238 pages. Within Japan, this was truly a collaborative effort, and was an achievement all the more remarkable because Japanese respondents were given very little time to prepare the report and have it translated into French and English (Fukisawa, 1912).

There was much to be learned from the Japanese report, but the general method of collaboration adopted by ICMI meant that the important ideas in the Japanese report would rarely, if ever, be discussed by those outside Japan. The issue of whether the West had much to learn from the East was not to be taken very seriously in a largely Eurocentric organization like ICMI, which seemed to be convinced that the best mathematicians, and therefore, probably, the best teaching of mathematics, were to be found mainly within Europe or North America. ICMI was officially part of ICM, and during its early existence, in the first half of the 20th century, it remained firmly under the wing of mathematicians (Furinghetti, 2008).

New Mathematics, Royaumont, and a Revival of an Exclusive Club

Following the Soviet Union's launching in 1957 of Sputnik, the US government worked with the Organisation for European Economic Cooperation (OEEC) to organize an international conference, held in Royaumont, France, to develop new thinking in mathematics and mathematical education (Moon, 1986). Attendance at this conference was by invitation, with each participating country being asked to send three delegates: a mathematician, a mathematics educator, and a secondary school teacher (OECC, 1961a, 1961b, 1961c). In fact, most of those who attended the Royaumont Conference were university-based mathematicians or mathematics educators, with the leadership coming from mathematicians.

The Royaumont conference was attended by representatives from 18 countries, 16 of which were European, the other 2 being Canada and the USA. From an international mathematics education perspective, the structure and attendance closely resembled the first ICMI conferences held before World War I. African, Asian or Australasian nations were not invited to send delegates who would make presentations to the conference. We see this episode as evidence of a revival of the old-boy, mathematics-dominated, European/North American club that had controlled the only moderately successful earlier attempts at international collaboration in mathematics education (OECC, 1961a, 1961b, 1961c).

The "New" ICMI

The much more encouraging and largely successful activities of a revived and restructured ICMI from around 1970 is only touched upon here, because they are dealt with in other chapters in this section, especially Chapter 28. It should be noted, though, that the new ICMI has used its influence and financial muscle to develop numerous impressive collaborative research activities in mathematics education. For example, in 2006 a 596-page book entitled *Mathematics Education in Different Cultural Traditions—A Comparative Study of East Asia and the West* was published as an outcome of the 13th ICMI Study Conference, which was held in Hong Kong in 2002 (Leung, Graf, & Lopez-Real, 2006). This book, which provided, among many other things, important commentary on historical, cultural and contextual background factors, offered well-argued debate surrounding issues like whether the West has much to learn from the high performance of Confucian-background students in international comparative studies. The book had 40 chapters contributed by authors from all continents. Quite a few of the authors prodded existing structures and attempted to shape future directions. The genre of this text was in stark contrast to that of the early ICMI reports which focussed on comparing programs, curricula, textbooks, and examinations. Other publications in the ICMI Study Series dealt with issues such as assessment, gender equity, mathematics education as a research domain, the teaching of geometry, the teaching and learning of mathematics at university level, history in mathematics education, and the teaching and learning of algebra. These publications typically were preceded by associated conferences in which recognized contributors to the issues under consideration were invited.

ICMI regional conferences on mathematics education have been held in many parts of the world, including Africa, South America, East Asia and Southeast Asia. Often the language of a conference has been English, but on other occasions the main language has been French, or some other language. Since the late 1960s, international congresses on mathematics education (ICMEs) have been conducted, every 4 years. Of the 12 ICMEs held so far, six were held in Europe, two in Asia, three in North America, and one in Australia. The last four ICMEs were held in Korea (2012), Mexico (2008), Denmark (2004) and Japan (2000). Although there has still been a tilt towards European/American leadership, there has been a noticeable change from the early 1900s.

Unexpected Outcomes of IEA Activities

In 1958, a group of scholars—mostly educational psychologists, sociologists and psychometricians, but not necessarily mathematicians or mathematics educators—came together at a UNESCO office in Germany to discuss problems of school and student evaluation. These scholars were especially interested in the evaluation of student learning outcomes, and they decided to gather international data on school students' knowledge of key mathematical concepts, their attitudes towards mathematics, and the extent of their participation in mathematics (Postlethwaite, 1967, 1993). They had the bright idea of requiring participating nations to contribute to the costs of the studies. In 1967, this group became formally known as the International Association for the Evaluation of Educational Achievement (hereafter IEA). The ongoing activities of IEA are dealt with in other chapters of this *Handbook*, and here it suffices to draw attention to some unexpected outcomes from IEA activities that have influenced thinking about what constitutes high-quality international collaboration in mathematics education research.

The design of IEA studies required consultation with local educators from all participating nations. Careful attention was given to creating pencil-and-paper instruments that corresponded to curricula adopted in participating nations, and to the delicate issue of translating tests so that, as far as possible, equivalent tests in different languages would be produced (Ellerton & Clements, 2000, 2002). Participating nations had to agree to allow stratified random samples to be selected. Although the designs were, from a psychometrical perspective, satisfactory, mathematics educators were unhappy that they had not been more fully and carefully consulted and involved in what were, after all, international mathematics education research studies (Keitel & Kilpatrick, 1999).

Governments of participating nations, however, were content to leave FIMS, SIMS, and TIMSS largely to the acknowledged expertise of IEA psychometricians and psychologists, who were assisted by carefully chosen curriculum experts, including some well-regarded mathematics education researchers. Those administering education systems, and politicians, wanted to know how well their students performed in comparison with corresponding students in other nations, and were prepared to pay to make sure that the forms of statistical analyses that were used were sufficiently authoritative for legitimate comparisons to be made.

IEA and OECD International Studies

IEA's First International Mathematics Study (FIMS) targeted 13-year-old students and pre-university students. Participating nations were Australia, Belgium, England, Finland, France, Germany (FRG), Israel, Japan, the Netherlands, Scotland, Sweden, and the USA. Students from Japan performed very well, and those from Australia and Israel performed at least as well as, if not better than, European students (Husén, 1967). It was a matter of interest, and perhaps surprise, that it appeared to be the case that Japanese students were learning school mathematics better than their European and American counterparts. The international education community, including mathematicians and mathematics educators, wanted to know if that was indeed the case, and if so why?

The Second IEA Mathematics Study (SIMS) was conducted between 1977 and 1981. Participating nations were Belgium, Canada, England and Wales, Finland, France, Hong Kong, Hungary, Israel, Japan, Luxembourg, Netherlands, New Zealand, Nigeria, Scotland, Swaziland, Sweden, Thailand, and the USA. SIMS examined the intended, implemented and attained curricula in mathematics, at two levels: (a) 13-year-olds, and (b) students in their last year of secondary education. Student performance was measured and reported separately for arithmetic, algebra, geometry, measurement, and statistics (Burstein, 1992; Robitaille & Garden, 1989; Travers & Westbury, 1989). Analyses of SIMS data showed that at the 13-year-old level, students from Japan and Hong Kong achieved the highest means. The concern raised by FIMS was now magnified, and Western mathematicians, educators, and parents began to demand that something be done to improve the situation (Clements, 2003).

IEA's Third International Study of Science and Mathematics (TIMSS) was conducted in the early 1990s. Over 500,000 students, in more than 15,000 schools in 45 participating nations, were involved, with students being mainly at three levels, Grade 4, Grade 8, and end-of-secondary school. Supposedly equivalent tests were developed in more than 30 different languages, and strict sampling procedures were followed within most of the participating nations. In addition, a TIMSS video study, which focussed on eighth-grade mathematics classes in three nations—Germany, Japan and the USA—was conducted (Hiebert, Stigler, & Manaster, 1999).

TIMSS analyses indicated that at both the fourth- and eighth-grade levels, the four best-performing nations were the four participating Confucian-background nations: Singapore, Korea, Japan, and Hong Kong. The mean for US fourth-grade students was slightly above the international average, and the mean for students from England and Wales was slightly below the mean. At the eighth-grade level, the means for US students and for students from England and Wales were below the international mean. The lowest national mean score was from South Africa.

The outstanding TIMSS results from Confucian-heritage nations struck a chord with educators and politicians around the world, who renewed their efforts to discover the Asian nations' "secret" (Menon, 2000). For example, curricular analyses suggested that US mathematics curricula were "a mile wide and an inch deep" (Schmidt et al., 2001), and video analyses (Hiebert et al., 1999) suggested that Japanese teachers taught in qualitatively different ways from their US and German counterparts. Japanese teachers did not rely wholly on drill and practice, as some had previously believed, but regularly engaged students in challenging problem solving and problem creation exercises. Such was the publicity given to TIMSS's 1995 findings that IEA has subsequently been funded, by numerous governments, to conduct new international comparative studies. IEA has continued to use the acronym TIMSS (now standing for "Trends in Mathematics and Science Study") and many Asian, Australasian and some African nations have participated in TIMSS assessments over the past decade (Stacey, 2010). Confucian-heritage nations have always topped international performance league tables.

The success of TIMSS encouraged the Organisation for Economic Co-operation and Development (OECD) to inaugurate its Programme for International Student Assessment (PISA) studies, which compared nations on problem solving and applications of mathematics and science. Forty-one nations participated in the second PISA study, conducted in 2003, which focussed on mathematical problem solving. Once again, students from Confucian-background nations excelled, although students from some other nations (e.g., Finland) also performed well. In a 2009 PISA study, about 75 nations participated, including Japan, Korea, Australia, New Zealand, Indonesia, Thailand, Singapore, Chinese Taipei, and three parts of China (Hong Kong, Shanghai, Macao) (Stacey, 2010). From the outset, the PISA studies have seen a greater involvement of mathematics education researchers than the IEA studies (Stacey, 2010). To help readers to appreciate the extent to which collaborative actions underlie the working structure of PISA, the following extract is quoted from the Foreword of the report of PISA 2009 results (OECD, 2010):

> This report is the product of a collaborative effort between the countries participating in PISA, the experts and institutions working within the framework of the PISA Consortium, and the OECD Secretariat. The report was drafted by Andreas Schleicher, Francesca Borgonovi, Michael Davidson, Miyako Ikeda, Maciej Jakubowski, Guillermo Montt, Sophie Vayssettes and Pablo Zoido of the OECD Directorate for Education, with advice as well as analytical and editorial support from Marilyn Achiron, Simone Bloem, Marika Boiron, Henry Braun, Nihad Bunar, Niccolina Clements, Jude Cosgrove, John Cresswell, Aletta Grisay, Donald Hirsch, David Kaplan, Henry Levin, Juliette Mendelovitz, Christian Monseur, Soojin Park, Pasi Reinikainen, Mebrak Tareke, Elisabeth Villoutreix and Allan Wigfield. Volume II also draws on the analytic work undertaken by Jaap Scheerens and Douglas Willms in the context of PISA 2000. Administrative support was provided by Juliet Evans and Diana Morales. The PISA assessment instruments and the data underlying the report were prepared by the PISA Consortium, under the direction of Raymond Adams at the Australian Council for Educational Research (ACER) and Henk Moelands from the Dutch National Institute for Educational Measurement (CITO). The expert group that guided the preparation of the reading assessment framework and instruments was chaired by Irwin Kirsch. The development of the report was steered by the PISA Governing Board, which is chaired by Lorna Bertrand (United Kingdom), with Beno Csapo (Hungary), Daniel McGrath (United States) and Ryo Watanabe (Japan) as vice chairs. Annex C of the volumes lists the members of the various PISA bodies, as well as the individual experts and consultants who have contributed to this report and to PISA in general. (p. 3)

IEA and OECD have generated some healthy, collaborative research studies within the domain of mathematics education research. One thing that these studies have made clear is that high-quality intended and implemented school mathematics curricula have often involved participants from both inside and outside of Europe and North America.

Mathematics education researchers, everywhere, quickly recognized that it was their responsibility to play a leading role in discussions about what the implications of the IEA and PISA results might be, locally, nationally and internationally. They also decided that more worthwhile discussion and associated actions were likely to emerge from collaborative research where all participants were regarded as equals. At the 29th Conference of the International Group for the Psychology of Mathematics Education, for example, there was a plenary panel session on "What Do Studies Like PISA Mean to the Mathematics Education Community?" Acknowledged leading mathematics education researchers provided the plenary addresses (see, Jones, 2005; Kieran, 2005; Neubrand, 2005; Shimizu, 2005; Williams, 2005), but all present were encouraged to participate in the keen discussions which these addresses precipitated.

Any claims that mathematics education researchers comprised a closed, largely European/North American club were no longer sustainable. Despite important criticisms of international comparative studies (e.g., Keitel & Kilpatrick, 1999), leaders of many national educational systems now wanted to learn from such studies. The lesson here is this: The mathematics education research community must recognize that even unsatisfactory approaches to collaboration can often generate structures and results that can underpin future, more successful collaborative research. The international mathematics education research community should continually seek to tweak existing research efforts so that better research will be conducted.

The IEA studies certainly generated much discussion and debate in relation to the quality of mathematics education offered in different nations. Inevitably, the performance league table raised questions about the merits and demerits of the pedagogies utilized in developed nations like the USA, Germany, and Britain in which students' performances were regarded as dismal (Stevenson & Stigler, 1992; Stigler & Hiebert, 1999). However, a range of methodological issues dealing with the nature and conduct of the surveys, the comparability of the populations tested, and the quality of the data obtained raised questions about the validity and applicability of the findings (Torney-Purta, 1987).

Then, when it was found that the attitudes towards mathematics, and the mathematics self-concepts, of many students in the high-performing Confucian-background nations were poor, important questions were raised that had not really been addressed in the IEA reports (Ellerton & Clements, 2010; Leung, 2006).

From Local Effort to a Cooperative Form of International Endeavour: The Case of RECSAM

We now turn to a completely different kind of collaboration, one which is only partly to do with mathematics education research. Since the beginning of the 20th century, mathematics education researchers within Asia, Africa, and Australasia have worked hard at becoming well-regarded members of the international mathematics education community. Their efforts have been facilitated by rapid improvements in transport (from rail and ship to automobile and to air transport), and escalating use of modern communication technologies. It is important to document how educators in these nations have overcome the tyranny of distance and have succeeded in developing a "local to national to international" progression in the quality of their research. We now discuss, very briefly, the case of the Regional Centre for Education in Science and Mathematics (RECSAM), a case which, we claim, has featured high-quality collaborative procedures from the outset.

RECSAM, located in Penang, Malaysia, has worked according to a model for international cooperation in which a wide range of participants from co-operating nations have successfully contributed to local, national and international research in mathematics education over a long period of time.

At the beginning of the 20th century, some Southeast Asian nations (e.g., Brunei Darussalam) did not have any formal schools—primary or secondary—and less than one percent of Southeast Asian children had ever attended a secondary school (Horwood & Clements, 2000). When schools were set up, they tended to be based on colonialist models (Asante, 2006; Clements, Grimison, & Ellerton, 1989). Over the past 50 years, however, Southeast Asian nations have taken giant strides toward achieving the goal of providing a quality mathematics education for all (Horwood & Clements, 2000; Singh & Lim, 2005).

RECSAM began in 1966 as a co-operative venture of the Southeast Asian Ministers of Education (SEAMEO). At first, SEAMEO was administered by Lao PDR, Malaysia, the Philippines, Singapore, Thailand, and Vietnam, but subsequently other nations—Brunei Darussalam, Cambodia, Indonesia, Myanmar, and Timor-Leste—have become full members. SEAMEO's achievement in establishing a cross-national science and mathematics education centre, which has been active 52 weeks of each year for 45 years, is noteworthy. Although, between 1966 and 2012, Southeast Asian nations differed in terms of religion, race, language, and development, at RECSAM those differences have always been celebrated as a positive feature of the diversity and challenge inherent in mathematics education in the region.

From the outset, SEAMEO-RECSAM personnel agreed to a mission statement by which it would provide for the needs of SEAMEO member countries in the development of expertise in science and mathematics education. Well-qualified science and mathematics education specialists have always been seconded to RECSAM from participating nations, with most serving as full-time, resident instructors and workshop leaders at RECSAM for at least 3 years. The language of instruction at the Centre has always been English. International consultants, chosen by RECSAM, have assisted RECSAM instructors in workshop- and research-based programs for practising teachers, who have stayed at RECSAM, for periods of up to 8 months. This work has been financed partly by international aid money, but mainly by those nations sending the teachers. The curricula for programs have been developed by RECSAM personnel (Clements & Ellerton, 1996).

In the 1960s, SEAMEO-RECSAM embarked on activities in five main areas, namely: (a) training, (b) research and development, (c) consultancy work, (d) designing, offering and evaluating conferences, seminars and workshops pertaining to science and mathematics education, and (e) publication. Its subsequent efforts in each of these areas have been noteworthy (Clements & Ellerton, 1996).

RECSAM's facilities are extensive. They include residence halls; spacious dining and sports amenities; a large administration block; science, mathematics and computer laboratories; a library; a printing facility; and numerous lecture theatres and classrooms. The decision that the Centre should be largely self-financing has presented a continuing challenge, but the aim has been achieved. International seminars are regularly conducted, and for many years, RECSAM has continued to publish its refereed *Journal of Science and Mathematics Education in Southeast Asia*. It has also regularly published summaries of Southeast Asian mathematics education research (see, e.g., RECSAM, 1991; Roadrangka & Liau Tet Loke, 1993). It was a proud moment for Southeast Asia when Singaporean students gained the highest mean scores in the 1995 TIMSS. The world suddenly learned that Southeast Asian education authorities must be doing something right.

Clements and Ellerton (1996) and Ellerton and Clements (2000) have argued that the findings of mathematics education research conducted in Western nations might not apply in Southeast Asian nations. Western theories might "represent an essentially Eurocentric view of education, in general, and of mathematics education in particular" (p. vii). Thus, for example, summaries of "cognitive stages of development," "hypothetical learning trajectories," and rubrics for "levels" supposedly "validated" in Western nations might not be helpful in SEAMEO nations. A valid and reliable test in one culture might not be valid and reliable in another. When a test instrument written in one language is translated into another, the difficulty of items can change dramatically, and claims of "equivalent" tests across languages could be spurious. The other side of the coin is that the rest of the world might learn something by studying mathematics curricula, teaching practices, and factors influencing learning in Southeast Asian nations. Similarly, mathematics textbooks written specifically for students in one nation are unlikely to be suitable for students in other nations.

RECSAM has provided, and continues to provide, an effective working, collaborative model whereby mathematics education scholars from different countries, often with very different cultural backgrounds, can walk and talk with each other on a daily basis. Seen in this way, collaborative research in perhaps its purest form has been going on for decades, and has helped to improve the teaching and learning of mathematics in all SEAMEO nations.

The Australian Association of Mathematics Teachers (AAMT) and the Mathematics Education Research Group of Australasia (MERGA)

Australia became a federated nation in 1901, when six states were brought together to form one nation. Each of these states was responsible for defining and administering its own education system. Because of the isolation of Australia from other continents (Blainey, 2001), and because of the states' large areas and relatively small populations, most teachers of mathematics remained out of the national and international education mainstreams. However, state mathematical associations were soon formed (e.g., the Mathematical Association of Victoria began in 1907), and in the early 1920s the Australian Mathematical Society (AMS) was formed.

The Australian Association of Mathematics Teachers (AAMT) was formed in the 1940s, and this brought Australian school teachers of mathematics into greater contact with each other. For the first three decades of its existence, AAMT held occasional conferences which were attended by mathematicians, mathematics educators, and mathematics teachers from the various states. In the mid-1940s, AAMT began publishing its journal *The Australian Mathematics Teacher*. As of 2012, AAMT is a composite association comprising 12 local and state mathematics teacher associations. Like the National Council of Teachers of Mathematics (NCTM) in the USA, it publishes teacher education journals, and conducts national conferences. It represents mathematics teachers on relevant education issues. Most of its members teach mathematics in schools rather than in universities or colleges (Ellerton & Clements, 1994).

The Mathematics Education Research Group of Australasia (MERGA) grew out of the Mathematics Education Research Group of Australia, which was established in 1976. MERGA conducts an annual conference, and publishes two refereed international journals [*Mathematics Educational Research Journal (MERJ)*, and *Mathematics Teacher Education and Development (MTED)*]. In recent years, an increasing number of international mathematics educators have reported their research in these journals.

Most MERGA members are mathematics educators working in Australia or New Zealand. So, in one sense, MERGA only just qualifies as an international organization. There is another sense, though, in which MERGA, like NCTM, is an international organization. Its members have been, and continue to be, active in international circles, being particularly well represented at annual conferences of the International Group for the Psychology of Mathematics Education. A life-member, and former president of MERGA, Gilah Leder of Australia, was awarded the Felix Klein Award for 2009 by ICMI; a former editor of *MERJ*, Bill Barton, of New Zealand, served as president of ICMI between 2008 and 2012; Lyn English, an Australian member of MERGA, was the founding editor (and continues as editor) of the American-based journal *Mathematical Thinking and Learning*. Mathematics Education researchers from all over the world attend the annual MERGA conferences, and submit articles to *MERJ* or to *MTED*. Evidence that MERGA is seeking to develop a more fully collaborative international profile is attested to by the fact that, in 2012, the annual MERGA Conference was not only held in Singapore, but was jointly organized by the Singaporean mathematics education community.

MERGA is a shining example of how a body that was established mainly with national agendas in mind was able to tweak its structure so that it evolved from being a national body to an international body which provides its members with the opportunity to collaborate with mathematics education researchers from all over the world.

The International Group for the Psychology of Mathematics Education (PME)

Other chapters in this *Handbook* outline the ongoing work of PME. The focus in this chapter is on forms of collaboration encouraged by PME. At the third ICME conference, held in Karlsruhe, Germany, in 1976, it was decided to establish PME, and the first PME annual conference was held in the Netherlands in 1977. Between 1977 and 2012, 36 annual PME conferences were held—19 in Europe, 6 in Asia, 6 in North America, 2 in South America, 2 in Australia, and 1 in Africa. Of the 6 conferences held in Asia, 2 were in Israel, 2 in Japan, 1 in Korea and 1 in Taiwan. From the outset, those attending PME conferences were mainly professional mathematics educators.

The last five PME conferences have been held in Mexico (2008), Greece (2009), Brazil (2010), Turkey (2011) and Taiwan (2012). By contrast, the first five PME conferences were held in the Netherlands (1977), Germany (1978), the UK (1979), the USA (1980), and France (1981). The geography of those data suggests that a sea change occurred within PME in its thinking about the responsibilities of an organization with the word *international* in its title. Both ICME and PME have adopted policies whereby a percentage of all conference registration fees is allocated to assist persons from less affluent regions of the world to attend.

Internationalism in mathematics education is now being interpreted among mathematics educators as implying that all mathematics educators in all parts of the world should be involved if they so wish. The emphasis at annual PME conferences is less on mathematics per se than on psychological and sociological issues associated with the intersection between mathematics and education, especially issues pertaining to the teaching and learning of mathematics.

One of the perennial areas of controversy among PME members and supporters has been the raison d'être of the organization. Every now and then there has been an attempt to change "Psychology of Mathematics Education" (PME) to something like "Research in Mathematics Education" (RME), the main reason being that many mathematics education researchers who want to attend PME conferences are not comfortable and do not identify themselves easily with the term *psychology*. Some of these scholars regard themselves more as sociologists, linguists, historians, and so on, or more generally as mathematics educators. These scholars tend to argue that retaining the term *psychology* in the name of the organization denies it the "inclusivity" that many would like to see it have, and tends to restrict the type of collaborative research that can be fostered by PME. However, despite numerous debates over the years on the matter, the original name of the organization has remained inviolate. Many of those who worked hard to establish the organization in its early days do not believe that it is fair for subsequent PME members to want to change the image of the organization by changing its name.

Although PME does much more than organize annual conferences, nevertheless the fact that PME annual conferences have been successfully held in many parts of the world has contributed in a large way to the growing internationalization of mathematics education research. Having an international PME conference hosted in one's own country helps members to believe that they belong. Marcelo Borba (2010), a PME member from Brazil, stated that although PME has gradually become "a more inclusive group," "the challenge for PME is how to become more international" (p. 2). Zahra Gooya (2010), a PME member from Iran, called for "PME to be a truly international community" (p. 4), and commented that "with the help and support from the greater community" (p. 4), countries in all regions of the world should be invited to host PME conferences. So perhaps even more tweaking is needed!

PME conferences are deliberately arranged so that they provide a forum for international cooperation. Not only are there numerous plenary forums at which acknowledged leaders discuss research associated with recent changes or ideas, but there are also working groups which provide opportunities for small-group discussion on members' areas of research interest. Often, international publications arise out of these working groups.

Thus for example, for several years during the 1990s, the second author of this chapter chaired a PME mathematics teacher development working group. A natural outcome of the often-animated discussions in this group was a 256-page edited collection of research articles titled *Mathematics Teacher Development: International Perspectives* (Ellerton, 1999). There were 16 contributing authors from 14 different countries.

The book provided robust commentary on the following seven controversial issues relating to pre-service and professional development mathematics teacher education programs:

1. Is it feasible to expect mathematics teachers and teacher educators to keep up with, and harness, the potential of the new technologies? If yes, then what are the responsibilities of their present or likely future employers to provide them with adequate time and expert training to engender the confidence and competence necessary to use and teach with the new technologies?
2. What do mathematics teachers and teacher educators need to do in order to make mathematics more meaningful for an ever-widening spectrum of students?
3. With the domain of mathematics expanding rapidly, and in directions which are often quite different from those which confronted mathematics teachers when they received their formal training in mathematics, what can mathematics teachers and mathematics teacher educators do to keep abreast of contemporary mathematics developments? Or, should that not really be a concern of most teachers of mathematics, who should only be concerned with mastering the content, and methods of teaching, the traditional basics?
4. More generally, to what extent should mathematics specialists (in contrast with mathematics education specialists) be solely responsible for teaching the content of mathematics to students enrolled in pre- and inservice mathematics education programs? Should mathematicians teach the mathematical content and leave the pedagogical aspects to mathematics teacher educators?
5. What are the implications of constructivist theories for mathematics teachers and mathematics teacher educators?
6. Should mathematics curricula and mathematics teacher education curricula be the same the world over? Should these be treated as if they are culture-free domains?
7. How should mathematics teacher educators respond to calls for standards-based mathematics curriculum and assessment regimes, and competency-based mathematics teacher education programs?

These issues were heavily researched in the 1990s, with PME contributors taking the lead. These research agendas are still alive, with collaborating researchers from different nations working together (Schmidt et al., 2011).

PME has attempted to extend its reach to all continents. The first PME annual conference to be held in Africa was PME-22, which took place in 1998. The 4-volume proceedings of PME-22, edited by two South African mathematics educators (Olivier & Newstead, 1998), included 5 plenary addresses, 117 refereed research reports, 84 short oral communications, and 34 poster presentations (calculations based on data reported in Olivier and Newstead, 1998). Altogether, 51 South Africans made presentations at the Conference—including 2 plenary addresses and 15 refereed research reports. This remains by far the greatest number of papers from an African nation ever accepted at any single PME conference. Somewhat disappointingly, however, only 13 of the remaining 215 presentations at PME-22 were made by African scholars—and 10 of those were by scholars from Mozambique. The only other African presenters came from Botswana, Swaziland, and the Cameroons. At the 2009 PME Conference held in Thessaloniki, Greece, there were only eight presenters from nations within the continent of Africa, and all eight were from South Africa (calculated from data reported in Tzekaki, Kaldrimidou, and Sakonidis, 2009). Clearly, the locations of prestigious conferences can influence the growth of a mathematics education teaching and research culture within a nation.

It would be wrong, however, to give the impression that South Africa is the only nation in Africa contributing to international movements in mathematics education. The ICMI West African Report for 2010 can be found at http://mathunion.org/icmi, and the African Mathematical Union (AMU) is active across Africa. AMU was founded in Morocco in 1976, and although this is primarily a union of African mathematicians, it has a working education committee called the Commission of Mathematics Education in Africa. AMU seeks to establish international partnership agreements with organizations such as IMU, ICMI, and UNESCO, and in 2011 an international symposium on mathematics education was held in Tunisia. In 2009 and 2010, ICMI, UNESCO, AMU, and the French Embassy collaborated to present exhibitions on "Experiencing Mathematics" in Senegal, Benin and Burkina-Fasco.

Undoubtedly, the international mathematics education community still has much to do before many African mathematics educators will feel welcome in the international mathematics education community (Howson & Kahane, 1990; Persens, 2006). In 1996, Sitsofe Anku, an African mathematics educator who was then working in Singapore, stated that "current PME members will not know what they are missing unless they learn to listen to those who have been excluded (by whatever reasons) for so long" (Anku, 1996, p. 8). According to Anku, "an atmosphere of trust and encouragement" (p. 8) will be needed. Anku's comment reminded his readers that once upon a time many of the world's leading mathematics education researchers did not realize that mathematics education researchers had much to learn from educators and researchers in nations like Japan, China, and Singapore.

A 2010 Internet report indicated that Sitsofe Anku was the President of the Ghana Mathematics Society (http://www.ghanaweb.com/GhanaHomePage/NewsArchive/artikel.php?ID=176590) and, in May 2010, he was a presenter at the ICMI-sponsored Third African Congress on Mathematics held in Botswana. In September 2011, the eleventh Mathematics Education into the 21st Century Project International Conference was held at Rhodes University in South Africa. It would be interesting to analyze attendance data for those conferences to see if they attracted more African mathematics education scholars from Northern, Western, Eastern and Central Africa than did PME-22, which was held in South Africa in 1998. Meanwhile, the international mathematics education research community would do well to involve African mathematics scholars actively in collaborative research efforts.

The Learner's Perspective Study

The book *Mathematics Classrooms in Twelve Countries: The Insider's Perspective* (Clarke, Keitel, & Shimizu, 2006) presents data from the Learner Perspective Study (hereafter LPS), from mathematics classrooms in 12 countries: Australia, China, Czech Republic, Germany, Israel, Japan, Korea, the Philippines, Singapore, South Africa, Sweden, and the USA. Each author of a county report was a participating researcher, and local participants chose their own perspectives and modes of operation within the agreed LPS research framework.

The LPS framework did have tight procedural specifications that provided structural uniformity in LPS research, which is now taking place in many nations. In each participating nation, local researchers identify three highly competent teachers teaching in demographically diverse urban government schools. The data generated comprise a sequence of ten successive lessons that each of the three teachers give to their regular classes. This approach is intended to allow each participating teacher's "normal" development of a topic, as well as student growth in understanding, to be monitored.

Three cameras (and operators) are required in each LPS lesson to capture developments in the whole class, as well as the teacher's contributions and the students' actions. In addition, teachers are interviewed about what they do in class, and students are interviewed about their perceptions of what happens in their lessons, their memories of events being heightened by video recall techniques during interviews. Students and teachers also respond to questionnaires.

Although the teachers who participated in the research described in existing LPS publications agreed to work within LPS requirements, they were not fully involved in the formulation of those requirements. So, in that sense, this collaborative research did not involve an equal partnership. There is also the issue of authenticity, whether teachers who are aware that their every word and action will be captured on video can be expected to generate anything that resembles representative, "normal" classroom data. Thus, although Clarke, Keietel, and Shimizu (2006) claimed that they presented "detailed portrayals of the practices of individual well-taught mathematics classrooms over sequences of ten lessons" (p. 6), doubts can arise about the extent of the authenticity of the recorded lessons and how much one would have to look beyond the actual classes and teachers studied. On the other hand, Clarke, Keitel and Shimizu took pains to emphasize that it was *not* intended that the lessons would be representative of mathematics lessons in the various participating countries. Rather, the criterion for selection of participating teachers was that their lessons would represent high-quality teaching. That distinction seems rather incongruous, however, given that the sub-title of one of the LPS reports is "Comparing Mathematics Classrooms around the World" (Clarke, Emanuelsson, Jablonka, & Mok, 2006).

LPS reports took pains to highlight that their modes of analyses contrasted sharply with those used in the 1995 Video Study of the Third International Mathematics and Science Study, and in the subsequent 1999 TIMSS Video Study. In those studies, statistically representative samples of classrooms were used in participating countries (Japan, Germany and the USA in 1995, and Australia, the Czech

Republic, Hong Kong SAR, Japan, the Netherlands, Switzerland and the USA in 1999), and lessons focussed on the teacher and on teacher practices were videotaped (Hollingsworth, Lokan, & McCrae, 2003; Stigler & Hiebert, 1997, 1999).

It is interesting to reflect on the type of collaboration embodied in the LPS design. It is possible that the researchers' and teachers' selection process for choosing classrooms and teachers for the study introduced biases into the data and the analyses. That raises issues associated with the design of collaborative research that is intended to generate high-quality data. With LPS one sees a delicate balance between allowing researchers to choose what they consider to be the "most typical representation of school mathematics" in their country and what that choice then means to international readers when they interpret the report for their own purposes.

In comparison with other international classroom studies, one of the most outstanding aspects of LPS was the freedom given to researchers in the different countries to use analytical tools of their own choice. Most likely, this freedom would have heightened the researchers' enthusiasm to discuss and compare their approaches with other LPS researchers. Also, the researchers would have felt that their contributions to the LPS study were genuinely those of independent researchers, rather than those of minor players within a large, formula-driven study.

Table 26.1, from Ellerton (2008, p. 131), provides a country-by-country listing of chapters in Clarke, Keietel, and Shimizu (2006). Note that in Chapter 1, seven generally-worded LPS research questions are listed, and chapters are grouped under those questions based on what the editors felt would be most appropriate for that report. This breakdown, however, does not assist readers interested in a particular topic or country. The general phrasing of the research questions indicates a need to maximize the opportunities for each group of researchers to select and apply interpretations consonant with their own unique contexts. Whether the loss of opportunities for answering more specific research questions applied to each of the different contexts is greater than the flexibility offered by the more generally-phrased research questions is open to debate.

Entries in Table 26.1 draw attention to the huge conceptual differences driving LPS research, on the one hand, and the early ICMI work on the other. The concept of collaboration embodied in LPS research means that many participants have the opportunity to report on the way they view the data. In the early ICMI comparative studies, most of the reports were prepared by detached experts who attempted to synthesize data that had been gathered. The IEA studies (FIMS, SIMS, and TIMSS) were such that it was expected that tight generalizations would be made, to permit the creation on international league tables in performance. Thus, careful attention had to be given to obtaining stratified random samples, and to generating supposedly culture- and linguistic-free comparisons. There can be little doubt that IEA did create the impression, even among mathematicians and statisticians, that legitimate generalizations could be made from their findings. But this impression came at a cost.

The research summarized in Table 26.1 speaks volumes about the directions mathematics education research needed to take. Despite claims made by those with extensive access to IEA data about the teaching power to be found in Japanese mathematics classrooms (see, e.g., Stevenson & Stigler, 1992; Stigler

Table 26.1
Summary of Chapters

Country	Chapter/Pages	Title	Authors
Australia	Chapter 15 pp. 221–236	Autonomous looking-in to support creative mathematical thinking: Capitalizing on activity in Australian LPS classrooms	Gaye Williams
China	Chapter 6 pp. 87–97	Teacher-dominating lessons in Shanghai: The insiders' story	Ida Ah Chee Mok
	Chapter 16 pp. 237–246	A tale of two cities: A comparison of six teachers in Hong Kong and Shanghai	Ida Ah Chee Mok and Francis Lopez-Real
	Chapter 18 pp. 263–274	Repetition or variation: Practising in the mathematics classrooms in China	Rongjin Huang, Ida Ah Chee Mok, F. K. S. Leung
Czech Republic	Chapter 19 pp. 275–288	Constitution of the classroom environment: A case study	Helena Binterová, Alena Hošpesova and Jarmila Novotná
Germany	Chapter 3 pp. 37–57	'Setting a task' in German schools: Different frames for different ambitions	Christine Keitel
	Chapter 11 pp. 167–182	Students' verbal actions in German mathematics classes	Astrid Begehr
Israel	Chapter 14 pp. 209–220	The Israeli classroom: A meeting place for dichotomies	Michael N. Fried and Miriam Amit
Japan	Chapter 4 pp. 59–73	The role of seatwork in three Japanese classrooms	Keito Hino
	Chapter 12 pp. 183–194	Discrepancies in perceptions of mathematics lessons between the teacher and the students in a Japanese classroom	Yoshinori Shimizu
	Chapter 20 pp. 289–306	Mathematical norms in Japanese classrooms	Yasuhiro Sekiguchi
Korea	Chapter 17 pp. 247–261	Mathematics lessons in Korea: Teaching with systematic variation	Kyungmee Park and Frederick Koon Shing Leung

(continued)

Table 26.1
(continued)

Country	Chapter/Pages	Title	Authors
The Philippines	Chapter 9 pp. 131–149	"Ganas"—A motivational strategy: Its influence on learners	Soledad Asuncion Ulep
	Chapter 13 pp. 195–208	Students' private discourse in a Philippine classroom: An alternative to the teacher's classroom discourse?	Florenda Lota Gallos
South Africa	Chapter 8 pp. 117–130	Fine-tuning a language of description for mathematics items which incorporate the everyday	Godfrey Sethole, Busi Goba, Jill Adler and Renuka Vithal
Singapore	Chapter 7 pp. 99–115	Mathematics teaching in two Singapore classrooms: The role of the textbook and homework	Berinderjeet Kaur, Low Hooi Kiam and Seah Lay Hoon
	Chapter 10 pp. 151–165	Case studies of Singapore secondary mathematics classrooms: The instructional approaches of two teachers	Seah Lay Hoon, Berinderjeet Kaur and Low Hooi Kiam
Sweden	Chapter 21 pp. 307–322	Same from the outside, different on the inside: Swedish mathematics classrooms from students' points of view	Jonas Emanuelsson and Fritjof Sahlström
USA	Chapter 5 pp. 75–85	Mathematics education reform in three US classrooms	Terry Wood, Soo Yeon Shin and Phu Down

& Hiebert, 1999), we believe that IEA research failed to answer the key question of why Confucian-heritage nations performed so well at the elementary and secondary levels.

Although Clarke, Keitel, and Shimizu (2006) state "that research into classrooms, and into learning in classrooms, in particular, must address the interactive and mutually dependent character of teaching and learning" (p. 6), they fail to put to rest concerns about the effect of the intrusion of three video cameras, as well as researchers/observers, on the very interaction and mutual dependence that is the object of their study. In Chapter 11, for example, Begehr (2006) noted that the lesson (in a German school) discussed in her chapter, was the fifth lesson of a total of 14 recorded in the school, "so that students and teacher had already had the opportunity to develop a certain level of familiarity in dealing with the three video cameras" (p. 174). So, what was reported was *not* what normally happened, but rather what happened with the presence of three video cameras in the classroom. It is instructive, nevertheless to reflect carefully on the locations of the many countries involved in LPS research (see the first column of Table 26.1). LPS research has welcomed the long-overdue participation of mathematics education scholars from many of the emerging, and of some of the "forgotten," nations of the world.

In their book *Teaching Mathematics in Australia,* TIMSS Video Study researchers Hollingsworth, Lokan, and McCrae (2003) included a short discussion on the influence of videotaping in classrooms. They asked teachers who were part of the TIMSS 1999 Video Study whether the presence of a single camera affected their teaching of the lesson. They reported that teachers in Australia, the USA and Switzerland thought that their lessons were "about the same," whereas teachers in the Czech Republic, Hong Kong and the Netherlands felt that their lessons were "worse than usual." There are two strong reasons why such self-report data should be questioned as evidence to support the use of video cameras in classroom research. First, not all teachers in all countries surveyed thought that their lesson would have been essentially the same if no camera had been present. And second, the teachers' self-report comments may not necessarily have reflected what happened in their classrooms. For example, teachers' self-report data indicated that the majority of the teachers involved in the TIMSS 1999 study in Australia and the USA believed that their lessons were in accord with contemporary *Standards*-based ideas about teaching and learning mathematics (Hollingsworth et al., 2003). However, there is ample evidence to suggest that teachers' perceptions of what went on in their classrooms were at variance with what independent researchers observed (see, e.g., McIntosh, 2003; Wood, Shin, & Doan, 2006).

Clearly, LPS leaders are committed to the methodology they have developed, including the use of three video cameras, and video-stimulated interview data. But that commitment of itself does not dispel doubts about the effects of the cameras on the validity of the data generated. Concern about the use of three video cameras will not simply disappear by continued assertions that their presence does not really affect the nature or quality of the data. We believe that the researchers would be wise to give serious consideration to working through and reporting alternative and less-intrusive ways of gathering classroom data.

According to the LPS design, researchers in each country were expected to explore each research question through relevant data sets leading to the possibilities of interesting cross-cultural comparisons. Clearly, however, with the non-random sampling employed in each country, generalizations about features of the mathematics classrooms in any one country, or identification of cross-cultural differences on a wide scale, could not validly be undertaken. This limitation is acknowledged by the LPS researchers themselves. Nevertheless, in *Making Connections: Comparing Mathematics Classrooms Around the World*, Clarke, Emanuelsson et al. (2006) sought to derive comparisons between classroom practices in different countries based on LPS findings.

In her review of LPS, Ellerton (2008) stated that there is a sense in which this set of studies could be considered as a pilot study on a grand scale. Ellerton viewed LPS as a study which (a) explored its techniques of data collection; (b) identified areas of interest and significance for further study within each country; (c) explored similarities and differences of learning and teaching environments in culturally different settings, and (d) provided initial responses to the various general research questions which were posed. She commented that if LPS were viewed as a pilot study, it would be easier to forgive areas of weakness. It would also be easier to celebrate strong findings in each of the chapters and to acknowledge that these represent exciting possibilities for further research. According to Ellerton, "pilot studies often appear to be fragmented in some way—and that description could easily be applied to the diverse findings presented in the various chapters" (p. 132).

High-quality collaborative international research involves more than the mere participation of large numbers of people from different nations. It seems that higher degrees of understanding of the significance of the data will occur if as many participants as possible are involved in the interpretation of the data and the reporting of findings.

International Aid Organizations and the Concept of International Community

Since the 1950s, international aid organizations, like UNESCO, UNICEF, the World Bank, the African Development Bank, and the Asian Development Bank, have often negotiated contracts with nations to assist educators in so-called developing nations to modernize their curricula and update their approaches to teaching mathematics. The typical model has been for expert consultants from so-called advanced nations to form a team to work with local educators and administrators. Naturally, however, these experts have usually tended to recommend the "latest" curricula and teaching approaches used in their own nations (Berman, 1992; Carnoy, 1974; Clements & Ellerton, 1996; Karp, 2008; Kitchen, 1995). Thus, for example, British new math approaches were still being introduced into Africa in the late 1960s and into Malaysia as late as the 1980s (Lee, 1982).

Often, local educators were financed by aid money to undertake higher degrees in "advanced" countries, and when they returned to their nations, armed with their graduate degrees, they took steps to introduce curricula and methods to which they had been exposed during their studies abroad. Such an approach sometimes subconsciously bypassed culturally appropriate ideas of local educators (Ellerton & Clements, 1989; Kitchen, 1995), and looked backwards to colonialist times. What was needed most was to look forward to current reform efforts which employed more democratic, more collaborative methods for working toward improvement. The intentions of most participants in most aid programs may have been noble (Karp, 2008), but too often the top-down model for interventions has been inappropriate. What was needed was better quality collaborative designs in which all interested parties had a voice.

Given the politically-fraught nature of this kind of work, and the financial entanglements that can arise, it would not be prudent to pursue this difficult theme further, here. It should be emphasized, though, that the results of education aid have not been entirely negative. For example, UNICEF's (2010) work among people living in remote and mountainous regions of Vietnam has carefully and sensitively taken local culture into account, and has helped to improve the education prospects of many children.

The main point, though, is that the "local to national to international" trend in mathematics education has not always generated improved curricula and teaching and learning approaches in countries receiving aid. Stronger forms of collaborative research are needed, for there have been too many examples in which outsiders have controlled the interventions, and they have been keen to impose their own nations' preferred models and materials on local cultures (Bishop, 1988). Such a model of collaboration needs to be consigned to history.

However, one of the messages of this chapter is that work featuring inadequate models of collaboration should not always be simply discarded. Creative tweaking can often pick out what has been good in the past, discard the bad, and pave the way for much more successful, profitable collaboration.

Quality of Collaboration for International Mathematics Education Research: A Rubric

In the preceding discussion we have identified and isolated particular aspects of collaboration that were applied in the development of certain international mathematics education research programs or organizations. During the discussions, we implied that certain programs or organizations had structures which varied in relation to the quality of collaboration that occurred. We implied that these structural features not only influenced the quality of relationships between researchers but also the research that was produced in the associated research exercises. Much of these discussions were implicit in the ways we viewed the work of the organizations and programs under consideration.

We thought it might be of interest if we attempted to formalize our thinking on this matter. By way of disclaimer, let us state from the outset that what follows is speculative, being based as it is on our own assumptions and conclusions. If it offends any reader we sincerely apologize. It is our hope that others will tighten the ideas that we now present.

Having considered the work done in all the programs and organizations we identified the following seven aspects that might be assessed in relation to "degree of collaboration" for each:

- How well defined were the purposes of the program or organization, in other words, its raison d'être?
- To what extent was the name of the program or organization consistent with its raison d'être?
- To what extent was the governance of the program or organization democratic, in the sense that all participants had a voice that might influence future directions?
- To what extent did the program or organization stimulate wider creative discussion, internationally, on issues raised in the work of the program or organization?
- To what extent did the program or organization influence policies, nationally and internationally, in relation to mathematics education practices?
- To what extent did the program or organization influence practices, nationally and internationally, in relation to mathematics education?
- To what extent did the program or organization influence research directions, nationally and internationally, in relation to mathematics education?

We decided to offer our joint assessment of the quality of collaboration evident in each of the main programs or organizations under consideration in this chapter, as mirrored within the scope of the above seven aspects that relate to collaborative research. We developed and used a 5-point rubric in which "0" indicates "hardly any," 1 indicates "a little bit," 2 indicates "a reasonable amount," 3 indicates "a fair bit," and 4 indicates "a great deal" (see Table 26.2).

We recognize and acknowledge the high level of subjectivity involved in the above exercise, and would remind the reader, again, that Table 26.2 is nothing more than *our* joint assessments. We hope, though, that the table stimulates reflection, not only with respect to the nine programs or organizations involved and the seven rubric properties for collaboration that we identified, but also with respect to collaborative research in which readers have some involvement of their own. It is through such reflection that all researchers can strive to make their collaborative efforts more effective in each of the different domains identified.

The entries in Table 26.2 reflect our belief that although the early efforts of ICMI (before World War I) established a basis for future collaboration in international mathematics education research, the quality of the early collaboration left much to be desired. However, the quality of collaboration in each of the later efforts—those of ICMI, IEA, PISA, RECSAM, MERGA, PME, LPS and international aid programs or organizations has been uniformly good, with each having succeeded in influencing the intended, implemented and attained school mathematics curricula within or across nations, as well as influencing the direction of international mathematics education research.

Table 26.2
Quality of Collaboration for International Mathematics Education Research

Rubric Property	Program or Organization							
	Early ICMI (Before 1915)	Modern ICMI (Since the 1970s)	IEA (FIMS, SIMS, TIMSS)	PISA	RECSAM	MERGA	PME	LPS	INT'L AID
Clarity of *raison d'être*?	2	4	4	4	4	4	3	4	4
Consistency with *raison d'être*?	2	4	4	4	4	4	3	4	3
Democratic governance?	1	2	2	3	4	4	4	3	3
Stimulated Discussions on Issues?	2	4	3	4	4	4	4	4	2
Influenced Policies?	2	4	4	3	3	2	4	3	4
Influenced Practices?	1	3	4	3	3	2	3	3	4
Influenced Research Directions?	1	3	3	4	3	4	4	4	3
TOTALS	11	24	24	25	25	24	25	25	23

A Final Comment

In closing this chapter, we want to dispel the notion that international collaborative research in mathematics education necessarily involves large teams of specialist researchers from different nations. That idea tends to suggest that effective international collaborative research cannot be conducted unless there is large enabling research funding. We do not think that such a point of view is appropriate in this era of globalization.

To be fair to those involved in the early collaborative attempts we have discussed in this chapter, we would point out that collaborative international research is more easily facilitated in the present, globalized world than ever before. But therein lies a challenge: What can we learn from current and previous projects about historical consequences of unbalanced collaboration? What makes collaboration across national borders productive, efficient and effective?

Just because researchers can communicate quickly, freely and openly, does not guarantee that they each understand each other's cultures. Nor does it guarantee that they can relate to and interpret the research contexts in the respective nations in which they work. Cultural sensitivity is essential, as is mutual support and encouragement. As we gaze into the crystal ball, looking to the future of collaborative mathematics education research ventures across national boundaries, some visions are in sharp focus—like the need for genuine attempts at collaboration, for constructive criticism or review of the work of others from other nations, and for support for those whose first language is not English (or the language in which the research is being conducted or reported).

Other visions may not yet be as clear, but represent possibilities that might facilitate collaboration—for example, one can ponder how future revolutions in technology might create even closer links with researchers in other nations. But there are shadows across all of the visions. Do the researchers with whom one would like to work all have easy access to the Internet? Will researchers' focussed visions stay focussed, and will others become sharply focussed? Which visions will come to productive fruition?

The crystal ball is likely to be shattered if one disregards, by design or by accident, any of the essential ingredients for collaboration that we have identified in this chapter. In a very real sense, the crystal ball of collaboration across national boundaries is in our collective hands.

References

Anku, S. (1996, November). First impressions of PME. *PME Newsletter,* 7.

Asante, M. K. (2006). Foreword. In G. J. Sefa Dei & A. Kempf (Eds.), *Anti-colonialism and education: The politics of resistance* (pp. ix–x). Rotterdam, The Netherlands: Sense Publishers.

Begehr, A. (2006). Students' verbal actions in German mathematics classes. In D. Clarke, C. Keitel, & Y. Shimizu (Eds.), *Mathematics classrooms in twelve countries: The insider's perspective* (pp. 167–182). Rotterdam, The Netherlands: Sense Publishers.

Berman, E. H. (1992). Donor agencies and third world educational development, 1945–1985. In R. F. Arnove, P. G. Altbach, & G. P. Kelly (Eds.), *Emergent issues in education: Comparative perspectives* (pp. 57–74). Albany, NY: State University of New York Press.

Bishop, A. J. (1988). *Mathematical enculturation*. Dordrecht, The Netherlands: Kluwer.

Blainey, G. (2001). *The tyranny of distance: How distance shaped Australia's history*. Sydney, Australia: Macmillan.

Borba, M. (2010, March/April). PME an inclusive international community. *PME Newsletter*, 2.

Burstein, L. (Ed.). (1992). *The IEA study of mathematics III: Student growth and classroom processes*. Oxford, UK: Pergamon Press.

Carnoy, M. (1974). *Education as cultural imperialism*. New York, NY: McKay.

Clarke, D., Emanuelsson, J., Jablonka, E., & Mok, I. A. C. (Eds.). (2006). *Making connections: Comparing mathematics classrooms around the world*. Rotterdam, The Netherlands: Sense Publishers.

Clarke, D., Keitel, C., & Shimizu, Y. (Eds.). (2006). *Mathematics classrooms in twelve countries: The insider's perspective*. Rotterdam, The Netherlands: Sense Publishers.

Clements, M. A. (2003). An outsider's view of North American school mathematics curriculum trends. In G. M. A. Stanic & J. Kilpatrick (Eds.), *A history of school mathematics* (pp. 1509–1580). Reston, VA: National Council of Teachers of Mathematics.

Clements, M. A., & Ellerton, N. F. (1996). *Mathematics education research: Past, present and future*. Bangkok, Thailand: Asia-Pacific Centre of Educational Innovation for Development/UNESCO.

Clements, M. A., Grimison, L., & Ellerton, N. F. (1989). Colonialism and school mathematics in Australia 1788–1988. In N. F. Ellerton & M. A. Clements (Eds.), *School mathematics: The challenge to change* (pp. 50–78). Geelong, Australia: Deakin University.

Curbera, G. P. (2009). *Mathematicians of the world, unite! The International Congress of Mathematicians: A human endeavor*. Wellesley, MA: A. K. Peters.

Donoghue, E. F. (1987). *The origins of a professional mathematics education program at Teachers College*. Doctoral dissertation, Teachers College, Columbia University. Retrieved from http://www.proquest.com/en-US/.

Ellerton, N. F. (Ed.). (1999). *Mathematics teacher education: International perspectives*. Perth, Australia: Meridian Press.

Ellerton, N. F. (2008). An outsider's examination of the Insider's perspective—A review of mathematics classrooms in twelve countries: The insider's perspective, edited by D. Clarke, C. Keitel, & Y. Shimizu. *Mathematics Education Research Journal, 20*(1), 127–131.

Ellerton, N. F., & Clements, M. A. (1989). *Teaching post-secondary mathematics at a distance: A report to the Commonwealth Secretariat*. Geelong, Australia: Deakin University.

Ellerton, N. F., & Clements, M. A. (1994). *The national curriculum debacle*. Perth, Australia: Meridian Press.

Ellerton, N. F., & Clements, M. A. (2000). The translation of pencil-and-paper mathematics tests: The question of equivalence. In M. A. Clements, H. Tairab, & K. Y. Wong (Eds.), *Science, mathematics, and technical education in the 20th and 21st centuries* (pp. 113–122). Gadong, Brunei Darussalam: Universiti Brunei Darussalam.

Ellerton, N. F., & Clements, M. A. (2002). Translating pencil-and-paper mathematics tests from one language to another: The myth of equivalence. In H. S. Dhindsa, I. P.-A. Cheong, C. P. Tendencia, & M. A. Clements (Eds.), *Realities in science, mathematics and technical education* (pp. 245–254). Gadong, Brunei Darussalam: Universiti Brunei Darussalam.

Ellerton, N. F., & Clements, M. A. (2010). Hidden weaknesses in mathematics education settings. In I. Hideki (Ed.), *Development of mathematical literacy in the lifelong learning society* (pp. 286–301). Hiroshima, Japan: Japan Society for the Promotion of Science.

Fukisawa, R. (Ed.). (1912). *Summary report of the teaching of mathematics in Japan*. Tokyo, Japan: Department of Education.

Furinghetti, F. (2008). Mathematics education in the ICMI perspective. *The International Journal for the History of Mathematics, 3*(2), 47–56.

Gooya, Z. (2010, March/April). PME an international community? Not yet! *PME Newsletter*, 3–4.

Hiebert, J., Stigler, J. W., & Manaster, A. B. (1999). Mathematical features of lessons in the TIMSS video study. *Zentralblatt für Didaktik der Mathematik, 31*(6), 196–201.

Hollingsworth, H., Lokan, J., & McCrae, B. (2003). *Teaching mathematics in Australia: Results from the TIMSS 1999 video study*. Melbourne, Australia: Australian Council for Educational Research.

Horwood, J., & Clements, M. A. (2000). A mirror to the past: Mathematics education in Australia and Southeast Asia, past, present and future. In M. A. Clements, H. Tairab, & K. Y. Wong (Eds.), *Science, mathematics, and technical education in the 20th and 21st centuries* (pp. 133–143). Gadong, Brunei Darussalam: Universiti Brunei Darussalam.

Howson, G., & Kahane, J.-P. (Eds.). (1990). *The popularization of mathematics*. Cambridge, UK: Cambridge University Press.

Husén, T. (Ed.). (1967). *A comparison of twelve countries: International Study of Achievement in Mathematics* (Vol. 1–2). Stockholm, Sweden: Almquist & Wiksell.

Jones, G. A. (2005). What do studies like PISA mean to the mathematics education community? In H. L. Chick & J. L. Vincent (Eds.), *Proceedings of the 29th Conference of the International Group for the Psychology of Mathematics Education* (Vol. 1, pp. 71–74). Melbourne, Australia: International Group for the Psychology of Mathematics Education.

Kaiser, G., Luna, E., & Huntley, I. (Eds.). (1999). *International comparisons in mathematics education*. London, UK: Routledge.

Karp, A. (2008). Interview with Geoffrey Howson. *International Journal for the History of Mathematics Education, 3*(1), 47–67.

Katz, J. S., & Martin, B. R. (1997). What is research collaboration? *Research Policy, 26*(1), 1–18.

Keitel, C., & Kilpatrick, J. (1999). The rationality and irrationality of international comparative studies. In G. Kaiser, G. Luna, & I. Huntley (Eds.), *International comparisons in mathematics education* (pp. 241–256). London, UK: Routledge.

Kieran, C. (2005). Some results for the PISA 2003 international assessment of mathematics learning: What makes items difficult for students? In H. L. Chick & J. L. Vincent (Eds.), *Proceedings of the 29th Conference of the International Group for the Psychology of Mathematics Education* (Vol. 1, pp. 83–86). Melbourne, Australia: International Group for the Psychology of Mathematics Education.

Kitchen, R. S. (1995). Mathematics pedagogy in the 3rd world. The case of a Guatemalan teacher. *ISGEm Newsletter, 10*(2), 1–4.

Lee, C. S. (1982). Reform in mathematics education in Malaysia. *Journal of Science and Mathematics Education in Southeast Asia, 5*(2), 34–40.

Leung, F. K. S. (2006). Mathematics education in East Asia and the West: Does culture matter? In F. K. S. Leung, K. D. Graf, & F. Lopez-Real (Eds.), *Mathematics education is different cultural traditions—A comparative study of East Asia and the West: The 13th ICMI Study* (pp. 21–46). New York, NY: Springer.

Leung, F. K. S., Graf, K.-D., & Lopez-Real, F. J. (Eds.). (2006). *Mathematics education is different cultural traditions—A comparative study of East Asia and the West: The 13th ICMI Study*. New York, NY: Springer.

McIntosh, A. (2003). A typical Australian Year 8 mathematics lesson? In H. Hollingsworth, J. Lokan & B. McCrae (Eds.), *Teaching mathematics in Australia: Results from the TIMSS 1999 Video Study* (pp. 106–108). Camberwell, Australia: Australian Council for Educational Research.

Menon, R. (2000). Should the United States emulate Singapore's education system to achieve Singapore's success in TIMSS? *Mathematics Teaching in the Middle School, 5*(6), 345–347.

Moon, B. (1986). *The "new maths" curriculum controversy: An international story*. London, UK: Falmer Press.

Mosteller, F., & Boruch, R. (2002). *Evidence matters: Randomized trials in education research*. Washington, DC: Brookings Institution Press.

Neubrand, M. (2005). The PISA study: Challenge and impetus to research in mathematics education. In H. L. Chick & J. L. Vincent (Eds.), *Proceedings of the 29th Conference of the*

International Group for the Psychology of Mathematics Education (Vol. 1, pp. 79–82). Melbourne, Australia: International Group for the Psychology of Mathematics Education.

Olivier, A., & Newstead, K. (Eds.). (1998). *Proceedings of the 22nd Conference of the International Group for the Psychology of Mathematics Education*. Stellenbosch, South Africa: University of Stellenbosch.

Organisation for Economic Co-operation and Development. (2010). *PISA 2009 results: What students know and can do* (Vol. I). Paris, France: Author.

Organisation for European Economic Cooperation. (1961a). *New thinking in school mathematics*. Paris, France: Author.

Organisation for European Economic Cooperation. (1961b). *School mathematics in OEEC countries*. Paris, France: Author.

Organisation for European Economic Cooperation. (1961c). *Synopses for school mathematics*. Paris, France: Author.

Persens, J. (2006, August), *Mathematics achievement in Africa*. Paper presented at an international workshop on "Setting a Collaborative Research Agenda: Mathematics Education in the U.S. and Africa," Dakar, Senegal.

Postlethwaite, T. N. (Ed.). (1967). *School organisation and student achievement: A study based on achievement in mathematics in twelve countries*. Stockholm, Sweden: Almquist & Wiksell.

Postlethwaite, T. N. (1993). Torsten Husén (1916–). *Prospects: The Quarterly Review of Comparative Education, 23*(3/4), 677–686.

Reeve, W. D. (1929). *Significant changes and trends in the teaching of mathematics throughout the world since 1910* (Fourth Yearbook of the National Council of Teachers of Mathematics). New York, NY: Columbia University, Teachers College, Bureau of Publications.

Regional Centre for Education in Science and Mathematics (RECSAM). (1991). *Research abstracts*. Penang, Malaysia: Author.

Roadrangka, V., & Liau Tet Loke, M. (1993). *A summary of research reports for the years 1986–1992*. Penang, Malaysia: SEAMEO-RECSAM.

Robitaille, D. F., & Garden, R. A. (Eds.). (1989). *The IEA study of mathematics II: Context and outcomes of school mathematics*. Oxford, UK: Pergamon Press.

Schmidt, W. H., Blömeke, S., & Tatto, M. T. (2011). *Teacher education matters: A study of middle school mathematics teacher preparation in six countries*. New York, NY: Teachers College Press.

Schmidt, W. H., McKnight, C. C., Houang, R. T., Wang, H.-C., Wiley, D. E., Cogan, L. S., & Wolfe, R. G. (2001). *Why schools matter: A cross-national comparison of curriculum and learning*. San Francisco, CA: Jossey-Bass.

Schubring, G. (2008a). Editorial. *International Journal for the History of Mathematics, 3*(2), 1–2.

Schubring, G. (2008b). The origins and the early history of ICMI. *International Journal for the History of Mathematics, 3*(2), 3–33.

Schwarz, B., Dreyfus, T., & Hershkowitz, R. (Eds.). (2009). Introduction. In B. Schwarz, T. Dreyfus, & R. Hershkowitz. *Transformation of knowledge through classroom interaction* (pp. 1–8). Oxford, UK: Routledge & Kegan Paul.

Sefa Dei, G. J., & Kempf, A. (Eds.). (2006). *Anti-colonialism and education: The politics of resistance*. Rotterdam, The Netherlands: Sense Publishers.

Shimizu, Y. (2005). From a profile to the scrutiny of student performance: Exploring the research possibilities by the international achievement studies. In H. L. Chick & J. L. Vincent (Eds.), *Proceedings of the 29th Conference of the International Group for the Psychology of Mathematics Education* (Vol. 1, pp. 75–78). Melbourne, Australia: International Group for the Psychology of Mathematics Education.

Sierpinska, A., Kilpatrick, J., Balacheff, N., Howson, A. G., Sfard, A., & Steinbring, H. (1993). What is research in mathematics education and what are its results? *Journal for Research in Mathematics Education, 23*(3), 274–279.

Singh, P., & Lim, C. S. (Eds.). (2005). *Improving teaching and learning of mathematics: From research to practice*. Shah Alam, Malaysia: Pusat Penerbitan Universiti.

Stacey, K. (2010). Mathematics and scientific literacy around the world. *Journal of Science and Mathematics Education in Southeast Asia, 33*(1), 1–16.

Stevenson, H. W., & Stigler, J. W. (1992). *The learning gap*. New York, NY: Summit Books.
Stigler, J., & Hiebert, J. (1997). Understanding and improving classroom instruction: An overview of the TIMSS video study. *Phi Delta Kappan, 79*(1), 14–21.
Stigler, J., & Hiebert, J. (1999). *The teaching gap*. New York, NY: Free Press.
Torney-Purta, J. (1987). The role of comparative education in the debate on excellence. In R. Lawson, V. Rust, & S. Shafer (Eds.), *Education and social concern: An approach to social foundations* (pp. 80–89). Ann Arbor, MI: Prakken Publications.
Travers, K. J., & Westbury, I. (Eds.). (1989). *The IEA study of mathematics I: Analysis of mathematics curricula*. Oxford, UK: Pergamon Press.
Tzekaki, M., Kaldrimidou, M., & Sakonidis, H. (Eds.). (2009). *In search for theories in mathematics education: Proceedings of the 33rd Conference of the International Group for the Psychology of Mathematics Education*. Thessaloniki, Greece: International Group for the Psychology of Mathematics Education.
UNICEF. (2010). *An analysis of the situation of children in Viet Nam 2010*. Hanoi, Vietnam: Author.
Williams, J. (2005). The foundation and spectacle of (the leaning) tower of PISA. In H. L. Chick & J. L. Vincent (Eds.), *Proceedings of the 29th Conference of the International Group for the Psychology of Mathematics Education* (Vol. 1, pp. 87–90). Melbourne, Australia: International Group for the Psychology of Mathematics Education.
Wood, T., Shin, S. Y., & Doan, P. (2006). Mathematics education reform in three US classrooms. In D. J. Clarke, C. Keitel, & Y. Shimizu (Eds.), *Mathematics classrooms in twelve countries: The insider's perspective* (pp. 75–86). Rotterdam, The Netherlands: Sense Publishers.
Young, J. W. A., Osgood, W. F., Smith, D. E., & Taylor, E. H. (Eds.). (1915). *Mathematics in the middle commercial and industrial schools of various countries represented in the International Commission on the Teaching of Mathematics*. Washington, DC: Government Printing Office.

Chapter 27
Influence of International Studies of Student Achievement on Mathematics Teaching and Learning

Vilma Mesa, Pedro Gómez, and Ui Hock Cheah

Abstract In this chapter, we present findings regarding the ways in which the results of international studies of student achievement have influenced the teaching and learning of mathematics in the classroom. We put forward a model of curriculum composed of four levels (global, intended, implemented, and attained) and four dimensions (conceptual, cognitive, formative, and social). This model allows us to describe the differences between two major international studies of student achievement—the Trends in the International Mathematics and Science Study (TIMSS) and the Programme for International Student Assessment (PISA)—and to situate the influences of these studies on classroom practice. Our search revealed that the question of how these studies have directly affected practice has not been systematically addressed. Although we found that there have been some influences of the international studies on classroom practice—for example, in the language used in public documents, in the localization of curriculum design, and in the impact of using imported textbooks—research on these influences has been conducted mostly in isolation, without any coherent plan. We use our curriculum model to propose a research agenda on three major issues: the impact of the notion of competency and the use of the studies' frameworks; curriculum control, design, and management; and teacher preparation and development and textbook use.

V. Mesa (✉)
University of Michigan, Ann Arbor, MI, USA
e-mail: vmesa@umich.edu

P. Gómez
Universidad de Granada, Granada, Spain

Universidad de Los Andes, Bogotá, Colombia

U. H. Cheah
SEAMEO-RECSAM, Penang, Malaysia

Our charge for this chapter is to present an account of efforts that have been made to take advantage of the information that international studies of student achievement offer to affect practice: the day-to-day of teaching and learning mathematics in classrooms. The chapter is complementary to Dossey and Wu's Chapter 31, in this *Handbook*, which speaks about the impact that these studies have had at the policy level.

Comparative education in mathematics is an old enterprise (e.g., Cairn, 1935; Young, 1900). Only after the 1960s did efforts to investigate how students from different countries perform in mathematics become more systematic and collaborative, and begin to involve a larger number of nations and educational systems (Bottani, 2006; Husén, 1967; Robitaille & Travers, 1992; Schmidt & McKnight, 1995; Travers & Westbury, 1989). Although there are many arguments for and against participating in studies that compare and contrast student attainment across countries (see, e.g., Bracey, 1998; Freudenthal, 1975; Husén, 1983; Keitel & Kilpatrick, 1998; Kilpatrick, 1971; Robitaille & Travers, 1992), there is an anticipation that the community will benefit from these studies, from the "research findings, the methods used in research, and [their] theoretical constructs" (Ferrini-Mundy & Schmidt, 2005, p. 169).

Indeed, Robitaille and Travers (1992) argued that these studies (p. 707, emphasis added).

> "*can* serve as valuable sources of data and information against which educators in a given country *can* compare and contrast the curriculum, the teaching practices, and the outcomes attained by students in their own system. The possible impact of alternative curricular offerings, teaching strategies, administrative arrangements, and the like *can* be estimated efficiently by examining their implementation in other jurisdictions, even when the countries are quite dissimilar culturally or economically. … Achievement comparisons *can* also provide indications about what is possible … what *can* be accomplished. They *should* serve as a spur and incentive for improvement."

As the emphasized words suggest, these are hypothetical expectations that researchers have formulated about comparative studies. This chapter presents our efforts in assessing the extent to which these studies have indeed influenced practice, directly or indirectly; whether alternative "curricular offerings, teaching strategies, administrative arrangements, and the like" have indeed occurred, and if they have, whether they have resulted in the changes or real effects on learning and attitudes toward mathematics of students and on the teaching of school mathematics in the participating countries and elsewhere. We discuss the ways in which these studies have been "catalysts" for research that informs certain levels of practice (Ferrini-Mundy & Schmidt, 2005) and the ways in which they have been sensitive to cultural variation (Clarke, 2003).

The chapter is presented in three sections. We start by presenting a conceptualization of curriculum that allows us to organize our findings regarding ways in which the Trends in International Mathematics and Science Study (TIMSS) and the Programme for International Student Assessment (PISA) have exerted influences at different levels. This section is followed by illustrative examples of these influences, which leads to a final section in which we propose research ideas that would move our community toward a better understanding of the actual impact that these studies can have on classroom processes.

Readers who wish to learn more of the histories of TIMSS and PISA are referred to Chapter 31 by Dossey and Wu. A third group of studies, those conducted by

UNESCO, focus on Latin America and Africa, and are part of the "Education for All" initiative, which seeks to have all the world's primary-aged children enrolled in school by 2015. The studies collect information on third- and sixth-grade students' attainment in mathematics, science, and reading. In addition, there are questionnaires for students (attitudes and background), teachers (content coverage and pedagogy), principals, and parents. A Sub-Saharan study also tested teachers' knowledge of the content on which the students were tested. There are very few reports on the results of these studies (among them are Bonnet, 2008; Lee & Zuze, 2010; Saito, 2010; Valdés et al., 2008), and those that exist are mostly descriptive, although some seek to interpret variability using the contextual variables collected. The Centre for Innovation in Mathematics Teaching at the University of Exeter has conducted two other longitudinal international studies of mathematical achievement for Innovation in Mathematics Teaching in the UK, the Kassel Project (http://www.cimt.plymouth.ac.uk/projects/kassel/default.htm), and the International Project on Mathematical Attainment (http://www.cimt.plymouth.ac.uk/projects/ipma/default.htm). The data have been given to heads of the departments of participating schools, but there is no information on their effects in the classroom. For examples of reports in Singapore, see Kaur and Yap (2009) and Kaur, Koay, and Yap (2009).

Conceptualizations of Curriculum

An interesting feature of international collaboration is the need to clarify terms and concepts in order to make the work transparent. This was the case in writing this chapter. As our writing progressed, it became increasingly clear that we were using different definitions for *curriculum*. From the Latin *currere*—to run—the word can refer to the sequences of courses that a student can take, the topics that are covered in a given grade, or the content, skills, competencies, and habits of mind that a person needs to acquire through schooling in order to participate successfully in society. The classical distinction between intended, implemented, and attained curricula (Travers & Westbury, 1989) was useful to describe how either notion of curriculum is transformed, but it did not differentiate other aspects that play significant roles in defining a curriculum. Thus, in this section, we propose a definition that will help us situate the influences that we found. We depart from a definition of curriculum encompassing only content or competencies by defining curriculum as a teaching and learning plan that can be described at different levels and that has different dimensions. We start with a description of how curriculum is understood by the studies that are the central to this chapter, namely, TIMSS and PISA.

Curriculum in the International Studies

The International Association for the Evaluation of Educational Achievement (IEA) has conceptualized curriculum as a tripartite model consisting of the intended, implemented, and attained curriculum (see Figure 27.1). The *intended* curriculum

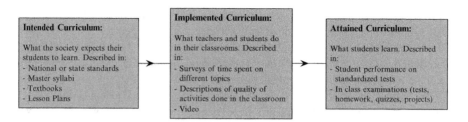

Figure 27.1. Different versions of the curriculum; adapted from Travers and Westbury (1989).

corresponds to the goals for learning mathematics that students are expected to attain, goals that may be established by national organizations (bureaus) or central ministries, states, departments, districts, or schools, and explained through course outlines, official syllabi, and textbooks. Some systems differentiate goals according to types of students (e.g., Gymnasium, Hauptschule, and Realschule in Germany), others produce the textbooks that all children should use (e.g., Cyprus), and others use achievement tests and diagnostic assessments that children take at different stages of their schooling (e.g., South Korea) to define learning expectations. The *implemented* curriculum corresponds to what is actually taught to students in classrooms by teachers; it refers to the interpretations of the intended curriculum made by teachers, who are directly responsible for helping students learn. The *attained* curriculum corresponds to what students have learned (as assessed by standardized tests, including international tests, and classroom assessment) as a consequence of the teaching received.

This view of curriculum presupposes different degrees of expertise in curriculum design at each level and assumes that teachers and students have little agency in designing the curriculum, accepting and agreeing with the information from the previous level. In particular, teachers are expected to use what is given to them (goals for society, goals for schooling, textbooks, official syllabi) to make decisions about what is best for the students they have. This view of curriculum also suggests that the studies are expected to exert a major influence at the policy level; that is, that the results will be used to shape intentions of the whole system that in turn will influence what will happen in the classrooms and with students, as those intentions get transformed into actions. As a side note, we acknowledge that this model excludes the possibility that after implementation, changes can be made to the intention of the curriculum. This local view of curriculum transformation is not accounted for in this model.

The mathematics framework developed for the 1995 Third International Mathematics and Science Study (TIMSS) consisted of three aspects: *content domains* (numbers; measurement; geometry; proportionality; functions, relations and equations; data representations, probability, and statistics; elementary analysis; and validation and structure), *processes or performance expectations* (knowing, using routine procedures, investigating and problem solving, mathematical reasoning, and communicating), and the *affective outcomes or perspectives* of school mathematics and science (attitudes, careers, participation, increasing interest, and habits of mind). The participating countries agreed upon the content domains, whereas the performance category was "aligned to the US National Assessment of Educational Progress's (NAEP's) concepts of mathematical abilities and mathematical power, both with roots in the National Council of Teachers of Mathematics' (NCTM's) standards" (Mullis, 1999, p. 15). Note that this definition encompassed more than content, although content was a major component of the framework.

The Organisation for Economic Co-operation and Development (OECD) departed from this content-focussed approach when setting out its assessment framework for its Programme for International Student Assessment (PISA). The first paragraph in the framework stated:

> The aim of the OECD/PISA assessment is to develop indicators of the extent to which the educational systems in participating countries have prepared 15-year-olds to play constructive roles as citizens in society. *Rather than being limited to the curriculum content students have learned,* the assessments focus on determining if students can use what they have learned in the situations they are likely to encounter in their daily lives. (OECD, 2003, p. 24, emphasis added)

The PISA mathematics framework used three components to describe a domain to be assessed in relation to the problems that students were expected to solve: (a) the situations or contexts in which the problems were located; (b) the mathematical content that had to be used to solve the problems, organized by certain overarching ideas, and, most importantly, (c) the competencies that had to be activated in order to use mathematics to solve real-world problems. *Content* was organized into four overarching ideas: quantity, space and shape, change and relationships, and uncertainty. Mathematical competence was described in terms of eight specific *competencies*: (a) thinking and reasoning, (b) argumentation, (c) communication, (d) modelling, (e) problem posing and solving, (f) representation, (g) using symbolic, formal and technical language and operations, and (h) use of aids and tools. The cognitive activities encompassed by these competencies were structured in three *competency clusters*: (a) reproduction, (b) connection, and (c) reflection.

The PISA framework introduced the idea of *mathematical literacy* in order to emphasize a functional view of school mathematics, and defined the term as the tools that should enable students to make well-founded judgments and be useful in students' lives as citizens. Thus, conceptually, PISA sought to assess the extent to which schools had prepared students for participation in the society, whereas TIMSS assessed the extent to which students showed proficiency with particular mathematical content at specific points in their school lives. We take these differences into account by situating the two studies as attending to two different aspects of what we will propose as curriculum.

The Concept of Curriculum in This Chapter

So far, we have identified three levels for the curriculum: the intended, implemented, and attained. Within the intended level, we distinguish several sublevels: the first is the education system that is particular to each individual country and in turn

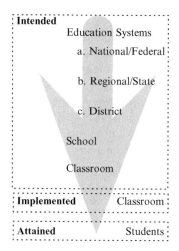

Figure 27.2. Levels of the curriculum considered in this chapter.

can be differentiated by national or federal, regional and state, or district mandates or guidelines; second, at the school level, we include the plans that schools use, perhaps adapted or adopted directly from federal, national, or regional mandates or guidelines; and third is the classroom level which refers to the plans that teachers create for teaching particular lessons, using their institutional, district, state, or national guidelines or other sources.

In contrast to these multiple sublevels, the implemented curriculum manifests mainly in the classroom, whereas the attained curriculum manifests mainly through student performance on class assessments and on standardized tests when these are available (see Figure 27.2, in which a gray arrow, in the background, shows the flow of suggested influences). Other aspects of the curriculum (e.g., "hidden," "null") are also identifiable at the school level—however, we shall concentrate only on the curriculum within the classroom, as that was the task for this chapter.

Concurrently with these levels, the curriculum is composed of four dimensions: conceptual, cognitive, formative, and social (Rico, 1997), each of which deals with four fundamental and interrelated questions: "What is knowledge, what is learning, what is teaching, and what is useful knowledge" (p. 386). We acknowledge, of course, that other conceptualizations of curriculum can be embedded within this definition (see, e.g., Beyer & Liston, 1996; Lattuca & Stark, 2009).

The *conceptual* dimension refers to content and topics that are specific to a given discipline; it defines those elements particular to a discipline (e.g., mathematics, the sciences) that are a synthesis of historical and cultural traditions; this dimension is informed by epistemology and the history of mathematics and defines larger cultural aims. The *cognitive* dimension refers to learning and the learner, and deals with understanding what learning is, how it happens, and how different people learn; it also has particular manifestations depending on a given discipline; it is directly informed by learning theories and defines specific expectations, development, and learning aims. The *formative* dimension refers to teaching and the teacher; it deals with aspects such as what teaching is and, in particular, what mathematics teaching is; it specifies practices that are believed to be useful for teaching (e.g., planning, differentiating instruction), and it provides the basis for generating programs for future and practicing teachers; it is informed by pedagogical theories and defines formative aims. The *social* dimension refers to the value that a society places on the utility and usefulness of the mathematical knowledge; it deals with questions such as:

> Which instruments are used to judge the mathematical capacity of an individual? What social mechanisms support that judgment? How and with what criteria are teachers' capacity and curriculum materials judged? [And] which criteria are used to assess the effectiveness of a curriculum? (Rico, 1997, p. 385)

This dimension is informed by sociology and other disciplines, and it defines social aims.

Hence, curriculum can be conceived as involving levels—from the national educational system through to the classroom—and four dimensions, as described above. At any given level, each dimension of curriculum acquires a specific meaning. At the classroom level, which is of particular interest for this chapter, the conceptual dimension of the curriculum refers to the mathematics topics that configure the content of a given grade or teaching unit; the learning goals of such grade or teaching unit are the expression of the cognitive dimension of the curriculum; the formative dimension of the curriculum refers to the teaching methodology set up for the grade or teaching unit; finally, the assessment instruments and criteria selected for the grade or teaching unit configure the social dimension of the curriculum at the classroom level.

For the purposes of this chapter, we include in our model two additional elements. First, there is a global level, which in the abstract refers to the possibility of having a curriculum that transcends individual systems and that could operate, in fact, as a global curriculum—a curriculum that is common to many education systems. Although not curricula themselves, the frameworks of the international studies can be seen as part of a global level because they represent the agreements across several education systems and nations towards a common set of content and learning expectations that will be used to assess students (Clarke, 2003). Second, in the attained level we focus on and distinguish between students' performance as assessed via national standardized tests and via tests prepared for international studies.

Because each of these dimensions manifests at different levels—at the level of classroom practice, at the administrative level of a particular school, and at the larger level of an educational system—we have combined these definitions with the tripartite version of the curriculum to generate a matrix that situates the different manifestations of curriculum, and have added several levels in order to represent better the different influences that we identified (see Figure 27.3).

Figure 27.3 allows us to situate different documents and aspects of the international comparison studies. In spite of their differences in emphasis, we situate the frameworks of TIMSS and PISA at the global level; and because both TIMSS and PISA are concerned with what students have learned and suggest content (TIMSS) and competencies (PISA) that are considered relevant, these studies relate to the conceptual and cognitive dimensions only. NCTM's (2000) *Principles and Standards*

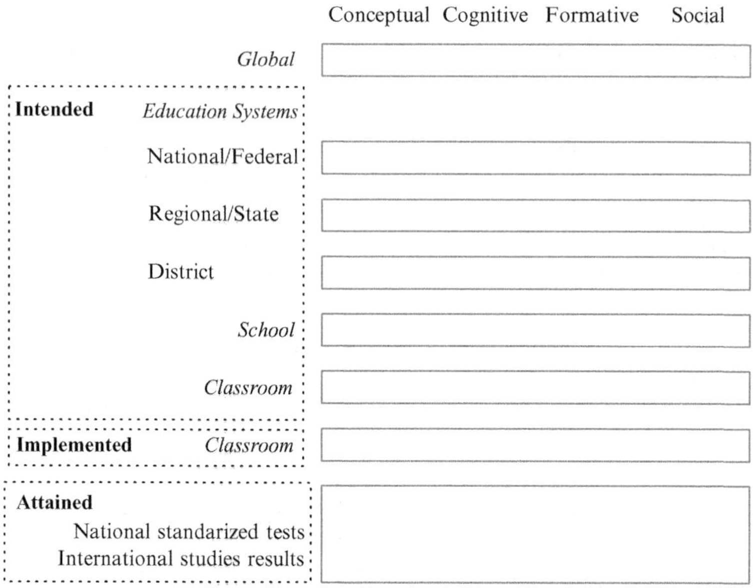

	Conceptual	Cognitive	Formative	Social
Global				

Intended
- Education Systems
- National/Federal
- Regional/State
- District
- School
- Classroom

Implemented
- Classroom

Attained
- National standarized tests
- International studies results

Figure 27.3. Levels and dimensions of the curriculum as understood in this chapter.

for School Mathematics and official syllabi in individual countries with centralized curricula can also fall into the conceptual and cognitive dimensions. NCTM (1991, 1995) also published professional teaching standards and assessment standards. These documents explicitly describe what is knowledge, what is learning, what is teaching, and what is useful knowledge, thus spanning all the dimensions of the curriculum at the national level.

The *Common Core State Standards*, recently released in the USA following an agreement among 48 governors, are an attempt to define content and learning outcomes for the country, and are situated at the conceptual and cognitive dimensions of the national level. At the time this chapter was being prepared, 40 states and the District of Columbia had adopted these *Standards*.

Some countries (e.g., Spain) include specific mandatory norms at the national level for each of these dimensions, whereas others (e.g., Colombia) only give guidelines and suggestions at the cognitive level (learning expectations). In Asian nations, many ministries of education oversee the development of mandatory national curricula which, although attending primarily to content, have over the last 10 years placed greater emphasis on mathematical processes. In some countries (e.g., Cyprus, South Korea), there are national teachers' guides that include ideas about how students can learn and about how topics should be taught. In some countries (e.g., the USA), there might be similar information, but this appears only in teachers' editions of student textbooks and might be totally unregulated.

With this definition of curriculum as backdrop, we turn now to the main task for this chapter.

Influences of Comparative Studies

To address the question driving this chapter, we searched numerous sources seeking to locate reports, documents, and articles related to the topic. We approached key informants in academia and in ministries or bureaus of education in several countries seeking information about possible studies into the impact of international comparative studies at the classroom level. These informants pointed to knowledgeable researchers and other relevant sources. Library searches provided links to dissertations, conference presentations, reports, books, and journal articles.

In the first group of studies that we found the majority, provided primary and secondary analyses of the results of international studies. Studies in this group mined the richness of the data sets in order to establish connections between the variables collected at country, school, classroom, teacher, and student level with the scores obtained. Because international comparison studies have been increasing in their sophistication, more powerful analyses have been conducted. These studies did not document changing practices or study possible changes to practice; rather, they sought to understand and explain the sources of differences in scores within and between countries. For this reason, we did not include these studies. We also excluded studies that mentioned students' performance (either high or low) to justify attending to a specific issue of educational interest, but we kept studies in which substantive elements of the studies (the frameworks, the test items, or the findings) were used.

We discuss here the only study that was very close in nature to the charge that we had for this chapter, and which anticipates our findings. That study, by Saracho (2006), described in detail some of the policies adopted by countries to improve their PISA results as consequence of their participation in PISA 2000 and 2003. It was commissioned by the OECD in Latin America and conducted by the Mexican foundation IDEA [Implementation, Design, Evaluation and Analysis of Public Policies]. Saracho used documents, secondary sources, and news, together with phone interviews and questionnaires answered by specialists, public officers, and university professors in the countries involved. The 14 countries analyzed (Austria, Belgium, Brazil, England, France, Germany, Ireland, Luxembourg, Mexico, The Netherlands, Norway, Poland, Spain, and Sweden) were selected according to the availability of information, the "size" of their reactions to the PISA 2000 results, and their relevance to the Mexican case.

Even though the Saracho (2006) report focussed on policy reactions, it considered issues related to the influence of PISA in the classroom and provided an overview of how each country reacted to the publication of the 2000 and 2003 results. However, he stated explicitly that he did not find any evidence of the impact of PISA on schools or classrooms. Nevertheless, he documented some reactions that could, indirectly, influence the teaching and learning of mathematics in the classroom, and where pertinent, we have included these findings in our review.

A second group of documents used elements of the international comparison of achievement studies—their frameworks, the released items, or the videos—to induce some change and to test the impact of those changes. Included in this group are studies that capitalized on the logic of "data-driven" analyses, by which changes at the local level can be initiated and evaluated using information that is of interest to the participants. We included these studies in the review even though they do not assess the impact of the changes.

A third group of studies reported actions taken as a result of looking at what successful countries do and consequently adopting salient elements believed to be directly associated with that success. The three most prominent cases are the shift of the locus of control for curriculum development from higher to lower levels, the incorporation of Japanese lesson study, and the adoption of Singapore mathematics textbooks. We included these studies, and in the case of the textbooks, we included available information about their impact in the classroom.

Thus, what we offer is an inventory of projects, activities, and programs whose impetus can be traced to either PISA or TIMSS and could have a direct or indirect connection to classroom practice. We believe that the inventory is not comprehensive, as many reports may not have been available to us. The majority of the initiatives have not been formally studied to assess their impact: How are they used? By how many people? What do teachers and students perceive about their effectiveness? And how are they related to student and teacher performance? Nevertheless, we believe that they merit consideration for their potential to generate substantive research in the future. Because both PISA and TIMSS have tests on science, when available we included works conducted in science education because they suggest possible uses that may have been given in mathematics, although we have not been successful in finding documentation of these uses.

We provide selected examples, classified by the main categories of influences that we were able to document. Within each category, we include details of how it was exemplified in individual countries as a consequence of either PISA or TIMSS. Readers will notice that most of the examples related to PISA focus on two regions, Europe and Latin America, and that most of the examples related to TIMSS focus on South East Asia and the USA. This reflects the regional impact of these studies. Dossey and Wu (Chapter 31) note that in terms of reactions to the studies, some countries (usually the high-performing ones) assume a congratulatory approach, and in general may not worry about making specific changes in their own systems; other countries assume an indifferent position, taking the results as yet another indicator of performance, with little interest in making changes; a third group of countries use their students' performance as a justification for engaging in activities that would alter practice. These last countries provide the examples of influences that we report here.

Our conceptualization of curriculum has allowed us to better understand and situate the influences that we identified in our search. We start with the model presented in Figure 27.3 and use numbered arrows to depict those influences (see Figure 27.4). We have identified two sets of influences by looking at where they originate: at the global level (Influences 1 and 2) or at the attained level of the international studies (Influences 3 to 7). Each influence starts at a particular dimension of these levels and ends in more than one level and dimension, illustrating that a specific aspect of the international studies can influence different features of a country's mathematics curriculum. The first set of influences start at a global level, with

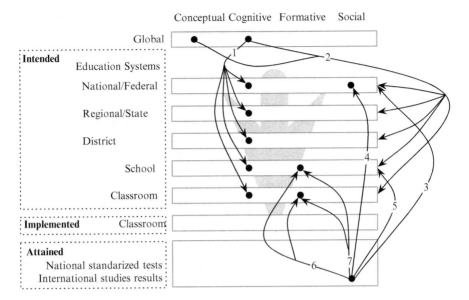

Figure 27.4. Influences of international studies on different aspects of the curriculum.

the frameworks proposed by the international studies. The two influences that can be tracked down to each level of the intended curriculum are as follows:

1. Competencies and standards in official documents; and
2. the adaptation and adoption of the frameworks and the items used in these studies.

Influence 1 starts at the cognitive dimension of the global level; it represents how the frameworks that have been used by the international studies, concretely PISA's, have had an effect on the cognitive dimension at all the other levels of the intended curriculum via the notion of *competencies*. In Spain, for example, a number of documents have been written at all the levels of the intended curriculum, targeting student learning of these competencies. Influence 2 starts at both the conceptual and cognitive dimension of the global level to show how the content and learning expectations in these studies have been used as benchmarks for standards and attainment, influencing various levels of the intended curriculum.

The second set of influences corresponds to those that we could track as consequence of the results of students' performance in the tests. The influences originate at the attained level, but they refer mainly to the results of tests generated by the international studies, rather than to the results of tests created by individual education systems, and go "up" to the intended level. We identified five different types of influences:

3. Localization of curriculum design (by which we refer to what is sometimes known as the pedagogical aspect of school autonomy);
4. National changes in assessment processes;

5. The implementation of cycles of data analysis to design, enact, and assess changes;
6. Professional development and teacher education; and
7. Use of textbooks from other countries.

In Figure 27.4, the concept of localization of curriculum design is represented with Arrow 3, from the social dimension of the attained level (results of the international studies) to the intended national/federal level expressing the influence the results of the studies have had, in some countries, in giving more freedom to schools to design dimensions of the mathematics curriculum. Arrow 4, on the other hand, represents influences of the results of the tests on assessment processes at the intended national level, and Arrow 5 represents the influence of the results of the tests on curriculum on all the dimensions of curriculum at the intended school level. Arrows 6 and 7 represent influences of the results of the tests on teaching methods both at the school and classroom intended levels.

The reader may, by now, have noticed that the arrows start at strategic points from which the influences could be expected according to the conceptualization of the international studies—a global, overarching level—and that the arrows end at various places in the intended level, none of them reaching the implemented level. This is in itself an important finding for us. As far as we know, there have been no systematic investigations on how these studies have reached the classroom.

In the next two sections, we describe these two sets of influences. Readers should note that the sections are *uneven* in the amount of information provided. This unevenness is a consequence of the types of documentation that we were able to secure; the differences in *depth* reveal, indirectly, the amount of information available about each type of influence.

From the Global Level to the Intended Curriculum

In this section, we describe the influences we established in Figure 27.4 that originate in the global level of our curriculum model and affect the intended curriculum. We briefly describe the studies and documents that illustrated the ways in which these influences were operating.

Competencies and standards in official documents (Influence 1). An important influence of these studies in several countries (mostly Latin American and European) has been on the language used in public and political discussions of education. The notion of competencies has gained currency, presumably because it is the word used in the PISA frameworks. The word competencies has been included in several national documents and in a European directive (Education Council, 2006), where it refers to competencies for lifelong learning. Competencies are meant to go beyond specific school content or skills:

> The assessment of student performance in selected school subjects took place with the understanding, though, that students' success in life depends on a much wider range of competencies. ... A competency is more than just knowledge and skills. It involves the ability

to meet complex demands, by drawing on and mobilising psychosocial resources (including skills and attitudes) in a particular context. For example, the ability to communicate effectively is a competency that may draw on an individual's knowledge of language, practical IT [information technology] skills and attitudes towards those with whom he or she is communicating. (OECD, 2003, pp. 3–4)

In reality, and perhaps unsurprisingly, a noticeable change in official rhetoric might have little impact on actual classroom practices, as a number of concrete cases suggest. In Europe, the 2006 directive for the development of competencies for lifelong learning (Education Council, 2006) recommended that "Member States develop the provision of key competences for all as part of their lifelong learning strategies, including their strategies for achieving universal literacy" (p. L 394/11). Since then, most European countries have introduced the idea of competency in their curricula, giving rise to what can be called a *competency-clash*:

> Not only do different notions of competence clash with each other, but the new educational "gospel" of standards, competencies, and outcome-orientation is at odds with [the countries'] traditions, such as content-based curricula and input-orientation (i.e., regulating the structures, processes and conditions of teaching and learning). (Sloane & Dilger, 2005, cited by Ertl, 2006, p. 628)

Spanish authorities explained the country's PISA results by arguing that because the Spanish curriculum was centred in content, and Spanish teaching focussed on teaching such content (Arias, 2006), Spanish students were not well prepared for the PISA tests. One action taken to address this concern was to distribute translations of the PISA's executive summary and framework to all schools nationwide. As a result, the idea of competency entered the Spanish educational discourse. Rico (2011) has noted that "the PISA assessment model has been determinant for the Spanish educational system" (p. 4). In fact, the new national curriculum seeks that students improve their command of basic competencies and introduces *competency* as the basis for curriculum innovation (Blanco & Rico, 2011; Maestro, 2006; Rico & Lupiáñez, 2008). This was also the case in the most autonomous regions in Spain (Ferreras, 2006; Graña, 2006), with many deciding to participate in future PISA tests with a representative sample or a census of their students (Gómez, 2006), making the need to prepare students on those competencies more pressing. Naturally this shift in language has also had an impact on the preparation of teachers and the design of textbooks in Spain.

PISA 2000 marked a milestone in Germany because the country ranked 21st among the 31 countries that participated in the study. The reactions to these results were known as the "PISA shock." No other country had a bigger reaction to the results (Gruber, 2006). Besides the media reaction, there was also a reaction in academic and political circles. Many articles and books were written on the results and on possible strategies to improve them; new teacher preparation and development courses were offered, and several educational policies were implemented. In particular, new standards were introduced for middle schools based on the idea of competency, together with standardized tests assessing students' performance on those standards (Neumann, Fischer, & Kauertz, 2010).

Colombia, which has participated in TIMSS since the first study in 1995, and will take part in PISA for the first time in 2012, adopted the notion of standards in the late

1990s, along with other Latin-American countries (Palamidessi, 2006). In 2006, the National Ministry of Education published its *Basic Standards of Competencies* (Ministerio de Educación Nacional [MEN], 2006). This document introduced some *general processes* that were similar in nature to PISA's competencies and set standards for pairs of grades, organized by types of mathematical knowledge—for instance, communicating; modelling; and formulating, treating, and solving problems (MEN, 2006, pp. 51–55). The Colombian standards are supposed to contribute to the development of competencies. This document has been distributed to most schools and teachers in the country. The Ministry has organized conferences and teacher-training events with the purpose of explaining the standards and the idea of competency. But, "teachers do not use the standards" (Monica López, personal communication, March 16, 2010). Because schools are autonomous in designing and developing curricula based on these standards, and most public schools do not provide or require textbooks, the responsibility of curriculum design and development has been passed to teachers, who usually produce so-called teaching guides for implementing instruction in their classrooms. A study by Gómez and Restrepo (2012) found that the word *competency* seldom appears in the school planning for any given grade. Furthermore, when the word appears, it is interpreted in many ways, usually differently from what the *Basic Standards of Competencies* document intended and most notably to refer to learning goals for a specific content topic and grade level, which is not consistent with the original meaning of competency.

In contrast, the use of standards and competencies does not seem to be a trend in Asia. Singapore, Hong Kong, and Japan, three of the Asian countries with outstanding results in TIMSS and PISA, do not refer to standards or competencies in the mathematics curriculum. Singapore's mathematics curriculum is based on its problem-solving curricular framework that focusses on five key components: skills, attitudes, concepts, metacognition, and processes. This framework has been used since it as first proposed 1990. Since then, there have been some changes to the curriculum in order to keep abreast of the shift in global trends towards a knowledge-based economy. For example, in 1998, there was an increased focus placed on thinking and processes as well as a trimming of some content (Kaur, 2003). However, there was no specific reference to TIMSS or PISA. There have, however, been numerous secondary analyses using the TIMSS data, which indicate that TIMSS has been used as an international benchmark to gauge the success of the mathematics curriculum in Singapore (e.g., Kaur, 2002, 2005, 2009; Kaur & Pereira-Mendoza, 2000a, 2000b).

Likewise, Hong Kong's curriculum does not refer directly to standards or competencies but rather focusses on the main aims of developing interest, communication, lifelong abilities, numeracy, spatial skills and understanding, and the acquisition of basic skills (Curriculum Development Council, 2000). In Japan, the mathematics curriculum has been undergoing a process of revision since 2005. One of the main purposes of the revision is to address weaknesses indicated by PISA results—such as students having difficulty writing problems that require thinking, decision making, expressing, and lacking motivation to learn (Ministry of Education Culture Sports and Science, 2010). The new revised course of study in the Japanese curriculum aims to prepare students through the acquisition of basic knowledge and skills and the development of abilities to think and to express ideas mathematically.

Although concerns arising from PISA results were considered, the new curriculum was not structured according to standards and competencies.

We found one Asian country, Indonesia, which has explicitly mentioned standards and competencies. The Indonesian mathematics standards use competencies to map the curriculum. The curriculum lists core competencies and corresponding student outcomes that can be used to indicate the achievement of these competencies (Departemen Pendidikan Nasional, 2003). There were no clear references to PISA frameworks in the mathematics curriculum, but because Indonesia has regularly participated in PISA studies since 2000, the use of standards and competencies in the curriculum provides some evidence of the effects of PISA on the curriculum.

Adaptation and adoption of frameworks and items (Influence 2). Besides the influences noted above, in terms of the language used and how it has affected curriculum standards, we have anecdotal evidence that the TIMSS frameworks have influenced the redesign of U.S. curriculum guidelines at the state level. The extent of this influence is unknown, although one dissertation (Landry, 2010) looked at the alignment between the TIMSS or PISA frameworks and the content standards in several states in relation to the high or low performance of countries and states. Landry found that, from the content point of view, most of the standards in most of the states in the United States cover a wide variety of topics that were repeated year after year—a finding that is consistent with results of the curricular analysis for TIMSS conducted by Schmidt and colleagues (Schmidt, McKnight, Valverde, Houang, & Wiley, 1996), which showed that in the United States topics tend to enter and stay longer in the curriculum than in other countries. Landry also found that high levels of alignment to the curriculum in high-performing countries and their assessments did not absolutely equate to high performance at the state level.

We have found that, in science education, researchers have used sections of the published questionnaires to create new instruments that are used in pre-/post-test designs to establish effectiveness of interventions (Lee, Deaktor, Enders, & Lambert, 2008; Lee, Deaktor, Hart, Cuevas, & Enders, 2005; Shymansky, Yore, & Anderson, 2004). However, we could not locate similar papers produced by members of the mathematics education community.

From the Attained to the Intended Curriculum

In this section, we discuss the second set of influences: documents and reports that speak about the ways in which results of the international studies—especially the scores that students obtained—have influenced different levels of the intended curriculum.

Localization of curriculum design (Influence 3). Most countries that perform well in PISA give local authorities the freedom to adapt curriculum (Schleicher & Shewbridge, 2008, p. 20). In PISA, school autonomy for curriculum is measured with an index of school responsibility for curriculum and assessment. This index is derived from categories that school principals classify as being the responsibility of schools—establishing student-assessment policies, choosing textbooks, determining which courses are offered, and the content of those courses (OECD, 2010). For instance, in reading literacy, school autonomy has a statistically significant positive relationship with student performance (OECD, 2005). However, when these findings are controlled for student and school-level factors, the relationship between school autonomy and student performance is weak.

Nonetheless, these results have led countries to push for more school autonomy, assuming that such change might have a larger effect on students' learning than other aspects of schooling. In Germany, for example, the decision was taken not only to give more authority to the states, with the central bureau prescribing a few core ideas (rather than all the curricular content), but also to give schools the autonomy for finishing the curriculum (Ertl, 2006), with both "external and internal assessments (of schools and students)" (p. 626) being required. Such a shift has resulted in teachers in some states becoming responsible for developing the whole curriculum for their schools, thus localizing the design of the curriculum. Although it is too early to establish the impact of such reform in Germany (Kotthoff & Pereyra, 2009), the perception is that the process is problematic, because German teachers are not qualified to assume responsibilities as curriculum designers and developers. Similar claims about the positive impact of giving more autonomy to schools to design curriculum have been made in Australia (Liberal Party of Australia, 2010).

Spain is also moving towards localization of curriculum design. In early March 2011, two major autonomous regions in Spain, Madrid and Catalonia, decided to transfer the authority of deciding 35% of the curriculum to schools—pending approval by the autonomous region (Alcaide & Álvarez, 2011; Bassells, 2011)—as a means to improve teaching. According to Alcaide and Álvarez (2011):

> The autonomy of schools for looking for better solutions for their context is being recognized recently as one of the most recurrent strategies for improving teaching; that is what can be assumed, for instance, from several OECD studies, like the PISA report. (p. 38, trans. by authors of this chapter)

The movement towards localization of curriculum design is not necessarily a consequence of the studies. Colombia is an example of curriculum localization that occurred prior to its participation in the international studies. The 1994 Colombian Education Law established that "the autonomy is a consequence of the will for differentiating each educational community, paying attention to different needs and expectations; it seeks that each educational institution educate citizens that can solve the problems of their own environment" (MEN, 1994, article 77, trans. by authors of this chapter).

Assessment (Influence 4). A common justification for low performance on the PISA examinations has been that students have little experience with standardized tests. According to Saracho (2006), although the influence of PISA has been limited, mainly because of the absence of a mechanism to inform schools about individual results, schools and teachers are feeling more inclined to agree about the need for accountability of results and are more willing to have standardized test results disaggregated at the school and classroom level, something that "was an alien notion until recently" (p. 27, see also Dossey and Wu, Chapter 31).

The case of Poland illustrates an indirect influence of the results of PISA on the definition of a standardized test, MATURA, which is administered at the end of the secondary school and serves both as a diagnostic tool of the performance of the education system and as an admission test for postsecondary education. The annual character of the test allows for tracking the effectiveness of changes and the identification of areas that need further change. The Polish results in PISA were useful in overcoming the public's and the schools' skepticism about the usefulness of the proposed test (Saracho, 2006).

Danish students obtained results that were close to the OECD average, which raised concerns in Denmark given its high expenditure in education. A government study named the lack of a standardized assessment culture in the country and low levels of satisfaction with the education system as possible reasons for the outcomes (OECD, 2004). In Brazil, another country whose PISA scores were low, the government recently introduced a new system of periodic standardized assessment of students' performance (Saracho, 2006).

Some Spanish regions (e.g., Andalusia) have introduced, by decree, diagnostic assessments in schools based on competencies (Junta de Andalucía, 2007). These diagnostic assessments have no influence on students' grades and are designed and implemented by schools. Their purpose is to establish the level at which students have developed basic competencies and help schools make decisions about their curriculum. Likewise, the Department of Education of the Autonomous Region of Navarra is promoting self- and external-assessment in schools. The Department anticipates that external assessments will encourage new teaching practices and that the PISA framework can be used to enrich the curriculum. In order to ensure coherence between the school assessments and the PISA tests, the Department has defined, for each knowledge domain, specific guidelines for improving reading competencies at all grade levels. It has also published standards in language, mathematics, and science based on the PISA framework. Regarding competency in mathematics problem-solving, the guidelines proposed that teachers use, during instruction, problems modelled on the PISA items, a suggestion in the formative dimension at the regional level. A Web page has been set up with information concerning PISA with proposals for improvement in each area (Ferreras, 2006). This is a first step in the data-driven approach to reform that we discuss next.

Data-driven approaches to reform (Influence 5). A report by the US National Research Council (1999) suggested that a way to capitalize on the results of these large-scale studies of achievement at the ground level (namely, schools and classrooms) is by using local results to initiate a process of self-reflection. This report was a direct outgrowth of the TIMSS study, and its focus was on the training of professional developers who could direct and assist schools in initiating the data-driven, self-reflection process. The process starts with teachers (schools or districts) getting acquainted with TIMSS, its design and its findings, analyzing the implications in their own contexts, and finding a particular focus of attention (e.g., student

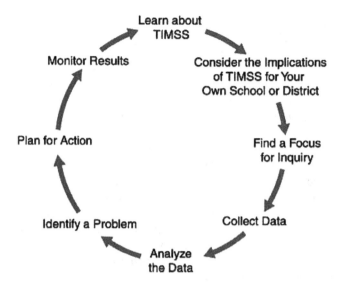

Figure 27.5. The data-driven inquiry process (Source: National Research Council, 1999, p. 398).

achievement or curriculum alignment). Teachers (schools or districts) can analyze data about student achievement in the TIMSS test or carry out a content analysis (guided by the TIMSS content analysis process) by which they can determine the extent to which the content is aligned with the framework that guided the design of the TIMSS test. This analysis should lead to the identification of a specific problem to work on: an area in which there is low student performance combined with an analysis of the content coverage for that area, and may result in a suggestion for a change (e.g., emphasize or de-emphasize instruction on certain topics). The change is monitored in order to determine its effectiveness, and this analysis starts the cycle again (see Figure 27.5). An important step in this cycle is benchmarking: testing a large sample of students to generate a baseline for later comparisons (similar to the Andalusian diagnostic tests). According to the report, several efforts emerged from this initial work (e.g., the Chicago area's First in the World Consortium, the Lake Shore school district in Michigan, and an urban school in Patterson, New Jersey). We found some documentation of these efforts in newspaper articles (e.g., Dunne, 2000) and descriptive reports (Kimmelman et al., 1999), but little in terms of assessment of their outcomes.

We found one dissertation that studied how a group of teachers immersed in a professional learning community in one school took advantage of this cycle of data analysis to improve their practice and collected data on the impact of this process. In her dissertation, Figueroa (2008) used information from the analysis of lessons from the TIMSS Video Study (Stigler, Gonzalez, Kawanaka, Knoll, & Serrano, 1999; Stigler & Hiebert, 1997, 1999) to generate a modified version of a TIMSS lesson plan that was piloted and used by several elementary teachers in a school in Arizona.

In Figueroa's (2008) study, the teachers engaged in a 2-year process of learning about the process of creating the lessons, presenting them, getting feedback, and redesigning the lessons for a new application using a multi-step approach that had 17 stages expanding on each of the 8 steps shown in Figure 27.5. Several elements of a lesson study process were present, including time that allowed the teachers to learn the method, seek resources, plan lessons, and observe each other's teaching.

In this very prescribed process, the observer teachers were trained to take notes on the percentage of students who could remember the strategies taught, were engaged in the problem-solving process, articulated the strategy, applied the strategy, were assessed for mastery/nonmastery of the learning strategy, and could relate the activities to the learning strategy. An evaluation of the impact of this two-year initiative followed a pre-/post-test design with a group of 65 fourth-graders in the school. The analysis revealed a statistically significant increase in the post-test scores with respect to the pre-test scores and a statistically significant association between time of test and categories of performance in the state test. But although Figueroa (2008) suggested that the method could be useful in helping districts improve their students' scores on state tests, unfortunately the study's design was problematic because it did not control for students' prior knowledge, for other students' characteristics, or for the quality of implementation of the lessons. Furthermore, the process relied on external funding, and there were no indications of attempts to sustain the effort.

In the USA, a large study led by William Schmidt and Joan Ferrini-Mundy from Michigan State University, titled "Promoting Rigorous Outcomes in Mathematics and Science Education" (PROM/SE), might provide information about the effectiveness, for achieving reform, of the approach which was adopted. PROM/SE (2006a, 2006b, 2008, 2009a, 2009b, 2009c) is funded by an 8-year grant that uses assessment of students and teachers for the purpose of improving standards and content coverage, and simultaneously building capacity among teachers and administrators. It involves over 300,000 students and 18,000 teachers in two states, Michigan and Ohio. The reports produced to date apply the logic model used to design TIMSS, capitalizing on many of their analytical strategies to deal with those data. In addition, the reports illustrate vividly the kind of educational system that exists in the USA—a system characterized by extreme variation that leads to substantial inequalities, not only between districts but within schools as well. This variation appears to be strongly determined by the differential access to resources (economic, cultural, and intellectual) of the community.

Teacher education and professional development (Influence 6). An example of influence on teacher education comes from Germany, as a reaction to unsatisfactory results in PISA (Ertl, 2006). The purpose of the SINUS project (*Steigerung der Effizienz des mathematisch-naturwissenschaftlichen Unterrichts*, trans. "Increasing the Efficiency of Science and Mathematics Instruction," http://sinus.uni-bayreuth.de/2956/) is to implement better learning environments at more than 1,000 schools. The project defined a strategy that was tested and later disseminated to schools

(Lindner, 2008). The strategy was based on curriculum design, through teachers' meetings, implementation in the classroom, and sharing of experiences. The purpose was to develop and share a new teaching and learning model that broke the German tradition, which was characterized by a strong emphasis on practising rules and algorithms, little attention to competencies such as modelling, compartmentalization of subjects, teaching methods that induce students to be passive, and an inappropriate mixture of learning and assessment (Blum, 2004, p. 1). Teams of three to ten teachers met six to eight times a year for 2–3 years. The meetings dealt specifically with subject issues and teaching methods, and teachers were expected to produce curriculum designs that would be implemented in their classrooms. Once the learning environments were established, the team evaluated the experience and improved the original design (Blum, 2004). The program had positive effects as shown in a large-scale comparison schools tested in PISA 2003 (Ostermeier, Prenzel, & Duit, 2010) This process is similar to that of lesson study (described in the next section), and also to some aspects of the data-driven approach.

Poland used the PISA results for validating its educational reform. Poland's approach to reform included several policies related to teacher education, development, and promotion: encouraging the improvement of teachers' social and economic status, introducing transparent mechanisms for promotion, improving teacher knowledge and competencies for the classroom, and offering permanent opportunities for teacher development. The improvement of Poland's PISA results seem to give credence to the need and importance of maintaining these policies (Barber & Mourshed, 2007; Saracho, 2006).

In addition, we found at least three ways in which professional development and teacher education have benefited from the results of the international studies. First, many initiatives have capitalized on the availability of videos of mathematics lessons in several countries collected through the TIMSS video study. Second, the Japanese lesson study has been used to engage teachers in improving mathematics lessons over time. And third, materials have been produced that assist teachers to explain the meaning of the changes in the curriculum and offer suggestions about using new textbooks.

Videos

The potential to generate change by observing practice has been the basis for developing professional development programs that make the analysis of video an important component (Kersting, Givvin, Sotelo, & Stigler, 2010; Roth & Givvin, 2008). It is undeniable that the availability of the TIMSS videos has been useful to many activities in which instruction is analyzed, both in programs of preparation of future teachers and in professional development, although the practice had been in place before the availability of these videos. However, there is no documentation about how these videos have been used in any of these settings.

Most of the findings of the analysis of the TIMSS video study are descriptive, characterizing the nature of instruction in different countries and documenting what has been termed the *lesson signature* (Givvin, Hiebert, Jacobs, Hollingsworth, & Gallimore, 2005; Stigler & Hiebert, 1999). This lesson signature reveals the particular ways in which lessons are deployed in each country and allows observers to understand other practices. Watching substantially different ways to organize instruction helps make visible features that are often taken for granted in one's own culture. Only recently, a Web site (http://timssvideo.com/videos/Mathematics) presenting full lessons from different countries participating in TIMSS (in Australia, the Czech Republic, Hong Kong, Japan, the Netherlands, Switzerland, and the USA for mathematics, and in Australia, Czech Republic, Japan, the Netherlands, and the USA for science) has been made public. In this Web site, viewers can watch and listen to a variety of English-subtitled eighth-grade lessons and download the lesson plans, a map of the class, and a one-page visual description of the lesson. There are four mathematics lessons per country, covering a wide range of topics from geometry, measurement, and algebra.

The lessons illustrate the many differences between countries in terms of instruction. A main purpose of this site is to offer video study readers a way to corroborate the main findings about instruction in these countries. The site also anticipates the likelihood that it will generate new ideas that will assist teachers as they prepare lessons. Naturally, teachers, teacher educators, and professional developers use these lessons in a variety of ways, and it may be informative to keep track of those uses and their connections to changes in classroom instruction.

Several practical suggestions have been derived from the analysis of the video component of TIMSS. For example, in a leadership journal targeting principals, Roth and Givvin (2008) summarized four main findings from the video study, two for mathematics and two for science, and made recommendations for taking action. In the case of mathematics, Roth and Givvin emphasized that in all countries except Japan, there was a strong emphasis on solving problems with the intention of learning procedures, and that "teachers in higher-achieving countries implement making connections to problems differently from teachers in the United States" (p. 24). They indicated that US teachers tended to simplify a problem rather than allow students to struggle with it, to make links across ideas and concepts, to generalize, or to conjecture. They offered suggestions for principals, in terms of having teachers participate in professional development opportunities that would increase their mathematical content knowledge, give them opportunities to observe how some teachers challenge "students to think about mathematics," and to help them "break the pattern of simplifying problems" so that they would become more likely to reinforce the idea that "students should struggle with important mathematics" (p. 24).

Lesson Study

The Japanese lesson study is a process by which teachers collectively plan lessons that are then implemented in the classroom; teachers observe and take notes, and the observations and notes are used for improving the lesson. The lessons become "research lessons" which are refined over time. This strategy has a very long tradition in the Japanese educational system. Lesson study was brought to the attention of US researchers in the late 1980s (Lesson Study Research Group, n.d.), thus predating both TIMSS and PISA. The high performance of Japanese students in TIMSS prompted questions about Japanese instruction, and in particular about the possibility of using lesson study as a professional development strategy that could spur changes in instruction in the USA (Hiebert & Stigler, 2000). A Web site is maintained by Teachers College at Columbia University (http://www.teacherscollege.edu/lessonstudy/index.html) for archival purposes and contains many documents, research papers, and manuals about lesson study. According to that site, by May 2004 nearly 2,300 US teachers in 32 states and 335 schools had been engaged in some form of lesson study. Most of the documentation on this site is concerned with understanding the method itself: What are the challenges for implementation? And what are the potential outcomes of using it in professional development (Lewis, Perry, & Hurd, 2004; Lewis & Tsuchida, 1998)? The papers on the site highlight the need for collaboration and for school reorganization (Watanabe, 2003) and, in particular, the need of allowing teachers time to meet and study (Liptak, 2002).

In Australia, the New South Wales (NSW) Department of Education and Training initiated a lesson study project in 2001 in which there was an attempt to adapt lesson study principles into a professional development program. Initially only three secondary schools took part, but by the end of the project in 2004, the number of schools in the project had grown to 200. The analyses of data derived from 117 teacher reports from 81 schools gathered in 200 to evaluate the lesson study program—showed that it had succeeded in changing teacher practices and beliefs. The teacher participants, when comparing the focus lessons to their normal practice, reported using more practical activities, concrete materials and technology, adopting new teaching procedures, focussing more on intellectually challenging mathematics, and collaborating more with fellow teachers. In a follow-up survey conducted 6 months later, 63% of the respondents ($n=64$) reported continuing to use the lesson study model of planning, evaluating, and refining to develop further lessons (White & Southwell, 2003). Lesson study has continued to be promoted in NSW government schools across many subject discipline areas—although mathematics remains the main subject of study. Examples of lessons are displayed on: http://www.curriculumsupport.education.nsw.gov.au/secondary/mathematics/prolearn/windows/public_lesson_study.htm.

Over the past decade, lesson study has increasingly been used as a professional development program in many countries. Some of this increase in dissemination has been due to the efforts of the Japan International Cooperation Agency (JICA), which provides international aid to developing economies. Through the collaborative efforts of JICA and lesson study experts, lesson study has been introduced to Cambodia, Colombia, Egypt, Ghana, Honduras, Indonesia, Kenya, Laos, the Philippines, and Thailand (Hattori, 2007; Inprasitha, 2007; Kimura, 2007; Koseki, 2007; López & Toro-Álvarez, 2008; Odani, 2007; Okubo, 2007; Saito, 2007; Shimizu, 2007; Yoshida, 2007). In addition, the Asia-Pacific Economic Cooperation (APEC) has, since 2006, hosted a project in conjunction with the University of Tsukuba, Japan, and Khon Kaen University, Thailand, to popularize lesson study. The project grew out of a recognition that lesson study constitutes an important approach towards developing human resources, especially the expertise of teachers in mathematics classroom instruction. As of 2011, five cycles of conferences had been held. Reports of the activities related to lesson study in the APEC countries can be found on the conference Web site: http://www.criced.tsukuba.ac.jp/math/apec/.

Much of the available literature documents challenges that can arise when the lesson study approach is used in professional development programs. Its use in Japan has very much been intertwined with the culture. Compared with teachers in Australia, for example, Japanese teachers hold less to the notion that the classroom is a private professional space (White & Lim, 2008). Thus, although the lesson study approach to professional development may be versatile, there are cultural challenges that need to be addressed in its adoption and implementation.

Hiebert and Stigler (2000) proposed that in order to change instruction, teachers need to "learn in context," that is, in their actual practice, and locate "substantive decisions for improving teaching within the schools and classrooms where teaching occurs" rather than having those decisions made "up the bureaucratic ladder" (p. 9). After reminding educators that although systemic and cultural change occurs slowly it does happen, they proposed using the process of lesson study as an ongoing professional development program that would be carried out in schools—where teaching happens and students learn.

Lewis (2011) reported the use of toolkits and Japanese textbooks in lesson study groups in the USA. The toolkits were used to provide support for elementary teachers to teach various mathematical concepts. The initial findings of a randomized controlled trial on the topic of fractions showed that, when compared with teachers in other professional development programs, there were significant improvements in both students' and teachers' knowledge when lesson study was used (Perry, Lewis, Friedkin, & Baker, 2011).

The spread, implementation, and success of lesson study as a professional development strategy can be traced back to the first TIMSS Video Study. The link is indirect, but it was the TIMSS Video Study that first drew worldwide attention to lesson study as a viable teacher professional development approach that had proved to be very successful in Japan. As with other cross-national adaptations, there are always limitations and challenges because of different cultural settings. Through the hard work, efforts, and creativity of teachers and researchers, lesson study has proven to be a sustainable teacher development approach, as is shown in the case of the USA (Perry et al., 2011).

Texts for Teachers

One way in which countries have attempted to reach teachers is by making more information available for their use, typically translations of frameworks and reports. We found in addition, however, two cases—Mexico and the USA—in which there were textbooks developed with the intention of training teachers in specific aspects of practice.

Mexico has participated in PISA since the first PISA study, and its results have motivated many reactions within the country (Rizo, 2006). One reaction has been in relation to informing and training teachers for PISA. The Web page of the INEE (National Institute for Educational Assessment) contains many documents (see http://www.inee.edu.mx) of different types: national reports, results analysis, research protocols, and materials for teachers. *PISA in the Classroom: Mathematics* (Aguilar & Loejo, 2008) explains the PISA project and the notion of mathematical competency. The core of the book contains several curriculum designs for the classroom on quantity, change and relations, and probability. These designs contain tasks that follow the PISA framework.

The remarkable and consistently high performance of Singapore in the international tests has led many to inquire about their curricular organization. Their mathematics textbooks, being in English, have certainly facilitated their incorporation into classroom practice in many nations. In the USA, for example, there is a professional development package that includes booklets, videos, and lessons that teachers can use in their classrooms if they wish to capitalize on the "Singapore method" for solving problems. With titles such as: *Place Value, Computation & Number Sense* (Chen, 2010); *8-Step Model Drawing: Singapore's Best Problem-Solving MATH Strategies* (Hogan & Forsten, 2007); *Problem-Solving Secrets from the World's Math Leader* (Hogan, 2005); and *Step-by-Step Model Drawing: Solving Word Problems the Singapore Way* (Forsten, 2010a, 2010b; Walker, n.d.), the books seek to illustrate how Singapore textbooks organize content across strands, how models are used, and in what ways can such a presentation and way of thinking reach all students and increase students' understanding of mathematical notions. There are also publications urging teachers to include parents in the process (Chen, 2008). All these efforts have been made at the elementary level, however, and, except for testimonials on the back covers of the actual books, there are no reports about the impact of the use of Singapore methods in the classroom.

Textbooks (Influence 7). Some countries use their own curriculum frameworks to design textbooks based on texts written in other nations, and others directly import them into their classrooms. In Spain, for example, textbooks are now based on the PISA conceptual framework (Lupiáñez, 2009) and, as in Mexico, a number of books for teachers have been published which explain the PISA framework and relate the idea of competency to curriculum (see, e.g., Rico & Lupiáñez, 2008). Recent studies have shown that the majority of tasks proposed by the textbooks are at a low level of competency, the reproduction cluster, with very few tasks from the connection cluster, and almost none from the reflection cluster (González, Monterrubio, Delgado, & Codes, 2011).

The consistently high performance of Singapore and Hong Kong students has led to the development of programs that encourage US districts and teachers to adopt and use their textbooks. Singapore textbooks have received more attention. A search using terms such as "Singapore," "Hong Kong," "adoption," "mathematics," "textbooks" produced about 20 results, with only three referring to Hong Kong. One study documented results of a pilot study investigating the impact of using Singapore's textbooks on student achievement. The study was conducted in different sites in the USA (Ginsburg, Leinwand, Anstrom, & Pollock, 2005). The main conclusion was that "under favorable conditions, Singapore mathematics textbooks can produce significant boosts in achievement, but introducing textbooks alone is insufficient to achieve improvement" (p. 127).

The results were conditioned by two factors: the mobility of the student population and the amount of professional development received. Schools with relatively small and stable populations showed large improvements over the 2-year period; similarly, there was a correlation between "improvements in the Singapore pilot schools and the intensity of the schools' participation in professional training, suggesting that teacher acceptance and commitment to the new Singapore mathematics program may be key to its success" (p. 127). A second important finding related to how confident the teachers using the Singapore textbooks were, which was corroborated with classroom observation. The researchers found that:

- Elementary teachers felt underprepared to teach with the materials and admitted that they had not understood many of the concepts they were supposed to teach.
- Nearly 40% of the teachers struggled with implementing the curriculum.
- About one-fifth of the teachers who received intense training successfully implemented the curriculum, whereas only 7% of teachers who received less training were successful.

In general, the teachers noticed that the Singapore textbooks offered a deeper treatment of mathematical topics, and that if a textbook returned to a topic it was treated with greater depth. They liked the numerous multistep problems included in the textbook and the visual explanations because they made abstract concepts concrete.

Nevertheless, the teachers identified challenges in bringing the Singapore textbooks and methods into US classroomssuch as teachers' lack of understanding of the method and their lack of knowledge of strategies to deal with students who have a weak background or who have not been exposed to the Singapore curriculum before. Other problems included the lack of real-world examples, the use of unfamiliar terms, and the unclear alignment between the Singapore content coverage and the state-mandated standards, all of which might require using supplemental material. In addition, because the Singapore curriculum assumed a spiral progression, a successful implementation would require an adoption as early as kindergarten. Finally, the Singapore textbooks did not revisit topics later on, something that is very typical in US textbooks, and thus there were no provisions for teaching or reteaching notions that had not been mastered.

The study concluded with suggestions for further studies in relation to five issues: (a) comparison of content coverage and sequencing at the state level with those of Singapore; (b) extensive textbook analysis to identify features that could be used in US textbooks (the emphasis on pictorial representations appears to be a feature that would benefit special education or limited English students); (c) comparison of performance on assessment by Singaporean students with national tests, such as NAEP, and by US students with Singaporean examinations; (d) changes in how teachers are tested with the PRAXIS teacher certification test, suggested by an examination of the alignment of this test with the entrance tests used in Singapore to select students for Singapore's National Institute of Education; and (e) extending the piloting of the Singapore textbooks in other schools, using the versions prepared for the American market.

Summary

It appears to us that the influences of TIMSS and PISA on teaching and learning within participating nations has been varied. Clearly there has been more reaction in countries that did not show outstanding student performance with either TIMSS or PISA, although there were countries (e.g., Denmark) for which an average performance was also a trigger for reflection. Results on the international studies seem to have stimulated action in the USA, Spain, and Germany, and certainly effects have been felt in other countries as well. It is possible that the availability of more organized and advanced facilities in the USA, Spain and Germany has made it possible for reform initiatives, arising from TIMSS and PISA results, to be planned and implemented with some success. In the USA, it is a story of many efforts, borne out of specific interests, rather than from common concerted collaborations aimed at addressing specific issues. There are many initiatives, but they seem to be carried out without any agreed-upon plan or goal. They appear to be left to individual states, researchers, schools, and practitioners. In countries with a more centralized organization, the efforts appear more coherent, with possibly Germany being the country in which efforts were most focussed.

There are two salient themes at the policy level that have important, and perhaps immediate, implications for the classroom: the increased interest in centrally defined standardized testing and the increased autonomy given to schools for designing curriculum that fits their local conditions. Central agencies are moving away from prescribing what should be learned, when, and how; limiting their prescriptions to a few core content topics and competencies; and giving schools and teachers the responsibility to complete the design of their curriculum. The appearance of standardized tests is the mechanism by which bureaus can control and verify that core content and competencies are indeed being implemented and achieved. One peril of this approach might be the convergence towards a narrower list of content and competencies, which might over time determine what schools and teachers will "design" locally for their classrooms. These initiatives seem to flourish in countries where there are no centralized curricula, such as the USA. These initiatives have not been documented in countries that fully prescribe a national curriculum, although Singapore is starting to experiment with the idea of giving more autonomy to schools while maintaining the central control of examinations.

It is unclear what the impact of the studies in the classrooms can be for countries in which student performance is high. In Singapore, for example, the Centre for International Comparative Studies (CICS) was set up in 2009 within the National Institute of Education. The aim of the CICS was to encourage further comparative analyses, using the results from international studies to provide stakeholders with findings that could predict the factors that affect student achievement. Being situated within the National Institute of Education, which conducts teacher development, the CICS could see that the findings were more immediately used and thus influence the content of teacher education courses. But although the TIMSS may play a significant role in mathematics education in Singapore, local educators have also been quick to point out that "local stakeholders are more likely to use as main indicators the performance of students in the public examinations" (Wong, Lee, Kaur, Foong, & Ng, 2009, p. 7), noting that local public examinations are high stakes, especially in the Asian region. Student performance is used as a criterion for university entrance and for awarding scholarships, which is very highly valued in Asian countries. It would therefore be safe to say that teachers in these countries are more likely to refer to items from the public examinations than from TIMSS in planning their lessons.

Proposing a Research Agenda to Investigate These Effects

The reader will have realized by now that we have been unable to document the ways in which the international comparative studies of student performance in mathematics have exerted an influence on learning and attitudes toward mathematics of students in the participating countries and elsewhere. Indeed, we have been able only tangentially and indirectly to find traces of such influence on the teaching of school mathematics. We did not find evidence of influence at the student level. So we have only partially fulfilled our task for this chapter.

Ferrini-Mundy and Schmidt (2005), referring to TIMSS, invited the mathematics education community to capitalize on the many elements of the studies to further our understanding by using the results, research methods, and theoretical constructs that were generated. As has been the case with other endeavours, the research community has been the main direct beneficiary of these studies. Because of their complexity, studies of this magnitude require substantial know-how at all levels: from design to application and from data management to analysis and reporting. The collaboration between participating countries becomes the vehicle by which new researchers could learn new techniques and generate research agendas that would use the data that had been obtained. It seems to us, however, that although there have been large and tangible benefits for the research community, the studies have not had much effect at the classroom level. For example, researchers have not explored sufficiently how different mathematics education stakeholders have interpreted the visions behind the international studies. The lack the coherence among those visions could pose constraints in the manner in which they get expressed in individual teachers' practices in the classroom (McNab, 2000).

Nor has the research community looked carefully enough at the effects of international studies from a global perspective. Yore, Anderson, and Chiu (2010) claimed, in relation to the PISA study, that

> many in the mathematics and science education research communities lament the lack of influence that research results have on the education profession, schools, and teaching. Academic research done in isolation of end-users—with the faint hope that teachers, politicians, and bureaucrats will access and utilise these results to inform curriculum, assessment, and instruction and to influence public policy—has not worked. (p. 593)

We believe that as researchers of student achievement in international contexts, we need to take responsibility for improving the knowledge transfer of the results from these studies into the places in which it matters—the schools, the classrooms, and the students—and to do so in a concerted and planned way. Isolated work runs the risk, as we see here, of not having an impact at the ground level, where all the policies, mandates, and guidelines are enacted. In what follows, we propose and reflect on several areas that we believe merit attention from the research community and can have direct impact on classroom practice. Our intention is to provide some coherence to the work that needs to be done to understand the impact that results of international studies might have at the classroom level.

Some Research Questions

In the following, we propose work that is geared towards understanding how important curricular ideas (e.g., competencies) and processes (e.g., localization of curriculum design) get interpreted and used differently at each of the intended levels, and how those interpretations mediate what teachers do for planning and enacting lessons. We assume that the influences we have described here (represented with gray arrows in Figure 27.6) are real, and propose to study how they might influence teachers' work, and hence, students' learning and attitudes (represented by a black continuing arrow in Figure 27.6).

The arrows make a connection to the classroom, both in the intended and the implemented levels of curriculum, because it is at these levels where teaching is planned and implemented and learning takes place. We first propose questions related to the idea of competencies and about the use of frameworks: that is, we consider how the influences represented by Arrows 1 and 2 affect the level of intention for the classroom and the implementation processes. The next set of questions refer to influences represented by Arrows 3, 4, and 5, regarding three related aspects of curriculum control, design, and management—namely, the localization of curriculum design, the emergence of standardized testing, and the use of data-driven approaches in local reform. Finally, we consider influences from Arrows 6 and 7, regarding teacher preparation and development and textbook use. Figure 27.6 draws attention to two levels of our curriculum framework: the intended classroom level,

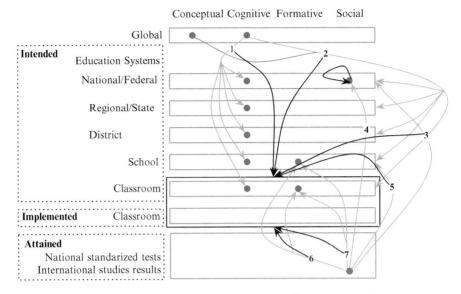

Figure 27.6. Proposed areas for further research on the impact of international studies in classrooms.

and the implemented. We mark this attention by the dark line surrounding these two levels. This is why all the arrows, except 4, point to this area in the drawing. In what follows, we identify these questions by the corresponding numbers in Figure 27.6.

Competencies and Frameworks

Our first set of questions refers to the impact of the notion of competencies in the curriculum and the use of the studies' frameworks. Concretely, we ask:

1. How is the idea of competency that guides the PISA framework being interpreted and transformed at the national, regional and district levels of curriculum? And how do schools and teachers interpret and use this idea when they plan and develop the curriculum in the classroom?
2. How are the frameworks of the international studies interpreted and implemented at national, regional and district levels of curriculum? And how are these interpretations and implementations used by schools and teachers at the classroom levels?

There is a growing interest in the notion of competencies—particularly in Europe—as a way of establishing long-term learning expectations by students that will empower them for participating in society. It will be important to see how this notion plays out in countries in which content has been traditionally more valued than competencies and, in particular, to understand how competencies are playing a role in moves towards localization of curriculum design. Because some initial information suggests that teachers' interpretations might be at odds with original intent of competencies, understanding the way in which teachers interpret and use the notion of competencies in their daily work (for planning and for delivering instruction) will be very important. We envisage analyses of the transformation of the idea of competencies at different levels of the curriculum; how it is interpreted at the national, regional, or local levels; and how these interpretations influence teachers' work. This knowledge would help researchers in organizing professional development strategies that can be more effective in making the idea of competencies more transparent, and thus generating potential benefits for planning and implementing instruction in the classroom.

The different nature of the TIMSS and PISA frameworks also calls for contrasts of their influence on classroom processes. The frameworks are fundamentally the studies' backbones; yet, although we know that they have had some impact, there is little documentation on the nature of such influence. We do not know how these frameworks are interpreted at the different levels of the curriculum, and in particular, whether and how teachers use them for planning and teaching, or whether the different natures of these frameworks are enacted differently in practice. If such differences were documented, then education systems could decide whether an approach that is more content-oriented, such as with TIMSS, would be better suited for its needs than an approach like PISA's, which values the development of life-long skills.

Curriculum Control, Design, and Management

Our second set of questions refers to studying the process of localization of curriculum design, the role of standardized tests, and the ways in which data-driven approaches support change. We ask:

3. What is the influence of localization of curriculum design on the work of teachers in schools and what is its impact on the teaching and learning of mathematics? Why does it seem that this policy works in some countries and not in others?
4. How are national standardized tests being aligned with tests used in international studies? And how is this influence reflected on what is taught and learned in classrooms?
5. In which ways do data-driven approaches influence the teaching and learning of mathematics?

We start by proposing a study of the impact of the localization of curriculum design, by which we mean the movement towards having schools and teachers produce the curriculum they will teach. Because greater responsibility is placed on schools and teachers for generating curriculum and because of concerns in some countries that teachers are not ready to assume this work, it is very important to document how such a process happens. Therefore, the studies that we propose are descriptive: What does it mean for school and teachers to produce the curriculum? What resources do they use? What are the characteristics of such a curriculum? Are there variations across communities that take into account the local contexts in which the curriculum is developed? What is the impact on students' learning and attitudes? Such studies would provide important information about how teachers are interpreting the task of curriculum design, the role that textbook companies play in the process, and the quality of the curriculum proposed. As mentioned before, there is some indication that in some countries teacher education programs are not preparing teachers to do this work (e.g., Gómez & Restrepo, 2012) and that schools have little resources to engage in this process. A second key aspect is to understand more fully why the localization of curriculum design is so successful in some countries and why it is harder to carry it out in others. The current status of this process presents itself as ripe for such studies, as some countries have a long tradition of localized curriculum design (e.g., Finland) and some are entering this process (e.g., Colombia, Germany, and Spain).

Next, we ask about the alignment of national standardized tests and the tests used in the international studies. Because of the increased push to localize the process of curriculum design in some countries, and because of growing concerns that educators are moving slowly into a "global curriculum" defined by the content and competencies assessed in these tests, it would be important to determine the rationale behind and the extent to which newly created (and already existing) standardized tests align with the international tests. For some education systems, this alignment is not an issue, because the international tests do not have the same prominence as their national tests. In countries in which new tests are being created, and in countries in which the tests are being revised, the issue is fundamental. A close alignment would suggest a strong influence from the international studies on the education system.

Studies such as those outlined above would require effective collaboration between researchers in charge of the international studies in each country and their counterparts in the bureaus of education or assessment agencies. In the cases where this alignment is strong, it would be important to study the extent to which the tests are becoming an important influence on what teachers do in planning and implementing lessons. We need to know the extent to which teachers feel it is important to prepare their students for the tests, Do they feel pressured to do so? What do teachers do to prepare their students for these tests?

Finally, we ask about the extent to which data-driven approaches that use the results from the TIMSS and PISA studies can indeed influence classroom practice. Earlier, we described the PROM/SE project, an effort to bring the strategy to a large number of US schools to generate processes of administrative, curricular, and pedagogical change. This project is not directly tied to an international study, but it is tied to results of local standardized tests (which might become more aligned to international studies tests!). It would be important to see the extent to which such high investment of resources in collecting information about student learning, studying ways to modify practise, implement and study the changes, and revise plans, is feasible for organizations like schools, whose primary mission is to teach children.

These three issues are closely connected: The localization of curriculum requires accountability through national or common standardized tests, and the tailoring of the curriculum requires a sustained study approach in order to obtain high results on the standardized tests. A danger associated with this close relationship is the convergence towards an impoverished view of content and competencies both in the definition of the curriculum and in the definition of the tests (Moss, 2004). In other words, the quality of the results will depend substantially on the quality of the international studies tests and the national standardized tests. That is because experience in many countries that have national standardized tests has shown that what is taught and partially learned is essentially what the agents (schools, teachers, and students) know will be assessed. National and international standardized tests more and more are not just assessing the students; the tests also assess teachers, schools, districts, and regions, and thus, students act as informants for assessing educational systems at all levels.

Teacher Preparation and Development and Textbook Use

Perhaps the most important area for research relates to the preparation of the teaching force, its continued development, and the resources that are being made available to teachers. In particular, we ask:

6. In which ways have international studies influenced the design and implementation of teacher education and professional development programs? And how do these new programs get reflected in what is taught and learned in the classroom?
7. How have textbooks from other countries been adapted and used? And what impact have these textbooks had on classroom processes?

A first step to answer questions about teacher preparation and development is documenting how widespread the use of international studies frameworks and research findings is in this area; this documentation can provide information about the ways in which the design and development of programs for teacher preparation and development have been influenced by the studies, and in which places each study has been more influential.

Beyond documenting the use of the international studies frameworks, it would be important to find out how teacher educators and teachers use resources such as lesson study, videos, or texts geared to teachers in planning and delivering their programs. It is unclear from our conversations with various teacher educators the extent to which these resources are being used. This area is particularly important because it is the closest to studying the teaching and learning processes as they are transformed from intentions to enactment in teacher education programs. Such studies, however, require continuous involvement with teachers, from the moment teachers begin to participate in a teacher education program or in a professional development course, to the process of planning, enacting, and assessing a lesson. Collecting information about all the aspects of this learning process, including measures of student learning, would be necessary to understand the impact of these resources in the classroom.

Finally, there are the textbooks that have been "imported" into many classrooms and used in many different ways. If the Ginsburg et al. (2005) study is illustrative, we can anticipate many barriers to seamless adoption. Yet more interesting will be to understand how teachers use these textbooks as additional resources for planning and teaching their lessons. Because the artefacts that teachers use get transformed over several iterations (Gueudet & Trouche, 2009), longitudinal studies are fundamentally important if we are to begin to understand such transformations and how they affect the ways lessons are designed and enacted.

Concluding Remark

The intricate connection that exists between these three areas—competencies and frameworks; curriculum control, design, and management; and teacher preparation and development and use of textbooks—is evident in our proposed research agenda. The international studies are playing an important, although subtle, role in shaping the landscape of curriculum design, implementation, and attainment. The lack of concerted efforts to trace their influence at the classroom level should be a cause for concern. To paraphrase an old saying, "All roads lead to the classroom." In the end, therefore, the substantial investments of money and effort that the international studies of student achievement require ought to have one single purpose: to improve teaching in order to improve learning. Collecting evidence in a systematic way may prove beneficial to achieving this ultimate goal.

Acknowledgements The authors thank Ms. Heejoo Suh for her invaluable bibliographic research skills and the three reviewers who raised important points that helped us clarify our argument.

References

Aguilar, M., & Loejo, A. (2008). *PISA en el aula: Matemáticas* [PISA in the classroom: Mathematics]. México, DF, México: Instituto Nacional para la Evaluación de la Educación (INEE).

Alcaide, S., & Álvarez, P. (2011, March 4). *Madrid ofrece a los colegios decidir un tercio del programa escolar* [Madrid offers to let schools decide a third of the school program] (p. 38). El País.

Arias, R. M. (2006). La metodología de los estudios PISA [Methodology in the PISA studies]. *Revista de Educación* (Extraordinario), 111–129.

Barber, M., & Mourshed, M. (2007). *How the world's best-performing school systems come out on top*. New York, NY: McKinsey.

Bassells, F. (2011, March 4). *Cataluña se prepara para que cada centro fije el contenido y el orden de las materiasl* [Catalonia gets ready for each center to set subject matter contents and sequencing] (p. 38). El País.

Beyer, L., & Liston, D. (1996). *Curriculum in conflict: Social visions, educational agendas, and progressive school reform*. New York, NY: Teachers College Press.

Blanco, L., & Rico, L. (2011, January 8). ¿Qué hacer con los datos de PISA? [What to do with PISA data?]. El País. Retrieved from http://www.elpais.com/articulo/sociedad/hacer/datos/PISA/elpepusoc/20110108elpepusoc_9/Tes.

Blum, W. (2004, July). *Opportunities and problems for "quality mathematics teaching": The SINUS and DISUM projects.* Paper presented as a regular lecture at the International Congress on Mathematical Education (ICME) 10. Retrieved from http://www.icme10.dk/proceedings/pages/regular_pdf/RL_Werner_Blum.pdf.

Bonnet, G. (2008). Do teachers' knowledge and behaviour reflect their qualifications and training? Evidence from PASEC and SACMEQ country studies. *Prospects, 38*(3), 325–244.

Bottani, N. (2006). La más bella del reino: El mundo de la educación en alerta con la llegada de un príncipe encantador [The fairest of them all: The world of education on alert with the arrival of a prince charming]. *Revista de Educación* (Extraordinario), 75–90.

Bracey, G. W. (1998). TIMSS, rhymes with "dims," as in "witted". *Phi Delta Kappan, 79*(9), 686–687.

Cairn, W. D. (1935). Advanced preparatory mathematics in England, France and Italy. *American Mathematical Monthly, 42*(1), 17–34.

Chen, S. (2008). *The parent connection for Singapore math: Tools to help them "get it" & get behind it.* Peterborough, NH: Crystal Springs Books.

Chen, S. (2010). *Singapore math: Place value, computation, & number sense.* Peterborough, NH: Crystal Springs Books.

Clarke, D. J. (2003). International comparative studies in mathematics education. In A. J. Bishop, M. A. Clements, C. Keitel, J. Kilpatrick, & F. K. S. Leung (Eds.), *Second international handbook of mathematics education* (pp. 143–184). Dordrecht, The Netherlands: Kluwer.

Curriculum Development Council. (2000). *Mathematics curriculum guide* (P1–P6). Hong Kong, China: Author. Retrieved from http://www.edb.gov.hk/index.aspx?nodeID=4907&langno=1.

Departemen Pendidikan Nasional. (2003). *Standar kompetensi: Mata pelajaran matematika* [Competency standards: Subject of mathematics]. Jakarta, Indonesia: Author.

Dunne, D. W. (2000, July 21). Why are Chicago-area students tops in the world in math and science? *Education World.* Retrieved from http://www.educationworld.com/a_curr/curr251.shtml.

Education Council. (2006). Recommendation of the European Parliament and the Council of 18 December 2006 on key competencies for lifelong learning. *Official Journal of the European Union, L*(394), 10–18.

Ertl, H. (2006). Educational standards and the changing discourse on education: The reception and consequences of the PISA study in Germany. *Oxford Review of Education, 32*(5), 619–634.

Ferreras, V. (2006). Participación de Navarra en la evaluación PISA [Participation of Navarra in the PISA assessment]. *Revista de Educación* (Extraordinario), 531–542.

Ferrini-Mundy, J., & Schmidt, W. (2005). International comparative studies in mathematics education: Opportunities for collaboration and challenges for researchers. *Journal for Research in Mathematics Education, 36*(3), 164–175.

Figueroa, M. (2008). *Examination of a new method to teach elementary students mathematics* (Doctoral dissertation). Northern Arizona University.

Forsten, C. (2010a). *Solving word problems the Singapore math model-drawing strategy.* Peterborough, NH: Crystal Springs Books [CD-ROM].

Forsten, C. (2010b). *Step-by-step model drawing: Solving word problems the Singapore way.* Peterborough, NH: Crystal Springs Books.

Freudenthal, H. (1975). Pupils' achievement internationally compared. *Educational Studies in Mathematics, 6,* 127–186.

Ginsburg, A., Leinwand, S., Anstrom, T., & Pollock, E. (2005). *What the United States can learn from Singapore's world-class mathematics system and what Singapore can learn from the United States: An exploratory study.* Washington, DC: American Institutes for Research.

Givvin, K. B., Hiebert, J., Jacobs, J. K., Hollingsworth, H., & Gallimore, R. (2005). Are there national patterns of teaching? Evidence from the TIMSS 1999 video study. *Comparative Education Review, 43*(3), 311–343.

Gómez, J. A. (2006). La Rioja hacia la evaluación censal de PISA [Rioja towards a census assessment with PISA]. *Revista de Educación* (Extraordinario), 521–529.

Gómez, P., & Restrepo, A. (2012, August). *Procesos de planificación en matemáticas y autonomía escolar* [Planning processes in mathematics and school autonomy]. Paper presented at the III Congreso Internacional y VIII Nacional de Investigación en Educación, Pedagogía y Formación Docente, Bogotá, Colombia.

González, M. T., Monterrubio, M. C., Delgado, M. L., & Codes, M. (2011, June). *Tipos de situaciones de las actividades de análisis matemático en los libros de texto de Educación Secundaria en España* [Types of situations in mathematical analysis activities in secondary education textbooks in Spain]. Paper presented at the XIIIth Inter American Conference on Mathematics Education.

Graña, J. (2006). El estudio PISA en Galicia: Pasado, presente y futuro [The PISA study in Galicia: Past, present and future]. *Revista de Educación* (Extraordinario), 515–520.

Gruber, K. H. (2006). The German "PISA-shock": Some aspects of the extraordinary impact of the OECD's PISA study on the German education system. In H. Ertl (Ed.), *Cross-national attraction in education. Accounts from Germany* (pp. 195–208). Didcot, UK: Symposium Books.

Gueudet, G., & Trouche, L. (2009). Towards new documentation systems for mathematics teachers? *Educational Studies in Mathematics, 71*(3), 199–218.

Hattori, K. (2007). Lesson study in South Africa. In M. Isoda, M. Stephens, Y. Ohara, & T. Miyakawa (Eds.), *Japanese lesson study in mathematics: Its impact, diversity and potential for educational improvement* (pp. 226–229). Singapore: World Scientific.

Hiebert, J., & Stigler, J. W. (2000). A proposal for improving classroom teaching: Lessons from the TIMSS video study. *Elementary School Journal, 101*(1), 3–20.

Hogan, B. (2005). *Singapore math: Problem-solving secrets from the world's math leader*. Peterborough, PA: Crystal Springs Books.

Hogan, B., & Forsten, C. (2007). *8-step model drawing: Singapore's best problem-solving MATH strategies*. Peterborough, PA: Crystal Springs Books.

Husén, T. (Ed.). (1967). *International study of achievement in mathematics* (Vols. 1 & 2). New York, NY: Wiley.

Husén, T. (1983). Are standards in U.S. schools really lagging behind those in other countries? *Phi Delta Kappan, 64*(7), 455–461.

Inprasitha, M. (2007). Lesson study in Thailand. In M. Isoda, M. Stephens, Y. Ohara, & T. Miyakawa (Eds.), *Japanese lesson study in mathematics: Its impact, diversity and potential for educational improvement* (pp. 188–193). Singapore: World Scientific.

Junta de Andalucía. (2007). Resolución de 10 de diciembre de 2010, de la Dirección General de Ordenación y Evaluación Educativa, por la que se establecen determinados aspectos de la Orden de 27 de octubre de 2009, por la que se regulan las pruebas de la evaluación de diagnóstico, para su aplicación en el curso 2010-2011 [Resolution of December 10, 2010, of the Directorate General of Educational Planning and Evaluation on establishing certain aspects of the Order of October 27, 2009, which regulates the testing for diagnostic evaluation applied during 2010–2011]. *BOJA, 12*, 22–23.

Kaur, B. (2002). TIMSS-R: Mathematics achievement of eighth graders from Southeast Asian countries. *Journal of Science and Mathematics Education in Southeast Asia, 25*(2), 66–92.

Kaur, B. (2003, March 31). *Evolution of Singapore's secondary school mathematics curricula*. Paper presented to the National Academy of Sciences. Retrieved from http://www7.nationalacademies.org/mseb/workshop_background_materials_kauer_spores.pdf.

Kaur, B. (2005). Schools in Singapore with high performance in mathematics at the eighth grade level. *Mathematics Educator, 9*(1), 29–38.

Kaur, B. (2009). Performance of Singapore students in Trends in International Mathematics and Science Studies (TIMSS). In K. Y. Wong, P. Y. Lee, B. Kaur, P. Y. Foong, & S. F. Ng (Eds.), *Mathematics education: The Singapore journey* (pp. 439–463). Singapore: World Scientific.

Kaur, B., Koay, P. L., & Yap, S. F. (2009). International project on mathematical attainment. In K. Y. Wong, P. Y. Lee, B. Kaur, P. Y. Foong, & S. F. Ng (Eds.), *Mathematics education: The Singapore journey* (pp. 494–511). Singapore: World Scientific.

Kaur, B., & Pereira-Mendoza, L. (2000a). Singapore primary school TIMSS data: Data representation, analysis and probability and patterns, relations and functions. *Mathematics Educator, 5*(1/2), 180–193.

Kaur, B., & Pereira-Mendoza, L. (2000b). TIMSS: Performance of Singapore secondary students (Part B): Proportionality, measurement, fractions and number sense. *Journal of Science and Mathematics Education in Southeast Asia, 23*(1), 54–70.

Kaur, B., & Yap, S. F. (2009). Kassel project on the teaching and learning of mathematics: Singapore participation. In K. Y. Wong, P. Y. Lee, B. Kaur, P. Y. Foong, & S. F. Ng (Eds.), *Mathematics education: The Singapore journey* (pp. 479–480). Singapore: World Scientific.

Keitel, C., & Kilpatrick, J. (1998). The rationality and irrationality of international comparative studies. In G. Kaiser, E. Luna, & I. Huntley (Eds.), *International comparisons in mathematics education* (pp. 242–257). London, UK: Falmer.

Kersting, N. B., Givvin, K. B., Sotelo, F. L., & Stigler, J. W. (2010). Teachers' analyses of classroom video predict student learning of mathematics: Further explorations of a novel measure of teacher knowledge. *Journal of Teacher Education, 61*(1–2), 172–181.

Kilpatrick, J. (1971). Some implications of the International Study of Achievement in Mathematics for mathematics educators. *Journal for Research in Mathematics Education, 2*(2), 164–171.

Kimmelman, P., Kroeze, D., Schmidt, W., van der Ploeg, A., McNeely, M., & Tan, A. (1999). *A first look at what we can learn from high performing school districts: An analysis of TIMSS data from the First in the World Consortium* (ERIC Document Reproduction Service No. ED 433 243). Jessup, MD: U.S. Department of Education.

Kimura, E. (2007). Lesson study in Honduras. In M. Isoda, M. Stephens, Y. Ohara, & T. Miyakawa (Eds.), *Japanese lesson study in mathematics: Its impact, diversity and potential for educational improvement* (pp. 230–235). Singapore: World Scientific.

Koseki, K. (2007). Lesson study in Indonesia. In M. Isoda, M. Stephens, Y. Ohara, & T. Miyakawa (Eds.), *Japanese lesson study in mathematics: Its impact, diversity and potential for educational improvement* (pp. 214–215). Singapore: World Scientific.

Kotthoff, H.-G., & Pereyra, M. A. (2009). La experiencia del PISA en Alemania: Recepción, reformas recientes y reflexiones sobre un sistema educativo en cambio [The experience of PISA in Germany: Response, recent reforms and reflections on a changing educational system]. *Profesorado. Revista de Currículum y Formación de Profesorado, 13*(2), 1–24.

Landry, S. D. (2010). *Degrees of alignment among K-12 mathematics content standards of instruction: An analysis of high-performing and low-performing data sets* (Doctoral dissertation). Texas Christian University, Fort Worth, TX.

Lattuca, L. R., & Stark, J. (2009). *Shaping the college curriculum: Academic plans in action*. San Francisco: Jossey-Bass.

Lee, O., Deaktor, R., Enders, C., & Lambert, J. (2008). Impact of a multiyear professional development intervention on science achievement of culturally and linguistically diverse elementary students. *Journal of Research in Science Teaching, 45*(6), 726–744.

Lee, O., Deaktor, R., Hart, J. E., Cuevas, P., & Enders, C. (2005). An instructional intervention's impact on the science and literacy achievement of culturally and linguistically diverse elementary students. *Journal of Research in Science Teaching, 42*(8), 857–887.

Lee, V. E., & Zuze, T. L. (2011). School resources and academic performance in Sub-Saharan Africa. *Comparative Education Review, 55*(3), 369–397.

Lesson Study Research Group. (n.d.). Timeline of U.S. lesson study. Retrieved from Teachers College, Columbia University http://www.teacherscollege.edu/lessonstudy/timeline.html.

Lewis, C. (2011, February 20). *Lesson study with Japanese curriculum materials: A randomized controlled trial*. Paper presented at the APEC Tsukuba International Conference on Lesson Study, Tsukuba, Japan.

Lewis, C., Perry, R., & Hurd, J. (2004). A deeper look at lesson study. *Educational Leadership, 61*(5), 6–11.

Lewis, C., & Tsuchida, I. (1998). A lesson is like a swiftly flowing river: How research lessons improve Japanese education. *American Educator, 22*(4), 12–17, 50–52.

Liberal Party of Australia. (2010, December 7). *PISA findings back school autonomy. Liberal.* Retrieved from http://www.liberal.org.au/Latest-News/2010/12/07/PISA-findings-back-school-autonomy.aspx.

Lindner, M. (2008). New programmes for teachers' professional development in Germany: The programme SINUS as a model for teachers' professional development. *Interacções, 4*(9), 149–155.

Liptak, L. (2002). It's a matter of time. *Research for Better School Currents, 5*(2), 6–7.

López, L. S., & Toro-Álvarez, C. (2008). Formación de docentes en la enseñanza de las matemáticas a través de la resolución de problemas en la red de comprensión lectora y matemáticas–CCyM, segunda etapa [Teacher preparation in mathematical thinking through problem solving in reading and mathematics comprehension, CCyM, network, second stage]. *Universitas Psychologica, 7*(3), 753–765.

Lupiáñez, J. L. (2009). *Expectativas de aprendizaje y planificación curricular en un programa de formación inicial de profesores de matemáticas de secundaria* [Learning expectations and curricular planning in a secondary mathematics teacher preparation program] (Doctoral dissertation). Universidad de Granada, Spain.

Maestro, C. (2006). La evaluación del sistema educativo [The evaluation of the education system]. *Revista de Educación* (Extraordinario), 315–336.

McNab, D. (2000). Raising standards in mathematics education: Values, vision, and TIMSS. *Educational Studies in Mathematics, 42*(1), 61–80.

Ministerio de Educación Nacional [MEN]. (1994). *Ley 115 de Febrero 8 de 1994. Por la cual se expide la ley general de educación* [Law 115 of February 8, 1994. In virtue of which the general law of education is issued]. Bogotá, Colombia: Author.

Ministerio de Educación Nacional [MEN]. (2006). *Estándares básicos de competencias en lenguaje, matemáticas, ciencias y ciudadanas* [Basic standards of competencies in language, mathematics, science and citizenship]. Bogotá, Colombia: Author.

Ministry of Education Culture Sports and Science. (2010). *Elementary school teaching guide for the Japanese course of study: Mathematics (Grade 1–6)* (M. Isoda, Trans.). Tsukuba, Japan: Center for Research on International Cooperation in Educational Development (CRICED), University of Tsukuba.

Moss, P. (2004). The risks of coherence. In M. Wilson (Ed.), *Towards coherence between classroom assessment and accountability* (103rd Yearbook of the National Society for the Study of Education, Part 2, pp. 217–238). Chicago, IL: University of Chicago Press.

Mullis, I. (1999). *Attaining excellence: TIMSS as a starting point to examine mathematics assessment. An in-depth look at geometry and algebra.* Washington, DC: U.S. Department of Education, Office of Educational Research and Improvement.

National Council of Teachers of Mathematics (NCTM). (1991). *Professional standards for teaching mathematics.* Reston, VA: Author.

National Council of Teachers of Mathematics (NCTM). (1995). *Assessment standards for school mathematics.* Reston, VA: Author.

National Council of Teachers of Mathematics (NCTM). (2000). *Principles and standards for school mathematics.* Reston, VA: Author.

National Research Council. (1999). *Global perspectives for local action: Using TIMSS to improve U.S. mathematics and science education—Professional development guide.* Washington, D.C.: National Academy Press.

Neumann, K., Fischer, H., & Kauertz, A. (2010). From PISA to educational standards: The impact of large-scale assessments on science education in Germany. *International Journal of Science and Mathematics Education, 8*(3), 545–563.

Odani, K. (2007). Lesson study in Cambodia. In M. Isoda, M. Stephens, Y. Ohara, & T. Miyakawa (Eds.), *Japanese lesson study in mathematics: Its impact, diversity and potential for educational improvement* (pp. 206–209). Singapore: World Scientific.

Okubo, K. (2007). Lesson study in Egypt. In M. Isoda, M. Stephens, Y. Ohara, & T. Miyakawa (Eds.), *Japanese lesson study in mathematics: Its impact, diversity and potential for educational improvement* (pp. 216–217). Singapore: World Scientific.

Organisation for Economic Co-operation and Development (OECD). (2003). *The PISA 2003 assessment framework: Mathematics, reading, science and problem solving knowledge and skills*. Paris, France: Author.
Organisation for Economic Co-operation and Development (OECD). (2004). *Learning for tomorrow's world: First results from PISA 2003*. Paris, France: Author.
Organisation for Economic Co-operation and Development (OECD). (2005). *School factors related to quality and equity: Results from PISA 2000*. Paris, France: Author.
Organisation for Economic Co-operation and Development (OECD). (2010). *Education at glance 2010*. Paris, France: Author.
Ostermeier, C., Prenzel, M., & Duit, R. (2010). Improving science and mathematics instruction: The SINUS Project as an example for reform as teacher professional development. *International Journal of Science Education, 32*(3), 303–327.
Palamidessi, M. (2006). *Desarrollos curriculares para la educación básica en el Cono Sur: prioridades de política y desafío de la práctica* [Curriculum development for basic education in the Southern Cone: Policy priorities and challenges to practice]. Geneva, Switzerland: United Nations Educational, Scientific and Cultural Organization (UNESCO).
Perry, R., Lewis, C., Friedkin, S., & Baker, E. K. (2011). *Improving the mathematical content base of lesson study: Interim summary of results*. Retrieved from http://www.lessonresearch.net/IES%20Abstract_01.03.11.pdf.
Promoting Rigorous Outcomes in Mathematics and Science Education [PROM/SE]. (2006a). *Knowing mathematics: What can we learn from teachers* (Research Report, Vol. 2). East Lansing, MI: Michigan State University.
Promoting Rigorous Outcomes in Mathematics and Science Education [PROM/SE]. (2006b). *Making the grade: Fractions in your schools* (Research Report, Vol. 1). East Lansing, MI: Michigan State University.
Promoting Rigorous Outcomes in Mathematics and Science Education [PROM/SE]. (2008). *Dividing opportunities: Tracking in high school science* (Research Report, Vol. 4). East Lansing, MI: Michigan State University.
Promoting Rigorous Outcomes in Mathematics and Science Education [PROM/SE]. (2009a). *Content coverage and the role of instructional leadership* (Research Report, Vol. 7). East Lansing, MI: Michigan State University.
Promoting Rigorous Outcomes in Mathematics and Science Education [PROM/SE]. (2009b). *Opportunities to learn in PROM/SE classrooms: Teachers' reported coverage of mathematics content* (Research Report, Vol. 6). East Lansing, MI: Michigan State University.
Promoting Rigorous Outcomes in Mathematics and Science Education [PROM/SE]. (2009c). *Variation across districts in intended topic coverage: Mathematics* (Research Report, Vol. 5). East Lansing, MI: Michigan State University.
Rico, L. (Ed.). (1997). *Bases teóricas del currículo de matemáticas en educación secundaria* [Theoretical basis for mathematics curriculum in secondary education]. Madrid, Spain: Síntesis.
Rico, L. (2011). El estudio PISA y la evaluación de la competencia matemática [The PISA study and the assessment of mathematical competency]. *Matematicalia, 7*(1), 1–11.
Rico, L., & Lupiáñez, J. L. (2008). *Competencias matemáticas desde una perspectiva curricular* [Mathematical competencies from a curricular perspective]. Madrid, Spain: Alianza Editorial.
Rizo, F. M. (2006). PISA en América Latina: lecciones a partir de la experiencia de México de 2000 a 2006 [PISA in Latin America: Lessons from the Mexican experience from 2000 to 2006]. *Revista de Educación* (Extraordinario), 153–167.
Robitaille, D. F., & Travers, K. J. (1992). International studies of achievement in mathematics. In D. A. Grouws (Ed.), *Handbook of research on mathematics teaching and learning* (pp. 687–709). Reston, VA: National Council of Teachers of Mathematics.
Roth, K., & Givvin, K. B. (2008). Implications for math and science instruction from the TIMSS video study. *Principal Leadership, 8*(9), 22–27.

Saito, M. (2010). Have gender differences in reading and mathematics achievement improved? *Southern and Eastern Africa Consortium for Monitoring Educational Quality (SACMEQ) Policy Issues Series, 4*, 1–3.

Saito, N. (2007). Lesson study in Laos. In M. Isoda, M. Stephens, Y. Ohara, & T. Miyakawa (Eds.), *Japanese lesson study in mathematics: Its impact, diversity and potential for educational improvement* (pp. 210–213). Singapore: World Scientific.

Saracho, A. (2006). *Casos de estudio de reacciones a PISA 2000 y 2003: Alemania, Polonia, Brasil y otros* [Case studies of reactions to PISA 2000 and 2003: Germany, Poland, Brazil and others]. México, DF: Fundación IDEA.

Schleicher, A., & Shewbridge, C. (2008). *What makes school systems perform? Seeing school systems through the prism of PISA*. Paris, France: Organisation for Economic Co-operation and Development (OECD).

Schmidt, W. H., & McKnight, C. C. (1995). Surveying educational opportunity in mathematics and science: An international perspective. *Educational Evaluation and Policy Analysis, 17*(3), 337–353.

Schmidt, W. H., McKnight, C. C., Valverde, G., Houang, R. T., & Wiley, D. E. (1996). *Many visions, many aims* (Vol. 1: A cross-national investigation of curricular intentions in school mathematics). Dordrecht, The Netherlands: Kluwer.

Shimizu, S. (2007). Lesson study in the Philippines. In M. Isoda, M. Stephens, Y. Ohara, & T. Miyakawa (Eds.), *Japanese lesson study in mathematics: Its impact, diversity and potential for educational improvement* (pp. 202–205). Singapore: World Scientific.

Shymansky, J. A., Yore, L. D., & Anderson, J. O. (2004). Impact of a school district's science reform effort on the achievement and attitudes of third- and fourth-grade students. *Journal of Research in Science Teaching, 41*(8), 771–790.

Sloane, P. F. E., & Dilger, B. (2005). The competence clash—Dilemmata bei der Übertragung des 'Konzepts der nationalen Bildungsstandards' auf die berufliche Bildung, Berufs- und Wirtschaftspädagogik [The competence clash—Dilemmas in conveying the "concept of national education standards" to vocational, professional, and business education]. *Berufs- und Wirtschaftspädagogik, 8*, 1–32. Retrieved from http://www.bwpat.de/ausgabe8/sloane_dilger_bwpat8.pdf.

Stigler, J. W., Gonzalez, P., Kawanaka, T., Knoll, S., & Serrano, A. (1999). *The TIMSS videotape classroom study: Methods and findings from an exploratory research project on eighth-grade mathematics instruction in Germany, Japan, and the United States (NCES 1999-074)*. Washington, DC: U.S. Department of Education, National Center for Education Statistics.

Stigler, J. W., & Hiebert, J. (1997). Understanding and improving classroom mathematics instruction: An overview of the TIMSS video study. *Phi Delta Kappan, 78*(1), 14–21.

Stigler, J. W., & Hiebert, J. (1999). *The teaching gap*. New York, NY: Free Press.

Travers, K. J., & Westbury, I. (Eds.). (1989). *The IEA Study of mathematics I: Analysis of mathematics curricula*. Oxford, UK: Pergamon.

Valdés, H., Treviño, E. G., Acevedo, C., Castro, M., Carrillo, S., Costilla, R., et al. (2008). *Student achievement in Latin America and the Caribbean: Results of the Second Regional Comparative and Explanatory Study (SERCE)*. Santiago, Chile: Laboratorio Latinoamericano de Evaluación de la Calidad de la Educación, United Nations Educational, Scientific and Cultural Organization (UNESCO).

Walker, L. (n.d.). *Model drawing for challenging word problems: Finding solutions the Singapore way*. Peterborough, NH: Crystal Springs Books.

Watanabe, T. (2003). Lesson study: A new model of collaboration. *Academic Exchange Quarterly, 7*(4), 180–184.

White, A., & Lim, C. S. (2008). Lesson study in Asia Pacific classrooms: Local responses to a global movement. *ZDM: International Journal for Mathematics Education, 40*(6), 915–939.

White, A., & Southwell, B. (2003). *Lesson study project: Evaluation report*. Ryde, Australia: Department of Education and Training, Professional Support and Curriculum Directorate.

Wong, K. Y., Lee, P. Y., Kaur, B., Foong, P. Y., & Ng, S. F. (2009). Introducing the landscape of Singapore mathematics. In K. Y. Wong, P. Y. Lee, B. Kaur, P. Y. Foong, & S. F. Ng (Eds.), *Mathematics education: The Singapore journey* (pp. 1–9). Singapore: World Scientific.

Yore, L., Anderson, J., & Chiu, M.-H. (2010). Moving PISA results into the policy arena: Perspectives on knowledge transfer for future considerations and preparations. *International Journal of Science and Mathematics Education, 8*(3), 593–609.

Yoshida, M. (2007). Lesson study in Ghana. In M. Isoda, M. Stephens, Y. Ohara, & T. Miyakawa (Eds.), *Japanese lesson study in mathematics: Its impact, diversity and potential for educational improvement* (pp. 222–225). Singapore: World Scientific.

Young, J. W. A. (1900). *The teaching of mathematics in the higher schools of Prussia*. New York, NY: Longmans, Green.

Chapter 28
International Organizations in Mathematics Education

Bernard R. Hodgson, Leo F. Rogers, Stephen Lerman, and Suat Khoh Lim-Teo

Abstract Although the history of internationalization in mathematics education goes back more than a century, the last few decades have witnessed a notable acceleration in the establishment of bodies aiming at grouping together members of the community. The purpose of this chapter is to survey international or multinational organizations created to support and enhance reflection and action about the teaching and learning of mathematics at various levels of education systems, worldwide or in some specific regions of the world. The oldest and best-known international organization in mathematics education is the International Commission on Mathematical Instruction, but there are many others established over the years, serving different purposes and covering various aspects of the field. Focussing on those connected to research in mathematics education, this chapter highlights the diversity thus encountered in connection with the aims of these organizations, their functioning, or the specific niche they occupy in the mathematics education landscape.

Introduction

A number of papers have appeared in the past few decades related to various aspects of internationalism in mathematics education. Three examples, taken from recent handbooks on mathematics education or the history of school mathematics,

---

B. R. Hodgson (✉)
Université Laval, Québec, Canada
e-mail: Bernard.Hodgson@mat.ulaval.ca

L. F. Rogers
University of Oxford, Oxford, UK

S. Lerman
London South Bank University, South London, UK

S. K. Lim-Teo
National Institute of Education, Singapore

are Jacobsen (1996), Robitaille and Travers (2003), and Atweh, Clarkson, and Nebres (2003). In the first paper, Jacobsen offered a survey from the perspective of international cooperation in mathematics education, the emphasis being on the role with respect to developing countries of institutions such as UNESCO or ICMI and its affiliates (see Appendix A for a list of acronyms used in this chapter). Robitaille and Travers emphasized international connections notably from the vantage point of comparative studies such as the Trends in International Mathematics and Science Study (TIMSS), recognized as contributing significantly to international debate and collaboration in mathematics education. In the third paper, concerned with international and global contexts, Atweh et al. first described what they understand by *globalization* and *internationalization*, the former term being connected to aspects essentially beyond one's control (e.g., a rising "global curriculum"), whereas the latter allows for a greater autonomy in participation. They then examined several cases of internationalization and globalization in mathematics education, ICMI and TIMSS standing among the examples proposed for the former. The tandem internationalization/globalization was also discussed in Atweh et al. (2008).

The present survey addresses the issue of internationalization in mathematics education by examining organizations created to support and enhance reflection and action about the teaching and learning of mathematics at various levels of education systems, worldwide or in some specific regions of the world, and by discussing how these organizations have affected, and continue to affect, mathematics education around the world. Except for a few, all organizations surveyed here came into existence since the mid-1970s, and since 2000 for six of them. (A chronological listing of these 30 or so organizations is given in Appendix B.)

ICMI, the International Commission on Mathematical Instruction, was established in 1908 and is the earliest example of an international institution related to mathematics education. Although its own activities were at times jeopardized—in particular around the two World Wars—ICMI remained for almost its first 50 years the only international organization aiming at fostering the development of mathematics education in all its aspects. The Commission has a long and rich history that offers a privileged way of understanding aspects of the evolution of mathematics education over the past century. ICMI was at times being influenced by or accompanying the evolution in mathematics education, and at times even fostering this evolution. From the 1950s, other international players came into the picture. To name a few: CIEAEM in Europe in 1950, then CIAEM in Latin America in 1961, and later the first Study Groups affiliated to ICMI. The domain of mathematics education was maturing and expanding, and new international organizations were created to reflect and better address the new needs.

An obvious difficulty encountered when approaching the topic of this chapter is the choice of organizations to be included. Of necessity, we need to be selective. Although some bodies unequivocally ought to be part of such a survey—the reader may probably easily identify a number of those—the matter becomes much less evident as the list expands. On what ground is selection to be made? And how is the information to be structured?

Our choice was to focus on organizations connected to research in mathematics education, while trying to remain eclectic as regards the kind of bodies, size, organizational infrastructure, or type of activities. We have understood the word *international* to be as inclusive as possible, from organizations aiming at reaching people worldwide to others concerned with collaboration among a few countries, sometimes on a regional basis (our use of the term *multinational* is often related to these latter cases). Some of these organizations are "general purpose," so to say, and aim at covering mathematics education in general (e.g., ICMI, CIEAEM), but sometimes with a focus on specific regions of the world (e.g., CIAEM, MERGA, AFRICME), whereas others are connected to specific areas or interests (e.g., ICTMA, Delta, MCG). A common concern among many of these organizations is to find a proper balance between high quality standards for the presentation and publication of research work, and inclusiveness of the community, particularly of newcomers to the field. Examples of initiatives related to this perennial issue are offered below (e.g., MERGA, ERME, MES, Delta).

This chapter thus addresses the history of the development of the mathematics education research community across the world over a not-inconsiderable period of time, and particularly over the last 35 years or so. As such, that development is not a matter of educational research, and certainly not of didactical research, but of historical appraisal. There are cases where a group has set itself up in opposition to existing groups for theoretical or ideological reasons (e.g., Mathematics, Education and Society, MES) or in support of existing groups (e.g., PME-NA), but we see these as social and historical phenomena partly motivated by local issues rather than entirely research-driven phenomena.

We attempt to do justice to the range of organizations that come under the umbrella of mathematics education research groups by charting their emergence, achievements to date, state of their current existence, the reasons that they were formed, and their current role and function. A most striking aspect of this work is the diversity thus encountered in connection with the aims of these organizations, their functioning, or the specific niche they occupy in the mathematics education landscape.

The chapter is organized around categories of mathematics education organizations active on the international scene. The first section is devoted exclusively to ICMI, an institution clearly of prime importance, both in itself and with respect to its multiple relations to other organizations, and which will play a pivotal role throughout the chapter. We then discuss five organizations among the first that were established after ICMI, from 1950 to the mid-1970s, namely CIEAEM, CIAEM, PME, HPM, and MERGA. Such a selection conveys a certain historical or chronological coherence, and also represents important cases already manifesting a rich diversity of contexts, aims, and functioning. The next section is based on the notion of affiliation to ICMI and displays other facets of diversity in mathematics education, as it can currently be seen among the ICMI community. The organizations discussed are ERME, IOWME, ICTMA, WFNMC, and MCG. We finally consider, in the last major section, various contexts in which given sub-communities emerged on the basis of regional or thematic interests. Ten different cases are surveyed, including the so-called ICMI Regional Conferences, organizations focussed on a given topic, like IASE, or regional structures of a general scope, such as NoRME. Among the organizations then discussed are some functioning on a regular basis but in a somewhat loose environment, without a formal body in the background (MES, Delta).

It is unavoidable, in such a paper, to use a plethora of acronyms, as we have already done. As noted above, Appendix A lists all acronyms used here, both for the organizations that we survey and for associated elements (related institutions, journals, conferences). In each case, we indicate the year of establishment and the URL to access the relevant Web site.

The Internationalization of the Mathematics Education Community: The Birth (and Rebirth) of ICMI

The International Commission on Mathematical Instruction (ICMI) undoubtedly distinguishes itself among international organizations by both its age and its global impact on the field of mathematics education. Established in 1908, ICMI is the oldest body specifically related to the teaching and learning of mathematics in a multinational perspective, and is reckoned to be the first international organization concerned with the teaching of a given scientific discipline. (A few international conferences on teaching, however, were held before the inception of ICMI, such as the congress organized in Paris on the occasion of the 1889 International Exposition, as reported by Buisson, 1911.) Over the years, the mission and influence of ICMI have evolved in such a way that the celebration of its centennial was seen as an opportunity "to investigate how key and perennial issues of mathematics education have developed during the existence of ICMI as shaped and/or reflected by ICMI activities" (Schubring, 2008a, p. 1).

A substantial literature deals with aspects of the history of ICMI, many of these documents being of recent vintage. Of particular interest are the following: the survey Howson (1984) published for the 75th anniversary of ICMI; ICMI-related sections from the book by Lehto (1998) on the history of the International Mathematical Union (IMU); the papers of Furinghetti (2003) and Schubring (2003), written for the centennial symposium of the journal *L'Enseignement Mathématique* (L'EM); the paper that Schubring (2008b) presented at the ICMI centennial celebration (see Menghini, Furinghetti, Giacardi, & Arzarello, 2008, for the proceedings of that symposium); and various ad hoc papers such as Bass and Hodgson (2004), Furinghetti (2008a), Arzarello, Giacardi, Furinghetti, and Menghini (2008), Hodgson (2009), and Hodgson (2011)—the comments in this section borrow from the latter. To these sources must be added the ICMI History Web site (http://www.icmihistory.unito.it/), an important on-going project devoted to the history of ICMI. Furinghetti and Giacardi (2010) offer a "walk" through the Web site's biographical gallery of key figures in ICMI history, as well as a schematic presentation of five periods structuring the first century of ICMI: (a) foundation in 1908 and early period up to the First World War; (b) crisis, dissolution, and ephemeral rebirth between the two World Wars; (c) rebirth in 1952 as a commission of IMU; (d) "renaissance" in the late 1960s and consolidation; (e) increased autonomy from IMU and new trends in ICMI action in recent decades. The place of women in the life of the Commission up to the 1970s is discussed by Fulvia Furinghetti (2008b). Historical information on ICMI is also found in this volume in Chapter 25, by Alexander Karp. Before examining what ICMI is today, we concentrate briefly on a few salient points from its history.

The Establishment of ICMI

The birth of ICMI happened in a social and intellectual environment where the idea of internationalism was gaining prominence. Such a context was in no way specific to mathematics education (see Chapter 25) and can be seen as related to a larger tendency of associating

> the world of teaching to the "*great movement of scientific solidarity*" which was emerging at the end of the 19th century, notably through the organization of international meetings such as the first International Congress of Mathematicians held in Zurich in 1897. (Coray & Hodgson, 2003, pp. 11–12)

A discussion of internationalism in relation to mathematics can be found in Parshall (2009).

Aspects such as the origins of ICMI; the links with the previous creation of the journal L'EM, the official organ of ICMI since its inception; and the roles played by Henri Fehr, Charles-Ange Laisant, and David Eugene Smith, to name a few key figures, have been discussed, for instance, in Coray and Hodgson (2003), Furinghetti (2003), and Schubring (2003, 2008b). It was during the fourth International Congress of Mathematicians (ICM), held in Rome in 1908, that ICMI was founded through a resolution asserting the importance of initiating a comparative study of the methods and programs of mathematics teaching in secondary schools and appointing to that effect a committee, under the presidency of the eminent German mathematician Felix Klein. ICMI was mainly known in those days via its French or German acronyms, CIEM and IMUK, or under its first English acronym, ICTM (Lehto, 1998; Schubring, 2003).

As noted by Lehto (1998), the first years of existence of ICMI were marked by much activity so far as curricular comparisons were concerned. However, international tensions provoked by the First World War, and the resulting decrease in international scientific contacts, brought ICMI to a quasi-stagnation (see Schubring, 2008b). It was only after the Second World War, in a context where the scientific community wanted to escape the difficulties encountered in the aftermath of the previous war, that the rebirth of the Commission occurred. In 1951, IMU formally came into existence (for a second time), and at its first general assembly, in March 1952, it was agreed that ICMI should be attached to the union as its education commission (see Sections 4.3, 5.1, and 5.4 of Lehto, 1998)—there had been a first incarnation of IMU during the period 1920–1932 (see Chapter 2 of Lehto, 1998)

ICMI as a Permanent Commission of IMU

Bass (2008) used the expression "Klein era" to describe a first phase in the history of ICMI, from its inception to the Second World War. The main actors in ICMI life were then mathematicians who, like the first ICMI president, had developed "a substantial, but peripheral interest in education … plus some secondary teachers of high mathematical culture" (Bass, 2008, p. 9). Furinghetti (2008a) commented that at the beginning of the second life of ICMI as a permanent IMU commission, the emphasis of its activities quickly went beyond the mere comparative analysis of school curricula, as done previously, in order to meet the challenges provoked by the "developments of society and schools" and the resulting increased "complexity of the educational problems" (p. 49). Bass (2008) called this new phase the "Freudenthal era," from the eighth president of ICMI. This was a time that witnessed substantial changes in reflections about the teaching and learning of mathematics; namely, a shift from mathematics education considered as a "national business" concerned with curricular comparisons, to a "personal business" centred on learners and teachers (Furinghetti, 2008a, p. 50).

The presidency of Hans Freudenthal, from 1967 to 1970, was a turning point in this renewal of ICMI, as two major events then occurred, essentially through his personal initiative: the launching of a new journal (*Educational Studies in Mathematics*, ESM) and a new series of congresses (the International Congress on Mathematical Education, ICME), both specifically devoted to the then-emerging discipline of mathematics education—see Hanna and Sidoli (2002) for a history of ESM. These spectacular undertakings of Freudenthal reflected, and at the same time nurtured, the ongoing development of a new field of research, with its actors often new as well. These were no longer mathematicians with an occasional interest in educational matters, but professional researchers in the teaching and learning of mathematics (known as *didacticians* in most languages except English). In the words of Bass (2008), "This period witnessed the emergence of mathematics education (didactics) as an international academic discipline, and of the corresponding scholarly community, for which ICMI was a major resource and agent" (p. 10).

Much more could be said about these remarkably active moments in the life of ICMI and the following post-Freudenthal years (Hodgson, 2009). Two points are especially worth mentioning. This deep evolution of ICMI under Freudenthal's influence did not happen without some tension with IMU—more details on this are given in Lehto (1998) and Hodgson (2009). Also, whereas the first years of the rebirth of ICMI as a commission of IMU still showed a strong preponderance of European countries, as in its early days (see Schubring, 2003, p. 56), ICMI soon became more worldwide-oriented. Attention was particularly given to the objectives of spreading ICMI actions in Asia, Latin America, or Africa, notably as regards the needs of developing countries, and of fostering the emergence of regional networking. It was in such a context, to take one example, that ICMI, via its president Marshall H. Stone, played a crucial role in the founding in the early 1960s of the *Comité Interamericano de Educación Matemática* (CIAEM; see below).

ICMI Today

In contrast with most other bodies discussed in this chapter, the members of ICMI are not individuals but countries, following the IMU model. ICMI currently has 92 members, 80 of which are de facto members through their IMU membership. The annual budget of ICMI is provided by IMU as part of the dues collected from its member countries. As a commission of IMU, ICMI belongs to the family of the

International Council for Science (ICSU), which entails that ICMI is to abide by the ICSU statutes, one of which establishes the fundamental principle of universality of science, based on nondiscrimination and equity.

ICMI is today a major institution in the field of mathematics education and it can arguably be considered as "perhaps the international organization that has the most direct effect on mathematics education" (Atweh et al., 2003, p. 192). A peculiarity of ICMI is its position "at the interface between mathematics and mathematics education," to borrow from the title of Artigue (2008b). Such an interface could be considered in relation to these fields of knowledge per se and the various ways they connect, as well as with regard to the mathematicians and mathematics educators interacting at that interface. Still another aspect is related to structures; that is, to the existence of ICMI as a commission of IMU. This connection defines ICMI legally and specifies the global context behind its actions.

This existence of ICMI at the interface with the community of mathematicians as represented by IMU provides a rich, albeit at times uneasy, framework for fulfilling its mission. Beyond the Freudenthal episode mentioned above, Artigue (2008b) spoke of an "increasing distance" (p. 188) between the two bodies in the 1980s and 1990s which eventually led to somewhat strained relations, in particular around the 1998 ICM (see also Hodgson, 2009, pp. 85–86). In spite of "voices asking ICMI to take its independence" (Artigue, 2008b, p. 189) from IMU, a decision was then made by the ICMI Executive Committee (EC) to build on "the strength of the epistemological links between mathematics and mathematics education" (p. 190), and to renew and reinforce the contacts and collaboration with IMU. Artigue and Hodgson report that the combined efforts of both bodies eventually resulted in robust and productive links between ICMI and IMU. A spectacular outcome, totally unexpected at the beginning of this century, was the approval at the 2006 IMU General Assembly (GA) of a new election procedure for ICMI transferring to the ICMI GA the decisive vote on the composition of the ICMI EC. Such a development points to the maturity not only of ICMI as an organization and the community it serves, but also of mathematics education as a scientific domain.

The two most widely familiar activities of ICMI are the series of ICME congresses, organized quadrennially, and the ICMI Study program, launched in the mid-1980s. The ICMEs are general conferences dealing with all aspects of mathematics education and are, as such, important vehicles for enabling mathematics educators (including, importantly, teacher educators and teachers) to come together and experience each other's ideas, cultural differences, and problems. Each ICMI Study, on the other hand, is devoted to a specific theme and aims at developing a state-of-the-art view of the topic at stake. Whereas the ICMEs typically attract some two to three thousand participants, (yet more than 3,600 people attended ICME-12 in Seoul in 2012) the studies are of a much smaller scale, with about 100 attendees taking part in an invitation-only working conference. Some of the recent ICMI Studies have been organized jointly with other bodies (IASE and ICIAM). ICMI is also involved in some ad hoc activities, such as the Pipeline Study or the Klein Project, both organized jointly with IMU, or the exhibition Experiencing Mathematics!, initiated and supported by UNESCO. (More details on these initiatives are available on the ICMI Web site: http://mathunion.org/icmi/)

Another strand of ICMI actions rests on the role of ICMI as a sort of umbrella organization offering a niche and support to various bodies. One example is the series of ICMI Regional Conferences, launched in the mid-1970s, gathering education communities in Latin America, Asia, Africa, and the Francophone world. Another is connected to a network of organizations affiliated to ICMI. These include *study groups* such as PME or HPM, affiliated in the 1970s, and also *multinational mathematical education societies* such as CIEAEM and CIAEM, established a while ago but affiliated to ICMI only recently. This is an evolving network, as it now comprises 10 bodies, 5 of whom got affiliation to ICMI in the years 2009–2011, thus reflecting the vitality and the diversity of the field of mathematics education nowadays. These regional conferences and affiliates are discussed separately below.

Other facets of the ICMI role can be seen by considering specific issues addressed at various moments. One such example with a long history, one that recently received renewed attention as a major aim of current ICMI actions, relates to outreach initiatives towards developing countries, as discussed by Artigue (2008b) and Hodgson (2009). The successful integration of colleagues from developing countries into the ICMI network requires a necessary evolution from the traditional "North–South" model towards "more balanced views and relationships" (Artigue, 2008b, p. 195). The Capacity and Networking Project aimed at developing countries was launched in 2011 jointly by ICMI, IMU, and ICIAM, in partnership with UNESCO. The objective of this initiative is to foster teacher development, both in mathematics and as professionals, and to help create and sustain networks of mathematicians, teachers, and mathematics educators in each region. A prerequisite for the acceptability of a given proposal is evidence of existing collaboration between local mathematicians and mathematics educators.

Although the publication of a research journal of its own is not part of its modus operandi, ICMI, as mentioned above, has strong historical connections with two major journals: L'EM, its official organ, and ESM, a journal deeply linked to the research community served by ICMI.

ICMI has clearly played, and continues to play, a leading role in mathematics education considered internationally. The number and scope of its activities are quite remarkable, especially bearing in mind its limited budget. Still, some issues can be raised when reflecting on its actions and mission. Atweh et al. (2003), for instance, stressed the financial difficulty for educators from developing countries to attend the ICMEs, in spite of the 10% "Solidarity Tax" now raised on registration fees since ICME-8 to support participation from less affluent countries. They also questioned the format of the forums provided by such international conferences and pointed to linguistic and cultural barriers leading, in ICMI's activities, to a domination "by educators and issues from Anglo-European countries" (p. 195). Artigue (2008b) discussed a series of "crucial challenges" that ICMI must face when looking at the future. These concern mainly thematic, cultural, and regional underpinnings of the mission of ICMI, for instance: improving and extending ICMI outreach and the accessibility to its activities, furthering new relationships between centres and peripheries, understanding cultural diversity, and benefiting from it.

Analogous comments are found in Hodgson (2008, 2009) concerning the challenges faced by ICMI in its renewed partnership with IMU. But as pointed out in the concluding words of Artigue (2008b), there are reasons to be confident about the capacity of ICMI to adapt to the new challenges. ICMI is "still at the interface, but today at a much wider interface, fostering exchanges and collaboration between the diverse communities which, all over the world, can contribute to the improvement of mathematics education" (p. 197). Observing that ICMI, as an institution, has in the past been able to "progress modestly and slowly, but … with coherence" (p. 197), Artigue concluded that this evolution has not come to an end and that ICMI "will go on moving and improving" (p. 197).

An Emerging Diversity of Interests and Structures in Mathematics Education: The 1950s to the 1970s

The previous section highlighted the long-standing presence of ICMI in the international mathematical education landscape and the scope and broad-ranging nature of its actions and mission, as well as the evolution it went through over the years. For more than 40 years, ICMI was essentially the sole international organization active in the field—although as already mentioned, it was itself at times in dormancy. When ICMI was reconstituted in 1952 as a commission of IMU, it was no longer the only player, as a new international organization was being established, partly, it must be stressed, in reaction to ICMI. It is thus with an agenda substantially different from that of the "traditional" ICMI that CIEAEM was launched in the early 1950s. That arrival provoked a context that eventually fostered a substantial evolution in ICMI itself, in connection with the Freudenthal era discussed above.

Ten years passed before the inception of another international organization in mathematics education, on that occasion with a substantial contribution from ICMI. In 1961, the first CIAEM conference was held, with the aim of fostering the development of mathematics education in Latin America. Such a regional structure proved to be a very fruitful model through which more attention could be paid to local needs as well as to linguistic specificities.

The next important movements in mathematics education infrastructure, considered internationally, happened during the 1970s. These were connected with the new activities, and more importantly the new spirit, resulting from the evolution of ICMI and reflect an increase in both strength and diversity inside mathematics education during the so-called post-Freudenthal era of ICMI, as discussed by Hodgson (2009). The birth of the first study groups affiliated to ICMI, in 1976, is directly connected to the ICME congresses launched in 1969 by Freudenthal. We discuss in this section the cases of PME and HPM, and examine another regional structure, MERGA, launched in Australasia in the second half of the 1970s.

This section thus presents five organizations that are somewhat "early," chronologically speaking, as they are among the first groups, besides ICMI, to come into existence, all having been established before 1980. But they also exemplify significant aspects of the structuring of the mathematics education community, as one can already identify in these five cases a variety of important models and scopes. They prefigure in a way the emergence in the following years of quite a few other organizations and their richness, both in number and diversity.

The five bodies that we discuss in this section are all linked to ICMI as affiliate organizations. But we delay a discussion of this notion of affiliation to the next section.

CIEAEM

In their discussion of the context that eventually led, in the late 1960s, to what they call the ICMI Renaissance, Furinghetti, Menghini, Arzarello, and Giacardi (2008) stressed the crucial role played in that connection by a new player in the field of mathematics education, the *Commission Internationale pour l'Étude et l'Amélioration de l'Enseignement des Mathématiques* (CIEAEM/International Commission for the Study and Improvement of Mathematics Teaching, ICSIMT). According to their comments, the "old ICMI," with its focus on curricular comparisons, was not seen as providing a context suitable for pertinent reflections on arising educational issues, so that the need was felt for a new arena.

Already in 1950, a small group of people concerned with the improvement of mathematics teaching first met at the initiative of Caleb Gattegno. This group was the nucleus of a community eventually assembled under CIEAEM, officially founded 2 years later during the fourth meeting of the group. From the outset, CIEAEM gathered people from various fields (mathematicians, educationalists, psychologists, epistemologists, secondary schoolteachers), mostly from Europe initially. French was the main language of communication inside CIEAEM in its early years, and it is still today one of its two official languages, alongside English. Among the early members were many distinguished scholars: Gustave Choquet, Jean Dieudonné, Hans Freudenthal, André Lichnerowicz, Georges Papy and Jean Piaget, to name a few, as well as eminent secondary schoolteachers like Emma Castelnuovo and Lucienne Félix. The group aimed at modernizing mathematics teaching and achieving a complete reconstruction of school mathematics "from kindergarten to university." Its thinking was influenced by the Bourbakist ideas of abstraction and structure, as well as by the importance of the link, advocated by Gattegno and others, between the learner's mental activity, mathematical knowledge, and the pedagogy of the classroom. CIEAEM soon influenced the evolution of mathematics education through the emphasis it placed on students and on the teaching process, in contrast to educational work typical of the time, as well as by the presence among its principles of a fundamental mathematics component, eventually crystallized under the "new mathematics."

Examples of innovative ideas promoted by CIEAEM were mentioned by Furinghetti et al. (2008), including the relevance of psychology in mathematics education, the key role of concrete materials, and the importance of empirical research.

These authors also claimed that "the collaboration between people of different backgrounds," central to the early actions of CIEAEM, fostered "the emergence of a new figure, the researcher in mathematics education" (p. 135), which in turn was a catalyst for a context favourable to "the emergence of mathematics education as a field of research" (p. 140). With regard to ICMI, all these developments were clearly instrumental in paving the way for the Freudenthal era discussed above.

Elements of the CIEAEM history are found in Félix (1986), Grugnetti (1996), and Bernet and Jaquet (1998), as well as in *Manifesto 2000,* which CIEAEM (2000) published to mark its 50th anniversary. This last document also describes the aims and functioning of CIEAEM. Information on the annual conferences is found in Bernet and Jaquet, and on the CIEAEM Web site (http://www.cieaem.net). In addition to Furinghetti et al. (2008), the role and influence of CIEAEM has been discussed in recent papers (e.g., Atweh et al., 2003; Furinghetti, 2008a, 2008b).

One contrast between CIEAEM and ICMI is the informal character of the group; for instance, it was only in 1996 that CIEAEM adopted a constitution. Moreover, membership is at the individual level and by cooption only. Although the annual conference of CIEAEM, its main activity, now attracts up to 300 or 400 participants, the membership remains rather small, of the order of 50 from about 15 countries. In spite of such a limited size, CIEAEM can be seen today as occupying a specific niche in the mathematics education landscape through the framework it proposes and the atmosphere of discussion and debates it aims at fostering in its conferences.

Central to CIEAEM's (2000) philosophy is the recognition, as asserted in the *Manifesto 2000,* of the importance of "creating links between scientific knowledge and craft wisdom and reinforcing the collaboration of mathematics education research and practice" (p. 2). This aspect, presented as "what distinguishes the organisation from others," (p. 2) gets reflected, it is claimed, "in all of its work and at all the meetings" (p. 2). Consequently the annual conference is intended as a study and working event where working groups, the "heart of the conference," aim at fostering contacts between researchers and teachers. The presence of schoolteachers among the CIEAEM community is still seen today as a distinctive feature and has been presented by Furinghetti (2008b) as having been instrumental in promoting the international visibility of women, especially in periods when, "as teachers, [they] previously had few opportunities to participate in the international debate" (p. 532).

The relationship between CIEAEM and ICMI has varied considerably over the years, as discussed in Furinghetti et al. (2008). Formal relations were reinvigorated after a long period of indifference, when CIEAEM became the second multinational organization affiliated to ICMI under the expanded affiliation scheme adopted by ICMI in 2009.

CIAEM

The *Comité Interamericano de Educación Matemática* (CIAEM/Inter-American Committee on Mathematics Education, IACME) is arguably the main multinational organization in mathematics education in Latin America (Ruiz, 2010). This regional organization was founded in 1961 by a group of mathematicians and mathematics educators from the three Americas under the leadership of the distinguished American mathematician Marshall H. Stone, at that time ICMI president. The context of the creation of CIAEM has been discussed in Barrantes and Ruiz (1998) and Ruiz and Barrantes (2011). The great interest by Stone in Latin America and his specific role in building CIAEM and bringing it international support through his personal reputation were stressed by Barrantes and Ruiz.

The first *Conferencia Interamericana de Educación Matemática* (CIAEM/ Interamerican Conference on Mathematics Education, IACME) was organized in 1961 under the auspices of ICMI and the Organization of American States (OAS). This was the time of the new math movement, with a strong influence of the "Bourbaki ideology" (Barrantes & Ruiz, 1998, p. 3), and the main objective was to bring together educators from the Americas to evaluate and reformulate mathematics curricula, with special attention to Latin America. A second conference was held 5 years later to analyze the progress made in the reforms identified during the first congress. It was during the first congress that the CIAEM committee was formed with Stone as president, originally as a pro tempore committee and later formalized during the second conference.

The list of the CIAEM conferences, now regularly held quadrennially, is found in Barrantes and Ruiz (1998) as well on the CIAEM Web site (http://www.ciaem-iacme.org). Jacobsen (1996) pointed to the support provided by UNESCO in publishing the proceedings for many of these conferences. The 13th conference, coinciding with the 50th anniversary of CIAEM, was held in 2011 and attracted more than 1,800 participants from 30 countries, substantially more than the typical CIAEM conference.

In early documents, CIAEM described itself as "a non-governmental body affiliated with the International Union of Mathematicians [sic] through the International Commission on Mathematical Instruction" (Barrantes & Ruiz, 1998, p. 25). A "Memorandum on Affiliation of IACME to ICMI" was later adopted by the ICMI Executive Committee (ICMI, 1975), but its actual impact on the relationship between the two bodies is not so clear, as the intensity of the contacts between ICMI and CIAEM varied over the years. In 2009, CIAEM became the first multinational mathematics education society linked to ICMI under the expanded affiliation scheme then adopted, thus launching a new era of collaboration going much beyond the mere recognition of CIAEM conferences as ICMI regional activities, as typically done earlier. It may be noted that the president of CIAEM at the time of this writing, Ángel Ruiz, is also vice-president of ICMI.

CIAEM does not have a formal notion of membership and relies for its networking on a set of representatives from different countries, mostly in Latin America. The mission of CIAEM is centred mainly around the series of CIAEM conferences, which can arguably be seen as a major contributor to the shaping of mathematics education in Latin America through its influence on researchers, teacher educators, and teachers. If CIAEM was born with an aim of "bridging" the Americas, and especially the countries in Latin America, despite substantial geographical or economic obstacles, it has evolved as a body providing as well links with the international community, as argued by Barrantes and Ruiz (1998). The recently reinvigorated relationship of CIAEM with ICMI can be seen as relevant in that connection. Another organization related to Latin America, CLAME, is discussed below.

PME

This section presents a modified version of the brief history of PME by Nicol and Lerman (2008). The International Group for the Psychology of Mathematics Education (IGPME, or PME) was recognized as officially affiliated to ICMI in 1976 at ICME-3. The impetus to develop an organization with a psychological focus on mathematics education began much earlier when, at the first International Congress on Mathematical Education in 1969, Efraim Fischbein, as reported in Fischbein (1990), was invited by ICMI president Hans Freudenthal to chair and organize a round table on psychology and mathematics education. A cognitive psychologist, Fischbein was very keen to take up Freudenthal's call to improve mathematics education in schools by going beyond philosophical discussions of mathematics teaching and learning to planning and implementing empirical scientific research in the field.

Participants attending this first round table were very enthusiastic to continue the discussion on psychological aspects of mathematics education. A working group dedicated to the psychology of mathematics education was offered at the second ICME in 1972. Hundreds of participants attended that workshop, recognizing, as Fischbein (1990) did, "that the psychological problems of mathematical learning and reasoning are scientifically exciting and at the same time genuinely relevant for mathematics education" (p. 4). Four years later at ICME-3, participants decided to organize a permanent group that would meet yearly to discuss and explore issues related to the psychology of mathematics education. So began PME, with Fischbein as first president and Richard Skemp as second president 4 years later in 1980. The first annual meeting was held in 1977 in Utrecht.

As PME developed, its focus broadened to include new ways of thinking about learning mathematics. There were periods where particular ideas were prominent in PME research, including the ideas of instrumental/relational thinking, Realistic Mathematics Education, constructivism as a theory of knowing, visualization, alternative forms of assessment, and others. These and other research agendas challenged previous ways of thinking about mathematical activity and provided new implications for instruction.

With time, there was growing discussion on the scientific direction of PME, with some members advocating a broadening of the focus of PME to go beyond psychological considerations to include also the process of teaching and teacher education, the epistemology of mathematics from a teaching/learning perspective, and equity and sociocultural issues of teaching and learning mathematics. At the 2005 General Meeting, with recognition that the group's stated aims over the previous 10 years had moved beyond the purely psychological aims of the early years, the membership voted to amend the statement of major goals to include the study of aspects of teaching and learning that drew upon disciplines other than psychology, such as sociology and anthropology. Nevertheless, the group kept its acronym PME for historical reasons and because it was so well established within the field with that name.

A strong vision of the field of mathematics education, as reflected through the research done by the PME community, can be developed from what Hershkowitz and Breen (2006) call the two "PME milestone publications" (p. ix); namely, the "research synthesis" of the first decade of PME work by Nesher and Kilpatrick (1990), and the "handbook of research" of Gutiérrez and Boero (2006), published on the occasion of the celebration of 30 years of existence of PME. These two volumes show how the trends, scope, and collected research in mathematics education among the PME community have expanded.

PME conferences, the main activity of the group, are held every year and consist of a range of forms of presentation, including plenary lectures, a plenary panel, research forums, working groups, discussion groups, and poster sessions. Membership during the beginning years of PME consisted mainly of mathematicians, mathematics educators, and psychologists from Europe and North America. Gradually, membership grew so that by the mid-1980s representation also included participants from other countries. Currently, members and conference participants represent more than one-third of the countries around the world. PME has between 700 and 800 individual members, and membership is open to all persons in active research interested or involved in furthering the group's direction.

From the early days of PME, national groups with a similar orientation were set up, some being affiliated to PME. For instance, PME-NA, the North American chapter of PME, started in 1979 and has met annually ever since.

The development of mathematics education as a research field over the last decades has been greatly influenced by the work done in the PME community. PME continues today to be a vibrant organization with international members contributing to mathematics education worldwide.

HPM

The idea of introducing historical components into mathematics education has a long history, as is shown by the preamble to a resolution from the third International Congress of Mathematicians held in Heidelberg in 1904:

> Considering that the history of mathematics nowadays constitutes a discipline of undeniable importance, that its benefit—from the directly mathematical viewpoint as well as from the pedagogical one—becomes ever more evident, and that it is, therefore, indispensable to accord it the proper position within public instruction. (Krazer, 1905, p. 51; trans. in Schubring, 2000, p. 91)

The establishment of the International Study Group on the Relations between the History and Pedagogy of Mathematics (ISGHPM, or HPM) recognized the importance of history of mathematics in educational issues. This movement was initiated during the 1972 ICME by Phillip S. Jones and Leo F. Rogers through a working group devoted to a similar theme, with the encouragement and support of Kenneth O. May, and led to the launching of HPM as a permanent study group and its affiliation to ICMI at ICME-3 in 1976. Information on the origins of HPM is given by Fasanelli and Fauvel (2006).

The aims of HPM concern mathematics education at all levels. They include the production of materials for teachers promoting awareness of history of mathematics as relevant for education and as a significant part of the development of cultures, and the furtherance of "a deeper understanding of the way mathematics evolves, and the forces which contribute to this evolution" (Rogers, 1978, pp. 26–27).

Since its inception, HPM has grown in influence and has produced a wide range of publications, from accounts of classroom experiences to research papers in a number of languages. The conviction that history of mathematics relates to our cultural and social background underlies the development of many lines of investigation and has brought a more socially aware style to the writing of the history of mathematics. Different cultures, be they past or present, have different histories, and the awareness of these necessarily has implications for education.

Very soon after its official establishment, HPM began contributing to the International Congresses of Mathematicians in connection with history of mathematics in university education, and stimulated meetings on history and pedagogy of mathematics at joint meetings of the MAA/AMS in the USA. In 1984, the North Americas Section of HPM was formed and has usually held its meetings alongside the annual meetings of NCTM.

HPM has also organized and collaborated with activities in Europe. On the foundation of the IREM network in France in the early 1970s, a number of universities decided to research the history and epistemology of mathematics in education. A considerable number of publications have been produced by the IREMs, supported by the inter-IREM Commission on Epistemology and History of Mathematics, established in 1975. From 1993, and supported by the IREMs, the European Summer University on History and Epistemology in Mathematics Education (ESU) has met almost triennially and has been the main venue for HPM in Europe. Further collaboration with European colleagues led to the regular occurrence of a Working Group on history in mathematics education at ERME congresses, starting at CERME 6 in 2009.

Among the activities of HPM are "satellite meetings" organized regularly in conjunction with ICME congresses. This tradition started with ICME-5 in 1984 and has resulted in a number of publications, listed on the HPM Web site (http://www.clab.edc.uoc.gr/HPM/).

Publications related to HPM activities also include the volume by Fauvel and van Maanen (2000) resulting from the Tenth ICMI Study. Background was presented in Fasanelli and Fauvel (2006), who claimed:

> ICMI's support for and promotion of this Study can thus be seen as recognition of how the HPM Study Group had encouraged and reflected a climate of greater international interest in the value of history of mathematics for mathematics educators, teachers and learners. (p. xxiii)

To disseminate information among its community, HPM has published a regular newsletter since 1972. Starting with Issue 45 (2000), it is available from the HPM Web site, where links to Web sites with similar interests are also found.

MERGA

The idea of an Australian national group in mathematics education research developed in the mid-1970s, and its first conference took place in 1977, where participants voted to establish the Mathematics Education Research Group of Australia (MERGA). The beginnings of MERGA are told in Clements (2007) and Mousley (2009). According to Mousley, this was the first national mathematics education research group formed anywhere in the world. A decade after its inception, the group became the Mathematics Education Research Group of Australasia, thus better reflecting its regional and multinational scope. At its 2011 annual meeting, the ICMI Executive Committee officially approved the affiliation of MERGA to ICMI as a multinational organization involved in mathematics education. MERGA describes its main aims as promoting quality research on the teaching and learning of mathematics at all levels, with a focus on Australasia, providing means for sharing of research results through publications and conferences, and fostering the implementation of research findings, particularly in the preparation of teachers. MERGA's (2002) policy statement maintained, in particular, that the conduct of research in mathematics education must be sensitive to the diverse cultural backgrounds in a given educational environment, and that support should be offered to "early researchers."

Concerns have been expressed in that connection about achieving a balance between high standards for research publications, and enabling researchers (both young and experienced) to improve via feedback and support from the MERGA community. As an accepted paper is typically needed to gain financial assistance for attending a conference, Mousley (2009) reported on different reviewing procedures recently implemented, including one distinguishing between papers meeting very basic criteria and accepted for presentation only (with publication of an abstract), and others for presentation and publication. The effects of such changes remain to be seen.

MERGA's annual conferences have been held regularly since 1977, the 35th being hosted in 2012 in Singapore–the first occurrence of a MERGA conference held outside Australia or New Zealand, thus demonstrating the increasing regional scope of the organization.

Besides its annual conference proceedings (available on its Web site http://www.merga.net.au), MERGA publishes quadrennial reviews of mathematics education research in Australasia, the most recent being Forgasz, Barkatsas, Bishop, Clarke, Keast, Wee, and Sullivan (2008), as well as two refereed journals.

One of these journals is *Mathematics Teacher Education and Development* (MTED), first published in 1999, the outcome of the amalgamation with MERGA in

1997 of the Mathematics Education Lecturers' Association (MELA), started in 1973 as an organization of lecturers in teachers' colleges. Klein, Putt, and Stillman (1999) stressed the need that was identified "to maintain an avenue for the dissemination of innovative practices in mathematics education and discussion of issues which affected lecturers and their roles" (p. 1), which led to the founding of MTED (Mousley, 2009). The authorship has evolved, so that Way, Anderson, and Bobis (2010) claimed, in a recent editorial of MTED, that "international authors [now] account for about 34% of the total content. … While the journal maintains a focus on publishing material of interest and application to the Australasian context, encouraging a global perspective on mathematics education is also important" (p. 1).

A similar concern is found in the other MERGA refereed journal, the *Mathematics Education Research Journal* (MERJ), launched in 1989 and described as "an Australasian-based international mathematics education journal" (Forgasz, 2004, p. 1). By publishing "high-quality papers" presenting research on mathematics education at all levels, MERJ aims to attract an international readership but, as indicated on its Web site (http://www.merga.net.au/publications/merj.php), "papers exploring specifically Australasian issues are welcome." Forgasz (2004) invited authors "to be inclusive of the Australasian and the wider readership [and to] consider … illustrating how findings from the studies discussed in articles are relevant to the Australasian as well as broader mathematics education contexts" (p. 2). MERJ is clearly among highly regarded research journals in mathematics education. However, its ambition to address the total Australasian context is still partly problematic, in particular with respect to the origins of its editorial board members, reviewers, and authors.

Diversity Inside the ICMI Community: The ICMI Affiliates

The five organizations of the last section all belong to the network of ICMI affiliates. Two of these, PME and HPM, were in fact the very first study groups affiliated to ICMI in 1976, and were established having that status at the very beginning. Typically, however, a body would not become an ICMI affiliate right at the time of its inception, but only somewhat later. Artigue (2008b) maintained that the creation of the first ICMI Affiliated Study Groups provided evidence of the "increasing number of communities that tended to be institutionalized inside the mathematics education world" (p. 189) in the 1970s, thus pointing to a growing diversity in the field and the way it was being structured. In addition to PME and HPM, four other study groups became affiliated to ICMI over the years: IOWME (1987), WFNMC (1994), ICTMA (2003), and MCG (2011), each being of a thematic nature and focussing on the study of a specific aspect of mathematics education.

Although implemented in practice since the mid-1970s, it is only with the 2002 revision of the ICMI terms of reference that the notion of affiliation was formally introduced within the structure of ICMI (see Hodgson, 2002). In 2009, it was enlarged to include not only study groups devoted to a specific field of interest and study in mathematics education, but also multinational mathematical education societies.

There are currently four organizations affiliated to ICMI under the expanded affiliation scheme (three of which were established before 1980 and discussed in the previous section): CIEAEM, CIAEM, ERME, and MERGA. Except for CIEAEM, these multinational societies are all of a regional nature. Both CIEAEM and CIAEM have a very long history intermingling regularly with that of ICMI, as already indicated.

Through the notion of affiliation, ICMI acts as an umbrella organization for bodies active in mathematics education and having an existence of their own. These affiliates are not created by ICMI, nor are they financially supported by it. But ICMI promotes their activities through its various channels of information and also ensures that specific slots are dedicated to them on the program of the ICME congresses. This is a way for ICMI to enlarge the scope of its actions and encourage the development of the field, and for the organizations to have their credibility supported internationally by a well-respected body. The Guidelines to the 2009 ICMI Terms of Reference (accessible on the ICMI Web site http://www.mathunion.org/icmi) also point to affiliation as facilitating "jointly sponsored activities" involving ICMI and its affiliates.

The ICMI Affiliate Organizations, whether study groups or multinational societies, thus provide niches where specific segments of the international mathematics education community may feel at home. The 10 current affiliates represent a great diversity, not only in size and in the type or frequency of their activities but also in the facets of mathematics education that each of them aims to foster.

ERME

Discussions about forming a specifically European society for research in mathematics education were initiated in the 1990s. A small conference was organized in Germany in 1995, and from this, the idea found a wider audience during ICME-8 in 1996. Representatives from 16 European countries met in 1997 in Osnabrück, Germany, to establish the European Society for Research in Mathematics Education (ERME). The spirit of the new society can be seen from the following comments by one of the founding members, Jaworski (2008):

> In true European spirit, we decided that we wanted a society that would bring together researchers from across Europe, particularly including colleagues from Eastern Europe, fostering communication, cooperation and collaboration. We wanted a conference that would explicitly provide such opportunity. We especially wanted to encourage and contribute to the education of young researchers. Thus ERME was born and began to take shape. (p. 43)

The society held its first congress in 1998, when Guy Brousseau (1999) and Jeremy Kilpatrick (1999) gave the keynote addresses.

ERME aims at promoting communication, cooperation, and collaboration (the "three Cs," Arzarello, 2009) in mathematics education research in Europe, especially through its main activities: the congresses, covering a wide spectrum of themes to profit from the diversity in European research, and the summer schools, where experienced researchers work together with beginners.

A chief aim of the Congress of the European Society for Research in Mathematics Education (CERME), held every other year since the second in 2001, is to move away from individual research presentations towards collaborative group work involving scholarly debate (Jaworski, 2008). The intensive working groups are a distinctive feature of CERME. Although criticisms are sometimes heard about the requirement to stay within a single group for the whole conference, the consensus shows that this format benefits the quality of debates, so that the CERME congresses "have remained faithful to the initial conception," as reported by Jaworski in the general introduction to Durand-Guerrier, Soury-Lavergne, and Arzarello (2010). Jaworski added: "Many participants have said in evaluation of the events that the opportunity to spend serious time in one group allowed them to really get to know researchers from other countries, and that this contributed significantly to the depth of thinking that was possible" (p. xx).

Jaworski (2008) mentioned issues facing the ERME community in its efforts to support and develop the language of communication used in its scientific work, so as not to disadvantage those for whom English is a second language, and also to achieve a balance between scientific quality and the will to be inclusive. This quality/inclusiveness dichotomy remains a point at issue, and various schemes are being tried, in particular in the review process, to include as many participants as possible by helping them to bring their paper to acceptable standards (Arzarello, 2009).

Another strand of the ERME mission is the community of Young European Researchers in Mathematics Education (YERME), established at CERME 2 (Krainer, 2002). As stated by Arzarello and Tirosh (2009), "The main idea of [YERME] initiatives is to support young researchers in their first years of work, particularly during and immediately after their PhDs" (p. 43), in a friendly and cooperative style and with support from highly qualified experts. A YERME day is held just prior to CERME, and since 2002 a YERME Summer School (YESS) has been held in alternation with the CERME congresses.

Early in 2010, ERME became the second multinational mathematical education society linked to ICMI under the expanded scheme of affiliate organization.

IOWME

In 1976, a meeting was arranged during the course of the ICME-3 congress to discuss the issue of "Women and Mathematics." A vivid description of this event is found in Shelley (1984), where some on-site adverse reactions to the meeting were reported. It was agreed at that meeting to launch IOWME. A resolution was approved concerning "the poor representation of women at all levels [in the Congress]: in delivering main papers, on panels, as reporters, and in the planning of this 1976 Congress" (p. 20). This resolution was presented at the final session of ICME-3.

Further contacts with ICMI, described by Shelley (1984), led to the affiliation of IOWME to ICMI in 1987. Part of the motivation for this affiliation, as reported in Mendick (2008), was to gain official recognition from ICMI, so to foster consultation between ICMI and IOWME and facilitate the input of the Group in the planning of events such as the ICME congresses.

The IOWME leadership is headed by an *international convenor,* supported by a team of *national coordinators,* and the aims of the group are centred on the relationship between gender and mathematics, emphasizing development and dissemination of research related to the participation of women in the mathematical sciences and factors influencing that participation. On its Web site (http://extra.shu.ac.uk/iowme/), IOWME describes itself as "an international network of individuals and groups who share a commitment to achieving equity in education," which includes the links between gender and the teaching and learning of mathematics.

The main activities of IOWME are related to components of the programs of ICME congresses, other sessions sometimes being held in association with PME conferences. The group was also involved in the ICMI Study on gender and mathematics education that resulted in books by Grevholm and Hanna (1995) and Hanna (1996).

ICTMA

The acronym ICTMA represents three different (but closely related) entities: the International Community of Teachers of Mathematical Modelling and Applications, established in 1983; the International Conferences on the Teaching of Mathematical Modelling and Applications, which are the biennial congresses organized by this community since its inception; and finally the International Study Group for Mathematical Modelling and Applications, the name under which the community chose to be known for the purpose of its affiliation to ICMI in 2003, keeping the same acronym already in use for two decades.

ICTMA aims at promoting applications and modelling at all levels in mathematics education, being concerned with these issues from different perspectives: research, teaching and practice. Galbraith (2004) presented fostering work with a "clear application/modelling content, contextualized within an educational framework appropriate to the issue being addressed" (p. 67) as the mission of ICTMA. He stressed this "double aspect" in distinction to a strictly mathematical focus, or a mathematics education context disconnected from applications and modelling. He concluded with the following assertion about the ICTMA community:

> A distinctive aspect of ICTMA is the interface it provides for collaboration between those whose main activity lies within mathematics, but who have an informed interest in educational issues, and those whose institutional affiliations are within education, but who have a commitment to promoting the application of quality mathematics. (p. 68)

The theme of the 14th ICMI Study was directly related to the Group's interests. The resulting volume (Blum, Galbraith, Henn, & Niss, 2007), prepared with substantial input from members of the ICTMA community, reflected to a large extent the vision of the group. The biennial ICTMA conferences, on the other hand, have resulted in a series of books which are listed on the group's Web site (http://www.ictma.net/).

WFNMC

Founded reportedly by "an international band of enthusiasts" (Taylor, 2009, p. 11), the World Federation of National Mathematics Competitions was established in 1984 on the occasion of a mathematics competitions session at ICME-5, mainly through the inspiration of Peter O'Halloran. Fundamentally, the main idea was to share experiences among those involved in mathematics competitions in different parts of the world.

The name of the Federation might imply goals related only to competitions, but the spirit of its actions concerns how competitions may contribute to the improvement of mathematics education in general. The WFNMC constitution describes its aims as promoting excellence in mathematics education and supporting those interested in the development of mathematics education through mathematics contests. A policy statement issued in 2002 presented competition activities for students at all levels as central to the interests of WFNMC, and listed related activities that the Federation aims to support—such as enrichment courses, mathematics clubs, mathematics camps, and the development of resources to meet the needs of talented students. Kenderov (2009) suggested that affiliation of WFNMC to ICMI, in 1994, can be seen as "a recognition for what it does for mathematics education" (p. 19).

The field of interests of WFNMC was taken as a starting point for the 16th ICMI Study (Barbeau & Taylor, 2009). A crucial issue for ICMI was to ensure that the study was not restricted to, say, "olympiad-type" competitions and included a wider reflection. An appropriate description of "mathematical challenges" was agreed upon that coincided with the general aims of WFNMC (ICMI, 2004):

> A challenge occurs when people are faced with a problem whose resolution is not apparent and for which there seems to be no standard method of solution. So they are required to engage in some kind of reflection and analysis of the situation, possibly putting together diverse factors. Those meeting challenges have to take the initiative and respond to unforeseen eventualities with flexibility and imagination. (p. 33)

The word *challenge* denotes here a relationship between a question or situation and an individual or a group, so that what is a challenge for some may not be for others. Barbeau and Taylor (2008) reported that about one-third of the contributors to the study were "competition types" (p. 82), with others representing a range of activities and interests in mathematics education.

The Federation meets every other year, as the WFNMC conferences take place quadrennially and are intertwined with meetings held in between on the occasion of the quadrennial ICMEs.

MCG

In contrast to the ICMI Affiliated Study Groups discussed above, all of which had their roots in the 1970s or 1980s, the International Group for Mathematical Creativity and Giftedness (MCG) began in 1999, when the first conference of a series known as the International Conferences on Creativity in Mathematics Education and the Education of Gifted Students (CMEG) was organized by Hartwig Meissner. The MCG Group was officially established at the 6th CMEG conference in 2010 and became affiliated to ICMI in 2011.

MCG aims to encourage research and dissemination of information on how creativity and giftedness can be identified, nurtured, and supported. It recognizes the role and needs of teachers as well as the ways educational systems are able to react to situations in order to develop the full potential of all students. The CMEG conferences are held typically every two years, and members of its community are important contributors to the working groups devoted to this theme at the ICME congresses.

The MCG community may be regarded as numerically modest. Nonetheless, the establishment of the Group, as well as its affiliation to ICMI, are indications of the importance in mathematics education of the specific needs MCG aims at addressing. Gifted students may not form the most visible or pressing segment of the student population, but they are there, and neglecting their case would lead to a substantial loss of human resources. On the other hand, all students should be enticed to fulfil their full potential, whatever the level of such potential may be. Making sure teachers are more and more sensitive to such needs and that they are well equipped to face them can only be to the benefit of society.

Regional and Thematic Communities: A Selection of Ten Organizations from the Mid-1980s to Today

The proliferation of international organizations related to mathematics education in the 1970s was not an accidental phenomenon doomed to exhaust rapidly. As a matter of fact, as is clearly seen from Appendix B, the following decades witnessed the establishment of several new organizations, each being aimed at addressing quite specific needs of the community.

This section will concentrate on a selection of 10 international organizations created between the mid-1980s and today. By regrouping these bodies together into a single section, we do not mean to suggest a strong similarity among them, although they all result either from a regional approach or a thematic focus. Our aim is rather to use this context to stress again the remarkable diversity achieved in the structuring over the years of the mathematics education community and its many sub-communities.

It is possible to identify some strands among the organizations discussed in this section. We start by discussing three communities meeting in the context of regional conferences organized under the auspices of ICMI. An interesting phenomenon to be observed is that to a large extent these groups, in spite of the regularity of their scientific meetings, have not felt the need to formalize the general infrastructure behind their activities via the establishment of a bona fide bureau or suchlike. Two additional examples of a similar informal situation, but with bodies centred on a thematic perspective, are also given. We conclude the section by discussing five organizations established in recent decades, some thematically oriented and the others regionally, but which rely on a fully fledged formal setting for their actions.

ICMI Regional Conferences

The ICMI policy of supporting regional conferences was officially launched in the 1970s, although its origin may be traced back to the establishment of CIAEM in the early 1960s. In his 1971–1974 presidential report, James Lighthill (1975) wrote: "ICMI adopted a new policy of holding Regional Symposia to facilitate wider discussion of mathematical education outside those areas of Europe and America where international meetings on the subject have mainly been held hitherto" (p. 330). This outreach initiative can also be seen as a way of addressing a "crucial challenge" for ICMI: the need to foster among the education community the development of a greater "sensitivity to the cultural and contextual dimensions of mathematics education" (Artigue, 2008b, p. 191). A region, however it may be delineated, can often be considered as a reasonable environment for dealing with such issues. Basic guidelines and criteria were adopted by ICMI about granting the status of ICMI Regional Conference to a conference under planning, including having a genuinely international conference, albeit maybe at a regional level, and ensuring quality standards through a broadly representative international scientific committee responsible for the program.

Over the years, meetings of different natures were recognized by ICMI as regional conferences organized under its auspices, sometimes on an ad hoc basis. Four current series of conferences correspond to stable regional networks that emerged within ICMI circles: the CIAEM conferences, EARCOME, EMF, and AFRICME (in chronological order). The first was discussed above, and we now consider the other three.

EARCOME. The East Asia Regional Conference in Mathematics Education (EARCOME) had its immediate roots in two ICMI Regional Conferences hosted by China in 1991 and 1994 which attracted participants mostly from China, Japan, and South Korea. It was then agreed to meet outside China and be more inclusive of other Northeast Asian countries. The first of these new conferences was held in Korea in 1998 under the name ICMI-EARCOME.

But the EARCOME as it now stands also results from combination with an older series, the Southeast Asian Conference on Mathematical Education (SEACME), begun in 1978 and hosted by almost every Southeast Asian country before it was subsumed under the EARCOME series at EARCOME-2 in 2002. Lim-Teo (2008) provided more information on the evolution of these conference series.

As noted by Nebres (2008), the SEACME series was initiated at the recommendation of Yukiyoshi Kawada, ICMI secretary-general from 1975 to 1978. The inaugural 1978 SEACME in Manila was a resounding success in the learning that took place, not only from the conference itself but also through series of activities leading up to the conference and follow-up actions afterwards. In the words of Nebres, this "led to a burst of mathematics education activity in Southeast Asia" (p. 149).

Each SEACME conference was organized by the host nation, which chose its own theme, invited speakers, and encouraged attendance by school mathematics teachers. It could be seen as "a national conference with regional and perhaps some international participation" (Lim-Teo, 2008, p. 248). She further commented: "The host country benefited … through providing their teachers and other participants the opportunity to learn from the regional and international speakers and participants" (p. 248). Lee (1992) stressed the importance of the pre- and post-conference activities connected to the first SEACME conference, "something which a regional conference could accomplish that no international conference could do" (p. 28). He also asserted that those involved "found that [they] could learn a lot from each other perhaps even more so than from the developed countries" (p. 28).

Although the SEACME host nations had roughly common contexts such as post-colonial independence which in turn generated educational issues, the North-Eastern Asian nations shared substantial similarity in their longer histories of civilization/structured governance, their greater homogeneity within each country, and a common Confucian-heritage culture with long educational traditions. In the first ICMI-China conference, in 1991, an oft-heard theme was the need for educational reform with more constructivist approaches to education. This conference was from a political perspective a crucial milestone in the ICMI relations with the North Asian nations, and a breakthrough in terms of cross-cultural understanding and cooperation.

Whereas the ICMI-China Regional Conferences and the SEACME initial conferences were "national" conferences with some international participation, the EARCOME series moved to a format rather similar to "Western" conferences, with a substantial number of foreign and local keynote speakers and paper presentations organized along various strands. This was a natural outcome of globalization and not unwelcome, since it enabled the academic discourse to be enriched with wider perspectives and diversity, to stay current and relevant. Foreign participation also increased. Nevertheless, the local–regional–international balance was maintained, and local participation remained strong, with many paper presentations being from local participants, discussing local issues and research and with parallel workshops for development of local teachers.

The role of EARCOME conferences can profitably be considered from the vantage point of *centres* and *peripheries*, as found in Artigue (2008b) or Nebres (2008). Discussing the development of mathematics education activities in East/Southeast Asia from the late 1970s, Nebres described countries from the region as remaining "in the periphery in the sense that, say in ICMEs, their unique voices are only heard in special sessions on mathematics education in developing countries" (p. 150). The EARCOME series has considerably helped the "periphery voices" in East/Southeast Asia to be heard more clearly and more coherently, both in the region and internationally.

EMF. The ICMI Regional Conferences *Espace mathématique francophone* (EMF, Francophone Mathematical Space), instigated in 2000, have as a peculiarity that they rest on a notion of region defined in linguistic rather than geographical terms, French being the common language among participants. This initiative originates from the *Commission Française pour l'Enseignement des Mathématiques* (CFEM), the French subcommission of ICMI. The EMF conferences are held triennially, and the principle of alternation of sites between developed and developing countries has hitherto been respected (France, Tunisia, Québec, Senegal, and Switzerland).

In a context where a majority of the participants in international mathematics education forums have English, today's lingua franca, as a second–or even third–language, it is definitely appropriate, from an ICMI perspective, to support collaboration and interactions in specific linguistic contexts, when the language at stake is shared in many regions around the world. Being "more and more sensitive to linguistic issues" is presented by Artigue (2008b) as a way for ICMI to increase its outreach, "all the more as the discussion of educational issues within a given language requires much more fluency than mathematical discussion" (p. 194). Such an argument was at the origin of the EMF network, whose success among the French-speaking community suggests that this may correspond globally to a genuine need. As the community gathered, EMF has felt no need to officialize its network via the establishment of an executive or a bureau. The presence of CFEM as a stable body closely supporting the network has greatly facilitated the passing of the baton from one conference to the next, something always an issue in such informal settings. At the EMF 2012 conference, it was decided to establish an "EMF Executive Bureau" composed of eight people representing North and South countries of the Francophonie and in charge of the general functioning of the EMF conferences. This body is responsible in particular for the selection of the site of a given conference as well as the transition from one scientific committee to the next.

AFRICME. Our final example of ICMI Regional Conferences is the Africa Regional Congress of ICMI on Mathematical Education (AFRICME), the first of which was organized in 2005 in South Africa largely at the initiative of Jill Adler, then ICMI vice-president. For ICMI, this project was most timely, as its executive committee had been reflecting for some time on ways to increase ICMI's role and impact in regions of the world where it was hardly present—Africa clearly fitted that description. The next AFRICME congresses were held in Kenya in 2007 and Botswana in 2010.

The main aim of AFRICME is to stimulate interactions among mathematicians, mathematics educators, and teachers across African countries, with a focus on the needs and specificities of the region. Because of the global African situation, a special emphasis is placed on issues pertinent to mathematics education in developing countries, but without neglecting the importance of keeping the reflections congruent with a more global framework, so as not to allow their impact and value to be trivialized or marginalized.

The AFRICME network is centred in Anglophone Africa, and efforts are being made to reinforce its links and collaboration with a subnetwork of the EMF community based in African Francophone countries.

Other Informal Structures

Most of the international organizations discussed in the previous sections had some form of "legal" existence connected to their status as constituted bodies, sometimes even as formally incorporated associations: one such example is MERGA, whose official name, as indicated in its constitution, is "The Mathematics Education Research Group of Australasia Incorporated."

But the organizational and legal frameworks supporting the missions of certain other groups are much more informal. A group may even exist essentially via a set of activities (typically a conference occurring periodically), without the presence of a formalized body in the background. There is usually a clearly identified and lively community supporting the group's activities, but for various reasons the need has never been felt for the creation of an organization with a constitution, a bureau, and so on. As a consequence, many such organizations do not have a centralized Web site, and Internet links move from one congress to the next. Many of these have not developed a formal notion of membership (with accompanying dues)—a group serves a given community via the organization of specific activities.

The three ICMI Regional Conferences we have just described provide good examples of this kind of informality. We next discuss two other groups independent of ICMI that support regular cycles of conferences but do not have formal structures defining their existence.

MES. In 1998, the first meeting of what has become a regular series of mathematics education conferences was set up explicitly as a challenge to the International Group for the Psychology of Mathematics Education (PME). Important aspects of the research field, it was felt, were not represented at PME, because of its insistence on a psychological focus. This meant that papers with a sociological, political, or philosophical focus were rejected by reviewers and therefore did not appear in PME's discussions and proceedings.

The original acronym of the new conferences was MEAS, standing for Mathematics Education And Society, but from the second meeting this was changed to MES. In his plenary talk at the first meeting, Alan Bishop marked MES's debt to the so-called Fifth Day Special Programme on Mathematics, Education and Society at ICME-6 (Keitel, Damerow, Bishop, & Gerdes, 1989).

The group has no formal management structure, no elections, no president or chair, and no standing committee. At each meeting, there is a general gathering at which a future meeting is proposed. A committee is then formed by volunteers with experience in organizing prior meetings and by new members.

Following the inaugural meeting in the UK, further meetings have been held in Portugal (2000), Denmark (2002), Australia (2005), Portugal (2008) and Germany (2010)—see Gellert, Jablonka, and Morgan (2010). There have been some discussions about whether the change in the constitution of PME in 2005, opening beyond psychology the theoretical perspectives that can inform research in mathematics education represented at PME conferences, should lead to the end of MES, but it has now a life and traditions of its own. The 7th MES meeting has been fixed for South Africa in 2013.

Meetings include plenary lectures, which are discussed subsequently by small groups. These groups stay together for those discussions throughout the meetings. Points from those discussions are fed back to the presenters at an open meeting. Other forms of interaction include presentations of reviewed papers and symposia.

Great emphasis is placed on participation from under-represented countries and social groups within countries, and a strong theme of all meetings is a concern for social justice issues. There is a tension between the inclusive ethos of the whole group's raison d'être, on the one hand, and the need for peer review of submitted papers and publication of proceedings respecting quality standards so that participants can secure funding from their institutions, on the other. Participants would probably describe the meetings as always challenging and inspiring, and supportive of research and researchers in the under-researched areas of working for human rights and social justice through mathematics education.

Some past proceedings of MES conferences are available at the Web site of the last meeting (http://www.ewi-psy.fu-berlin.de/en/v/mes6).

Delta. Delta is an informal collaboration network among Southern Hemisphere countries, focussing on the teaching and learning of mathematics and statistics at the undergraduate level. Its action is based on biennial conferences, the main organizer of the next conference acting as the chair of an informal steering committee for the community gathered through these conferences. Although the Delta conferences may attract educators involved in research about undergraduate mathematics and statistics education, their main aim is to address the community of research mathematicians and mathematics lecturers who are possibly not involved themselves in formal research, according to standard educational paradigms, but who are nevertheless committed to improving their own teaching. A central idea of Delta is to provide a forum in which mathematicians feel comfortable in discussing issues related to tertiary mathematics teaching and learning without being intimidated by what some may consider educational jargon or constructs. Many participants at the conferences are thus mathematicians wishing to report about a teaching experience or experiment that would normally not classify as bona fide research in mathematics education, but may still be helpful in inspiring those who want to reflect on their teaching.

Some of the contributions submitted to the Delta conferences are research reports by experts publishable as peer-reviewed material in standard education research journals. Selected papers from recent conferences have, for instance, appeared in special issues of the *International Journal of Mathematical Education in Science and Technology* (iJMEST). But in order to achieve inclusiveness, the conference organizers propose channels for other types of publications. Reports on pedagogical experiments can appear in the conference proceedings, whereas other contributions may take the form of posters. In all cases, mathematicians are provided with an opportunity to discuss their own practice in an environment where the teaching and learning issues are at the core of the discussion. This is the case at the Delta conferences to a much larger extent than at typical congresses of mathematical societies, even when an educational strand appears on the program.

Although Delta welcomes participants from all continents, they are predominantly from the Southern Hemisphere. As a matter of fact, all conferences since the first in 1997 have been held in Southern Hemisphere countries.

In contrast to most bodies discussed in this chapter, the name *Delta* does not constitute a set of initials but simply refers to the famous "delta" of higher mathematics, thus reflecting the idea of continuous change in university mathematics and statistics education which is at the core of the group interests (Oates & Engelbrecht, 2009).

Stable Structures with a Thematic or Regional Scope

In contrast to those just discussed, the five organizations discussed in this section offer activities that take place in environments supported by well-established infrastructures. We start by considering two thematic groups whose scope of action is built around a particular facet of the teaching and learning of mathematics: the concerns of ISGEm lie in the cultural basis of much mathematics learning, whereas those of IASE have relatively recently become more significant in mathematics education. We conclude the section by discussing three regional organizations of a more general nature which arose because of needs felt by specific language groups: NoRME, CLAME, and FISEM.

ISGEm. The term *ethnomathematics* was coined by Ubiratan D'Ambrosio (1985) in the late 1970s in the context of his research concerning

> the mathematics which is practised among identifiable cultural groups, such as national-tribal societies, labor groups, working children, professional classes, and so on, [that is] practices which are typically mathematical ... [but] done in radically different ways than those which are commonly taught in the school system. (pp. 44–45)

He linked the ethnomathematical approach to mathematical practices related to counting, ordering, classifying, measuring, inferring, or modelling that can be seen as existing "outside the school":

> This is a very broad range of human activities which, throughout history, have been expropriated by the scholarly establishment, formalized and codified and incorporated into what we call academic mathematics. But which remain alive in culturally identified groups and constitute routines in their practices. (p. 45)

Research in ethnomathematics quickly became an important strand in the field of mathematical education, as shown by the ICME-5 plenary of talk D'Ambrosio, (1986), or by the Fifth Day Special Programme on Mathematics, Education and Society at ICME-6 (Keitel et al., 1989). This last activity, it has been claimed, happened in a context of "a growing awareness of the importance of ethnomathematical activities as a means to overcome Eurocentrism and cultural oppression in mathematical learning" (p. 1). Reflections by D'Ambrosio on ethnomathematics, its reception by the community, and its place in mathematics education were offered in the ICME-10 plenary interview by Michèle Artigue (2008a).

At the annual NCTM meeting in 1985, a group of mathematical educators, including D'Ambrosio, founded the International Study Group on Ethnomathematics (ISGEm).

In 1990 it became an affiliate of NCTM. Quadrennial International Conferences on Ethnomathematics (ICEm) are organized under the auspices of ISGEm.

The ISGEm Web site (http://isgem.rpi.edu/) presents the main centres of activity and interests of the group in eight broad areas of ethnicity and geography: Africa, Pacific Islands, Asia, Native and African America, European, Latino, and the Middle East. Information is available in English, Spanish, Italian, and Portuguese, and there are 21 diverse subgroups of special interests classified, in addition to ethnicity/geography, by utility (including ethnomathematics in the classroom) and by social categories (including social studies of professional mathematics, multicultural mathematics, ethnomathematics in peace and social justice, and indigenous knowledge systems).

ISGEm has three chapters: in North America (NASGEm), Brazil (BR.ISGEm), and Southern Africa (SAEmSG). In addition to a newsletter, it has published, since 2006, via its North American Chapter, the *Journal of Mathematics and Culture*, a peer-reviewed journal examining "the intersections between mathematics and culture in both western and non-western societies" and with particular interest in "pedagogical applications of ethnomathematics" (from the NASGEm Web site: http://nasgem.rpi.edu/).

IASE. The International Association for Statistical Education (IASE) is an organization whose formal existence takes place in a context analogous to that linking ICMI to its mother organization IMU, as IASE is the Education Section of the International Statistical Institute (ISI).

Its precursor was the ISI Education Committee, launched in 1948. Vere-Jones (1995) observed that "although statistical education had been a concern of the ISI since its inception in 1885, it was the setting up of the Education Committee which marked the beginning of a systematic education programme" (p. 4) among ISI activities. Actions of the ISI Education Committee include the Round Table Conferences, launched in 1968 and now held as satellite meetings associated with the ICME congresses, and the International Conferences on Teaching Statistics (ICOTS), held quadrennially since 1982 and arguably the major international event in statistical education. Of particular significance, added Vere-Jones, was the resulting change in the focus of the ISI education program, "from the relatively narrow one of training statistical staff for developing countries to the broadest consideration of statistical education" (p. 10). This shift stimulated a greater emphasis on research in education and eventually the birth of IASE, in 1991, as a new section of ISI in charge of educational matters.

IASE has approximately 500 individual members, mainly lecturers and professors of statistics, applied and government statisticians, education researchers, and some teachers. It aims at supporting the development of effective and efficient educational services through international contacts among individuals and organizations. Phillips (2002) presented IASE as the main international organization devoted to the improvement of statistical education at all levels. Besides being the educational arm of ISI, it also provides a forum for the furtherance of research in statistical education in its own right.

Ottaviani and Batanero (1999) discussed the role of IASE in the promotion and development of statistical education research, notably through ICOTS and the round table conferences, which play a central role by providing regular forums where research problems, methodologies, and results are presented and discussed. Of particular interest was the round table organized in 2008 as a satellite conference to the ICME-11 congress. This was a conference for the study of "Statistics Education in School Mathematics: Challenges for Teaching and Teacher Education," organized jointly by IASE and ICMI (the 18th ICMI Study). This collaborative project, a success from the point of view of both organizations, has resulted in the study volume edited by Carmen Batanero, Gail Burrill and Chris Reading (2011).

The *Statistics Education Research Journal* (SERJ), launched in 2002, is a freely accessible peer-reviewed electronic journal published twice a year, and is a joint publication of IASE and ISI. The aims of SERJ are "to advance research-based knowledge that can help to improve the teaching, learning, and understanding of statistics or probability at all educational levels and in both formal (classroom-based) and informal (out-of-classroom) contexts" (from SERJ Web site http://www.stat.auckland.ac.nz/~iase/publications.php?show=serj).

The IASE Web site (http://www.stat.auckland.ac.nz/~iase/) provides links to resources around the world in statistics education. These are grouped by categories, including research, learning, assessment, curriculum guidelines, journals, software, and organizations.

In their paper on the role of IASE, Ottaviani and Batanero (1999) summarized recent and current trends as heralding a substantial growth in statistical education, interpreted in the broadest sense. IASE, they claimed, has a central role through its program of activities and the research fostered among its community. This role concerns not only statistical education in the school environment, but also the understanding of the fundamental concepts of statistics in society at large, as well as in other discipline areas or among professional bodies.

NoRME. The Nordic Society for Research in Mathematics Education (NoRME), established in 2008, rests on a tradition of regional cooperation in mathematics education going back to the 1960s, when a Nordic committee was created to collaborate in developing curricula in "modern" mathematics. This led to a series of conferences for mathematics teachers circulating among the Nordic countries. A more recent initiative organized between 1988 and 1993, the Danish project Mathematics Education and Democracy, "paved the way for the continued Nordic collaboration in the 1990s and the 21st century" (Niss, personal communication, June 2011). Nissen and Blomhøj (1993), a widely known publication stemming from this project, was connected to a symposium described as "one of the [important] starting points for the Nordic collaboration in the field" (Blomhøj, Valero, & Häggström, 2009, p. 2). Nordic colleagues also attended meetings of international organizations like PME and CIEAEM, and among the outcomes of all these activities was the launching of both the journal *Nordic Studies in Mathematics Education* (Nomad or *Nordisk Matematikdidaktik*) in 1993, and a series of Nordic research conferences in mathematics education (NORMA) in 1994.

NoRME was founded at a meeting held in 2008, during the fifth NORMA conference. Grevholm (2009a) presented the background to that event, stressing the need felt, after ICME-10, to create an umbrella organization linking the various organizations supporting mathematics education research in the region. Members of NoRME are thus national or regional societies, and the Nomad association. Among arguments supporting the creation of NoRME, Grevholm emphasized the importance of offering a "home" for the journal Nomad, of supporting the NORMA conferences, of continuing the collaborative activities for doctoral studies carried out by the Nordic Graduate School in Mathematics Education (NoGSME), and in general the need to strengthen and widen regional cooperation. Launched in 2004 with a time-limited funding from the Nordic Research Board, NoGSME formally existed for a mere period of 6 years but during that period was highly influential–both Grevholm (personal communication, June 2011) and Niss (personal communication, June 2011) claim that NoRME can be seen as an offspring of NoGSME.

Besides the more than 50 doctoral dissertations successfully defended in its 40 institutions from Nordic and Baltic countries, other activities of NoGSME have included workshops, seminars, and summer schools, as discussed by Grevholm (2009b) in her final report on the graduate school. A responsibility inherited by NoRME is thus "to administer the legacy ... handed down from NoGSME" (Grevholm & Rønning, 2010, p. 97), that is, to ensure the survival of the NoGSME network and the kind of activities that built up during those years of collaboration (Grevholm, 2009b).

As stated in its constitution, the raison d'être of NoRME is to support and raise the quality of Nordic and Baltic research in mathematics education, especially through regional collaboration. A recent Nomad editorial emphasized the need for the journal to maintain and develop its regional identity, as "the authors or the contents of the published papers have connections to, or specific relevance for, the mathematics education milieus in the region" (Blomhøj, Rønning, & Häggström, 2010, p. 1). Although they welcomed the increasing number of submissions in English, the editors argued for "the publishing of research papers in the Scandinavian languages" as a way for the journal to be more closely connected to the regional mathematics teacher education community. This expectation can be seen as an interesting illustration of the possible difficulties, at least linguistically speaking, in balancing regional action and international scope. MERGA publications, as was discussed above, also need to find a proper regional/international balance. Still, they do not face the additional issue of English being today's lingua franca.

CLAME. Besides CIAEM, other organizations are concerned with mathematics education in Latin America. One is the *Comité Latinoamericano de Matemática Educativa* (CLAME, or Latin American Committee on Mathematics Education), established to support the attendance of a series of annual conferences on teacher education and mathematics education research in Central America and the Caribbean. Launched in 1987, these meetings provided a forum for colleagues who, although geographically close, did not have regular opportunities for exchange of views using Spanish as a language of communication. CLAME was formally created at the 1996 conference. Since 1997, the conferences, typically attended by some 1,500 participants, have become the *Reunión Latinoamericana de Matemática Educativa* (RELME). CLAME is thus both an outcome and a catalyst of an increasing professionalization of the Spanish-speaking mathematics education community in Latin America. It has individual members and collaborates, through its Web site (http://www.clame.org.mx/), with national mathematics education societies and graduate schools in the region.

CLAME supports two important publications: the *Acta Latinoamericana de Matemática Educativa* (ALME), an annual collection of reviewed papers resulting from the RELME conferences, and *Revista Latinoamericana de Investigación en Matemática Educativa* (Relime), a research journal having three issues each year. Launched in 1997, Relime aims at fostering the publication of quality research contributing to the development of a Latin American school of mathematics education of an international level principally connected with the regional culture and educational systems. Papers appear mainly in Spanish, but they also appear in Portuguese, French, and English.

FISEM. Another recently established body, with a larger geographical scope, is the *Federación Iberoamericana de Sociedades de Educación Matemática* (FISEM), aiming at the coordination of efforts among Spanish- and Portuguese-speaking countries. From the mid-1990s, leaders of various Ibero-American mathematics education bodies discussed the federation of their societies. In 2003 they created FISEM, which now covers national organizations from 13 different countries. Among the FISEM projects is the electronic journal *UNIÓN*, launched in 2005 and aimed at teachers from all levels in the Ibero-American landscape. Another project fostering "cross-fertilization between Latin American, Spanish and Portuguese mathematics educators" (Jacobsen, 1996, p. 1242), is the *Congreso Iberoamericano de Educación Matemática* (CIBEM).

The focus of CIBEM conferences is to contribute to the development of mutual knowledge about the teaching and learning of mathematics within the cultural framework of Ibero-America (including Spain and Portugal), as well as those countries where Spanish and Portuguese are spoken (Sánchez Vázquez & García Blanco, 1991). The first CIBEM congress took place in 1990 at the instigation of the Spanish community, and in particular of the then recently established FESPM (*Federación Española de Sociedades de Profesores de Matemáticas*), and in collaboration with CIAEM. The inception of the CIBEM series also received support from UNESCO, which published the proceedings of the first conference. The CIBEM conferences are now taking place quadrennially under the auspices of FISEM, alternating with the CIAEM conferences.

Coda: A Glimpse at Other Facets of Diversity Among Organizations

A survey like the present one is bound to lead to difficult decisions about the organizations to be included. We next mention a few bodies clearly of interest to our topic but not given fuller consideration in this chapter because of space limitations.

These offer additional illustrations of the richness and diversity of contexts and structures among institutions concerned with international aspects of mathematical education.

This chapter being devoted to international organizations, the question arises how this internationalism is to be enacted. Could, for instance, a body whose existence is primarily of a national nature be included? One approach is to assess the extent to which its activities contribute substantially beyond the country's boundary. To take a concrete case, what about the National Council of Teachers of Mathematics (NCTM), possibly the world's largest organization in mathematics education, with more than 90,000 members and 230 Affiliates throughout the United States and Canada (http://www.nctm.org/about/content.aspx?id=174)?

Although NCTM is primarily concerned with the U.S. scene, the positions that it regularly takes on mathematics education issues, for instance via documents such as its *Principles and Standards for School Mathematics* (NCTM, 2000), undoubtedly have an impact of an international nature. But more to the point for our discussion, NCTM launched in 1970 the *Journal for Research in Mathematics Education* (JRME), considered today among the most influential journals in the field internationally. From that perspective, NCTM definitely belongs to the scope of this survey.

NCTM is not the only national organization publishing a research journal of international stature. Other examples of such tandems, each with its own specificity, are the following:

- The British Society for Research into Learning Mathematics (BSRLM), in the United Kingdom, and its journal *Research in Mathematics Education* (RME);
- The Canadian Mathematics Education Study Group (CMESG) and *For the Learning of Mathematics* (FLM); and
- The *Association pour la Recherche en Didactique des Mathématiques* (ARDM), in France, which publishes *Recherches en Didactique des Mathématiques* (RDM).

Another possible question about international organizations is the kind of infrastructure on which they rely. We have seen in this survey a great variety among models for the existence, more or less formal, of the various institutions. In some cases, a certain community would even gather on a regular basis without having a well-defined body supporting it. Membership in some organizations was aimed at individuals and, in other cases, at associations or even countries. Another variation of the same vein is the case of a selective group with membership by invitation only. Such is the situation of the BACOMET group (for *BA*sic *CO*mponents of *M*athematics *E*ducation for *T*eachers), launched at the end of the 1970s and which ceased its activities about 10 years ago. According to Silver and Kilpatrick (1994), its working style and organizational scheme are of interest. Its membership, international and representing a high degree of scholarship, was always small (about 15–20 people at a given time) and varied over the years, being based on an invitation to collaborate on a specific project. Those involved in a given project were also generally responsible for setting up a team for the next one. BACOMET was active for a little more than two decades, during which time it carried out five projects and produced four books, published between 1986 and 2005. Further information on BACOMET is given by Biehler (2005).

Another community with a substantial level of activity in the 1980s and early 1990s but which, like BACOMET, has now ceased to function, organizationally speaking, is the International Study Group on Theory of Mathematics Education (TME). As opposed to other bodies discussed in this paper, TME never developed into a full-fledged organization, and the notion of an actual membership did not become an issue. TME remained a kind of ad hoc group connected to a vision and a program of its founder, Hans-Georg Steiner, and corresponding to the community of those attending the conferences organized by Steiner around the concept of "theories of mathematics education." The influence of the TME group is visible through a landmark book devoted to the development of "didactics of mathematics as a scientific discipline" (Biehler, Scholz, Strässer, & Winkelmann, 1994), whose genesis is directly connected to the activities of TME.

Our final example concerns the International Council on Mathematics in Developing Countries (ICOMIDC), initiated during the 1983 International Congress of Mathematicians with the objective of fostering both mathematics teaching and research in mathematics in less affluent countries. Its brief existence was discussed by Lehto (1998), who linked the beginnings of ICOMIDC to the lack of satisfaction at that time in some circles with outreach activities towards less affluent countries under the auspices of IMU. Lehto also pointed to some of the difficulties that led to the disappearance of ICOMIDC, some being financial but the most serious being those of a political nature.

The interested reader is referred to Hodgson and Rogers (2011, 2012) for further details about the organizations mentioned in this section.

Conclusion

Writing a survey demands selectivity, if only because of space limitations. We are aware that some organizations playing a substantial role internationally in mathematics education are absent from this chapter. In some cases, this omission results from choices we made about the kind of bodies to be discussed. For instance, we decided not to review per se a specialized governmental institution like UNESCO, in spite of the major role that it has played, and continues to play, in educational matters, including mathematical education. (UNESCO was briefly mentioned above, notably in connection with ICMI, and is treated extensively by Jacobsen (1996). See UNESCO (2011) for a recent UNESCO publication, written by Michèle Artigue, on mathematics in basic education.) The same remark applies for major international education assessment organizations such as OECD's PISA, or the IEA's TIMSS, in spite of their strong influence on the school curriculum and on political decisions within individual countries. We also chose not to consider institutions directly resulting from policy decisions of governments, such as the IREM network in France, despite their possible impact on mathematics education internationally (however, IREMs have been mentioned above in connection with HPM). Our choice was to concentrate on autonomous, academic organizations supporting research in mathematics education and whose actions are close to the community of mathematicians, mathematics educators (including researchers), those involved in the preparation of schoolteachers, and teachers actually involved in the teaching and learning of mathematics.

It may happen that some newcomers to research in mathematics education develop the impression that it all began sometime in the 1970s with organizations like PME. By looking back at the "prehistory" of research in mathematics teaching and learning, which witnessed the inception of bodies such as ICMI, CIEAEM, and CIAEM, to name three famous forerunners, one does grasp a better understanding of the background that led to the emergence of didactics of mathematics as a bona fide research domain and as an international academic discipline. It does not belong to this paper to review the evolution of mathematics education research per se, and the interested reader should consult the papers by Niss (2004), by Kilpatrick (1992, 1999, 2008)—plus the reaction of Dorier (2008) to the last—and the ICMI Study 8 volume (Sierpinska & Kilpatrick, 1998).

A large proportion of international organizations in mathematics education are of a rather recent vintage (see Appendix B). Particularly striking on that account is the decade starting in 1976, the year of ICME-3. The proliferation then happening can be seen as a reflection of, and also a stimulus to, both the growth of research in mathematics education and the emergence of a new community supporting that research. It is also a testimony to an increasing diversity of interests within this community, to a growing sensitivity to cultural and contextual characteristics, and to the development of new, or renewed, perspectives and approaches on issues at stake in various contexts for the teaching and learning of mathematics. A typical example is the creation of the first three ICMI Affiliated Study Groups—PME, HPM, and IOWME—each corresponding to a particular strand in the mathematics education landscape. The presence of such subcommunities wanting to become institutionalized within the mathematical education world can be interpreted as a sign of the vitality of the field and the diversity of its global community.

Clearly, what strikes anyone looking at the organizations that constitute the mathematics education research community today is their sheer proliferation. A "one size fits all" approach to the problems of mathematics teaching and learning is obviously not viable. The whole educational enterprise has become more complex, more specialized. This has been accompanied by the development of various subcommunities, often quite naturally, in support of those involved in that endeavour. The institutionalization of those communities in an international setting has materialized through a variety of models, as seen above. The resulting web of organizations is still evolving and expanding so as better to reflect the changing contexts for mathematics education.

One can map some of these developments, as we have tried to do in this chapter, as owing to representations of local activity; some to the growth of theoretical perspectives, sometimes competing; others to the establishment of special interest groups, or other patterns that might be identified. We can suggest, for example, that there was an early development of thematic interest groups in the period 1975–1990

(HPM, IOWME, PME, ICTMA, WFNMC, and ISGEm). Perhaps under the influence of localizing theoretical perspectives, whether ethnomathematical, liberationist, inspired by critical theory, or of the death of universalist goals within postmodernism, there has been a growth in organizations with regional emphases (CIAEM, MERGA, EARCOME, and EMF, to name a few). Such regional structures may be related to special sensitivities about culture and identity, whereas others, like AFRICME, are explicitly addressing specific needs for solidarity and outreach. Interestingly, a majority of the organizations established over the last 15 years or so are of a regional nature, including ERME and NoRME.

Individuals are likely to belong to several groups, since many of these organizations have overlapping values and aims. One might say they are competing for the same ecological niche in a context where the resources (both human and material) are limited. For instance, one could think that PME, created in the mid-1970s, may have attracted some of those previously participating in the CIEAEM community. Others may have preferred to remain with CIEAEM, possibly because of the specific role played by schoolteachers in that community. Similar questions could be raised about CIAEM and CLAME, for example. Or one may see a competition about actual participation in conferences, between, for example, general "all-encompassing" events like the ICME congresses, and more specialized activities like the satellite conferences organized at each ICME by groups such as PME and HPM.

Is this, possibly astonishing, proliferation of organizations a good thing or a bad thing? This may seem a strange question to ask, given that there is no "United Nations" of mathematics education to engage with such a phenomenon. However, questions are certainly being asked by some academics about the potential danger of the proliferation of theoretical perspectives in our field (Sriraman & English, 2010). These questions are raised both in terms of possible incommensurability between perspectives, making communication difficult, and in the sense that with competing perspectives the community will never be able to build the kind of body of scientific knowledge that one sees developed and developing in medicine, for example, to take another field that is engaged in both practice and theory. But, then, education is sometimes characterized as being linked to the social sciences rather than the traditional sciences, and thus this situation may be inevitable. We can expect the proliferation to continue. It will be of great interest to historians of the field in years to come to see what patterns and directions are discernible in those developments.

Acknowledgements The authors wish to express their gratitude to all the colleagues who brought their support for collecting information on the organizations discussed in this survey, and in particular to: Michèle Artigue, Ferdinando Arzarello, Nicolas Balacheff, Bill Barton, Carmen Batanero, Christer Bergsten, Annie Bessot, Rolf Biehler, Alan Bishop, Morten Blomhøj, Jean-Luc Dorier, Viviane Durand-Guerrier, Johann Engelbrecht, Fulvia Furinghetti, Livia Giacardi, Merrilyn Goos, Barbro Grevholm, Lucia Grugnetti, Corinne Hahn, Geoffrey Howson, Barbara Jaworski, Jean-Pierre Kahane, Gabriele Kaiser, Jeremy Kilpatrick, Lee Peng Yee, Lin Fou-Lai, Hartwig Meissner, Cynthia Nicol, Mogens Niss, Marie-Jeanne Perrin, Hilary Povey, André Rouchier, Ángel Ruiz, Marie-Hélène Salin, Gert Schubring, Elaine Simmt, Walter Spunde, Bryan Wilson, Carl

Winsløw. Of course the authors take full responsibility for any remaining inaccuracies or oversights.

Part of this work was done while the first author was a visiting researcher at the *Centre de Recherches Mathématiques* (CRM), in Montréal.

Appendix A: Glossary of Acronyms

The following list gives the acronym (or set of initials) for the organisations, conferences or journals mentioned in this paper. In each case, the year of establishment is indicated, as well as the Web site address. For periodic conferences, the Web site of the last or the next one may be indicated, when no common site was found. General information on the ICMI Regional Conferences and the ICME congresses is accessible via the ICMI Web site. (These URLs were valid at the time of publication.)

AFRICME — Africa Regional Congress of ICMI on Mathematical Education/2005. Recent conference: AFRICME-3, Gaborone, Botswana, 2010. http://www.mat.uc.pt/~jaimecs/icmi/AFRCME3_2ndCall.doc

ALME — *Acta Latinoamericana de Matemática Educativa* [Latin American Acts of Mathematics Education]—published yearly by CLAME in connection with the RELME conferences. http://www.clame.org.mx/alme.htm

AMS — American Mathematical Society/1888. http://www.ams.org/

ARDM — *Association pour la Recherche en Didactique des Mathématiques* [Association for Research on the Didactics of Mathematics]/1992. http://www.ardm.eu

BACOMET — Basic Components of Mathematics Education for Teachers/1979. http://lama.uni-paderborn.de/personen/rolf-biehler/projekte/bacomet.html

BR.ISGEm — *Seção Brasileira do International Study Group on Ethnomathematics* [Brazilian Chapter of ISGEm]

BSRLM — British Society for Research into Learning Mathematics/1978 as BSPLM (British Society for the Psychology of Learning Mathematics), renamed BSRLM in 1985. http://www.bsrlm.org.uk/

CERME — Congress of the European Society for Research in Mathematics Education—see ERME/1998. Recent conference: CERME 7, Rzeszów, Poland, 2011. http://www.cerme7.univ.rzeszow.pl/

CFEM — *Commission Française pour l'Enseignement des Mathématiques* [French Subcommission of ICMI]/1975. http://www.cfem.asso.fr/

CIAEM — *Comité Interamericano de Educación Matemática* [Inter-American Committee on Mathematics Education]/1961. http://www.ciaem-iacme.org

CIAEM — *Conferencia Interamericana de Educación Matemática* [Interamerican Conference on Mathematics Education]—see the CIAEM Committee/1961. Recent conference: XIII CIAEM, Recife, Brasil, 2011. http://www.cimm.ucr.ac.cr/ocs/index.php/xiii_ciaem/

CIBEM	*Congreso Iberoamericano de Educación Matemática* [IberoAmerican Congress on Mathematics Education]/1990. Next conference: VII CIBEM, Montevideo, Uruguay, 2013. *http://www.cibem7.semur.edu.uy/*
CIEAEM	*Commission internationale pour l'étude et l'amélioration de l'enseignement des mathématiques* [International Commission for the Study and Improvement of Mathematics Teaching]/1950. *http://www.cieaem.org/*
CIEM	*Commission internationale de l'enseignement mathématique*—see ICMI
CLAME	*Comité Latinoamericano de Matemática Educativa* [Latin American Committee on Mathematics Education]/1996. *http://www.clame.org.mx/*
CMEG	International Conference on Creativity in Mathematics Education and the Education of Gifted Students—see MCG/1999. Recent conference: CMEG-7, Busan, Korea, 2012. *http://www.mcg7.org/*
CMESG	Canadian Mathematics Education Study Group/1977. *http://cmesg.math.ca*
Delta	Southern Hemisphere Conference on Undergraduate Mathematics and Statistics Teaching and Learning/1997. Recent conference: 8th Delta Conference, Rotorua, New Zealand, 2011. *http://www.delta2011.co.nz/*
EARCOME	East Asia Regional Conference in Mathematics Education/1998. Recent conference: EARCOME 5, Tokyo, Japan, 2010. *http://www.sme.or.jp/earcome/*
EM	See L'EM
EMF	*Espace mathématique francophone* [French Mathematical Space]/2000. Recent conference: EMF 2012, Geneva, Switzerland, 2012. *http://www.emf2012.unige.ch/*
ERME	European Society for Research in Mathematics Education/1998. *http://www.mathematik.uni-dortmund.de/~erme/*
ESM	*Educational Studies in Mathematics* – journal launched under the auspices of ICMI/1968. *http://www.springer.com/journal/10649*
ESU	European Summer University on the History and Epistemology in Mathematics Education/1993. Recent conference: Sixth ESU, Vienna, Austria, 2010. *http://educmath.ens-lyon.fr/Educmath/dossier-manifestations/archives/esu-6/*
FESPM	*Federación Española de Sociedades de Profesores de Matemáticas* [Spanish Federation of Associations of Mathematics Teachers]/1988. *http://www.fespm.es/*
FISEM	*Federación Iberoamericana de Sociedades de Educación Matemática* [Iberoamerican Federation of Societies of Mathematics Education]/2003. *http://www.fisem.org/*
FLM	*For the Learning of Mathematics*—journal published by CMESG/1980. *http://flm.math.ca*

HPM	International Study Group on the Relations between the History and Pedagogy of Mathematics/1976 (Affiliated to ICMI in 1976). http://www.clab.edc.uoc.gr/HPM/. HPM-Americas/1984. http://www.hpm-americas.org/
IACME	Inter-American Committee on Mathematics Education—see the CIAEM Committee
IASE	International Association for Statistical Education/1991. http://www.stat.auckland.ac.nz/~iase/
ICEm	International Conference on Ethnomathematics—see ISGEm/1998. Recent conference: ICEm-4, Towson, MD, USA, 2010. http://pages.towson.edu/shirley/ICEM-4.htm
ICIAM	International Council for Industrial and Applied Mathematics/1987. www.iciam.org
ICM	International Congress of Mathematicians—see IMU/1897. Next conference: ICM 2014, Seoul, Korea, 2014. http://www.icm2014.org/
ICME	International Congress on Mathematical Education—see ICMI/1969. Recent conference: ICME-12, Seoul, Korea, 2012. http://www.icme12.org/
ICMI	International Commission on Mathematical Instruction/1908. http://www.mathunion.org/icmi
ICOMIDC	International Council on Mathematics in Developing Countries/1983. (Originally International Committee on Mathematics in Developing Countries)
ICOTS	International Conference on Teaching Statistics—see IASE/1982. http://icots.net/
ICSIMT	International Commission for the Study and Improvement of Mathematics Teaching—see CIEAEM
ICSU	International Council for Science (Formerly International Council of Scientific Unions)/1931. http://www.icsu.org/
ICTM	International Commission on the Teaching of Mathematics—see ICMI
ICTMA	International Community of Teachers of Mathematical Modelling and Applications—International Study Group for Mathematical Modelling and Applications/1983 (Affiliated to ICMI in 2003). http://www.ictma.net/
IEA	International Association for the Evaluation of Educational Achievement/1967. http://www.iea.nl/
ICTMA	International Conference on the Teaching of Mathematical Modelling and Applications—see ICTMA community/1983. http://www.ictma.net/conferences.html
IGPME	See PME
iJMEST	*International Journal of Mathematical Education in Science and Technology*/1970. http://www.tandf.co.uk/journals/tmes
IMU	International Mathematical Union/1920, 1951. http://www.mathunion.org/

IMUK	*Internationale mathematische Unterrichtskommission* – see ICMI
IOWME	International Organisation of Women and Mathematics Education/1976 (Affiliated to ICMI in 1987). *http://extra.shu.ac.uk/iowme/*
IREM	*Instituts de Recherche sur l'Enseignement des Mathématiques* [Research Institutes on Mathematical Education]/1969. *http://www.univ-irem.fr/*
ISGEm	International Study Group on Ethnomathematics/1985. *http://isgem.rpi.edu/*
ISGHPM	See HPM
ISI	International Statistical Institute/1885. *http://www.isi-web.org/*
JRME	*Journal for Research in Mathematics Education*—published by NCTM/1970. *http://www.nctm.org/publications/jrme.aspx*
L'EM	*L'Enseignement Mathématique* [Mathematics Teaching]—official organ of ICMI since the inception of the Commission/1899. *http://www.unige.ch/math/EnsMath/*
MAA	Mathematical Association of America/1915. *http://www.maa.org/*
MCG	International Group for Mathematical Creativity and Giftedness/2010 (Affiliated to ICMI in 2011). *http://www.igmcg.org/*
MEAS	Mathematics Education and Society—see MES
MELA	Mathematics Education Lecturers' Association (Australia)/1973. (Amalgamated with MERGA in 1997)
MERGA	Mathematics Education Research Group of Australasia/1977. *http://www.merga.net.au*
MERJ	*Mathematics Education Research Journal*—published by MERGA/1989. *http://www.merga.net.au/publications/merj.php*
MES	Mathematics Education and Society/1998. *http://mes.crie.fc.ul.pt/*. *http://www.ewi-psy.fu-berlin.de/en/v/mes6*
MTED	*Mathematics Teacher Education and Development*—journal published by MERGA/1999. *http://www.merga.net.au/node/42*
NASGEm	North American Study Group on Ethnomathematics—Chapter of ISGEm. *http://nasgem.rpi.edu/*
NCTM	National Council of Teachers of Mathematics/1920. *http://www.nctm.org/*
NoGSME	Nordic Graduate School in Mathematics Education/2004. *http://www.nogsme.no/*
Nomad	*Nordic Studies in Mathematics Education* (Nordisk Matematikdidaktik)— journal published by NoRME /1993. *http://ncm.gu.se/nomad*
NORMA	Nordic Research Conferences in Mathematics Education—see NORME/1994. Recent conference: NORMA 11, Reykjavik, Iceland, 2011. *http://vefsetur.hi.is/norma11/*
NoRME	Nordic Society for Research in Mathematics Education/2008. *http://www.norme.me/*
OAS	Organization of American States/1889. *http://www.oas.org/*

OECD	Organisation for Economic Co-operation and Development/1961. http://www.oecd.org/
PISA	Programme for International Student Assessment—see OECD/2000. http://www.pisa.oecd.org/
PME	International Group for the Psychology of Mathematics Education/1976 (Affiliated to ICMI in 1976). http://igpme.org/
PME x	PME Annual Conference—see PME/1977. http://igpme.gandi-site.net/#/past-conferences/3807862
PME-NA	North American Chapter of the International Group for the Psychology of Mathematics Education—see PME/1979. http://www.pmena.org/
RDM	*Recherches en Didactique des Mathématiques* [Research in Mathematics Education]—journal published by ARDM/1980. http://rdm.penseesauvage.com/
Relime	*Revista Latinoamericana de Investigación en Matemática Educativa* [Latin American Journal of Research in Mathematics Education]—published by CLAME/1997. http://www.clame.org.mx/relime.htm
RELME	*Reunión Latinoamericana de Matemática Educativa* [Latin American Meeting on Mathematics Education]—see CLAME/1987. http://www.clame.org.mx/relme.htm
RME	*Research in Mathematics Education*—journal published by BSRLM/1999. http://www.tandf.co.uk/journals/rrme/
SAEmSG	Southern African Ethnomathematics Study Group—Chapter of ISGEm. http://www.rpi.edu/~eglash/isgem.dir/texts.dir/SAEmSG.htm
SEACME	Southeast Asian Conference on Mathematical Education—see EARCOME/1978
SERJ	*Statistics Education Research Journal* – published by IASE/2002. http://www.stat.auckland.ac.nz/~iase/publications.php?show=serj
TIMSS	Trends in International Mathematics and Science Study/1995. http://www.iea.nl/timss2011.html
TME	International Study Group on Theory of Mathematics Education/1984
UNESCO	United Nations Educational, Scientific and Cultural Organization/1946. http://www.unesco.org/
WFNMC	World Federation of National Mathematics Competitions/1984 (Affiliated to ICMI in 1994). http://www.wfnmc.org/
YERME	ERME community of Young European Researchers in Mathematics Education—see ERME/2001. http://www.mathematik.uni-dortmund.de/~erme/index.php?slab=yerme
YESS	YERME Summer School—see YERME/2002. YESS-6, Faro, Portugal, 2012. http://www.ese.ualg.pt/yess6/

Appendix B: List of Organisations

The following table gives the mathematics education organisations included in this survey, listed according to the year of their inception.

Year	Organisation
1908	ICMI
1950	CIEAEM
1961	CIAEM
1976	HPM
	IOWME
	PME
1977	MERGA
1978	BSRLM
1979	BACOMET
1983	ICOMIDC
1984	WFNMC
1983	ICTMA
1985	ISGEm
1991	IASE
1992	ARDM
1996	CLAME
1997	Delta
1998	EARCOME
	ERME
	MES
2000	EMF
2001	YERME
2003	FISEM
2005	AFRICME
2008	NoRME
2010	MCG

References

Artigue, M. (2008a). Plenary interview session (interviewees: U. D'Ambrosio, G. Hanna, J. Kilpatrick and G. Vergnaud). In M. Niss (Ed.), *Proceedings of the Tenth International Congress on Mathematical Education* (pp. 105–122). Roskilde, Denmark: IMFUFA.

Artigue, M. (2008b). ICMI: A century at the interface between mathematics and mathematics education. In M. Menghini, F. Furinghetti, L. Giacardi, & F. Arzarello (Eds.), *The first century of the International Commission on Mathematical Instruction (1908–2008). Reflecting and shaping the world of mathematics education* (pp. 185–198). Rome, Italy: Istituto della Enciclopedia Italiana.

Arzarello, F. (2009). Recent and future activities of the European Society for Research in Mathematics Education (ERME). *EMS Newsletter, 72*, 43–44.

Arzarello, F., Giacardi, L., Furinghetti, F., & Menghini, M. (2008). *Celebrating the first century of ICMI (1908–2008): Some aspects of the history of ICMI* (Regular lecture presented at ICME-11). Unpublished manuscript.

Arzarello, F., & Tirosh, D. (2009). ERME activities for young researchers. *EMS Newsletter, 73*, 43–44.

Atweh, B., Calabrese Barton, A., Borba, M. C., Gough, N., Keitel, C., Vistro-Yu, C., & Vithal, R. (Eds.). (2008). *Internationalisation and globalisation in mathematics and science education.* Dordrecht, The Netherlands: Springer.
Atweh, B., Clarkson, P., & Nebres, B. (2003). Mathematics education in international and global contexts. In A. J. Bishop, M. A. Clements, C. Keitel, J. Kilpatrick, & F. K. S. Leung (Eds.), *Second international handbook of mathematics education* (pp. 185–229). Dordrecht, The Netherlands: Kluwer Academic Publishers.
Barbeau, E. J., & Taylor, P. J. (2008). ICMI Study: Mathematical challenges. *Mathematics Competitions, 21*(2), 81–86.
Barbeau, E. J., & Taylor, P. J. (Eds.). (2009). *Challenging mathematics in and beyond the classroom: The 16th ICMI Study* (New ICMI Study Series, Vol. 12). New York, NY: Springer.
Barrantes, H., & Ruiz, A. (1998). *The history of the Inter-American Committee on Mathematics Education* (Bilingual Spanish and English edition). Bogotá, Colombia: Academia Colombiana de Ciencias Exactas, Físicas y Naturales.
Bass, H. (2008). Moments in the life of ICMI. In M. Menghini, F. Furinghetti, L. Giacardi, & F. Arzarello (Eds.), *The first century of the International Commission on Mathematical Instruction (1908–2008). Reflecting and shaping the world of mathematics education* (pp. 9–24). Rome, Italy: Istituto della Enciclopedia Italiana.
Bass, H., & Hodgson, B. R. (2004). The International Commission on Mathematical Instruction– What? Why? For whom? *Notices of the American Mathematical Society, 51*(6), 639–644.
Batanero, C., Burrill, G., & Reading, C. (Eds.). (2011). *Teaching statistics in school mathematics—Challenges for teaching and teacher education. A joint ICMI/IASE Study: The 18th ICMI Study* (New ICMI Study Series, Vol. 14). New York, NY: Springer.
Bernet, T., & Jaquet, F. (1998). *La CIEAEM au travers de ses 50 premières rencontres: Matériaux pour l'histoire de la Commission* [CIEAEM through its first 50 meetings: Materials for the history of the Commission]. Neuchâtel, Switzerland: Commission Internationale pour l'Étude et l'Amélioration de l'Enseignement des Mathématiques.
Biehler, R. (2005). *BACOMET.* Retrieved from http://lama.uni-paderborn.de/personen/rolf-biehler/projekte/bacomet.html.
Biehler, R., Scholz, R. W., Strässer, R., & Winkelmann, B. (Eds.). (1994). *Didactics of mathematics as a scientific discipline* (Mathematics Education Library, Vol. 13). Dordrecht, The Netherlands: Kluwer.
Blomhøj, M., Rønning, F., & Häggström, J. (2010). Ledare/Editorial: Nomad—A regional journal in mathematics education research. *Nordic Studies in Mathematics Education, 15*(2), 1–3.
Blomhøj, M., Valero, P., & Häggström, J. (2009). Ledare/Editorial: Quality criteria in mathematics education research. *Nordic Studies in Mathematics Education, 14*(1), 1–5.
Blum, W., Galbraith, P. L., Henn, H.-W., & Niss, M. (Eds.). (2007). *Modelling and applications in mathematics education: The 14th ICMI Study* (New ICMI Study Series, Vol. 10). New York, NY: Springer.
Brousseau, G. (1999). Research in mathematics education: observation and ... mathematics. In I. Schwank (Ed.), *European research in mathematics education I* (Proceedings of the first conference of the European Society for Research in Mathematics Education, Vol. 1, pp. 34–48). Osnabrück, Germany: Forschungsinstitut für Mathematikdidaktik.
Buisson, F. (1911). *Nouveau dictionnaire de pédagogie et d'instruction primaire. Entrée: Congrès d'instituteurs congrès pédagogiques* [New dictionary of pedagogy and primary education. Entry: Teachers' and pedagogical congresses]. Paris, France: Hachette. Retrieved from http://www.inrp.fr/edition-electronique/lodel/dictionnaire-ferdinand-buisson/document.php?id=2434.
Clements, M. A. (2007). The beginnings of MERGA. In J. Watson & K. Beswick (Eds.), *Mathematics: Essential research, essential practice.* Proceedings of the 30th annual conference of the Mathematics Education Research Group of Australasia, Hobart (Vol. 1, p. 2). Adelaide, Australia: Mathematics Education Research Group of Australasia.
Commission Internationale pour l'Étude et l'Amélioration de l'Enseignement des Mathématiques. (2000). *50 years of CIEAEM: Where we are and where we go? Manifesto 2000 for the Year of Mathematics.* Retrieved from http://www.cieaem.net/50_years_of_c_i_e_a_e_m.htm.

Coray, D., & Hodgson, B. R. (2003). Introduction. In D. Coray, F. Furinghetti, H. Gispert, B. R. Hodgson, & G. Schubring (Eds.), *One hundred years of L'Enseignement Mathématique: Moments of mathematics education in the twentieth century* (Monographie 39, pp. 9–15). Geneva, Switzerland: L'Enseignement Mathématique.

D'Ambrosio, U. (1985). Ethnomathematics and its place in the history and pedagogy of mathematics. *For the Learning of Mathematics, 5*(1), 44–48.

D'Ambrosio, U. (1986). Socio-cultural bases for mathematical education. In M. Carss (Ed.), *Proceedings of the Fifth International Congress on Mathematical Education* (pp. 1–6). Boston, MA: Birkhäuser.

Dorier, J.-L. (2008). Reaction to J. Kilpatrick's plenary talk "The development of mathematics education as an academic field." In M. Menghini, F. Furinghetti, L. Giacardi, & F. Arzarello (Eds.), *The first century of the International Commission on Mathematical Instruction (1908–2008). Reflecting and shaping the world of mathematics education* (pp. 40–46). Rome, Italy: Istituto della Enciclopedia Italiana.

Durand-Guerrier, V., Soury-Lavergne, S., & Arzarello, F. (Eds.). (2010). *Proceedings of the Sixth Congress of the European Society for Research in Mathematics Education (CERME 6)*. Lyon, France: Institut National de Recherche Pédagogique.

Fasanelli, F., & Fauvel, J. (2006). The International Study Group on the relations between the history and pedagogy of mathematics: The first twenty-five years, 1976–2000. In F. Furinghetti, S. Kaisjer, & C. Tzanakis (Eds.), *Proceedings of HPM 2004 & ESU 4: ICME 10 satellite meeting of the HPM Group & Fourth European Summer University on the history and epistemology in mathematics education* (pp. x–xxviii). Iraklion, Greece: University of Crete.

Fauvel, J., & van Maanen, J. (Eds.). (2000). *History in mathematics education: The ICMI Study* (New ICMI Study Series, Vol. 6). Dordrecht, The Netherlands: Kluwer Academic Publishers.

Félix, L. (1986). *Aperçu historique (1950–1984) sur la Commission Internationale pour l'Étude et l'Amélioration de l'Enseignement des Mathématiques* [Historical overview (1950–1984) of the International Commission for the Study and Improvement of Mathematics Teaching] (2nd ed.). Bordeaux, France: IREM de Bordeaux.

Fischbein, E. (1990). Introduction. In P. Nesher & J. Kilpatrick (Eds.), *Mathematics and cognition: A research synthesis by the International Group for the Psychology of Mathematics Education* (ICMI Study Series, pp. 1–13). Cambridge, UK: Cambridge University Press.

Forgasz, H. (2004). Editorial. *Mathematics Education Research Journal, 16*(2), 1–2.

Forgasz, H., Barkatsas, A., Bishop, A., Clarke, B., Keast, S., Wee, T. S., & Sullivan, P. (Eds.). (2008). *Research in mathematics education in Australasia 2004–2007*. Rotterdam, The Netherlands: Sense Publishers.

Furinghetti, F. (2003). Mathematical instruction in an international perspective: The contribution of the journal *L'Enseignement Mathématique*. In D. Coray, F. Furinghetti, H. Gispert, B. R. Hodgson, & G. Schubring (Eds.), *One hundred years of L'Enseignement Mathématique: Moments of mathematics education in the twentieth century* (Monographie 39, pp. 19–46). Geneva, Switzerland: L'Enseignement Mathématique.

Furinghetti, F. (2008a). Mathematics education in the ICMI perspective. *International Journal for the History of Mathematics Education, 3*(2), 47–56.

Furinghetti, F. (2008b). The emergence of women on the international stage of mathematics education. *ZDM: The International Journal of Mathematics Education, 40*(4), 529–543.

Furinghetti, F., & Giacardi, L. (2010). People, events, and documents of ICMI's first century. *Actes d'Història de la Ciència i de la Tècnica, 3*(2), 11–50.

Furinghetti, F., Menghini, M., Arzarello, F., & Giacardi, L. (2008). ICMI renaissance: The emergence of new issues in mathematics education. In M. Menghini, F. Furinghetti, L. Giacardi, & F. Arzarello (Eds.), *The first century of the International Commission on Mathematical Instruction (1908–2008). Reflecting and shaping the world of mathematics education* (pp. 131–147). Rome, Italy: Istituto della Enciclopedia Italiana.

Galbraith, P. L. (2004). Report by the International Study Group for Mathematical Modelling and Applications (ICTMA)–Activities 2000–2004. *ICMI Bulletin, 54,* 66–69.

Gellert, U., Jablonka, E., & Morgan, C. (2010). Introduction. In U. Gellert, E. Jablonka, & C. Morgan (Eds.), *Proceedings of the Sixth International Mathematics Education and Society Conference* (Vol. 1, pp. 1–5). Berlin, Germany: Freie Universität Berlin.

Grevholm, B. (2009a). Creation of a Nordic Society for Research in Mathematics Education, NoRME. In C. Winsløw (Ed.), *Nordic research in mathematics education. Proceedings from NORMA08* (pp. 387–389). Rotterdam, The Netherlands: Sense Publishers.

Grevholm, B. (2009b). Nordic collaboration in mathematics education research. *Nordic Studies in Mathematics Education, 14*(4), 89–100.

Grevholm, B., & Hanna, G. (Eds.). (1995). *Gender and mathematics education: An ICMI Study*. Lund, Sweden: Lund University Press.

Grevholm, B., & Rønning, F. (2010). Nordic collaboration in mathematics education research—continued. *Nordic Studies in Mathematics Education, 15*(1), 97–102.

Grugnetti, L. (1996). Un regard historique sur la CIEAEM [A historical look at CIEAEM]. In C. Keitel, U. Gellert, E. Jablonka, & M. Müller (Eds.), *Mathematics (education) and common sense. Proceedings of the 47th CIEAEM meeting* (pp. 9–13). Berlin, Germany: Freie Universität Berlin.

Gutiérrez, A., & Boero, P. (Eds.). (2006). *Handbook of research on the psychology of mathematics education: Past, present and future*. Rotterdam, The Netherlands: Sense Publishers.

Hanna, G. (Ed.). (1996). *Towards gender equity in mathematics education: An ICMI Study* (New ICMI Study Series, Vol. 3). Dordrecht, The Netherlands: Kluwer Academic Publishers.

Hanna, G., & Sidoli, N. (2002). The story of ESM. *Educational Studies in Mathematics, 50*(2), 123–156.

Hershkowitz, R., & Breen, C. (2006). Foreword—Expansion and dilemmas. In A. Gutiérrez & P. Boero (Eds.), *Handbook of research on the psychology of mathematics education: Past, present and future* (pp. ix–xii). Rotterdam, The Netherlands: Sense Publishers.

Hodgson, B. R. (2002). New terms of reference for ICMI. *ICMI Bulletin, 51*, 8–12.

Hodgson, B. R. (2008). Some views on ICMI at the dawn of its second century. In M. Menghini, F. Furinghetti, L. Giacardi, & F. Arzarello (Eds.), *The first century of the International Commission on Mathematical Instruction (1908–2008). Reflecting and shaping the world of mathematics education* (pp. 199–203). Rome, Italy: Istituto della Enciclopedia Italiana.

Hodgson, B. R. (2009). ICMI in the post-Freudenthal era: Moments in the history of mathematics education from an international perspective. In K. Bjarnadóttir, F. Furinghetti, & G. Schubring (Eds.), *"Dig where you stand": Proceedings of the conference on On-going research in the history of mathematics education* (pp. 79–96). Reykjavik, Iceland: The University of Iceland.

Hodgson, B. R. (2011). Collaboration et échanges internationaux en éducation mathématique dans le cadre de la CIEM: Regards selon une perspective canadienne [ICMI as a space for international collaboration and exchange in mathematics education: Some views from a Canadian perspective]. In P. Liljedahl, S. Oesterle, & D. Allan (Eds.), *Proceedings of the 2010 annual meeting of the Canadian Mathematics Education Study Group/Groupe canadien d'étude en didactique des mathématiques* (pp. 31–50). Burnaby, Canada: CMESG/GCEDM.

Hodgson, B. R., & Rogers, L. F. (2011). Aspects of internationalism in mathematics education: National organizations with an international influence. *International Journal for the History of Mathematics Education, 6*(2), 87–97.

Hodgson, B. R., & Rogers, L. F. (2012). On international organizations in mathematics education. *International Journal for the History of Mathematics Education, 7*(1), 17–27.

Howson, A. G. (1984). Seventy-five years of the International Commission on Mathematical Instruction. *Educational Studies in Mathematics, 15*(1), 75–93.

International Commission on Mathematical Instruction. (1975). Memorandum on affiliation of IACME to ICMI. *ICMI Bulletin, 5*, 6.

International Commission on Mathematical Instruction. (2004). The sixteenth ICMI Study: Challenging mathematics in and beyond the classroom—Discussion document. *ICMI Bulletin, 55*, 32–46.

Jacobsen, E. (1996). International co-operation in mathematics education. In A. J. Bishop, M. A. Clements, C. Keitel, J. Kilpatrick, & C. Laborde (Eds.), *International handbook of mathematics education* (pp. 1235–1256). Dordrecht, The Netherlands: Kluwer.

Jaworski, B. (2008). The European Society for Research in Mathematics Education (ERME). *EMS Newsletter, 73*, 43–44.

Keitel, C., Damerow, P., Bishop, A., & Gerdes, P. (Eds.). (1989). *Mathematics, education and society* (Science and Technology Education Document Series, Vol. 35). Paris, France: UNESCO.

Kenderov, P. S. (2009). A short history of the World Federation of National Mathematics Competitions. *Mathematics Competitions, 22*(2), 14–31.

Kilpatrick, J. (1992). A history of research in mathematics education. In D. A. Grouws (Ed.), *Handbook of research on mathematics teaching and learning* (pp. 3–38). New York, NY: Macmillan.

Kilpatrick, J. (1999). Ich bin europäisch [I am European]. In I. Schwank (Ed.), *European research in mathematics education I*. (Proceedings of the first conference of the European Society for Research in Mathematics Education, Vol. 1, pp. 49–68). Osnabrück, Germany: Forschungsinstitut für Mathematikdidaktik.

Kilpatrick, J. (2008). The development of mathematics education as an academic field. In M. Menghini, F. Furinghetti, L. Giacardi, & F. Arzarello (Eds.), *The first century of the International Commission on Mathematical Instruction (1908–2008). Reflecting and shaping the world of mathematics education* (pp. 25–39). Rome, Italy: Istituto della Enciclopedia Italiana.

Klein, M., Putt, I., & Stillman, G. (1999). Editorial: Sabre rattling or genuine change. *Mathematics Teacher Education and Development, 1*, 1–3.

Krainer, K. (2002). From ERME to YERME—News from the European Society for Research in Mathematics Education. *ICMI Bulletin, 51*, 75–76.

Krazer, A. (Ed.). (1905). *Verhandlungen des dritten Internationalen Mathematiker-Kongresses 1904* [Proceedings of the Third International Congress of Mathematicians 1904]. Leipzig, Germany: B. G. Teubner.

Lee, P. Y. (1992). A glimpse of SEACME. *ICMI Bulletin, 33*, 27–29.

Lehto, O. (1998). *Mathematics without borders: A history of the International Mathematical Union*. New York, NY: Springer.

Lighthill, J. (1975). Report on the period 1971–1974. *L'Enseignement Mathématique, 21*, 329–330.

Lim-Teo, S. K. (2008). ICMI activities in East and Southeast Asia: Thirty years of academic discourse and deliberations. In M. Menghini, F. Furinghetti, L. Giacardi, & F. Arzarello (Eds.), *The first century of the International Commission on Mathematical Instruction (1908–2008). Reflecting and shaping the world of mathematics education* (pp. 247–252). Rome, Italy: Istituto della Enciclopedia Italiana.

Mathematics Education Research Group of Australasia. (2002). *Research policy: Statement on research in mathematics education (Rev. 2002)*. Retrieved from http://www.merga.net.au/node/11.

Mendick, H. (Ed.). (2008). *Her-stories of IOWME*. Retrieved from http://www.icmihistory.unito.it/iowme.php.

Menghini, M., Furinghetti, F., Giacardi, L., & Arzarello, F. (Eds.). (2008). *The first century of the International Commission on Mathematical Instruction (1908–2008): Reflecting and shaping the world of mathematics education*. Rome, Italy: Istituto della Enciclopedia Italiana.

Mousley, J. (2009). *A history of MERGA*. Retrieved from http://www.merga.net.au/node/75.

National Council of Teachers of Mathematics. (2000). *Principles and standards for school mathematics*. Reston, VA: National Council of Teachers of Mathematics.

Nebres, B. F. (2008). Centers and peripheries in mathematics education. In M. Menghini, F. Furinghetti, L. Giacardi, & F. Arzarello (Eds.), *The first century of the International Commission on Mathematical Instruction (1908–2008). Reflecting and shaping the world of mathematics education* (pp. 149–163). Rome, Italy: Istituto della Enciclopedia Italiana.

Nesher, P., & Kilpatrick, J. (Eds.). (1990). *Mathematics and cognition: A research synthesis by the International Group for the Psychology of Mathematics Education* (ICMI Study Series). Cambridge, UK: Cambridge University Press.

Nicol, C., & Lerman, S. (2008). *A brief history of the International Group for the Psychology of Mathematics Education (PME)*. Retrieved from http://www.icmihistory.unito.it/pme.php.

Niss, M. (2004). Key issues and trends in research on mathematical education. In H. Fujita, Y. Hashimoto, B. R. Hodgson, P. Y. Lee, S. Lerman, & T. Sawada (Eds.), *Proceedings of the Ninth International Congress on Mathematical Education* (pp. 37–57). Dordrecht, The Netherlands: Kluwer Academic Publishers.

Nissen, G., & Blomhøj, M. (Eds.). (1993). *Criteria for scientific quality and relevance in the didactics of mathematics*. Roskilde, Denmark: Roskilde University, IMFUFA.

Oates, G., & Engelbrecht, J. (2009). Foreword: Mathematics in a dynamic environment. *International Journal of Mathematical Education in Science and Technology, 40*(7), 847–849.

Ottaviani, M.-G., & Batanero, C. (1999). The role of the IASE in developing statistical education. In *Proceedings of the Sixth Islamic Countries Conference on Statistical Sciences (ICCS-VI), Lahore, Pakistan, August 27–31, 1999* (Vol. 11, pp. 171–186). Retrieved from http://www.stat.auckland.ac.nz/%7Eiase/about/ottabatpak.pdf.

Parshall, K. H. (2009). The internationalization of mathematics in a world of nations, 1800–1960. In E. Robson & J. Stedall (Eds.), *The Oxford handbook of the history of mathematics* (pp. 85–104). Oxford, UK: Oxford University Press.

Phillips, B. (2002). *The IASE: Background, activities, and future* (Invited paper at the Sixth International Conference on Teaching Statistics, ICOTS 6). Retrieved from http://www.stat.auckland.ac.nz/%7Eiase/publications/1/5d2_phil.pdf.

Robitaille, D. F., & Travers, K. J. (2003). International connections in mathematics education. In G. M. A. Stanic & J. Kilpatrick (Eds.), *A history of school mathematics* (pp. 1491–1508). Reston, VA: National Council of Teachers of Mathematics.

Rogers, L. F. (1978). International Study Group on the relations between the history and the pedagogy of mathematics. *ICMI Bulletin, 10*, 26–27.

Ruiz, A. (2010). Editorial: Affiliate organizations. *ICMI News, 15*. Retrieved from http://www.mathunion.org/icmi/publications/icmi-news/.

Ruiz, A., & Barrantes, H. (2011). En los orígenes del CIAEM [On the origins of CIAEM]. *Cuadernos de Investigación y Formación en Educación Matemática, 7*, 13–46.

Sánchez Vázquez, G., & García Blanco, M. (1991). Introducción. In M. García Blanco (Ed.), *Memorias del primer Congreso Iberoamericano de Educación Matemática* (Science and Technology Education Document Series No. 42, pp. v–vi). Paris, France: UNESCO.

Schubring, G. (2000). History of mathematics for trainee teachers. In J. Fauvel & J. van Maanen (Eds.), *History in mathematics education: The ICMI Study* (New ICMI Study Series, Vol. 6, pp. 91–142). Dordrecht, The Netherlands: Kluwer Academic Publishers.

Schubring, G. (2003). *L'Enseignement Mathématique* and the first international commission (IMUK): The emergence of international communication and cooperation. In D. Coray, F. Furinghetti, H. Gispert, B. R. Hodgson, & G. Schubring (Eds.), *One hundred years of L'Enseignement Mathématique: Moments of mathematics education in the twentieth century* (Monographie 39, pp. 47–65). Geneva, Switzerland: L'Enseignement Mathématique.

Schubring, G. (2008a). Editorial. *International Journal for the History of Mathematics Education, 3*(2), 1–2.

Schubring, G. (2008b). The origins and early history of ICMI. *International Journal for the History of Mathematics Education, 3*(2), 3–33.

Shelley, N. (1984). A brief history of IOWME. *ICMI Bulletin, 17*, 19–22.

Sierpinska, A., & Kilpatrick, J. (Eds.). (1998). *Mathematics education as a research domain, a search for identity: An ICMI Study* (New ICMI Study Series, Vol. 4). Dordrecht, The Netherlands: Kluwer Academic Publishers.

Silver, E. A., & Kilpatrick, J. (1994). E pluribus unum: Challenges of diversity in the future of mathematics education research. *Journal for Research in Mathematics Education, 25*(6), 734–754.

Sriraman, B., & English, L. (2010). *Theories of mathematics education: Seeking new frontiers*. New York, NY: Springer.

Taylor, P. J. (2009). WFNMC 25 years on: Some experiences. *Mathematics Competitions, 22*(1), 10–18.

UNESCO. (2011). *Les défis de l'enseignement des mathématiques dans l'éducation de base* (Education Sector ED.2010/WS/37). [English translation: *Challenges in basic mathematics education* (2012).] Paris, France: UNESCO.

Vere-Jones, D. (1995). The coming of age of statistical education. *International Statistical Review, 63*(1), 3–23.

Way, J., Anderson, J., & Bobis, J. (2010). Editorial: The importance of teacher knowledge and teacher thinking. *Mathematics Teacher Education and Development, 12*(1), 1–2.

Chapter 29
Toward an International Mathematics Curriculum

Jinfa Cai and Geoffrey Howson

Abstract This chapter revisits the notion of an international curriculum, analyzing the various forces that might push countries toward one and reasons why countries should develop their own distinct curricula. We first describe the term *curriculum* to set the stage for our later discussion. We then discuss, in turn, common influences for curriculum change, common learning goals, common driving forces of public examinations, common emphases and treatments, and common issues for future curriculum development. Although the tendency for countries to include a more-and-more internationally-accepted core selection of topics in their national curricula is to a great extent both to be welcomed and expected, this move has had a potential negative effect on curriculum development. Significant work also remains to be done to explore the way in which new technology (especially digital technology) could affect both the mathematics included in the curriculum and how it could more effectively contribute to the teaching and learning of mathematics in general.

The main purpose of the Second International Commission on Mathematics Instruction (ICMI) Study "School Mathematics in the 1990s" was to provoke and stimulate discussion about the school curriculum and directions it might take. In the study report, Howson and Wilson (1986) noted the manner in which a canonical school mathematics curriculum, influenced by industrial and commercial needs, "has been adopted practically everywhere" (p. 19), although there were still considerable differences of opinion on how the word *geometry* should be interpreted (p. 38). This claim was made despite the drastic variations of socio-economic

J. Cai (✉)
University of Delaware, Newark, DE, USA
e-mail: jcai@math.udel.edu

G. Howson
The University of Southampton, Southampton, UK

circumstances to be found between countries, as the report showed by its comparison of the situations in Japan and Mexico. In Japan, almost all students enrolled in elementary school, and nearly 95% of them completed secondary education. In contrast, in Mexico, about 60% of children enrolled in elementary school, and only about 3% of them completed secondary education. However, the two countries had quite similar mathematical topics at each grade level. Furthermore, the coverage of certain mathematical topics in each grade level in Japan and Mexico was determined by adopting syllabuses developed elsewhere, where students resembled neither students in Japan nor in Mexico. This argument and example clearly showed a trend toward an international curriculum at that time.

However, caution was expressed about the results of such a trend. The benefits of countries learning from each other when developing a mathematics curriculum were acknowledged, but the study report warned about the direct adoption of a curriculum developed elsewhere, because "local circumstances" matter. Curriculum developers were advised to "pay more attention to their own actual circumstances and needs than to consideration of international 'standards' and issues of comparability ... since the goals of school mathematics will not be identical everywhere" (Howson & Wilson, 1986, p. 21). This ICMI study took place not long after the world had seen a wave of curriculum development in mathematics ("the new math(s)" or "modern mathematics"), yet the curricula, or to be more exact the mathematical syllabuses, of the various countries had already assumed a near common standard form. Therefore, the caution was timely.

Today, more than 25 years since the ICMI report was published, mathematics educators are still experiencing technological revolutions, even greater than those in the 1980s. In fact, the world has been changing dramatically, and these changes are happening much faster than many anticipated. The change in the past 25 years has gone way beyond a technological revolution.

Historically, across the nations, changing the curriculum has been viewed and used as an effective way to change classroom practice and to influence student learning to meet the needs of the ever-changing world (Cai, Nie, & Moyer, 2010; Howson, Keitel, & Kilpatrick, 1981; Senk & Thompson, 2003). In fact, curriculum has been called a changing agent for educational reform (Ball & Cohen, 1996; Darling-Hammond, 1993). In the preface to the 72nd Yearbook of the National Council of Teachers of Mathematics (NCTM), Reys, Reys, and Rubenstein (2010) correctly pointed out that one thing does not change about the mathematics curriculum: The school mathematics curriculum remains a central issue in efforts to improve students' learning. The curriculum plays a significant role in mathematics education because it effectively determines what students learn, when they learn it, and how well they learn it. But what aspects of the curriculum should be changed, and how might a curriculum be improved to meet the needs of the ever-changing world? Educators, researchers, and policy makers around the globe are constantly seeking answers to that question (Usiskin & Willmore, 2008).

Given the sensitive and critical roles of the curriculum in a changing world, in this chapter we revisit the notion of an international curriculum, analyzing the various forces that might push countries toward one and reasons why countries should develop their own distinct curricula. This chapter has seven major sections, and in the first we try to define what a curriculum is. A curriculum can be interpreted in different levels, and the term *curriculum* may have different meanings. We plan to describe the term *curriculum* to set the stage for our later discussion. We then consider, in turn, common influences for curriculum change, common learning goals, common driving forces of public examinations, common emphases and treatments, and common issues for future curriculum development, before offering a summing-up conclusion.

What Is a Curriculum?

In mathematics, we usually want to start with a definition of a concept. To set the stage for discussing an international curriculum, it is natural to define what a curriculum is. However, in searching for a definition of *curriculum,* we quickly found that it is almost impossible to give a universally acceptable definition. The notion of curriculum can be discussed at different levels and in different ways, and there are different conceptions about curriculum (Jackson, 1992). After surveying over 1,100 curriculum books, Cuban (1992) found "each with different versions of what 'curriculum' means; many of the definitions conflict" (p. 221).

Although there is no consensus about the actual definition of *curriculum,* two things are quite clear. The first is that one can talk about the curriculum from different levels. The International Association for the Evaluation of Educational Achievement [IEA] (Travers & Westbury, 1989) distinguished between three levels of curriculum (intended, implemented, and attained), and these distinctions have been widely accepted and used in mathematics education. This categorization highlights the differences in what a society would like to have taught, what is actually taught, and what students have actually learned (National Research Council, 2004; Pinar, 2003; Senk & Thompson, 2003; van den Akker, Kuiper, & Hameyer, 2003).

The intended curriculum refers to the formally written documents that set system-level expectations for the learning of mathematics. It usually includes goals and expectations set at the educational system level along with official syllabi or curriculum standards, and, in some countries, approved textbooks. The intended curriculum, then, is concerned with the *system level*. The implemented curriculum refers to school and classroom processes for the teaching and learning of mathematics as interpreted and implemented by teachers, according to their experience and beliefs for particular classes. It, then, operates at the *classroom level*. The classroom is, of course, central to students' learning since it is there that students acquire most of their mathematical knowledge and form their attitudes to the subject (Robitaille & Garden, 1989). Regardless of how well a curriculum is designed, its ultimate value depends on how it is implemented in the classroom. The attained curriculum refers to what is actually learned by students and is manifested in their achievements and attitudes. It is at the *student level*. It deals with those aspects of the intended curriculum that are taught by teachers and learned by students. In addition to the intended, implemented, and attained curriculum, researchers have also talked about the *ideal* curriculum, the *hidden* curriculum, relating to the values stressed and their social and political implications, and, perhaps most importantly, the *tested* curriculum (Burkhardt, Fraser, & Ridgway, 1990).

The second problem is that the term *curriculum* can be used as both a product and a process. A curriculum is a product: a set of instructional guidelines and materials for students' acquisition of certain culturally valued knowledge and skills. In many countries, for example, the curriculum is based on a so-called syllabus, which is usually referred to as a summary of the mathematical topics and skills which students need to know at a particular grade level (Hershkowitz et al., 2002). A curriculum can also be viewed as a process. In this sense the curriculum is not a physical thing, like textbooks, but rather the interaction of teachers, students, and knowledge. In this view, teachers are an "integral part of the curriculum constructed and enacted in classrooms" (Clandinin & Connelly, 1992, p. 363). In other words, the curriculum is what actually happens in the classroom. In this sense, a curriculum is a particular form of specification about the practice of teaching. It is not a set of topics covered in the classroom. Instead, it is a way of translating any educational idea into a hypothesis testable in practice (Smith, 1996/2000). The teachers' role is as curriculum maker—to engage in the process of developing a coherent sequence of learning situations, together with appropriate materials, the implementation of which has the potential to bring about intended changes in learners' knowledge (Clandinin & Connelly, 1992; Hershkowitz et al., 2002).

In this chapter, for discussion purposes, we do not distinguish between several related terms: *curriculum, curriculum materials, standards, syllabus,* and *textbooks.* Rather, we view the curriculum from different levels (intended, implemented, and attained), and as both a product and a process.

Toward Common Influences for Curriculum Change

Curriculum development is the process of developing a coherent sequence of learning situations, materials, and student assessment procedures, which has the potential to bring about desired changes in students' learning (Hershkowitz et al., 2002). In the first *International Handbook of Mathematics Education,* Clarke, Clarke, and Sullivan (1996) discussed two broad areas of influence on commonality of contents in mathematics curriculum. The first area is related to the uniformity of contents based on the notion of a canonical curriculum (Howson & Wilson, 1986). The second is related to common external influences. These two broad areas are still the driving forces towards an international curriculum.

During the past 25 years, however, two unique influences have reinforced a move toward an international curriculum: (a) globalization, and (b) internationalization. In this section, we specifically discuss these two influences.

Globalization

Globalization refers to a process by which regional economies, societies, and cultures have become integrated through a global network of communication, collaboration, transportation, and trade. Indeed, the increasing global connectivity, integration, and interdependence in so many aspects of our life require educators to think again about mathematics education. People of the world are inextricably linked in such a way that local happenings are shaped by events happening around the globe. Mathematics and science education communities have been responding to this challenge by developing frameworks for understanding globalization and analyzing its impact on education (Atweh et al., 2007). Yet globalization has not brought homogeneity. Vast differences still exist between, and in some cases within, countries concerning resources available for education, let alone curriculum renewal, and these differences, of course, necessitate different responses (Skovsmose & Valero, 2008; Vithal & Volmink, 2005).

Such globalization requires us to rethink not only the content topics and sequencing of topics in the school mathematics curriculum but also the goals of school mathematics. More than ever, the mathematics curriculum needs to be designed so as to assist students not only to develop the abilities to think critically and to solve problems but also to foster cross-cultural communication and collaboration, and nurture creativity and innovation.

Internationalization

In economics, internationalization refers to a process of planning and implementing products and services so that they can easily be adapted to specific local contexts or markets. In mathematics education, internationalization is not a new phenomenon. It takes several different forms: the most apparent being the use in colonial days of textbooks, or "local adaptations" of these, originally written with students from very different backgrounds in mind. Yet it has also meant studying in other countries, the formation of international organizations such as ICMI (formed in 1908), and visits to study other countries' schools. For example, Arnold (1868/2008), commissioned by a UK government worried by the increasing industrial and military power of continental rivals, provided an account of the then vastly superior educational systems of France, Prussia, and Switzerland. More recently, emphasis has switched to international comparative studies, and in particular to the Trends in International Mathematics and Science Study (TIMSS) and the Programme for International Student Assessment (PISA), which seek to assess students' knowledge and then to rank countries by their students' performances. The results of these, but not their processes and effects, are readily and frequently quoted and acted upon by politicians. Yet, the TIMSS studies, in particular, have, thrown much light on classroom practices and textbooks worldwide and have highlighted the fact that the curricula developed in individual countries greatly influence the test results of their students. The actual test results of TIMSS and PISA also have importance for educators. However, to quote one warning:

> The main conclusion from this study, therefore, is that little is to be gained from studying the ranking lists of countries to be found in Mullis et al. (1998). Indeed, their simplicity may well prove a tempting trap into which politicians might fall. There are some clear warnings in such tables, of which heed should be taken—but they, by themselves, do not provide "value for use." A country wishing to use TIMSS data to improve its mathematics teaching must carry out a careful study of responses to individual items; to the validity and importance they assign to these; and to the way that students' successes or failures can be linked, not only to specific topic areas given the country's curriculum content and pedagogy, but, perhaps more importantly, to the varying cognitive demands of the individual items. (Howson, 2002, p. 123)

Comparative studies can, indeed, not only provide information on students' achievements but also, and less controversially, help to identify effective aspects of educational practice. Postlethwaite (1988) identified four objectives of comparative studies: (a) identifying what is happening in different countries that might help improve education systems and outcomes; (b) describing similarities and differences in educational phenomena between systems of education and interpreting why these exist; (c) estimating the relative effects of variables that are thought to be determinants of educational outcomes (both within and between systems of education); and (d) identifying general principles concerning educational effects. These four objectives have been sought in many comparative studies, especially in the designs of IEA studies such as TIMSS (Medrich & Griffith, 1992).

The US Board on International Comparative Studies in Education identified the following six reasons why the USA should participate in international studies (Bradburn & Gilford, 1990):

1. Improving understanding of education systems;
2. Providing information on the students' achievement in relation to the much broader range of the world's education systems;
3. Identifying the factors that do and do not promote educational achievement;
4. Enhancing the research enterprise itself;
5. Recording the diversity of educational practice; and
6. Promoting issue-centred studies.

These are reasons that understandably are very similar to those to be found in Postlethwaite (1988).

Internationalization and globalization are powerful influences for curriculum development, instructional design, and educational policy, and compel countries and educators critically to examine themselves with respect to intended, implemented, and attained curricula. These two forces have, then, provided mechanisms for better understanding concerning how different education systems address similar problems, and accordingly allow educational policy makers, researchers, and practitioners to look beyond experiences evident in their own systems and thus to reflect upon issues in curriculum and instruction which could facilitate educational improvement.

Yet these factors in themselves should not necessarily cause a movement towards an international curriculum. Although countries might well simply pick and choose what, of what they see, would seem most suitable for them, this does not always appear to have been the case. Vithal and Volmink (2005) discuss this problem with post-apartheid South Africa in mind. They remark on the different forces which shaped curriculum development in the Western countries in the 1960s, the new mathematics movement and its counterpart back-to-basics, behaviourism, structuralism, formalism, problem-solving and integrated curriculum approaches (Howson et al., 1981), all of which left their mark on the South African curriculum. Note that some of these forces sprang from mathematical considerations, others from psychological ones. Then came constructivism, which they describe as *strong epistemology* but with a *weak pedagogy*, and ethnomathematics, which by its very nature did not fit comfortably with an international curriculum, and perhaps was thought of as the acceptance of something weaker, or second-rate. Now outcome-based education is in the ascendancy. But how are the "outcomes" to be decided?

Globalization cannot solve the problems of a country where large inequalities exist in access to mathematical education, provision of resources, and opportunities to learn: Alternatives have to be found (Skovsmose, 2003). Yet the Organisation for Economic Co-operation and Development [OECD] (1999), which through its name emphasizes economic rather than education considerations, has, through its presentation of the results of PISA, promoted comparisons that national governments have interpreted as a need to improve mathematics education for the sole purpose of creating a qualified workforce that will be competitive in a globalized economy. This interpretation can militate against any curriculum development that seeks to meet an economically hard-pressed country's true educational and mathematical needs and also serves to restrict the way in which curriculum design and curricular aims are viewed in more economically-fortunate countries. This aspect of internationalization is, then, a strong influence towards a common international curriculum—but one that would be restricted both mathematically and socially.

Toward Common Learning Goals

Education is a goal-directed activity, and a curriculum specifies learning goals. It is assumed that the effectiveness of curriculum and instruction is related to the goal of high achievement for all students (National Academy of Education, 1999). In the complex endeavour of schooling, teachers encounter many unexpected events. Although a teacher has an overall plan, she or he naturally cannot exactly follow the detailed script for action. With a clear learning goal or goals for a lesson, the teacher can make immediate decisions to address the unexpected and guide students toward the learning goals. Determination of learning goals for each lesson requires that teachers know mathematics, curriculum emphasis, students as learners, and pedagogical strategies. It also depends on teachers' beliefs about mathematics and conceptions about teaching mathematics. If mathematics is viewed as a collection of isolated facts and skills, then teachers may just focus on rules, procedures, rote memorizations, and practice. Instead, if mathematics is viewed as a way of thinking with wide applications, then teachers must teach it with that in mind.

Across the nations, mathematics is in the central place in school curricula. We can justify the need to study mathematics in school from different perspectives (Christiansen, Howson, & Otte, 1986; Romberg, 2002). There was a reasonable convergence of views until the 1960s; a view which was widely based on the assumption that different courses were needed for the students in different types of schools. Then, the coming of the new mathematics brought new ideas on what the aims of mathematics teaching should be, particularly for students in schools with "high ability," and this led to a wide divergence of views on what the school curriculum should contain. In recent years, some of the reform material has been accepted into the curriculum and some has been rejected, leading towards more commonly accepted learning goals in school mathematics. In addition to developing traditionally accepted mathematical knowledge and skills through mathematics instruction, increasing emphasis has been placed on developing students' higher-order thinking skills. Although there are no commonly accepted definitions of such skills, the frequently cited list to be found in Resnick (1987) might help. According to Resnick, higher-order thinking:

1. Is *non-algorithmic*. That is, the path of action is not fully specified in advance.
2. Tends to be *complex*. The total path is not "visible" (mentally speaking) from any single vantage point.
3. Often yields *multiple solutions*, each with costs and benefits, rather than unique solutions.
4. Involves *nuanced judgment* and interpretation.
5. Involves the application of *multiple criteria*, which sometimes conflict with one another.
6. Often involves *uncertainty*; not everything that bears on the task at hand is known.
7. Involves *self-regulation* of the thinking process.
8. Involves *imposing meaning*, finding structure in apparent disorder.
9. Is *effortful*; considerable mental work is involved in the kinds of elaborations and judgments required.

This list clearly shows that higher-order thinking skills involve the abilities to think flexibly so as to make sound decisions in complex and uncertain problem situations. In addition, such skills involve self-monitoring one's own thinking—metacognitive skills. In particular, ideally, mathematics instruction should provide students with opportunities to: (a) think about things from different points of view, (b) step back to look at things again, and (c) consciously think about what they are doing and why they are doing it. Resnick's list does not include the ability to collaborate with others, but being able to work together with others is also an essential higher-order thinking skill. Collaborative work encourages students to think together about ideas and problems as well as to challenge each other's ideas and ask for clarification.

This desirable aim of developing such skills is related to the view that mathematics education should be seen as something more than just contributing to the intellectual development of individual students. Certainly, preparing them to live as informed and functioning citizens in contemporary society, and providing them with the potential to take their places in the fields of commerce, industry, technology, and science are important objectives (Robitaille & Garden, 1989). In addition, mathematics education should seek to teach students about the nature of mathematics. Mathematics, then, is viewed no longer as simply a prerequisite subject but rather as a fundamental aspect of *literacy* for a citizen in contemporary society (Mathematics Sciences Education Board [MSEB], 1993; NCTM, 1989). Ideally, with this view of school mathematics, teachers need to move toward mathematical thinking, reasoning, and problem solving and away from merely memorizing procedures; move toward conjecturing, inventing, generalizing, proving, and problem posing and away from an emphasis on mechanistic answer-finding; and move toward connecting mathematics, its ideas, and its applications and away from treating mathematics as a body of isolated concepts and procedures.

Here it is valuable to pause, however, and contemplate how many these qualities and of Resnick's higher-order skills are tested—indeed, are capable of being tested—by TIMSS, PISA, and the examination systems to be found in individual countries. Until such skills are featured in the examined curriculum, there is little chance of their being widely accepted into the implemented curriculum. Indeed, it could be claimed that as a result of recent pressures and changes within school systems, educators are further away from attaining such goals than ever. Are these suitable goals for all students, goals at which countries should continue to aim, or should the emphasis be on improving current practice in less demanding ways?

This question of goals is also related to the needs of an ever-changing world. Today, possessing a large amount of knowledge and information is not sufficient. Instead, the most important qualities that teachers can help their students develop are the abilities to think independently and critically, to learn, and to be creative, as well as to learn how to learn. In his best-selling book, *The World Is Flat*, Friedman (2005) pointed out that "there may be a limit to the number of good factory jobs in the world, but there is no limit to the number of idea-generated jobs in the world" (p. 230)—again an economically generated reason for the teaching of mathematics. Education in general and mathematics education in particular have the responsibility for nurturing students' creativity and critical thinking skills not only for their lifelong learning but also for their general benefit and pleasure. Again, it is essential to ask whether or not educators are any nearer to achieving those goals.

In the USA, NCTM specified five goals for students in its monumental standards document published in 1989:

1. To learn to value mathematics;
2. To learn to reason mathematically;
3. To learn to communicate mathematically;
4. To become confident of their mathematical abilities; and
5. To become mathematical problem solvers.

NCTM also specified major shifts to achieve these goals in teaching mathematics, including movement toward:

- Classrooms as mathematical communities—away from classrooms as simply collections of individuals;
- Logic and mathematical evidence as verification—away from the teacher as the sole authority for right answers;
- Mathematical reasoning—away from merely memorizing procedures;
- Conjecturing, inventing, and problem solving—away from an emphasis on mechanistic answer-finding;
- Connecting mathematics, its ideas, and its applications—away from treating it as a body of isolated concepts and procedures.

In China, there have been notable movements in mathematics education in recent years. The most recent curriculum reform in China began in 2001 and has focussed on developing new curriculum standards, textbooks, teaching methods, and assessment systems. In 2001, the Chinese Ministry of Education published *Curriculum Reform Guidelines for the Nine-Year Compulsory Education*. The main objectives of the new curriculum reform included the following:

1. Shifting from overemphasizing knowledge transmission to placing more emphasis on students' active participation and to developing such mathematical abilities as collecting and processing new information, gaining new knowledge independently, analyzing and solving problems, and communicating and cooperating with others;
2. Shifting the curriculum structure from an overemphasis on separate school subjects to emphasizing more on the integration of school mathematics; and
3. Shifting from complicated and outdated curriculum content to curriculum content that reflects students' life and the new developments of modern science and technology. (Basic Education Curriculum Material Development and Chinese Ministry of Education, 2001)

Ni, Li, Cai, and Hau (2012) have analyzed the goals in the new mathematics curriculum standards in China. These goals include helping students to (a) acquire important knowledge and the basic problem-solving skills in mathematics that are important for their lifelong learning; (b) apply knowledge of mathematics and related skills to observe, analyze, and solve problems in daily life and in other subjects by using mathematical methods; and (c) to appreciate the close relationship between mathematics, nature, and society. As Ni et al. pointed out, these new goals not only require students to acquire basic mathematical knowledge and skills; they also aim to provide them with the opportunities to reason about evidence and explanation, evaluate knowledge claims, use acquired knowledge and skills to solve real-life problems, and develop interest and confidence in learning and using mathematics.

In the UK, the Cockroft (1982) Report advocated problem solving as a means of developing mathematical thinking as a tool for daily living. Developing problem-solving ability lies at the heart of doing mathematics because it is the means by which mathematics can be applied to a variety of unfamiliar situations. In the 2000 revised English National Curriculum, the key skills of communication, application of number, information technology, working with others, improving one's own learning and performance, problem solving and other skills such as thinking skills, financial capability, enterprise and entrepreneurial skills and work-related learning were elaborated (Department for Education and Employment and Qualifications and Curriculum Authority, 2000). This list, however, was in response to a first attempt to establish a National Curriculum some 10 years earlier. Since 2000, the mathematics curriculum has seen yet more changes, and still more are promised. This does not mean that the early, overall aims were misjudged, but rather that it is relatively easy to set out well-intentioned objectives for a curriculum. However, the ensuing difficulties of determining the details of such a curriculum, of establishing appropriate assessment procedures, and of implementing change in classrooms have yet to be fully comprehended by politicians. Moreover, it is the politicians' view of educational aims in general that will, to a large extent, determine how mathematics curricula will develop. This is a matter in which mathematics educators have a great role to play, but their voices will not always prove to be the dominant ones.

Through analyzing curriculum documents, Wong (2004) compared the goals of school mathematics in various countries/regions, including Western nations (Australia, France, Germany, the UK, and the USA) and far Eastern ones (Mainland China, Hong Kong, Taiwan, Singapore, and Japan). In his reflection, he listed "higher-order thinking skills" as the most important common goal across the nations/regions in his analysis. However, governmental reasons for developing students' higher-order thinking skills are, clearly, to increase economic and political competitiveness. Sriraman and Törner (2008) analyzed European didactic traditions. They found that there are some commonalties in curricular changes in the European countries and also concluded that a "good mathematics education was regarded as favoring industrial and economic, hence political, competitiveness" (p. 680). Given such common goals, they asked if a common European curriculum were possible. Since such learning goals are even more commonly shared, this could lead to a call for an international curriculum.

But could one have such a curriculum at even a "system" level, for how could this be implemented across classrooms in even, say, Europe? Good education and, in particular, curriculum development depends upon sound financial support—how could this be achieved when many countries are on the brink of bankruptcy? What would happen in countries with totally different educational systems: comprehensive in many countries, tripartite in Germany, multilateral in the Netherlands. In England, the "public"—that is, independent private—schools do not have to follow the National Curriculum and increasingly do not use the state examinations, believing that these are insufficiently ambitious and do not cater for the needs of the more able student. Howson (1991) surveyed the curricula of the then 12 European Union countries, plus Japan and Hungary, and found not only interesting national deviations, but also differences within Germany. The "canonical" international curriculum is only the basic, shared common curriculum that is tested in international tests. Each country has its own individual add-ons and countries vary greatly concerning the hours devoted to mathematics teaching, the age at which they wish to introduce topics, and the length of time they consider it desirable to devote to a topic immediately after it is introduced (see, e.g., Howson, 1991, 1995; Howson, Harries, & Sutherland, 1999). It would appear that an "international curriculum," equally suited to the educational traditions (see, e.g., Leung, Graf, & López-Real, 2006), finances and aspirations of all countries, is a chimera that should not be chased. Moreover, the difficulties of actually effecting curriculum development even within a single country are so great that, if such an international curriculum could be established worldwide, it would prove almost impossible to change and develop it in any significant manner.

Toward a Common Influence of Public Examinations

The division into the intended, implemented, and attained curriculum is a useful one. It is clear where the responsibility lies for the curriculum that is intended, but the notion that the teacher is responsible solely for the implemented curriculum is far from the case in countries with public examinations, for it is those who set the examinations who carry considerable responsibility for what content is taught and the level of demands made for the analysis and synthesis of mathematical content and capabilities. Moreover, in some countries, for example, Denmark and England, officially-controlled testing now occurs more frequently throughout a student's school career than was the case 20 or so years ago. Again, the variety of means of examining at their disposal (e.g., oral, computer marked, and coursework) can either strengthen or limit examiners' abilities to examine important factors in mathematics learning. All these factors have had an immediate feedback in the classroom and have become a major influence on the implemented curriculum and classroom instruction. That is, if the testing does not include the assessment of certain topics, it is unlikely that classroom instruction will cover them.

For example, in the last 50 years arguments in favor of including mathematical modelling in the curriculum have frequently been advanced (Blum, Galbraith, Henn, & Niss, 2006). Recently, the Common Core State Standards in the USA clearly emphasized the importance of mathematical modelling (National Governors Association Center for Best Practices, & Council of Chief State School Officers, 2010). Yet how can mathematical modelling be adequately tested in a relatively short, timed examination? This was attempted in England by the School Mathematics Project in the 1960s, but the constraints imposed by the examination and the demands made on students by asking them to think under examination conditions about genuine modelling, as opposed to solving more involved yet still routine "real-life" problems, proved too great. Coursework could provide the answer, but problems then arise concerning plagiarism, and in a public examination, taken by tens of thousands of students, those marking the scripts have to have firm guidelines laid down for the awarding of marks. There is an inevitable recession to examinations containing only questions having a single correct answer. Yet public examinations in mathematics receive comparatively little attention from researchers. The research emphasis on assessment tends to veer towards teacher assessment—an important aspect of mathematics teaching—but, it could be argued, secondary to the overall control of the curriculum frequently exercised by the public examination. Brown (2006) explained how it was intended to assess primary school students' attainment under the National Curriculum then just imposed in England. Again, though admirable in its aims, this arrangement proved too costly and time-exhausting to implement. Simpler, less-informative, and essentially less-educational, methods replaced it. If by *curriculum* one means *syllabus content,* then there was no great loss, but if *curriculum* is taken to include the generation and cultivation of means of thought and expression, then little was left in the assessment and one suspects, accordingly, in the implemented curriculum.

There are particular concerns in China, which is one of the many countries where, to a great extent, the scores on examinations can determine one's opportunity of additional education and even one's future career (Cai & Nie, 2007). Many Chinese parents (and even teachers) believe that obtaining higher scores in examinations means being intellectually elite (Zhang, Tang, & Liu, 1991). At the same time, most students view examinations as competitions and filters for better opportunities. To a great degree, therefore, one of the main goals of classroom instruction is to prepare for examinations and to ensure high scores in examinations. The vast majority of problems in any examinations are related to basic knowledge and skills. Thus, the principal purpose of instruction in problem solving is interpreted in terms of helping students grasp basic knowledge and skills, so that, when examined, they can receive higher scores. Starting in the early 1990s, however, it was recommended to include some modelling (open-ended and real-life) problems in both the College Entrance Examinations and Senior High School Entrance Examinations.

Two reasons were advanced for integrating modelling problems into the Chinese College and Senior High School Entrance Examinations (Cai & Nie, 2007). First, the current mathematics curriculum reform in China emphasizes the development of students' abilities to pose, analyze, and solve problems. This new emphasis requires a corresponding response within assessment and evaluation. Second, given the nature of examination-driven instruction in China, the inclusion of such questions could be used as the driving force to integrate more modelling problems into school mathematics. That is, examinations were being used to influence classroom instruction in a positive way. Yet even this presents inherent problems, for using examinations as a means to influence teaching not only succeeds in driving teachers to teach what is examined (for that is in a sense desirable), but to teach *only* those aspects of modelling it is possible to examine within traditional examination constraints—and that will not necessarily achieve all the stated curricular aims.

Public examinations play, then, a great role in determining what actually happens in the classroom, and their influence can feed back into the intended curriculum. This process is almost certainly happening as a result of what are now "public examinations" shared between many countries; namely, TIMSS and PISA. Both of these were launched with the objectives of assisting the attainment of educational objectives, and to a certain extent these have been achieved, but perhaps more successfully away from the actual tests, for example, in the work on comparative curriculum analysis and the study, using video, of classroom methods in various countries, carried out by TIMSS.

A detailed critique of TIMSS can be found in Keitel and Kilpatrick (1998), and the whole problem of comparisons in mathematics education is considered in Kaiser, Luna, and Huntley (1998), the book in which Keitel and Kilpatrick's critique is included. More recently criticisms have also been raised concerning PISA and, in particular, the extent to which its "real-life" problems do, in fact, reflect real life for all those students tested and, in particular, the range of participating countries in which this is the case (Jablonka, 2007).

It is essential, then, that any country wishing to use TIMSS and PISA data to improve its mathematics teaching must carry out a careful study of responses to individual items. It is these, and their validity, that will carry "value for use." This, however, is not the view that ministers of education in many countries have taken: To them, the country's rank in the "league table" is all important. The result is that in some countries the mathematics curriculum is in danger of being circumscribed by what can be effectively examined by PISA and TIMSS. This change will ease the way "toward an international curriculum," but the losses could be huge.

Toward a Common Emphasis

Over the past 30 years there appears to have been some common emphases affecting school mathematics. Here we discuss two of these: the teaching of statistics and probability, and the teaching of algebra.

Statistics and probability began to enter the secondary school curriculum of some countries during the reforms of the 1960s. In the late 1980s and early 1990s, there were calls to increase the teaching of these subjects. Jones, Langrall, and Mooney (2007) analyzed curriculum documents from Australia, the USA, and the UK, and found remarkable similarities of big ideas for probability content across the three nations: nature of chance and randomness, sample space, probability measurement, and probability distribution. Jones et al. concluded that the presence of fundamental ideas in the three national curriculum documents "adds further grist to the argument that researchers across the international arena were influenced in comparable ways" (p. 915). A similar analysis of Chinese curriculum documents was recently undertaken and interestingly, there were similar emphases to be found in them.

Initially probability and statistics were thought of as secondary school mathematics (Jones & Tarr, 2010; Jones et al., 2007; Li, 2004). However, in recent years, statistical work on data display (e.g., bar charts, pictograms, pie charts) and analysis has been introduced and emphasized in elementary schools. For example, in the USA, the National Assessment of Educational Progress (NAEP) showed an increased emphasis on "data analysis" from 1996 to 2000 to 2003 (Kloosterman & Walcott, 2010). As part of the NAEP survey, teachers were asked to report curricular emphasis on particular strands. In 1996, only 8% of the fourth-grade teachers reported heavy emphasis of data analysis, but the percentage increased to 18% in 2000, and 23% in 2003. Elementary school work on probability has been more limited and has not proved so successful—perhaps because of the teachers' lack of understanding.

A second visible development is the early introduction of algebraic ideas (Cai & Knuth, 2011; Stacey, Chick, & Kendal, 2004). An important curricular emphasis, common around the globe, is the development of students' algebraic thinking in earlier grades. This is not a particularly new idea; in China and Russia, for example, algebraic concepts were introduced to elementary school students in the 1950s and 1960s. In other countries (e.g., in Europe and in North America), the discussion of integrating algebraic ideas into mathematics curricula in the earlier grades started in the 1970s. In the past decade, however, there has been an increased emphasis on and a wider acceptance of the advantages of developing students' algebraic ideas and thinking in earlier grades, and this new degree of acceptance is reflected in a number of influential policy documents.

There is a common theme in different countries that the curriculum is designed to help students see algebra in the context of arithmetic (Britt & Irwin, 2011; Cai, et al., 2010; Russell, Schifter, & Bastable, 2011; Subramaniam & Banerjee, 2011). For example, in the Chinese curriculum, elementary school students are asked to solve problems using both an arithmetical approach (no variables involved) and an algebraic one (involving variables), in the belief that this practice will help students develop ways of thinking about problem solving better than would be the case when such tasks were separated by some years.

In Singapore, the elementary mathematics curriculum provides a wide variety of experiences to help younger children develop algebraic thinking, and this development is made possible by using "model methods" or "pictorial equations" to analyze parts and wholes, generalize and specify, and do and undo. The model method is diagram- or model-drawing (Cai, Ng et al., 2011; Kho, 1987). It was believed that if children were provided with the means to visualize a word problem—be it a simple arithmetical word problem or an algebraic one—the structural underpinning of the problem would be made more apparently overt (Kho, 1987). In the earlier grades, pictures of real objects are initially used to model problem situations, but then the pictures are replaced by the more abstract rectangles. Elementary students solve word problems using the "model method" to construct pictorial equations that represent the relationship in word problems as a cohesive whole, rather than as distinct parts. The aim of using the model method described above is to provide a smooth transition from working with unknowns in a less abstract form to the more abstract use of letters in formal secondary-school algebra.

In India, the key aspect of early algebra learning is focussing on symbolic arithmetic as a preparation for algebra. Students work with numerical expressions (without letter variables), with the goal of building on the operational sense acquired through the experience of arithmetic. The aim is not just to compute the value of an expression, but to understand the structure of the expressions (Subramaniam & Banerjee, 2011). In the USA, similarly, students are provided with opportunities to develop algebraic thinking in the context of arithmetic (Russell et al., 2011).

These two examples serve to illustrate not only common international moves in curriculum development but also how these can spring from different forces. Knowledge of probability and statistics is of great importance for a well-educated citizen today; by providing this, educators are meeting national needs. Yet the changes in the teaching of algebra have been dictated by pedagogical reasons: the realization that the teaching of algebra, which is not intended to meet the same citizenship needs as that of probability and statistics, can be improved by the subject's early introduction. Here, then, we see examples of how two major pressures help generate curriculum development.

Toward Common Issues in the Mathematics Curriculum

In this section, we specifically focus our discussion on three common issues that need to be addressed in curriculum development: (a) nurturing creativity and thinking skills; (b) developing conceptual understanding and procedural skills; and (c) mathematics for all and mathematics for the gifted. The first two of these issues have been chosen because of their importance notwithstanding the difficulty that would seem to lie in their successful implementation. The last is a longstanding problem that changes in the organization of school systems in many countries in the last half of a century have tended to intensify.

It should be understood, however, that many other issues arise and can be approached in alternative ways. Thus, for example, Zalman Usiskin (2010) discussed several general issues in curriculum development, including pure versus applied mathematics, deduction versus induction versus statistical influence, algorithms versus problem solving, fluency versus flippancy, culture free versus culture dependent, and hard versus easy. While Usiskin situated his discussion in the context of curricula developed in the USA, these issues would, for example, also appear applicable when discussing an international curriculum.

Nurturing Creativity and Thinking Skills

We base the discussion of creativity and thinking skills on comparative research relating to the USA and some Asian countries. Early studies appear to suggest that although Chinese students perform better than their US counterparts on tasks requiring knowledge routinely learned in school, they may not be better creative thinkers than US students (Cai, 2000, 2001). For example, when solving process-open tasks, US students actually had higher mean scores than their Chinese counterparts. In fact, educators and government officials tend to believe that the USA does a better job of nurturing students' creativity than Asian countries do. Some Chinese leaders have openly criticized education to the age of 16 for giving students knowledge but not the ability to think creatively. In India, Prime Minister Manmohan Singh stated that two-thirds of the nation's universities and 90% of its degree-granting colleges failed to perform at expected norms and that university curricula typically failed to give emphasis to the educational skills and attitudes required by employers or job seekers (Bharucha, 2008). Yet, in the USA, several recent reports call for the nation to learn from Asia because it is believed that Asian countries such as China, India, and Singapore provide much more effective mathematics and science education, thus posing a major threat to the global competitiveness of the USA (Asian Society, 2006).

In contrast, some Asian-born scholars believe that Asian countries should learn from the USA about science, technology, engineering and mathematics (STEM) education (e.g., Bharucha, 2008; Zhao, 2008) because, it is argued, the USA does a better job of nurturing creativity. It is well-documented that classroom instruction in Asian countries is very traditional in many ways. It is often content-based, examination-driven, and teacher-centered (Fan, Wong, Cai, & Li, 2004). Classroom instruction is usually conducted in a whole-classroom setting, with a large class of 50–60 students and with little interaction among the students. In contrast, classroom instruction in the USA is usually conducted in small classes that encourage class participation. Students are encouraged to take intellectual risks and challenge accepted wisdom. Therefore, the learning environment in the USA may be perceived as nurturing students' creativity better than that in China, India, or other Asian countries, since students in the USA are nurtured better to tolerate deviation from tradition and the norm. Yet, the classrooms may simply reflect different cultures and traditions. Societies may even view creativity and the ways to nurturing creativities quite differently (Gardner, 1989).

Although it is generally believed that students in the USA are more creative than are those in Asian countries, the field lacks empirical studies to assess directly Chinese, Indian, and US students' *creativity,* however one chooses to interpret that word in general and, in particular, in mathematics. The real question is the following: Can a curriculum be designed and developed to nurture creative thinking skills, and how can those skills best be examined and assessed? This is, of course, a question faced in all countries, not simply in the USA and Asia. There is no obvious answer, although it is easy to determine methods and trends that militate against the achievement of such goals.

Developing Conceptual Understanding and Procedural Skills

The learning of mathematics involves both understanding an idea conceptually and being able to perform related procedures fluently. Experience and research have ably demonstrated that expertise in carrying out routine applications does not imply expertise in complex and novel problem solving. Routine applications can often be solved using procedural knowledge; in contrast, complex and novel problem solving usually requires that the solver uses conceptual knowledge and a synthesis of ideas in order to find a solution.

In a longitudinal study of curricular effect on students' learning, Cai, Moyer, Wang, and Nie (2011) found differential effects of "reformed curricula" on students' conceptual understanding and procedural skills in their LieCal Project (Longitudinal Investigation of the Effect of Curricula on Algebra Learning). The LieCal Project examined the similarities and differences between a reformed curriculum in the USA, called the Connected Mathematics Program (CMP), and more traditional curricula, called non-CMP curricula. CMP is a complete middle-school mathematics curriculum (Grades 6–8) and differences between the CMP curriculum and more traditional curricula are analyzed by Cai and his associates (Cai et al., 2010; Nie et al., 2009). The researchers investigated not only the ways and circumstances under which the CMP and non-CMP curricula promoted or hindered student achievement gains, but also the characteristics of the reform and traditional curricula that contributed to those gains. The longitudinal analyses showed that students did not sacrifice basic mathematical skills if they were taught using a reformed mathematics curriculum like CMP, but across the three middle-school years, students using the CMP curriculum showed significant gains over the non-CMP students on assessment items measuring conceptual understanding and problem solving (Cai, Moyer et al., 2011).

Another recent longitudinal study investigated whether or not the current curriculum reform in Mainland China brought about desirable student learning outcomes in elementary mathematics (Ni, Li, Cai, & Hau, in press; Ni, Li, Li, & Zhang, 2011). Improved performance was observed in the students of both groups over time on measures of computation, simple problem solving, and complex problem solving, which included process-constrained and process-open questions. However, although the reform group performed better than the non-reform group did on the complex problem-solving tasks, they did not do as well as the non-reform group on computation and simple problem solving. The findings from Ni, Li, Cai, and Hau suggested that, when using reformed curricula, students' conceptual understanding came at the expense of the development of basic mathematical skills; but is the tradeoff worth it? Ni, Li, Cai, and Hau argued that this tradeoff was indeed worthwhile and provided two reasons for their conclusion. The first reason was related to the new goals for the new curriculum: to improve students' competence in solving non-routine mathematics problems, an area of weakness in previous studies of Chinese students' mathematics achievement. This goal had been achieved. The second reason was that the students receiving the new curriculum still performed adequately with basic computations. Thus, a more balanced development in mathematics achievement had been attained.

How can educators ensure that the development of students' conceptual understanding does not come at the expense of the development of basic mathematical skills? Can students learn algorithms and master basic skills as they engage in explorations of mathematically intriguing problems? These are two of the fundamental questions that mathematics educators need to consider in curriculum development and research. However, this is not a new issue. A committee was established in the UK in 1946 to report on the teaching of primary school mathematics, and it determined that "[we] plead for attempts to develop mathematical ideas through the study of broad environmental topics and through the investigation of situations and phenomena at first hand" (Mathematical Association, 1955, p. 20). The committee also stressed that "practice without the power of mathematical thinking leads nowhere; the power of mathematical thinking without practice is like knowing what to do but not having the skills or tools to do it; but the power of mathematical thinking supported by practice and rote learning will give the best opportunity for all children to enjoy and pursue mathematics" (p. 4).

It is possible to develop and implement a curriculum for fostering students' conceptual understanding and problem solving. Much effort is needed by the international community to explore the ways of designing a curriculum that develops both basic mathematical skills and mathematical thinking.

Mathematics for All and Mathematics for the Gifted

Many countries are undergoing a mathematics education reform. These countries all face tensions and debates over issues involved in such a reform effort. One of these relates to the debate about mathematics for all and mathematics for the gifted. In some countries, such as the USA and Singapore, gifted education is regulated under state or national laws to develop systematically and strategically the potential of these students. In other countries, such as China, there is no law or policy pertaining to the education of gifted students.

In the USA, developing and discovering scientific talent is a national strategic goal (National Science Board, 2010). In a recent report, the National Science Foundation pointed out:

> The Board's 2-year examination of this issue made clear one fundamental reality: *the U.S. education system too frequently fails to identify and develop our most talented and motivated students who will become the next generation of innovators.* ... The possibility of reaching one's potential should not be met with ambivalence, left to chance, or limited to those with financial means. Rather, the opportunity for excellence is a fundamental American value and should be afforded to all. (p. 5)

Similarly, in China, developing a talent pool with creativity and innovation is the overarching goal of the China's next 5-year plan for the nation.

The tension about education for all and for the gifted is a problem faced in all countries and one that has been exacerbated over the last 50 years by the growth of comprehensive education—in opposition to selective education, which apportioned students to different types of school at a comparatively early age. Although it is important to make mathematics accessible for all students, meeting the special needs of mathematically gifted students is also critical. It cannot be assumed that mathematically talented students can succeed on their own or will wish to continue with the study of mathematics if they are not provided with challenges appropriate to their capabilities. There is a clear need, then, for all countries to address the needs of gifted students within the context of mathematics for all. One strategy for countries is to have different curricula for different students. Where this strategy has been ruled out, an important question arises: Should gifted students be accelerated through the grades, or should there be provision for their mathematics learning to be enhanced in some manner while they remain with their age group? Questions of equity and quality are indeed extremely important ones and are explored more fully in an international context in Atweh et al. (2011) and in Chapter 1.

Conclusion

Education is commonly seen as the key to a nation's economic growth and prosperity and to its ability to compete in the global economy. As already described, the curriculum plays a vital role in students' education, and there is an increasing tendency for countries to include a more-and-more internationally accepted core selection of topics in their national curricula. This tendency is to a great extent both to be welcomed and expected. Students do need to gain mathematical literacy in the global arena, and what constitutes such literacy does not differ greatly across nations. This move has, however, had a potential negative effect on curriculum development, for overmuch attention is now being given to what can be covered in national examinations and international assessments such as TIMSS and PISA tests. That is, a wider view of mathematics learning, other than simply the acquisition of an acceptable degree of "literacy," can become ignored. Indeed, a wider view of education, itself, is under threat. A concentrated basic curriculum, simply targeted at all students attaining good marks in tests of mathematical literacy, could be harmful for teachers, students, and mathematics education in general.

Nearly 30 years ago, at a talk given at a 1983 symposium organized by the ICMI at the International Congress of Mathematicians in Warsaw (see Damerow, Dunkley, Nebres, & Werry, 1984), Jan de Lange asked the following question: "Does 'mathematics for all' mean 'no mathematics for all'?" His answer was no. We can ask a similar question nowadays: "Would an international mathematics curriculum mean no mathematics for all?" Again, the answer is no, but it could mean that vital elements of mathematics become neglected or ignored. Would there, for example, be geometry (other than mensuration) for all? Indeed, geometry (which has been rarely mentioned so far in this chapter) presents particular problems (Mammana & Villani, 1998). There is, in fact, no agreement on what should comprise a geometry curriculum within, for example, China, or England (Royal Society, 2001). Seeking international agreement on a geometry curriculum could be impossible, for since the "fall" of Euclid, countries have largely gone their own way. The only way ahead would seem to be through experiments in a number of separate countries in the hope that some satisfactory approach that can be more universally adopted emerges. However, it is clear that what at present constitutes the common core curriculum offers little insight into what mathematics can offer other than an extremely useful, indeed essential, tool kit.

The arithmetical and algebraic content now to be found in that curriculum has not changed significantly in the last century. It is, then, to be expected that there is common agreement both relating to its acceptance and, to some extent, ways in which it should be taught. However, it should be noted that even so, the introduction of concepts and topics within a nation's curriculum can differ considerably: "It is essentially a core mathematical *syllabus* that is, more or less, universally shared." For example, Cai and his associates (Cai et al., 2010; Nie et al., 2009) analyzed the way in which the concepts of variables, equations, and equation solving were introduced in reform and traditional mathematics curricula in the USA. They found totally different approaches to these concepts, with the reform curriculum focussing on a functional approach and the traditional curriculum focussing on a structural approach. Statistics at an elementary level has also reached a canonical degree of acceptance, although there are differences in how far one might develop the teaching of that and probability at a higher school level. However, many important questions remain relating to a possible international curriculum to which we appear to be heading. Is there the mathematics that will catch the imagination or astonish, nurture creativity, prepare for logical proof and a search for generality? Such mathematics may well not be suitable for, or within the grasp of, all, but what will be the consequences for the future development of mathematics of not offering it to the most gifted? How, then, does one plan a curriculum to cope with students of widely differing mathematical abilities?

Significant work also remains to be done to explore the way in which new technology could affect both the mathematics included in the curriculum and how it could more effectively contribute to the teaching and learning of mathematics in general (Atkinson & Mayo, 2010; Clark-Wilson, Oldknow, & Sutherland, 2011). The influence of technology has still to be felt on the actual mathematics taught in schools. Even the pioneering work on how technology can assist in the teaching of well-established mathematical content, for example, in geometry, has had little take-up. With the growing emphasis on processes within national curricula (e.g., problem solving, communication), the impact of technology in terms of the kind of dynamic learning environment it can create merits more attention (Hoyles & Lagrange, 2010). Most importantly, the impact of technology should also be examined with respect to the delivery format of a curriculum. With the advance of digital technology and accessibility of technologies, perhaps a digital form of an international *core* curriculum can be a reality in the near future (Atkinson & Mayo, 2010; Clark-Wilson et al., 2011; Kim, 2011), but here it is important to distinguish between a citizenship-based and politician-pleasing core and a sounder, more mathematically-inspired, complete curriculum. Already the national curricula of a growing number of countries can be found on the ICMI Web site, and these should provide guidance to countries wishing to renew their mathematics curricula.

In summary, curriculum development aimed at other ends than simply improving the teaching and learning of accepted curriculum content—although that in itself is a worthy aim—depends upon the freedom of individual countries to experiment within the confines of their curriculum. The absence of that freedom, through the emergence of an all-embracing international curriculum, would seriously prevent any future developments in the curriculum as a whole. Here, it is significant to point out that following the curricular reforms of the 1960s, and despite the enormous technological advances that have occurred since then, little, if any, new mathematics has entered the school curriculum and little, apart from logarithms for the gifted, has left it. Indeed, the most noteworthy attempt to spell out a model curriculum for the early 21st century, the Kahane Commission in France, was carried out as a hypothetical exercise intended to justify and define the place of mathematics in the school curriculum, but whose findings, it was accepted, could never be implemented—see Merle (2003) for a brief description of the findings.

It is also essential that countries retain the freedom to decide upon the ages at which topics are introduced to students for this will vary much according to local social and environmental circumstances. Let countries then rejoice in what they share in common with other countries—that essential *core* of mathematical literacy—and may comparative mathematics education flourish. But may they also be aware that the present international "common core" cannot constitute a satisfactory curriculum in itself and that new initiatives within the curriculum must come from within individual countries and will only be accepted into an international core over the years. Most importantly, and a lesson from the 1960s, is that the school mathematics curriculum is not something that can be forever extended without some paring down of what already has a place in it. Moreover, in recognizing that technology is at the core of virtually every aspect of our daily lives and work, mathematics educators must endeavour both to develop the content of syllabuses bearing in mind the new possibilities it presents, and also employ it to provide engaging and powerful learning experiences for students whatever topics in mathematics they may be studying.

Acknowledgements The first author would like to acknowledge the support of grants from the National Science Foundation (ESI-0454739 and DRL-1008536). When the first draft of this chapter was written, the first author was invited to serve as a Program Director at the Division of Research on Learning, the US National Science Foundation. The support of the US National Science Foundation is greatly appreciated, but any opinions expressed herein are those of the authors and do not necessarily represent the views of the National Science Foundation. We are grateful for the assistance of Dr Bikai Nie in preparing the reference list. We are also grateful for the insightful comments provided by Jeremy Kilpatrick, Gilah Leder, and Paola Valero on an earlier version of this chapter.

References

Arnold, M. (2008). *Schools and universities on the continent* (Original work published 1868). Whitefish, MT: Kessinger.

Asian Society. (2006). *Math and science education in a global age: What the U.S. can learn from China*. New York, NY: Author.

Atkinson, R. D., & Mayo, M. (2010). *Refueling the U.S. innovation economy: Fresh approaches to science, technology, engineering and mathematics (STEM) education*. Washington, DC: The Information Technology and Innovation Foundation.

Atweh, B., Calabrese Barton, A., Borba, M., Gough, N., Keitel, C., Vistro-Yu, C., & Vithal, R. (Eds.). (2007). *Internationalisation and globalisation in mathematics and science education*. Dordrecht, The Netherlands: Springer.

Atweh, B., Graven, M., Secada, W., & Valero, P. (Eds.). (2011). *Mapping equity and quality in mathematics education*. Dordrecht, The Netherlands: Springer.

Ball, D. L., & Cohen, D. K. (1996). Reform by the book: What is—or might be—the role of curriculum materials in teacher learning and instructional reform? *Educational Researcher, 25*(9), 6–8, 14.

Basic Education Curriculum Material Development Center, Chinese Ministry of Education. (2001). *National mathematics curriculum standards at the compulsory education level (draft for consultation)* [in Chinese]. Beijing, China: Beijing Normal University.

Bharucha, J. (2008, January 25). America can teach Asia a lot about science, technology, and math. *Chronicle of Higher Education, 54*(20). Retrieved from http://chronicle.com/weekly/v54/i20/20a03301.htm.

Blum, W., Galbraith, P., Henn, H-W., & Niss, M. (Eds.). (2006) *Applications and modelling in mathematics education* (New ICMI Studies Series No. 10). New York, NY: Springer.

Bradburn, M. B., & Gilford, D. M. (1990). *A framework and principles for international comparative studies in education*. Washington, DC: National Academies Press.

Britt, M. S., & Irwin, K. C. (2011). Algebraic thinking with and without algebraic representation: A pathway for learning. In J. Cai & E. Knuth (Eds.), *Early algebraization: A global dialogue from multiple perspectives* (pp. 137–159). New York, NY: Springer.

Brown, G. (2006, January 14). *The future of Britishness*. Speech to the Fabian Society New Year Conference. Retrieved from http://www.fabians.org.uk/events/speeches/the-future-of-britishness.

Burkhardt, H., Fraser, R., & Ridgway, J. (1990). The dynamics of curriculum change. In I. Wirszup & R. Streit (Eds.), *Developments in school mathematics education around the world* (Vol. 2, pp. 3–30). Reston, VA: National Council of Teachers of Mathematics.

Cai, J. (2000). Mathematical thinking involved in U.S. and Chinese students' solving process-constrained and process-open problems. *Mathematical Thinking and Learning, 2*, 309–340.

Cai, J. (2001). Improving mathematics learning: Lessons from cross-national studies of U.S. and Chinese students. *Phi Delta Kappan, 82*(5), 400–405.

Cai, J., & Knuth, E. (Eds.). (2011). *Early algebraization: A global dialogue from multiple perspectives*. New York, NY: Springer.

Cai, J., Moyer, J. C., Wang, N., & Nie, B. (2011). Examining students' algebraic thinking in a curricular context: A longitudinal study. In J. Cai & E. Knuth (Eds.), *Early algebraization: A global dialogue from multiple perspectives* (pp. 161–186). New York, NY: Springer.

Cai, J., Ng, S. F., & Moyer, J. C. (2011). Developing students' algebraic thinking in earlier grades: Lessons from China and Singapore. In J. Cai & E. Knuth (Eds.), *Early algebraization: A global dialogue from multiple perspectives* (pp. 25–42). New York, NY: Springer.

Cai, J., & Nie, B. (2007). Problem solving in Chinese mathematics education: Research and practice. *ZDM—International Journal on Mathematics Education, 39*, 459–473.

Cai, J., Nie, B., & Moyer, J. (2010). The teaching of equation solving: Approaches in *Standards*-based and traditional curricula in the United States. *Pedagogies: An International Journal, 5*(3), 170–186.

Christiansen, B., Howson, G., & Otte, M. (Eds.). (1986). *Perspectives on mathematics education*. Dordrecht, The Netherlands: Reidel.

Clandinin, D. J., & Connelly, F. M. (1992). Teacher as curriculum maker. In P. Jackson (Ed.), *Handbook of research in curriculum* (pp. 363–401). New York, NY: Macmillan.

Clarke, B., Clarke, D. M., & Sullivan, P. (1996). The mathematics teacher and curriculum development. In A. J. Bishop, K. Clements, C. Keitel, J. Kilpatrick, & C. Laborde (Eds.), *International handbook of mathematics education* (pp. 1207–1233). Dordrecht, The Netherlands: Kluwer.

Clark-Wilson, A., Oldknow, A., & Sutherland, R. (2011). *Digital technologies and mathematics education*. London, UK: Joint Mathematical Council of the United Kingdom.

Cockroft, W. H. (1982). *Mathematics counts: Report of the committee of inquiry into the teaching of mathematics in school*. London, UK: Her Majesty's Stationery Office.

Cuban, L. (1992). Curriculum stability and change. In P. W. Jackson (Ed.), *Handbook on research on curriculum* (pp. 216–247). New York, NY: Macmillan.

Damerow, P., Dunkley, M. E., Nebres, B. F., & Werry, B. (Eds.). (1984). *Mathematics for all* (Science and Technology Education Document Series No. 20). Paris, France: UNESCO.

Darling-Hammond, L. (1993, June). Reframing the school reform agenda. *Phi Delta Kappan, 74*(10), 752–761.

Department for Education and Employment and Qualifications and Curriculum Authority (DfEE/QCA). (2000). *Curriculum guidance for the foundation stage*. London, UK: Author.

Fan, L., Wong, N.-Y., Cai, J., & Li, S. (Eds.). (2004). *How Chinese learn mathematics: Perspectives from insiders*. Singapore: World Scientific.

Friedman, T. L. (2005). *The world is flat: A brief history of the twenty-first century*. New York, NY: Farrar, Straus, & Giroux.

Gardner, H. (1989). *To open minds*. New York, NY: Basic Books.

Hershkowitz, R., Dreyfus, T., Ben-Zvi, D., Friedlander, A., Hadas, N., Resnick, T., Tabach, M., & Schwarz, B. (2002). Mathematics curriculum development for computerized environments. A designer-researcher-teacher-learner activity. In L. D. English, M. Bartolini Bussi, G. A. Jones, R. A. Lesh, & D. Tirosh (Eds.), *Handbook of international research in mathematics education: Directions for the 21st century* (pp. 657–694). Mahwah, NJ: Erlbaum.

Howson, A. G. (1991). *National curricula in mathematics*. Leicester, UK: Mathematical Association.

Howson, A. G. (1995). *Mathematics textbooks: A comparative study of Grade-8 texts*. Vancouver, Canada: Pacific Educational Press.

Howson, A. G. (2002). Advanced mathematics: Curricula and student performance. In D. E. Robitaille & A. E. Beaton (Eds.), *Secondary analysis of the TIMSS data* (pp. 113–123). New York, NY: Kluwer.

Howson, A. G., Harries, T., & Sutherland, R. (1999). *Primary school mathematics textbooks*. London, UK: Qualifications and Curriculum Authority.

Howson, A. G., Keitel, C., & Kilpatrick, J. (1981). *Curriculum development in mathematics*. Cambridge, UK: Cambridge University Press.

Howson, A. G., & Wilson, B. (1986). *School mathematics in the 1990s*. Cambridge, UK: Cambridge University Press.

Hoyles, C., & Lagrange, J.-B. (Eds.). (2010). *Mathematics education and technology—Rethinking the terrain: The 17th ICMI Study*. New York, NY: Springer.

Jablonka, E. (2007). The relevance of modelling and applications: Relevant to whom and for what purpose? In W. Blum, P. Galbraith, H.-W. Henn, & M. Niss (Eds.), *Modelling and applications in mathematics education: The 14th ICMI Study* (pp. 193–200). Berlin, Germany: Springer.

Jackson, P. (1992). *Handbook of research on curriculum*. New York, NY: Macmillan.

Jones, G. A., Langrall, C. W., & Mooney, E. S. (2007). Research in probability: Responding to classroom realities. In F. K. Lester Jr. (Ed.), *Second handbook of research on mathematics teaching and learning* (pp. 909–956). Charlotte, NC: Information Age.

Jones, D., & Tarr, J. E. (2010). Recommendations for statistics and probability in school mathematics over the past century. In B. J. Reys, R. E. Reys, & R. Rubenstein (Eds.), *K–12 mathematics curriculum: Issues, trends, and future directions* (72nd yearbook of the National Council of Teachers of Mathematics, pp. 65–76). Reston, VA: NCTM.

Kaiser, G., Luna, E., & Huntley, I. (Eds.). (1998). *International comparisons in mathematics education*. London, UK: Falmer.

Keitel, C., & Kilpatrick, J. (1998). The rationality and irrationality of international comparative studies. In G. Kaiser, E. Luna, & I. Huntley (Eds.), *International comparisons in mathematics education* (pp. 241–256). London, UK: Falmer.

Kho, T. H. (1987). Mathematical models for solving arithmetic problems. *Proceedings of the Fourth Southeast Asian Conference on Mathematical Education (ICMI–SEAMS). Mathematical education in the 1990s* (pp. 345–351). Singapore: National Institute of Education.

Kim, S. (2011, August 10). South Korea powers toward an all-digital scholastic network. *Education Week, 30*(37), 13.

Kloosterman, P., & Walcott, C. (2010). What we teach is what students learn: Evidence from the National Assessment. In B. Reys, R. Reys, & R. Rubenstein (Eds.), *K–12 mathematics curricu-*

lum: Issues, trends, and future directions (72nd yearbook of the National Council of Teachers of Mathematics, pp. 89–102). Reston, VA: NCTM.

Leung, F. K. S., Graf, K.-D., & López-Real, F. J. (Eds.). (2006). *Mathematics education in different cultural traditions*. New York, NY: Springer.

Li, J. (2004). Teaching approach: Theoretical or experimental? In L. Fan, N. Y. Wong, J. Cai, & S. Li (Eds.), *How Chinese learn mathematics: Perspectives from insiders* (pp. 443–461). Singapore: World Scientific.

Mammana, C., & Villani, V. (Eds.). (1998). *Perspectives on the teaching of geometry for the 21st century*. Dordrecht, The Netherlands: Kluwer.

Mathematical Association. (1955). *The teaching of mathematics in primary schools*. London, UK: Bell.

Mathematical Sciences Education Board. (1993). *Measuring what counts: A conceptual guide for mathematics assessment*. Washington, DC: National Academy Press.

Medrich, E. A., & Griffith, J. E. (1992). *International math and science assessment: What have we learned?* (U.S. Department of Education Report No. NCES92011). Washington, DC: Office of Educational Research and Improvement and National Center for Education Statistics.

Merle, M. (2003). Defining mathematical literacy in France. In B. L. Madison & L. A. Steen (Eds.), *Quantitative literacy: Why numeracy matters for schools and college* (pp. 221–223). Princeton, NJ: National Council on Education and the Disciplines.

Mullis, I. V. S., Martin, M. O., Beaton, A. E., Gonzalez, E. J., Kelly, D. L., & Smith, T. A. (1998). *Mathematics and science achievement in the final years of secondary school: IEA's Third International Mathematics and Science Study (TIMSS)*. Chestnut Hill, MA: Center for the Study of Testing, Evaluation, and Educational Policy, Boston College.

National Academy of Education. (1999). *Recommendations regarding research priorities: An advisory report to the National Educational Research Policy and Priorities Board*. New York, NY: Author.

National Council of Teachers of Mathematics. (1989). *Curriculum and evaluation standards for school mathematics*. Reston, VA: Author.

National Governors Association Center for Best Practices, & Council of Chief State School Officers. (2010). *Common Core State Standards: Mathematics*. Retrieved from http.//www.corestandards.org/assets/CCSSI_MathStandards.pdf.

National Research Council. (2004). *On evaluating curricular effectiveness: Judging the quality of K–12 mathematics evaluations*. Washington, DC: National Academies Press.

National Science Board (NSB). (2010). *Preparing the next generation of STEM innovators: Identifying and developing our nation's human capital*. Arlington, VA: National Science Foundation.

Ni, Y., Li, Q., Cai, J., & Hau, K.-T. (in press). Has curriculum reform made a difference in classroom? An evaluation of the new mathematics curriculum in Mainland China. In B. Sriraman, J. Cai, K.-H. Lee, L. Fan, Y. Shimuzu, L. C. Sam, & K. Subramanium (Eds.), *The first sourcebook on Asian research in mathematics education: China, Korea, Singapore, Japan, Malaysia and India*. Information Age: Charlotte, NC.

Ni, Y., Li, Q., Li, X., & Zhang, Z.-H. (2011). Impact of curriculum reform: An analysis of student mathematics achievement in Mainland China. *International Journal of Educational Research, 50*(2), 100–116.

Nie, B., Cai, J., & Moyer, J. C. (2009). How a standards-based mathematics curriculum differs from a traditional curriculum: With a focus on intended treatments of the ideas of variable. *ZDM—International Journal on Mathematics Education, 41*(6), 777–792.

Organisation for Economic Co-operation and Development. (1999). *Measuring student knowledge and skills: A new framework for assessment*. Paris, France: Author.

Pinar, W. (Ed.). (2003). *International handbook of curriculum research*. Mahwah, NJ: Erlbaum.

Postlethwaite, T. N. (Ed.). (1988). *The encyclopedia of comparative education and national systems of education*. Oxford, UK: Pergamon.

Resnick, L. (1987). *Education and learning to think*. Washington, DC: National Academy Press.

Reys, B. J., Reys, R. E., & Rubenstein, R. (Eds.). (2010). *Mathematics curriculum: Issues, trends, and future directions* (72nd yearbook of the National Council of Teachers of Mathematics). Reston, VA: NCTM.

Robitaille, D. F., & Garden, R. A. (1989). *The IEA Study of Mathematics II: Contexts and outcomes of school mathematics.* New York, NY: Pergamon.

Romberg, T. (2002). Thirty years of mathematics education research. *Wisconsin Center for Educational Research Highlights, 14*(3), 1–3.

Royal Society. (2001). *Teaching and learning geometry* (pp. 11–19). London, UK: Author.

Russell, S. J., Schifter, D., & Bastable, V. (2011). Developing algebraic thinking in the context of arithmetic. In J. Cai & E. Knuth (Eds.), *Early algebraization: A global dialogue from multiple perspectives* (pp. 43–69). New York, NY: Springer.

Senk, S. L., & Thompson, D. R. (Eds.). (2003). *Standards-based school mathematics curricula: What are they? What do students learn?* Mahwah, NJ: Erlbaum.

Skovsmose, O. (2003). *Uncertainty and responsibility: Notes about aporia.* Aalborg, Denmark: Department of Education and Learning, Aalborg University.

Skovsmose, O., & Valero, P. (2008). Democratic access to powerful mathematical ideas. In L. D. English (Ed.), *Handbook of international research in mathematical education: Directions for the 21st century* (2nd ed., pp. 415–438). Mahwah, NJ: Erlbaum.

Smith, M. K. (2000). Curriculum theory and practice. *Encyclopedia of informal education.* (Original work published 1996). Retrieved from http://www.infed.org/biblio/b-curric.htm.

Sriraman, B., & Törner, G. (2008). Political union/mathematical education disunion: Building bridges in European didactic traditions. In L. English, M. Bartolini Bussi, G. A. Jones, R. Lesh, B. Sriraman, & D. Tirosh (Eds.), *Handbook of international research in mathematics education* (2nd ed., pp. 656–690). New York, NY: Routledge, Taylor & Francis.

Stacey, K., Chick, H. M., & Kendal, M. (Eds.). (2004). *The future of the teaching and learning of algebra.* Dordrecht, The Netherlands: Kluwer.

Subramaniam, K., & Banerjee, R. (2011). The arithmetic-algebra connection: A historical-pedagogical perspective. In J. Cai & E. Knuth (Eds.), *Early algebraization: A global dialogue from multiple perspectives* (pp. 87–107). New York, NY: Springer.

Travers, K. J., & Westbury, I. (Eds.). (1989). *The IEA Study of Mathematics I: Analysis of mathematics curricula.* Oxford, UK: Pergamon.

Usiskin, Z. (2010). The current state of the school mathematics curriculum. In B. J. Reys, R. E. Reys, & R. Rubenstein (Eds.), *K–12 mathematics curriculum: Issues, trends, and future directions* (72nd yearbook of the National Council of Teachers of Mathematics, pp. 25–40). Reston, VA: NCTM.

Usiskin, Z., & Willmore, E. (Eds.). (2008). *Mathematics curriculum in Pacific Rim countries: China, Japan, Korea, and Singapore.* Charlotte, NC: Information Age Publishing.

van den Akker, J. J. H., Kuiper, W., & Hameyer, U. (Eds.). (2003). *Curriculum landscape and trends.* Dordrecht, The Netherlands: Kluwer Academic Publishers.

Vithal, R., & Volmink, J. (2005). Mathematics curriculum research: Roots, reforms, reconciliation and relevance. In R. Vithal, J. Adler, & C. Keitel (Eds.), *Researching mathematics education in South Africa: Perspectives, practices and possibilities* (pp. 3–27). Cape Town, South Africa: Human Sciences Research Council.

Wong, N. Y. (2004). The CHC learner's phenomenon: Its implications on mathematics education. In L. Fan, N. Y. Wong, J. Cai, & S. Li (Eds.), *How Chinese learn mathematics: Perspectives from insiders* (pp. 503–534). Singapore: World Scientific.

Zhang, D., Tang, R., & Liu, H. (1991). *Pedagogy of mathematics* [in Chinese]. Nanchang, China: Jiangxi Educational Press.

Zhao, Y. (2008, February). What knowledge has the most worth? Reconsidering how to cultivate skills in U.S. students to meet the demands of global citizenry. *School Administrator, 65*(2), 20–27. Retrieved from http://www.aasa.org/SchoolAdministratorArticle.aspx?id=6032&terms=zhao.

Chapter 30
Methods for Studying Mathematics Teaching and Learning Internationally

Mogens Niss, Jonas Emanuelsson, and Peter Nyström

Abstract The focus of this chapter is issues related to methods for studying mathematics teaching and learning internationally. The chapter identifies three sorts of overarching purposes and goals of international studies, namely to uncover and analyze, across a group of countries: differences in students' learning outcomes, achievements and attitudes; differences in curricula, teaching approaches, resources and the environments of mathematics education; and possible links between the latter and the former. The chapter provides detailed accounts of the designs, methods, methodologies, and instruments that have been used in two kinds of studies—large-scale international comparative studies, such as TIMSS and PISA, and so-called focal studies concentrating on more specific *problématiques* or themes. The last part of the chapter offers reflections on the nature of international comparative studies with an emphasis on their strengths and potentials as well as on their challenges and limitations. One fundamental question in this context is the extent to which the results of such studies can be meaningfully interpreted, especially in view of the massive interest amongst politicians, administrators, media, and the general public, who often do not pay sufficient attention to the characteristics and conditions of the studies.

M. Niss (✉)
Roskilde University, Roskilde, Denmark
e-mail: mn@ruc.dk

J. Emanuelsson
University of Gothenburg, Gothenburg, Sweden

P. Nyström
Umeå University, Umeå, Sweden

Introduction: The Relationship Between Study Issues and Methodology

Since the creation of the International Commission on Mathematical Instruction (ICMI) in 1908 (Schubring, 2008), there has been an interest in considering mathematics teaching and learning from an international perspective. Until the 1960s, the focus was on describing and comparing mathematics curricula across different countries, or on proposing—from normative points of view—new curriculum approaches or components (such as the notion of function in the early decades of the 20th century or the so-called new math or modern mathematics movement from the mid-1950s to the mid-1970s). When the international congresses on mathematical education (the ICMEs) came into being (the first one was held in Lyon, France, in 1969), the majority of the contributions in the early ICMEs were designed to exchange information, views, and experiences amongst delegates from different countries about the actual or potential structures of mathematics curricula, the orchestration of teaching, teaching materials and resources, teaching experiments, and—to a lesser extent—student reactions to the "diets" they were offered.

Even though it dates back to the beginning of the 20th century, the sharing of information, ideas, and experiences has never ceased to be of interest. For example, the so-called International Seminar at the Park City Mathematics Institute (PCMI), held under the auspices of the Princeton Institute for Advanced Study every summer in Park City, Utah, USA, has provided a platform for such exchange since 2001.

The goal of all these endeavours has been to allow participants to learn from each other in terms of ideas, approaches, materials for teaching, and the reported outcomes thereof. Even though selecting, collecting, and presenting the factual information involved in these activities may well have been difficult and time consuming in places, it would not be reasonable to say that these endeavours amount to *studying* mathematics teaching and learning internationally in a scholarly or scientific sense. Studying something is closely linked to trying to come to grips with essential features of or issues related to the objects, situations, or systems to be studied; in other words, seeking answers to pertinent questions by way of some investigation, a disciplined inquiry. Studying something is usually focussed on uncovering and explaining relationships, with particular regard to mechanisms, correlations, and causalities. Therefore, any discussion of the choice and implementation of the methods to be put to use in a study must take its point of departure in the issues and questions that the investigation is designed to address. So, what are the issues addressed and the questions asked in studying mathematics teaching and learning internationally? And what are individuals' and agencies' (or even countries') purposes of engaging in such studies? This is related to the question asked by Clarke (2003) with regard to *comparative research*: "Who are the stakeholders of international comparative research?" (p. 151).

In the sections that follow, we provide more specific and detailed answers to these questions as far as the most important international studies are concerned, of which the first seems to be the so-called FIMS—First International Mathematics Study—which was carried out in 1964 (see below). However, at an overall level it is fair to claim that most international studies are designed to deal with three major *problématiques*: The first is to uncover and analyze *differences in students' learning outcomes, achievement, and attitudes* across a group of countries. The second is to uncover and analyze *differences in curricula, teaching approaches, resources for teaching, classroom cultures, teachers' educational and other backgrounds, and more general cultural and socio-economic environments of mathematics education.* The third, and often the most significant, is to *link* the former *problématique* to the latter; in particular, in order to come to understand, if possible, the former as a function of the latter. It goes without saying that the methodological deliberations and issues arising in this context (should) depend heavily on the quantitative and qualitative characteristics of the students considered; on the specific learning outcomes, kinds of achievement, and sorts of attitude in focus; on the cultural, societal, economic, and institutional conditions of the countries involved; and on those aspects of teaching approaches and resources, classroom cultures, and teacher backgrounds that are selected to be of interest in the investigation. Clarke (2003) adds a twist to the third *problématique*; namely, what he calls "evaluative comparisons: not just to document similarities and differences, but attaching value to performances judged as superior by some criterion" (p. 152).

Against this background, one may well raise the more general question of the extent to which it makes sense, and is methodologically feasible, to detect, investigate, and interpret differences and to make comparisons across and among countries with particular regard to mathematics education, when multitudes of cultural, societal, and economic and other factors exert predominant influences on the systems in which mathematics education takes place. We return to this issue later in this chapter.

In dealing with issues concerning study methods, a number of words almost automatically enter the stage: *design, method, methodology, instrument, technique,* and *procedure,* among others. Transparency in deliberations and exposition requires some clarification of what these terms are supposed to mean. If we take our point of departure in the idea that scholarly and scientific studies are undertaken in order to answer certain more or less clearly delineated questions (Niss, 2010), we propose the following definitions in the present context.

By the term *design* of a study, we understand the entire *collection of approaches* (whether conceptual, theoretical, or empirical) employed *to provide answers* to the set of questions that drive the study; in other words, the overall *layout* of the study. Each approach is focussed on answering a subset of the questions (but several approaches may be used, e.g., in combination, to answer the same question) and hence gives rise to issues of *methodology.* By *methodology,* we understand the set of deliberations, reflections, and analyses involved in choosing, implementing, and assessing one or more *methods* with a potential to answer a certain class of questions. Typically this involves comparing, contrasting, and relating different actual and potential methods with particular regard to their potentialities, limitations, and tractability in the given context and under the circumstances present. So, we use the term *method* to designate a package of specific undertakings by which a certain class of questions may be answered, and the term *methodology* to include all meta-level considerations about methods. Adopting a particular method as a means for answering certain questions presupposes the belief that the method actually can, or at least has the potential to, provide valid answers to the questions. A method may be established and well-described, but it may also be in a process of inception or under construction for a certain purpose. Implementing a method normally involves putting a number of *instruments* to use. Typically an instrument—say, a questionnaire—is not restricted to be part of a particular method but will be available for use in several different methods. Finally, using an instrument often requires the activation of various more or less specific *techniques*, some of which may take the form of standardized *procedures*, whereas others may be more loosely defined. In the following sections, these rather general definitions are given flesh and blood when we deal with concrete studies.

This chapter is structured as follows: In the next two sections, we attempt to provide factual presentations, without much commentary, of the studies under consideration in the chapter, including their goals, designs, and methods. In the last section, we offer our more analytic reflections on key issues related to those and other studies.

Different Kinds of Studies and Their Goals

In gross terms we deal with two kinds of internationally-oriented studies of mathematics teaching and learning. The first kind consists of *large-scale international comparative studies*, where the term *large-scale* refers to at least two features—the involvement of a multitude of countries and of large numbers of students. Sometimes *large-scale* also means "many dimensions," such as student achievement and affect, socio-economic background variables, structure of education systems, curriculum organization, approaches to teaching, and teacher backgrounds. Studies of the second kind, let us agree to call them *focal studies*, have a narrower focus—for example, problem solving, curriculum structure, textbooks, classroom interaction—and typically involve just a few countries. Large-scale studies—which almost by definition require huge efforts and human and material resources, including funding, and are time consuming—tend to attract a lot of public interest and debate, especially if league tables are included in the reporting, whereas focal studies rather attract the attention of mathematics educators and researchers, and occasionally of politicians dealing with education.

Large-Scale Studies

We begin by listing the international large-scale studies that are taken into consideration in this chapter. Because of the resources required to undertake large-scale studies, there are not so many of them. Although comparative international studies of education at large have a long history (Kaiser, 1999a), as previously mentioned the first large-scale comparative international study of *mathematics* was the FIMS. It was produced and published by the IEA, the International Association for the Evaluation of Educational Achievement, which was created by a group of educationists in 1958 and established as a legal entity based in the Netherlands in 1967. The study was designed and conducted during the years 1961–1964, and students' achievements in mathematics in 12 countries were tested in 1964 (Freudenthal, 1975). The outcomes were reported in 1967 (Husén, 1967). Freudenthal (1975) made the following comments on the aims of FIMS:

> The overall aim is, with the aid of psychometric techniques, to compare outcomes in different educational *systems*. The fact that these comparisons are cross-national should not be taken as an indication that the primary interest was, for instance, national means and dispersions in school achievement at certain age and school levels. ...
>
> The main objective of the study is to investigate the "outcomes" of various school systems by relating as many as possible of the relevant input variables (to the extent that they could be assessed) to the output assessed by international test instruments. (p. 131)

Two populations of students took part in the study, one consisting of 13-year-olds, and one consisting of students at the final year of upper secondary school.

It is worth noticing in the above quotation that the ultimate goal of FIMS was to compare different educational systems and that students' achievements in mathematics were used as *the* indicator of the outcomes of these different systems.

The next comparative IEA study, SIMS, the Second International Mathematics Study, was decided upon in 1976 (Travers & Weinzweig, 1999), and data were collected during 1980–1982 (Robitaille & Travers, 1992). The final reports were published some years later (Robitaille & Garden, 1989; Travers & Westbury, 1990). SIMS was considerably more complex than FIMS. First and foremost, the goal was broader: "The overall objective was to produce an international portrait of mathematics education, with a particular emphasis on the mathematics classroom" (Travers & Weinzweig, 1999). More specifically, the emphasis was on an in-depth study of the curriculum:

> The curriculum in many countries is mandated at the national or system level. This is spelled out in curriculum guides and presented in the approved textbooks. Teachers are then expected to translate these guides into actual classroom instruction. There is an implicit assumption that students will learn the material presented in the classroom. How well do teachers translate what has been mandated? How close a match is there between what actually goes on in the classroom and what has been mandated? How much and what do the students learn? (p. 20)

Thus the focus of this study was on mathematics education as an end in itself, not as a means to a different end as was the case with FIMS. Based on the intentions indicated in the quotation, SIMS introduced a distinction which since then has become standard in mathematics education: the distinction between the *intended* curriculum, the *implemented* curriculum and the *attained* curriculum (a curriculum-oriented version of Bauersfeld's (1979) older distinction between the matter "meant," the matter "taught," and the matter "learned"). The student populations targeted in the study were roughly the same as the ones in FIMS; namely, 13-year-olds and those students at the final year of upper secondary school whose program had mathematics as a substantial component. Seventeen countries took part in SIMS, and also the Canadian provinces Ontario and British Columbia. Of the 17 countries, the French- and Flemish-speaking parts of Belgium entered the study as separate entities.

TIMSS, The Third International Mathematics and Science Study, conducted in 1995 under the auspices of the IEA, represented further growth of scale and complexity in comparison with SIMS. The focus on the intended, the implemented, and the attained curriculum and the relationships between them was maintained in TIMSS. Beaton and Robitaille (1999) listed four "research questions" that underlay the study design. First, as to the intended curriculum, the question concerns the ways in which countries vary in the intended learning goals for mathematics and how these goals are influenced by the characteristics of the educational systems, the schools and the students, the ways in which the curriculum is articulated, and the locus of curricular decision-making. Next, when it comes to the implemented curriculum, the question concerns (possible) differences between the implemented and the intended curriculum and the multitude of factors that may be responsible for observed differences. Factors that influence the attained curriculum form the concern of the third question, including students' homework, investment of effort, classroom behaviour, attitudes and aspirations with regard to education, and self-concept, as well as parents' economic status and expectations for their children. The fourth and final question addresses the relationships between the three curriculum aspects and the social and educational contexts, including "arrangements for teaching and learning, and outcomes of the educational process" (p. 34).

The student populations addressed in TIMSS were three, roughly comprising 9-year-olds, who were not included in FIMS or SIMS, 13-year-olds, and the students in the final year of upper secondary schooling. Forty-five countries took part in the study with at least one of these three populations. A huge body of reports were published about TIMSS in the late 1990s (c.f., http://timss.bc.edu), including one on mathematics achievement in the primary school years (1997), one on mathematics achievement in the middle-school years (1996) and one on mathematics and science achievement in the final year of secondary schooling (1998), in addition to various survey and technical reports (e.g. Martin & Kelly, 1996; Martin, Gregory & Stemler, 2000; and Martin, Mullis & Christowsky, 2004). Moreover, three so-called TIMSS monographs on curriculum frameworks for mathematics and science, research questions and study design, and textbooks, respectively, were published as well.

A follow up on TIMSS, called TIMSS-Repeat (TIMSS-R), was conducted in 1999. It focussed on the 13-year-olds only (Population 2 in TIMSS), but slightly changed the definition of the group. The four general research questions posed in TIMSS (1995) were also in focus in TIMSS-R: What kinds of mathematics and science are students expected to learn? Who provides the instruction? How is instruction organized? What have students learned?

Since then, taking advantage of the fact that the acronym TIMSS has become a brand in itself, IEA decided, rather than to insert still new first letters, to change the acronym to Trends in International Mathematics and Science Study, with the year in which it was conducted added to the acronym. Under that heading, subsequent studies were conducted in 2003, 2007, 2008, and 2011. Accordingly, previous studies were renamed to TIMSS 1995 and TIMSS 1999. The change from *third* to *trends* also reflects a new focus on trends in the IEA studies. The definition of

TIMSS target populations (Populations 1–3) has developed from a focus on age to a focus on grade level. By attempting to compare students' achievements after the same amount of schooling, the researchers assume the results will be directly useful for educational purposes.

In 1964 FIMS targeted not only compulsory schooling but also post-compulsory secondary education. As previously described, TIMSS 1995 contained such an element as well, and around 2005 initiatives were taken to establish a study enabling comparison with upper secondary school results from 1995. These initiatives led to TIMSS Advanced 2008, aimed at assessing the advanced mathematics (and physics) achievement of students in the final year of secondary schooling, which in most countries is the 12th year (Garden et al., 2006). For advanced mathematics, the target population was defined as those students in the final year of secondary schooling who have taken courses in advanced mathematics.

During the writing of this chapter, TIMSS 2011 was well under way. This study aimed at Populations 1 and 2 with similar definitions to those found in TIMSS 2007. A unique characteristic of this TIMSS cycle is that the IEA study PIRLS (Progress in International Reading Literacy Study) was done simultaneously in Grade 4. This created opportunities for research aiming at investigating and understanding relationships between language and mathematics.

TIMSS always took its point of departure in student achievement vis-à-vis school curricula. In contrast, the Organisation for Economic Co-operation and Development (OECD) decided in the late 1990s to mount a series of international comparative studies that focussed on the outcomes of schooling for students leaving compulsory education in most countries, settling on students of age 15, irrespective of the curricula according to which they have been taught. The purpose was to study education systems' ability to equip the youth in the participating countries with the capabilities needed for citizenship in a broad sense, but with particular regard to reading, mathematics, and science. This undertaking was given the name Programme for International Student Assessment, better known as PISA (for an in-depth comparison between TIMSS and PISA, see de Lange, 2007). The first study was to take place in 2000, and then every three years a new study would be conducted. The introduction to the initiating publication of PISA, *Measuring Student Knowledge and Skills: A New Framework for Assessment* (OECD, 1999) reads:

> How well are young adults prepared to meet the challenges of the future? Are they able to analyse, reason and communicate their ideas effectively? Do they have the capacity to continue learning throughout life? Parents, students, the public and those who run education systems need to know. ...
>
> OECD/PISA will produce policy-oriented and internationally comparable indicators of student achievement on a regular and timely basis. The assessments will focus on 15-year-olds, and the indicators are designed to contribute an understanding of the extent to which education systems in participating countries are preparing their students to become lifelong learners and to play constructive roles as citizens in society. (p. 9)

Furthermore,

> PISA is the most comprehensive and rigorous international effort to date to assess student performance and to collect data on the student, family and institutional factors that can help to explain differences in performance. (p. 14)

The international consortium chosen by the OECD to be in charge of conducting the study was the Australian Council for Educational Research (ACER). It was decided to adopt a cyclical study structure, such that for each round—cycle-one of the three domains reading, mathematics, and science would be the major domain, and the other two would be minor domains. Thus, reading was the major domain in 2000, mathematics in 2003, science in 2006, reading again in 2009, and so on. Mathematics will be the major domain again in 2012.

The fact that the purpose of PISA is to uncover the capabilities for citizenship and lifelong learning that students gain from schooling in different countries, implies that the focus of the study is, and has been from the very beginning, expressed in terms of *literacy*, including mathematical literacy. The first definition of *mathematical literacy* was as follows:

> Mathematical literacy is an individual's capacity to identify and understand the role that mathematics plays in the world, to make well-founded mathematical judgments and to engage in mathematics in ways that meet the needs of that individual's current and future life as a constructive, concerned and reflective citizen. (OECD, 1999, p. 43)

Very minor changes were made to this definition in the frameworks for PISA 2003, 2006, and 2009. However, as a result of changes in the composition and management of PISA instigated by the OECD in 2009, the U.S. organization Achieve became associated with the consortium with the specific task to oversee the development of a new framework for PISA mathematics in 2012. As part of this process, a new definition of mathematical literacy was agreed upon. Its purpose was to spell out, in an explicit way, the main components involved in identifying and understanding the role of mathematics and in engaging with it:

> *Mathematical literacy* is an individual's capacity to formulate, employ, and interpret mathematics in a variety of contexts. It includes reasoning mathematically and using mathematical concepts, procedures, facts, and tools to describe, explain, and predict phenomena. It assists individuals to recognise the role that mathematics plays in the world and to make the well-founded judgments and decisions needed by constructive, engaged and reflective citizens. (OECD, 2010b)

In 2000 (OECD, 2001), 32 countries participated in PISA, including 28 OECD countries. In 2002, another 13 countries joined the first cycle. In the 2003 round, in which mathematics was the major domain, 30 OECD countries and 11 non-OECD countries participated (OECD, 2004). In 2006, the 30 OECD countries were joined by 27 other countries or "economies" (OECD, 2007), whereas 34 OECD countries and 31 other countries or "economies" took part in PISA 2009 (OECD, 2010a). In addition to the outcomes reports just referenced, OECD PISA has published hosts of other reports, some of which are technical reports, whereas others focus on specific themes or issues (see http://www.pisa.oecd.org).

Focal Studies

When it comes to what we here call international focal studies, there are quite a few of them. Some are accompanying or following up on large-scale studies, whereas others are independent studies. A study of the former kind is the so-called Survey of Mathematics and Science Opportunities (SMSO), a four-year study on instructional practices in six countries (France, Japan, Norway, Spain, Switzerland, and the USA), "charged with developing the research instruments and procedures that would be used in the Third International Mathematics and Science Study (TIMSS)" (Cogan & Schmidt, 1999, p. 69) with particular regard to 9- and 13-year-old students. Although SMSO was conducted prior to TIMSS itself, the so-called TIMSS Video Study of eighth-grade classrooms in Germany, Japan, and the USA, and the so-called Case Study Project of TIMSS concerning the same three countries, were supplementary additions to TIMSS proper, even though they were funded by the US Department of Education, National Center for Education Statistics (Kawanaka, Stigler, & Hiebert, 1999; Stevenson, 1999). Germany and Japan were chosen because they were, at the time, seen as major economic competitors with the USA, and because Japan was consistently obtaining scores at the top end of international comparison tests (Kawanaka et al., 1999; Stevenson, 1999). Another related study (Schmidt et al., 1997) surveyed the curricular intentions in school mathematics in a number of countries.

One driving force behind the development of the TIMSS Video Study was the ambition to go beyond international comparisons of students' achievements as measured by tests. IEA wanted also to consider so-called contextual factors (Stigler, Gallimore, & Hiebert, 2000). Previously, information on teaching processes had relied solely on the responses of teachers and students to questionnaires.

The overall goal of the Video Study was to provide a rich account of what happens inside Grade 8 classrooms in the three countries, and in that context:

> To develop objective observational measures of classroom instruction to serve as quantitative indicators at a national level of teaching practices in the three countries.
>
> To compare actual mathematics teaching methods in the US and the other countries with those recommended in current reform documents and with teachers' perceptions of those documents.
>
> To assess the feasibility of applying videotape methodology in future wider-scale national and international surveys of classroom instructional practices. (Kawanaka et al., 1999, p. 87)

The Video Study was later extended to include eight countries in the TIMSS-R video survey study.

The Case Study Project was included in TIMSS "in the hope that [the findings] would provide in-depth information about beliefs, attitudes and practices of students, parents and teachers that would complement and amplify information obtained through the questionnaires used in the main TIMSS study" (Stevenson, 1999, p. 106). The research topics chosen were meant to "be of interest to US policymakers who deal with elementary and secondary schooling" (p. 107), and comprised "national standards, teachers' training and working conditions, attitudes towards dealing with differences in ability and the place of school in adolescents' lives" (p. 107).

So, the common task of the Video Study and the Case Study of TIMSS was to zoom in on factors in Germany, Japan, and the USA that might potentially serve to explain the differences in outcomes of mathematics (and science) education, including students' achievements, in these countries.

In the beginning of 2000, the Learner's Perspective Study (LPS) was launched. Initially research groups from four countries—Australia, Germany, Japan and the USA—participated. The study was mainly funded by Australian means (Clarke, Keitel, & Shimizu, 2006). There were different rationales behind the original study. One of the more important ambitions was to be able to situate Australian mathematics teaching in relation to results from the first TIMSS video survey study (Stigler & Hiebert, 1999). Later the study was extended by research groups from several additional countries joining the project. At the time of writing this chapter the number of participating groups amounts to 15 (see the Web site of the project http://www.lps.iccr.edu.au). As a result, the original project has gradually been expanded and can today rather be seen as a network of researchers with a common interest in classrooms studies in an international context.

A broad range of research questions are addressed within the LPS. Since the project is a conglomerate of research groups belonging to different traditions, there is no unifying set of questions. Clarke, Keitel, & Shimizu (2006) put forward a set of seven overarching questions ranging from addressing issues of the presence of coherent and culturally-specific student and teacher practices, over relationships between these practices, to variability within classrooms and countries as well as among classrooms and countries. The questions also reflect ambitions of the project to provide information about the practices studied.

It is also worth mentioning that in comparison with the large-scale international studies described in this chapter, the LPS stands out by not being anchored in an international organization such as IEA, OECD, or ICMI. Instead, it is based on researcher-driven interests. Hence, LPS is an example of scholarly stakeholders working in the field of international comparative studies.

The US–Japan Cross-cultural Research on Students' Problem-Solving Behaviours is an early example of another independent focal study with the researchers themselves as the stakeholders, emphasizing problem solving. It began by joint US–Japan seminars in 1987 instigated by Jerry Becker and T. Miwa, and was subsequently developed into a research project, the purpose of which was "to collect descriptive data pertaining to the performance of Japanese and US students on certain kinds of problem-solving behaviours," and "contrasts in these behaviours between students in the two countries were also sought" (Becker, Sawada, & Shimizu, 1999, p. 121). The students under consideration were 4th, 6th, 8th, and 11th graders in the two countries.

A comparative study—called the Kassel Project—of secondary mathematics teaching in England and Germany was carried out in the 1990s. One of the rationales stated for this study (Kaiser, 1999b) was that European countries will, to an increasing extent, receive each others' students. Therefore it will be important to know what students know and to develop a mutual understanding of the different education systems in the European countries. The goals were to provide

> an examination of the differences in the mathematical achievement of English and German students.
>
> an analysis of the differences in the ways of teaching and learning mathematics in both countries. Based on this, the teaching methods will be questioned, and ideas gathered on how to improve the different ways of teaching mathematics. (p. 141)

An entirely different kind of comparative study is found in the 13th ICMI Study *Mathematics Education in Different Cultural Traditions: A Comparative Study of East Asia and the West* (Leung, Graf, and Lopez-Real, 2006). In this study, which is actually a collection of different theoretical and empirical contributions, numerous aspects of observed differences between the Confucian tradition and approach to mathematics education, which is predominant in East Asia, and the Western traditions are investigated. In contradistinction to what is common to several other international comparative studies, where the overall idea is, in some way or another, to provide lessons for learning from each other, the 13th ICMI study had a different, if not outright opposite, rationale:

> The globalisation processes are producing reactions from mathematics educators in many countries who are concerned that regional and local differences in educational approach are being eradicated. This is not just a mathematical ecology argument, about being concerned that the rich global environment of mathematical practices is becoming quickly impoverished. It is also an argument about education, which recognises the crucial significance of any society's cultural and religious values, socio-historical background and goals for the future, in determining the character of that society's mathematics education. (p. 6).

In other words, this study can be seen as an attempt to counteract (Western) cultural and educational imperialism with regard to mathematics education. It did so by comparing and contrasting the contexts of mathematics education, the curricula, teaching and learning and, finally, values and beliefs in Confucian and Western cultures and traditions.

Several other focal studies might have been mentioned, for example, Collaborative Studies on Innovations for Teaching and Learning Mathematics in Different Cultures in APEC Member Economies (cf. http://www.criced.tsukuba.ac.jp/math/apec and http://www.crmekku.ac.th), but they would not fundamentally expand the set of purposes already encountered in the international studies mentioned.

Designs and Methods Adopted in International Studies

Based on the distinctions introduced in the first section, we concentrate here on presenting and discussing the *designs* (i.e., the set of approaches adopted to answer the questions that drive a given study) and the *methods* chosen and implemented for pursuing these approaches. Moreover, we consider the most important *instruments* involved in these methods.

The IEA Studies

The *design* adopted for FIMS consisted of three approaches to answering the question driving the study (Robitaille & Travers, 1992). As the fundamental idea in FIMS was to measure and compare outcomes of education systems by way of student achievement in mathematics, the overarching and most important approach was to *construct achievement tests*. This was closely linked to the second approach, *choosing the student populations* in participating countries whose achievements were to represent countries' school achievements at large. That constituted the second approach. The third approach to answering the primary question was to *ask students, parents, and teachers* about attitudes, demographics, socio-economic backgrounds, and so on.

Considerations about which *student populations* to involve in FIMS led to the definition of three student populations to be tested, but results were reported for only two of these: A younger population, consisting of students close to the very end of compulsory schooling in most countries (Postlethwaite, 1971), roughly speaking consisting of 13-year-olds, and an older population, consisting of students at the end of secondary schooling. Both populations were divided into two subpopulations, but the details are omitted here. Methods for identifying samples of these populations in the participating countries were employed nationally according to general guidelines, which included stratified random probability sampling.

As to the *achievement tests*, the method adopted was to construct them in accordance with a matrix structure: "topics" by "cognitive behaviour levels." Although the topics varied across the populations, the five cognitive behaviour levels were the same for all populations (Husén, 1967): (a) knowledge and information: recall of definitions, notation, concepts; (b) techniques and skills: solutions; (c) translation of data into symbols or schema and vice versa; (d) comprehension: capacity to analyze problems, to follow reasoning; and (e) inventiveness: reasoning creatively in mathematics. The sets of test items constructed with this matrix structure in mind were then administered to students in all participating countries after having been filtered through elaborate piloting procedures. More specifically, each student in a given population was required to do the same three-to-four one-hour item booklets—forming the test *instruments*—such that each student had to complete a total of 50 to 70 items (Postlethwaite, 1971; Robitaille & Travers, 1992). Most of the items had a multiple response format, but a couple of open-ended items were included in each booklet. Included in the item booklets were also some scale-based questions concerning student attitudes to mathematics and its learning (Postlethwaite, 1971). More specifically, these questions concerned "mathematics as a process," "difficulties of learning mathematics," "the place of mathematics in society," "school and school learning," and "man and his environment."

Finally, the method to probe into institutional characteristics, socio-economic background variables, career perspectives, teacher backgrounds, and so on, was to make use of four types of questionnaires—each forming a sociological *instrument*—student questionnaires, teacher questionnaires, school questionnaires, and a national case study questionnaire.

Given its focus on portraying mathematics education at large, and curricula in particular, SIMS had a somewhat different *design*, which was based on an overall framework distinguishing between the intended, the implemented, and the attained curriculum. This framework gave rise to three *different approaches* to answering questions concerning the constitution of each type of curriculum across participating countries. However, the basic—and more overarching—approach was to decide on the student populations whose curricula were to be investigated in the study.

Again, as part of the design of SIMS *target populations* had to be chosen. It seems as though the basic approach leading to the selection of these populations was to keep the definitions of FIMS, whenever possible, but also to attempt to solve some of the delineation problems encountered with FIMS, especially with students in the older population. In most countries, the actual samples of students representing each population studied were selected by using probabilistic sampling methods at a national level.

As to the method adopted in the identification of *the intended curriculum* in participating countries, a matrix-based specification in terms of a content dimension and a cognitive behaviour dimension, similar to but not identical with that employed in FIMS, was chosen (Travers & Weinzweig, 1999). Subdivided content strands were identified for the two populations (five for the younger and nine for the older population). As regards the cognitive behaviour dimension, SIMS deviated from FIMS in making use of a more hierarchical classification: computation, comprehension, application, and analysis. Considerable effort was made to avoid ambiguity, for example, by describing the resulting cells in the matrices by detailed examples of what the SIMS committee had in mind such that countries' respondents were able to tell whether a certain cell was part of their curriculum or not. Moreover, countries' respondents were asked to indicate the degree of importance of each cell for the curriculum at issue in their country. In other words, the instruments employed in this method were content-by-cognitive behaviour grids, together with illustrations and comments, which country respondents were asked to fill out and return accompanied by importance degrees assigned to each cell.

When it came to investigating *the implemented curriculum* in the SIMS countries, that is, the second approach in the design of the study, the method employed was to ask teachers to fill in detailed questionnaires—the instruments—about their classrooms, their teaching methods during the school year, their attitudes and beliefs, and the place and role of each cell in the above-mentioned grids. For "each topic, a detailed description of a large variety of teaching methods that could be utilized in the teaching of that topic" was provided (Travers & Weinzweig, 1999, p. 22).

Finally, the core approach in the design was to capture *the attained curriculum* in participating countries. As in FIMS, the method to investigate this curriculum first of all consisted in written student achievement tests containing items referring to the content-by-cognitive behaviour grid mentioned above. The number of items belonging to each cell was determined by the importance assigned to that cell by participating countries. The final pool of items also contained some anchor items in order to detect possible changes for the 11 countries that participated in both FIMS and SIMS. The actual instrument employed consisted of multiple item booklets, such that each student answered one or two booklets, at least one of which was from a set of rotated booklets. This rotation was introduced in order to ensure a broad coverage of grid cells across countries (Travers & Weinzweig, 1999). Moreover, the instrument also included, for each item, a student and a teacher question, asking whether the content implicated in the item had been taught or not, and if so when.

The *design* of the Third International Mathematics and Science Test (TIMSS) was a continuation of that of SIMS. For TIMSS, the design was focussed on answering what Beaton and Robitaille (1999) called Research Questions 1–4, using the three-part model of intended, implemented, and achieved curricula. Methods used to describe and evaluate the different curriculum levels were similar to those of SIMS, but there were also some differences. In TIMSS 1995, a set of performance items was used as a supplement to the core paper-and-pencil tests given to students. Furthermore, the construction of the tests was based on a framework specifying three dimensions in a mathematics curriculum: content, performance expectations, and perspectives (Robitaille et al., 1993). The content dimension listed the mathematical content areas to be covered, performance expectations defined competencies such as knowing and communicating, and perspectives covered other aspects such as attitudes and habits of mind. The target populations in TIMSS 1995 were similar, though not identical, to those of FIMS and SIMS: Population 1 (9-year-olds), Population 2 (13-year-olds), and Population 3 (students in their final year of secondary schooling). All participating countries were required to enter Population 2, whereas the other two were optional. A two-stage random-sampling procedure was used as the method for identifying samples representing the sample populations in each participating country. In Populations 1 and 2, entire classrooms were sampled, whereas in Population 3, individual students were selected.

TIMSS 1999 is often described as a repetition of TIMSS 1995, using basically the same *design*. The framework for constructing tests in TIMSS 1999 was the same as for TIMSS 1995. Thus the mathematical content covered was the same. The goal with TIMSS 1999 was "more modest in scope, focussing on one target population only." Nevertheless, it "yielded valuable information on the curricular intentions of participating countries" (Martin, Gregory, & Stemler, 2000). Even though the design was essentially unchanged, some important changes in the *methods* employed were introduced in TIMSS 1999, which proved significant for the development of successive TIMSS cycles. As far as the *achievement test* approach is concerned, additional items were developed since two-thirds of the items from TIMSS 1995 had been released and consequently had to be replaced by similar items in order to cover the framework. In so doing, TIMSS 1999 introduced the focus on trends which later became a "trademark" of TIMSS. In earlier studies, some items had been reused, but there had not been a focus on the trend aspect as such. Next, substantial and influential changes in the third approach, *the questionnaires*, were implemented. A curriculum questionnaire to be answered by the National Research Coordinator of each participating country, summarizing features of the school system on a national level, was introduced. Similar questionnaires were used in all subsequent TIMSS cycles. Whereas the TIMSS 1999 school questionnaire was very similar to the 1995 version, several changes were made to the teacher questionnaires for the 1999 cycle, mainly because the previous ones were considered too lengthy. In the student questionnaire, questions dealing with student self-concept in mathematics, Internet access, and its use for mathematical activities were added. It is an interesting fact that outcomes of the TIMSS Video Study helped frame a set of questions about activities in mathematics classes in TIMSS 1999.

TIMSS 2003 confirmed the focus on trends introduced in TIMSS 1999. Furthermore, the transition of definitions of participating populations from age to years of schooling was taken one step further. In addition to a basic definition based on age, the population definition stated that the identified grade level was intended to represent 4 and 8 years of schooling (Martin et al., 2004). In the first three cycles of TIMSS (1995, 1999, and 2003), *student achievement* in mathematics in addition to an overall result was reported in content domains (e.g., algebra, geometry). At the time, several other international studies (e.g., PIRLS—also conducted by IEA—and PISA) had introduced reporting of student achievement in different cognitive domains. TIMSS participating countries also expressed a need for comparative information about cognitive aspects of how students performed in mathematics (and science). An international group of mathematics experts was gathered to develop categories that could be the basis for meaningful reporting of achievement in cognitive domains. Previous definitions of four cognitive domains had been used in the development of items for the TIMSS assessments, but the existing model led to some overlap across these domains. The expert group worked to develop mutually exclusive cognitive domains for reporting the TIMSS 2003 results (Mullis, Martin, & Foy, 2005) leading to the definition of three cognitive domains: knowing facts, procedures and concepts; applying knowledge and understanding; and reasoning. These domains, supported by categorization of items from TIMSS 2003 and reanalysis of TIMSS 2003 data with respect to these categories, were published in 2005 (Mullis, Martin, & Foy, 2005).

Further refinement of the assessment framework was done in the early stages of TIMSS 2007 as published in the TIMSS 2007 assessment frameworks (Mullis, Martin, Ruddock et al., 2005). Based on the development project mentioned above, the number of content domains and cognitive domains was decreased. The revision of the framework was at least partly a consequence of a decision made that, beginning with TIMSS 2007, frameworks were to be updated with every cycle of the study, thereby permitting the frameworks, the achievement tests, and the procedures to evolve gradually into the future. Another small but still significant change from 2003 to 2007 is found in the definition of the study populations. An important feature of the research design that TIMSS represents is that these populations must be defined rather precisely and can be viewed as "a collection of units to which the survey results apply" (Olson et al., 2008, p. 78). A subset of the target population was sampled for participation in the study, and a lot of effort was put into identifying the sample in such a way that results from the sample can be generalized to the entire target population.

TIMSS Advanced 2008 focussed on a population which had not been targeted in IEA studies since TIMSS 1995—that is, students at the end of upper secondary education (Grade 12) who had taken courses in advanced mathematics. Apart from that, the basic *design* was essentially the same as for TIMSS 2007, the aim being to study the intended, the implemented, and the achieved curriculum. The *methods* used were also similar to those of TIMSS 2007. The assessment framework guiding the development and construction of instruments defined three broad mathematical content domains (algebra, calculus, and geometry) and three cognitive domains (knowing, applying, and reasoning) (Garden et al., 2006).

PISA: Programme for International Student Achievement

PISA 2000 and studies which followed upon it, were not research studies as such even though they have given rise to several research questions, some of which have been pursued in follow-up studies. Instead, PISA is a survey designed to assess students' "ability to complete tasks relating to real life, depending on a broad understanding of key concepts, rather than assessing the possession of specific knowledge" (OECD, 2001, p. 19). Thus the *design* of PISA 2000 was focussed on charting students' performance with regard to reading (the major domain), mathematical and scientific (the minor domains) *literacy* (see the definition of mathematical literacy above), and relating such performance to student and school background factors. Correspondingly, four approaches were pursued: constructing an assessment *framework* for literacy (OECD, 1999), constructing and administering *achievement tests*, and constructing and administering *background questionnaires*. Further, an approach to *ranking participating countries* according to various performance variables was part of the design as well. The basic decision to assess 15-year-olds in participating countries was taken much before the other design decisions.

The *method* undertaken in constructing the framework was to ask an expert group for each domain, to devise such a framework. As far as mathematics is concerned, the framework contained three dimensions: a content dimension, which for PISA 2000 had two components "change and relations" and "space and shape"; a process dimension (called "competency clusters") "reproduction," "connections," and "reflection"; and a situation dimension focussing on the spheres in which students live, that is, private/personal, school, work and sports, local community and society, and scientific spheres of life. These dimensions then formed the platform for constructing the test items. The items were devised to be literacy items and were, moreover, to be cast in one of three paper-and-pencil response formats: multiple choice, closed constructed, and open constructed response. A total of 64 items, chosen as a result of extensive field-testing, comprised the test.

The methods involved in identifying educational background factors and relating them to student performance consisted in devising two questionnaires: a student and a school questionnaire. Responses to those questionnaires were then correlated by way of several statistical analyses to student performance so as to explain a multitude of performance variations. Also, the methods employed in ranking countries by way of certain ranking measures were probabilistic and statistical in nature, based, more specifically, on the so-called Rasch model. In particular, the methods in item response theory were utilized.

The *instruments* adopted consisted of the actual student tests and questionnaire and a school questionnaire to be completed by the principals of the schools whose students were included in the sample. Each student was given one out of nine item booklets, containing items from the three domains (reading, mathematics, science) without any indication of which domain they belonged to. This rotation principle implied that different students were completing different booklets. Each student was given two hours to complete the booklet. The questionnaire that each student was asked to complete was a 30-minutes questionnaire containing questions about students' and parents' economic, cultural, and social status; student characteristics and family backgrounds; and learning strategies and attitudes (OECD, 2001). The school principals' questionnaire—which also was meant to take 20 to 30 minutes to complete—contained questions concerning school policies and practices, classroom practices, school resources and type of school.

In PISA 2003, mathematics was the major domain, the aims and overall design were not much different from those of PISA 2000, except in one respect: *trends* from PISA 2000 to PISA 2003 were sought. As before, the primary aim of the OECD/PISA assessment was "to determine the extent to which young people have acquired the wider knowledge and skills in reading, mathematical and scientific literacy" that they would need in adult life (OECD, 2003, p. 12).

The *framework* part of the design was unchanged along the main lines. But there were minor changes in the content, process, and situations dimensions. Two new content categories, "overarching ideas," were added to the ones in PISA 2000; namely, "quantity" and "uncertainty," thus forming a total of four. The situation and context categories were slightly modified as well. As to the mathematical process dimension, the notion of eight mathematical competencies as developed in the Danish KOM-project (Niss & Hoejgaard, 2011; Niss & Jensen, 2002) was introduced to underpin the competency clusters that were utilized in PISA 2000.

In the *achievement test*, a rotated design was employed, with a total of 85 mathematics items included in the pool, 20 of which were also used in PISA 2000. These are called "link items." Student and school *questionnaires* were included as in PISA 2000, and also contained questions concerning students' self-concept, learning strategies, and affects specifically concerning mathematics. Again, the items were selected and the questionnaires finalized after substantial field trialling.

The method adopted for *charting trends* in mathematics performance from PISA 2000 to PISA 2003 was to establish common PISA 2000–2003 performance scales. This was done by using the detected changes of difficulty in the 20 link items from 2000 to 2003 to construct a transformation of scores so as to fit a common scale (OECD, 2004), having 500 score points as the OECD average. With that in hand, PISA 2000 and PISA 2003 subscales for the two content categories which were common to both cycles, "space and shape" and "change and relationships," were constructed. It was then possible to see that the OECD average in space and shape grew from 494 to 496 score points, whereas in change and relationships, scores grew from 488 to 499. The 2003 score for quantity was 501, and for uncertainty 502. It did not make sense to make an overall comparison of mathematics performance from 2000 to 2003, since the combined average score was set to be 500.

In PISA 2006 and PISA 2009, mathematics was again a minor domain. Therefore, only minor changes were made to the *design* of the study as far as mathematics and student and school questionnaires are concerned. In 2006 only 48 items were used. As these were also included in 2003, they were all link items. Each participating student received a randomly selected booklet. With regard to detection of trends the PISA 2003 scale with an average OECD score of 500 was used as the benchmark (OECD, 2007), and again the link items were used to create a transformation that allowed for comparison between the two assessments. The OECD mathematics score for 2006 was 498, which was not significantly different from the 500 in 2003. In 2009 the total testing time in mathematics was reduced and only 35 items were included in the test. The OECD average score in mathematics 2009 was 496, which was not significantly different from 2006.

Various changes were incorporated in PISA 2012, when mathematics was again the major domain, but it is premature to go into details with these changes. For current information consult OECD (2010b). More changes are likely to occur from 2015 as a new contractor will be in charge of the future development of frameworks.

The TIMSS Video and Case Studies

In the TIMSS Video Study, the *design* adopted was chosen so as to reduce the conceptual and terminological ambiguities within and across cultures that could arise from using questionnaires, as well as to avoid dependence on coding schemes fixed beforehand and the impossibility of critical scrutiny of documentation of live observations (Kawanaka et al. 1999):

> We needed data that could be analyzed and re-analyzed objectively by researchers working from a variety of perspectives. The idea of using videotapes began to emerge, and the final decision was made to collect direct information on classroom processes by videotaping instructional practices. (p. 88)

So, approaching the reality of classrooms by *videotaping* them was, of course, the fundamental approach in the study. This decision allowed researchers to engage in many iterations and related discussions between observations and post hoc coding of the observations. Teachers' views of the representativeness of the lessons videotaped and their goals were sought as well, by means of *questionnaires*. The next key approach in the design was *analyzing and coding the data* generated by the videotapes, and the final approach was to devise ways to *represent and depict mathematics classroom reality* in a manner that would make sense to researchers outside the project.

Each of these approaches gave rise to its own set of *methodological issues* and decisions. First, how to *sample the classrooms* that were to be videotaped, and when and for how long should they be videotaped? Another important issue to decide upon was what to aim cameras at and hence what type of classroom activities to document. It was decided to focus on the middle TIMSS population only (eighth grade) in Germany, Japan, and the USA. The classrooms sampled were a subsample of the national random probability samples in TIMSS 1995. Eventually 100 German, 50 Japanese, and 81 US classrooms were included in the study. Classrooms were videotaped in 1994–1995 (Stigler et al., 1999) evenly across the school year in Germany and in the USA, but less so in Japan, where the sample was skewed towards a time of the school year when geometry was predominant in the curriculum (Kawanaka et al., 1999).

When seeking a method for *coding the tapes*, Kawanaka et al. (1999) had three dimensions in focus: the work environment in the classroom, the nature of the work students are engaged in, and the methods teachers use for engaging students in work. The coding schemes were developed with the aim to construct objective and reliable categories and codes that allowed for capturing, representing, and quantifying characteristic features and patterns in the classrooms of the three countries.

In putting the method of videotaping into practice, the actual *instrument* employed was to film one complete lesson per classroom by one camera, representing the perspective of an ideal(ized) student, typically focussing on the teacher. Prior to that event, participating teachers were given a common set of information and instructions, and afterwards they completed the questionnaires mentioned above (Stigler et al., 1999). All videotapes were digitized, and lessons were translated into English and transcribed, linking the transcript to the video by time codes (Kawanaka et al., 1999). The final instrument for coding was very elaborate. It focussed on what was called "lesson tables."

> These lesson tables were skeletons of each lesson that showed, on a time-indexed chart, how the lesson was organized through alternating segments of classwork and seatwork, what pedagogical activities were used ..., what tasks were presented and the solution strategies for the tasks that were offered by the teacher and by the students. (p. 96)

The tables included several components: organization of the class; outside interruption; organization of interaction; activity segments; mathematical content referring to units (Stigler et al., 1999) and to mathematical topics (numbers; measurement; geometry; proportionality; functions, relations and equations; data representation, probability and statistics; elementary analysis; validation and structure; other). Also, a very detailed coding of classroom discourse, based on a rather fine-grained division of public talk and private talk, respectively, was undertaken. Coding schemes were refined along the road when warranted by the analysis of the videos and intercoder reliability checks (Kawanaka et al., 1999). In addition to being guides to the entire video of a classroom, the lesson tables also served as separate reporting outcomes which could themselves be coded. Statistical analyses were conducted to capture and describe patterns for comparison across the three countries.

The *design* of the TIMSS Case Study encompassed three approaches to seeking in-depth answers to the initiating question "about the beliefs, attitudes and practices of students, parents and teachers" in Germany, Japan, and the USA (Stevenson, 1999). The *first approach* was to identify the topics on which information was to be sought. The method adopted was to select, after consultation with the funding agencies, four such topics: national standards, teachers' training and working environment, dealing with differences in students' ability, and, finally, the place of secondary school in adolescents' lives. One of the *instruments* put to use in relation to this method was to attach a number (15 to 35) of predetermined tags, in terms of key concepts and words, to each topic so as to facilitate subsequent computerized retrieval of the tagged instances. It was further decided not to form a particular set of hypotheses from the outset but to let them be generated from the data collected. The *second approach* was to identify the units from which information should be collected. The method then was to concentrate on one primary and two secondary sites in each of the three countries, all chosen to be representative in demographic and socioeconomic terms. Each site would contain several schools. The *third*—key—*approach* concerned the ways in which researchers were to gather information. Here the method was to make each researcher responsible for one of the four topics and to conduct a number of so-called encounters (i.e., interviews, observations, conversations) of a minimum duration of one hour. Moreover, each researcher was to produce and circulate weekly field notes—another instrument—to the other researchers. A total of more than 960 encounters were conducted in the three countries. In addition, 250 hours of observation of mathematics and science classes were carried out. All interviews were to be conducted according to a predetermined semi-structured format, which involved yet another instrument. Whenever possible, the encounters were tape-recorded, which constituted the final instrument involved in implementing the third approach.

The Learner's Perspective Study

The design used within the TIMSS Videotape Study was extended for use in the Learner's Perspective Study (LPS), and measures were taken to improve the possibilities to capture not only teachers' activities during lessons but also the students' learning processes. The capturing of students' learning processes—the *first approach* in the *design*—was operationalized by adding some features to the design of the TIMSS Videotape Study. An important such feature, which differs from earlier major studies with comparative possibilities, was that sequences of lessons rather than singular ones were documented. A minimum of 10 consecutive lessons were recorded at each site. The main characteristic of the method adopted in this approach is the use of video documentation of teachers' and students' work in eighth-grade mathematics classrooms. Three cameras were used in each classroom: one stationary camera equipped with a wide-angle lens capturing as much of the classroom as possible, a second one pointing to a group of so-called focus students, and finally a manually operated camera following and documenting the activities of the teacher. Depending on the seating plan, one to four focus students' work was video- and audio-recorded in each lesson.

In each city, three teachers' classrooms were selected for recording. The relatively small number of classrooms investigated is a trade-off with the comparatively large number of consecutive lessons documented. The sampling of participating teachers, classrooms, and hence students was not made randomly but was based on the selection of "competent" teachers as defined by the local community in each city and country. The focus students were interviewed in a stimulated recall interview—the *second approach* in the design—after the lesson. This decision was informed by the aim to explore learners' practices and allow them to generate reconstructive accounts of classroom events. Three times during a lesson sequence the teachers, too, were interviewed in a subsequent stimulated recall session. The actual recordings of the focus student and the teacher cameras were used as recall stimulus in the interviews (Clarke, 1998, 2001, 2003, n.d.). The interviewees were invited to comment on each recorded lesson in terms of what they found significant in the classroom activities. They were in control of the replay of the videos and could freely choose when to use the fast forward (or rewind) buttons and when to stop and comment on the recordings.

Documenting sequences of lessons allows for analyses of single lessons but also analyses that stretch beyond those, hence making it possible to address questions on how both teaching and learning unfold over a longer period of time. When it comes to analyzing the data—*the third approach*—there is no framework common to all the participating research groups in the network. However, the overall approach is informed by a Vygotskian point of view where teaching and learning are seen as mutually constitutive processes.

Complementarity is a distinguishing characteristic of the research design on four levels (Clarke, Emanuelsson, Jablonka, & Mok, 2006):

> (a) At the level of data, the accounts of the various classroom participants are juxtaposed; (b) At the level of primary interpretation, complementary interpretations are developed by the research team from the various data sources related to particular incidents, settings, or individuals; (c) At the level of theoretical framework, complementary analyses are generated from a common data set through the application by different members of the research team of distinct analytical frameworks; and (d) At the level of culture, complementary characterizations of practice and meaning are constructed for the classrooms in each culture (and by the researchers from each culture) and these characterizations can then be compared and any similarities or differences identified for further analysis, particularly from the perspective of potential cross-cultural transfer. (pp. 12–13)

All video materials were transcribed and translated into English. The transcripts, together with digitized videos, were included in a database which also contained seating plans describing students' positions during class and so-called lesson plans; that is, rough summaries of each lesson. Survey materials such as short teacher questionnaires, performance tests compiled from released items from TIMSS studies, scanned copies of the focus students' work, and textbooks were also part of the integrated datasets constructed by each participating research group.

The US–Japan Problem-Solving Study

In order to compare and contrast Japanese and US students' abilities, behaviours, and views concerning problem solving in mathematics, the design of the US–Japan Cross-cultural Research on Students' Problem-Solving Behaviors (Becker, 1992; Becker et al., 1999) included the following *four approaches*. First, the subjects to be studied had to be specified. Next, the ways in which they were to be studied had to be determined. More specifically, it was decided to put the students selected to work on certain tasks, and they as well as their teachers were asked to complete questionnaires pertinent to the problems solved and to mathematics at large. Finally, student problem responses were coded by means of certain predetermined categories, and the questionnaire answers were analyzed.

As to the *first approach*, the subjects to be studied formed a number of populations in the two countries. The method was to sample students—with their teachers—in 4th, 6th, 8th, and 11th grades from large rural, small urban and large urban schools in Japan and the USA in the school year 1989–1990. The selection of the schools seems to have been made on pragmatic grounds, namely from districts near the researchers' own institutions. At least two classes participated in each region in each country. Neither the schools nor the classes were randomly selected (Becker, 1992; Becker et al., 1999). The number of students involved in the study was several hundred from each population in both the USA and Japan.

The method employed to implement the *second approach* was to give all but the 11th-grade students two problems to solve. The problems had been used and investigated by researchers in previous studies, and their final formulation and place in problem work booklets—the *instrument* employed—had been tried out in a pilot study (Becker, 1992). The US 11th-graders also got an extra problem to solve. Each student was given exactly 15 minutes to solve each of the two problems, except that the US 11th graders got an additional 10 minutes to solve the third problem. For all problems, students were asked to solve them in as many different ways as possible—on separate answer sheets handed out to them—within the given time frame. This introduces an unusual feature in task-based studies, which usually only ask for single solutions.

As to the *third approach*, students were asked to fill out a questionnaire—forming one *instrument* in this approach—after having worked on the problems. The questionnaire contained questions concerning students' degrees of interest, difficulty, and familiarity with the problems they had just solved, and their attitudes and self-concept with regard to mathematics. Teachers were asked to fill out their questionnaires (another instrument) while the students were doing the problems. These questionnaires, in addition to seeking information about the school and the students, addressed the teacher's view of the problems posed and of the students' reactions to them (Becker et al., 1999).

The *final approach* was to analyze the data collected. Individual or pairs of researchers were responsible for analyzing the data for one problem (Becker et al., 1999). The focus of the analyses, which often made use of categories established by previous Japanese or American research, was on comparison of correctness of responses, solution strategies, and modes of explanation.

The Kassel Project

The so-called Kassel Project (Kaiser, 1999b), aiming at comparing essential features of secondary mathematics teaching and learning in England and Germany and at explaining observed differences, had a *design* which in important respects differed from the designs adopted in most international studies, even though the study—as is often the case—is a combined quantitative and qualitative one. The quantitative part of the study was focussed on longitudinal student achievement, and the qualitative part concentrated on capturing and charting key features of teaching and learning in the two countries.

In the achievement part of the study, the *first approach* was to identify two comparable lower secondary cohorts in 1993 (a sample of about 800 students in Year 8 in Germany and about 1,000 students in Year 9 in England) who were then followed and tested until Year 10 and, respectively, Year 11. Testing—the *second approach*—was conducted in 1993, 1994, and 1996 (Kaiser, 1999b). Tests were informal, nonstandardized, and based on an analysis of curricula. All test rounds covered three large topic areas: number, algebra, and functions and graphs with geometry.

The main difference from other studies lies in the qualitative part of the project. The *third approach* adopted was to conduct participant observer classroom observations in about 240 lessons in 17 schools in England and about 100 lessons in 12 German schools (Kaiser, 1999b). Based on the entire set of observations, idealized descriptions—*constituting the fourth approach*—of typical mathematics teaching in German and English classrooms were constructed so as to encompass the following three foci: mathematical theory (including introduction of new concepts and methods, importance of theory and rules, organization by subject structure or a spiral curriculum, the role of proofs, rules versus examples, the role of precise language and formal notations); the role of real-world examples; and teaching and learning styles (for further details of the method adopted, see Kaiser, 1997).

The 13th ICMI Study

The final study to be considered here is the 13th ICMI Study, *Mathematics Education in Different Cultural Traditions: A Comparative Study of East Asia and the West* (Leung et al., 2006). This study was not a uniform, coherent one, based on one single design, but rather an umbrella overarching a variety of specific studies with different foci and perspectives, all seeking to compare and contrast fundamental features of mathematics education in East Asia and the West.

We confine ourselves to outlining, in an aggregate manner, some of the most significant aspects of methodology involved in this study. One of the pertinent issues dealt with in the study was how it can be that East Asian students excel in international comparative mathematics achievement tests such as TIMSS and PISA while at the same time possessing negative attitudes and low self-concept towards mathematics and its study (Leung et al., 2006). One of the methods adopted to answer this question is to undertake historico-cultural investigations of the origins and development of the fundamental traditions in East Asian and Western countries, in particular with regard to the role of the teacher. This was done, for example, in Hirabayashi's and Ueno's chapters, as far as Japan is concerned, and in Wong's and Li Shiqi chapters concerning China. The Western tradition was depicted in Keitel's chapter. Analytic comparisons between Eastern and Western *curricula* were made in Bessot and Comiti's chapter on French and Vietnamese curricula, and in Wu and Zhang's chapter, whereas comparative analyses of Eastern and Western *textbooks* were presented in Li Yeping and Ginsburg's chapter and in Park and Leung's chapter. There was a focus on teachers' beliefs and values in the last part of the book. Perry, Wong, and Howard's chapter reported on a questionnaire-based study comparing Australian and Hong Kong primary and secondary teachers' beliefs about mathematics, mathematics learning, and mathematics teaching, whereas middle-school teachers' beliefs in the USA and China were studied through a combined questionnaire–interview–observation approach reported in An, Kulm, Wu, Ma, and Wang's chapter.

The ICMI study book also included chapters which surveyed the other comparative studies referred to in the present chapter. In summary, a fair sample of the spectrum of research methods employed in international studies of mathematics teaching and learning were represented in the ICMI volume.

Reflections on Designs and Methods

Before we summarize, analyze, and reflect on the designs and methods encountered in the studies presented in this chapter, we consider a more fundamental question which has been briefly touched upon above: To what extent are international comparative studies at all possible and meaningful? It goes without saying that the very carrying out of such studies presupposes that they appear as both meaningful and possible to those who conduct them. Otherwise they would not exist. As this is a deep and complex issue, which in a way deserves a chapter of its own, we have to confine ourselves to sketching some basic deliberations.

First, one should bear in mind that the task of this chapter is to present and analyze—from a methodological perspective—studies that actually exist. The primary task is not to assess and judge them. The agencies and people who instigate and conduct the studies—the primary stakeholders—do so for a purpose, and to them the most significant issues therefore are whether a given study serves its purpose and can be said to be methodologically sound so as to produce results that are useful, valid, and reliable relative to that purpose. What is likely to be less important to the stakeholders of a study is whether or not it is useful, valid, and reliable with respect to other sorts of purposes. So, any critique of a study should be more concerned with the extent to which it lives up to what it purports to be, than with its capability of responding adequately to something else.

There are two components involved in "international comparative studies," namely "international" and "comparative study." The fundamental component in the question about the possibility of international comparative studies seems to be the very notion of comparative study. Whenever entities (such as objects, situations, conditions, relationships, mechanisms, phenomena, or categories of contents) are subjected to any form of comparison, certain features of the entities are deemed irrelevant or less important and left out of consideration, yet others are chosen to be in focus. How then can one be sure that the entities left out of consideration do not—behind the curtain, so to speak—exert a significant influence on the features actually considered in the comparison? In disciplined inquiry in general, and in science in particular, this may well be seen as the most essential question of all. Since it is usually extremely difficult to guarantee that no hidden variables have an impact on the entities being compared, and hence on the outcomes of a comparative study, the most important thing is to subject the study to open discussion, critical scrutiny, alternative studies, methodological debate, and so on, much of which will concentrate on the balance between the factors left out, or kept in, in the comparisons undertaken.

When comparative studies deal with human beings and human behaviour, the issues just mentioned become aggravated. For instance, this is the case when we compare nth-grade students in different schools in the same town, in different parts of the same country, in different socioeconomic, ethnic, cultural, or religious groups, and so on. Going beyond the borders of one country to involve other countries, continents, cultural traditions, and so on, implies further complexity. It introduces changes of degree or orders of magnitude, but not fundamental changes. Needless to say, even more openness, care, analysis, scrutiny, and alternative views or investigations are needed in international comparative studies than in other kinds of comparative studies. But it would be unreasonable to claim that whereas comparisons between nth-grade students in two schools in neighbourhood N of municipality M in county C in country S are perfectly possible and meaningful, the possibility and meaningfulness of comparisons disappear when national, regional, continental, or cultural borders are being crossed.

We now offer a number of more specific observations concerning the designs and methodologies of international studies of the teaching and learning of mathematics. The first observation worth making is that a large fraction of the studies have investigated *student achievement on written tasks* as a key component of their design, not only when assessment of achievement *is* actually the primary subject of study but also when the purpose of the study is to come to grips with something else. Since the time allocated per test item is usually very limited, ranging from 1 to 2 minutes, to 15 minutes, only those kinds of achievement which can come to fruition within such a time frame are represented in the tests. It is remarkable that student achievement on tasks is taken to epitomize *mathematical competence at large*. This fact is indeed worth discussing, not only because of the constrained spectrum of forms of achievement which can find their way into the test but also because mathematical competence possesses many more significant dimensions than the ability to do well in achievement tests. It can, of course, be very well justified to include achievement tests in a given study, and sophisticated test items may have been developed for the study according to the highest international standards. So, achievement tests are not a problem in and of themselves. However, a problem occurs when no other probes into mathematical competence are taken into account and employed.

The problem is aggravated when media, politicians, and other outsiders to mathematics education oversimplify things even further by equating mathematical competence with success on achievement tests. It is not unusual to encounter the following line of argumentation: As the test results are numbers that speak for themselves, you are not allowed to interpret what they tell us, let alone to argue against them.

This way of reasoning is not very different from saying: The thermometer in my hand yields a result you can't argue with. It displays the gravitational force on the spot where I'm standing! It may be seen as surprising that no international studies have devised methods to investigate aspects of students' mathematical competence that cannot be accessed by tightly time-constrained achievement tests. It is conceivable that future international studies would benefit considerably from the development of new kinds of gauges of mathematical competence. There are, however, huge challenges in adopting more complex assessment situations in the wide variety of school contexts found among the many countries participating in large-scale international studies such as TIMSS and PISA.

In cross-national achievement studies, the fact that all the items included in the tasks have to be meaningful and reasonable in participating countries leads to a fair amount of harmonization of items, item types, response formats, and score coding. This is true both of curriculum-referenced studies, such as TIMSS, where items at least to some extent have to be related to the curricula students have been exposed to, and of literacy or competency referenced studies, where some basic degree of familiarity with contexts and situations needs to be ascertained, as is the case in, for example, PISA. It poses particular challenges to test mathematics embedded in extra-mathematical contexts in a manner that is not *too* dependent on cultural, technological, or socio-economic contexts. All this being said, items in international studies are typically highly thoughtfully and carefully constructed, developed, piloted, field-trialled, score coded, and rated, sometimes with an impressive degree of sophistication. Also, sophisticated item analysis methods that allow for studying achievement conditioned on a variety of (sub-)population characteristics and other background variables are put into use in many studies, especially large-scale ones such as TIMSS and PISA. Against this background, the various pools of items from international studies are goldmines for research and practice, as are the multitude of achievement databases, many of which have already been subjected to several correlation studies. However, unfortunately this happens too seldom, and the existing item pools and databases deserve to be put to use in new research.

The next observation is that even though student achievement tests are a major component in several international studies, tests never stand alone. They are *always accompanied by other approaches and instruments*, such as student or teacher questionnaires and interviews, classroom observations, analyses of written materials such as curriculum documents, teacher education programs, textbooks, and assessment instruments. There are three main reasons for making use of such other approaches in relation to achievement studies: to provide a means for interpreting and understanding what students had in mind in their solution processes and how these were related to what and how they had been taught, to provide causal or correlational explanations of students' achievement or of related observed phenomena in terms of background factors and variables, or to provide an entirely different sort of information from that sought in the achievement tests; for instance, about students' attitudes, beliefs, and career perspectives. It goes without saying that the approaches listed above are not only utilized in connection with achievement studies, they also can, and often do, stand alone as independent approaches.

As an independent approach, or in addition to other approaches, *questionnaires* to students, teachers, school principals, or other target groups primarily serve two purposes. Sometimes the primary purpose is to gather information of intrinsic, separate interest in the study. At other times, it is to constitute a platform to follow up on or lead into other approaches, say, classroom observations or interviews. Questionnaire questions come in different types. Some questions ask for factual, unambiguous, objective information such as student sex and age, number of students in a class, types of school programs, and the like. Other questions may ask for multiple-choice responses representing predetermined, but not necessarily well-defined, entities, while other questions may ask the respondent to describe objects, phenomena, or situations in his or her own words, and still others may concern affective or attitudinal matters.

It is generally acknowledged that questionnaires give rise to many methodological issues, at least as far as nontrivial, nonfactual questions are concerned. One such issue is that the response categories offered in multiple-choice questions may often not be understood or accepted by respondents, for instance, because of ambiguity or problematic demarcation lines between response options. This becomes a special concern when questionnaire responses are subjected to subsequent quantitative aggregation. A related issue concerns questions in which respondents are asked to estimate the frequency of the occurrence of certain kinds of experiences or acts, where it may simply be difficult to remember things well enough to provide reliable answers. Another issue is to do with questions that ask respondents to write comments or statements which are likely to be difficult to interpret by researchers. In some contexts, respondents may tend to figure out which answers are "good" or "right," or would impress or please those who administer the questionnaires, and then respond accordingly. Moreover, there may well be socio-cultural biases in the occurrence of this tendency. (Similar arguments are posed in relation to video observation and interviews considered below.) However, designed with reflection and care and treated with caution, questionnaires can be powerful instruments, both in quantitative surveys and in in-depth qualitative investigations.

As with questionnaires, *interviews* can be a method to gather information of independent research interest, and they may constitute an approach accompanying other approaches. To the extent interviews are used in large-scale studies, they are typically used for the latter purpose, as a method to probe deeper into issues or phenomena which have emerged through other means, such as achievement tests, questionnaires, or classroom observations. It may be that students' comments and reflections on their solutions to problems are sought, in order to shed light on their background knowledge, strategies, or solution processes. Or it may be that elaboration on students' responses to attitudinal questions in a questionnaire is needed, either as a means for ascertaining investigators' interpretation of the responses or as a way to resolve possible inconsistencies in the responses. Or it may be that the reasons for teachers' observed acts and decisions in classrooms need or deserve further elucidation. Usually interviews employed in large-scale studies address a much smaller subject sample than does the study itself. Therefore, such interview data are rarely aggregated in a quantitative form but remain qualitative data, possibly subjected to some sort of classification. Since the interviews typically serve specific purposes, seeking certain kinds of information, they often—but not always—take place according to some protocol, either a completely structured protocol, not admitting deviation from predefined questions, or a semi-structured protocol that admits tangential excursions to follow up on the responses obtained while returning to the main track afterwards.

When interviews are used in focal studies, all the features just mentioned apply as well, but additional features become relevant. Most importantly, in some studies interviews are given the predominant role. This is typically the case when respondents' comments, experiences, or views are sought on a broad spectrum of topics or issues—for instance, when the aim is to obtain a multi-faceted and integral picture of the respondents selected. In such cases, loosely structured interviews may come into play; that is, in the shape of more freely flowing conversations in which the route taken by the interviewer depends on what happens along the road.

The conducting of interviews poses many challenges, as do their recording, registration, analysis, and sometimes coding. It is often demanding to "get what one is after," because interviews are a form of human interaction and hence subjected to implicit or explicit socio-cultural boundary conditions, which are likely to differ from country to country. It may not only be difficult to obtain a fair degree of homogeneity across countries, but also challenging for the interviewer to steer the conversation according to the interview protocol and pose follow-up questions while paying close attention to the social relationship with the interviewee and perhaps managing the equipment, taking field notes, and so on. Recording, registering, transcribing, coding, or otherwise analyzing the interviews conducted are enormously time-consuming and intellectually demanding activities, especially when it comes to selecting what to store and to interpreting what respondents said. No wonder that a huge body of research literature exists on interviews as a research method.

Comparative *classroom studies*, too, have given rise to huge bodies of methodological considerations, many of which pay special attention to the instruments, procedures, and techniques involved in conducting such studies. As is the case with interviews, classroom studies can take place with varying degrees of structuring, ranging from unstructured studies in which observers, whether participant or neutral observers, focus on what appears to them to be significant along the road, to semi-structured studies, in which researchers concentrate their attention (or intervention) on certain predetermined topics or issues but are also ready to follow up on interesting opportunities or sidetracks that emerge during classroom sessions, through to completely structured studies, where researchers record and classify instances of certain sorts of phenomena or situations in predefined categories and neglect everything else. Since the mathematics classroom is an immensely complex organism, the set of potential objects of study is immensely complex as well. Forms and content of classroom interaction and communication between the teachers and the whole class, student groups or individual students, or among students, may be one possible focus point. Student activities and the teacher's role in orchestrating them may be another focus point, as may student behaviour in particular respects, for example problem solving, hypothesis formation, or explanation of solutions.

Also the nature of the mathematics actually being dealt with in the classroom by teacher or students may be of interest to researchers.

The main reason that the technicalities of classroom research preoccupy researchers is that a classroom session is by definition of a transient nature, so measures that make it possible to register and fix the significant components of the session, either for documentation or for later analysis, are crucial for the entire undertaking. Field notes or written forms to be filled out by the researcher during class, audio or video recordings of whole sessions or episodes, are some of the instruments typically used alone or in combination in such research. Providing detailed information about the procedures followed and the circumstances under which the instruments have been employed is an important documentation task. The concurrent or post hoc coding of the classroom entities identified, and the grounds on which the coding has been performed, are equally important tasks, as is the tracing of them in the analysis. This is not the place to go into details. It is worth noting, though, that some of the international methodological and technical standards for classroom study research in mathematics education have to a large extent been established and moved forward by the international comparative studies, especially as regards to the handling of large samples.

Comparative *curriculum and textbook analyses* are conducted on written documents, and the methods employed therefore involve text analysis. However, apart from general aspects of such analysis and analysis of curricula in relation to education systems—what students get what sort of education, where, with whom, and taught by whom—curriculum and textbook analyses in mathematics education have strong mathematical components in terms of content, exposition, processes, competencies, tasks, activities, and so on, which can be analyzed in a multitude of different, and sometimes even conflicting, ways. Therefore, frameworks for curriculum and textbook analyses in mathematics represent important methodological challenges and decisions, the outcomes of which have a decisive impact on the nature and results of the research conducted. Here, too, many of the international comparative studies considered have contributed to setting and improving significant aspects of the standards of research internationally.

Tasks for students are essential in teaching and learning of school mathematics and in international achievement tests, to such a degree that the nature of the tasks given to students to a large extent codetermines the outcomes of international studies. Against this background, task construction and task analysis become key methodological issues. It is interesting to observe that already in FIMS a matrix-based framework (content-by-cognitive behaviour level) for selecting and analyzing test items was put into practice. In other words, test items were classified not according to more or less traditional content strands only, but according to other dimensions as well. This was the case with all subsequent large-scale studies, including TIMSS. The schemes adopted by PISA were the most elaborate of all. Item classification according to different dimensions gives rise to a variety of correlational item analysis studies of a statistical type.

In addition to the tasks employed in comparative studies, it is also interesting to study the kinds of tasks utilized in mathematics education in different countries. It is therefore somewhat surprising that only few publications of this kind exist. An exception is the book by Shimizu, Kaur, and Clarke (2010).

Concluding Comments

This chapter has attempted two things: (a) to provide a detailed account of the purposes and goals of a number of important large-scale or focal studies of mathematics teaching and learning internationally and of the designs and methods employed to conduct these studies; (b) to analyze and reflect on those designs and methods.

We have found that most of the studies have adopted a *multi-faceted design*, in which combinations of *different approaches* have been used to answer different subsets of the set of questions that gave rise to the study. These approaches are as follows: frameworks to conceptualize the domain being studied, especially as regards mathematics as a subject; construction and administering of student achievement tests; analysis of intrinsic item characteristics; analysis of student responses; student, teacher, or school questionnaires; sampling of the populations studied; interviews with students, teachers, or parents, and methods for analyzing the outcomes; observation (participant or neutral) of real classrooms and methods for recording and analyzing the resulting data; analysis of curricula as part of education systems and as separate entities, textbooks, and assessment tasks; and analytic reflections on the traditions and cultural environments of mathematics education.

Together with these approaches comes a variety of different methods, each of which is implemented by the use of various specific instruments. The methods and the instruments in turn involve a multitude of different procedures and techniques that we have had to leave aside in this chapter, even though quite a few of them are interesting in their own right.

It is a remarkable fact that most, if not all, of the studies considered have contributed to substantial progress in the development of the research designs, approaches, methods, and instruments applied in the studies. Among other things, this progress is due to the fact that several studies have had many human and material resources at their disposal, primarily because the stakeholders of the studies often attribute large amounts of prestige and impact to the outcomes and the politico-administrative uses of the studies.

This phenomenon implies that several sorts of research not meant to deal with international comparisons of one kind or the other can benefit greatly from the contributions to research methodology offered by the international studies.

We have found, however, that the studies display certain limitations as well. This is particularly true of the approaches to gauge student achievement in mathematics, where time and format constraints exclude essential aspects of mathematical competence from being taken into consideration in the studies. This is an issue on which substantial new developments are sorely needed.

Another limitation has been that the overall cultural, economic, and structural contexts and boundary conditions of the education systems at large, and of schooling in particular, have only rarely entered the studies in a direct manner. Such factors influence the classroom reality in ways that go beyond the reality being produced by participants in practice only. Here, too, new approaches directly linking classroom reality to the surrounding contexts are needed.

A chapter such as this one cannot end without comments on the fact that international comparative studies attract a massive interest amongst politicians, media, and the general public. There is a clear tendency of these parties to summarize things in a manner that is "clear, brief, and wrong." This is on the boundary of involving misuse of the studies, but it is a misuse that is difficult to counteract by those involved in them. However, it would probably contribute to more balanced and fact-based debates if researchers undertook to engage in them to a larger extent than is typically seen.

Acknowledgments The authors would like to thank the reviewers, Michael Fried and Maitree Inprasitha, and above all the section editor, Jeremy Kilpatrick, for significant observations and constructive suggestions which have helped to improve this chapter.

References

Bauersfeld, H. (1979). Research related to the mathematical learning process. In B. Christiansen & H. G. Steiner (Eds.), *New trends in mathematics teaching* (Vol. 4, pp. 199–213). Paris, France: UNESCO.

Beaton, A. E., & Robitaille, D. F (1999). An overview of the Third International Mathematics and Science Study. In G. Kaiser, E. Luna, & I. Huntley (Eds.), *International comparisons in mathematics education* (Studies in Mathematics Education Series No. 11, pp. 30–47). London, UK: Falmer.

Becker, J. P. (Ed.). (1992). *Report of U.S.-Japan cross-national research on students problem solving behaviors*. Carbondale: Southern Illinois University at Carbondale. http://eric.ed.gov./PDFS/ED351204.pdf.

Becker, J. P., Sawada, T., & Shimizu, Y. (1999). Some findings of the US-Japan cross-cultural research on students' problem solving behaviours. In G. Kaiser, E. Luna, & I. Huntley (Eds.), *International comparisons in mathematics education* (Studies in Mathematics Education Series No. 11, pp. 121–139). London, UK: Falmer.

Clarke, D. J. (1998). Studying the classroom negotiation of meaning: Complementary accounts methodology. In A. Teppo (Ed.), *Qualitative research methods in mathematics education* (Journal for Research in Mathematics Education Monograph No. 9, pp. 98–111). Reston, VA: National Council of Teachers of Mathematics.

Clarke, D. J. (Ed.). (2001). *Perspectives on practice and meaning in mathematics and science classrooms*. Dordrecht, the Netherlands: Kluwer Academic Publishers.

Clarke, D. J. (2003). International comparative research in mathematics education. In A. J. Bishop, M. A. Clements, C. Keitel, J. Kilpatrick, & F. K. S. Leung (Eds.), *Second international handbook of mathematics education* (pp. 145–186). Dordrecht, the Netherlands: Kluwer Academic Publishers.

Clarke, D. J. (n.d.). *The learners' perspective study: Research design*. Retrieved from 2011-04-10. http://extranet.edfac.unimelb.edu.au/DSME/lps/assets/lps.pdf.

Clarke, D. J., Emanuelsson, J., Jablonka, E., & Mok, I. A. C. (Eds.). (2006). *Making connections: Comparing mathematics classrooms around the world*. Rotterdam, the Netherlands: Sense Publishers.

Clarke, D. J., Keitel, C., & Shimizu, Y. (2006). The Learner's perspective study. In D. J. Clarke, C. Keitel, & Y. Shimizu (Eds.), *Mathematics classrooms in twelve countries: The insider's perspective* (pp. 1–14). Rotterdam, the Netherlands: Sense Publishers.

Cogan, L. S., & Schmidt, W. H. (1999). An examination of instructional practices in six countries. In G. Kaiser, E. Luna, & I. Huntley (eds.) *International comparisons in mathematics education* (Studies in Mathematics Education Series No. 11, pp. 68–85). London, UK: Falmer.

De Lange, J. (2007). Large-scale assessment and mathematics education. In F. K. Lester Jr. (Ed.), *Second handbook of research on mathematics teaching and learning* (pp. 1111–1142). Charlotte, NC: Information Age.

Freudenthal, H. (1975). Pupils' achievement internationally compared: The IEA. *Educational Studies in Mathematics, 6*, 127–186.

Garden, R. A., Lie, S., Robitaille, D. F., Angell, C., Martin, M. O., Mullis, I. V. S., et al. (2006). *TIMSS advanced 2008 assessment frameworks.* Boston, MA: TIMSS & PIRLS International Study Center, Lynch School of Education, Boston College.

Husén, T. (Ed.). (1967). *International study of achievement in mathematics* (Vols. 1 & 2). Stockholm, Sweden: Almqvist & Wiksell.

Kaiser, G. (1997). Vergleichende Untersuchungen zun mathematikunterricht im englischen und deutschen Schulwesen [Comparative studies of mathematics instruction in English in the German school system]. *Journal für Mathematik-Didaktik, 18*(2/3), 127–170.

Kaiser, G. (1999a). International comparisons in mathematics education under the perspectives of comparative education. In: G. Kaiser, E. Luna, & I. Huntley (Eds.), *International comparisons in mathematics education* (Studies in Mathematics Education Series No. 11, pp. 3–15). London, UK: Falmer.

Kaiser, G. (1999b). Comparative studies on teaching mathematics in England and Germany. In G. Kaiser, E. Luna, & I. Huntley (Eds.), *International comparisons in mathematics education* (Studies in Mathematics Education Series No. 11, pp. 140–150). London, UK: Falmer.

Kawanaka, T., Stigler, J. W., & Hiebert, J. (1999). Studying mathematics classrooms in Germany, Japan and the United States: Lessons from the TIMSS Videotape Study. In G. Kaiser, E. Luna, & I. Huntley (Eds.), *International comparisons in mathematics education* (Studies in Mathematics Education Series No. 11, pp. 86–103). London, UK: Falmer.

Leung, F. K. S., Graf, K. D., & Lopez-Real, F. J. (Eds.). (2006). *Mathematics education in different cultural traditions: A comparative study of East Asia and the West: The 13th ICMI Study.* New York, NY: Springer.

Martin, M. O., Gregory, K. D., & Stemler, S. E. (Eds.). (2000). *TIMSS 1999 technical report.* Boston, MA: International Study Center, Lynch School of Education, Boston College.

Martin, M. O., & Kelly, D. L. (Eds.). (1996). *TIMSS technical report: Vol. I. Design and development.* Boston, MA: Center for the Study of Testing, Evaluation, and Educational Policy, Boston College.

Martin, M. O., Mullis, I. V. S., & Chrostowski, S. J. (Eds.). (2004). *TIMSS 2003 Technical report.* Boston, MA: TIMSS & PIRLS International Study Center, Lynch School of Education, Boston College

Mullis, I. V. S., Martin, M. O., & Foy, P. (2005). *IEA's TIMSS 2003 international report on achievement in the mathematics cognitive domains. Findings from a developmental project.* Boston, MA: TIMSS & PIRLS International Study Center, Lynch School of Education, Boston College.

Mullis, I. V. S., Martin, M. O., Ruddock, G. J., O'Sullivan, C. Y., Arora, A., & Erberber, E. (2005). *TIMSS 2007 assessment frameworks.* Boston, MA: TIMSS & PIRLS International Study Center, Lynch School of Education, Boston College.

Niss, M. (2010). What is quality in a PhD dissertation in mathematics education? *Nordisk Matematik(k)Didaktik(k)/Nordic Studies in Mathematics Education, 15*(1), 5–23.

Niss, M., & Hoejgaard, T. (Eds). (2011). *Competencies and mathematical learning. Ideas and inspiration for the development of mathematics teaching and learning in Denmark* (English edition, October 2011, IMFUFAtekst no. 485). Roskilde, Denmark: Roskilde University, Department of Science, Systems and Models.

Niss, M., & Jensen, T. H. (Eds.). (2002). *Kompetencer og matematiklæring, Ideer og inspiration til udvikling af matematikundervisning i Danmark* [Competences and mathematics education: Ideas and inspiration for the development of mathematics education in Denmark] (Uddannelsesstyrelsens temahæfteserie nr. 18). Copenhagen, Denmark: Ministry of Education.

Olson, J. F., Martin, M. O., & Mullis, I. V. S. (Eds.). (2008). *TIMSS 2007 technical report*. Boston, MA: TIMSS & PIRLS International Study Center, Lynch School of Education, Boston College.

Organisation for Economic Co-operation and Development. (1999). *Measuring student knowledge and skills: A new framework for assessment*. Paris, France: Author.

Organisation for Economic Co-operation and Development. (2001). *Knowledge and skills for life: First results from the OECD Programme for International Student Assessment (PISA) 2000*. Paris, France: Author.

Organisation for Economic Co-operation and Development. (2003). *The PISA 2003 assessment framework: Mathematics, reading, science and problem solving knowledge and skills*. Paris, France: Author.

Organisation for Economic Co-operation and Development. (2004). *Learning for tomorrow's world: First results from PISA 2003*. Paris, France: Author.

Organisation for Economic Co-operation and Development. (2005). *School factors related to quality and equity. Results from PISA 2000*. Paris, France: Author.

Organisation for Economic Co-operation and Development. (2006). *Assessing scientific, reading and mathematical literacy. A framework for PISA 2006*. Paris, France: Author.

Organisation for Economic Co-operation and Development. (2007). *Science competencies for tomorrow's world*. Paris, France: OECD.

Organisation for Economic Co-operation and Development. (2010a). *What students know and can do: Student performance in reading, mathematics and science*. Paris, France: Author.

Organisation for Economic Co-operation and Development. (2010b). *PISA 2012 mathematics framework. To OECD, November 30, 2010* [Draft subject to possible revisions after the field trial]. Retrieved from http://www.pisa.oecd.org/dataoecd/8/38/46961598.pdf.

Postlethwaite, T. N. (1971). International Association for the Evaluation of Educational Achievement (IEA): The mathematics study. *Journal for Research in Mathematics Education, 2*, 69–103.

Robitaille, D. F., & Garden, R. A. (1989). *The IEA Study of Mathematics II: Contexts and outcomes of school mathematics*. Oxford, UK: Pergamon.

Robitaille, D. F., Schmidt, W. H., Raizen, S. A., McKnight, C. C., Britton, E., & Nicol, C. (1993). *Curriculum frameworks for mathematics and science* (TIMSS Monograph No. 1). Vancouver, Canada: Pacific Educational Press.

Robitaille, D. F., & Travers, K. J. (1992). International studies of achievement in mathematics. In D. A. Grouws (Ed.), *Handbook of research on mathematics teaching and learning* (pp. 687–707). New York, NY: Macmillan.

Schmidt, W. H., McKnight, C., Valverde, G. A., Houang, R. T., & Wiley, D. E. (1997). *Many visions, many aims: A cross-national investigation of curricular intentions in school mathematics*. Dordrecht, the Netherlands: Kluwer Academic Publishers.

Schubring, G. (2008). The origin and early incarnations of ICMI. In: M. Menghini, F. Furinghetti, L. Giacardi, & F. Arzarello (Eds.), *The first century of the International Commission on Mathematical Instruction (1908–2008). Reflecting and shaping the world of mathematics education* (pp. 113–130). Rome, Italy: Instituto della Enciclopedia Italiana.

Shimizu, Y., Kaur, B., & Clarke, D. J. (2010). *Mathematical tasks around the world*. Rotterdam, The Netherlands: Sense Publishers.

Stevenson, H. W. (1999). The Case Study Project of TIMSS. In G. Kaiser, E. Luna, & I. Huntley (Eds.), *International comparisons in mathematics education* (Studies in Mathematics Education Series No. 11, pp. 104–120). London, UK: Falmer.

Stigler, J. W., Gallimore, R., & Hiebert, J. (2000). Using video surveys to compare classrooms and teaching across cultures: Examples and lessons from the TIMSS and TIMSS-R video studies. *Educational Psychologist, 35*, 87–100.

Stigler, J. W., Gonzales, P., Kawanaka, T., Knoll, S., & Serrano, A. (1999, February). *The TIMSS Videotape Classroom Study: Methods and findings from an exploratory research project on eighth-grade mathematics instruction in Germany, Japan, and the United States* (National

Center for Educational Statistics, Research and Development Report). Washington, DC: U.S. Department of Education. Retrieved from http://nces.ed.gov.

Stigler, J., & Hiebert, J. (1999). *The teaching gap: Best ideas from the world's teachers for improving education in the classroom*. New York, NY: Free Press.

Travers, K. J., & Weinzweig. A. I. (1999). The Second International Mathematics Study. In G. Kaiser, E. Luna, & I. Huntley (Eds.), *International comparisons in mathematics education* (Studies in Mathematics Education Series No. 11, pp. 19–29). London, UK: Falmer.

Travers, K. J., & Westbury, I. (1990). *The IEA Study of Mathematics II: Analysis of mathematics curricula*. Oxford, UK: Pergamon.

Chapter 31
Implications of International Studies for National and Local Policy in Mathematics Education

John A. Dossey and Margaret L. Wu

Abstract This chapter examines large-scale comparative studies of mathematics education focussed on student achievement in an attempt to explain how such investigations influence the formation and implementation of policies affecting mathematics education. In doing so, we review the nature of comparative studies and policy research. Bennett's (1991) formulation of policy development and implementation is used in examining national reactions to the results of international studies. Focus is given to the degree to which mathematics educators and others have played major roles in determining related policy outcomes affecting curriculum and the development and interpretations of the assessment instruments and processes themselves.

International Studies of Mathematics Education

Writing in 1999, Martin Carnoy stated:

> The *quality* of national educational systems is increasingly being compared internationally. This has placed increased emphasis on mathematics and science curricula, English as a foreign language and communication skills. Testing and standards are part of a broader effort to increase accountability by *measuring* knowledge production and using such measures to assess education workers (teachers) and managers. Yet, the way testing is used to "improve quality" is heavily influenced by the *political* context and purposes of the evaluation system. Again, to develop effective policies for education improvement, the ideological–political content of a testing programme has to be clearly separated from its educational management content. (p. 16)

J. A. Dossey (✉)
Illinois State University, Normal, IL, USA
e-mail: jdossey@ilstu.edu

M. L. Wu
Victoria University, Melbourne, Victoria, Australia

Carnoy's insightful comments are as true today as they were in 1999. The appearance of large-scale international comparative studies of mathematics education, starting with those of the 1960s, have engendered three significant changes to the mathematics education landscape at national, state/provincial, and local levels. The first is an ever present reliance on large-scale sample survey data as a policy-making base for curricular and mathematics instruction decisions. The second is the heightened role occupied by nonmathematical organizations, governmental and nongovernmental, in the decision-making structure for what is important and what is not important in mathematics education programs, their design, and their implementation. The third is the use of comparative assessments as a lever for encouraging the convergence of curricular plans toward "national" or "global" models.

These changes have brought with them a reliance on values based in numerical indicators emanating from surveys, rather than from a body of mathematics education research based on a series of related research studies, be they quantitative or qualitative. Although expert panels who assisted in the design of these international assessments have involved knowledgeable mathematicians and mathematics educators, the resulting structure of the questionnaires and assessments used in the final collection of data has often then been modified to fit time, legal, or policy-based constraints which have distanced, in many cases, the data from classroom practice. Notwithstanding these disconnects, the global outcomes of international comparative studies of school mathematics have emerged as powerful arbiters of educational policy discussions of student competence, teacher quality, the path to school improvement, and the structure of schooling itself. At national, state/provincial, and local levels, assessment systems similar to the international assessments have been instituted by legislative acts as primary monitors of trends. Decision making and the institution of "educational crises" have become major media events stemming from the release of participant rankings in league-like tables of student achievement results or teacher qualifications.

The first major international comparative work in mathematics education was initiated by the International Commission on the Teaching of Mathematics during the 4 years following the organization's founding at the Fourth International Congress of Mathematicians in Rome in 1908. The study was created with the expressed purpose of conducting a comparative study on the methods and plans of teaching mathematics at the secondary and other levels of schooling. The study, spanning the years from 1908 through 1914, produced 187 volumes, containing 310 reports from eighteen countries (ICMI, 2011a). Excerpts based on data from the study can be found in the *Teaching of Arithmetic* and *Mathematics in the Elementary Schools of the United States* (Bidwell & Clason, 1970; Smith, 1909; United States Bureau of Education, 1911). In 1954, the Commissions' parent body, the International Mathematics Union, restructured and renamed the commission as the International Commission on Mathematics Instruction (ICMI). Along with the shift in the name, there was an implicit shift from the study of "the teaching of mathematics" to "mathematics instruction" in the activities of the organization (ICMI, 2011b).

In 1967, a *New York Times* article provided, in a manner similar to a sport's league standings table in a newspaper, the order of finish of national student achievement performances in the First International Mathematics Study (FIMS). This public release and the media's presentation focussing on standings signalled the emergence of a new way of examining and evaluating nation/state or provincial/local mathematics education programs. The influence of this approach to policy-building and blame-directing has only increased over time with the quadrennial release of data from the Trends in International Mathematics and Science Study (TIMSS) of the International Association for the Evaluation of Educational Achievement (IEA) and the triennial release of data from the Programme for International Student Assessment (PISA) developed by the Organisation for Economic Co-operation and Development (OECD). Mixed among the output from these massive studies are findings emanating from United Nations Educational, Scientific and Cultural Organisation (UNESCO), the World Bank, educational and economic think tanks, national assessments, doctoral dissertations, and education consortia and bureaus in individual countries.

These comparisons have grown over time to include comparisons of student achievement, teaching, teacher preparation, the context of mathematics education, and specialized topics included within or related to the mathematics curriculum such as problem solving, modelling, statistics, textbook contents, and information technology. A full discussion of all of these findings and their policy implications is beyond the scope of this chapter, which will focus on large-scale international studies of student achievement and the impact that they have had and continue to have on education policy in mathematics education.

International Association for the Evaluation of Educational Achievement (IEA)

The IEA was conceived in 1958 at an UNESCO meeting of sociologists, educational psychologists, and psychometricians. The IEA today, consists of a linked body of ministries of education and similar nationally-representative structures. Mathematically, the IEA became an important entity with the release of the findings of the First International Mathematics Study (FIMS) in 1967. This 12-nation study, based on data collected in 1964, focussed on 13-year-olds and students in the pre-university year of schooling. Policy relevant constructs emerging from the study were the importance of student opportunity-to-learn and equity issues as they affected academic performance. Other issues focussed on particular national differences in the education of teachers (Husén, 1967; IEA, 2011).

Seventeen years later, in 1981–1982, the IEA returned to mathematics assessment with the 20-nation Second International Mathematics Study (SIMS). This assessment featured a sharpened design based on a mathematics content framework and substantially more input from the mathematics community. The SIMS design featured pre- and post-measures about student opportunities to learn and perform in mathematics for 13-year-olds and students in the final year of secondary school. This study aroused increased interest in students' opportunity-to-learn, while heightening the key roles of curriculum and number of topics students are exposed to in a given year of study. In-depth questionnaires were used to probe teachers' coverage of key topics in the teaching of prealgebra at the middle-school level and content in precalculus and calculus at the end of secondary school (Burstein, 1993; McKnight et al., 1987; Robitaille & Garden, 1989; Travers & Westbury, 1989).

In 1995, the IEA returned to mathematics with the Third International Mathematics and Science Study (TIMSS). This time, the study cohort contained 45 countries, and the focal populations included 9- and 13-year-olds, as well as students in the final year of secondary education. In addition to focussing on major in- and out-of-school determinants of educational outcomes of schooling, TIMSS also conducted a special substudy comparing the mathematics curricula in the countries participating. The careful design and implementation of the design for the TIMSS 1995 study has provided an anchor for the subsequent IEA cycle of trend studies in mathematics, science, and reading. These ongoing data collections also highlight the semi-permanent status of such studies (now reconceived and renamed under the same acronym: the Trends in International Mathematics and Science Studies). The first assessments in the new formulation of TIMSS were carried out in 1999, 2003, and 2007. In 2011, more than 60 countries and jurisdictions participated in TIMSS 2011. Results from these studies are available online at the TIMSS study centre (http://timss.bc.edu/) and in a series of research monographs (Robitaille & Beaton, 2002; Schmidt, McKnight, Cogan, Jakwerth, & Houang, 1999; Schmidt, McKnight, Valverde, Houange, & Wiley, 1997; Schmidt, McKnight, & Raizen, 1997; Schmidt et al., 1996, 2001; Valverde, Bianchi, Wolfe, Schmidt, & Houang, 2002).

Results from TIMSS increased interest in the teacher preparation policies and practices around the world. At writing, the IEA is involved in the Teacher Education and Development Study in Mathematics (TEDS-M). This IEA study is focussed on how teacher preparation policies, programs, and practices contribute to the capability of teachers to teach mathematics in elementary and lower secondary schools (Grades 4 and 8). The framework, data, and findings from this study are available at the TEDS-M study centre (http://teds.educ.msu.edu/).

Programme for International Student Assessment (PISA)

In 1997, the Organisation for Economic Co-operation and Development, a group of democratic countries sharing economic-related information, decided to initiate a program of literacy assessments for 15-year-olds in the domains of mathematics literacy, science literacy, and reading in the mother tongue. PISA conducted its first survey in 2000, with subsequent surveys following in a triennial cyclic pattern, with the three domains rotating in their degree of overall emphasis within each passing assessment cycle. As a result, mathematics was the major focus of the assessment in 2003, in 2012, and will again be slated for 2021. In the intervening assessment cycles, mathematics is assessed only for trend reporting, with one of the other domains taking the role of primary focus. PISA also includes measures of general or cross-curricular competencies such as problem solving, measured in 2003 and 2012, and financial literacy measured in 2012.

Unlike TIMSS's focus on curricular-based knowledge, PISA focusses on measuring students' mathematical literacy, envisioned as students' ability to apply mathematical knowledge and skills and their developed capabilities to analyze, reason, and communicate effectively as they examine, interpret, and solve problems in contextualized settings. In PISA 2009, 34 OECD member countries and 41 partner countries participated. PISA is the only international education survey to measure the knowledge and skills of 15-year-olds, an age at which students in most countries are nearing the end of their compulsory time in school. Although PISA's results provide a picture of students' capabilities, they provide less direct relationships to the schooling students have received. At the same time, they may provide a better picture of the future capabilities of nations' students to cope with everyday applications of mathematics and science. These results allow countries and economies to compare best practices and to further develop their own improvements—ones appropriate for their school systems (McGaw, 2008; OECD, 2004a, 2004b).

In addition to this difference in aim, the PISA governing board is made up of representatives of national governments or members of their national ministries of education. Although some of these individuals are researchers, many are policy and legislative leaders with responsibilities related to reporting on the output of their nation's schools and status of the implementation of the approved curricula for mathematics.

Both the TIMSS and PISA assessments have had their share of proponents and detractors from within and outside of the educational world. From the foci of the assessments' content and the publication of the assessment frameworks to what students are expected to do in responding to the items and finally to the statistical analysis and reporting of the data, the studies have created a great amount of interest in student learning, performance, and achievement (Hopmann & Brinek, 2007; Kang, 2009; Kilpatrick, 2009; Murphy, 2010; Prais, 2003; Sjøberg, 2007). Supporting this interest, countries and professional societies have released special national studies, and the contractors carrying out the assessments have provided released items and other sample materials available along with other supporting documentation (Kilpatrick, 2009; materials on OECD/PISA Web site: http://www.pisa.oecd.org and on the TIMSS Web site: http://timss.bc.edu).

Other International Assessments of Mathematics Education

The Educational Testing Service (Lapointe, Mead, & Askew, 1992; Lapointe, Mead, & Phillips, 1988) conducted the International Assessment of Educational Progress (IAEP) with 13-year-olds in 1988, and 9- and 13-year-olds in 1991 with an expressed purpose of comparing participating countries' performances with that of US states through a statistical linking of the National Assessment of Educational Progress (NAEP) items common to NAEP and IAEP. This analysis showed wide variation in the performance of US states, with some performing statistically as well as the Asian nations, whereas others performed at the level of developing countries (Pashley & Phillips, 1993). Similar results were found in a special follow-up study to the TIMSS test conducted in 1998 (Kimmelman et al., 1999; Mullis et al., 2001).

Another international comparative education project was the Kassel project. It was initiated in 1993 by England, Germany, and Scotland, and later joined by Australia, Brazil, the Czech Republic, Finland, Greece, Holland, Hungary, Japan, Norway, Poland, Russia, Singapore, Thailand, Ukraine, and the USA. This project is focussed on collecting longitudinal samples of pupil work from the participating countries. As such, it differs from the preceding studies in that it focusses on individual student work over time rather than cross-sectional samples of student work. The analysis of the growth trajectories in these students is then used to ferret out key factors that lead to successful progress in mathematics within the participating countries (Blum & Kaiser, 2004; Burghes, Kaur, & Thompson, 2004).

Three other international comparative studies of note are the International Project on Mathematical Attainment (IPMA) study, the Southern and Eastern Africa Consortium for Monitoring Educational Quality (SACMEQ) studies, and the First International Comparative Study of Language and Mathematics in Latin America.

The IPMA study focussed on student progress from the first year of primary school through secondary school, with data collected concerning student achievement, methods of teaching, resources available to teachers and students, and the nature of the curriculum studied. Countries participating for all or part of the study were Brazil, China, the Czech Republic, England, Estonia, Finland, Greece, Hungary, Ireland, Japan, Poland, Russian, Singapore, South Africa, Ukraine, the USA, and Vietnam. Reports from the study are available at (http://www.cimt.plymouth.ac.uk) and through a summary volume (Burghes, Geach, & Roddick, 2004; IPMA, 2011).

The SACMEQ series of studies report on student performance in reading and mathematics. The sponsoring organization consists of a consortium of the ministries of education from Botswana, Kenya, Lesotho, Malawi, Mauritius, Mozambique, Namibia, Seychelles, South Africa, Tanzania (mainland and Zanzibar), Uganda, Zambia, and Zimbabwe. Starting in 1995, there have been three cycles of assessment, with individual nation reports of recommendations derived from an overall data set representative of the member nations. The SACMEQ results report on the achievement of Standard 6 students (12–14 years of age). Cycle 1 reports were released in 2001, Cycle II reports in 2005, and Cycle III reports in 2010. These reports are available at the consortium Web site at (http://www.sacmeq.org/index.htm). SACMEQ began with support from UNESCO and has grown into a self-sufficient organization through joint support and the development of internal capacity (Greaney & Kelleghan, 2008; SACMEQ, 2011).

The First International Comparative Study of Language and Mathematics in Latin America was a project of the Latin American Laboratory for Assessment of the Quality of Education (LLECE) and involved a consortium of nations consisting of Argentina, Bolivia, Brazil, Chile, Colombia, Costa Rica, Cuba, The Dominican Republic, Honduras, Mexico, Paraguay, Peru, and Venezuela. The ministries of education of these Latin American and Caribbean nations were brought together in 1994 through the coordinating efforts of the UNESCO Regional Office for Latin

America and the Caribbean to develop a study focussed on information on students' achievements and associated factors that would be useful in establishing and implementing education policies within countries. The OECD has assisted UNESCO in the actual collection, analysis, interpretation, and reporting of the LLECE data. The focal content areas for the assessment were the language and mathematics knowledge and skills of third and fourth graders in the participating countries. In addition, information on a significant number of background and contextual variables was obtained from the schools and students (Casassus, Froemel, Palafox, & Cusato, 1998; LLECE, 2001). Information on the study and reports can be found at the consortium's Web site at http://www.llece.org/public/content/view/8/3/lang,en.

International Studies and Educational Policy

Reach of Educational Policy

Comparative international educational research in its purest form involves empirical work aimed at the revision of existing theories of the relationships within or between educational systems or between variables describing educational systems and economic indices or demographic data (Carnoy, 2006). If this is the case, how do the international comparative assessments of mathematics education fit this model? One might argue that their purpose is to describe student achievement at national levels. However, such a response would be short sighted. In reality, their purpose appears to be the creation of a platform for illustrating and relating students' achievement to salient policy variables such as distribution of achievement across racial and cultural groups, the relative performance of different genders in mathematical situations, the distribution of resources and teachers across geographical units, the relationships between the flow of students through the academic mathematics pipeline, and the relationship of various levels of output to national needs and labor projections. Within education, the output of such studies is of direct interest to curriculum experts, teacher educators, and those involved in professional development programs, and textbook writers and publishers of mathematical learning materials, as in Kilpatrick, Mesa, and Sloane (2007). Other interested consumers include governmental and policy experts, parents, and the public in general. As such, the results of national and educational comparative studies is a huge lever for those involved in educational policy, especially those interested in educational reform (Kellaghan, Greaney, & Murray, 2009).

TIMSS and PISA Assessment Frameworks

Given the role that the IEA and PISA results play in serving as levers in international and national discussions of educational policy, one might examine their geneses and stated purposes. IEA studies result from a cooperative group of research bodies, some of which are governmental and some not. In either case, the bodies are research oriented first and policy oriented second. The TIMSS studies are closely linked to instructional processes in classrooms and the curricula of the participating nations. The mission of the TIMSS assessments is

> to provide high quality information on student achievement outcomes and on the educational contexts in which students achieve. ... [In doing so, TIMSS is dedicated] to providing countries with information to improve teaching and learning in these curriculum areas. Conducted every four years on a regular cycle, TIMSS assesses achievement in mathematics and science at the fourth and eighth grades. The achievement data are collected together with extensive background information about the availability of school resources and the quality of curriculum and instruction. (Mullis et al., 2009, pp. 2, 7)

The mission statement for TIMSS places learning outcomes, teaching, and learning contexts at the forefront, with an implied goal of linking achievement to curricula and instructional practices.

PISA studies, on the other hand, assess how well 15-year-old students are prepared to deal with contextualized situations where mathematics might provide assistance in finding resolutions. PISA refers to this capability as *mathematical literacy* and defines it as follows:

> An individual's capacity to formulate, employ, and interpret mathematics in a variety of contexts. It includes reasoning mathematically and using mathematical concepts, procedures, facts, and tools to describe, explain, and predict phenomena. It assists individuals to recognise the role that mathematics plays in the world and to make the well-founded judgments and decisions needed by constructive, engaged and reflective citizens. (OECD, 2010c, p. 4)

Although PISA does not reject curricular links in developing students' literacy, the assessment's primary purpose is the determination and description of students' capabilities to formulate, implement, and solve mathematical problems.

The linking of TIMSS to teaching and learning and PISA to literacy does not say that they ignore the other's main focus. To do so would be a denial of the intrinsic link between the two goals and the huge overlap of outcomes that are examined by both programs. Many of the curricular and instructional research objectives in TIMSS are driven by policy considerations, and many policy objectives in PISA result in research themes linking PISA findings directly to school programs. In fact, over the past decade, the two large-scale assessment programs have moved toward one another in goals and in the nature of the items used in their assessments. In addition, their role in policy decisions also increased as nations, states and provinces, and local school districts have looked for guidance in forming curricular plans and selecting instructional approaches and materials.

Role of International Studies in Shaping Policy

The IEA international mathematics assessments came of age in the 1980s just in time to fill the increased desire within UNESCO and, later, within OECD for a set of indicators of student performance. Indicators, viewed as variables taking on values which describe inputs, processes, or outputs from the educational enterprise of some defined country or defined grouping of countries provided a way of quantifying education. Over time, such indicators became the source of policy, and at the same time, their values provided another lever for policy change. This recasting of indicators in quantifiable form further spread the influence of indicator systems, especially those that had linked assessments, as a source of policy initiatives.

Reports portraying indicators from such studies have fuelled governmental and nongovernmental agency reports on educational outcomes for the past 50 years. The OECD indicators had their birth in the OECD's International Indicators of Education Systems project (INES) in the late 1980s, a movement that coincided with a shift from research-based assessments of student performance within a nation to national assessments of educational progress. This shift was very evident with the maturing National Assessment of Educational Progress (NAEP) in the USA and the initiation of similar, but newer, assessment programs in New Zealand, Portugal, Spain, Sweden, and the UK. Central to the growth for the demand for data on education was the UNESCO (1990) World Declaration on Education for All recognizing education as a human right and relating it to the physical and economic health of nations. Its foci on learning, equity, and supportive environments and resources for education promoted the need for more policy-based items as part of the background and demographic sections of national and international assessments (Moskowitz, & Stephens, 2004; Rutkowski, 2008: UNESCO, 2011).

In 1991, OECD began the publication of annual indices of indicators in its *Education at a Glance* series. This provided easy reference for policy analysts to countries' profiles, as well as their comparative performances relative to other countries. The indicators and supporting data exhibit a wide range of outcomes discussing student performance when parental education, social-economic status, and other factors are considered and when national performances are adjusted for national economic indicators (OECD, 2010c).

At the same time, the influence of PISA was growing within OECD nations. Several non-member OECD nations participated in the 2003, 2006, 2009, and 2012 PISA cycles. This participation multiplied the influence PISA indicators have had on just the member states by including another group of developed nations and an even larger group of developing nations. These indicators do not just inform the leaders of these countries, they assist in the framing of policies and the direction of reforms. Nóvoa and Yariv-Masal (2003) noted:

> Such researches produce a set of conclusions, definitions of "good" or "bad" educational systems, and required solutions. Moreover, the mass media are keen to diffuse the results of these studies, in such a manner that reinforces a need for urgent decisions, following lines of action that seem undisputed and uncontested, largely due to the fact that they have been internationally asserted. (p. 424)

Results from PISA 2000 and 2003 indicators supported the development of national goals for secondary-school curricula in Flemish-speaking regions of Belgium, strengthening mathematics program implementation by increasing the numbers of secondary-school mathematics advisors in New Zealand, and allocating more resources for the mathematical and science education of prospective primary school teachers in Sweden (Owen, Stephens, Moskowitz, & Gil, 2004). When the favourable results of PISA were announced in Finland, rather than reflect positively on them, the Finnish government moved to harmonize the system and to allocate more time to core subjects, potentially removing some of the advantage Finnish schools might be providing their students (Välijärvi, Linnakylä, Kupari, Reinikainen, & Arffman, 2002). But, perhaps the most notable effect was the reception of the PISA 2000 results in Germany. The below median performance in overall literacy ranking stunned the nation, and German educational authorities called for urgent reform measures to right the ship and get Germany back on course. This outcry for reform was based on the argument that the PISA test measures outcomes associated with the emerging world and its workplace requirements (Ertl, 2006). Again, a governmental reaction, in this case was based partially on economics and partially in shock.

Parallel to the growth of indicators and assessments, national educational ministry personnel worldwide were becoming active in the administration, development, analysis, and reporting of findings of their own national assessments and national reports of TIMSS and PISA. This shift of ministry officials into becoming players rather than consumers was aimed more at policy issues than research issues. In several countries, the shift was also accompanied by a shifting from curriculum questions and research-oriented issues to sharpening background assessments to answer other less-curricular-centred issues that were more pressing nationally from a policy standpoint. This movement, in some cases, weakened the focus on content alone in favour of more general policy-centred questions within assessments. Within Europe, first, and then on a broader stage, concern began to arise about the shifting use of indicators from being a focal point for understanding the educational enterprise to potentially being used to shape and control the educational systems of nations from a normative standpoint (Grek et al., 2011; Lester, Dossey, & Lindquist, 2007).

With the advent of multinational industries and the easy international exchange of knowledge and data, there has arisen a demand for evidence-based research findings to quantify the adequacy of state or provincial and national educational systems. This demand is a side product of the advent of reform programs in mathematics based on standards (Chatterji, 2002; Steiner-Khamsi, 2006, 2007). Further, the demand for information in the form of comparative data has grown to the point where data resulting from the output of large-scale comparative studies can be viewed as forming international knowledge banks which enter into the processes of borrowing and lending. This view of the outcome data serving as a knowledge bank was first broached at an educational meeting at the World Bank in 1996 (Eaton & Kortum, 1996; Jones, 2004; Jones & Coleman, 2005).

The rise of such international knowledge banks has brought with it two new major policy influences. The establishment of international means and indices for outcomes has resulted in funder/donor pressures to move country means above international averages on targeted indices. This has been especially true in developing countries. In many cases, such targeted indices are only marginally related to national or local goals or to the culture of a given country. Such pressures are most prevalent in developing countries, where ministries themselves find continued funding associated with a "harmonization" of their programs and expectations to share the funders' knowledge and approaches and to work together with partner countries to "converge" and improve their programs (World Bank, 2005). With this pressure from large donors, financial lenders, or policy groups such as OECD comes a change in their behaviour. They begin to make a shift from being a lender of capital to becoming a lender of educational policy.

The second shift related to policy is directly tied to the emergence of the large knowledge banks of studies such as those associated with the IEA and OECD studies. This shift comes from within the affected countries themselves. As particular indices are seen as being associated with positive movement and successful transformation of curricula, national and local politicians and policy makers use the indices of the knowledge banks as fulcrums for change. Politicians and policy makers turn to the existence of such study-based indices as external justification for the policy points they are promoting as needed changes in their national or local programs (Cussó & D'Amico, 2005; Grek, 2009; Peters, 2002; Phillips & Ochs, 2003).

Luhmann (1990) and Schriewer (1990) argued that the very existence of rankings provided by assessments such as the IEA and OECD studies provide a perceived base of scientific rationality for policy proposals and their public explanation. In fact, it has been argued that this very perception of the large-scale studies answering questions about curriculum has led to the lack of other research on curriculum reforms (Vithal, Adler, & Keitel, 2005). The use of indices as a basis for monitoring and leveraging change in countries, especially lower-performing nations, often leads to the declaration of crises and the increase in educational policy "borrowing" from league-leading nations. Such adoption of other countries' policies is made without careful consideration of the internal system supports which have made the policy successful, the cultural differences between the programs of the lending and borrowing nations' educational programs, and the impact such changes will make on the internal coherence of the curriculum of the "borrowing" nation (Nguyen, Elliott, Terlouw, & Pilot, 2009; Phillips & Ochs, 2003; Ripley, 2011; Thomas, 2001).

Policy Convergence

The concept of policy convergence was first introduced by Kerr (1983) as "the tendency of societies to grow, more alike, to develop similarities in structures, processes and performance" (p. 3). Over time, many have noted this tendency and attributed it to a number of causes. Bennett (1991) examined the topic at a level less general than "societal convergence" in his examination of policy convergence. He claimed that policy convergence should be examined as a movement from varied positions to a common point over time. In examining the forces that lead to convergence, Bennett (1991) posited a taxonomy of four processes that result in policy convergence in times of change:

1. *Emulation.* This approach to convergence involves the utilization of evidence about another's programs to modify one's own programs. As it is the adoption of a blueprint, emulation can explain some policy changes, but not outcomes themselves.
2. *Elite networking and policy communities.* This form of policy convergence is based in the actions of a transnational group of policy makers sharing a common focus on a policy issue. Unlike emulation, there is shared engagement in working on and adopting similar policies. There may even be a group charged with discussing a set of issues around the topic central to an emergent policy; in other cases, such groups may be self-appointed.
3. *Harmonization.* This approach to convergence of policy involves interdependence of the policy-making bodies and the existence of a super-body responsible for shaping and monitoring the common policy. However, the harmonization provides a movement together without the need for external controls or oversight. The European Community (EC) and OECD were held up as examples of such linked policy-making bodies. Harmonization requires a balance of relinquished autonomy with a hope for a gain in unproductive diversity in cross-national policies.
4. *Penetration.* Convergence through penetration occurs when the policy-adopting bodies are forced to implement an externally developed policy. Examples of penetration exist when nations are forced to implement an international standard or be closed out of a market. In some cases, this may be the result of harmonization strengthened to a regulatory system that defines who can participate or benefit in a given market of human activity—telecommunications, intercontinental aviation, and measurement standards.

It is the latter, and more coercive, types of convergence that are causing concern among educational policy experts. External loan institutions (e.g., UNESCO, World Bank, International Monetary Fund) have the leverage of expected improved outcomes for continued funding. Internal politicians and policy makers have the leverage of the public press to achieve convergence through public opinion and political power. Such uses of IEA and OECD data are being questioned in many quarters of the comparative education and mathematics education communities (Alexiadou, 2007; Carnoy, 2006; Grek, 2009). The results of research have also raised doubts about whether curricular convergence-focussed activities result in increased outputs (Grier & Grier, 2007; Mayer-Foulkes, 2010). Studies of interventions aimed at convergence have often shown that although convergence to the mean occurred with several variables, significant, unwanted, and unexpected increases occurred in the variance of both the focus and related variables. Such patterns could be very counterproductive in educational settings struggling to improve across the board.

Impact of Educational Policy in Different Countries

An examination of national reactions to the release of IEA or OECD data presents an opportunity to study the actual impact of large-scale international comparative studies on mathematics education programs at a national level. Reactions to the release need to be monitored from public, media, and policy levels, as the degree of knowledge and potential leverage differ greatly among these bodies as one moves from nation to nation.

In high-performing countries, at least as characterized by the study league-tables, the reaction has often been one of satisfaction, raised even to the point of self-congratulation. This reaction is often accompanied by reference to performance on particular indices comparing their performance with that of other countries. In low-performing countries, there are public calls for reform, which often take one of two forms. One is a call for a return to the basics; the other is a call for changes leading to harmonizing the national program with that of other countries having higher league-values in indices of comparison that are viewed as desirable. In other instances, the release in lower-performing countries is a governmental one calling for change in policy with specific reference to a greater federal role. Sometimes, action has been called for to fold the perceived needs into supporting an even broader political agenda involving governmental roles and the roles of public–private education within the country. A third reaction is one of indifference, suggesting that the results are just one way in which one could evaluate the outcomes of the national system. Such reactions might result in no action, the institution of a study to look into the results more deeply, or starting a small-scale study or group of projects examining alternatives without a great deal of fanfare.

Mini-Case Studies of Policy Influence of International Studies on Mathematics Education

The following mini-case studies of national performance at the Grade 8 (13-year-old) levels of the IEA studies and the PISA 15-year-old literacy studies present brief histories of the reactions and policy decisions surrounding the release of results from these large-scale international comparative studies. Occasional comments will be made concerning issues tied to either Grade 4 or 12 aspects of the IEA program. The first country examined is the USA, a country where changes have traversed the full span of Bennett's levels of convergence because of reactions to performance in international studies and recommendations from governmental studies and professional organizations. Next, Germany and Finland are examined for the differences they experienced in student results and reactions to public opinions. Finally, some comments are made concerning Singapore and past and present movements in mathematics education there.

USA

Perhaps the region having the most public reactions to study releases has been the USA, where each national and international assessment has had the same level of reaction in the media as a major sporting event. These reactions have even triggered national commissions whose reports have also created waves of interest and policy-related actions.

The USA has participated in every IEA mathematics study at some level. Our study of reactions to large-scale international assessments in the USA is presented in three phases: reaction to the 1967 release of FIMS, reaction to the 1987 release of SIMS, and reaction to the 1995 release of TIMSS and successive releases of IEA Trend studies and to OECD PISA studies. The USA performed significantly below the IEA average in FIMS and SIMS, at the international average in 1995, and above the international average in 1999, 2003, and 2007 (Beaton et al., 1996; Husén, 1967; Mullis et al., 2000, 2004, 2008; Robitaille & Garden, 1989).

In the OECD PISA studies of 15-year-olds' performances, the USA performed no differently from the PISA mean in the 2000 assessment and then significantly below the international mean in the 2003, 2006, and 2009 assessments (OECD, 2001, 2004a, 2007, 2010a).

Phase 1. Reactions to the 1967 release of the first IEA mathematics study (FIMS) began with the *New York Times* coverage. Although the notion of international comparative studies was new to the public, curricular and instructional scholars, and policy practitioners, each group immediately saw the data and findings as potential policy levers.

The USA was, at this time, about 10 years into the development of new curricular programs, which originated in the mid-1950s because of a perceived lack of skills of students entering scientific, technology, engineering, and mathematics study at the collegiate level. The most notable of these was the School Mathematics Study Group (SMSG). Although the initial impetus for these new curricular programs had been the unpreparedness of entering university students in the broad sciences, the 1957 launching of the Sputnik satellite by the Soviet Union was quickly given the credit for their creation. Financial support for these programs, and others, was provided by the National Science Foundation (NSF) and materials were quickly brought to field tests in the schools of the nation (NCTM, 1961, 1964).

These programs, which were backed by many in mathematics education and the mathematics community, focussed on the development of new textbook series and supporting materials. School mathematics was to have a greater focus on its underlying structure and the relationship between this structure and the algorithms that had dominated the content of the traditional programs. Instructionally, there was a shift from teacher presentation to an approach making greater use of guided discovery and manipulative materials to illustrate and motivate mathematics learning. Paralleling this work, the projects, universities, and school districts instituted a number of professional development projects for teachers aimed at strengthening their understanding of and capabilities to teach the newer curricula. Parents were also factored into the change equation with workshops held in conjunction with school parent–teacher organizations. The press labelled the entire reform effort the *new math*.

National reports were issued by the College Entrance Examination Board (CEEB—later to become the College Board) and a group of mathematicians and mathematics educators looking into the future. The CEEB (1959) Report of the Commission on Mathematics presented a review of secondary mathematics programs and made a call for mathematics for all students before turning to its main point—revising the secondary program for college-capable students. In particular, the report provided a call for a balanced treatment of concepts and skills with a stress on deductive reasoning throughout the secondary-school program. It also suggested attention be given to structure, use of sets and functions as a unifying feature, combined with a functional approach to trigonometry (Jones & Coxford, 1970).

Once the overall program of reform was well underway, the movement was not without its critics. Foremost among these was Morris Kline (1961) of New York University and a list of other mathematicians ("On the mathematics curriculum of the high school," 1962). This group of mathematicians was concerned about the undue emphasis on mathematical structure in the reforms, the lack of ties to the real world, and the lack of reasons for studying mathematics beyond mathematics itself. Another line of attack came from the Executive Director of the National Association of Secondary School Principals (P. Elicker, personal communication, January 23, 1962), calling for cutting off of federal funds for SMSG, as it was creating a national curriculum which would usurp the state and local rights to educational policy. This trickle of dissatisfaction from some vocal voices in the mathematics community, coupled with the voices of teachers stressed by dealing with new curricula, and joined by the dissatisfaction of parents unable to assist their children with homework viewed as unfamiliar and abstract, set the stage for change.

The release of the results from IEA FIMS was accompanied by a headline on the first page of the March 12, 1967, *New York Times* which read: "United States Gets Low Marks in Math." This headline, and the accompanying report that the USA had finished 11th out of 12 at the 13-year-old level and 13th out of 13 at the final year secondary-school level, cast a significant blow to the reform movement in school mathematics in the USA. The trickle of dissatisfaction turned to a torrent, with critics pointing to a downplaying of the "basics" or arithmetic facts and computational algorithm proficiency as the culprit. A crisis was proclaimed, the FIMS results served as the lever, and the result was a backlash against curricular reform in school mathematics.

Jeremy Kilpatrick (1971) presented a thoughtful analysis of the FIMS study, noting especially the tradeoffs that a researcher makes in moving from very small samples to a large sample where the notion of the context of student learning is lost. Students' opportunity-to-learn stood out as a salient, researchable topic to pursue in secondary and follow-up studies. It was clear that US students had far less exposure to advanced topics, and more review of previously studied topics, at both the 13-year-old level and at the final year of the secondary-school level. Further, student performance on advanced topics indicated that they were potentially teachable and learnable at the levels where assessments were given.

The following decade brought work on redefining the basics, based in many cases on curricula from other countries. This work resulted in a gradual expectation for a greater focus on algebraic and geometric content in the middle school. More importantly, it led to the NCTM developing, with wide feedback, its 1980 *An Agenda for Action* which laid out a new broad conception of the basics in school mathematics and moved problem solving to a pre-eminent position in the curriculum. The *Agenda* endorsed appropriate uses of technology in the curriculum and recommended that assessment of students be expanded beyond the traditional algorithmic-based approaches. Further, the *Agenda* called on teachers to exhibit greater levels of efficiency and effectiveness in their instruction.

Employing Bennett's (1991) model for convergence, one might indicate that the period from 1967 to 1980 was a period of reflection and emulation. Although there were smaller cycles of focus on manipulatives and the appearance of hand calculators during the interval, the policy focus was on defining a new way forward in school mathematics based on looking at others, learning from the first findings of the fledgling National Assessment of Educational Progress which released its first findings in the early 1970s, and developing a more policy-oriented outlook in the mathematics and mathematics education community. As there was no national department of education in the federal government at this point, the focus was on opening a conversation and providing a model, the *Agenda*, that the profession could examine and debate. The emergence of the professional community and its contacts at the first international mathematics education congresses with leaders from other nations and the emergence of the research community in mathematics education during this same period began to lead toward the formation of elite networks of policy-minded individuals in the mathematics and mathematics education communities.

Phase 2. The *Agenda* ushered in a decade of work which ultimately resulted in the development and release of NCTM's *Curriculum and Evaluation Standards for School Mathematics* in the spring of 1989 (McLeod, Stake, Schappelle, Mellissinos, & Gierl, 1996). Across the 1980s, prior to and immediately following the release of the influential *Nation at Risk* report (National Commission on Excellence in Education, 1983) calling for reform in US education, the NCTM, along with other major mathematical groups, had been moving toward drafting a statement of what students should know and be able to do as a result of their mathematics education. This process was guided by an emerging group that Bennett (1991) would term an *elite*. Formed by educationally-oriented members of the mathematics community and leaders of the National Council of Teachers of Mathematics, this group worked to form a community of teachers, researchers, and scholars fuelled by the notion of improving school mathematics and making the reform stick.

The report of the 1986 NAEP, *The Mathematics Report Card: Are We Measuring Up?* (Dossey, Mullis, Lindquist, & Chambers, 1988), noted growth in students' mathematics achievement since previous NAEP assessments. But the report also noted that students were frequently unable to work straightforward problems involving concepts of which they should have full command at their grade level. Since this was the first report of US student achievement after the *Nation at Risk* report and the release of the results of the 19-nation IEA's Second International Mathematics Study (SIMS) in the January 1987 publication of *The Underachieving Curriculum* (McKnight et al., 1987), the nation could have shifted immediately into a crisis mode (K. J. Travers, personal communication, July 4, 2011).

However, the US mathematics community and mathematics education community, in conjunction with the National Research Council, had formed the Mathematical Sciences Education Board (MSEB) in 1985 to coordinate the nation's response to the underperformance in mathematics education. The MSEB was structured to be broadly representative of the mathematics community from elementary school teachers to distinguished university professors, representatives of state and local boards of education, employers from the scientific and technological sectors, and representatives of teacher, parent, and policy groups. In January 1989, the MSEB released *Everybody Counts,* which set the stage for what US mathematics education programs needed to do, based on research and comparative studies, to reach the goal of mathematics for all and the goal of an increased flow of qualified students at all levels along the mathematics pipeline. This document served as a policy precursor for the release of the NCTM *Curriculum and Evaluation Standards for School Mathematics* in March 1989. This release was met with positive comments and a lack of crisis focus. States signed on to the standards, and within three years all but a few states had changed their curricular frameworks to parallel the recommendations of the NCTM standards.

This was a tremendous step forward for mathematics education policy in the USA. Although the nation now had a Department of Education, direction of schools was still vested in the state departments of education, which, to a large degree, abrogated their responsibilities for curriculum to the leaders of over 15,000 separate school systems spread across the country. This vast and dispersed responsibility for mathematics education at the local level has been a major and defining feature of US mathematics education. The appearance of the *Curriculum and Evaluation Standards* (NCTM, 1989), and the year-long public vetting of the draft with special attention paid to state departments of education, led to 46 states setting or modifying, within three years, their written state curricula to parallel the recommendations of the final standards document. Further, professional development materials and training sessions were provided to educate leaders to talk about the standards and work with state and local school districts in implementing the standards at the local level. With this effort and the formulation of the MSEB and its work with NCTM, the policy community focussed on convergence of the mathematics education curriculum and attempts at convergence moved to the harmonisation level of Bennett's (1991) taxonomy. Leaders of the mathematics, mathematics education, and policy communities met regularly to shape and monitor activities aimed at strengthening US school mathematics. Although not everyone supported the standards-based movement, there was focussed change afoot.

Phase 3. Subsequent releases of the IEA documents from TIMSS and from the IEA Trends studies in 1999, 2003, and 2007 were viewed as signals of distances to go, but not as imminent crises. The same could be said for the release of the OECD

PISA studies of 2000, 2003, 2006, and 2009. The scores show consistent underperformance at the OECD level. Although each release was met with media proclamations which spoke of doom and despair over the state of education, especially in reading, mathematics, or science, little direct action was taken at the local level. Teachers were involved in professional development, and the updating of the NCTM standards with recommendations shaped ever closer to grade/age-level expectations appeared as the *Principles and Standards for School Mathematics* (NCTM, 2000).

In Washington, DC, the situation was different. With the change of administrations in 2000, President George Bush pushed for and won legislative approval for his *No Child Left Behind* (NCLB) law that created a mandatory national testing program which held schools accountable for achieving specific and increasing levels of performance. Those levels were keyed to a new NAEP framework for mathematics that called for increased focus on algorithmic skills and a lessening of attention to measurement, geometry, and probability as targets for the NAEP assessments. In addition, the legislation moved the NAEP to an annual testing program for all students in Grades 3 through 8 and at one level in secondary school. The law further instituted a requirement that all states ensure that the schools under their aegis bring their students up to the "proficient" level of performance by 2014 (Olson, 2004). Intervals defining *below basic, basic, proficient,* and *advanced* levels of performance were defined psychometrically via achievement-level-setting procedures working with the individual NAEP items, student percentages, and Item Response Theory (IRT) parameter information (Pellegrino, Jones, & Mitchell, 1999). This focus on accountability by achievement levels had been growing across the 1990s parallel to the implementation of the NCTM standards, but NCLB brought it front and centre.

With the institution of the NCLB law, the federal NAEP testing program, and its framework, one had the essence, at least, of penetration in the policy community. However, at the time of this writing, the lasting impact of this legislation and its punitive aspects for schools that fail to achieve raising their students to the *proficient* level by 2014 is uncertain, as legislative forces are afoot to change NCLB. The path to convergence that had its roots in the IEA release of the FIMS data, the growth of the policy community within the mathematics and mathematics education community through NCTM, MSEB, and the many state mathematics teacher groups, and the success of the standards showed a pattern of harmonization. However, the impact of the NCLB, the insertion of a NAEP assessment system not harmonized with the NCTM (1989, 2000) standards, but having punitive outcomes for noncompliance illustrates the power of the existence of policy groups which have the ability to force convergence through penetration.

At the time of this writing, US schools are working through another policy-induced change to the mathematics curriculum and state level assessment systems. In 2010, the Council of Chief State School Officers (CCSSO) and Achieve, an organization formed in 1996 by the state governors and corporate leaders and focussed on educational reform, released their *Common Core State Standards for School Mathematics* (CCSSM) (see CCSSO & NGA, 2010; Porter, McMaken, Hwang, &

Yang, 2011). This set of recommendations was immediately adopted by 40 states as their state-level standards for school mathematics for K-12 public schools. As such, the CCSSM provides the framework for expected mathematics outcomes and becomes the state-level proxies for meeting the NCLB goals for student progress toward proficiency. Although it is too early to judge the impact of this rapid insertion of new materials into the mathematics reform and policy mix in the USA, it clearly shows that convergence by insertion is the order of the day, with the impetus for structural change originating outside the professional mathematics education community. Time will tell the outcome of the NCTM (1989, 2000)-standards-led move to convergence of the K-12 mathematics education curriculum in the schools of the USA and the influence of the CCSSM movement on the trajectory the NCTM standards engendered.

Germany and Finland

Although Germany and Finland are close geographically, their experiences with the PISA assessments and policy reactions are quite dissimilar. Both countries place a high value on public education but toward different ends. Neither had been consistent participants in the TIMSS 13-year-old (eighth grade) level assessments from 1995 forward. Germany had an eighth-grade ranking of 23rd out of 41 countries in 1995, whereas Finland had an eighth-grade ranking of 14th out of 38 countries in 1999. In PISA, both countries participated in each assessment from 2000 forward. The countries' performances can be viewed in terms of place ranking out of the number of participating countries or by their PISA mathematical literacy score. Using this notation (ranking, literacy score), Germany's results for the four assessments were as follows: 2000 (21/41, 490), 2003 (19/40, 503), 2006 (17/48, 504), and 2009 (16/65, 513). Finland's results were: 2000 (5/41, 536), 2003 (2/40, 544), 2006 (1/48, 540), and 2009 (6/65, 565).

Germany. Germany's students' performances in 2000 through 2006 were met with public outcries, and the nation was caught up in rethinking its educational structure, what other factors might have influenced the scores, and a myriad of other possibilities (Miserable Noten für Dekutsche schüler, 2001; OECD, 2002; Stanat et al., 2002). Finland, on the other hand, had high performances, and its citizens were hardly aware of the PISA assessment program or their students' achievements. The different reactions are reflective of the countries' cultures and their approaches to educating their children. However, the policy reactions are somewhat surprising.

Prior to this time, Germany's education expectations were organized at the state level, with each state developing and monitoring school outcomes within their own *Länder*. The reaction was swift to the 2000 and 2003 findings. By 2003, there was a report outlining recommended standards and assessments by which these expectations would be monitored (Klieme et al., 2003). This report was passed through the Standing Committee of *Länder* Ministers in December 2003 and became the law of the land for implementation with the 2004–2005 school year. This was change, and unanimous change, at an unprecedented pace for German education. Unlike other reforms, the trade unions and businesses and industry quickly endorsed the changes as well (Ertl, 2006). As a result, new curricular guidelines and texts had to be developed and teachers provided with professional development relative to the implementation of the new goals. Individual states in Germany still had the authority to react to the strictures of the new standards in their own fashion. Educators in Germany felt that the changes within the mathematics curricular recommendations moved the curriculum closer to an empirical and practice-focussed conception than to the more didactical–cultural conception that had defined German education (Bohl, 2004). The conception of OECD literacy as an outcome was not central to German schooling prior to the reactions to the national PISA outcomes. This, combined with the notion of developing the competencies associated with the individual disciplines sampled by PISA, has furthered the stress in moving from traditional approaches to schooling (Sloane & Dilger, 2005).

The process and changes that resulted in the convergence observed in Germany was a significantly compressed version of that observed in the USA. In the USA, the transitions occurred over a period of 40–50 years in moving from the uncoordinated curricula of the early 1950s to the adoption of standards-based outcomes by the states in the late 1990s. In Germany, these transitions were compressed into little over a 4-year span. Given that many of the mathematics educators in Germany were well linked to others in the international mathematics education community and that the notions of *competencies* defining outcomes were part of the experience in Germany's neighbouring country of Denmark (Niss, 1999; OECD, 2003), clearly, communication was already in place between the leaders of the curricular areas in German education and other international policy players at the start of the period of reaction to the PISA results. However, the reflective convergence that usually accompanies change resulting from harmonization was sharply curtailed by the quick institution of new standards by the ministers of culture and education in 2003. Germany is a case where the *Länder* ministries and educational administrators were handed the new standards almost as a fait accompli to be inserted into a new nationwide mathematics curricular structure.

Not all sectors of the education establishment were happy with the decisions made by the ministers and the move to standards-based outcomes. Ertl (2006) noted that the

> Federal Ministry's post-PISA agenda seems to be firmly focused on raising national educational standards by pursuing measures that will improve Germany's low ranking in the PISA league table. It places less emphasis on the solution of the other major problems identified by PISA, the strong connections between the socio-economic background of students and their education achievement. (p. 630)

Finland. The situation in Finland, contrasted with that in Germany, shows another country where education was valued, but the philosophical view of the process was different. Finland did not participate in TIMSS 1995 but did participate in 1999, where their Grade 8 equivalent students performed significantly higher than the IEA average performance in mathematics for this level (Mullis et al., 2000).

Finland's performances in the OECD PISA assessments have been stellar, with its students attaining the highest non-Asian country performance in each of the PISA assessments from 2003 through 2009 (OECD, 2001, 2004a, 2007, 2010b).

Finland's student achievements in the TIMSS and OECD assessments have garnered considerable kudos in the international education community and the public press. This focus has brought attention to the differences in both curricular programs and quality of instructional staff found in Finnish schools. Many have asked what factors led to their consistently high achievement on the PISA mathematical literacy assessments. To answer that question, one can start with the fact that Finland has a National Board of Education (FNBE) which oversees the educational enterprise of the nation. Starting with the 1985 mathematical curricular framework, the FNBE started a movement away from the comprehensive school with a strong core curriculum in mathematics for Grades 1–9. Although the board still provided a framework with four mathematical strands (number concepts, expressions and equations, geometry, and applied mathematics), the focus shifted from an emphasis on basic concepts and structure to one emphasizing problem solving, applications, and everyday uses of mathematics. This change was accompanied by professional development for teachers on teaching through problem solving and the use of projects to involve students in using their mathematics to solve problems from everyday settings. Follow-up research indicated that this movement was a partial success, but it succeeded in moving teachers to teaching only about problem solving, not through problem solving.

To further aid teachers in the transition, the FNBE and the municipalities provided teachers with more professional development, publishers produced problem booklets keyed to grade levels, and special emphasis was given to Japanese-style "open-problems." This change moved the agenda on problem solving and realistic applications of mathematics further. The biggest change which might have affected the PISA results, was the release of a revised framework for mathematics by the FNBE in 1994. This action decentralized the curricular oversight by removing the listing of specific content and turned the task of developing the mathematics curriculum over to the local schools' teachers. The FNBE did provide guidance that teachers should still examine the traditional content critically and thin the curriculum of material that did not have any use in the further development of mathematics. The FNBE also stated that Grades 1–6 should master the basic concepts and be capable of performing calculations on paper, mentally, and through the use of a hand-calculator (Kupiainen & Pehkonen, 2008).

In 1999, Finnish education officials provided schools with a marking guide scaled from 4 (reject) to 10 (excellent) with advice to move students to at least the 8 (good) level. Although there is no national assessment used to place each student in an achievement level bracket for mathematics, Finland does have an assessment given to a representative sample of ninth graders. These papers are analyzed, published, and discussed. Further, individual schools can buy copies of these tests to be given locally and then compare their results, and marks, with those given on the national sample of tests. This information helps provide a degree of uniformity to outcomes at a national level. There is also an assessment given to a sample of sixth-graders every fifth year.

When Finnish educators reflect on what has enabled their system to perform so well, they cite the following factors: their comprehensive educational structure with heterogeneous grouping of students, the societal focus of the schools with free healthcare and cohesive group-focussed structure, the use of specialist teachers of mathematics at lower grades in many schools, the focus on equity and Co-operation rather than competition, the focus on problems and the use of mathematics, widespread student belief that they can solve problems, and the strong and supported corps of teachers (Kupiainen & Pehkonen, 2008; Malaty, 2006; Rautalin & Alasuutari, 2009; Sahlberg, 2010).

Teachers in Finland had a more advanced education than their peers in most countries, and this education is balanced between content knowledge and content-based pedagogical knowledge. This advanced preparation for their teaching and for the professional ways in which they approach the tasks confronting them has resulted in teaching being one of the most respected careers in Finland. This confidence in teachers as a whole has allowed them to plan and implement curricula and assessment programs fitting to their individual schools.

Other nations might note the heavy focus on equity and Co-operation—not choice and competition—in Finnish schools. Also, when teachers are provided government-paid educational preparation and are given significant recognition and public backing for their work, teaching becomes a desired profession by well-qualified individuals. Although Finland is reticent to say "Do this and you, too, can have high scores," their Ministry of Education has reflected on the differences in Finnish education and tried to provide some background that might explain the cultural differences and practices as reasons for their performances (Hautamäki et al., 2008).

As in other countries, there is some concern about the high PISA scores from the mathematics community in Finland. Citing students' recent low performance on graduation tests, members of university faculty argue that PISA provides a view of everyday mathematics and note the value of such knowledge, but also argue that such knowledge does not include advanced concepts and skills in algebra and other core subjects necessary for study and gainful employment after secondary school (Astala et al., 2005).

That said, there is still concern about the influence of outside forces on Finnish education (Grek et al., 2011; Rautalin & Alasuutari, 2009). The development of the Finnish system of education and the changes made between 1985 and 2000 were based on within-country self-study and the selective importation and emulation of practices seen to work in other countries. These imports were carefully woven into the curricular and professional development work provided for teachers. The OECD PISA results are seen with some distrust, as they come with a cloak of data and information, but bear the impact of scientific truth. Researchers notice that statistical comparisons can often lead to the emulation of some practice of a country placed above the average of other countries (Rautalin & Alasuutari, 2009). Such comparisons and interpretations then become levers for change. In fact, in Finland, the outcome that Finnish student achievement levels had the least variance as a system in the PISA assessments was read as suggesting that perhaps there should be more attention paid to the top students, perhaps they could achieve even more. Although the Finnish take the homogeneity as one of their strengths, the numerical interpretation can be used to suggest a failing. In reaction to the 2000 results, Finnish officials decided to add more emphasis to the curriculum and instituted a call for more core subject influence (Välijärvi et al., 2002). Hence, the numerical results suggested a possible weakness, and hence, even in the face of superior achievement, changes antithetical to the historical culture of Finnish education were made.

In Finland, we again see a country growing out of its own educational history through emulation to develop a system drawing on the best practices of other counties and schools within its own borders. The FNBE directives on curriculum made changes across the 1980s and 1990s consistent with programmatic changes in other countries that seemed to fit Finnish schools, but did so by modifying those practices to the culture of Finnish education. Although this approach led to harmonization through curricular guidance, the Finnish Ministry in 1985 backed off a bit in decentralizing the education system to provide more local control of curriculum within broader guidelines. It was only with the numerical results dealing with the homogeneity of results across the Finnish student body in the PISA assessment that one saw outside influence reach Finland in the form of indicator influence inducing local policy. Although the influence did not have the impact of penetration noted in Germany, this is an instance of convergence of education structure as a result of international assessment and indicator results.

Singapore

Asian student performance has dominated the achievement charts as their countries have held the majority of top rankings in the international large-scale assessments of mathematics performance since their inceptions in the middle 1980s. Asian students' stellar performances have originated from Hong Kong, Japan, Korea, Macau, Singapore, Taipei, and most recently Shanghai. Despite their high rankings in international assessments, Asian countries have not been complacent with their current education systems.

Singapore did not participate in the FIMS or SIMS studies; rather, it made its entry with the TIMSS 1995 study. In 1995 through 2007, Singapore's students performed in the top group of countries and had the highest means, with the exception of the 2007 study, when Chinese Taipei had the highest numerical position but not significantly higher than that of Singapore (Beaton et al., 1996; Mullis et al., 2000, 2004, 2008). In the OECD studies, Singapore, a non-OECD country, has participated in only the 2009 assessment. The Singapore students finished second numerically but not significantly lower than the students of Shanghai-China (OECD, 2010a).

In Singapore, the gap between the intended and the implemented and achieved curricula is small. This alignment results from a close monitoring of teacher progress and student achievement. There is a strong and articulated program of professional development that parallels the curriculum, providing important, grade-specific suggestions in the same time frame where teachers can immediately implement them in their classrooms (Kaur, 2009). The Singapore Ministry of Education noted three problems emanating from the TIMSS findings. The first was that students did not perform well on mathematics that they had not specifically learned and practised. The second was student difficulty in transferring learned knowledge to different contexts. The third dealt with comprehension problems rooted in language issues which arose when unfamiliar words appeared in problems. All three of these issues have found their ways into curricular reform for mathematics in Singapore.

Singapore leaders feel that the results of the studies can provide fresh perspectives and benchmark their performance relative to other countries. However, there is some fear that the high performance levels may lead to feelings of complacency relative to local standards. Singapore feels that such participation provides opportunities to participate in other international comparative projects which have the possibilities of enriching their programs. In particular, they have participated in the Kassell project, the multinational IPMA, and a bilateral project with Brunei Darussalam. Singapore mathematics educators also participate in study tours to other countries and attend conferences of professional mathematics groups internationally. All of these efforts are viewed as adding new vistas to their program's possibilities (Wong, Lee, Kaur, Yee, & Fong, 2009).

Unlike other top-performing Asian countries, Singapore students not only performed well in mathematics, they also displayed a positive attitude towards learning mathematics. The high performance of Singapore students attracted the attention of many Western mathematics educators. The Singapore mathematics curriculum and textbooks have been the focus of a number of studies aimed at identifying factors contributing to the high performance of Singapore students (American Institutes for Research, 2005). Such focussed cross-cultural studies are examples of many small-scale international comparative studies initiated as a result of TIMSS and PISA findings. These again are illustrative of attempted emulation and harmonization processes under Bennett's (1991) model of policy convergence. But this time, other countries want to learn from Singapore's success story.

Concerns Regarding the Impact of International Studies

From the case studies, it is clear that international studies have had, and continue to have, a strong impact on policies for a number of countries. Although such an impact may lead to positive outcomes for mathematics education, there could also be consequences from international studies that are damaging to mathematics education. A critical review of the impact of international studies is essential. The following presents a discussion of concerns regarding the possible impact of international studies.

Concerns Regarding Statistical Precision of the Results

It is not unusual for policy makers to draw quick conclusions by looking at the change in country rankings from one assessment cycle to another. For example, if the ranking (or the country mean score) is worse than for the previous cycle, there may be an immediate outcry about the decline of mathematics standards in the country. This outcry in turn could lead to policy changes. What the policy makers have often missed is that there is always a margin of error in any reported measure. Although those who conduct international studies take great pains in articulating the confidence level surrounding performance measures, these margins of error are often ignored. A policy change may be totally unwarranted, as the change in country ranking could simply be the result of random fluctuation due to the sampling of students (Wu, 2010a, 2010b).

Concerns regarding inferences on causal relationships. International studies such as PISA and TIMSS are cross-sectional sample surveys. Such survey designs are not powerful in establishing causal relationships. Even though student and school background characteristics are captured and correlated with achievement measures, positive correlations do not establish causal relationships. A positive correlation between students' interest in mathematics and test scores in mathematics may be expected. But it is difficult to conclude whether higher interest in mathematics leads to higher achievement, or in fact, higher achievement raises interest. Similarly, better school resources could be positively correlated with higher achievement. But there could be mediating variables such as student socio-economic status (SES) that explain both student achievement and school resources. For example, private schools may have better resources and higher achievement scores, but both could be due to the higher SES of students in private schools. In general, translating survey results into policy measures relies on many assumptions and hypotheses. Some policy changes in response to international study results may be completely off the track.

Concerns regarding using mean scores only. Often the main focus on results of international studies is the country mean score. Although the mean score summarizes overall performance, it could be the case that a country has a large group of low achievers because of geographical remoteness or immigrant composition. That is, the lower mean score could be the result of specific factors rather than an inefficient education system across the board. Policy changes need to take into consideration a myriad of indicators and not just the ranking and mean score of a country. The emphasis on ranking and mean scores, often fuelled by the media, could lead to inappropriate policy changes (Hutchison & Schagen, 2007).

Concerns regarding policy convergence. Although there is a great deal of benefit arising from collaboration, whether internationally or between local communities, there are also a number of concerns in "borrowing" from other education systems, be it the curriculum, assessment, or a management approach. In Bennett's (1991) model, policy convergence in the form of *emulation* appears to be the most flexible, and *penetration* appears to be the most rigid. An authoritarian approach to enforcing standards may work well, provided the standards are sound. The mini-case studies in this paper show that there are significant differences between education systems across the world, and that there are different success models. Finland has clearly showed that a decentralized system with little emphasis on standardized testing can lead to high education attainments, whereas East Asian countries with highly centralized and examination-based education systems are also top performers. What works for one country may not work for another country because of cultural differences and local conditions. This variability is also the case for policy convergence within a country. A national curriculum brings uniformity across states or provinces but stifles diversity and innovation. If education systems are regarded as business models, then the importance of diversity and competition cannot be ignored, as educators have learned from the political and economic arena. When a borrowed system does not fit well within an education community, the consequence could range from a waste of resources to serious damages to the education system (Vithal et al., 2005).

Concerns regarding the use of assessment to drive teaching and learning. Assessments of students should be undertaken as an evaluation of the outcomes of education. Assessments should be designed around teaching and learning, and not the other way round, where teaching and learning are designed around assessment. This direction of design is important as there are important differences between teaching and assessment. One may design an authentic task in assessment where multiple skills are required to solve a problem, but to teach those skills, basic building blocks of skills need to be taught separately, and often in a context-free mode. Only when students have mastered individual skills can they combine the skills and apply them. That is, the way mathematics is taught may be at variance with the way mathematics is assessed. As international studies like PISA and TIMSS are assessments, the adaptation or adoption of the PISA and TIMSS assessment frameworks as curriculum frameworks may not be desirable. For example, as PISA focusses only on problem solving and application in everyday settings, it would be an error for curriculum designers not to include skills involving abstract mathematics as well as basic foundations of mathematics which are often context-free. There is a particular concern when, in order to improve a country's international test scores, the curriculum is changed to match the assessment frameworks of the international assessments.

Additional comments relative to design, interpretation, difficulties in conducting cross-cultural studies, and the drawing of inferences were the focus of a symposium held by the Board on Comparative Studies in Education at the National Research Council in Washington, DC, in 2000 (Porter & Gamoran, 2002).

Retrospective

It is generally acknowledged that international studies such as TIMSS and PISA have an enormous impact on educational policy debates, if not on the policies themselves (Figazzolo, 2009). However, it is not always straightforward to identify the impact of international studies on policies since many policy changes are influenced by international assessments in subtle and indirect ways. Sometimes policy changes evolve over a long period of time, moving slowly and thoughtfully through each of the steps to lasting educational reform. In such cases, it is difficult to attribute a specific lever that triggers a policy implementation. In other cases, media-induced crises lead to rapid, and often thoughtless, reforms lacking foundations in either research or practice. In this chapter, we have reviewed national and local reactions to international studies that are quite public, as well as political and economic processes whose implications are less overt but nevertheless important in influencing policies. Below, we provide a summary of different kinds of policy implications of international studies.

First, results of international studies have been used as policy levers. This trend has been increasing with recent data releases, most notably in the USA, Germany, and Japan. In some cases, the results are used simply as an opportunistic justification for some policies that have already been rolled out. In other cases, new policies have been devised in direct response to the poor performance of students. The policy changes range from changing curriculum content to providing resources to schools.

Second, international studies have been used as performance measures to gauge the success or otherwise of a policy. For example, a policy might be linked to an international study through the setting of a target level of a country's performance in the study. More recently, national achievement measures have been used as economic incentives or indicators by international funding organizations working with developing countries.

Third, international studies have provided a wealth of data and, with that, opportunities for mathematics education researchers to carry out in-depth analyses ranging from classroom climate, gender equity, to curriculum design. Many of these studies are funded by policy bodies with a view that these analyses may influence policies down the line, even if there may be no immediate policy changes based on the research findings. The authors of a number of chapters in this handbook have discussed the link between mathematics education research and policy implementation.

Fourth, international studies such as TIMSS and PISA have led to further transnational dialogs between researchers in assessment, curriculum, and instruction. In Bennett's (1991) model of policy convergence, these are examples of emulation and harmonization.

Fifth, international studies have increased an awareness of the use of student performance measures, and, in some cases, led to the establishment of national sample-based or full-cohort standardized tests. Such enforced tests are examples of Penetration under Bennett's (1991) model where, by law, achievement targets from the tests are set.

International studies have had both positive and negative impacts. On the one hand, it is encouraging to see increased discussion and debate on curriculum content in mathematics, teaching strategies in the classrooms, assessment methodologies, and a rethinking of the values and goals of education more generally. The discussions have certainly stimulated a great deal of reflection, evaluation and constructive criticisms. These have been positive outcomes from international studies. On the other hand, there have also been hasty reactions to study results and rash policy decisions based on unfounded inferences. In particular, the media and some policy makers have been prone to brush aside caveats clearly stated in the study reports, to ignore the degree of confidence one can have in the measures, and to launch into actions that have been typically politically motivated. There are often policy measures that are quick fixes to improve test scores rather than for long-term investment for a better education. These are examples of the negative impact of international studies.

Hopefully, the outcomes of international studies have fostered curricular considerations and productive changes, a careful reflection on cross-cultural comparative methodology, and steps to the improvement of student learning of mathematics worldwide. Researchers and policy-inclined individuals in the mathematics education community need to ask what should be and what are the policy ramifications associated with the TIMSS and PISA assessments, as well as those associated with other international and national assessments of mathematics education. What are the benefits that can be obtained from a careful analysis of the tests, curricula, instructional patterns, opportunity-to-learn, instructional materials and other resources, teacher preparation and professional development and support programs, and related research findings? What are the positive and negative effects resulting from borrowing and promoting the TIMSS and PISA frameworks for developing countries and inducing the insertion of these frameworks into national curricular framework discussions? These questions shape an agenda for mathematics education and policy researchers to examine in the coming decade, as the role of international assessments will surely continue to grow in the number and range of nations participating and in the sources of important indicators chosen (Jones, 2005). For mathematics educators to dismiss the powerful force such assessment programs have on educational policy decisions worldwide would be a dangerous mistake from cultural, mathematical, and educational perspectives.

References

Alexiadou, N. (2007). The Europeanisation of education policy: Researching a changing governance and "new" modes of coordination. *Research in Comparative and International Education*, 2(2), 102–116.

American Institutes for Research. (2005). *What the United States can learn from Singapore's world-class mathematics system (and what Singapore can learn from the United States): An exploratory study*. Washington, DC: AIR. Retrieved from http://www.air.org/reports-products/index.cfm?fa=viewContent&content_id=598

Astala, K., Kivelä, S. K., Koskela, P., Martio, O., Näätänen, M., & Tarvainen, K. (2005, December). The PISA survey tells only a partial truth of Finnish children's mathematical skills. *Matilde*, 29, 9.

Beaton, A. O., Mullis, I. V. S., Martin, M. O., Gonzalez, E. J., Kelly, D. L., & Smith, T. A. (1996). *Mathematics achievement in the middle school years: IEA's Third International Mathematics and Science Study*. Chestnut Hill, MA: Center for the Study of Testing, Evaluation, and Educational Policy, Boston College.

Bennett, C. J. (1991). What is policy convergence and what causes it? *British Journal of Political Science, 21*(2), 215–233.

Bidwell, J. K., & Clason, R. G. (Eds.). (1970). *Readings in the history of mathematics education.* Washington, DC: National Council of Teachers of Mathematics.

Blum, W., & Kaiser, G. (2004). *Kassel Project in Germany.* Budapest, Hungary: Wolters–Kluwer.

Bohl, T. (2004). Empirische Unterrichtsforschung und Allgemeine Didaktik. Entstehung, Situation und Konsequenzen eines prekären Spannungsverhältnisses im Kontext der PISA-Studie [Empirical research on teaching and general didactics: A precarious tension and consequences in the context of the PISA study]. *Die Deutsche Schule, 96*(4), 414–425.

Burghes, D., Geach, R., & Roddick, M. (Eds.). (2004). *IPMA.* Budapest, Hungary: Wolters–Kluwer.

Burghes, D., Kaur, B., & Thompson, D.R. (Eds.). (2004). *Kassel Project—Final Report.* Budapest, Hungary: Wolters–Kluwer.

Burstein, L. (Ed.). (1993). *The IEA Study of Mathematics III: Student growth and classroom processes.* Oxford, UK: Pergamon.

Carnoy, M. (1999). *Globalization and educational reform: What planners need to know.* Paris, France: United Nations Education, Scientific, and Cultural Organisation.

Carnoy, M. (2006). Rethinking the comparative and the international. *Comparative Education Review, 50*(4), 551–570.

Casassus, J., Froemel, J. E., Palafox, J. C., & Cusato, S. (1998). *First international comparative study of language, mathematics, and associated factors in third and fourth grades.* Santiago, Chile: Latin American Laboratory for Evaluation of the Quality of Education.

Chatterji, M. (2002). Models and methods for examining standards-based reforms and accountability initiatives: Have the tools of inquiry answered pressing questions on improving schools? *Review of Educational Research, 72*(3), 345–386.

College Entrance Examination Board. (1959). *Program for college preparatory mathematics.* New York, NY: Author

Council of Chief State School Officers & National Governors Association. (2010). Common core state standards for mathematics. Retrieved from http://www.corestandards.org/assets/CCSSI_Math%20Standards.pdf.

Cussó, R., & D'Amico, S. (2005). From development comparativism to globalization comparativism: Towards more normative international education statistics. *Comparative Education, 41*(1), 1–16.

Dossey, J. A., Mullis, I. V. S., Lindquist, M. M., & Chambers, D. L. (1988). *The mathematics report card: Are we measuring up?* Princeton, NJ: Educational Testing Service.

Eaton, J., & Kortum, S. (1996). Trade in ideas: Patenting and productivity in the OECD. *Journal of International Economics, 40*(3–4), 251–278.

Ertl, H. (2006). Educational standards and the changing discourse on education: The reception and consequences of the PISA study in Germany. *Oxford Review of Education, 32*(5), 619–634.

Figazzolo, L. (2009). *Impact of PISA 2006 on the education policy debate.* Education International. Retrieved from http://www.ei-ie.org/research/en/documentation.php.

Greaney, V., & Kellaghan, T. (2008). *Assessing national achievement levels in education.* Washington, DC: International Bank for Reconstruction and Development/The World Bank.

Grek, S. (2009). Governing by numbers: The PISA "effect" in Europe. *Journal of Educational Policy, 24*(1), 23–37.

Grek, S., Lawn, M., Lingard, B., Rinne, R., Segerholm, C., & Simola, H. (2011). National policy brokering and the construction of the European Education Space in England, Sweden, Finland and Scotland. In J. Ozga, P. Dahler-Larsen, C. Segerholm, & H. Simola (Eds.), *Fabricating quality in education: Data and governance in Europe* (pp. 47–65). London, UK: Routledge.

Grier, K., & Grier, R. (2007). Only income diverges: A neoclassical anomaly. *Journal of Developmental Economics, 84*, 25–45.

Hautamäki, J., Harjunen, E., Hautamäki, A., Karjalainen, T., Kupiainen, S., Laaksonen, S., Lavonen, J., Pehkonen, E., Rantanen, P., Scheinin, P., Halinen, I., & Jakku-Sihvonen, R. (2008). *PISA 2006 Finland: Analyses, reflections and explanations.* Helsinki, Finland: Ministry of Education.

Hopmann, S. T., & Brinek, G. (2007). Introduction: PISA according to PISA—Does PISA keep what it promises? In S. T. Hopmann, G. Brinek, & M. Retzel (Eds.), *PISA according to PISA—Does PISA keep what it promises* (pp. 9–19). Vienna, Austria: LIT-Verlag.

Husén, T. (Ed.). (1967). *International Study of Achievement in Mathematics: A comparison of twelve countries* (2 vols.). Stockholm, Sweden: Almqvist & Wiksell.

Hutchison, D., & Schagen, I. (2007). Comparisons between PISA and TIMSS—Are we the man with two watches? In T. Loveless (Ed.), *Lessons learned—What international assessments tell us about math achievement* (pp. 227–261). Washington, DC: Brookings Institute Press.

International Association for the Evaluation of Educational Achievement. (2011). *Brief history of the IEA*. Retrieved from http://www.iea.nl/brief_history_of_iea.html.

International Commission on Mathematical Instruction. (2011a). *A historical sketch of ICMI*. Retrieved from http://www.mathunion.org/icmi/about-icmi/a-historical-sketch-of-icmi/.

International Commission on Mathematical Instruction. (2011b). *The first century of the International Commission on Mathematical Instruction (1908–2008)*. Retrieved from http://www.icmihistory.unito.it/timeline.php.

International Project on Mathematical Attainment. (2011). *CIMT research projects and publications*. Retrieved from http://www.cimt.plymouth.ac.uk/menus/research.htm.

Jones, P. W. (2004). Taking the credit: Financing and policy linkages in the educational portfolio of the World Bank. In G. Steiner-Khamsi (Ed.), *The global politics of educational borrowing and lending* (pp. 188–200). New York, NY: Teachers College Press.

Jones, G. A. (2005). What do studies like PISA mean to the mathematics education community? In H. L. Chick & J. L. Vincent (Eds.), *Proceedings of the 29th Conference of the International Group for the Psychology of Mathematics Education* (Vol. 1, pp. 71–74). Melbourne, Australia: PME.

Jones, P. W., & Coleman, D. (2005). *The United Nations and education: Multilateralism, development and globalization*. London, UK: Routledge/Falmer.

Jones, P. S., & Coxford, A. F., Jr. (Eds.). (1970). *A history of mathematics education in the United States and Canada* (Thirty-second Yearbook of the National Council of Teachers of Mathematics). Washington, DC: National Council of Teachers of Mathematics.

Kang, H. J. (2009). A cross-cultural curriculum study on U.S. elementary mathematics textbooks. In S. L. Swars, D. W. Stinson, & S. Lemons-Smith (Eds.), *Proceedings of the 31st Annual Meeting of the North American Chapter of the International Group for the Psychology of Mathematics Education* (Vol. 5, pp. 379–386). Atlanta, GA: Georgia State University.

Kaur, B. (2009). Performance of Singapore students in Trends in International Mathematics and Science Studies (TIMSS). In K. Y. Wong, P. P. Lee, B. Kaur, F. P. Yee, & N. S. Fong (Eds.), *Mathematics education: The Singapore journey* (pp. 439–463). Singapore: World Scientific.

Kellaghan, T., Greaney, V., & Murray, T. S. (2009). *Using the results of a national assessment of educational achievement*. Washington, DC: World Bank.

Kerr, C. (1983). *The future of industrial societies: Convergence or continuing diversity?* Cambridge, MA: Harvard University Press.

Kilpatrick, J. (1971). Some implications of the International Study of Achievement in Mathematics for mathematics educators. *Journal for Research in Mathematics Education, 2*(2), 164–171.

Kilpatrick, J. (2009, February–March). TIMSS 2007 mathematics: Where are we? *MAA Focus*, 4–7.

Kilpatrick, J., Mesa, V., & Sloane, F. (2007). U.S. algebra performance in an international context. In T. Loveless (Ed.), *Lessons learned: What international assessments tell us about math achievement* (pp. 85–126). Washington, DC: Brookings Institution Press.

Kimmelman, P., Kroze, D., Schmidt, W., van der Ploef, A., McNeely, M., & Tan, A. (1999). *A first look at what we can learn from high-performing school districts: An analysis of TIMSS data from the First in the World Consortium*. Washington, DC: National Institute on Student Achievement, Curriculum, and Assessment.

Klieme, E., Avenarius, H, Blum, W., Döbrich, P., Gruber, H., Prenzel, M., Reiss, K., Riquarts, K., Rost, J., Tenorth, H., & Vollmer, H. (2003). *Zur Entwicklung nationaler Bildungsstandards* [Toward the development of national standards for education]. Berlin, Germany: Bundesministerium für Bildung und Forschung.

Kline, M. (1961, October). Math teaching assailed as peril to U.S. scientific progress. *New York University Alumni News*, n.p.

Kupiainen, S., & Pehkonen, E. (2008). PISA 2006 mathematical literacy assessment. In J. Hautamäki, E. Harjunen, A. Hautamäki, T. Karjalainen, S. Kupiainen, S. Laaksonen, J. Lavonen, E. Pehkonen, P. Rantanen, P. Scheinin, I. Halinen, & R. Jakku-Sihvonen (Eds.), *PISA 2006 Finland: Analyses, reflections and explanations* (pp. 117–143). Helsinki, Finland: Ministry of Education.

Lapointe, A. E., Mead, N. A., & Askew, J. M. (1992). *Learning mathematics*. Princeton, NJ: Educational Testing Service.

Lapointe, A. E., Mead, N. A., & Phillips, G. W. (1988). *A world of differences: An international assessment of mathematics and science*. Princeton, NJ: Educational Testing Service.

Latin American Laboratory for Assessment of the Quality of Education. (2001). *First international comparative study of language, mathematics, and associated factors for third and fourth grade primary school students*. Santiago, Chile: Latin American Laboratory for Assessment of the Quality of Education.

Lester, F., Jr Dossey, K. J. A., & Lindquist, M. M. (2007). Challenges and opportunities in the analysis of NAEP mathematics results. In P. Kloosterman & F. K. Lester Jr. (Eds.), *Results and interpretations of the 2003 mathematics assessment of the National Assessment of Educational Progress* (pp. 311–331). Reston, VA: National Council of Teachers of Mathematics.

Luhmann, N. (1990). *Essays on self-reference*. New York, NY: Columbia University Press.

Malaty, G. (2006, December). What are the reasons behind the success of Finland in PISA? *Matilde, 29*, 4–8.

Mathematical Sciences Education Board. (1989). *Everybody counts*. Washington, DC: National Academy Press.

Mayer-Foulkes, D. (2010). *Divergences and convergences in human development* (Human Development Research Paper 2010/20). New York, NY: United Nations Development Programme.

McGaw, B. (2008). The role of the OECD in international comparative studies of achievement. *Assessment in Education: Principles, Policy & Practice, 15*(3), 223–243.

McKnight, C. C., Crosswhite, F. J., Dossey, J. A., Kifer, E., Swafford, J. O., Travers, K. J., & Cooney, T. J. (1987). *The underachieving curriculum: Assessing U.S. school mathematics from an international perspective*. Champaign, IL: Stipes Publishing.

McLeod, D. B., Stake, R. E., Schappelle, B. P., Mellissinos, M., & Gierl, M. J. (1996). Setting the standards: NCTM's role in the reform of mathematics education. In S. A. Raizen & E. D. Britton (Eds.), *Bold ventures: Case studies of U.S. innovations in mathematics education* (Vol. 3, pp. 13–132). Dordrecht, The Netherlands: Kluwer.

Miserable Noten für deutsche Schüler [Abysmal marks for German students]. (2001, December 4). *Frankfurter Allgemeine*.

Moskowitz, J. H., & Stephens, M. (Eds.). (2004). *Comparing learning outcomes: International assessments and educational policy*. London, UK: Routledge/Falmer.

Mullis, I. V. S., Martin, M. O., Foy, P., Olson, J. F., Preuschoff, C., Erberber, E., Arora, A., & Galia, J. (2008). *TIMSS 2007 international mathematics report: Findings from IEA's Trends in International Mathematics and Science Study at the fourth and eighth grades*. Chestnut Hill, MA: TIMSS & PIRLS International Study Center, Lynch School of Education, Boston College.

Mullis, I. V. S., Martin, M. O., Gonzalez, E. J., & Chrostowski, S. J. (2004). *TIMSS 2003 international mathematics report*. Chestnut Hill, MA: TIMSS & PIRLS International Study Center, Lynch School of Education, Boston College.

Mullis, I. V. S., Martin, M. O., Gonzalez, E. J., Gregory, K. D., Garden, R. A., O'Connor, K. M., Chrostowski, S. J., & Smith, T. A. (2000). *TIMSS 1999 international mathematics report*. Chestnut Hill, MA: International Study Center, Lynch School of Education, Boston College.

Mullis, I. V. S., Martin, M. O., Gonzalez, E. J., O'Connor, K. M., Chrostowski, S. J., Gregory, K. D., Garden, R. A., & Smith, T. A. (2001). *Mathematics benchmarking report: TIMSS*

1999–Eighth grade. Chestnut Hill, MA: TIMSS International Study Center, Lynch School of Education, Boston College.

Mullis, I. V. S., Martin, M. O., Ruddock, G. J., O'Sullivan, C. Y., & Preuschoff, C. (2009). *TIMSS 2011 assessment frameworks*. Chestnut Hill, MA: TIMSS & PIRLS International Study Center, Boston College.

Murphy, S. (2010). The pull of PISA: Uncertainty, influence, and ignorance. *Interamerican Journal of Education for Democracy, 3*(1), 28–44.

National Commission on Excellence in Education. (1983). *A nation at risk: The imperatives of education reform*. Washington, DC: Government Printing Office.

National Council of Teachers of Mathematics. (1964). *The revolution in school mathematics*. Washington, DC: Author.

National Council of Teachers of Mathematics. (1961). *An analysis of new mathematics programs*. Washington, DC: Author.

National Council of Teachers of Mathematics. (1980). *An agenda for action*. Reston, VA: Author.

National Council of Teachers of Mathematics. (1989). *Curriculum and evaluation standards for school mathematics*. Reston, VA: Author.

National Council of Teachers of Mathematics. (2000). *Principles and standards for school mathematics*. Reston, VA: Author.

Nguyen, M., Elliott, J., Terlouw, C., & Pilot, A. (2009). Neocolonialism in education: Cooperative learning, Western pedagogy in an Asian context. *Comparative Education, 45*(1), 109–130.

Niss, M. A. (1999). Kompetencer og uddannelsesbeskrivelse [Competencies and description of education]. *Uddannelse, 9*, 21–29.

Nóvoa, A., & Yariv-Masal, T. (2003). Comparative research in education: A mode of governance or a historical journey? *Comparative Education, 39*(4), 423–438.

Olson, L. (2004, September 22). No Child Left Behind Act changes weighed. *Education Week, 24*(4), 31, 34.

On the mathematics curriculum of the high school. (1962). *American Mathematical Monthly, 69*(3), 189–193.

Organisation for Economic Co-operation and Development. (2001). *Knowledge and skills for life: First results from PISA 2000*. Paris, France: Directorate for Education, OECD.

Organisation for Economic Co-operation and Development. (2002). *PISA in the news in Germany: Dec 2001–Jan 2002*. Paris, France: Directorate for Education, OECD.

Organisation for Economic Co-operation and Development. (2003). *The PISA 2003 assessment framework: Mathematics, reading, science and problem solving knowledge and skills*. Paris, France: Directorate for Education, OECD.

Organisation for Economic Co-operation and Development. (2004a). *Learning for tomorrow's world: First results from PISA 2003*. Paris, France: Directorate for Education, OECD.

Organisation for Economic Co-operation and Development. (2004b). *What makes school systems perform? Seeing school systems through the prism of PISA*. Paris, France: Directorate for Education, OECD.

Organisation for Economic Co-operation and Development. (2007). *PISA 2006: Science competencies for tomorrow's world* (Vol. 1: Analysis). Paris, France: Directorate for Education, OECD.

Organisation for Economic Co-operation and Development. (2010a). *PISA 2009 results: What students know and can do–Student performance in reading, mathematics and science* (Vol. 1). Paris, France: Directorate for Education, OECD.

Organisation for Economic Co-operation and Development. (2010b). *Draft of PISA 2012 Mathematics Framework*. Paris, France: Directorate for Education, OECD.

Organisation for Economic Co-operation and Development. (2010c). *Education at a glance 2010: OECD indicators*. Paris, France: Directorate for Education, OECD.

Owen, E., Stephens, M., Moskowitz, J., & Gil, G. (2004). Toward education improvement: The future of international assessment. In J. H. Moskowitz & M. Stephens (Eds.), *Comparing learning outcomes: International assessments and educational policy* (pp. 3–23). London, UK: Routledge/Falmer.

Pashley, P. J., & Phillips, G. W. (1993). *Toward world class standards: A research study linking international and national assessments*. Princeton, NJ: Educational Testing Service.

Pellegrino, J. W., Jones, L. R., & Mitchell, K. J. (Eds.). (1999). *Grading the nation's report card: Evaluating NAEP and transforming the assessment of educational progress*. Washington, DC: National Academy Press.

Peters, M. (2002). Education policy research and the global knowledge economy. *Educational Philosophy and Theory, 34*(1), 91–102.

Phillips, D., & Ochs, K. (2003). Processes of policy borrowing in education: Some explanatory and analytic devices. *Comparative Education, 39*(4), 451–464.

Porter, A. C., & Gamoran, A. (Eds.). (2002). *Methodological advances in cross-national surveys of educational achievement*. Washington, DC: National Academy Press.

Porter, A., McMaken, J., Hwang, J., & Yang, R. (2011, April). Common core standards: The new U.S. intended curriculum. *Educational Researcher, 40*(3), 103–116.

Prais, S. J. (2003). Cautions on OECD's recent educational survey (PISA). *Oxford Review of Education, 29*(2), 139–163.

Rautalin, M., & Alasuutari, P. (2009). The uses of the national PISA results by Finnish officials in central government. *Journal of Education Policy, 24*(5), 539–556.

Ripley, A. (2011). The world's schoolmaster: How a German scientist is using test data to revolutionize global learning. *Atlantic Magazine, 308*(1), 109–110.

Robitaille, D. F., & Beaton, A. E. (Eds.). (2002). *Secondary analysis of the TIMSS data*. Dordrecht, The Netherlands: Kluwer.

Robitaille, D. F., & Garden, R. A. (Eds.). (1989). *The IEA Study of Mathematics II: Contexts and outcomes of school mathematics*. Oxford, UK: Pergamon.

Rutkowski, D. (2008). Towards an understanding of educational indicators. *Policy Futures in Education, 6*(4), 470–481.

Sahlberg, P. (2010, December 27). Learning from Finland: How one of the world's top educational performers turned around. *Boston Globe*, A9.

Schmidt, W. H., Jorde, D., Cogan, L. S., Barrier, E., Gonzalo, I., Moser, U., Shimizu, K., Sawada, T., Valverde, G., McKnight, D., Prawat, R., Wiley, D. E., Raizen, S., Britton, E. D., & Wolfe, R. G. (1996). *Characterizing pedagogical flow: An investigation of mathematics and science teaching in six countries*. Dordrecht, The Netherlands: Kluwer.

Schmidt, W. H., McKnight, C., Cogan, L. S., Jakwerth, P. M., & Houang, R. T. (1999). *Facing the consequences: Using TIMSS for a closer look at U.S. mathematics and science education*. Dordrecht, The Netherlands: Kluwer.

Schmidt, W. H., McKnight, C. C., Houang, R. T., Wang, H. C., Wiley, D. E., Cogan, L. S., & Wolfe, R. G. (2001). *Why schools matter: A cross-national comparison of curriculum and schooling*. San Francisco, CA: Jossey-Bass.

Schmidt, W. H., McKnight, C., & Raizen, S. (1997). *A splintered vision: An investigation of U.S. science and mathematics education*. Dordrecht, The Netherlands: Kluwer.

Schmidt, W. H., McKnight, C., Valverde, G. A., Houange, R. T., & Wiley, D. E. (1997). *Many visions, many aims* (Vol. 1: A cross-national investigation of curricular intentions in school mathematics). Dordrecht, The Netherlands: Kluwer.

Schriewer, J. (1990). The method of comparison and the need for externalization: Methodological criteria and sociological concepts. In J. Schriewer & B. Holmes (Eds.), *Theories and methods in comparative education* (pp. 25–83). Frankfurt am Main, Germany: Lang.

Sjøberg, S. (2007). PISA and "real life challenges": Mission impossible? In S. T. Hopmann, G. Brinek, & M. Retzel (Eds.), *PISA according to PISA—Does PISA keep what it promises?* (pp. 203–224). Vienna, Austria: LIT-Verlag.

Sloane, P. F. E., & Dilger, B. (2005) The competence clash—Dilemata bei der übertragung des "Konzepts der nationalen Bildungsstandards" auf die berufliche Bildung [The competence clash: Dilemmas in the transmission of the "concept of national education standards" to vocational training]. *Berufs- und Wirtschaftspädagogik*. Retrieved from http://www.bwpat.de.

Smith, D. E. (1909). *The teaching of arithmetic*. Boston, MA: Ginn.

Southern and Eastern Africa Consortium for Monitoring Educational Quality. (2011). SACMEQ: 1995–2010. Retrieved from http://www.sacmeq.org/.

Stanat, P., Artelt, C., Baumert, J., Klieme, E., Neubrand, M., Prenzel, M., Schiefele, U., Schneider, W., Schümer, G., Tillmann, K., & Weiß, M. (2002). *PISA 2000: Overview of the study—Design, method and results.* Berlin, Germany: Max Planck Institute for Human Development.

Steiner-Khamsi, G. (2006). The economics of policy borrowing and lending: A study of late adopters. *Oxford Review of Education, 32*(5), 665–678.

Steiner-Khamsi, G. (2007). International knowledge banks and the production of educational crises. *European Educational Research Journal, 6*(3), 285–292.

Thomas, J. (2001). Globalization and the politics of mathematics education. In B. Atweh, B. Nebres, & H. Forgasz (Eds.), *Sociocultural research on mathematics education: An international perspective* (pp. 95–112). Mahwah, NJ: Erlbaum.

Travers, K. J., & Westbury, I. (Eds.). (1989). *The IEA Study of Mathematics II: The analysis of mathematics curricula.* Oxford, UK: Pergamon.

UNESCO. (1990). *World declaration on education for all.* Retrieved July 15, 2011 from http://www.unesco.org/education/efa/ed_for_all/background/jomtien_declaration.shtml.

UNESCO. (2011). *The hidden crisis: Armed conflict and education.* Paris, France: United Nations Educational, Scientific, and Cultural Organization.

United States Bureau of Education. (1911). *Mathematics in the elementary schools of the United States.* Washington, DC: Government Printing Office.

Välijärvi, J., Linnakylä, P., Kupari, P., Reinikainen, P., & Arffman, I. (2002). *The Finnish success in PISA—And some reasons behind it.* Jyväskylä, Finland: Finnish Institute for Educational Research, University of Jyväskylä.

Valverde, G. A., Bianchi, L. J., Wolfe, R. G., Schmidt, W. H., & Houang, R. T. (2002). *According to the book: Using TIMSS to investigate the translation of policy into practice through the world of textbooks.* Dordrecht, The Netherlands: Kluwer.

Vithal, R., Adler, J., & Keitel, C. (Eds.). (2005). *Researching mathematics education in South Africa: Perspectives, practices and possibilities.* Cape Town, South Africa: Human Sciences Research Press.

Wong, K. Y., Lee, P. P., Kaur, B., Yee, F. P., & Fong, N. S. (Eds.). (2009). *Mathematics education: The Singapore journey.* Singapore: World Scientific.

World Bank. (2005, October 12). *Education for all—Fast track initiative. Fact sheet: About aid effectiveness.* Washington, DC: World Bank.

Wu, M. (2010a). *Comparing the similarities and differences of PISA 2003 and TIMSS* (OECD Education Working Paper No. 32). Paris, France: Directorate for Education, OECD.

Wu, M. (2010b). Measurement, sampling, and equating errors in large-scale assessments. *Educational Measurement: Issues and Practice, 29*(4), 15–27.

Brief Biographical Details of Authors

Nadja Maria Acioly-Régnier has a bachelor and masters degree in Psychology from the Federal University of Pernambuco, Brazil. She did her Diploma of Advanced Studies and PhD in Psychology at the University René Descartes, Paris V Sorbonne. She received a Habilitation to supervise research projects at the Université Claude Bernard Lyon 2. Currently she is a permanent Lecturer at the Institut Universitaire de Formation des Maîtres—Université Claude Bernard Lyon 1, and a researcher of the "Equipe d'Accueil Mixte," Laboratoire Santé Individu Société, France. She works in the areas of Psychology and Educational Sciences, focusing on culture and cognition, intercultural psychology, professional didactic, scientific and mathematical concept development in formal and informal learning contexts. She is married to Jean-Claude Régnier, with whom she conducts research in the area of obstacles in learning mathematical, statistical and scientific concepts.

Cecilia Agudelo-Valderrama is a Commissioner of CONACES—one of Colombia Ministry of Education's advisory groups on matters related to the quality of higher education programs. She has been Senior Lecturer in mathematics education at various Faculties of Education in Colombia, and has many years of experience as a teacher of mathematics at the primary and secondary levels. Her research has focused on mathematics teachers' conceptions, which includes their attitudes to change, and their professional learning. During the last two years she has been collaborating with colleagues at Monash University, Australia, in order to research the impact of a "connected mathematics and science teaching approach" on the practice of a group of Colombian mathematics and science teachers.

Arthur Bakker is Assistant Professor at the Freudenthal Institute for Science and Mathematics Education, Utrecht University, in the Netherlands. His PhD thesis (2004) was about design research in statistics education, with a focus on technology, the history of statistics, and Peircean semiotics (diagrammatic reasoning). He was advisor and curriculum author in the *TinkerPlots* project (1999–2003), directed by Clifford Konold (USA), and did postdoctoral studies in the *Techno-mathematical*

Literacies in the Workplace project (2004–2007), which was co-directed by Celia Hoyles and Richard Noss (UK). With Keino Gravemeijer, he co-directed the project *Boundary Crossing Between School and Work for Developing Techno-Mathematical Competencies in Vocational Education* (2007–2011).

Dani Ben-Zvi is Senior Lecturer in Mathematics Education and in Learning, Instruction and Teacher Education, and head of the Educational Technologies Graduate Program, in the Faculty of Education at the University of Haifa, Israel. He is a recognized international scholar in statistics education, focusing on developing students' statistical reasoning that involves creating and evaluating data-based claims that are used as a means of increasing credibility of arguments and of making decisions under uncertainty. He designs and studies innovative educational technologies and technology-enhanced learning communities as a means of making complex domains—such as statistics—more accessible to learners.

Rolf Biehler is Professor for Didactics of Mathematics at Paderborn University, Germany. He is co-director of the KHDM (German Centre for Mathematics Education Research in Tertiary Education) and member of the board of directors of the German Centre for Continuous Professional Development of Mathematics Teachers (DZLM). His interests are research into the teaching and learning of probability and statistics, the design and use of e-learning and software tools for teaching and learning, and mathematics education at the tertiary level. He is author of over 100 publications in mathematics education and currently the editor-in-chief of *Journal für Mathematik-Didaktik*, published by Springer.

Alan J. Bishop is Emeritus Professor at Monash University, Australia. He was Professor of Education at Monash University between 1992 and 2002 .He edited (1978–1990) the international research journal, *Educational Studies in Mathematics*, published by Kluwer, and has been an Advisory Editor since 1990. He is Managing Editor of the book series Mathematics Education Library, also published by Kluwer (1980–present). He has authored or edited several influential books, reports, articles and chapters on mathematics education, and was Chief Editor of the first two International Handbooks of Mathematics Education (1996 and 2003) published by Kluwer (now Springer).

Marcelo C. Borba is Professor in the graduate program in mathematics education and of the Mathematics Department of UNESP (State University of Sao Paulo), campus of Rio Claro, Brazil. He researches the use of digital technology in mathematics education, online distance education and qualitative research methodology. From 2008 through 2011 he served on important education committees within the main research funding agency of Brazil. He is a member of the editorial board of *Educational Studies in Mathematics*, and is currently an associate editor of ZDM. He has given invited talks in countries such as Argentina, Canada, Colombia, Denmark, Germany, Italy, Mozambique, Mexico, New Zealand and the USA. He has been a member of program committees for several international conferences, and has written several books, book chapters and papers published in Portuguese and in English. During the past 10 years he has edited a collection of books in Brazil which includes 24 books to date.

Tony Brown is Professor of Mathematics Education at Manchester Metropolitan University. His research is primarily concerned with mathematics education through the lens of contemporary social theory. He has written seven books, including three titles for Springer's Mathematics Education Library series, most recently *Becoming a Mathematics Teacher* and *Mathematics Education and Subjectivity*. The journal *Educational Studies in Mathematics* has published eight of his papers. He has also had a long-standing interest in professionally-oriented research, typically carried out by practitioners working on doctoral studies. He recently organized a *Mathematics Education and Contemporary Theory* conference, and an associated special issue of *Educational Studies in Mathematics*.

Jinfa Cai is Professor of Mathematics and Education at the University of Delaware. He is interested in how students learn mathematics and solve problems, and how teachers can provide and create learning environments so that students can make sense of mathematics. He has received a number of awards, including a US National Academy of Education Spencer Fellowship, an American Council on Education Fellowship, an International Research Award, and a Teaching Excellence Award. He has been a visiting professor at various institutions, including Harvard University. He has served as a Program Director at the US National Science Foundation and is currently serving as the senior co-chair of the American Educational Research Association's Special Interest Group on Research in Mathematics Education.

Olive Chapman is Professor of Mathematics Education and Associate Dean of Undergraduate Programs Education in the Faculty of Education at the University of Calgary. She is Editor-in-Chief of the *International Journal of Mathematics Teacher Education*. Her research interests include prospective and practising mathematics teachers' thinking—their beliefs, conceptions, perspectives, practical knowledge, mathematical sense-making; learning, and change; mathematics knowledge for teaching; mathematical thinking, problem solving, problem posing and contextual/word problems; inquiry-based mathematics pedagogy, and inquiry-based discourse to facilitate mathematical thinking. She teaches mathematics education courses at the undergraduate and graduate levels and supervises graduate students in mathematics education.

Ui Hock Cheah has been Senior Specialist in Mathematics at the Regional Centre for Education in Science and Mathematics (RECSAM), Penang, Malaysia, since 2004. He holds a masters degree from Deakin University and a PhD from Universiti Sains Malaysia. He began his career as a secondary mathematics teacher before moving into teacher education. He has a long history of engagement in mathematics education research and the professional development of teachers, and has been heavily involved in the implementation and designing of professional development programs for Malaysian teachers. Since 2006 he has been the chief editor of *Journal of Science and Mathematics Education in Southeast Asia*.

David Clarke is Professor at the University of Melbourne and Director of the International Centre for Classroom Research (ICCR). Over the last 15 years, his research activity has centred on capturing the complexity of classroom practice through a program of international video-based classroom research. Other significant research has addressed teacher professional learning, metacognition, problem-based learning, and assessment (particularly the use of open-ended tasks for assessment and instruction in mathematics). Current research activities involve multi-theoretic research designs, cross-cultural analyses, discourse in and about classrooms internationally, curricular alignment, and the challenge of research synthesis in education. He has written books on assessment and on classroom research and has published his research in over 150 book chapters, journal articles and papers in conference proceedings.

Philip Clarkson has taught at the Australian Catholic University since 1985, where he is Professor of Education. This followed nearly five years as Director of a Research Centre at the Papua New Guinea University of Technology, and prior to that as a lecturer at Monash University and tertiary colleges in Melbourne. He began his professional life as a secondary school teacher. He has served as President of the Mathematics Education Research Group of Australasia (MERGA) and was foundation editor of the *Mathematics Education Research Journal*. His research interests are wide ranging, from evaluation of schools, education systems and research programs, through to various areas of mathematics education.

M. A. (Ken) Clements is Professor in the Department of Mathematics at Illinois State University. After teaching in schools for 10 years, he taught in three Australian universities (Monash, Deakin, and Newcastle), and at Universiti Brunei Darussalam (1997–2004). He has served as a consultant in India, Malaysia, PNG, South Africa, Thailand and Vietnam, and has been an editor for the three international handbooks on mathematics education (1996, 2003, 2012). He has written or edited 25 books and has authored many peer-reviewed articles. In 1996 he co-authored, with Nerida Ellerton, a UNESCO book on mathematics education research. He is honorary life member of both the Mathematics Education Research Group of Australasia (MERGA) and the Mathematical Association of Victoria (MAV).

John A. Dossey is Distinguished Professor of Mathematics Emeritus at Illinois State University. His research has focused on comparative studies of student achievement (NAEP, TIMSS, PISA) and the mathematics curriculum. He has served as President of the National Council of Teachers of Mathematics (NCTM), Chair of the Conference Board of Mathematical Sciences (CBMS), and on numerous other professional commissions, committees, policy boards, and editorial boards. He has authored/coauthored textbook series for grades 6–12, collegiate texts for preservice and inservice teachers of mathematics, research in mathematics education, and collegiate texts for discrete mathematics and mathematical modeling, as well as numerous monographs and papers.

Paul Drijvers is Associate Professor in Mathematics Education at the Freudenthal Institute of Utrecht University, the Netherlands. His major research interests are the use of ICT in mathematics education, algebra education, teacher professional development, and curriculum development.

Nerida F. Ellerton has been Professor in the Department of Mathematics at Illinois State University since 2002. She holds two doctoral degrees—one in chemistry and the other in mathematics. After teaching in schools, she took an academic appointment at Deakin University, and between 1991 and 1992 was Director of Deakin's National Centre for Research and Development in Mathematics Education. She was Professor of Mathematics Education at Edith Cowan University (1993–1997), editor of the *Mathematics Education Research Journal* (1993–1997), and Dean of Education at the University of Southern Queensland (1997–2001). She has led research projects in 10 nations, has written or edited 12 books, and has had more than 150 articles published in refereed journals, conference proceedings, or books. She is currently studying toward a Diploma in Book Conservation with the American Academy of Bookbinding. With Ken Clements, she wrote the book *Rewriting the History of School Mathematics in North America 1607–1861*, which Springer published in 2012.

Jonas Emanuelsson is Senior Lecturer in the Department of Pedagogical, Curricular and Professional Studies at the University of Gothenburg, Sweden. He specializes in classroom studies of teaching and learning in mathematics and science. His recent work includes international comparative studies conducted in the context of the Learner's Perspective Study (LPS). His major research interests focus on teachers' and students' co-construction of learning processes in classroom interactions. Analyses are aimed at understanding both the content dealt with, and the organisation of interaction. Previous work examined formative assessment as an aspect of classroom interaction.

Jeff Evans is Emeritus Reader in Adults' Mathematical Learning at Middlesex University, London. He teaches and offers consulting on social research methodology and statistics. His research interests include: mathematical thinking and the emotions; adult numeracy and demands arising from everyday life and work; cultural images of mathematics; social survey methods and measurement of adult skills and attitudes; and methods for fruitfully combining quantitative and qualitative methods. He is on the Editorial Board of *Educational Studies in Mathematics*, and also reviews for several other educational journals.

Joan Ferrini-Mundy served as Director of the Division of Science and Mathematics Education at Michigan State University during the period 1999–2006, with appointments in Mathematics and Teacher Education. She was an MSU University

Distinguished Professor in Mathematics Education and was previously Associate Dean for Science and Mathematics Education in the College of Natural Science. Her research interests are in calculus learning and mathematics education reform, K–14. She chaired the writing group for *Principles and Standards for School Mathematics*, the 2000 revision of NCTM *Standards*. Her research projects address the development of leadership in mathematics and science education, school-district level improvement in mathematics and science education, and knowledge of algebra for secondary school teaching. She has recently worked on secondment to the National Science Foundation, where she served as Director of the Division of Research on Learning in Formal and Informal Settings. She currently is a member of the US government Senior Executive Service, serving as Assistant Director, National Science Foundation, for Education and Human Resources.

Cristina Frade is a Lecturer at Universidade Federal de Minas Gerais [UFMG], Brazil. She previously taught mathematics in schools for 25 years, and now teaches mathematics education and supervises postgraduate students. She coordinates a research group at UFMG with projects funded by the university, state and national research agencies. Her research and publication areas include: the tacit-explicit dimension of mathematics practice in and out of school contexts; socio-cultural theories of learning; psychology; culture and affect in mathematics education.

Fulvia Furinghetti is Professor of Mathematics Education in the Department of Mathematics at the University of Genoa. Her research interests include: beliefs, images of mathematics in society, proof, problem solving, use of history of mathematics in teaching, teacher professional development, and history of mathematics education. She organized the celebrations of the centenary of the journal *L'Enseignement Mathématique* and of ICMI, and edited the proceedings. In addition, she developed the website on the history of the first 100 years of ICMI. In 2000–2004 she chaired HPM (the International Study Group on the relations between History and Pedagogy of Mathematics, which is affiliated to ICMI).

George Gadanidis is Associate Professor at the Faculty of Education, University of Western Ontario, Canada. His research explores the intersection of mathematics education, technology and the arts. Classroom documentaries of his work can be viewed at www.researchideas.ca

Uwe Gellert is Professor of Mathematics Education within the Faculty of Education and Psychology at Freie Universität Berlin, Germany. His research interests include social inequalities in mathematics education, cross-cultural studies, microanalysis of classroom interaction, sociological perspectives on mathematics, and mathematics teacher education. He is active in the *Mathematics Education and Society* group (MES) and in the *Commission Internationale pour l'Étude et l'Amélioration de l'Enseignement des Mathématiques* (CIEAEM). His collaborative international research projects take him regularly to several South American countries.

Pedro Gómez is a mathematics education researcher at the University of Granada (Spain) and visiting Professor at the Universidad de los Andes in Bogotá (Colombia). He is editor of PNA, a mathematics education research journal. His research interests focus on mathematics teacher education.

Merrilyn Goos is Professor of Education and Director of the Teaching and Educational Development Institute at the University of Queensland. Her research interests include secondary school students' mathematical thinking, the professional learning of mathematics teachers, the role of digital technologies in mathematics teaching and learning, numeracy education in school and non-school contexts, school reform, and assessment practices in university courses. She is currently President of the Mathematics Education Research Group of Australasia (MERGA), and an Associate Editor of *Educational Studies in Mathematics*.

Zahra Gooya is Associate Professor of Mathematics Education in the Mathematics Department of Shahid Beheshti University in Iran. Her experience has included being a member of writing teams for five secondary national mathematics textbooks, running more than 11 national training sessions for mathematics teachers regarding the new textbooks, editor of the "Roshd" Mathematics Teacher Education Journal (in Farsi), an influential scholar for the establishment of the masters program of "Mathematics Education" in Iran. Since 2006 she has been an elected member of the Iranian Mathematics Society. In recent years, she has been researching the professional development and professional learning of mathematics teachers.

Ghislaine Gueudet is Professor at IUFM Bretagne within the University of Brest, France. She is co-director of the CREAD (Center for Research on Education, Learning and Didactics). Her research concerns all the resources intervening in the teacher's activity, with a particular focus on Internet resources. With Luc Trouche, she introduced an approach to the study of mathematics teachers' documentation work, considering the interactions between teachers and ressources, and the consequences of these interactions for teacher professionnal development.

Lulu Healy is a lecturer in the post-graduate program in Mathematics Education, at Bandeirante University of São Paulo. She has been working in Brazil since 2002, following a research post in mathematics education at the Institute of Education, University of London. Her research interests focus on the use of digital technologies in the teaching and learning of mathematics, and especially on the design of innovative ways of doing and expressing school mathematics. She is particularly interested in the challenges associated with the building of a more inclusive school mathematics and in understanding the mathematical practices of learners with disabilities. She currently coordinates the research group *Technology and Means for Expressing Mathematics* and directs the research program, *Towards an Inclusive School Mathematics*, which investigates relationships between sensory experiences and mathematical cognition and designs and evaluates mathematical learning scenarios for students with a diverse range of educational needs.

M. Kathleen Heid is a Distinguished Professor of Mathematics Education at Pennsylvania State University. She has served as Editor of the *Journal for Research in Mathematics Education*, as a member of the Board of Directors for the National Council of Teachers of Mathematics, as a member of the Board of Governors for the Mathematical Association of America, and as co-PI for NSF-funded projects and Centers. She has co-edited *Research on Technology and the Teaching and Learning of Mathematics*, has co-authored technology-intensive mathematics textbooks for high school students, and has conducted research on the impact of technology on the learning of mathematics.

Rosa Becerra Hernández is Professor in the Pedagogical University Experimental Libertador (UPEL) in Caracas, Venezuela. In addition to her PhD in Education from UPEL (2006), she holds an MA from the University of the Pacific, USA (1983), where she was an Outstanding Credential Candidate. From 2003 to 2005, she served as a member of Venezuela's National Program to Promote Research. She currently is the coordinator of the Research Center of Mathematics and Physics, the coordinator of the Undergraduate Mathematics Program, and the coordinator of the Master of Education Program in Mathematics Education at UPEL. In 2001, she was awarded the Benefit National Award for University Academics (CONABA).

Bernard R. Hodgson is *Professeur titulaire* in the Department of Mathematics and Statistics at Université Laval, where he has been a faculty member since 1975. His research and teaching interests include mathematical logic and theoretical computer science, mathematical education and its history, and history of mathematics. He was an invited regular lecturer at the International Congress of Mathematicians (1990 and 1998) and at ICME-7 (1992), and a plenary lecturer at ICME-12 (2012). He has served as Vice-President of the Canadian Mathematical Society, President of the Canadian Mathematics Education Study Group, and Secretary-General (1999–2009) of the International Commission on Mathematical Instruction (ICMI).

Geoffrey Howson is Emeritus Professor of Mathematical Curriculum Studies at Southampton University, England. He has written, edited, or contributed to many books and papers on mathematics, education, and mathematics education. Among other commitments, he has served as Assistant Director of the Centre for Curriculum Renewal and Educational Development Overseas, President of the Mathematical Association, Chairman of the School Mathematics Project, a founder Director of BACOMET, and Secretary-General of ICMI.

Celia Hoyles is Professor of Mathematics Education at the Institute of Education, University of London, and is Director of the National Centre for Excellence in the Teaching of Mathematics. Her research interests include proof conceptions, the mathematical skills needed in modern workplaces, and the design of digital environments for mathematics. In 2004, she was awarded an OBE for services to mathematics education, and also the first ICMI Hans Freudenthal medal. Between 2004 and 2007 she served as the UK Government's Chief Adviser for Mathematics.

In 2011, she was awarded the first Royal Society Kavli Education Medal for distinguished contribution to science or mathematics education.

Eva Jablonka is Professor of Mathematics Education at Luleå University of Technology, Sweden. She earned her PhD at the University of Technology in Berlin in 1996, and worked for many years at Freie Universität in Berlin. As a member of the Learner's Perspective Study (LPS), she spent one year in Australia helping to the set up the International Centre for Classroom Research at the Universty of Melbourne. She has lectured in a variety of undergraduate and graduate programs in mathematics education. Her research areas include sociological theories of mathematics and mathematics education, mathematical modelling and mathematical literacy, and comparative empirical classroom studies. Currently, she is engaged in an international study on the emergence of disparity in achievement in mathematics classrooms.

Barbara Jaworski is Professor of Mathematics Education in the Mathematics Education Centre at Loughborough University, in the UK. Her experience has included being a teacher of mathematics at both secondary and university levels, a teacher educator for prospective secondary teachers, and extensively working with practising teachers in several countries including the UK, Norway and Pakistan. She has edited the *Journal of Mathematics Teacher Education* and has been President of the European Society for Research in Mathematics Education. Her research has focused on mathematics teaching at all levels and its development. She has been particularly interested in creating and researching communities of inquiry involving teachers and educators in which both groups are researchers and bring complementary knowledge to the partnership.

Alexander Karp is Associate Professor of Mathematics Education at Teachers College, Columbia University. He received his Ph.D. in mathematics education from Herzen Pedagogical University in St. Petersburg, Russia, and also holds a degree from the same University in history and education. Currently, his scholarly interests span several areas, including gifted education, mathematics teacher education, the theory of mathematical problem solving, and the history of mathematics education. He is the managing editor of the *International Journal for the History of Mathematics Education* and the author of over 100 publications, including over 20 books.

Christine Keitel is Professor of Mathematics Education at the Freie University, Berlin, where she teaches prospective primary and secondary school teachers of mathematics. In the 1970s she worked as a research fellow at Max-Planck-Institute for Educational Research in Berlin on theoretical and practical approaches to curriculum development. In 1980, she became Director of a practice-oriented teacher education project at the Institute for Didactics of Mathematics in Bielefeld. She then moved to the Technical University of Berlin (TUB) and passed her Habilitation/venia legendi in the Mathematics Department of that University. In 1990 she obtained her professorship at Freie University.

Her main research interests are mathematics as a social practice; philosophy and sociology of mathematics and the sciences; mathematics for all; mathematical literacy; mathematics education and technology; social justice and mathematics education: gender, ethnicity and class and the politics of schooling; history and the current state of mathematics education around the world; comparative studies on mathematics classroom practice and learners' perspectives; political and social dimensions of research on mathematics classroom practice, internationalization and globalization of scientific collaboration; difficulties faced by students and teachers in mathematics classrooms; mathematics education and values.

In 1994 and in 1999–2001 she was guest professor in South Africa/Durban, and in Melbourne and Queensland in 2002 and 2004. In 1999 she was awarded an Honorary Doctorate in Sciences by the University of Southampton, and the A. v. Humboldt-Scholarship Award for Research and Capacity Building in South Africa. In 2009 she was awarded an Honorary Doctorate in Sciences by the University of Shumen, Bulgaria. Between 1999 and 2001 and also between 2007 and 2010 she was Vice-President of Freie University responsible for transformation of study orders into Bachelor and Master programs.

Carolyn Kieran is Professor Emerita of Mathematics Education at the Université du Québec à Montréal. For many years her research has focused on the learning and teaching of algebra, more recently involving the roles played by computing tools. She has served as President of the International Group for the Psychology of Mathematics Education and as member of the International Program Committees for ICME-11, ICMI Study Group 12 on Algebra, and ICMI Study 22 on Task Design. Recent publications include chapters in the *Handbook of Research on the Psychology of Mathematics Education* and the *Second Handbook of Research on Mathematics Teaching and Learning*.

Jeremy Kilpatrick is Regents Professor of Mathematics Education at the University of Georgia. He holds A.B. and M.A. degrees from the University of California, Berkeley, and M.S. and Ph.D. degrees from Stanford University. Before joining the Georgia faculty, he taught at Teachers College, Columbia University. He is a National Associate of the National Academy of Sciences and a Fellow of the American Educational Research Association, received a 2003 Lifetime Achievement Award from the National Council of Teachers of Mathematics, and received the 2007 Felix Klein Medal honoring lifetime achievement in mathematics education from the International Commission on Mathematical Instruction.

Konrad Krainer is Professor at the University of Klagenfurt, head of the Austrian Educational Competence Centre for Instructional and School Development (Klagenfurt and Vienna), and the leader of the nation-wide IMST project. He was associate editor of JMTE, co-editor of the *International Handbook of Mathematics Teacher Education*, was a founding member of the ERME-Board (e.g., establishing a summer school for young researchers), and is a member of the Education Committee of the EMS. He gave plenary lectures at ICME 10 (co-presenter) and

PME 35. His research interests are mathematics teacher education, school development, and educational system development.

Troels Lange is Senior Lecturer in Mathematics for young children in teacher education at Malmö University, Sweden. After teaching mathematics and science in a high school program for adults, he taught mathematics education in Danish teacher-education programs. He has recently worked in Australian primary teacher education. In 2009 he defended his PhD thesis, at Aalborg University, Denmark, on children's perspectives on having difficulties with learning mathematics. Since then, he has published a number of peer-reviewed articles. He is now involved in research projects investigating preschools' approach to mathematics education.

Stephen Lerman was a secondary school teacher of mathematics in London and abroad, including five years as a head of Mathematics. Since then he has been in mathematics teacher education and research, and he is now Professor of Mathematics Education at London South Bank University. He is a former President of the International Group for the Psychology of Mathematics Education (IGPME) and Chair of the British Society for Research in Learning Mathematics (BSRLM). His interests are in theories of learning and socio-cultural theory. His recent work has drawn on sociological theory to inform studies of disadvantage in school classrooms.

Allen Leung is Associate Professor in Mathematics Education at the Hong Kong Baptist University. He obtained his PhD in mathematics at the University of Toronto and has been Assistant Professor at the University of Hong Kong and the Hong Kong Institute of Education. His main research areas are the pedagogical and epistemological aspects of dynamic geometry environments, the use of tools in mathematics classrooms, the development of the Theory of Variation in mathematics pedagogy, and the development of lesson and learning studies. He is an IPC member of the 22nd ICMI Study on task design.

Frederick Koon-Shing Leung is Professor in Mathematics Education within the Faculty of Education of the University of Hong Kong. Born and raised in Hong Kong, he obtained B.Sc., Cert. Ed. and M.Ed. qualifications from the University of Hong Kong, and Ph.D. from the University of London Institute of Education. His major research interests are in the comparison of mathematics education in different countries, and in the influence of culture on teaching and learning. He is principal investigator of a number of major research projects, including the Hong Kong component of the Trends in International Mathematics and Science Study (TIMSS), the TIMSS Video Study, and the Learner's Perspective Study (LPS). He was a member of the Executive Committee of the International Commission on Mathematical Instruction (ICMI) and the Standing Committee of the International Association for the Evaluation of Academic Achievement (IEA). He was awarded a Senior Fulbright Scholarship in 2003, and is a honorary professor of Beijing Normal University, Southwest University, and Zhejiang Normal University in China.

Jun Li is Associate Professor within the Department of Mathematics at the East China Normal University. She has a special interest in studying students' understanding of mathematics, especially in the field of statistics and probability. She is also interested in curriculum issues, teacher training, the use of technology in classrooms, and the influences of culture on mathematics education. She is a member of the writing group of the Standards of Mathematics Curriculum for Senior High Schools issued by the Ministry of Education of China, and is also author of a mathematics textbook being used in junior high schools in China.

Abigail Fregni Lins (Bibi Lins) is a permanent Lecturer in the Mathematics Department and in the Graduate Master Program in Mathematics Education at the State University of Paraíba, Brazil. She was born in the capital city of São Paulo, Brazil, and obtained her BSc in Pure Mathematics at the Catholic University of São Paulo-PUCSP in 1985, her MPhil in Number Theory at the University of Nottingham in 1992, and a PhD in Mathematics Education at the University of Bristol in 2003. She has taught at undergraduate and graduate levels in national and international universities. Her professional interests are in pure mathematics, and in mathematics education with emphasis on the use of technologies in the teaching and learning of mathematics. She is President of the Brazilian Mathematics Education Society in the State of Paraíba and is involved in various national research projects.

Chap Sam Lim is Professor of Mathematics Education in the School of Educational Studies at Universiti Sains Malaysia (USM), Penang, Malaysia. She taught mathematics in secondary schools for 8 years and lectured at a teacher-training college for one and a half years before taking an appointment at USM in 1993. In 1999 she was awarded a PhD by Exeter University, in the UK. She gained the Asian Scholar award from the Asian Scholarship Foundation (2004–2005), and was Fulbright Scholar in 2008 and 2009. She has written or edited 7 books and more than 50 peer-reviewed articles focusing on cross-cultural study, public images of mathematics, teaching mathematics in a second language, and lesson study as a form of professional development for mathematics teachers.

Lim-Teo, Suat Khoh is Associate Professor at the National Institute of Education, Singapore. She has held various appointments there, including Head of Mathematics, Associate Dean for pre-service programs and Dean of Faculty Affairs. She was also a past President of Singapore's Association of Mathematics Educators, a past Vice-President of the Singapore Mathematical Society, and has served on local and international conference committees. She was an IPC member for ICME10 (Denmark, 2004) and IPC chair for EARCOME 2002. She has been a mathematics teacher educator for more than 20 years and her current research focus is on teacher education.

Katie Makar is Senior Lecturer in mathematics education at the University of Queensland in Australia. She has authored over 40 peer-reviewed publications, presented her research on six continents and led several national projects researching the teaching of mathematics through inquiry, statistical argumentation, and informal statistical inference at the school level. She has conducted consulting work for government agencies on innovative teaching of mathematics, promoting statistical reasoning and teacher professional standards, and teaches preservice secondary mathematics teachers as well as courses in mathematics education, interdisciplinary curriculum and statistics. She obtained her PhD in 2004, and before that she taught secondary mathematics for 15 years in the USA and in Asia.

José Manuel Matos did his undergraduate studies in applied mathematics at the University of Lisbon, Portugal. For some years he worked as a certified teacher of mathematics in secondary schools in Portugal, and later he obtained his masters degree at Boston University and his doctorate at the University of Georgia, his dissertation being in mathematics education. He currently works at the New University of Lisbon where he teaches undergraduate and graduate courses on education and mathematics education. His research interests are currently focused on learning, curricular issues, and the history of mathematics education.

Claire Margolinas is "Maître de Conférence" and member of the *Acté* research center at Blaise Pascal University, France. Since 1993 she has taught mathematics education and mathematics for secondary and primary teachers at Auvergne University's Institute for Teacher Education (IUFM: Institut Universitaire de Formation des Maîtres). Between 2003 and 2006 she was editor of the journal *Recherches en Didactique des Mathématiques* and she is currently a member of the editorial board of *Educational Studies in Mathematics.* She is known for her work about teachers' situations.

Tamsin Meaney is Professor of Mathematics for young children at Malmö University, Sweden. She has worked as a mathematics educator in Australia, Kiribati, and New Zealand. Before moving to Sweden, Tamsin worked as a teacher educator at the University of Otago, Dunedin, New Zealand, and at Charles Sturt University, Wagga Wagga, Australia, Much of her research interests have focused on issues associated with language in mathematics, particularly in relationship to the use of Indigenous languages for the teaching of mathematics. This includes a long-running research project within a Māori-immersion school in Rotorua, which has been the theme of her recent book, published by Springer.

Marta Menghini is Associate Professor in the Department of Mathematics of Sapienza University in Rome. Her research interests are as follows: history of mathematics teaching, particularly of geometry teaching; relations between historical development, foundations, cognitive aspects and curriculum in the learning of mathematics; influences of research in geometry on the teaching of geometry in the last century; the approach to definitions in geometry, intuitive geometry and the use of concrete materials. She was the chief organizer of the international symposium, held in Rome in March 2008, marking the occasion of the centennial of ICMI, and she edited the proceedings of that symposium.

Vilma Mesa is Assistant Professor of Education at the University of Michigan. She is currently investigating the role that resources play in developing teaching expertise in undergraduate mathematics, specifically at community colleges and in inquiry-based learning classrooms. She has conducted several analyses of textbooks and evaluation projects on the impact of innovative mathematics teaching practices for students in science, technology, engineering, and mathematics. She has a B.S. in computer science and a B.S. in mathematics from the University of Los Andes in Bogotá, Colombia, and holds a master's degree and a Ph.D. in mathematics education from the University of Georgia.

Elena Nardi is a Greek-born researcher in mathematics education. She is currently Reader at the University of East Anglia in the UK. Her research is mainly on the teaching and learning of mathematics at university level, cognitive, social and affective issues of student engagement with mathematics, and, secondary mathematics teacher knowledge and beliefs. She is joint Editor-in-Chief of the Routledge journal *Research in Mathematics Education*, the official journal of the *British Society for Research into the Learning of Mathematics* (BSRLM), and her monograph *Amongst Mathematicians: Teaching and Learning Mathematics at University Level* was published by Springer in 2008. E-mail: e.nardi@uea.ac.uk

Mogens Niss is Professor of Mathematics and Mathematics Education at Roskilde University, Denmark. Trained as a pure mathematician, his research interests turned towards mathematical modelling and mathematical education. Today, his main field of research is mathematics education, especially the justification problem in mathematics education; applications and modelling in the teaching and learning of mathematics; assessment; the nature of mathematics education research as a scientific discipline; and mathematical competencies in mathematics education. He has been member of several committees, including the Executive Committee of ICMI, 1987–1998 (Secretary General 1991–1998), the ICMI Awards Committee (chair 2008–2011), and the Education Committee of the European Mathematical Society.

Jarmila Novotná is Professor at Charles University in Prague and researcher in *LACES*, at Université Bordeaux Segalen. She was member of the IPC of the Fifteenth ICMI Study *Professional Education and Development of Teachers of Mathematics* and of the ICME 10 Survey Team, *The Professional Development of Mathematics Teachers*. She coordinated the PME 27/PMENA25 plenary Panel: *Navigating Between Theory and Practice*. She is a member of the Educational Committee of the European Mathematical Society. She is known for her work related to mathematics teacher training and to the transfer into practice of research results.

Peter Nyström is Assistant Professor in the Department of Applied Educational Science at Umeå University, in Sweden. He is currently serving as the mathematics expert for TIMSS in Sweden, and since 2004 has led a group that has been developing national tests in mathematics for upper secondary schools in Sweden. He is an active participant in, and a member of the board of, the Umeå Research Centre for Mathematics Education (UMERC). Although his research is mainly in the field of educational assessment with a special focus on mathematics and science, he has also taken interest in issues such as curriculum, ability grouping, and the teaching of mathematics.

Neil A. Pateman is Professor in the Department of Curriculum Studies at the University of Hawai'i at Mānoa. His research interests are teaching and learning in mathematics classrooms, and socio-cultural dimensions of mathematics education. He is currently working with others on a project to recover traditional mathematical knowledge in Pacific island entities with a view to developing curriculum for schools in those entities in order to keep the traditional practices alive.

Arthur B. Powell is Associate Professor, Department of Urban Education at Rutgers, the State University of New Jersey. He has taught in Brazil, China, and Mozambique, as well as in the USA, has authored or edited five books, and has published numerous articles. His research areas include ethnomathematics, critical mathematics, analysis of curriculum materials, subordination of teaching to learning in mathematics, and collaborative mathematical problem solving with technology. He created the NGO, Elevating Learning above Teaching (ELAT), to support professional development projects for teachers of elementary schools in Haiti. In 2003, he co-founded the Bronx Charter School for Better Learning (BCSBL) in New York City, and there he conducts professional development workshops for teachers.

David Lindsay Roberts is Adjunct Professor at Prince George's Community College, in Maryland (USA). He has an M.A. in mathematics from the University of Wisconsin-Madison and a PhD. in the history of science from Johns Hopkins University. His research focusses on the history of mathematics education in the nineteenth and twentieth centuries. He has received the Lester R. Ford Award for expository writing on mathematics from the Mathematical Association of America, and was co-author, with Peggy Aldrich Kidwell and Amy Ackerberg-Hastings, of *Tools of American Mathematics Teaching, 1800–2000*, published by the Johns Hopkins University Press in 2008.

Ornella Robutti is Associate Professor in Mathematics Education in the Department of Mathematics of the University of Torino, in Italy. She obtained her Masters degree in Mathematics in 1984 and her Masters degree in Physics in 1989 at the University of Torino. She is involved in teacher education in national projects in Italy (m@t.abel, PON), and is in charge of the GeoGebra Institute of Torino and of the project DIFIMA. She is the author of many publications in the field of mathematics education as well as high school books on physics and mathematics. Her main field of research is teaching and learning processes in mathematics with the support of technologies.

Leo Rogers is a founder member of the British Society for the History of Mathematics and founder of the International Study Group on the History and Pedagogy of

Mathematics (HPM). He has taught in Primary and Secondary schools in England, and as a trainer of teachers, worked with pupils and teachers in a number of European Community curriculum and research projects. His principal interests are the historical, philosophical and cultural aspects of mathematics as they relate to the development of curricula, mathematical pedagogies, and individual learning. When not involved with education, he dances the Argentine Tango.

Ana Isabel Sacristán has been a full-time researcher in mathematics education at the Center for Research and Advanced Studies (Cinvestav) in Mexico City, Mexico, since 1989. She received her PhD from the Institute of Education, University of London in 1997. Her main area of research is the use of digital technologies in the teaching and learning of mathematics and, more recently, professional development of teachers aimed at assisting them to incorporate digital technologies into their practice. She has participated in several national and international projects and committees in these areas, including as a member of the program committee of the ICMI 17 study group.

Bernard Sarrazy is Professor of Educational Sciences at the University of Bordeaux Segalen, France. He was director of the *Département des Sciences de l'Education* for 2 years and at present is director of the *Laboratoire Cultures, Education, Sociétés* (LACES). In his research he focusses on the study of phenomena of mathematics education at the intersections of post-structuralist anthropology and the theory of didactical situations. He pays special attention to anthropological and social determinations of the relationships of pupils towards the didactical contract.

J. Michael Shaughnessy, who is Professor, Department of Mathematics and Statistics at Portland State University, has recently served a 4-year term as President of the National Council of Teachers of Mathematics. He has taught mathematics courses and directed professional development programs for teachers from K–12 through university level. He has authored over 70 scholarly articles, books, and book chapters on issues in mathematics education. His principal research interests have been the teaching and learning of statistics, probability, and geometry. He was a member of the Mathematics Department at Oregon State University from 1976 to 1993, and subsequently joined the Department of Mathematics and Statistics at Portland State University.

Yoshinori Shimizu is Professor of Mathematics Education within the Graduate School of Comprehensive Human Sciences, University of Tsukuba. He taught at Tokyo Gakugei University, one of the largest national institutions for teacher education in Japan, for 15 years before joining University of Tsukuba in 2005. His primary research interests include international comparative studies of mathematics classroom instruction and student assessment. He was a member of the Mathematics Expert Group for OECD/PISA 2003, 2006, and 2009, and was one of the founders of Learner's Perspective Study (LPS), an international comparative study on mathematics classrooms. He is the Japanese team leader for LPS.

Nathalie Sinclair is Associate Professor in the Faculty of Education at Simon Fraser University in Canada. Before that, she worked in the Department of Mathematics at Michigan State University. Her primary research interests focus on the consequences of embodied cognition in mathematics thinking and learning. She studies the role of the aesthetic in the development of mathematics as a discipline and in the understandings of both research mathematicians and school learners, and investigates the ways in which digital technologies, and dynamic geometry software in particular, change the way people think, move and feel mathematically.

Parmjit Singh is Associate Professor within the Faculty of Education, University Technology Mara, Malaysia. He obtained his PhD in Mathematics Education from Florida State University in 1998, and his research interests focus on children's learning and development in mathematics with a specific interest in cognitive processes. He has been a principal investigator on several funded research projects such as Gaps in Children's Mathematical Learning, Multiplicative Thinking Structures, Problem Solving among College Students, Children's Numeracy, and Key Performance Indicators (KPI) for Public Universities in Malaysia. He has given numerous paper presentations and gained research publications in local, national, and international arenas and has authored several books aimed at the primary, secondary and tertiary levels of mathematics education.

Bharath Sriraman is an Indian-born academic editor, mathematician, and educator best known for his contributions to theory development in mathematics education and in gifted education. He is Professor of Mathematics at The University of Montana—with a secondary appointment in the Department of Central Asian Studies where he offers courses in Indo-Iranian studies. He travels and collaborates extensively with colleagues at institutions in Norway, Sweden, Denmark, and Iceland, in addition to Cyprus, Germany, Turkey, Iran, Australia, and Canada, and supervises doctoral students from these nations. He is the founder and editor-in-chief of *The Mathematics Enthusiast*. E-mail: sriramanb@mso.umt.edu

Kaye Stacey is Foundation Professor of Mathematics Education at the University of Melbourne. She works as a researcher and teacher educator, training teachers for both primary and secondary schools and supervising graduate research. Her research interests centre on mathematical thinking and learning, problem solving and the mathematics curriculum, particularly the challenges and opportunities that arise in adapting to the new technological environment. Her research work is renowned for its high engagement with schools. Her doctoral thesis from the University of Oxford, was in number theory. She has a Centenary Medal from the Australian government for outstanding services to mathematics education.

Mike Thomas is Professor in the Mathematics Department at The University of Auckland, New Zealand. His research interests are in using technology to improve learning, developing theories of advanced mathematical thinking, the learning and teaching of calculus and undergraduate mathematics, school and university teaching, and connections between mathematics education and cognitive neuroscience. He has given invited research seminars in a number of countries and is on the editorial boards of *Mathematics Education Research Journal* and the *International Journal of Mathematical Education in Science and Technology*. He has recently leading a survey team for the 2012 International Congress on Mathematical Education (ICME), on the mathematical difficulties inherent in the transition from school to university.

Luc Trouche Professor at the French Institute for Education (Ecole Normale Supérieure de Lyon, France), is head of the research department. His work was previously dedicated to the didactical study of ICT integration in mathematics education, considering the interplay between instrumentation and conceptualization processes. He introduced the notion of orchestration to model the didactical management of available artefacts in a classroom. This led him, in a joint work with Ghislaine Gueudet, to introduce a documentary approach of didactics, analyzing teacher development as an interplay between practice, individual and collective, and resources.

David Wagner is Associate Dean and a Mathematics Education Associate Professor in the Faculty of Education of the University of New Brunswick, Canada. He is most interested in human interaction in mathematics and mathematics learning and the relationship between such interaction and social justice. This inspires his research, which has focussed on identifying positioning structures in mathematics classrooms by analyzing language practices, on ethnomathematical conversations in Mi'kmaw communities, and on working with teachers to interrogate authority structures in their classrooms. He currently serves as managing editor on the board of directors of the journal *For the Learning of Mathematics* and as a member of the Nonkilling Science and Technology Research Committee. He has published articles in various journals, hosted conferences, led working groups at a number of conferences, taught in five countries and co-edited a book in Springer's Mathematics Education Library series.

Margaret Walshaw is Professor in the School of Curriculum and Pedagogy at Massey University, New Zealand. She is Co-Director of the Centre of Excellence for Research in Mathematics Education at her University, and as Research Director coordinates, for her college, the professional doctorate in education. Her overriding research interest is in making connections between new theories of the social and mathematics education. In this interest she has written and edited several books and published in a wide range of journals.

Tine Wedege is Professor in the Faculty of Education and Society, Malmö University, Sweden, where she teachers mathematics teacher education. During 2005–2010 she was also Professor in the Department of Mathematical Sciences, Norwegian University of Science and Technology, Norway. Until 2005, she was Associate Professor at Roskilde University, Denmark, where she defended her doctoral thesis in 2000. She has written and/or edited more than 100 scientific publications, and is a member of the editorial committee of *Nordic Studies in Mathematics Education*. Internationally, she has been active in the Adults Learning Mathematics (ALM) research forum since 1994, and has been on the editorial board of ALM's international journal.

Allan Leslie White is Associate Professor in Mathematics Education, within the School of Education at the University of Western Sydney (UWS). Before UWS he taught for over 20 years in primary and secondary schools in three Australian states and during that time held positions of Principal, Head of Mathematics, Boarding Director. He has worked with intellectually handicapped and emotionally disturbed students, and has a long history of working throughout South East Asia in the areas of mathematics education, research, and teacher professional development. He is currently working in, or has worked in, Brunei Darussalam, Cook Islands, Indonesia, Japan, Malaysia, New Zealand, The Philippines, Singapore, Thailand, and Taiwan.

Dylan Wiliam is Emeritus Professor of Educational Assessment at the Institute of Education, University of London where, from 2006 to 2010 he was its Deputy Director. In a varied career, he has taught in urban public schools, directed a large-scale testing program, and served a number of roles in university administrations, including Dean of a School of Education. For the last 15 years, his scholarly work has focussed on supporting teachers to develop their use of assessment in support of learning—sometimes called formative assessment or assessment for learning.

Gaye Williams is a Lecturer at Deakin University, and an Australian Research Council Post-doctoral Fellow hosted by the International Centre for Classroom Research at the University of Melbourne. She is also a member of the Learner's Perspective Study (LPS). Her research foci are influenced by her many years as a secondary mathematics teacher, provider of professional learning in primary and secondary schools, and by her participation in curriculum development and assessment projects at local, state and national levels. She has visited classrooms in Australia, Japan, Germany, the USA, the Philippines, South Africa, and Sweden, and analysed lesson videos from these countries and from China. Collaborations with local researchers have enriched her understanding of intercultural contexts.

Julian Williams is Professor of Mathematics Education within the University of Manchester's School of Education. He began his professional career teaching mathematics in comprehensive schools. His work increasingly focussed on research and research supervision in mathematics education. His interests have always been not only in modelling and problem solving but also in learning, assessment and teaching mathematics, situated and embodied intuition, workplace and school–college/university transitions, and widening participation in mathematics. His recent theoretical work deals with cultural-historical psychology/activity theory, semiotics, identity and Bourdieusian sociology. He is currently attempting to synthesize cultural psychology in the educational field as a localization of capitalism.

Margaret Wu has a background in statistics and educational measurement. She has worked as a researcher in educational institutes, as well as a mathematics teacher in schools. Margaret has been involved in a number of international studies including PISA and TIMSS, taking on the roles of data analyst and test developer. She grew up in Taiwan but has lived in Australia for many years. She has a keen personal interest in international comparative studies, and her main area of research is in the development of item response models. She is a co-author of an item response modeling software program, ConQuest.

Keiko Yasukawa is Lecturer in Adult Education at the University of Technology, Sydney. Keiko has taught in mathematics, engineering and adult education departments in Australian universities. She now teaches and coordinates adult literacy and numeracy teacher education courses, and conducts research in the areas of critical mathematics/numeracy, adult literacy and numeracy pedagogies and policy, and workplace literacy and numeracy.

Rose Mary Zbiek is Professor of Mathematics Education and in charge of the Mathematics Education Program at The Pennsylvania State University. She is Series Editor for the Essential Understanding book series published by the National Council of Teachers of Mathematics. A past Chair of the International Committee for Computer Algebra in Mathematics Education, her research interests include the use of technology in learning and teaching mathematics and the development and employment of teacher understanding of mathematics in technology-intensive environments. In addition, she has developed secondary school mathematics curriculum materials which capitalize on mathematics technology.

Names of Reviewers

Every *Handbook* chapter was reviewed by at least two independent reviewers, appointed by the editorial team. In addition, each chapter was read critically by the appropriate section editor and by Ken Clements. The editors thank the following persons who served as reviewers.

Section	Name of Reviewer	Reviewer's Address
A	Abraham Arcavi	Weizmann Institute of Science, Israel
D	Michèle Artigue	Université Paris Diderot, France
A	Bill Atweh	Curtin University, Australia
A	Takuya Baba	The University of Hiroshima, Japan
B	Patricia Baggett	New Mexico State University, USA
B	Heinrich Bauersfeld	Universität Bielefeld, Germany
B	Andy Begg	Auckland University of Technology, New Zealand
D	Kristín Bjarnadóttir	Iceland University, Iceland
D	Jo Boaler	Stanford University, USA
C	Peter Boon	University of Utrecht, The Netherlands
C	Nigel Calder	The University of Waikato, New Zealand
A	Marta Civil	The University of Arizona, USA
A	Barbara Clarke	Monash University, Australia
B	Doug Clarke	The Australian Catholic University, Australia
D	Patricia Cline Cohen	The University of California, Santa Barbara
C	Damien Debell	Boston College, USA
C	Michael de Villiers	Kennesaw State University, USA
A	Jan Draisma	Eindhoven University of Technology, The Netherlands
A	Renaud d'Enfert	Université Paris Sud 11, France
A	Paul Ernest	Formerly, The University of Exeter, UK
B	Ruhama Even	Weizmann Institute of Science, Israel
A	Helen Forgasz	Monash University, Australia
C	Josep Fortuny	Universitat Autònma de Barcelona, Spain
D	Michael Fried	Institute for Applied Research, Israel
C	Rossella Garuti	Università di Genova, Italy
C	Vince Gieger	Australian Catholic University, Australia
A	Simon Goodchild	The University of Agder, Norway
D	Kian-Sam Hong	Universiti Malaysia Sarawak, Malaysia
D	Maitree Inpasitha	Khon Kaen University, Thailand
C	Gabrielle Kaiser	University of Hamburg, Germany
B	Tom Kieren	The University of Alberta, Canada
C	Jean-B. Lagrange	Université Paris 7, France
C	Zsolt Lavicza	The University of Cambridge, UK

Section	Name of Reviewer	Reviewer's Address
D	Gilah Leder	Monash University, Australia
A	Stephen Lerman	London South Bank University, UK
B	Salvador Llinares	University of Alicante, Spain
B	Francis Lopez-Real	The University of Hong Kong
A	Juergen Maasz	The University of Linz, Austria
D	George Malaty	The University of Joensuu, Finland
A	Ramakrishnan Menon	Georgia Gwinnett College, USA
B	Fayez M. Mina	Ain Shams University, Egypt
D	Candia Morgan	The University of London, UK
A	Judit Moschkovich	The University of California, Santa Cruz
A	Christopher Ormell	Formerly, The University of East Anglia, UK
D	Minoru Ohtani	Kanazawa University, Japan
A	Birgit Pepin	University College Sør-Trøndelag, Norway
B	Andrea Peter-Koop	Carl von Ossietzky Universität Oldenburg, Germany
C	Rui Pimenta	Instituto Politécnico do Porto, Portugal
B	Joao Pedro da Ponte	Universidade de Lisboa, Portugal
C	Renate Retkute	The University of Nottingham, UK
C	Teresa Rojano	CINVESTAV, Instituto Politécnico Nacional, Ministry of Education, Mexico
C	Kenneth Ruthven	The University of Cambridge, UK
B	Wee Tiong Seah	Monash University, Australia
A	Mamokgethi Setati	The University of South Africa, South Africa
B	Yasuhiro Sekaguchi	The University of Alicante, Spain
B	Anna Sfard	The University of Haifa, Israel
B	Mihaela Singer	The University of Ploiesti, Romania
A	Ole Skovsmose	Aalborg University, Denmark
B	Judy Sowder	San Diego State University, USA
C	Peter Sullivan	Monash University, Australia
D	David Tall	The University of Warwick, UK
B	Zalman Usiskin	The University of Chicago, USA
B	Pongchawee Vaiyavutjamai	Chiang Mai University, Thailand
D	Paola Valero	Aalborg University, Denmark
B	Marja van den Heuvel-Panhuizen	Utrecht University, The Netherlands
D	Renuka Vithal	University of Kwazulu-Natal, South Africa
B	Jeannette Vogelaar	UNESCO, France
A	Anne Watson	Oxford University, UK
C	Ngai-Ying Wong	Chinese University of Hong Kong, Hong Kong
D	Ban Har Yeap	Marshall Cavendish Institute, Singapore
B	Orit Zaslavsky	New York University, Steinhardt, USA

Author Index

A
Aaron, Wendy Rose, 55
Abrahamson, Louis, 746
Abramov, Alexander, 807
Abrantes, Paulo, 506
Abu Zahari bin Abu Bakar, 12
Acevedo, Carmen Gloria, 863
Achiron, Marilyn, 839
Acioly-Régnier, Nadja, 101, 107–111
Ackerberg-Hastings, Amy, 27, 530–533, 535, 536, 538, 539
Adam, Shehenaz, 170, 178, 182
Adams, Barbara L., 181, 183, 193
Adams, Jennifer, 79
Adams, Raymond, 839
Adler, Jill, 21, 58, 80, 81, 120, 149, 328–331, 364, 399, 400, 412, 432, 479, 815, 816, 821, 850, 1019, 1034
Adorno, Theodor, 45
Agalianos, Angelos, 757, 765
Agatston, Patricia Walton, 706
Agudelo-Valderrama, Cecilia, 336, 393, 394, 417, 418
Agudo, J. Enrique, 654
Aguilar, Miguel, 884
Aguirre, Julia M., 89
Ahlfors, Lars Valerian, 285
Ahmed, Ayesha, 729
Ainley, Janet, 572, 671
Aitchison, John J. W., 206
Akamatsu, C. Tane, 79, 82
Akiba, Motoko, 244, 250
Alasuutari, Pertti, 1030
Alatorre, Silva, 209
Albers, Donald J., 802
Alcaide, Solidad, 876
Aldon, Gilles, 539
Alexanderson, Gerald L., 802
Alexiadou, Nafsika, 1020
Allal, Linda, 118, 728
Alldredge, J. Richard, 652
Allen, Ann Taylor, 534
Almond, Russell G., 722, 734
Alonso, F., 620
Alonso, Orlando B., 807
Alrø, Helle, 61, 184, 185
Alston, Alice S., 92, 364, 375, 377, 415
Altbach, Philip G, 852
Althusser, Louis, 45, 56, 468–470
Altrichter, Herbert, 346, 364, 368, 382
Altschuld, James W., 249
Alvarez, A., 876
Álvarez, P., 876
Ameis, Jerry A., 538
Amit, Miriam, 849
An, Shuhua, 814
Ander-Egg, Ezequiel, 348
Anderson, Judy O., 399, 875, 888, 917
Anderson, Maria H., 532
Andersson, Annica, 186–189, 191, 192
Andrade, Fernanda, 108
Andrew-Ihrke, Dora, 170
Andrews, Jane, 90
Andrich, David, 148
Angell, Carl, 981, 989
Anku, Sitsofe, 846
Annetta, Leonard A., 705
Ansell, Ellen, 77
Anstrom, Terry, 885, 893
Anthony, Glenda, 45, 120, 176
Antonini, Alberto Severo, 541
Anyon, Jean, 468, 470
Apel, Naomi, 644, 667, 668
Apple, Michael, 45, 352

Appleby, Yvon, 222
Araújo, A., 220, 224, 226
Araújo, Cláudia Roberta, 107, 108
Araújo, J. L., 118
Arbaugh, Fran, 336, 416
Arcavi, Abraham, 575, 620, 622
Arffman, Inga, 1018, 1031
Arias, Rosario Martinez, 873
Arifin, Achmad, 11
Arnold, Matthew, 953
Arnold, Stephen, 14
Arnove, Robert F, 852
Arora, Alka, 150, 989, 1022, 1031
Artelt, Cordula, 1027
Artigue, Michèle, 303, 305, 306, 310, 313, 321, 447, 518, 539, 599, 602–605, 613, 630, 631, 762, 770, 907–909, 917, 923–925, 928
Artiles, Alfredo J., 87
Arzarello, F., 49, 51, 281, 287, 574–576, 579, 580, 587, 591, 592, 622, 699, 904, 910, 911, 918, 919
Asante, Molefi Kete, 841
Ascher, Marcia, 112, 177, 529
Ascher, Robert, 112, 177, 529
Ash, Katie, 532
Ashburner, John, 81
Ashfield, Jean, 532
Ashton, Helen S., 731, 732
Asiala, Mark, 309
Askew, Janice M., 1013
Askey, Richard, 489
Asp, Gary, 744
Assude, Teresa, 762
Astala, Kari, 1030
Atkinson, Robert D., 969
Atweh, Bill, 74, 121, 173, 382, 693, 798, 800, 813, 818, 902, 907, 908, 911, 953, 968
Ausejo, Elana, 806
Awtry, Thomas, 362
Azevedo, A., 220, 224, 226

B

Baccaglini-Frank, A., 573
Back, Jenni, 120
Bacon, Harold M., 285
Bacon, Lili, 375
Bailey, David H., 574
Baker, David P., 213, 798, 819, 820
Baker, Elizabeth K., 214, 217, 219, 221, 222, 234, 883
Bakhtin, Mikhail, 46

Bakker, Arthur, 205, 212, 215, 216, 551, 554, 643, 644, 649, 660, 667, 668, 670, 675, 679–681
Balacheff, Nicolas, 296, 583, 602, 621, 828
Balatti, Josephine, 179
Baldino, Roberto R., 73, 121
Ball, Deborah, 121, 328–331, 364, 395, 399, 400, 402, 419, 432, 466, 629–632, 771, 815, 816, 821, 950
Ballard, Keith, 90
Banerjee, Rakhi, 963
Bao, Jiansheng, 377–379, 382, 386
Bapoo, Abdool, 80, 81
Barabash, Marita, 697, 698
Barbeau, Edward J., 921
Barber, Michael, 880
Barbosa, Jonei Cerqueira, 119
Barkatsas, Anastasios, 743, 916
Barkatsas, T., 402
Barnard, Henry, 15
Baron, Georges-Louis, 89
Baron, Stephen, 477
Barrantes, H., 912, 913
Barrantes, Manuel, 333
Barrier, E., 1012
Barta, J., 171, 179
Bartolini Bussi, Maria G., 77, 287, 534, 539–541, 553, 557, 573
Barton, Angela Calabrese, 798, 902, 953
Barton, Bill, 53, 61, 106, 112, 121, 177
Barton, David, 222
Baruch, Rachel, 697, 698
Barwell, Richard, 51–53, 336
Bass, Hyman, 629, 631, 804, 904–906
Bassells, F., 876
Bassey, Michael, 403
Bastable, Virginia, 963
Batanero, Carmen, 930
Batarce, Marcelo Salles, 121
Battey, Daniel, 89
Battista, M. T., 147, 574, 575, 577, 578, 586
Baudrillard, Jean, 50
Bauersfeld, Heinrich, 285, 291, 294, 295, 979
Baulac, Yves, 571
Baumert, Jürgen, 336, 1027
Bazzini, Luciana, 350, 364
Beach, King D., 210, 224
Beaton, Albert E., 509, 954, 980, 987, 1012, 1022, 1031
Beatty, Ruth, 699
Becker, Jerry, 30, 984, 995, 996
Becker, Richard A., 646
Bednarz, Nadine, 373–375, 381, 384, 385, 628
Beevers, Cliff E., 731, 732

Author Index

Beevers, J., 731
Begehr, Astrid, 849, 851
Begle, Edward G., 26, 284, 288, 296, 297, 811, 812
Behr, Merlyn J., 306, 307
Bejar, Isaac I., 723
Bekdemir, Mehmet, 344
Belfiore, Mary Ellen, 223
Bell, Charles B., 285
Bellemain, Franck, 571
Bellman, Richard E., 285
Beman, William, 28
Benavot, Aaron, 9, 11, 16
Benjamin, Harold R. W., 9
Benke, Gertraud, 382
Benn, R. Roseanne, 209, 210
Bennett, Colin J., 1009, 1019–1021, 1023–1025, 1032, 1033, 1035
Bennett, Randy Elliot, 736
Bennison, Anne, 630, 632
Ben-Yehuda, Miriam, 316
Ben-Zvi, Dani, 644, 645, 649, 651, 653, 667, 668, 670, 671, 675, 679, 680, 682, 952
Berger, Margot, 293
Berger, Peter, 45
Berglund, Lars, 528
Bergsten, Christer, 42–44
Berliner, David C., 72, 453
Berman, Edward H., 852
Berman, Simon L., 531
Bernáldez, M., 762
Berne, Jennifer, 437
Bernet, Thé, 911
Bernstein, Basil, 45, 104, 127, 128, 213, 311, 312, 314, 553
Berry, J. M., 76, 83, 709
Berry, Robert Q., 76
Bers, Lipman, 285
Bertrand, Lorna, 839
Bessot, Annie, 17, 437
Beyer, Landon, 866
Bhanot, Ruchi, 519
Bharucha, Jamshed, 965
Bhattacherjee, Anol, 654
Bianchi, Leonard J., 1012
Bibby, Tamara, 59
Biber, George Edward, 15
Bicudo, Maria A. V., 217, 219
Biddy, Tamara, 176, 181, 192, 195
Bidwell, James K., 1010
Biehler, Rolf, 296, 644, 645, 647–651, 656–658, 660, 663, 667, 671, 672, 674–676, 680, 934

Bigott, Luis, 348, 350
Biguenet, A., 284, 286, 293
Biklen, Sariknapp K., 333
Binet, Alfred, 18
Binkley, Marilyn, 737
Binterová, Helena, 849
Birchal, T. S., 130, 131
Birkhoff, Garrett, 285
Bishop, Alan J., 7, 9, 21, 33, 57, 71, 75, 79, 104, 106, 113–118, 146, 170, 171, 177, 179, 210, 248, 266–268, 295–297, 402, 461, 572, 792, 805, 812, 817, 853, 916, 926, 928
Björnssen, Julius, 735, 737
Bjuland, Raymond, 371, 408
Black, Laura, 52, 468, 556
Black, Max, 551
Black, Paul J., 724, 747
Black, Stephen, 223, 225
Blainey, Geoffrey, 842
Blanco, García M., 932
Blanco, Lorenzo J., 333, 873
Blanton, Maria, 473, 573
Blessing, Axel M., 773
Block, Jeanne Humphrey, 19
Bloem, Simone, 839
Blömeke, Sigrid, 21, 845
Blomhøj, Morten, 930, 931
Blum, Werner, 336, 550, 551, 557, 880, 920, 960, 1014
Blume, Glendon W., 540, 599, 601, 602, 628, 629, 631
Blumer, Herbert, 45
Blunden, Andy, 553
Boal, Augusto, 713
Boaler, Jo, 57, 109, 115, 116, 120, 205, 468, 470
Bobis, Janet, 399, 409, 917
Boero, Paolo, 575, 583, 914
Boerst, Timothy, 147
Bogdan, Robert C., 333
Bohl, Jeffrey V., 464, 474
Bohl, Thorsten, 1030
Böhm, Josef, 599, 616
Boiron, Marika, 839
Bond, M., 207, 223, 224, 226
Boneh, Tal, 735
Bonfil, Guidlermo, 346
Bonne, Linda, 402
Bonnet, Gabrielle, 863
Boohan, Richard, 662
Boon, Marko, 745
Boon, Peter, 592, 769
Boondao, Sakorn, 218–219, 234

Boorstin, Jon, 712, 713
Booth, Shirley, 578
Borba, Marcelo C., 113, 121, 177, 287, 592, 696, 699, 701–703, 706–708, 711–714, 798, 844, 902, 953
Borda, Orlando Fals, 346–348
Bordo, Susan, 476, 477
Borel, Émile, 29, 286, 535, 540
Borgioli, Gina M., 87
Borgonovi, Francesca, 839
Boruch, Robert, 830
Borwein, J. David B., 574
Bosch, Marianna, 117, 118, 313
Bosley, Jennifer H., 631, 632
Bottani, Norberto, 860
Bottge, Brian, 736
Bourbaki, Nicolas, 308, 309
Bourdage, Nicole, 375
Bourdieu, Pierre, 45, 104, 127, 128
Bowen, A., 327–328
Bowers, Janet, 115, 658
Boyask, Ruth, 477
Boylan, Mark, 337
Braams, Bastian J., 487
Bracey, Gerald W., 245, 253, 862
Bradburn, Norman M., 954
Brady, Kate, 173
Braga, S. M., 109
Brantlinger, Andrew M., 184
Braun, Henry, 839
Breaux, G. A., 758
Breen, Chris, 49, 346, 400–402, 439, 914
Breiteig, Trygve, 371, 408
Brenner, Margaret E., 111, 171, 172
Brenner, Margaret W., 18
Brickman, William, 799, 802
Bright, George W., 336, 436
Brinek, Gertrude, 1013
Brito, Arlete de Jésus, 808
Britt, Murray S., 963
Britton, Edward D., 988, 1012
Britzman, Deborah, 46, 477
Brock, William H., 27, 535
Bromme, Rainer, 364
Brooks, Edward, 18, 25
Brossard, Alain, 104
Brousseau, Guy, 4, 33, 117, 208, 295, 296, 396, 431, 433, 438, 440, 442, 444, 451, 453, 918
Brown, Annette M., 56, 57, 556
Brown, Gordon, 961
Brown, John Seely, 115
Brown, L., 345, 439, 449
Brown, Margaret, 205

Brown, Raymond, 171, 741, 742, 764
Brown, Sally A., 405
Brown, Tony, 49, 57, 120, 203, 221, 459, 464, 467, 469, 470, 765
Brownell, William A., 10, 17, 19, 31
Bruckheimer, Maxim, 575
Bruner, Jerome S., 22, 23, 103, 115, 284, 292, 293, 434, 552, 806
Brunner, Martin, 336
Bruns, Barbara, 12, 13
Bryans, Martha B., 472
Bshouty, Daoud, 181
Buchberger, Bachmair B., 602
Buerk, Dorothy, 222
Buisson, Ferdinand, 904
Bukarau, Jared, 382
Bull, Rebecca, 79
Bulmer, Michael, 745
Bunar, Nihad, 839
Burghes, David, 1014
Búrigo, Elisabete Zardo, 808
Burkhardt, Hugh, 551, 952
Burnham, William H., 17, 18
Burrill, Gail, 644, 645, 671, 675, 930
Burstein, Leigh, 838, 1012
Burton, Leone, 124, 210, 331, 347
Burton, Warren, 531
Bussi, Bartolini Maria G., 77, 287, 534, 539–541, 553, 557, 573
Buswell, Guy T., 17, 31
Buteau, Chantal, 762
Butts, R. Freeman, 530
Buxton, Laurie, 210

C
Cabral, Tania C., 73
Cahalan, Cara, 736
Cahnmann, Melissa S., 179
Cai, Jinfa, 153, 465, 473, 949, 950, 958, 961, 963–966, 969
Cairn, William D., 862
Cajori, Florian, 17, 292, 531
Cakir, Murat Perit, 705, 708
Calder, Nigel, 767
Callingham, Rosemary A., 679, 745
Campbell, Stephen, 321
Campedelli, Luigi, 284, 286, 293
Campos, Celso R., 183
Canady, Herman George, 103
Cannon, Joanna, 150
Cantoni, Gina, 170
Cantoral, Ricardo, 348
Carlsen, Martin, 52

Carnoy, Martin, 852, 1009, 1015, 1020
Carpenter, Thomas P., 436
Carr, Wilfred, 404, 450
Carraher, David W., 104, 107, 154, 172, 179, 205, 209, 211
Carraher, Terezhinha N., 107, 296
Carreira, Susana, 109, 185–187, 218, 224
Carrillo, Sandra, 863
Carter, Glenda, 473
Carter, Kevin, 339
Caruso, J. Borreson, 695
Carvalho, Amelia, 756
Carver, Robert P., 739
Casassus, Juan, 1015
Cassundé, Maria A., 107
Castells, M., 714, 715
Castelnuovo, E., 284, 286, 293
Castro, Manuel, 863
Cavanagh, Michael, 436
Cazes, Claire, 699, 730, 763
Cedillo, Tenoch, 621
Cekmez, Erdein, 575
César, Margarida, 146, 171
Chambers, Donald L., 1024
Chambers, John M., 646
Chan, Yip Cheung, 572
Chance, Beth L., 644, 648, 649, 651, 675, 740
Chang, Peichin, 52
Chantarasonthi, U-Savadee, 218–219, 234
Chaplin, George, 130
Chapman, Chris, 695
Chapman, K., 259
Chapman, Olive, 327, 341, 394, 405, 449
Charles, Elizabeth S., 705
Charlot, B., 806
Chatterji, Madhabi, 1018
Chazan, Daniel, 466, 628
Cheah, Ui Hock, 861
Chen, S., 884
Cheng, Yin Cheong, 467
Cheung, Wai Ming, 467
Chevallard, Yves, 117, 118, 313, 321, 434, 784
Chew, Cheng Meng, 258
Chiappini, Giampaolo P., 626
Chick, Helen L., 725, 735, 963
Chiew, Chin Mon, 415
Childs, Carla P., 104
Chinnappan, Mohan, 341, 631, 632, 698
Chiu, M.-H. (Ming), 888
Chodkiewicz, Andrew, 219, 226
Chokshi, Sonal, 150
Choong, K. F., 257
Choppin, Jeffrey, 53
Choquet, Gustave, 284, 286, 292, 293, 805

Christiansen, Bent, 285, 289, 291, 295, 296, 956
Christiansen, Iben M., 53, 120
Christie, Michael J., 259
Christou, Constantine, 575
Chronaki, Anna, 53, 56, 120, 121
Chrostowski, Stephen J., 989, 1014, 1022, 1028, 1031
Chui, Angel Miu-Ying, 156, 158, 160
Churchhouse, Robert F., 287, 694, 757
Cirade, Gisèle, 784
Cirillo, Michelle, 52, 53, 369, 380, 383, 384
Civil, Marta, 79, 115, 172, 179, 180, 188, 192, 204, 219, 346
Clandinin, D. Jean, 339, 952
Clark, Stacy, 181
Clarke, Barbara, 402, 409, 916, 952
Clarke, David J., 21, 72, 120, 149, 155–159, 161, 409, 412, 444, 445, 459, 466, 467, 470, 476, 487, 509, 847, 848, 851, 852, 862, 867, 976, 977, 984, 985, 995, 1003
Clarke, Doug M., 952
Clarkson, Phillip C., 21, 80, 113, 115, 693, 694, 696, 699, 798, 800, 813, 818, 902, 907, 908, 911
Clark-Wilson, Alison, 632, 746, 969
Claro, Susana, 519
Clason, Robert G., 1010
Clay, Ellen, 701
Clements, M. A.(Ken), 10, 13–17, 21, 27, 29, 31, 57, 266, 267, 292, 296, 396, 410, 411, 531, 572, 811, 812, 837, 838, 840–842, 852, 853, 916
Clements, Niccolina, 839
Clerc, Benjamin, 772
Clycq, Noel, 106
Coballes-Vega, Carmen, 366
Cobb, George W., 644, 645, 740
Cobb, Paul, 42, 54, 115, 118, 146, 147, 153, 176, 249, 267, 401, 409, 494, 551, 553, 558, 560, 658, 778
Coben, Diana, 204, 205, 207, 208, 214, 215, 217, 219, 221, 222, 226, 232, 234, 728
Coburn, Cynthia, 509
Cochran-Smith, Marilyn, 384, 404, 439
Cockburn, Ann, 344–345
Cockroft, Wilfred H., 958
Codes, M., 884
Coe, Michael D., 529
Cogan, Leland S., 19, 20, 838, 983, 1012
Cohen, David K., 465, 469, 771, 950
Cohen, Patricia Cline, 31, 207, 531, 799
Colburn, Warren, 15, 16

Cole, Michael, 76, 77, 103, 104, 107, 109, 123, 212, 473, 566
Coleman, David, 1018
Coles, Alf, 345, 439
Collins, Allan, 115
Collins, Patricia Hill, 45
Colwell, Dhamma, 205
Comiti, Claude, 17
Confrey, Jere, 147, 401, 494, 550
Connell, Raewyn, 469
Connell, William, 25
Connelly, F. Michael, 339, 952
Connelly, Julianne, 818
Conner, Anna Marie, 576
Cooks, Jamal, 189
Cooney, Thomas J., 330, 1012, 1025
Cooper, Barry, 127, 128, 178
Cooper, Patricia, 394, 806
Coray, Daniel, 798, 905
Corbin, Juliet M., 333, 353
Cordasco, Francesco, 530
Cordy, Michelle, 707, 708
Corey, Douglas L., 382
Cornu, Bernard, 287, 756, 757
Cortes, Viviana, 51
Cosgrove, Jude, 839
Cossentino, Jacqueline, 536
Costa, W. G., 112, 125
Costilla, R., 863
Cottrill, Jim, 309
Coulange, Lalina, 437
Coupland, Mary, 124
Coupland, Nikolas, 315
Couture, Christina, 375
Coxford, Arthur F. Jr., 1023
Cramer, Kathleen, 306, 307
Crawford, Kathryn, 364, 400
Creemers, Bert, 364
Crespo, Sandra, 343
Cresswell, John W., 333, 334, 339, 839
Crinion, Jenny T., 81
Croon, Lucille, 72
Cross, Adney E., 808
Crosswhite, F. Joe, 1012, 1025
Cryer, Jonathan D., 651
Csapo, Beno, 839
Cuban, Larry, 951
Cubberley, Ellwood P., 10
Cuevas, Peggy Dickinson, 875
Cuisenaire, E.G., 536
Cuoco, Al, 628
Curbera, Guillermo P., 833
Cusato, Sandra, 1015
Cussó, Roser, 1019

D

Da Rocha Falcão, J. T., 107, 108
Daher, Wajeer, 519
Daly, Alan J., 507, 508
D'Ambrosio, Beatrice, 338
D'Ambrosio, Ubiratán, 73, 74, 104, 111, 112, 121, 147, 154, 177, 178, 205, 209–211, 248, 251, 291, 295, 296, 792, 808, 928
Damerow, Peter, 8, 9, 295, 805, 926, 928, 968
D'Amico, Sabrina, 1019
Dana-Picard, Thierry, 572, 621
Daniels, Harry, 54
Darling-Hammond, Linda, 950
David, Maria M., 120, 121
Davidenko, Susana, 179
Davidson, Michael, 839
Davidson, P. S., 530, 536
Davis, Brent, 51, 60
Davis, Ernest Kof, 179
Davis, Gary, 608
Davis, Joy C., 31, 809
Davis, Jon D., 342
Davis, Philip J., 52, 349, 468, 556
Davis, Robert B., 362
Davydov, Vasily V., 554–556
Dawes, Mark, 771
Dawson, Sandy, 396
Day, Christopher, 471
de Abreu, Guida, 54, 114, 115, 146, 147, 154, 170, 171, 179
de Andalucía, Junta, 877
De Andrade, S., 58, 59
de Costra, R. M., 575
De Freitas, Elizabeth, 52, 53, 57, 477
De Geest, Els, 773
de la Villa, Augustin, 620
De Lange, Jan, 26, 466, 981
De Moor, Ed. W. A., 292
De Silva, G. V. S., 346
de Villiers, Michael, 574–576, 582
Deaktor, Rachael, 875
Dean, Chrystal, 778
DeBell, Matthew, 695
Debien, Josianne, 366
DeBlois, Lucie, 336, 394
DeBoer, George E., 507
Defoe, Tracy A., 223
Delgado, Lorenzo M., 884
delMas, Robert C., 645, 675
Delors, Jacques, 13
delos Santos, Alan, 631, 632
DeLuca Fernandez, Sonia, 84, 87
Demana, Franklin, 539

Derrida, Jacques, 50
Derry, Jan, 667, 679, 680
Descartes, René, 352
Desgagné, Serge, 375
Desrosiers, C., 342
Devine, N., 465
Dewey, John, 24, 26, 706
Di Bucchianico, Alessandro, 745
Diaconis, Percy, 647
Dick, Thomas P., 600, 631
Dienes, Zoltan Paul, 22, 23, 293, 307, 536
Dieudonné, Jean, 284, 286, 292, 293
Díez-Palomar, Javier, 219
Dilger, Bernadette, 873, 1028
DiSessa, Andy A., 401, 679
Dixon-Román, Ezekiel, 56, 75
Doerr, Helen, 306, 307, 558
Dohrmann, Christian, 773
Doig, Brian, 148
Doku, Philip Atteh, 809
Dolciani, Mary P., 531
Domite, Maria de Carmo, 217, 219
Donne, Julie, 666, 667
Donoghue, Eileen F., 802, 821, 833
Donovan, M. Suzanne, 512
Doorman, L.Michiel, 30, 592, 658, 769
Dorier, Jean-Luc, 935
Dossey, John A., 1009, 1012, 1018, 1024, 1025
Douady, Régine, 433
Dowker, Ann, 86
Dowling, Paul, 45, 56, 57, 121, 125, 127, 128, 187, 330
Down, Phu, 850
Drake, Corey, 338
Drake, Pat, 56, 57
Dreyfus, Tommy, 157, 831, 952
Drijvers, Paul, 564, 572, 592, 599, 602, 608, 609, 612, 613, 738, 740, 753, 763, 764, 766, 768–770
Dubet, François, 122, 123, 127, 131
Dubinsky, Ed, 309
Duckworth, Eleanor, 92
Dufour-Janvier, Bernadette, 375
Duggan, Mary Anne, 698
Duguid, Paul, 115
Duit, Reinders, 880
Duncan, Allan G., 626, 633
Dunkel, Harold B., 26
Dunkley, Merv E., 8, 805, 968
Dunne, Diane Weaver, 878
Dunne, Máiréad, 127, 128
Durand-Guerrier, Viviane, 919
Durkheim, Émile, 44
Dussel, Enrique, 346

Duval, Raymond, 574, 605
Dyson, Alan, 87

E

Eaton, Jonathan, 1018
Ebbutt, David, 333
Edwards, Laurie, 49, 51, 699
Edwards, Michael Todd, 602
Efron, Bradley, 647
Ehrenfeucht, Aniela, 807
Eisenberg, Michael B., 695
El Sawi, Mohamed, 808, 809
Ellemor-Collins, David, 85
Ellerton, Nerida F., 14–17, 21, 27, 29, 256, 292, 531, 811, 812, 827, 837, 840–842, 845, 848, 852, 853
Elliott, Julian G., 403, 729, 1019
Ellis, Amy B., 466
Ellison, Nicole B., 695
Ellsworth, Elizabeth, 46
Ely, Robert, 303, 305
Emanuelsson, Jonas, 149, 155, 157, 160, 161, 847, 850, 852
Emerson, George B., 16
Enders, Craig K., 875
Engel, Arthur, 287
Engelbrecht, Johann, 697, 699, 701, 708, 714, 928
Engeström, Yrjö, 44, 210, 212, 555, 558
English, Lyn D., 42, 57, 296, 303, 304, 307, 312, 321, 397, 398, 936
Ensor, Paula, 54, 56
Erberber, Ebru, 150, 989, 1022, 1031
Erez, Michael M., 577
Erickson, Fredrick, 328, 333, 345
Ericsson, K. Anders, 159
Ernest, Paul, 56, 104–106, 121, 124, 125, 130, 184, 210, 246, 247, 330, 710, 818, 819
Ernie, K., 648
Ertl, Hubert, 873, 876, 879, 1018, 1028
Esmonde, Indigo, 179
Estrella, Gucci, 519
Etterbeck, Wallace, 156, 158, 160
Evans, Jeff, 50, 76, 109, 203, 204, 209, 210, 212, 213, 218, 220–222, 224–226, 232
Evans, Juliet, 839
Evans, Michael, 631
Even, Ruhama, 364, 394, 396, 400, 402, 419–421
Everson, Michelle G., 644, 652
Eysenck, Michael W., 732

F

Fairclough, Norman, 46, 170, 173
Fan, Lianghuo, 466, 473, 965
Fan, Y., 378, 386
Faragher, Rhonda, 561, 633
Faria, D., 113, 114
Farsi, Daryoush, 536
Fasanelli, Florence, 915
Fauvel, John, 915
Favilli, Franco, 146, 181
Fehr, Henri, 280
Feiman-Nemser, Sharon, 436
Feldman, Allan, 346, 364, 368, 382
Félix, Lucienne, 284, 911
Fennema, Elizabeth, 124, 245, 336, 436, 464
Fenstermacher, Gary D., 399
Fernandes, Anna Paula, 78, 92, 220, 224, 226
Fernandez, Clea, 150, 375, 376, 382, 414
Ferrara, Francesca, 699
Ferreras, Villanueva, 873, 877
Ferri, R. Borromeo, 550, 557
Ferrini-Mundy, Joan, 486, 489, 758, 862, 879
Fey, James T., 287, 599, 602, 628, 629, 806
Fiddes, David, 731
Fielding-Wells, Jill, 682
Fierro, Cecilia, 350
Figazzolo, Laura, 1034
Figueroa, M., 878, 879
Fink, Karl, 15
Finnigan, Kara S., 507, 508
Finzer, William F., 650, 654–656
Fischbein, Efraim, 293, 294, 913
Fischer, Hans E., 873
Fitzallen, Noleine, 632, 666
FitzSimons, Gail E., 113, 115, 204, 208, 209
Fletcher, Trevor J., 284, 286, 293
Floden, Robert E., 486
Flores, Alfinio, 652
Florio, S., 345
Flynn, Peter, 741, 744, 745
Folinsbee, Sue, 223
Folta, Elizabeth, 705
Fong, Ng Swee, 1032
Fonseca, Maria C. F. R., 112, 213, 217, 219, 225, 756
Foong, P. Y., 887
Ford, Donna Y., 106, 126
Ford, Timothy G., 485
Forgasz, Helen J., 71, 74, 75, 79, 124, 245, 402, 630–632, 743, 762, 798, 916, 917
Forman, Ellice, 77, 315, 316
Forsten, Char, 884
Fortoul, Bertha, 350

Foucault, Michel, 46, 474
Fowler, Frances C., 507
Foy, Pierre, 150, 989, 1022, 1031
Frackowiak, Richard S., 81
Frade, Cristina, 101, 108, 109, 113–117, 120
Franci, Raffaella, 14, 15
François, Karen, 32
Franke, Megan L., 336, 436
Frankenstein, Marilyn, 72–74, 112, 115, 121, 183, 184, 210, 248, 251
Frant, Janete B., 575
Fraser, Rosemary, 952
Frederickson, Ann, 667, 679
Freeman, R. Edward, 365
Freilich, Julius, 531
Freire, Paulo, 11, 210, 346, 348
Freudenthal, Hans, 30, 285, 288, 461, 540, 550, 558, 679, 805, 817, 862, 979
Fried, Katalin, 807
Fried, Michael N., 303, 305, 849
Friedkin, Shelley, 883
Friedlander, Alex, 767, 952
Friedman, Thomas L., 957
Friel, Susan N., 667
Froebel, Friedrich, 534–536
Froemel, Juan Enrique, 1015
Fromm, Erich, 45
Frost, Laurie A., 124
Frykholm, Jeffrey A., 336
Fuglestad, Anne Berit, 371, 408, 447, 780, 781
Fuhrmann, Susan H., 490
Fukisawa, Rikitaro, 834, 835
Fukuyama, Yoshihiro Francis, 820
Fullan, Michael, 363, 472
Furinghetti, Fulvia, 273, 279–281, 287–290, 798, 801, 834, 835, 904–906, 910, 911

G

Gadamer, Hans-Georg, 170
Gadanidis, George, 700, 703, 707–709, 711–713
Gaisman, Maria Trigueros, 591
Gal, Iddo, 208, 228, 229, 644, 645
Galant, Jaamiah, 54, 56
Galbraith, Peter L., 120, 550, 551, 557, 920, 960
Galia, Joseph, 150, 1022, 1031
Gallagher, Ann, 736
Gallannaugh, Frances, 87
Galligan, Linda, 118
Gallimore, Ronald, 156, 158, 160, 881, 983
Gallo, Melina L., 223
Gallos, Florenda Lota, 21, 149, 412, 850

Gamerman, Ellen, 152
Gamoran, Adam, 1034
Garcia, A., 620
Garcia, Eugene E., 80
Garcia, F., 620
Garcia, Ronald, 480
Garden, Robert A., 838, 951, 957, 979, 981, 989, 1012, 1014, 1022, 1028, 1031
Gardner, Howard, 104, 105, 706, 965
Garegae, Kgomotso Gertrude, 256, 257, 260
Garfield, Joan, 644, 645, 649, 651, 652, 675, 740
Garfinkel, Harold, 45
Garnier, Helen, 156, 158, 160
Garuti, Rossella, 575
Gascón, J., 117, 118, 313
Gates, Peter, 8, 9, 126, 127, 508, 511, 586
Gattegno, Caleb, 284, 286, 292, 293, 536
Gay, Geneva, 90
Gay, John, 109
Geach, Russell, 1014
Geary, D., 85
Gebre, Alemayehu H., 213
Gebremichael, A., 480
Gee, J. P., 114, 170
Geertz, C., 104
Geiger, V., 561, 633, 699, 707
Geist, P. K., 464
Gelis, J.-M., 626
Gellert, Uwe, 44, 55, 61, 177, 184, 327, 336, 337, 349, 352, 419, 926
Gerber, Sue, 697, 699
Gerdes, Paulus, 74, 113, 211, 295, 805, 926, 928
Gervasoni, Ann, 84, 86, 93
Gholson, Maisie L., 247, 248
Giacardi, Livia, 29, 279, 281, 287, 289, 290, 904, 910, 911
Gierl, Mark J., 488, 1024
Gifford, Sue, 85
Gigerenzer, Gerd, 215, 234
Gil, Einat, 644, 667, 668
Gil, Guillermo N., 1018
Gil, J., 333
Gil, Ignatio N., 782
Gilford, Dorothy M., 954
Gillborn, David, 106, 126
Gillespie, John, 230
Ginsburg, Alan, 885, 893
Ginsburg, Herbert P., 86
Ginsburg, Margery B., 9
Giri, Jason, 604, 629
Giroux, Henry A., 349
Gispert, H., 798

Givvin, K. B., 156, 158, 160, 880, 881
Glas, Eduard, 27
Glaser, Barney G., 333
Glaymann, Maurice, 280
Glevey, Kwame E., 126, 127
Glick, Joseph A., 109
Goba, Busi, 850
Godfrey, David, 607, 608
Goetz, Judith P., 333
Goffmann, Erving, 45
Goldenberg, Michael Paul, 464, 465
Goldstine, Harvey H., 539
Gomes, Nilma Lino, 106
Gómez, Pedro, 348, 861, 873, 874, 891
Gonzales, Patrick, 156, 158, 160, 992, 993
Gonzalez, Eugenio J., 954, 1014, 1022, 1028, 1031
González, Martinez, 884
Gonzalez, Patrick, 878
Gonzalez, René, 80
Gonzalo, Ignacio, 1012
Good, Richard, A., 599, 602, 628, 629
Goodchild, Simon, 371, 373, 402, 404, 408
Goos, Merrilyn, 120, 439, 449, 473, 561, 630, 632
Gooya, Zahra, 393, 398, 411, 412, 416, 421, 844
Gorgorió, Nuria, 81, 106, 115, 121, 170, 171, 179, 186
Gough, Noel, 798, 902, 953
Gould, Peter, 409
Gould, Robert, 644
Gould, Stephen Jay, 103
Gouseti, Anastasia, 503
Gouzévitch, Dmtri, 538
Gouzévitch, Irina, 538
Graeber, Anna, 806
Graf, Klaus D., 153, 160, 810, 820, 836, 960, 985, 997, 998
Graham, Alan, 605
Gramsci, Antonio, 45
Graña, José, 873
Granger, Robert C., 510
Grant, Carl A., 88
Grant, Timothy, 736
Gravemeijer, Koeno P. E., 30, 118, 396, 550, 551, 592, 649, 658, 769, 806
Graven, Mellony, 119, 968
Graves, Barbara, 439
Graves, Frank P., 15
Gray, Eddie M., 608
Greaney, Vincent, 1014, 1015
Greene, Charles E., 10, 17, 19
Greenfield, Patricia M., 104

Greenhough, Pamela M., 90
Greeno, James G., 109, 115, 120
Greenwood, Davydd J., 92, 346
Greer, Brian, 74, 75, 105, 121, 312
Gregory, Kelvin D., 988, 1014, 1022, 1028, 1031
Grek, Sotira, 232, 1018–1020, 1030
Grevholm, Barbro, 124, 371, 408, 920, 931
Grier, Kevin, 1020
Grier, Robin, 1020
Griffin, Patrick, 76
Griffith, Jeanne E., 954
Griffiths, H. Brian, 284
Griffiths, Rachel, 16
Grimison, Lindsay, 16, 841
Grisay, Aletta, 839
Grosjean, François, 80
Groulx, Lionel, 105
Grouws, Douglas A., 266, 296
Grove, Myrna J., 10
Gruber, Karl Heinz, 873
Grugeon, Brigitte, 778, 782
Grugnetti, Lucia, 911
Gu, Quing, 471
Guala, Elda, 583
Guberman-Glebov, Raisa, 697, 698
Gueudet, Ghislaine, 572, 699, 772, 775, 781, 782, 893
Guile, David, 551
Guin, Dominique, 287, 305, 599, 603, 621, 631, 632
Gunawardena, A. J., 810
Gurtner, Jean-Luc, 623
Guskey, Thomas R., 409
Gustafsson, Lars, 208
Gutierrez, Angel, 575, 699, 914
Gutiérrez, Rochelle, 56, 70, 73, 75, 76, 84, 88, 106
Gutstein, Eric, 74, 90, 186–188
Guven, Bilgehan, 575
Gvozdenko, Eugene, 725, 732, 733, 735

H

Habermas, Jürgen, 45, 349
Hache, S. Bastien, 772
Hadas, Nurit, 572, 575, 952
Hadi, Sutarto, 414
Hagège, Claude, 81
Hager, Paul, 209
Häggström, Johan, 930, 931
Hahn, Corinne, 218, 220, 223, 226
Haines, Christopher R., 557
Halinen, Irmeli, 1030
Hall, Carol, 215, 728
Hall, Jennifer, 667
Halliday, Michael, 46, 50, 51
Halmos, Mfiria, 807
Hameyer, Uwe, 951
Hammer, David, 679
Hancock, Chris, 648, 658
Hanley, Una, 57, 58
Hanna, Gila, 277, 289, 551, 574, 814, 906, 920
Hannah, John, 629
Hannula, A., 219
Hannula, Markku S., 120
Hansen, Barbara A., 509
Hansen, Eric G., 734
Hansen, Hans C., 29
Harding, Ansie, 697, 699, 701, 708, 714
Harding, David, 13
Harding, Kelly J., 346, 347
Harel, Guershon, 248, 306, 307, 530, 605
Hargreaves, David H., 471
Hargreaves, M., 735, 736
Harjunen, Elina, 1030
Harkness, Shelly S., 338
Harlos, Carol Ann, 697, 699
Harradine, Anthony, 628, 659, 666, 679
Harré, Rom, 46
Harries, Tony, 178, 960
Harris, Mary, 56, 57
Harris, Pam, 21, 259
Hart, Juliet E., 875
Hart, Laurie C., 334, 344, 364, 375, 377, 415
Hartono, H., 414
Hasan, Ruqaiya, 46
Hasenbank, Jon, 603
Haslina bte Hj Mahmud, 410
Hassi, Marja-Liisa, 219
Hatano, Giyoo, 399
Hattie, John A., 363, 734
Hattori, Katsunori, 882
Hau, Kit-Tai, 958, 966
Hauser, Peter C., 79
Hautamäki, Airi, 1030
Hautamäki, Jarkko, 254, 260, 1030
Hayter, John, 23
Hazin, I. Falcão, 108
He, X., 378
Head, Alison J., 695
Healy, Lulu, 69, 78, 92, 93, 572, 575, 776, 780, 781
Heater, Brenda, 339
Hegedus, Stephen J., 52, 599, 626, 776
Heiberger, Richard M., 651
Heid, M. Kathleen, 247, 248, 540, 597, 599, 601, 602, 623, 624, 626, 628, 629, 631

Heinz, Karen, 464
Henn, Hans-Wolfgang, 550, 551, 920, 960
Hennessy, Sara, 768
Henningsen, Inge, 225
Herbel-Eisenmann, Beth A., 51–53, 369, 371, 380, 382–384, 386
Herber, George Mead, 45
Herbert, Carrie, 404
Herbert, Sandra, 604, 629
Herbst, Patricio G, 42, 52, 574
Hernández, Rosa Becerra, 327
Hernandez-Martinez, Paul, 52, 468, 556
Hernandez-Sánchez, M., 763
Herrera, Tony A., 697–699
Hersant, Magali, 699
Hersh, R., 349
Hershkowitz, Rina, 157, 572, 575, 831, 914, 952
Hiebert, James, 146, 150, 155, 156, 158, 160, 411, 412, 414, 436, 819, 838, 840, 848, 851, 878, 881–883, 983, 984, 992, 993
Higgins, Steven, 532
Higgins, T. L., 667, 680, 681
Higginson, William, 710
Hill, Heather C., 395, 629, 631
Hillel, Joel, 623
Hino, Keito, 464, 849
Hirsch, Donald, 839
Hivon, Laurent, 699, 767, 770
Hjh Rosmawati bte Hj Abu Bakar, 410
Hjh Rozaimah bte Hj Abdul Wahid, 410
Hjh Tini bte Hj Sani, 410
Hjh Hafizah bte Hj Salat, 410
Hjh Kamsiah bte Hj Ismail, 410
Hjh Mardiah, 410
Hjh Ramnah bte Pg Hj Abdul Rajak, 410
Hjh Shimawati, 410
Hjh Yunaidah bte Hj Yunus, 410
Hjh Zarinah bte Hj Jamudin, 410
Hoang, Van Sit, 20
Hobart, Michael E., 529
Hodge, Lynn L., 146, 153
Hodge, Rachel, 222
Hodgson, Bernard R., 274, 798, 812, 904–909, 917, 934
Hodgson, Thomas, 603
Hodkinson, Phil, 209
Hoehn, L., 615
Højgaard, Tomas, 991
Hofmann, T., 644
Hogan, Bob, 884
Hogan, Maureen P., 181
Hohenwarter, Marcus, 771

Hokstad, Leif Martin, 519
Holcomb, John P., 644, 740
Holder, Deborah, 214, 217, 219, 221, 222, 234
Hole, B., 763
Holland, Dorothy, 551
Hollar, Jeannie C., 602
Hollebrands, Karen Flanagan, 572, 575, 576, 599, 631, 644, 699
Holliday, Amanda, 91
Hollingsworth, Hilary, 156, 158, 160, 848, 851, 881
Holman, Henry, 15
Holton, Derek, 518, 628
Homer, Matthew, 723, 735, 736
Homse, Lucas Correa, 187, 191, 192
Hong, Ye Yoon, 604, 630–632
Honig, Meridith I, 509
Hoon, Seah Lay, 850
Hoover, Matthew, 763
Hopmann, Stefan T., 1013
Hopp, Carolyn, 124
Horkheimer, Max, 45
Horstman, Theresa, 708
Horton, Nicholas J., 671, 675, 679
Horwood, John, 10, 13, 16, 17, 840, 841
Hošpesova, Alena, 118, 382, 849
Houang, Richard T., 489, 838, 1012
Hough, Sarah, 341
Hourbette, Danièle, 89
Houssart, Jenny, 225
Howson, Geoffrey, 276, 282–284, 287, 289, 292, 294, 460, 757, 801, 805, 806, 828, 846, 904, 946, 950, 952, 954, 956, 960
Hoya, S., 620
Hoyles, Celia, 56, 57, 205, 212, 215, 216, 396, 486, 502, 511, 518, 543, 551, 554, 560, 572, 575, 590–592, 602, 644, 667, 694, 696, 699, 767, 770, 969
Høyrup, Jens, 14, 15
Hua, Yinchin, 537
Huang, Rongjin, 158, 377–379, 382, 386, 849
Huberman, A. Michael, 333
Huckstep, Peter, 395
Hughes, Janet, 707–709, 711–713
Hughes, R. Martin, 90
Hull, Glynda, 223
Hunkin-Finau, Salusulumalo S., 260
Hunt, D. Neville, 651
Hunter, Judy, 223
Hunter, Robert, 413
Huntley, Ian, 830
Hurd, Jacqueline, 415, 418, 882
Hurford, Amy, 308, 557

Husén, Torsten, 244, 792, 794, 837, 862, 979, 987, 1011, 1022
Husman, Jenefer, 698
Husserl, Edmund, 45
Hutcheson, Graeme, 468
Hutchison, Dougal, 1033
Hutton, B. Meriel, 215, 728
Huzzard, Tony, 54
Hwang, Jun, 1026–1027
Hyde, Janet S., 124
Hymes, Dell, 345

I

Idris, Yazid, 695
Ikeda, Miyako, 839
Ilavarasan, P. Vigneswara, 693
Illeris, Knud, 221, 222
Inagaki, Kayoko, 399
Ingram, Naomi, 120
Ingvarson, Lawrence, 252
Inprasitha, Maitree, 882
Ireland, Seth, 666
Irvine, Sidney, 722
Irwin, Kathryn C., 963
Ismael, Abdulcarimo, 180
Isoda, Masami, 414, 534, 540, 541
Ivanic, Roz, 222

J

Jablonka, Eva, 21, 41–44, 55, 61, 149, 155, 157, 161, 177, 184, 211, 304, 311–314, 319, 349, 352, 847, 852, 926, 962
Jablonski, Nina G., 130
Jabobini, Octavio R., 183
Jackiw, Nicholas, 571, 578, 654, 656
Jackson, Kara, 494
Jackson, Nancy S., 223
Jackson, Paul, 951
Jacobs, Jennifer K., 156, 158, 160, 881
Jacobsen, Edward Carl, 812, 902, 912, 932, 934
Jacobsen, Rebecca, 505
Jahn, Ana-Paula, 92, 93, 576, 584, 773, 774
Jahnke, Hans N., 29, 551, 574
Jakku-Sihvonen, Ritka, 1030
Jakubowski, Maciej, 839
Jakwerth, Pamela M., 1012
Jancarík, Antonin, 447
Jankvist, Uffe T., 303, 305
Janvier, Bernadette, 628
Jaquet, François, 911
Jardine, Rick, 703

Jarvin, Linda, 536, 537
Jarvis, Dan, 778, 782
Jaworski, Adam, 315
Jaworski, Barbara, 91, 120, 328, 364, 371–373, 381, 383, 385, 393, 394, 400–406, 408, 414, 417, 419, 421, 440, 918, 919
Jeeves, Malcolm A, 307
Jenkins, Michael, 575, 577
Jennings, Todd, 88
Jensen, Jens H., 991
Jeronnez, Louis, 536
Jiménez-Molotla, Jesús, 760
Jockusch, Elizabeth, 362
Johnson, Erin P., 531
Johnson, Martin L., 91, 551
Johnson, R. Burke, 334
Johnston, Bill, 205, 219, 226
Johnston-Wilder, Sue, 605, 757, 765
Jones, Doug, 962
Jones, Graham A., 839, 962, 1036
Jones, Ian S., 731
Jones, Keith D., 572, 574–577, 584, 771
Jones, Lee R., 1026
Jones, Peter L., 631
Jones, Phillip S., 1023
Jones, Phillip W., 1018
Jones-Newton, Kristie, 343
Jonker, Vincent, 766
Jordan, Alexander, 336
Jorde, Doris, 1012
Jorgensen, Robyn, 146, 154, 225
Joseph, George G., 553
Judson, P. T., 601
Julie, Cyril, 758, 759, 762
Jun, Li, 101
Jungwirth, Helga, 204, 226

K

Kaasila, Raimo, 339
Kafai, Yasmin, 89
Kahane, Jean-Pierre, 287, 757, 846
Kahneman, Daniel, 675
Kaiser, Gabriele, 550, 557, 830, 962, 978, 984, 996, 997, 1014
Kalas, Ivan, 699, 767, 770
Kaldrimidou, Maria, 846
Kamens, David H., 9, 11, 16
Kang, Helen J., 1013
Kanu, Yatta, 346
Kao, L. L., 89
Kaplan, David, 644, 740, 839
Kapur, J. N., 810, 811

Kaput, James N., 599, 626
Karim, Abdul Muchtar, 414
Kariya, Takehiko, 122
Karjalainen, Tommi, 254, 260, 1030
Karp, Alexander, 31, 274, 799, 800, 803, 805, 807, 809, 817, 852, 853
Karpinski, Louis C., 14, 17
Karshmer, Arthur, 536
Katsap, Ada, 182, 192
Katz, J. Sylvan, 830
Kauertz, Alexander, 873
Kaur, Berinderjeet, 158, 850, 874, 887, 1003, 1014, 1031, 1032
Kawanaka, Takako, 156, 158, 878, 983, 992, 993
Kawasaki, Teresinha F., 694
Kazak, Sibel, 671, 672
Keady, P., 261
Keane, Mark T., 732
Keast, Stephen, 402, 916
Keitel, Christine, 7, 21, 22, 27, 29, 57, 106, 121, 124, 150, 155–159, 266, 267, 295, 296, 349, 412, 444, 572, 798, 805, 818, 820, 837, 839, 847–849, 851, 902, 926, 928, 950, 953, 962, 984, 1019, 1034
Kellaghan, Thomas, 1014, 1015
Kelly, Anthony, 92, 763
Kelly, B., 539
Kelly, Dana L., 954, 1022, 1031
Kelly, Ronald R., 79
Kemmis, Stephen, 404
Kempf, Arlo, 828
Kendal, Margaret, 632, 963
Kendall, Mike, 756
Kenderov, Petar S., 921
Kent, Phillip, 205, 212, 215, 216, 551, 554, 644, 667
Keogh, John J., 226
Kerr, Clark, 1019
Kerr, Stephen, 708
Kersting, Nicole B., 156, 158, 160, 880
Khalil, Hassan K., 680
Khaneboubi, Mehdi, 89
Kho, Tek Hong, 963
Khuzwayo, Herbert, 161
Kiam, Low Hooi, 850
Kidron, Ivy, 572, 624
Kidwell, Peggy Aldrich, 27, 530–533, 535, 536, 538, 539
Kieran, Carolyn, 315, 316, 361, 367, 572, 602, 608, 609, 612, 613, 621, 623, 696
Kieran, Tom, 464
Kifer, Edward, 1012, 1025

Kilpatrick, Jeremy, 7, 25, 29, 57, 104, 150, 207, 266, 267, 274, 282, 284, 288, 293, 296, 312, 346, 395, 572, 791, 792, 794, 805, 812, 820, 828, 837, 839, 862, 913, 914, 918, 933, 935, 950, 962, 1013, 1015, 1023
Kim, M., 671, 672
Kim, Sam, 969
Kim, Yongnam, 20
Kimmelman, Paul, 878, 1014
Kimura, Eiichi, 882
Kinach, Barbara M., 436
King, Karen, 118
King, Samuel, 746
Kingston, Neal M., 735
Kington, Alison, 471
Kinsel, M., 464
Kirsch, Irwin, 839
Kirshner, David, 115, 362
Kitchen, Richard S, 852, 853
Kivelä, Simo K., 1030
Klein, David, 487
Klein, Felix, 277, 280, 281, 283, 535, 536, 802, 817
Klein, Mary, 917
Klesath, Marta, 705
Klieme, Eckhard, 1027
Kline, Morris, 285, 1023
Klineberg, Otto, 103
Klingner, Janette K., 87
Kloosterman, Peter, 124, 245, 962
Klothou, Anna, 52
Klusmann, Uta, 336
Knijnik, Gelsa, 74, 112, 113, 121, 125, 177, 178, 211, 217–219, 233, 479
Knobel, Michèle, 704
Knoll, Steffen, 156, 158, 878, 992, 993
Knorr, Wilbur Richard, 533
Knuth, Eric, 147, 573, 963, 966
Koay, Phong Lee, 874
Koc, Yusuf, 518, 777–779, 782
Koehler, Matthew J., 447, 448, 652, 768, 778, 782
Koistinen, Laura, 480
Kolezab, Eugenia, 118
Kolmogorov, Andrey, 807, 811
Kolpakowski, T., 679
Konold, Clifford, 648, 650, 659, 660, 666–668, 671, 672, 679–681
Kor, Liew Kee, 258
Korabinski, Athol A., 731, 732
Korb, Katrina, 735
Kortenkamp, Ulrich, 773
Kortum, Samuel, 1018

Koseki, Kiyoshi, 882
Koskela, Pekka, 1030
Kotsopoulos, Donna, 705
Kotthoff, Hans-Georg, 876
Kotzmann, E., 349
Kowalski, Robin, 706
Kozol, Jonathan, 72
Krainer, Konrad, 328–331, 336, 346, 361, 364, 365, 375, 382, 384, 387, 394, 399, 400, 402, 403, 432, 449, 503, 780, 815, 816, 821, 919
Kramer, Rita, 536
Krauss, Stefan, 336
Krazer, Adolf, 914
Kreis, Yves, 773
Kress, Gunther, 46, 709
Kroeze, David, 878, 1014
Krutetskii, Valdim, 554
Krygowska, A. Z., 8, 288, 289
Krzywacki, Heidi, 480
Kuhn, Jonathan R. D., 652
Kuiper, William, 951
Kunter, Mareike, 336
Kuntz, Gerard, 772
Kupari, Pekka, 1018, 1031
Kupiainen, Sirkku, 254, 260, 1029, 1030
Kwon, Jung-Min, 736
Kynigos, Chronis, 575, 699, 780, 781

L

Laaksonen, Seppo,, 1030
Laborde, Collette, 266, 287, 296, 518, 571, 572–576, 584, 590, 591, 602, 605, 607, 613, 630, 699
Laborde, J. M., 571
Labov, William, 84
Lacan, Jacques, 46
Laclau, Ernesto, 468
Lacroix, Guy I., 342
Lacy, Doreen, 181
Lagrange, Jean-B., 446, 511, 518, 599, 602, 605, 613, 626, 630, 694, 696, 767, 776, 778, 782, 969
Lahaye, Louise, 121
Lakoff, George, 551, 561
Lambdin, Diana, 518, 777–779, 782
Lambert, Julie, 875
Lampe, Cliff, 695
Lampert, Magdalene, 557
Landry, S. D., 875
Lang, Harry G., 79
Lange, Troels, 169
Langrall, Cynthia W., 962

Lankshear, Colin, 704
Lanz, Carlos, 350
Lapointe, Anthony E., 1013
Laridon, Paul, 178
LaRoque, Perry, 736
Lassak, Marshall, 591
Latour, Bruno, 558
Lattuca, Lisa R., 866
Lave, Jean, 44, 74, 104, 105, 107, 108, 110, 115, 116, 119, 205, 209–212, 214, 558
Lavicza, Zsolt, 771
Lavonen, Jari, 480, 1030
Lavy, Ilana, 316
Law, John, 337
Lawn, Martin, 1018, 1030
Lawson, Michael J., 341
Leavy, Aisling, 342
LeBaron, John, 757, 758
Lebethe, Agathe, 437–440, 442, 449
LeCompte, Margaret D., 33, 333
Leder, Gilah C., 124, 245
Lee, Arthur M. S., 543, 544, 730, 763
Lee, Chuan Seng, 852
Lee, Hyunjoo, 20, 644
Lee, Lesley, 628
Lee, Ming Chee, 775
Lee, Oklee, 875
Lee, Peng Yee, 887, 924, 1032
Lee, Valerie, 863
Lee, Yi-Fang, 249, 552
Leffler, James C., 509
Lehrer, Richard, 30, 401, 550, 551, 575, 577, 667, 671, 672
Lehto, Ollie, 802, 904–906, 934
Leigh-Lancaster, David, 631, 763
Leikin, Roza, 336, 394, 817
Leinwand, Steven, 885, 893
Lemut, Enrica, 575
Leonard, Jacqueline, 247, 248
Leontiev, Alexei, 76, 108, 212, 552, 553, 555, 556, 558
Lerman, Stephen, 43, 50, 58, 74, 104, 105, 109, 120, 146, 147, 218, 224, 296, 304, 306, 311–314, 321, 679, 699, 703
Lesh, Richard, 92, 306–308, 310, 550, 557, 558, 763, 776
Leslie, Elsie, 465
Lessard, Claude, 121
Lesser, Lawrence, 652
Lester, Frank K. Jr., 57, 267, 401, 486, 1018
LeTendre, Gerald K., 153, 244, 250, 798, 819, 820

Leu, Yu-Chyn, 21
Leung, Allen Yuk Lun, 52, 525, 528, 543, 544, 572, 574, 577, 578, 580, 758, 759, 762
Leung, Frederick Koon Shing, 21, 57, 153, 160, 266, 296, 517–519, 572, 810, 811, 820, 836, 840, 849, 960, 985, 997, 998
Levenson, Esther, 118, 344
Levin, Henry, 839
Levin, Morten, 92, 346
Levy, Mark R., 693
Lévy, Pierre, 702, 712, 714
Lewis, Benjamin Merrill, 382
Lewis, Catherine, 153, 414, 415, 418, 882, 883
Li, J., 962
Li, Qiong, 341, 958, 966
Li, S., 473, 965
Li, Xiaobao, 966
Li, Yeping, 9, 150, 378, 386
Liang, Hai-Ning, 700
Liau Tet Loke, Michael, 841
Libbrecht, Paul, 773
Lichnerowicz, André, 284, 286, 292, 293
Lidz, Carol S., 729
Lie, Svein, 981, 989
Lietzmann, Walther, 804
Lighthill, James, 923
Liljedahl, Peter, 337, 345
Lim, Chap Sam, 243, 256, 258, 414, 415, 841, 883
Lim, Geok Kuan, 410
Lima, Priscila Coelho, 219
Limber, Susan, 706
Lim-Teo, Suat Khoh, 901, 923
Lin, Fou Lai, 328–331, 399, 432, 466, 519, 701, 815, 816, 821
Linchevski, Liora, 316
Lindenskov, Lina, 84, 86, 93, 206
Lindlom-Ylanne, Sari, 148
Lindner, Martin, 880
Lindquist, Mary M., 1018, 1024
Lingard, Bob, 231, 232, 399, 401, 1018, 1030
Linn, Robert Lee, 328
Linnakylä, Pirjo, 1018, 1031
Lins, Abigail Fregni, 525
Lins Lessa, Mônica Maria, 108
Lipka, Jerry, 170, 181, 183, 193
Lipták, László, 882
Liston, Daniel P., 866
Little, Judith W., 88, 364
Liu, Haibin, 961
Livingston, Carol V., 125
Livne, Nava L., 734

Livne, Oren E., 734
Llinares, Salvador, 341, 364, 365, 449, 701, 779
Lloyd, Gwendolyn M., 338, 464, 465, 474
Lluis, E., 807, 808
Lobato, Joanne, 466
Loejo, Antonio, 884
Lohwater, Arthur J., 807
Loisy, Catherine, 782
Lokan, Jan, 848, 851
Lombardo, Denise F, 183
Long, Howard H., 103
Long, Pamela D., 14
Loong, Esther, 698
López, Luz S., 882
Lopez, Monica, 874
Lopez, Mottier L., 118
Lopez-Real, Francis J., 153, 160, 810, 820, 836, 849, 960, 985, 997, 998
Loveless, Tom, 148–150, 152
Lowe, Jim, 633
Lozano, Maria D., 591, 761
Lu, Allison, 771
Lubienski, Sheryl T., 327–328
Luckmann, Thomas, 45
Luhmann, Niklas, 45, 1019
Luna, Eduardo, 830, 962
Lundeberg, Mary Anne, 366
Lupiáñez, José L., 873, 884, 889
Luria, Aleksandr R., 107, 120
Lynch, Trevor, 364
Lytle, Susan L., 384, 404, 439

M
Ma, Liping, 794
Maasz, Juergen, 204, 226
Macedo, Suzana, 107
Machado, Milene C., 114, 115
Mackrell, Kate, 575, 576
Maclure, J. Stuart, 23
MacMullen, Edith N., 15
Macrae, Sheila, 205
Madden, Sandra R., 644
Maestro, Carmen, 873
Magalhães, V. P., 107
Magne, Olof, 84, 85, 87
Maguire, Terry, 226
Maher, Carolyn A., 92
Maher, Frinde, 251
Main, Susan, 519
Makar, Katie, 385–386, 648, 670, 675, 680, 682
Maksheev, Z., 803
Malaty, George, 1030
Males, Lorraine, 384

Malheiros, Ana Paula dos Santos, 696, 701
Maloney, Alan ,550
Malsch, Fritz, 804
Maltempi, Marcus V., 696, 701
Mammana, Carmelo, 969
Mammana, C., 969
Manaster, Alfred B., 156, 158, 160, 838
Manaster, Carl, 156, 158, 160
Manen, Max Van, 45
Manly, Myrna, 228, 229
Mantoan, Maria Teresa, 87
Maplesoft, 731
Maratas, Ilhan K., 575
Marcuse, Herbert, 45
Margolinas, Claire, 431, 434, 437
Maria bte Abdullah, 410
Mariotti, Maria-Alessandra, 77, 539, 572, 573, 575, 577, 579, 584, 590, 592, 699
Marrades, R., 575, 699
Marschark, Marc, 79
Martignone, Claudia, 541
Martin, Ben R., 830
Martin, Danny, 56, 70, 75, 76, 92, 126, 222, 224, 247, 248
Martin, J. R. R., 46
Martin, Lyndon, 120
Martin, Michael O., 150, 954, 981, 988, 989, 1014, 1016, 1022, 1028, 1031
Martin, Romain, 737
Martinovic, Dragana, 697, 708
Martio, Olli, 1030
Martland, Jim, 91
Marton, Ference, 578
Marx, Karl, 44, 50
Maschietto, Michela, 539
Masingila, Joanna O., 177, 179
Mason, John, 31, 296, 337, 605, 607, 628
Massarwe, Khayriah, 181
Matos, João F., 44, 88, 91, 108
Matos, José Manuel, 273
Matras, Mary A., 601, 602
Matthews, Lou E., 184
Mattos, Adriana Cesar, 50, 121
Matz, Marilyn, 362
Maxara, Carmen, 656–658, 675
Mayer, Connie, 79, 82
Mayer-Foulkes, David, 1020
Mayes, Robert, 602, 628
Maynard, Rebecca A., 510
Mayo, Merrilea, 969
McClain, Kay, 118, 316, 551, 553
McClain, Oren L., 76
McCrae, Barry, 848, 851
McCullough, B. D., 651

McCusker, Sean, 680
McDermott, Ray, 87
McDonnell, Lorraine M., 509
McDonough, Elizabeth, 757, 758
McGaw, Barry, 1013
McGee, David, 14
McGlaughlin, Ale, 126
McGrath, Daniel, 839
McGuire, George, 731
McIntyre, Donald, 394, 403, 405
McKnight, Curtis C.,, 19, 20, 362, 489, 838,
McLeod, Douglas B., 488, 1024
McMaken, Jennifer, 1026–1027
McMullin, Lin, 621
McMurchy-Pilkington, Colleen, 125
McMurry, Charles A., 24, 25
McNab, Donald, 888
McNamara, Olwen, 57, 120, 469
McNeely, Maggie, 878, 1014
McNeil, Nicole M., 536, 537
McVittie, Janet, 120
Mead, Nancy A., 1013
Meagher, Michael, 572
Meaney, Tamsin, 169
Mechelli, Andrea, 81
Medina, Elsa, 644, 649, 651
Medrich, Elliott A., 954
Mehta, Niranjan, 346
Meira, Luciano, 116, 117, 120
Meissner, Hartwig, 767
Meletiou-Mavrotheris, M., 667
Mellin-Olsen, Stieg, 56, 57, 75, 121, 248, 251, 295, 460
Mellissinos, Ma, 88, 1024
Menasian, James, 601, 602
Menchaca, Maria, 103
Mendelovitz, Juliette, 839
Mendick, Heather, 75, 76, 919
Menghini, Marta, 281, 287, 904, 910, 911
Menon, Ramakrishnan, 21, 838
Mercat, Christian, 773
Mercier, Alain, 305
Merle, Michel, 970
Mesa, Vilma, 52, 793, 794, 861, 1015
Messina, Graciela, 348
Meyfarth, Thorsten, 678
Micheletti, Chiara, 575, 576, 587
Miles, Matthew B., 333
Milgram, R. James, 489
Miliukov, Paul N., 820
Mill, John Stuart, 349
Miller, Catherine D., 650
Miller, Jen, 532
Millroy, Wendy, 108

Mills, Charles Wright, 45
Mingat, Alain, 12, 13
Mishra, Punya, 447, 448, 768, 778, 782
Mislevy, Robert J., 722, 723
Mitchell, Karen Janice, 1026
Mittag, Hans-Joachim, 652
Miyakawa, Takeshi, 414
Moelands, Henk, 839
Mogari, David, 178
Mohammad Ariffin bin Hj Bakar, 410
Mok, Ida Ah Chee, 149, 155, 157, 161, 847, 849, 852
Monaghan, John, 604, 630, 632, 780, 781
Monge, Gaspard, 535
Monkman, Karen, 397, 399
Monroe, Eula E., 810, 811
Monroe, Paul, 23
Monroe, Will S., 15
Monseur, Christian, 839
Montague, Marjorie, 87
Monterrubio, M. Consuelo, 884
Montessori, Maria, 15, 536
Montt, Guillermo, 839
Moon, Bob, 19, 284, 835
Mooney, Edward S., 962
Moore, Darlinda, 644, 645, 648
Moore, Eliakim Hastings, 28, 278, 282, 535
Moore, Joyce L., 109
Moore, Rob, 227, 235
Morales, Diana, 839
Moreira, Darlinda, 179
Moreira, Leonor, 185–187
Moreira, Plinio C., 121
Moreno, Constanza, 79
Moreno-Armella, Luis, 77
Morgan, Candia, 44, 50, 51, 76, 109, 173, 213, 218, 224, 926
Morin, Edgar, 451
Morris, Anne K., 340, 436
Morrone, Anastasia S., 338
Moschkovich, Judit N., 53, 76, 79, 80, 90, 111
Moser, Urs, 1012
Mosimege, Mogege, 178, 180
Moskowitz, Jay H., 1017, 1018
Moss, Joan, 575–577, 584
Moss, Pamela A., 892
Mosteller, Frederick, 830
Motard, L., 284, 286, 293
Mouffe, Chantal, 467
Mourshed, Mona, 880
Mousley, Judith, 346, 400, 518, 777–779, 782, 916, 917
Mousoulides, Nicholas, 308, 575
Mouwitz, Lars, 208

Moya, Andrés, 350
Moyer, John C., 950, 963, 966, 969
Moyer, Patricia S., 536
Mukhopadhyay, Swapna, 74, 75
Mullen, Jana, 220, 224, 226
Mullis, Ina V. S., 150, 865, 954, 981, 989, 1014, 1016, 1022, 1024, 1028, 1031
Munn, Penny, 85
Murata, Aki, 364, 375–377, 386, 410, 415
Murray, T. Scott, 1013, 1015
Mwakapenda, Willy, 809

N

Näätänen, Marjatta, 1030
Nabonnand, Philippe, 277
Nachlieli, Talli, 52
Nardi, Elena, 303, 316, 320
Nascimento, Jorge C., 108
Nasir, Na'Ilah Suad, 176, 189, 249
Nathan, Mitchell J., 436
Nebres, Ben F., 8, 9, 13, 74, 693, 798, 800, 805, 902, 907, 908, 911, 923, 924, 968
Nédélec, Dreslard D., 442
Nelson, Barbara, 464
Nelson, Stephen R., 509
Nelson, Tamara, 416
Nelson-Barber, Sharon, 74, 75, 90
Nemirovsky, Ricardo, 628
Nesher, Perla, 914
Neubrand, Michael, 336, 394, 839, 1027
Neumann, Knut, 873
Neuwirth, Eric, 651
Newman, Denis, 76
Newman, M. Anne, 410
Newmarch, Barbara, 214, 217, 219, 221, 222, 234
Newstead, Karen, 845
Ng, Swee F., 887, 963
Nguyen, Phuong-Mai, 1019
Ni, Yujing, 958, 966
Nicéas, Lenice, 107
Nichols, Sharon L., 453
Nicholson, Ann, 735
Nicholson, James, 680
Nicol, Cynthia, 121, 913, 988
Nicolet, Jean Louis, 284, 286, 293
Nie, Bikai, 465, 473, 950, 961, 963, 966, 969
Nieto, L. J. B., 449
Niss, Mogens, 221, 401, 550, 551, 920, 935, 960, 977, 991, 1028
Nissen, Gunhild, 930
Nixon, Althed Scott, 89

Nkhwalume, Alakanani Alex, 124
Nkopodi, Nkopodi, 180
Noffke, Susan E., 346
Nolan, Kathleen, 57, 477
Nolasco, Margarita, 346
Noppeney, Uta, 81
Norjah bte Hj Burut, 410
Norman, Donald A., 713
Norton, Pam, 631
Norwood, Karen, 602
Noss, Richard, 56, 57, 205, 212, 215, 216, 226, 503, 518, 543, 551, 554, 560, 572, 590–592, 602, 644, 667, 699, 757, 765, 767, 770
Nóvoa, António, 1017
Novotná, Jarmila, 118, 328–331, 399, 431, 432, 437–440, 442, 447, 449, 451, 815, 816, 821, 849
Noyes, Andrew, 209
Nunes, Terezhinha, 79, 82, 107, 146, 154, 172, 205, 209, 211
Núñez, Rafael, 561
Nussbaum, Miguel, 519

O

Oates, Greg, 631, 928
O'Brien, Mia, 385–386
O'Callaghan, Brian R., 602
Ochs, Kimberly, 1019
O'Connor, Carla, 84, 87
O'Connor, Kathleen M., 1014, 1022, 1028, 1031
Odani, Kenji, 882
O'Day, Jennifer, 487
O'Doherty, John, 81
O'Donoghue, John, 204, 208, 226
Oepnjuru, George, 213
Ogborn, John, 662
Ograjenšek, Irena, 644
O'Hagan, Joan, 234
Ohara, Yutaka, 414
Ohuche, R. Ogbonna, 809
Okubo, Kazuyoshi, 882
Olaya, Rojas A., 348
Oldknow, Adrian, 969
Olesen, Henning Salling, 204, 207
Olive, John, 648
Oliveira, Claudia Hosé, 177
Olivera, Mercedes L., 146, 346
Olivero, Frederica, 573–576, 579–582, 584, 587
Olivier, Alwyn, 845
O'Loughlin, Noreen, 342
OLPC Foundation, 759
Olson, John F., 150, 989, 1022, 1031
Olson, Lynn, 1026
Oner, Diler, 575
Onwuegbuzie, Anthony J., 334
Or, C. M., 52, 574
Ormell, Christopher P., 29, 30
O'Rode, Nancy L., 341
O'Rourke, John, 519
Orr, Eleanor W., 83, 84
Orton, R., 306, 307
Osana, Helen P., 342, 575, 577
Osborne, Jonathan, 507
Osgood, William F., 835
Osler, Audrey, 131
Ostermeier, Christian, 880
O'Sullivan, Christine Y., 989, 1016
Ottaviani, Maria-Gabriella, 930
Otte, Michael, 463, 956
Otten, Samuel, 380
Oughton, Helen, 223, 226
Owen, Eugene, 1018

P

Page, David P., 18
Pagliaro, Claudia M., 79
Pais, Alexandre, 57, 73, 87, 178, 181
Paiti, Margaret, 465
Palafox, Juan Carlos, 1015
Palamidessi, Mariano, 874
Palhares, Pedro, 112, 113
Palmiter, Jeanette, 601
Pampaka, Maria, 52, 468
Panizzon, Debra, 364, 384
Paola, Domingo, 574–576, 579, 580, 587, 699
Paparistodemou, Efi, 667
Papert, Seymour, 287, 757, 765
Papy, Frédérique, 287
Parada, Sandra, 782
Paraide, Patricia, 179
Pardala, Antoni, 807
Park, Kyungmee, 849
Park, Soojin, 839
Parker, Thomas, 487
Parshall, Karen H., 905
Parson, Talcott, 45
Parsons, Jim B., 346, 347
Pashley, Peter J., 1014
Passeron, Jean-Claude, 104, 127, 128
Passos, Claudio C. M., 118
Pateman, Neil A., 243
Paterson, Judy E., 630
Patrick, Rachel, 183, 192, 193
Patton, Michael Q., 333
Pea, Roy, 728, 737

Pegg, John, 364, 384
Pehkonen, Erkki, 1029, 1030
Pelgrims Ducrey, Greta, 728
Pellegrino, James W., 1026
Pellerey, Michelle, 285
Peltenburg, Marjolijn, 729
Pena, Sergio D. J., 130, 131
Pennington, M. Nan, 698
Penuel, William R., 52, 519, 746
Pepin, Birgit, 771
Pereira-Mendoza, Lionel, 874
Pereyra, Miguel A., 876
Perks, Pat, 419–421
Perret-Clermont, Anne-Nelly, 104
Perry, John, 27, 28, 535
Perry, Rebecca, 415, 418, 882, 883
Persens, Jan, 846
Peskette, J. W., 284, 286, 293
Pestalozzi, Johann, 534, 535
Peters, Michelle, 1019
Peterson, Blake E., 382
Peterson, Jodi, 507
Petrosino, Anthony, 436
Pfannkuch, Maxine, 645, 646, 671, 675, 678, 679
Phillips, Brian, 929
Phillips, David, 1019
Phillips, Eileen, 710
Phillips, Gary W., 1013, 1014
Philp, Hugh, 83
Piaget, Jean, 108, 109, 284, 286, 292, 293, 306–309, 315, 316, 434, 461, 806
Picker, Susan H., 709
Pierce, Robyn, 600, 604, 628–630, 632, 633, 743, 768
Pilot, Albert, 1019
Pimm, David, 50, 51, 53, 710, 757, 765
Pinar, William F., 951
Pineau, Elysse Lamm, 713
Pinto, Marcia M. F., 694
Pinxten, Rik, 32
Pirie, Susan, 120
Pitombeira de Carvalho, João, 804
Pittalis, Marios, 575
Pitta-Pantazi, Demetra, 575
Planas, Núria, 80, 81, 106, 115, 121, 170, 171, 179, 186, 346
Plofker, Kim, 528
Pluvinage, François, 287, 757
Podworny, Stefan, 678
Poirier, Louise, 375
Pollack, Robert, 18
Pollak, Henry O., 31, 551, 806
Pollatsek, Alexander, 668

Pollitt, Alastair, 729
Pollock, Elizabeth, 885, 893
Pollock, Maggie, 725
Pólya, George, 550, 556, 807, 817
Ponte, João Pedro, 396, 701
Pool, Peter, 723, 735, 736
Popkewitz, T. Neville, 116
Popper, Karl, 306
Porter, Andrew C., 1026–1027, 1034
Posadas, Linda S., 758, 759, 762
Posch, Peter, 346, 364, 368, 382
Posner, Jill K., 108, 109
Post, Thomas, 306, 307, 763
Postlethwaite, T. Neville, 837, 954, 986
Potari, Despina, 209
Potter, John, 503
Powell, Arthur B., 69, 74, 75, 80–83, 88, 91, 92, 112, 115, 121, 126, 184, 248, 251, 286
Pozzi, Stefano, 215, 560
Prais, Sig J., 1013
Pratt, Dave, 661, 671, 699
Pratt, Nick, 120
Prawat, Richard S., 1012
Preissle, Judith, 33
Premkumar, G. Prem, 654
Prenzel, Manfred, 880, 1027
Prescott, Anne, 436
Presmeg, Norma C., 114, 115, 146, 147, 170, 171, 176, 177, 256, 258
Prestage, Stephanie, 419–421
Preuschoff, Corinna, 150, 1016, 1022, 1031
Price, Beth, 725, 735
Price, Cathy J., 81
Price, Michael, 27, 286, 535
Prieto, Annalyse K. C., 187, 191, 192
Pring, Richard, 405
Pritchard, C., 697
Pritchard, Ruth, 402
Prömmel, Andreas, 676
Prusak, Anna, 120, 316
Prus-Wisniowska, Ewa, 179
Pryor, John, 220, 225
Puig Adam, Pedro, 284, 286, 293
Pulte, Helmut, 574
Purakam, O., 811
Putt, Ian, 917

Q

Quan, Wang Lin, 811
Quander, Judith, 147
Quinn, Naomi, 551
Quirk, William, 487

R

Rabardel, Para, 572, 591, 603, 631
Radford, Luis, 49, 51, 171, 172, 190, 470, 565, 592
Rafanan, Ken, 519
Rahman, Anisur, 346
Raizen, Senta A., 489, 988, 1012
Rakotomalala, Ramahatra, 12, 13
Ralston, Anthony, 287
Ramsey, Fred L., 653
Rands, Kathleen, 71
Rantanen, Pekka, 1030
Rappaport, I., 221
Rasmussen, Chris, 118
Rasmussen, P., 204, 207
Rautalin, Marjaana, 1030
Rav, Y., 574
Ravitch, Diane, 253
Ravn, Ole, 61
Reading, Chris, 930
Reason, Rea, 85
Recorde, Robert, 534
Redmond, Trevor, 633
Reed, Helen J., 107, 592, 769
Reed, Yvonne, 80, 81
Rees, Gareth, 477
Reeve, William D., 24, 804, 834
Regan, Meridith M., 671, 675, 679
Régnier, Jean-Claude, 538
Reid, Constance, 802
Reinikainen, Pasi, 839, 1018, 1031
Reis, Maria C., 213, 225
Reisner, Edward H., 19
Reiss, Kristina M., 467
Relethford, J. H., 130
Remillard, Janine T., 179, 464, 472–474
Renert, Moshe, 60, 352
Renshaw, Peter, 120, 399, 401
Resnick, Lauren, 956, 957
Resnick, Tzippora, 952
Restivo, Sal, 121
Restrepo, A., 874, 891
Reyes, Laurie Hart, 245
Reynolds, David, 364
Reys, Barbara J., 950
Reys, Robert E., 950
Rhodes, Valerie, 205
Richter, Jutta, 277
Rickhuss, Mike G., 602
Rico, Luis, 873, 886
Rico, Mercedes, 654
Ridgway, Jim, 680, 952
Rijpkema, Koo, 745
Rinne, Risto, 1018, 1030
Ripley, Amanda, 1019

Rismark, Marit, 519
Rivera, Antonio, 480
Rivkin, Steven, 253
Rizo, Filipe M., 884
Rizvi, Fazal, 231, 232
Roadrangka, Vantipa, 841
Robert, Aline, 305
Roberts, David Lindsay, 27, 28, 519, 525, 530–533, 535, 536, 538, 539, 812
Roberts, T., 178
Robertson, Roland, 399
Robinson, Carol, 746
Robinson, Sharon P., 8
Robitaille, David F., 509, 838, 862, 902, 951, 957, 979–981, 985–989, 1012, 1022
Robitzsch, Alexander, 729
Robutti, Ornella, 574–576, 579–582, 584, 587, 592, 622, 699
Rock, Donald A., 736
Roddick, Margaret, 1014
Rodrigues, José F., 220, 224, 226
Rodrigues, Marco Aurélio Borela, 187, 191, 192
Rodriguez, Gerardo, 620
Roesken, Bettina, 120
Rogalski, Janine, 305
Rogers, Alan, 207, 213
Rogers, Leo F., 915, 934
Rogoff, Barbara, 44, 76, 109, 154
Rohlen, Thomas P., 153
Rojano, Teresa, 628, 760, 761, 767
Rojas, Yury M., 187
Romberg, Thomas A., 813, 956
Rønning, Frode, 931
Rorty, Richard, 129
Rosa, Milton, 679
Rosas, Lesvia, 350
Roschelle, Jeremy, 519, 599, 746
Rosen, Gershan, 437–440, 442, 449, 451
Rosenberg, Diana, 808
Rossi-Becker, Joanne, 124
Rossman, Allan J., 648
Roth, Kathleen J., 880, 881
Roth, Wolf-Michael, 212, 552, 565
Rowe, David, 215, 728
Rowe, Ken, 252
Rowland, Tim, 395, 404
Rozina bte Awg Hj Salim, 410
Rubenson, Kjell, 206, 232
Rubenstein, Rheta, 950
Ruddock, Graham J., 989, 1016
Rueda, Enrique, 736
Ruiz, Angel, 911–913
Russell, Susan Jo, 680, 963
Ruthven, Kenneth, 29, 286, 287, 305, 402, 603, 623, 757, 762, 765, 768

Author Index

Rutkowski, David, 1017
Ryan, Julie, 558
Ryan, Marilyn, 124
Ryve, Andreas, 51, 314, 315, 319

S

Sabra, Hussein, 772
Sacristán, Ana I., 758–760, 762, 763, 767, 770, 782
Safford, Katherine, 204
Sahlberg, Pasi, 151, 152, 1030
Sahlström, Fritjof, 157, 160, 850
Saito, Mioko, 863
Saito, Noboru, 882
Saiz, Mariana, 480
Sakonidis, Haralambos, 52, 846
Saldanha, Luis, 602, 612
Saleh, Fatimah, 258
Salin, Marie-Hélène, 442
Säljö, Roger, 146, 373
Saló i Nevado, Läia, 219
Salsburg, David, 677
Salway, Leida, 90
Sammons, Pam, 471
Sanchez, Hector, 654
Sánchez Vázquez, G., 932
Sandlin, Jennifer A., 644
Sandoval, Ivonne, 782
Sangiorgi, O., 808
Sangwin, Chris J., 730, 763
Santagata, Rossella, 340, 446, 779
Santos, Madelina, 108, 767
Santos, Silvanoc C., 702
Sapere, Patricia, 79
Saracho, Alberto, 869, 876, 877, 880
Sarmiento-Klapper, J. W., 704, 705
Sarrazy, Bernard, 431, 451, 453
Sato, Manabu, 374
Sauble, Irene, 31
Saunders, Lesley, 471
Sawada, Toshio, 984, 995, 996, 1012
Saxe, Geoffrey B., 107–109, 179
Schafer, Daniel W., 653
Schaffer, Gene, 364
Schagen, Ian, 1033
Schappelle, Bonnie P., 488, 1024
Scharlemann, Martin, 489
Schauble, Leona, 401, 671, 672
Scheerens, Jaap, 839
Scheinin, Patrik, 1030
Scherer, Petra, 346
Scheuermann, Friedrich, 735, 737
Schiefele, Ulrich, 1027
Schiffman, Zachary S., 529
Schifter, Deborah, 963
Schilling, Steven G., 395
Schleicher, Andreas, 228, 839, 875
Schliemann, Analúcia D., 104, 107, 108, 154, 172, 179, 205, 209, 211, 212
Schlöglmann, Wolfgang, 204, 226
Schmidt, William H., 19–21, 487, 489, 838, 845, 862, 878, 879, 983, 988, 1012, 1014
Schmitt, Mary Jane, 228, 229
Schneider, Barbara, 485
Schneider, Wolfgang, 1027
Schoenfeld, Alan H., 104, 267, 303, 305, 307, 467, 556, 557, 630
Scholz, Roland W., 296, 934
Schön, Donald A, 364
Schrage, Michael, 692, 704
Schriewer, Jürgen, 1019
Schubring, Gert, 24, 274, 277, 281, 283, 798–802, 804, 833, 834, 904–906, 914, 976
Schümer, G., 1027
Schütz, Alfred, 45
Schwarz, Baruch, 157, 572, 575, 831, 952
Schwingendorf, Keith E., 309
Scribner, Jay P., 244, 250
Scribner, Sylvia, 104, 107, 109
Scucuglia, Ricardo, 709, 713
Seago, Nanette, 336, 394
Seah, Wee Tiong, 21, 113, 115, 179, 402, 916
Secada, Walter G., 72, 88, 245, 968
Sedig, Kamran, 700, 703
Sedlmeier, Peter, 675
Seeger, Falk, 121, 146, 318
Seel, Norbert M., 315
Sefa Dei, George J., 828
Segerholm, Christine, 1018, 1030
Séguin, Edouard, 15
Seidman, Edward, 510
Sekiguchi, Yasuhiro, 464, 849
Seligman, M. E. P., 161
Seliktar, Miriam, 124
Selleck, Richard J. W., 16, 25
Sells, Lucy W., 104
Selwyn, Neil, 503, 695
Semenov, Alexei, 758, 759, 762
Senk, Sharon L., 950, 951
Sennett, Richard, 235
Sensevy, Gérard, 305
Serrano, Ana, 156, 158, 878, 992, 993
Serrano, W., 352
Servais, Willy, 284, 286, 293
Setati, Mamokgethi, 53, 80, 81, 120, 255, 256, 346
Sethole, Godfrey, 21, 149, 850

Sewell, Tony, 126
Seymour, Dale, 530, 536
Sfard, Anna, 52, 60, 74, 120, 315–320, 328, 558, 828
Sharat-Amir, Yael, 679
Sharp, Donald W., 109
Sharygin, I. F., 818
Shaughnessy, J. Michael, 361, 681
Shavelson, Richard J., 479
Shearer, Brenda A., 366
Sheets, Charlene, 599, 601, 602, 628, 629
Sheffield Hallam University, 503
Shelley, Nancy, 919
Sherin, Bruce, 679
Sherin, Miriam G., 436, 779
Sherman, Julia, 245
Shevkin, L., 818
Shewbridge, Claire, 875
Shibata, Rokuji, 537
Shimizu, Katsuhiko, 1012
Shimizu, Shizumi, 882
Shimizu, Yoshinori, 21, 145, 149–151, 155–160, 412, 414, 444, 464, 839, 847–849, 851, 984, 995, 996, 1003
Shin, Jongho, 20
Shin, Soo Yeon, 850
Shiohata, Mariko, 220, 225
Shirley, Lawrence, 113
Shiu, Christine, 14
Shockey, Todd L., 177
Shorrocks-Taylor, Diane, 735, 736
Shuell, Thomas J., 697, 699
Shulman, Lee S., 394–396, 447, 631
Shumar, Wesley, 705
Shute, Valerie J., 734
Shymansky, James A., 875
Siang, Kim Teng, 538
Sidoli, Nathan, 277, 906
Sierpinska, Anna, 118, 207, 296, 312, 313, 828, 935
Sila'ila'i, Emilie, 465
Silva, D., 351
Silva, Vanisio L., 112, 125
Silver, Edward A., 42, 120, 933
Silverman, Frederick L., 182, 192
Silverman, Jason, 701
Simmel, Georg, 45
Simmt, Elaine, 51
Simola, Hanu, 1018, 1030
Simon, Martin, 396, 397, 405, 414, 415, 421, 464
Simons, Herbert A., 159
Simpson, Adrian, 608

Sinclair, Nathalie, 320, 571, 574–577, 582, 584, 590, 591, 710
Singh, Parmjit, 827, 841
Singh, S., 125
Sintsov, D., 803, 804
Siu, M. K., 282
Sjøberg, S., 1013
Skemp, Richard, 293, 294
Skinner, Burrhus F., 293
Sklar, Jeffrey C., 652
Skott, Jeppe, 336
Skovsmose, Ole, 12, 46, 56, 61, 74, 118, 121, 172, 176, 178, 183–185, 194, 210, 248, 296, 349, 373, 479, 953, 955
Slavit, David, 416
Sloan, Deborah, 121
Sloane, Finbarr, 1015
Sloane, Peter F. E., 873, 1028
Slovic, Paul, 675
Smees, Rebecca, 471
Smid, Harm J., 285
Smith, David Eugene, 17, 23, 28, 280, 537, 801, 804, 835, 1010
Smith, Dorothy E., 45
Smith, David R., 109
Smith, Heather J., 532
Smith, Louis M., 120, 147, 156, 158, 160, 339
Smith, Margaret S., 120, 147, 156, 158, 160
Smith, Mark K., 952
Smith, Marshall S., 486, 487, 509, 510
Smith, Matthew L., 486, 509, 510
Smith, Marvin E., 334, 344
Smith, Ryan C., 576
Smith, Shannon D., 695
Smith, Stephanie Z., 334, 344
Smith, Teresa A., 954, 1014, 1022, 1028, 1031
Snell, J. Laurie, 644, 651
Snow, Catherine E., 512
Sølvberg, Asrid M., 519
Som, Nicholas A., 652
Somekh, Bridget, 346, 364, 368, 382, 479
Son, Ji-Won, 343
Sonenberg, Elizabeth, 735
Sotelo, Francisco L., 880
Soury-Lavergne, Sophie, 773, 774, 782, 919
Southwell, Beth, 31, 415, 882
Sowell, Evelyn J., 536
Speer, For, 336
Speer, Natasha M., 336
Spencer, Herbert, 25
Spencer, Patricia E., 79
Spitzer, Sandy M., 436
Spivak, Gayatri C., 346

Author Index

Sriraman, Bharath, 42, 57, 74, 77, 105, 121, 283, 303–305, 307, 308, 310, 312, 314, 318, 319, 321, 550, 936, 959
Stacey, Kaye,, 600, 628–633, 721, 725, 735, 743,765 768, 838, 839, 963
Stadler, E., 320
Stahl, Alan M., 14
Stahl, Gerry, 698, 704, 705, 708
Stake, Robert E., 333, 488, 1024
Stanat, Petra, 1027
Stanic, George M. A., 24, 25, 245
Stanley, Christine A., 76
Star, Jon R., 340
Stark, Joan, 866
Starkey, Hugh, 131
Staub, Natalie, 213
Steen, Lynn A., 31, 72, 221, 727
Steffe, Leslie P., 296, 464
Stehlíková, Naďa, 445
Steinberg, Linda S., 722
Steinbring, Heinz, 296, 346, 350, 364, 396, 828
Steiner, Hans Georg, 291, 296, 350
Steiner-Khamsi, Gita, 1018
Steinfield, Charles, 695
Steinle, Vicki, 725, 735
Stemhagen, Kurt, 292
Stemler, Steven E., 988
Stengers, Isabelle, 452
Stenhouse, Lawrence, 364, 403
Stentoft, Diana, 59
Stephan, M., 30, 118, 414, 1017, 1018
Stephens, Maria, 1017–1018
Stephens, Max, 414
Sternberg, Robert J., 104, 105, 146
Stevens, Peter A. J., 106
Stevenson, Harold W., 20, 840, 848, 983, 993
Stevenson, Ian, 575, 576
Stewart, Sepideh, 629, 630
Stigler, James W., 146, 150, 155, 156, 158, 160, 340, 411, 412, 414, 446, 779, 819, 838, 840, 848, 851, 878, 880–883, 983, 984, 992, 993
Stillman, Gloria, 179, 550, 557, 917
Stinson, David W., 74, 76
Stobart, Gordon, 471
Stocker, David, 188
Stockero, Shari L., 340
Stohl Lee, Hollylynne, 92
Stokoe, William C., 81
Stone, John C., 31
Sträßer, Rudolf, 220, 223, 296, 558, 560, 572, 575, 591, 699, 934
Straughn, Celka, 706
Strauss, Anselm L., 333, 353

Streefland, Leen, 558
Street, Brian, 213
Strickland, Sharon K., 340
Stringfield, Sam, 364
Strømme, Alex, 519
Stromquist, Neil P., 397, 399
Stroup, Walter M., 763
Struik, Dirk J., 275
Styles, Irene, 148
Stylianides, Andreus J., 573
Stylianou, Despina A., 573
Stzajn, Paola, 88, 91
Subramaniam, K., 963
Sukthankar, Neela, 124
Sullivan, Peter, 55, 328, 394, 400, 402, 916, 952
Sundefeld, Marcia L. M. M., 187, 191, 192
Suri, H, 72
Sutherland, Rosamund, 960, 969
Sutton, John, 147
Swafford, Jane O., 1012, 1025
Swain, Jon, 214, 217, 219, 221, 222, 234
Swanson, Dalene, 479
Swantz, Marja-Lissa, 346, 347
Swars, Susan L., 334, 344
Swetz, Frank J., 14, 17, 806
Swinnerton, Bronwyn, 723, 735, 736
Sykes, Gary, 485
Symanzik, Jurgen, 652
Szablewski, Jackie, 222
Sztajn, Paola, 468, 470

T

Tabach, Michael, 952
Taimina, Daina, 534, 540, 541
Tait, Kenneth, 735, 736
Tall, David O., 608
Tan, Alexandra, 878, 1014
Tan, Daphne M. M., 695
Tan, Hazel, 743
Tan, Kok Eng, 258
Tan, Magdalene X. J., 695
Tang, K. N., 258
Tang, R., 961
Tardif, Maurice, 121
Tareke, Mebrak, 839
Tarr, James E., 962
Tarvainen, Kyosti, 1030
Tashakkori, Abbas, 334
Tatar, Deborah, 599
Tate, William F., 87
Tatsis, Konstantinos, 118, 120
Tatto, Maria T., 21, 845
Tay, K., 695

Taylor, Chris, 477
Taylor, Edward H., 835
Taylor, Peter J., 921
Taylor, Steve, 630
Teddlie, Charles B., 334, 364
Teese, Richard, 55
Teixeira, Paulo J. M., 118
Terlouw, Cees, 1019
Terman, Nancy, 341
Tesch, Renata, 33
Thanh, Nguyen Chi, 758, 759, 762
Thanheiser, Eva, 343
Thom, René, 285
Thomas, G., 91
Thomas, J., 1019
Thomas, Michael O. J., 409, 518, 604, 606–609, 628–632
Thomas, Peter G., 602
Thomas, William B., 103
Thompson, Alba G., 336
Thompson, Denisse R., 950, 951, 1014
Thorndike, Edward L., 19
Thorsten, M., 157, 460
Threlfall, John, 723, 735, 736
Thunder, Kateri, 76
Thwaites, Anne, 395
Thwaites, Bryan, 285, 287
Tichá, Marie, 382
Tillmann, Klaus-Jürgen, 1027
Timmerman, Christianne, 106
Timmons-Brown, Stephanie, 91
Tintori, Stefania, 181
Tirosh, Dina, 118, 294, 328, 344, 394, 396, 400, 402, 420, 421, 919
Tobias, Sheila, 210
Tobies, Renate, 28
Todd, Peter M., 215, 234
Todes, Daniel P., 19
Tolstoy, Leo N., 819
Tomlin, Alison, 188, 193, 213
Toomey, Ron, 696
Törner, Günter, 283, 305, 314, 467, 959
Torney-Purta, Judith, 840
Toro-Álvarez, Catalina, 882
Torres, Rosa M., 20
Tout, Dave, 228, 229
Towers, Jo, 120
Towne, Lisa, 472, 479
Tran, Si Nguyen, 20
Travers, Kenneth J., 838, 862–864, 902, 951, 979, 985–987, 1012, 1025
Treffers, Adri, 558
Treviño, Ernesto G., 863
Trewin, Dennis, 644

Trgalovà, Jana, 773, 774
Trigueros, Maria, 760, 761, 763, 770
Trouche, Luc, 287, 305, 518, 539, 540, 590, 592, 599, 602, 603, 605, 613, 621, 630–632, 680, 681, 699, 738, 753, 767, 768, 770, 772, 775, 781, 782, 893
Tsai, K. H., 775
Tsai, Yi-Miau, 336
Tsamir, Pessia, 118, 294, 344
Tsang, Betty, 489
Tsatsaroni, Anna, 43, 76, 213, 225, 232
Tseng, Ellen, 156, 158, 160
Tseng, Vivian, 507, 510
Tsuchida, Ineko, 414, 882
Tucker, Bradley J., 342
Tukey, John W., 646, 671
Tuomi-Gröhn, Terttu, 210
Tusting, Karin, 222
Tversky, Amos, 675
Tyack, David B., 530
Tymoczko, Thomas, 791
Tzekaki, Marianna, 846
Tzur, Ron, 464

U

Ueno, Kenji, 29
Ulep, Soledad Asuncion, 21, 149, 160, 412, 850
Umaki, Sandra, 465
Umland, Kristin, 303, 305, 306, 321
Upex, Stephen, 12
Upitis, Rena, 710
Ursini, Sonia, 760
Usiskin, Zalman, 599, 950, 964
Uworwabayeho, Alphonse, 346

V

Vacc, Nancy N., 336, 436
Vadcard, Lucile, 576, 582
Vahey, Phil, 599
Valdés, Hector, 863
Valencia, Enrique, 346
Valencia, Richard, 102, 106, 122, 126
Valero, Paola, 44, 56, 57, 59, 61, 73, 75, 87, 105, 112–113, 115, 121, 189, 347, 348, 930, 953, 968
Välijärvi, Jouni, 1018, 1031
Valls, Julia, 341, 779
Valsiner, Jaan, 103
Valverde, Gilbert A., 489, 1012
van Berkum, Emiel, 745

van den Akker, Jan J. H., 951
van den Heuvel-Panhuizen, Marja, 30, 396, 729
van der Kooij, Henk, 209
van der Ploeg, Arie, 878, 1014
Van der Veer, Rene, 103
Van Egmond, Warren, 14
Van Es, Elizabeth, 779
van Groenestijn, Mieke, 228, 229
Van Houtte, M., 106
Van Leeuwen, Theo, 709
van Lint, Jacobus H., 287, 757
van Maanen, Jan, 915
van Oers, Bert, 30, 550, 551
van Oosterum, Boers M. A. M., 602
van Velthoven, Wim, 649
van Zoest, Laura R., 464, 474
Vandebrouck, Fabrice, 699
VanLehn, Kurt, 735
Varga, Tamas, 807
Vayssettes, Sophie, 839
Vazquez, S. C., 80, 81
Velleman, Paul, 647, 651
Vere-Jones, David, 929
Vergel, Rodolfo, 417, 418
Vergnaud, Gérard, 108, 110, 191, 210, 212
Vérillon, Pierre, 572, 603, 631
Vermandel, Alfred, 296
Verner, Igor, 181
Verschaffel, Lieven, 30, 550, 551
Verzani, John, 648, 651
Viadero, Debra, 8
Vicki Zack, 439
Vico, Giambattista, 353
Viirman, Olive, 320
Villani, Vinicio, 969
Villarreal, Monica E., 592, 696, 699, 702, 712, 713
Villoutreix, Elisabeth, 839
Vincent, Jill L., 118, 626
Visnovska, Jana, 778
Vistro-Yu, Catherine, 8, 9, 508, 511, 586, 798, 902, 953
Vithal, Renuka, 21, 128, 129, 149, 172, 176, 178, 347, 412, 798, 850, 902, 953, 955, 1019, 1034
Vogeli, Bruce R., 807
Voigt, Jörg, 55, 121, 146
Volmink, John, 953, 955
von Jezierski, Deter, 538
Voss, Tamar, 336
Vukasinovic, Natasha, 652
Vygotsky, Lev, 77, 78, 105, 107–109, 120, 146–149, 212, 434, 461, 536, 551–555, 572, 573, 728, 806

W

Wadsworth, Leigh M., 698
Wager, Anita A., 74
Wagner, David, 41, 51–53, 188
Wagner, Jon, 372
Waits, B. K., 539
Wake, Geoff D., 52, 210, 220, 224, 226, 468, 556, 559
Walcott, Crystal, 962
Walker, Judith, 206
Walker, Lorraine, 884
Walkerdine, Valerie, 116, 127, 210, 213, 466
Wall, Kate, 532
Walls, Fiona, 57, 58
Walsh, Michael, 345
Walshaw, Margaret, 41, 57, 75, 120, 176
Wander, Roger, 600, 628, 630, 632, 633
Wanderer, Fernanda, 177
Wang, Tsung-Yi, 838, 1012
Wang, Ning, 966
Wang, Qiyun, 695
Wang, Tsung-Yi, 775
Ward, Jenny, 91
Warman, Arturo, 346
Wartofsky, Marx W., 550, 566
Waschescio, Ute, 121, 146
Watanabe, Ryo, 839
Watanabe, Tad, 882
Waters, Michael, 693
Watson, Anne, 115, 119, 120, 147, 558, 773, 814
Watson, Jane M., 644, 645, 666, 667, 679
Watts, Michael, 333
Way, Jenni, 399, 917
Waywood, Andrew, 32, 266, 267
Wearne, Diane, 156, 158, 160
Weaver, J. Fred, 293
WebAssign, 729, 730, 737
Weber, Keith, 92
Weber, Max, 44
Webster, Joan Parker, 181
Wedege, Tine, 177, 203, 204, 206, 208, 222
Weekes, Debbie, 126
Weeks, Keith, 215, 728
Weigand, H.-G., 599
Weigel, Margaret, 706
Weiner, Lois, 106
Weinzweig, Aurum I., 979
Weiß, Manfred, 1027
Weisner, Thomas S., 510
Weissglass, Julian, 341
Welch, Wayne W., 486
Wells, Gordon, 404
Wenger, Etienne, 44, 110, 116, 119, 169, 177, 189–191, 211, 397, 404, 558

Werry, Beven, 8, 805, 968
Wertsch, James V., 77
West, Roscoe L., 10, 17, 19
West, Webster, 651
Westbrook, Susan, 473
Westbury, Ian, 8, 9, 838, 862–864, 951, 979, 987, 1012
Westwood, Peter S., 26, 31, 32
Weyl, Herman, 306
White, Allan Leslie, 31, 150, 393, 409–411, 414, 415, 882, 883
White, Bruce, 698
White, Dorothy, 147
White, Jeffrey L., 249, 250
Whitescarver, Keith, 536
Whitney, Hassler, 290
Whitson, James A., 115
Whitty, Geoff, 471, 757, 765
Wiener, Hermann, 286
Wigfield, Allan, 512, 839
Wight, Charles A., 734
Wignaraja, Ponna, 346
Wijers, Monica, 766
Wikipedia, 486
Wilcox, Brian L., 510
Wild, Chris J., 645, 646, 671, 675, 678, 679
Wild, David G., 731
Wilensky, Uri, 699, 763, 767, 770
Wiley, David E., 489, 838, 1012
Wiliam, Dylan, 486, 721, 723, 724, 733, 738, 747, 765
Wilkins, Jessie L. M., 343
Willey, Ruth, 91
William T. Grant Foundation, 486, 507
Williams, Gaye, 120, 145, 159–161
Williams, Julian, 52, 210, 220, 224, 226, 465, 468, 549, 556, 558, 559, 839
Williamson, David M., 723
Willis, Sue, 124, 173, 174, 194
Willmore, Edwin, 950, 964
Willmore, Thomas, 286
Willms, Douglas, 839
Wilson, Barry J., 651, 808, 950, 952
Wilson, G., 697
Wilson, Melvin, 330
Wilson, Suzanne M., 437, 464, 465
Wilson, W. Stephen, 487, 530
Winbourne, Peter, 109, 115, 119, 147, 558
Winkelmann, Bernard, 296, 934
Winter, Jan C., 90
Winters, Tina M., 472
Wirszup, Isaak, 812
Wise, Lauress L., 472

Witmann, Erik C., 401
Wittgenstein, Ludwig, 443, 451
Wodewotzki, Maria Lucia, 183
Wolfe, Richard G., 838, 1012
Wolley, Norman, 728
Wong, Khoon Yoong, 887, 1032
Wong, Ka-Lok, 544, 730, 763
Wong, Ngai-Ying, 20, 473, 518, 959, 965
Wood, L., 124
Wood, Ruth, 532
Wood, Terry, 91, 147, 328, 364, 394, 400–402, 780, 850
Woodward, John, 87
Woolley, Norman, 215
Woolman, David C., 17
Wooton, William, 284
World Bank, 1011, 1018–1020
World Class Arena, 726, 727
Wright, Cecile, 126
Wright, EricOlin, 45
Wright, Robert J., 85, 409
Wu, C.-J., 21
Wu, Hung-Hsi, 489
Wu, Margaret L., 9, 21, 26, 1009, 1031

X
Xu, Guo-Rong, 43

Y
Yackel, Erna, 54, 118, 319, 551, 553
Yahya, Kurnia, 768, 778, 782
Yamaguti, M., 296, 757
Yanez, Evelyn, 170, 181
Yang, Rui, 1026–1027
Yap, Sook Fwe, 874
Yariv-Masal, Tali, 1017
Yasukawa, Keiko, 44, 203, 205, 219, 221, 223, 225, 226
Yates, Frank, 646
Yee, Foong Pui, 1032
Yee, Wan Ching, 90
Yerushalmy, Michal, 577, 628, 696
Yin, Robert K., 333
Yore, Larry D., 875, 888
Yoshida, Minoru, 150, 375, 376, 382, 414, 882
Young, Jacob W. A., 28, 835, 862
Young, Michael, 45, 218, 235
Young-Loveridge, Jennifer, 409
Youngson, Martin A., 731, 732
Yurita, Violeta, 320, 590

Author Index

Z

Zack, Vicki, 346, 400, 437–440, 442, 449, 451
Zang, Li-Fang, 146
Zangeneh, Bijan, 411
Zannoni, Claudia, 340, 446, 779
Zaslavsky, Claudia, 809
Zaslavsky, Orit, 419, 421
Zawojewski, Judith, 306, 307, 550, 558
Zbiek, Rose Mary, 518, 520, 599, 602, 623, 624, 626, 628, 629, 631
Zeichner, Ken, 345
Zemel, Alan, 705, 708
Zeuli, John S., 366
Zevenbergen, Robyn, 56, 75, 112–113, 115, 124, 223, 699
Zhang, Dake, 9, 961
Zhao, Y., 965
Zhu, Yan, 466
Zieffler, Andrew, 644, 740
Zikopoulos, Marianthi, 814
Zimmerman, Jonathan, 10
Ziqiang, Zhu, 810, 811
Žižek, Slavoj, 46, 468
Zoido, Pablo, 839
Zolkower, Betina, 53
Zoltan Dienes, P., 536
Zulatto, Rúbia B. A., 701, 702
Zuze, Tia L., 863
Zwick Rebecca, 652

Subject Index

A
Abacus, 537–538, 540
Abbaco tradition
    Pestalozzi's challenge, 15–16
    Treviso Arithmetic, 14
Accountability, 876, 892
Acculturation, 111, 113–115
ACE. *See* Advanced Certificate in Education (ACE)
Action research, 305, 328, 345–353, 398, 399, 412
    and collaborative research, 382
    and professional development of teachers, 368, 380, 384, 397
    and university-based researchers, 382, 384
Active Mathematics in Classrooms (AMIC), Brunei Darussalam, 410
Activity theory, 553, 556, 558, 560, 561, 565
ActivStats, 652
The Adult Literacy and Lifeskills Survey (ALL), 228
Adults' mathematics education (AME), 4
    adult definition, 204
    aspects of
        adult learner, 221
        aims/goals, 221
        curriculum and pedagogy, 221
    awkward realities, 232
    contextuality, 206
    critical perspectives
        invisibility of mathematics, 225–226
        landless peasants, 217
        mathematics learning, engagement, 226
        motivation to study mathematics, 214
        numeracy for nurses, 215
        social practice, 224–225
        techno-mathematical literacies, 215–216
        transfer of mathematical learning, 224
    FAME
        basic education, 219
        higher education, 217–218
    field of study
        CHAT, 212
        discursive perspectives, 213
        emergence and identity, 207–209
        ethnomathematics perspectives, 211
        human-capital approach, 233
        international research field, 210
        knowledge/skill requirements, 209
        pedagogy, 222–224
        situated perspectives, 211–212
        utilitarian perspectives, 210
    IFAME, 220–221
    learners quality, 232
    LLL concept, 206
    main problems, 204
    NFAME
        parents, 219–220
        workplace, 220
    numeracy, 215
    OECD, 228
    policy context
        change, 227
        international survey risk, 227–228
        international surveys of adults' skills, 227–228
        PIAAC (*see* Programme for the International Assessment of Adult Competencies (PIAAC))
Advanced Certificate in Education (ACE), 450
Advisory Committee on Mathematics Education (ACME), 496

Subject Index

Advisory Council for Adult and Continuing Education (ACACE), 210, 496, 497, 507
Affect, 106–111, 115, 117, 129
Africa, 863
  school mathematics
    African Mathematics Program (Entebbe Project), 809
    Africanizing mathematics education, 809
    colonial influence, 809
    conference of African Ministers of Education, 1961, 809
    religious propaganda, 808
African-American students, 83
African Mathematical Union (AMU), 846
Africa Regional Congress of ICMI on Mathematical Education (AFRICME), 925
Algebra, 396, 417
  and CAS
  curricula, 599, 600, 602, 629
  instruction
    computational transposition, 602
    Computer-Intensive Algebra project, 602
    developers and mathematics educators, 602
    graphing and spreadsheet program, 602
    instrumental genesis, 603
    mathematical knowledge, 603
    meta-analytical approach, 602
    re-balanced approach, 601
    symbolic calculation programs, 601
    tool-assisted procedures, 603
  international mathematics curriculum, 963–964
  pedagogy, 600, 631–633
  representations, 598, 600, 605, 606, 610, 611, 620–626, 628, 633
  transformations, 599, 603, 605, 606, 610, 623
American Institutes for Research, 1032
Anthropological theory of didactics (ATD), 304, 305, 311, 313, 314, 321
Anti-colonialist theories, 828
APEC. *See* Asia-Pacific Economic Cooperation (APEC)
Applets, 652
Argentina, 759
Arithmetic for all
  abbaco tradition, 14–16
  ability concept, 18
  colonialist assumptions, 16–17
  "Committee of Fifteen" report, 18
  face-to-face teaching, 17
  legislation, 17

Asia-Pacific Economic Cooperation (APEC), 882, 883
Asian Development Bank, 818
Assessment
  authentic, 782
  online, 763
Assessment tasks for teacher education, 340, 341, 352
Assessment and technology
  and the assessment of basic skills, 740
  assessment with support, 728–729
  scoring issues and procedures, 723, 724
Association for Research on the Didactics of Mathematics (ARDM), 933
Australia, 72, 86, 92, 102, 124, 384, 399, 402, 408, 412, 415, 416, 461, 464, 472, 474, 476, 536, 631, 666, 741, 743, 805, 809, 817, 876, 881–883
Australian Association of Mathematics Teachers (AAMT), 842
Australian Council for Educational Research (ACER), 759, 982
Australian Curriculum, Assessment and Reporting Authority (ACARA), 644–645
Authentic assessment, 727–728, 747, 748

B

Babylonian, 526
Basic components of mathematics education for teachers (BACOMET), 933–934
Bayesian net, 734
Begle, E.G., 811, 812
  and SMSG, 811
Benchmarking, 878
Better: Evidence-Based Education journal, 502
Bilingual learners, 80, 81
Blackboards, 531, 532, 544
Black English Vernacular (BEV), 80, 83
BMDP, 647
Bootstrapping, 648
Borba, M.C.
  and "*Pass-the Pen*", 701–703, 706–708
  and *Geometricks*, 701, 702
Borderland discourse, 114
Brazil, 72, 78, 92, 102, 107, 112, 125, 131, 696, 713, 804, 816, 869, 877
British Council, 810
British educational system, 801
British Society for Research into Learning Mathematics (BSRLM), 933
Brunei Darussalam, 410
Business-Higher Education Forum (BHEF), 249, 250

C

CAA. *See* Computer-aided assessment (CAA)
Cabri, 587
Cabri Géomètre, 571, 760
Calculator
  calculation tools, 539
  controversies, 539
  graphical, 696
  history of, 526, 529, 537, 540
  scientific, 696
  use in assessment, 724, 738–744, 746, 748
Cambodia, 882
Cambridge University, 801
Canada, 72, 369, 530, 536, 653, 667, 816
Canadian, 373–375
Canadian Mathematics Education Study Group (CMESG), 933
Caribbean school mathematics, 807, 820
CAS. *See* Computer algebra system (CAS)
Casino problem, 674
Catalonia, 876
Cell phones, 692
Centre de Recherche sur l'Enseignement des Mathématiques (CREM), 574
Centre for Innovation in Mathematics Teaching (University of Exeter), 863
Centre for International Comparative Studies (CICS), 887
Centre for Observation and Research in Mathematics Education (COREM), 432, 438, 442–444
Centre for Research on International Cooperation in Educational Development (CRICED), 541
CHAT. *See* Cultural-historical activity theory (CHAT)
Chicago, 1893 meeting of mathematicians, 802
Chile, 348
China, 369, 377, 378, 386, 461, 463–465, 473, 476, 528, 537, 538, 810, 811, 814, 815, 820
  cultural revolution, 810
Chinese, 377–380
  abacus *(suanpan)*, 537
  set square, 527
Chinese-language Google Scholar, 817
CIEAEM. *See* Commission for the Study and Improvement of Mathematics Teaching (CIEAEM)
Cipherbook, 531
Classrooms
  influence of the internet on, 692, 693, 696–699, 705, 711, 713–715
  new forms of discourse, 699

Cockcroft Report, 209
Cognition, 105–112, 115, 116
Cognitive psychology, 147, 160, 161
Collaboration
  asynchronous, 692, 701, 704, 705
  hybrid, 701
  remote, 703
  synchronous, 701
Collaboration in mathematics education, 799–801
Collaborative learning, 767, 768, 770
Collection of technologies (COT), 628
Collective learning, 221
College Board, 738, 740, 743, 744
College Entrance Examination Board (CEEB), 1023
Colombia, 346–348, 417, 418, 807, 868, 873, 874, 876, 882, 891
Colonialism, 256
  and neo-colonialism, 346
  and school mathematics, 810
Commission Internationale pour l'Étude et l'Amélioration de l'Enseignement des Mathématiques, 910
Commission for the Study and Improvement of Mathematics Teaching (CIEAEM), 805, 807, 816, 817
Committee of Inquiry into the Teaching of Mathematics in Schools, 738
Commognition, 318
Common Core State Standards, 960
Common Core State Standards Initiative (CCSSI)
  K–12 STEM, 491
  Race to the Top program, 493
Common Core State Standards (National Governors Association (NGA), USA), 868
Common Core State Standards for School Mathematics (CCSSM), 1026–1027
Common Core State Standards (USA), 488, 491–493, 508
Community of practice (CoP), 119, 372, 404, 410
Comparative education, 862
  history of, 862
Compass, 533, 541
Competencies
  influenced by PISA, 872, 874
  influenced by TIMSS, 874
Computational transposition, 602
Computer-aided assessment (CAA), 730

Computer algebra system (CAS), 539, 544,
    730, 731
  and algebra curricula, 602
  algebra instruction, 601–604
  and assessment of learning, 722, 730, 739
  basic utilities, 598
  and calculus, 601, 624, 629
  and examinations, 630, 631
  extending procedures
    Böhm's task, 618
    Diophantine equations, 614
    factorization, 612–513
    generated table, 618
    symbolic approaches, 616–617
    tabular and graphical approaches,
       616–617
    Tartaglia–Cardano method, 614
    TTK framework, 613
  graphical capabilities, 598
  history of, 598–601
  implementation issues
    external assessment, 630
    pedagogical technology knowledge, 632
    RIPA, 631
    student attitudes, 629
    technology, 630
  implications of, 603, 604, 629–633
  mathematical concepts
    algebraic transformations, 606–607
    continuity, 610–611
    epistemic value, 604
    equation and equivalence, 607–610
    formulas and equation solution, 606
    generalization, 604
    pragmatic value, 604
    representational versatility, 606
    versatile thinking, 606
  mathematical model
    activity theory, 565
    blackbox, 564
    confrontation, 562
    "false," 561
  pedagogical opportunities, 600
  personal, 532, 539, 544
  reasoning opportunities
    dynamically linked representations,
       626–628
    graphical representations, 623–624
    integrated technology environments,
       626, 628
    symbolic representations, 620–622
  and research, 598, 600–605, 609, 613, 626,
       628, 630, 631, 633–634
  school algebra, 599
  school curriculum, role of, 628–629
  slow uptake of, 632
  software, 532, 544
  symbolic manipulations, 598
  symbolic procedures, 600
  tabular capabilities, 598, 616
  theoretical discourse for, 603
  and technique, 602–604, 606, 609, 612,
       613, 624, 625, 629, 631, 633
  technological resources, 598
  technology, 564
  use in assessment, 763
  use in Victoria, Australia, 763
Computer-based assessment
  and automated scoring, 723, 724, 729,
       731, 734
  and on-line learning systems, 723, 725
  compared with paper-based assessment,
       735–737
  diagnostic feedback, 735
  difficulties with, 731
  effects of sliders, 725
  enhancing item presentation, 724, 726
  partial credit, 726, 730–732
  smart tests, 725, 735
  world-class tests, 726
Computers in education
  game playing, 694
  historical perspectives on, 704
  ICMI studies on, 694
Computer-Intensive Algebra project, 602
Computer technology, 551
  and mathematics education, 813
Concept maps, 341–342
Confirmatory data analysis (CDA), 646, 647
Confucian approach, 153
Confucian heritage and mathematics, 20, 21
Connecting data, 666, 671–674
  by modelling, 671–674
Consciousness to speech method, 15
Constructionism, 765
Constructivism, 310, 401, 464–466, 479,
       759, 765
Continuing professional development (CPD),
       495, 499
Continuity, 610–611
CoP. *See* Community of Practice (CoP)
Copybook, 531
COREM. *See* Centre for Observation
       and Research in Mathematics
       Education
Costa Rica, 808
Council of Chief State School Officers
       (CCSSO), 1026

Subject Index

Council of Chief State School Officers
    & National Governors
    Association, 1026
Counting board, 526
Counting On (CO) program, 409
Count Me In Too (CMIT),
    Australia, 408
Critical friends, 383
Critical mathematics education
    Brazil, 183
    four aims, 184
    Frankenstein's suggestion, 183–184
    Freire, Paolo, 183
    Gutstein's problems, 187
    inquiry cooperation model, 184, 185
    "mathemacy," 184
    mathematical model, 185
    Sandra's comment, 188
    societal contexts, 186
    statistical learning, 187
    "Terrible Small Numbers" project, 184
    two geographical groups, 183
Critical theory, 72, 75, 87, 104, 123, 828
Cube root block, 534, 535
Cuisenaire rods, 536
Cultural-historical activity theory (CHAT),
    212, 553, 566
Cultural and historical perspectives, 151–152
Cultural learners, cognition and affect
    academic failure children, 108
    binomial interaction, 111
    empirical data, 107–108
    focus consciousness, 109
    poor economic backgrounds, 107
Culture, 106, 109, 117, 118, 121, 125–127,
    129, 130, 808–810, 813, 818–820
    cognitive development, 108
    cultural influences on mathematics
        education, 153
    deficit model, 102–104
    frame, 110
    and mathematics curricula
        ethnomathematics, 111–113
        mathematical enculturation and
            acculturation, 113–115
        semiotic interactions, 115
        universal character, 116
Curriculum, 119, 122, 125–128
    attained, 863, 864, 866
    change, 464, 471
    classroom level, 866, 867
    cognitive dimensions, 866, 871
    conceptual dimensions, 866–867
    control, 870, 888, 890–893
    definition, 863
    degrees of expertise, 864
    design, 864, 871, 872, 874–876, 880, 884,
        888, 890–893
    development, 463, 464, 470–474, 476
    dimensions of
        cognitive, 866, 868, 871
        conceptual, 866, 868
        formative, 866, 868
        social, 866, 868
    education system, 865–866
    formative dimensions, 866–867
    global level, 867, 868, 872–875, 891
    IEA, 863–864
    implemented, 863–866, 888
    influences
        assessment, 876–877
        classroom levels, 869
        data-driven approaches, 877–879
        dimension, 870
        global level, 872–875
        localization, 875–876
        mathematics curriculum
            dimensions, 872
        school level, 872
        teacher education
            (*see* Teacher education)
        textbooks, 884–886
        types, 871–872
    intended, 863–866, 872–875
    localization of, 871, 872, 875–876, 888,
        890–892
    locus of control, 870
    management, 888, 890–893
    mathematics, 111–117
    national, 873, 876
    national standardized tests, 867
    OECD, 865
    policy, 890
    school-based development, 873
    school level, 866
    social dimensions, 867
    standards, 868
Curriculum Development Council, 874
Cyprus, 667, 864, 868
Czech Republic, 881

D

Data collection methods, 315, 342, 351
DataDesk, 647
DataScope, 648
Deaf and hard-of-hearing (DHH)
    learners, 79

Deficit models, 101–132, 223. *See also* Socio-cultural perspectivescognitive, 85
    linguistic minority students, 79
    mathematics learning, 89
Delta conferences, 927–928
Denmark, 764, 820, 877, 886
Departemen Pendidikan Nasional, 875
Department for Education (DfE), 467
Design
    modes, 771–775
    networks, 773
Design experiments, 486
Design research, 401, 409
Deutsche Telekom Foundation, 504
Developing mathematics educators, 270
    ACE course, 450
    cooperation, teachers and researchers
        teacher A's reflections, 440–441
        teacher B's reflections, 441
    didactical culture and social anticipation, 453–454
    ICT, 446–447
    mathematical education, 452–453
    observation
        COREM, 442
        decision-making, 442–443
        learning and teaching, 434–435
        learning mathematics, 433–434
        mathematics activities, 442
        ordinary classrooms, 444–445
        pupil–teacher–mathematics, interactions, 437
        student teachers, 436
        types, 443
    practical skills, 449
    pupils interactions, 452
    specific knowledge, 449
    teachers *vs.* researchers
Developmental psychology, 147, 154
DGE. *See* Dynamic geometry environment (DGE)
Didactical contract, 118
Didactical cultures, 432, 453–454
Didacticians, 371–373, 381, 383, 385, 396, 408
Didactic systems, 435
Digital-Math Environment, 732
Digital technology. *See* Technology
Digital Windows into Mathematics, 703, 706–707
Diophantine equations, 614
Discourse, 102, 104, 118, 127, 129
    analysis
        antagonistic relationship, 53
        classroom, 51, 52
        commognition, 52
        "discursive psychology," 52
        Educational Studies in Mathematics, 53
        Ryve's analysis, 51
        SFL tools, 51, 52
        social psychology perspective, 52
        "the mathematics register," 51
    borderland, 114
    deficit, 105, 106
    mathematics, 120
    social justice, 121–123
Discovery learning, 806
Discrimination in school mathematics, 104
Discursive approach to mathematics education research, 303, 314–316
    communicational framework, 317
    endorsed narratives and routines, 319
    focal analysis, 317
    methodological and epistemological grounds, 315
    preoccupation analysis, 317
    verbalization skills, 320
    visual mediators, 319
    word use, 319
Document camera, 532
Dragging tool
    cognitive implications, 576, 582, 590
        dummy locus, 579
        equilateral triangle, 578
        spatio-graphical field, 580
        spectral dragging, 580
        theoretical field, 580
    declarative, 578
    demonstrative, 578
    drag-and-drop, 653, 654, 662
    dragging points, 654–656
    elementary school level, 584–587
    epistemological implications
        declarative and demonstrative process, 578
        figural aspects and geometric proofs, 577
        rhombus and quadrilaterals, 577
        Shape Maker, 577
        spatial and temporal type, 578
        theory of variation, 578
        transformational-saliency hypothesis, 577
        unconscious visual transformation, 577
    in high school mathematics, 572, 574
    and invariance, 656
Drexel University's Mathematics Forum, 696
Drijvers, P., 753, 763, 764, 766, 768–770
    four assessment policies of, 763–764

Subject Index 1099

Dutch *Realistic Mathematics Education* curriculum, 466
Dynamic geometry environment (DGE), 320
   in classroom contexts, 584
   didactical issues, 576, 592
   didactic implications, 590–591
   dragging tool
      cognitive implications, 579–580
      elementary school level, 584–587
      epistemological implications, 576–579
   in elementary schools, 584–587
   epistemological issues, 572, 573, 575–578, 581–584, 587, 590
   and finding invariants, 577, 578, 580, 582, 584
   in high schools, 584, 587–590
   locus and tracing, 584
   measurement tool, 580
      cognitive implications, 582–583
      epistemological implications, 581–582
      high school setting, 587–590
   non-Euclidean geometry, 576
   and proof, 571–592
   protocols, 573, 584
   proving process, 576
   research, 571–573, 579, 580, 583–592
   role of, 572
   semiotic issues, 572–573, 579, 592
   tools, 573, 576, 581, 583–584, 590–592
Dynamic geometry and the internet, 699

E

Early childhood mathematics, 434
Early Numeracy Research Project (ENRP), Australia, 408
East Asia Regional Conference in Mathematics Education (EARCOME), 923–924
Ecological validity, 318
EDA. *See* Exploratory data analysis (EDA)
Educational Studies in Mathematics (ESM), 289, 311, 315, 330, 332, 334, 398
Educational technology, 552
Educational Testing Service (ETS), 730, 1013
Education Council, 872, 873
Education policy, 486, 492, 506, 511
Education technology. *See* Technology
Egypt, 820, 882
EMAT project, 759–761
Enciclomedia program (Mexico), 759, 761–762, 773, 784
Enculturation, 111, 113–115
England, 72

English National Curriculum, 502
Equations
   and equivalence, 607–610
   conditional, 608
   identical, 608
   surface structure view of, 607
Equity, 120
   issues with assessment, 739, 744
Equity and access, 4–5
   aboriginal Australians in homeland communities, 259–260
   American Sāmoa, 260
   Botswana, 256–257
   liberal perspective
   Malaysia, 257–259
   marginalization, 248
   NCTM, 247
   OTL mathematics, 244
   PISA, 245
   policy formulation model, 260, 261
   South Africa, 255–256
   for students, 249
   students performance, 253
   teaching mathematics
      BHEF, 250
      diversity, 250
      English-speaking school graduates, 255
      language policy, 255
      mainstream view, 251
      politically-motivated attacks, 254
      poverty, 253
      productive discourse, 254
      VAM approach, 252
   "The Politics of," 246–247
ESM. *See* Educational Studies in Mathematics (ESM)
Ethnicity, 125–127
Ethnomathematics, 74, 86, 105, 111–115, 117, 125, 808
   classroom, 179
   Hindu-Arabic system, 179
   intercultural studies, 147, 154
   learners' views, 180
   Papua New Guinea, 179
   permeability, 178
   research program, 177
   Western mathematics, 177–178
ETS. *See* Educational testing service (ETS)
Euclid, 282, 283
Euclidean geometry, 284, 287, 306, 307, 535, 576
European Society for Research in Mathematics Education (ERME), 918–919
European Union (EU), 206, 230, 232
Every Childs Counts, 498

1100 Subject Index

Excel, 651, 681, 760
Exploratory data analysis (EDA)
   graphical data exploration, 647
   historical overview of, 646

F
Facebook
   and education, 692, 695, 715
   and the relationship revolution, 692
   in North Africa, 692
   in the Middle East, 692
   research on usage of, 695
Factoring, 612–614, 616, 617
Fathom, 648
   aggregate representations, 680
   drag-and-drop graphing, 654
   dragging points, 654
   dynamic nature, 679
   link multiple representations, 656
   microworlds, 656
   personal features, 680
   visual nature, 679
Feminist scholarship, 828
Financial literacy program, 219
Finland, 891, 1014, 1018, 1021,
      1027–1031, 1033
Finland has a National Board of Education
   (FNBE), 1029
Finland, international benchmark rankings,
   151–152
Finnish education system, 152
First International Mathematics Study (FIMS),
   792, 831, 837, 976, 1011
Formal adult mathematics education (FAME)
   basic education, 219
   higher education, 217–218
Formative assessment, 724, 732, 734, 746
For the Learning of Mathematics (FLM), 933
France, 758, 764, 771–773, 775, 779
French Mathematical Space (EMF), 924–925
Freudenthal, H., 461, 550, 558, 805, 817
   Freudenthal era, 804
Function (mathematical concept), 803
Functions, 602, 605–608, 610–613, 615, 616,
      618, 621, 623, 624, 626, 628, 629
Funds of Knowledge project, 179–180
Further Mathematics (FM) Network, 498

G
Garagae observation, 257
Gender, 123–125
Gendered identities, 70, 71

Generalization, 605, 611–613, 615, 616,
      620, 628
Geometer's Sketchpad, 571
Geometric argumentation, 584
Geometric models, 535
Geometric solids, 534
Geometry, 411
   dynamic, 531, 541, 543, 544
   education, 282, 291
   Euclidean, 535
   learning tools for, 526
   teaching tools for, 535, 538
   virtual environments, 544
Geometry in school curricula
   After Euclid, 801
   in primary schools, 807
Germany, 396, 400, 412, 535, 541, 667, 681,
      801, 804, 864, 869, 873, 876, 879,
      886, 891, 1014, 1018, 1019,
      1027–1031, 1035
   and PISA shock, 873
Ghana, 179, 882
Globalization, 397–399, 693, 798
   and mathematics education, 798, 818
Glocalization, 399
Good education, 246
GPIMEM, 702
Graphical representations, 605, 611–613, 615,
      616, 620, 628
Graphing calculators, 651, 652, 696
Grounded theory, 333, 335, 336, 400
Guided measuring, 583
Gutiérrez's criteria, 73–74

H
Habilidades Digitales para Todos (Mesxico)
   program, 759, 762
Handbook of Mathematics Teacher
   Education, 400
Hart inversor, 533
Heller Reports, 746
Higher Education Opportunity
   Act, 491
History of mathematics, 2, 105
   arithmetic for all
      abbaco tradition, 14–16
      ability concept, 18
      colonialist assumptions, 16–17
      "Committee of Fifteen" report, 18
      face-to-face teaching, 17
      legislation, 17
   basic mathematics for all
      Asian–American students, 21

Confucian-heritage nations, 21
quality-of-teaching factor, 20
TIMSS, 19, 20
beyond arithmetic, 21–22
The Curriculum Revolution, 20th century
Herbartianists, 25, 26
humanists and *developmentalists*, 25
national reports, 24
seven threads, 24–25
social efficiency educators and *meliorists*, 25
examinations
Mathematics Applicable Project, 29–30
"mathematics for all"
"algebra for all," 8
curriculum design, 8–9
modern mathematics era, 8
Saber-Tooth Curriculum, 9
scholarly discussion, 10
secondary school, 23–24
"small-m" and "big-M" forms, 9
UNESCO, 8
vertical and horizontal curriculum relationships, 9
non-trivial mathematical modelling, 26–27
numeracy for all, 31–32
Perryism, 28
Perry, John concept, 27–28
RME program, 30, 31
schooling for all
distance education, 14
elementary forms, 10
Harding report, 13
Indonesia, 11, 12, 14
Jomtien commitment, 12
Nebres comments, 13
populations percentage, 10, 11
primary and secondary, 12
principle, 12
religious knowledge, 10
UNESCO, 12, 13
SMSG, 26
History and Pedagogy of Mathematics (HPM), 914–916
Honduras, 882
Hong Kong, 412, 466, 758, 874, 881, 884, 885
Hungarian mathematics educators, 818
Hungary, 816

I

Iberoamerican Federation of Societies of Mathematics Education (FISEM), 932
ICME. *See* International Congress on Mathematics Education (ICME)
ICMI. *See* International Commission on Mathematical Instruction (ICMI)
ICT. *See* Information and communication technologies (ICT)
Identity, 105, 110, 115–117, 120
IGPME. *See* International group of psychology in mathematics education (IGPME)
Improvement, 462, 463, 475
ideological bases for, 466–472
IMU. *See* International Mathe
Inclusive liberalism. *See* Lifelong learning (LLL) concept
India, 528, 810
Indonesia, 875, 882
Inequality, 123–126, 129–131
Inferential statistics
and *Fathom Dynamic Statistical* software, 674
hypothesis testing, 674–677
pathways to, 674–678
Influences of international studies
cycles of data analysis, 872
localization, 871, 872
national changes, 871
on school autonomy, 871
professional development and teacher education, 872
use of textbooks from other countries, 872
Informal adult mathematics education (IFAME), 204
Informal Mathematics Learning Project, 92
Information and communication technologies (ICT), 432, 446–447, 695
preceded by information technology, 695
Inquiry community, 404, 405
Institute of International Education, 814
Institutional contexts, 270
academic networking and research community, 475–476
curriculum development and evidence-based policy
Chinese curriculum, 472
conceptualization and cultural specificity, 473
curriculum decision and mathematical ideas, 471
policy makers and teacher biographies, 471
teaching approach, 472
ideology
culturally dependent, 467
curriculum content, 467

Institutional contexts (*cont.*)
    effective practice/program success, 470
    group affiliations, 468
    neo-Marxist theory, 467
    problem-solving approaches, 467
    reform agenda, 469
    research design, 468
    societal structures, 469
    teaching styles, 467
    value systems, 470
  mathematics education and research
    classrooms, 460
    collective conceptions, 461
    communities, of people, 462
    English usage, 462
    international competition, 460
    international tests, 461
    knowledge, 461–462
    mathematical activity/performance, 461
    OECD, 460
    pedagogical attitudes, 461
    PISA, 460
    school mathematics, 460
    TIMSS, 460
  publication networks, 474–475
  reform
    child-centred approaches, 466
    constructivism, 464
    inquiry methods, 464
    instructional approaches, 465
    Japanese mathematics educators, 464
    "math wars," 464
    modern conceptions, 464
    veteran/traditional teachers, 465
  training and education researchers, 476–478
Instrumental genesis
  and instrumentalization, 603, 632
  and instrumentation, 603, 632
Instrumental orchestration, 768
Interaction patterns, 443
Interactive forms of classroom discourse, 770
Interactive whiteboard, 699
Inter-American Committee on Mathematics Education (IACME), 911–913
Intercultural contexts
  challenges ahead, 162–163
  dichotomies
    cultural traditions, 153–154
    in-school and out-of-school, 154, 155
    International Comparative Research Studies, 155–156
  international benchmark rankings
    Finland, 151–152
    lesson study, 150–151
    pedagogical practices, 150
    public talk, 150
  international "video survey" studies, 158–159
  learners, and learning processes, 145–146
  LPS
    "between-desk instruction," 161
    collaborative negotiation, 157
    data generation techniques, 157
    needs, 157
    new mathematical ideas, 162
    participation patterns, 161
    protocol, common research, 157
    qualitative video study, 158
    research, 157, 160–161
    research teams, 157–158
    Williams' study, 160–161
  qualitative and quantitative studies
    experimental research, 148
    large-scale quantitative studies, 148–149
    resourced countries, 149
    TIMSS and PISA, 148
  social interaction and meaning making, 159–160
  and theoretical perspectives, 146–148
  TIMSS 1999 Video Study, 156
Intergeo project (Europe), 773–776
Intermediate-level professionals, 215–216
The International Adult Literacy Survey (IALS), 228
International Association for the Evaluation of Educational Achievement (IEA), 1011
  activities, outcomes, 836–837
  design of, 837
  FIMS, 837
  First International Mathematics Study, 794
  goal, 793
  international studies, 1011–1012
  PISA, 838–839
  SIMS, 837–838
  TIMSS, 838, 864
  tripartite model for curriculum, 863
International Association for the Evaluation of Educational Achievement (IEA), 507
  and Trends in Mathematics and Science Study (TIMSS), 509
International Association for Statistical Education (IASE), 929–930
International collaboration
  AAMT, 842
  action research, 829
  British educational system, 801
  channels of communication, 811

Subject Index

contemporary period
  brain drain phenomenon, 813–814
  centralization, 813
  CIEAEM conferences, 816, 817
  foreign-born population, United States, 814
  foreign students, United States, 814, 815
  JRME, 815, 816
  mathematics teacher education, 815, 816
  national curriculum, 813
definition, 829
design research, 829
education administrators, 830
globalization, 798
ICMI, 801, 803, 833–836
IEA
  activities, 837
  design of, 837
  FIMS, 837
  PISA, 838–839
  SIMS, 837–838
  TIMSS, 838
IMU, 802
inferential statistical approaches, 828
International Aid Organizations, 852–853
international contacts, 803
Iron Curtain, 798
large-scale studies, 831
lesson study, 829
linguistic factors, 828
LPS
  analytical tools, 848
  authenticity, 847
  cameras and operators, 847
  country-by-country listing, 848–850
  design, 852
  leaders, 851
  non-random sampling, 852
  reports, 847–848
  requirements, 847
  research, 847, 848
  self-report data, 851
  structural uniformity, 847
  videotaping, 851
mathematics curricula, quality, 828
MERGA, 843
mixed-methods research, 829
national differences, 819–820
national systems, 800
pedagogical and methodological approach, 803
PME
  AMU, 846
  annual conference, 843–844
  international community, 844
  mathematics teacher education programs, 845
postmodern approaches, 828
programs/organizations, 832
psychometricians, 830
quality, 853–855
RECSAM, 840–842
Royaumont conference, 835
Russian mathematics education, 799, 800
school curriculum, 803
social and cultural forces, 832
Soviet Union
  Africa, 808–809
  Asia/Australia, 809–811
  Eastern Europe, 806–807
  Freudenthal era, 804
  Latin America and Caribbean, 807–808
  Western Europe and North America, 805–806
teaching and learning, quality, 828
UNESCO, 812
International Commission on Mathematical Instruction (ICMI), 268, 274, 730, 801, 803, 833–836, 904, 912, 950, 976, 1010
  activities, 907
  affiliation, 903, 917–918
  age, 904
  annual budget, 906
  Commission Internationale de l'Enseignement Mathématique (CIEM), 798
  de facto members, 906
  early history of, 801
  establishment of, 905
  history, 902, 904
  permanent IMU commission, 905–906
  post-Freudenthal era, 909
  regional conferences
    AFRICME, 925
    EARCOME, 923–924
    EMF, 924–925
  role of, 908
  studies, 17, 694, 763
International Commission for the Study and Improvement of Mathematics Teaching (CIEAEM), 910–911
International Commission on the Teaching of Mathematics, 1010
International Community of Teachers of Mathematical Modelling and Applications (ICTMA), 550, 552, 557, 920

International Congress on Mathematical
    Education (ICME), 290, 328–330,
    348, 804–805
International Council on Mathematics in
    Developing Countries
    (ICOMIDC), 934
*Internationale Matematik Unterricht
    Kommission* (IMUK), 798
International Group of Psychology in
    Mathematics Education (IGPME),
    43–47, 304, 311, 329, 348, 400,
    449, 572, 813, 913–914
Internationalization in mathematics education,
    814–188
International Mathematical Union (IMU), 274,
    802, 804, 1010
International mathematics curriculum
  algebra teaching, 963–964
  conceptual understanding and procedural
      skills, 965–967
  creativity and thinking skills, 957, 964–965
  curriculum, definition, 951–952
  gifted students, 967–968
  globalization, 953
  internationalization, 953–955
  learning goals
    in China, 958
    determination, 955
    higher-order thinking skills, 956
    NCTM goal, 957–958
    school curriculum, 956
    in UK, 958–959
  public examinations
    in China, 961
    classroom, role in, 961
    coursework, 960
    mathematical modelling, 960
    TIMSS and PISA, 961–962
  role of, 950
  statistics and probability, 962–963
International Organisation of Women and
    Mathematics Education (IOWME),
    919–920
International perspectives
  categories, 792
  of policy, 793
  of practice, 793
  of profession, 794
International Project on Mathematical
    Attainment (IPMA) study, 863, 1014
International Society for Technology in
    Education, 754, 756
International studies
  cross-sectional sample surveys, 1033

data, 1035
educational policy, 1015–1016
educational policy, impact of, 1021
Educational Testing Service, 1013
Germany and Finland, 1027–1031
IEA, 1011–1012
INES, 1017
influences of textbooks, 868, 872
influences on classroom practices, 870
influences on curriculum, 871, 872, 875
influences on teacher education, 872,
    879, 892
international knowledge banks, 1018
IPMA study, 1014
Kassel project, 1014
LLECE, 1014–1015
mean score, 1033
national achievement measures, 1035
OECD indicators, 1017
PISA, 1012–1013, 1035
policy convergence, 1019–1020,
    1033–1034
SACMEQ series, 1014
Singapore, 1031–1032
teaching and learning, 1034
TIMSS, 1015–1016, 1035
USA, 1022–1027
International studies of student achievement.
    See Programme for International
    Student Assessment; Third
    International Mathematics and
    Science Study
International Study Group on
    Ethnomathematics (ISGEm),
    928–929
Internet, 398, 651–653, 754, 755, 764, 777, 783
  advantage, 693
  algebra, 696
  applets, 696
  blended learning environment, 701
  collaboration
    classroom setting, 706
    Digital Windows into Mathematics, 703
    geometricks, 701–702
    humans-with-media, 702
    information revolution, 704
    problematic dynamics, 706
    student interactions, 705
    teacher-centred communication, 705
    virtual math teams, 704
    visual and textual realizations, 705
  different uses of, 697, 700–702, 704,
      705, 707
  dynamic geometry software, 699

Subject Index

education resources, 696
Facebook, 692
fundamental foci, 692
geometric figures, 696
graphical calculators, 696
human–computer interaction design, 700
influence on mathematics education, 693–700
managing and monitoring learning, 714
mathematics education sites, 692
multimodality, 692, 693, 700–701, 706–709, 714, 715
networking, 695, 715
number of users of, 694, 703, 707
numerical and alpha databases, 694
online
  activities, 695
  help sites, 697
  education, 692, 695, 696, 701, 704–706, 713, 714
  mathematics courses, 692
paper-and-pencil medium, 699
pedagogical designs, 692
performance
  classroom mathematics, 711–712
  cognitive ecology, 712
  images of mathematicians, 709–711
  in mathematics, 713
  multimedia authoring tools, 709
  teaching-as-performance, 713
  technological design principles, 713
popularization of, 694, 695
quality of teaching, 694
and real-world data, 699
researchers and curriculum developers, 693–694
and school mathematics, 704
software, 696
spreadsheets, 696
teacher role, 698
and technology, 693–696, 702, 712, 715
and textbooks, 691, 692, 714
Twitter, 692
virtual environments, 692
Wikipedia, 695
YouTube, 692
Iran, 398, 410, 411, 416, 421, 820
Iron Curtain, 798
Israel, 412, 420
Italy, 816
Items
  evidence accumulation, 747
  evidence identification, 747
  multiple-choice, 723, 727, 729, 741, 742

operation, 747
preparation, 747
presentation, 747

J
Japan, 364, 369, 376, 377, 379, 382, 383, 399, 409, 412, 464, 466, 473, 476, 535, 537, 538, 541, 799, 819, 874, 881–883
  lesson study, 870, 880, 881
Japanese, 375–377
Japanese Society of Mathematics Education, 574
Jesuit Educational Institutions, 799
Journal for Research in Mathematics Education (JRME), 327–332, 334, 398, 815, 816
Journal of Mathematics Teacher Education (JMTE), 327–332, 334, 337, 364, 400

K
Kassel Project (UK), 863
K–12 education
  CCSSI, 491
  US government agencies, 511
Keli (in China), 377–379
Kenya, 882
Kinematical Models for Design Digital Library (KMDDL), 541
KMK, 645
Korea, 464–466, 476, 819, 864, 868

L
LaboMEP, 772
Landless People Movement of Brazil, 178
Landscapes of investigation, 118
Language factors in mathematics learning
  code-switching, 257, 258
  language of instruction, 254–255
  policies, 255
Lao Country Paper, 20
Laos, 882
Laptops, 754, 759, 762
Latin America, 125, 346–349, 352, 758, 773, 799, 805, 807–808, 814, 815, 817, 818, 863, 869, 870, 872, 874
Latin American Committee on Mathematics Education (CLAME), 931–932
Latin American Laboratory for Assessment of the Quality of Education (LLECE), 1014–1015

LCM. *See* Learning Communities Mathematics Project (LCM)
Learner's Perspective Study (LPS), 407, 412, 432, 438, 444–445, 793, 984
  analytical tools, 848
  authenticity, 847
  "between-desk instruction," 161
  cameras and operators, 847
  collaborative negotiation, 157
  country-by-country listing, 848–850
  data generation techniques, 157
  design, 852
  leaders, 851
  needs, 157
  new mathematical ideas, 162
  non-random sampling, 852
  participation patterns, 161
  protocol, common research, 157
  qualitative video study, 158
  reports, 847–848
  requirements, 847
  research, 157, 160–161, 847, 848
  research teams, 157–158
  self-report data, 851
  structural uniformity, 847
  videotaping, 851
  Williams' study, 160–161
Learning communities in mathematics (LCM) (Norway), 371, 408
Learning Communities Mathematics Project (LCM), 371, 373
Learning spaces, 765–768, 770
Learning trajectories, 768
L'Ensignement Mathématique (journal), 801, 803
Lesson study (LS), 364, 367, 375–377, 379–383, 385, 386, 396, 402, 409, 410, 414–416, 421
  cultural challenges with, 883
  and the Japan International Cooperation Agency (JICA), 882
Liberal Party of Australia, 876
Lifelong learning (LLL) concept, 206, 872, 873
Literacy and Numeracy Studies: An International Journal in the Education and Training of Adults, 208
Logo (Seymour Papert), 757, 760
LPS. *See* Learner's Perspective Study (LPS)

M
Making Mathematics Count (MMC), 497–498
Malaysia, 399, 414–416
Mandarin numeration, 82, 83

Manipulatives, 536, 537
Manor Program (Israel), 420
Making mathematics count (UK), 496, 497
Mathematical communication, 268
Mathematical Creativity and Giftedness (MCG), 921–922
Mathematical learning spaces
  collaborative learning, 767
  mental learning space, 766
  MobileMath game, 766–767
  out-of-the-classroom/out-of-school learning, 766
  for teachers, 767
  ubiquitous learning, 766
Mathematical literacy, 723, 724, 865
Mathematical model
  computer algebra systems
    activity theory, 565
    blackbox, 564
    confrontation, 562
    "false," 561
  cultural-historical perspectives
    academic scientific model, 553
    activity theory, 553
    formal language, 554
    human labor, 552
    schools and academies, 553
    scientific concepts, 553
    scientific/theoretical thinking, 554
    Wason's reasoning task, 555
    zone of proximal development, 555
  cultural models, 551
  definition, 550–551
  metacognitive process, 550
  physical models, 551
  problem solving
    Freudenthal tradition, 557
    ICTMA conferences, 557
    internalisation, 558
    metacognitive reflection, 557
    realistic and authentic school problems, 558
    "set up a simple model," 557
    social motivation, 558
    verbalism, 557
  traditional, 550
  workplace technology
    activity theory, 560
    Dan's formula, 559
    everyday workplace, 560
    internalisations and externalisations, 560
    mathematical thought, 560
    pedagogical models, 560
    social and cultural context, 561

Subject Index 1107

spreadsheet formula, 559
timeline, 559
Mathematical Sciences Education Board
    (MSEB), 1025
Mathematical teaching spaces, 768–770
Mathematicians, 434, 435, 449, 452, 453
    and education policy, 496, 497
    and mathematics education, 497
    and mathematics education research, 435
Mathematics
    A-level, 504
    and science, 486, 488, 490, 491, 495, 496
    anxiety, 210
    assessment
        authentic assessment, 727–728
        automated scoring, 723
        CAS, 730
        classroom connectivity, 746
        communications infrastructure, 722
        computer-and paper-based assessment,
            723, 735–737
        computer-based testing, 722
        distributed cognition, 728
        dynamic geometry diagram, 730
        equity principle, 739, 743–745
        examination-based assessment, 738
        expanding assessment, 725–727
        feedback, 733–735
        human-scored partial credit
            assessment, 732
        Internet technology, 745
        learning principle, 739–743
        Maple visualization tools, 731
        mathematical assistant, 724, 732
        mathematically-able software, 728, 739
        mathematics analysis tools, 723
        mathematics principle, 739–743
        m-rater scoring engine, 730
        on-line assessment system, 730, 737
        pen-and-paper algorithms, 738
        reflections, 747–748
        spreadsheets and statistics
            programs, 745
        standardized *vs.* dynamic
            assessment, 729
        StatLab program, 745
        steps method, 731, 732
        student model, 722
        student-task interaction, 732–733
        teacher-friendly tools, 725
        tutor model, 722
        Vygotsky's distinction, 728
        WebAssign, 729, 730
    curricula, 489, 490, 506, 519, 522, 523

    laboratories, 535, 538
    statistics learning, 221
    textbooks, 529–531, 535, 541
Mathematics analysis software (MAS), 628
Mathematics education
    academic field, 269
    curricular reforms, twentieth century
        calculating machines, 287
        drawing instruments, 286
        elements, 283
        Euclidean method, 282
        geometry, 283
        German reform movement, 283
        manipulatives, 286
        modern/new math(s), 283–286
        practical mathematics, 282
        teaching aids, 286
    curriculum (*see* International mathematics
        curriculum)
    development of, 270
    educational policy, 270
    educational problems, 288
    empirical research, 287
    field of study
        epistemology, 267
        professional and academic, 269
        research, 267–268
        theoretical field, 269
    IMU and ICMI, 274
    institutional contexts, 270
    as an international discipline, 804
    international handbook
        epistemology, 267
        NCTM, 267
        research, 267
    international studies, national and local
        policy
        cross-sectional sample surveys, 1033
        data, 1035
        educational policy, 1015–1016, 1021
        Educational Testing Service, 1013
        Germany and Finland, 1027–1031
        IEA, 1011–1012
        INES, 1017
        international knowledge banks, 1018
        IPMA study, 1014
        Kassel project, 1014
        LLECE, 1014–1015
        mean score, 1033
        national achievement measures, 1035
        OECD indicators, 1017
        PISA, 1012–1013, 1035
        policy convergence, 1019–1020,
            1033–1034

Mathematics education (*cont.*)
  SACMEQ series, 1014
  Singapore, 1031–1032
  teaching and learning, 1034
  TIMSS, 1015–1016, 1035
  USA, 1022–1027
 mathematical communication, 268
  conferences, 290
  journals, 276, 289–290
  mathematics teachers association, 278
 mathematics didactics, 289
 plan, 289
 policy and practice, 270
 psychology and
  active learning, 292
  behavioural psychological theories, 293
  Decroly's method, 292
  German kindergarten movement, 292
  human and social interactions, 292
  instrumental understanding, 293
  practice schools, 291
  pupil behaviour, 292
  social interactions, 293
  teaching methods, 292
 research, 798, 811–818
 social, cultural and political dimensions, 294–296
 social justice, 3
 teachers, key stakeholders, 269
 teachers learning, 270
 theoretical field, 269
 unbounded
  ICMI, 280–281
  L'Enseignement Mathématique, 279–280
Mathematics Education Journal (Iran), 398
Mathematics Education Advisory Board, 490, 491, 494
Mathematics education policy, 270
 Advisory Committee on Mathematics Education, 495–496
 curriculum reform, in Portugal, 506
 definition, 485–486
 educational reform, in Mexico, 506
 in England
  policy research, 494–496
  teachers professional development, 498–499
  Every Childs Counts, 498
  Making Mathematics Count, 497–498
  mathematics A-level, 504
  mathematics education research community, 506–507
  National Centre for Excellence in the Teaching of Mathematics, 498–501
 policymakers, 508–510
 stakeholders, 507–508
 in USA
  Common Core State Standards, 491–493
  Higher Education Opportunity Act of 2008, 491
  National Council of Teachers of Mathematics, 488–489
  National Mathematics Standards, 487–488
Mathematics education researchers, 432, 435, 437
Mathematics education research (MER), 2, 310
 adoption and assimilation of theories
 attitudes of mathematicians towards, 509
 attitudes of policy-makers towards, 509
  good practice identification, 47
  heresy/development, 48
  networking theory, 48–49
  research contexts, 48
  "translation" and well-developed theory, 48
 communication and cognition interaction, 59–60
 discourse analysis
  antagonistic relationship, 53
  classroom, 51, 52
  commognition, 52
  "discursive psychology," 52
  Educational Studies in Mathematics, 53
  Ryve's analysis, 51
  SFL tools, 51, 52
  social psychology perspective, 52
  "the mathematics register," 51
 ethnomathematics, 60, 61
 evolutionary perspective, 60
 Fairclough's work, 50
 "incommensurability," 42
 intertextuality, 42
 Luis Radford's cultural theory of objectification, 49
 Morgan's comment, 50
 new perspectives, 50, 51
 policies, 486–488, 490, 495, 498, 506, 508, 509, 511, 519, 522, 523
 postmodern approaches
  Adler and Lerman's view, 58
  De Andrade report, 58–59
  discursive analysis, 59
  impact and take-up, 57
  Lacan work, 57
  objectivity, 58
  post-Freudian psychoanalytic theor, 59

Subject Index   1109

"pure mathematics classrooms," 59
  teaching and learning, 57
questions and methodologies, 50
Skovsmose's work, 61
social, political and cultural dimensions, 42
sociology
  Bourdieu and Bernstein's work, 55
  curriculum hierarchy, 55
  ethnomethodology, 54, 55
  "ideological state apparatus," 56
  power dynamics, 56
  socio-cultural perspectives, 54
  symbolic interactionism and
    phenomenology, 54
  Vygotsky, 54
theoretical perspective, 42, 60
trends and advances
  classroom, 46
  constructivism, 47
  explicit theory, 46
  global document search function, 44
  journals, 43
  non-traditional frames, 43
  PME, 43–47
  psychology, 43
  sub-fields, 43
  theoretical frameworks, 47
  Vygotskian and neo-Vygotskian
    theories, 44
Mathematics Education Research
  Group of Australia (MERGA), 843,
  916–917, 925
Mathematics, Education and Society (MES),
  926–927
Mathematics knowledge for teaching, 406, 418
Mathematics learning
  collaboration, 91–93
  conflicts with mainstream mathematics
    education, 71
  deficiency, 70
  deficit models, 3
  disadvantage, 2, 70
  equity and disability
    cognitive and attitudinal factors, 87
    Dowker's view, 86
    "dyscalculia," 85
    Geary's study, 85
    groups, 84
    othering process, 87
    special educational needs, 84
    underperformance, 86
  language, 5
  language and mediation, 79–80
  linguistic resources, 82–84

multilinguals and cognitive resources,
  80–82
multiple resources examination, 77
reflection, 93–94
sensory, material and semiotic
    interplay
  blind learners, 78
  DHH learners, 79
  visuo-spatial representations, 79
  Vygotsky's work, 77
societal discourses, 70
sociocultural directions, 3
socio–political approaches
  cognitive behaviours, 74
  discursive positioning, 76
  "dominant mathematics," 75
  enculturation and emancipation, 75
  knowledge, 74, 75
  social turn, 74
  "western mathematics," 75
static characteristic, 70
teacher education
  and cultural sensitivity, 90–91
  equity, 88–90
views on equity
  equity definition, 72
  ethnicity and race, 73
  "fairness" and "justice," 72
  micro and macro criteria, 73, 74
  social–political status, 73
Mathematics problem, 557
  definition of, 557
Mathematics teachers
  and curriculum development, 523
  professional development, 487, 488, 490,
    498, 501
  roles in policy development, 504
Mathematics teacher education, 434, 435,
    445–446, 449, 451
  action research, 345
  classroom practice, 344
  emancipatory action research, 346
  participatory action research
    development, 347–349
    implementation and, 350–352
  research, 352
  surveying research methods
    data construction and data analysis,
      335–336
    development methodologies, 336–337
    geographic origin, 331–332
    ICME10, 329
    JMTE, 330–331
    participating teachers, 331

Mathematics teacher education (*cont.*)
    qualitative and quantitative methods, 334–335
    referenced research volumes, 333
    techniques
        concept maps, 341–342
        narratives/stories, 338–339
        questionnaires/surveys, 343–344
        tests and tasks, 342–343
        videos, 340–341
Mathematics teacher education research, 352
Mathematics teacher educators (MTEs), 406, 410, 412, 415, 421
Mathematics teacher inquiry, 413, 417, 418, 421
Mathenpoche, 772
Math in Cultural Contexts project, 178
Matrices, 619, 620
Maya, 528
Measurement
    with *Cabri-Géomètre*, 571
    guided, 583, 589
    Measure probatoire, 582, 583
    Mesure exploratoire, 582
    perceptual, 583
Measurement tool. *See* Dynamic geometry environment (DGE)
Mental learning space, 766, 770
    out-of-school learning, 766
Merit and meritocracy, 101, 122
Meta analyses comparing effects of modes of assessment, 735
Metacognition, 558
Mexico, 346, 348, 759–762, 771, 773, 784, 869, 883, 884
Microworlds, 649–651, 656, 663–666, 673
    definition of, 649
Ministère de l'Éducation Nationale, 758
Ministère de l'Éducation Nationale, de l'Enseignement Supérieur et de la Recherche (MENESR), 758
Ministerio de Educación Nacional (MEN), 874
Ministry of Education, People's Republic of China, 574
Minitab, 648
Minitools, 649
Miniwatts Marketing Group, 251
Miserable Noten für deutsche Schüler [Abysmal marks for German students], 1027
Mixed-methods research, 334, 338, 343
MMC. *See Making Mathematics Count (MMC)*
MobileMath game, 766–767
Modelling, 549–566

definitions of, 549–566
Modes of belonging
    context transition, 189
    critical mathematics education (*see* Critical mathematics education)
    ethnomathematics (*see* Ethnomathematics)
    imagination, alignment, and engagement, 189, 190
    mathematics classrooms, 189
    trajectories formation, 190
    Wenger's method, 189
Mongolian mathematics education, 818
Moscow Mathematics Compendium, 803
MTE. *See* Mathematics teacher educator (MTEs)

N

Nanyang Technological University, 695
Narrative analysis, 338
National Academy of Sciences, 489
National Assessment of Educational Progress (NAEP) (USA), 865, 886, 1013, 1017, 1024, 1026
National Center for Educational Statistics (NCES), 245
National Centre for Excellence in the Teaching of Mathematics (NCETM), 499–504, 508, 510
National Centre for Excellence in the Teaching of Mathematics (NCETM) (UK), 773
    Funded Projects Scheme, 502
    independent evaluation studies, 503
    mathematics A-level, 504
    portal's homepage, 501
    professional learning framework, 500
    self-evaluation tools, 502
    Teacher Enquiry Bulletins, 502
National Commission on Excellence in Education, 1024
National Council of Teachers of Mathematics (NCTM), 488–491, 508
National Council of Teachers of Mathematics (NCTM) (USA), 72, 247, 267, 277, 286, 365, 464, 475, 488–489, 574, 645, 696, 704, 755, 865, 933, 1024
    an agenda for action, 1024
    principles and standards (2000), 867
    standards documents, 1025
    standards movement, 491
    4th yearbook (1924), 804
National Education Association, 18
National Governors Association Center for Best Practices, 492, 960

Subject Index

National Library of Virtual Manipulatives for mathematics education (Utah University), 697
National Mathematics Advisory Panel, 490, 491, 494
National mathematics curricula, 465
National Mathematics Curriculum Standards (NMCS) (Japan), 377
National Mathematics Standards, 487–488
National Research Council, 472, 510, 877, 878
National Research Council Mathematical Sciences Education Board, 739
National Science Foundation, 967
NCLB. See No Child Left Behind (NCLB)
NCTM. See National Council of Teachers of Mathematics (NCTM) (USA)
NCETM. See National Centre for Excellence in the Teaching of Mathematics (NCETM) (UK)
Neo-Marxism, 467
The Netherlands, 102, 125, 805, 806, 812
Networking, 598, 599
    and professional geneses, 782–783
    of teachers, 698
Neuroscience and mathematics education, 321
Newman error analysis, 422
New Math (Maths/Mathematics), 284, 531, 536, 806, 807, 809, 811, 1027
    difficulties with, 809
News media, 692
New Zealand, 102, 125, 208, 402, 408, 413
New Zealand Ministry of Education, 645
NMCS. See National Mathematics Curriculum Standards (NMCS) (USA)
No Child Left Behind (NCLB), 1026, 1027
No Child Left Behind Federal legislation, 488
Non-Euclidean geometry, 576
Non-formal mathematics education (NFAME), 204
Nordic Society for Research in Mathematics Education (NoRME), 930–931
Norway, 369, 371, 373, 408
Norwegian, 371–373
Numeracy development projects (NDP), 408
Numeracy for all, 31–32
Numeracy for nurses, 728

O

Observation, 432–438, 442–446, 450, 452
OECD. See Organisation for Economic Co-operation and Development (OECD)
One Laptop per Child (OLPC) computer, 759

Online mathematics education
    advantages of, 693, 698, 699
    disadvantages of, 693
    and the internet, 696, 701, 706
    multimodal forms of, 706
    networks of teachers, 698
    professional development of teachers, 696, 698
    roles of teachers, 698
On the mathematics curriculum of the high school, 1023
Opportunity to learn (OTL) mathematics, 244
Organisation for Economic Cooperation and Development (OECD), 32, 206, 506, 743, 793, 838, 865, 873, 876, 877, 981, 1011–1013, 1015–1017, 1019–1022, 1025–1031
    2012 International PISA Survey, 724
    and PISA, 150, 151, 509, 724
Organisation for European Economic Cooperation (OEEC), 284, 835
Overhead projectors, 532
Oxford University, 801

P

Pantograph, 541–543
Parents, 863, 884, 890
Park City Mathematics Institute (PCMI), 976
Partial credit, 726, 730–732
    "steps" method, 731, 732
Participation, 105, 110, 115–117, 119–121, 124, 125, 130
Participatory action research, 328, 346–353
    ICME 10 and PME 28, 348
    implementation
        elements, 350
        investigation, 351
        planning workshops, 351
        social framing, 350
    silence policy, 349
    socio-political role, 349
PCK. See Pedagogical content knowledge (PCK)
Pedagogical approaches
    alignment, 190, 193
    boundary crossings, 170
    complex transition process, 172
    context transition process, 4, 194
    critical mathematics education
        four aims, 184
        Frankenstein's suggestion, 183–184
        Gutstein's problems, 187
        inquiry cooperation model, 184
        "mathemacy," 184

Pedagogical approaches (*cont.*)
    mathematical model, 185
    Sandra's comment, 188
    societal contexts, 186
    statistical learning, 187
    "Terrible Small Numbers" project, 184
    two geographical groups, 183
  dispositions to learn, 172
  engagement, 190–192
  ethnomathematics
    classroom, 179
    Hindu-Arabic system, 179
    learners' views, 180
    Papua New Guinea, 179
    permeability, 178
    research program, 177
    Western mathematics, 177–178
  imagination, 190, 192–193
  knowledge transfer, 172
  learning, 195
  ongoing process, 171
  reflection, 172–173
  social justice
    disadvantage, 173, 174
    drawing process, 173
    educational task, 173
    interviews, 176
    investigations, 173–175
    learners' perceptions, 172
    mathematics curriculum, 176
    symbolic violence, 176
Pedagogical content knowledge (PCK), 394, 396, 402
Pedagogical technology knowledge (PTK), 632
Perceptual measuring, 583
Peru, 759
Pew Hispanic Center, 814
The Philippines, 882
PISA. *See* Programme for International Student Assessment (PISA)
PME. *See* International Group of Psychology in Mathematics Education (PME)
Poland, 819, 869, 877, 880
Policy and mathematics education
  analysis, 487
  and assessment, 486, 489
  and mathematics education research, 486–487, 494–496, 506–510
  formation, 496
  influences on
    in Germany, 504
    in Mexico, 506
    in Portugal, 508
    in UK, 507
    in USA, 491, 492
  policy-makers, 507–510
  stakeholders, 487, 505, 507–508
Policy convergence, 1019–1020, 1032–1035
Policy dimensions, 755, 784
Polya, G., 550, 556, 817
  how to solve it, 807
Polynomials, 613–618
Postmodern approaches
  Adler and Lerman's view, 58
  De Andrade report, 58–59
  discursive analysis, 59
  impact and take-up, 57
  Lacan work, 57
  objectivity, 58
  post-Freudian psychoanalytic theory, 59
  "pure mathematics classrooms," 59
  teaching and learning, 57
PowerPoint, 532
Praxis, 121
Practitioner model, 768
Principles of assessment
  equity principle, 739, 743–745
  learning principle, 739–745
  mathematics principle, 739–741, 744, 745
Problem posing, 865
ProbSim, 648
Professional development (PD) of teachers, 396, 450, 632, 872, 880
  and lesson study, 880
Professional learning of teachers, 403, 407, 409
Programme for the International Assessment of Adult Competencies (PIAAC)
  aspects of survey validity, 230–231
  conception of numeracy, 229–230
  human capital approach, 229
  OECD, 228
  survey design and administration, 230
Programme for International Student Achievement, 72, 990
Programme for International Student Assessment (PISA), 32, 397, 460, 466, 467, 480, 696, 792, 793, 813, 838–839, 1011–1013, 1015–1018, 1021, 1022, 1026–1036
  classroom level, 869
  competencies, 871–874, 889–890
  content coverage and pedagogy, 863
  curriculum (*see* Curriculum)
  data-driven analyses, 870
  Europe and Latin America, 870
  frameworks, 865, 871, 872, 889–890
  hypothetical expectations, 862

Subject Index

international studies, 1012–1013, 1035
mathematical content, 865
mathematical literacy, 865
policy reactions, 869
primary and secondary analyses, 869
public examinations, 961–962
schools and classrooms, 869
situations/contexts, 865
students learning and attitudes, 888, 889
Sub-Saharan study, 863
teacher preparation and development, 892–893
tests on science, 870
textbook use, 892–893
themes, 886
Programme for International Student Assessment Governing Board, 724
Promoting Rigorous Outcomes in Mathematics and Science Education (PROM/SE), 879, 891
Proof measuring, 583
PropWin, 676
Prosperity for all in the global economy: World class skills: The Leitch review of skills, 495
Providing feedback, 724, 732–735
Psychologists, 433, 442
and mathematics education research, 432
Psychology and mathematics education
active learning, 292
AMU, 846
annual conference, 843–844
behavioural psychological theories, 293
Decroly's method, 292
German kindergarten movement, 292
human and social interactions, 292
instrumental understanding, 293
international community, 844
mathematics teacher education programs, 845
practice schools, 291
pupil behaviour, 292
social interactions, 293
teaching methods, 292
PTK. *See* Pedagogical technology knowledge (PTK)
Public examinations, 887
in China, 961
classroom, role in, 961
coursework, 960
mathematical modelling, 960
TIMSS and PISA, 961–962
Pythagoras theorem, 527, 614, 791

Q
Quadrilaterals, 587
Qualifications and Curriculum Authority, 645
Qualitative Cognition-Focused Research, 147
Qualitative research, 329, 333, 334
Quantitative literacy for all. *See* Numeracy for all
Quantitative research, 334
Questionnaires, 335, 343–344
Quipus, 529

R
Race, 102, 104, 114, 125, 130, 131
Race to the Top (USA), 493
Radford, Luis, 171, 172, 190
Realistic Mathematics Education (RME) program, 30, 31, 466, 806
Reflections on teaching, 372, 380
Reform mathematics, 463
Regional Education Centre for Science and Mathematics (RECSAM), 832, 840–842, 854, 855
Requirements of software tools, 649–650
Research
agenda, 887–888, 893
questions, 888–889
Research in Mathematics Education (RME), 933
and reform, 461, 463–465, 469–472
and teaching, 463–469, 471–473, 475, 476, 480, 481
Response times, 732, 733, 747
Response types, 722
Riemann surface, 536
RME. *See* Realistic Mathematics Education (RME) program
Rotato, 727
Royaumont Conference, 1959, 806
Russia, 537, 538, 799, 800, 803, 815, 818
berlin wall, 812
cold war, 799
history of school mathematics in, 811
iron curtain, 805
and mathematics education, 818
socialist mathematics, 806–807
sputnik, 805
Union of Soviet Socialist Republics (USSR), 814, 815
Russian-language Google Scholar, 817–818
Russian mathematics education, 799, 800
Rwanda, 759

S

Salamis Tablet, 526
School
   curriculum, computer algebra systems, 628–629
   failure, 102, 103, 107, 108
Schooling for all
   distance education, 14
   elementary forms, 10
   Harding report, 13
   Indonesia, 11, 12, 14
   Jomtien commitment, 12
   Nebres comments, 13
   populations percentage, 10, 11
   primary and secondary, 12
   principle, 12
   religious knowledge, 10
   UNESCO, 12, 13
School Mathematics Project (SMP), 806
School Mathematics Study Group (SMSG), 26, 284, 806, 811, 1022
Science, technology, engineering, and mathematics (STEM), 509
Second International Mathematics Study (SIMS), 831, 837
Self-evaluation tools (SETs), 502
Semiotics, 105, 108, 110, 115, 120
Sempoa, 537
Sesamath Association, 771–772
Set square, 527
Shapemaker, 577
Singapore, 461, 464, 466, 473, 476, 863, 870, 874, 884–887, 1014, 1021, 1031–1032
   National Institute of Education, 886, 887
SINUS project (Germany), 879
Situated cognition, 110
Slate, 525–544
Sliders, 655, 665, 666, 680
Slide rule, 529, 537–539
   history of, 529, 537
Smith, D.E.
   end the formation of ICMI, 801
   ties with Germany, 804
SMP. *See* School Mathematics Project (SMP)
SMSG. *See* School Mathematics Study Group (SMSG)
Social class, 127–129
Social factors, 109, 310–313
Social justice
   ethnicity, 125–127
   gender, 123–125
   meritocratic model, 121
   social class, 127–129

Social turn, 104, 105, 119, 296
Socio-cultural perspectives, 103–106, 111, 120, 121, 129, 131
   classroom dynamic
      classroom designs, 118–119
      social negotiations, 117–118
      teaching modes, 119–121
   cultural learners, cognition and affect
      academic failure children, 108
      binomial interaction, 111
      empirical data, 107–108
      focus consciousness, 109
      poor economic backgrounds, 107
   culture and mathematics curricula
      ethnomathematics, 111–113
      mathematical enculturation and acculturation, 113–115
      semiotic interactions, 115
      universal character, 116
   deficit model
      assumptions, 104
      characteristics, 102
      research designs, 103
   social justice
      ethnicity, 125–127
      gender, 123–125
      meritocratic model, 121
      social class, 127–129
Socio-economic status, 244, 245
Sociolinguistics, 105
Sociology, 105
Software
   Fathom (*see Fathom*)
   TinkerPlots (*see TinkerPlots*)
Software development
   for algebra, 696
   for functions, 696
Soroban, 537
South Africa, 72, 102, 119, 125, 128, 412, 758
Southern and Eastern Africa Consortium for Monitoring Educational Quality (SACMEQ) studies, 1014
Spain, 806, 868, 869, 871, 873, 876, 884, 886, 891
Spatio-graphical field, 574–576, 580–583, 587, 589, 590
Specific Mathematics Assessments that Reveal Thinking (SMART), 735
Spectral dragging, 580
Spot-and-show orchestration, 769–770
Spreadsheets, 598, 602, 615, 622, 648–651, 681
SPSS, 647
Square root of 2, 526

Subject Index 1115

Squared paper, 535
Sri Lanka, 810
Stakeholders, mathematics education policy, 507–508
Stakeholders in research on teaching, 363–366, 368, 369, 387, 388
Standardized tests, 864, 866, 867, 873, 876, 877, 886, 888, 890–892
Standard American English (SAE), 80, 83
Standards movement (USA), 491, 508
Statistical distributions, 666, 668
Statistical reasoning
  future possibilities, 681–682
  support for, 644, 646–650, 655, 659, 662, 666, 668, 669, 672, 674, 675, 678–680
  as travelling (metaphor), 678
Statistical software
  ActiveStats, 652
  BMDP, 647
  CyberStats, 652
  DataDesk, 647, 651, 652
  Excel, 649, 651, 664, 681
  Fathom, 648, 650–654, 674
  Google Spreadsheet, 651
  R package, 651
  SAS, 651
  S package, 647
  SPSS, 647, 651
  StatCrunch, 651
Statistics
  data-based enquiry, 645
  data collection and exploration, 645
  Fathom (see Fathom)
  education
    data and chance, 644, 666, 671–674
    inferential statistics, 647, 671
    probability and statistics, 671
    randomization, 647, 672
    simulation, 649
    with TinkerPlots, 646, 648–650, 653, 671, 682
  history of
    children's learning, 648–649
    confirmatory data analysis, 646
    exploratory data analysis, 646
    software tool requirements, 649–650
    statistical programming language *S*, 646
    textbooks, 648
  inquiry-based pedagogies, 645
  intellectual method, 644
  literacy skills, 645
  "reformed" approach, 645
  statistical inference, 677

technological tools
  applets and stand-alone applications, 652
  data and materials repositories, 653
  educational software, 653
  graphing calculators, 652
  multimedia materials, 652
  software packages, 650
  spreadsheets, 651
  TinkerPlots (see TinkerPlots)
  traditional approaches, 645
  workplace, 643
STEM education, 491, 509
Strategic Education Research Partnership (SERP), 506
Student-centred pedagogies, 465
Student teachers, 436, 445–447
Studying mathematics teaching and learning internationally
  comparative research, 976
  designs and methods
    classroom studies, 1002–1003
    components, 998–999
    cross-national achievement studies, 1000
    curriculum and textbook analyses, 1003
    independent approach, 1000–1001
    interviews, 1001–1002
    questionnaires, 1001
    student achievement tests, 1000
    tasks, students, 1003
  FIMS, 976–977
  focal studies
    contextual factors, 983
    goals, 984
    LPS, 984
    13th ICMI study, 985
    videotape study, 983
  ICMEs, 976
  IEA studies
    assessment framework, 989
    attained curriculum, 987
    construct achievement tests, 986
    implemented curriculum, 987
    intended curriculum, 987
    student achievement, 989
    target populations, 987
    TIMSS 1999, 988
  international studies, 977
  Kassel project, 996–997
  large-scale studies
    FIMS, 979
    OECD, 982
    SIMS, 979–980
    TIMSS, 980–981

Studying mathematics teaching (*cont.*)
    LPS, 994–995
    multi-faceted design, 1004
    PISA, 990–992
    13th ICMI study, 997–998
    TIMSS video and case studies
        classrooms, 992
        coding schemes, 993
        first approach, 993
        second approach, 993–994
        third—key—approach, 994
    US–Japan problem-solving study, 995–996
Suanpan, 537
Sudan, 808
Summative assessment, 732, 734
Sur l'épreuve pratique de mathématiques au baccalauréat en France, 764
Survey of Mathematics and Science Opportunities (SMSO), 983
Survey research, 329–337
Sweden, 208, 412
Switzerland, 536
Symbolic representations, 620–623
Symbol sense, 622, 625
Systemic functional linguistics (SFL) tools, 51, 52

T
TableTop, 648
Taiwan, 466, 810, 816
Tartaglia–Cardano method, 614
Task–Technique–Theory (TTT), 613
Teacher
    autonomy, 465
    biographies, 472
    craft knowledge of, 394, 398, 412
    inservice, 418
    preservice, 418–421
    reflections of, 363, 364
    as researchers, 394, 398, 400–403, 405, 409, 410, 412, 413, 415, 417, 418
Teacher education, 254, 327–353, 432, 434, 435, 442, 445–450, 454, 872, 879–880, 887, 891, 892
    associative *vs.* institutional, 778
    feedback mechanisms, 782
    individual *vs.* collective, 778
    networking and professional geneses, 782–783
    online discussions, 779–780
    Pairform@nce program, 781
    and professional development
        German tradition, 880
        lesson study, 881–883
        Poland approach, 880
        SINUS project, 879
        text for teachers, 883–884
        videos, 880–881
    Second International Handbook of Mathematics Education, 777
    technology integration, 780
    video clubs, 779
    video resources, 777
Teacher education and development study in mathematics (TEDS-M), 432, 1012
Teacher education and technology
    inservice, 760, 780–782
    preservice, 760, 778–780
    trends, 778, 780, 792
Teacher Enquiry Bulletins, 502
Teachers, key stakeholders, 269
    action research dimension
        classroom materials, 383
        collaborative, 380–382
        inquiry-based activity, 380–382
        lesson study, 382
        professional and scientific artefacts, 384
        reflective, 380–382
        videos, 383
    alumni, 367
    blackboard, 362
    collaboration, 365
    design research activities, 363
    developed and developing countries
        Canadian, 373–375
        Chinese, 377–380
        Japanese, 375–377
        Norwegian, 371–373
        USA, 369–371
    diagnostic test, 362
    identity, 362
    intervention research, 364
    management strategy, 365
    professional knowledge, 367
    recipients, 366–367
    research and professional development
        classrooms, 386
        lesson-study activity, 386
        practitioner constructs, 385
        strategies, 385
    research enterprise, 365
    scientific community, 366
    scientific knowledge, 368
    students investigation, 363
    teacher-researcher collaboration, 365
Teachers learning, 270
    active teacher participation, 400

Subject Index   1117

classroom activity, 405
collaboration, 345
craft knowledge, 394
design research, 401
global influences, 397–402
inquiry community, 404
insider researchers, 403
knowledge and learning, 403
large-scale projects
    AMIC workshop, 410
    CFPR, 408
    CMIT program, 409
    CO program, 409
    facilitated model, 409
    LCM, 408
    LPS, 412
    LS model, 409–410
    MTEs, 408–409
    numeracy programs, 409
    TBM, 408
local influences, 399
mathematical knowledge, 395
MTE knowledge, 406
outsider researchers, 403
participating teachers, 396
pedagogical methods, 399
personal reflections, 393–394
preservice programs
    knowledge, levels, 419
    teacher-educator program, 419, 420
professional development, 396
small-scale professional learning projects
    communication and participation, 413
    data collection, 417
    insiders and outsiders, 417
    LS, 414
    schools and university, 417
    TIMSS video study, 414
teaching–learning development, 403
theory and theory development, 400
Teachers as researchers, 382–384, 439
Teaching better mathematics (TBM) project, Norway, 408
Teaching Mathematics with Technology (EMAT) program (Mexico), 759
*Technological Pedagogical and Content Knowled*ge (TPACK), 768
Technological tools
    DataScope, 648, 650
    InspireData, 649
    Minitab, 648, 651
    ProbSim, 648, 650
    TableTop, 648–650, 658
    TinkerPlots, 650, 651, 653

Technology, 549–566
access to, 754–756, 758, 759, 762, 764–769, 777, 783, 784
calculation tools
    abacus, 537–538
    calculator, 539
    slide rule, 538–539
Chinese set square, 527
competence among teachers, 779
computer, 551
computer algebra systems, 630
demonstration
    colored cubes and rods, 536
    cube root block, 534–535
    Cuisenaire rods, 536
    cylinders, 536
    geometric models, 535
    geometric solids, 534
    geometry instruction, 533
    Hart inversor, 533
    history of, 533
    internet, 537
    laboratory method, 535
    manipulatives, 536–537
    physical objects, 536
    Riemann surfaces, 536
and digital textbooks, 771
education, 552
fundamental symbolic tools, 528
guiju, 527
He Tu, 527–528
information display tool, 531–532
information storage tool, 529–531
integration, 754–759, 763–765, 774, 778, 781–783
Luò Shû, 527–528
and mathematics education, 526
Mayan calendar wheels, 528–529
modelling and technology
    (*see* Mathematical model)
quipus, 529
Salamis Tablet, 526
in school mathematics
    evolution of, 598, 599, 628
    limitations of, 598, 611, 616
    research into, 598, 602–604
square root of 2, 526
statistics (*see* Statistics)
three-policy dimensions, 784
and traditional textbooks, 771
in twenty-first century
    Kinematical Models for Design Digital Library, 541
    pantograph, 541

Technology (*cont.*)
    techno-pedagogic task design, 544
    webbing, 543
    two-policy dimensions, 755
Technology-driven developments and policy implications
    access/support dimension, 754, 755
    assessment policies, 763–764
    constructionist, 765
    constructivist, 765
    digital resources, profusion
        individual/collective, 776
        Intergeo project, 773–774
        policy questions, 771
        production paradigm, 776
        resources quality, 775–776
        Sesamath Association, 771–772
    digital technology, 757
    individual students and teachers learning, 754
    mathematical learning spaces
        collaborative learning, 767
        mental learning space, 766
        MobileMath game, 766–767
        out-of-the-classroom/out-of-school learning, 766
        for teachers, 767
        ubiquitous learning, 766
    mathematical teaching spaces, 768–770
    National curricula recommendations
        EMAT project, 759–761
        Enciclomedia, 761–762
        Habilidades Digitales para Todos program, 762
        OLPC computer, 759
        pedagogical and implementation strategies, 759
    new educational paradigms, 756–758
    teacher education
        associative *vs.* institutional, 778
        feedback mechanisms, 782
        individual *vs.* collective, 778
        networking and professional geneses, 782–783
        online discussions, 779–780
        Pairform@nce program, 781
        Second International Handbook of Mathematics Education, 777
        technology integration, 780
        video clubs, 779
        video resources, 777
    three policy dimensions, 784
    top-down/bottom-up dimension, 754
    two policy dimensions, 755

TEDS-M. *See* Teacher education and development study in mathematics (TEDS-M)
Textbook companies and assessment, 731
Textbooks, 598, 605
    effects of international studies on, 892
    importing textbooks, 884, 893
    Singapore mathematics textbooks, 884, 885
Textbooks and the Internet, 691, 692, 714
Text messaging, 695
Thailand, 882
Theoretical field, 574, 580–583, 587, 590
Theories of mathematics education, 303–321
    Anna Sfard's commognitive framework
        communicational framework, 317
        endorsed narratives and routines, 319
        focal analysis, 317
        methodological and epistemological grounds, 315
        preoccupation analysis, 317
        verbalization skills, 320
        visual mediators, 319
        word use, 319
    APOS theory, 309
    ATD, 313
    authentic task situations, 312
    didactical transposition, 313
    grand theory, 310
    international community, 305
    MER, 310
    neo-Piagetian theory, 313
    operational definition, 308
    physical and natural sciences, 308
    Piaget's theory, 309
    PISA, 312
    psychological learning theory, 311
    scientific research, 310
    six-stage theory, 307
    strengths and weaknesses, 312
    structural mechanism, 308
Theory development, 304, 306, 309, 310
Theory of didactical situations (TDS), 305, 313
Theory of Mathematics Education (TME), 296, 401, 934
Theory of multiple intelligences, 105
Third International Mathematics and Science Study, 72, 244, 460, 461, 466, 467, 473, 475, 480, 810, 813, 819, 831, 838, 980, 1012
    affective outcomes/perspectives, 864
    classroom level, 869
    competencies, 871, 874, 889–890
    content coverage and pedagogy, 863
    content domains, 864

Subject Index 1119

curriculum (*see* Curriculum)
frameworks, 867, 871, 889–890
hypothetical expectations, 862
primary and secondary analyses, 869
processes/performance expectations, 864
South East Asia and United States, 870
students learning and attitudes, 888, 889
Sub-Saharan study, 863
teacher preparation and development, 892–893
tests on science, 870
textbook use, 892–893
video study, 878
TIMSS. *See* Third International Mathematics and Science Study; Trends in International Mathematics and Science Study
TinkerPlots
aggregate representations, 680
basic actions, 659
Connections Project, 686
data connection, 671
data exploration with, 666–671
dynamic nature, 679
microworlds, 650
in middle schools, 658, 666, 681
multivariate representations and analysis, 660–661
organizing and representing data, 658–660
personal features, 680
in primary schools, 658, 668
random samples, 680
reorganizing virtual data cards, 659
simulation, 650, 658, 661–663, 671, 672
visual nature, 679
Tools
of calculation, 529, 537–539
of demonstration, 529, 532–537
of information display, 529, 531–532, 539, 540
of information storage, 529–532, 540
for reasoning, 626–628
Towards an Inclusive Mathematics Education, 92, 93
TPACK. *See* Technological Pedagogical and Content Knowledge (TPACK)
Training model, 88
Transition, 114
Transition studies, 320
Trends in International Mathematics and Science Study, 19, 20, 397, 414, 1011, 1012
international studies, 1015–1016, 1035
public examinations, 961–962

Treviso Arithmetic, 14
Triarchic theory of intelligence, 105
Twitter, 692, 715
cTWO, 645

U

Understanding, 105, 107, 111, 113, 117, 120, 123, 125, 128, 131
UNESCO. *See* United Nations Educational, Scientific and Cultural Organisation (UNESCO)
UNICEF, 852, 853
Unione Matematica Italiana, 574
United Kingdom (UK), 102, 127, 128, 399, 461, 464, 467, 469, 472, 480, 535, 801, 805, 806, 808, 811, 813, 816, 1017
United Nations Development Programme, 227
United Nations Educational, Scientific and Cultural Organisation (UNESCO), 12, 13, 754, 756, 808, 809, 812, 863, 934, 1011, 1014–1017, 1020
United States of America (USA), 72, 102, 124–126, 365, 369–371, 377, 380, 383, 384, 386, 399, 412, 415, 421, 473, 530, 598, 739, 804, 806, 808, 811–816, 818, 819, 868, 870, 879, 881–886, 1014, 1017, 1021–1028, 1035
international studies, 1022–1027
mathematics education policy
Common Core State Standards, 491–493
Higher Education Opportunity Act of 2008, 491
National Council of Teachers of Mathematics, 488–489
National Mathematics Standards, 487–488
National Research Council, 739
United States Bureau of Education, 1010
United States Department of Education, 473
University mathematicians, 320
University of Modena (Italy), 541
laboratory of mathematical machines, 541
Uruguay, 759, 819
USA. *See* United States of America (USA)
U.S. Census Bureau, 814
U.S. Commissioner of Education, 10, 11, 15, 16, 19
U.S. Department of Education, 488, 489, 491
US National Council of Teachers of Mathematics, 464
Utah University, 696
Utilitarian perspective, 246

V

Validation measuring, 583
Validity issues, 723, 728, 734, 736, 737, 739, 747
Value-added measures (VAM) approach, 252
Venezuela, 349, 351
Verbalization in mathematics education, 320
Victoria (Australia), 763
Victorian Certificate of Education (VCE), 744, 746
Victorian Curriculum and Assessment Authority (VCAA), 740, 743, 746
Video analysis, 340, 341
Video-recordings of mathematics classrooms, 408
Video studies
  qualitative, LPS, 158
  TIMSS 1999 Video Study, 156
Videotaped lessons, 381
Virtual learning environments, 543, 544
Virtual pantograph, 543
Vygotsky-inspired activity theories, 828

W

Wandering measuring, 582
Wason's reasoning task, 555
Web, 525–544
Webbing, 543
Western mathematics educators, 807, 817
WFNMC. *See* World Federation of National Mathematics Competitions (WFNMC)
Whiteboards, 532
  interactive, 532
Wikileaks, 692
Wikipedia, 692, 695, 708
Workplace mathematics, 215
World Bank, 818, 1011, 1018–1020
World Class Arena, 727
World Federation of National Mathematics Competitions (WFNMC), 813, 921
World Wars, 799, 801–805, 807, 810
  and mathematics education, 801–804, 807
World Wide Web (WWW), 694, 704, 713

Y

Yolngu Matha, 259
YouTube, 692, 700, 709

Z

Zone of Proximal Development (ZPD), 120, 555, 561, 564, 565